Grundlehren der mathematischen Wissenschaften 290

A Series of Comprehensive Studies in Mathematics

Springer
New York
Berlin
Heidelberg
Barcelona
Hong Kong
London
Milan
Paris
Singapore
Tokyo

J.H. Conway N.J.A. Sloane

Sphere Packings, Lattices and Groups

Third Edition

With Additional Contributions by
E. Bannai, R.E. Borcherds, J. Leech,
S.P. Norton, A.M. Odlyzko, R.A. Parker,
L. Queen and B.B. Venkov

With 112 Illustrations

Springer

J.H. Conway
Mathematics Department
Princeton University
Princeton, NJ 08540 USA
conway@math.princeton.edu

N.J.A. Sloane
Information Sciences Research
AT&T Labs – Research
180 Park Avenue
Florham Park, NJ 07932 USA
njas@research.att.com

Mathematics Subject Classification (1991): 05B40, 11H06, 20E32, 11T71, 11E12

Library of Congress Cataloging-in-Publication Data
Conway, John Horton
Sphere packings, lattices and groups. — 3rd ed. / J.H. Conway, N.J.A. Sloane.
p. cm. — (Grundlehren der mathematischen Wissenschaften ; 290)
Includes bibliographical references and index.

1. Combinatorial packing and covering. 2. Sphere. 3. Lattice theory. 4. Finite groups. I. Sloane, N.J.A. (Neil James Alexander). 1939– . II. Title. III. Series.
QA166.7.C66 1998
511′.6—dc21 98-26950

Printed on acid-free paper.

Springer-Verlag Berlin Heidelberg New York
a part of Springer Science+Business Media
springeronline.com

Text prepared by the authors using TROFF.

9 8 7 6 5 4 3 2

ISBN 978-1-4419-3134-4

Preface to First Edition

The main themes. This book is mainly concerned with the problem of **packing spheres** in Euclidean space of dimensions 1,2,3,4,5,.... Given a large number of equal spheres, what is the most efficient (or densest) way to pack them together? We also study several closely related problems: the **kissing number problem**, which asks how many spheres can be arranged so that they all touch one central sphere of the same size; the **covering problem**, which asks for the least dense way to cover n-dimensional space with equal overlapping spheres; and the **quantizing problem**, important for applications to analog-to-digital conversion (or data compression), which asks how to place points in space so that the average second moment of their Voronoi cells is as small as possible. Attacks on these problems usually arrange the spheres so their centers form a *lattice*. Lattices are described by *quadratic forms*, and we study the **classification of quadratic forms**. Most of the book is devoted to these five problems.

The miraculous enters: the E_8 and Leech lattices. When we investigate those problems, some fantastic things happen! There are two sphere packings, one in eight dimensions, the E_8 *lattice*, and one in twenty-four dimensions, the *Leech lattice* Λ_{24}, which are unexpectedly good and very symmetrical packings, and have a number of remarkable and mysterious properties, not all of which are completely understood even today. In a certain sense we could say that the book is devoted to studying these two lattices and their properties.

At one point while working on this book we even considered adopting a special abbreviation for "It is a remarkable fact that", since this phrase seemed to occur so often. But in fact we have tried to avoid such phrases and to maintain a scholarly decorum of language.

Nevertheless there are a number of astonishing results in the book, and perhaps this is a good place to mention some of the most miraculous. (The technical terms used here are all defined later in the book.)

— The occurrence of the Leech lattice as the unique laminated lattice in 24 dimensions (Fig. 6.1).
— Gleason's theorem describing the weight enumerators of doubly-even self-dual codes, and Hecke's theorem describing the theta series of even unimodular lattices (Theorems 16 and 17 of Chap. 7).
— The one-to-one correspondence between the 23 deep holes in the Leech lattice and the even unimodular 24-dimensional lattices of minimal norm 2 (Chapters 16, 23, 26).
— The construction of the Leech lattice in $II_{25,1}$ as $w^{\perp}/w$, where

$$w = (0,1,2,3,\ldots,24\,|\,70)$$

is a vector of zero norm (Theorem 3 of Chap. 26). (We remind the reader that in Lorentzian space $\mathbf{R}^{25,1}$, which is 26-dimensional space equipped with the norm $x \cdot x = x_1^2 + \cdots + x_{25}^2 - x_{26}^2$, there is a unique even unimodular lattice $II_{25,1}$ — see §1 of Chap. 26.)
— The occurrence of the Leech lattice as the Coxeter diagram of the reflections in the automorphism group of $II_{25,1}$ (Chap. 27).

Some further themes. Besides the five problems that we mentioned in the first paragraph, there are some other topics that are always in our minds: *error-correcting codes, Steiner systems* and *t-designs*, and the theory of *finite groups*. The E_8 and Leech lattices are intimately involved with these topics, and we investigate these connections in considerable detail. Many of the *sporadic simple groups* are encountered, and we devote a whole chapter to the *Monster* or *Friendly Giant* simple group, whose construction makes heavy use of the Leech lattice.

The main applications. There are many connections between the geometrical problems mentioned in the first paragraph and other areas of mathematics, chiefly with *number theory* (especially quadratic forms and the geometry of numbers).

The main application outside mathematics is to the *channel coding problem*, the design of signals for data transmission and storage, i.e. the *design of codes for a band-limited channel* with white Gaussian noise. It has been known since the work of Nyquist and Shannon that the design of optimal codes for such a channel is equivalent to the sphere packing problem. *Theoretical* investigations into the information-carrying capacity of these channels requires knowledge of the best sphere packings in large numbers of dimensions. On the other hand *practical* signalling systems (notably the recently developed Trellis Coded Modulation schemes) have been designed using properties of sphere packings in low dimensions, and certain modems now on the market use codes consisting of points taken from the E_8 lattice!

There are beautiful applications of these lattices to the numerical evaluation of *n*-dimensional integrals (see § 4.2 of Chap. 3).

Of course there are connections with *chemistry*, since crystallographers have studied three-dimensional lattices since the beginning of the subject. We sometimes think of this book as being a kind of higher-dimensional analog of Wells' "Structural Inorganic Chemistry" [Wel4]. Recent work on *quasi-crystals* has made use of six- and eight-dimensional lattices.

Higher-dimensional lattices are of current interest in *physics*. Recent developments in *dual theory* and *superstring theory* have involved the E_8 and Leech lattices, and there are also connections with the Monster group.

Other applications and references for these topics will be found in Chap. 1.

Who should buy this book. Anyone interested in sphere packings, in lattices in n-dimensional space, or in the Leech lattice. Mathematicians interested in finite groups, quadratic forms, the geometry of numbers, or combinatorics. Engineers who wish to construct n-dimensional codes for a band-limited channel, or to design n-dimensional vector quantizers. Chemists and physicists interested in n-dimensional crystallography.

Prerequisites. In the early chapters we have tried to address (perhaps not always successfully) any well-educated undergraduate. The technical level fluctuates from chapter to chapter, and generally increases towards the end of the book.

What's new here. A number of things in this book have not appeared in print before. We mention just a few.
— Tables of the densest sphere packings known in dimensions up to one million (Tables 1.2, 1.3).
— Tables and graphs of the best coverings and quantizers known in up to 24 dimensions (Chap. 2).
— A table of the best coding gain of lattices in up to 128 dimensions (Table 3.2).
— Explicit arrangements of n-dimensional spheres with kissing numbers that grow like $2^{0.003n}$ (Eq. (56) of Chap. 1), the best currently known.
— Explicit constructions of good spherical codes (§2.5 of Chap. 1).
— Explicit constructions of n-dimensional lattice coverings with density that grows like $2^{0.084n}$ (Eq. (13) of Chap. 2), again the best known.
— New formulae for the theta series of many lattices (e.g. for the Leech lattice with respect to an octahedral deep hole, Eq. (144) of Chap. 4).
— Tables giving the numbers of points in successive shells of these lattices (e.g. the first 50 shells of the Leech lattice, Table 4.14).
— A new general construction for sphere packings (Construction A_f, §3 of Chap. 8).
— New descriptions of the lattice packings recently discovered by McKay, Quebbemann, Craig, etc. (Chap. 8).
— Borcherds' classification of the 24-dimensional odd unimodular lattices and his neighborhood graph of the 24-dimensional even unimodular lattices (Chap. 24).
— A list of the 284 types of shallow hole in the Leech lattice (Chap. 25).
— Some easy ways of computing with rational and integral quadradic forms, including elementary systems of invariants for such forms, and a handy notation for the genus of a form (Chap. 15).
— Tables of binary quadratic forms with $-100 \leqslant \det \leqslant 50$, indecomposable ternary forms with $|\det| \leqslant 50$, genera of forms with $|\det| \leqslant 11$, genera of p-elementary forms for all p, positive definite forms

with determinant 2 up to dimension 18 and determinant 3 up to dimension 17 (Chap. 15).
– A simple description of a construction for the Monster simple group (Chap. 29).

Other tables, which up to now could only be found in journal articles or conference proceedings, include:
– Bounds for kissing numbers in dimensions up to 24 (Table 1.5).
– The Minkowski-Siegel mass constants for even and odd unimodular lattices in dimensions up to 32 (Chap. 16).
– The even and odd unimodular lattices in dimensions up to 24 (Table 2.2 and Chaps. 16, 17).
– Vectors in the first eight shells of the E_8 lattice (Table 4.10) and the first three shells of the Leech lattice (Table 4.13).
– Best's codes of length 10 and 11, that produce the densest packing known (P_{10c}) in 10 dimensions and the highest kissing number known (P_{11c}) in 11 dimensions (Chap. 5).
– Improved tables giving the best known codes of length 2^m for $m \leqslant 8$ (Table 5.4) and of all lengths up to 24 (Table 9.1).
– Laminated lattices in dimensions up to 48 (Tables 6.1, 6.3).
– The best integral lattices of minimal norms 2, 3 and 4, in dimensions up to 24 (Table 6.4).
– A description of E_8 lattice vectors in terms of icosians (Table 8.1).
– Minimal vectors in McKay's 40-dimensional lattice M_{40} (Table 8.6).
– The classification of subsets of 24 objects under the action of the Mathieu group M_{24} (Fig. 10.1).
– Groups associated with the Leech lattice (Table 10.4).
– Simple groups that arise from centralizers in the Monster (Chap. 10).
– Second moments of polyhedra in 3 and 4 dimensions (Chap. 21).
– The deep holes in the Leech lattice (Table 23.1).
– An extensive table of Leech roots, in both hyperbolic and Euclidean coordinates (Chap. 28).
– Coxeter-Vinberg diagrams for the automorphism groups of the lattices $\mathrm{I}_{n,1}$ for $n \leqslant 20$ (Chap. 28) and $\mathrm{II}_{n,1}$ for $n \leqslant 24$ (Chap. 27).

The contents of the chapters. Chapters 1-3 form an extended introduction to the whole book. In these chapters we survey what is presently known about the packing, kissing number, covering and quantizing problems. There are sections on quadratic forms and their classification, the connections with number theory, the channel coding problem, spherical codes, error-correcting codes, Steiner systems, t-designs, and the connections with group theory. These chapters also introduce definitions and terminology that will be used throughout the book.

Chapter 4 describes a number of important lattices, including the cubic lattice $\mathbf{Z}^n$, the root lattices A_n, D_n, E_6, E_7, E_8, the Coxeter-Todd lattice K_{12}, the Barnes-Wall lattice Λ_{16}, the Leech lattice Λ_{24}, and their duals. Among other things we give their minimal vectors, densities, covering radii, glue vectors, automorphism groups, expressions for their theta series, and tables of the numbers of points in the first fifty shells. We also include a brief discussion of reflection groups and of the technique of gluing lattices together.

Chapters 5-8 are devoted to techniques for constructing sphere packings. Many of the constructions in Chaps. 5 and 7 are based on error-correcting codes; other constructions in Chapter 5 build up packings by layers. Layered packings are studied in greater detail in Chap. 6, where the formal concept of a *laminated lattice* Λ_n is introduced. Chapter 8 uses a number of more sophisticated algebraic techniques to construct lattices.

Chapter 9 introduces analytical methods for finding bounds on the best codes, sphere packings and related problems. The methods use techniques from harmonic analysis and linear programming. We give a simplified account of Kabatiansky and Levenshtein's recent sphere packing bounds.

Chapters 10 and 11 study the Golay codes of length 12 and 24, the associated Steiner systems $S(5,6,12)$ and $S(5,8,24)$, and their automorphism groups M_{12} and M_{24}. The MINIMOG and MOG (or Miracle Octad Generator) and the Tetracode and Hexacode are computational tools that make it easy to perform calculations with these objects. These two chapters also study a number of related groups, in particular the automorphism group Co_0 (or $\cdot 0$) of the Leech lattice. The Appendix to Chapter 10 describes all the sporadic simple groups.

Chapter 12 gives a short proof that the Leech lattice is the unique even unimodular lattice with no vectors of norm 2. Chapter 13 solves the kissing number problem in 8 and 24 dimensions — the E_8 and Leech lattices have the highest possible kissing numbers in these dimensions. Chapter 14 shows that these arrangements of spheres are essentially unique.

Chapters 15-19 deal with the classification of integral quadratic forms. Chapters 16 and 18 together give three proofs that Niemeier's enumeration of the 24-dimensional even unimodular lattices is correct. In Chap. 19 we find all the extremal odd unimodular lattices in any dimension.

Chapters 20 and 21 are concerned with geometric properties of lattices. In Chap. 20 we discuss algorithms which, given an arbitrary point of the space, find the closest lattice point. These algorithms can be used for vector quantizing or for encoding and decoding lattice codes for a bandlimited channel. Chapter 21 studies the Voronoi cells of lattices and their second moments.

Soon after discovering his lattice, John Leech conjectured that its covering radius was equal to $\sqrt{2}$ times its packing radius, but was unable to find a proof. In 1980 Simon Norton found an ingenious argument which shows that the covering radius is no more than 1.452... times the packing radius (Chap. 22), and shortly afterwards Richard Parker and the authors managed to prove Leech's conjecture (Chap. 23).

Our method of proof involves finding all the "deep holes" in the Leech lattice, i.e. all points of 24-dimensional space that are maximally distant from the lattice. We were astonished to discover that there are precisely 23 distinct types of deep hole, and that they are in one-to-one correspondence with the Niemeier lattices (the 24-dimensional even unimodular lattices of minimal norm 2) — see Theorem 2 of Chap. 23. Chapter 23, or the Deep Holes paper, as it is usually called, has turned out to be extremely fruitful, having stimulated the remaining chapters in the book, also Chap. 6, and several journal articles.

In Chap. 24 we give 23 constructions for the Leech lattice, one for each of the deep holes or Niemeier lattices. Two of these are the familiar constructions based on the Golay codes. In the second half of Chap. 24 we introduce the *hole diagram* of a deep hole, which describes the environs of the hole. Chapter 25 (the Shallow Holes paper) uses the results of Chap. 23 and 24 to classify *all* the holes in the Leech lattice.

Considerable light is thrown on these mysteries by the realization that the Leech lattice and the Niemeier lattices can all be obtained very easily from a single lattice, namely $II_{25,1}$, the unique even unimodular lattice in Lorentzian space $\mathbf{R}^{25,1}$. For any vector $w \in \mathbf{R}^{25,1}$, let

$$w^{\perp} = \{x \in II_{25,1} : x \cdot w = 0\} .$$

Then if w is the special vector

$$w_{25} = (0,1,2,3,\ldots,23,24 \,|\, 70) ,$$

$w^{\perp}/w$ is the Leech lattice, and other choices for w lead to the 23 Niemeier lattices.

The properties of the Leech lattice are closely related to the geometry of the lattice $II_{25,1}$. The automorphism groups of the Lorentzian lattice $I_{n,1}$ for $n \leqslant 19$ and $II_{n,1}$ for $n = 1$, 9 and 17 were found by Vinberg, Kaplinskaja and Meyer. Chapter 27 finds the automorphism group of $II_{25,1}$. This remarkable group has a reflection subgroup with a Coxeter diagram that is, speaking loosely, isomorphic to the Leech lattice. More precisely, a set of fundamental roots for $II_{25,1}$ consists of the vectors $r \in II_{25,1}$ satisfying

$$r \cdot r = 2 , \quad r \cdot w_{25} = -1 ,$$

and we call these the *Leech roots*. Chapter 26 shows that there is an isometry between the set of Leech roots and the points of the Leech lattice. Then the Coxeter part of the automorphism group of $II_{25,1}$ is just the Coxeter group generated by the Leech roots (Theorem 1 of Chap. 27).

Since $II_{25,1}$ is a natural quadratic form to study, whose definition certainly does not mention the Leech lattice, it is surprising that the Leech lattice essentially determines the automorphism group of the form.

The Leech roots also provide a better understanding of the automorphism groups of the other lattices $I_{n,1}$ and $II_{n,1}$, as we see in Chap. 28. This chapter also contains an extensive table of Leech roots. Chapter 29 describes a construction for the Monster simple group, and the final chapter describes an infinite-dimensional Lie algebra that is obtained from the Leech roots, and conjectures that it may be related to the Monster.

The book concludes with a bibliography of about 1550 items.

The structure of this book. Our original plan was simply for a collection of reprints. But over the past two years the book has been completely transformed: many new chapters have been added, and the original chapters have been extensively rewritten to bring them up to date, to reduce duplicated material, to adopt a uniform notation and terminology, and to eliminate errors.

We have however allowed a certain amount of duplication to remain, to make for easier reading. Because some chapters were written at different times and by different authors, the reader will occasionally notice differences in style from chapter to chapter.

The arrangement of the chapters presented us with a difficult problem. We feel that readers are best served by grouping them in the present arrangement, even though it means that one or two chapters are not in strict logical order. The worst flaw is that the higher-dimensional part of the laminated lattices chapter (Chap. 6) depends on knowledge of the deep holes in the Leech lattice given in Chap. 23. But the preceding part of Chapter 6, which includes all the best lattice packings known in small dimensions, had to appear as early as possible.

Chapters 1-4, 7, 8, 11, 15, 17, 20, 25 and 29 are new. Chapter 5 is based on J. Leech and N.J.A.S., *Canad. J. Math.* **23** (1971) (see reference [Lee10] in the bibliography for full details); Chap. 6 on J.H.C. and N.J.A.S., *Annals of Math.* **116** (1982) [Con32]; Chap. 9 is based on N.J.A.S., *Contemp. Math.* **9** (1982), published by the American Mathematical Society [Slo13]; Chap. 10 on J.H.C., in *Finite Simple Groups*, edited by M. B. Powell and G. Higman, Academic Press, N.Y. 1971 [Con5]; Chap. 12 on J.H.C., *Invent. Math.* **7** (1969), published by Springer-Verlag, N.Y. [Con4]; Chap. 13 on A. M. Odlyzko and N.J.A.S., *J. Combin. Theory* **A26** (1979), published by Academic Press, N.Y. [Odl5]; Chap. 14 on E. Bannai and N.J.A.S., *Canad. J. Math.* **33** (1981) [Ban13]; Chap. 16 on J.H.C. and N.J.A.S., *J. Number Theory* **15** (1982) [Con27], and *Europ. J. Combin.* **3** (1982) [Con34], both published by Academic Press, N.Y.; Chapter 18 on B. B. Venkov, *Trudy Mat. Inst. Steklov* **148** (1978), English translation in *Proc. Steklov Inst. Math.*, published by the American Mathematical Society [Ven1]; Chap. 19 on J.H.C., A. M. Odlyzko and N.J.A.S., *Mathematika*, **25** (1978) [Con19]; Chap. 21 on J.H.C. and N.J.A.S., *IEEE Trans. Information Theory*, **28** (1982) [Con28]; Chap. 22 on S. P. Norton, *Proc. Royal Society London*, **A380** (1982) [Nor4]; Chap. 23 on J.H.C., R. A. Parker and N.J.A.S., *Proc. Royal Society London*, **A380** (1982) [Con20]; Chap. 24 on J.H.C. and N.J.A.S., *Proc. Royal Society London*, **A381** (1982) [Con30]; Chap. 26 on J.H.C. and N.J.A.S., *Bulletin American Mathematical Society*, **6** (1982) [Con31]; Chap. 27 on J.H.C., *J. Algebra*, **80** (1983), published by Academic Press, N.Y. [Con13]; Chap. 28 on J.H.C. and N.J.A.S., *Proc. Royal Society London*, **A384** (1982) [Con33]; Chap. 30 on R. E. Borcherds, J.H.C., L. Queen and N.J.A.S., *Advances in Math.* **53** (1984), published by Academic Press, N.Y. [Bor5].

Our collaborators mentioned above are:

Eiichi Bannai, Math. Dept., Ohio State University, Columbus, Ohio 43210;

Richard E. Borcherds, Dept. of Pure Math. and Math. Statistics, Cambridge University, Cambridge CB2 1SB, England;

John Leech, Computing Science Dept., University of Stirling, Stirling FK9 4LA, Scotland;

Simon P. Norton, Dept. of Pure Math. and Math. Statistics, Cambridge University, Cambridge CB2 1SB, England;

Andrew M. Odlyzko, Math. Sciences Research Center, AT&T Bell Laboratories, Murray Hill, New Jersey 07974;

Richard A. Parker, Dept. of Pure Math. and Math. Statistics, Cambridge University, Cambridge CB2 1SB, England;

Larissa Queen, Dept. of Pure Math. and Math. Statistics, Cambridge University, Cambridge CB2 1SB, England;

B. B. Venkov, Leningrad Division of the Math. Institute of the USSR Academy of Sciences, Leningrad, USSR.

Acknowledgements. We thank all our collaborators and the publishers of these articles for allowing us to make use of this material.

We should like to express our thanks to E. S. Barnes, H. S. M. Coxeter, Susanna Cuyler, G. D. Forney, Jr., W. M. Kantor, J. J. Seidel, J.-P. Serre, P. N. de Souza, and above all John Leech, for their comments on the manuscript. Further acknowledgements appear at the end of the individual chapters. Any errors that remain are our own responsibility: please notify N. J. A. Sloane, Information Sciences Research, AT&T Labs - Research, 180 Park Avenue, Florham Park, NJ 07932-0971, USA (email: njas@research.att.com). We would also like to hear of any improvements to the tables.

We thank Ann Marie McGowan, Gisele Wallace and Cynthia Martin, who typed the original versions of many of the chapters, and especially Mary Flannelly and Susan Tarczynski, who produced the final manuscript. B. L. English and R. A. Matula of the Bell Laboratories library staff helped locate obscure references.

N.J.A.S. thanks Bell Laboratories (and especially R. L. Graham and A. M. Odlyzko) for support and encouragement during this work, and J.H.C. thanks Bell Laboratories for support and hospitality during various visits to Murray Hill.

We remark that in two dimensions the familiar hexagonal lattice

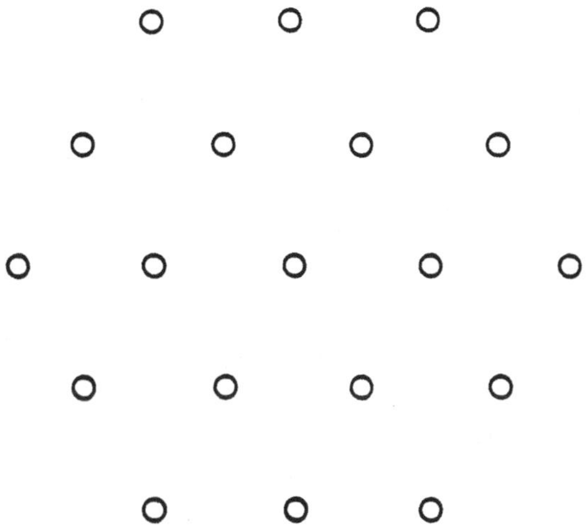

solves the packing, kissing, covering and quantizing problems. In a sense this whole book is simply a search for similar nice patterns in higher dimensions.

Preface to Third Edition

Interest in the subject matter of the book continues to grow. The Supplementary Bibliography has been enlarged to cover the period 1988 to 1998 and now contains over 800 items. Other changes from the second edition include a handful of small corrections and improvements to the main text, and this preface (an expanded version of the preface to the Second Edition) which contains a brief report on some of the developments since the appearance of the first edition.

We are grateful to a number of correspondents who have supplied corrections and comments on the first two editions, or who have sent us copies of manuscripts.[1] We thank in particular R. Bacher, R. E. Borcherds, P. Boyvalenkov, H. S. M. Coxeter, Y. Edel, N. D. Elkies, L. J. Gerstein, M. Harada, J. Leech, J. H. Lindsey, II, J. Martinet, J. McKay, G. Nebe, E. Pervin, E. M. Rains, R. Scharlau, F. Sigrist, H. M. Switkay, T. Urabe, A. Vardy, Z.-X. Wan and J. Wills. The new material was expertly typed by Susan K. Pope.

We are planning a sequel, tentatively entitled *The Geometry of Low-Dimensional Groups and Lattices*, which will include two earlier papers [Con36] and [Con37] not included in this book, as well as several recent papers dealing with groups and lattices in low dimensions ([CSLDL1]–[CSLDL8], [CoSl91a], [CoSl95a], etc.).

A Russian version of the first edition, translated by S. N. Litsyn, M. A. Tsfasman and G. B. Shabat, was published by Mir (Moscow) in 1990.

Recent developments, comments, and additional corrections. The following pages attempt to describe recent developments in some of the topics treated in the book. The arrangement roughly follows that of the chapters. Our

[1]We also thank the correspondent who reported hearing the first edition described during a talk as "the bible of the subject, and, like the bible, [it] contains no proofs". This is of course only half true.

coverage is necessarily highly selective, and we apologize if we have failed to mention some important results. In a few places we have also included additional comments on or corrections to the text.

Three books dealing with lattices have recently appeared: in order of publication, these are Ebeling, *Lattices and Codes* [Ebe94], Martinet, *Les Réseaux Parfaits des Espaces Euclidiens* [Mar96] and Conway and Fung, *The Sensual (Quadratic) Form* [CoFu97].

An extensive survey by Lagarias [Laga96] discusses lattices from many points of view not dealt with in our book, as does the Erdős-Gruber-Hammer [ErGH89] collection of **unsolved problems concerning lattices**. The encyclopedic work on **distance-regular graphs** by Brouwer, Cohen and Neumaier [BrCN89] discusses many mathematical structures that are related to topics in our book (see also Tonchev [Ton88]). See also the works by Aschbacher [Asch94] on **sporadic groups**, Engel [Eng86], [Eng93] on **geometric crystallography** [Eng93], Fejes Tóth and Kuperberg [FeK93], and Fejes Tóth [Fej97] on **packing and covering**, Gritzmann and Wills [GriW93b] on **finite packings and coverings**, and Pach and Agarwal [PacA95] on **combinatorial geometry**.

An electronic **data-base of lattices** is now available [NeSl]. This contains information about some 160,000 lattices in dimensions up to 64. The theta-series of the most important lattices can be found in [SloEIS]. The computer languages PARI [BatB91], KANT [Schm90], [Schm91] and especially MAGMA [BosC97], [BosCM94], [BosCP97] have extensive facilities for performing lattice calculations (among many other things).

Notes on Chapter 1: Sphere Packings and Kissing Numbers

Hales [Hal92], [Hal97], [Hal97a], [Hal97b] (see also Ferguson [Ferg97] and Ferguson and Hales [FeHa97]) has described a series of steps that may well succeed in proving the long-standing conjecture (the so-called "**Kepler conjecture**") that no packing of three-dimensional spheres can have a greater density than that of the face-centered cubic lattice. In fact, on August 9, 1998, just as this book was going to press, Hales announced [Hal98] that the final step in the proof has been completed: the Kepler conjecture is now a theorem.

The previous best upper bound known on the density of a three-dimensional packing was due to Muder [Mude93], who showed that the density cannot exceed $0.773055\ldots$ (compared with $\pi/\sqrt{18} = 0.74048\ldots$ for the f.c.c. lattice).

A paper by W.-Y. Hsiang [Hsi93] (see also [Hsi93a], [Hsi93b]) claiming to prove the Kepler conjecture contains serious flaws. G. Fejes Tóth, reviewing the paper for *Math. Reviews* [Fej95], states: "If I am asked whether the paper fulfills what it promises in its title, namely a proof of Kepler's conjecture, my answer is: no. I hope that Hsiang will fill in the details, but I feel that the greater part of the work has yet to be done." Hsiang [Hsi93b] also claims to have a proof that no more than 24 spheres can touch an equal sphere in four dimensions. For further discussion see [CoHMS], [Hal94], [Hsi95].

K. Bezdek [Bez97] has made some partial progress towards solving the **dodecahedral conjecture**. This conjecture, weaker than the Kepler conjecture, states that the volume of any Voronoi cell in a packing of unit spheres in $\mathbb{R}^3$ is at least as large as the volume of a regular dodecahedron of inradius 1. See also Muder [Mude93].

A. Bezdek, W. Kuperberg and Makai [BezKM91] had established the Kepler conjecture for packings composed of parallel strings of spheres. See also Knill [Knill96].

There was no reason to doubt the truth of the Kepler conjecture. However, A. Bezdek and W. Kuperberg [BezKu91] show that there are packings of congruent ellipsoids with density $0.7533\ldots$, exceeding $\pi/\sqrt{18}$, and in [Wills91] this is improved to $0.7585\ldots$.

Using spheres of two radii $0 < r_1 < r_2$, one obviously obtains packings in 3-space with density $> 0.74048\ldots$, provided r_1/r_2 is sufficiently small. In [VaWi94] it is shown that that this is so even when $r_1/r_2 = 0.623\ldots$.

There are infinitely many three-dimensional nonlattice packings (the Barlow packings, see below) with the same density as the f.c.c. lattice packing. In [Schn98] it is shown that large finite subsets of the f.c.c. lattice are denser (in the sense of parametric density, see below) than subsets of any other Barlow packing.

What are all the best sphere packings in low dimensions?

In [CoSl95a] we describe what may be *all* the best packings of nonoverlapping equal spheres in dimensions $n \leq 10$, where "best" means both having the highest density and not permitting any local improvement. For example, it appears that the best five-dimensional sphere packings are parameterized by the 4-colorings of the one-dimensional integer lattice. We also find what we believe to be the exact numbers of "uniform" packings among these, that is, those in which the automorphism group acts transitively. These assertions depend on certain plausible but as yet unproved postulates.

There are some surprises. We show that the Korkine-Zolotarev lattice Λ_9 (which continues to hold the density record it established in 1873) has the following astonishing property. Half the spheres can be moved bodily through arbitrarily large distances without overlapping the other half, only touching them at isolated instants, and yet the density of the packing remains the same at all times. A typical packing in this family consists of the points of

$$D_9^{\theta+} = D_9 \cup D_9 + \left(\left(\frac{1}{2}\right)^8, \frac{1}{2}\theta\right),$$

for any real number θ. We call this a "fluid diamond packing," since $D_9^{0+} = \Lambda_9$ and $D_9^{1+} = D_9^+$ (cf. Sect. 7.3 of Chap. 4). All these packings have the same density, the highest known in 9 dimensions. Agrell and Eriksson [AgEr98] show D_9^+ is a better 9-dimensional quantizer than any previously known. In [CoSl95a] we also discuss some new higher-dimensional packings, showing for example that there are extraordinarily many 16-dimensional packings that are just as dense as the Barnes-Wall lattice Λ_{16}.

Mordell-Weil Lattices

One of the most exciting developments has been Elkies' ([Elki], [Elki94], [Elki97]) and Shioda's [Shiod91d] construction of lattice packings from the Mordell-Weil groups of elliptic curves over function fields. These lattices have a greater density than any previously known in dimensions from about 80 to 4096, and provide the following new entries for Table 1.3 of Chap. 1:

n	54	64	80	104	128
$\log_2 \delta \geq$	15.88	24.71	40.14	67.01	97.40
reference	[Elki]	[Elki]	[Shi7]	[Shi7]	[Elki]

n	256	512	1024	2048	4096
$\log_2 \delta \geq$	294.80	797.12	2018.2	4891	11527
reference	[Elki]	[Elki]	[Elki]	[Elki]	[Elki]

In this Introduction we will use MW_n to denote an n-dimensional Mordell-Weil lattice. For further information about this construction see Shioda [Shiod88]–[Shiod91e], Oguiso and Shioda [OgS91], Dummigan [Dum94]–[Dum96], Gow [Gow89], [Gow89a], Gross [Gro90], [Gro96], Oesterlé [Oes90], Tiep [Tiep91]–[Tiep97b].

Several other new record packings will be mentioned later in this Introduction. Because of this, it seems worthwhile to include two new tables. (The latest versions of these two tables are also available electronically [NeSl].)

A new table of densest packings

Table I.1 gives the densest (lattice or nonlattice) packings and Table I.2 the highest kissing numbers presently known in dimensions up to 128. These tables update and extend Table 1.2 and part of Table 1.3 of Chapter 1, to which the reader is referred for more information about most of these packings. Others are described later in this Introduction. There are several instances in Table I.1 where the highest known density is achieved by a nonlattice packing: these entries are enclosed in parentheses.

At dimension 32, Q_{32} denotes the Quebbemann lattice constructed on page 220 (although the Mordell-Weil lattice MW_{32} or Bachoc's lattice $\mathcal{B}_{32}$ [Baco95], [Baco97] have the same density). The lattices $Q_{33}, \ldots, Q_{40}$ were

Table I.1(a) Densest packings presently known in dimensions $n \leq 128$. The table gives the center density δ, defined on page 13.

n	δ (lattice)	δ (nonlattice)	Lattice (nonlattice)
0	0		Λ_0
1	$1/2 = 0.50000$		$\Lambda_1 \cong A_1 \cong \mathbb{Z}$
2	$1/2\sqrt{3} = 0.28868$		$\Lambda_2 \cong A_2$
3	$1/4\sqrt{2} = 0.17678$		$\Lambda_3 \cong A_3 \cong D_3$
4	$1/8 = 0.12500$		$\Lambda_4 \cong D_4$
5	$1/8\sqrt{2} = 0.08839$		$\Lambda_5 \cong D_5$
6	$1/8\sqrt{3} = 0.07217$		$\Lambda_6 \cong E_6$
7	$1/16 = 0.06250$		$\Lambda_7 \cong E_7$
8	$1/16 = 0.06250$		$\Lambda_8 \cong E_8$
9	$1/16\sqrt{2} = 0.04419$		Λ_9
10	$1/16\sqrt{3} = 0.03608$	$(5/128 = 0.03906)^*$	Λ_{10} $(P_{10c})^*$
11	$1/18\sqrt{3} = 0.03208$	$(9/256 = 0.03516)^*$	K_{11} $(P_{11a})^*$
12	$1/27 = 0.03704$		K_{12}
13	$1/18\sqrt{3} = 0.03208$	$(9/256 = 0.03516)^*$	K_{13} $(P_{13a})^*$
14	$1/16\sqrt{3} = 0.03608$		Λ_{14}
15	$1/16\sqrt{2} = 0.04419$		Λ_{15}
16	$1/16 = 0.06250$		Λ_{16}
17	$1/16 = 0.06250$		Λ_{17}
18	$1/8\sqrt{3} = 0.07217$	$(3^9/4^9 = 0.07508)^*$	Λ_{18} $(\mathcal{B}^*_{18}$ [BiEd98]$)^*$
19	$1/8\sqrt{2} = 0.08839$		Λ_{19}
20	$1/8 = 0.12500$	$(7^{10}/2^{31} = 0.13154)^*$	Λ_{20} $(\mathcal{B}^*_{20}$ [Vard95]$)^*$
21	$1/4\sqrt{2} = 0.17678$		Λ_{21}
22	$1/2\sqrt{3} = 0.28868$	$(0.33254)^*$	Λ_{22} $(\mathcal{A}^*_{22}$ [CoSl96]$)^*$
23	$1/2 = 0.50000$		Λ_{23}
24	1		Λ_{24}
25	$1/\sqrt{2} = 0.70711$		Λ_{25}
26	$1/\sqrt{3} = 0.57735$		Λ_{26}, T_{26} (see Notes on Chap. 18)

*Nonlattice packing.

Table I.1(b) Densest packings presently known in dimensions $n \leq 128$. The table gives the center density δ, defined on page 13.

n	δ (lattice)	δ (nonlattice)	Lattice (nonlattice)
27	$1/\sqrt{3} = 0.57735$	$(1/\sqrt{2} = 0.70711)^*$	$\mathcal{B}_{27}$ ($\mathcal{B}^*_{27}$ [Vard98])*
28	$2/3 = 0.66667$	$(1)^*$	$\mathcal{B}_{28}$ ($\mathcal{B}^*_{28}$ [Vard98])*
29	$1/\sqrt{3} = 0.57735$	$(1/\sqrt{2} = 0.70711)^*$	$\mathcal{B}_{29}$ ($\mathcal{B}^*_{29}$ [Vard98])*
30	$3^{13.5}/2^{22} = 0.65838$	$(1)^*$	$\mathcal{Q}_{30}$ ($\mathcal{B}^*_{30}$ [Vard98])*
31	$3^{15}/2^{23.5} = 1.20952$		$\mathcal{Q}_{31}$
32	$3^{16}/2^{24} = 2.56578$		$\mathcal{Q}_{32}$ and others
33	$3^{16.5}/2^{25} = 2.22203$		$\mathcal{Q}_{33}$ [Elki94], [Elki]
34	$3^{16.5}/2^{25} = 2.22203$		$\mathcal{Q}_{34}$ [Elki94], [Elki]
35	$2\sqrt{2} = 2.82843$		B_{35} (p. 234)
36	$2^{18}/3^{10} = 4.43943$		Ks_{36} [KsP92]
37	$4/\sqrt{2} = 5.65685$		$\mathcal{D}_{37}$
38	8		$\mathcal{D}_{38}$
39	$3^{16}/2^{20}\sqrt{14} = 10.9718$		From P_{48p}, see p. 167
40	$3^{17}/2^{22.5} = 21.7714$		From P_{48p}, see p. 167
41	$3^{17}/2^{21.5} = 43.5428$		From P_{48p}
42	$3^{18}/2^{22} = 92.3682$		From P_{48p}
43	$3^{19}/2^{22.5} = 195.943$		From P_{48p}
44	$3^{20}/2^{23} = 415.657$	$(17^{22}/2^{43}3^{24} = 472.799)^*$	From P_{48p} ($\mathcal{A}_{44}$ [CoSl96])*
45	$3^{21}/2^{23.5} = 881.742$	$(17^{22.5}/2^{44}3^{24} = 974.700)^*$	From P_{48p} ($\mathcal{A}_{45}$ [CoSl96])*
46	$3^{21.5}/2^{23} = 2159.82$	$(13^{23}/3^{46.5} = 2719.94)^*$	From P_{48p} ($\mathcal{A}_{46}$ [CoSl96])*
47	$3^{23}/2^{24} = 5611.37$	$(35^{23.5}/2^{70}3^{24} = 5788.81)^*$	From P_{48p} ($\mathcal{A}_{47}$ [CoSl96])*
48	$3^{24}/2^{24} = 16834.1$		P_{48n}, P_{48p}, P_{48q}
54	$2^{15.88}$		MW_{54}
56	$1.5^{28} = 2^{16.38}$		$L_{56,2}(M)$, $\tilde{L}_{56,2}(M)$ [Nebe98]
64	$3^{16} = 2^{25.36}$		Ne_{64} [Nebe98], [Nebe98b]
80	$2^{40.14}$		MW_{80}
128	$2^{97.40}$		MW_{128}

*Nonlattice packing.

constructed by Elkies [Elki] by laminating $\mathcal{Q}_{32}$ (or MW_{32} or $\mathcal{B}_{32}$) above certain half-lattice points. Each putative deep hole of norm 4 (two-thirds the minimal norm of the lattice) is surrounded by 576 lattice points. For $n \leq 8$ we obtain a lattice $\mathcal{Q}_{32+n}$ with center density $\delta = 2^{-24}3^{16+0.5n}/\sqrt{\lambda_n}$ and kissing number $261120 + 576\tau_n$, where λ_n and τ_n are respectively the determinant and kissing number of Λ_n (cf. Tables 6.1, 6.3). At the present

Table I.2(a) Highest kissing numbers τ presently known for packings in dimensions $n \le 128$.

n	τ (lattice)	τ (nonlattice)	Lattice (nonlattice)
0	0		Λ_0
1	2		$\Lambda_1 \cong A_1 \cong \mathbb{Z}$
2	6		$\Lambda_2 \cong A_2$
3	12		$\Lambda_3 \cong A_3 \cong D_3$
4	24		$\Lambda_4 \cong D_4$
5	40		$\Lambda_5 \cong D_5$
6	72		$\Lambda_6 \cong E_6$
7	126		$\Lambda_7 \cong E_7$
8	240		$\Lambda_8 \cong E_8$
9	272	(306)*	Λ_9 $(P_{9a})^*$
10	336	(500)*	Λ_{10} $(P_{10b})^*$
11	438	(582)*	$\Lambda_{11}^{\max}$ $(P_{11c})^*$
12	756	(840)*	K_{12} $(P_{12a})^*$
13	918	(1130)*	K_{13} $(P_{13a})^*$
14	1422	(1582)*	Λ_{14} $(P_{14b})^*$
15	2340		Λ_{15}
16	4320		Λ_{16}
17	5346		Λ_{17}
18	7398		Λ_{18}
19	10668		Λ_{19}
20	17400		Λ_{20}
21	27720		Λ_{21}
22	49896		Λ_{22}

* The kissing number in a nonlattice packing may vary from sphere to sphere – we give the largest value (see Table 1.2)

[CoSl95] describes an especially interesting imperfect 11-dimensional lattice, which we call **anabasic**: it is generated by its minimal vectors, but no set of 11 minimal vectors forms a basis.

Several important books have appeared that deal with the construction of very good codes and very dense lattices from algebraic curves and algebraic function fields (cf. §1.5 of Chap. 1 and §2.11 of Chap. 3): Goppa [Gop88], Tsfasman and Vladuts [TsV91] and Stichtenoth [Stich93], Lachaud et al. [Lac95]. Garcia and Stichtenoth [GaSt95] have given a fairly explicit construction for an infinite sequence of good codes over a fixed field $GF(q^2)$. Elkies [Elki97a] has worked on finding

Table I.2(b) Highest kissing numbers τ presently known for packings in dimensions $n \le 128$.

n	τ (lattice)	τ (nonlattice)	Lattice (nonlattice)
23	93150		Λ_{23}
24	196560		Λ_{24}
25	196656		Λ_{25}
26	196848		Λ_{26}
27	197142		Λ_{27}
28	197736		Λ_{28}
29	198506		Λ_{29}
30	200046		Λ_{30}
31	202692		Λ_{31}
32	261120	(276032)*	Q_{32} and others ([EdRS98])
33	262272	(294592)*	Q_{33} ([EdRS98])
34	264576	(318020)*	Q_{34} ([EdRS98])
35	268032	(370892)*	Q_{35} ([EdRS98])
36	274944	(438872)*	Q_{36} ([EdRS98])
37	284160	(439016)*	Q_{37} ([EdRS98])
38	302592	(566652)*	Q_{38} ([EdRS98])
39	333696	(714184)*	Q_{39} ([EdRS98])
40	399360	(991792)*	Q_{40} ([EdRS98])
44	2708112	(2948552)*	MW_{44} ([EdRS98])
48	52416000		P_{48n}, P_{48p}, P_{48q}
64	138458880	(331737984)*	Ne_{64} [Nebe98b] ([EdRS98])
80	1250172000	(1368532064)*	L_{80} [BacoN98] ([EdRS98])
128	218044170240	(8863556495104)*	MW_{128} [Elki] ([EdRS98])

* The kissing number in a nonlattice packing may vary from sphere to sphere – we give the largest value (see Table 1.2)

explicit equations for modular curves of various kinds that attain the Drinfeld-Vladuts bound. See also Manin and Vladuts [MaV85], Stichtenoth and Tsfasman [StTs92], Tsfasman [Tsf91], [Tsf91a]. The problem of decoding codes constructed from algebraic geometry is considered by Feng and Rao [FeRa94], Justesen et al. [JuL89], [JuL92], Pellikaan [Pel89], Skorobogatov and Vladuts [SkV90], Sudan [Suda96], [Suda97], Vladuts [Vlad90] (see also Lachaud et al. [Lac95]).

Quebbemann [Queb88] uses class field towers and Alon et al. [AlBN92], Sipser and Spielman [SipS96] and Spielman [Spiel96] use expander graphs to construct asymptotically good codes. The codes in [SipS96] and [Spiel96] can be decoded in linear time.

Several important papers have appeared dealing with the construction of dense lattices in high-dimensional space using algebraic number fields and global fields (cf. §1.5 of Chap. 1 and §7.4 of Chap. 8) — see Quebbemann [Queb89]–[Queb91a] and Rosenbloom and Tsfasman [RoT90], as well as Tsfasman and Vladuts [TsV91].

The results in [Rus1] (see p. 19 of Chap. 1) have been generalized and extended in [ElkOR91], [Rus89]–[Rus92].

Yudin [Yud91] gives an upper bound for the number of disjoint spheres of radius r in the n-dimensional torus, and deduces from this a new proof of Levenshtein's bound (Eq. (42) of Chap. 1).

When discussing **Hermite's constant** γ_n on page 20 of Chap. 1, we should have mentioned that it can also be defined as the minimal norm of an n-dimensional lattice packing of maximal density, when that lattice is scaled so that its determinant is 1. See also Bergé and Martinet [BerM85].

Let $\mu(\Lambda)$ denote the minimal nonzero norm of a vector in a lattice Λ. Bergé and Martinet [BerM89] call a lattice **dual-critical** if the value of $\mu(\Lambda)\mu(\Lambda^*)$ is maximized (where Λ^* is the dual lattice). They prove that A_1, A_2, A_3, A_3^*, D_4, A_4, A_4^* are the only lattices in dimensions $n \leq 4$ on which the product $\mu(\Lambda)\mu(\Lambda^*)$ attains a local maximum. That A_1, A_2, D_4 (and E_8) are the only dual-critical lattices in dimensions 1, 2, 4 and 8 respectively follows because these are the densest lattices, and they are unique and isodual. See also [CoSl94], [Mar96], [Mar97].

Fields [Fie2], [Fiel80] and Fields and Nicolich [FiNi80] have found the **least dense lattice packings** in dimensions $n \leq 4$ that minimize the density under local variations that preserve the minimal vectors.

For recent work on **quasilattices** and **quasicrystals** (cf. §2.1 of Chap. 1), see for example [BaaJK90], [Bru86], [Hen86], [MRW87], [OlK89], [RHM89], [RMW87], [RWM88], [Wills90a], [Wills90b]. Baake et al. [BaaJK90] show that many examples of quasiperiodic tilings of the plane arise from projections of root lattices.

Spherical codes

There has been considerable progress on the construction of spherical codes and the Tammes problem (cf. §2.3 of Chap. 1). In particular, Hardin, Smith and Sloane have computed extensive tables of spherical codes in up to 24 dimensions. For example, we have found conjecturally optimal packings of N spherical caps on a sphere in n dimensions for $N \leq 100$ and $n \leq 5$; and coverings and maximal volume codes in three dimensions for $N \leq 130$ and in four dimensions for $N \leq 24$. A book is in preparation [HSS]. Many of these tables are also available electronically [SloHP]. The papers by Hamkins and Zeger [HaZe97], [HaZe97a] show how to construct good spherical codes by (a) "wrapping" a good lattice packing around a sphere or (b) "laminating" (cf. Chap. 6) a good spherical code in a lower dimension. Other recent papers dealing with the construction of spherical codes are Dodunekov, Ericson and Zinoviev [DEZ91], Kolushov and Yudin [KolY97], Kottwitz [Kott91], Lazić, Drajić and Senk [LDS86], [LDS87], Melissen [Mel97].

A series of papers by Boyvalenkov and coauthors [Boy93]–[BoyN95] has investigated (among other things) the best polynomials to use in the linear programming bounds for spherical codes (cf. Chaps. 9, 13). This has led to small improvements in the (rather weak) upper bounds on the kissing number in dimensions 19, 21 and 23 given in Table 1.5 [Boy94a]. Coverings of a sphere by equal spherical caps are also discussed in [Tar5], [Tar6].

Drisch and Sonneborn [DrS96] give a table of upper bounds on kissing numbers (cf. Table 1.5 of Chap. 1) that extends to dimension 49.

Concerning the **numbers of lattice points** in or on various regions, (cf. §2.4 of Chap. 1), see [ArJ79], [Barv90], [BoHW72], [DuS90], [Dye91], [ElkOR91], [GoF87], [GriW93a], [Kra88], [Levi87], [MaO90], [Sar90].

Gritzmann and Wills [GriW93b] give a survey of recent work on **finite packings** and **coverings** (cf. §1.5 of Chap. 1), with particular emphasis on "sausage problems" and "bin-packing." The term "sausage problem" arises from the "sausage catastrophe," first observed by Wills in 1983 [Wills83]: which arrangement of N equal three-dimensional spheres has the smallest convex hull? It appears that for N up to about 55 a sausage-like linear arrangement of spheres is optimal, but for all larger N except 57, 58, 63 and 64 a three-dimensional cluster is better. In [Wills93] the notion of **parametric density** was introduced, permitting a joint theory of finite and infinite packings, with numerous applications. There are many interesting papers on the best packings of N balls when N is large and the connection with the **Wulff shape**, etc.: see Arhelger et al. [ABB96], Betke and Böröczky [BetB97], Böröczky and Schnell [BorSch97], [BorSch98], [BorSch98a], Dinghas [Ding43], von Laue [Laue43], Schnell [Schn98], Wills [Wills90]–[Wills98a]. For the sausage catastrophe in dimension 4, see [GaZu92].

Several recent papers have studied the problems of packing N points in a (a) circle [Fej97], [GraLNO], [Mel94], [Mel97], (b) square [Fej97], [Golb70], [Mel97], [MolP90], [NuOs97], [PeWMG], [Scha65], [Scha71], [Vall89], (c) triangle [Fej97], [GraL95], [Mel97], and (d) torus (Hardin and Sloane, unpublished).

Other papers dealing with finite packings and coverings are [BetG84], [BetG86], [BetGW82], [Chow92], [DaZ87], [FGW90], [FGW91], [GrW85], [Wills90]–[Wills90b].

The techniques that we use in [HSS] to find spherical codes have also proved successful in constructing **experimental designs** for use in statistics, and have been implemented in a general-purpose experimental design program called *Gosset* [HaSl93], [HaSl96a]. The name honors both the amateur mathematician Thorold Gosset (1869 – 1962) (cf. p. 120 of Chap. 8) and the statistician William Seally Gosset (1876 – 1937).

We have also used the same optimization methods to find good (often optimal) packings of lines through the origin in $\mathbb{R}^n$ (that is, antipodal spherical codes), and more generally packings of m-dimensional subspaces of $\mathbb{R}^n$. These **Grassmannian packings** are described in [CoHS96], [ShS98], [CHRSS].

The material in the Appendix to Chapter 1 on **planetary perturbations** was obtained in conversations among E. Calabi, J. H. Conway and J. G. Propp about Conway's lectures on "Games, Groups, Lattices and Loops" at the University of Pennsylvania in 1987 — see [Prop88].

For the application of lattices in **string theory** (cf. §1.4 of Chap. 1), see for example Gannon and Lam [GaL90].

Finally, we cannot resist calling attention to the remark of Frenkel, Lepowsky and Meurman, that **vertex operator algebras** (or conformal field theories) are to lattices as lattices are to codes (cf. [DGM90]–[DGM90b], [Fre1]–[Fre5], [Godd89], [Hoehn95], [Miya96], [Miya98]).

Notes on Chapter 2: Coverings, Lattices and Quantizers

Miyake [Miya89, Section 4.9] gives an excellent discussion of the classical result that the **theta function** of an integral lattice is a **modular form** for an appropriate subgroup of $SL_2(\mathbb{Z})$.

The papers [CSLDL6], [CSLDL8] and Chap. 3 of [CoFu97] describe how the **Voronoi cell** of a lattice (cf. §1.2 of Chap. 2) changes as that lattice is continuously varied. We simplify the usual treatment by introducing new parameters which we call the vonorms and conorms of the lattice. [CSLDL6] studies lattices in one, two and three dimensions, ending with a proof of the theorem of Fedorov [Fed85], [Fed91] on the five types of three-dimensional lattices. The main result of [CSLDL6] (and Chap. 3 of [CoFu97]) is that each three-dimensional lattice is uniquely represented by a projective plane of order 2 labeled with seven numbers, the conorms of the lattice, whose minimum is 0 and whose support is not contained in a proper subspace. Two lattices are isomorphic if and only if the corresponding labelings differ only by an automorphism of the plane.

These seven "conorms" are just 0 and the six "Selling parameters" ([Sel74], [Bara80]). However, this apparently trivial replacement of six numbers by seven numbers whose minimum is zero leads to several valuable improvements in the theory:

(i) The conorms vary continuously with the lattice. (For the Selling parameters the variation is usually continuous but requires occasional readjustments.)

(ii) The definition of the conorms makes it apparent that they are invariants of the lattice. (The Selling parameters are almost but not quite invariant.)
(iii) All symmetries of the lattice arise from symmetries of the conorm function. (Again, this is false for the Selling parameters.)

[CSLDL8] (summarized in the Afterthoughts to Chap. 3 of [CoFu97]) uses the same machinery as [CSLDL6] to give a simple proof of the theorem of Delone (= Delaunay) [Del29], [Del37], as corrected by Stogrin [Sto73], that there are 52 types of **four-dimensional lattices**. We also give a detailed description of the 52 types of four-dimensional parallelotopes (these are also listed by Engels [Eng86]). Erdahl and Ryskov [ErR87], [RyE88] show that there are only 19 types of different Delaunay cells that occur in four-dimensional lattices. (See also [RyE89].)

We call a lattice that is geometrically congruent to its dual **isodual** ([CoSl94]). We have used the methods of [CSLDL6] to determine the densest three-dimensional isodual lattice [CoSl94]. This remarkable lattice, the **m.c.c.** (or **mean-centered cuboidal**) **lattice**, has Gram matrix

$$\frac{1}{2}\begin{bmatrix} 1+\sqrt{2} & 1 & 1 \\ 1 & 1+\sqrt{2} & 1-\sqrt{2} \\ 1 & 1-\sqrt{2} & 1+\sqrt{2} \end{bmatrix} . \qquad (1)$$

In a sense this lattice is the geometric mean of the f.c.c. and b.c.c. lattices. (Consider the lattice generated by the vectors $(\pm u, \pm v, 0)$ and $(0, \pm u, \pm v)$ for real numbers u and v. If the ratio u/v is respectively 1, $2^{1/2}$ or $2^{1/4}$ we obtain the f.c.c., b.c.c. and m.c.c. lattices.) The m.c.c. lattice is also the thinnest isodual covering lattice. It is of course nonintegral. The m.c.c. lattice also recently appeared in a different context, as the lattice corresponding to the period matrix for the hyperelliptic Riemann surface $w^2 = z^8 - 1$ [BerSl97].

A **modular** lattice is an integral lattice that is geometrically similar to its dual (the term was introduced by Quebbemann [Queb95]; see also [Queb97]). In other words, an n-dimensional integral lattice Λ is *modular* if there exists a similarity σ of $\mathbb{R}^n$ such that $\sigma(\Lambda^*) = \Lambda$, where Λ^* is the dual lattice. If σ multiplies norms by N, Λ is said to be N-*modular*, and so has determinant $N^{n/2}$. A unimodular lattice is 1-modular. A modular lattice becomes isodual when rescaled so that its determinant is 1. For example, the sporadic root lattices E_8, F_4 ($\cong D_4$), G_2 ($\cong A_2$) are respectively 1-, 2- and 3-modular. In the last two cases the modularity maps short roots to long roots. The densest lattice packings presently known in dimensions 1, 2, 4, 8, 12, 16, 24, 48 and 56 are all modular.

Root lattices

In [CoSl91a] the Voronoi and Delaunay cells (cf. §1.2 of Chap. 2) of the lattices A_n, D_n, E_n and their duals are described in a simple geometrical way. The results for E_6^* and E_7^* simplify the work of Worley [Worl],

[Wor2], and also provide what may be new space-filling polytopes in dimensions 6 and 7. Pervin [Per90] and Baranovskii [Bara91] have also studied the Voronoi and Delaunay cells of E_6^*, E_7^*. Moody and Patera [MoP92], [MoP92a] have given a uniform treatment of the Voronoi and Delaunay cells of root lattices that also applies to the hyperbolic cases.

If a lattice Λ has covering radius R (cf. §1.2 of Chap. 2) then closed balls of radius R around the lattice points just cover the space. Sullivan [Sul90] defines the **covering multiplicity** $CM(\Lambda)$ to be the maximal number of times the interiors of these balls overlap. In [CoSl92] we show that the least possible covering multiplicity for an n-dimensional lattice is n if $n \leq 8$, and conjecture that it exceeds n in all other cases. We also determine the covering multiplicity of the Leech lattice and of the lattices I_n, A_n, D_n, E_n and their duals for small values of n. Although it appears that $CM(I_n) = 2^{n-1}$ if $n \leq 33$, it follows from the work of Mazo and Odlyzko [MaO90] that as $n \to \infty$ we have $CM(I_n) \sim c^n$, where $c = 2.089\ldots$. The results have applications to numerical integration.

The covering problem

Several better coverings of space by spheres have been found in low dimensions, giving improvements to Table 2.1 and Fig. 2.4 of Chap. 2. The lattice $A_n[s]$ mentioned on p. 116 of Chap. 4 is generated by the vectors of the translate $[s] + A_n$, where s is any divisor of $n+1$, and is the union of the r translates $[i] + A_n$ for $i = 0, s, 2s, \ldots, (r-1)s$, where $r = (n+1)/s$. This is the lattice A_n^{+r}, in the notation of [CSLDL1], or A_n^r in Coxeter's notation [Cox10]. It has determinant $(n+1)/r^2$ and minimal norm $rs/(n+1)$.

Baranovskii [Bara94] shows that A_9^{+5} has covering radius $\sqrt{24}/5$ and thickness

$$\Theta = (2^{13} 3^{4.5} / 5^{8.5}) V_9 = 1.3158\ldots V_9 = 4.3402\ldots ,$$

whereas A_9^*, the old record-holder, has thickness

$$\Theta = (3^{4.5} 11^{4.5} / 2^{13} 5^4) V_9 = 1.3306\ldots V_9 = 4.3889\ldots .$$

Furthermore, W. D. Smith [Smi88] has shown that Λ_{22}^* and Λ_{23}^* are better coverings than A_{22}^* and A_{23}^*. Smith finds that the thickness Θ of Λ_{22}^* is at most

$$2\sqrt{3}\left(\frac{\sqrt{17}}{3}\right)^{22} V_{22} = 27.8839\ldots$$

and the thickness of Λ_{23}^* is at most

$$2\left(\frac{\sqrt{31}}{4}\right)^{23} V_{23} = 15.3218\ldots .$$

For other work on the covering radius of lattices see [CaFR95].

Integer coordinates for integer lattices

[CSLDL5] is concerned with finding descriptions for integral lattices (cf. §2.4 of Chap. 2) using integer coordinates (possibly with a denominator). Let us say that an n-dimensional (classically) integral lattice Λ is **s-integrable**, for an integer s, if it can be described by vectors $s^{-1/2}(x_1, \ldots, x_k)$, with all $x_i \in \mathbb{Z}$, in a Euclidean space of dimension $k \geq n$. Equivalently, Λ is s-integrable if and only if any quadratic form $f(x)$ corresponding to Λ can be written as s^{-1} times a sum of k squares of linear forms with integral coefficients, or again, if and only if the dual lattice Λ^* contains a eutactic star of scale s. [CSLDL5] gives many techniques for s-integrating low-dimensional lattices (such as E_8 and the Leech lattice). A particular result is that any one-dimensional lattice can be 1-integrated with $k = 4$: this is Lagrange's four-squares theorem. Let $\phi(s)$ be the smallest dimension n in which there is an integral lattice that is not s-integrable. In 1937 Ko and Mordell showed that $\phi(1) = 6$. We prove that $\phi(2) = 12$, $\phi(3) = 14$, $21 \leq \phi(4) \leq 25$, $16 \leq \phi(5) \leq 22$, $\phi(s) \leq 4s + 2$ (s odd), $\phi(s) \leq 2\pi e s(1 + o(1))$ (s even) and $\phi(s) \geq 2 \ln \ln s / \ln \ln \ln s(1 + o(1))$.

Plesken [Plesk94] studies similar embedding questions for lattices from a totally different point of view. See also Cremona and Landau [CrL90].

Complexity

For recent results concerning the complexity of various lattice- and coding-theoretic calculations (cf. §1.4 of Chap. 2), see Ajtai [Ajt96], [Ajt97], Downey et al. [DowFV], Hastad [Has88], Jastad and Lagarias [HaL90], Lagarias [Laga96], Lagarias, Lenstra and Schnorr [Lag3], Paz and Schnorr [PaS87], Vardy [Vard97].

In particular, Vardy [Vard97] shows that computing the minimal distance of a binary linear code is NP-hard, and the corresponding decision problem is NP-complete. Ajtai [Ajt97] has made some progress towards establishing analogous results for lattices. Downey et al. [DowFV] show that computing (the nonzero terms in) the theta-series of a lattice is NP-hard.

For **lattice reduction algorithms** see also [Schn87], [Val90], [Zas3]. Most of these results assume the lattice in question is a sublattice of $\mathbb{Z}^n$. In this regard the results of [CSLDL5] mentioned above are especially relevant. Ivanyos and Szántó [IvSz96] give a version of the LLL algorithm that applies to indefinite quadratic forms.

Mayer [Maye93], [Maye95] shows that every Minkowski-reduced basis for a lattice of dimension $n \leq 6$ consists of strict Voronoi vectors (cf. [Rys8]). He also answers a question raised by Cassels ([Cas3], p. 279) by showing that in seven dimensions (for the first time) the Minkowski domains do not meet face to face.

Barvinok [Barv92a] has described a new procedure for finding the minimal norm of a lattice or the closest lattice vector to a given vector that makes use of theta functions (cf. [Barv90], [Barv91], [Barv92]).

The **isospectral problem** for planar domains was solved by Gordon, Webb and Wolpert in 1992 ([GWW92], [Kac66] see also [BuCD94]). A particularly simple solution appears in Chap. 2 of [CoFu97]. This has aroused new interest in other isospectrality problems, for instance that of finding the largest n such that any positive definite quadratic form of rank n is determined by its representation numbers. Equivalently, what is the smallest dimension in which there exist two inequivalent lattices with the same theta series? We shall call such lattices *isospectral*: they are the subject of Chap. 2 of [CoFu97]. As mentioned in §2.3 of Chap. 2, Witt [Wit4], Kneser [Kne5] and Kitaoka [Kit2] found isospectral lattices in dimensions 16, 12 and 8 respectively. Milnor [Mil64] pointed out the connection with the isospectral manifold problem.

In 1986 one of the present authors observed that pairs of isospectral lattices in 6 and 5 dimensions could be obtained from a pair of codes with the same weight enumerator given by the other author [Slo10]. These lattices are mentioned on p. 47, and have now been published in [CoSl92a]. One lattice of the six-dimensional pair is a scaled version of the cubic lattice I_6. The five-dimensional pair have Gram matrices

$$\begin{bmatrix} 2 & 0 & 0 & 2 & 2 \\ 0 & 2 & 0 & 2 & 0 \\ 0 & 0 & 2 & 0 & 2 \\ 2 & 2 & 0 & 8 & 4 \\ 2 & 0 & 2 & 4 & 8 \end{bmatrix}, \quad \begin{bmatrix} 2 & 1 & 0 & 2 & 2 \\ 1 & 2 & 0 & 2 & 2 \\ 0 & 0 & 6 & 4 & 4 \\ 2 & 2 & 4 & 8 & 4 \\ 2 & 2 & 4 & 4 & 8 \end{bmatrix} \qquad (2)$$

and determinant 96 (and are in different genera).

The first pair of isospectral 4-dimensional lattices was found in 1990 by Schiemann [Schi90], by computer search, and we have been informed by Schulze-Pillot (personal communication) that Schiemann has since found at least a dozen such pairs. Another pair has been given by Earnest and Nipp [EaN91]. The main result of [CoSl92a] is to give a simple 4-parameter family of pairs of isospectral lattices, which includes many of the known examples (including Schiemann's first pair) as special cases. The typical pair of this family is

$$\langle 3w - x - y - z, \quad w + 3x + y - z, \quad w - x + 3y + z, \quad w + x - y + 3z \rangle$$

and

$$\langle -3w - x - y - z, \quad w - 3x + y - z, \quad w - x - 3y + z, \quad w + x - y - 3z \rangle \ ,$$

where w, x, y, z are orthogonal vectors of distinct lengths, and the pointed brackets mean "lattice spanned by."

Schiemann [Schi97] has now completed the solution to the original problem by showing that any three-dimensional lattice is determined by its theta series. Thus n-dimensional isospectral lattices exist if and only if n is at least 4. For more about these matters see [CoFu97].

Lattice quantizers

Coulson [Coul91] has found the mean squared error G for the perfect (and isodual) six-dimensional lattice $P_6^5 \cong A_6^{(2)}$ (defined in §6 of Chap. 8 and studied in [CSLDL3]). He finds $G = 0.075057$, giving an additional entry for Table 2.3 of Chap. 2. Viterbo and Biglieri [ViBi96] have computed G for the lattices of Eqs. (1) and (2), the Dickson lattices of page 36, and other lattices.

Agrell and Eriksson [AgEr98] have found 9- and 10-dimensional lattices with $G = 0.0716$ and 0.0708, respectively, and show that the nonlattice packings D_7^+ and D_9^+ have $G = 0.0727$ and 0.0711, respectively. These values are all lower (i.e. better) than the previous records.

Notes on Chapter 3: Codes, Designs and Groups

Lattice codes. Several authors have studied the error probability of codes for the Gaussian channel that make use of constellations of points from some lattice as the signal set – see for example Banihashemi and Khandani [BanKh96], de Buda [Bud2], [Bud89], Forney [Forn97], Linder et al. [LiSZ93], Loeliger [Loel97], Poltyrev [Polt94], Tarokh, Vardy and Zeger [TaVZ], Urbanke and Rimoldi [UrB98].

Urbanke and Rimoldi [UrB98], completing the work of several others, have shown that lattice codes bounded by a sphere can achieve the capacity $\frac{1}{2}\log_2(1 + P/N)$ (where P is the signal power and N is the noise variance), using minimal-distance decoding. This is stronger than what can be deduced directly from the Minkowski-Hlawka theorem ([Bud2], [Cas2], [Gru1a], [Hla1], [Rog7]), which is that a rate of $\frac{1}{2}\log_2(P/N)$ can be achieved with lattice codes.

There has been a great deal of activity on **trellis codes** (cf. §1.4 of Chap. 3) — see for example [BDMS], [Cal91], [CaO90], [Forn88], [Forn88a], [Forn89a], [Forn91], [FoCa89], [FoWe89], [LaVa95], [LaVa95a], [LaVa96], [TaVa97], [VaKs96].

Another very interesting question is that of finding **trellis representations** of the standard codes and lattices: see Forney [Forn94], [Forn94a], Feigenbaum et al. [FeFMMV], Vardy [Vard98a] and many related papers: [BanBl96], [BanKh97], [BlTa96], [FoTr93], [KhEs97], [TaBl96], [TaBl96a].

We have already mentioned recent work on Goppa codes and the construction of codes and lattices from algebraic geometry (cf. §2.11 of Chap. 3) under §1.5 of Chap. 1.

Tables of codes

Verhoeff's table [Ver1] of the best binary linear codes (cf. §2.1 of Chap. 3) has been greatly improved by Brouwer [BrVe93], [Bro98]. For other tables of codes see [BrHOS], [BrSSS], [Lits98], [SchW92].

For recent work on the **covering radius of codes** see the book by Cohen et al. [CHLL] as well as the papers [BrLP98], [CaFR95], [CLLM97], [DaDr94], [EtGr93], [EtGh93], [EtWZ95], [Habs94]–[Habs97], [LeLi96], [LiCh94], [Stru94], [Stru94a], [Tiet91], [Wee93], [Wille96].

Spherical t-designs

The work of Hardin and Sloane on constructing experimental designs mentioned under Chap. 1 has led to new results and conjectures on the existence of spherical 4-designs (cf. §3.2 of Chap. 3). In three dimensions, for example, we have shown that spherical 4-designs containing M points exist for $M = 12$, 14 and $M \geq 16$, and we conjecture that they do not exist for $M = 9, 10, 11, 13$ and 15 [HaSl92]. Similarly, we conjecture that in four dimensions they exist precisely for $M \geq 20$; in five dimensions for $M \geq 29$; in six dimensions for $M = 27$, 36 and $M \geq 39$; in seven dimensions for $M \geq 53$; and in eight dimensions for $M \geq 69$ [HaSl92]. The connections between experimental designs and spherical designs have become much clearer thanks to the work of Neumaier and Seidel [NeS92].

Other recent papers dealing with spherical designs and numerical integration on the sphere are [Atk82], [Boy95], [BoyDN], [BoyN94], [BoyN95], [KaNe90], [Kea87], [Kea1], [LySK91], [NeS88], [Neut83], [NeSJ85], [Rezn95], [Sei90], [Yud97]. Several of J. J. Seidel's papers (including in particular the joint papers [Del15], [Del16]) have been reprinted in [Sei91].

Finite matrix groups

For a given value of n there are only finitely many nonisomorphic finite groups of $n \times n$ integer matrices. This theorem has a long history and is associated with the names Jordan, Minkowski, Bieberbach and Zassenhaus (see [Mil5], [Bro10]). For $n = 2$ and 3 these groups were classified in the past century, because they are needed in crystallography (see also [AuC91], [John91]). The maximal finite subgroups of $GL(4, \mathbb{Z})$ were given by Dade [Dad1], and the complete list of finite subgroups of $GL(4, \mathbb{Z})$ by Bülow, Neubüser and Wondratschek [Bül1] and Brown et al. [Bro10]. The maximal irreducible finite subgroups of $GL(5, \mathbb{Z})$ were found independently by Ryskov [Rys4], [Rys5], and Bülow [Bül2]. That work was greatly extended by Plesken & Pohst [Ple5], who determined the maximal irreducible subgroups of $GL(n, \mathbb{Z})$ for $n = 6, 7, 8, 9$, and by Plesken [Ple3], who dealt with $n = 11, 13, 17, 19, 23$.

In these papers the subgroups are usually specified as the automorphism groups of certain quadratic forms. In [CSLDL2] we give a geometric description of the maximal irreducible subgroups of $GL(n, \mathbb{Z})$ for $n = 1, \ldots, 9, 11, 13, 17, 19, 23$, by exhibiting lattices corresponding to these quadratic forms (cf. §4.2(i) of Chap. 3): the automorphism groups of the lattices are the desired groups. By giving natural coordinates for these lattices and determining their minimal vectors, we are able to make their symmetry groups clearly visible. There are 176 lattices, many of which have not been studied before (although they are implicit in the above references and in [Con16]).

The book by Holt and Plesken [HoPl89] contains tables of perfect groups of order up to 10^6, and includes tables of crystallographic space groups in dimensions up to 10.

Nebe and Plesken [NePl95] and Nebe [Nebe96], [Nebe96a] (see also [Plesk96], [Nebe98a]) have recently completed the enumeration of the maximal finite irreducible subgroups of $GL(n, \mathbb{Q})$ for $n \leq 31$, together with the associated lattices. This is an impressive series of papers, which contains an enormous amount of information about lattices in dimensions below 32.

Notes on Chapter 4: Certain Important Lattices and Their Properties

Several recent papers have dealt with gluing theory (cf. §3 of Chap. 4) and related techniques for combining lattices: [GaL91]–[GaL92a], [Gers91], [Sig90], [Xu1]. Gannon and Lam [GaL92], [GaL92a] also give a number of new theta-function identities (cf. §4.1 of Chap. 4).

Scharlau and Blaschke [SchaB96] classify all lattices in dimensions $n \leq 6$ in which the root system has full rank.

Professor Coxeter has pointed out to us that, in the last line of the text on page 96, we should have mentioned the work of Bagnera [Bag05] along with that of Miller.

For recent work on quaternionic reflection groups (cf. §2 of Chap. 4) see Cohen [Coh91].

Hexagonal lattice A_2

The number of inequivalent sublattices of index N in A_2 is determined in [BerSl97a], and the problems of determining the best sublattices from the points of view of packing density, signal-to-noise ratio and energy are considered. These questions arise in cellular radio. See also [BaaPl95].

Kühnlein [Kuhn96] has made some progress towards establishing Schmutz's conjecture [Schmu95] that the distinct norms that occur in A_2 are strictly smaller than those in any other (appropriately scaled) two-dimensional lattice. See also Schmutz [Schmu93], Schmutz Schaller

[Schmu95a], Kühnlein [Kuhn97]. A related idea (the Erdős number) is discussed in the Notes on Chapter 15.

Leech lattice

A very simple construction for the Leech lattice Λ_{24} was discovered by Bonnecaze and Solé ([BonS94], see also [BonCS95]): lift the binary Golay code to $\mathbb{Z}_4$ (the ring of integers mod 4), and apply "Construction A_4". The details are as follows.

The [23, 12, 7] Golay code may be constructed as the cyclic code with generator polynomial $g_2(x) = x^{11} + x^9 + x^7 + x^6 + x^5 + x + 1$, a divisor of $x^{23} - 1$ (mod 2). By Hensel-lifting this polynomial (using say Graeffe's root-squaring method, cf. [HaKCSS], p. 307) to $\mathbb{Z}_4$ we obtain

$$g_4(x) = x^{11} + 2x^{10} - x^9 - x^7 - x^6 - x^5 + 2x^4 + x - 1 ,$$

a divisor of $x^{23} - 1$ (mod 4). By appending a zero-sum check symbol to the cyclic code generated by $g_4(x)$, we obtain a self-dual code of length 24 over $\mathbb{Z}_4$. Applying Construction A_4 (cf. Chapter 5), that is, taking all vectors in $\mathbb{Z}^{24}$ which when read mod 4 are in the code, we obtain the Leech lattice.

In this version of the Leech lattice the 196560 minimal vectors appear as 4.16.759 of shape $2^2 1^8 0^{14}$, 2.24.2576 of shape $2^1 1^{12} 0^{11}$, 32.759 of shape $1^{16} 0^8$ and 48 of shape $4^1 0^{23}$.

The general setting for this construction is the following ([BonCS95], Theorem 4.1). Define the Euclidean norms of the elements of $\mathbb{Z}_4$ by $N(0) = 0$, $N(\pm 1) = 1$, $N(2) = 4$, and define $N(u)$, $u = (u_1, \ldots, u_n) \in \mathbb{Z}_4^n$, by $N(u) = \sum N(u_i)$. Then if C is a self-dual code over $\mathbb{Z}_4$ in which the Euclidean norm of every codeword is divisible by 8, Construction A_4 produces an n-dimensional even unimodular lattice.

J. Young and N. J. A. Sloane showed (see [CaSl97]) that the other eight doubly-even binary self-dual codes of length 24 can also be lifted to codes over $\mathbb{Z}_4$ that give the Leech lattice (see also Huffman [Huff98a]).

[CaSl95] considers the codes obtained by lifting the Golay code (and others) from $\mathbb{Z}_2$ to $\mathbb{Z}_4$ to $\mathbb{Z}_8$ to ..., finally obtaining a code over the 2-adic integers $\mathcal{Z}_2$. For more about codes over $\mathbb{Z}_4$ see the Notes on Chapter 16.

For other recent results on the Leech lattice and attempts to generalize it see Bondal et al. [BKT87], Borcherds [Borch90], Conway and Sloane [CoSl94a], Deza and Grishukhin [DezG96], Elkies and Gross [ElkGr96], Harada and Lang [HaLa89], [HaLa90], Koike [Koik86], Kondo and Tasaka [Kon2], [KoTa87], Kostrikin and Tiep [KoTi94], Ozeki [Oze91], Seidel [Sei90b].

Lindsay [Lin88] describes a 24-dimensional 5-modular lattice associated with the proper central extension of the cyclic group of order 2 by the

Hall-Janko group J_2 (cf. Chap. 10). The density of this lattice is about a quarter of that of the Leech lattice.

Napias [Napa94] has found some new lattices by investigating cross-sections of the Leech lattice, the 32-dimensional Quebbemann lattice and other lattices.

Shadows and parity (or characteristic) vectors

The notion of the "shadow" of a self-dual code or unimodular lattice, introduced in [CoSl90], [CoSl90a], has proved useful in several contexts, and if we were to rewrite Chapter 4 we would include the following discussion there. We will concentrate on lattices, the treatment for codes being analogous.

Let Λ be an n-dimensional odd unimodular (or Type I) lattice, and let Λ_0 be the even sublattice, of index 2. The dual lattice Λ_0^* is the union of four cosets of Λ_0, say

$$\Lambda_0^* = \Lambda_0 \cup \Lambda_1 \cup \Lambda_2 \cup \Lambda_3 \ ,$$

where $\Lambda = \Lambda_0 \cup \Lambda_2$. Then we call $S := \Lambda_1 \cup \Lambda_3 = \Lambda_0^* \setminus \Lambda$ the *shadow* of Λ. If Λ is even (or Type II) we define its shadow S to be Λ itself. The following properties are easily established [CoSl90].

If $s \in S$ and $x \in \Lambda$, then $s \cdot x \in \mathbb{Z}$ if $x \in \Lambda_0$, $s \cdot x \in \frac{1}{2}\mathbb{Z} \setminus \mathbb{Z}$ if $x \in \Lambda_1$. In fact the set $2S = \{2s : s \in S\}$ is precisely the set of *parity vectors* for Λ, that is, those vectors $u \in \Lambda$ such that

$$u \cdot x \equiv x \cdot x \pmod 2 \quad \text{for all} \quad x \in \Lambda \ .$$

Such vectors have been studied by many authors, going back at least as far as Braun [Brau40] (we thank H.-G. Quebbemann for this remark). They have been called *characteristic vectors* [Blij59], [Bor1], [Elki95], [Elki95a], [Mil7], *canonical elements* [Ser1], and *test vectors*. We recommend "parity vector" as the standard name for this concept.

The existence of a parity vector u also follows from the fact that the map $x \rightarrow x \cdot x \pmod 2$ is a linear functional from Λ to $\mathbb{F}_2$. The set $2S$ of all parity vectors forms a single class $u + 2\Lambda$ in $\Lambda/2\Lambda$. If Λ is even this is the zero class.

We also note that for any parity vector u, $u \cdot u \equiv n \pmod 8$.

Gerstein [Gers96] gives an explicit construction for a parity vector. Let $v_1, \ldots, v_n$ be a basis for Λ and $v'_1, \ldots, v'_n$ the dual basis. Then $\sum c_i v_i$ is a parity vector if and only if $c_i \equiv v'_i \cdot v'_i \pmod 2$ for all i.

Elkies [Elki95], [Elki95a] shows that the minimal norm $p(\Lambda)$ of any parity vector for Λ satisfies $p(\Lambda) \leq n$, and $p(\Lambda) = n$ if and only if $\Lambda = \mathbb{Z}^n$. Furthermore, if $p(\Lambda) = n - 8$ then $\Lambda = \mathbb{Z}^{n-r} \oplus M_r$, where M_r is one of the fourteen unimodular lattices whose components are E_8, D_{12}, E_7^2 A_{15}, D_8^2, $A_{11}E_6$, D_6^3, A_9^2, $A_7^2 D_5$, D_4^5, A_5^4, A_3^7, A_1^{22}, O_{23} (using the notation of Chapter 16).

The shadow may also be defined for a more general class of lattices. If Λ is a *2-integral* lattice (i.e. $u \cdot v \in \mathbb{Z}_2$, the 2-adic integers, for all $u, v \in \Lambda$), and $\Lambda_0 = \{u \in \Lambda \ : \ u \cdot u \in 2\mathbb{Z}_2\}$ is the even sublattice, we define the shadow $S(\Lambda)$ of Λ as follows [RaSl98a]. If Λ is odd, $S(\Lambda) = (\Lambda_0)^* \setminus \Lambda^*$, otherwise $S(\Lambda) = \Lambda^*$. Then

$$S(\Lambda) = \{v \in \Lambda \otimes \mathbb{Q} : 2u \cdot v \equiv u \cdot u \pmod{2\mathbb{Z}_2} \quad \text{for all} \quad u \in \Lambda\} \ .$$

This includes the first definition of shadow as a special case. The theta series of the shadow (for both definitions) is related to the theta series of the lattice by

$$\Theta_{S(\Lambda)}(z) = (\det \ \Lambda)^{1/2} \left(\frac{e^{\pi i/4}}{\sqrt{z}}\right)^{\dim \Lambda} \Theta_\Lambda \left(1 - \frac{1}{z}\right) \ . \tag{3}$$

It is also shown in [RaSl98a] that if Λ has odd determinant, then for $u \in S(\Lambda)$,

$$u \cdot u \equiv \frac{1}{4} \text{ oddity } \Lambda \pmod{2\mathbb{Z}_2} \tag{4}$$

(compare Chap. 15). In particular, if Λ is an odd unimodular lattice with theta series

$$\Theta_\Lambda(z) = \sum_{r=0}^{[n/8]} a_r \theta_3(z)^{n-8r} \Delta_8(z)^r \tag{5}$$

(as in Eq. (36) of Chap. 7), then the theta series of the shadow is given by

$$\Theta_S(z) = \sum_{r=0}^{[n/8]} \frac{(-1)^r}{16^r} a_r \theta_4(2z) \theta_2(z)^{n-8r} \ . \tag{6}$$

For further information about the shadow theory of codes and lattices see [CoSl90], [CoSl90a], [CoSl98], [Rain98], [RaSl98], [RaSl98a]. See also [Dou95]–[DoHa97].

Coordination sequences

Crystallographers speak of "coordination number" rather than "kissing number." Several recent papers have investigated the following generalization of this notion ([BaaGr97], [BrLa71], [GrBS], [MeMo79], [O'Ke91], [O'Ke95]). Let Λ be a (possibly nonlattice) sphere packing, and form an infinite graph Γ whose nodes are the centers of the spheres and which has an edge for every pair of touching spheres. The **coordination sequence** of Γ with respect to a node $P \in \Gamma$ is the sequence $S(0)$, $S(1)$, $S(2), \ldots$, where $S(n)$ is the number of nodes in Γ at distance n from P (that is, such that the shortest path to P contains n edges).

If Λ is a lattice then the coordination sequence is independent of the choice of P. In [CSLDL7], extending the work of O'Keeffe [O'Ke91],

[O'Ke95], we determine the coordination sequences for all the root lattices and their duals. Ehrhart's reciprocity law ([Ehr60]–[Ehr77], [Stan80], [Stan86]) is used, but there are unexpected complications. For example, there are points Q in the 11-dimensional "anabasic" lattice of [CoSl95], mentioned in the Notes to Chapter 1, with the property that $2Q$ is closer to the origin than Q (in graph distance).

We give two examples. For a d-dimensional lattice Λ it is convenient to write the generating function $S(x) = \sum_{n=0}^{\infty} S(n)x^n$ as $P_d(x)/(1-x)^d$, where we call $P_d(x)$ the *coordinator polynomial*. For the root lattice A_d it turns out that

$$P_d(x) = \sum_{k=0}^{d} \binom{d}{k}^2 x^k \ ,$$

and for E_8 we have

$$P_8(x) = 1 + 232x + 7228x^2 + 55384x^3 + 133510x^4 + 107224x^5 + 24508x^6 + 232x^7 + x^8 \ .$$

Thus the coordination sequence of E_8 begins 1, 240, 9120, 121680, 864960, ... For further examples see [BattVe98], [CSLDL7], [GrBS], and [SloEIS]. We do not know the coordination sequence of the Leech lattice.

In [CSLDL7] we also show that among all the Barlow packings in three dimensions (those obtained by stacking A_2 layers, cf. [CoSl95a]) the hexagonal close packing has the greatest coordination sequence, and the face-centered cubic lattice the smallest. More precisely, for any Barlow packing,

$$10n^2 + 2 \le S(n) \le [21n^2/2] + 2 \quad (n > 0) \ .$$

For any $n > 1$, the only Barlow packing that achieves either the left-hand value or the right-hand value for all choices of central sphere is the face-centered cubic lattice or hexagonal close-packing, respectively. This interesting result was conjectured by O'Keeffe [O'Ke95]; it had in fact already been established (Conway & Sloane 1993, unpublished notes). There is an assertion on p. 801 of [Hsi93] that is equivalent to saying that any Barlow packing has $S(2) = 44$, and so is plainly incorrect: as shown in [CoSl95a], there are Barlow packings with $S(2) = 42$, 43 and 44. [CSLDL7] concludes with a number of open problems related to coordination sequences.

Notes on Chapter 5: Sphere Packing and Error-Correcting Codes

The **Barnes-Wall lattices** ([Bar18], §6.5 of Chap. 5, §8.1 of Chap. 8) are the subject of a recent paper by Hahn [Hahn90].

On p. 152 of Chap. 5 we remarked that it would be nice to have a list of the best **cyclic codes** of length 127. Such a list has now been supplied by Schomaker and Wirtz [SchW92]. Unfortunately this does not improve the $n = 128$ entry of Table 8.5. Perhaps someone will now tackle the cyclic codes of length 255.

The paper by Ozeki mentioned in the postscript to Chap. 5 has now appeared [Oze87].

*Construction B**

The following construction is due to A. Vardy [Vard95], [Vard98] (who gives a somewhat more general formulation). It generalizes the construction of the Leech lattice given in Eqs. (135), (136) of Chap. 4 and §4.4 of Chap. 5, and we refer to it as Construction B* since it can also be regarded as a generalization of Construction B of §3 of Chap.5

Let $\mathbf{0} = 0\ldots0$ and $\mathbf{1} = 1\ldots1$, and let $\mathcal{B}$ and $\mathcal{C}$ be (n, M, d) binary codes (in the notation of p. 75) such that $c \cdot (\mathbf{1} + b) = 0$ for all $b \in \mathcal{B}$, $c \in \mathcal{C}$. Let Λ be the sphere packing with centers

$$\mathbf{0} + 2b + 4x, \qquad \mathbf{1} + 2c + 4y,$$

where x (resp. y) is any vector of integers with an even (resp. odd) sum, and $b \in \mathcal{B}$, $c \in \mathcal{C}$. (We regard the components of b and c as *real* 0's and 1's rather than elements of $\mathbb{F}_2$.) In general Λ is not a lattice.

The most interesting applications arise when d is 7 or 8, in which case it is easily verified that Λ has center density $M\,7^{n/2}/4^n$ (if $d = 7$ and $n \geq 20$) or $M/2^{n/2}$ (if $d = 8$ and $n \geq 24$).

Vardy [Vard95], [Vard98] uses this construction to obtain the nonlattice packings $\mathcal{B}_{20}^*$ and $\mathcal{B}_{27}^* - \mathcal{B}_{30}^*$ shown in Table I.1. In dimension 20 he uses a pair of $(20, 2^9, 7)$ codes, but we will not describe them here since the same packing will be obtained more simply below. For dimensions 28 and 30 he takes $\mathcal{B} = \mathcal{C}^{\perp}$ to be the $[28, 14, 8]$ or $[30, 15, 8]$ double circulant codes constructed by Karlin (see [Mac6], p. 509). Both codes contain $\mathbf{1}$, are not self-dual, but are equivalent to their duals.

For $n = 27$ we shorten the length 28 code to obtain a $[27, 13, 8]$ code $\mathcal{A}$ and set $\mathcal{B} = \mathbf{1} + \mathcal{A}$, $\mathcal{C} =$ even weight subcode of $\mathcal{A}^{\perp}$. Similarly for $n = 29$.

Once n exceeds 31, we may use Construction D (see Chap. 8, §8) instead of Construction B*, obtaining a lattice packing from an $[n, k, 8]$ code. In particular, using codes with parameters $[37, 31, 8]$ and $[38, 22, 8]$ (Shearer [Shea88]) we obtain the lattices $\mathcal{D}_{37}$ and $\mathcal{D}_{38}$ mentioned in Table I.1.

As far as is known at the present time, codes with parameters $[32, 18, 8]$, $[33, 18, 8]$, $[34, 19, 8]$, $\ldots$, $[38, 23, 8]$, $[39, 23, 8]$ might exist. If any one of these could be constructed, a new record for packing density in the corresponding dimension would be obtained for using Construction D.

Bierbrauer and Edel [BiEd98] pointed out that Construction B* also yields a new record 18-dimensional nonlattice packing, using the $[18, 9, 8]$ quadratic residue code and its dual. This packing has center density $(3/4)^9 = 0.07508\ldots$.

An alternative construction of Vardy's 20-dimension packing was given in [CoSl96]. This construction, which we call the **antipode construction**, also produces new records (denoted by $\mathcal{A}_n$ in Table I.1) in dimensions 22 and 44–47. It is an analogue of the "anticode" construction for codes ([Mac6], Chap. 17, Sect. 6). The common theme of the two constructions is that instead of looking for well-separated points, which is what most constructions do, now we look for points somewhere else that are close together and factor them out.

Let Λ be a unimodular lattice (the construction in [CoSl96] is slightly more general) of minimal norm μ in an n-dimension Euclidean space W. Let U, V be respectively k- and l-dimensional subspaces with $W = U \oplus V$, $n = k + l$, such that $\Lambda \cap U = K$ and $\Lambda \cap V = L$ are k- and l-dimensional lattices. Then the projections $\pi_U(\Lambda)$, $\pi_V(\Lambda)$ are the dual lattices K^*, L^*. Suppose we can find a subset $S = \{u_1, \ldots, u_s\} \subseteq K^*$ such that $\text{dist}^2(u_i, u_j) \le \beta$ for all i, j. Then

$$\mathcal{A}(S) = \{\pi_V(w) : w \in \Lambda,\ \ \pi_U(w) \in S\}$$

is an l-dimensional packing of minimal norm $\mu - \beta$ and center density equal to $\delta = s(\mu - \beta)^{l/2} 2^{-l} / \sqrt{\det L}$.

In dimension 20 we take $\Lambda = \Lambda_{24}$, $L = \Lambda_{20}$, $K = \sqrt{2} D_4$, $K^* = 2^{-1/2} D_4^*$,

$$S = 2^{-1.5}\{0\ 0\ 0\ 0,\quad 1\ 1\ 1\ 1,\quad 2\ 0\ 0\ 0,\quad 1\ 1\ 1\ -1\},\quad \beta = \tfrac{1}{2}\ ,$$

which produces Vardy's packing $\mathcal{A}_{20} \cong \mathcal{B}_{20}^*$ of center density $7^{10}/2^{31} = 0.13154$ and kissing number 15360.

In dimension 22 we take $\Lambda = \Lambda_{24}$, $K = \sqrt{2} A_2$, $K^* = 2^{-1/2} A_2^*$, $S =$ three equally spaced vectors in K^* with $\beta = \tfrac{1}{3}$, obtaining a packing $\mathcal{A}_{22}$ with center density $2^{-23} 3^{-10.5} 11^{11} = 0.33254$ and kissing number 41472. To obtain explicit coordinates for this packing, take those vectors of Λ_{24} in which the first three coordinates have the form $2^{-1.5}(a, a, a, \ldots)$, $2^{-1.5}(a + 2, a, a, \ldots)$ or $2^{-1.5}(a, a - 2, a, \ldots)$ and replace them by their respective projections $2^{-1.5}(a, a, a, \ldots)$, $2^{-1.5}(a + \tfrac{2}{3}, a + \tfrac{2}{3}, a + \tfrac{2}{3}, \ldots)$ and $2^{-1.5}(a - \tfrac{2}{3}, a - \tfrac{2}{3}, a - \tfrac{2}{3}, \ldots)$.

In dimensions 44–47, we take $\Lambda = P_{48p}$, for example, and as on p. 168 find subspaces U in $\mathbb{R}^{48}$ such that K^* is respectively $3^{-1/2} A_1^*$, $3^{-1/2} A_2^*$, $3^{-1/2} A_3^*$, $3^{-1/2} D_4^*$. In these four lattices we can find $s = 2, 3, 4, 4$ points, respectively, for which $\beta = \tfrac{1}{6}, \tfrac{2}{9}, \tfrac{1}{3}, \tfrac{1}{3}$, obtaining the packings $\mathcal{A}_{44}$–$\mathcal{A}_{47}$ mentioned in Table I.1.

Notes on Chapter 6: Laminated Lattices

In 1963 Musès ([Cox18], p. 238; [Mus97], p. 7) discovered that the highest possible kissing number for a lattice packing in dimensions $n = 0$ through 8 (but presumably for no higher n) is given by the formula

$$n\left(n+\left\lceil \frac{2^n}{12} \right\rceil\right) \tag{7}$$

where $\lceil x \rceil$ is the smallest integer $\geq x$ (cf. Table 1.1).

All laminated lattices Λ_n in dimensions $n \leq 25$ are known, and their kissing numbers are shown in Table 6.3. In dimensions 26 and above, as mentioned on pp. 178–179, the number of laminated lattices seems to be very large, and although they all have the same density, we do not at present know the range of kissing numbers that can be achieved.

In the mid-1980's the authors computed the kissing numbers of one particular sequence of laminated lattices in dimensions 26–32, obtaining values that can be seen in Table I.2. Because of an arithmetical error, the value we obtained in dimension 31 was incorrect. Musès [Mus97] independently studied the (presumed) maximal kissing numbers of laminated lattices (finding the correct value 202692 in dimension 31) and has discovered the formulae analogous to (7).

In the Appendix to Chapter 6, on page 179, third paragraph, it would have been clearer if we had said that, for $n \leq 12$, the integral laminated lattice $\Lambda_n\{3\}$ of minimal norm 3 consists of the projections onto $v^\perp$ of the vectors of Λ_{n+1} having even inner product with v, where $v \in \Lambda_{n+1}$ is a suitable norm 4 vector. For $n \leq 10$, $K_n\{3\}$ is defined similarly, using K_{n+1} instead of Λ_{n+1}. Also $\Lambda_n\{3\}^\perp$, $K_n\{3\}^\perp$ denote the lattices orthogonal to these in $\Lambda_{23}\{3\}$.

A sequel to Plesken and Pohst [Ple6] has appeared — see [Plesk92].

Notes on Chapter 7: Further Connections Between Codes and Lattices

Upper bounds. The upper bounds on the minimal norm μ of a unimodular lattice and the minimal distance d of a binary self-dual code stated in Corollary 10 of Chapter 7 have been strengthened. In [RaSl98a] it is shown that an n-dimensional unimodular lattice has minimal norm

$$\mu \leq 2\left\lfloor \frac{n}{24} \right\rfloor + 2\ , \tag{8}$$

unless $n = 23$ when $\mu \leq 3$. The analogous result for binary codes (Rains [Rain98]) is that minimal distance of a self-dual code satisfies

$$d \leq 4\left\lfloor \frac{n}{24} \right\rfloor + 4\ , \tag{9}$$

unless $n \equiv 22 \pmod 4$ when the upper bound must be increased by 4.

These two bounds are obtained by studying the theta series (or weight enumerator) of the shadow of the lattice (or code) — see Notes on Chapter 4, especially equations (5), (6).

In [RaSl98] and [RaSl98a] it is proposed that a lattice or code meeting (8) or (9) be called **extremal**. This definition coincides with the historical usage for even lattices and doubly-even codes, but for odd lattices extremal has generally meant $\mu = [n/8] + 1$ and for singly-even codes that $d = 2[n/8] + 2$. In view of the new bounds in (8) and (9) the more uniform definition seems preferable. A lattice or code with the highest possible minimal norm or distance is called **optimal**. An extremal lattice or code is *a priori* optimal.

By using (5) and (6) it is often possible to determine the exact values of the highest minimal norm or minimal distance — see Table I.3, which is (essentially) taken from [CoSl90] and [CoSl90a]. The extremal code of length 62 was recently discovered by M. Harada.

Table I.3 Highest minimal norm (μ_n) of an n-dimensional integral unimodular lattice, and highest minimal distance (d_{2n}) of a binary self-dual code of length $2n$.

n	μ_n	d_{2n}	n	μ_n	d_{2n}
1	1	2	19	2	8
2	1	2	20	2	8
3	1	2	21	2	8
4	1	4	22	2	8
5	1	2	23	3	10
6	1	4	24	4	12
7	1	4	25	2	10
8	2	4	26	3	10
9	1	4	27	3	10
10	1	4	28	3	12
11	1	6	29	3	10
12	2	8	30	3	12
13	1	6	31	3	12
14	2	6	32	4	12
15	2	6	33	3	12
16	2	8	34	3 – 4	12
17	2	6	35	3 – 4	12 – 14
18	2	8	36	4	12 – 16

In the years since the manuscript of [CoSl90a] was first circulated, over 50 sequels have been written, supplying additional examples of codes in the range of Table I.3. In particular, codes with parameters $[70, 35, 12]$ (filling a gap in earlier versions of the table) were found independently by W. Scharlau and D. Schomaker [ScharS] and M. Harada [Hara97]. Other self-dual binary codes are constructed in [BrP91], [DoGH97a], [DoHa97], [Hara96], [Hara97], [KaT90], [PTL92], [Ton89], [ToYo96], [Tsa91], but these are just a sampling of the recent papers (see [RaSl98]).

For ternary self-dual (and other) codes see [Hara98], [HiN88], [Huff91], [KsP92], [Oze87], [Oze89b], [VAL93].

The classification of Type I self-dual binary codes of lengths $n \leq 30$ given in [Ple12] (cf. p. 189 of Chap. 7) has been corrected in [CoPS92] (see also [Yor89]).

Lam and Pless [LmP90] have settled a question of long standing by showing that there is no $[24, 12, 10]$ self-dual code over $\mathbb{F}_4$. The proof was by computer search, but required only a few hours of computation time. Huffman [Huff90] has enumerated some of the extremal self-dual codes over $\mathbb{F}_4$ of lengths 18 to 28.

We also show in [CoSl90], [CoSl90a], [CoSl98] that there are precisely five Type I optimal (i.e. $\mu = \mu_n$) lattices in 32 dimensions, but more than 8×10^{22} optimal lattices in 33 dimensions; that unimodular lattices with $\mu = 3$ exist precisely for $n \geq 23$, $n \neq 25$; that there are precisely three Type I extremal self-dual codes of length 32; etc.

Nebe [Nebe98] has found an additional example of an extremal unimodular lattice (P_{48n}) in dimension 48, and Bachoc and Nebe [BacoN98] contruct two extremal unimodular lattices in dimension 80. One of these (L_{80}) has kissing number 1250172000 (see Table I.2). The existence of an extremal unimodular lattice in dimension 72 (or of an extremal doubly-even code of length 72) remains open.

Several other recent papers have studied extremal unimodular lattices, especially in dimensions 32, 40, 48, etc. Besides [CoSl90], [CoSl98], which we have already mentioned, see Bonnecaze et al. [BonCS95], [BonS94], [BonSBM], Chapman [Chap 96], Chapman and Solé [ChS96], Kitazume et al. [KiKM], Koch [Koch86], [Koch90], Koch and Nebe [KoNe93], Koch and Venkov [KoVe89], [KoVe91], etc. Other lattices are constructed in [JuL88].

For doubly-even binary self-dual codes, **Krasikov and Litsyn** [KrLi97] have recently shown that the minimal distance satisfies

$$d \leq 0.166315\ldots n + o(n), \quad n \to \infty \; . \tag{10}$$

No comparable bound is presently known for even unimodular lattices.

For a comprehensive survey of self-dual codes over all alphabets, see Rains and Sloane [RaSl98].

[RaSl98a] also gives bounds, analogous to (8), for certain classes of **modular lattices** (see Notes to Chapter 2), and again there is a notion of **extremal** lattice. (Scharlau and Schulze-Pillot [SchaS98] have proposed a somewhat different definition of extremality for modular lattices.) Bachoc [Baco95], [Baco97] has constructed a number of examples of extremal modular lattices with the help of self-dual codes over various rings. Further examples of good modular lattices can be found in Bachoc and Nebe [BacoN98], Martinet [Mar96], Nebe [Nebe96], Plesken [Plesk96], Tiep [Tiep97a], etc.

For generalizations of the theorems of Assmus-Mattson and Venkov (cf. §7 of Chap. 7), see [CaD92], [CaD92a], [CaDS91], [Koch86], [Koch90], [KoVe89].

Several papers are related to **multiple theta series** of lattices. Peters [Pet90], extending the work of Ozeki [Oze4], has investigated the second-order theta series of extremal Type II (or even) unimodular lattices (cf. §7 of Chap. 7). [Pet89] studies the Jacobi theta series of extremal lattices. See also Böcherer and Schulze-Pillot [BocS91], [BocS97].

The connections between multiple weight enumerators of self-dual codes and Siegel modular forms have been investigated by Duke [Duke93], Ozeki [Oze76], [Oze4], [Oze97] and Runge [Rung93]–[Rung96]. Borcherds, Freitag and Weissauer [BorchF98] study multiple theta series of the Niemeier lattices.

Ozeki [Oze97] has recently introduced another generalization of the weight enumerator of a code C, namely its *Jacobi polynomial.* For a fixed vector $v \in \mathbb{F}^n$, this is defined by

$$Jac_{C,v}(x, z) = \sum_{u \in C} x^{wt(u)} z^{wt(u \cap v)} .$$

These polynomials have been studied in [BanMO96], [BanO96], [BonMS97]. They have the same relationship to Jacobi forms [EiZa85] as weight enumerators do to modular forms.

P_{48q} **and** P_{48p}. G. Nebe informs us that the automorphism groups of P_{48q} and P_{48p} are in fact $SL_2(47)$ and $SL_2(23) \times S_3$. We have modified page 195 accordingly. In the first paragraph on page 195, it would have been clearer if we had said that the vectors of these two lattices have the same *coordinate* shapes.

Notes on Chapter 8: Algebraic Constructions for Lattices

As we discuss in §7 of Chap. 8, there are several constructions for lattices that are based on **algebraic number theory**. The article by Lenstra [Len92] and the books by Bach and Shallit [BacSh96], Cohen [CohCNT] and Pohst and Zassenhaus [PoZ89] describe algorithms for performing algebraic number theory computations. (See also Fieker and

Pohst [FiP96].) The computer languages KANT, PARI and MAGMA (see the beginning of this Introduction) have extensive facilities for performing such calculations.

The papers [BoVRB, BoV98] give algebraic constructions for lattices that can be used to design signal sets for transmission over the Rayleigh fading channel.

Corrections to Table 8.1

There are four mistakes in Table 8.1. The entry headed $\omega_{IK} \to (GJH)$ should read

$$\begin{matrix} -1 & -2 & 0 & 0 & 0 \\ \sigma & 0 & 0 & 0 & 2 \\ 0 & 0 & 0 & 0 & 0 \\ -\tau & -2 & 0 & 0 & 2 \end{matrix},$$

and the entry headed $\omega_{JI} \to (GKH)$ should read

$$\begin{matrix} -1 & -2 & 0 & 0 & 0 \\ -\tau & -2 & 0 & 0 & 2 \\ \sigma & 0 & 0 & 0 & 2 \\ 0 & 0 & 0 & 0 & 0 \end{matrix}.$$

Further examples of new packings

Dimensions 25 to 30. As mentioned at the beginning of Chapter 17, the 25-dimensional unimodular lattices were classified by Borcherds [Bor1]. All 665 lattices (cf. Table 2.2) have minimal norm 1 or 2.

In dimension 26, Borcherds [Bor1] showed that there is a unique unimodular lattice with minimal norm 3. This lattice, which we will denote by S_{26}, was discovered by J. H. Conway in the 1970's.

The following construction of S_{26} is a modification of one found by Borcherds. We work inside a Lorentzian lattice P which is the direct sum of the unimodular Niemeier lattice A_4^6 and the Lorentzian lattice $I_{2,1}$ (cf. Chaps. 16 and 24). Thus $P \cong I_{26,1}$. Let $\rho = (-2, -1, 0, 1, 2)$ denote the Weyl vector for A_4, so that $\rho' = \rho \oplus \rho \oplus \rho \oplus \rho$ is the Weyl vector for A_4^6, of norm 60, and let $v' = (4, 2 \mid 9) \in I_{2,1}$. Then S_{26} is the sublattice of P that is perpendicular to $v = \rho' \oplus v' \in P$. S_{26} can also be constructed as a complex 13-dimensional lattice over $Q[(1+\sqrt{5})/2]$ ([Con16], p. 62).

Here are the properties of S_{26}. It is a unimodular 26-dimensional lattice of minimal norm 3, center density $\delta = 3^{13}2^{-26} = .0237\ldots$ (not a record), kissing number 3120 (also not a record), with automorphism group of order $2^8.3^2.5^4.13 = 18720000$, isomorphic to $Sp_4(5)$ (cf. [Con16], p. 61; [Nebe96a]). The minimal norm of a parity vector is 10, and there are 624 such vectors. The group acts transitively on these vectors. The theta series begins

$$1 + 3120q^3 + 102180q^4 + 1482624q^5 + \cdots .$$

We do not know the covering radius.

There is a second interesting 26-dimensional lattice, T_{26}, an integral lattice of determinant 3, minimal norm 4 and center density $1/\sqrt{3}$. This is best obtained by forming the sublattice of T_{27} (see below) that is perpendicular to a norm 3 parity vector. T_{26} is of interest because it shares the record for the densest known packing in 26 dimensions with the (nonintegral) laminated lattices Λ_{26}. The kissing number is 117936 and the group is the same as the group of T_{27} below.

Bacher and Venkov [BaVe96] have classified all unimodular lattices in dimensions 27 and 28 that contain no roots, i.e. have minimal norm ≥ 3. In dimension 27 there are three such lattices. In two of them the minimal norm of a parity vector is 11. These two lattices have theta series

$$1 + 2664q^3 + 101142q^4 + 1645056q^5 + \cdots$$

and automorphism groups of orders 7680 and 3317760, respectively. The third, found in [Con16], we shall denote by T_{27}. It has a parity vector of norm 3, theta series

$$1 + 1640q^3 + 119574q^4 + 1497600q^5 + \cdots$$

and a group of order $2^{13}3^5 7^2 13 = 1268047872$, which is isomorphic to the twisted group $2 \times ({}^3D_4(2) : 3)$ ([Con16], p. 89; [Nebe96a]). That this is the unique lattice with a parity vector of norm 3 was established by Borcherds [Bor1].

The following construction of T_{27} is based on the descriptions in [Con16], p. 89 and [Bor1]. Let V be the vector space of 3×3 Hermitian matrices

$$y = \begin{bmatrix} a & C & \bar{B} \\ \bar{C} & b & A \\ B & \bar{A} & c \end{bmatrix} = (a, b, c \mid A, B, C), \quad a, b, c \quad \text{real} ,$$

over the real Cayley algebra with units $i_\infty = 1$, $i_0, \ldots, i_6$, in which i_∞, $i_{n+1} \to i$, $i_{n+2} \to j$, $i_{n+4} \to k$ generate a quaternion subalgebra (for $n = 0, \ldots, 6$). V has real dimension $3 + 8 \times 3 = 27$. We define an inner product on V by $Norm(y) = \sum Norm(y_{ij})$. The lattice T_{27} is generated by the 3×3 identity matrix and the 819 images of the norm 3 vectors

$$\underset{(3)}{\begin{bmatrix} 1 & 0 & 0 \\ 0 & -1 & 0 \\ 0 & 0 & -1 \end{bmatrix}}, \quad \underset{(48)}{\begin{bmatrix} -1 & 0 & 0 \\ 0 & 0 & 1 \\ 0 & 1 & 0 \end{bmatrix}}, \quad \underset{(768)}{\begin{bmatrix} 0 & \bar{s} & \bar{s} \\ s & -\tfrac{1}{2} & \tfrac{1}{2} \\ s & \tfrac{1}{2} & -\tfrac{1}{2} \end{bmatrix}}$$

where $s = (i_\infty + i_0 + \cdots + i_6)/4$, under the group generated by the maps taking $(a, b, c \mid A, B, C)$ to $(a, b, c \mid iAi, iB, iC)$, $(b, c, a \mid B, C, A)$ and $(a, c, b \mid \bar{A}, -\bar{C}, -\bar{B})$, respectively, where $i \in \{\pm i_\infty, \ldots, \pm i_6\}$. These 820 norm 3 vectors and their negatives are all the minimal vectors in the lattice.

The identity matrix and its negative are the only parity vectors in T_{27} of norm 3. Taking the sublattice perpendicular to either vector gives T_{26}, which therefore has the same group as T_{27}.

In 28 dimensions Bacher and Venkov [BaVe96] show that there are precisely 38 unimodular lattices with no roots. Two of these have a parity vector of norm 4 and theta series

$$1 + 1728q^3 + 106472q^4 + \cdots ,$$

while for the other 36 the minimal norm of a parity vector is 12 and the theta series

$$1 + 2240q^3 + 98280q^4 + \cdots .$$

One of these 36 is the exterior square of E_8, which has group $2 \times G.2$, where $G = O_8^+(2)$ (whereas E_8 itself has group $2.G.2$). One of these 36 lattices also appears in Chapman [Chap97].

Bacher [Bace96] has also found lattices $\mathcal{B}_{27}$, $\mathcal{B}_{28}$, $\mathcal{B}_{29}$ in dimensions 27–29 which are denser than the laminated lattices Λ_{27}, Λ_{29}, and are the densest lattices presently known in these dimensions (although, as we have already mentioned in the Notes to Chapter 5, the densest packings currently known in dimensions 27 to 31 are all nonlattice packings).

$\mathcal{B}_{28}$ can be obtained by taking the even sublattice S_0 of S_{26}, which has determinant 4 and minimal norm 4, and finding translates $r_0 + S_0$, $r_1 + S_0$, $r_2 + S_0$ with $r_0 + r_1 + r_2 \in S_0$ and such that the minimal norm in each translate is 3. We then glue S_0 to a copy of A_2 scaled so that the minimal norm is 4, obtaining a lattice $\mathcal{B}_{28}$ with determinant 3, minimal norm 4, center density $1/\sqrt{3}$ and kissing number 112458. This is a nonintegral lattice since the r_i are not elements of the dual quotient S_0^*/S_0. $\mathcal{B}_{29}$ is obtained in the same way from T_{27}, and has determinant 3, minimal norm 4, center density $1/\sqrt{3}$ and kissing number 109884.

Dimensions 32, 48, 56

Nebe [Nebe98] studies lattices in dimension $2(p-1)$ on which $SL_2(p)$ acts faithfully. For $p \equiv 1 \pmod 4$ these are cyclotomic lattices over quaternion algebras. The three most interesting examples given in [Nebe98] are a 32-dimensional lattice with determinant 17^4, minimal norm $\mu = 6$, center density $\delta = 2^{-16}3^{16}17^{-2} = 2.2728\ldots$, kissing number $\tau = 233376$; a 56-dimensional lattice with det $= 1$, $\mu = 6$, $\delta = (3/2)^{28} = 85222.69\ldots$, $\tau = 15590400$; and a 48-dimensional even unimodular lattice with minimal norm 6 that is not isomorphic to either P_{48p} or P_{48q}, which we will denote by P_{48n}. Its automorphism group contains a subgroup $SL_2(13)$ whose normalizer in the full group is an absolutely irreducible group $(SL_2(13) \otimes SL_2(5)).2^2$.

Dimensions 36 and 60

Kschischang and Pasupathy [KsP92] combine codes over $\mathbb{F}_3$ and $\mathbb{F}_4$ to obtain lattice packings Ks_{36}, Ks_{60} with center densities δ given by

$$\log_2 \delta = 2.1504 \qquad \text{(in 36 dimensions)},$$

$$\log_2 \delta = 17.4346 \qquad \text{(in 60 dimensions)},$$

respectively, thus improving two entries in Table 1.3. Their construction is easily described using the terminology of §8 of Chap. 7. If $\mathcal{E}$ denotes the Eisenstein integers, there are maps

$$\pi_4 : \mathcal{E} \to \mathcal{E}/2\mathcal{E} \to \mathbb{F}_4 \ ,$$

$$\pi_3 : 2\mathcal{E} \to 2\mathcal{E}/2\theta\mathcal{E} \to \mathbb{F}_3 \ ,$$

where $\theta = \sqrt{-3}$. If C_1 is an $[n, k_1, d_1]$ code over $\mathbb{F}_4$, and C_2 is an $[n, k_2, d_2]$ code over $\mathbb{F}_3$, we define Λ to be the complex n-dimensional Eisenstein lattice spanned by the vectors of $(2\theta\mathcal{E})^n$, $\pi_4^{-1}(C_1)$ and $\pi_3^{-1}(C_2)$. The real version of Λ (cf. §2.6 of Chap. 2) is then a $2n$-dimensional lattice with determinant $2^{2n-4k_1}3^{3n-2k_2}$ and minimal norm $= \min\{12, d_1, 4d_2\}$. In the case $n = 18$ Kschischang and Pasupathy take C_1 to be the $[18, 9, 8]$ code S_{18} of [Mac4] and C_2 to be the $[18, 17, 2]$ zero-sum code; and for $n = 30$ they take C_1 to be the $[30, 15, 12]$ code Q_{30} of [Mac4] and C_2 to be a $[30, 26, 3]$ negacyclic code. Christine Bachoc (personal communication) has found that Ks_{36} has kissing number 239598.

Elkies [Elki] points out that the densities of **Craig's lattices** (§6 of Chap. 8) can be improved by adjoining vectors with fractional coordinates. However, these improvements do not seem to be enough to produce new record packings, at least in dimensions up to 256. The Craig lattices (among others) have also been investigated by Bachoc and Batut [BacoB92].

Lattices from representations of groups

Of course nearly every lattice in the book could be described under this heading. We have already mentioned the lattices obtained by Nebe and Plesken [NePl95] and Nebe [Nebe96], [Nebe96a] from groups of rational matrices. See also Adler [Adle81].

Scharlau and Tiep [SchaT96], [SchaT96a] have studied lattices arising from representations of the symplectic group $Sp_{2n}(p)$. Among other things, [SchaT96] describes "p-analogues" of the Barnes-Wall lattices.

The concept of a "globally irreducible" lattice was first investigated by Thompson [Tho2], [Tho3], [Thomp76] in the course of his construction of the sporadic finite simple group Th. The construction involves a certain even unimodular lattice 248-dimensional lattice TS_{248} with minimal norm 16 (the **Thompson-Smith lattice**), with $Aut(TS_{248}) = 2 \times Th$. (For more about this lattice see also Smith [Smith76], Kostrikin and Tiep [KoTi94].)

This lattice shares with Z, E_8 and the Leech lattice the property of being **globally irreducible**: $\Lambda/p\Lambda$ is irreducible for every prime p.

However, Gross [Gro90] remarks that over algebraic number rings such lattices are more common. He gives new descriptions of several familiar lattices as well as a number of new families of unimodular lattices. Further examples of globally irreducible lattices have been found by Gow [Gow89], [Gow89a]. See also Dummigan [Dum97], Tiep [Tiep91]–[Tiep97b].

Thompson and Smith actually constructed their lattice by decomposing the Lie algebra of type E_8 over $\mathbb{C}$ into a family of 31 mutually perpendicular Cartan subalgebras. Later authors have used other Lie algebras to obtain many further examples of lattices, including infinite families of even unimodular lattices. See Abdukhalikov [Abdu93], Bondal, Kostrikin and Tiep [BKT87], Kantor [Kant96], and especially the book by Kostrikin and Tiep [KoTi94].

Lattices from tensor products

Much of the final chapter of Kitaoka's book [Kita93] is concerned with the properties of tensor products of lattices. The minimal norm of a tensor product $L \otimes M$ clearly cannot exceed the product of the minimal norms of L and M, and may be less. Kitaoka says that a lattice L is of **E-type** if, for any lattice M, the minimal vectors of $L \otimes M$ have the form $u \otimes v$ for $u \in L$, $v \in M$. (This implies $\min(L \otimes M) = \min(L) \min(M)$.) Kitaoka elegantly proves that every lattice of dimension $n \leq 43$ is of E-type.

On the other hand the Thompson-Smith lattice TS_{248} is not of E-type. (Thompson's proof: Let $L = TS_{248}$, and consider $L \otimes L \cong Hom(L, L)$. The element of $L \otimes L$ corresponding to the identity element of $Hom(L, L)$ is easily seen to have norm 248, which is less than the square of the minimal norm of L.) Steinberg ([Mil7], p. 47) has shown that there are lattices in every dimension $n \geq 292$ that are not of E-type.

If an extremal unimodular lattice of dimension 96 (with minimal norm 10) could be found, or an extremal 3-modular lattice in dimension 84 (with minimal norm 16), etc., they would provide lower-dimensional examples of non-E-type lattices.

Coulangeon [Cogn98] has given a generalization of Kitaoka's theorem to lattices over imaginary quadratic fields or quaternion division algebras. Such tensor products provide several very good lattices. Bachoc and Nebe [BacoN98] take as their starting point the lattice L_{20} described on p. 39 of [Con16]. This provides a 10-dimensional representation over $\mathbb{Z}[\alpha]$, $\alpha = (1+\sqrt{-7})/2$ for the group $2.M_{22}.2$. (L_{20} is an extremal 7-modular lattice with minimal norm 8 and kissing number 6160.) Bachoc and Nebe form the tensor product of L_{20} with A_2^2 over $\mathbb{Z}[\alpha]$ and obtain a 40-dimensional extremal 3-modular lattice with minimal norm 8, and of L_{20} with E_8 to obtain an 80-dimensional extremal unimodular lattice L_{80} with minimal norm 8 and kissing number 1250172000 (see Notes on Chapter 1).

Lattices from Riemann surfaces

The period matrix of a compact Riemann surface of genus g determines a real $2g$-dimensional lattice. Buser and Sarnak [BuSa94] have shown that from a sphere packing point of view these lattices are somewhat disappointing: for large g their density is much worse than the Minkowski bound, neither the root lattices E_6, E_8 nor the Leech lattice can be obtained, and so on. Nevertheless, for small genus some interesting lattices occur [BerSl97], [Quin95], [QuZh95], [RiRo92], [Sar95], [TrTr84].

One example, the m.c.c. lattice, has already been mentioned in the Notes on Chapter 1. The period matrix of the Bring curve (the genus 4 surface with largest automorphism group) was computed by Riera and Rodríguez, and from this one can determine that the corresponding lattice is an 8-dimensional lattice with determinant 1; minimal norm 1.4934... and kissing number 20 (see [NeSl]).

Lattices and codes with no group

Etsuko Bannai [Bann90] showed that the fraction of n-dimensional unimodular lattices with trivial automorphism group approaches 1 as $n \to \infty$. Some explicit examples were given by Mimura [Mimu90]. Bacher [Bace94] has found a Type I lattice in dimension 29 and a Type II lattice in dimension 32 with trivial groups $\{\pm 1\}$. Both dimensions are the lowest possible.

Concerning codes, Orel and Phelps [OrPh92] proved that the fraction of binary self-dual codes of length n with trivial group approaches 1 as $n \to \infty$. A self-dual code with trivial group of length 34 (conjectured to be the smallest possible length) is constructed in [CoSl90a], and a doubly-even self-dual code of length 40 (the smallest possible) in [Ton89]. See also [BuTo90], [Hara96], [Huff98], [LePR93], and [Leo8] (for a ternary example).

Notes on Chapter 9: Bounds for Codes and Sphere Packings

Samorodnitsky [Samo98] shows that the Delsarte linear programming bound for binary codes is at least as large as the average of the Gilbert-Varshamov lower bound and the McEliece-Rodemich-Rumsey-Welch upper bound, and conjectures that this estimate is actually the true value of the pure linear programming bound.

Krasikov and Litsyn [KrLi97a] improve on Tietavainen's bound for codes with $n/2 - d = o(n^{1/3})$.

Laihonen and Litsyn [LaiL98] derive a straight-line upper bound on the minimal distance of nonbinary codes which improves on the Hamming, linear programming and Aaltonen bounds.

Levenshtein [Lev87], [Lev91], [Lev92] and Fazekas and Levenshtein [FaL95] have obtained new bounds for codes in finite and infinite polynomial association schemes (cf. p. 247 of Chap. 9).

Table 9.1 has been revised to include several new bounds on $A(n,d)$. A table of lower bounds on $A(n,d)$ extending to $n \leq 28$ (cf. Table 9.1 of Chap. 9) has been published by Brouwer et al. [BrSSS] (see also [Lits98]). The main purpose of [BrSSS], however, is to present a table of lower bounds on $A(n,d,w)$ for $n \leq 28$ (cf. §3.4 of Chap. 9).

Notes on Chapter 10: Three Lectures on Exceptional Groups

Curtis [Cur89a], [Cur90] discusses further ways to generate the Mathieu groups M_{12} and M_{24} (cf. Chaps. 10, 11).

Hasan [Has89] has determined the possible numbers of common octads in two Steiner systems $S(5,8,24)$ (cf. §2.1 of Chap. 10). The analogous results for $S(5,6,12)$ were determined by Kramer and Mesner in [KrM74].

Figure 10.1 of Chap. 10 classifies the binary vectors of length 24 into orbits under the Mathieu group M_{24}. [CoSl90b] generalizes this in the following way. Let C be a code of length n over a field $\mathbb{F}$, with automorphism group G, and let C_w denote the subset of codewords of C of weight w. Then we wish to classify the vectors of $\mathbb{F}^n$ into orbits under G, and to determine their distances from the various subcodes C_w. [CoSl90b] does this for the first-order Reed-Muller, Nordstrom-Robinson and Hamming codes of length 16, the Golay and shortened Golay codes of lengths 22, 23, 24, and the ternary Golay code of length 12.

For recent work on the subgroup structure of various finite groups (cf. Postscript to Chap. 10) see Kleidman et al. [KlL88], [KlPW89], [KlW87], [KlW90], [KlW90a], Leibeck et al. [LPS90], Linton and Wilson [LiW91], Norton and Wilson [NoW89], Wilson [Wil88], [Wil89]. The "modular" version of the ATLAS of finite groups [Con16] has now appeared [JaLPW].

On page 289, in the proof of Theorem 20, change "$x \cdot y = y \cdot y = 64$" to "$x \cdot x = y \cdot y = 64$". On page 292, 8th line from the bottom, change "$\{i\}$" to "$\{j\}$"

Borcherds points out that in Table 10.4 on page 291 there is a third orbit of type 10 vectors, with group $M_{22}.2$. (Note that [Con16], p. 181, classifies the vectors up to type 16).

Notes on Chapter 11: The Golay Codes and the Mathieu Groups

For more about the MOG (cf. §5 of Chap. 11) see Curtis [Cur89].

The cohomology of the groups M_{11}, M_{12} and J_1 has been studied in [BenCa87], [AdM91] and [Cha82], respectively.

Notes on Chapter 13: Bounds on Kissing Numbers

Drisch and Sonneborn [DrS96] have given an upper bound on the degree of the best polynomial to use in the main theorem of §1 of Chap. 13.

Notes on Chapter 15: On the Classification of Integral Quadratic Forms

A recent book by Buell [Bue89] is devoted to the study of binary quadratic forms (cf. §3 of Chap. 15). See also Kitaoka's book [Kita93] on the arithmetic theory of quadratic forms (mentioned already in the Notes on Chapter 8). Hsia and Icaza [HsIc97] give an effective version of Tartakovsky's theorem.

For an interpretation of the "oddity" of a lattice, see Eq. (4) of the Notes on Chapter 4.

Tables

Nipp ([Nip91] has constructed a table of reduced positive-definite integer-valued four-dimensional quadratic forms of discriminant ≤ 1732. A sequel [Nip91a] tabulates five-dimensional forms of discriminant ≤ 256. These tables, together with a new version of the Brandt-Intrau [Bra1] tables of ternary forms computed by Schiemann can also be found on the electronic *Catalogue of Lattices* [NeSl].

Universal forms

The **15-theorem**. Conway and Schneeberger [Schnee97], [CoSch98] (see also [CoFu97]) have shown that for a positive-definite quadratic form with integer matrix entries to represent all positive integers it suffices that it represent the numbers 1, 2, 3, 5, 6, 7, 10, 14, 15. It is conjectured (the **290-conjecture**) that for a positive-definite quadratic form with integer values to represent all positive integers it suffices that it represent the numbers 1, 2, 3, 5, 6, 7, 10, 13, 14, 15, 17, 19, 21, 22, 23, 26, 29, 30, 31, 34, 35, 37, 42, 58, 93, 110, 145, 203, 290.

The 15-theorem is best-possible in the sense that for each of the nine critical numbers c there is a positive-definite diagonal form in four variables that misses only c. For example $2w^2 + 3x^2 + 4y^2 + 5z^2$ misses only 1, and $w^2 + 2x^2 + 5y^2 + 5z^2$ misses only 15. For the other c in the 290-conjecture the forms are not diagonal and sometimes involve five variables. For example a form that misses only 290 is

$$\begin{bmatrix} 2 & \frac{1}{2} & 0 & 0 & 0 \\ \frac{1}{2} & 4 & \frac{1}{2} & 0 & 0 \\ 0 & \frac{1}{2} & 1 & 0 & 0 \\ 0 & 0 & 0 & 29 & 14\frac{1}{2} \\ 0 & 0 & 0 & 14\frac{1}{2} & 29 \end{bmatrix} .$$

For other work on universal forms see Chan, Kim and Raghavan [ChaKR], Earnest and Khosravani [EaK97], [EaK97b], Kaplansky [Kapl95].

M. Newman [Newm94] shows that any symmetric matrix A of determinant d over a principal ideal ring R is congruent to a tridigonal matrix

$$\begin{bmatrix} c_1 & d_1 & 0 & 0 & & & \\ d_1 & c_2 & d_2 & 0 & & & \\ 0 & d_2 & c_3 & d_3 & & & \\ & & & \cdots & & & \\ & & & & d_{n-2} & c_{n-1} & d_{n-1} \\ & & & & 0 & d_{n-1} & c_n \end{bmatrix}$$

in which d_i divides d for $1 \le i \le n-2$. In particular, the Gram matrix for a unimodular lattice can be put into tridiagonal form where all off-diagonal entries except perhaps the last one are equal to 1.

[CSLDL1] extends the classification of positive definite integral lattices of small determinant begun in Tables 15.8 and 15.9 of Chap. 15. Lattices of determinants 4 and 5 are classified in dimensions $n \le 12$, of determinant 6 in dimensions $n \le 11$, and of determinant up to 25 in dimensions $n \le 7$.

The four 17-dimensional even lattices of determinant 2 (cf. Table 15.8) were independently enumerated by Urabe [Ura89], in connection with the classification of singular points on algebraic varieties. We note that in 1984 Borcherds [Bor1, Table −2] had already classified the 121 25-dimensional even lattices of determinant 2. Even lattices of dimension 16 and determinant 5 have been enumerated by Jin-Gen Yang [Yan94], and other lattice enumerations in connection with classification of singularities can be found in [Tan91], [Ura87], [Ura90], [Wan91]. For the connections between lattices and singularities, see Eberling [Ebe87], Kluitmann [Klu89], Slodowy [Slod80], Urabe [Ura93], Voigt [Voi85].

Kervaire [Kerv94] has completed work begun by Koch and Venkov and has shown that there are precisely 132 indecomposable even unimodular lattices in **dimension 32** which have a "complete" root system (i.e. the roots span the space). Only 119 distinct root systems occur.

Several recent papers have dealt with the construction and classifications of lattices, especially unimodular lattices, over rings of integers in number fields, etc. See for example Bayer-Fluckiger and Fainsilber [BayFa96], Benham et al. [BenEHH], Hoffman [Hof91], Hsia [Hsia89],

Hsia and Hung [HsH89], Hung [Hun91], Takada [Tak85], Scharlau [Scha94], Zhu [Zhu91]–[Zhu95b].

Some related papers on class numbers of quadratic forms are Earnest [Earn88]–[Earn91], Earnest and Hsia [EaH91], Gerstein [Gers72], Hashimoto and Koseki [HaK86].

Hsia, Jochner and Shao [HJS], extending earlier work of Friedland [Fri89], have shown that for any two lattices Λ and M of dimension > 2 and in the same genus (cf. §7 of Chap. 15), there exist isometric primitive sublattices Λ' and M' of codimension 1.

Fröhlich and Thiran [FrTh94] use the classification of Type I lattices in studying the quantum Hall effect.

Erdős numbers

An old problem in combinatorial geometry asks how to place a given number of distinct points in n-dimensional Euclidean space so as to minimize the total number of distances they determine ([Chu84], [Erd46], [ErGH89], [SkSL]). In 1946 Erdős [Erd46] considered configurations formed by taking all the points of a suitable lattice that lie within a large region. The best lattices for this purpose are those that minimize what we shall call the *Erdős number* of the lattice, given by

$$E = Fd^{1/n},$$

where d is the determinant of the lattice and F, its *population fraction*, is given by

$$F = \lim_{x\to\infty} \frac{P(x)}{x}, \quad \text{if} \quad n \geq 3,$$

where $P(x)$ is the population function of the corresponding quadratic form, i.e. the number of values not exceeding x taken by the form.[2] The Erdős number is the population fraction when the lattice is normalized to have determinant 1. It turns out that minimizing E is an interesting problem in pure number theory.

In [CoSl91]] we prove all cases except $n = 2$ (handled by Smith [Smi91]) of the following proposition:

> The lattices with minimal Erdős number are (up to a scale factor) the even lattices of minimal determinant. For $n = 0, 1, 2, \ldots$ these determinants are
> $$1,\ 2,\ 3,\ 4,\ 4,\ 4,\ 3,\ 2,\ 1,\ 2,\ 3,\ 4,\ 4,\ 4,\ \ldots,$$
> this sequence continuing with period 8.

---

[2]For $n \leq 2$ these definitions must be modified. For $n = 0$ and 1 we set $E = 1$, while for $n = 2$ we define F by $F = \lim_{x\to\infty} x^{-1} P(x)\sqrt{\log x}$.

For $n \le 10$ these lattices are unique:

$$A_0, A_1, A_2, A_3 \simeq D_3, D_4, D_5, E_6, E_7, E_8, E_8 \oplus A_1, E_8 \oplus A_2,$$

with Erdős numbers

$$1,\ 1,\ 2^{-3/2}3^{1/4} \prod_{p \equiv 2(3)} \left(1 - \frac{1}{p^2}\right)^{-1/2} = 0.5533,$$

$$\frac{11}{24}4^{1/3} = 0.7276, \quad \frac{4^{1/4}}{2} = 0.7071, \quad \frac{4^{1/5}}{2} = 0.6598, \quad \frac{3^{1/6}}{2} = 0.6005,$$

$$\frac{2^{1/7}}{2} = 0.5520, \quad \frac{1}{2}, \quad \frac{2^{1/9}}{2} = 0.5400, \quad \frac{3^{1/10}}{2} = 0.5581,$$

(rounded to 4 decimal places), while for each $n \ge 11$ there are two or more such lattices. The proof uses the p-adic structures of the lattices (cf. Chap. 15). The three-dimensional case is the most difficult. The crucial number-theoretic result needed for our proof was first established by Peters [Pet80] using the Generalized Riemann Hypothesis. The dependence on this hypothesis has been removed by Duke and Schulze-Pillot [DuS90].

Notes on Chapter 16: Enumeration of Unimodular Lattices

Recent work on the classification of lattices of various types has been described in the Notes on Chapter 15.

The **mass formula** of H. J. S. Smith, H. Minkowski and C. L. Siegel (cf. §2 of Chap. 16) expresses the sum of the reciprocals of the group orders of the lattices in a genus in terms of the properties of any one of them. In [CSLDL4] we discuss the history of the formula and restate it in a way that makes it easier to compute. In particular we give a simple and reliable way to evaluate the 2-adic contribution. Our version, unlike earlier ones, is visibly invariant under scale changes and dualizing. We then use the formula to check the enumeration of lattices of determinant $d \le 25$ given in [CSLDL1]. [CSLDL4] also contains tables of the "standard mass," values of the L-series $\Sigma(\frac{n}{m})m^{-s}$ (m odd), and genera of lattices of determinant $d \le 25$.

Eskin, Rudnick and Sarnak [ERS91] give a new proof of the mass formula using an "orbit-counting" method. Another proof is given by Mischler [Misch94].

The classification of the Niemeier lattices is rederived by Harada, Lang and Miyamoto [HaLaM94]. Montague [Mont94] constructs these lattices from ternary self-dual codes of length 24, and Bonnecaze et al. [BonGHKS] construct them from codes over $\mathbb{Z}_4$ using Construction A_4.

Codes over $\mathbb{Z}_4$

The glue code for the Niemeier lattice with components A_4^6 (cf. Table 16.1 of Chap. 16) is a certain linear code (the **octacode**) of length 8 over the ring $\mathbb{Z}_4$ of integers mod 4. This code may be obtained from the code generated by all cyclic shifts of the vector (3231000) by appending a zero-sum check symbol. The octacode contains 256 vectors and is self-dual with respect to the standard inner product $x \cdot y = \sum x_i y_i \pmod 4$. It is the unique self-dual code of length 8 and minimal Lee weight 6 [CoSl93].

In [FoST93] it is shown that if the octacode is mapped to a binary code of twice the length using the **Gray map**:

$$0 \to 00, \quad 1 \to 01, \quad 2 \to 11, \quad 3 \to 10, \tag{11}$$

then we obtain the (nonlinear) Nordstrom-Robinson code (mentioned in §2.12 of Chapter 2).

Furthermore, if we apply "Construction A_4" (see Notes on Chapter 4) to the octacode, we obtain the E_8 lattice [BonS94], [BonCS95].

The octacode may also be obtained by Hensel-lifting the Hamming code of length 7 from $GF(2)$ to $\mathbb{Z}_4$. If the same process is applied to an arbitrary binary Hamming code of length $2^{2m+1} - 1$ $(m \geq 1)$ we obtain the (nonlinear) Preparata and Kerdock codes [HaKCSS]. If this process is applied to the binary Golay code of length 23, then as already mentioned in the Notes on Chapter 4, we obtain a code over $\mathbb{Z}_4$ that lifts by Construction A_4 to the Leech lattice [BonS94], [BonCS95].

Many other nonlinear binary codes also have a simpler description as $\mathbb{Z}_4$ codes. Consider, for example, the $(10, 40, 4)$ binary nonlinear code found by **Best** [Bes1], which leads via Construction A to the densest 10-dimensional sphere packing presently known (Chapter 5, p. 140). This code now has the following simple description [CoSl94b]: take the 40 vectors of length 5 over $\mathbb{Z}_4$ obtained from

$$(c - d, b, c, d, b + c), \quad b, c, d \in \{+1, -1\} \ ,$$

and its cyclic shifts, and apply the Gray map (11). Litsyn and Vardy [LiV94] have shown that the Best code is unique.

In some cases the $\mathbb{Z}_4$-approach has also led to (usually nonlinear) binary codes that are better than any previously known: see Calderbank and McGuire [CaMc97], Pless and Qian [PlQ96], Shanbhag, Kumar and Helleseth [ShKH96], etc.

For other recent papers dealing with codes over $\mathbb{Z}_4$ see [DoHaS97], [Huff98a], [Rain98a], [RaSl98].

Bannai et al. [BanDHO97] investigate self-dual codes over $\mathbb{Z}_{2k}$ and their relationship with unimodular lattices.

Notes on Chapter 17: The 24-Dimensional Odd Unimodular Lattices

The chapter has been completely retyped for this edition, in order to describe the enumeration process more clearly, and to correct several errors in the tables.

Neighbours

Borcherds uses Kneser's notion of neighboring lattices [Kne4], [Ven2], defined as follows. Two lattices L and L' are *neighbors* if their intersection $L \cap L'$ has index 2 in each of them. We note the following properties.

(i) Let L be a unimodular lattice. Suppose $u \in L$, $\frac{1}{2}u \notin L$ and $u \cdot u \in 4\mathbb{Z}$. Let $L_u = \{x \in L : x \cdot u \equiv 0 \pmod 2\}$ and $L^u = \langle L_u, \frac{1}{2}u\rangle$. Then L^u is a unimodular neighbor of L, all unimodular neighbors of L arise in this way, and u, u' produce the same neighbor if and only if $\frac{1}{2}u \equiv \frac{1}{2}u' \bmod L_u$.

(ii) Let L be an even unimodular lattice. Suppose $u \in L$, $\frac{1}{2}u \notin L$ and $u \cdot u \in 4\mathbb{Z}$. Then L^u is an even unimodular lattice if and only if $u \cdot u \in 8\mathbb{Z}$. If $\frac{1}{2}u \equiv \frac{1}{2}u' \bmod L$ then $L^u = L^{u'}$ and $\frac{1}{2}u \equiv \frac{1}{2}u' \pmod{L_u}$.

(iii) Define the *integral part* of an arbitrary lattice M to be

$$[M] = M \cap M^* = \{l \in M : l \cdot M \subseteq \mathbb{Z}\} . \tag{12}$$

If L is unimodular, and $u \in L$, $\frac{1}{2}u \notin L$, $u \cdot u \in 4\mathbb{Z}$, then the neighbor L^u is also equal to

$$[\langle L, \tfrac{1}{2}u\rangle] . \tag{13}$$

The proofs are straightforward, and are written out in full in [Wan96]. (Note that a neighbor of a unimodular lattice need not be unimodular: it must have determinant 1, but need not be integral. Overlooking this point caused some inaccuracies in the statement of these properties in the second edition.)

For example, if $L = \mathbf{Z}^8$ and $x = (\frac{1}{2})^8$, we obtain $B = E_8$.

As a second example, let L be the Niemeier lattice A_1^{24}, which we take in the form obtained by applying Construction A to C_{24} (see §1 of the previous chapter). Any $a \in A_1^{24}$ has the shape $8^{-1/2}\ (2c + 4z)$ for $c \in C_{24}$, $z \in \mathbf{Z}^{24}$. There are just two inequivalent choices for x. If $x = 8^{-1/2}(2^{24})$ then L^u is the odd Leech lattice O_{24}, with minimal norm 3 (see the Appendix to Chap. 6). If $x = 8^{-1/2}(-6, 2^{23})$ then L^u is the Leech lattice Λ_{24} itself. In fact A_1^{24} is the unique even neighbor of Λ_{24} (see Fig. 17.1).

Eq. (5) can be used to construct new lattices even if x does not satisfy (1). For example many years ago Thompson [Tho7] showed that the Leech lattice can be obtained from D_{24} by

$$\Lambda_{24} = [\langle D_{24}, x\rangle] \tag{14}$$

where

$$x = \left(\frac{0}{47}, \frac{1}{47}, \ldots, \frac{23}{47}\right) . \tag{15}$$

But since the intersection $\Lambda_{24} \cap D_{24}$ has index 47 in each of them, Λ_{24} and D_{24} are not neighbors in our sense.

Notes on Chapter 18: Even Unimodular 24-Dimensional Lattices

An error in Venkov's proof was found and corrected by Wan [Wan96].

There are also two tiny typographical errors on page 435. In line 13, change "x_2" to "x_4", and in line 18 from the bottom, change "$x_1 = (0, 0, 1, 1, -2, -2)$" to "$x_1 = (0, 0, 1, 1, 2, -2)$".

Scharlau and Venkov [SchaV94] use the method of Chapter 17 to classify all lattices in the genus of BW_{16}.

Notes on Chapter 20: Finding Closest Lattice Point

The following papers deal with techniques for finding the closest lattice point to a given point (cf. Chap. 20), and with the closely related problem of "soft decision" decoding of various binary codes. Most of these are concerned with the Golay code and the Leech lattice. [AdB88], [Agr96], [Allc96], [Amr93], [Amr94], [BeeSh91], [BeeSh92], [Forn89], [FoVa96], [LaLo89], [RaSn93], [RaSn95], [RaSn98], [SnBe89], [Vard94], [Vard95a], [VaBe91], [VaBe93].

Notes on Chapter 21: Voronoi Cells of Lattices and Quantization Errors

Because of an unfortunate printer's error, the statement of one of the most important theorems on Voronoi cells (Theorem 10 of Chap. 21) was omitted from the first edition. This has now been restored at the top of page 474. Concerning the previous theorem (Theorem 9), Rajan and Shende [RaSh96] have shown that the conditions of the theorem are necessary and sufficient for the conclusion, and also that the only lattices whose relevant vectors are precisely the minimal vectors are the root lattices (possibly rescaled). Geometrically, this assertion is just that the root lattices are the only lattices whose Voronoi cells have inscribed spheres.

Viterbo and Biglieri [ViBi96] give an algorithm for computing the Voronoi cell of a lattice. A recent paper by Fortune [Fort97] surveys techniques for computing Voronoi and Delaunay regions for general sets of points.

Notes on Chapter 23: The Covering Radius of the Leech Lattice

Deza and Grishukhin [DezG96] claim to give a simpler derivation of the covering radius of the Leech lattice, obtained by a refinement of Norton's argument of Chapter 22. However, as the reviewer in *Math. Reviews* points out (MR98b: 11074), there is a gap in their argument. A correction will be published in *Mathematika*.

Notes on Chapter 25: The Cellular Structure of the Leech Lattice

On page 520, in Fig. 25.1(b) the shaded node closest to the top of the page should not be shaded.

Notes on Chapter 27: The Automorphism Group of the 26-Dimensional Lorentzian Lattice

Borcherds [Borch90] has generalized some of the results of Chap. 27, by showing that, besides the Leech lattice, several other well-known lattices, in particular the Coxeter-Todd lattice K_{12} and the Barnes-Wall lattice BW_{16}, are related to Coxeter diagrams of reflection groups of Lorentzian lattices.

Borcherds (personal communication) remarks that some of the questions on the last page of the chapter can now be answered.

(ii) There is a proof of the main result of this chapter that does not use the covering radius of the Leech lattice on page 199 of [Borch95a], but it is very long and indirect.

(iv) The group is transitive for $\mathrm{II}_{33,1}, \ldots$. This follows from Corollary 9.7 of [Borch87].

(v) The answer is yes: see Notes on Chapter 30.

Kondo [Kon97] has made use of the results in this chapter and in [Borch87] in determining the full automorphism group of the Kummer surface associated with a generic curve of degree 2.

Notes on Chapter 28: Leech Roots and Vinberg Groups

A lattice Λ in Lorentzian space $\mathbb{R}^{n,1}$ is called **reflexive** if the subgroup of Aut(Λ) generated by reflections has finite index in Aut(Λ). Esselmann [Ess90] (extending the work of Makarov [Mak65], [Mak66], Vinberg [Vin1]–[Vin13], Nikulin [Nik3], Prokhorov [Pro87] and the present authors, see Chap. 28) shows that for $n = 20$ and $n \geq 22$ reflexive lattices do not exist, and that for $n = 21$ the example found by Borcherds (the even sublattice of $I_{21,1}$ of determinant 4) is essentially the only one.

Allcock ([Allc97], [Allc97a]) found several new complex and quaternionic hyperbolic reflection groups, the largest of which behave rather like the reflection group of $II_{25,1}$ described in Chapter 27.

Two other recent papers dealing with the Lorentzian lattices and related topics are Scharlau and Walhorn [SchaW92], [SchaW92a].

Notes on Chapter 30: A Monster Lie Algebra?

The Lie algebra of this chapter is indeed closely related to the Monster simple group. In order to get a well behaved Lie algebra it turns out to be necessary to add some imaginary simple roots to the "Leech roots." This gives the *fake* Monster Lie algebra, which contains the Lie algebra of this chapter as a large subalgebra. See [Borch90a] for details (but note that the fake Monster Lie algebra is called the Monster Lie algebra in this paper). The term "Monster Lie algebra" is now used to refer to a certain "Z/2Z-twisted" version of the fake Monster Lie algebra. The Monster Lie algebra is acted on by the Monster simple group, and can be used to show that the Monster module constructed by Frenkel, Lepowsky, and Meurman satisfies the moonshine conjectures; see [Borch92a].

For other recent work on the Monster simple group (Chap. 29) and related Lie algebras see [Borch86]–[BorchR96], [Con93], [Fre5], [Iva90], [Iva92], [Iva92a], [Lep88], [Lep91], [Nor90], [Nor92].

For further work on "Monstrous Moonshine" see Koike [Koik86], Miyamoto [Miya95], the conference proceedings edited by Dong and Mason [DoMa96], and the Cummins bibliography [Cumm].

Errata for the Low-Dimensional Lattices papers [CSLDL1] – [CSLDL7]

[CSLDL1]. On page 37, in line 8 from the bottom, the second occurrence of $p \equiv 1$ should be $p \equiv -1$.

[CSLDL2]. On page 41, in Table 1, the first occurrence of $Q_{23}(4)^{+4}$ should be $Q_{23}(4)^{+2}$.

[CSLDL3]. On page 56, in Table 1, the entry for P_5^3 should have $\mu = 2$ (not 4). On page 62, at about line 10, "$78k_c$" should be "$78c_k$". On page 68, in the entry for p_7^{12}, the third neighbor should be changed from "p_7^{28}" to "p_7^{4}". In the entry for p_7^{13}, the six neighbors should read, in order, "p_7^{11}, p_7^{5}, p_7^{1}, p_7^{17}, p_7^{1}, p_7^{1}". On page 70, in the entry for p_7^{17}, the second neighbor should be changed from "p_7^{15}" to "p_7^{7}".

[CSLDL4]. In line 1 of page 273, change "7" to "3".

List of Symbols

		page
a_n	ordinary Coxeter-Dynkin diagram	481
A	Gram matrix of lattice	5
A_n	n-dimensional zero-sum root lattice	108
$A_n^{(m)}$	Craig's lattice	223
A_n	alternating group of degree n	82
$A.B$, AB, $A:B$, $A{\cdot}B$	extensions of groups	92
$Aut(\Lambda)$	automorphism group of lattice	90
B_n	lattice constructed from BCH codes	235
BW_n	n-dimensional Barnes-Wall lattice	234
C_n	cyclic group of order n	109
$\mathscr{C}_4$	[4, 2, 3] tetracode over $\mathbf{F}_3$	81, 320
$\mathscr{C}_6$	[6, 3, 4] hexacode over $\mathbf{F}_4$	82, 300
$\mathscr{C}_{12}$	[12, 6, 6] Golay code over $\mathbf{F}_3$	85, 321
$\mathscr{C}_{24}$ or $\mathscr{C}$	[24, 12, 8] Golay code over $\mathbf{F}_2$	84, 303
c.w.e.	complete weight enumerator	78
d	minimal distance of code (usually)	76
d_n	ordinary Coxeter-Dynkin diagram	481
D_n	n-dimensional checkerboard root lattice	117
det	square of volume of fundamental region	5
e_6, e_7, e_8	ordinary Coxeter-Dynkin diagrams	481
E_6, E_7, E_8	Gosset's root lattices	120
$\mathscr{E}$	Eisenstein integers	52
$\mathbf{F}_q$	Galois field of order q	75
$\mathscr{G}$	Gaussian integers	52
g_0, g_1, g_2	orders of glue groups	100
G_0, G_1, G_2	glue groups	100
$G(P)$	second moment of polytope	34
$G(\Lambda)$	second moment of Voronoi cell of lattice	35
$\mathscr{H}$	Hurwitzian integral quaternions	53
HiS	Higman-Sims group	292
HJ or J_2	Hall-Janko group	293
I	identity matrix	77
I_n	The Euclidean integer lattice (often called $\mathbf{Z}_n$); also the genus to which this belongs	42

II_n	the genus of even unimodular Euclidean lattices in n dimensions	48
$I_{n,1}$, $II_{n,1}$	odd and even unimodular Lorentzian lattices, or the genera to which they belong	385
$J_1, J_2, \ldots$	Janko groups	273, 298
K_{12}	12-dimensional Coxeter-Todd lattice	127
M	generator matrix for code or lattice;	4, 77
	also the Monster simple group	297, 554
M_{40}	McKay's 40-dimensional lattice	221
n	dimension of lattice or length of code	3, 75
$N(x) = x \cdot x$	norm of vector	41
N_m or $N(m)$	number of vectors of norm m	102
O_{23}	shorter Leech lattice	179
O_{24}	odd Leech lattice	179
P_{48p}, P_{48q}	extremal 48-dimensional lattices	195
$q = e^{\pi i z}$	variables for theta series	45
$\mathbf{Q}$	the rational numbers	366
Q_{32}, Q_{64}	Quebbemann's lattices	220, 213
R	covering radius	33
$\mathbf{R}$	the real numbers	3
$\mathbf{R}^n$	n-dimensional Euclidean space	3
$\mathbf{R}^{n,1}$	Lorentzian space	522
S_n	symmetric group of degree n	82
$S(t, k, v)$	Steiner system	88
Suz	Suzuki group	293
V_4	Klein 4-group (rare)	118
V_n	volume of n-dimensional unit sphere	9
$V(u)$	Voronoi cell around u	33
W_C	weight enumerator of code	78
$\mathbf{Z}$	the integers	8
$\mathbf{Z}^n$	n-dimensional cubic lattice	106
γ_n	Hermite's constant	20
$\delta = \Delta / V_n$	center density	13
Δ	density of packing	6
Δ_{24}	generating function for Ramanujan numbers	51
$\eta(\Lambda)$	lattice obtained from Construction E	238
$\theta = \Theta / V_n$	normalized thickness	32

$\theta = \omega - \bar{\omega} = \sqrt{-3}$	an Eisenstein integer	55
$\theta_1, \theta_2, \theta_3, \theta_4$	Jacobi theta functions	102
Θ	thickness of covering	31
Θ_Λ	theta series of lattice	45
Λ_n	n-dimensional laminated lattice	163
Λ_{16}	16-dimensional Barnes-Wall lattice	129
Λ_{24}	24-dimensional Leech lattice	131, 286
$\Lambda(C)$	packing obtained from Construction A	182
μ	minimal norm of lattice	42
ρ	packing radius of lattice	10
$\sigma_k(n)$	sum of k-th powers of divisors of n	51
τ	kissing number	21
$\tau(m)$	Ramanujan number	51
ϕ_1, ϕ_2	theta series of hexagonal lattice	103
ψ_k	theta series of $\mathbf{Z} + k^{-1}$	103
ω	a cube root of unity	53
$\cdot 0 = Co_0$	automorphism group of Leech lattice	287
$\cdot 1 = Co_1, \cdot 2 = Co_2, \ldots$	simple groups	290
$\cdot \infty = Co_\infty$	automorphism group of Leech lattice including translations	290
$\cdot m, \cdot lmn,$	subgroups of $\cdot 0$	290
Λ^*	dual lattice	10
$[n, k, d]$	parameters of a code	76
$[x]$	integer part of x	154
$[0], [1], \cdots$	glue vectors	100

We draw the reader's attention to the fact that some errors have been noted in the Preface to the Third Edition.

Contents

1

Sphere Packings and Kissing Numbers

J. H. Conway and N. J. A. Sloane

The first three chapters describe some of the questions that motivated this book. In this chapter we discuss the problems of packing spheres in Euclidean space and of packing points on the surface of a sphere. The kissing number problem is an important special case of the latter, and asks how many spheres can just touch another sphere of the same size. We summarize what is known about these topics and also introduce terminology that will be used throughout the book.

1. The Sphere Packing Problem

1.1 Packing ball bearings. The classical sphere packing problem, still unsolved even today, is to find out how densely a large number of identical spheres (ball bearings,[1] for example) can be packed together. To state this another way, consider a large empty region, such as an aircraft hangar, and ask what is the greatest number of ball bearings that can be packed into this region. If instead of ball bearings we try to pack identical wooden cubes, children's building blocks, for example, the answer becomes easy. The cubes fit together with no wasted space in between, we can fill essentially one hundred percent of the space (ignoring the small amount of space that may be left over around the walls and at the ceiling), and so the number of cubes we can pack is very nearly equal to the volume of the hangar divided by the volume of one of the cubes.

But spheres do not fit together so well as cubes, and there is always some wasted space in between. No matter how cleverly the ball bearings are arranged, about one quarter of the space will not be used. One familiar arrangement is that shown in Figs. 1.1a and b, in which the centers of the spheres form the *face-centered cubic* (or *fcc*) lattice. (This lattice is discussed further in Chap. 4, where there is a detailed account of

---

(1) Or more pedantically, "bearing balls".

the most important packings in low dimensions.) In this packing the spheres occupy $\pi/\sqrt{18} = .7405...$ of the total space. By arranging the ball bearings in this way the number that can be packed into the hangar is about .7405... times the volume of the hangar, divided by the volume of one ball bearing. We therefore say that the face-centered cubic lattice packing has *density* .7405... .

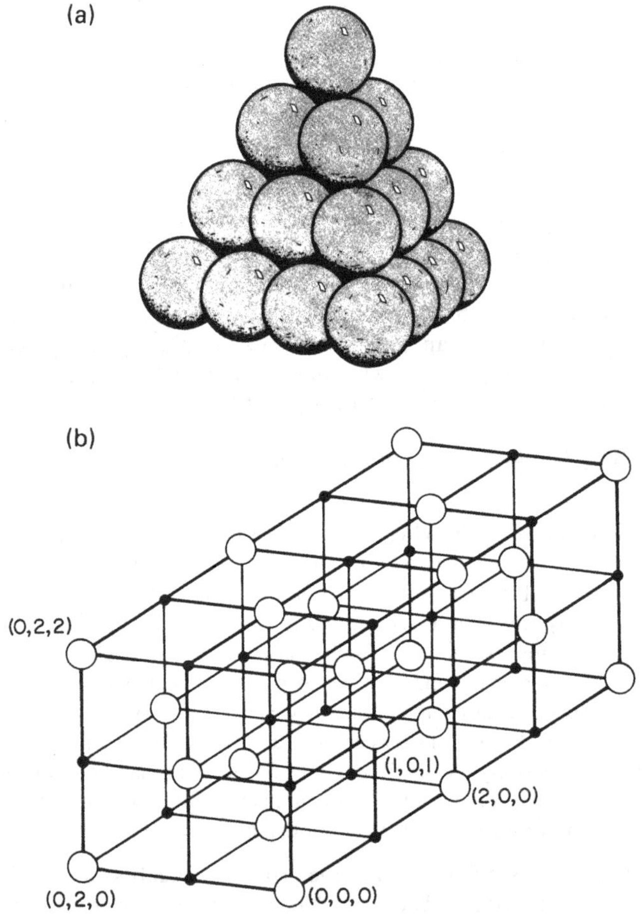

Figure 1.1. Two views of the most familiar sphere packing, in which the centers form the face-centered cubic (or fcc) lattice. This is the packing usually found in fruit stands, in piles of cannon balls on war memorials, or in crystalline argon. The spheres occupy 0.7405... of the space, and each sphere touches 12 others. (b) Shows only the centers of the spheres (the open circles). These centers are obtained by taking those points of the cubic lattice $\mathbf{Z}^3$ whose coordinates add up to an even number.

The classical version of the sphere packing problem now asks: is this the greatest density that can be attained? Unfortunately this is an unsolved problem, one of the famous open problems in mathematics. For many years the best bound known was Rogers' 1958 result (Eq. (39) below, [Rog2], [Rog7]) that no packing of ball bearings can have density greater than .7796... . Recently Lindsey [Lin9] has tightened Rogers' bound, replacing .7796... by .7784... . (Muder [Mud1] and Lindsey [Lin9a] have obtained further small improvements to the upper bound. See also [Fej9, p. 298].) As Rogers [Rog2] remarks, "many mathematicians believe, and all physicists know" that the correct answer is .7405... . The situation is complicated, however, by the fact that there are partial packings that are denser than the face-centered cubic lattice over larger regions of space than one might have supposed [Boe1].

The general sphere-packing problem, to which we return in §1.4, asks for the densest packing of equal spheres in n-dimensional space. This problem has considerable practical importance even for dimensions greater than 3.

We pause to assure the reader that there is nothing mysterious about n-dimensional space. A point in real n-dimensional space $\mathbf{R}^n$ is simply a string of n real numbers

$$x = (x_1, x_2, x_3, \ldots, x_n).$$

A sphere in $\mathbf{R}^n$ with center $u = (u_1, \ldots, u_n)$ and radius ρ consists of all the points $x = (x_1, x_2, \ldots, x_n)$ satisfying

$$(x_1 - u_1)^2 + (x_2 - u_2)^2 + \cdots + (x_n - u_n)^2 = \rho^2.$$

We can describe a sphere packing in $\mathbf{R}^n$ just by specifying the centers u and the radius. Everything is done with coordinates, and there is no need to draw pictures. There has been a great deal of nonsense written in science fiction about the mysterious fourth dimension. One should certainly not think that the fourth dimension represents time. In mathematics 4-dimensional space just consists of points with four coordinates instead of three (and similarly for any number of dimensions). For a concrete model the reader may imagine a telegraph wire over which numbers are being sent in sets of four: (1.8, 2.9, −1.3, 2.0), (1.1, −0.8, 0.5, 3.1), ... Each set of four numbers is a point in 4-dimensional space. Later in this chapter we'll do some easy calculations in 4-dimensional space.

1.2 Lattice packings. The packing shown in Fig. 1.1 is called a *lattice packing* because it has the properties that 0 is a center and that if there are spheres with centers u and v then there are also spheres with centers $u + v$ and $u - v$. In other words the set of centers forms an additive group. In crystallography these lattices are usually called *Bravais lattices* ([Hah1], [Kit4], [Ple1], [Rys5]). We can find three centers v_1, v_2, v_3 (or in general n centers $v_1, v_2, \ldots, v_n$ for an n-dimensional lattice) such that the set of all centers consists of the sums $\sum k_i v_i$ where the k_i are integers.

The vectors $v_1, \ldots, v_n$ are then called a *basis* for the lattice. The parallelotope consisting of the points

$$\theta_1 v_1 + \cdots + \theta_n v_n \quad (0 \leqslant \theta_i < 1)$$

is a *fundamental parallelotope.* Figure 1.2 shows a two-dimensional lattice and the fundamental parallelotope determined by the basis v_1, v_2. A fundamental parallelotope is an example of a *fundamental region* for the lattice, that is, a building block which when repeated many times fills the whole space with just one lattice point in each copy. Figure 1.3c below shows a hexagonal fundamental region for a two-dimensional lattice.

There are many different ways of choosing a basis and a fundamental region for a lattice Λ. But the volume of the fundamental region is uniquely determined by Λ, and the square of this volume is called the *determinant* or *discriminant* of the lattice. (Some authors use this term for the unsquared volume.) There is a simple formula for the determinant. Let the coordinates of the basis vectors be

$$\begin{aligned} v_1 &= (v_{11}, v_{12}, \ldots, v_{1m}), \\ v_2 &= (v_{21}, v_{22}, \ldots, v_{2m}), \\ &\cdots \\ v_n &= (v_{n1}, v_{n2}, \ldots, v_{nm}), \end{aligned}$$

where $m \geqslant n$. (Sometimes it is convenient to use $m > n$ coordinates to describe an n-dimensional lattice.) The matrix

$$M = \begin{bmatrix} v_{11} & v_{12} & \cdots & v_{1m} \\ v_{21} & v_{22} & \cdots & v_{2m} \\ & \cdots & \cdots & \\ v_{n1} & v_{n2} & \cdots & v_{nm} \end{bmatrix} \tag{1}$$

is called a *generator matrix* for the lattice, and the lattice vectors consist of all the vectors

$$\xi M, \tag{2}$$

where $\xi = (\xi_1, \ldots, \xi_n)$ is an arbitrary vector with integer components ξ_i. The matrix

$$A = MM^{tr}, \tag{3}$$

where tr denotes transpose, is called a *Gram matrix* for the lattice. The (i,j)th entry of A is the inner product $v_i \cdot v_j$. (Given two vectors $u = (u_1, \ldots, u_m)$, $v = (v_1, \ldots, v_m)$, their *inner* or *scalar product* $u_1 v_1 + \cdots + u_m v_m$ will be denoted in this book either by $u \cdot v$ or (u,v).) The determinant of Λ is then the determinant of the matrix A,

$$\det \Lambda = \det A. \tag{4}$$

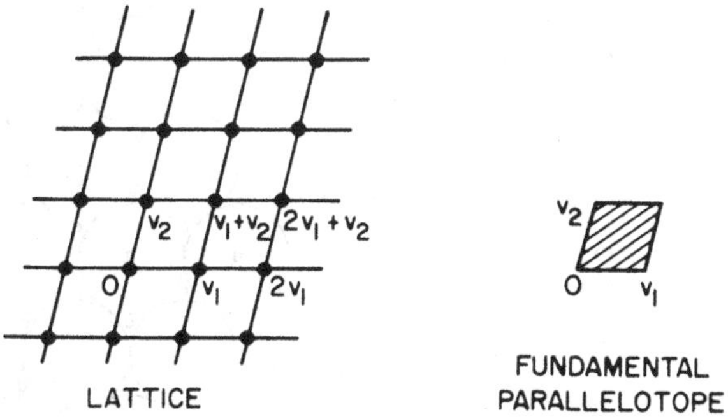

Figure 1.2. The plane divided into fundamental parallelotopes of a 2-dimensional lattice.

If M is a square matrix this reads

$$\det \Lambda = (\det M)^2. \tag{5}$$

Consider for example the familiar planar hexagonal lattice shown in Fig. 1.3a. An obvious generator matrix is

$$M = \begin{bmatrix} 1 & 0 \\ 1/2 & 1/2\sqrt{3} \end{bmatrix}, \tag{6}$$

for which the Gram matrix is

$$A = MM^{tr} = \begin{bmatrix} 1 & 1/2 \\ 1/2 & 1 \end{bmatrix}, \tag{7}$$

and $\det \Lambda = \det A = 3/4$. But for many purposes it is better to use the generator matrix

$$M' = \begin{bmatrix} 1 & -1 & 0 \\ 0 & 1 & -1 \end{bmatrix}, \tag{8}$$

with Gram matrix

$$A' = M'M'^{tr} = \begin{bmatrix} 2 & -1 \\ -1 & 2 \end{bmatrix}, \tag{9}$$

and now $\det \Lambda = \det A' = 3$. Eq. (8) uses three coordinates to define a two-dimensional lattice, but the points all lie in the plane $x + y + z = 0$. Eqs. (6) and (8) both describe the hexagonal lattice in Fig. 1.3a, but in different coordinates and on a different scale. Formally we say that the lattices defined by (6) and (8) are *equivalent* (see §1.4). The advantages of (8) over (6) are that the coordinates are nicer (the $\sqrt{3}$'s have disappeared), and the symmetries of the lattice are more easily seen. It is

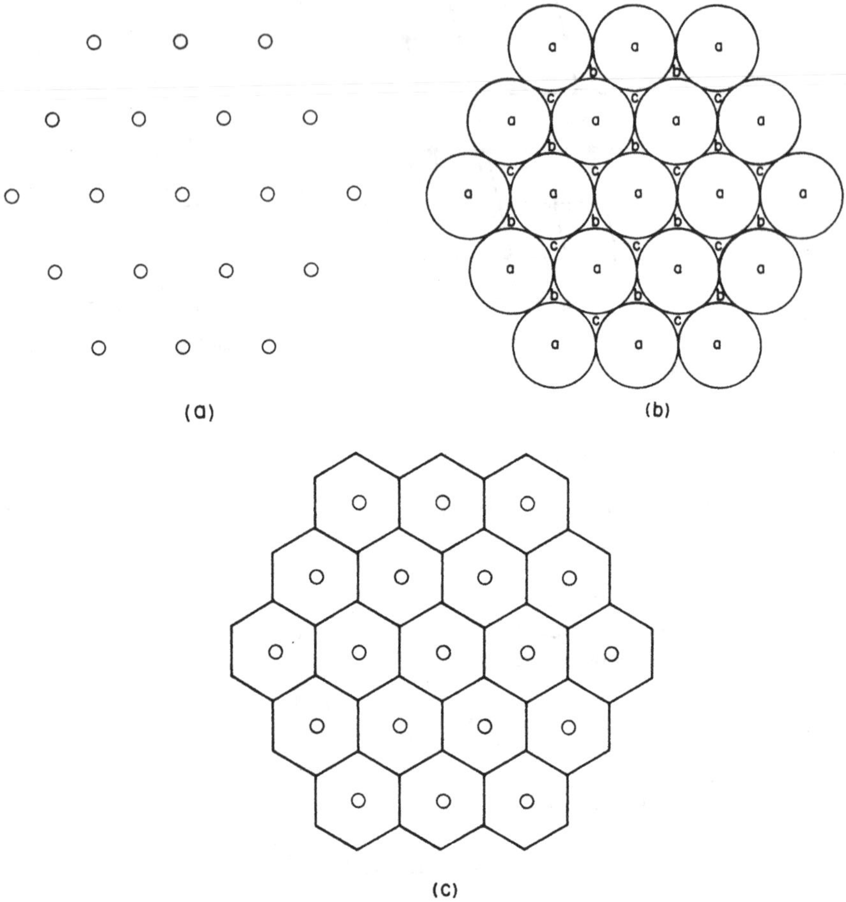

Figure 1.3. (a) The hexagonal lattice in the plane. (b) The corresponding sphere-packing (in this case a circle-packing). The points labeled a are the lattice points; the points labeled b and c are the "deep holes" in this lattice, the points of the plane furthest from the lattice. (c) The *Voronoi cells* are regular hexagons. (d) An enlargement of (b) and (c), showing the *packing radius* ρ of the lattice (the radius of the circles in (b)), and the *covering radius* R, the distance from a lattice point to a deep hole. For this lattice $R = 2\rho/\sqrt{3}$.

clear from (8) that the three coordinates can be permuted in any way without changing the lattice. (8) is the two-dimensional representative (A_2) of the infinite sequence of *root lattices* $A_1, A_2, A_3, \ldots$ (see Chap. 4, §6.1).

We can now give a precise definition of the density Δ of a lattice packing:

$$\Delta = \text{proportion of the space that is occupied by the spheres} \qquad (10)$$

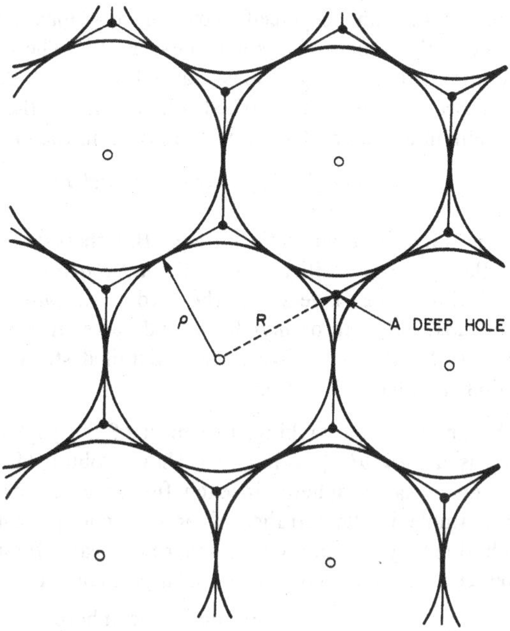

(d)

Figure 1.3 (cont.)

$$= \frac{\text{volume of one sphere}}{\text{volume of fundamental region}}$$

$$= \frac{\text{volume of one sphere}}{(\det \Lambda)^{1/2}} . \tag{11}$$

If the hexagonal lattice is defined by (6) we may take the spheres (actually circles in this case) to have radius $\rho = 1/2$, obtaining the packing shown in Fig. 1.3b, with density

$$\Delta = \frac{\pi}{\sqrt{12}} = 0.9069\ldots . \tag{12}$$

Using (8) leads to the same value: the density is independent of the choice of coordinates, as we should expect it to be.

1.3 Nonlattice packings. Of course most packings are not lattices. In three dimensions, for example, there are nonlattice packings that are just as dense as the face-centered cubic lattice. This happens because the fcc can be built up by layers, beginning with one layer of spheres placed in the hexagonal lattice arrangement of Fig. 1.3b, with the centers at the points marked 'a'. There are then two (equivalent) ways to place the second layer: the spheres can be placed above the positions marked '*b*' or above those marked '*c*'. Suppose we place them at '*b*'. The third layer can now be placed either above the positions marked '*a*', i.e. directly above the first layer, or above the positions marked '*c*'. Similarly there are two choices for every remaining layer. Choosing the layers in the order

$$...abcabcabc... \quad \text{or} \quad ...acbacbacb...$$

produces the face-centered cubic lattice. But there are uncountably many other (nonlattice) possibilities, such as ...*acbabacbca*..., all with the same density. Taking the layers in the order ...*ababab*... produces the *hexagonal close packing* or *hcp* (as found for example in helium crystals [Kit4], [Wel4]). We shall give a more detailed study of such laminating constructions in Chaps. 5 and 6.

The hexagonal close packing, like most of the packings encountered in this book, is a *periodic* packing, one that is obtained by placing a fixed configuration of say s spheres in each fundamental region of a lattice Λ. Another example is the tetrahedral or diamond packing (Chap. 4, §6.4) that is obtained by placing $s = 8$ spheres in each fundamental region of the simple cubic lattice. The density of a periodic packing is given by

$$\Delta = \frac{s \cdot \text{volume of one sphere}}{(\det \Lambda)^{1/2}} . \tag{13}$$

In the diamond packing the spheres have radius $\sqrt{3}/8$, so the density is only

$$\Delta = \frac{\pi\sqrt{3}}{16} = 0.3401... . \tag{14}$$

The density of an arbitrary (not necessarily periodic) packing is found by taking a very large spherical region, of radius R say, calculating the proportion of this region that is occupied by the spheres of the packing, and letting $R \to \infty$ (see [Rog7, Chap. 1]).

1.4 *n*-dimensional packings. The first main problem of the book, then, is to find the densest (lattice or nonlattice) packing of equal nonoverlapping, solid spheres (or balls) in n-dimensional space. In one dimension this is trivial, for a one-dimensional ball is a line segment, a rod for example, and we can obtain a density $\Delta = 1$ by centering these rods at the integer points on a line (Fig. 1.4). These points form the one-dimensional lattice $\mathbf{Z}$.

The answer is also known in two dimensions: the highest density that can be attained is $\Delta = \pi/\sqrt{12} = 0.9069...$, as in the hexagonal lattice packing of Fig. 1.3b. This result has a long history — see especially Thue's 1910 paper [Thu1] and the proof given by Fejes Tóth in 1940 [Fej3]. For a concise and elegant proof see [Fej10, pp. 58-61]. Other references are [Fol1], [Gra5], [Rog7, p. 11], [Seg1].

In three dimensions, as we remarked in §1.1, the answer is unknown. But if we consider only *lattice* packings then the answer is known: Gauss

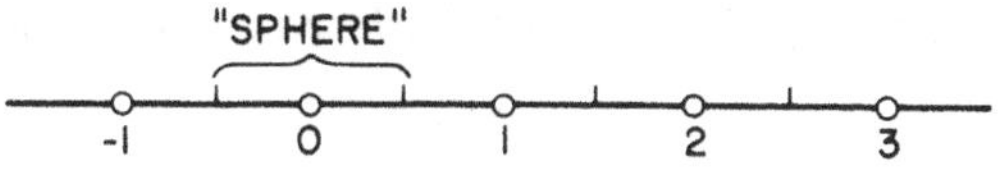

Figure 1.4. The 1-dimensional lattice $\mathbf{Z}$, the integers.

showed in 1831 [Gau2] that the face-centered cubic lattice is the densest lattice packing in three dimensions. Proofs of this result may be found in [Cas2, Chap. II, Th. III], [Con42], [Cox14, §18.4], [Dem1] or [Mor5].

What about dimensions higher than 3? To demonstrate how easy it is work in 4-dimensional space, we describe the densest sphere packing known in four dimensions. This is the *checkerboard lattice* D_4 (Chap. 4, §7.2), in which the centers of the spheres are all the points (u_1, u_2, u_3, u_4) where the u_i are integers that add to an even number. Thus $(0,0,0,0)$ is allowed, and $(1,1,0,0)$, but not $(1,0,0,0)$. The sphere centered at $(0,0,0,0)$ has 24 spheres around it, centered at the points $(\pm 1, \pm 1, 0, 0)$, with any choice of signs and any ordering of the coordinates. (There are $\frac{1}{2} 4 \cdot 3 = 6$ positions for the 0's and then 4 choices for the signs, for a total of 24.) Any two distinct centers must differ by at least 1 in at least two coordinates, or by at least 2 in one coordinate, so the minimal distance between the centers is $\sqrt{2}$. Then if the spheres are taken to have radius $\rho = \sqrt{2}/2 = 1/\sqrt{2}$, they will not overlap, and each sphere will touch 24 others.

It is clear from the definition that D_4 contains the centers $(2,0,0,0)$, $(0,2,0,0), \ldots, (0,0,0,2)$, $(1,1,0,0)$, $(1,0,1,0), \ldots, (0,0,1,1)$, and conversely that each center of D_4 is an integer combination of these vectors. Actually it is enough to use $(2,0,0,0)$, $(1,1,0,0)$, $(1,0,1,0)$ and $(1,0,0,1)$, and

$$M = \begin{bmatrix} 2 & 0 & 0 & 0 \\ 1 & 1 & 0 & 0 \\ 1 & 0 & 1 & 0 \\ 1 & 0 & 0 & 1 \end{bmatrix} \tag{15}$$

is a generator matrix for D_4. [E.g. $(0,2,0,0) = 2(1,1,0,0) - (2,0,0,0)$.] Therefore D_4 has determinant $\det D_4 = (\det M)^2 = 4$.

In order to compute the density of D_4 we need to know the volume of an n-dimensional sphere of radius ρ: this is

$$V_n \, \rho^n , \tag{16}$$

where V_n, the volume of a sphere of radius 1, is given by

$$V_n = \frac{\pi^{n/2}}{(n/2)!} = \frac{2^n \pi^{(n-1)/2} ((n-1)/2)!}{n!} . \tag{17}$$

(The second form avoids the use of $(n/2)!$ when n is odd. See §2C of Chap. 21, or [Som1, p. 136].) Also

$$\log_2 V_n = -\frac{n}{2} \log_2 \frac{n}{2\pi e} - \frac{1}{2} \log_2 (n\pi) - \epsilon , \tag{18}$$

where $0 < \epsilon < (\log_2 e)/(6n)$. The surface area of a sphere of radius ρ is

$$nV_n \rho^{n-1}. \tag{19}$$

From (11) and (16) the density of a lattice packing Λ is given by

$$\Delta = \frac{V_n \rho^n}{(\det \Lambda)^{1/2}}, \tag{20}$$

where ρ is the radius of the spheres. We take ρ to be half the minimal distance between lattice points; ρ is called the *packing radius* of Λ.

For the D_4 lattice we have $V_4 = \pi^2/2$, $\rho = 1/\sqrt{2}$, so the density is

$$\Delta = \frac{\pi^2}{16} = 0.6169\ldots . \tag{21}$$

This is the highest density known in four dimensions, and Korkine and Zolotareff proved in 1872 [Kor1] that this is the highest possible density for a 4-dimensional *lattice* packing. By using Mordell's inequality (Eq. (19) of Chap. 6) Korkine and Zolotareff's result can be deduced immediately from the optimality of the fcc lattice packing ([Mor4], [Opp1]).

D_4 may be generalized in an obvious way to an n-dimensional packing D_n (take n integer coordinates with an even sum). In Chap. 4 we summarize the properties of the lattices D_n, and of other important lattices that occur throughout this book, including the n-dimensional lattices A_n (for $n \geqslant 1$), the 6-, 7- and 8-dimensional lattices E_6, E_7 and E_8, and the 24-dimensional Leech lattice Λ_{24}. (The subscript usually gives the dimension.)

If one lattice can be obtained from another by (possibly) a rotation, reflection and change of scale we say they are *equivalent*, or *similar*, written $\cong$. Two generator matrices M and M' define equivalent lattices if and only if they are related by

$$M' = c\,U M B, \tag{22}$$

where c is a nonzero constant, U is a matrix with integer entries and determinant ± 1, and B is a real orthogonal matrix (with $BB^{tr} = I$). The corresponding Gram matrices are related by

$$A' = c^2 U A U^{tr}. \tag{23}$$

If U has determinant ± 1 and $c = 1$ then M and M' are *congruent* lattices (they are *directly congruent* if $\det U = +1$). For example the fcc lattice occurs in both the A_n and D_n sequences as $A_3 \cong D_3$.

Any n-dimensional lattice L_n has a *dual* (or *reciprocal*, or *polar*) lattice, L_n^*, given by

$$L_n^* = \{x \in \mathbf{R}^n : x \cdot u \in \mathbf{Z} \text{ for all } u \in L_n\}. \tag{24}$$

For example the dual of the fcc is the *body-centered cubic* or *bcc lattice* $A_3^* \cong D_3^*$ (see Chap. 4, §6.8). If A is a Gram matrix for L_n, L_n^* has Gram matrix A^{-1}, and $\det L_n^* = (\det L_n)^{-1}$. If M is a square generator matrix for L_n, then $(M^{-1})^{tr}$ is a generator matrix for L_n^*.

Why do we care about finding dense packings of n-dimensional spheres? There are many reasons.

(i) This is an interesting problem in pure *geometry*. Hilbert mentioned it in 1900 in his list of open problems [Hil1], [Mil5]. There is an enormous literature (see the bibliography), from which we select a few items of particular interest: [Bar1], [Cox18], [Fej1], [Fej9], [Fej10], [Gru1], [Hah1], [Ham2], [Hil2], [Mil7], [Rog7], [Sch16], [Sig1], [Slo14], [Wel4].

Furthermore the best packings turn out to have connections, sometimes totally unexpected, with other branches of mathematics. For example the best lattice packings in up to eight dimensions belong to the families A_n, D_n and E_n, and the corresponding Coxeter-Dynkin diagrams (see Chap. 4) turn up in several apparently unrelated areas — see Hazewinkel et al.'s survey article "The ubiquity of Coxeter-Dynkin diagrams (an introduction to the $A-D-E$ problem)" [Haz1]. Similarly in 24 dimensions the Leech lattice Λ_{24} has mysterious connections with hyperbolic geometry, Lie algebras, and the Monster simple group — see Chaps. 23-30 of this book, and the "Monstrous Moonshine" articles [Con11], [Con17], [Fon1], [Kac2]-[Kac6], [Koi1], [Kon1], [Kon2], [Mas2], [Mas3], [Nor5], [Smi13], [Tho5], [Tho6]. We expect that one day someone will write an article on "The ubiquity of the Leech lattice".

(ii) There are direct applications of lattice packings to *number theory*, for example in solving Diophantine equations, and to "the geometry of numbers" — see [Cas2], [Gru1], [Gru1a], [Han3], [Hla1], [Hla3], [Kel1], [Min4]-[Min6]. We shall say more about this in Chap. 2, §2.

(iii) There are important practical and theoretical applications of sphere packings to problems arising in *digital communications*, as we shall see in Chap. 3. Just to give the flavor, here is a typical question from the design of spread-spectrum communications for mobile radio (cf. [Coo1], [Maz1]): how many spheres of radius 0.25 can be packed into a sphere of radius 1 in 100-dimensional space?

(iv) Two- and three-dimensional packings have many applications. For example the circles in a two-dimensional packing may represent optical fibers as seen in the cross-section of a cable [Kin1]. Three-dimensional packings have applications in *chemistry* and *physics* [Ber11], [Hoa1], [Hoa2], [Kit4], [O'Ke1], [Slo17], [Slo19], [Teo1]-[Teo3], [Wel2]-[Wel5], [Zim1]; *biology* [Rit1], [Tam5]; *antenna design* [Str1]; choosing directions for X-ray *tomography* [She3], [She4], [Smi7]; and in performing *statistical analysis* on spheres [Wat24].

(v) n-dimensional packings may be used in the *numerical evaluation of integrals*, either on the surface of a sphere in $\mathbf{R}^n$ or in its interior. See

[Bab2], [Bou2], [Del6], [Del16], [Fro1], [Hla2], [Hua1], [Kea1], [Nie1], [Roo2], [Sha0], [Sob3], [Str3], [Zar1] and especially [Goe5], [McL0], [Sln1], [Sln2], [Sob2]. We shall discuss this topic further in §3.2 of Chap. 3. A related application that has not yet received much attention is the use of these packings for solving n-dimensional *search* or *approximation* problems — cf. [Air1], [Dob1], [Ren1], [Sob1].

(vi) Recent developments in physics (in *dual theory* and *superstring theory*) have involved the E_8 and Λ_{24} lattices and the related Lorentzian lattices in dimensions 10 and 26 discussed in Chaps. 26 and 27 ([Cha3], [God1], [Gre1], [Jac2], [Sch3], [Sch15], [Thi1], [Thi2]).

1.5 Sphere packing problem-summary of results. In this section we summarize what is presently known about the sphere packing problem. Table 1.1 shows the results in dimensions 1 to 8, 12, 16 and 24, comparing the densest known packings with the answers to some of the other questions discussed in the first two chapters.

The entries in Table 1.1 enclosed in boxes are known to be optimal. The boxes in the first row indicate that the densest possible sphere packings are known only in 1 and 2 dimensions! The entries to the left of the double lines are known to be optimal *among lattices*. In particular the densest possible lattice packings are known in dimensions $n \leqslant 8$. We have already mentioned dimensions 1-4. Korkine and Zolotareff [Kor3] showed that D_5 is the densest lattice packing in 5 dimensions, and guessed E_6, E_7, E_8, Λ_9, Λ_{10} in the next five dimensions. Blichfeldt [Bli2]-[Bli4] established the optimality of E_6, E_7 and E_8. Blichfeldt's proof is quite difficult, but has been confirmed by Watson [Wat5] and Vetchinkin [Vet2]. Mordell's inequality enables the optimality of E_8 to be deduced from that of E_7, but no simple proof of the optimality of E_7 is known, nor has Blichfeldt's work been extended to higher dimensions. The optimality of E_6 also follows

Table 1.1. Records for packings, kissing numbers, coverings and quantizers. (Box: optimal. To left of double line: known to be optimal among lattices.) For $n \leqslant 8$ the entry in the first row is $\cong \Lambda_n$.

DIMENSION	1	2	3	4	5	6	7	8	12	16	24
DENSEST PACKING	Z	A_2	A_3	D_4	D_5	E_6	E_7	E_8	K_{12}	Λ_{16}	Λ_{24}
HIGHEST KISSING NUMBER	Z 2	A_2 6	A_3 12	D_4 24	D_5 40	E_6 72	E_7 126	E_8 240	P_{12a} 840	Λ_{16} 4320	Λ_{24} 196560
THINNEST COVERING	Z	A_2	A_3^*	A_4^*	A_5^*	A_6^*	A_7^*	A_8^*	A_{12}^*	A_{16}^*	Λ_{24}
BEST QUANTIZER	Z	A_2	A_3^*	D_4	D_5^*	E_6^*	E_7^*	E_8	K_{12}	Λ_{16}	Λ_{24}

from Barnes's classification of all 6-dimensional lattices whose density is a local maximum [Bar6], [Bar7]. It is known that the densest lattice packings in dimensions 1 to 8 are unique. (See Barnes [Bar6], [Bar7] and Vetchinkin [Vet2] for the uniqueness of E_6, E_7, E_8.)

The symbol Λ_n in Table 1.1 represents a *laminated lattice*. These will be defined and studied in detail in Chap. 6. In particular the laminated lattices are the densest lattices in up to 8 dimensions, and we have the following equivalences:

$$\begin{array}{ll} \Lambda_1 \cong Z \cong A_1 \cong A_1^*, & \Lambda_2 \cong A_2 \cong A_2^*, \\ \Lambda_3 \cong A_3 \cong D_3, & \Lambda_4 \cong D_4 \cong D_4^*, \\ \Lambda_5 \cong D_5, & \Lambda_6 \cong E_6, \\ \Lambda_7 \cong E_7, & \Lambda_8 \cong E_8 \cong E_8^*. \end{array} \tag{25}$$

$\Lambda_{16} \cong BW_{16}$ is the Barnes-Wall lattice (§10 of Chap. 4, [Bar18]), Λ_{24} is the Leech lattice (§11 of Chap. 4, [Lee5]) and K_{12}, the only entry in Table 1.1 not yet mentioned, is the Coxeter-Todd lattice (§9 of Chap. 4, [Cox29]). It seems likely that K_{12}, Λ_{16} and Λ_{24} are the densest lattices in dimensions 12, 16 and 24 respectively, although this has not been proved. In fact it is a reasonable guess that all the entries in Table 1.1 are optimal.

The densities of these packings are shown in Fig. 1.5 and in Table 1.2, which is an updated version of a table in [Lee10], and includes most of the best packings known in up to 24 dimensions. As well as the density Δ itself, Table 1.2 shows the *center density* δ, given by

$$\delta = \frac{\Delta}{V_n}, \tag{26}$$

which is a much simpler number. For Λ_{24} for example (an extreme case), $\Delta = \pi^{12}/479001600 = 0.001930...$, whereas $\delta = 1$. The table gives the exact values only for δ. If the spheres have radius 1, δ is the number of centers per unit volume. For a lattice packing, from (20),

$$\delta = \rho^n (\det \Lambda)^{-1/2}. \tag{27}$$

The center densities in Fig. 1.5 have been rescaled to make them easier to see. The vertical axis gives

$$\log_2 \delta + \frac{1}{96} n(24-n).$$

Fig. 1.5 and Table 1.2 also give the best upper bound on the center density presently known. For $n = 3$ this is Lindsey's bound mentioned in §1.1, while for $n \geqslant 4$ it is Rogers' bound [Rog2] as calculated by Leech [Lee5]. Leech's figures are *truncated* to 5 decimal places, but apart from this the last digits of all decimal expansions in this book have been *rounded off* to the closest digit. In the column headed Type in Table 1.2, B indicates that both a lattice and a nonlattice packing with this density and kissing number are known, L indicates that only a lattice packing is known, N that only a nonlattice packing is known and A indicates a local

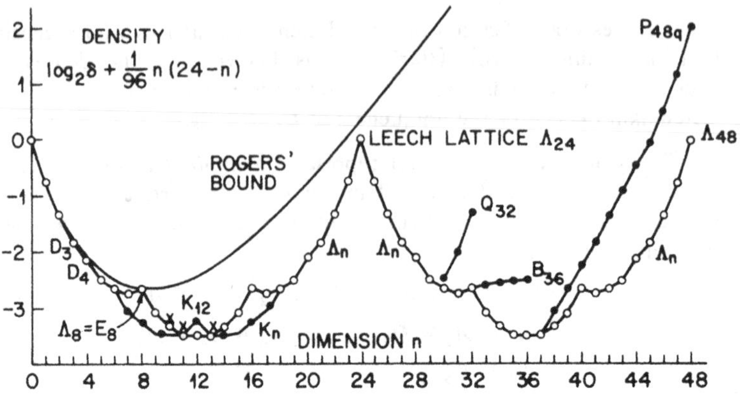

Figure 1.5. The densest sphere packings known in dimensions $n \leq 48$. The vertical axis gives $\log_2 \delta + n(24-n)/96$, where δ is the center density. The Λ_n are laminated lattices, the K_n are described in Chap. 6, K_{12} is the Coxeter-Todd lattice, the crosses are nonlattice packings (Chap. 5, §§2.6,4.3), and Q_{32}, B_{36} and P_{48q} are described in Table 1.3a. The upper bound is Rogers' bound (39), (40). (See also Table I.1 of the Introduction.)

arrangement (or cluster) of spheres touching one sphere. The previous column indicates where in this book these packings are described.

We see from Fig. 1.5 and Table 1.2 that the laminated lattices Λ_n are the densest packings known in dimensions $n \leqslant 29$, except for dimensions 10-13. In dimension 12, as already mentioned, the Coxeter-Todd lattice K_{12} is the densest known, and in dimensions 10, 11 and 13 there are non-lattice packings, to be described in Chap. 5, that are denser than any known lattices. In dimensions 30-32 Quebbemann's lattices (§4 of Chap. 8, [Que5], [Que6]) surpass Λ_n.

Table 1.3 gives a selection of the densest packings known from dimensions 24 to about 10^6. All these packings are lattices. The lattices in Table 1.3 are all quite explicit: without too much difficulty one could write down generator matrices (the amount of effort needed being polynomial in the dimension n).

In contrast, Minkowski gave a *nonconstructive* proof in 1905 that there exist lattices with density satisfying

$$\Delta \geqslant \frac{\zeta(n)}{2^{n-1}}, \tag{28}$$

where $\zeta(n) = \Sigma_{k=1}^{\infty} k^{-n}$ is the Riemann zeta-function. Thus

$$\log_2 \Delta \geqslant -n + 1, \text{ as } n \to \infty. \tag{29}$$

This implies (from (27)) that there are lattices with center density at least as large as the following:

n	128	256	512	1024	4096	65536	1048576
$\log_2\delta$	63	249	750	2006	12102	324603	7290660

Table 1.2. Sphere packings in up to 24 dimensions. (See Tables I.1, I.2 of the Introduction for recent improvements.)

n	Name of packing	Density Δ	Center density δ Attained	Center density δ Bound	Kissing number Max	Kissing number Ave	Ch., §	Type
0	Λ_0	1	1	1	1	1	6,2	L
1	$\Lambda_1 \cong A_1$	1	$1/2 = 0.5$	0.5	2	2	4,6.1	L
2	$\Lambda_2 \cong A_2$	0.90690	$1/2\sqrt{3} = 0.28868$	0.28868	6	6	4,6.2	L
3	$\Lambda_3 \cong D_3$	0.74048	$1/4\sqrt{2} = 0.17678$	0.1847	12	12	4,6.3	B
4	$\Lambda_4 \cong D_4$	0.61685	$1/8 = 0.125$	0.13127	24	24	4,7.2	L
5	$\Lambda_5 \cong D_5$	0.46526	$1/8\sqrt{2} = 0.08839$	0.09987	40	40	4,7.1	B
6	$\Lambda_6 \cong E_6$	0.37295	$1/8\sqrt{3} = 0.07217$	0.08112	72	72	4,8.3	B
7	$\Lambda_7 \cong E_7$	0.29530	$1/16 = 0.0625$	0.06981	126	126	4,8.2	B
8	$\Lambda_8 \cong E_8$	0.25367	$1/16 = 0.0625$	0.06326	240	240	4,8.1	L
9	Λ_9	0.14577	$1/16\sqrt{2} = 0.04419$	0.06007	272	272	6,4	B
	P_{9a}	0.12885	$5/128 = 0.03906$		306	235 3/5	5,2.6	N
10	Λ_{10}	0.09202	$1/16\sqrt{3} = 0.03608$	0.05953	336	336	6,4	B
	P_{10a}	0.09463	$19/512 = 0.03711$		372	353 9/19	5,2.6	N
	P_{10b}	0.08965	$9/256 = 0.03516$		500	340 1/3	5,2.6	N
	P_{10c}	0.09962	$5/128 = 0.03906$		372	372	5,2.6	N
11	Λ_{11}^{max}	0.05888	$1/32 = 0.03125$	0.06136	438	438	6,4	B
	K_{11}	0.06043	$1/18\sqrt{3} = 0.03208$		432	432	6,2	L
	P_{11a}	0.06624	$9/256 = 0.03516$		566	519 7/9	5,2.6	N
	P_{11b}				580		5,4.3	A
	P_{11c}				582		5,2.6	A
12	Λ_{12}^{max}	0.04173	$1/32 = 0.03125$	0.06559	648	648	6,4	B
	K_{12}	0.04945	$1/27 = 0.03704$		756	756	4,9	L
	L_{12}	0.04456	$3^7/2^{16} = 0.03337$		704	704	5,3.3	N
	P_{12a}	0.04694	$9/256 = 0.03516$		840	770 2/3	5,2.6	N
13	Λ_{13}^{max}	0.02846	$1/32 = 0.03125$	0.07253	906	906	6,4	B
	K_{13}	0.02921	$1/18\sqrt{3} = 0.03208$		918	918	6,2	B
	P_{13a}	0.03201	$9/256 = 0.03516$		1130	1060 2/3	5,4.3	N
	P_{13b}				1066		5,4.3	A
14	Λ_{14}	0.02162	$1/16\sqrt{3} = 0.03608$	0.08278	1422	1422	6,4	B
	P_{14a}				1484		5,4.3	A
	P_{14b}				1582		5,4.3	A
15	Λ_{15}	0.01686	$1/16\sqrt{2} = 0.04419$	0.09735	2340	2340	6,4	B
	P_{15a}				2564		5,4.3	A
16	Λ_{16}	0.01471	$1/16 = 0.0625$	0.11774	4320	4320	4,10	B
17	Λ_{17}	0.008811	$1/16 = 0.0625$	0.14624	5346	5346	6,6	B
18	Λ_{18}	0.005928	$1/8\sqrt{3} = 0.07217$	0.18629	7398	7398	6,6	B
19	Λ_{19}	0.004121	$1/8\sqrt{2} = 0.08839$	0.24308	10668	10668	6,6	B
20	Λ_{20}	0.003226	$1/8 = 0.125$	0.32454	17400	17400	6,6	B
21	Λ_{21}	0.002466	$1/4\sqrt{2} = 0.17678$	0.44289	27720	27720	6,6	B
22	Λ_{22}	0.002128	$1/2\sqrt{3} = 0.28868$	0.61722	49896	49896	6,6	L
23	Λ_{23}	0.001905	$1/2 = 0.5$	0.87767	93150	93150	6,6	L
24	Λ_{24}	0.001930	$1 = 1.0$	1.27241	196560	196560	4,11	L

A comparison with Table 1.3 shows that in dimensions above 1000 these lattices are denser than any presently known. Unfortunately the proof of (28) (see [Cas2, Chap. 6], [Lek1, §19], [Rog7, Chap. 4]) uses an averaging argument, and does not say how lattices satisfying (28) may be

Table 1.3. Sphere packings in more than 24 dimensions. (See Tables I.1, I.2 of the Introduction for recent improvements.)

n	Name	$\log_2 \delta$	Bound	Kissing no.	Ch., §
32	Λ_{32}	0	5.52	208320	6,7
	BW_{32}	0		146880	8,8.2f
	C_{32}	1		249280	8,8.2h
	Q_{32}	1.359		261120	8,4
36	P_{36p}	−1	8.63	42840	5,5.5
	Λ_{36}	1		234456	5,5.3
	B_{36}	2			8,8.2d
48	Λ_{48}	12	15.27		6,7
	P_{48p}	14.039		52416000	5,5.7
	P_{48q}	14.039		52416000	5,5.7
60	P_{60p}	16.548	27.85	3908160	5,5.5
64	BW_{64}	16	31.14	9694080	8,8.2f
	Q_{64}	18.719		2611200	8,2
	P_{64c}	22			8,8.2e
80	$\eta(E_8)$	36	49.90		8,10c
96	$\eta(P_{48q})$	52.078	70.96		8,10g
104	$\eta(E_8)$	60	80.20		8,10c
128	BW_{128}	64	118.6	1260230400	8,8.2f
	P_{128b}	85			5,6.6
	$\eta(E_8)$	88			8,10c
136	$\eta(E_8)$	100	129.4		8,10c
150	$A_{150}^{(15)}$	113.06	153.2		8.6
180	$\eta(\Lambda_{20})$	133	206.7		8,10c
	$A_{180}^{(17)}$	154.12			8,6

found. Many generalizations and extensions of (28) have been found, although no essential improvement is known for large n. In its general form this bound is known as the *Minkowski-Hlawka theorem*.

We still do not know how to construct packings that are as good as (28). In 1959 Barnes and Wall found an infinite sequence of lattices BW_n in dimensions $n = 2^m$ with

$$\frac{1}{n} \log_2 \Delta \sim -\frac{1}{4} \log_2 n, \text{ as } n \to \infty \tag{30}$$

(§8.2f of Chap. 8, [Bar18]); in 1971 [Lee10] (see §6.7 of Chap. 5) constructed nonlattice packings with

$$\frac{1}{n} \log_2 \Delta \sim -\frac{1}{2} \log_2 \log_2 n; \tag{31}$$

Table 1.3 (cont.)

n	Name	$\log_2 \delta$	Bound	Ch., §
192	$\eta(\Lambda_{24})$	156	230.0	8,10e
	$A_{192}^{(18)}$	171.44		8,6
256	BW_{256}	192	357.0	8,8.2f
	B_{256}	250		8,8.2g
	$A_{256}^{(23)}$	270.89		8,6
508	$A_{508}^{(41)}$	742.66	948.1	8,6
512	B_{512}	698	957.4	8,8.2g
520	$A_{520}^{(42)}$	767.46	980.1	8,6
1020	$A_{1020}^{(74)}$	1922	2406	8,6
1030	$A_{1030}^{(74)}$	1947	2439	8,6
2052	$A_{2052}^{(135)}$	4755	5871	8,6
4096	$\eta(\Lambda_{16})$	11344	13750	8,10c
4098	$A_{4098}^{(246)}$	11279	13758	8,6
8184	$\eta(\Lambda_{24})$	26712	31547	8,10e
8190	$A_{8190}^{(454)}$	26154	31573	8,6
8208	$\eta(\Lambda_{24})$	26808	31655	8,10e
16380	$A_{16380}^{(844)}$	59617	71325	8,6
16392	$\eta(\Lambda_{24})$	61608	71387	8,10e
32784	$\eta(\Lambda_{24})$	139488	159154	8,10e
65520	$\eta(\Lambda_{24})$	311364	350788	8,10e
65544	$\eta(\Lambda_{24})$	311496	350932	8,10e
131088	$\eta_3(\Lambda_{24})$	664962	767395	8,10f
262152	$\eta_3(\Lambda_{24})$	$1.475 \cdot 10^6$	$1.666 \cdot 10^6$	8,10f
524304	$\eta_3(\Lambda_{24})$	$3.178 \cdot 10^6$	$3.594 \cdot 10^7$	8,10f
1048584	$\eta_3(\Lambda_{24})$	$6.918 \cdot 10^6$	$7.711 \cdot 10^7$	8,10f

and in 1973 [Slo1] (see §6.8 of Chap. 5) gave nonlattice packings for which

$$\frac{1}{n} \log_2 \Delta \gtrsim -6. \tag{32}$$

Lattices with

$$\frac{1}{n} \log_2 \Delta \sim -\frac{1}{2} \log_2 \log_2 n \tag{33}$$

were found in [Bar15] (see Chap. 8, §8.2g) and by Craig (see Chap. 8, §6), and lattices with

$$\frac{1}{n} \log_2 \Delta \sim -\log_2^* n \tag{34}$$

were given in [Bos3] (see Chap. 8, §10h). Here $\log_2^* n$ is the smallest value of k for which the kth iterated base 2 logarithm of n is less than 1. The lattices constructed in Chap. 8 are however still reasonably dense in quite large dimensions. For example there are lattices in dimensions $n \leqslant 98328$ with

$$\frac{1}{n} \log_2 \Delta = -1.2454... + \epsilon,$$

where $|\epsilon| < 12\, n^{-1} \log_2(n/6)$, in dimensions $n \leqslant 10^{51}$ with

$$\frac{1}{n} \log_2 \Delta = -2.0006... + \epsilon',$$

and in dimensions $n \leqslant 10^{9870}$ with

$$\frac{1}{n} \log_2 \Delta = -2.2005... + \epsilon''. \tag{35}$$

Again all these packings are quite explicit. Litsyn and Tsfasman [Lit3], [Lit5] have recently announced that nonlattice packings with

$$\frac{1}{n} \log_2 \Delta \gtrsim -1.31, \tag{36}$$

and lattices with

$$\frac{1}{n} \log_2 \Delta \gtrsim -2.30 \tag{37}$$

may be constructed using the methods of Chap. 8 applied to codes obtained from algebraic curves (cf. Chap. 3, §2.11).

As we shall see in §7.4 of Chap. 8, the infinite class field towers found by Golod and Shafarevich [Gol9], Martinet [Mar2] and others produce infinite sequences of lattices satisfying, in the most favorable case known,

$$\frac{1}{n} \log_2 \Delta \sim -2.218... \quad \text{as } n \to \infty. \tag{38}$$

(This was pointed out by Litsyn and Tsfasman [Lit3].) However there does not seem to be any practicable method known for explicitly finding these lattices.

Remark. Although we have described Minkowski's result and others as "nonconstructive", there are algorithms that would theoretically enable one to write down generating vectors in a bounded time, and so in the logical sense these results are effective. However the typical such algorithm involves a search through an extremely large population and would require a super-exponential running time. In contrast our "explicit" constructions can be used to write down generators very quickly. We deliberately refrain from making precise definitions, since we feel that these distinctions are best made informally.

Very recently it has been shown in [Rus1] that packings in dimension $n = p^2$, where p is a prime, with

$$\frac{1}{n} \log_2 \Delta \gtrsim -1.000000007719...$$

(the right-hand side is $-(1 + 2(\log_2 e)/e^{2\pi^2} + \cdots))$ can be found by searching over a specific finite set of error-correcting codes of length n. Although this result is also "nonconstructive", the search is over a much smaller population than in Minkowski's theorem.

Minkowski's result (28) guarantees that dense lattices exist. In the other direction, Rogers showed in [Rog2], [Rog7] that the density of *any* n-dimensional packing satisfies

$$\Delta \leqslant \sigma_n , \tag{39}$$

where σ_n is defined as follows. Let S be a regular n-dimensional simplex of edge length 2. Spheres of radius 1 centered at the vertices of S are disjoint, and σ_n is the ratio of the volume of the part of S covered by the spheres to the volume of S. Eq. (39) implies that no two-dimensional packing can be denser than the hexagonal lattice, and in three dimensions it gives the bound 0.7796... mentioned in §1.1. Another form of (39) found by Leech [Lee10] is

$$\log_2 \delta \leqslant \frac{1}{2} n \log_2 (\frac{n}{4e\pi}) + \frac{3}{2} \log_2 n - \log_2 \frac{e}{\sqrt{\pi}} + \frac{5.25}{n+2.5} , \tag{40}$$

the last term being approximate. For large n (39) gives

$$\frac{1}{n} \log_2 \Delta \lesssim -0.5 . \tag{41}$$

In 1979 Levenshtein [Lev7] showed that

$$\Delta \leqslant \frac{j(\frac{1}{2}n)}{\Gamma(\frac{1}{2}n+1)^2 4^n} , \tag{42}$$

where $j(t)$ is the smallest positive zero of the Bessel function $J_t(x)$. For large $n = 2t$ the approximate formula ([Bos2], [Olv1])

$$\begin{aligned} j(t) \sim t &+ 1.8557571\, t^{1/3} + 1.033150\, t^{-1/3} - 0.003971\, t^{-1} \\ &- 0.0908\, t^{-5/3} + 0.043\, t^{-7/3} \end{aligned} \tag{43}$$

can be used in (42). Asymptotically (42) gives

$$\frac{1}{n} \log_2 \Delta \lesssim -0.5573... . \tag{44}$$

In 1979 Kabatiansky and Levenshtein [Kab1] obtained an even stronger bound, which for large n gives

$$\frac{1}{n} \log_2 \Delta \lesssim -0.5990 . \tag{45}$$

The precise form of their bound and an outline of its proof are given in Chap. 9.

Other bounds are given in [Lev4], [Lev5], [Lev9], [Rog7], [Sid1], [Sid3], [Ura1]. Rogers' bound (39), (40) is the best for $n \leqslant 42$, and above that the Kabatiansky-Levenshtein bound takes over. The upper bounds in Table 1.3 are obtained from these two bounds.

Table 1.4 compares the various densities in dimension 65536, showing the best packings known, the density that we know from Minkowski's result is possible, and three upper bounds on the highest attainable density.

In summary, from (29) and (45) the densest (lattice or nonlattice) packings satisfy

$$-1 \lesssim \frac{1}{n} \log_2 \Delta \lesssim -0.5990 \tag{46}$$

for large n. (Roughly speaking, when the dimension increases by 1, the density of the best packing is divided by a number between 2 and 1.51.)

Many writers on number theory use *Hermite's constant* γ_n, given by

$$\gamma_n = 4\, \delta_n^{2/n}, \tag{47}$$

where δ_n is the center density of the densest lattice packing in $\mathbf{R}^n$ [Lek1, p. 294]. γ_n is known for $n \leqslant 8$ (it can be obtained from Tables 1.1, 1.2), and for large n (29) and (45) imply

$$\frac{1}{2\pi e} \lesssim \frac{\gamma_n}{n} \lesssim \frac{1.744}{2\pi e}. \tag{48}$$

In view of Eq. (28)-(34) of Chap. 3, γ_n measures the highest attainable *coding gain* of an n-dimensional lattice.

Table 1.4. Comparison of center densities δ in dimension 65536.

Type	Name	$\log_2\delta$	Chap.
Constructions	BW_{65536}	180224	8,§8.2f
	B_{65536}	290998	8,§8.2g
	$\eta(\overline{\Lambda}_{32})$	295120	8,§10c
	$A_{65536}^{(2954)}$	297740	8,§6
	$\eta(\Lambda_{24})^*$	311364	8,§10e
Existence	Minkowski	324603	1,(28)
Upper bounds	Kab.-Lev.	350885	9
	Levenshtein	353768	1,(42)
	Rogers	357385	1,(40)

* in dimension 65520

Although this book only considers the problem of packing a very large number of spheres, there is a considerable body of literature dealing with the best packings of N spheres for small values of N. See for example [Ben3], [Boe1], [Fej2a], [Fej2b], [Gol8], [Gra4a], [Gri12]-[Gri14], [Gro0], [Weg1], [Wil0].

2. The Kissing Number Problem

2.1 The problem of the thirteen spheres. The second main problem in this book, closely related to the packing problem, is fondly known as the *kissing number* problem. In three dimensions this asks how many billiard balls can be arranged so that they all just touch, or "kiss", another billiard ball of the same size. Don't imagine this on a billiard table – instead consider one billiard ball fixed in space, and mentally try to arrange the others around it. (For this problem billiard balls seem easier to "handle" than ball bearings, and besides "kiss" is a billiards term!) This question was the subject of a famous discussion between Isaac Newton and David Gregory in 1694. Newton believed the answer was 12, as found in the fcc lattice for example (see Fig. 1.1a), while Gregory thought that 13 might be possible.

The correct answer to this question, which is often called the problem of the thirteen spheres, is now known to be 12. Some proofs appeared in the nineteenth century in [Ben1], [Hop1] and [Gün1]. Schütte and van der Waerden gave a detailed proof in 1953 [Sch12]. The best proof now available is Leech's [Lee2], but although elementary and straightforward, it cannot be called trivial. (See also [Boe1], [Was1].)

The difficulty arises because the arrangement is not unique; in fact there are infinitely many ways to arrange 12 billiard balls around another. For example, if the 12 balls are placed at positions corresponding to the vertices of a regular icosahedron concentric with the central ball, the twelve outer balls do not touch each other and may all be moved freely. (In fact *any* permutation of the twelve spheres can be achieved by rolling the outer spheres around the inner one – see the Appendix). The icosahedral arrangement of 12 atoms around a central atom occurs in certain alloys and other compounds [Bri1], [Teo2], [Wel4, p. 71], and in the recently discovered nonperiodic structures known as *quasilattices* or *quasicrystals* (see for example [Con16a], [Els1], [Els2], [Gra6], [Grü3], [Lev10], [Lev11], [Nel1], [She0]), while the arrangement of 12 atoms around one found in the fcc lattice (forming the vertices of a cuboctahedron) is very common [Wel4, p. 71].

2.2 Kissing numbers in other dimensions. More generally we may define the *kissing number* (usually denoted by τ) of a sphere packing in any dimension to be the number of spheres that touch one sphere. For a lattice packing τ is the same for every sphere, but for an arbitrary packing τ may vary from one sphere to another. Other names for τ that have been used are *Newton number* (after the originator of the problem), *contact number*, *coordination number* or *ligancy* (the last two being chemist's terms).

The n-dimensional version of the kissing number problem then asks for the greatest value of τ attained by any packing of n-dimensional spheres. In one-dimensional space the answer is 2, and in two dimensions it is 6 (it is easy to show that although six pennies can be arranged around another penny on a table-top without overlapping, as in Fig. 1.3b, seven can not.) In three dimensions, as we have mentioned, the answer is 12.

It is somewhat surprising that we also know the answers in 8 and 24 dimensions ([Odl5] ≅ Chap. 13, [Lev7]), but in no other dimension above 3. In fact these numbers (240 and 196560 respectively) are technically easier to establish than the result in three dimensions. This is partly because in these dimensions the arrangements are unique: the only way to place 240 8-dimensional spheres around a central sphere is the arrangement found in the E_8 lattice, and similarly in 24 dimensions the arrangement must be that found in the Leech lattice in one of its two mirror-image forms ([Ban13] ≅ Chap. 14).

On the other hand in four dimensions, for example, the highest attainable kissing number is at present only known to be either 24 or 25. 24 occurs in the D_4 lattice, while the best bound known is 25. The kissing numbers of various packings in dimensions up to 24 can be found in Tables 1.1 and 1.2, while Table 1.3 includes the kissing numbers of some lattices in dimensions 24 to 128. (Beyond that the kissing numbers in Table 1.3 are not known.) The best bounds currently known on τ_n, the highest attainable kissing number in dimension n, are shown in Table 1.5 for $1 \leqslant n \leqslant 24$. The lower bounds in this table are taken from Table 1.2, and the upper bounds will be derived in Chap. 13.

So far we have considered the kissing numbers of arbitrary arrangements of spheres. If we restrict ourselves to lattice packings rather more is known: Watson [Wat7] showed that the laminated lattices Λ_n have the highest possible kissing numbers for lattices in dimensions $n \leqslant 9$. For $n \leqslant 8$ the highest kissing numbers presently known for arbitrary packings coincide with what can be attained by lattices. However in dimension 9 a nonlattice packing (P_{9a}) exists in which some spheres touch 306 others (see Table 1.2 and §2.6 of Chap. 5), whereas the highest possible kissing number for a lattice is 272 [Wat7] (see also [Wat6], [Wat8], [Wat9], [Wat13]). So as regards the maximal kissing number, 9 is the first dimension in which nonlattice packings are known to be superior to lattice packings.

It is worth commenting on our present state of knowledge about 11-dimensional packings, which is summarized in the following table (part of Table 1.2):

Name	Type	δ	τ_{max}
P_{11a}	nonlattice	0.03516	566
P_{11c}	nonlattice	low	582
K_{11}	lattice	0.03208	432
Λ_{11}^{max}	lattice	0.03125	438

This suggests that in general we should expect the lattice and nonlattice versions of the packing and kissing problems to have four different answers. The moral is that the kissing number question is a local problem, while the sphere packing question is a *global* problem! Of course the packings in this table are not known to be optimal, and it is still possible that there is a single lattice that is superior to all four.

Table 1.5. Range of possible values for τ_n, the highest attainable kissing number in dimension n. The third column gives the degree of the polynomial used in Chap. 13 to obtain the upper bound.

n	τ_n	Deg.
1	2	
2	6	
3	12	
4	24-25	9
5	40-46	10
6	72-82	10
7	126-140	10
8	240	6
9	306-380	11
10	500-595	11
11	582-915	11
12	840-1416	11
13	1130-2233	12
14	1582-3492	12
15	2564-5431	12
16	4320-8313	13
17	5346-12215	13
18	7398-17877	13
19	10668-25901	13
20	17400-37974	13
21	27720-56852	13
22	49896-86537	14
23	93150-128096	14
24	196560	10

In higher dimensions (returning now to the general kissing number problem) rather less is known. Kabatiansky and Levenshtein showed that in n dimensions the kissing number is bounded above by

$$\tau \leqslant 2^{0.401n(1+o(1))} \tag{49}$$

(see [Kab1] and Chap. 9, Eq. (55)), and Wyner [Wyn1] showed that arrangements of spheres exist with

$$\tau \geqslant 2^{(1-0.5\log_2 3)n(1+o(1))} = 2^{0.2075\ldots n(1+o(1))}. \tag{50}$$

But Wyner's result, like Minkowski's Eq. (28), is nonconstructive. Up to now the best constructive result was due to Leech, who showed that the Barnes-Wall lattice BW_n in dimension $n = 2^m$ has kissing number

$$\tau = \prod_{i=1}^{m} (2^i + 2) \tag{51}$$

$$\sim 4.768\ldots \cdot 2^{m(m+1)/2} = 4.768\ldots \cdot 2^{0.5\log_2 n(\log_2 n+1)}$$

(§6.5 of Chap. 5, §8.2f of Chap. 8, [Bar15], [Bar18], [Lee4], [Bos3]). The Barnes-Wall lattices, already mentioned in Eq. (30), provide one possible way to continue the sequence of lattices D_4, E_8, $\cdots$ to all dimensions 2^m. However, as we shall see in Chaps. 5, 8, these lattices are based on Reed-Muller codes and are therefore not especially good packings in high dimensions. Λ_{16}, the third lattice in the sequence, has the highest density and kissing number known in dimension 16 (see Chap. 4, §10), but the fourth lattice, which is equivalent to the laminated lattice Λ_{32} (Chap. 6) is inferior to Quebbemann's 32-dimensional lattice Q_{32} (§4 of Chap. 8, [Que5], [Que6]). For large n the kissing numbers of the Barnes-Wall lattices are very small compared to (50). An explicit construction for much higher kissing numbers will be given in §2.5.

2.3 Spherical codes. The kissing number problem can be stated in another way: how many points can be placed on the surface of a sphere in n-dimensional space $\mathbf{R}^n$ so that the angular separation between any two points is at least 60°? To see that this is exactly the same problem, consider any arrangement of nonoverlapping spheres touching a central sphere. Then the angular separation between the points where the outer spheres touch the central sphere is at least 60° (Fig. 1.6).

Thus the kissing number problem can be regarded as a packing problem, analogous to that considered in §1, but packing the points on the surface of a sphere in $\mathbf{R}^n$ (rather than in $\mathbf{R}^n$ itself). This leads to an important generalization. We denote the surface of the sphere by Ω_n:

$$\Omega_n = \{(x_1, \cdots, x_n) \in \mathbf{R}^n : \sum x_i^2 = 1\}. \tag{52}$$

We call a finite subset X of Ω_n a *spherical code* (by analogy with binary codes — see Chap. 3, §2), and say that X has *minimal angle* ϕ if ϕ is the largest number for which

$$x \cdot y \leqslant \cos\phi \quad \text{for any } x, y \in X, x \neq y.$$

(The inequality is $\leqslant$ rather than $\geqslant$, since the closer two points are on Ω_n, the larger is their inner product. When $x = y$, $x \cdot y = 1$. When y is the *antipodal* point to x, i.e. $y = -x$, then $x \cdot y = -1$.)

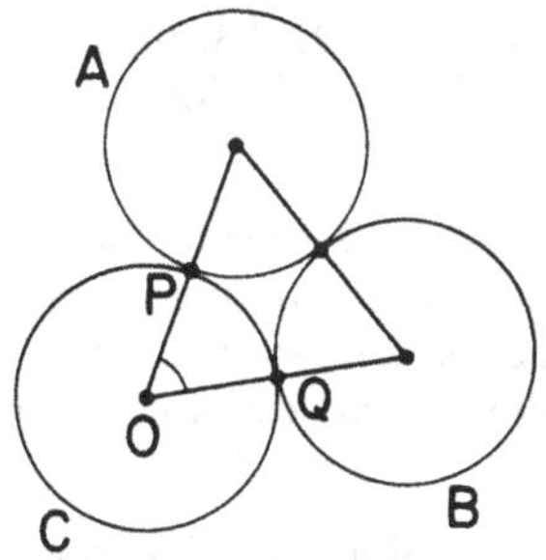

Figure 1.6. If spheres A and B touch sphere C, then the angular separation between the contact points P and Q, $\measuredangle POQ$, is at least 60°. For if A touches B, the three centers form an equilateral triangle.

We now generalize the kissing number problem by asking, for given n and ϕ, what is the maximal number $A(n,\phi)$ of points in a spherical code in Ω_n of minimal angle ϕ? $A(n,\pi/3)$ is the kissing number problem, and we have seen that $A(2,\pi/3) = 6$, $A(3,\pi/3) = 12$, $A(4,\pi/3) = 24$ or 25, etc.

The analogy with the sphere packing problem becomes clearer if we consider equal spherical *caps* placed on the sphere and centered at the points of the spherical code. Then $A(n,\phi)$ is the maximal number of spherical caps of angular diameter ϕ that can be placed on Ω_n without overlapping.

Spherical codes have many applications — see the references at the end of §1.4, and also Chap. 3, §1.2. The three-dimensional problem, the determination of $A(3,\phi)$, has been extensively studied. The inverse form is often used: what is the largest angular separation that can be attained in a spherical code on Ω_3 containing M points? This is sometimes called *Tammes' problem*, after the Dutch botanist who was led to this question by studying the distribution of pores on pollen grains [Tam5]. Equivalently we can ask: where should M inimical dictators build their palaces on a planet so as to be as far away from each other as possible?

Examples. The best code with $M = 4$ points consists of the vertices of a regular tetrahedron, with minimal angle $\phi = \pi - \cos^{-1} 1/3 \approx 109.47^\circ$. For $M = 6$, 8, 12 and 24 the best codes are given by a regular octahedron, a square antiprism, a regular icosahedron and a snub cube, with $\phi = 90^\circ$, 74.859°, $\tan^{-1} 2 \approx 63.435^\circ$ and 43.7° respectively [Fej10], [Rob1], [Sch11]. But for $M = 20$ the answer is *not* the regular dodecahedron, for which $\phi \approx 41.810^\circ$, since van der Waerden [Wae1] found a much less symmetric code with $\phi \approx 47.431^\circ$. The best codes known for larger values of M are also often not the obvious (highly symmetric) candidates.

The best three-dimensional spherical codes presently known can be found in [Bör3], [Bru1], [Cla1], [Cox15, p. 325], [Dan1], [Fej2], [Fej10, p. 168], [Gol3]-[Gol7], [Hab1], [Lee11], [Mel1], [Rob1], [Rob2], [Sch11], [Str2], [Szé2], [Tar1]-[Tar4], [Why1]. Three-dimensional codes with large numbers of points have been constructed in [Bau1], [Lub1], [Lub2].

See also [Arn1], [Bec1], [Ber9], [Ber10], [Bla1], [Grü2], [Lee3], [Lin1], [Mel1], [Mey1], [Moh1], [Mue1], [Slo13a], [Wel4]. For four-dimensional spherical codes see [Cox24], [Mac0], [Nun1], [War4].

There are two simple and general methods for constructing spherical codes that are not restricted to three dimensions, one based on sphere packings and the other on error-correcting codes. These constructions, like the three- and four-dimensional codes just mentioned, all give lower bounds on $A(n,\phi)$.

2.4 The construction of spherical codes from sphere packings. Let Λ be a sphere packing in $\mathbf{R}^n$, and let us place the origin of coordinates at a convenient point P. (Usually P is taken to be the center of a sphere or a point equidistant from a certain number of centers. For example, in Fig. 1.3b we might choose P to be a point marked 'a' or 'b', or midway between two points marked 'a'.) Suppose there are N spheres in Λ with centers at distance u from P. Then these centers, when rescaled by dividing them by u, form an n-dimensional spherical code of size N. In other words we take a *shell* of points around P as the spherical code [Slo12].

The distance between the centers of the spheres is at least 2ρ (where ρ is the radius of a sphere), so the minimal angle in this code is at least

$$2 \sin^{-1}(\rho/u). \tag{53}$$

The number of points in this code is given by the theta series of the packing with respect to P (Chap. 2, §2.3).

Examples. Consider the lattice D_4 (§1.4), with the origin P at a lattice point and $\rho = 1/\sqrt{2}$. The first shell has $u = \sqrt{2}$ and contains 24 centers (the kissing number), and the corresponding spherical code consists of the points $2^{-1/2}(\pm 1, \pm 1, 0, 0)$, with minimal angle $2\sin^{-1}\frac{1}{2} = 60°$. The second shell has $u = 2$ and also contains 24 points, and the spherical code consists of $(\pm 1, 0, 0, 0)$ and $\frac{1}{2}(\pm 1, \pm 1, \pm 1, \pm 1)$. Again the minimal angle is $60°$, so in this case (53) does not give the exact value of ϕ. In fact this second code is a rotation of the first. Both examples show that $A(4, 60°) \geqslant 24$.

On the other hand if we move P to $(1,0,0,0)$ (a "deep hole" in D_4 — see Fig. 1.3d) we obtain a different sequence of spherical codes. The first shell now contains 8 points at distance 1 from P, forming a code of minimal angle $90°$ (the vertices of a regular 4-dimensional generalized octahedron). Thus $A(4, 90°) \geqslant 8$. The numbers of points in these two families of spherical codes are the coefficients of the theta series $\frac{1}{2}(\theta_3(2z)^4 + \theta_4(2z)^4)$ and $\frac{1}{2}\theta_2(2z)^4$ respectively (Chap. 4, §7).

2.5 The construction of spherical codes from binary codes. Let C be a binary error-correcting code (Chap. 3, §2) of length n and minimal distance d. A spherical code is obtained by changing the 1's to -1's and the 0's to $+1$'s in every codeword, and dividing by $\sqrt{n}$. The resulting points lie on Ω_n and the minimal angle is given by

$$\phi = \cos^{-1}(1 - \frac{2d}{n}), \tag{54}$$

$$\frac{d}{n} = \sin^2 \frac{\phi}{2}. \tag{55}$$

Examples. The code containing all 2^n binary vectors of length n produces the spherical code consisting of the vertices $n^{-1/2}(\pm 1, \cdots, \pm 1)$ of an n-dimensional cube. Any other spherical code obtained by the construction is a subset of this. The Golay code $\mathscr{C}_{24}$ (Chap. 3, §2.8.2) with $n = 24$, $d = 8$ produces a spherical code containing 4096 points on Ω_{24} with $\phi = \cos^{-1} 1/3 \approx 70.529°$. Thus $A(24, \cos^{-1} 1/3) \geqslant 4096$.

By using certain codes constructed by Justesen, Weldon, and Sugiyama et al. (Chap. 3, §2.6), we may obtain infinite sequences of spherical codes, with $n \to \infty$, having a fixed minimal angle ϕ and containing 2^{cn} points (where c depends only on ϕ). Justesen's codes are the simplest, but have $d/n \leqslant 0.110...$, and only produce spherical codes with $\phi \leqslant 38.73°$. For larger angles the codes of Weldon and Sugiyama et al. must be used. The kissing number problem for example requires $\phi = 60°$, and from the Sugiyama et al. codes we obtain an explicit sequence of arrangements of spheres in $\mathbf{R}^n$ with kissing number

$$\tau = 2^{0.003n(1+o(1))}. \tag{56}$$

(This is not very good when compared with Wyner's existence result (50), but at least grows exponentially with n.)

2.6 Bounds on $A(n,\phi)$. Again there is a nonconstructive lower bound [Sha6], [Wyn1]:

$$A(n,\phi) \geqslant 2^{-n \log_2 \sin\phi (1+o(1))}. \tag{57}$$

(Eq. (50) is the special case $\phi = 60°$.) In the other direction there are a number of bounds. We start with bounds that apply when ϕ is large, and work down. Rankin [Ran4] found the exact value for $\phi \geqslant 90°$:

$$A(n,\phi) = 1, \quad \text{for } \pi < \phi \leqslant 2\pi, \tag{58}$$

$$A(n,\phi) = [1 - \sec\phi], \quad \text{for } \sec^{-1}(-n) \leqslant \phi \leqslant \pi, \tag{59}$$

$$A(n,\phi) = n + 1, \quad \text{for } \pi/2 < \phi \leqslant \sec^{-1}(-n), \tag{60}$$

$$A(n,\pi/2) = 2n. \tag{61}$$

The spherical codes corresponding to (60) and (61) are respectively the vertices of a regular simplex and a regular cross-polytope or generalized octahedron. Rankin also showed that

$$A(n,\phi) \leqslant (\tfrac{1}{2}\pi n^3 \cos\phi)^{1/2} (\sqrt{2} \sin \tfrac{1}{2}\phi)^{-n} (1 + o(1)). \tag{62}$$

Coxeter [Cox16] conjectured and Böröczky [Bör2] proved that

$$A(n,\phi) \leqslant \frac{2F_{n-1}(\alpha)}{F_n(\alpha)}, \tag{63}$$

where $\sec 2\alpha = \sec \phi + n - 2$ and $F_n(\alpha)$ is Schläfli's function defined by

$$F_n(\alpha) = \frac{2^n U}{n \cdot n!\, V_n}, \tag{64}$$

where U is the "area" of a regular spherical simplex of angle 2α contained in Ω_n. For large n, (63) is stronger than (62) by a factor approaching $2/e$. For the kissing number problem, setting $\phi = 60^\circ$ in (63) gives

$$\frac{1}{n} \log_2 \tau \lesssim \frac{1}{2},$$

which is not as good as the Kabatiansky-Levenshtein bound (49). Kabatiansky and Levenshtein [Kab1] also show that for fixed ϕ with $0 < \phi < \pi/2$, and large n,

$$\frac{1}{n} \log_2 A(n,\phi) \lesssim \frac{1+\sin\phi}{2\sin\phi} \log_2 \frac{1+\sin\phi}{2\sin\phi} - \frac{1-\sin\phi}{2\sin\phi} \log_2 \frac{1-\sin\phi}{2\sin\phi} \tag{65}$$

and, for $0 < \phi < \phi^*$,

$$\frac{1}{n} \log_2 A(n,\phi) \lesssim -\frac{1}{2} \log_2 (1-\cos\phi) - 0.0990, \tag{66}$$

where $\phi^* \approx 63^\circ$ is the root of a certain equation. When $\phi = 60^\circ$, (66) yields (49). Eqs. (65) and (66) follow from linear programming — see Chap. 9, §3.5(iv). The linear programming method also produces many tight estimates of $A(n,\phi)$ in particular cases — see for example Tables 1.4 and 9.2. For $\cos\phi < 1/\sqrt{n}$, Delsarte et al. [Del16] showed that

$$A(n,\phi) \leqslant \frac{n(1-\cos\phi)\,(2+(n+1)\cos\phi)}{1-n\cos^2\phi}, \tag{67}$$

and Astola [Ast1] obtained

$$A(n,\phi) \lesssim n(2.2+\log_e(1+n\alpha)) \tag{68}$$

for $\cos\phi = o(n^{-2/3})$, and

$$A(n,\phi) \lesssim \tfrac{1}{2}\, n \log_e (n\cos\phi) \tag{69}$$

provided that $n \cos\phi$ is unbounded as $n \to \infty$.

Bounds on $A(n,\phi)$ also lead to bounds on the density Δ of an n-dimensional sphere packing via the inequality

$$\Delta \leqslant (\sin \tfrac{1}{2}\phi)^n A(n+1,\phi), \text{ for } 0 < \phi \leqslant \pi \tag{70}$$

(see Chap. 9, Th. 6). For example Eq. (45) was deduced from (66) and (70).

Sphere-packing problems in hyperbolic space have been studied in [Bez1], [Bez3], [Bör1], [Cox18, pp. 173-177], [Fej1, p. 325], [Fej4]-[Fej7], [Fej11].

Appendix. Planetary Perturbations

Consider a cluster of 12 unit spheres, the "planets", touching a given one, the "sun", arranged as if they were in the fcc lattice. We shall show that there is so much "play" that it is possible to achieve any permutation of the 12 planets by cleverly rolling them around the sun in such a way that they never overlap.

Figure 1.7 shows one way to prove this. Figure 1.7a shows the 12 points of contact when the planetary spheres are in the original cuboctahedral arrangement of the fcc lattice. This cuboctahedron can be rotated about a square face to achieve the permutation

$$\pi_1 = (0,10,3,11)(1,9,4,2)(5,6,7,8) .$$

If the planets are perturbed by rolling them in the directions indicated by the arrows, they can be brought continuously into the icosahedral arrangement of Fig. 1.7b. (A precise description of this perturbation can be obtained from [Cox20, §8.4].) The distances have now increased so that the 12 planets do not touch each other. In this configuration call any planet, say number 1, the South pole, and its opposite (number 4), the North pole, and regard each other planet as the "satellite" of the nearest of these two polar ones, as indicated in Fig. 1.7c. Now move each satellite towards its polar parent planet until they touch (as suggested by the arrows in Fig. 1.7c). The five satellites of the North polar planet can now revolve around it to yield the permutation

$$\pi_2 = (3,6,5,9,10) .$$

The different choices for the poles give twelve 5-cycles like π_2. We shall show that these twelve 5-cycles generate the alternating group A_{12}. In Chap. 11, §18 it is shown that their pairwise quotients generate the Mathieu group M_{12}. Since M_{12} is quintuply transitive there is an element of it which transforms π_2 into any desired 5-cycle, and the set of *all* 5-

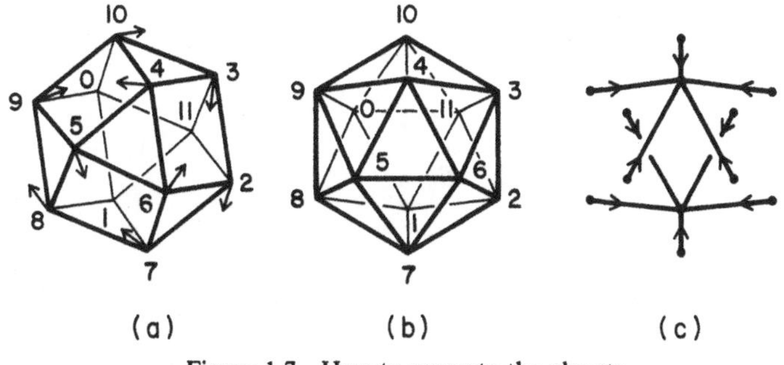

Figure 1.7. How to permute the planets.

cycles is well known to generate A_{12}. Since we can also achieve the odd permutation π_1, we can achieve the full symmetric group S_{12}, as claimed.

By a slight perturbation of the satellite motions we can obtain the twelve 5-cycles like π_2 (which generate A_{12}) even when the planets have radius slightly larger than the sun. We do not know whether the whole group S_{12} can be achieved for such a radius.

2

Coverings, Lattices and Quantizers

J. H. Conway and N. J. A. Sloane

This second chapter continues the description of the questions motivating this book. We first discuss the problem of finding the best covering of space by overlapping spheres, a kind of dual to the packing problem. Then we introduce the language of quadratic forms, show that lattices and quadratic forms are really the same, and explain the connections with number theory. One of the central issues is the classification of integral quadratic forms or lattices. The last section describes the problem of designing good quantizers or analog-to-digital converters. For each problem we summarize what is presently known about its solution.

1. The Covering Problem

1.1 Covering space with overlapping spheres. We have already met the packing and kissing number problems. The third main topic of this book is a kind of dual to the packing problem, and asks for the most economical way to *cover* n-dimensional Euclidean space with equal overlapping spheres. Figure 2.1 shows two different ways to cover the plane with overlapping circles. In (a) the centers of the circles belong to the square lattice $\mathbf{Z}^2$, and in (b) to the hexagonal lattice. Clearly (b) is a more efficient covering than (a), since there is less overlap among the circles.

To make this precise we define the thickness Θ of a covering in the same way as the density Δ of a packing. Suppose an arrangement of spheres of radius R covers $\mathbf{R}^n$. If the centers form a lattice Λ then the *thickness* is defined by formulae similar to Eqs. (10), (11), (20) of Chap. 1:

$$\Theta = \text{average number of spheres that contain a point of the space}$$

$$= \frac{\text{volume of one sphere}}{(\det \Lambda)^{1/2}} = \frac{V_n R^n}{(\det \Lambda)^{1/2}} . \tag{1}$$

Θ is also called the *density* or the *sparsity* of the covering. None of these terms is completely satisfactory, and *thickness* seems the most descriptive. The thickness of an arbitrary covering is defined in the same way as the density of an arbitrary packing [Rog7, Chap. 1]. We always have $\Delta \leqslant 1 \leqslant \Theta$. The *normalized thickness* (or *center density*) θ is given by

$$\theta = \frac{\Theta}{V_n} \tag{2}$$

(compare (26) of Chap. 1). The lattice coverings in Fig. 2.1 have thickness $\Theta = \pi/2 = 1.5708...$ and $\Theta = 2\pi/3\sqrt{3} = 1.2092...$ respectively.

Then the *covering problem* asks for the thinnest covering of n-dimensional space by spheres, i.e. for the covering with minimal thickness.

Kershner showed in 1939 [Ker3] that no arrangement of circles can cover the plane more efficiently than the hexagonal lattice arrangement shown in Fig. 2.1b. For a concise proof see [Fej10, pp. 58-61]. (See also [Aki1], [Aki2].) But, just as for packings, the optimal coverings are not known in higher dimensions.

In three dimensions the best covering known, which Bambah [Bam1] showed is optimal among three-dimensional lattice coverings, is the body-centered cubic lattice mentioned in §1.4 of Chap. 1. This is at first sight surprising, since the densest lattice *packing* is a different lattice, the fcc lattice. In fact the bcc is the dual of the fcc lattice. To understand this situation we must look further into the structure of these lattices, and introduce some additional terminology.

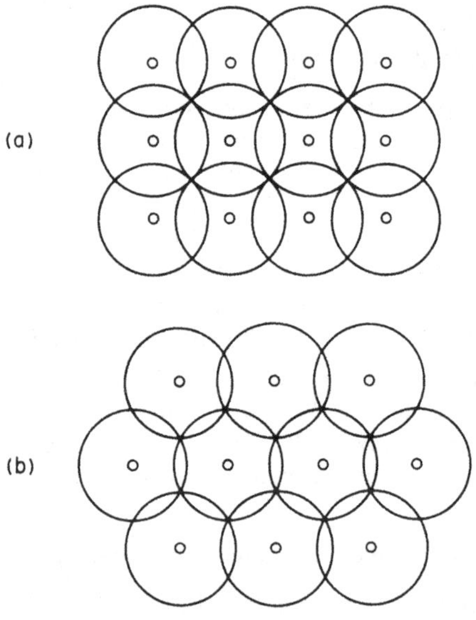

Figure 2.1. Covering the plane with circles. In (a) the centers belong to the square lattice $\mathbf{Z}^2$, in (b) they belong to the hexagonal lattice. (b) is a more efficient or *thinner* covering.

1.2 The covering radius and the Voronoi cells. Consider any discrete collection of points $\mathcal{P} = \{P_1, P_2, \ldots\}$ in $\mathbf{R}^n$. The least upper bound for the distance from any point of $\mathbf{R}^n$ to the closest point P_i is called the *covering radius* of $\mathcal{P}$, usually denoted by R. Thus

$$R = \sup_{x \in \mathbf{R}^n} \inf_{P \in \mathcal{P}} \operatorname{dist}(x, P). \tag{3}$$

(If the upper bound does not exist we set $R = \infty$.) Then spheres of radius R centered at the points of $\mathcal{P}$ will cover $\mathbf{R}^n$, and no smaller radius will do.

Around each point $P_i \in \mathcal{P}$ is its *Voronoi cell*, $V(P_i)$, consisting of those points of $\mathbf{R}^n$ that are at least as close to P_i as to any other P_j. Thus

$$V(P_i) = \{ \; x \in R^n \; : \; \operatorname{dist}(x, P_i) \leqslant \operatorname{dist}(x, P_j) \text{ for all } j \; \}. \tag{4}$$

If P_1, $P_2, \ldots$ represent schools, the Voronoi cells are the school districts! Other names are *nearest neighbor regions, Dirichlet regions, Brillouin zones* and *Wigner-Seitz cells* (the last two are physicists' terms). The Voronoi cells of the hexagonal lattice, for example, are the regular hexagons shown in Fig. 1.3c. The Voronoi cells of many other lattices are described in Chaps. 4 and 21.

The interiors of the Voronoi cells are disjoint, although they have faces in common. Each face lies in the hyperplane midway between two neighboring points P_i. The Voronoi cells are convex polytopes whose union is the whole of $\mathbf{R}^n$. (It is true, but not entirely obvious, that the intersection of two abutting Voronoi cells is an entire face of each of them.) If $\mathcal{P}$ is a lattice Λ, all the Voronoi cells are congruent and have volume equal to $(\det \Lambda)^{1/2}$.

The vertices of the Voronoi cells are especially interesting. They include the points of $\mathbf{R}^n$ whose distance from $\mathcal{P}$ is a local maximum: these are called the *holes* in $\mathcal{P}$ (cf. Fig. 1.3d). If there is a point whose distance from $\mathcal{P}$ is an absolute maximum it is called a *deep hole* and its distance from $\mathcal{P}$ is the covering radius R. (The children who live at deep holes are those that have the longest walk to school.) Holes that are not deep are called *shallow*. The deep holes in the hexagonal lattice are the points marked b and c in Fig. 1.3b. For this lattice $R = 2\rho/\sqrt{3}$, where ρ is the packing radius. If $\mathcal{P}$ is a lattice, the vertices of the Voronoi cells are exactly the holes, but in general they may include other points.

For a lattice packing Λ, with Voronoi cells congruent to a polytope V say, the packing radius ρ is the *inradius* of V (the radius of the largest inscribed sphere), while the covering radius R is the *circumradius* of V (the radius of the smallest circumscribed sphere). We now see the difference between the packing and covering problems. For a good packing we try to maximize ρ, i.e. we wish to choose the centers of the spheres so that the inradius of the Voronoi cells is as large as possible (for a nonlattice packing we consider the smallest inradius of any Voronoi cell). On the other hand for a good covering we try to minimize R, i.e. we wish to choose the centers so that the circumradius of the Voronoi cells is as small as possible (for a nonlattice covering we consider the largest circumradius of any Voronoi cell).

Let us return to the three-dimensional question. The Voronoi cell for the fcc lattice is a rhombic dodecahedron (Fig. 2.2a), one of the semi-regular polyhedra [Cun1, p. 114], [Fej9], [Hol1], [Loe1, p. 42], [Wel4, p. 73]. If we choose the scale so that the lattice has determinant 1 (and the Voronoi cell has volume 1) then the inradius and circumradius of the Voronoi cell are $\rho = 2^{-5/6} = 0.5612...$ and $R = 2^{-1/3} = 0.7937...$ respectively. On the other hand the Voronoi cell for the bcc lattice is a truncated octahedron (Fig. 2.2b), one of the Archimedean polyhedra [Cun1, p. 98], [Fej9], [Hol1], [Loe1, p. 129], [Wel4, p. 73], [Wen1, p. 21]. When this lattice is scaled so as to have determinant 1 we find $\rho = 2^{-5/3}\,3^{1/2} = 0.5456...$ and $R = 2^{-5/3}\,5^{1/2} = 0.7043...$. Thus although the fcc lattice is the better packing, the bcc lattice is indeed a better covering. There is another difference between these two lattices. In the bcc lattice, as in the planar hexagonal lattice, there is only one kind of hole (all holes are deep), but in the fcc lattice there are two kinds (shallow and deep holes). See Fig. 2.2 and also Chap. 4, §§6.3, 6.7. This phenomenon is particularly striking in the Leech lattice, where there are 23 kinds of deep hole and 284 kinds of shallow hole (Chaps. 23 and 25).

There are two other important concepts associated with the Voronoi cell $V(P_i)$ around a point P_i. The *normalized second moment* of $V(P_i)$ is defined to be

$$G = G(V(P_i)) = \frac{1}{n}\,\mathrm{Vol}(V(P_i))^{-1-\frac{2}{n}} \int_{V(P_i)} \|x - P_i\|^2 dx .$$

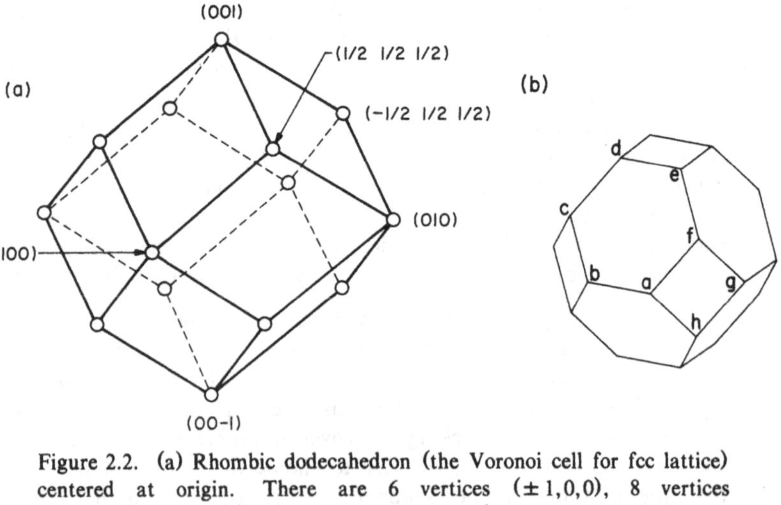

Figure 2.2. (a) Rhombic dodecahedron (the Voronoi cell for fcc lattice) centered at origin. There are 6 vertices $(\pm 1,0,0)$, 8 vertices $(\pm\frac{1}{2}, \pm\frac{1}{2}, \pm\frac{1}{2})$. (b) Truncated octahedron (the Voronoi cell for bcc lattice) centered at origin. There are 24 vertices $(\pm 1, \pm\frac{1}{2}, 0)$. For example $a = (0,\frac{1}{2},1)$, $b = (0,1,\frac{1}{2})$, $c = (\frac{1}{2},1,0)$, $d = (1,\frac{1}{2},0)$, $e = (1,0,\frac{1}{2})$, $f = (\frac{1}{2},0,1)$, $g = (0,-\frac{1}{2},1)$, $h = (-\frac{1}{2},0,1)$.

If the points P_i form a lattice Λ, then G will be denoted by $G(\Lambda)$. Apart from the scale factors, G is the average squared distance that a child who lives in $V(P_i)$ has to walk to school. (The second scale factor makes G a dimensionless quantity.) G appears in Davenport's construction of efficient coverings (§1.3), in studying quantizers (§3.2), and in other applications.

The points $P_{i1}, \ldots, P_{ir}$ such that the hyperplane between P_i and P_{ij} contains a face of $V(P_i)$ with positive area are called *Voronoi-relevant* (or just *relevant*) for P_i. These are the points actually needed to define the Voronoi cell. For a lattice Λ the *(Voronoi-)relevant vectors* $u \in \Lambda$ are those that are relevant for the origin, i.e. that are needed to define $V(0)$.

The following references discuss Voronoi cells: [Aki1], [Aki2], [Ash3], [Bar9], [Cha2], [Con28], [Con38], [Dir1], [Ede1], [Fej9], [How1], [Jon4], [Kel1], [Kit4], [Koc1], [Lan2], [Loe1], [Mau1], [Pre2], [Shu3], [Ven5], [Vor1], [Wan1], [Wig1]. Algorithms for computing Voronoi cells are mentioned in §1.4.

Before leaving this section we introduce the Delaunay cells associated with the points $\mathscr{P}$. There is a *Delaunay cell* for each point that is the vertex of a Voronoi cell; it is the polytope which is the convex hull of the points of $\mathscr{P}$ closest to that point [Cox26], [Del0], [Del1], [Gal0], [Rog7]. (The early works of Boris Nikolaevich Delone were published under the name Delaunay, and that spelling has come to be associated with these polytopes.)

The Delaunay cells form a partition of $\mathbf{R}^n$ into convex regions that is a kind of dual to the partition into Voronoi cells. Consider for example the set $\mathscr{P}$ indicated by the small circles in Fig. 2.3. The Voronoi cells are triangles such as XYZ, and there are two types of Delaunay cells: squares,

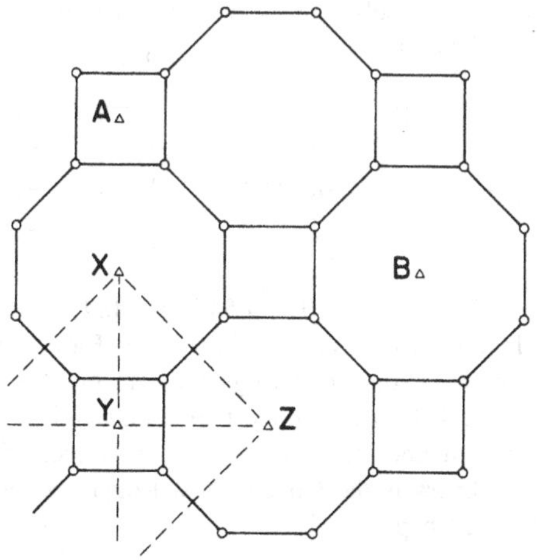

Figure 2.3. For the points $\mathscr{P}$ indicated by small circles the Delaunay cells are squares (A) and octagons (B); the Voronoi cells are triangles (e.g. XYZ).

centered at shallow holes such as A, and octagons centered at deep holes such as B.

In the fcc lattice the Delaunay cells form a tessellation of alternating tetrahedra and octahedra. For the bcc lattice the Delaunay cells are all semiregular tetrahedra (see p. 116 and [Con43g]).

1.3 Covering problem-summary of results. In this section we summarize what is presently known about the covering problem (see Tables 1.1, 2.1 and Fig. 2.4). As mentioned in §1.1, the thinnest possible coverings are only known in dimensions 1 and 2. The thinnest *lattice* coverings are known in dimensions 1 through 5, and in each case the optimal lattice is A_n^* (the dual of A_n, and sometimes referred to as "Voronoi's principal lattice of the first type" — see Chap. 4, §6.6). The results are due to Bambah in 1954 for $n = 3$ [Bam1], Delone (or Delaunay), Ryskov in 1963 for $n = 4$ [Del4], and Ryskov and Baranovskii in 1975 for $n = 5$ [Rys12], [Rys13]. (See also [Bam2], [Bar5], [Bar14], [Ble1], [Del2], [Dic4]-[Dic6], [Few1], [Gam1], [Gam2], [Kau1], [Woo1].)

The covering associated with A_n^* has thickness

$$\Theta = V_n \sqrt{n+1} \left\{ \frac{n(n+2)}{12(n+1)} \right\}^{n/2} , \tag{5}$$

and is in fact the thinnest covering known[(1)] in all dimensions $n \leqslant 23$. (It is surprising that A_8^*, with $\Theta = 3.6658...$, is better than E_8, which has $\Theta = 4.0587...$.) But A_{24}^* has thickness 63.269..., and is inferior to the Leech lattice Λ_{24}, for which $\Theta = 7.9035...$ (Chap. 4, §11, and [Con20] = Chap. 23).

In three dimensions, A_3^* (the bcc lattice) is not just the thinnest lattice covering, it is the only *locally* optimal lattice covering. In $\mathbf{R}^4$ Dickson [Dic5] found that there are precisely three locally optimal lattice coverings, namely A_4^* and the lattices with Gram matrices

$$Di_{4a} : \begin{bmatrix} 2 & \alpha & -1 & -1 \\ \alpha & 2 & -1 & -1 \\ -1 & -1 & 2 & 1-\alpha \\ -1 & -1 & 1-\alpha & 2 \end{bmatrix}, \quad Di_{4b} : \begin{bmatrix} 3-\gamma & \gamma & -1 & -1 \\ \gamma & 3-\gamma & -1 & -1 \\ -1 & -1 & 2+2\beta & -\beta \\ -1 & -1 & -\beta & 2+2\beta \end{bmatrix} \tag{6}$$

where $\alpha = (5 - \sqrt{13})/2$ and $\beta \approx 0.544$, $\gamma \approx 0.499$ are the roots of certain polynomials. In dimensions $n = 5, 7, 9, \cdots$ Barnes and Trenerry [Bar17] found locally optimal lattice coverings that are only slightly worse than A_n^*. Table 2.1 and Fig. 2.4 show the thinnest coverings known in dimensions $n \leq 24$, together with the thicknesses of certain other lattices of interest. The lower bound on the thickness is the Coxeter-Few-Rogers bound (16), calculated via (17) and Eq. (40) of Chap. 1.

---

(1) It is now known that Λ_{23}^* is a better covering than A_{23}^*, and there are other improvements to Table 2.1 and Fig. 2.4 in dimensions just below 24.

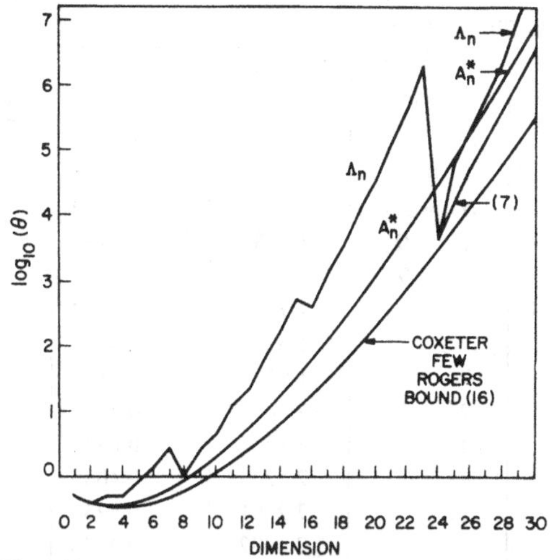

Figure 2.4. The thickness of various lattice coverings in dimensions $n \le 24$. The values for Λ_{13} to Λ_{15} and Λ_{17} to Λ_{23} are lower bounds. (They are computed using the *subcovering radius* of Chap. 6.) See Notes on Chapter 2 for recent improvements.

In all dimensions $n = 24\ell + m \geqslant 24$ (for $0 \leqslant m \leqslant 23$) the lattice $\Lambda_{24} \oplus \Lambda_{24} \oplus \cdots \oplus \Lambda_{24} \oplus A_m^*$, with ℓ copies of Λ_{24}, has thickness

$$\Theta = V_n \sqrt{m+1} \left(\frac{m+2}{m+1} \right)^{m/2} \left(\frac{n}{12} \right)^{n/2}, \tag{7}$$

and is a thinner covering than A_n^* [Bam3]. Even thinner coverings for $n \geqslant 25$ are given in [Con43]. However all the preceding coverings have the defect that, as $n \to \infty$,

$$\frac{1}{n} \log_2 \Theta \sim \log_2 \sqrt{\frac{2\pi e}{12}} = 0.2546\ldots . \tag{8}$$

In 1952 Davenport [Dav2] found a method that yields thinner coverings than (8) when n is large. Davenport's construction may be described as follows. We start with a fixed k-dimensional lattice Λ having generator matrix M say, and construct the km-dimensional lattice $\mathcal{L}$ with the generator matrix shown in Fig. 2.5. Davenport showed that if ℓ and m are large the thickness of $\mathcal{L}$ is bounded above by a quantity which is essentially independent of ℓ. For ℓ fixed (and large), and $m, n \to \infty$, the thickness of $\mathcal{L}$ satisfies

$$\frac{1}{n} \log_2 \Theta \le \log_2 \sqrt{2\pi e G(\Lambda)}, \tag{9}$$

where $G(\Lambda)$ is defined in §1.2. If $\Lambda = \mathbf{Z}$, $G(\Lambda) = 1/12$ (see Eq. (88) below), and (9) reduces to (8). Davenport took $\Lambda = A_k$, calculated $G(A_k)$

Table 2.1. Coverings in up to 24 dimensions. (See Notes on Chapter 2 for recent improvements.)

n	Name of covering	Thickness Θ	Center density θ Attained	Bound
0	Λ_0	1	1	1
1	$A_1^* \cong \mathbf{Z}$	1	0.5	0.5
2	$A_2^* \cong A_2$	1.2092	0.3849	0.3849
3	$A_3^* \cong D_3^*$	1.4635	0.3494	0.3419
	$A_3 \cong D_3$	2.0944	0.5	
4	A_4^*	1.7655	0.3578	0.3360
	Di_{4b}	1.89	0.382	
	Di_{4a}	1.93	0.391	
	$D_4^* \cong D_4$	2.4674	0.5	
	A_4	3.1780	0.644	
5	A_5^*	2.1243	0.4036	0.3581
	BT_5	2.2301	0.4237	
	D_5^*	2.4982	0.4746	
	D_5	4.5977	0.8735	
	A_5	5.9218	1.125	
6	A_6^*	2.5511	0.4937	0.4087
	D_6^*	4.3603	0.8437	
7	A_7^*	3.0596	0.6476	0.4949
	BT_7	3.2441	0.6865	
	E_7^*	4.1872	0.8862	
	D_7^*	4.5687	0.9670	
8	A_8^*	3.6658	0.9032	0.6319
	E_8	4.0587	1	
	D_8^*	8.1174	2	
9	A_9^*	4.3889	1.331	0.8460
	BT_9	4.6569	1.412	
	D_9^*	8.6662	2.627	
10	A_{10}^*	5.2517	2.059	1.183
11	A_{11}^*	6.2813	3.334	1.721
12	A_{12}^*	7.5101	5.624	2.597
	K_{12}	17.7834	13.318	
13	A_{13}^*	8.9768	9.858	4.055
14	A_{14}^*	10.727	17.90	6.537
15	A_{15}^*	12.817	33.60	10.86
16	A_{16}^*	15.311	65.06	18.56
17	A_{17}^*	18.288	129.7	32.57
18	A_{18}^*	21.841	265.9	58.63
19	A_{19}^*	26.082	559.4	108.1
20	A_{20}^*	31.143	1207	204.0
21	A_{21}^*	37.185	2666	393.5
22	A_{22}^*	44.395	6023	775.2
23	A_{23}^*	53.000	13908	1558
24	Λ_{24}	7.9035	4096	3193
	A_{24}^*	63.269	32789	

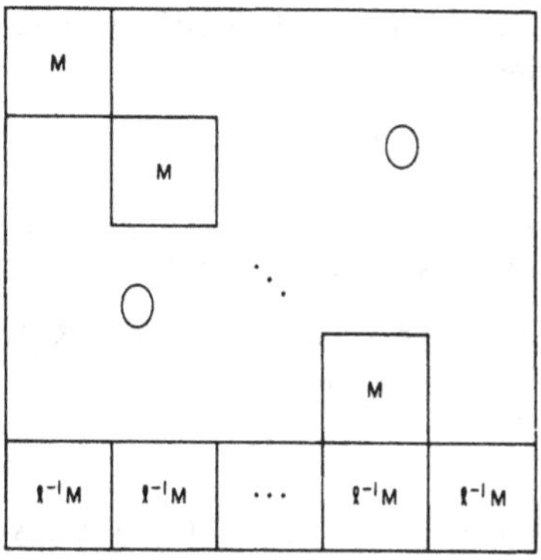

Figure 2.5. Davenport's construction of thin lattice coverings. This lattice contains the direct sum of m copies of the starting lattice Λ, whose generator matrix is M.

(see Chap. 21, Eq. (26)), finding that it is minimized when $k = 8$. This produces lattices $\mathscr{L}$ with

$$\frac{1}{n} \log_2 \Theta \lesssim 0.2012\ldots . \tag{10}$$

Ryskov [Rys1] showed that lattices obtained from Davenport's construction (taking $\Lambda = A_2$) are thinner coverings than A_n^* for all $n \geqslant 200$.

Watson [Wat2] used $\Lambda = E_8$ and obtained

$$\frac{1}{n} \log_2 \Theta \lesssim 0.1460\ldots . \tag{11}$$

(Watson's calculation of $G(E_8)$ does not quite agree with the value 929/12960 we find in Chap. 21, but the discrepancy is so small that it has no effect on (11).)

Finally, by taking Λ to be the Leech lattice Λ_{24}, and using our numerical estimate [Con38]

$$G(\Lambda_{24}) = 0.065771 \pm 0.000074, \tag{12}$$

we obtain coverings with thickness satisfying

$$\frac{1}{n} \log_2 \Theta \lesssim 0.084\ldots . \tag{13}$$

These appear to be the most efficient coverings known.

On the other hand these are all extremely poor coverings when compared with the existence results. For Rogers showed in [Rog3] that lattice coverings exist with

$$\Theta \leqslant c \; n (\log_e n)^a \tag{14}$$

for some constant c, where $a = \frac{1}{2} \log_2(2\pi e)$, and in [Rog1] that (possibly nonlattice) coverings exist with

$$\Theta \leqslant n \log_e n + n \log_e \log_e n + 5n , \tag{15}$$

for $n \geqslant 3$ (see also [Erd2], [Gri11], [Rog7]). As usual these results are nonconstructive, in the sense explained in Chap. 1.

In the other direction Coxeter, Few and Rogers [Cox26] showed that any n-dimensional covering satisfies

$$\Theta \geqslant \tau_n . \tag{16}$$

This is a companion to Eq. (39) of Chap. 1. Again let S be a regular simplex of edge 2. Spheres of radius $\{2n/(n+1)\}^{1/2}$ centered at the vertices of S just cover S, and τ_n is the ratio of the sum of the volumes of the intersections of these spheres with S to the volume of S. Thus

$$\tau_n = \left\{ \frac{2n}{n+1} \right\}^{n/2} \sigma_n \tag{17}$$

$$\sim \frac{n}{e\sqrt{e}} \text{ as } n \to \infty . \tag{18}$$

In summary, from (15)-(18), the thinnest coverings satisfy

$$\frac{n}{e\sqrt{e}} \lesssim \Theta \leqslant n \log_e n + n \log_e \log_e n + 5n . \tag{19}$$

Remarks. (i) These bounds are considerably closer together than the corresponding bounds for packings (Chap. 1, Eq. (46)); on the other hand the best constructions known are further from the bounds. (ii) There are no known examples of nonlattice coverings that are thinner than the best lattice coverings. (iii) One may also consider the covering problem on the sphere Ω_n (analogous to Chap. 1, §2.3). There are some results in [Fej1, p. 325], [Rog6], [Wyn2], but not much is known about this problem. (iv) Davenport's construction can be adapted to produce binary codes with low covering radius [Gra4].

1.4 Computational difficulties in packings and coverings. Given a generator matrix for an n-dimensional lattice Λ we can find its determinant from Eq. (4) of Chap. 1. But anything else is hard! For an arbitrary lattice Λ the best algorithms known for finding either the packing radius ρ or the covering radius R require a number of steps that grows exponentially with n [Die1], [Fin1], [Hel3], [Hel4], [Kan2], [Poh1]. The problem of finding the covering radius is known to be in the class of NP-hard problems [Emd1], and finding the packing radius is conjectured to be in this class. Evidence for this conjecture is supplied by the fact that certain closely related coding theory problems are NP-hard [Ber8]. Finding the covering radius of a code is also known to be NP-hard [McL2].

Lenstra, Lenstra and Lovász [Len1] describe an algorithm (the "LLL algorithm") which finds lattice vectors of moderately small length in a number of steps that grows only as a polynomial function of n. The short vectors in Λ have length 2ρ; the LLL algorithm finds lattice vectors that are guaranteed to have length $\leqslant 2^{(n+1)/2}\rho$. For modest values of n the algorithm does much better than this and very often finds a vector of minimal length. It has been used on lattices of dimension as large as 100 in connection with breaking certain encryption schemes [Lag4], [Odl4]. The LLL algorithm actually converts the given basis for the lattice into a "reduced basis". One of the reduced basis vectors is then a short vector in the lattice. Many other *reduction algorithms* are known; the classical Minkowski reduction algorithm is briefly described in Chap. 15, §10.1. Reduction algorithms are discussed in [Aff1], [Bab1], [Bar11]-[Bar13], [Dic7], [Don1], [Gru1], [Hel3], [Hel4], [Lag3], [Min1], [Nov1], [Rys2], [Rys3], [Rys8], [Rys14], [Sch4], [Sto1], [Tam1]-[Tam4], [Wae3], [Wae5]. (For three-dimensional lattices see also the crystallographic references mentioned at the end of this section.) Babai [Bab1] has shown that the LLL algorithm can be used to find an upper bound on the covering radius that is $\leqslant 3^n R$.

There is of course no practicable algorithm known for finding the best lattice packings or coverings of a given dimension. Other algorithms related to the classification of lattices and quadratic forms are discussed in Chap. 15, §11.

The situation improves when the lattices have some algebraic structure. For the root lattices (Chap. 4, §2) and some related lattices we are able to calculate the exact value of R and find the Voronoi cells in Chap. 21. In some other special cases R can be found by Simon Norton's method ([Nor4] ≅ Chap. 22, [Con37], [Con38], [Con43]). Otherwise if n is small one may try to find the Voronoi cells first and then obtain R as the distance of the furthest vertex of a Voronoi cell from the lattice.

Algorithms for computing the Voronoi cells of arbitrary sets of points are described in [Aki2], [Bow2], [Bro3], [Dev1], [Dev2], [Fin5], [Lee1], [Pre2], [Sei4], [Wat1]. References [Con29] ≅ Chap. 20, [Con35], [Con38], [Con40], [For2] describe algorithms for reasonably nice lattices that, given an arbitrary point $x \in \mathbf{R}^n$, identify which Voronoi cell x belongs to, i.e. find the closest lattice point to x. Such algorithms are important for the applications described in §3 below and §1 of Chap. 3. There are many algorithms used by crystallographers for classifying periodic structures in three dimensions — see for example [Ahm1], [Bur1], [Bur2], [Fer1], [Hal5], [Koc2], [Pat1], [San1], [Say2], [Zim2]. See also [Coh4a], [Spe1].

2. Lattices, Quadratic Forms and Number Theory

2.1 The norm of a vector. When studying lattice vectors $x = (x_1, x_2, \ldots, x_n) \in \Lambda$ it is generally easier to work, not with their length, but with their *squared length* or norm. This is denoted by

$$N(x) = x \cdot x = (x,x) = \sum x_i^2 .$$

The minimal squared distance between distinct lattice vectors, or more simply the *minimal norm* of Λ, is

$$\min \{ N(x-y) : x,y \in \Lambda,\ x \neq y \} \tag{20}$$

$$= \min \{ N(x) \ : \ x \in \Lambda,\ x \neq 0 \}. \tag{21}$$

If the minimal norm is μ, the packing radius of Λ is given by

$$\rho = \tfrac{1}{2} \sqrt{\mu}. \tag{22}$$

Formulae (20) and (21) are equivalent definitions for the minimal norm of a *lattice*, for if x and y are distinct vectors in Λ then $x' = x - y$ is a nonzero vector of Λ. (20) and (22) — but not necessarily (21) — are also valid for nonlattice packings.

2.2 Quadratic forms associated with a lattice. Quadratic forms provide an alternative language for studying lattices, especially useful for investigating arithmetical properties. In this section we show how quadratic forms and lattices are related. Some of the material is rather detailed, and the reader may like to turn at once to §3 where we discuss the quantization problem. The classification of quadratic forms and lattices is briefly discussed in §2.4; we shall say much more in Chaps. 15-18.

Let Λ be a lattice in n-dimensional space $\mathbf{R}^n$, having basis vectors $v_1, \ldots, v_n$ (forming the rows of a generator matrix M). The generic lattice vector $x = (x_1, \ldots, x_n) \in \Lambda$ may be written (see Eq. (2) of Chap. 1) as

$$x = \xi_1 v_1 + \cdots + \xi_n v_n = \xi M, \tag{23}$$

where the ξ_i are integers and $\xi = (\xi_1, \ldots, \xi_n)$. The norm of this vector is

$$N(x) = N(\xi_1 v_1 + \cdots + \xi_n v_n) = \sum_{i=1}^{n} \sum_{j=1}^{n} \xi_i \xi_j \, v_i \cdot v_j$$

$$= \xi \, M M^{tr} \, \xi^{tr} = \xi \, A \, \xi^{tr} = f(\xi) \ \ (\text{say}), \tag{24}$$

where $A = MM^{tr}$ is a Gram matrix for Λ. Regarded as a function of the n integer variables $\xi_1, \ldots, \xi_n$, $f(\xi)$ is a *quadratic form* associated with the lattice.

For example, from the Gram matrix given in Eq. (7) of Chap. 1, one quadratic form associated with the hexagonal lattice is

$$\xi_1^2 + \xi_1 \xi_2 + \xi_2^2. \tag{25}$$

The n-dimensional cubic lattice $\mathbf{Z}^n$ (Chap. 4, §5) has generator matrix I_n (the identity matrix), and the corresponding quadratic form is

$$\xi_1^2 + \xi_2^2 + \cdots + \xi_n^2. \tag{26}$$

It is sometimes simpler to use the ξ-coordinates $(\xi_1, \ldots, \xi_n)$ for lattice vectors rather than the x-coordinates. For instance if we start with the quadratic form f we may define the lattice by setting the norm of $(\xi_1, \ldots, \xi_n)$ equal to $f(\xi)$. (Or we may factor A as MM^{tr} and use M as a generator matrix.)

We recall that congruent lattices are related by Eqs. (22) and (23) of Chap. 1 with $c = 1$ and $\det U = \pm 1$. Quadratic forms corresponding to congruent lattices are called *integrally equivalent*. Thus there is a one-to-one correspondence between congruence classes of lattices and integral equivalence classes of quadratic forms. For example the vectors v_1, v_2 in Fig. 2.6a and w_1, w_2 in Fig. 2.6b also span lattices that are congruent to the hexagonal lattice of Fig. 1.3a. The Gram matrices for these bases are

$$\begin{bmatrix} 1 & -1/2 \\ -1/2 & 1 \end{bmatrix} \quad \text{and} \quad \begin{bmatrix} 3 & 3/2 \\ 3/2 & 1 \end{bmatrix} \quad \text{respectively,}$$

and so the corresponding quadratic forms

$$\xi_1^2 - \xi_1\xi_2 + \xi_2^2 \quad \text{and} \quad 3\xi_1^2 + 3\xi_1\xi_2 + \xi_2^2 \tag{27}$$

are integrally equivalent to each other and to (25).

If Λ is a lattice of full rank, i.e. a lattice in n-dimensional space that is spanned by n independent vectors, then M has rank n, A is a positive definite matrix, and the associated quadratic form is called a positive definite form.

Indefinite quadratic forms (those for which A is an indefinite matrix) may also be studied via lattices, although these are no longer lattices in Euclidean space $\mathbf{R}^n$. One of the most interesting cases occurs when A has signature $(n-1, 1)$, so that the corresponding lattice is in Lorentzian (or hyperbolic) space. We shall say much more about this in Chaps. 15, 26 and 27.

In view of the equivalence between lattices and quadratic forms, all the geometrical problems described in Chaps. 1 and 2 have corresponding formulations in terms of quadratic forms. For example, finding the

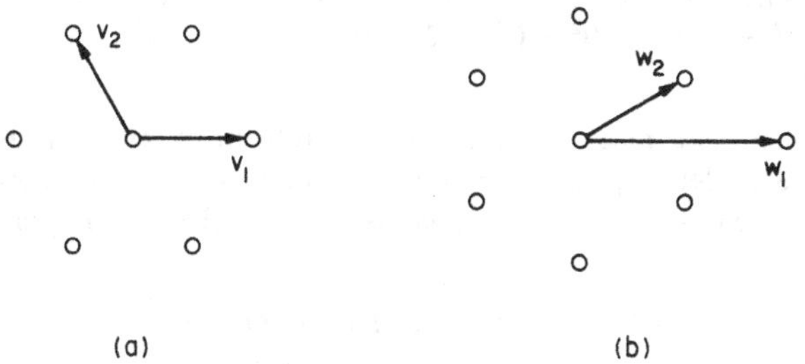

Figure 2.6. Two congruent versions of the hexagonal lattice.

minimal norm of a lattice is equivalent to finding the minimal value of $f(\xi)$ over all integral $\xi \neq 0$, the so-called *homogeneous minimum* of the quadratic form [Cas2]. Similarly the *inhomogeneous minimum* of the form is the square of the covering radius of the lattice. A locally optimal lattice packing, i.e. a lattice Λ whose density Δ does not increase when Λ is perturbed slightly, corresponds to an *extreme* form [Bac2], [Bar4], [Bar6]-[Bar10], [Con42], [Cox10], [Kne2], [Rys15]. We know all the extreme forms in at most 6 variables [Bar6], [Bar7], [Hof1]. Extreme forms in 7 variables have been studied in [Bar9], [Con42], [Lar1], [Sco1], [Sco2], [Shu4], [Sta1], [Sta2]. A quadratic form corresponding to a densest lattice packing is called an *absolutely extreme* form. The results described in Chap. 1, §1.5 mean that the absolutely extreme forms are known for $n \leqslant 8$.

Much of the research into lattices has been carried out in the language of quadratic forms. However in this book we usually prefer to study lattices geometrically, since the choice of a particular set of basis vectors is often arbitrary and conceals the symmetry of the situation. For example the simplest quadratic form for the E_8 lattice is

$$\begin{aligned} &\xi_1^2 - \xi_1\xi_2 + \xi_2^2 - \xi_2\xi_3 + \xi_3^2 - \xi_3\xi_4 + \xi_4^2 - \xi_4\xi_5 \\ &\quad + \xi_5^2 - \xi_5\xi_6 + \xi_6^2 - \xi_6\xi_7 + \xi_7^2 - \xi_5\xi_8 + \xi_8^2 , \end{aligned} \tag{28}$$

obtained from the basis given in Chap. 4, Fig. 4.7. The definition given in Chap. 4, Eq. (97), reveals much more of the symmetry.

Nevertheless quadratic forms are important for studying certain aspects of lattices, and there is an extensive literature — see the references in §1 of Chap. 15. Quadratic forms are particularly useful for investigating arithmetical aspects of lattices, as we now show.

2.3 Theta series and connections with number theory. A very old problem asks for the number of ways of writing an integer m as a sum of four squares, or in other words for the number of quadruples of integers (x_1, x_2, x_3, x_4) such that

$$x_1^2 + x_2^2 + x_3^2 + x_4^2 = m . \tag{29}$$

For example when $m = 2$ there are 24 solutions, consisting of all permutations of $(\pm 1, \pm 1, 0, 0)$. (We agree to count $2 = 1^2 + 1^2 + 0^2 + 0^2$, $2 = 1^2 + (-1)^2 + 0^2 + 0^2$, $2 = 1^2 + 0^2 + 1^2 + 0^2$ etc. as distinct solutions.)

There is a nice way to state this problem in terms of lattices. For any lattice Λ, let N_m be the number of vectors $x \in \Lambda$ of norm m, i.e. with $x \cdot x = m$. From (24), N_m is also the number of integral vectors ξ such that

$$\xi A \xi^{tr} = m , \tag{30}$$

or in other words the number of times the quadratic form associated with Λ *represents* the number m. Eq. (30) is an example of a *Diophantine equation* of degree 2 [Mor6]. Now $x_1^2 + x_2^2 + x_3^2 + x_4^2$ is the quadratic form associated with the four-dimensional cubic lattice $\mathbf{Z}^4$. So the number of ways of writing m as a sum of four squares is equal to the number of vectors of norm m in the lattice $\mathbf{Z}^4$.

The calculation of these numbers is facilitated by introducing the *theta series* of a lattice Λ, which is

$$\Theta_\Lambda(z) = \sum_{x \in \Lambda} q^{x \cdot x} \tag{31}$$

$$= \sum_{m=0}^{\infty} N_m \, q^m , \tag{32}$$

where $q = e^{\pi i z}$. For many purposes we can just think of Θ_Λ as a formal power series in an indeterminate q, although for deeper investigations we must take $q = e^{\pi i z}$, where z is a complex variable. In this case $\Theta_\Lambda(z)$ is easily seen to be a holomorphic function of z for $\text{Im}(z) \geqslant 0$ [Gun1, p. 71], [Ser1, p. 109].

We can also use (31) to define the theta series of a nonlattice packing Λ. The commonest examples of this occur when Λ is a translate of a lattice, or a union of translates (e.g. Eq. (35)).

For a periodic packing $\mathscr{P}$, consisting of a union of say s translates

$$u_j + \Lambda, \qquad j = 1, \ldots, s ,$$

of a lattice Λ (compare Eq. (13) of Chap. 1), we define the *average theta series* to be

$$\Theta_{\mathscr{P}}(z) = \frac{1}{s} \sum_{j=1}^{s} \sum_{k=1}^{s} \sum_{x \in \Lambda} q^{N(x + u_j - u_k)}$$

$$= \Theta_\Lambda(z) + \frac{2}{s} \sum_{j<k} \sum_{x \in \Lambda} q^{N(x + u_j - u_k)} \tag{33}$$

[Odl6]. If $\mathscr{P}$ is *distance invariant* (cf. [Mac6, p. 40]), i.e. has the property that the number of points of $\mathscr{P}$ at any distance d from $x \in \mathscr{P}$ is independent of x, then $\Theta_{\mathscr{P}}(z)$ reduces to

$$\Theta_{\mathscr{P}}(z) = \sum_{j=1}^{s} \sum_{x \in \Lambda} q^{N(x + u_j - u_1)} . \tag{34}$$

For example, the theta series of the integers $\mathbf{Z}$ is

$$\Theta_{\mathbf{Z}}(z) = \sum_{m=-\infty}^{\infty} q^{m^2} = 1 + 2q + 2q^4 + 2q^9 + 2q^{16} + \cdots ,$$

which is the Jacobi theta function $\theta_3(z)$ (Chap. 4, §4.1). The translate $\mathbf{Z} + \frac{1}{2} = \{\ldots, -1\frac{1}{2}, -\frac{1}{2}, \frac{1}{2}, 1\frac{1}{2}, \ldots\}$ has theta series

$$\Theta_{\mathbf{Z}+1/2}(z) = \sum_{m=-\infty}^{\infty} q^{(m+1/2)^2} = 2q^{1/4} + 2q^{9/4} + 2q^{25/4} + \cdots , \tag{35}$$

which is the Jacobi theta function $\theta_2(z)$.

We now see one of the advantages of using the norm $N(x) = x \cdot x$ in Eq. (31). Because $N((x_1, \cdots, x_n)) = x_1^2 + \cdots + x_n^2 = N(x_1) + \cdots + N(x_n)$, we have

$$\Theta_{\mathbf{Z}^n}(z) = \Theta_{\mathbf{Z}}(z)^n = \theta_3(z)^n .$$

So the answer to the problem at the start of this section is simply the coefficient of q^m in the expansion of $\theta_3(z)^4$ in powers of q.

To obtain the theta series of the lattice D_n (which consists of the points of $\mathbf{Z}^n$ whose coordinates sum to an even number, Chap. 4, §7.1) we introduce the theta function θ_4. θ_3 and θ_2 may be regarded as assigning unit masses to the integer and half-integer points on the real line respectively. To get θ_4 we assign masses of +1 to the even integers and −1 to the odd integers, or formally

$$\theta_4(z) = \sum_{m=-\infty}^{\infty} (-q)^{m^2} = 1 - 2q + 2q^4 - 2q^9 + 2q^{16} - \cdots . \tag{36}$$

Then

$$\Theta_{D_n}(z) = \frac{1}{2} \{\theta_3(z)^n + \theta_4(z)^n\} . \tag{37}$$

We cannot resist giving two more examples. It is not difficult to show that the translate $(½, \ldots, ½) + D_n$ has theta series $½\theta_2(z)^n$. The packing

$$D_n^+ = D_n \cup (½, \ldots, ½) + D_n \tag{38}$$

is the diamond packing in three dimensions and the E_8 lattice in eight dimensions. (See Chap. 4, §7.3. D_n^+ is a lattice if and only if n is even, and D_4^+ is congruent to $\mathbf{Z}^4$.) Thus we have the appealing formulae ([Slo17])

$$\Theta_{\text{diamond}}(z) = ½ \left(\theta_2(z)^3 + \theta_3(z)^3 + \theta_4(z)^3 \right), \tag{39}$$

$$\Theta_{E_8}(z) = ½ \left(\theta_2(z)^8 + \theta_3(z)^8 + \theta_4(z)^8 \right). \tag{40}$$

Many other theta series will be given in later chapters. We calculate the theta series of the hexagonal lattice in Chap. 4, §6.2, to illustrate one general method of obtaining the theta series from the quadratic form. A more general theta function that includes θ_2, θ_3 and θ_4 as special cases is the Jacobi theta function $\theta_3(\xi \mid z)$ (Chap. 4, Eq. (6)). However most of the lattices we encounter can be handled using θ_2, θ_3, θ_4 and other simpler functions defined in §4.1 of Chap. 4.

We note that the theta series of a lattice tells us the packing radius ρ, the kissing number τ and the density Δ, since

$$\Theta_\Lambda(z) = 1 + \tau q^{4\rho^2} + \cdots , \tag{41}$$

$$\Delta = \lim_{r \to \infty} \left(\frac{\rho}{r} \right)^n \sum_{m \leq r^2} N_m . \tag{42}$$

We also obtain the theta series of the dual lattice Λ^* by a simple substitution:

$$\Theta_{\Lambda^*}(z) = (\det \Lambda)^{1/2} \left(\frac{i}{z} \right)^{n/2} \Theta_\Lambda \left(-\frac{1}{z} \right) \tag{43}$$

— see Chap. 4, Eq. (19). Here for the first time we must use z rather than q. There is also a generalization of (43) which applies to nonlattice packings that are the union of translates of a lattice [Odl6]. Two illustrations are given in Example 6 of Chap. 7.

The actual values of the theta series are also of importance in chemistry (for example in calculating the *Madelung constant*, see [Bor6], [Gel2], [Gla7], [Gla8], [Zuc1]), and in communication theory. The error probability when a lattice is used as a code for a particular Gaussian channel can be estimated by the value of $\Theta_\Lambda(q)$ for a particular choice of q — see Eq. (35) of Chap. 3.

The theta series is determined by the lattice, but not conversely. Witt [Wit4] observed that in 16 dimensions D_{16}^+ and the direct sum $E_8 \oplus E_8$ are inequivalent lattices of determinant 1 with the same theta series. It is easy to verify this, and the verification illustrates how identities between theta functions have lattice-theoretic proofs. First, since D_4^+ is congruent to $\mathbf{Z}^4$, we have the identity $\frac{1}{2}(\theta_2(z)^4 + \theta_3(z)^4 + \theta_4(z)^4) = \theta_3(z)^4$, i.e.

$$\theta_3(z)^4 = \theta_2(z)^4 + \theta_4(z)^4 . \tag{44}$$

We must show that the theta series of D_{16}^+ and $E_8 \oplus E_8$ are equal, i.e. that

$$\frac{1}{2} \left(\theta_2(z)^{16} + \theta_3(z)^{16} + \theta_4(z)^{16} \right) = \frac{1}{4} \left(\theta_2(z)^8 + \theta_3(z)^8 + \theta_4(z)^8 \right)^2$$

(from (37)-(40)). This follows immediately by eliminating $\theta_3(z)$ using (44) and expanding both sides.

Similarly Kneser [Kne5] observed that D_{12} and $D_4 \oplus E_8$ are inequivalent 12-dimensional lattices of determinant 4 with the same theta series. (This identity also follows immediately from (44).) Later Kitaoka [Kit2] gave an example of two 8-dimensional lattices of determinant 81 with the same theta series, and examples are now known in dimension 5. In dimension 2, on the other hand, the theta series determines the lattice uniquely (see for example [Wat20]). (Some related references are [Con43i], [Hsi5], [Hsi7], [Kit1], [Li1].)

2.4 Integral lattices and quadratic forms. The class of *integral* lattices includes many important examples. In this book we shall call a lattice or quadratic form *integral* if the inner product of any two lattice vectors is an integer, or in other words if the Gram matrix A has integer entries. (The term *classically integral* is sometimes used — see the discussion of integrality notions in Chap. 15.) Equivalently, a lattice Λ is integral if and only if

$$\Lambda \subseteq \Lambda^* ; \tag{45}$$

and for many geometrical purposes this is the best form of the definition. Most of the lattices encountered so far in this chapter are integral when suitably scaled. Our general policy (for example when selecting basis vectors in Chap. 4) is to choose the scale so as to make the determinant as small as possible while making the lattice integral. For example the hexagonal lattice A_2 as defined by Eq. (8) of Chap. 1 is integral, although the equivalent lattice defined by (6) is not.

An important group associated with an integral lattice Λ is its *dual quotient group* Λ^*/Λ, which has order det Λ. For example A_2^*/A_2 is a cyclic group of order 3. The three cosets of A_2 in A_2^* are indicated by the letters a, b, c in Fig. 1.3b.

Note that an integral lattice Λ has the property

$$\Lambda \subseteq \Lambda^* \subseteq \frac{1}{\det \Lambda} \Lambda . \tag{46}$$

(For if $x \in \Lambda^*$, $x = \xi (M^{-1})^{tr} = \xi (M^{-1})^{tr} M^{-1} M = (\det \Lambda)^{-1} \xi \operatorname{adj}(A) M$ $= (\det \Lambda)^{-1} \xi' M$, where $\xi, \xi' \in \mathbf{Z}^n$ and $\operatorname{adj}(A)$ is the adjoint of A.)

An integral lattice with $|\det \Lambda| = 1$, or equivalently with $\Lambda = \Lambda^*$, is called *unimodular* or *self-dual*. (We have written $|\det \Lambda|$ rather than det Λ so that the definition also applies to Lorentzian lattices.) If Λ is integral then $x \cdot x$ is necessarily an integer for all $x \in \Lambda$. If $x \cdot x$ is an *even* integer for all $x \in \Lambda$, then Λ is called *even*; otherwise *odd*. Even unimodular lattices (also called *Type* II lattices) are especially interesting. E_8 and Λ_{24} are even unimodular lattices, while $\mathbf{Z}$, $\mathbf{Z}^2$, $\mathbf{Z}^3$,... are odd unimodular (or *Type* I) lattices.

It is known that if a unimodular lattice has the property that the norm of every lattice vector is a multiple of some positive integer c, then c is either 1 or 2 [O'Mel, p. 324].

The classification of odd and even unimodular lattices is an important problem in number theory and other parts of mathematics [Bri2], [Dol1], [Haz1], [Hir4], [Nik2], and the known results are summarized in Table 2.2. Even unimodular lattices exist if and only if the dimension is a multiple of 8 (Cor. 18 of Chap. 7), while odd unimodular lattices exist in all dimensions. E_8 is the unique even unimodular 8-dimensional lattice, and $E_8 \oplus E_8$ and D_{16}^+ are the only two such 16-dimensional lattices [Wit4]. The even unimodular 24-dimensional lattices were enumerated by Niemeier [Nie2], who found that there are 24 such lattices, 23 with minimal norm 2 and one, the Leech lattice Λ_{24}, with minimal norm 4 — see Table 16.1 of Chap. 16. Three separate verifications of Niemeier's result will be given in Chaps. 16 and 18.

Table 2.2. The number of n-dimensional unimodular lattices. a_n is the number of unimodular lattices of dimension n containing no vectors of norm 1. If $n \equiv 0 \pmod 8$, a_n is written as $d_n + e_n$, where d_n is the number of odd lattices and e_n is the number of even lattices. Similarly b_n is the total number of unimodular lattices of dimension n (including those with vectors of norm 1).

n	0	1	2	3	4	5	6	7	8
a_n	0 + 1	0	0	0	0	0	0	0	0 + 1
b_n	0 + 1	1	1	1	1	1	1	1	1 + 1
n	9	10	11	12	13	14	15	16	17
a_n	0	0	0	1	0	1	1	1 + 2	1
b_n	2	2	2	3	3	4	5	6 + 2	9
n	18	19	20	21	22	23	24	25	26
a_n	4	3	12	12	28	49	156 + 24	368	?
b_n	13	16	28	40	68	117	273 + 24	665	?

Kneser [Kne4] (see also [Her0], [Kol], [Mor3], [Smi6]) enumerated the odd unimodular lattices in dimensions $n \leqslant 16$, the authors ([Con27], [Con34], Chap. 16) extended this to $n \leqslant 23$, and Borcherds ([Bor1], [Bor3] = Chap. 17) recently enumerated the 24- and 25-dimensional odd lattices. If Λ is a unimodular lattice so is $\mathbf{Z}^i \oplus \Lambda$ for all i, so it is simplest to record how many lattices are *not* of the form $\mathbf{Z} \oplus \Lambda'$ for some Λ', i.e. do not contain a vector of norm 1. The numbers (a_n) of lattices of this type are given in the first row of Table 2.2, while the second row (b_n) gives the total number. If n is not a multiple of 8 the table simply gives the number of odd unimodular lattices; if $n \equiv 0 \pmod 8$ the numbers are written to show the division into odd and even lattices. For $1 \leqslant n \leqslant 8$ there is only one odd unimodular lattice, $\mathbf{Z}^n$, while for $n = 9, 10, 11, \ldots$ there is also $E_8 \oplus \mathbf{Z}^{n-8}$. In dimension 12 a third odd lattice (D_{12}^+) appears, and so there are three odd unimodular 12-dimensional lattices, $\mathbf{Z}^{12}$, $E_8 \oplus \mathbf{Z}^4$ and D_{12}^+. This explains the first twelve columns of Table 2.2. The enumeration is extended to dimension 24 in Chaps. 16 and 17. See also the remarks on *extremal* self-dual lattices in Chap. 7.

The Minkowski-Siegel *mass formulae* provide a powerful check that these enumerations are correct. Let $|\mathrm{Aut}(\Lambda)|$ be the order of the automorphism group of a lattice Λ (defined in §4.1 of Chap. 3). The mass formula gives an explicit constant a_0 such that

$$\sum \frac{1}{|\mathrm{Aut}(\Lambda)|} = a_0, \qquad (47)$$

where the sum is over all inequivalent lattices from a given genus (for example all n-dimensional odd unimodular lattices). A slightly more general formula gives an explicit formula for

$$\sum_{\Lambda} \frac{\Theta_\Lambda(z)}{|\mathrm{Aut}(\Lambda)|}.$$

We illustrate by using the mass formula to check that the first nine columns of Table 2.2 are correct. First we consider even unimodular lattices. In dimension 8 the constant $a_0 = 1/696729600$ is given in Chap. 16, Table 16.4, and indeed

$$\frac{1}{|\mathrm{Aut}(E_8)|} = \frac{1}{696729600}$$

(see Chap. 4, §8.1). The values of a_0 for odd unimodular lattices are given in Chap. 16, Table 16.2. For $1 \leqslant n \leqslant 8$, $a_0 = 1/(2^n n!)$, confirming that $\mathbf{Z}^n$ is the unique odd unimodular lattice (since $|\mathrm{Aut}(\mathbf{Z}^n)| = 2^n n!$, Chap. 4, §5). However, for $n = 9$, $a_0 = 17/2786918400$, and indeed there are two 9-dimensional lattices, $E_8 \oplus \mathbf{Z}$ and $\mathbf{Z}^9$, and

$$\frac{1}{|\mathrm{Aut}(E_8 \oplus \mathbf{Z})|} + \frac{1}{|\mathrm{Aut}(Z^9)|}$$

$$= \frac{1}{2 \cdot 696729600} + \frac{1}{2^9 \cdot 9!} = \frac{17}{2786918400}. \tag{48}$$

We shall say more about these formulae in Chap. 16. The most dramatic application of the mass formulae is the verification of Niemeier's list of 24-dimensional even unimodular lattices given in Chap. 16, §2.

The mass formulae also provide a lower bound on the number of inequivalent lattices of each dimension (since $|\mathrm{Aut}(\Lambda)| \geqslant 2$ in (47)). It follows from Chap. 16, Table 16.4, for example that there are at least 80000000 distinct even unimodular 32-dimensional lattices, so it is extremely unlikely that Niemeier's work will ever be extended to 32 or higher dimensions.

This book also has results on the classification of (positive definite) quadratic forms of other small determinants (2,3,...) in Chap. 15, and contains some information about indefinite forms (Chaps. 15, 26, 27).

2.5 Modular forms. Further connections between lattices and number theory arise because the theta series of an integral lattice is a *modular form*. We do not state the general theorem here; some important special cases will be found in Chap. 7, Theorems 7 and 17. There is a brief account of the general theory of the connections between quadratic forms and modular forms in Cassels [Cas3, p. 382], and numerous other references: [Gun1], [Har4], [Hec2], [Hec3], [Kit3], [Kno1], [Lan9], [Ogg1], [Pet4], [Ran6], [Sch6], [Sch7], [Vig1].

The main consequence of this theory is that there are only a limited number of possibilities for the theta series when the determinant is small. From this it is sometimes possible to derive explicit expressions for the coefficients N_m in the theta series, or to obtain accurate numerical estimates for the coefficients. We shall give five samples of such results; many others will be found in the above references.

(i) In dimensions $n \leqslant 7$, as we have mentioned, the lattice $\mathbf{Z}^n$ with quadratic form (26) is the unique unimodular lattice. From this it follows that when n is even there are simple formulae for the coefficients of the theta series, or in other words for the number of ways of writing a number as a sum of n squares. For example the final answer to our original questions is that the number of ways of writing m as a sum of four squares is equal to

$$8 \sum_{d \mid m,\, 4 \nmid d} d \tag{49}$$

(cf. Chap. 4, Eq. (49)).

(ii) There are many other beautiful identities that arise in connection with the particular modular forms known as *Eisenstein series*, whose coefficients involve the arithmetical function $\sigma_k(m)$, the sum of the kth powers of the divisors of m. For example the theta series of E_8, Eq. (40), is also equal to

$$\begin{aligned}\Theta_{E_8}(z) &= 1 + 240 \sum_{m=1}^{\infty} \sigma_3(m) q^{2m} \\ &= 1 + 240\, q^2 + 2160 q^4 + 6720 q^6 + \cdots \end{aligned} \tag{50}$$

– see Chap. 4, §8.1 and Chap. 7.

(iii) Let us define

$$\begin{aligned}\Delta_{24}(z) &= q^2 \prod_{m=1}^{\infty} (1-q^{2m})^{24} = \sum_{m=0}^{\infty} \tau(m) q^{2m} \quad \text{(say)} \\ &= q^2 - 24q^4 + 252q^6 - 1472q^8 + \cdots, \end{aligned} \tag{51}$$

in which the function τ is called *Ramanujan's function.* It now follows from modular form theory that there is a simple expression for the coefficients of the theta series of the Leech lattice Λ_{24}. Let

$$\begin{aligned}\Theta_{\Lambda_{24}}(z) &= \sum_{m=0}^{\infty} N_m q^m \\ &= 1 + 196560 q^4 + 16773120 q^6 + \cdots . \end{aligned} \tag{52}$$

Then, for $m > 0$,

$$N_{2m} = \frac{65520}{691} (\sigma_{11}(m) - \tau(m)) \tag{53}$$

– see Chap. 4, §11.

(iv) The second term on the right-hand side of (53) is much smaller than the first. In fact it is known from Deligne's work [Deg1], [Deg2], [Kat3] that

$$|\tau(m)| \leqslant m^{11/2}\, d(m), \tag{54}$$

where $d(m)$ is the number of divisors of m, and so

$$\tau(m) = O(m^{11/2+\epsilon}) \tag{55}$$

for every positive ϵ. Since $\sigma_{11}(m) > m^{11}$,

$$N_m \approx \frac{65520}{691}\sigma_{11}\left(\frac{m}{2}\right) \tag{56}$$

is a very good approximation. This is typical of the estimates that can be obtained for the coefficients of theta series (see for example [Ran6, §4.5], [Ran7], and §1.4 of Chap. 3), at least for dimensions $n \geqslant 4$.

(v) The theta series of any even unimodular lattice can be written as a polynomial in $\Delta_{24}(z)$ and the theta series of E_8 (Th. 17 of Chap. 7). Consider for example the 48-dimensional lattices P_{48p} and P_{48q} mentioned in Table 1.3. Knowing only that they are even unimodular lattices of minimal norm 6 we can immediately deduce from this theorem that both have theta series equal to

$$\begin{aligned}&\Theta_{E_8}(z)^6 - 1440\,\Theta_{E_8}(z)^3\Delta_{24}(z) + 125280\,\Delta_{24}(z)^2 \\ &= 1 + 0q^2 + 0q^4 + 52416000q^6 + 39007332000q^8 + \cdots .\end{aligned} \tag{57}$$

The kissing number 52416000 is obtained with almost no extra work! Further information about this and similar examples will be found in Chaps. 7 and 19.

2.6 Complex and quaternionic lattices. The lattices defined so far in this chapter are subgroups of real n-dimensional space $\mathbf{R}^n$ that are generated by n linearly independent vectors. They are closed under the operations of addition, subtraction, and multiplication by integers $m \in \mathbf{Z}$, or in other words are $\mathbf{Z}$-*modules*. There are several possible generalizations, for example by using complex or quaternionic vectors instead of real vectors. These more general lattices are of interest in their own right, and sometimes lead to simpler constructions of real lattices. They are introduced in this section; further examples will be found for example in Chaps. 7, 8 and in [Cas3, Chap. 7], [Cha1], [Con36], [Hsi6a], [Kar1], [Lew1], [O'Me1], [Ong1], [Ong2], [Otr1], [Que4], [Que6], [Rog9], [Rog11], [Ser2], [Smi11].

Let $\mathbf{C}$ denote the field of complex numbers $\{z = x + iy : x, y \in \mathbf{R}\}$, where $i^2 = -1$, and let $\mathbf{H}$ denote the (skew) field of quaternions $\{z = u + iv + jw + kx : u, v, w, x \in \mathbf{R}\}$, where

$$\begin{aligned}&i^2 = j^2 = k^2 = -1, \\ &ij = -ji = k,\ jk = -kj = i,\ ki = -ik = j\end{aligned}$$

([Bir1], [Cox8], [Cox17], [Her1]). The *conjugate* $\bar{z}$ of z is defined to be $x - iy$ (in $\mathbf{C}$) or $u - iv - jw - kx$ (in $\mathbf{H}$), and the *norm* of z is $N(z) = z\bar{z}$, which is $x^2 + y^2$ or $u^2 + v^2 + w^2 + x^2$. Similarly the norm of a vector $z = (z_1, \ldots, z_n)$ is

$$N(z) = z_1\bar{z}_1 + \cdots + z_n\bar{z}_n .$$

Besides changing the field we also change what we mean by an integer. There are many possibilities, but here we shall just consider three *rings of integers* to be used instead of $\mathbf{Z}$. These are the *Gaussian* and *Eisenstein integers*

$$\mathcal{G} = \{ a+ib \ : \ a,b \in \mathbf{Z} \} \subset \mathbf{C}, \tag{58}$$

$$\mathcal{E} = \{ a+\omega b \ : \ a,b \in \mathbf{Z} \} \subset \mathbf{C}, \tag{59}$$

where $\omega = (-1+i\sqrt{3})/2$ is a complex cube root of unity [Cox10, p. 421], [Cox21, p. 145], [Fei2], [Har5, p. 179], [Lan0, III, pp. 5, 16], and the *Hurwitz quaternionic integers* [Hur1], [Cox18]

$$\mathcal{H} = \{ a+ib+jc+kd \ : \ a,b,c,d \in \mathbf{Z} \text{ or } a,b,c,d \in \mathbf{Z} + \tfrac{1}{2} \} \subset \mathbf{H}. \tag{60}$$

Let J be one of the rings of integers $\mathbf{Z}$, $\mathcal{G}$, $\mathcal{E}$, $\mathcal{H}$, and let K be the corresponding field ($\mathbf{R}$, $\mathbf{C}$, $\mathbf{C}$, $\mathbf{H}$ respectively). A *J-lattice* is defined as follows. Take n vectors $v_1, \ldots, v_n \in K^n$ that are linearly independent over K. Then the J-lattice Λ generated by these vectors consists of all linear combinations

$$\xi_1 v_1 + \cdots + \xi_n v_n , \tag{61}$$

where $\xi_1, \ldots, \xi_n \in J$. Since J is a ring, Λ is closed under addition, subtraction and multiplication (on the left, in the quaternionic case) by elements of J, i.e. is a J-module.

Aside. Other choices for J that lead to interesting examples include the rings of algebraic integers $\{ a+b\theta : a,b \in \mathbf{Z} \}$, where θ is one of

$$\sqrt{-2},\ \sqrt{-5},\ \frac{-1+\sqrt{-7}}{2},\ \frac{-1+\sqrt{-11}}{2},\ \frac{1+\sqrt{5}}{2}, \tag{62}$$

or the ring of cyclotomic integers $\mathbf{Z}[\zeta]$ where $\zeta = e^{2\pi i/m}$ (see Chap. 8, [Bay1], [Con36], [Cos1], [Cra2]-[Cra5], [Que6]). However, one must be careful. If J is not a principal ideal domain (e.g. when $\theta = \sqrt{-5}$) Λ may be n-dimensional but not have n generators, i.e. may not be a *free* J-module; these are called *non-principal lattices*. If J contains an irrational real number (as in the last example in (62)) then Λ is dense in K^n, i.e. is not a (discrete) lattice. Nevertheless these examples can still be used to produce lattice packings by redefining the norm or the associated quadratic form (see e.g. §4 of Chap. 8). By restricting ourselves in this section to the special rings mentioned above we avoid these complications.

Let Λ be a J-lattice. Just as in the case of real lattices we may now define a generator matrix M (Eq. (1) of Chap. 1) and Gram matrix $A = M\overline{M}^{tr}$ (cf. Chap. 1, Eq. (3)). The main difference from the real case is that the inner product of vectors $x = (x_1, \ldots, x_n)$, $y = (y_1, \ldots, y_n)$ is now defined to be the hermitian inner product

$$x \cdot \bar{y} = x_1\bar{y}_1 + \cdots + x_n\bar{y}_n . \tag{63}$$

The determinant of Λ is given by Eq. (4) of Chap. 1, the minimal norm μ by Eq. (21) above, the packing radius ρ by Eq. (22), the covering radius by a formula analogous to Eq. (3), the theta series by

$$\Theta_\Lambda(z) = \sum_{x \in \Lambda} q^{N(x)}, \quad q = e^{\pi i z}, \tag{64}$$

(cf. Eq. (31)), and the *dual* lattice by

$$\Lambda^* = \{ x \in K^n \ : \ x \cdot \bar{u} \in J \text{ for all } u \in \Lambda \} \tag{65}$$

(cf. Chap. 1, Eq. (24)). Λ is *integral* if $\Lambda \subseteq \Lambda^*$, and *unimodular* or *self-dual* if $\Lambda = \Lambda^*$.

If Λ is a complex J-lattice in $\mathbf{C}^n$ there is a corresponding real lattice Λ_{real} in $\mathbf{R}^{2n}$ consisting of the vectors

$$(\text{Re}(z_1), \text{Im}(z_1), \ldots, \text{Re}(z_n), \text{Im}(z_n)) \tag{66}$$

for $(z_1, \ldots, z_n) \in \Lambda$. Also

$$\det \Lambda_{\text{real}} = \partial^n \, 4^{-n} (\det \Lambda)^2, \tag{67}$$

where ∂, the absolute value of the discriminant of J, is equal to 3 if $J = \mathcal{E}$ or 4 if $J = \mathcal{G}$. The theta series of the dual lattice is

$$\Theta_{\Lambda^*}(z) = (\det \Lambda)^2 \left(\frac{2i}{z\sqrt{\partial}} \right)^n \Theta_\Lambda \left(-\frac{4}{z\partial} \right). \tag{68}$$

If Λ is a quaternionic $\mathcal{H}$-lattice in $\mathbf{H}^n$ the corresponding real lattice Λ_{real} in R^{4n} is obtained by replacing every component $u + iv + jw + kx$ of every vector in Λ by (u, v, w, x). In this case

$$\det \Lambda_{\text{real}} = 4^{-n} (\det \Lambda)^4. \tag{69}$$

In all cases Λ_{real} has the same packing and covering radii as Λ, and we define the density, kissing number and thickness of Λ to be equal to the corresponding values for Λ_{real}.

In the case where Λ is an $\mathcal{E}$-lattice (where $\mathcal{E}$ is the ring of Eisenstein integers), multiplication of vectors in Λ by ω is a *fixed-point-free automorphism* of order 3 of Λ and Λ_{real}; only the zero vector is fixed. Conversely, a real lattice in $\mathbf{R}^{2n}$ with a fixed-point-free automorphism of order 3 can be regarded as an $\mathcal{E}$-lattice in $\mathbf{C}^n$. If Λ has generator matrix $M = X + iY$ (X, Y real), then Λ_{real} has generator matrix

$$\begin{bmatrix} X & Y \\ \tfrac{1}{2}(X + \sqrt{3}Y) & \tfrac{1}{2}(Y - \sqrt{3}X) \end{bmatrix}. \tag{70}$$

Similarly if Λ is a $\mathcal{G}$-lattice (where $\mathcal{G}$ is the ring of Gaussian integers), multiplication by i is an automorphism σ of Λ and Λ_{real}, σ has order 4, and the powers σ, σ^2, σ^3 have no fixed points except 0. If Λ has generator matrix $M = X + iY$ (X, Y real), then Λ_{real} has generator matrix

$$\begin{bmatrix} X & Y \\ -Y & X \end{bmatrix}. \tag{71}$$

The situation is a little more complicated if Λ is an $\mathcal{H}$-lattice (where $\mathcal{H}$ is the ring of Hurwitz integers). $\mathcal{H}$ contains 24 integers of norm 1, or *units*, namely

$$\begin{gathered} \pm 1, \pm i, \pm j, \pm k, \pm \omega, \pm \omega^i, \\ \pm \omega^j, \pm \omega^k, \pm \bar{\omega}, \pm \bar{\omega}^i, \pm \bar{\omega}^j, \pm \bar{\omega}^k, \end{gathered} \tag{72}$$

where

$$\begin{aligned} \omega &= \tfrac{1}{2}(-1 + i + j + k), \\ \bar{\omega} &= \tfrac{1}{2}(-1 - i - j - k), \end{aligned} \tag{73}$$

and

$$\begin{aligned}
\omega^i &= i^{-1}\omega i = j\omega = \omega k = \bar{\omega} + i = \tfrac{1}{2}(-1+i-j-k), \\
\omega^j &= j^{-1}\omega j = k\omega = \omega i = \bar{\omega} + j = \tfrac{1}{2}(-1-i+j-k), \\
\omega^k &= k^{-1}\omega k = i\omega = \omega j = \bar{\omega} + k = \tfrac{1}{2}(-1-i-j+k), \\
\bar{\omega}^i &= i^{-1}\bar{\omega} i = -k\bar{\omega} = -\bar{\omega} j = \omega - i = \tfrac{1}{2}(-1-i+j+k), \\
\bar{\omega}^j &= j^{-1}\bar{\omega} j = -i\bar{\omega} = -\bar{\omega} k = \omega - j = \tfrac{1}{2}(-1+i-j+k), \\
\bar{\omega}^k &= k^{-1}\bar{\omega} k = -j\bar{\omega} = -\bar{\omega} i = \omega - k = \tfrac{1}{2}(-1+i+j-k).
\end{aligned} \tag{74}$$

As a multiplicative group this set of units is isomorphic to $2A_4$, where A_4 is the alternating group of degree 4. Let L_π denote left multiplication of vectors in Λ by π, where π is one of the 24 units in $\mathcal{H}$. If $\pi \neq 1$, L_π is a fixed-point-free automorphism of Λ and Λ_{real}. Conversely, if L is a real lattice in $\mathbf{R}^{4n}$ which is a left $(2A_4)$-module, and every nontrivial element of $2A_4$ acts in a fixed-point-free manner, then L can be regarded as an $\mathcal{H}$-lattice in $\mathbf{H}^n$. If Λ has generator matrix $M = U + iV + jW + kX$ (U, V, W, X real), then Λ_{real} has generator matrix

$$\begin{bmatrix} U' & V' & W' & X' \\ -V & U & -X & W \\ -W & X & U & -V \\ -X & -W & V & U \end{bmatrix}, \tag{75}$$

where $\omega M = U' + iV' + jW' + kX'$.

Examples. If $\Lambda = \mathcal{G}$ in $\mathbf{C}^1$, Λ_{real} is the square lattice $\mathbf{Z}^2$; if $\Lambda = \mathcal{E}$, Λ_{real} is the hexagonal lattice (Fig. 1.3a); and if $\Lambda = \mathcal{H}$, Λ_{real} is D_4 (Chap. 4, §7.2).

The densest six-dimensional lattice E_6 has a simple description as an Eisenstein lattice. If Λ is the $\mathcal{E}$-lattice in $\mathbf{C}^3$ with generator matrix

$$\begin{bmatrix} \theta & 0 & 0 \\ 0 & \theta & 0 \\ 1 & 1 & 1 \end{bmatrix} \tag{76}$$

where $\theta = \omega - \bar{\omega} = \sqrt{-3}$, then Λ_{real} is E_6.

Similarly the densest eight-dimensional lattice E_8 has a simple description as a Hurwitzian lattice. If Λ is the $\mathcal{H}$-lattice in $\mathbf{H}^2$ with generator matrix

$$\begin{bmatrix} 1+i & 0 \\ 1 & 1 \end{bmatrix} \tag{77}$$

then Λ_{real} is E_8.

The Leech lattice can be constructed in all three ways, as a $\mathcal{G}$-, $\mathcal{E}$- or $\mathcal{H}$-lattice — see §1 of Chap. 6 and Example 12 of Chap. 7.

Finally, we briefly discuss the existence of *even* complex lattices. The *norm* of Λ, denoted by $\mathcal{N}(\Lambda)$, is the subgroup of the real numbers generated by $\{N(x) : x \in \Lambda\}$. We have already seen that if Λ is a real self-dual lattice then $\mathcal{N}(\Lambda)$ is either $\mathbf{Z}$ or $2\mathbf{Z}$ [O'Mel, pp. 227,324]. The same proof shows that this result also holds for complex lattices. Lattices for which $\mathcal{N}(\Lambda) = 2\mathbf{Z}$ are called *even*. Even self-dual $\mathcal{G}$-lattices exist in $\mathbf{C}^n$ if and only if n is a multiple of 4; there are no even self-dual $\mathcal{E}$-lattices [Ger5, Th. 3.6]. There are however even self-dual lattices over other rings of integers — see for example §4 of Chap. 8.

3. Quantizers

3.1 Quantization, analog-to-digital conversion, and data compression. As mentioned at the end of §1.4 of Chap. 1, although this book primarily discusses geometrical problems, many of them have direct application to other branches of science. In this section and in §1 of Chap. 3 we describe two of the principal applications, to vector quantizers and to the design of signals for use over a noisy channel.

The real world is full of nasty numbers like 0.79134989..., while the world of computers and digital communication deals with nice numbers like 0 and 1. A *quantizer*, or *analog-to-digital converter*, is a device for converting nasty numbers into nice numbers. Quantizers are found in digital measuring instruments or recording devices, and in digital communication systems (including most medium- or long-distance telephone networks, those that use pulse code modulation for example — see §1.1 of Chap. 3). We must add, however, that most existing quantizers are one-dimensional; the n-dimensional quantizers described here are not yet in general use.

Figure 2.7 shows a hypothetical two-dimensional quantizer. Eleven representative points $P_1, \ldots, P_{11}$ have been chosen in advance. The input

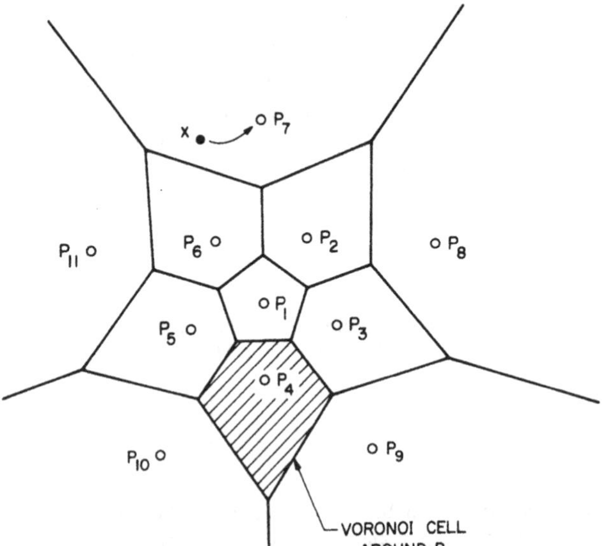

Figure 2.7. An example of a (hypothetical) two-dimensional quantizer. Eleven representative points $P_1, \ldots, P_{11}$ have been chosen in advance. Any point x in the plane is then "rounded off", or quantized, to the closest point P_i. The illustration also shows the Voronoi cells around each P_i. For example, any point x falling in the Voronoi cell of P_4 (the shaded region) is rounded off to P_4.

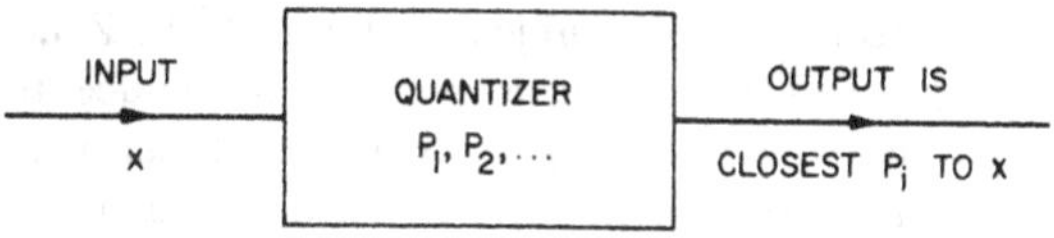

Figure 2.8. An n-dimensional quantizer.

to the quantizer is a pair of real numbers (x_1,x_2), which we regard as a point x in the plane, and the output is the closest one of $P_1, \ldots, P_{11}$ to x. Any point x in the plane is thus replaced by (or *rounded-off* to) the closest point P_i. This process is also called *data-compression*; the output could equally well be the index i of the closest P_i.

The formal definition of an n-dimensional quantizer is as follows (Fig. 2.8). M points $P_1, \ldots, P_M$ are chosen in $\mathbf{R}^n$. The input x is an arbitrary point of $\mathbf{R}^n$; the output is the closest P_i to x. If the closest P_i is not unique, one of the closest P_i is chosen at random to be the output.

The action of this quantizer may also be described by saying that the space $\mathbf{R}^n$ is partitioned into the Voronoi cells (§1.2) $V(P_1)$, $V(P_2)$,... around the P_i; if the input x belongs to $V(P_i)$, the output is P_i (see Fig. 2.7).

Quantizing introduces errors, the magnitude of the error being measured by the Euclidean distance from x to the closest P_i. The P_i should therefore be chosen according to the probability distribution of x so as to be representative or "typical" points. More precisely we shall attempt to choose the P_i so as to minimize the *mean squared error* (or m.s.e.), the average value of $N(x-P_i(x))$, where $P_i(x)$ denotes the closest point to x.

If x has probability density function $p(x)$, the *average mean squared error per dimension* is

$$E = \frac{1}{n} \int_{\mathbf{R}^n} N(x-P_i(x))\, p(x)dx\,, \tag{78}$$

which, since the Voronoi cells partition $\mathbf{R}^n$, can be written as

$$E = \frac{1}{n} \sum_{i=1}^{M} \int_{V(P_i)} N(x-P_i)p(x)dx\,. \tag{79}$$

Remarks. (i) (78) and (79) assume the probability that the input x lies *exactly* on the boundary of a Voronoi cell to be negligibly small. (ii) The factor $1/n$ is to ensure a fair comparison between quantizers of different dimensions. (iii) The mean squared error criterion is only one of many possible distortion measures; it has the advantage of being widely used and is mathematically the simplest. For applications to speech or picture processing the correct choice of distortion measure is a difficult problem. See for example [Ata1]-[Ata3], [Fla1], [Sch10].

The main theorem in this subject is due to P. L. Zador, who showed in 1963 ([Zad1], [Zad2]) that it is possible to reduce the average mean squared error per dimension by using higher-dimensional quantizers. In other words it is more efficient to wait until several numbers have arrived at the quantizer, and then to quantize them all at once by regarding them as specifying a point in some high-dimensional space. It pays to procrastinate.

Unfortunately Zador's result, like others we have mentioned, is nonconstructive, and the problem of explicitly finding good multidimensional quantizers is still unsolved. But the lattices shown in Table 1.1 perform very well if the input x has a uniform distribution, as we shall see.

We now give a more precise account of Zador's result. Given the dimension n, the number of points M, and the probability density function $p(x)$ of the input, we wish to find the infimum

$$E(n,M,p) = \inf_{\{P_i\}} E, \tag{80}$$

i.e. the smallest error attainable by any set of M points $P_1, \ldots, P_M \in \mathbf{R}^n$. Zador ([Zad1], [Zad2]; see also [Buc2], [Buc3], [Ger2], [Yam1]) showed under quite general assumptions about $p(x)$ that

$$\lim_{M \to \infty} M^{2/n} E(n,M,p) = G_n \left(\int_{R^n} p(x)^{n/(n+2)} dx \right)^{\frac{n+2}{n}}, \tag{81}$$

where G_n only depends on n. Zador also showed that

$$\frac{1}{(n+2)\pi} \Gamma\left(\frac{n}{2}+1\right)^{2/n} \leqslant G_n \leqslant \frac{1}{n\pi} \Gamma\left(\frac{n}{2}+1\right)^{2/n} \Gamma\left(1+\frac{2}{n}\right). \tag{82}$$

For large n the upper and lower bounds in (82) agree, giving

$$G_n \to \frac{1}{2\pi e} = 0.058550\ldots \text{ as } n \to \infty. \tag{83}$$

Since the probability density function $p(x)$ only appears in the last term of (81), we may choose any convenient $p(x)$ when attempting to find G_n. The simplest choice, and one which is of considerable interest in its own right, is to assume that x is uniformly distributed over a large region of $\mathbf{R}^n$ (a large ball, for example). With a uniformly distributed input the mean squared error is minimized if each point P_i lies at the centroid of its Voronoi region $V(P_i)$ [Ger2]. We are most interested in the case when M, the number of points, is very large. This enables us to ignore boundary effects, and leads to simpler answers. If the input is uniformly distributed, (79) and (81) become

$$E = \frac{\frac{1}{n} \sum_{i=1}^{M} \int_{V(P_i)} N(x-P_i) dx}{\sum_{i=1}^{M} \int_{V(P_i)} dx}, \tag{84}$$

$$G_n = \lim_{M \to \infty} \frac{E(n,M,p)}{\left[\frac{1}{M} \sum_{i=1}^{M} \int_{V(P_i)} dx \right]^{2/n}}, \tag{85}$$

where $E(n,M,p) = \inf E$. Equations (84) and (85) can be interpreted as saying that, for optimal quantization of a uniformly distributed input using a large number of points, G_n is the *average mean squared error per dimension, scaled so as to be a dimensionless quantity* (independent of the choice of scale of the P_i).

3.2 The quantizer problem. From now on we assume that the input x has a uniform distribution over a large ball in $\mathbf{R}^n$, and that M, the number of points P_i, is large. The *quantizer problem* is to choose points $P_1, \ldots, P_M$ in $\mathbf{R}^n$ so as to minimize

$$\frac{1}{n} \cdot \frac{\frac{1}{M} \sum_{i=1}^{M} \int_{V(P_i)} N(x-P_i)dx}{\left\{ \frac{1}{M} \sum_{i=1}^{M} \mathrm{Vol}(V(P_i)) \right\}^{1+\frac{2}{n}}}, \tag{86}$$

where $V(P_i)$ is the Voronoi cell of P_i. Then (from (84), (85)) G_n is equal to the limit as $M \to \infty$ of the infimum of (86) over all choices of the P_i. By the previous section the solution to the quantizer problem gives both the optimal quantizer for uniformly distributed inputs and, via (81), the error probability for an optimal quantizer when the input has an arbitrary probability density function $p(x)$.

If all the Voronoi cells $V(P_i)$ are congruent (as in the case when the P_i form a lattice), (86) can be simplified. Suppose the Voronoi cells are congruent to a polytope Π. If we place the origin of coordinates at the centroid of Π, (86) becomes

$$\frac{\frac{1}{n} \int_{\Pi} x \cdot x \, dx}{\mathrm{Vol}(\Pi)^{1+\frac{2}{n}}} = G(\Pi), \tag{87}$$

the normalized second moment of Π defined in §1.2. If the points P_i form (a large segment of) a lattice Λ, with Voronoi cells congruent to a polytope Π, we write $G(\Lambda) = G(\Pi)$. The *lattice quantizer problem* is to find an n-dimensional lattice Λ for which $G(\Lambda)$ is a minimum.

3.3 Quantizer problem-summary of results. It is not difficult to show that the solution to the one-dimensional quantizer problem is to place the points P_i uniformly along the real line, so as to form (a large segment of) the integer lattice $\mathbf{Z}$. The Voronoi cells are intervals of length 1, and we take Π to be the interval $[-\frac{1}{2}, \frac{1}{2}]$. From (87) we have

$$G_1 = \frac{\int_{-1/2}^{1/2} x^2 dx}{\int_{-1/2}^{1/2} dx} = \frac{1}{12} = 0.083333\ldots . \tag{88}$$

This is a familiar result in quantization theory: if uniformly distributed numbers are quantized one at a time, the average squared error is 1/12. Zador's thesis shows that this error can be reduced by using higher-dimensional quantizers.

For the two-dimensional problem Fejes Tóth ([Fej8]; see also [Fej10], [Ger2], [New1]) showed that the points should form the hexagonal lattice; correspondingly

$$G_2 = G(\text{regular hexagon}) = \frac{5}{36\sqrt{3}} = 0.0801875... \tag{89}$$

(see Chap. 21, §2I). As usual the problem is unsolved in higher dimensions.

The solution to the *lattice* quantizer problem is known in three dimensions [Bar16]: the best lattice quantizer is the bcc lattice, and

$$G(\text{bcc}) = G(\text{truncated octahedron}) = \frac{19}{192 \cdot 2^{1/3}} = 0.078543... . \tag{90}$$

It is shown in [Bar16] that the bcc lattice is the only lattice at which $G(\Pi)$ is even a local minimum. For comparison

$$G(\text{fcc}) = G(\text{rhombic dodecahedron}) = 2^{-11/3} = 0.078745... \tag{91}$$

(Chap. 21, Eq. (26)), but this can be reduced by perturbing the lattice slightly. In higher dimensions even the best lattice quantizers are not known.

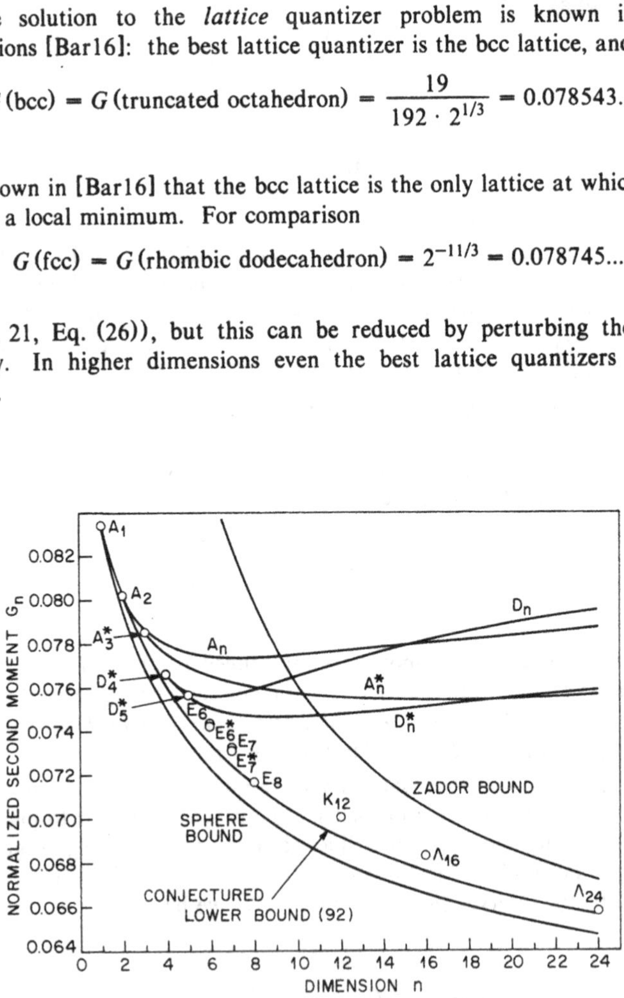

Figure 2.9. The best quantizers known in dimensions $n \leqslant 24$.

Table 2.3. Bounds for G_n, the mean squared error of the optimal n-dimensional quantizer. (92) is our conjectured lower bound. (See also Introduction to Third Edition.)

n	Name of quantizer	Mean squared error G	Lower bound (92)
0	Λ_0	0	0
1	$\mathbf{Z}$	0.083333	0.083333
2	$A_2^* \cong A_2$	0.080188	0.080188
3	$A_3^* \cong D_3^*$	0.078543	0.077875
	$A_3 \cong D_3$	0.078745	
4	$D_4^* \cong D_4$	0.076603	0.07609
	A_4^*	0.077559	
	A_4	0.078020	
5	D_5^*	0.075625	0.07465
	D_5	0.075786	
	A_5^*	0.076922	
	A_5	0.077647	
6	E_6^*	0.074244	0.07347
	E_6	0.074347	
7	E_7^*	0.073116	0.07248
	E_7	0.073231	
8	$E_8^* \cong E_8$	0.071682	0.07163
	D_8^*	0.074735	
	D_8	0.075914	
	A_8^*	0.075972	
	A_8	0.077391	
12	$K_{12}^* \cong K_{12}$	0.070100 ± 0.000024	0.06918
16	$\Lambda_{16}^* \cong \Lambda_{16}$	0.068299 ± 0.000027	0.06759
24	$\Lambda_{24}^* \cong \Lambda_{24}$	0.065771 ± 0.000074	0.06561

Tables 1.1, 2.3 and Fig. 2.9 show the best quantizers known in dimensions $n \leqslant 24$. For some of these lattices $G(\Lambda)$ can be determined exactly, as shown in Chap. 21. However, for E_6^*, E_7^*, K_{12}, Λ_{16} and Λ_{24}, $G(\Lambda)$ was evaluated in [Con38] by Monte Carlo integration, making use of algorithms (see §1.4 and Chap. 20) that find the closest lattice point to an arbitrary point of $\mathbf{R}^n$.[2]

Table 2.3 and Fig. 2.9 also show Zador's bounds (82). The lower bound in (82) on G_n is the normalized second moment of an n-dimensional sphere – see Chap. 21, Eq. (8).

[2] Worley [Wor1], [Wor2] has since found exact values for E_6^* and E_7^* – see §3.D of Chap. 21.

In [Con39] we conjecture a new lower bound on G_n which is significantly stronger than Zador's lower bound. The conjecture is that G_n is at least

$$\frac{n+3-2H_{n+2}}{4n(n+1)} \left\{ (n+1)(n!)^4 F_n^2 \left(\frac{1}{2} \cos^{-1} \frac{1}{n} \right) \right\}^{\frac{1}{n}}, \qquad (92)$$

where $H_m = 1^{-1} + 2^{-1} + \cdots + m^{-1}$ $(m = 1, 2, \ldots)$ is a harmonic sum, and $F_n(\alpha)$ is defined in Eq. (64) of Chap. 1. This bound is also given in Table 2.3 and Fig. 2.9. Although no formal proof has been found, there are plausible geometrical reasons for believing that the bound is correct. The conjectured new bound is analogous to the Rogers bound for the sphere packing problem (Eq. (39) of Chap. 1), the Coxeter-Few-Rogers bound (16) for the covering problem, and the Coxeter-Böröczky bound for spherical codes (Eq. (63) of Chap. 1).

It is worth remarking that the best n-dimensional quantizers presently known are always the duals of the best packings known. However, we do not expect that this will always be the case. It would be nice to have more data. What is $G(\Lambda)$ for Dickson's four-dimensional lattices mentioned in §1.3, or for the Barnes-Trenerry lattices also mentioned there?

For further information about quantizers see [Ado1], [Ber1]-[Ber3], [Buc1]-[Buc3], [Dav4], [Dun10], [Eli1], [Gal2], [Ger1]-[Ger3], [Gis1], [Gra7], [Jay1], [Lin2], [Max1], [Ziv1], [Ziv2].

3

Codes, Designs and Groups

J. H. Conway and N. J. A. Sloane

This is the last of the three introductory chapters. We begin by describing the second main application of sphere packings, the design of signals for data transmission or storage systems. The remaining sections are devoted to topics that, although not our primary concern in this book, are always in our minds: error-correcting codes, Steiner systems, t-designs and finite groups.

1. The channel coding problem

1.1 The sampling theorem. The second main application is to the design of signals for use in data transmission or storage systems. A typical system is shown in Fig. 3.1. Messages are produced by an *information source*, such as a human speaker, an orchestra or a computer. The *source encoder* converts the messages to digital form. This process often involves quantization or analog-to-digital conversion, as described in the previous chapter. The messages are to be conveyed to the *destination* via a *channel*, which may be a data *transmission* channel, such as a copper telephone wire, microwave radio link, or optical fiber, or a data *storage* device such as a magnetic or optical disk. There is *noise* on the channel, so what is received (or retrieved) may differ slightly from what was transmitted (or stored).

The source produces messages at a certain rate, and we wish to communicate them to the destination reliably, efficiently and cheaply. The communication system in Fig. 3.1 accomplishes this by choosing a set of special signals called the *code*, the members of which are designed to be easily distinguishable from each other even in the presence of noise. Only signals from the code will be used on the channel.

The *channel encoder* takes the output from the source encoder and replaces it by one of the code signals, which is then transmitted over the channel (or stored in the recording device). The *channel decoder* reverses

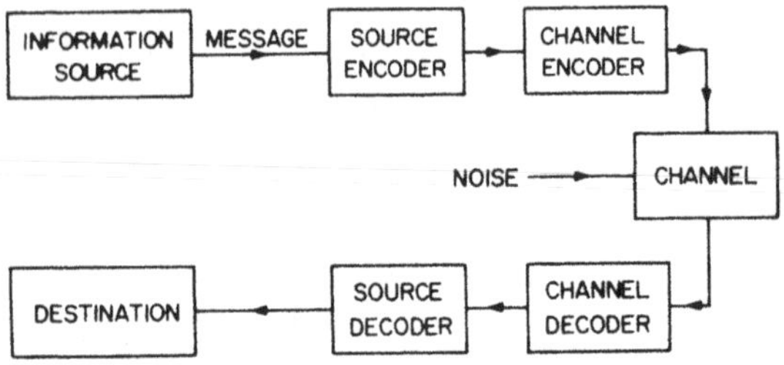

Figure 3.1. Block diagram of data transmission or storage system.

this process by taking the received (or retrieved) signal and making a (hopefully correct) estimate of the code signal; the *source decoder* then converts this back to the original message.

From now on we shall just describe the data transmission problem, always keeping in mind that there are identical results for data storage systems.

The *sampling theorem* plays a crucial role in many communication systems. The sampling theorem (Fig. 3.2) states that if $f(t)$ is a signal (i.e. a function of time) containing no components of frequency greater than W cycles per second, then $f(t)$ is completely specified by its samples

$$\ldots, f\left(-\frac{1}{2W}\right), f(0), f\left(\frac{1}{2W}\right), f\left(\frac{2}{2W}\right), \ldots \tag{1}$$

taken every $1/(2W)$ seconds (see [Dym1], [Hig1], [Jer1], [Lan1], [Lee0], [Pet2], [Sha2], [Ste1], [Woz1]). In fact there is an explicit formula, the *cardinal series*, expressing $f(t)$ in terms of its sample values:

$$f(t) = \sum_{k=-\infty}^{\infty} f\left(\frac{k}{2W}\right) \frac{\sin 2\pi W(t-k/2W)}{2\pi W(t-k/2W)}. \tag{2}$$

Since

$$\int_{-\infty}^{\infty} \frac{\sin 2\pi W(t-k/2W)}{2\pi W(t-k/2W)} \frac{\sin 2\pi W(t-l/2W)}{2\pi W(t-l/2W)} dt$$

is equal to 0 if $k \neq l$, or to $1/(2W)$ if $k = l$, the *energy* in $f(t)$ is given by

$$\int_{-\infty}^{\infty} f(t)^2 dt = \frac{1}{2W} \sum_{k=-\infty}^{\infty} f\left(\frac{k}{2W}\right)^2. \tag{3}$$

The sampling theorem is of both practical and theoretical importance in digital communication systems. The theorem is easily applied in practice, since from Eq. (2) the signal $f(t)$ can be recovered from its samples by

(a)   f(t)

O

$\frac{1}{2W}$

T

(b)

O

T

(c)

$$\left(f(0), f\left(\frac{1}{2W}\right), f\left(\frac{2}{2W}\right), \cdots, f\left(\frac{n-1}{2W}\right) \right)$$

(d)

•

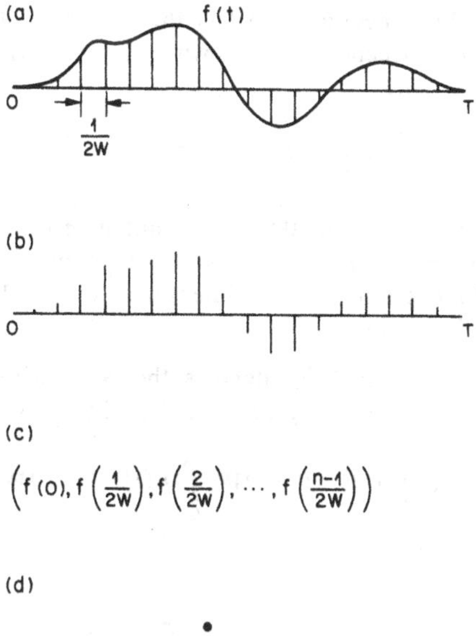

Figure 3.2. The sampling theorem is the basis for many digital communication systems. (a) A function $f(t)$ which does not change too quickly, in particular one which does not contain any frequency components greater than W cycles per second, is sampled every $1/(2W)$ seconds. (b) The sample values $f(0)$, $f(1/2W)$, $f(2/2W)$,... are transmitted. The theorem states that $f(t)$ can be reconstructed exactly from these samples. This is the principle underlying pulse code modulation (PCM). (c) In T seconds there will be n samples, where $n = 2TW$. These samples $(f(0), f(1/2W), \ldots, f((n-1)/2W))$ are the coordinates of a point in n-dimensional space. (d) Thus the complicated function $f(t)$ has been represented by a single point in n-dimensional space.

feeding them into a simple electrical circuit (a low pass filter). Moreover the recovery is stable, in the sense that small errors in the sample values produce only correspondingly small errors in the recovered signal.

For example the sampling theorem is the basis for the digital transmission system known as *pulse code modulation*, or PCM ([Bell], [Ben4], [Oli1]). Amplitude modulation (AM) and frequency modulation (FM) are no doubt familiar to the reader. PCM is less well-known but just as important, being widely used in medium-distance telephone calls. The speaker's voice is sampled at the appropriate rate (usually 8000 times a second), and each sample is rounded off or quantized to one of 256 levels, and then expressed as an 8-bit binary number. The binary numbers are transmitted and at the receiving end the speaker's voice is reconstructed from the samples. Since the sampling theorem is a true theorem, the listener is unaware that this has happened.

The sampling theorem also has theoretical implications. If the signal $f(t)$ lasts for T seconds[(1)], as shown in Fig. 3.2, then there are $n = 2TW$ samples, for instance

$$(f(0), f(\frac{1}{2W}), f(\frac{2}{2W}), \ldots, f(\frac{n-1}{2W})) . \tag{4}$$

But these are the coordinates of a point in n-dimensional space. So the sampling theorem makes it possible to represent this complicated function $f(t)$ by a single point $(\cdot)$ in n-dimensional space. This shows the power of mathematical notation!

Furthermore, if $F \in \mathbf{R}^n$ denotes the point with coordinates (4), then from (3) the norm of F is proportional to the energy in $f(t)$:

$$N(F) = F \cdot F = 2W \int_0^T f(t)^2 dt = 2WTP = nP , \tag{5}$$

where

$$P = \frac{1}{T} \int_0^T f(t)^2 dt \tag{6}$$

is the *average power* in the signal. If G represents another signal $g(t)$, then

$$N(F-G) = \frac{n}{T} \int_0^T (f(t) - g(t))^2 dt , \tag{7}$$

$$F \cdot G = \frac{n}{T} \int_0^T f(t)\, g(t) dt . \tag{8}$$

1.2 Shannon's theorem. The observation that signals of finite bandwidth can be represented by points in Euclidean space is a crucial ingredient in the research that Claude E. Shannon carried out between 1940 and 1960 into the mathematical foundations of communication theory [Sha1]-[Sha8]. As David Slepian [Sle5] has said, "probably no single work in this century has more profoundly altered man's understanding of communication than Shannon's 1948 paper *A Mathematical Theory of Communication.*"

Before describing some of Shannon's results let us mention the two idealized models for the channel in Fig. 3.1 that are the most important. One is the *binary symmetric channel*, in which only sequences of 0's and 1's are transmitted and received. In this case the code is called a binary

---

(1) Strictly speaking it is impossible for a function to have simultaneously a finite bandwidth and to be nonzero for only a finite interval $0 \leqslant t \leqslant T$. To make this precise we should consider bandlimited signals which have *almost all* of their energy in the range $0 \leqslant t \leqslant T$. See [Dym1], [Lan3], [Lan4], [Sle6], [Sle7].

error-correcting code; we discuss such codes in §2. The other is the *Gaussian white noise channel*; the problem of designing codes for this channel is related to the problem of packing spheres in Euclidean space.

The Gaussian white noise channel transmits continuous signals. All frequencies above a cutoff frequency of W cycles per second (called the *bandwidth* of the channel) are attenuated completely; frequencies below W are passed without attenuation. In the course of transmitting the signal, the channel adds white Gaussian noise to it.

This channel has a very simple description based on the sampling theorem. A transmitted signal $f(t)$ is represented by a point $F = (f_1, \ldots, f_n)$ in n-dimensional Euclidean space. During transmission this point is perturbed by the addition of a noise vector $Y = (y_1, \ldots, y_n)$ whose components are independent Gaussian random variables with mean 0 and variance σ^2. (Thus σ^2 is the average power of the noise.) The received signal is represented by the vector $F + Y$.

For this channel the code is a set of points in $\mathbf{R}^n$. If there are M code points, each representing a signal of bandwidth W and duration T seconds, the *rate* of the code is defined to be

$$R = \frac{1}{T} \log_2 M \quad \text{bits/sec.} \tag{9}$$

The decoder finds the closest code point to the received vector and from it reconstructs the signal. If the noise is small, the closest code point will be the transmitted point and the signal $f(t)$ will be recovered correctly. But if the noise is large the received vector may be closer to some other code point and will be decoded incorrectly; this is a *decoding error*.

Thus we may reduce the effect of the noise by placing the code points further apart. On the other hand by (5) this requires signals of greater energy and increases the cost. One of Shannon's basic theorems is the surprising result that it is possible to attain essentially error-free transmission over this channel while using only signals of finite power, provided the rate of the code does not exceed a critical threshold called the *capacity* of the channel. There are similar results for many other channels, including the binary symmetric channel.

The precise statement of Shannon's theorem for the Gaussian channel is as follows. For any rate R less than the *capacity*

$$C = W \log_2 \left(1 + \frac{P}{\sigma^2} \right), \tag{10}$$

by making T and hence $n = 2WT$ sufficiently large we can find a code of rate R (Eq. (9)) and average power at most P (Eq. (6)) for which the probability of a decoding error is arbitrarily small. Conversely such codes do not exist for rates $R \geqslant C$.

We shall sketch a proof of a weaker result, to show the connection with sphere packing; for a rigorous treatment of the full theorem see [Gal1],

[McEl1], [Sha2], [Sha6], [Sle5], [Wol3], [Wyn3]-[Wyn5].

Since the signals F have average power $\leqslant P$, by (5) they lie in a sphere of radius $\sqrt{nP}$ around the origin. By making n sufficiently large, we can assume that the noise vector Y, which has n components of variance σ^2, has norm $N(Y) \leqslant n(\sigma^2+\epsilon)$, where ϵ is arbitrarily small. In other words, with high probability the received vector $F + Y$ lies in a small sphere of radius $\leqslant \{n(\sigma^2+\epsilon)\}^{1/2}$ centered at F. The average power of the received vector is $\leqslant P + \sigma^2 + \epsilon$, and in particular $F + Y$ lies in a sphere around the origin of radius $\{ n(P+\sigma^2+\epsilon) \}^{1/2}$.

Therefore, if we take the code to consist of the centers of spheres in a dense packing of spheres of radius $\{n(\sigma^2+\epsilon)\}^{1/2}$ in a large sphere of radius $\{ n(P+\sigma^2+\epsilon) \}^{1/2}$, we obtain a code in which the transmitted vector will be decoded correctly with probability close to 1. The number of code points is, from Eq. (10) of Chap. 1,

$$M = \Delta \left(\frac{P+\sigma^2+\epsilon}{\sigma^2+\epsilon} \right)^{n/2} < \Delta \left(\frac{P+\sigma^2}{\sigma^2} \right)^{n/2}, \tag{11}$$

where Δ is the density of the sphere packing. By making n large we can make M arbitrarily close to (11). The rate of this code is, from (9), (11),

$$R \approx W \log_2 \left(1 + \frac{P}{\sigma^2} \right) + \frac{2W}{n} \log_2 \Delta .$$

Therefore, by making T and n sufficiently large, and using a sphere packing of density satisfying $(\log_2 \Delta)/n \geqslant -1$ (which we know exists by Chap. 1, Eq. (46)), we can achieve essentially error-free transmission at any rate

$$R < W \log_2 \left(1 + \frac{P}{\sigma^2} \right) - 2W . \tag{12}$$

More sophisticated arguments are needed to establish that rates up to the capacity (10), and no higher, can be achieved without error — see the references cited above. Shannon's argument, which uses an averaging method, is one of the best known non-explicit proofs in coding theory. (See also the remarks in §1.5 of Chap. 1.)

The proof does show that, in order to achieve rates close to (10), n and T must be large. More precise versions of the theorem investigate how the probability P_e of a decoding error drops as the dimension n increases, for a particular rate R. Slepian [Sle1] has examined the tradeoffs between all seven parameters W, T, n, R, P, σ^2 and P_e.

The rigorous proof of (10) also shows that, for small values of σ, finding an optimal code of maximal power P is closely related to finding the densest packing of spheres in $\mathbf{R}^n$, as we saw in our heuristic derivation of (12). De Buda and Kassem [Bud2], [Bud3], [Kas2] have shown that codes obtained from lattice sphere packings are essentially as good as any codes, for all values of σ.

A second, very similar problem asks for the best code in which all signals have the same energy (a *constant energy* code). For small values of σ the solution to this problem is closely related to the problem of packing spherical caps on the n-dimensional sphere Ω_n, i.e. to the problem of constructing spherical codes (see §2.3 of Chap. 1).

1.3 Error probability. We can state the channel coding problem more precisely by giving an expression for the error probability P_e. Suppose the code consists of M code points $C_1, \ldots, C_M$ in $\mathbf{R}^n$, and let $V(C_k)$ be the Voronoi cell for C_k. Given that C_k is transmitted, the decoder makes the correct decision if and only if the noise vector Y is in $V(C_k)$, an event of probability

$$\frac{1}{(\sigma\sqrt{2\pi})^n} \int_{V(C_k)} e^{-x\cdot x/2\sigma^2}\, dx \,. \tag{13}$$

Assuming that all code points are equally likely to be used, the error probability for this code is

$$P_e = 1 - \frac{1}{M} \sum_{k=1}^{M} \frac{1}{(\sigma\sqrt{2\pi})^n} \int_{V(C_k)} e^{-x\cdot x/2\sigma^2}\, dx \,. \tag{14}$$

If all the Voronoi cells are congruent, to some polytope Π say, this simplifies to

$$P_e = 1 - \frac{1}{(\sigma\sqrt{2\pi})^n} \int_{\Pi} e^{-x\cdot x/2\sigma^2}\, dx \,. \tag{15}$$

Then one version of the Gaussian *channel coding problem* is the following. Given the dimension n, the number of code points M, and a *power constraint*

$$N(C_k) \leqslant nP, \qquad k = 1, \ldots, M \,, \tag{16}$$

find a code $C_1, \ldots, C_M \in \mathbf{R}^n$ satisfying (16) for which the error probability P_e given by (14) is minimized. The *constant energy* problem replaces (16) by the requirement that $N(C_k) = nP$ for all k, or equivalently that all $C_k \in \Omega_n$.

The *lattice* version of the Gaussian channel coding problem is to find, for a given value of σ, that n-dimensional lattice of determinant 1 for which (15) is minimized, where Π (of volume 1) is the Voronoi cell of the lattice.

The difference between the four main lattice problems is now clear. Let Λ be a lattice with Voronoi cell Π, of volume 1. For the packing problem (§1 of Chap. 1) we maximize the in-radius of Π, for the covering problem (§1 of Chap. 2) we minimize the circumradius, for the quantizing problem (§3 of Chap. 2) we minimize the second moment $G(\Pi)$, and for the channel coding problem we minimize P_e (Eq. (15)).

In general the answers to the channel coding problems will depend on σ. Unfortunately, unlike the expression for quantizer error given in

Eq. (87) of Chap. 2, the integrals in (14), (15) cannot be evaluated exactly even in simple cases (except in one or two dimensions, or if Π is a cube, when P_e can be expressed in terms of the error function (18)). Because of the difficulty in evaluating the integral, most investigators have replaced (14) by a simpler expression.

Probably the most useful estimate is the following *union bound* (cf. [Wel7], [Woz1]). Consider the Voronoi cell $V(C_k)$ of the kth code point C_k. Suppose the walls of $V(C_k)$ are determined by r code points $C_{k1}, \ldots, C_{kr}$ (i.e. these are the ***relevant*** points for C_k in the notation of §1.2 of Chap. 2). The complement of $V(C_k)$ is contained in the union of the corresponding half-spaces. Therefore the error probability given that C_k is transmitted, $P_e^{(k)}$, is bounded by

$$P_e^{(k)} \leqslant \sum_{j=1}^{r} \frac{1}{\sigma\sqrt{2\pi}} \int_{\rho_j}^{\infty} e^{-x^2/2\sigma^2} dx ,$$

where $\rho_j = ½ \, \| C_k - C_{kj} \|$. We write this as

$$P_e^{(k)} \leqslant \sum_{j=1}^{r} ½ \operatorname{erfc} \left(\frac{\rho_j}{\sigma\sqrt{2}} \right), \tag{17}$$

where

$$\operatorname{erfc}(x) = \frac{2}{\sqrt{\pi}} \int_{x}^{\infty} e^{-t^2} dt \tag{18}$$

is the *error function* [Abr1, p. 297]. An upper bound on (18) is obtained by including the contribution from the half-spaces corresponding to *all* $C \neq C_k$. Then averaging over k we get

$$P_e \leqslant \frac{1}{M} \sum_{C} \sum_{C' \neq C} ½ \operatorname{erfc} \frac{\| C - C' \|}{\sigma\sqrt{8}} . \tag{19}$$

As $\sigma \to 0$ the right-hand side of (19) approaches P_e. This is the case when $P >> \sigma$, the case of high *signal-to-noise ratio*. Let $\rho = \min \rho_j$ be half the minimal distance between code points, and let $\hat{\tau}$ be the average number of code points at distance 2ρ from a code point. (This is not the same as the average kissing number.) Then for small σ the solution to the coding problem is found by minimizing

$$\frac{1}{2} \hat{\tau} \operatorname{erfc} \left(\frac{\rho}{\sigma\sqrt{2}} \right). \tag{20}$$

So even when σ is small it is not quite correct to say that the coding problem coincides with the sphere packing problem, since the latter maximizes ρ but ignores $\hat{\tau}$. At the other extreme, when σ is large, we can approximate $\exp\{-x \cdot x / (2\sigma^2)\}$ by $1 - x \cdot x / 2\sigma^2 + \cdots$ in (14), and the coding problem coincides with the quantizing problem.

1.4 Lattice codes for the Gaussian channel. We now consider the lattice channel coding problem in more detail. In one and two dimensions the optimal lattice is independent of σ. (In one dimension there is only one lattice, and in two dimensions the optimality of the hexagonal lattice for all σ follows by taking $f(x) = (\sigma\sqrt{2\pi})^{-1} \exp\{-x^2/2\sigma^2\}$ in the theorem on page 81 of [Fej10].)

We now assume σ is small. From (20) we may approximate the error probability P_e by

$$P_e' = \frac{\tau}{2} \operatorname{erfc}\left(\frac{\rho}{\sigma\sqrt{2}}\right). \tag{21}$$

Let the code consist of all lattice points C with $N(C) \leqslant nP$. The number of points, M, is given by

$$M \cdot V_n \rho^n = \Delta \cdot V_n (nP)^{n/2} (1+o(1)), \tag{22}$$

(where the term $o(1) \to 0$ as $M \to \infty$), so

$$\rho = \sqrt{nP} \left(\frac{\Delta}{M}\right)^{1/n} (1+o(1)). \tag{23}$$

The average norm of the code points is

$$\frac{n}{n+2} \cdot nP(1+o(1)) \tag{24}$$

(see [Cal7, Th. 4]). For this analysis it is better to define the *rate* of the code to be

$$R = \frac{1}{n} \log_2 M \quad \text{bits/dimension} \tag{25}$$

(rather than (9)). Then

$$\rho = \frac{\sqrt{nP}}{2^R} \Delta^{1/n} (1 + o(1)). \tag{26}$$

We also define the *normalized signal-to-noise ratio* of the code to be

$$S = \frac{P}{2^{2R}\sigma^2} = \frac{P}{M^2\sigma^2}. \tag{27}$$

By combining (21), (26), (27) we obtain our final estimate for the error probability of a Gaussian channel code with normalized signal-to-noise ratio S, obtained from an n-dimensional lattice of density Δ and kissing number τ:

$$P_e'' = \frac{\tau}{2} \operatorname{erfc}\left(\sqrt{\frac{nS}{2}}\, \Delta^{1/n}\right). \tag{28}$$

This estimate becomes more accurate as S increases. For large values of x

$$\operatorname{erfc}(x) \sim \frac{1}{x\sqrt{\pi}} e^{-x^2} \tag{29}$$

and therefore

$$\log P_e'' \sim \text{constant} - \left\{\frac{1}{2} n \Delta^{2/n} \log e\right\} S . \tag{30}$$

We have calculated the error probability of various lattices using (28). Figure 3.3 shows those lattices in dimensions 2, 4, 8, 24 and 48 that have the smallest known values of P_e'', for $P_e'' < 10^{-3}$. In this range the best lattice codes presently known coincide with the densest known lattice sphere packings. Furthermore in this range $\log_{10} P_e''$ is a linear function of S, with slope given by (30).

From (28)-(30) we see that, as long as the densest lattice sphere packings in $\mathbf{R}^n$ all have the same kissing number, then as $\sigma \to 0$ (or $S \to \infty$) the lattice coding problem coincides with the lattice sphere packing problem. But if there are two or more equally dense lattices with different kissing numbers, then the lattice having the smallest kissing number is to be preferred.

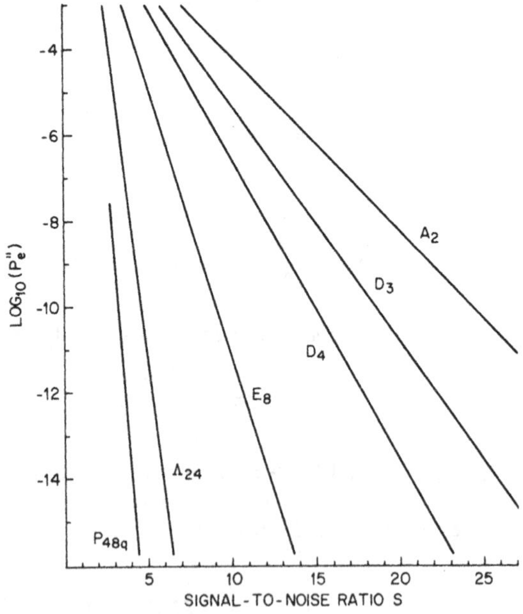

Figure 3.3. The smallest known error probability P_e'' (Eq. (28)) for any n-dimensional lattice, as a function of the normalized signal-to-noise ratio S (Eq. (27)), for various dimensions $n \leqslant 48$.

We have also calculated P_e "exactly" from (15) for various low dimensional lattices using Monte Carlo integration. The results (as yet unpublished) agree with those obtained from (28).

A cruder measure of the efficiency of a code is provided by the ratio μ/E, where μ is the minimal squared distance between code points and E is their average norm (or energy). The *(nominal) coding gain* of a code with parameters μ_1, E_1 over another with parameters μ_2, E_2 is defined to be

$$10 \log_{10} \left(\frac{\mu_1}{E_1} / \frac{\mu_2}{E_2} \right) \text{ db}. \tag{31}$$

For example, if we take our code to consist of all points C of norm at most nP in a lattice Λ, then

$$\frac{\mu}{E} = \frac{4(n+2)}{n} \left(\frac{\Delta}{M} \right)^{\frac{2}{n}} (1+o(1)) \tag{32}$$

from (23), (24). For comparison we use a roughly equal number of points arranged in an n-dimensional cubical array centered at the origin (an essentially one-dimensional code). For this code

$$\frac{\mu}{E} = \frac{12}{nM^{2/n}} (1+o(1)). \tag{33}$$

The nominal coding gain obtained from the lattice Λ is therefore

$$10 \log_{10} \left\{ \frac{1}{3}(n+2)\Delta^{\frac{2}{n}} \right\}. \tag{34}$$

The coding gains of various lattices in up to 128 dimensions are shown in Table 3.1. The table also gives the value of $\gamma = 4\,\delta^{2/n}$ (where $\delta = \Delta/V_n$) for each lattice. γ is a lower bound to Hermite's constant (Eq. (47) of Chap. 1), and in view of (28)-(34) is another measure of the coding gain of a lattice.

As pointed out by Forney [For1], the estimate (19) for P_e can be related to the theta series of the lattice. Since [Mit6, p. 291]

$$\operatorname{erfc}(a) \leqslant e^{-a^2} \quad (a \geqslant 0),$$

we have

$$P_e \leqslant \frac{1}{2}\, \Theta_\Lambda \left(q = e^{-1/(8\sigma^2)}\right) - \frac{1}{2}, \tag{35}$$

and the right-hand side approaches P_e as $\sigma \to 0$.

Welti [Wel6] and Welti and Lee [Wel7] have used (20) to estimate P_e for other four-dimensional codes obtained from lattices. Other approximate formulae for the error probability of lattice codes have been used in [Bla3] (cf. [Slo11]) and [Bud2].

Table 3.1. Nominal coding gain (Eq. (34)) for various lattices in Tables 1.2, 1.3, and the values of the center density δ and $\gamma = 4\delta^{2/n}$, a lower bound on Hermite's constant.

n	Lattice	δ	Gain (db)	γ
2	A_2	$1/(2\sqrt{3})$	0.825	1.155
4	D_4	$1/8$	1.961	1.414
6	E_6	$1/(8\sqrt{3})$	2.832	1.665
8	E_8	$1/16$	3.739	2
12	K_{12}	$1/27$	4.514	2.309
16	Λ_{16}	$1/16$	5.491	2.828
24	Λ_{24}	1	7.116	4
48	P_{48q}	$(3/2)^{24}$	9.037	6
64	P_{64c}	2^{22}	9.397	6.442
80	$\eta(E_8)$	2^{36}	10.070	7.464
128	$\eta(E_8)$	2^{88}	11.557	10.375

The channel coding problems described in this section are idealized mathematical problems. In practice the situation is more complicated and many other factors must be considered. Shannon's theorem implies that T (the duration of the signals) and n (the dimension) must be increased in order to reduce the probability of error, but this both delays the encoding and decoding processes and makes them more complicated. For a code to be used in practice, efficient encoding and decoding algorithms must exist. Furthermore other types of distortion besides Gaussian noise must be considered (intersymbol interference, for example, or noise which occurs in bursts).

A great many constructions of practical channel codes have been proposed: for example permutation codes [Bil2], [Sle2], "group codes" (see §4.2), and the codes described by [Cam4], [Ein1], [Ein2], [Fos1], [Fos2], [Gao1], [Ger4], [Gil1], [Gin1], [Kan1], [Ker2], [Kri1], [Laz1], [Say1], [Zet1], [Zet2]. See also the references on the construction of spherical codes given in §2.3 of Chap. 1.

In recent years, geometrical channel codes of the type described in this section have been combined with error correcting codes (§2), especially with convolutional codes, to produce what are called *trellis code modulation* (TCM) schemes. These are in effect very high dimensional codes, but have practical encoding and decoding algorithms. See [And1], [Cal2]-[Cal8], [Fal1], [For2]-[For3], [Ung1], [Wei1], [Wil20].

2. Error-correcting codes

2.1 The error-correcting code problem. The other idealized model for the channel in Fig. 3.1 is the *binary symmetric channel.* For this channel the input and output symbols are 0's and 1's, and there is some fixed probability $p < \frac{1}{2}$ that when a 0 or 1 is transmitted, the other symbol is received; with probability $1 - p > \frac{1}{2}$, the correct symbol is received. A *binary code*[(2)] C of *length* n is a set of binary vectors (called *codewords*) with n coordinates, or in other words is a subset of $\mathbf{F}_2^n$, where $\mathbf{F}_2 = \{0,1\}$ is the Galois field of order 2.

Similarly a *q-ary code* is a subset of $\mathbf{F}_q^n$, where $\mathbf{F}_q$ is the Galois field of order q and q is a prime or prime-power. Besides $\mathbf{F}_2$, the ternary field $\mathbf{F}_3 = \{0,1,-1\}$ and the field of order 4, $\mathbf{F}_4 = \{0,1,\omega,\omega^2\}$ (where $\omega^2 = \overline{\omega} = \omega + 1$, $\omega^3 = 1$) are especially interesting. A q-ary code is used on the *q-ary symmetric channel*, where the input and output symbols are labeled with the elements of $\mathbf{F}_q$.

As in §1.2, we wish to choose the codewords so they are easy to distinguish from each other even when errors have occurred. To make this precise we define the *Hamming distance* between two vectors

$$u = (u_1, \cdots, u_n), \quad v = (v_1, \cdots, v_n),$$

$u_i, v_i \in \mathbf{F}_q$, to be the number of coordinates where they differ:

$$d(u,v) = |\{ i : u_i \neq v_i \}|.$$

The *(Hamming) weight* $wt(u)$ of a vector u is the number of nonzero coordinates u_i; therefore

$$d(u,v) = \mathrm{wt}(u-v). \tag{36}$$

The *minimal distance* d of a code is

$$d = \min \{ d(u,v) : u,v \in C, u \neq v \}. \tag{37}$$

If a code has minimal distance d, the "Hamming spheres" of radius

$$\rho = [\tfrac{1}{2}(d-1)] \tag{38}$$

around the codewords are disjoint (so ρ is the packing radius of the code, cf. Eq. (22) of Chap. 2), and therefore the code can correct ρ errors. A code of length n, containing M codewords and with minimal distance d is said to be an (n,M,d) code.

---

(2) Strictly speaking this is a *block* code. *Convolutional* codes (see for example [Lin0], [Vit1]) are an important rival family of codes. Although essential for many practical applications, convolutional codes are not directly related to the geometrical problems discussed in this book.

A *linear* (or *group*) code C is a linear subspace of $\mathbf{F}_q^n$: the set of codewords is closed under vector addition and coordinatewise multiplication by elements of $\mathbf{F}_q$. The *dimension* k (also written dim C) is the dimension of this subspace, and there are q^k codewords. The *rate* of the code is

$$R = \frac{1}{n} \log_2 M = \frac{k}{n} \log_2 q \quad \text{bits/symbol}. \tag{39}$$

A linear code of length n, dimension k and minimal distance d is said to be an $[n,k,d]$ code (or sometimes an $[n,k]$ code). The minimal distance of a linear code is the minimal nonzero weight of any codeword:

$$d = \min \{ \mathrm{wt}(u) : u \in C, u \neq 0 \}. \tag{40}$$

In a good code n is small (to reduce delays), M is large (to make efficient use of the channel), and d is large (to correct many errors). Naturally these goals are incompatible. The *error-correcting code problem* is the following: given n and d, find $A(n,d)$, the maximal number of codewords in any (n,M,d) code. There is a similar problem for linear codes. In general these problems are unsolved. However many upper and lower bounds on $A(n,d)$ have been found, and a large number of constructions of codes are known. A table of $A(n,d)$ for $n \leqslant 24$ is given in Table 9.1 of Chap. 9. Other tables may be found in [Bes3], [Che1], [Gra2], [Hel7], [Mac6], [Pet3], [Pro1], [Slo13], [Ver1], [Zin2], [Zin4].

We will see in Chaps. 5 and 7 that there are strong connections between codes and sphere packings, and between linear codes and lattice packings. This is not surprising, since the error-correcting problem is also a sphere packing problem, although in $\mathbf{F}_2^n$ or $\mathbf{F}_q^n$ rather than $\mathbf{R}^n$.

A *constant weight* code is one in which all codewords have equal weight. Binary constant weight codes are a discrete analog of the spherical codes described in §2.3 of Chap. 1.

Since we have written at length about error-correcting codes in [Mac6], and numerous other references are available ([Ass4], [Ber4], [Ber6], [Bla2], [Bla5], [Bla9], [Lin0], [Lin11], [Pet3], [Vit1]), the discussion in this chapter will be brief. In the next section we establish some further notation, and then give a number of examples.

The covering problem of Chap. 2 also has a binary analog: one wishes to find the smallest number of overlapping Hamming spheres that will cover $\mathbf{F}_2^n$. Equivalently, let the *covering radius* of a code C be

$$\max_x \min_c d(x,c) \quad (x \in \mathbf{F}_2^n, c \in C) . \tag{41}$$

Then the coding version of the covering problem is, for a given length and covering radius, to find the smallest possible number of codewords. We shall not discuss such *covering codes* in this book, but refer the reader to [Coh3], [Coh4], [Dow4], [Gra4], [Kil1], [Kil2], [Slo15], [Slo16]. Much less is known about the binary analog of the quantizing problem.

2.2 Further definitions from coding theory. Many of the following definitions are analogs of the corresponding definitions for sphere packings. Let B be an $n \times n$ *monomial* matrix, containing exactly one nonzero element from $\mathbf{F}_q$ in each row and column. The set of all such matrices forms the full *monomial group* over $\mathbf{F}_q$, of order $(q-1)^n n!$. The matrix B sends a code C over $\mathbf{F}_q$ into the *equivalent* code

$$C' = CB = \{ uB : u \in C \}. \tag{42}$$

If C' is the *same* code as C then B belongs to the *automorphism group* $\mathrm{Aut}(C)$ of C (the corresponding notion for sphere packings is defined in §3). Thus the number of codes equivalent to C is

$$\frac{(q-1)^n n!}{|\mathrm{Aut}(C)|}. \tag{43}$$

An $[n,k]$ linear code C may be specified by a *generator matrix*. This is a $k \times n$ matrix M such that C consists of all linear combinations (with coefficients from $\mathbf{F}_q$) of the rows of M. It is always possible to find a code equivalent to C which has a generator matrix of the form

$$G = [I_k \mid A], \tag{44}$$

where I_k is a $k \times k$ identity matrix and A is a $k \times (n-k)$ matrix.

Conjugation in the field $\mathbf{F}_q$, where $q = p^a$, p prime, is defined by $x \to \bar{x} = x^p$, for $x \in \mathbf{F}_q$. A similar notation is used for vectors: $\bar{u} = (\bar{u}_1, \cdots, \bar{u}_n)$, matrices, etc. The *dual* (or *orthogonal*) code is defined to be

$$C^* = \{ x \in \mathbf{F}_q^n : x \cdot \bar{u} = 0 \ \text{ for all } \ u \in C \} \tag{45}$$

(cf. Eq. (24) of Chap. 1, Eq. (65) of Chap. 2). Then $\dim C^* = n - \dim C$. A code is *self-dual* if $C = C^*$; this implies that n is even and $\dim C = n/2$. If C has generator matrix (44) then C^* has generator matrix

$$[-\bar{A}^{tr} \mid I_{n-k}], \tag{46}$$

and C is self-dual if and only if

$$A\bar{A}^{tr} = -I_{n/2}. \tag{47}$$

There are four especially interesting families of self-dual codes, those in which the weights of the codewords are multiples of a constant $c > 1$. In a binary self-dual code every codeword has even weight. In some binary self-dual codes the weight of every codeword is a multiple of 4: these are called *Type* II, or *doubly-even* codes; the other binary self-dual codes are of *Type* I. In a ternary self-dual code the weight of every codeword is a multiple of 3. These are called *Type* III codes. Finally in a self-dual code over $\mathbf{F}_4$ every codeword has even weight. These are called *Type* IV codes.

It was shown by Gleason and Pierce that these are essentially the only possibilities: two proofs of this theorem are given in [Slo10, §6.1]. Codes of type I, II, III or IV exist if and only if n is a multiple of 2, 8, 4 or 2 respectively (see Chap. 7). Similar results for lattices were given in §§2.4, 2.6 of Chap. 2. We shall describe what is known about the enumeration of these codes in Chap. 7.

Let C be an (n, M, d) code over $\mathbf{F}_q$. We let $A_i(c)$ denote the number of codewords at Hamming distance i from a codeword $c \in C$. The numbers $\{ A_i(c) \}$ are called the *weight distribution* of C with respect to c. Of course $A_0(c) = 1$, $A_i(c) \geqslant 0$, and $\Sigma_i A_i(c) = M$. The $A_i(c)$ also satisfy certain less obvious inequalities found by Delsarte, which play a crucial role in the theory. These are described in Chap. 9.

For linear codes (and some nonlinear codes) $A_i(c)$ is independent of c and will be denoted by A_i. The *(Hamming) weight enumerator* of a linear code C is

$$W_C(x,y) = \sum_{u \in C} x^{n-\mathrm{wt}(u)} y^{\mathrm{wt}(u)} \tag{48}$$

$$= \sum_{i=0}^{n} A_i x^{n-i} y^i . \tag{49}$$

This is a homogeneous polynomial of degree equal to the length of the code, and strongly resembles the theta series of a lattice (Eqs. (31), (32) of Chap. 2).

The *MacWilliams identity* [Mac2], [Mac6, Chap. 5] gives the weight enumerator of the dual code C^*:

$$W_{C^*}(x,y) = \frac{1}{M} W_C(x+(q-1)y, x-y) \tag{50}$$

– compare Eq. (43) of Chap. 2. There is a generalization to nonlinear codes [Mac6, Chap. 5, §5], [Mac7].

The Hamming weight enumerator classifies codewords according to the number of nonzero coordinates. More detailed information is supplied by the *complete weight enumerator* (c.w.e.), which gives the number of codewords of each composition. For example the c.w.e. of a ternary code C is

$$\mathrm{c.w.e.}_C(x,y,z) = \sum_{u \in C} x^{n_0(u)} y^{n_1(u)} z^{n_{-1}(u)}, \tag{51}$$

where $n_i(u)$ is the number of times $i \in \mathbf{F}_3$ occurs in u. There is also a version of the MacWilliams identity for c.w.e.'s.

There are *mass formulae* for codes, just as for lattices (cf. Eq. (47) of Chap. 2) giving explicit constants a_0 such that

$$\sum_C \frac{1}{|\mathrm{Aut}(C)|} = a_0, \tag{52}$$

where the sum is over all inequivalent codes of a certain type (for example, all Type II codes of length n) – see [Mac6], [Slo10]. A slightly more general version gives an explicit formula for

$$\sum_C \frac{W_C(x,y)}{|\mathrm{Aut}(C)|}. \tag{53}$$

An (n,M,d) code over $\mathbf{F}_q$ can be *extended* to an $(n+1,M,d')$ code by appending a *zero-sum check* digit

$$c_n = -\sum_{i=0}^{n-1} c_i \tag{54}$$

to each codeword $c_0\, c_1 \dots c_{n-1}$. If $q = 2$ and d is odd the extended code has $d' = d + 1$.

2.3 Repetition, even weight and other simple codes. We begin with some easy examples of codes.

(2.3.1) The $[n,0,n]$ *zero* code of length n contains just the codeword $00\dots0$.

(2.3.2) The $[n,n,1]$ *universe* code $\mathbf{F}_q^n$ is the dual of (2.3.1).

(2.3.3) The $[n,1,n]$ *repetition* code contains all codewords $a\,a\dots a$, $a \in \mathbf{F}_q$.

(2.3.4) The $[n,n-1,2]$ *zero-sum* code contains all vectors such that $\sum c_i = 0$; it is the dual of (2.3.3). When $q = 2$ this is called the *even weight* code, sometimes denoted by $\mathcal{E}_n$, since it consists of all binary vectors containing an even number of 1's.

2.4 Cyclic codes. A code is *cyclic* if whenever $c_0 c_1 \dots c_{n-1}$ is a codeword so is $c_{n-1} c_0 c_1 \dots c_{n-2}$. Unless stated otherwise a cyclic code is assumed to be linear.

Let $q = p^a$ where p is prime, and let C be an $[n,k,d]$ cyclic code over $\mathbf{F}_q$. It is convenient to represent a codeword $c = c_0 c_1 \dots c_{n-1} \in C$ by the polynomial $c(x) = c_0 + c_1 x + \cdots + c_{n-1} x^{n-1}$ in the ring $R_n(x)$ of polynomials modulo $x^n - 1$ with coefficients from $\mathbf{F}_q$. Then $xc(x)$ represents a cyclic shift of c, and so a linear cyclic code is represented by an ideal in $R_n(x)$ [Mac6, Chap. 7]. This can be generated by a single polynomial $g(x)$, a *generator polynomial* for the code. It is easily seen that $g(x)$ divides $x^n - 1$ over $\mathbf{F}_q$ and that the dimension of C is $k = n - \deg g(x)$.

Examples. For the universe code $g(x) = 1$, for the zero sum code $g(x) = x - 1$, and for the repetition code $g(x) = 1 + x + \cdots + x^{n-1}$.

(2.4.1) The [7, 4, 3] *Hamming* code $\mathcal{H}_7$ is the binary cyclic code of length 7 with generator polynomial $g(x) = 1 + x + x^3$; it has generator matrix

$$\begin{bmatrix} 1 & 1 & 0 & 1 & 0 & 0 & 0 \\ 0 & 1 & 1 & 0 & 1 & 0 & 0 \\ 0 & 0 & 1 & 1 & 0 & 1 & 0 \\ 0 & 0 & 0 & 1 & 1 & 0 & 1 \end{bmatrix}, \qquad (55)$$

where the rows correspond to $g(x)$, $xg(x)$, etc. $\mathscr{H}_7$ has weight distribution $A_0 = 1$, $A_3 = A_4 = 7$, $A_7 = 1$, which we abbreviate to

$$0^1\ 3^7\ 4^7\ 7^1 .$$

The automorphism group $\mathrm{Aut}(\mathscr{H}_7)$ is Klein's group of order 168. The dual code $\mathscr{H}_7^*$ is a [7, 3, 4] code with weight distribution $0^1\ 4^7$, the codewords forming the vertices of a regular simplex.

(2.4.2) The [8, 4, 4] *extended Hamming* code $\mathscr{H}_8$ is formed by appending a zero-sum check digit to $\mathscr{H}_7$. If the coordinates in (55) are labeled 0, 1, 2, ..., 6, and the zero-sum digit is labeled ∞, then $\mathscr{H}_8$ contains 14 codewords of weight 4, those with 1's in coordinates

$$\infty 124,\ \infty 235,\ \infty 346,\ \infty 450,\ \infty 561,\ \infty 602,\ \infty 013 \qquad (56)$$

and their complements. An alternative generator matrix is

$$\begin{bmatrix} 0 & 0 & 0 & 0 & 1 & 1 & 1 & 1 \\ 0 & 0 & 1 & 1 & 0 & 0 & 1 & 1 \\ 0 & 1 & 0 & 1 & 0 & 1 & 0 & 1 \\ 1 & 1 & 1 & 1 & 1 & 1 & 1 & 1 \end{bmatrix}. \qquad (57)$$

$\mathscr{H}_8$ has weight distribution $0^1\ 4^{14}\ 8^1$; equivalently its weight enumerator is

$$\psi_8 = x^8 + 14x^4y^4 + y^8 ,$$

and $\mathrm{Aut}(\mathscr{H}_8)$ is the affine group of order $8 \cdot 7 \cdot 6 \cdot 4 = 1344$. This important code shares many properties with the E_8 lattice. For example $\mathscr{H}_8$ is the smallest nontrivial example of a self-dual Type II code.

Returning to the general theory, let n be relatively prime to q, and let α be a primitive n-th root of unity, so that

$$x^n - 1 = \prod_{i=0}^{n-1} (x - \alpha^i) .$$

If m is the multiplicative order of q modulo n, then $\alpha \in \mathbf{F}_{q'}$ where $q' = q^m$. Also

$$g(x) = \prod_{i \in K} (x - \alpha^i) , \qquad (58)$$

for some set $K \subseteq \{0, 1, \ldots, n-1\}$, where since $g(x)$ divides $x^n - 1$ over $\mathbf{F}_q$, $i \in K$ implies $qi \in K$ [Mac6, Chap. 7]. The α^i for $i \in K$ are called the *zeros* of the code. There are good lower bounds on the minimal distance of a cyclic code in terms of its zeros — see van Lint and Wilson [Lin13].

(2.4.3) For example, the binary *Hamming code* $\mathcal{H}_n$ with parameters $n = 2^m - 1$, $k = n - m$, $d = 3$ is obtained by taking $K = \{1, 2, 4, \ldots, 2^{m-1}\}$, so that $g(x)$ is the minimal polynomial of α.

(2.4.4) By appending a zero-sum check digit we obtain the *extended Hamming code* $\mathcal{H}_{2^m}$ with parameters $[2^m, 2^m - m - 1, 4]$.

2.5 BCH and Reed-Solomon codes. The *Bose-Chaudhuri-Hocquenghem* (*BCH*) code of length n and *designed distance* d over $\mathbf{F}_q$ is the cyclic code whose generator polynomial $g(x)$ has roots exactly $\alpha, \alpha^2, \alpha^3, \ldots, \alpha^{d-1}$ and their conjugates [Ber4], [Lin11], [Mac6, Chaps. 7-9], [Pet3]. The actual minimal distance between codewords is at least the designed distance and may exceed it.

BCH codes of length $n = q^m - 1$ are called *primitive*. For example primitive binary BCH codes of length $n = 2^m - 1$ and designed distance 3 are Hamming codes.

BCH codes over $\mathbf{F}_q$ of length $n = q - 1$ are usually called *Reed-Solomon* codes; they were the first to be discovered and are of great practical and theoretical importance [Bla2], [Mac6, Chap. 10], [Ree1]. The minimal distance of a Reed-Solomon code is equal to its designed distance. Thus a Reed-Solomon code over $\mathbf{F}_q$ with parameters $[n = q - 1, k, d = n - k + 1]$ has generator polynomial of the form

$$g(x) = (x - \alpha^b)\,(x - \alpha^{b+1}) \cdots (x - \alpha^{b+d-2}) , \tag{59}$$

where α is a primitive element of $\mathbf{F}_q$. Usually b is taken to be 0 or 1. These codes exist for all q and $1 \leqslant k \leqslant n$.

It is possible to extend a Reed-Solomon code by adding 1, 2 or 3 additional coordinates [Mac6, Chaps. 10, 11]. For any q and k there exist $[q, k, q - k + 1]$ and $[q + 1, k, q - k + 2]$ extended Reed-Solomon codes over $\mathbf{F}_q$; and there also exist $[2^m + 2, 3, 2^m]$ and $[2^m + 2, 2^m - 1, 4]$ triply-extended Reed-Solomon codes over $\mathbf{F}_q$ when $q = 2^m$.

For any linear code $d \leqslant n - k + 1$ (the *Singleton bound* [Mac6, Chap. 1]). Codes for which $d = n - k + 1$ are called *maximal distance separable* (*MDS*) codes. The preceding Reed-Solomon and extended Reed-Solomon codes are all MDS codes, as are the codes (2.3.2)-(2.3.4).

The following are two especially interesting examples of MDS codes; they are also extended Reed-Solomon codes and are self-dual.

(2.5.1) The [4, 2, 3] ternary *tetracode* $\mathcal{C}_4$ has generator matrix

$$\begin{bmatrix} 1 & 1 & 1 & 0 \\ 0 & 1 & -1 & 1 \end{bmatrix}, \tag{60}$$

weight enumerator

$$\psi_4 = x^4 + 8xy^3 , \tag{61}$$

complete weight enumerator

$$x\{ x^3+(y+z)^3 \} \tag{62}$$

and automorphism group $2.S_4$, where S_n is the symmetric group of order $n!$.

(2.5.2) The [6, 3, 4] *hexacode* $\mathscr{C}_6$ is a code over $\mathbf{F}_4$ with generator matrix

$$\begin{bmatrix} 0 & 0 & 1 & 1 & 1 & 1 \\ 0 & 1 & 0 & 1 & \omega & \bar{\omega} \\ 1 & 0 & 0 & 1 & \bar{\omega} & \omega \end{bmatrix}. \tag{63}$$

The following matrix defines an equivalent code:

$$\begin{bmatrix} 1 & 0 & 0 & 1 & \omega & \omega \\ 0 & 1 & 0 & \omega & 1 & \omega \\ 0 & 0 & 1 & \omega & \omega & 1 \end{bmatrix}. \tag{64}$$

Using the definition (63), the 64 *hexacodewords* are obtained from the five words

$$\begin{array}{ccc} 01 & 01 & \omega\bar{\omega} \\ \omega\bar{\omega} & \omega\bar{\omega} & \omega\bar{\omega} \\ 00 & 11 & 11 \\ 11 & \omega\omega & \bar{\omega}\bar{\omega} \\ 00 & 00 & 00 \end{array} \tag{65}$$

by freely permuting the three pairs, reversing any even number of pairs, and multiplying by any power of ω. The five words have 36, 12, 9, 6, 1 images respectively. The Hamming weight enumerator of $\mathscr{C}_6$ is

$$x^6 + 45x^2y^4 + 18y^6, \tag{66}$$

and $\mathrm{Aut}(\mathscr{C}_6) = 3.A_6$, where A_n is the alternating group of order $\frac{1}{2}n!$. (By making use of the field automorphism that interchanges ω with $\bar{\omega}$ we obtain the slightly larger group $3.S_6$.) This code is the basis for the MOG construction of the Leech lattice — see §11 of Chap. 4 and Chap. 11.

2.6 Justesen codes. For large n, BCH codes are weak: for any fixed rate R, the ratio $d/n \to 0$ as $n \to \infty$. In contrast, Justesen codes are an explicit family of binary linear codes for which both R and d/n are bounded away from zero as n increases. We sketch their construction; further details can be found in [Jus1], [Mac6, Chap. 10]. The starting point is a Reed-Solomon code C over $\mathbf{F}_q$, $q = 2^m$, with length $N = 2^m - 1$, dimension K and minimal distance $D = N - K + 1$. Let α be a primitive element of $\mathbf{F}_q$. If $c = (c_0 c_1 \ldots c_{N-1})$, $c_i \in \mathbf{F}_q$, is an arbitrary codeword of C, let c' be the vector

$$c' = (c_0, c_0, c_1, \alpha c_1, \ldots, c_{N-1}, \alpha^{N-1} c_{N-1}) \tag{67}$$

of length $2N$ over $\mathbf{F}_q$.

Now $\mathbf{F}_q$ (for $q = 2^m$) is a vector space of dimension m over $\mathbf{F}_2$, so we may set up a one-to-one correspondence between $\mathbf{F}_q$ and $\mathbf{F}_2^m$. Let c'' be obtained from c' by replacing each component by the corresponding binary m-tuple. Then c'' is a binary vector of length $n = 2mN$; the set of all c'' for $c \in C$ is a Justesen code. The binary dimension is $k = mK$ and the rate is $R = k/n = K/2N < ½$.

Since C is a Reed-Solomon code, a large number of the c_i are nonzero. Because each c_i is multiplied by a different power of α, the binary $2m$-tuples corresponding to nonzero pairs $(c_i, \alpha^i c_i)$ are distinct. This is enough to guarantee that a significant fraction of the components of c'' are nonzero; one can show (see the above references) that the minimal distance d of the Justesen code satisfies

$$\frac{d}{n} \gtrsim 0.110(1-2R) \tag{68}$$

for $0 < R < ½$. Justesen codes of higher rate may be obtained by deleting the appropriate number of coordinates from c''.

Justesen's construction has been generalized by Weldon [Wel1] and Sugiyama et al. [Sug1]-[Sug3] to yield codes with higher rates and minimal distances.

2.7 Reed-Muller codes. Let the *binary weight* $W(i)$ of a non-negative integer i be the number of ones in the binary expansion of i. For $1 \leqslant r \leqslant m-2$, the rth order binary *punctured Reed-Muller* code of length $n = 2^m - 1$ is the cyclic code whose generator polynomial has as roots those α^i such that $1 \leqslant i \leqslant 2^m - 2$ and $1 \leqslant W(i) \leqslant m - r - 1$.

The rth order *Reed-Muller* (*RM*) code of length 2^m is formed by appending a zero-sum check digit to the rth order punctured Reed-Muller code, for $1 \leqslant r \leqslant m-2$. An rth order RM code has parameters

$$[n = 2^m,\ k = \sum_{i=0}^{r} \binom{m}{i},\ d = 2^{m-r}]. \tag{69}$$

The 0th, $(m-1)$th and mth order RM codes are the repetition, even-weight and universe codes respectively. The $(m-2)$th order RM codes are extended Hamming codes. The rth and $(m-r-1)$th order RM codes are duals, for all r. Thus $\mathcal{H}_8$ is a first-order Reed-Muller code.

Reed-Muller codes may also be defined in terms of Boolean functions — see for example [Mac6, Chap. 13]. Quite a lot is known about the weight distributions of RM codes; for example the number of codewords of minimal weight $d = 2^{m-r}$ is

$$A_d = 2^r \cdot \prod_{i=0}^{m-r-1} \frac{2^{m-i}-1}{2^{m-r-i}-1} \tag{70}$$

[Mac6, Chaps. 13-15].

It is clear from the definitions that BCH and RM codes are "nested". More precisely, the BCH code of designed distance d is contained in that of designed distance $d - 1$, and the rth order RM code is contained in the $(r+1)$th order RM code. The rth order RM code is a subcode of the code obtained by appending a zero-sum check digit to the BCH code of designed distance $2^{m-r} - 1$.

2.8 Quadratic residue codes. Let p, n be primes such that p is a square modulo n. For $p = 2$, the most important case, this means that n is a prime of the form $8m \pm 1$. The *quadratic residue* code of length n over $\mathbf{F}_p$ is the cyclic code whose generator polynomial has roots $\{ \alpha^i : i \neq 0$ is a square modulo $n \}$ ([Lin11], [Mac6, Chap. 16]; see also [Leo5], [Leo6]). In other words the roots are α^i, where i is a quadratic residue modulo n. The dimension of this code is $(n+1)/2$. The extended quadratic residue code is obtained by appending a zero-sum check digit. If n is of the form $4a - 1$ then the extended code is self-dual. In particular, extended binary quadratic residue codes of length $8m$ are self-dual Type II codes.

For example when $q = 2$, $n = 7$ we obtain the Hamming codes $\mathscr{H}_7$ and $\mathscr{H}_8$ again.

(2.8.1) The [23, 12, 7] *binary Golay* code $\mathscr{C}_{23}$ is the quadratic residue code of length 23.

(2.8.2) The [24, 12, 8] (extended) Golay code $\mathscr{C}_{24}$ is obtained by appending a zero-sum check digit to $\mathscr{C}_{23}$. One of many possible generator matrices is shown in Fig. 3.4. $\mathscr{C}_{24}$ has weight distribution

$$0^1 \, 8^{759} \, 12^{2576} \, 16^{759} \, 24^1, \tag{71}$$

or equivalently its weight enumerator is

$$x^{24} + 759x^{16}y^8 + 2576x^{12}y^{12} + 759x^8y^{16} + y^{24}. \tag{72}$$

												0	1	2	3	4	5	6	7	8	9	10	
1												1	0	1	0	0	0	1	1	1	0	1	1
	1											1	1	0	1	0	0	0	1	1	1	0	1
		1										0	1	1	0	1	0	0	0	1	1	1	1
			1									1	0	1	1	0	1	0	0	0	1	1	1
				1								1	1	0	1	1	0	1	0	0	0	1	1
					1							1	1	1	0	1	1	0	1	0	0	0	1
						1						0	1	1	1	0	1	1	0	1	0	0	1
							1					0	0	1	1	1	0	1	1	0	1	0	1
								1				0	0	0	1	1	1	0	1	1	0	1	1
									1			1	0	0	0	1	1	1	0	1	1	0	1
										1		0	1	0	0	0	1	1	1	0	1	1	1
											1	1	1	1	1	1	1	1	1	1	1	1	0

Figure 3.4. A generator matrix for the binary Golay code $\mathscr{C}_{24}$. Blank entries are zero. The pattern of 0's and 1's in the right-hand half is defined by the quadratic residues modulo 11.

The automorphism group of $\mathscr{C}_{24}$ is the Mathieu group M_{24} (see Chap. 10). As we shall see, the code $\mathscr{C}_{24}$ shares many properties with the Leech lattice.

(2.8.3) When $q = 2$, $n = 47$ we obtain [47, 24, 11] and [48, 24, 12] codes.

(2.8.4) The [11, 6, 5] *ternary Golay* code $\mathscr{C}_{11}$ is the quadratic residue code of length 11 over $\mathbf{F}_3$.

(2.8.5) The [12, 6, 6] (extended) *ternary Golay* code $\mathscr{C}_{12}$ is obtained by appending a zero-sum check digit to $\mathscr{C}_{11}$. One of many possible generator matrices is

$$\begin{bmatrix} 1 & 0 & 0 & 0 & 0 & 0 & 0 & 1 & 1 & 1 & 1 & 1 \\ 0 & 1 & 0 & 0 & 0 & 0 & -1 & 0 & 1 & -1 & -1 & 1 \\ 0 & 0 & 1 & 0 & 0 & 0 & -1 & 1 & 0 & 1 & -1 & -1 \\ 0 & 0 & 0 & 1 & 0 & 0 & -1 & -1 & 1 & 0 & 1 & -1 \\ 0 & 0 & 0 & 0 & 1 & 0 & -1 & -1 & -1 & 1 & 0 & 1 \\ 0 & 0 & 0 & 0 & 0 & 1 & -1 & 1 & -1 & -1 & 1 & 0 \end{bmatrix} . \tag{73}$$

$\mathscr{C}_{12}$ has weight enumerator

$$x^{12} + 264x^6y^6 + 440x^3y^9 + 24y^{12} , \tag{74}$$

and complete weight enumerator (assuming the all-ones word is present)

$$f_{12} = x^{12} + y^{12} + z^{12} + 22(x^6y^6+y^6z^6+z^6x^6)$$

$$+ 220(x^6y^3z^3+x^3y^6z^3+x^3y^3z^6) . \tag{75}$$

The automorphism group of $\mathscr{C}_{12}$ is $2.M_{12}$. The Golay codes are studied in more detail in Chaps. 10 and 11. See also [Gol1], [Gol2], [Mac6, Chap. 20].

(2.8.6) When $q = 3$ and $n = 23$ or 47 we obtain [24, 12, 9] and [48, 24, 15] codes. The c.w.e. of the latter is

$$\sum_{i+j+k=48} A_{ijk}\, x^i\, y^j\, z^k ,$$

where the A_{ijk} for $i \geqslant j \geqslant k$ are given in Table 3.2. For the [24, 12, 9] code see [Mal2]. For later use we remark that the codewords of maximal weight in these [12, 6, 6], [24, 12, 9] and [48, 24, 15] ternary codes consist exactly of the rows of a Hadamard matrix and its negative (cf. §2.13).

2.9 Perfect codes. A *perfect* code is defined by the condition that its packing radius is equal to its covering radius. The packing radius of an (n,M,d) code C over $\mathbf{F}_q$ is $\rho = [(d-1)/2]$, and C is perfect if and only if the Hamming spheres of radius ρ around the codewords are simultaneously a packing and a covering of $\mathbf{F}_q^n$. (There is no connection with the concept of a perfect quadratic form [Cas3, p. 281].)

Perfect codes were essentially classified by Tietäväinen and van Lint (see [Lin11], [Mac6, Chap. 6]). The list is as follows.

(i) Certain trivial codes (a code with one codeword, the universe code, or a binary repetition code of odd length).

(ii) Hamming codes, i.e. linear codes with $\rho = 1$ over any field $\mathbf{F}_q$, with parameters $n = (q^m - 1)/(q-1)$, $k = n - m$ and $d = 3$, for $m \geqslant 2$.

(iii) Nonlinear codes with the same parameters as Hamming codes (these codes have not been completely enumerated).

(iv) The binary [23, 12, 7] Golay code $\mathscr{C}_{23}$ and the ternary [11, 6, 5] Golay code $\mathscr{C}_{11}$.

Table 3.2. Complete weight enumerator for both the quadratic residue and Pless [48, 24, 15] ternary codes.

A_{ijk}	i	j	k	Number of terms
1	48	0	0	3
17296	33	12	3	6
190256	33	9	6	6
190256	30	15	3	6
4280760	30	12	6	6
11225104	30	9	9	3
951280	27	18	3	6
38621968	27	15	6	6
209281600	27	12	9	6
94	24	24	0	3
2092816	24	21	3	6
138220984	24	18	6	6
1343397616	24	15	9	6
2777932180	24	12	12	3
210423136	21	21	6	3
3333094864	21	18	9	6
12362073856	21	15	12	6
20145351688	18	18	12	3
36133419520	18	15	15	3

2.10 The Pless double circulant codes. Pless ([Ple9], [Ple11], [Bla7], [Bla8], [Ito1], [Mac6, Chap. 16] [Mal2]) has constructed a $[2p+2, p+1]$ ternary code S_{2p+2} whenever p is a prime $\equiv -1$ (modulo 6). Let $B = (b_{ij})$ be a $(p+1) \times (p+1)$ matrix with rows and columns labeled ∞, 0, 1, ..., $p-1$, where

$$b_{\infty\infty} = 0,\ b_{\infty i} = 1,\ b_{ii} = 0,$$

$$b_{i\infty} = 1 \quad \text{if} \quad p \equiv 1 \pmod 4,\ b_{i\infty} = -1 \quad \text{if} \quad p \equiv -1 \pmod 4,$$

$$b_{ij} = 1 \quad \text{if} \quad j - i \quad \text{is a square mod } p \ (i \neq j),$$

$$b_{ij} = -1 \quad \text{if} \quad j - i \quad \text{is a nonsquare mod } p \ (i \neq j).$$

Then S_{2p+2} has generator matrix $[I\,B]$. For example the generator matrix for $p = 5$ is shown in (73), and S_{12} coincides with the Golay code $\mathscr{C}_{12}$. The next five examples are [24, 12, 9], [36, 18, 12], [48, 24, 15], [60, 30, 18] and [84, 42, $\leqslant$ 21] codes. The c.w.e. of S_{48} is given in Table 3.2. The c.w.e.'s of S_{24}, S_{36} and S_{60} are given in [Mal2]. S_{24} and S_{48} have the same c.w.e.'s as the corresponding quadratic residue codes but different automorphism groups.

The following properties of these codes will be used later. (i) S_{2p+2} is a self-dual (Type III) code containing the all-ones vector. (ii) The numbers of -1's, 0's or $+1$'s in any codeword are all multiples of 3. (iii) The codewords of maximal weight contain among them the rows of a Hadamard matrix and its negative (cf. §2.13), and for S_{12}, S_{24} and S_{48} these are the only codewords of maximal weight. (iv) For S_{12}, S_{24} and S_{48} the codewords containing only 0's and $+1$'s are of the types 0^n, 1^n and $0^{n/2}1^{n/2}$.

2.11 Goppa codes and codes from algebraic curves. In recent years it has been shown that very good codes may be obtained from algebraic curves, although we shall not make use of these codes in this book. See for example [Gop1]-[Gop5], [Hir2], [Iha1], [Kat2], [Lac1], [Len2], [Lit4], [Man1], [Mor7], [Ser3], [Tsf1], [Tsf2], [Vla1]-[Vla3], [Zin3].

2.12 Nonlinear codes. Many nonlinear codes are known which contain more codewords than the best linear codes of the same length and minimal distance. Some examples are:

(i) Codes obtained from Hadamard (§2.13) or conference matrices [Mac6, pp. 44, 55], [Slo18].

(ii) Nonlinear single-error-correcting codes, such as Golay's (9, 20, 4) code, Best's (10, 40, 4) code and Julin's (12, 144, 4) code. For these and other single-error-correcting codes see Chap. 5, [Bes1], [Gol2], [Jul1], [Kat1], [Lit1], [Mac6, Chap. 2], [Rom1], [Slo20].

(iii) The (16, 256, 6) Nordstrom-Robinson code [Mac6, p. 73], [Nor1], [Sem1], and the Kerdock and Preparata codes and their generalizations [Kan4], [Ker1], [Mac6, Chap. 15], [Pre1].

2.13 Hadamard matrices. A *Hadamard matrix* H of order n is an $n \times n$ matrix of $+1$'s and -1's such that

$$H\,H^{tr} = nI\,. \tag{76}$$

Multiplying any row or column by -1 changes H into another Hadamard matrix. By this means we can change the first row and column of H into $+1$'s. Such a Hadamard matrix is called *normalized*. If a Hadamard matrix of order n exists then n is 1, 2 or a multiple of 4. Conversely it is believed that Hadamard matrices exist whenever the order is a multiple of

4, although this has not been proved. A large number of constructions are known, and the smallest order for which a Hadamard matrix has not yet been constructed is (in 1987) 428. Connections with coding theory are described in [Mac6, Chap. 2, §3]. For this purpose one often changes H to a binary matrix by replacing +1's by 0's and −1's by 1's. For further information see [Hal1]-[Hal3], [Har8], [Hed1], [Ito2], [Leo1], [Lev1], [Lin10], [Lon1], [Lon2], [Mac6], [Saw1], [Ton1], [Tur1], [Wal2].

3. t-Designs, Steiner systems and spherical t-designs

3.1 t-Designs and Steiner systems. t-designs are a class of constant weight codes that satisfy conditions similar to (but weaker than) those satisfied by perfect codes. Let X be a v-set (i.e. a set with v elements) whose elements are called *points*. A *t-design* is a collection of distinct k-subsets (called *blocks*) of X with the property that any t-subset of X is contained in exactly λ blocks. This is also called a $t-(v,k,\lambda)$ design. If $\lambda = 1$ a t-design is called an $S(t,k,v)$ *Steiner system*. The terminology of the subject comes from the original application of such designs in agricultural experiments, although they are now used in all branches of science [Dia3].

If $X = \{1,2, \ldots, v\}$, each block $B \subseteq X$ may be represented by its *indicator* or *characteristic vector* $(c_1, \ldots, c_v)$, where $c_i = 1$ if $i \in B$, $c_i = 0$ if $i \notin B$. In this way a t-design becomes a binary code of length v in which every codeword has weight k. Usually there is no need to distinguish between a block and its indicator vector.

The parameters of a t-design must satisfy certain arithmetical conditions. Let $P_1, \ldots, P_t \in X$ be distinct points. For $1 \leqslant i \leqslant t$ the number of blocks containing $P_1, \ldots, P_i$ is

$$\lambda_i = \lambda \binom{v-i}{t-i} \Big/ \binom{k-i}{t-i}, \tag{77}$$

and the total number of blocks is

$$b = \lambda \binom{v}{t} \Big/ \binom{k}{t}. \tag{78}$$

Each point belongs to $r (=\lambda_1)$ blocks, where

$$bk = vr, \tag{79}$$

$$\lambda_2 (v-1) = r(k-1) \quad \text{if } t \geq 2. \tag{80}$$

Given t, k and v one wishes to find a design with the smallest λ, i.e. with the smallest number of blocks. Of course a necessary condition for a design to exist is that the right-hand sides of (77) and (78) are integers. In some cases these conditions are also sufficient (e.g. for Steiner systems $S(2,3,v)$, $S(2,4,v)$, $S(2,5,v)$ and $S(3,4,v)$), but not always. In general the precise conditions under which t-designs exist are not known. For further information about t-designs see [Ass2]-[Ass4], [Cam2], [Hal3],

[Han1], [Han2], [Hug2], [Hug3], [Jac1], [Lan5], [Lin3], [Mac6, Chap. 2], [Rag1], [Ton4], [Wil18], [Wil19].

Many t-designs may be constructed by taking the blocks to be the codewords of a particular weight, often the minimal weight, in a good binary error-correcting code (see especially the Assmus-Mattson theorem, Theorem 22 of Chap. 7). Conversely, good codes can sometimes be constructed as the linear span of the blocks of a t-design. We may also obtain t-designs from nonbinary codes by taking the blocks to be the locations of the nonzero components of the codewords of a particular Hamming weight; i.e. the blocks are the distinct *supports* of the codewords of a particular weight. In the nonbinary case it is much harder to recover the code from the design.

Examples. (i) The seven codewords of weight 3 in the Hamming code $\mathcal{H}_7$ form a Steiner system $S(2,3,7)$, and the 14 codewords of weight 4 in $\mathcal{H}_8$ form an $S(3,4,8)$.

(ii) More generally the codewords of weight 3 in any Hamming code $\mathcal{H}_n$, $n = 2^m - 1$, form an $S(2,3,2^m - 1)$, and the codewords of weight 4 in any extended Hamming code of length 2^m form an $S(3,4,2^m)$.

(iii) The supports of the codewords of weight 5 in the ternary Golay code $\mathcal{C}_{11}$ form an $S(4,5,11)$, and the supports of the words of weight 6 in $\mathcal{C}_{12}$ form an $S(5,6,12)$. The 132 blocks of this design are called *hexads* (see Chap. 11).

(iv) The codewords of weight 7 in the binary Golay code $\mathcal{C}_{23}$ form an $S(4,7,23)$, and the codewords of weight 8 in $\mathcal{C}_{24}$ form an $S(5,8,24)$. These 759 blocks (or the corresponding codewords) are called *octads* (see Chap. 11).

Many other examples will be found in the references cited above. Until recently $t = 5$ was the largest t for which a t-design was known ([Ass3], [Den1], [Mil4], [Wit2], [Wit3]), but in 1983 Magliveras and Leavitt constructed 6-designs [Mag1], [Mag2], and Teirlinck [Tei1] has now shown that t-designs exist for all t.

3.2 Spherical t-designs. Just as t-designs are a special class of constant weight codes, so spherical t-designs are a special class of spherical codes (cf. §2.3 of Chap. 1). The original motivation for studying these objects came from the numerical evaluation of multi-dimensional integrals. The integral of a polynomial function over the sphere Ω_n may be approximated by its average value at the code points; if the code is a spherical t-design, the approximation is exact for polynomials of degree $\leqslant t$. Formally, a *spherical t-design X* is a finite subset of Ω_n with the property that

$$\int_{\Omega_n} f(\xi) d\omega(\xi) = \frac{1}{|X|} \sum_{x \in X} f(x) \tag{81}$$

for all polynomials f of degree $\leqslant t$, where $\omega(\xi)$ is Lebesgue measure on the sphere, scaled so that the total measure of Ω_n is 1. (It would require too much of a digression to explain why these are called t-designs — see [Del11], [Del16].)

Many examples of spherical t-designs may be found by using the vectors of a particular norm, often the minimal norm, in a good lattice packing. For example the 240 minimal vectors in the E_8 lattice (when re-scaled so as to lie on Ω_8) form a spherical 7-design, and the 196560 minimal vectors in the Leech lattice form a spherical 11-design. The appropriate value of t for such designs may sometimes be determined by using theorems of Sobolev (see §4.2) or Venkov (see Chap. 7, §7).

Given n and t one wishes to find a spherical t-design with the smallest number of points. Delsarte, Goethals and Seidel [Del16] show that the number of points satisfies

$$|X| \geqslant \binom{n+s-1}{n-1} + \binom{n+s-2}{n-1}, \quad \text{for } t = 2s, \tag{82}$$

$$|X| \geqslant 2\binom{n+s-1}{n-1}, \quad \text{for } t = 2s+1. \tag{83}$$

If equality holds in either (82) or (83) the spherical t-design is said to be *tight*. Very few tight t-designs exist, and in particular Bannai and Damerell [Ban7], [Ban8] have shown that, for $n \geqslant 3$, there are no tight $(2s)$-designs with $2s \geqslant 6$ and no tight $(2s+1)$-designs with $2s+1 \geqslant 9$ except for the just-mentioned 11-design obtained from the Leech lattice. (The latter is also known to be unique — see Chap. 14.) For generalizations of these results to other spaces see Chap. 9 and [Ban9], [Ban10], [Hog1]-[Hog3], [Neu2].

On the other hand Seymour and Zaslavsky [Sey1] have shown that (non-tight) spherical t-designs exist in Ω_n for all values of n and t. For further information about spherical t-designs see [Ban1]-[Ban13], [Del11], [Del15], [Del16], [Dun5], [Goe5], [Goe6], [Hon1], [Sob2], [Sob3].

4. The connections with group theory

4.1 The automorphism group of a lattice. First of all, lattices give rise to groups, their symmetry groups. The *automorphism group* (or *symmetry group*) $\mathrm{Aut}(\Lambda)$ of a lattice Λ is the set of distance-preserving transformations (or *isometries*) of the space that fix the origin and take the lattice to itself. For a lattice in ordinary Euclidean space, $\mathrm{Aut}(\Lambda)$ is finite and the transformations in $\mathrm{Aut}(\Lambda)$ may be represented by orthogonal matrices. Let Λ have generator matrix M. Then an orthogonal matrix B is in $\mathrm{Aut}(\Lambda)$ if and only if there is an integral matrix U with determinant ± 1 such that

$$UM = MB \tag{84}$$

(cf. Eq. (22) of Chap. 1). This implies $U = MBM^{tr} A^{-1}$, where A is the Gram matrix. On the other hand the set of integral matrices U for which there exists an orthogonal matrix B satisfying (84) forms an *integral representation* of Aut(Λ).

For example, the automorphism group of the hexagonal lattice in Fig. 1.3a is a dihedral group of order 12, generated by a rotation through 60° and a reflection in a line joining the center of two spheres. The automorphism group of the simple three-dimensional cubic lattice contains all permutations and sign changes of the three coordinates, and has order $2^3\, 3! = 48$. A lattice and its dual have the same automorphism group.

On some occasions we consider the infinite group of *all* distance-preserving transformations of the underlying space (the *affine automorphisms*) that take the lattice to itself. This is obtained by adjoining the translations in lattice vectors to Aut(Λ).

The automorphism groups Aut(Λ) of the lattices in Table 1.1 in up to 8 dimensions are especially interesting: they have subgroups of low index that are reflection groups (or Coxeter groups) — see §2 of Chap. 4. In higher dimensions other remarkable groups appear. For example the automorphism group of the 24-dimensional Leech lattice is the group Co_0 (or $\cdot 0$) of order 8315553613086720000, which will be studied in considerable detail in Chaps. 10 and 11. The corresponding infinite group (obtained by adjoining translations by lattice vectors) is called Co_∞ (or $\cdot\infty$). Actually Co_0 is interesting for other reasons than its size — the 24-dimensional lattice D_{24} has a much larger group, of order

$$2^{24} \cdot 24! = 10409396852733332453861621760000 .$$

The interest of Co_0 lies in its connection with the classification of finite groups. Any finite group can be built up out of certain special groups called *simple* groups, and the classification of all finite simple groups is a project that was completed in 1982 after occupying the attention of a great many mathematicians for over fifty years (see for example [Gor2]-[Gor5]). When Co_0 was discovered in 1968 it produced three new simple groups (Co_1, Co_2 and Co_3 — see [Con2], [Con3] and Chap. 10). Furthermore the largest of the "sporadic" simple groups (and one of the last to be found), the Friendly Giant or Fischer-Griess "Monster" simple group, of order

$$808017424794512875886459904961710757005754368000000000$$

was constructed by R. L. Griess in 1981 using the Leech lattice ([Gri4], [Gri5]). A simplified construction is given in Chap. 29.

The automorphism group of a lattice in a Lorentzian space is usually infinite, the even unimodular lattice $II_{25,1}$ being a notorious case — see Chaps. 26-28. One of the surprises is that even those groups have been illuminated by studying the Leech lattice.

Error-correcting codes (§2) and t-designs (§3) also give rise to groups. The automorphism group of a code was defined in §2.2, and similarly the automorphism group of a t-design is the set of all permutations of the points that takes the set of blocks into itself.

We follow the ATLAS conventions [Con16, p. xx], and write $A \times B$ for a direct product; $A.B$ or AB for a group with a normal subgroup isomorphic to A, for which the corresponding quotient group is isomorphic to B; $A{:}B$ for the case of $A.B$ which is a split extension or semi-direct product; and $A{\cdot}B$ for the case when $A.B$ is not a split extension.

As general references for groups we mention [Coh1], [Coh2], [Con16], [Cox20], [Cox21], [Cox28], [Gor2]-[Gor4], [Hup1], [She2].

4.2 Constructing lattices and codes from groups. We may also take the opposite point of view, beginning with a finite group G, and attempting to construct objects of various kinds of which G is the automorphism group. Three of the most important constructions are the following.

(i) *The construction of lattices from integral representations.* Given a representation of a group G by integral matrices U, it is possible to construct a lattice invariant under G in the following way. We let the group act on quadratic forms $f(\xi) = \xi A \xi^{tr}$ by sending A to UAU^{tr}, $f(\xi)$ to $f(\xi U)$ (so that U multiplies the generator matrix of the lattice on the left, as in Eq. (22) of Chap. 1 and Eq. (84)). Then if $f(\xi)$ is an arbitrary positive definite quadratic form, the quadratic form

$$\sum_{U \in G} f(\xi U)$$

is invariant under G, as is the corresponding lattice. For instance, the group of order 3 generated by $\begin{pmatrix} 0 & 1 \\ -1 & -1 \end{pmatrix}$ permutes the three quadratic forms

$$\xi_1^2 + \xi_2^2, \ \ \xi_1^2 + 2\xi_1\xi_2 + 2\xi_2^2, \ \ 2\xi_1^2 + 2\xi_1\xi_2 + \xi_2^2$$

and so preserves their sum $4(\xi_1^2+\xi_1\xi_2+\xi_2^2)$, which is a quadratic form for the hexagonal lattice (Eq. (25) of Chap. 2). In this example the resulting lattice is actually invariant under a larger group than we started with (the full group has order 12, as we saw in the previous section). The study of the integral representations of a given group G is to a large extent the study of lattices invariant under that group. (See [Bro10]-[Bro13], [Bül1], [Con42], [Cra1], [Cra2], [Dad1], [Fei1], [Gud1], [Gus1], [Hem1], [Ple1]-[Ple5], [Rei2]-[Rei4], [Rog10], [Shu1], [Shu2], [Tho3].)

(ii) *The construction of codes from permutation representations.* Similarly, given a representation of group elements by permutations, we can work modulo 2 and obtain a representation of G on a vector space V over the field of order 2. The invariant subspaces (the subspaces of V taken into themselves by every group element) are then all the binary codes C for which G is a subgroup of Aut(C). Similar methods produce codes over arbitrary fields. This technique has been used in [Bro0]-[Bro2], [Cal9], [Kna1], [Par1]. Important information about these codes can be obtained from the theory of modular representations of groups.

Calderbank and Wales [Cal9] used this idea to construct a binary code of length 176 and dimension 22 whose automorphism group is the Higman-Sims group. Brooke [Bro0]-[Bro2] has found all binary codes obtainable in this way from the primitive permutation representations of many interesting groups. See also [Kna1], [Phe1].

(iii) *The construction of spherical codes and designs from orthogonal representations.* Given a representation of group elements by $n \times n$ orthogonal matrices B, we may construct a spherical code $X \subset \Omega_n$ by choosing an initial point $x_0 \in \Omega_n$ and setting

$$X = \{ x_0 B : B \in G \}, \tag{85}$$

the *orbit* of x_0. These codes are sometimes called *group* codes, and have been investigated in [Bil1], [Bil3], [Bla4], [Bla6], [Bla9], [Dow1]-[Dow3], [Ing1], [Kar3], [Sle3], [Sle4].

There is a beautiful connection between these codes and invariant theory, first discovered by Sobolev. Let G be a finite group of $n \times n$ real or complex matrices B. The elements of G act on polynomials $f(x) = f(x_1, \cdots, x_n)$ by $B \circ f(x) = f(Bx^{tr})$. Polynomials for which $f(x) = f(Bx^{tr})$ for all $B \in G$ are said to be *invariant* under G, and the set of all invariant polynomials forms a ring R^G.

Now suppose G is a group of real orthogonal matrices. Of course the squared length $\phi(x) = x_1^2 + \cdots + x_n^2$ is invariant under G. Sobolev showed that if the degree of the first invariant that is not a polynomial in $\phi(x)$ is $t+1$, then every orbit under G is a spherical t-design (see [Sob2] and the other references on spherical t-designs given at the end of §3.2).

Examples. The automorphism group of the E_8 lattice, namely the Weyl group of type E_8, is a reflection group (Chap. 4, §2). The ring of invariants R^G has a basis consisting of homogeneous polynomials of degrees

$$2, 8, 12, 14, 18, 20, 24, 30, \tag{86}$$

where the invariant of degree 2 is $\phi(x) = x_1^2 + \cdots + x_8^2$ [Cox28, Table 10, p. 141]. It follows from Sobolev's theorem that every orbit under this group is a spherical 7-design. In particular the set of 240 minimal vectors in E_8 is a spherical 7-design, as we remarked in §3.2.

The ring of invariants of the automorphism group of the Leech lattice has been investigated in [Huf5], where it shown that the first invariant not a polynomial in ϕ has degree 12. Therefore every orbit under this group (for example the set of 196560 minimal vectors) is a spherical 11-design. These examples are studied further in Chaps. 13, 14. Venkov's theorem (Chap. 7, §7) generalizes these examples. We shall see other applications of invariant theory in Chap. 7.

4

Certain Important Lattices and Their Properties

J. H. Conway and N. J. A. Sloane

This chapter describes the properties of a number of important lattices, including the cubic lattice $\mathbf{Z}^n$, the root lattices A_n, D_n, E_6, E_7, E_8, the Coxeter-Todd lattice K_{12}, the Barnes-Wall lattice Λ_{16}, the Leech lattice Λ_{24}, and their duals. Among other things we give their minimal vectors, densities, covering radii, glue vectors, automorphism groups, expressions for their theta series, and tables of the numbers of points in the first fifty shells. We also include a brief discussion of reflection groups and of the technique of gluing lattices together.

1. Introduction

In this chapter we introduce a number of important lattices that will be used throughout the book, namely $\mathbf{Z}^n (n \geqslant 1)$, $A_n (n \geqslant 1)$, $D_n (n \geqslant 3)$, E_6, E_7, E_8, K_{12}, Λ_{16}, Λ_{24} and their duals, and give a summary of the basic properties of each lattice. Proofs of most of these results will be given in later chapters. Some of the reasons for the importance of these lattices can be found in Table 1.1 of Chap. 1: they include many of the best packings, coverings and quantizers, as well as the lattices with the highest kissing numbers, in dimensions up to 24. It is likely that they will also be good candidates for the solution of other problems the reader may have in mind in these dimensions.

The "root lattices" $\mathbf{Z}^n$, A_n, D_n, E_n are treated first, in §§5-8. The terminology arises because these lattices are generated by the root systems of certain Lie algebras, although we shall say very little about this connection (cf. [Bou1], [Hum1]). These lattices are also closely related to reflection groups, as we discuss in §2. Another reason the lattices $\mathbf{Z}^n$, A_n, D_n, E_n are the first to arise in many investigations is that any integral lattice generated by vectors of norms 1 and 2 is a direct sum of these lattices (Witt's theorem — see §3). There is no corresponding theory for larger norms. "Gluing theory" (§3) describes how component lattices may be combined to produce lattices in higher dimensions.

The remaining lattices treated in this chapter (in §§9-11) are K_{12}, Λ_{16} and Λ_{24}, all of which occur inside the Leech lattice Λ_{24}. In fact $\mathbf{Z} \cong A_1$, A_2, $A_3 \cong D_3$, D_4, D_5, E_6, E_7, E_8, Λ_{16} and Λ_{24} are all laminated lattices (see Eq. (25) of Chap. 1 and Chap. 6), and thus are found inside Λ_{24}.

Two important lattices not included in this chapter are the 48-dimensional self-dual lattices P_{48p} and P_{48q}. A summary of their properties will be found in Example 9 of Chap. 7.

The theta series of a lattice (§2.3 of Chap. 2) tells how many points there are at each distance from the origin, i.e. gives the number of points in each spherical layer or shell. The theta series of the lattices in this book can be expressed in terms of simpler functions such as the Jacobi theta series θ_2, θ_3 and θ_4. These are defined in §4 together with various other terms.

We have included tables of the first 50 or so theta coefficients of most of these lattices. Such tables are useful in studying the lattices, investigating the connections with number theory (§2.3 of Chap. 2), and because configurations of lattice points often form excellent codes (§2.3 of Chap. 1, §1 of Chap. 3). Individual shells form *spherical* or *constant energy codes*, while the set of lattice points lying inside or on a shell forms a *bounded energy code* or *spherical cluster*.

We also give brief descriptions of the Voronoi cells (or nearest neighbor regions) around the lattice points (§1.2 of Chap. 2); further information will be given in Chap. 21. Although much of the material in this chapter may be found elsewhere (for example [Bou1], [Con27], [Con28], [Con37], [Cox20], [Int1], [Nie2], [Slo12], [Slo19], [Teo2], [Teo3]), some of the theta series and tables are published here for the first time. (Some related tables are given in [Fra1], [Mit5], [New2], [Pry1].) Further information about the packings in two and three dimensions (including more extensive tables, both for the packings given here and other translates, plus tables of the vectors in the first few layers) will be found in [Slo17], [Slo19], [Teo2], [Teo3].

2. Reflection groups and root lattices

Figure 4.1 shows a kaleidoscope whose three mirrors (or walls) cut the sphere in a spherical triangle having angles $\pi/2$, $\pi/3$ and $\pi/5$. The reflections in these walls generate a group of order 120, called the reflection group [3,5]. The whole surface of the sphere is divided into 120 triangles, one for each group element.

This is an instance of a *finite*, or *spherical*, *reflection group*. (Reflection groups are often also called *Coxeter groups*.) In general such a group (if irreducible) is generated by reflections in the walls of a spherical simplex, all of whose dihedral angles are submultiples of π. (The reducible reflection groups are direct products of irreducible ones.) The

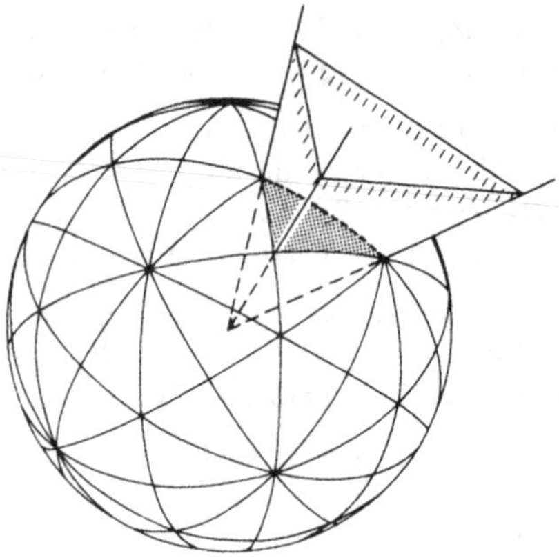

Figure 4.1. Kaleidoscope intersecting sphere in a spherical triangle. The whole surface of sphere is divided into 120 copies of this triangle.

infinite cone bounded by the reflecting walls or hyperplanes (i.e. the kaleidoscope) is a fundamental region for the reflection group. The group may be succinctly described by a *Coxeter-Dynkin diagram*, having one node for each wall, two nodes being joined by a line marked p when the corresponding walls are at an angle of π/p. Certain abbreviations are customarily adopted for lines labeled with small values of p, as shown in Fig. 4.2. For example the diagram for the preceding example is shown in Fig. 4.3.

If R_i is the reflection in the ith wall of the fundamental region then from the diagram we can read off a set of defining relations (or a *presentation*) for the group, namely

$$R_i^2 = (R_i R_j)^{p_{ij}} = 1 \quad (i,j=1,\ldots,n), \tag{1}$$

where π/p_{ij} is the angle between the ith and jth walls. A celebrated theorem of Coxeter [Cox4], [Cox28, §9.3] asserts that every finite group with a presentation of this form is a reflection group.

The classification of the finite reflection groups has a complicated history [Kan3]. The elegant complete enumeration of the finite reflection groups in terms of diagrams is due to Coxeter [Cox2], [Cox3], [Cox20]. However Mitchell [Mit1]-[Mit4] had already solved an essentially harder

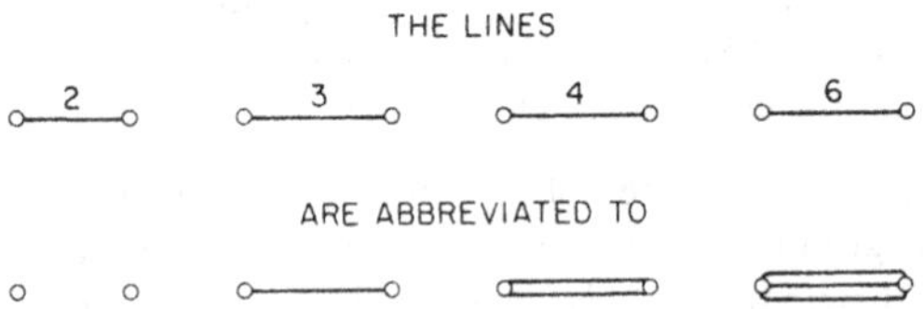

Figure 4.2. Conventions for Coxeter-Dynkin diagrams. (See also Chap. 27.)

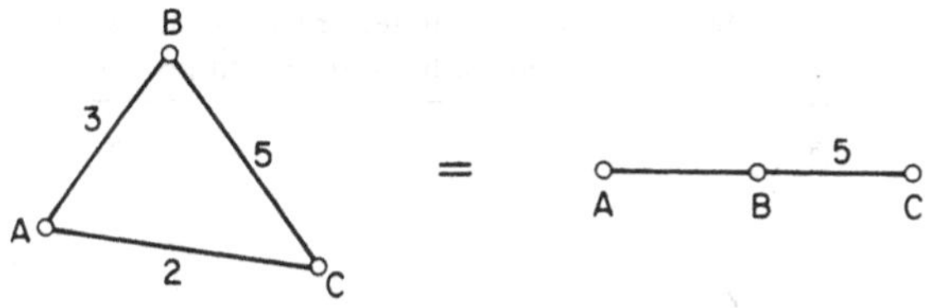

Figure 4.3. Diagram for reflection group [3,5].

problem. We shall concentrate on the *crystallographic* reflection groups, for which p takes only the values 2, 3, 4 and 6, since these are associated with lattices. They are given in the fourth column of Table 4.1. The only finite indecomposable non-crystallographic reflection groups are shown in Fig. 4.4.

We can specify each reflecting hyperplane by a vector perpendicular to it, called a *root vector* (or simply a *root*). The root vectors perpendicular to the walls of the fundamental region are the *fundamental roots* for the group, while the entire set of root vectors is called a *root system*. The entire root system is found by taking all images of the fundamental roots under the group. The lattice generated by the root system is called a *root lattice*, and the fundamental roots form an integral basis for the root lattice.

Conversely, given a lattice Λ, we define a *root* (*vector*) for Λ to be a vector $r \in \Lambda$ for which the associated reflection

$$x \rightarrow x - 2\,\frac{x \cdot r}{r \cdot r}\,r \tag{2}$$

is a symmetry of Λ. The group H generated by the reflections in the roots is the *reflection subgroup* of $\mathrm{Aut}(\Lambda)$. If Λ is integral and unimodular the roots are just the vectors of norm 1 or 2 in Λ, which are called *short* and *long* roots respectively. The reflecting hyperplanes corresponding to all the roots of Λ partition the space into fundamental regions for H. The roots corresponding to the walls of any one fundamental region are a set of *fundamental roots* for the lattice (and for H). If Λ is not a root lattice then in general the fundamental roots are not a basis for Λ.

All the indecomposable finite root systems, and so all the indecomposable crystallographic finite reflection groups, are shown in Table 4.1. The first column of the table is the usual name for the root system, the second column is our name for the root lattice generated by the roots, and the third column gives the determinant d of this lattice. The fourth column describes the fundamental roots, the reflections in which

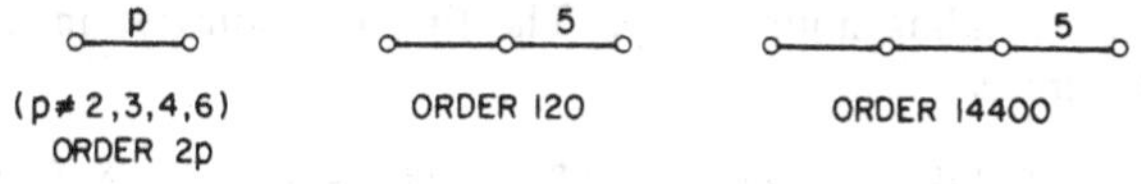

Figure 4.4. Indecomposable non-crystallographic reflection groups.

Table 4.1. Indecomposable finite root systems or equivalently, indecomposable crystallographic finite reflection groups.

ROOT SYSTEM	LATTICE	DET
A_n	A_n	$n+1$
B_n	Z^n	2
C_n	D_n	2
D_n	D_n	4
G_2	A_2	1
F_4	D_4	1
E_6	E_6	3
E_7	E_7	2
E_8	E_8	1

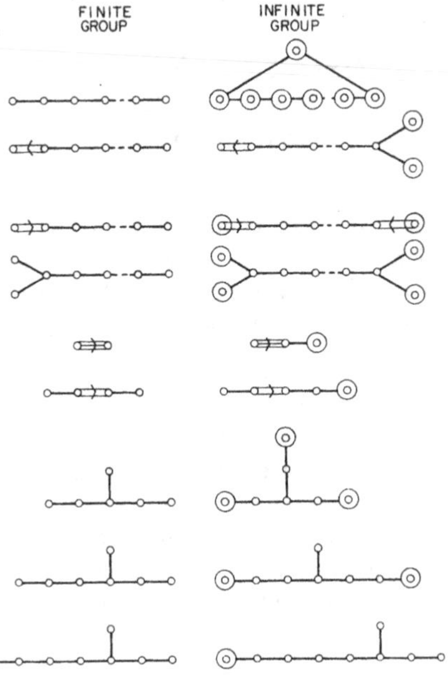

generate the corresponding finite reflection group. The right-most roots in the fourth column have norm 2. The lengths of the other roots are determined by the convention that a k-fold edge whose arrowhead (if any) points from r to s indicates that $N(r) = k\,N(s)$.

Note that the conventions of Fig. 4.2 imply that if roots r and s are joined by a k-fold edge then the angle θ between them is given by $4\cos^2\theta = k$ and $\cos\theta \leqslant 0$, i.e. by

k:	0	1	2	3
θ:	90°	120°	135°	150°

The final column contains the corresponding (infinite) *Euclidean reflection group*, sometimes called the *affine Weyl group*, obtained by adjoining the translations by root vectors. This is again a reflection group, whose diagram is obtained by adjoining one more node (called the *extending node*) to the diagram for the finite group. If we had started from the diagram for the infinite group, there would be exactly d nodes, called the *tips* or *special* nodes, whose removal would leave a diagram for the corresponding finite group. The tips are indicated in the diagrams by double circles.

The product of all the generating reflections is called a *Coxeter element*, and all the Coxeter elements have the same order h, the *Coxeter number*. The Coxeter number has many other interpretations. For example the total number of roots is nh. Thus for the lattices A_n, D_n and E_n (when all roots have the same length), nh is the kissing number.

Note on B_n, C_n, G_2 and F_4. We explain why only A_n, D_n and E_n (but not B_n, C_n, G_2 or F_4) appear in the rest of the book. The fundamental root systems for the Lie algebras B_n and C_n may be taken as

$$\begin{array}{cccc} & B_n & & \\ 1 & 0 & 0 & 0 \ldots \\ -1 & 1 & 0 & 0 \ldots \\ 0 & -1 & 1 & 0 \ldots \\ & \ldots & & \end{array} \qquad \begin{array}{cccc} & C_n & & \\ 2 & 0 & 0 & 0 \ldots \\ -1 & 1 & 0 & 0 \ldots \\ 0 & -1 & 1 & 0 \ldots \\ & \ldots & & \end{array}$$

Since these differ only by scalar factors, the corresponding *finite* reflection groups are identical. However the root lattices generated by these roots (which are respectively $\mathbf{Z}^n$ and D_n) are different, as are the infinite groups obtained by adjoining the translations in these lattices to the finite groups. Similar phenomena occur in the cases G_2 and F_4. It may be observed that every root lattice is generated by its roots of shortest length. Thus our list of lattices includes A_n, D_n and E_n, which are root lattices for which all roots have the same length, but not B_n, C_n, G_2 and F_4. In fact the root system for G_2 consists of the vectors of the first two norms (2 and 6) in the root lattice A_2. Similarly the root system for F_4 consists of the first two layers (of norms 2 and 4) in the root lattice D_4.

For further information about reflection groups, crystallographic groups (including the Bieberbach theorems) and related topics see [Aus1], [Bie1], [Bou1], [Bus1], [Cox20], [Cox28], [Del7], [Fri2], [Gro3], [Haz1], [Hil3], [Hum1], [Oli2], [Vin1]-[Vin15], [Zas1].

3. Gluing theory

Gluing theory is a way to describe the general n-dimensional integral lattice L that has a sublattice which is a direct sum

$$L_1 \oplus L_2 \oplus \cdots \oplus L_k$$

of given integral lattices $L_1, \ldots, L_k$ of total dimension n. The typical vector of L can be written

$$y = y_1 + y_2 + \cdots + y_k, \tag{3}$$

where each component y_i is in the subspace spanned by L_i, although not necessarily in L_i itself. What are the possibilities for $y_1, \ldots, y_k$?

The inner product of y_i with any vector of L_i is an integer, since it is the same as the inner product of y with that vector. This shows that y_i must be a member of the dual lattice L_i^*.

Plainly any y_i can be altered by adding a vector of L_i, so we may suppose that y_i is one of a standard system of representatives for the cosets of L_i in L_i^*. These representatives are called the *glue vectors* for L_i. It is usual to choose the glue vectors to be of minimal length in their cosets. The quotient group L_i^*/L_i is called the *dual quotient*, or *glue group* for L_i. As remarked in §2.4 of Chap. 2, its order is equal to det L_i.

So each possible lattice L is generated by $L_1 \oplus \cdots \oplus L_k$ together with certain vectors (3), where each y_i is a glue vector for L_i, and we need only check that the various vectors (3) have integral inner products with each other and are closed under addition modulo $L_1 \oplus \cdots \oplus L_k$. We describe such a lattice L informally by saying that the *components* $L_1, \ldots, L_k$ have been *glued* together by the glue vectors (3). It may happen that there is a glue vector (3) in which only one y_i is nonzero, when we say that the component L_i has *self-glue*, and that y is a *self-glue* vector. For example D_n is self-glued to form D_n^+.

There is an important special case.

Theorem 1. *If a unimodular lattice L is formed by gluing together two lattices L_1 and L_2 in such a way that there is no self-glue, i.e. if*

$$L_1 = (L_1 \otimes \mathbf{R}) \cap L, \quad L_2 = (L_2 \otimes \mathbf{R}) \cap L,$$

then the dual quotients L_1^*/L_1 and L_2^*/L_2 are isomorphic groups.

The isomorphism is given by $y_1 + L_1 \to y_2 + L_2$ whenever there is a glue vector $y = y_1 + y_2$.

There is a standard choice of glue vectors $[0], [1], \ldots, [d-1]$ for each root lattice of determinant d, and we use $[a_1 \cdots a_k]$ as an abbreviation for the glue vector (3) in which $y_1 = [a_1]$ for L_1, $y_2 = [a_2]$ for $L_2, \ldots$. (There is one glue vector for each tip node.)

In our applications of gluing theory it will usually be obvious that all automorphisms of L will permute the lattices $L_1, \ldots, L_k$ (often because $L_1 \oplus \cdots \oplus L_k$ will be the part of L generated by vectors of norms 1 and 2). In these circumstances there is a simple description of the automorphism group $G(L)$ of L.

The group of all permutations of the L_i that arise from automorphisms in $G(L)$ we shall call $G_2(L)$. It is isomorphic to the quotient group $G(L)/G_{01}$, where G_{01} consists of just those automorphisms that give the trivial permutation.

Let $G_0(L)$ be the normal subgroup of G_{01} consisting of those automorphisms which, for every i, send each glue vector y_i into a vector in the same coset $y_i + L_i$, i.e. which fix the glue vectors modulo the components. Then $G_{01}/G_0(L)$ is isomorphic to a permutation group acting on the glue vectors of each component: we call this permutation group $G_1(L)$. Thus the full group $G(L)$ is compounded of the groups $G_0(L)$, $G_1(L)$, $G_2(L)$, and has order

$$g(L) = g_0(L) g_1(L) g_2(L), \tag{4}$$

where $g_i(L)$ is the order of $G_i(L)$. Also $G_0(L)$ is the direct product of the groups $G_0(L_i)$. But in general $G_1(L)$ is only a subgroup of the direct product of the $G_1(L_i)$ and therefore must be computed directly for each L.

For example there is a unique 17-dimensional unimodular lattice with no vector of norm 1 (see Chap. 16). It is obtained from $A_{11} \oplus E_6$ by adjoining the glue vector

$$[2,1] = \left[\left(\frac{2}{12}^{10}, \frac{-10}{12}^{2}\right); \left(0, \frac{2}{3}^{2}, \frac{-1}{3}^{4}, 0\right)\right]$$

which is the concatenation of the glue vectors $y_1 = [2]$ for A_{11} and $y_2 = [1]$ for E_6. For this lattice G_0 is the product of the Weyl groups $G_0(A_{11})$ and $G_0(E_6)$; G_1 has order 2 (we can simultaneously negate y_1 and y_2); while G_2 is plainly trivial (no automorphism can interchange A_{11} with E_6).

Witt's theorem [Wit5], [Kne5] tells us that, for any integral lattice L, the sublattice generated by vectors of norms 1 and 2 is a direct sum of root lattices. Thus the root lattices are particularly appropriate choices for the L_i. For a root lattice X_n, G_0 is the Weyl group (the subgroup generated by the reflections in the minimal vectors of X_n), and G_1 is the symmetry group of the (ordinary) Coxeter-Dynkin diagram (the so-called graph automorphism group). The symmetry group of the extended Coxeter-Dynkin diagram has order $g_1 d$.

The outstanding application of gluing theory is provided by Niemeier's list of even unimodular 24-dimensional lattices — see §3 of Chap. 16. Many other examples are also given in Chaps. 16, 17. Gluing theory has also proved useful in coding theory [Con21], [Con24], [Leo7], [Leo8], [Ple18].

4. Notation; theta functions

In the following sections the subscript n on the symbol for a lattice L usually indicates its dimension, M is a generator matrix for the lattice, $A = MM^{tr}$ is a Gram matrix, det is the determinant (the square of the volume of a Voronoi cell), τ is the kissing number, h is the Coxeter number (see §2), ρ is the packing radius, R is the covering radius, $\Delta = V_n \rho^n/\sqrt{\det}$ is the density, $\delta = \Delta/V_n$ is the center density and $\Theta = V_n R^n/\sqrt{\det}$ is the thickness, where $V_n = \pi^{n/2}/(n/2)!$ is the volume of a sphere of radius 1 (see Chap. 1). Diamond brackets are sometimes used to indicate the generators of a lattice, as in $A_2 = \langle(1,-1,0), (0,1,-1)\rangle$. The symbol $\oplus$ indicates a direct sum.

The glue group is L^*/L, of order $\det L$, and the glue vectors (representatives for the cosets of L in L^*) are indicated by square brackets $[i]$. The automorphism group $G(L)$ has order $g = g_0 g_1$, where g_0, g_1 are the orders of the groups $G_0(L)$, $G_1(L)$ (see §2).

We use obvious abbreviations for the components of vectors. For example the glue vector

$$[1] = \left(\frac{1}{4}, \frac{1}{4}, \frac{1}{4}, \frac{1}{4}, \frac{1}{4}, \frac{1}{4}, -\frac{3}{4}, -\frac{3}{4}\right)$$

for E_7 would be abbreviated to

$$\left\{ \left(\frac{1}{4} \right)^6 \left(-\frac{3}{4} \right)^2 \right\}.$$

The norm of a vector x is its squared length $x \cdot x$. The minimal norm of L is $\{\min x \cdot x : x \in L, x \neq 0\}$, which is $4\rho^2$. The theta series of L is

$$\Theta_L(z) = \sum_{x \in L} q^{x \cdot x} = \sum N(m) q^m, \quad q = e^{\pi i z}, \ \mathrm{Im}(z) > 0, \tag{5}$$

where $N(m)$ (sometimes written N_m) is the number of vectors in L of norm m (see §2.3 of Chap. 2). Thus the coefficients in the expansion of $\Theta_L(z)$ in powers of q give the numbers of vectors in the successive shells of L around the origin. We also use Eq. (5) to define the theta series of a nonlattice packing L (for example the translate of a lattice).

4.1 Jacobi theta functions. The theta series of the lattices encountered in this book can be expressed in terms of the Jacobi theta function

$$\theta_3(\xi \mid z) = \sum_{m=-\infty}^{\infty} e^{2mi\xi + \pi i z m^2}, \quad \mathrm{Im}(z) > 0. \tag{6}$$

(References are given at the end of this section.) For many purposes it is enough to work with the simpler theta functions $\theta_1'(z)$, $\theta_2(z)$, $\theta_3(z)$, $\theta_4(z)$ given by

$$\theta_1'(z) = \theta_1'(0 \mid z) = \sum_{m=-\infty}^{\infty} (-1)^m \, (2m+1) q^{(m+1/2)^2}$$

$$= 2q^{1/4} - 6q^{9/4} + 10q^{25/4} - 14q^{49/4} + \cdots$$

$$= 2q^{1/4}(1 - 3q^2 + 5q^6 - 7q^{12} + 9q^{20} - \cdots), \tag{7}$$

$$\theta_2(z) = e^{\pi i z/4} \, \theta_3\left(\frac{\pi z}{2} \,\Big|\, z \right) = \sum_{m=-\infty}^{\infty} q^{(m+1/2)^2}$$

$$= 2q^{1/4} + 2q^{9/4} + 2q^{25/4} + \cdots$$

$$= 2q^{1/4}(1 + q^2 + q^6 + q^{12} + q^{20} + \cdots), \tag{8}$$

$$\theta_3(z) = \theta_3(0 \mid z) = \sum_{m=-\infty}^{\infty} q^{m^2}$$

$$= 1 + 2q + 2q^4 + 2q^9 + \cdots, \tag{9}$$

$$\theta_4(z) = \theta_3\left(\frac{\pi}{2} \middle| z\right) = \theta_3(z+1) = \sum_{m=-\infty}^{\infty} (-q)^{m^2}$$

$$= 1 - 2q + 2q^4 - 2q^9 + \cdots, \tag{10}$$

where $q = e^{\pi i z}$. Other useful functions are

$$\psi_k(z) = e^{\pi i z/k^2}\theta_3\left(\frac{\pi z}{k} \middle| z\right) = \sum_{m=-\infty}^{\infty} q^{(m+1/k)^2}, \tag{11}$$

for $k = \pm 1, \pm 2, \ldots$, and

$$\phi_0(z) = \theta_2(2z)\theta_2(6z) + \theta_3(2z)\theta_3(6z)$$
$$= \tfrac{1}{2}\{\theta_3(z/2)\theta_3(3z/2) + \theta_4(z/2)\theta_4(3z/2)\}$$
$$= 1 + 6q^2 + 6q^6 + 6q^8 + 12q^{14} + \cdots, \tag{12}$$
$$\phi_1(z) = \theta_2(2z)\theta_3(6z) + \theta_3(2z)\theta_2(6z)$$
$$= \tfrac{1}{2}\theta_2(z/2)\theta_2(3z/2)$$
$$= 2q^{1/2} + 2q^{3/2} + 4q^{7/2} + 2q^{9/2} + \cdots. \tag{13}$$

These functions are related by a labyrinth of identities. The deepest of these is due to Poisson (1827) and Jacobi (1828) — see [Wht1, p. 475], [Bel2, p. 4]:

$$\theta_3(\xi \mid z) = (-iz)^{-1/2} e^{\xi^2/\pi i z}\, \theta_3\left(\frac{\xi}{z} \middle| -\frac{1}{z}\right) \tag{14}$$

(where the square root is the principal value). This implies

$$\theta_1'(-1/z) = (z/i)^{3/2}\,\theta_1'(z), \tag{15}$$

$$\theta_2(-1/z) = (z/i)^{1/2}\,\theta_4(z), \tag{16}$$

$$\theta_3(-1/z) = (z/i)^{1/2}\,\theta_3(z), \tag{17}$$

$$\phi_0(-1/z) = \frac{z}{i\sqrt{3}}\,\phi_0(z/3), \tag{18}$$

and Jacobi's formula for the theta series of the dual lattice:

$$\Theta_{\Lambda^*}(z) = (\det \Lambda)^{1/2}\,(i/z)^{n/2}\,\Theta_\Lambda(-1/z). \tag{19}$$

One may regard Eq. (14)-(19), as well as the MacWilliams identity for weight enumerators of codes (Eq. (50) of Chap. 3), as consequences of the general version of the Poisson summation formula, which states that the sum of a function over a linear space is equal to the sum of the Fourier transform of the function over the dual space [Dym1, Chap. 2, §11.3], [Igu1, p. 44], [Loo1, p. 153], [Ser1, Chap. VII, §6].

Other useful identities, arranged roughly by degree, and to show the symmetry[(1)] between θ_2, θ_3 and θ_4, are:

$$\theta_2(z+1) = \sqrt{i}\,\theta_2(z),\ \ \theta_3(z+1) = \theta_4(z),\ \ \theta_4(z+1) = \theta_3(z), \tag{20}$$

$$\theta_2(-1/z) = (z/i)^{1/2}\,\theta_4(z),\ \ \theta_3(-1/z) = (z/i)^{1/2}\,\theta_3(z),$$

$$\theta_4(-1/z) = (z/i)^{1/2}\,\theta_2(z), \tag{21}$$

$$\theta_2(z) + \theta_3(z) = \theta_3(z/4),\ \ -\theta_2(z) + \theta_3(z) = \theta_4(z/4), \tag{22}$$

$$\theta_3(z) + \theta_4(z) = 2\theta_3(4z),\ \ \theta_3(z) - \theta_4(z) = 2\theta_2(4z), \tag{23}$$

$$\theta_2(z)\theta_3(z) = \tfrac{1}{2}\theta_2(z/2)^2,\ \ \theta_3(z)\theta_4(z) = \theta_4(2z)^2,$$

$$\theta_2(z)\theta_4(z) = \frac{1}{2\sqrt{i}}\,\theta_2\left(\frac{z+1}{2}\right)^2, \tag{24}$$

$$\theta_2(z)^2 + \theta_3(z)^2 = \theta_3(z/2)^2,\ \ -\theta_2(z)^2 + \theta_3(z)^2 = \theta_4(z/2)^2, \tag{25}$$

$$\theta_3(z)^2 + \theta_4(z)^2 = 2\theta_3(2z)^2,\ \ \theta_3(z)^2 - \theta_4(z)^2 = 2\theta_2(2z)^2, \tag{26}$$

$$\theta_2(z)\theta_2(3z) - \theta_3(z)\theta_3(3z) + \theta_4(z)\theta_4(3z) = 0, \tag{27}$$

$$\theta_2(z)\theta_3(3z) + \theta_3(z)\theta_2(3z) = \tfrac{1}{2}\theta_2(z/4)\,\theta_2(3z/4), \tag{28}$$

$$\theta_2(z)\theta_2(3z) + \theta_3(z)\theta_3(3z) =$$

$$\tfrac{1}{2}\{\theta_3(z/4)\,\theta_3(3z/4) + \theta_4(z/4)\theta_4(3z/4)\}, \tag{29}$$

$$\theta_2(z)\theta_3(z)\theta_4(z) = \theta_1'(z), \tag{30}$$

$$\theta_2(z)^4 + \theta_4(z)^4 = \theta_3(z)^4. \tag{31}$$

[(1)]Equations (20), (21) show that the substitutions $z \to z+1$, $z \to -1/z$ permute θ_2, θ_3, θ_4 up to trivial factors.

These functions may also be expanded as infinite products (Jacobi's triple product identities):

$$\theta_3(\xi|z) = \prod_{m=1}^{\infty} (1-q^{2m})(1+q^{2m-1}e^{2i\xi})\,(1+q^{2m-1}e^{-2i\xi}), \qquad (32)$$

$$\theta_1'(z) = 2q^{1/4} \prod_{m=1}^{\infty} (1-q^{2m})^3, \qquad (33)$$

$$\theta_2(z) = 2q^{1/4} \prod_{m=1}^{\infty} (1-q^{2m})\,(1+q^{2m})^2, \qquad (34)$$

$$\theta_3(z) = \prod_{m=1}^{\infty} (1-q^{2m})\,(1+q^{2m-1})^2, \qquad (35)$$

$$\theta_4(z) = \prod_{m=1}^{\infty} (1-q^{2m})\,(1-q^{2m-1})^2. \qquad (36)$$

Another important product is

$$\Delta_{24} = q^2 \prod_{m=1}^{\infty} (1-q^{2m})^{24} \qquad (37)$$

$$= \{\tfrac{1}{2}\theta_2(z)\theta_3(z)\theta_4(z)\}^8 = \{\tfrac{1}{2}\theta_1'(z)\}^8 \qquad (38)$$

$$= \sum_{m=0}^{\infty} \tau(m) q^{2m} = q^2 - 24q^4 + 252q^6 - 1472q^8 + \cdots, \qquad (39)$$

in which the coefficients $\tau(m)$ are called *Ramanujan numbers*. (The first 300 coefficients are given in [Leh2]. See also Theorem 19 of Chap. 7.) Some values of $\psi_k(z)$ (Eq. (11)) can be expressed in terms of other functions:

$$\psi_1(z) = \theta_3(z),\ \psi_2(z) = \theta_2(z),\ \psi_3(z) = \frac{1}{2}(\theta_3(\frac{z}{9}) - \theta_3(z)), \qquad (40)$$

$$\psi_4(z) = \frac{1}{2}\theta_2(\frac{z}{4}),\ \psi_6(z) = \frac{1}{2}(\theta_2(\frac{z}{9}) - \theta_2(z)), \qquad (41)$$

$$\psi_{-k}(z) = \psi_k(z). \qquad (42)$$

For further information see [Bel2], [Gro2], [Har4], [Igu1], [Kra3], [Mum1], [Rad1], [Ran6], [Rau1], [Rau2], [Tan1], [Wht1].

Rescaling. It is often necessary to rescale a lattice, changing Λ to $\Lambda' = c\Lambda = \{cx : x \in \Lambda\}$ for some constant $c \in \mathbf{R}$. The parameters of Λ and Λ' are related as follows: $M' = cM$, $A' = c^2 A$, $\det' = c^{2n} \det$, (minimal norm)$' = c^2$ (minimal norm), $\rho' = c\rho$, $R' = cR$; the kissing number τ, densities Δ and δ, and thickness Θ are unchanged; $(\Lambda')^* = c^{-1}\Lambda^*$, and

$$\Theta_{\Lambda'}(z) = \Theta_\Lambda(c^2 z). \qquad (43)$$

5. The n-dimensional cubic lattice $\mathbf{Z}^n$

The set of integers ..., -2, -1, 0, 1, 2, 3, ... is denoted by $\mathbf{Z}$, and

$$\mathbf{Z}^n = \{(x_1, \ldots, x_n) : x_i \in \mathbf{Z}\}$$

is the n-dimensional *cubic* or *integer lattice*. ($\mathbf{Z}^2$ is better called the *square lattice*, as seen in ordinary graph paper.) As generator matrix M we may simply take the identity matrix. Then det $= 1$, minimal norm $= 1$, kissing number $\tau = 2n$, and the minimal vectors are $(0, \ldots, \pm 1, \ldots, 0)$. The packing radius $\rho = 1/2$, the covering radius $R = \sqrt{n}/2 = \rho\sqrt{n}$, the density $\Delta = V_n 2^{-n}$ and the center density $\delta = 2^{-n}$. Thus $\mathbf{Z}$ has density $\Delta = 1$, but the densities of $\mathbf{Z}^2$, $\mathbf{Z}^3$ and $\mathbf{Z}^4$ are only $\pi/4 = 0.785...$, $\pi/6 = 0.524...$ and $\pi^2/32 = 0.308...$. A typical deep hole is $(½, ½, ..., ½)$, and the Voronoi cells are cubes. $\mathbf{Z}^n$ is self-dual. Its automorphism group consists of all permutations and sign changes of the coordinates, and has order $2^n n!$. (This is the Weyl group of B_n.)

The theta series of $\mathbf{Z}^n$ is $\theta_3(z)^n$, and the theta series of the translate $\mathbf{Z}^n + (0^a\, ½^{n-a})$ is

$$\theta_2(z)^{n-a}\, \theta_3(z)^a \tag{44}$$

(§2.3 of Chap. 2). Tables of the numbers of points in the first 50 shells of $\mathbf{Z}^2$, $\mathbf{Z}^3$ and their translates are given in Tables 4.2, 4.3. In Table 4.2 the columns headed $N^{(0)}$, $N^{(1)}$, $N^{(2)}$ give respectively the numbers of points of norms m, $m + 1/4$, $2m + 1/2$ in $\mathbf{Z}^2$ with the origin of coordinates at the

Table 4.2. Numbers of points in the first shells of the square lattice $\mathbf{Z}^2$ with respect to various choices for the origin. See text for full explanation.

m	$N^{(0)}$	$N^{(1)}$	$N^{(2)}$	m	$N^{(0)}$	$N^{(1)}$	$N^{(2)}$	m	$N^{(0)}$	$N^{(1)}$	$N^{(2)}$
0	1	2	4	17	8	0	0	34	8	4	8
1	4	4	8	18	4	4	8	35	0	0	0
2	4	2	4	19	0	0	0	36	4	8	16
3	0	4	8	20	8	2	4	37	8	4	8
4	4	4	8	21	0	8	16	38	0	4	8
5	8	0	0	22	0	4	8	39	0	4	8
6	0	6	12	23	0	0	0	40	8	0	0
7	0	4	8	24	0	4	8	41	8	0	0
8	4	0	0	25	12	4	8	42	0	6	12
9	4	4	8	26	8	0	0	43	0	4	8
10	8	4	8	27	0	4	8	44	0	0	0
11	0	4	8	28	0	4	8	45	8	4	8
12	0	2	4	29	8	4	8	46	0	8	16
13	8	4	8	30	0	2	4	47	0	0	0
14	0	0	0	31	0	8	16	48	0	4	8
15	0	4	8	32	4	0	0	49	4	4	8
16	4	8	16	33	0	0	0	50	12	0	0

Table 4.3. Numbers of points in the first shells of the cubic lattice $\mathbf{Z}^3$ with respect to various choices for the origin.

m	$N^{(0)}$	$N^{(1)}$	$N^{(2)}$	$N^{(3)}$	m	$N^{(0)}$	$N^{(1)}$	$N^{(2)}$	$N^{(3)}$
0	1	2	4	8	26	72	32	24	72
1	6	8	8	24	27	32	24	48	96
2	12	10	8	24	28	0	32	32	120
3	8	8	16	32	29	72	40	24	48
4	6	16	12	48	30	48	26	40	104
5	24	16	8	24	31	0	48	48	168
6	24	10	24	48	32	12	48	16	96
7	0	24	16	72	33	48	16	56	48
8	12	16	16	24	34	48	32	32	120
9	30	8	24	56	35	48	32	16	72
10	24	32	16	72	36	30	32	64	96
11	24	24	16	48	37	24	56	40	192
12	8	18	28	72	38	72	48	32	72
13	24	24	32	72	39	0	24	32	144
14	48	16	8	48	40	24	64	36	96
15	0	24	32	48	41	96	32	40	72
16	6	32	32	120	42	48	26	48	144
17	48	32	16	72	43	24	56	48	120
18	36	16	40	56	44	24	16	32	96
19	24	32	16	96	45	72	40	48	104
20	24	34	16	24	46	48	64	48	192
21	48	16	40	120	47	0	64	16	72
22	24	48	40	120	48	8	16	80	120
23	0	16	32	48	49	54	40	40	192
24	24	16	36	96	50	84	48	24	48
25	30	56	16	96	51	48	32	80	144

center of a lattice point, an edge, a square; or in other words the coefficients of q^m, $q^{m+1/4}$, $q^{2m+1/2}$ in the expansion of

$$\theta_3(z)^2, \quad \theta_3(z)\theta_2(z), \quad \theta_2(z)^2 \tag{45}$$

respectively. In Table 4.3 the columns headed $N^{(0)}$, $N^{(1)}$, $N^{(2)}$, $N^{(3)}$ give respectively the numbers of points of norms m, $m + 1/4$, $m + 1/2$, $2m + 3/4$ in $\mathbf{Z}^3$ with the origin of coordinates at the center of a lattice point, an edge, a square face, a cube; or in other words the coefficients of q^m, $q^{m+1/4}$, $q^{m+1/2}$, $q^{2m+3/4}$ in the expansion of

$$\theta_3(z)^3, \quad \theta_3(z)^2\theta_2(z), \quad \theta_3(z)\theta_2(z)^2, \quad \theta_2(z)^3 \tag{46}$$

respectively.

If we expand

$$\Theta_{\mathbf{Z}_n}(z) = \theta_3(z)^n = \sum_{m=0}^{\infty} r_n(m) q^m , \tag{47}$$

then the coefficient $r_n(m)$ is the number of ways of writing m as a sum of n squares. In particular for $m > 0$ we have

$$r_2(m) = 4\delta(m) , \tag{48}$$

where $\delta(m)$ is the difference between the numbers of divisors of m of the two forms $4a + 1$, $4a + 3$,

$$r_4(m) = \begin{cases} 8 \sum_{d \mid m} d , & \text{if } m \text{ is odd}, \\ 24 \sum_{d \mid m,\, d \text{ odd}} d , & \text{if } m \text{ is even}, \end{cases} \tag{49}$$

$$r_8(m) = 16 \sum_{d \mid m} (-1)^{m-d} d^3 . \tag{50}$$

Many similar, although more complicated, formulas are known for other values of n. For further information see [Bat1], [Cas3], [Dic2], [Fri1], [Gla1]-[Gla6], [Gro2], [Har3]-[Har5], [Mal6], [Mor2], [Rad1], [Ran6], [Smi5], [Val1]-[Val3].

If we define $r_4'(m) = r_4(m)/8$, then $r_4'(m)$ is *multiplicative*, i.e. satisfies

$$r_4'(\ell m) = r_4'(\ell) r_4'(m) \text{ whenever } \ell \text{ and } m \text{ are relatively prime.} \tag{51}$$

This property simplifies the calculation of $r_4'(m)$, for now it is enough to know the values when $m = p^a$ is a power of a prime. These are

$$r_4'(2^a) = 3, \quad (a \geqslant 1),$$
$$r_4'(p^a) = 1 + p + p^2 + \cdots + p^a , \quad (a \geqslant 0),$$

where p is an odd prime. A similar multiplicative property holds for the theta functions of certain other lattices. Bateman [Gro2, p. 131] has shown that $r_n(m)/2n$ is multiplicative if and only if n is 1, 2, 4 or 8. (This reflects the fact that there are unique factorization theorems for suitable rings of integers in the real, complex, quaternionic, and Cayley numbers.)

6. The n-dimensional lattices A_n and A_n^*

6.1 The lattice A_n. For $n \geqslant 1$,

$$A_n = \{ (x_0, x_1, \ldots, x_n) \in \mathbf{Z}^{n+1} : x_0 + \cdots + x_n = 0 \} ,$$

which uses $n + 1$ coordinates to define an n-dimensional lattice: A_n lies in the hyperplane $\sum x_i = 0$ in $\mathbf{R}^{n+1}$. The standard integral basis is marked on the Coxeter-Dynkin diagram in Fig. 4.5. The extending node in this basis is $(1,0,0,\ldots,0,-1)$, and the extended diagram is given in the final column of Table 4.1. From Fig. 4.5 we can read off the generator matrix

$$M = \begin{bmatrix} -1 & 1 & 0 & 0 & \cdots & 0 & 0 \\ 0 & -1 & 1 & 0 & \cdots & 0 & 0 \\ 0 & 0 & -1 & 1 & \cdots & 0 & 0 \\ \cdot & \cdot & \cdot & \cdot & \cdots & \cdot & \cdot \\ 0 & 0 & 0 & 0 & \cdots & -1 & 1 \end{bmatrix}. \tag{52}$$

Two possible Gram matrices are

$$\begin{bmatrix} 2 & -1 & 0 & \cdots & 0 & 0 \\ -1 & 2 & -1 & \cdots & 0 & 0 \\ 0 & -1 & 2 & \cdots & 0 & 0 \\ \cdot & \cdot & \cdot & \cdots & \cdot & \cdot \\ 0 & 0 & 0 & \cdots & 2 & -1 \\ 0 & 0 & 0 & \cdots & -1 & 2 \end{bmatrix}, \quad \begin{bmatrix} 2 & 1 & 1 & \cdots & 1 \\ 1 & 2 & 1 & \cdots & 1 \\ 1 & 1 & 2 & \cdots & 1 \\ \cdot & \cdot & \cdot & \cdots & \cdot \\ 1 & 1 & 1 & \cdots & 1 \\ 1 & 1 & 1 & \cdots & 2 \end{bmatrix}. \tag{53}$$

(The first of these corresponds to (52).) Also det $= n + 1$, minimal norm $= 2$, kissing number $\tau = n(n+1)$, minimal vectors are all permutations of $(1,-1,0,\ldots,0)$, Coxeter number $h = n + 1$, packing radius $\rho = 1/\sqrt{2}$, center density $\delta = 2^{-n/2}(n+1)^{-1/2}$, covering radius

$$R = \rho \left\{ \frac{2a(n+1-a)}{n+1} \right\}^{1/2}, \tag{54}$$

where a is the integer part of $(n+1)/2$, a typical deep hole being the glue vector $[a]$. The Voronoi cell is given in Chap. 21. Glue vectors:

$$[i] = \left[\frac{i}{n+1}, \ldots, \frac{i}{n+1}, \frac{-j}{n+1}, \ldots, \frac{-j}{n+1} \right], \tag{55}$$

with j components equal to $i/(n+1)$, and i components equal to $-j/(n+1)$, where $i + j = n + 1$ and $0 \leqslant i \leqslant n$. The norm of $[i]$ is $ij/(n+1)$. Glue group: cyclic C_{n+1}, with addition $[j] + [k] = [j+k]$. Automorphism group: G_0 is the Weyl group of A_n, which is the symmetric group S_{n+1} of all permutations of the coordinates, and G_1 is the group of

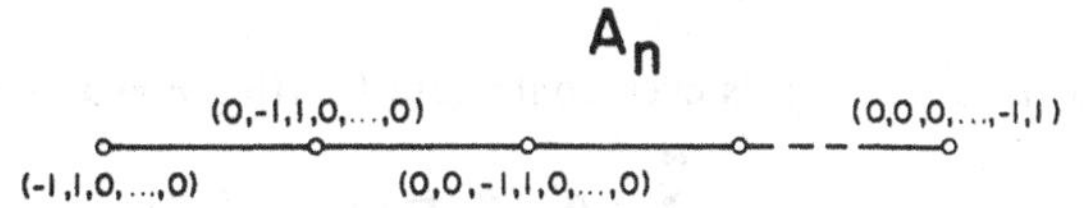

Figure 4.5. Diagram for A_n showing integral basis for root lattice.

order 2 generated by the negation of all coordinates (which interchanges $[i]$ and $[n+1-i]$), except that $G_1 = 1$ when $n = 1$. Theta series:

$$A_{n-1}: \quad \frac{\sum_{k=0}^{n-1} \theta_3\left(\frac{k\pi}{n} \,\middle|\, z\right)^n}{n\,\theta_3(nz)}, \tag{56}$$

$$A_{n-1} + [\ell]: \quad \frac{\sum_{k=0}^{n-1} \zeta^{-k\ell}\, \theta_3\left(\frac{k\pi}{n} \,\middle|\, z\right)^n}{n \sum_{m=-\infty}^{\infty} q^{n(m+\frac{\ell}{n})^2}}, \tag{57}$$

where $\zeta = e^{2\pi i/n}$.

6.2 The hexagonal lattice. We discuss the first few cases in more detail. Of course $A_1 \cong \mathbf{Z}$. A_2 is equivalent (or *similar*, see §1.4 of Chap. 1) to the familiar **hexagonal lattice** shown in Fig. 1.3a of Chap. 1, so called because the Voronoi cells are hexagons. The hexagonal lattice may be spanned by the vectors $(1,0)$ and $(-1/2, \sqrt{3}/2)$, and so an alternative generator matrix to (52) is

$$M = \begin{bmatrix} 1 & 0 \\ \frac{-1}{2} & \frac{\sqrt{3}}{2} \end{bmatrix}. \tag{58}$$

In this form det $= 3/4$, minimal norm $= 1$, $\tau = 6$, minimal vectors are $(\pm 1, 0)$ and $(\pm 1/2, \pm\sqrt{3}/2)$, $h = 3$, $\rho = 1/2$, density $\Delta = \pi/\sqrt{12} = 0.9069...$ $(\delta = 1/\sqrt{12})$, $R = 2\rho/\sqrt{3}$, typical deep holes being the points $(\pm 1/2, \pm 1/2\sqrt{3})$ (see Fig. 1.3d), and thickness $\Theta = 2\pi/3\sqrt{3} = 1.2092....$ Glue vectors:

$$[0] = (0,0), \quad [1] = \left(\frac{1}{2}, \frac{1}{2\sqrt{3}}\right), \quad [2] = \left(\frac{-1}{2}, \frac{-1}{2\sqrt{3}}\right).$$

Glue group $= C_3$. Automorphism group: $g_0 = 3!$, $g_1 = 2$, $g = 12$. The quadratic form associated with (58) is

$$x^2 - xy + y^2, \tag{59}$$

and the theta series is therefore

$$\Theta_{\text{hex}}(z) = \sum_{x,y=-\infty}^{\infty} q^{x^2 - xy + y^2} = \sum_{x,y=-\infty}^{\infty} q^{\left(x - \frac{1}{2}y\right)^2 + \frac{3}{4}y^2}.$$

The terms in which y is even contribute (putting $r = x - \frac{1}{2}y$, $s = \frac{1}{2}y$)

$$\sum_{r,s=-\infty}^{\infty} q^{r^2 + 3s^2} = \theta_3(z)\theta_3(3z),$$

and those with y odd contribute (putting $r = x - (y-1)/2$, $s = (y-1)/2$)

$$\sum_{r,s=-\infty}^{\infty} q^{\left(r+\frac{1}{2}\right)^2 + 3\left(s+\frac{1}{2}\right)^2} = \theta_2(z)\theta_2(3z).$$

Therefore

$$\Theta_{\text{hex}}(z) = \theta_3(z)\theta_3(3z) + \theta_2(z)\theta_2(3z) = \phi_0\left(\frac{z}{2}\right) \tag{60}$$

(see (12)). The series begins

$$\Theta_{\text{hex}}(z) = 1 + 6q + 6q^3 + 6q^4 + 12q^7 + \cdots, \tag{61}$$

and the first 50 nonzero terms are given in Table 4.4. If the origin is moved midway between two lattice points (i.e. to the middle of an edge, when each point is joined by edges to its six neighbors), the theta series is

$$\Theta_{\text{hex (edge)}}(z) = \frac{1}{2}\theta_2\left(\frac{z}{4}\right)\theta_2\left(\frac{3z}{4}\right) = \phi_1\left(\frac{z}{2}\right)$$

$$= 2q^{1/4} + 2q^{3/4} + 4q^{7/4} + 2q^{9/4} + \cdots. \tag{62}$$

If the origin is moved to a deep hole the theta series is

$$\Theta_{\text{hex}+[1]}(z) = \theta_2(z)\psi_6(3z) + \theta_3(z)\psi_3(3z)$$

$$= 3q^{1/3} + 3q^{4/3} + 6q^{7/3} + 6q^{13/3} + \cdots \tag{63}$$

Table 4.4. Theta series of planar hexagonal lattice.

m	$N(m)$	m	$N(m)$	m	$N(m)$	m	$N(m)$
0	1	28	12	67	12	109	12
1	6	31	12	73	12	111	12
3	6	36	6	75	6	112	12
4	6	37	12	76	12	117	12
7	12	39	12	79	12	121	6
9	6	43	12	81	6	124	12
12	6	48	6	84	12	127	12
13	12	49	18	91	24	129	12
16	6	52	12	93	12	133	24
19	12	57	12	97	12	139	12
21	12	61	12	100	6	144	6
25	6	63	12	103	12	147	18
27	6	64	6	108	6	148	12

(see [Slo19], [Teo3]). If we write $\Theta_{\text{hex}}(z) = \sum_{m=0}^{\infty} N(m) q^m$, then $N(m)$ is the number of times the quadratic form (59) represents m, and $N'(m) = N(m)/6$ is multiplicative (see (51)). Thus $N(m)$ is specified by the values

$$
\begin{aligned}
N'(3^a) &= 1, && \text{for all } a \geqslant 0, \\
N'(p^a) &= a + 1, && \text{for } p \equiv 1 \pmod 3, \\
N'(p^a) &= \begin{cases} 0, \\ 1, \end{cases} && \begin{array}{l} \text{for } p \equiv 2 \pmod 3,\ a \text{ odd}, \\ \text{for } p \equiv 2 \pmod 3,\ a \text{ even}, \end{array}
\end{aligned}
$$

where $p \neq 3$ is a prime.

We shall use the name A_2 only when the minimal norm is 2. From (60) we have

$$\Theta_{A_2}(z) = \phi_0(z) = 1 + 6q^2 + 6q^6 + \cdots . \tag{64}$$

6.3 The face-centered cubic lattice. Both A_3 and D_3 (§7) are equivalent to the **face-centered cubic lattice** (or fcc), illustrated in every chemistry textbook, and found in the pyramids of oranges on any fruit stand (Fig. 1.1). Fruiterers normally use square-based pyramids, but triangular-based pyramids lead to geometrically equivalent packings. The simplest definition is via D_3: the fcc consists of the points (x,y,z), where x, y and z are integers with an even sum. A generator matrix is

$$M = \begin{bmatrix} -1 & -1 & 0 \\ 1 & -1 & 0 \\ 0 & 1 & -1 \end{bmatrix}, \tag{65}$$

det = 4, minimal norm = 2, $\tau = 12$, and the minimal vectors consist of all permutations of $(\pm 1, \pm 1, 0)$, the twelve vertices of a regular cuboctahedron. Also $h = 4$, the packing radius $\rho = 1/\sqrt{2}$, the density $\Delta = \pi/\sqrt{18} = 0.7405...$ $(\delta = 2^{-5/2})$ and the covering radius $R = \rho\sqrt{2} = 1$. The Voronoi cell is a rhombic dodecahedron (Fig. 2.2a and Chap. 21). There are two kinds of holes in the fcc, corresponding to the two types of vertices of the Voronoi cell: deep or *octahedral* holes, such as $(0,0,1)$, surrounded by six lattice points, and shallow or *tetrahedral* holes, such as $(\frac12,\frac12,\frac12)$, surrounded by four lattice points. Glue vectors:

$$[0] = (0,0,0),\quad [1] = (\tfrac12,\tfrac12,\tfrac12),\quad [2] = (0,0,1),\quad [3] = (\tfrac12,\tfrac12,-\tfrac12).$$

Glue group: C_4. Automorphism group: $g_0 = 24$, $g_1 = 2$, $g = 48$. The fcc is the unique lattice with this density, although there are equally dense nonlattice packings (see §6.5). Theta series:

$$
\begin{aligned}
\Theta_{\text{fcc}}(z) &= \tfrac12 (\theta_3(z)^3 + \theta_4(z)^3) = \theta_3(4z)^3 + 3\theta_3(4z)\theta_2(4z)^2 \\
&= 1 + 12q^2 + 6q^4 + 24q^6 + \cdots .
\end{aligned} \tag{66}
$$

There is a complicated explicit formula for the coefficients [Dic2, Vol. II, p. 263], [Gro2]. The first 50 coefficients are given in Table 4.5. For the translates we have:

$$\Theta_{fcc+[1]}(z) = \tfrac{1}{2}\,\theta_2(z)^3 = 4q^{3/4} + 12q^{11/4} + 12q^{19/4} + \cdots, \quad (67)$$

$$\begin{aligned}\Theta_{fcc+[2]}(z) &= \tfrac{1}{2}\,(\theta_3(z)^3 - \theta_4(z)^3)\\ &= 6q + 8q^3 + 24q^5 + 30q^9 + \cdots. \quad (68)\end{aligned}$$

Table 4.5. Theta series of face-centered cubic lattice. The table gives $N(m)$ for $m = 10r + s$.

$10r\backslash s$	0	2	4	6	8	10	12	14	16	18
0+	1	12	6	24	12	24	8	48	6	36
20+	24	24	24	72	0	48	12	48	30	72
40+	24	48	24	48	8	84	24	96	48	24
60+	0	96	6	96	48	48	36	120	24	48
80+	24	48	48	120	24	120	0	96	24	108
100+	30	48	72	72	32	144	0	96	72	72

6.4 The tetrahedral or diamond packing. The *tetrahedral packing*, as found in *diamond* (the hardest substance known), consists of the points

$$\text{fcc} \;\cup\; [1] + \text{fcc}. \quad (69)$$

There is a picture in [Kit4, p. 25], for example. This is the packing D_3^+ (§7.3) and is *not* a lattice. All the points are equivalent, and the Voronoi region around any one of them has volume 1, the minimal norm = 3/4, kissing number $\tau = 4$, and the minimal vectors = (½,½,½), (½,−½,−½), (−½,½,−½), (−½,−½,½). Thus the four neighbors of each point form a regular tetrahedron. The packing radius $\rho = \sqrt{3}/4$, density $\Delta = \pi\sqrt{3}/16 = 0.3401...$ (Chap. 1, Eq. (14)), $\delta = 2^{-6}3^{3/2}$, covering radius $R = 2\rho = \sqrt{3}/2$, and the holes are tetrahedral holes such as (½,½,−½). Theta series ([Slo17]):

$$\begin{aligned}\Theta_{\text{diamond}}(z) &= \tfrac{1}{2}\,(\theta_2(z)^3 + \theta_3(z)^3 + \theta_4(z)^3)\\ &= 1 + 4q^{3/4} + 12q^2 + 12q^{11/4} + 6q^4 + \cdots. \quad (70)\end{aligned}$$

If the origin is moved to a hole the theta series is

$$\begin{aligned}\Theta_{\text{diamond(hole)}}(z) &= \tfrac{1}{2}\,(\theta_2(z)^3 + \theta_3(z)^3 - \theta_4(z)^3)\\ &= 4q^{3/4} + 6q + 12q^{11/4} + 8q^3 + \cdots. \quad (71)\end{aligned}$$

6.5 The hexagonal close-packing. As remarked in Chap. 1, A_1, A_2 and A_3 are the densest known packings in dimensions 1, 2 and 3 (and the densest possible *lattice* packings in those dimensions). In three dimensions however there are infinitely many nonlattice packings with the same density and kissing number as the fcc. The simplest of these is the *hexagonal close-packing* or hcp (§1.3 of Chap. 1), which is of considerable importance in chemistry. The hcp is not itself a lattice, but may be defined as the union of the lattice L spanned by the vectors $(1,0,0)$, $(1/2,\sqrt{3}/2,0)$ and $(0,0,\sqrt{8/3})$, and the translate $L + (1/2,1/\sqrt{12},\sqrt{2/3})$. Alternatively it is the union of the lattice spanned by the vectors

$$2^{-1/2}(1,1,0),\ 2^{-1/2}(1,0,1),\ 2^{-1/2}(-4/3,-4/3,-4/3)$$

and its translate by $2^{-1/2}(0,1,1)$. A third definition is that the hcp consists of the points

$$\pm\ (6i+1,\ 6j+1,\ 6k-2,\ 2\sqrt{3}(4\ell-1)), \tag{72}$$

where i, j, k, ℓ are integers with $i+j+k=0$ [Slo19]. With the first definition the Voronoi cells have volume $1/\sqrt{2}$, minimal norm $= 1$, $\tau = 12$, minimal vectors $= (\pm 1,0,0)$, $(\pm 1/2, \pm\sqrt{3}/2, 0)$, $(0, -1/\sqrt{3}, \pm\sqrt{2/3})$ and $(\pm 1/2, 1/\sqrt{12}, \pm\sqrt{2/3})$, $\rho = 1/2$, $\Delta = \pi/3\sqrt{2} = 0.7405\ldots$, and $R = \rho\sqrt{2} = 1/\sqrt{2}$. There are two kinds of holes, deep or octahedral holes such as $(0, 1/\sqrt{3}, 1/\sqrt{6})$ and shallow or tetrahedral holes such as $(1/2, 1/\sqrt{12}, 1/\sqrt{24})$. Theta series:

$$\Theta_{\mathrm{hcp}}(z) = \phi_0(z)\left\{\theta_3\left(\frac{8z}{3}\right) - \frac{1}{2}\theta_2\left(\frac{8z}{3}\right)\right\} + \frac{1}{2}\phi_0\left(\frac{z}{3}\right)\theta_2\left(\frac{8z}{3}\right)$$

$$= 1 + 12q + 6q^2 + 2q^{8/3} + \cdots \tag{73}$$

(Table 4.6). If the origin is moved to a deep hole the theta series is

$$\Theta_{\mathrm{hcp\,(hole)}}(z) = \theta_2(2z/3)\{\theta_2(z)\psi_6(3z) + \theta_3(z)\psi_3(3z)\}$$

$$= 6q^{1/2} + 6q^{3/2} + 6q^{11/6} + 12q^{5/2} + \cdots . \tag{74}$$

For further information about three-dimensional lattices see Chap. 1, [Bal1], [Cox18], [Cox20], [Fej8a]-[Fej10], [Int1], [Kit4], [Sma1], [Wel2]-[Wel4], [Wyc1], [Wyc2].

Table 4.6. Theta series of 3-dimensional hexagonal close-packing. The table gives $N(m)$ for $m = (10r+s)/3$.

$10r\backslash s$	0	1	2	3	4	5	6	7	8	9
0+	1	0	0	12	0	0	6	0	2	18
10+	0	12	6	0	0	12	0	12	6	6
20+	12	24	6	0	0	12	0	12	0	24
30+	12	12	2	12	6	24	6	12	0	24
40+	0	12	0	6	24	12	12	24	6	12
50+	0	24	0	24	18	12	12	24	0	12

6.6 The dual lattice A_n^*. The lattice dual to A_n is

$$A_n^* = \bigcup_{i=0}^{n} ([i]+A_n), \tag{75}$$

with generator matrix

$$M = \begin{bmatrix} 1 & -1 & 0 & \cdots & 0 & 0 \\ 1 & 0 & -1 & \cdots & 0 & 0 \\ \cdot & \cdot & \cdot & \cdots & \cdot & \cdot \\ 1 & 0 & 0 & \cdots & -1 & 0 \\ \frac{-n}{n+1} & \frac{1}{n+1} & \frac{1}{n+1} & \cdots & \frac{1}{n+1} & \frac{1}{n+1} \end{bmatrix}. \tag{76}$$

An equivalent definition uses either the Gram matrix

$$A = \begin{bmatrix} n & -1 & -1 & \cdots & -1 \\ -1 & n & -1 & \cdots & -1 \\ \cdot & \cdot & \cdot & \cdots & \cdot \\ -1 & -1 & -1 & \cdots & n \end{bmatrix}, \tag{77}$$

(cf. Eq. (49) of Chap. 8), with determinant $(n+1)^{n-1}$, or the associated quadratic form

$$n \sum_{i=1}^{n} x_i^2 - \sum_{i \neq j}^{n} x_i x_j, \tag{78}$$

often referred to as Voronoi's principal form of the first type [Rys12], [Rys13]. Using the definition (76), det $= 1/(n+1)$, minimal norm $= n/(n+1)$, $\tau = 2$ $(n=1)$ or $2n+2$ $(n \geqslant 2)$,

$$\rho = \frac{1}{2}\sqrt{\frac{n}{n+1}}, \quad \delta = \frac{n^{n/2}}{2^n (n+1)^{(n-1)/2}}, \tag{79}$$

$$R = \rho\sqrt{\frac{n+2}{3}} = \sqrt{\frac{n(n+2)}{12(n+1)}}, \tag{80}$$

a typical deep hole being

$$\frac{1}{2n+2}(-n, -n+2, -n+4, \ldots, n-2, n), \tag{81}$$

the thickness is

$$\Theta = V_n \sqrt{n+1} \left(\frac{n(n+2)}{12(n+1)} \right)^{n/2}, \tag{82}$$

and the Voronoi cells are permutohedra (see Chap. 21), with vertices all permutations of (81). $A_1^* \cong A_1 \cong \mathbf{Z}$, $A_2^* \cong A_2$, and A_3^* is the body-centered cubic lattice (see §6.7). A_n^* is the best covering of $\mathbf{R}^n$ for $n = 2$, the best lattice covering for $n \le 5$, and the best covering known for $n \le 21$ [Bam1]-[Bam3], [Del4], [Few1], [Gam1], [Gam2], [Rys7], [Rys12], [Rys13]. (See the Preface to the Second Edition.)

The typical lattice between A_n and A_n^* is $A_n[s]$, where s is any divisor of $n+1$. In Coxeter's notation [Cox10] this is A_n^r, where $rs = n+1$. Its theta series can be found by summing (57) for $\ell = 0, s, 2s, \ldots, n+1-s$.

6.7 The body-centered cubic lattice. Both A_3^* and D_3^* (§7) are equivalent to the *body-centered cubic lattice* (or bcc), also familiar from chemistry. The simplest definition is via D_3^*: the bcc consists of the points (x,y,z) where x, y and z are all even or all odd integers. A generator matrix is

$$M = \begin{bmatrix} 2 & 0 & 0 \\ 0 & 2 & 0 \\ 1 & 1 & 1 \end{bmatrix}, \qquad (83)$$

det = 16, minimal norm = 3, $\tau = 8$, minimal vectors are $(\pm 1, \pm 1, \pm 1)$, $\rho = \sqrt{3}/2$, density $\Delta = \pi\sqrt{3}/8 = 0.6802\ldots$ $(\delta = 3\sqrt{3}/32)$, and $R = \rho\sqrt{5/3} = \sqrt{5}/2$. There is only one type of hole in the bcc, namely tetrahedral holes such as $(0, 1/2, 1)$, surrounded by four lattice points. Also

$$\Theta = \pi\, 5\sqrt{5}\,/24 = 1.4635\ldots,$$

and the Voronoi cells are truncated octahedra (Fig. 2.2b, Chap. 21). Theta series:

$$\Theta_{\text{bcc}}(z) = \theta_2(4z)^3 + \theta_3(4z)^3 = 1 + 8q^3 + 6q^4 + 12q^8 + \cdots \qquad (84)$$

(see Table 4.7). If the origin is moved to a hole the theta series is

Table 4.7. Theta series of body-centered cubic lattice.

m	$N(m)$	m	$N(m)$	m	$N(m)$	m	$N(m)$
0	1	36	30	75	56	115	48
3	8	40	24	76	24	116	72
4	6	43	24	80	24	120	48
8	12	44	24	83	72	123	48
11	24	48	8	84	48	128	12
12	8	51	48	88	24	131	120
16	6	52	24	91	48	132	48
19	24	56	48	96	24	136	48
20	24	59	72	99	72	139	72
24	24	64	6	100	30	140	48
27	32	67	24	104	72	144	30
32	12	68	48	107	72	147	56
35	48	72	36	108	32	148	24

$$\Theta_{bcc(hole)}(z) = \tfrac{1}{2}\theta_2(z)\theta_2(2z)^2$$
$$= 4q^{5/4} + 4q^{13/4} + 8q^{21/4} + 12q^{29/4} + \cdots . \quad (85)$$

For further information see the references at the end of §6.5.

7. The n-dimensional lattices D_n and D_n^*

7.1 The lattice D_n. For $n \geqslant 3$,

$$D_n = \{(x_1, \cdots, x_n) \in \mathbf{Z}^n : x_1 + \cdots + x_n \text{ even}\},$$

or in other words D_n is obtained by coloring the points of $\mathbf{Z}^n$ alternately red and white with a checkerboard coloring, and taking the red points. D_n is sometimes called the *checkerboard lattice.* The standard integral basis is marked on the Coxeter-Dynkin diagram in Fig. 4.6. The extending node in this basis is $(0, \cdots, 0, 1, 1)$, and the extended diagram is given in the final column of Table 4.1. From Fig. 4.6 we obtain the generator matrix

$$M = \begin{bmatrix} -1 & -1 & 0 & \cdots & 0 & 0 \\ 1 & -1 & 0 & \cdots & 0 & 0 \\ 0 & 1 & -1 & \cdots & 0 & 0 \\ \cdot & \cdot & \cdot & \cdots & \cdot & \cdot \\ 0 & 0 & 0 & \cdots & 1 & -1 \end{bmatrix}. \quad (86)$$

Then det $= 4$, minimal norm $= 2$, kissing number $\tau = 2n(n-1)$, minimal vectors $=$ all permutations of $(\pm 1, \pm 1, 0, \ldots, 0)$, $h = 2n-2$, packing radius $\rho = 1/\sqrt{2}$, center density $\delta = 2^{-(n+2)/2}$, and covering radius $R = \rho\sqrt{2}$ $(n = 3)$ or $\rho\sqrt{n/2}$ $(n \geq 4)$. There are two kinds of hole, deep holes such as the glue vectors [2] (if $n \leq 4$) or [1] (if $n \geq 4$), and shallow holes such as the glue vectors [1] (if $n \leq 4$) or [2] (if $n \geq 4$). For $n = 4$ there is only one kind of hole. The Voronoi cell is given in Chap. 21, and the Delaunay cells in §4.2 of Chap. 5. Glue vectors:

[0] =	$(0, 0, \ldots, 0)$,	norm 0,
[1] =	$(\tfrac{1}{2}, \tfrac{1}{2}, \ldots, \tfrac{1}{2})$,	norm $n/4$,
[2] =	$(0, 0, \ldots, 1)$,	norm 1,
[3] =	$(\tfrac{1}{2}, \tfrac{1}{2}, \ldots, -\tfrac{1}{2})$,	norm $n/4$.

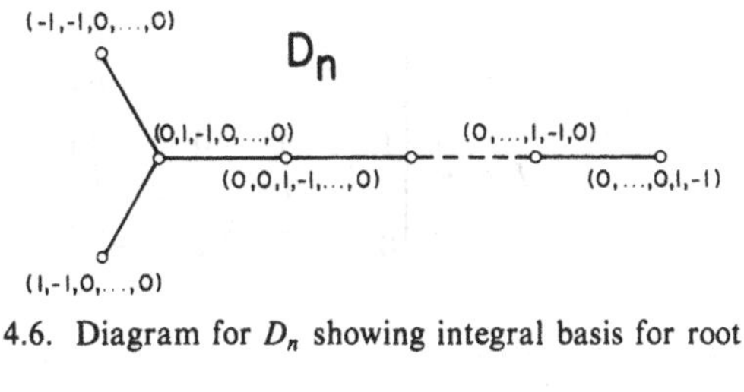

Figure 4.6. Diagram for D_n showing integral basis for root lattice.

Glue group: Klein V_4 ($[i]+[i]=0$) if n is even, cyclic C_4 ($[1]+[2]=[3]$) if n is odd. Automorphism group:

$$n=4: \quad g_0 = 2^3 \cdot 4!, \quad g_1 = 3! \quad \text{(all perms of [1], [2], [3])},$$

$$n \neq 4: \quad g_0 = 2^{n-1} \cdot n!, \quad g_1 = 2 \quad \text{(interchange [1] and [3])}.$$

G_0 is generated by all permutations together with sign changes of evenly many coordinates, and G_1 contains the sign change of the last coordinate and, for $n=4$ only, the Hadamard matrix

$$H_4 = \tfrac{1}{2}\begin{bmatrix} 1 & 1 & 1 & 1 \\ 1 & -1 & 1 & -1 \\ 1 & 1 & -1 & -1 \\ 1 & -1 & -1 & 1 \end{bmatrix}.$$

Theta series (see Chap. 2, Eq. (37)):

$$D_n : \tfrac{1}{2}(\theta_3(z)^n + \theta_4(z)^n) = \sum_{m=0}^{\infty} r_n(2m) q^{2m}, \tag{87}$$

using the notation of (47),

$$D_n + [1] \text{ or } D_n + [3] : \tfrac{1}{2}\theta_2(z)^n, \tag{88}$$

$$D_n + [2] : \tfrac{1}{2}(\theta_3(z)^n - \theta_4(z)^n). \tag{89}$$

$D_3 \cong A_3$ is the face-centered cubic lattice (§6.3). D_3, D_4 and D_5 are the densest possible lattice packings in dimensions 3, 4 and 5, and densest known packings in these dimensions, although for $n=3$ and 5 there are equally dense nonlattice packings (§6.5 and Chap. 5).

7.2 The four-dimensional lattice D_4. D_4, which was also described in §1.4 of Chap. 1, is one of the two most useful four-dimensional lattices (the other being A_4^*). D_4 is defined by (86) or equivalently by the generator matrix

$$M' = \begin{bmatrix} 1 & 0 & 0 & 0 \\ 0 & 1 & 0 & 0 \\ 0 & 0 & 1 & 0 \\ \tfrac{1}{2} & \tfrac{1}{2} & \tfrac{1}{2} & \tfrac{1}{2} \end{bmatrix}. \tag{90}$$

The matrix

$$T = \frac{1}{2}\begin{bmatrix} 1 & 1 & 0 & 0 \\ 1 & -1 & 0 & 0 \\ 0 & 0 & 1 & 1 \\ 0 & 0 & 1 & -1 \end{bmatrix} \tag{91}$$

maps the first version into the second. ((90) also defines the dual lattice D_4^* — see §7.4 — showing that $D_4 \cong D_4^*$.) Using the definition (86), det $= 4$, minimal norm $= 2$, $\tau = 24$, minimal vectors $=$ all permutations of $(\pm 1, \pm 1, 0, 0)$, $h = 6$, $\rho = 1/\sqrt{2}$, $\Delta = \pi^2/16 = 0.6169\ldots$ $(\delta = 1/8)$, $R = \rho\sqrt{2} = 1$, typical deep holes $= (\pm ½, \pm ½, \pm ½, \pm ½)$ and $(0, 0, 0, \pm 1)$, $\Theta = \pi^2/4 = 2.4674\ldots$. The Voronoi cell is the regular 4-dimensional polytope known as the 24-cell or $\{3, 4, 3\}$ (Chap. 21, [Cox20, §8.2]). The three nonzero cosets $D_4 + [i]$, $i = 1, 2, 3$, are equivalent. Theta series

$$\Theta_{D_4}(z) = ½(\theta_3(z)^4 + \theta_4(z)^4) = \theta_2(2z)^4 + \theta_3(2z)^4 \tag{92}$$

(cf. Eq. (89) of Chap. 7). D_4 is the unique lattice with this density.

Using the second definition, (90), D_4 consists of the vectors (a, b, c, d) with a, b, c, d all in $\mathbf{Z}$ or all in $\mathbf{Z} + ½$, and therefore may be regarded as the lattice of Hurwitz *integral quaternions* ([Cox8], [Cox18, p. 25], [Har5, §20.6], [Hur1]). This gives the lattice points the structure of a skew domain. As an Eisenstein lattice, D_4 is generated by (2,0) and $(1, \theta)$.

The first 50 coefficients of the theta series for D_4 (using the first definition, Eq. (86)) are shown in Table 4.8. These coefficients are given explicitly by $N(2m) = r_4(2m)$, using the second formula in Eq. (49). Also $N(2m)$ is the number of integral quaternions of norm m, and $(24)^{-1}N(2m)$ is a multiplicative function of m. In the notation of Schläfli and Coxeter, D_4 is the regular honeycomb $\{3,3,4,3\}$ (see [Cox20, §7.8]).

Table 4.8. Theta series of 4-dimensional lattice D_4. The table gives $N(m)$ for $m = 10r + s$.

$r\backslash s$	0	2	4	6	8	10	12	14	16	18
0	1	24	24	96	24	144	96	192	24	312
2	144	288	96	336	192	576	24	432	312	480
4	144	768	288	576	96	744	336	960	192	720
6	576	768	24	1152	432	1152	312	912	480	1344
8	144	1008	768	1056	288	1872	576	1152	96	1368
10	744	1728	336	1296	960	1728	192	1920	720	1440

7.3 The packing D_n^+. The covering radius of D_n increases with n, and when $n = 8$ it is equal to the minimal distance between the lattice points. So when $n \geqslant 8$ we can slide another copy of D_n in between the points of D_n, doubling the number of points (and the density) without reducing the distance between them! Formally, we define

$$D_n^+ = D_n \cup ([1] + D_n). \tag{93}$$

D_n^+ is a lattice packing if and only if n is even. D_3^+ is the tetrahedral or diamond packing (§6.4) and $D_4^+ \cong \mathbf{Z}^4$. When $n = 8$ this construction is especially important, the lattice D_8^+ being known as E_8 (§8.1). The

Voronoi cells of D_n^+ have volume 1, the minimal norm $= n/4$ $(n \leqslant 8)$ or 2 $(n \geqslant 8)$, center density $\delta = (\sqrt{n}/4)^n$ $(n \leqslant 8)$ or $2^{-n/2}$ $(n \geqslant 8)$, kissing number $\tau = 2^{n-1}$ $(n \leqslant 7)$, 240 $(n = 8)$, or $2n(n-1)$ $(n \geqslant 9)$, and the theta series is

$$\tfrac{1}{2}(\theta_2(z)^n + \theta_3(z)^n + \theta_4(z)^n). \tag{94}$$

7.4 The dual lattice D_n^*

$$D_n^* = D_n \cup ([1]+D_n) \cup ([2]+D_n) \cup ([3]+D_n).$$

Generator matrix

$$M = \begin{bmatrix} 1 & 0 & \cdots & 0 & 0 \\ 0 & 1 & \cdots & 0 & 0 \\ & & \cdots & & \\ 0 & 0 & \cdots & 1 & 0 \\ \tfrac{1}{2} & \tfrac{1}{2} & \cdots & \tfrac{1}{2} & \tfrac{1}{2} \end{bmatrix}, \tag{95}$$

det $= 1/4$, minimal norm $= 3/4$ $(n = 3)$ or 1 $(n \geq 4)$, $\tau = 8$ $(n = 3)$, 24 $(n = 4)$ or $2n$ $(n \geq 5)$, $\rho = \sqrt{3}/4$ $(n = 3)$, $\rho = \tfrac{1}{2}$ $(n \geq 4)$, $\delta = 3^{1.5}\, 2^{-5}$ $(n = 3)$ or $2^{-(n-1)}$ $(n \geq 4)$, $R = \rho\, n^{1/2}/\sqrt{2}$ (n even), $\rho\sqrt{5/3}$ $(n = 3)$ or $\rho(2n-1)^{1/2}/2$ (n odd ≥ 5). For the Voronoi cells see Ch. 21. Theta series:

$$\theta_2(z)^n + \theta_3(z)^n. \tag{96}$$

D_3^* is the body-centered cubic lattice (§6.7), and $D_4^* \cong D_4$.

8. The lattices E_6, E_7 and E_8

8.1 The 8-dimensional lattice E_8. The existence of an even unimodular 8-dimensional form was first established nonconstructively in [Smi6, p. 521], and an explicit construction was given in [Kor1]-[Kor3] and [Min0]. Gosset in 1900 [Gos1] seems to have been the first to study the lattice E_8 itself. $E_8 = D_8^+$ is a special case of the family of packings constructed in §7.3, and so might be called the 8-dimensional diamond lattice. In the *even coordinate system* E_8 consists of the points

$$\{(x_1, \ldots, x_8) : \text{ all } x_i \in \mathbf{Z} \text{ or all } x_i \in \mathbf{Z} + \tfrac{1}{2}, \quad \sum x_i \equiv 0 \pmod 2\}. \tag{97}$$

The *odd* coordinate system is obtained by changing the sign of any coordinate: the points are

$$\{(x_1, \ldots, x_8) : \text{ all } x_i \in \mathbf{Z} \text{ or all } x_i \in \mathbf{Z} + \tfrac{1}{2}, \quad \Sigma\, x_i \equiv 2x_8 \pmod 2\}. \tag{98}$$

The standard integral basis (in the odd coordinate system) is marked on the Coxeter-Dynkin diagram in Fig. 4.7. The extending node is $(\tfrac{1}{2}, -\tfrac{1}{2}^7)$,

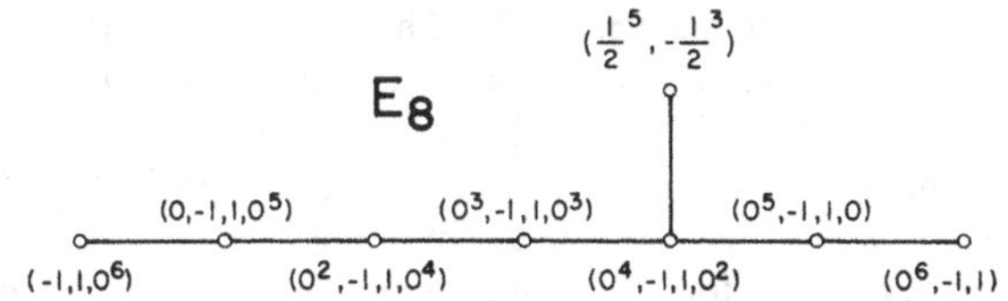

Figure 4.7. Diagram for E_8 showing integral basis for root lattice in the odd coordinate system. A basis in the even coordinate system can be obtained by changing the signs of any odd number of coordinates (for example the last three coordinates).

and the extended diagram is given in the final column of Table 4.1. A generator matrix (in the even coordinate system) is

$$M = \begin{bmatrix} 2 & 0 & 0 & 0 & 0 & 0 & 0 & 0 \\ -1 & 1 & 0 & 0 & 0 & 0 & 0 & 0 \\ 0 & -1 & 1 & 0 & 0 & 0 & 0 & 0 \\ 0 & 0 & -1 & 1 & 0 & 0 & 0 & 0 \\ 0 & 0 & 0 & -1 & 1 & 0 & 0 & 0 \\ 0 & 0 & 0 & 0 & -1 & 1 & 0 & 0 \\ 0 & 0 & 0 & 0 & 0 & -1 & 1 & 0 \\ \frac{1}{2} & \frac{1}{2} & \frac{1}{2} & \frac{1}{2} & \frac{1}{2} & \frac{1}{2} & \frac{1}{2} & \frac{1}{2} \end{bmatrix}, \qquad (99)$$

det = 1, minimal norm = 2, kissing number $\tau = 240$, and the minimal vectors are all vectors $(\pm 1^2, 0^6)$, together with those vectors $(\pm \frac{1}{2}^8)$ that have an even number of minus signs in the even coordinate system, or an odd number of minus signs in the odd coordinate system. The Coxeter number $h = 30$, packing radius $\rho = 1/\sqrt{2}$, density $\Delta = \pi^4/384 = 0.2537...$ $(\delta = 1/16)$, covering radius $= \rho\sqrt{2} = 1$, thickness $\Theta = \pi^4/24 = 4.0587...$, and the Voronoi cell is the reciprocal of the polytope 4_{21} (see Chap. 21 and [Cox20, §11.8]). There are two kinds of holes in E_8, deep holes such as $(0^7, 1)$ (halves of norm 4 vectors), surrounded by 16 points, and shallow holes such as $((1/6)^7, 5/6)$, surrounded by 9 lattice points (see Fig. 21.8). $E_8^* = E_8$, so there is no glue (or more precisely the only glue is $[0] = (0^8)$).

E_8 is the unique lattice with this density and minimal norm [Vet2]. Automorphism group: $g_0 = 2^{14}\, 3^5\, 5^2\, 7 = 696729600$, $g_1 = 1$. G_0 is the Weyl group $W(E_8)$, generated (in the even coordinate system) by all permutations of 8 letters, all even sign changes, and the matrix

$$\text{diag}\,\{H_4, H_4\} \qquad (100)$$

(cf. §7.1). See also [Cox20], [Cox28], [Edg3].

E_8 may also be obtained by applying Construction A to the Hamming code $\mathcal{H}_8$ (§2.5 of Chap. 5, Example 5 of Chap. 7). The minimal vectors (after rescaling) now consist of $2^4 \cdot 14 = 224$ vectors $((\pm \frac{1}{2})^4, 0^4)$, where the nonzero coordinates support a minimal weight codeword in $\mathcal{H}_8$, together with 16 vectors $(\pm 1, 0^7)$. This definition is the best for showing the identification of E_8 with the integral *Cayley numbers* or *octaves* ([Bli5], [Bli6], [Con16, p. 85], [Cox9], [Cox17], [Cox18, Chap. 2]). Let the coordinates be labeled ∞, 0, 1, 2, . . . , 6 as in §2.4.2 of Chap. 3. The *real Cayley algebra* is spanned by "numbers" $i_\infty = 1$, i_0, $i_1, \ldots, i_6$ with the property that for the quadruples $\{a,b,c,d\} = \{\infty,1,2,4\}, \ldots, \{\infty,0,1,3\}$ given in Eq. (56) of Chap. 3 (corresponding to the minimal vectors of $\mathcal{H}_8$ that contain ∞) the sets $\{i_a, i_b, i_c, i_d\} \cong \{1, i, j, k\}$ span a quaternion subalgebra. Thus $i_1 i_2 = i_4$, $i_2 i_1 = -i_4$, $i_1 i_3 = i_0$, etc. The *integral Cayley numbers* are the members of the E_8 sublattice spanned by vectors of the form $(\pm \frac{1}{2}^4, 0^4)$ for which the distinguished sets of four coordinates are those obtained from Eq. (56) of Chap. 3 by interchanging ∞ and 0, together with their complements. The 240 minimal vectors become the 240 unit (or norm 1) Cayley numbers.

Alternatively, E_8 can also be identified with the ring of *icosians*, as described is §2 of Chap. 8. Table 8.1 gives icosian names for the 240 minimal vectors. A quaternionic construction for E_8 is given in Eq. (77) of Chap. 2. Theta series:

$$\Theta_{E_8}(z) = \tfrac{1}{2}(\theta_2(z)^8 + \theta_3(z)^8 + \theta_4(z)^8)$$

$$= \theta_2(2z)^8 + 14\,\theta_2(2z)^4\theta_3(2z)^4 + \theta_3(2z)^8$$

$$= \sum_{m=0}^{\infty} N_m\, q^m = 1 + 240q^2 + 2160q^4 + \cdots, \tag{101}$$

$$N_m = 240\sigma_3\left(\frac{m}{2}\right), \tag{102}$$

where

$$\sigma_r(m) = \sum_{d \mid m} d^r \tag{103}$$

(Table 4.9). N_m is the number of integral Cayley numbers of norm $m/2$, and $240^{-1} N_{m/2}$ is multiplicative. If the origin is moved to a deep hole the theta series is

$$\Theta_{E_8(\text{hole})}(z) = \tfrac{1}{2}(\theta_2(z)^8 + \theta_3(z)^8 - \theta_4(z)^8)$$

$$= \sum_{m=1}^{\infty} N_m'\, q^m = 16q + 128q^2 + 448q^3 + 1024q^4 + \cdots, \tag{104}$$

$$N_m' = 16\{\sigma_3(m) - \sigma_3(m/2)\}, \tag{105}$$

with the convention that $\sigma_3(x) = 0$ if x is not an integer.

Table 4.10 gives coordinates for representative vectors in the first eight layers of E_8. The full list of vectors is obtained by applying arbitrary permutations and signs to the vectors in the table, except that if the vector is prefixed by an E (resp. D) then an even (resp. odd) number of minus

Table 4.9. Theta series of Gosset's lattices E_6, E_7 and E_8.

m	$N_m(E_6)$	$N_m(E_7)$	$N_m(E_8)$
0	1	1	1
2	72	126	240
4	270	756	2160
6	720	2072	6720
8	936	4158	17520
10	2160	7560	30240
12	2214	11592	60480
14	3600	16704	82560
16	4590	24948	140400
18	6552	31878	181680
20	5184	39816	272160
22	10800	55944	319680
24	9360	66584	490560
26	12240	76104	527520
28	13500	99792	743040
30	17712	116928	846720
32	14760	133182	1123440
34	25920	160272	1179360
36	19710	177660	1635120
38	26064	205128	1646400
40	28080	249480	2207520
42	36000	265104	2311680
44	25920	281736	2877120
46	47520	350784	2920320
48	37638	382536	3931200
50	43272	390726	3780240
52	45900	470232	4747680
54	59040	505568	4905600
56	46800	532800	6026880
58	75600	615384	5853600
60	51840	640080	7620480
62	69264	701568	7150080
64	73710	799092	8987760
66	88560	809424	8951040
68	62208	853776	10614240
70	108000	1006992	10402560
72	85176	1051974	13262640
74	98640	1031688	12156960
76	97740	1195992	14817600
78	122400	1286208	14770560
80	88128	1313928	17690400
82	151200	1469664	16541280
84	110700	1474704	20805120
86	133200	1547784	19081920
88	140400	1797768	23336640
90	157680	1776600	22891680
92	114048	1809360	26282880
94	198720	2104704	24917760
96	147600	2130968	31456320
98	176472	2123982	28318320
100	162270	2382156	34022160

signs are required (in the even coordinate system). The group G_0 acts transitively on the sets of vectors of norms 2, 4, 6, 10 and 12. (There are two orbits of norm 8 vectors: such a vector may or may not be twice a lattice vector. The vectors of the second orbit, divided by 3, give the shallow holes near 0.)

For further information about E_6, E_7 and E_8 see later chapters and [Bou1], [Cox10], [Cox18], [Cox20], [Cox28], [Haz1].

Table 4.10. The first 8 shells of E_8.

Norm	Number	Vectors
0	1	0^8.
2	240	1^20^6, $E(1/2)^8$.
4	2160	20^7, 1^40^4, $D(3/2)(1/2)^7$.
6	6720	21^20^5, 1^60^2, $E(3/2)^2(1/2)^6$.
8	17520	2^20^6, 21^40^3, 1^8, $D(3/2)^3(1/2)^5$, $E(5/2)(1/2)^7$.
10	30240	310^6, $2^21^20^4$, 21^60, $D(5/2)(3/2)(1/2)^6$, $E(3/2)^4(1/2)^4$.
12	60480	31^30^4, 2^30^5, $2^21^40^2$, $E(5/2)(3/2)^2(1/2)^5$, $D(3/2)^5(1/2)^3$.
14	82560	3210^5, 31^50^2, $2^31^20^3$, 2^21^6, $D(7/2)(1/2)^7$, $E(5/2)^2(1/2)^6$, $D(5/2)(3/2)^3(1/2)^4$, $E(3/2)^6(1/2)^2$.
16	140400	40^7, 321^30^3, 31^7, 2^40^4, 2^31^40, $E(7/2)(3/2)(1/2)^6$, $D(5/2)^2(3/2)(1/2)^5$, $E(5/2)(3/2)^4(1/2)^3$, $D(3/2)^7(1/2)$.

8.2 The 7-dimensional lattices E_7 and E_7^*. The vectors in E_8 perpendicular to any minimal vector $v \in E_8$ form the lattice E_7:

$$E_7 = \{x \in E_8 : x \cdot v = 0\} . \tag{106}$$

There are several possible coordinate systems. Using the even coordinate system for E_8 and taking $v = (½^8)$ we obtain

$$E_7 = \{(x_1, \ldots, x_8) \in E_8 : x_1 + \cdots + x_8 = 0\} ; \tag{107}$$

using the odd coordinate system and taking $v = (½^7, -½)$ we obtain

$$E_7 = \{(x_1, \ldots, x_8) \in E_8 : \Sigma x_i = 2x_8\} ; \tag{108}$$

and (in either coordinate system) $v = (0^6, 1, -1)$ leads to

$$E_7 = \{(x_1, \ldots, x_8) \in E_8 : x_7 = x_8\} . \tag{109}$$

A standard integral basis (using definition (107)) is marked on the Coxeter-Dynkin diagram in Fig. 4.8. The extending node is $(0^6, -1, 1)$. A generator matrix for E_7 as defined by (107) is

$$M = \begin{bmatrix} -1 & 1 & 0 & 0 & 0 & 0 & 0 & 0 \\ 0 & -1 & 1 & 0 & 0 & 0 & 0 & 0 \\ 0 & 0 & -1 & 1 & 0 & 0 & 0 & 0 \\ 0 & 0 & 0 & -1 & 1 & 0 & 0 & 0 \\ 0 & 0 & 0 & 0 & -1 & 1 & 0 & 0 \\ 0 & 0 & 0 & 0 & 0 & -1 & 1 & 0 \\ ½ & ½ & ½ & ½ & -½ & -½ & -½ & -½ \end{bmatrix} . \tag{110}$$

Still another definition (see §2.5 of Chap. 5) leads to the generator matrix

$$M' = \frac{1}{2} \begin{bmatrix} 2\,0 & 0\,0 & 0\,0 & 0 \\ 0\,2 & 0\,0 & 0\,0 & 0 \\ 0\,0 & 2\,0 & 0\,0 & 0 \\ 0\,0 & 0\,2 & 0\,0 & 0 \\ 1\,1 & 1\,0 & 1\,0 & 0 \\ 0\,1 & 1\,1 & 0\,1 & 0 \\ 0\,0 & 1\,1 & 1\,0 & 1 \end{bmatrix} . \tag{111}$$

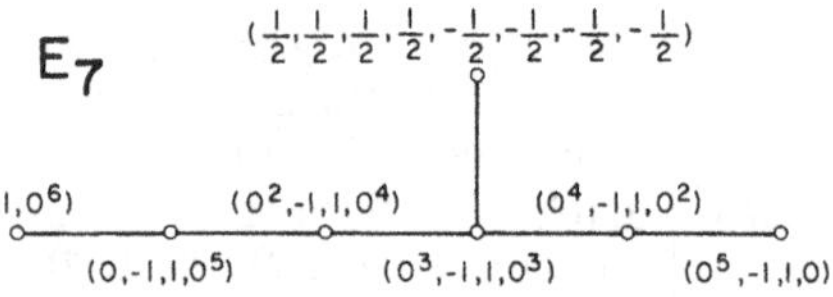

Figure 4.8. Diagram for E_7 showing integral basis for root lattice in the coordinate system (107).

For the definition via (107), (110) we have det = 2, minimal norm = 2, $\tau = 126$, the minimal vectors consist of 56 of the form $(1,-1,0^6)$ and 70 of the form $(½^4,-½^4)$, $h = 18$, $\rho = 1/\sqrt{2}$, $\Delta = \pi^3/105 = 0.2953...$ $(\delta = 1/16)$, $R = \rho\sqrt{3} = \sqrt{3/2}$, a typical deep hole being the glue vector [1], while the Voronoi cell is given in Chap. 21. Glue vectors:

$$[0] = (0,0,0,0,0,0,0,0)\,, \qquad \text{of norm } 0,$$
$$[1] = (1/4,1/4,1/4,1/4,1/4,1/4,-3/4,-3/4)\,, \qquad \text{of norm } 3/2.$$

Glue group = C_2. Automorphism group: G_0 is the Weyl group $W(E_7)$, of order $g_0 = 2^{10}\cdot 3^4 \cdot 5 \cdot 7 = 2903040$, $G_1 = 1$. E_7 is the unique lattice with this density [Vet2], although there are equally dense nonlattices (§4.2 of Chap. 5). Theta series (Table 4.9):

$$\Theta_{E_7}(z) = \theta_3(2z)^7 + 7\,\theta_3(2z)^3\,\theta_2(2z)^4$$
$$= 1 + 126\,q^2 + 756\,q^4 + 2072\,q^6 + \cdots\,, \tag{112}$$
$$\Theta_{E_7+[1]}(z) = \theta_2(2z)^7 + 7\,\theta_2(2z)^3\,\theta_3(2z)^4$$
$$= 56\,q^{3/2} + 576\,q^{7/2} + 1512\,q^{11/2} + 4032\,q^{15/2}$$
$$+ 5544\,q^{19/2} + 12096\,q^{23/2} + 13664\,q^{27/2} + \cdots\,. \tag{113}$$

Other constructions will be given in Chap. 5.

The dual lattice is

$$E_7^* = E_7 \cup ([1]+E_7)\,, \tag{114}$$

with generator matrix

$$M = \begin{bmatrix} -1 & 1 & 0 & 0 & 0 & 0 & 0 & 0 \\ 0 & -1 & 1 & 0 & 0 & 0 & 0 & 0 \\ 0 & 0 & -1 & 1 & 0 & 0 & 0 & 0 \\ 0 & 0 & 0 & -1 & 1 & 0 & 0 & 0 \\ 0 & 0 & 0 & 0 & -1 & 1 & 0 & 0 \\ 0 & 0 & 0 & 0 & 0 & -1 & 1 & 0 \\ -3/4 & -3/4 & 1/4 & 1/4 & 1/4 & 1/4 & 1/4 & 1/4 \end{bmatrix}, \tag{115}$$

det = ½, minimal norm = 3/2, $\tau = 56$, minimal vectors = $\pm(1/4^6,-3/4^2)$, $\rho = \sqrt{3/8}$, $R = \rho\sqrt{7/3} = \sqrt{7/8}$, a typical deep hole being $(7/8,(-1/8)^7)$, whose images are the vertices of the Voronoi cell (see [Wor2]). The theta series is the sum of (112) and (113).

8.3 The 6-dimensional lattices E_6 and E_6^*. The vectors in E_8 perpendicular to any A_2-sublattice V in E_8 form the lattice E_6:

$$E_6 = \{x \in E_8 : x \cdot v = 0 \ \text{ for all } v \in V\}\,. \tag{116}$$

Again there are several possible coordinate systems. Using the even coordinate system for E_8 and taking $V = \langle (1,0^6,1), (½^8) \rangle$ we obtain

$$E_6 = \{\,(x_1,\ldots,x_8) \in E_8 : x_1 + x_8 = x_2 + \cdots + x_7 = 0\,\}\,; \tag{117}$$

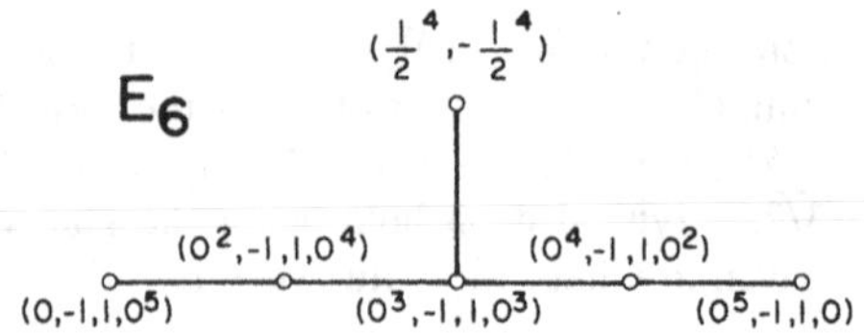

Figure 4.9. Diagram for E_6 showing integral basis for root lattice in the coordinate system (117).

while (in either coordinate system) $V = < (0^5, 1, -1, 0),\ (0^6, 1, -1) >$ leads to

$$E_6 = \{(x_1, \ldots, x_8) \in E_8 : x_6 = x_7 = x_8\}. \tag{118}$$

A standard integral basis (using definition (117)) is marked on the Coxeter-Dynkin diagram in Fig. 4.9. The extending node is $(-1;0^6;1)$. From Fig. 4.9 we obtain the generator matrix

$$M = \begin{bmatrix} 0 & -1 & 1 & 0 & 0 & 0 & 0 & 0 \\ 0 & 0 & -1 & 1 & 0 & 0 & 0 & 0 \\ 0 & 0 & 0 & -1 & 1 & 0 & 0 & 0 \\ 0 & 0 & 0 & 0 & -1 & 1 & 0 & 0 \\ 0 & 0 & 0 & 0 & 0 & -1 & 1 & 0 \\ \frac{1}{2} & \frac{1}{2} & \frac{1}{2} & \frac{1}{2} & -\frac{1}{2} & -\frac{1}{2} & -\frac{1}{2} & -\frac{1}{2} \end{bmatrix}. \tag{119}$$

E_6 also has a simple construction as a three-dimensional complex lattice, a lattice over the Eisenstein integers (§2.6 of Chap. 2). A generator matrix for this version of E_6 is

$$\begin{bmatrix} \theta & 0 & 0 \\ 0 & \theta & 0 \\ 1 & 1 & 1 \end{bmatrix}, \tag{120}$$

where $\theta = \omega - \bar{\omega} = \sqrt{-3}$.

Using the definitions (117) and (119), we have det = 3, minimal norm = 2, $\tau = 72$, the minimal vectors consist of 30 of the form $(0;1,-1,0^4;0)$, 40 of the form $\pm(\frac{1}{2};\frac{1}{2}^3,-\frac{1}{2}^3;-\frac{1}{2})$ and 2 of the form $\pm(1;0^6;-1)$, $h = 12$, $\rho = 1/\sqrt{2}$, $\Delta = \pi^3/48\sqrt{3} = 0.3729\ldots$ $(\delta = 1/8\sqrt{3})$, $R = \rho\sqrt{8/3}$, a typical deep hole being the glue vector [1], and the Voronoi cell is given in Chap. 21. Glue vectors:

$$\begin{aligned} [0] &= (0;0,0,0,0,0,0;0), && \text{of norm } 0, \\ [1] &= (0;-\tfrac{2}{3},-\tfrac{2}{3},\tfrac{1}{3},\tfrac{1}{3},\tfrac{1}{3},\tfrac{1}{3};0), && \text{of norm } \tfrac{4}{3}, \\ [2] &= -[1], && \text{of norm } \tfrac{4}{3}. \end{aligned}$$

Glue group = C_3. Automorphism group: G_0 is the Weyl group $W(E_6)$, of order $g_0 = 2^7 \cdot 3^4 \cdot 5 = 51840$, $G_1 = C_2$ (generated by negation). (See [Cox1], [Cox7], [Cox13].) E_6 is the unique lattice with this density

[Bar7], [Vet2], although there are equally dense nonlattices (§4.2 of Chap. 5). Theta series (Table 4.9):

$$\Theta_{E_6}(z) = \phi_0(z)^3 + {}^1\!/_4\{\phi_0(z/3) - \phi_0(z)\}^3 ,$$
$$= 1 + 72q^2 + 270q^4 + 720q^6 + \cdots , \quad (121)$$
$$\Theta_{E_6^*}(z) = {}^1\!/_3[\phi_0(z/3)^3 + {}^1\!/_4\{3\phi_0(z) - \phi_0(z/3)\}^3]$$
$$= 1 + 54q^{4/3} + 72q^2 + 432q^{10/3} + 270q^4$$
$$+ 918q^{16/3} + 720q^6 + 2160q^{22/3} + \cdots , \quad (122)$$
$$\Theta_{E_6+[1]}(z) = \Theta_{E_6+[2]}(z) = {}^1\!/_2\{\Theta_{E_6^*}(z) - \Theta_{E_6}(z)\}$$
$$= 27q^{4/3} + 216q^{10/3} + 459q^{16/3} + \cdots . \quad (123)$$

(Eq. (121) follows from definition (120), as shown in Chap. 7, and (122) is derived from (121) via (18).)

The dual lattice is

$$E_6^* = E_6 \cup ([1]+E_6) \cup ([2]+E_6), \quad (124)$$

with generator matrix

$$M = \begin{bmatrix} 0 & -1 & 1 & 0 & 0 & 0 & 0 & 0 \\ 0 & 0 & -1 & 1 & 0 & 0 & 0 & 0 \\ 0 & 0 & 0 & -1 & 1 & 0 & 0 & 0 \\ 0 & 0 & 0 & 0 & -1 & 1 & 0 & 0 \\ 0 & 2/3 & 2/3 & -1/3 & -1/3 & -1/3 & -1/3 & 0 \\ 1/2 & 1/2 & 1/2 & 1/2 & -1/2 & -1/2 & -1/2 & -1/2 \end{bmatrix} . \quad (125)$$

As a three-dimensional complex lattice E_6^* has generator matrix

$$M' = \begin{bmatrix} \theta & 0 & 0 \\ 1 & -1 & 0 \\ 1 & 0 & -1 \end{bmatrix} . \quad (126)$$

Using the definition (125), det $= 1/3$, minimal norm $= 4/3$, $\tau = 54$, the minimal vectors consist of 30 of the form $\pm(0;(-2/3)^2,(1/3)^4;0)$ and 24 of the form $\pm(\pm 1/2;(-1/6)^5,5/6;\mp 1/2)$, $\rho = 1/\sqrt{3}$, and the covering radius $R = \rho\sqrt{2} = \sqrt{2/3}$, corresponding to the deep hole $(0;1,1,1,-1,-1,-1;0)/3$, whose images are the vertices of the Voronoi cell (see [Wor1]). Theta series: Eq. (122).

9. The 12-dimensional Coxeter-Todd lattice K_{12}

So far all lattices mentioned were known in the nineteenth century, but now we move into the twentieth century. The automorphism group of K_{12} was discovered by Mitchell in 1914 [Mit3], while the lattice itself was first explicitly described by Coxeter and Todd in 1954 [Cox29]. Like E_6, K_{12}

has a simple description as a complex lattice. K_{12} is the real form of the 6-dimensional complex lattice over the Eisenstein integers that is generated by the vectors

$$\frac{1}{\sqrt{2}}(\pm\theta, \pm 1^5), \tag{127}$$

where $\theta = \omega - \bar{\omega} = \sqrt{-3}$ may be in any position and there are an even number of minus signs. An alternative definition (Example 10a of Chap. 7) uses the *hexacode* (§2.5.2 of Chap. 3, in the version defined by (64)) and corresponds to the generator matrix

$$M = \begin{bmatrix} 2 & 0 & 0 & 0 & 0 & 0 \\ 0 & 2 & 0 & 0 & 0 & 0 \\ 0 & 0 & 2 & 0 & 0 & 0 \\ 1 & \omega & \omega & 1 & 0 & 0 \\ \omega & 1 & \omega & 0 & 1 & 0 \\ \omega & \omega & 1 & 0 & 0 & 1 \end{bmatrix}. \tag{128}$$

(See also Example 11c of Chap. 7.) In the definition given by (127), K_{12} may be regarded as the set of all Leech lattice vectors of the form

$$\begin{array}{|cccc|} u_1 & u_2 & \cdots & u_6 \\ v_1 & v_2 & \cdots & v_6 \\ v_1 & v_2 & \cdots & v_6 \\ v_1 & v_2 & \cdots & v_6 \end{array} \tag{129}$$

(see Chap. 6). (129) corresponds to the K_{12} vector

$$(u_1 + \theta v_1, \ldots, u_6 + \theta v_6)/\sqrt{8}.$$

For example the Leech vector with $v_1 = u_2 = \cdots = u_6 = 2$, $u_1 = v_2 = \cdots = v_6 = 0$ corresponds to $(\theta, 1^5)/\sqrt{2}$. K_{12} may also be obtained as the set of Leech lattice vectors of the form

$$\begin{array}{|cccc|} 0 & 0 & \cdots & 0 \\ x_1 & x_2 & \cdots & x_6 \\ y_1 & y_2 & \cdots & y_6 \\ z_1 & z_2 & \cdots & z_6 \end{array} \tag{130}$$

where $x_i + y_i + z_i = 0$, $i = 1, \ldots, 6$ (Fig. 6.3). Using the definition (127), det = 729, minimal norm = 4, kissing number $\tau = 756$, the minimal vectors consist of 576 of the form $\omega^\nu(\pm\theta, 1^5)/\sqrt{2}$, $\nu = 0, 1, 2$, with an even number of minus signs, and 180 of the form $\omega^\nu(\pm 2^2, 0^4)/\sqrt{2}$, packing radius $\rho = 1$, density $\Delta = \pi^6/19440 = 0.04945\ldots$ ($\delta = 1/27$), and covering radius $R = \rho\sqrt{8/3}$, a typical deep hole being $(4/\theta, 0^5)/\sqrt{2}$. The Voronoi cell has 4788 faces, 756 corresponding to the minimal vectors and 4032 to those of the next layer [Con38, Th. 3]. As a complex lattice over the Eisenstein integers K_{12} is an integral unimodular lattice, and so (as a real lattice) $K_{12}^* \cong K_{12}$. The automorphism group of the real lattice has order

Table 4.11. Theta series of 12-dimensional Coxeter-Todd lattice K_{12}.

m	$N(m)$	m	$N(m)$
0	1	38	48009024
2	0	40	64049832
4	756	42	70709184
6	4032	44	102958128
8	20412	46	124782336
10	60480	48	142254252
12	139860	50	189423360
14	326592	52	237588120
16	652428	54	248250240
18	1020096	56	344391264
20	2000376	58	397510848
22	3132864	60	433936440
24	4445532	62	554879808
26	7185024	64	671393772
28	10747296	66	677557440
30	13148352	68	908374824
32	21003948	70	1018507392
34	27506304	72	1079894844
36	33724404		

$2^{10} \cdot 3^7 \cdot 5 \cdot 7 = 78382080$. The automorphisms preserving the complex lattice form Mitchell's complex reflection group of half this order (and isomorphic to $6 \cdot P\Omega_6^-(3) \cdot 2$ and to $6 \cdot PSU_4(3) \cdot 2$). Other names for Mitchell's group are $6 \cdot U_4(3) \cdot 2$, $6 \cdot HO(4, 3^2) \cdot 2$ [Dic1], $[2\ 1\ ;\ 3]^2$ (see [She1]), $[3\ 2\ 1]^3$ (see [Ben0]), and $W(K_6)$ [Coh1]. It is number 34 on Shephard and Todd's list [She2]. See also [Edg1], [Edg2]. Theta series:

$$\Theta_{K_{12}}(z) = \phi_0(2z)^6 + 45\phi_0(2z)^2\phi_1(2z)^4 + 18\phi_1(2z)^6$$
$$= 1 + 756q^4 + 4032q^6 + 20412q^8 + \cdots \quad (131)$$

(Table 4.11). For further information see later chapters and [Con37], [Con38], [Fei2], [Ham1], [Har6], [Har7], [Lin6], [Lin7], [Tod1], [Tod2].

10. The 16-dimensional Barnes-Wall lattice Λ_{16}

This lattice appears to have been first published by Barnes and Wall in 1959 [Bar18], and since then has been rediscovered by many authors. There are several constructions. Construction B of Chap. 5 applied to the first-order Reed-Muller code of length 16 leads to the generator matrix shown in Fig. 4.10, for which det = 256, minimal norm = 4, kissing number $\tau = 4320$, the minimal vectors consist of 480 of the form $2^{-1/2}(\pm 2^2, 0^{14})$ and 3840 of the form $2^{-1/2}(\pm 1^8, 0^8)$, where the positions of

$$
\frac{1}{\sqrt{2}}
\left[\begin{array}{cccccccccccccccc}
4 & & & & & & & & & & & & & & & \\
2 & 2 & & & & & & & & & & & & & & \\
2 & 0 & 2 & & & & & & & & & & & & & \\
2 & 0 & 0 & 2 & & & & & & & & & \mathbf{0} & & & \\
2 & 0 & 0 & 0 & 2 & & & & & & & & & & & \\
2 & 0 & 0 & 0 & 0 & 2 & & & & & & & & & & \\
2 & 0 & 0 & 0 & 0 & 0 & 2 & & & & & & & & & \\
2 & 0 & 0 & 0 & 0 & 0 & 0 & 2 & & & & & & & & \\
2 & 0 & 0 & 0 & 0 & 0 & 0 & 0 & 2 & & & & & & & \\
2 & 0 & 0 & 0 & 0 & 0 & 0 & 0 & 0 & 2 & & & & & & \\
2 & 0 & 0 & 0 & 0 & 0 & 0 & 0 & 0 & 0 & 2 & & & & & \\
1 & 1 & 1 & 1 & 0 & 1 & 0 & 1 & 1 & 0 & 0 & 1 & & & & \\
0 & 1 & 1 & 1 & 1 & 0 & 1 & 0 & 1 & 1 & 0 & 0 & 1 & & & \\
0 & 0 & 1 & 1 & 1 & 1 & 0 & 1 & 0 & 1 & 1 & 0 & 0 & 1 & & \\
0 & 0 & 0 & 1 & 1 & 1 & 1 & 0 & 1 & 0 & 1 & 1 & 0 & 0 & 1 & \\
1 & 1 & 1 & 1 & 1 & 1 & 1 & 1 & 1 & 1 & 1 & 1 & 1 & 1 & 1 & 1
\end{array}\right]
$$

Figure 4.10. Generator matrix for 16-dimensional Barnes-Wall lattice Λ_{16}. The last five rows are a generator matrix for the first-order Reed-Muller code of length 16.

the ± 1's form one of the 30 codewords of weight 8 in the first-order Reed-Muller code and there are an even number of minus signs. The packing radius $\rho = 1$, density $\Delta = \pi^8/16 \cdot 8! = 0.01471\ldots$ ($\delta = 1/16$), and covering radius $R = \rho\sqrt{3}$, a typical deep hole being $2^{-1/2}(1^6, 0^{10})$, where the six 1's are at a deep hole in the first-order Reed-Muller code (corresponding to a "bent function" of four variables [Mac6, Chap. 14]) — see §5 of Chap. 6. Theta series (Table 4.12):

Table 4.12. Theta series of 16-dimensional Barnes-Wall lattice Λ_{16}.

m	$N(m)$	m	$N(m)$
0	1	32	8593797600
2	0	34	11585617920
4	4320	36	19590534240
6	61440	38	25239859200
8	522720	40	40979580480
10	2211840	42	50877235200
12	8960640	44	79783021440
14	23224320	46	96134307840
16	67154400	48	146902369920
18	135168000	50	172337725440
20	319809600	52	256900127040
22	550195200	54	295487692800
24	1147643520	56	431969276160
26	1771683840	58	487058227200
28	3371915520	60	699846624000
30	4826603520	62	776820326400

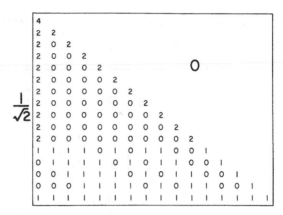

$$\Theta_{\Lambda_{16}}(z) = \tfrac{1}{2}\{\theta_2(2z)^{16}+\theta_3(2z)^{16}+\theta_4(2z)^{16}+30\theta_2(2z)^8\theta_3(2z)^8\}$$
$$= 1 + 4320q^4 + 61440q^6 + \cdots . \qquad (132)$$

Λ_{16} may be constructed from the Leech lattice Λ_{24}. There are involutory symmetries of Λ_{24} that fix a 16-space, and the portion of Λ_{24} that lies in such a space is a copy of Λ_{16}. The automorphism group of Λ_{16} is the centralizer of such an involution in the automorphism group of Λ_{24}, considered modulo the involution itself. The group has order $g = 2^{21} \cdot 3^5 \cdot 5^2 \cdot 7 = 89181388800$, and structure $2^{1+8} \cdot O_8^+(2)$, where $O_8^+(2)$ is the simple group of order $2^{12} \cdot 3^5 \cdot 5^2 \cdot 7 = 174182400$.

Λ_{16} may also be obtained by applying Construction C (or D) to either of the following sequences of binary codes: [16, 1, 16], [16, 11, 4], [16, 16, 1] or [16, 5, 8], [16, 15, 2] (see §3.4 of Chap. 5, Example 7 of Chap. 7, §8.2f of Chap. 8). Other constructions will be found in §4 of Chap. 6 and in §4 of Chap. 8, where Λ_{16} appears as a self-dual lattice over $\mathbf{Z}[e^{\pi i/4}]$. The last-mentioned construction implies that $\Lambda_{16}^* \cong \Lambda_{16}$.

11. The 24-dimensional Leech lattice Λ_{24}

This lattice was discovered by Leech in 1965 [Lee5]. We shall give several constructions in this book. The following definition is probably the simplest. We state it in three equivalent forms.

(i) The Leech lattice Λ_{24} is generated by all vectors of the form

$$\frac{1}{\sqrt{8}}(\mp 3, \pm 1^{23}), \qquad (133)$$

where the ∓ 3 may be in any position, and the upper signs are taken on a "$\mathcal{C}$-set", i.e. the set of coordinates where a codeword of the binary Golay code $\mathcal{C}_{24}$ is 1.

(ii) Equivalently, we label the coordinate positions $\infty, 0, 1, \ldots, 22$ and let Q denote the set of nonzero quadratic residues modulo 23. Then Λ_{24} is spanned by the vectors

$$\begin{aligned} &a(2^{12},0^{12}), \quad \text{23 vectors, supported on a translate} \\ &\qquad \{\{0\} \cup Q\} + i, \quad 0 \leqslant i \leqslant 22, \\ &a(-3,1^{23}), \quad \text{a single vector}, \\ &a(\pm 4^2,0^{22}), \quad 2 \cdot 24 \cdot 23 \ \text{vectors}, \end{aligned} \qquad (134)$$

where $a = 1/\sqrt{8}$ (see §3.2 of Chap. 10).

(iii) Equivalently, Λ_{24} consists of the vectors

$$a(\mathbf{0}+2c+4x), \qquad (135)$$

$$a(\mathbf{1}+2c+4y), \qquad (136)$$

where $a = 1/\sqrt{8}$, $\mathbf{0} = (0^{24})$, $\mathbf{1} = (1^{24})$, $c \in \mathcal{C}_{24}$ (regarding the components of c as *real* 0's and 1's rather than elements of $\mathbf{F}_2$) and $x, y \in \mathbf{Z}^{24}$ satisfy

$\Sigma x_i \equiv 0 \pmod 2$, $\Sigma y_i \equiv 1 \pmod 2$. (135) are called the *even* vectors and (136) the *odd* vectors. (See §4.4 of Chap. 5.)

Other constructions for the Leech lattice are given in §5.7 of Chap. 5 (from the [24,12,9] quadratic residue or Pless code over $\mathbf{F}_3$); §6 of Chap. 6 (as a laminated lattice); Example 12 of Chap. 7 and §3.6 of Chap. 10 (as a complex 12-dimensional lattice, using the [12, 6, 6] Golay code $\mathcal{C}_{12}$ over $\mathbf{F}_3$); §2 of Chap. 8 (as a three-dimensional icosian lattice, an analog to the Turyn construction of $\mathcal{C}_{24}$); §3 of Chap. 8 (from Construction A_f, a generalized version of Construction A); §5 of Chap. 8 (McKay's construction from a Hadamard matrix of order 12); §7.3b of Chap. 8 (Craig's construction as an ideal in the cyclotomic field $\mathbf{Q}(e^{2\pi i/39})$); §7.5 of Chap. 8 (Thompson's construction from the cyclotomic field $\mathbf{Q}(e^{2\pi i/23})$); Chap. 11 (the MOG definition – see below); Eq. (7) of Chap. 17 (from D_{24}); Chap. 17 and §5 of Chap. 18 (as a neighboring lattice to A_1^{24}); Chap. 24 (twenty-three constructions, one for each Niemeier lattice!); Theorems 1, 2 and 3 of Chap. 26 (three constructions of the Leech lattice as a Lorentzian lattice), and Theorem 1 of Chap. 27 (as the Coxeter diagram of the automorphism group of the Lorentzian lattice $\mathrm{II}_{25,1}$).

In a certain sense the construction given in §6 of Chap. 6 is the simplest of all, since it states that the Leech lattice may be built up by "induction", starting with the 1-dimensional lattice of integers and each time extending to a densest lattice in the next dimension. The lattices constructed in this way can be seen in Fig. 6.1 of Chap. 6; in 24 dimensions one obtains the Leech lattice. This is a "no-input" construction! Another is given in Chap. 27.

But for calculating with the Leech lattice it is best to use MOG (Miracle Octad Generator) coordinates, which are arranged in a 4×6 rectangle. The MOG coordinates for $\mathcal{C}$-sets are obtained from the codewords of the hexacode (§2.5.2 of Chap. 3) by replacing each digit by a column of length 4 as shown in Fig. 4.11. The $\mathcal{C}$-sets are obtained either by giving each digit an *odd* interpretation in any way such that the top row contains an odd number of points, or an *even* interpretation in any way such that the top row contains an even number of points. The Leech lattice is spanned by the vectors $(\pm 2^8, 0^{16})$, where the support is a $\mathcal{C}$-set and there are an even number of minus signs, together with the vector $(-3, 1^{23})$. MOG arrays are extensively studied in Chap. 11.

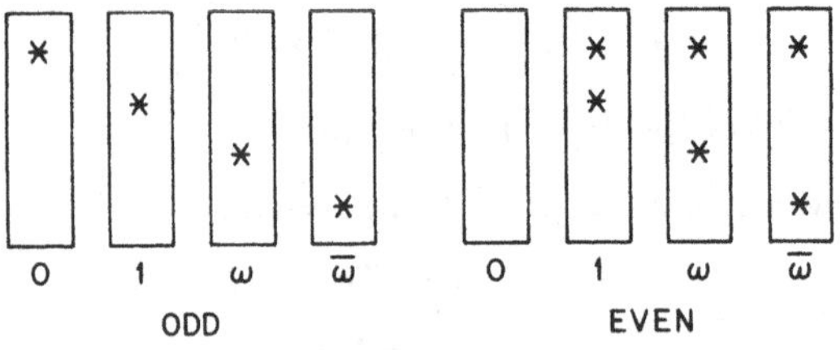

Figure 4.11. The two odd interpretations of a digit are the sets on the left or their complements. Similarly the two even interpretations are the sets on the right or their complements.

$\frac{1}{\sqrt{8}}$

8 0 0 0	0 0 0 0	0 0 0 0	0 0 0 0	0 0 0 0	0 0 0 0
4 4 0 0	0 0 0 0	0 0 0 0	0 0 0 0	0 0 0 0	0 0 0 0
4 0 4 0	0 0 0 0	0 0 0 0	0 0 0 0	0 0 0 0	0 0 0 0
4 0 0 4	0 0 0 0	0 0 0 0	0 0 0 0	0 0 0 0	0 0 0 0
4 0 0 0	4 0 0 0	0 0 0 0	0 0 0 0	0 0 0 0	0 0 0 0
4 0 0 0	0 4 0 0	0 0 0 0	0 0 0 0	0 0 0 0	0 0 0 0
4 0 0 0	0 0 4 0	0 0 0 0	0 0 0 0	0 0 0 0	0 0 0 0
2 2 2 2	2 2 2 2	0 0 0 0	0 0 0 0	0 0 0 0	0 0 0 0
4 0 0 0	0 0 0 0	4 0 0 0	0 0 0 0	0 0 0 0	0 0 0 0
4 0 0 0	0 0 0 0	0 4 0 0	0 0 0 0	0 0 0 0	0 0 0 0
4 0 0 0	0 0 0 0	0 0 4 0	0 0 0 0	0 0 0 0	0 0 0 0
2 2 2 2	0 0 0 0	2 2 2 2	0 0 0 0	0 0 0 0	0 0 0 0
4 0 0 0	0 0 0 0	0 0 0 0	4 0 0 0	0 0 0 0	0 0 0 0
2 2 0 0	2 2 0 0	2 2 0 0	2 2 0 0	0 0 0 0	0 0 0 0
2 0 2 0	2 0 2 0	2 0 2 0	2 0 2 0	0 0 0 0	0 0 0 0
2 0 0 2	2 0 0 2	2 0 0 2	2 0 0 2	0 0 0 0	0 0 0 0
4 0 0 0	0 0 0 0	0 0 0 0	0 0 0 0	4 0 0 0	0 0 0 0
2 0 2 0	2 0 0 2	2 2 0 0	0 0 0 0	2 2 0 0	0 0 0 0
2 0 0 2	2 2 0 0	2 0 2 0	0 0 0 0	2 0 2 0	0 0 0 0
2 2 0 0	2 0 2 0	2 0 0 2	0 0 0 0	2 0 0 2	0 0 0 0
0 2 2 2	2 0 0 0	2 0 0 0	2 0 0 0	2 0 0 0	2 0 0 0
0 0 0 0	0 0 0 0	2 2 0 0	2 2 0 0	2 2 0 0	2 2 0 0
0 0 0 0	0 0 0 0	2 0 2 0	2 0 2 0	2 0 2 0	2 0 2 0
-3 1 1 1	1 1 1 1	1 1 1 1	1 1 1 1	1 1 1 1	1 1 1 1

Figure 4.12. Generator matrix for the 24-dimensional Leech lattice Λ_{24} in standard MOG coordinates (the rows of the matrix are divided into six blocks of 4, corresponding to the columns of the MOG, read downwards).

The generator matrix for Λ_{24} given in Fig. 4.12 has been chosen to be compatible with the MOG coordinates. For Λ_{24}, det = 1, minimal norm = 4, and kissing number $\tau = 196560$, the minimal vectors consist of $2^7 \cdot 759 = 97152$ of the form $8^{-1/2}(\pm 2^8, 0^{16})$, where the positions of the ± 2's form one of the 759 codewords of weight 8 in $\mathscr{C}_{24}$ and there are an even number of minus signs, $24 \cdot 2^{12} = 98304$ of the form (133), and $2 \cdot 24 \cdot 23 = 1104$ of the form $8^{-1/2}(\pm 4^2, 0^{22})$. The vectors in the first three layers are shown in Table 4.13, with the signs suppressed (see also

Table 4.13. Vectors in the first three shells of the Leech lattice. $\Lambda(n)_i$ indicates the Leech vectors of norm $2n$ (or type n) and shape i; the signs have been suppressed.

Class	Shape	Number	Class	Shape	Number
$\Lambda(0)_1$	(0^{24})	1	$\Lambda(4)_{2+}$	$(2^{16}0^8)$	$2^{11} \cdot 759$
$\Lambda(2)_2$	$(2^8 0^{16})$	$2^7 \cdot 759$	$\Lambda(4)_{2-}$	$(2^{16}0^8)$	$2^{11} \cdot 759 \cdot 15$
$\Lambda(2)_3$	$(3\,1^{23})$	$2^{12} \cdot 24$	$\Lambda(4)_3$	$(3^5 1^{19})$	$2^{12}\binom{24}{5}$
$\Lambda(2)_4$	$(4^2 0^{22})$	$2^2\binom{24}{2}$	$\Lambda(4)_4$	$(4^4 0^{20})$	$2^4\binom{24}{4}$
$\Lambda(3)_2$	$(2^{12}0^{12})$	$2^{11} \cdot 2576$	$\Lambda(4)_{4+}$	$(4^2 2^8 0^{14})$	$2^9 \cdot 759 \cdot \binom{16}{2}$
$\Lambda(3)_3$	$(3^3 1^{21})$	$2^{12}\binom{24}{3}$	$\Lambda(4)_{4-}$	$(4\,2^{12}0^{11})$	$2^{12} \cdot 2576 \cdot 12$
$\Lambda(3)_4$	$(4\,2^8 0^{15})$	$2^8 \cdot 759 \cdot 16$	$\Lambda(4)_5$	$(5\,3^2 1^{21})$	$2^{12}\binom{24}{3} \cdot 3$
$\Lambda(3)_5$	$(5\,1^{23})$	$2^{12} \cdot 24$	$\Lambda(4)_6$	$(6\,2^7 0^{16})$	$2^7 \cdot 759 \cdot 8$
			$\Lambda(4)_8$	$(8\,0^{23})$	$2^1 \cdot 24$

§3.2 of Chap. 10). The first 15 layers are described on page 181 of [Con16]. The packing radius $\rho = 1$, density $\Delta = \pi^{12}/12! = 0.001930...$ $(\delta = 1)$, covering radius $R = \sqrt{2}\ \rho = \sqrt{2}$, there are 23 different types of deep hole (Chap. 23), one of which is $8^{-1/2}(4,0^{23})$ (this is an "octahedral" hole, surrounded by 48 lattice points), and 284 types of shallow hole (Chap. 25). The thickness $\Theta = (2\pi)^{12}/12! = 7.9035...$. The Voronoi cell has 16969680 faces, 196560 corresponding to the minimal vectors and 16773120 to those of the next layer (from [Con38] or Theorem 10 of Chap. 21). Λ_{24} is the unique integral lattice of determinant 1 and minimal norm 4 (Chap. 12). Its automorphism group is Co_0 (or $\cdot 0$), of order

$$2^{22}\, 3^9\, 5^4\, 7^2\, 11 \cdot 13 \cdot 23 = 8315553613086720000 \tag{137}$$

(see Chaps. 10, 11). Theta series:

$$\Theta_{\Lambda_{24}}(z) = \Theta_{E_8}(z)^3 - 720\,\Delta_{24}(z) \tag{138}$$

$$= \frac{1}{8}\{\theta_2(z)^8+\theta_3(z)^8+\theta_4(z)^8\}^3 - \frac{45}{16}\{\theta_2(z)\theta_3(z)\theta_4(z)\}^8 \tag{139}$$

$$= \frac{1}{2}\{\theta_2(z)^{24}+\theta_3(z)^{24}+\theta_4(z)^{24}\} - \frac{69}{16}\{\theta_2(z)\theta_3(z)\theta_4(z)\}^8 \tag{140}$$

$$= \sum_{m=0}^{\infty} N(m)q^m = 1 + 196560q^4 + 16773120q^6 + \cdots, \tag{141}$$

where (cf. Eq. (53) of Chap. 2)

$$N(m) = \frac{65520}{691}\left\{\sigma_{11}\left(\frac{m}{2}\right) - \tau\left(\frac{m}{2}\right)\right\} \tag{142}$$

and $\Delta_{24}(z)$, $\tau(n)$ are defined in (39), $\sigma_{11}(n)$ in (103). The first 50 coefficients $N(m)$ are given in Table 4.14. The decompositions into prime factors of the first 20 coefficients $N(m)$ are given in [Slo8, Table X]. The coefficients grow rapidly, and

$$N(m) \approx \frac{65520}{691}\,\sigma_{11}\left(\frac{m}{2}\right) \tag{143}$$

is a good approximation (see Eqs. (54), (56) of Chap. 2). If the origin is shifted to a deep hole of octahedral (or A_1^{24}) type the theta series is

$$\frac{1}{8}\{\theta_2(z)^8+\theta_3(z)^8-\theta_4(z)^8\}^3 + \frac{3}{16}\{\theta_2(z)\theta_3(z)\theta_4(z)\}^8 \tag{144}$$

$$= \sum_{m=2}^{\infty} N_m'\, q^m = 48q^2 + 4096q^3 + 97152q^4 + \cdots, \tag{145}$$

$$N_m' = \frac{16}{691}\left\{\sigma_{11}(m) - \sigma_{11}\left(\frac{m}{2}\right) - \tau(m) + \tau\left(\frac{m}{2}\right)\right\} \tag{146}$$

Table 4.14. Theta series of 24-dimensional Leech lattice Λ_{24}.

m	$N(m)$	m	$N(m)$
0	1	52	348188700764268000
2	0	54	527108117540659200
4	196560	56	786767288036446080
6	16773120	58	1156841376897024000
8	398034000	60	1680521645295642240
10	4629381120	62	2409208986562560000
12	34417656000	64	3417887115322439760
14	187489935360	66	4792384230947389440
16	814879774800	68	6658492791534948000
18	2975551488000	70	9154704673132462080
20	9486551299680	72	12486419179545402000
22	27052945920000	74	16869989277755228160
24	70486236999360	76	22632233269428619200
26	169931095326720	78	30102943984468992000
28	384163586352000	80	39789443408903042400
30	820166620815360	82	52181704975196160000
32	1668890090322000	84	68053817735463006720
34	3249631112232960	86	88114798569141227520
36	6096882661243920	88	113523923563982568000
38	11045500816896000	90	145290076878792867840
40	19428439855275360	92	185116016465911440000
42	33213186220032000	94	234407918373703925760
44	55431591273414720	96	295640558277712240320
46	90344564568760320	98	370725724086681600000
48	144355739339448000	100	463209975930723119280
50	226066364540190720		

with the convention that $\sigma_{11}(x)$ and $\tau(x)$ are 0 if x is not an integer. Other references dealing with the Leech lattice are [Bay1], [Bro4], [Lep2], [Lin6], [Lin8], [Mil1], [Tit6], [Tit7], [Vos1]. The "odd Leech lattice" O_{24} (Chap. 6, Appendix) was discovered by O'Connor and Pall in 1944 [O'Co1].

5

Sphere Packing and Error-Correcting Codes

John Leech and N. J. A. Sloane

Error-correcting codes are used to construct dense sphere packings in n-dimensional Euclidean space $\mathbf{R}^n$, and other packings are obtained by taking cross-sections or building up by layers. In this way we construct the densest packings known in all dimensions $n \leqslant 29$ and in some higher dimensions, as well as a number of other interesting packings.

1. Introduction

In this chapter we make systematic use of error-correcting codes to construct dense sphere packings. By use of cross-sections we then obtain packings in spaces of lower dimension, and by building up packings by layers we obtain packings in spaces of higher dimension. The packings constructed in this chapter include the densest packings presently known in all dimensions up to 29, as well as a number of other interesting packings. They are summarized in Tables 1.2, 1.3 and Fig. 1.5 of Chap. 1.

We begin by defining the coordinate array of a point in §1.1, and then §2 describes Construction A, which is most effective in up to 15 dimensions. §3 describes Construction B, effective in dimensions 8 to 24. §4 digresses to deal with packings built up from layers, while §5 gives some special constructions for dimensions 36, 40, 48 and 60. §6 deals with Construction C, which generalizes A and B and is especially effective in dimensions that are powers of 2. The constructions make use of the material on error-correcting codes given in §2 of Chap. 3.

Packings built up by layers are considered further in Chap. 6, while Chaps. 7 and 8 give additional properties and generalizations of Constructions A, B and C. This chapter is a revised and updated version of [Lee10].

1.1 The coordinate array of a point. The *coordinate array* of a point $x = (x_1, \ldots, x_n)$ with integer coordinates is obtained by writing the binary expansions of the coordinates x_i in columns, beginning with the least significant digit [Lee5, §1.42]. Complementary notation is used for negative numbers, as illustrated in the display below. The first, second, third, ... rows of the coordinate array are called the 1's, 2's, 4's,... rows respectively. The 1's row of the array comprises the 1's digits of the coordinates, and thus has 0's for even coordinates and 1's for odd coordinates. The 2's, 4's, 8's,... rows similarly comprise the 2's, 4's, 8's,... digits of the coordinates. For example, the coordinate array of the point (4, 3, 2, 1, 0, −1, −2, −3) is

$$\begin{bmatrix} 0 & 1 & 0 & 1 & 0 & 1 & 0 & 1 \\ 0 & 1 & 1 & 0 & 0 & 1 & 1 & 0 \\ 1 & 0 & 0 & 0 & 0 & 1 & 1 & 1 \\ 0 & 0 & 0 & 0 & 0 & 1 & 1 & 1 \\ & \cdots & & \cdots & & \cdots & & \end{bmatrix} \begin{matrix} 1's \text{ row} \\ 2's \text{ row} \\ 4's \text{ row} \\ 8's \text{ row} \\ \\ \end{matrix} \qquad (1)$$

The number of rows is potentially infinite, but all rows after the first few are identical.

2. Construction A

2.1 The construction. Let C be an (n,M,d) binary code. The following construction specifies a set of centers for a sphere packing in $\mathbf{R}^n$.

Construction A. $x = (x_1, \ldots, x_n)$ is a center if and only if x is congruent (modulo 2) to a codeword of C.

Thus a point x with integer coordinates is a center if and only if the 1's row of the coordinate array of x is in C. A lattice packing is obtained if and only if C is a linear code. Construction A is a generalization of the construction of Λ_4 as given in [Lee4, §1.1], and is studied further in Chaps. 7 and 8, where several generalizations are given.

2.2 Center density. On the unit cube at the origin,

$$\{ 0 \leqslant x_i \leqslant 1 \ : \ i = 1, \ldots, n \},$$

the centers are exactly the M codewords. All other centers are obtained by adding even integers to any of the coordinates of a codeword. This corresponds to shifting the unit cube by two in any direction. Thus all the centers may be obtained by repeating a building block consisting of a $2 \times 2 \times \cdots \times 2$ cube with the codewords marked on the vertices of the $1 \times 1 \times \cdots \times 1$ cube in one corner.

Each copy of the $2 \times 2 \times \cdots \times 2$ cube contributes M spheres of radius ρ (say), so the center density obtained from Construction A is

$$\delta = M \rho^n 2^{-n}. \qquad (2)$$

If two distinct centers are congruent to the same codeword their distance apart is at least 2. If they are congruent to different codewords then they differ by at least 1 in at least d places and so are at least $\sqrt{d}$ apart. Thus we may take the radius of the spheres to be

$$\rho = \tfrac{1}{2} \min\{2, \sqrt{d}\}. \tag{3}$$

2.3 Kissing numbers. Let S be a sphere with center x, where x is congruent to the codeword c. Candidates for centers closest to x are as follows. (a) There are $2n$ centers of the type $x + ((\pm 2)0^{n-1})$ at a distance of 2 from x. (b) Let $\{A_i(c)\}$ be the weight distribution of C with respect to c (§2.2 of Chap. 3). Since there are $A_d(c)$ codewords at a distance of d from c, there are $2^d A_d(c)$ centers of the type $x + ((\pm 1)^d 0^{n-d})$ at a distance of $\sqrt{d}$ from x. Therefore the number of spheres touching S, the kissing number of S, is

$$\tau(S) = \begin{cases} 2^d A_d(c) & \text{if } d < 4, \\ 2n + 16A_4(c) & \text{if } d = 4, \\ 2n & \text{if } d > 4. \end{cases} \tag{4}$$

2.4 Dimensions 3 to 6. Let C be the $[n, n-1, 2]$ linear code consisting of all codewords of even weight (§2.3.4 of Chap. 3). There are $\tfrac{1}{2}n(n-1)$ codewords of weight 2, and applying Construction A we obtain the checkerboard lattice D_n in $\mathbf{R}^n$ (§7.1 of Chap. 4), with $\rho = 1/\sqrt{2}$, $\delta = 2^{-(n+2)/2}$ and kissing number $\tau = 2n(n-1)$. The lattice points are alternate vertices of the cubic lattice $\mathbf{Z}^n$.

As mentioned in §1.5 of Chap. 1, for $n = 3, 4, 5$, D_n is the densest possible lattice packing in $\mathbf{R}^n$, and is denoted by Λ_n, since it is also a laminated lattice (see the next chapter). In $\mathbf{R}^3$ and $\mathbf{R}^5$ there are equally dense nonlattice packings (§4.2 below, §6.5 of Chap. 4, [Lee6]).

The densest six-dimensional lattice, $E_6 \cong \Lambda_6$, is not directly given by this version of Construction A, but will be obtained by stacking layers of Λ_5 in §4.2, as a section of Λ_7 in §4.5, and from a generalization of Construction A in §8 of Chap. 7.

2.5 Dimensions 7 and 8. From now on we apply Construction A to codes with minimal distance $d = 4$. The packing obtained is of spheres of unit radius, and has center density $\delta = M2^{-n}$ and kissing number $\tau = 2n + 16A_4(c)$.

Let H_n denote the binary matrix obtained from an $n \times n$ Hadamard matrix (§2.13 of Chap. 3) upon replacing the +1's by 0's and −1's by 1's. We assume that H_n has been normalized so that the first row and column are all 0's. Up to obvious equivalences there is a unique H_8, and if the first column is deleted the rows form the [7,3,4] code $\mathcal{H}_7$ (§2.4.1 of Chap. 3), with $A_4 = 7$. By applying Construction A we obtain $E_7 \cong \Lambda_7$, the densest lattice packing in $\mathbf{R}^7$, with $\delta = 2^{-4}$ and $\tau = 2 \cdot 7 + 16 \cdot 7 = 126$ (§8.2 of Chap. 4).

The rows of H_8 together with their complements form the [8,4,4] Hamming code $\mathscr{H}_8$, with $A_4 = 14$ (§2.4.2 of Chap. 3). This is also a first-order Reed-Muller (or RM) code. From this code we obtain $E_8 \cong \Lambda_8$, the densest lattice packing in $\mathbf{R}^8$, with $\delta = 2^{-4}$ and $\tau = 2\cdot 8 + 14\cdot 16 = 240$ (§8.1 of Chap. 4).

2.6 Dimensions 9 to 12. H_{12} is also unique up to equivalence. Let $\mathscr{D}$ be the 11×11 matrix consisting of the vector 11011100010 (with 1's at position zero and at the quadratic residues modulo 11) together with its cyclic shifts. Then H_{12} may be taken as

$$\begin{bmatrix} 0 & \mathbf{0}^{tr} \\ \mathbf{0} & \mathscr{D} \end{bmatrix}. \tag{5}$$

The sums modulo 2 of pairs of rows of H_{12} and the complements of such sums form the 132 vectors of a Steiner system $S(5, 6, 12)$ (see [Lee4] for example). Since no two of these vectors can overlap in more than four places, they form a (12, 132, 4) code, every codeword having weight six. This code may be increased to a (12, 144, 4) nonlinear code by adding six codewords of type $0^{10}\,1^2$ and six of type $0^2\,1^{10}$, the six 1^2 and the six 0^2 being disjoint sets. Several different versions of this code are possible, depending on the relationship between the positions of the 1's in the 12 "loose" codewords and the vectors of the Steiner system.

By shortening the (12, 144, 4) code we obtain (11, 72, 4), (10, 38, 4) and (9, 20, 4) codes. Codes equivalent to these four were first given by Golay [Gol2] and Julin [Jul1], and alternative constructions and generalizations are given in [Lit1], [Mac6], [Rom1], [Slo18], [Slo20].

Applying Construction A to the various versions of the (12, 144, 4) code we obtain nonlattice packings in $\mathbf{R}^{12}$ with center density $\delta = 144\cdot 2^{-12} = 2^{-8}\cdot 3^2 = 0.03516...$, which is less than that of K_{12} (§9 of Chap. 4), but in the most favorable versions some spheres touch as many as 840 others, as we now show. Let c be a codeword of weight 6. Any vector of weight 4 and length 12 is contained in exactly 8 vectors of weight 5, and therefore in exactly 4 codewords of weight 6 (since any vector of weight 5 is contained in a unique codeword of weight 6). So the number of codewords of weight 6 that are at a Hamming distance of 4 from c is $3\binom{6}{4} = 45$.

If the 12 loose codewords are chosen so that there are three words $0^{10}\,1^2$ whose 1's coincide with pairs of 1's from c, and three words $0^2\,1^{10}$ whose 0's coincide with pairs of 0's from c, then there are an additional 6 codewords at a Hamming distance of 4 from c and $A_4(c) = 45 + 6 = 51$. Based on these codes we obtain packings in $\mathbf{R}^{12}$ with a maximal kissing number of $2\cdot 12 + 16\cdot 51 = 840$ for those spheres whose centers are congruent to c.

There are still several inequivalent versions of these packings, depending on the choice of the remaining loose codewords, but they all have the same maximal kissing number of 840, and are collectively called P_{12a} in Table 1.2.

Other choices for the 12 codewords give packings with maximal kissing number of 824 or 808. It can be shown that, regardless of the choice of the 12 loose codewords, the average kissing number obtained is equal to $770^2/_3$. (This is true even for those packings where the maximal kissing number is less than 840.)

For $\mathbf{R}^{11}$ we set $x_{12} = 1$ in P_{12a}, and find the maximal kissing number to be 566 in the most favorable cases, collectively called P_{11a}, but it can be only 550 or 534 in other cases (if $0^{10}1$ is altered to 0^{11}). Except in this last case the average is $519^7/_9$. (P_{11a} is studied further in [Vet1].)

For $\mathbf{R}^{10}$, if we set $x_{11} = x_{12} = 1$ in P_{12a}, where x_{11}, x_{12} are a pair of both $0^{10}1^2$ and $1^{10}0^2$, we obtain packings P_{10b} with $\delta = 2^{-8} \cdot 3^2$ and a maximal kissing number of 500. If instead we set $x_{12} = 1$, $x_{10} = 0$ where x_{10}, x_{12} is not a pair of either form, we obtain packings with $\delta = 2^{-9} \cdot 19$. In the most favorable cases, called P_{10a}, the maximal kissing number is 372. The average kissing number is $353^9/_{19}$ for P_{10a} and $340^1/_3$ for P_{10b}.

For $\mathbf{R}^9$ we set $x_{10} = 0$, $x_{11} = x_{12} = 1$ in P_{12a} and obtain packings P_{9a} with $\delta = 2^{-7} \cdot 5$ and a maximal kissing number of 306; the average kissing number is $235^3/_5$.

An alternative derivation of P_{10b} may be given. The tetrads of a Steiner system $S(3, 4, 10)$ (see §3.1 of Chap. 3) form 30 codewords of length 10, weight 4 and Hamming distance at least 4 apart. For example the tetrads may be taken to be the cyclic permutations of 1110001000, of 1101100000 and of 1010100100. By including the zero codeword and five codewords of weight 8, several different (10, 36, 4) codes are obtained. Construction A then gives the packing P_{10b}.

Similarly, the packings P_{9a}, P_{10a}, P_{11a} may be obtained by applying Construction A to the (9, 20, 4), (10, 38, 4) and (11, 72, 4) codes. All these are nonlattice packings.

A (10, 40, 4) code was found by Best [Bes1]; it consists of the cyclic permutations of 1010000001, 1100101100, 0001010111 and 0111111010. Although nonlinear, the weight distribution with respect to each codeword is $0^1\ 4^{22}\ 6^{12}\ 8^5$. By applying Construction A a 10-dimensional nonlattice packing P_{10c} is obtained with $\delta = 2^{-7} \cdot 5$, in which each sphere touches $2 \cdot 10 + 16 \cdot 22 = 372$ others.

Best [Bes1], [Bes3] also found eleven different constant weight codes, each consisting of 35 codewords of length 11, weight 4 and distance 4; one such code is shown in Fig. 5.1. By applying Construction A to any one of these codes an 11-dimensional arrangement of spheres is obtained in which some spheres touch $2 \cdot 11 + 16 \cdot 35 = 582$ others. These are collectively denoted by P_{11c} in Table 1.2. The densities of these packings are quite low.

2.7 Comparison of lattice and nonlattice packings. It is still an open question whether there exists any nonlattice packing with density exceeding that of the densest lattice packing in its dimension. But the nonlattice packings P_{10c}, P_{11a} and P_{13a} of §4.3 below have density greater than that

1 1 1 1 1 1 1 1 1 1 1 1 0 0 0 0 0 0 0 0 0 0 0 0 0 0 0 0 0 0 0 0 0 0 0

1 1 1 0 0 0 0 0 0 0 0 0 1 1 1 1 1 1 1 1 1 0 0 0 0 0 0 0 0 0 0 0 0 0 0

0 0 0 1 1 1 0 0 0 0 0 0 1 1 1 0 0 0 0 0 0 1 1 1 1 1 1 0 0 0 0 0 0 0 0

0 0 0 1 0 0 1 1 0 0 0 0 1 0 0 1 1 1 0 0 0 1 1 0 0 0 0 1 1 1 1 0 0 0 0

0 0 0 0 1 0 0 0 1 1 0 0 1 0 0 0 0 0 1 1 1 0 0 1 1 0 0 1 1 1 0 1 0 0 0

1 0 0 0 0 1 0 0 1 0 1 0 0 0 0 1 0 0 1 0 0 1 0 1 0 1 0 1 0 0 1 0 1 1 0

0 1 0 0 0 1 1 0 0 0 0 1 0 0 0 0 1 0 0 1 0 0 1 0 1 0 1 1 0 0 0 1 1 1 0

0 0 1 0 0 0 0 1 1 0 0 1 0 1 0 0 1 0 0 0 1 0 0 0 1 1 0 0 1 0 1 0 1 0 1

0 0 1 0 0 0 1 0 0 1 1 0 0 0 1 0 0 1 1 0 0 1 0 0 0 0 1 0 1 0 0 1 1 0 1

1 0 0 1 0 0 0 0 0 1 0 1 0 1 0 0 0 1 0 1 0 0 0 1 0 0 1 0 0 1 1 0 0 1 1

0 1 0 0 1 0 0 1 0 0 1 0 0 0 1 1 0 0 0 0 1 0 1 0 0 1 0 0 0 1 0 1 0 1 1

Figure 5.1. The columns form a constant weight code found by Best, containing 35 words of length 11, weight 4 and minimal distance 4.

of the densest *known* lattice packings. It has already been mentioned in §2.2 of Chap. 1 that it follows from Watson's work [Wat7] that P_{9a} has a higher kissing number than is possible in any 9-dimensional lattice.

3. Construction B

3.1 The construction. Let C be an (n, M, d) binary code with the property that the weight of each codeword is even. A sphere packing in $\mathbf{R}^n$ is given by:

Construction B. $x = (x_1, \ldots, x_n)$ is a center if and only if x is congruent (modulo 2) to a codeword of C, and $\sum_{i=1}^{n} x_i$ is divisible by 4.

Thus a point x with integer coordinates is a center if and only if the 1's row of the coordinate array of x is a codeword $c \in C$ and the 2's row has either even weight if the weight of c is divisible by 4, or odd weight if the weight of c is divisible by 2 but not by 4. A lattice packing is obtained if and only if C is a linear code. Construction B is a generalization of the construction of Λ_8 as given in [Lee4, §1.1], and will be studied further in Chap. 7.

3.2 Center density and kissing numbers. The number of centers is half that of Construction A, so $\delta = M \rho^n 2^{-n-1}$.

Let S be a sphere with center x, where x is congruent to a codeword c. Candidates for centers closest to x are (a) the $2n(n-1)$ centers of type $x + ((\pm 2)^2 0^{n-2})$, and (b) the $2^{d-1} A_d(c)$ centers congruent to the codewords differing minimally from c. Therefore the kissing number of S is

$$\tau(S) = \begin{cases} 2^{d-1} A_d(c) & \text{if } d < 8, \\ 2n(n-1) + 128 A_8(c) & \text{if } d = 8, \\ 2n(n-1) & \text{if } d > 8; \end{cases} \quad (6)$$

and S has radius $\rho = \frac{1}{2} \min\{\sqrt{d}, \sqrt{8}\}$.

3.3 Dimensions 8, 9 and 12. In $\mathbf{R}^8$ we apply Construction B to the repetition code $\{0^8, 1^8\}$ to obtain the packing E_8 again. In $\mathbf{R}^9$ we use the code $\{0^9, 1^80\}$ and obtain the lattice packing Λ_9, with $\rho = \sqrt{2}$, $\delta = 2^{-4.5}$ and $\tau = 272$.

In $\mathbf{R}^{12}$ we use the (12, 24, 6) code formed from the rows of H_{12} and their complements to obtain the nonlattice packing L_{12} (see [Lee4]) with $\delta = 2^{-16} \cdot 3^7$ and $\tau = 704$. The [12, 2, 8] code $\{0^{12}, 0^41^8, 1^40^41^4, 1^80^4\}$ gives the lattice packing $\Lambda_{12}^{\max}$, with $\delta = 2^{-5}$ and $\tau = 648$.

3.4 Dimensions 15 to 24. In $\mathbf{R}^{16}$ the [16, 5, 8] first-order RM code gives the lattice packing Λ_{16} having $\delta = 2^{-4}$ and $\tau = 4320$ (§10 of Chap. 4). Shortening this code by equating a coordinate to zero we obtain the [15, 4, 8] simplex code, which gives the lattice packing Λ_{15} in $\mathbf{R}^{15}$ having $\delta = 2^{-4.5}$ and $\tau = 2340$.

Let C_i, $i = 0, 1, \ldots, 5$, denote the shortened code obtained from the [24, 12, 8] Golay code $\mathcal{C}_{24}$ by setting i coordinates equal to zero, and let a_i denote the number of codewords of weight 8 in C_i. Then $a_0 = 759$, $a_1 = 506$, $a_2 = 330$, $a_3 = 210$, $a_4 = 130$ and $a_5 = 78$ (see Table 10.1 of Chap. 10).

The sequence of lattice packings in $\mathbf{R}^{24-i}$, $i = 0, 1, \ldots, 5$, obtained from C_i has $\delta_i = 2^{-(i+2)/2}$ [Lee4, §2.4]. In $\mathbf{R}^{19}$, $\mathbf{R}^{20}$, $\mathbf{R}^{21}$ these are Λ_{19}, Λ_{20}, Λ_{21}, the densest packings known, but in $\mathbf{R}^{22}$ to $\mathbf{R}^{24}$ they can be improved, as will be shown in §§4.4, 4.5. We denote the 24-dimensional lattice by $h\Lambda_{24}$; this consists of half of the points of the lattice Λ_{24} constructed in §§4.4, the "even" part of that lattice.

Remark. Constructions A and B cannot be successful for large n. Consider for example Construction B applied to a code with $d = 8$. It is easy to show that $\log_2 \delta \leqslant n/2 - 3 \log_2 n + o(\log_2 n)$, as $n \to \infty$, and so, using (18) of Chap. 1, $\log_2 \Delta \lesssim -\frac{1}{2} n \log_2 (n/\pi e)$. But much denser packings exist — see §1.5 of Chap. 1. Also $A_8 \leqslant \binom{n}{8}$, and therefore $\tau = O(n^8)$ at most, whereas much higher values of τ are known — see §2.2 of Chap. 1 or §6.5 below.

4. Packings built up by layers

4.1 Packing by layers. The basic idea is very simple (cf. [Lee6], [Lee7]). Let Λ be a lattice sphere packing in $\mathbf{R}^n$ with finite covering radius R (Eq. (3) of Chap. 2), and let $D(\Lambda)$ be the set of deep holes in Λ. (The same construction may also be applied to suitable nonlattice packings.)

A *layer* of spheres in $\mathbf{R}^{n+1}$ is a set of spheres whose centers lie in a hyperplane, and whose cross section in the hyperplane is Λ. In this hyperplane there are two distinguished sets of points, the set C of centers of Λ, and the set D of the deep holes in Λ.

We shall try to build up a dense sphere packing in $\mathbf{R}^{n+1}$ by stacking such layers as closely as possible. We therefore place adjacent layers so that the set C of one layer is opposite to some or all of the set D of the next. Since the layers are lattice packed, if one point of C is opposite to a point of D, then so are all the points of C.

It may happen that D is more numerous than C, in which case several inequivalent packings may be produced in $\mathbf{R}^{n+1}$. These may be lattice or nonlattice packings or both. Examples will be found below and in succeeding sections. (So far we have not found an example where no such lattice packing can be formed, but this does not seem impossible. Indeed, there might even be a case where the deep holes fail to give either lattice or nonlattice packings.)

For example let Λ be the square lattice $\mathbf{Z}^2$ in $\mathbf{R}^2$. Then C comprises the vertices of the squares and D their centers. The layers are placed so that the vertices of the squares in one layer are opposite to the centers of the squares in the next. Since C and D are equally numerous, this arrangement is unique, and the face-centered cubic lattice packing Λ_3 in $\mathbf{R}^3$ is obtained (§6.3 of Chap. 4).

The situation is different if Λ is chosen to be Λ_2, the hexagonal lattice in $\mathbf{R}^2$. Now D consists of the points marked 'b' or 'c' in Fig. 1.3b, and is twice as numerous as C (the points 'a'). The spheres of a layer can be placed opposite to the points 'b' or to the points 'c' of the adjacent layer. As already discussed in §1.3 of Chap. 1, if the layers are stacked so that the spheres in the two layers adjacent to any layer are opposite points marked with different letters, then the lattice Λ_3 is obtained. But if the adjacent spheres on both sides of each layer are opposite points marked with the same letter, then the (nonlattice) *hexagonal close packing* (§6.5 of Chap. 4) is obtained, having the same density as Λ_3. If we require uniform packings there is no further choice, as all layers have to be fitted alike, and this construction gives precisely these two packings.

In general, the layers are placed just far enough apart that the smallest distance between centers in adjacent layers equals the smallest distance between centers in the same layer. There are four cases which can arise here. The spheres of one layer may (i) not reach, (ii) just touch, or (iii) penetrate, the central hyperplane of an adjacent layer, or (iv) it may be possible to fit the spheres of one layer into the spaces between the spheres of the adjacent layer and so merge the two layers.

Examples of case (i) are $n = 2$ (as above) and 3-5 (in §4.2). In case (ii), extra contacts may occur because of spheres on opposite sides of a layer touching. Examples are the case $n = 6$ in §4.2, and the packing P_{13a} in §4.3. In case (iii), the layers on each side of a given layer must be staggered to avoid overlapping. Examples are the construction of E_8 from layers of D_7 as given in [Lee7], and the local arrangements P_{14b}, P_{15a} in §4.3. Case (iv) doubles the density of the original packing. Examples of this in $\mathbf{R}^8$, $\mathbf{R}^{12}$, $\mathbf{R}^{24}$ and $\mathbf{R}^{48}$ are given in §4.4 and §5.6.

It is sometimes advantageous to stack layers which are not lattices. Examples are the local arrangements of spheres in $\mathbf{R}^{11}$, $\mathbf{R}^{14}$ and $\mathbf{R}^{15}$ and the nonlattice packing P_{13a} in $\mathbf{R}^{13}$ to be described in §4.3.

This laminating construction is investigated further in the next chapter.

4.2 Dimensions 4 to 7. We consider packings in $\mathbf{R}^{n+1}$ formed by stacking layers of the lattice D_n for $n \geqslant 3$. The Delaunay cells (§1.2 of Chap. 2) are of two kinds: each omitted vertex of $\mathbf{Z}^n$ (such as the glue vector [2] given in §7.1 of Chap. 4) is the center of an octahedral cell β_n, while the center of each cube of $\mathbf{Z}^n$ (such as the glue vectors [1] and [3]) is the center of a "half-cube" $h\gamma_n$ [Cox20, p. 155]. The latter cells are twice as numerous as the former, and we shall regard the cells $h\gamma_n$ as being colored alternately black and white.

For $n = 3$ the octahedral cells β_3 are larger than the tetrahedral cells $h\gamma_3$, and so we place the spheres of each layer opposite to octahedra of the adjacent layer. As the octahedra and spheres are equally numerous we arrive uniquely at the lattice packing $D_4 \cong \Lambda_4$.

For $n = 4$ the cells β_4 and $h\gamma_4$ are congruent, the lattice D_4 being the regular honeycomb {3, 3, 4, 3} [Cox20, p. 136]. Thus there is a threefold choice for placing each layer: the spheres of each layer may be placed opposite to the omitted vertices or the black cells or the white cells. In this case we obtain either the lattice packing $D_5 \cong \Lambda_5$ or three distinct uniform nonlattice packings, all having the same density and kissing numbers [Lee6].

For $n > 4$ the cells $h\gamma_n$ are larger than the cells β_n, so for maximal density in $\mathbf{R}^{n+1}$ we stack the layers with the spheres of each layer opposite to the cubes of the adjacent layers. At each stage the spheres may be placed opposite to the black or the white cubes. For $n = 5$ or 6 we obtain in this way two uniform packings, the lattice packing Λ_{n+1} and a nonlattice packing of equal density. For $n = 5$ both have the same kissing numbers. For $n = 6$ each sphere in the lattice packing $E_7 \cong \Lambda_7$ has two more contacts than in the nonlattice packing, because it touches spheres in layers two away from it. This is an example of case (ii) above. For $n = 7$ only the lattice packing $E_8 \cong \Lambda_8$ is produced. This is an example of case (iii) above, where the adjacent layers have to be staggered to avoid overlap.

4.3 Dimensions 11 and 13 to 15. In $\mathbf{R}^{11}$ we construct a local arrangement P_{11b} of 580 spheres touching one sphere (inferior to P_{11c} of §2.6), by stacking three partial layers. The central layer consists of the 500 centers of P_{10b} touching one sphere, with eleventh coordinate zero. The two outer layers consist of all points of the form

$$(c_1, \ldots, c_{10}, 0) - (\tfrac{1}{2}, \ldots, \tfrac{1}{2}, \pm \tfrac{1}{2}\sqrt{6}),$$

where $(c_1, \ldots, c_{10})$ runs through the (10, 40, 4) code, and contain 80 points. It does not seem possible to extend this to a dense space packing.

In $\mathbf{R}^{13}$ we take any one of the packings P_{12a} and its translates by multiples of the vector $((\tfrac{1}{2})^{12}, \pm 1)$, and obtain a family of nonlattice packings collectively called P_{13a}. Any of these packings has $\delta = 2^{-8} \cdot 3^2$ and a maximal kissing number of $840 + 2 \cdot 144 + 2 = 1130$, the last 2 being caused by a sphere touching spheres two layers away. The average kissing number is $1060\tfrac{2}{3}$.

The tetrads of a Steiner system $S(3, 4, 14)$ (see [Han1]) form 91 codewords of length 14, weight 4 and Hamming distance at least 4 apart. For example, if the fourteen coordinates are labeled 1 2 3 4 5 6 7 and 1′ 2′ 3′ 4′ 5′ 6′ 7′, the tetrads may be taken to be 1 2 3 6, 2′ 5′ 6′ 7′, 1 2 4 2′, 6 4′ 6′ 7′, 2 7 1′ 6′, 2 4 4′ 6′, 5 7 1′ 3′, 1 5 3′ 7′, 1 4 4′ 7′, 4 7 1′4′, 6 7 1′ 2′, 5 6 2′ 3′, 4 5 3′ 4′, and their transforms under the permutation (1 2 3 4 5 6 7) (1′ 2′ 3′ 4′ 5′ 6′ 7′). Construction A then gives a local arrangement P_{14a} of 1484 spheres touching one sphere. This can be improved, however, by cutting down to thirteen dimensions and rebuilding. The best section of $S(3, 4, 14)$ in $\mathbf{R}^{13}$ has 65 codewords, giving a local arrangement P_{13b} with $26 + 16 \cdot 65 = 1066$ contacts, inferior to P_{13a}.

We now form a local arrangement P_{14b} in $\mathbf{R}^{14}$ by stacking five partial layers. The central layer is P_{13b}. The adjacent layers are $(c, 0) - ((\frac{1}{2})^{13}, \frac{1}{2}\sqrt{3})$ and $(c', 0) - ((\frac{1}{2})^{13}, -\frac{1}{2}\sqrt{3})$, where c runs through the [13, 8, 4] shortened Hamming code and c' denotes $c + (1^2, 0^{11})$ reduced modulo 2. The two outer layers contain just two spheres each: $(\pm 1, 0^{12}, \pm\sqrt{3})$. Thus the central sphere touches $1066 + 2 \cdot 256 + 2 \cdot 2 = 1582$ others.

Similarly in $\mathbf{R}^{15}$ we form a local arrangement P_{15a} from five partial layers. The central layer is P_{14a}. The adjacent layers are $(c, 0) - ((\frac{1}{2})^{14}, \frac{1}{2}\sqrt{2})$ and $(c', 0) - ((\frac{1}{2})^{14}, -\frac{1}{2}\sqrt{2})$, where c runs through the [14, 9, 4] shortened Hamming code and $c' = c + (1^2, 0^{12})$ reduced modulo 2. The outer layers are $(x_1, \ldots, x_{14}, \pm\sqrt{2})$, with $x_1, \ldots, x_{14}$ all zero except for one pair x_{2i-1}, x_{2i} which are ± 1 in all four combinations. The central sphere touches $1484 + 2 \cdot 512 + 2 \cdot 28 = 2564$ others.

4.4 Density doubling and the Leech lattice Λ_{24}. In $\mathbf{R}^8$ two copies of D_8 can be fitted together without overlap to form the lattice packing $E_8 \cong \Lambda_8$, the second copy being a translation of the first by $((\frac{1}{2})^8)$.

In $\mathbf{R}^{24}$ it was shown in [Lee5, §2.31] that two copies of the 24-dimensional packing obtained in §3.4 from the Golay code $\mathscr{C}_{24}$ may be fitted together without overlap to form a packing Λ_{24} (now usually called the *Leech lattice*), with $\delta = 1$ and $\tau = 196560$. The second copy is a translation of the first by $(-1\frac{1}{2}, (\frac{1}{2})^{23})$, and this produces Λ_{24} in the form given in Eqs. (133)-(136) of Chap. 4, rescaled.

4.5 Cross sections of Λ_{24}. All of the densest known *lattice* packings in fewer than 24 dimensions occur as sections of Λ_{24}. There are two main sequences of sections, Λ_i and K_i, $i = 0, 1, \ldots, 24$, where the subscript indicates the dimension, and $\Lambda_i \cong K_i$ for $i \leqslant 6$ and $i \geqslant 18$. Because of the symmetry of Λ_{24}, there are several different ways of describing some of the sections.

The sequence of lattice packings Λ_i is defined as follows ([Lee4, §2.4; Lee5, §2.41]). Λ_{23}, Λ_{22} and Λ_{21} are obtained from Λ_{24} by equating any two, three or four coordinates to each other. Λ_{21} (again), Λ_{20} and Λ_{19} are obtained by equating any three, four or five coordinates to 0. We recall from §3.1 of Chap. 3 that codewords of weight 8 in $\mathscr{C}_{24}$ form the octads of a Steiner system $S(5, 8, 24)$. Because of the fivefold transitivity of the Steiner system, the choice of coordinates in forming $\Lambda_{19}, \ldots, \Lambda_{23}$ is arbitrary.

Λ_{19} (again), Λ_{18}, Λ_{17} and Λ_{16} are obtained by equating to 0 the sum of eight coordinates forming a Steiner octad and also any four, five, six or all of them.

The [16, 5, 8] RM code occurs as a subcode of $\mathscr{C}_{24}$, and we associate the remaining 16 coordinates with coordinate positions of this subcode. Any two intersecting octads of the [16, 5, 8] code split the coordinates into four tetrads. Λ_{15}, Λ_{14}, Λ_{13} and Λ_{12} are obtained by equating to 0 the sum of the coordinates forming a tetrad and none, any one, any two or all of them. Λ_{11}, Λ_{10} and Λ_9 are obtained from Λ_{12} by equating any two, any three or all four coordinates to each other in one of the remaining tetrads. Λ_9 (again) and Λ_8 are obtained by equating to 0 any three or all four coordinates in a tetrad.

$\Lambda_7,\ldots,\Lambda_4$ are obtained as sections of Λ_8 in the same way that $\Lambda_{11},\ldots,\Lambda_8$ are obtained from Λ_{12}, defining any four coordinates of the eight to be a tetrad. Finally, $\Lambda_3,\ldots,\Lambda_1$ are obtained from Λ_4 (calling the four coordinates a tetrad) in the same way that $\Lambda_{15},\ldots,\Lambda_{13}$ were obtained from Λ_{16}.

The sequence $\Lambda_1,\ldots,\Lambda_{24}$ includes the densest known lattice packings in $\mathbf{R}^1,\ldots,\mathbf{R}^{10}$ and $\mathbf{R}^{14},\ldots,\mathbf{R}^{24}$. But in dimensions 11 to 13 the lattices K_{11}, K_{12}, K_{13} of the K_n sequence are denser than Λ_n [Lee5], [Lee10]. These are described in the next chapter. K_{12} is the Coxeter-Todd lattice (§9 of Chap. 4), with $\delta = 3^{-3}$ and $\tau = 756$. K_{11}, K_{12} and K_{13} are the densest known lattice packings in these dimensions, although P_{11a} and P_{13a} are denser nonlattice packings. Also the nonlattice packings P_{10a}, P_{10c} are denser than Λ_{10}.

We remark that there is a close analogy between the packings $D_3,\ldots,D_8$ of §2.4 and the packings in $\mathbf{R}^{19},\ldots,\mathbf{R}^{24}$ of §3.4, in that the first three are the densest known, the last can be doubled in density, and the densest intermediate packings can be derived either as sections of the doubled packing or by building up in layers.

5. Other constructions from codes

5.1 A code of length 40. We construct a binary code of length 40 which will be used in §5.2 to obtain a sphere packing. Let C_1 be the [16, 11, 4] Hamming code. The 140 codewords of weight 4 form the blocks of a Steiner system $S(3, 4, 16)$. Let $\mathscr{C}_{24}$ be the [24, 12, 8] Golay code, with coordinates arranged so that $1^8 0^{16}$ is a codeword. There are 759 codewords of weight 8, forming the octads of a Steiner system $S(5, 8, 24)$.

Let C be the code of length 40 consisting of all codewords of the form (x,y,z) where $y \in C_1$, $z \in C_1$, and $(x,y+z) \in \mathscr{C}_{24}$.

Theorem 1. *C is a* [40, 23, 8] *linear code.*

Proof. We note that if the eight coordinates belonging to an octad are deleted from all the codewords of $\mathscr{C}_{24}$, the truncated codewords form two copies of C_1 (see [Ber5] or Chap. 11). In C, y and z can each be chosen in 2^{11} ways and then, by the previous remark, there are two ways of choosing x so that $(x, y+z) \in \mathscr{C}_{24}$. Therefore C contains 2^{23} codewords. Since

$$(x,y,z) = (x,y+z,0) + (0,z,z), \quad \mathrm{wt}(x,y,z) \geqslant \mathrm{wt}(x,y+z,0) \geqslant 8$$

by construction, and so C has minimal weight 8.

Theorem 2. *The number of codewords of minimal weight in C is* 2077.

Proof. Let $(x,y,z) = (x,y+z,0) + (0,z,z)$ be a codeword of weight 8. There are five cases. (1) If $y = z = 0$, then $x = 1^8$. (2) If $y \neq 0$, $z = 0$, then there are 758 codewords of the form $(x,y) \in \mathscr{C}_{24}$, $y \neq 0$. (3) If $y = 0$, $z \neq 0$, again there are 758 possibilities. (4) If $y = z \neq 0$, then $\mathrm{wt}(y) = 4$, $x = 0$, and there are 140 choices for y. (5) If $y \neq 0$, $z \neq 0$, $y+z \neq 0$, then $x = 0$, $\mathrm{wt}(y+z) = 8 = \mathrm{wt}(y) + \mathrm{wt}(z)$. Therefore y and z have disjoint sets of 1's. y can be chosen in 140 ways and for each of these there are three choices for z so that $(0, y+z) \in \mathscr{C}_{24}$, giving 420 codewords. The total of (1)-(5) is 2077.

Remark. If $(0,y,z) \in C$, then $(0,y+z) \in \mathscr{C}_{24}$ and so $y+z$ is in the [16, 5, 8] 1st order RM code and therefore (y,z) is in the [32, 16, 8] 2nd order RM code.

5.2 A lattice packing in $\mathbf{R}^{40}$. A lattice packing in $\mathbf{R}^{40}$ can be obtained from the code C of §5.1 as follows. The 40 coordinates are divided into 8 + 16 + 16 corresponding to the division of the codewords of C.

Then $X = (x,y,z)$ is a center if and only if the 2's row of X is in C, the 1's row is all 0's or is 0 on one of the sets 8, 16, 16 and 1 on the other two, and the 4's row has even weight if the set of 8 is even or has odd weight if the set of 8 is odd.

The centers closest to the origin are shown in Table 5.1. (The centers described in the first line of the Table have 1's row $0^8 1^{32}$, 2's row equal to any codeword $(0,y,z) \in C$, and 4's row chosen to make all the nonzero coordinates equal to ± 1, and by the remark at the end of § 5.1 the number of such centers is 2^{16}. The other types of centers are easily counted.) Thus we have a lattice packing Λ_{40} with $\rho = 2\sqrt{2}$, $\delta = 2^4$ and $\tau = 531120$.

Table 5.1. Minimal vectors (x,y,z) in Λ_{40}. In the lines marked (‡), one entry ± 1 must be replaced by ∓ 3.

x	y	z	Shape	Number
zero	odd	odd	$0^8(\pm 1)^{32}$	2^{16}
odd	zero	odd	$(\pm 1)^8 0^{16} (\pm 1)^{16}$ (‡)	$24 \cdot 2^{12}$
odd	odd	zero	$(\pm 1)^8 (\pm 1)^{16} 0^{16}$ (‡)	$24 \cdot 2^{12}$
even	even	even	$0^{32}(\pm 2)^8$	$2^7 \cdot 2077$
even	even	even	$0^{38}(\pm 4)^2$	$40 \cdot 39 \cdot 2$
				531120

5.3 Cross sections of Λ_{40}. Λ_{40} contains Λ_{24} as a section, and for $1 \leqslant n \leqslant 16$ has a section Λ_{40-n} in $\mathbf{R}^{40-n}$ of center density 16 times that of Λ_{24-n}. For instance in $\mathbf{R}^{36}$ there is a section Λ_{36} with $\delta = 2$ and $\tau = 234456$. Λ_{36} is obtained from Λ_{40} by equating to zero four coordinates from a tetrad contained in one of the sets of 16. Λ_{32} is obtained by equating to zero all coordinates from an octad contained in one of the sets of 16, and has $\delta = 1$ and $\tau = 208320$.

5.4 Packings based on ternary codes. Let C be an (n, M, d) ternary code. By analogy with the constructions based on binary codes we obtain sphere packings in $\mathbf{R}^n$ from the following constructions.

Construction A_3. $x = (x_1, \ldots, x_n)$ is a center if and only if x is congruent (modulo 3) to a codeword of C.

Construction B_3. In addition, $\sum_{i=1}^{n} x_i$ is divisible by 2.

From Construction A_3 we obtain a packing of spheres for which

$$\rho = \min\{3/2, 1/2\sqrt{d}\}, \qquad \delta = M\rho^n 3^{-n}, \tag{7}$$

and from Construction B_3, if d_e and d_o are respectively the minimal even and odd distances between codewords,

$$\rho = \min\{3/\sqrt{2}, \tfrac{1}{2}\sqrt{d_e}, \tfrac{1}{2}\sqrt{d_o+3}\}, \tag{8}$$

$$\delta = M\rho^n 2^{-1}3^{-n}. \tag{9}$$

Other constructions from ternary codes are described in §8 of Chap. 7.

5.5 Packings obtained from the Pless codes. By applying Construction B_3 to the Pless double circulant codes (§2.10 of Chap. 3) we obtain the lattice packings shown in Table 5.2.

Table 5.2. Lattice obtained by applying Construction B_3 to the Pless codes.

n	Code	Radius	δ	Name
12	[12, 6, 6]	$2^{-1/2}\,3^{1/2}$	2^{-7}	D_{12}
24	[24, 12, 9]	$3^{1/2}$	2^{-1}	$h\Lambda_{24}$
36	[36, 18, 12]	$3^{1/2}$	2^{-1}	P_{36p}
48	[48, 24, 15]	$2^{-1/2}\,3$	$2^{-25}\,3^{24}$	hP_{48p}
60	[60, 30, 18]	$2^{-1/2}\,3$	$2^{-31}\,3^{30}$	P_{60p}

By applying Construction A_3 to the [12, 6, 6] Golay code $\mathscr{C}_{12}$ we obtain the lattice D_{12}^2 with $\rho = \sqrt{3/2}$ and $\delta = 2^{-6}$. It is known that D_{12} can be doubled in density to give D_{12}^2, and in fact quadrupled to give Λ_{12} [Lee4, §2.1].

We show in §5.7 that the packings in $\mathbf{R}^{24}$ and $\mathbf{R}^{48}$ of Table 5.2 can also be doubled in density, to give Λ_{24} and a lattice P_{48p}. On the other hand we have not found doublings of the packings P_{36p} and P_{60p}, the technique of §5.7 failing because of the existence of codewords of maximal weight with odd numbers of each sign [Mal2]. (See Postscript, p. 156.)

In P_{36p} the centers closest to the origin are those of the type $((\pm 1)^{12}0^{24})$ and correspond to codewords of minimal weight, so $\tau = A_{12} = 42840$. In P_{60p}, $\tau = A_{18} + 2 \cdot 60 \cdot 59 = 3908160$.

5.6 Packings obtained from quadratic residue codes. As pointed out in §§2.8, 2.10 of Chap. 3, there are [24, 12, 9] and [48, 24, 15] ternary quadratic residue codes with the same weight distributions as the corresponding Pless codes. Applying Construction B_3, we obtain packings $h\,\Lambda_{24}$ and $h\,P_{48q}$. It will now be shown that these can be doubled in density to give Λ_{24} in $\mathbf{R}^{24}$ and another lattice P_{48q} in $\mathbf{R}^{48}$.

5.7 Density doubling in $\mathbf{R}^{24}$ and $\mathbf{R}^{48}$. Any of the lattices $h\,\Lambda_{24}$, $h\,P_{48p}$ and $h\,P_{48q}$ may be doubled in density, by adding a second copy which is a translation of the first by $(-2\frac{1}{2}, (\frac{1}{2})^{n-1})$.

To show this, we must verify that no point of the original lattice is closer to $(-2\frac{1}{2}, (\frac{1}{2})^{n-1})$ than the origin. Every point of the original lattice differs from this point by at least $\frac{1}{2}$ in every coordinate. If any coordinates are congruent to -1 (modulo 3) then at least three are, by §2.10 (ii) of Chap. 3, and so at least three coordinates differ by $1\frac{1}{2}$. On the other hand if all coordinates are congruent to 0 or 1 (modulo 3), then by §2.10 (iv) they are congruent to one of the types 0^n, 1^n or $0^{n/2}1^{n/2}$. Since the sum is even, at least one coordinate differs by $2\frac{1}{2}$.

In $\mathbf{R}^{24}$ the doubled packing has center density 1 and so must be Λ_{24} (regardless of whether it came from the Pless code or the quadratic residue code) since this lattice is unique (Chap. 12).

In $\mathbf{R}^{48}$ we obtain a lattice which we call P_{48p} by doubling that given by the [48, 24, 15] Pless code, and one which we call P_{48q} from the [48, 24, 15] quadratic residue code. Both P_{48p} and P_{48q} have $\delta = 2^{-24} \cdot 3^{24}$. The kissing numbers for P_{48q} and P_{48p} can be obtained from the weight distribution given in Table 3.2. The centers closest to the origin are shown in Table 5.3; their total is $\tau = 52416000$ (in agreement with the value found in Eq. (57) of Chap. 2). The cross-sections of P_{48p} are investigated in §2 of Chap. 6. The lattice P_{48q} was first found by Thompson [Tho7]

Table 5.3. Minimal vectors in either P_{48q} or P_{48p}.

Coordinates	Number
$(\pm 3)^2 0^{46}$	$2 \cdot 48 \cdot 47$
$(\mp 2)(\pm 1)^{14} 0^{33}$	$15 \cdot 415104$
$(\pm 1)^{18} 0^{30}$	20167136
$(\mp 2\frac{1}{2})(\pm \frac{1}{2})^{47}$	$48 \cdot 96$
$(\pm 1\frac{1}{2})^3 (\pm \frac{1}{2})^{45}$	$4 \cdot 6503296$

using the construction given in §7.5 of Chap. 8. We return to these lattices in Example 9 of Chap. 7, where we give their determinants and theta series, etc.

6. Construction C

6.1 The construction. Let $C_i = (n, M_i, d_i)$, $i = 0, 1, \ldots, a$, be a family of codes with $d_i = \gamma \cdot 4^{a-i}$, where $\gamma = 1$ or 2. A sphere packing in $\mathbf{R}^n$ is given by:

Construction C. A point x with integer coordinates is a center if and only if the 2^i's row of the coordinate array of x is in C_i, for $i = 0, 1, \ldots, a$.

This is a generalization of the constructions for $\mathbf{R}^n$, $n = 2^m$, given in [Lee4, §1.6]. Construction C follows the trend of A and B in successively imposing more restrictions on the coordinate array. In general a nonlattice packing is obtained. A modification of this construction (Construction D) given in §8 of Chap. 8 always produces lattice packings provided only that the codes C_i are linear and nested, i.e. satisfy $C_i \subseteq C_{i+1}$.

6.2 Distance between centers. If the first row in which two centers differ is the 2^i's row, then (i) if $i > a$ their distance apart is at least 2^{a+1}, and (ii) if $0 \leqslant i \leqslant a$ they differ by at least 2^i in at least d_i coordinates, and so are at least $(d_i \cdot 4^i)^{1/2} = \sqrt{\gamma} \cdot 2^a$ apart. Thus we may take the radius of the spheres to be $\rho = \sqrt{\gamma} \cdot 2^{a-1}$.

6.3 Center density. How many integer points x satisfy the conditions of Construction C? The fraction of x's with 1's row in C_0 is $M_0 2^{-n}$, of these a fraction $M_1 2^{-n}$ have 2's row in C_1, and so on. Thus the fraction of integer points which are accepted as centers is $M_0 M_1 \cdots M_a 2^{-(a+1)n}$, and so the center density is

$$\delta = M_0 M_1 \cdots M_a 2^{-(a+1)n} \rho^n = M_0 M_1 \cdots M_a \gamma^{n/2} 2^{-2n}. \tag{10}$$

If the codes are linear, and $\dim C_i = k_i$, then

$$\log_2 \delta = \sum_{i=0}^{a} k_i - 2n \quad \text{if } \gamma = 1, \tag{11a}$$

$$\log_2 \delta = \sum_{i=0}^{a} k_i - \frac{3n}{2} \quad \text{if } \gamma = 2. \tag{11b}$$

For large n we have (from (18) of Chap. 1)

$$\log_2 \Delta = \sum_{i=1}^{a} k_i - \frac{n}{2} \log_2 (8n/\pi e) + O(\log_2 n). \tag{12}$$

For example, we may obtain a 64-dimensional nonlattice packing by using the following codes from Table 5.4:

$$C_0:[64,1,64],\ C_2:[64,57,4],\ C_1:[64,28,16],\ C_3:[64,64,1] \tag{13}$$

(here $\gamma = 1$, $a = 3$). From (11a) the center density is $\delta = 2^{22}$. In §8.2e of Chap. 8 we obtain a lattice packing with the same density.

6.4 Kissing numbers. We shall calculate the number of spheres touching the sphere at the origin. From the discussion in §6.2 this means that we must find the number of centers of type $(\pm 2^r)^{d_r} 0^{n-d_r}$ for each $r = 0,1,\ldots,a$. A coordinate equal to $+2^r$ contributes to the coordinate array a column with a single one in the 2^r's row; while a coordinate equal to -2^r has ones in the 2^i's row for all $i \geqslant r$. The 1's in the 2^r's row form a codeword c (say) in C_r, and the minus signs must be at the locations of the ones in a codeword in $C_{r+1} \cap C_{r+2} \cap \cdots \cap C_a$. The 2^r's row can be chosen in A_{d_r} ways, and for each of these the number of ways of choosing the signs is equal to the number of codewords in $C_{r+1} \cap C_{r+2} \cap \cdots \cap C_a$ which are contained in the codeword c of C_r. Let the latter number be $N_r(c)$.

Then the number of centers of the desired type is $\sum N_r(c)$, and

$$\tau = \sum_{r=0}^{a} \sum N_r(c), \tag{14}$$

where $\sum$ denotes the sum over all $c \in C_r$. If $N_r(c) = N_r$ is independent of c, this becomes $\tau = \sum_{r=0}^{a} A_{d_r} N_r$.

6.5 Packings obtained from Reed-Muller codes. As an example of Construction C we take C_r to be the $(2r)$th order RM code of length $n = 2^m$, to obtain a packing P_{nr} in $\mathbf{R}^n$. This and similar packings were given in [Lee4].

If m is even, $\gamma = 1$ and $a = m/2$, while if m is odd, $\gamma = 2$ and $a = (m-1)/2$; in both cases the radius is $2^{(m-2)/2}$. Since RM codes are nested, $N_r(c)$ is equal to the number of codewords in the $(2r+2)$th order RM code of length 2^m which are contained in a codeword of minimal weight in the $(2r)$th order RM code, and this is the same as the number of codewords in the 2nd order RM code of length 2^{m-2r}. Thus

$$\log_2 N_r(c) = 1 + \binom{m-2r}{1} + \binom{m-2r}{2}. \tag{15}$$

Using the properties of RM codes given in §2.7 of Chap. 3, the center density and maximal kissing number are then found to be

$$\delta = 2^{-5n/4}\, n^{n/4}, \tag{16}$$

$$\tau = (2+2)\,(2+2^2)\cdots(2+2^m) \tag{17}$$

$$\sim 4.768\ldots\, 2^{m(m+1)/2}.$$

The details may be found in [Lee4]. (The relation between the *k-parity* of a vector as defined in [Lee4] and RM codes is that a binary vector of length 2^m has k-parity if and only if it is in the $(m-k)$th order RM code.)

The packing P_{nr} in $\mathbf{R}^n$, $n = 2^m$, is a lattice if and only if $n \le 64$. The Barnes-Wall lattice packing BW_n (§8.2f of Chap. 8) in dimension $n = 2^m$ coincides with P_{nr} for $n \le 32$ and has the same density and maximal kissing number for all n.

6.6 Packings obtained from BCH and other codes. Both very low and very high order RM codes contain the maximal possible number of codewords, but intermediate order RM codes are poor. For $n \geqslant 64$ there are BCH codes with more codewords, and these too are known not to be optimal. For length 64 extended cyclic codes are known which are better than BCH codes. Unfortunately for lengths greater than 99 cyclic non-BCH codes have not been extensively studied. Cyclic codes of length $n \leqslant 63$ are listed in [Che1], [Pet3, Appendix D], and Promhouse and Tavares [Pro1] have extended this table to $n = 99$. It would be nice to have a list of the best cyclic codes of length 127. In Table 5.4 we give a selection of the best codes presently known with length $n = 2^m$ $(5 \leqslant m \leqslant 8)$ and minimal distance $d = 2^i$. The Goethals codes are nonlinear (see [Goe1], [Goe2], [Mac6, Chap. 15, §7]); the others are linear. The trivial $[n, n, 1]$, $[n, n-1, 2]$ $[n, 1, n]$ codes have been omitted from the table.

Table 5.4. The best codes known of length 2^m.

n	$\log_2 M$	d	Name
32	26	4	Hamming
	16	8	2nd order RM
	17	8	[Che3]
	6	16	1st order RM
64	57	4	Hamming
	45	8	BCH
	46	8	extended cyclic
	47	8	Goethals
	24	16	BCH
	28	16	extended cyclic
	7	32	1st order RM
128	120	4	Hamming
	106	8	BCH
	78	16	BCH
	43	32	BCH
	8	64	1st order RM
256	247	4	Hamming
	231	8	BCH
	233	8	Goethals
	199	16	BCH
	139	32	BCH
	55	64	BCH
	9	128	1st order RM

For completeness we mention that the [64, 46, 8] code in Table 5.4 is obtained by adding a zero-sum check digit to the [63, 46, 7] cyclic code having generator polynomial

$$g(x) = M^{(1)}(x) M^{(5)}(x) M^{(9)}(x) M^{(21)}(x), \tag{18}$$

where $M^{(i)}(x)$ is the minimal polynomial of α^i, α = primitive element of $\mathbf{F}_{64}$ (see §2.4 of Chap. 3). Similarly the [64, 28, 16] code comes from the [63, 28, 15] cyclic code with generator polynomial

$$g(x) = M^{(21)}(x) \prod_{i=0}^{5} M^{(2i+1)}(x) \tag{19}$$

[Pet3, Appendix D].

Since the center density in Construction C is proportional to the number of codewords, by using the best codes from this table we obtain considerable improvements over the packings of the previous section. In $\mathbf{R}^{64}$ for example, as we saw in §6.3, we obtain a nonlattice packing with $\delta = 2^{22}$. For $m \geqslant 7$ we use extended BCH codes to obtain an infinite family of nonlattice packings P_{nb} in $\mathbf{R}^n$, $n = 2^m$. For $m = 7, 8, 9$ these have $\delta = 2^{85}, 2^{250}, 2^{698}$ respectively, giving improvements over both BW_n and P_{nr} by factors of $2^{21}, 2^{58}, 2^{186}$ respectively. Lattices with the same density as P_{nb} will be obtained in §8.2g of Chap. 8 using Construction D. For dimensions $n \geqslant 256$ Craig's lattices (§6 of Chap. 8) are denser still — see the discussion in §1.5 of Chap. 1.

The density of P_{nb} is estimated for all m in the next section.

By using Goethals' code in place of the BCH code with $d = 8$ we can increase the density of P_{nb} by a factor of 4 for $n = 2^m \geqslant 256$ when m is even. However, this does not affect the following analysis.

BCH codes are nested and so also are the extended cyclic codes of length 64 in Table 5.4. Therefore calculation of the kissing numbers in these packings requires knowledge of the number of codewords of minimal weight d in each code used and the number of codewords in the code with minimal distance $d/4$ that are wholly contained in such minimal weight codewords. This appears to be a difficult problem.

6.7 Density of BCH packings. This section contains lower and upper bounds and an asymptotic expansion (Theorem 3) for the center density of the packings in $\mathbf{R}^n$ obtained by using extended BCH codes in Construction C. Here "code" will mean "extended BCH code of length $n = 2^m$." Two packings are considered. Packing (a) uses the codes of (actual) Hamming distance $1, 4, 16, \ldots, 4^{[m/2]}$, and packing (b) uses the codes of (actual) Hamming distance $2, 8, 32, \ldots, 2 \cdot 4^{[(m-1)/2]}$. Then let P_{nb} denote the denser of the two packings.

By the remark at the end of §2.7 of Chap. 3, the code of designed distance 2^λ has actual distance 2^λ. Let this code have dimension k_λ. But there may be codes of designed distance less than 2^λ also having actual distance 2^λ. Let K_λ be the largest dimension of such a code. Then

$k_\lambda \leqslant K_\lambda < k_{\lambda-1}$. For example, for codes of length 128 it is known that $k_5 = 36$, $K_5 = 43$, $k_4 = 78$ [Kas1].

If δ_a and δ_b denote the center densities of the two packings, then from (11a), (11b) we have

$$\log_2 \delta_a = \sum_{i=0}^{[m/2]} K_{2i} - 2n, \quad \log_2 \delta_b = \sum_{i=0}^{[(m-1)/2]} K_{2i+1} - 3n/2.$$

An algorithm for calculating k_λ is given in [Ber4, §12.3]. The results of this algorithm may be stated as follows. Let numbers $a_{i,j}$ be defined by

$$a_{i,j} = 2^j - 1, \qquad \text{for } 1 \leqslant j \leqslant i,$$
$$a_{i,j} = a_{i,j-i} + a_{i,j-i+1} + \cdots + a_{i,j-1}, \qquad \text{for } 1 \leqslant i < j.$$

Then $k_0 = n$, $k_m = 1$ and $k_\lambda = m + a_{m-\lambda,m}$ for $0 < \lambda < m$. Let $A_i(x) = \sum_{j=1}^{\infty} a_{i,j} x^j$. From the definition of $a_{i,j}$,

$$A_i(x) = \frac{x + 2x^2 + \cdots + ix^i}{1 - x - x^2 - \cdots - x^i}.$$

Let $1 - x - x^2 - \cdots - x^i = \Pi_{\nu=1}^{i}(1 - t_\nu x)$. The partial fraction expansion of $A_i(x)$ is then

$$A_i(x) = \sum_{\nu=1}^{i} \frac{1}{1 - t_\nu x},$$

and so

$$a_{i,j} = \sum_{\nu=1}^{i} t_\nu^j.$$

The polynomial $G(y) = y^{i+1} - 2y^i + 1 = (y-1)(y^i - y^{i-1} - ... - y - 1)$ has roots $1, t_1, \ldots, t_i$. A sketch of $G(y)$ shows that one root, t_1 (say), is close to 2. In fact for $i > 1$, $G(2 - 2^{1-i}) < 0$ and $G(2 - 2^{-i}) > 0$, so

$$2 - 2^{1-i} < t_1 < 2 - 2^{-i}, \quad i > 1.$$

It has been shown by Mann [Man2], while calculating the number of codewords in the BCH codes of length $2^m - 1$ and distances 2^λ and $2^\lambda + 1$, that $|t_\nu| < 1$ for $\nu = 2, 3, \ldots, i$. Then

$$(2 - 2^{1-i})^m - i + 1 < a_{i,m} < (2 - 2^{-i})^m + i - 1, \quad i > 1.$$

Collecting these results we find that lower and upper bounds to $\log_2 \delta_a$ are respectively

$$2^m \sum_{i=1}^{[m/2]} (1 - 2^{2i-m})^m - 2^m + O(m^2),$$
$$2^m \sum_{i=1}^{[m/2]} (1 - 2^{2i-m-2})^m - 2^m + O(m^2).$$

The sum in the lower bound may be written as

$$\sum_{i=0}^{[m/2]-c\log m} + \sum_{i=[m/2]-c\log m}^{[m/2]-d\log m} + \sum_{i=[m/2]-d\log m}^{[m/2]} = \sum\nolimits_{\mathrm{I}} + \sum\nolimits_{\mathrm{II}} + \sum\nolimits_{\mathrm{III}},$$

where $c > 1$ and $d < 1$ are constants, and the logarithms are to base 4. Then

$$\sum\nolimits_{\mathrm{I}} > \sum_{i=0}^{[m/2]-c\log m} (1 - m4^i \cdot 2^{-m}) > \frac{m}{2} - \log_4 m + O(1),$$

since c may be made arbitrarily close to 1. Also $\sum_{\mathrm{II}} = o(\log_4 m)$, since c and d can be made arbitrarily close together. Finally, since $d < 1$,

$$0 < \sum\nolimits_{\mathrm{III}} < (1 - m^{-d})^m (d \cdot \log_4 m) + 1 = o(\log_4 m) .$$

Therefore

$$\log_2 \delta_a > m 2^{m-1} - 2^m \log_4 m + o(2^m \log_4 m),$$

$$= \tfrac{1}{2}n \log_2 n - \tfrac{1}{2}n \log_2 \log_2 n + o(n \log_2 \log_2 n),$$

since $n = 2^m$. Similar arguments apply to the upper bound and to bounds for $\log_2 \delta_b$. This proves:

Theorem 3. *Let δ be the center density of either of the packings in $\mathbf{R}^n$, $n = 2^m$, obtained by using extended BCH codes in Construction* C. *Then*

$$\log_2 \delta \sim \tfrac{1}{2}n \log_2 n - \tfrac{1}{2}n \log_2 \log_2 n \quad as \quad n \to \infty. \tag{20}$$

It follows using Eq. (18) of Chap. 1 that the actual density $\Delta = V_n \delta$ of these packings satisfies

$$\log_2 \Delta \sim -\tfrac{1}{2} n \log_2 \log_2 n . \tag{21}$$

6.8 Packings obtained from Justesen codes. A higher density than (21) can be achieved by replacing some of the BCH codes by Justesen codes (§2.6 of Chap. 3). Suppose $n = m2^m = 2^{2a}$, and take $\gamma = 1$. For $0 \leqslant i \leqslant 5$ let C_i consist of the 0 codeword alone, for $6 \leqslant i \leqslant \frac{5}{8}\log_2(2a)$ let C_i be a Justesen code of minimal distance 4^{a-i} shortened to have length n, and for $\frac{5}{8}\log_2(2a) < i \leqslant a$ let C_i be an extended BCH code of minimal distance 4^{a-i} as in §6.7. Then it can be shown [Slo1] that Construction C produces a nonlattice packing with

$$\frac{1}{n} \log_2 \Delta \gtrsim -6. \tag{22}$$

Remark. If we used codes meeting the Gilbert bound (Eq. (3) of Chap. 9) in Construction C we would obtain nonlattice packings with

$$\frac{1}{n} \log_2 \Delta \gtrsim -1.2919.... \tag{23}$$

(However, although such codes exist, it is not known how to construct them.) On the other hand, from the JPL bound (Eq. (1) of Chap. 9), the density of any packing obtained from Construction C is bounded above by

$$\frac{1}{n} \log_2 \Delta < -0.9042.... \tag{24}$$

Eqs. (23), (24) are to be compared with Eq. (46) of Chap. 1.

Postscript: a uniform construction for extremal Type II lattices in dimensions 24, 32, 40, 48, 56, 64. Ozeki [Ternary code construction of even unimodular lattices, preprint] asserts that the following method can be used to double the density of lattices obtained by applying Construction B_3 to a ternary self-dual C of length n, for all $n = 24, 32, 40, \ldots, 64$, just as we did for $n = 24$ and 48 in §5.7. One adjoins a vector v of the form $((\pm 1/2)^{n-1}, (\mp 5/2))$, $((\pm 1/2)^{n-2}, (-3/2)^2)$ or $((\pm 1/2)^{n-1}, -3/2)$, according as $n \equiv 0$, 8 or 16 (mod 24), where the signs are chosen so that $2v$ is in the undoubled lattice. He shows that if C has minimal distance 9 for $n = 24, 32, 40$, or distance 15 for $n = 48, 56, 64$ then the resulting lattice, suitably scaled, is an extremal Type II lattice (cf. Chap. 7, §7). The required codes are briefly discussed in §10 of Chap. 7. We remark that such codes of lengths 32 and 56 may be obtained by "subtracting" [Con24, §VII] the tetracode $\mathcal{C}_4$ (Chap. 3, (2.5.1)) from the double circulant codes S_{36} and S_{60} (Chap. 3, §2.10).

6

Laminated Lattices

J. H. Conway and N. J. A. Sloane

We study the densest lattice packings that can be built up in layers. Start with the 1-dimensional lattice Λ_1 of even integer points; at the nth step stack layers of a suitable $(n-1)$-dimensional lattice Λ_{n-1} as densely as possible, keeping the same minimal norm; the result is a laminated lattice Λ_n. In this chapter the density of Λ_n is determined for $n \leqslant 48$, all Λ_n are found for $n \leqslant 25$, and at least one Λ_n is found for $26 \leqslant n \leqslant 48$. The unique Λ_{24} is the Leech lattice. Denser lattices than Λ_n are now known for $n \geqslant 30$.

1. Introduction

A natural and familiar way to construct a lattice packing of spheres in n-dimensional Euclidean space $\mathbf{R}^n$ is the following. Starting with the one-dimensional lattice consisting of the even integer points, $2\mathbf{Z}$, we obtain a two-dimensional packing by drawing a row of circles of radius 1 centered at the even integer points along one axis, then a similar row next to it as close as possible, then another row at the same spacing, and so on. This produces the hexagonal lattice packing A_2. To go to three dimensions we place a layer of billiard balls in the hexagonal lattice packing, then place a similar layer next to it and as close as possible, then another layer at the same spacing, and so on, obtaining the face-centered cubic lattice packing $A_3 \cong D_3$. At each step, in going from i to $i+1$ dimensions, we stack layers (or laminae) consisting of copies of a suitable i-dimensional lattice Λ_i as close together as possible. The resulting lattice(s) — for in some higher dimensional spaces more than one lattice can be obtained in this way — we call *laminated lattices* (a more precise definition is given in §2). A typical n-dimensional laminated lattice will be denoted by Λ_n.

In this chapter we determine the density of Λ_n for $n \leqslant 48$ (see Fig. 1.5 of Chap. 1 and Table 6.1), find all Λ_n for $n \leqslant 25$ (see Fig. 6.1), and find at least one Λ_n for $26 \leqslant n \leqslant 48$. We see that laminated lattices are the densest packings known in dimensions $n \leqslant 29$ except for $n = 10$, 11, 12 and 13.

Table 6.1. The laminated lattice Λ_n has minimal norm 4, determinant λ_n and center density $\delta = \lambda_n^{-1/2}$. The distance between the laminae Λ_n in Λ_{n+1} is $\pi_n^{1/2}$, and the greatest subcovering radius of any Λ_n is h_n. Note that $\pi_n = 4 - h_n^2$ and $\lambda_n = \pi_{n-1} \lambda_{n-1}$. Also $h_n \le r_n$ (the greatest covering radius of any Λ_n), with equality if either $h_n \le \sqrt{3}$ or $r_n \le \sqrt{3}$.

n	λ_n	π_n	h_n^2	n	λ_n	π_n	h_n^2
0	1	4	0	24	1	2	2
1	4	3	1	25	2	3/2	5/2
2	12	8/3	4/3	26	3	4/3	8/3
3	32	2	2	27	4	1	3
4	64	2	2	28	4	1	3
5	128	3/2	5/2	29	4	3/4	13/4
6	192	4/3	8/3	30	3	2/3	10/3
7	256	1	3	31	2	1/2	7/2
8	256	2	2	32	1	1	3
9	512	3/2	5/2	33	1	3/4	13/4
10	768	4/3	8/3	34	3/4	2/3	10/3
11	1024	1	3	35	1/2	1/2	7/2
12	1024	1	3	36	1/4	1/2	7/2
13	1024	3/4	13/4	37	1/8	3/8	29/8
14	768	2/3	10/3	38	$3 \cdot 2^{-6}$	1/3	11/3
15	512	1/2	7/2	39	2^{-6}	1/4	15/4
16	256	1	3	40	2^{-8}	1/2	7/2
17	256	3/4	13/4	41	2^{-9}	3/8	29/8
18	192	2/3	10/3	42	$3 \cdot 2^{-12}$	1/3	11/3
19	128	1/2	7/2	43	2^{-12}	1/4	15/4
20	64	1/2	7/2	44	2^{-14}	1/4	15/4
21	32	3/8	29/8	45	2^{-16}	3/16	61/16
22	12	1/3	11/3	46	$3 \cdot 2^{-20}$	1/6	23/6
23	4	1/4	15/4	47	2^{-21}	1/8	31/8
24	1	2	2	48	2^{-24}	?	?

History. A detailed account of this laminating process in up to eight dimensions (and a description of the nonlattice packings that can be constructed in the same way) was given in §4 of the previous chapter and in [Lee7], where it is shown that $\Lambda_1, \ldots, \Lambda_8$ are unique and coincide with the densest lattice packings $A_1 \cong Z$, A_2, $A_3 \cong D_3$, D_4, D_5, E_6, E_7, E_8 in these dimensions (see §1.5 of Chap. 1). It appears that in 1946 Chaundy [Cha4] found the laminated lattices Λ_9 and Λ_{10}, but did *not* show these to be the densest possible lattices in 9 and 10 dimensions (which have still not been determined). E. S. Barnes has informed us that in unpublished work Chaundy also found some of the Λ_n in the next few dimensions.

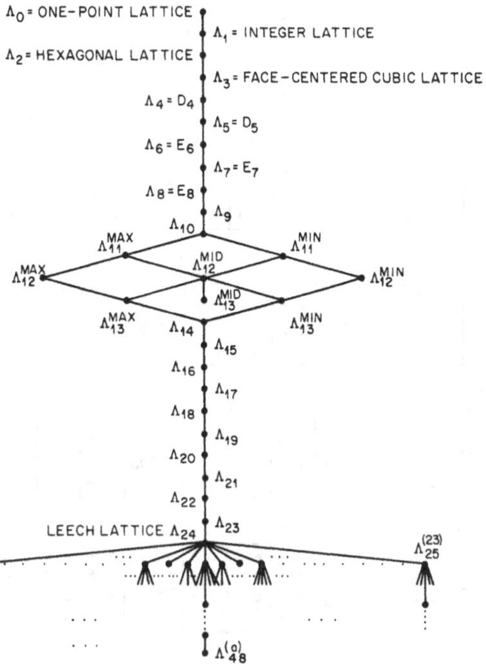

Figure 6.1. Inclusions among the laminated lattices Λ_n. All Λ_n for $n \leqslant 24$ are shown, while there are 23 Λ_{25}'s and probably a large number of Λ_{26}'s. At least one Λ_n is known for $26 \leqslant n \leqslant 48$.

We shall see that there are unique laminated lattices Λ_{15} and Λ_{16}, which are in fact isomorphic to lattices found by Barnes and Wall [Bar18] in 1959. The unique Λ_{24} is the Leech lattice, and the "main sequence" of cross-sections

$$\Lambda_0, \Lambda_1, \ldots, \Lambda_{10}, \Lambda_{11}^{\max}, \Lambda_{12}^{\max}, \Lambda_{13}^{\max}, \Lambda_{14}, \Lambda_{15}, \ldots, \Lambda_{24} \tag{1}$$

was described in [Lee4]-[Lee7] and the previous chapter, although there these lattices were not proved to give the densest laminations for $n > 8$. The members of this sequence are distinguished by having the highest kissing numbers among the Λ_n (see Table 6.3 below). In Fig. 6.2 we give particularly simple coordinates for these lattices as cross-sections of the Leech lattice (cf. [Lee5]) using the MOG notation of Chap. 4, §11. Their determinants are given in Table 6.1. The lattice $\Lambda_{12}^{\max}$ is Chaundy's J_{12} (see [Cox29]).

It was shown in [Lee4], [Lee5], [Lee10] that there is a second important sequence of sections of Λ_{24}, denoted by $K_0, K_1, \ldots, K_{24}$. These coincide with Λ_n for $n \leqslant 6$ and $18 \leqslant n \leqslant 24$, but for $n = 7, \ldots, 17$ are not laminated lattices. K_{12} was found by Coxeter and Todd ([Cox29]; see also §9 of Chap. 4), K_{11} by Barnes [Bar9], and K_{13} by Leech (unpublished). Figure 6.3 gives coordinates for the K_n as sections of Λ_{24},

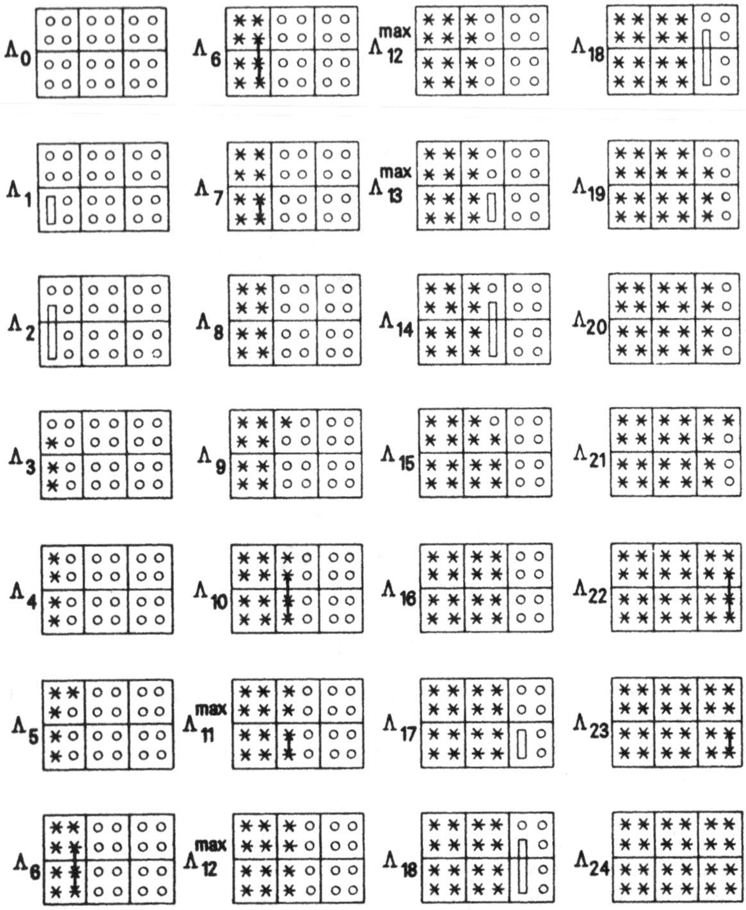

Figure 6.2. MOG coordinates for $\Lambda_0, \ldots, \Lambda_n$ as sections of the Leech lattice Λ_{24}. The figure shows Λ_0, $\Lambda_1 \cong \mathbf{Z} \cong A_1$, $\Lambda_2 \cong A_2$, $\Lambda_3 \cong A_3 \cong D_3$, $\Lambda_4 \cong D_4$, $\Lambda_5 \cong D_5$, $\Lambda_6 \cong E_6$, $\Lambda_7 \cong E_7$, $\Lambda_8 \cong E_8$, Λ_9, Λ_{10}, $\Lambda_{11}^{\max}$, $\Lambda_{12}^{\max}$, $\Lambda_{13}^{\max}$, Λ_{14},.... A small circle represents a zero coordinate, a hollow loop is a set of coordinates adding to zero, an asterisk is a free coordinate, and a line of asterisks is a set of equal coordinates.

and Table 6.2 gives their determinants. Some of the Λ_n for $n \leqslant 16$ and the K_n for $n \leqslant 13$ were rediscovered by Skubenko [Sku1], who appears to have been unaware of the earlier work on this subject. Particular lattices $\Lambda_{25}, \ldots, \Lambda_{40}$ are also described in the previous chapter. Leech has shown that, for $n \leqslant 21$, the layers can be stacked to give nonlattice packings that are as dense as Λ_n and K_n, except for Λ_1, Λ_2, Λ_4, Λ_8, K_{11} and K_{12}.

Sequels to this chapter. We have generalized this work in two ways in [Con36], by considering (a) laminated *complex* and *quaternionic* lattices (cf. §2.6 of Chap. 2), and (b) *integral* laminated lattices, where Λ_n is required to be an integral (real, complex or quaternionic) lattice with the

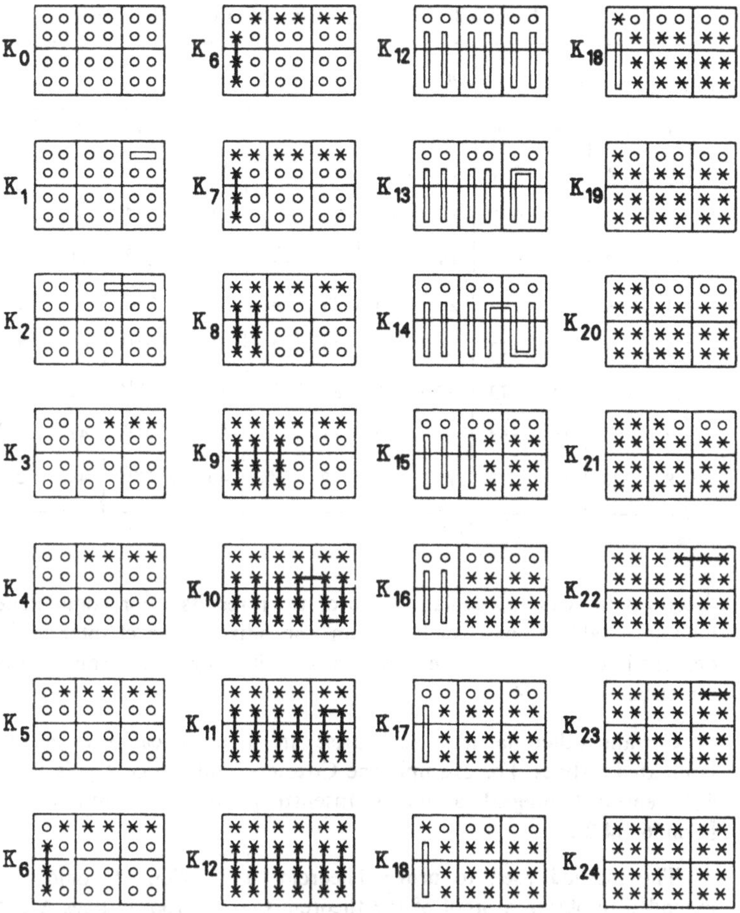

Figure 6.3. MOG coordinates for the lattices $K_0, \ldots, K_n$ as sections of the Leech lattice. The symbols have the same meaning as in Figure 6.2.

prescribed minimal norm. Let us temporarily denote the particular laminated lattices constructed in §7 below in dimensions 25-48 by $\Lambda_n^{(a)}$. Then we show in [Con36] that the lattices

$$\Lambda_0, \Lambda_4, \Lambda_8, \Lambda_{12}^{\max}, \Lambda_{16}, \Lambda_{20}, \Lambda_{24}, \Lambda_{28}^{(a)}, \Lambda_{32}^{(a)}, \ldots, \Lambda_{48}^{(a)}$$

can be given the structure of quaternionic lattices (over the Hurwitz integers), and that these are laminated quaternionic lattices. Furthermore they also have the structure of Gaussian or Eisenstein lattices (cf. §2.6 of Chap. 2), and as such are again laminated lattices. Integral laminated lattices with minimal norm M over a ring of integers J have a similar definition to laminated lattices, with the additional requirement that the layers must be stacked so as to form an integral lattice, i.e. so that all inner products $u \cdot \bar{v}$ belong to J. Real integral laminated lattices have been investigated by computer by Plesken and Pohst [Ple6], [Poh2], and the

Table 6.2. Determinants κ_n of lattices K_n.

n	κ_n	n	κ_n	n	κ_n	n	κ_n	n	κ_n	n	κ_n
0	1	8	576	16	576	24	1	32	9/4	40	$3^2 \cdot 2^{-10}$
1	4	9	864	17	384	25	2	33	$3^3 \cdot 2^{-4}$	41	$3 \cdot 2^{-10}$
2	12	10	972	18	192	26	3	34	$3^5 \cdot 2^{-8}$	42	$3 \cdot 2^{-12}$
3	32	11	972	19	128	27	4	35	$3^5 \cdot 2^{-9}$	43	2^{-12}
4	64	12	729	20	64	28	4	36	$3^6 \cdot 2^{-12}$	44	2^{-14}
5	128	13	972	21	32	29	4	37	$3^5 \cdot 2^{-11}$	45	2^{-16}
6	192	14	972	22	12	30	3	38	$3^5 \cdot 2^{-12}$	46	$3 \cdot 2^{-20}$
7	384	15	864	23	4	31	3	39	$3^3 \cdot 2^{-10}$	47	2^{-21}
8	576	16	576	24	1	32	9/4	40	$3^2 \cdot 2^{-10}$	48	2^{-24}

theorems in [Con36] explain some of their results. For completeness we include a table from [Con36] as an appendix to this chapter, giving the densest integral lattices presently known with minimal norms 2, 3 or 4 in dimensions up to 24.

Certain other lattices also arise in a canonical way as integral laminated lattices. For example the Coxeter-Todd lattice K_{12} is the unique 6-dimensional integral laminated Eisenstein lattice with minimal norm 2 [Con36, §4.2].

The method used to prove Theorem 10 of [Con36] could easily be extended to obtain a proof of the theorem that the root lattices Λ_0, A_1, A_2, $A_3 \cong D_3$, D_4, D_5, E_6, E_7 E_8 (see Theorem 2 below) are the unique laminated lattices $\Lambda_0, \ldots, \Lambda_8$. The only extra step needed is a proof that the deep holes in $\Lambda_0, \ldots, E_7$ belong to the dual lattices $\Lambda_0^*, \ldots, E_7^*$. This can be supplied using Norton's technique (see Chap. 22 and §5 below). In the present chapter we deduce this theorem from the results of Blichfeldt and Vetchinkin. The alternative proof would make our determination of the Λ_n for $n \leqslant 48$ independent of Blichfeldt and Vetchinkin's work (the inductive argument would then proceed one dimension at a time instead of in steps of eight dimensions). Of course Theorem 3 below would still depend on [Bli4] and [Vet2].

Lexicographic codes, which are a coding-theory analog of laminated lattices, are studied in [Con41].

Outline. The rest of this chapter is arranged as follows. The next few paragraphs establish some notation, and §2 states the main results. In §3 we collect some properties of $\Lambda_1, \ldots, \Lambda_8$ that will be used throughout the chapter and then §§4-7 are devoted to the proof of Theorem 1.

Notation. As in §1.4 of Chap. 1, if two lattices L and M differ only by a rotation, a scale factor and possibly a reflection, we say they are equivalent and write $L \cong M$. We shall use three different norms for lattices. The laminated lattices Λ_n will always have minimal norm 4, to correspond to a packing of unit spheres. The root lattices A_n, D_n and E_n may have minimal norm 2 or 4, depending on which coordinates are being used. Finally in MOG notation (see §11 of Chap. 4) the minimal norm is 32. If the minimal norm of a lattice is μ, its center density (in this chapter the adjective "center" will usually be omitted) is

$$\delta = (\mu/4)^{n/2}(\det L)^{-1/2}, \tag{2}$$

which becomes $\delta = (\det L)^{-1/2}$ for a laminated lattice. A *hole* (see §1.2 of Chap. 2) in a lattice $L_n \subseteq \mathbf{R}^n$ is a point of $\mathbf{R}^n$ whose distance from L_n is a local maximum. The greatest distance of any hole from L_n is the *covering radius* of L_n (Eq. (3) of Chap. 2), and a hole at this distance is called a *deep* hole. By a k-dimensional *section* of a lattice L_n we mean a k-dimensional lattice $M_k \subseteq \mathbf{R}^k \subseteq \mathbf{R}^n$ such that $M_k = L_n \cap \mathbf{R}^k$. The *dual* lattice $L_n^* = \{ x \in \mathbf{R}^n : x \cdot y \in \mathbf{Z} \text{ for all } y \in L_n \}$ (Eq. (24) of Chap. 1). If V is a subspace of $\mathbf{R}^n$ the *orthogonal* subspace $V^\perp = \{ x \in \mathbf{R}^n : x \cdot y = 0 \text{ for all } y \in V \}$.

2. The main results

Definition of laminated lattice Λ_n. Let Λ_0 be the one-point lattice. For $n \geqslant 1$ we take all n-dimensional lattices with minimal norm 4 that have at least one sublattice Λ_{n-1}, and select those of minimal determinant. Any such lattice is a laminated lattice Λ_n. When necessary we use superscripts (e.g. $\Lambda_{12}^{\max}$, $\Lambda_{12}^{\min}$) to distinguish between different Λ_n's.

The following geometrical ideas will be useful. The projection of a given laminated lattice Λ_n onto the 1-dimensional subspace $(\mathbf{R}\Lambda_{n-1})^\perp$, where Λ_{n-1} is one of its sublattices, is a 1-dimensional lattice of minimal norm π_{n-1}, say. Then Λ_n may be regarded as a union

$$\cdots \bigcup \Lambda_{n-1}^{(-1)} \bigcup \Lambda_{n-1}^{(0)} \bigcup \Lambda_{n-1}^{(1)} \bigcup \cdots$$

of translates of Λ_{n-1}, $\Lambda_{n-1}^{(i)}$ having the coordinate $i\sqrt{\pi_{n-1}}$ in $(\mathbf{R}\Lambda_{n-1})^\perp$. The smallest norm, s say, of vectors in $\Lambda_n^{(1)}$ is at least 4 (and it will turn out that in all the cases we use, s actually *is* 4). If v is a vector in $\Lambda_{n-1}^{(1)}$ of norm s and the distance of v from $\Lambda_{n-1}^{(0)}$ is h_{n-1}, then

$$\pi_{n-1} = s - h_{n-1}^2 \geqslant 4 - h_{n-1}^2. \tag{3}$$

So h_{n-1}, which we shall call the *subcovering radius*, is a lower bound for the covering radius r_{n-1} of Λ_{n-1}. It is easy to see that h_{n-1} and r_{n-1} are equal if either is $\leqslant \sqrt{3}$. (It seems likely that h_{n-1} is always equal to r_{n-1}, although we cannot prove this.) If the determinant of Λ_n is denoted by λ_n, we have

$$\lambda_n = \pi_{n-1}\lambda_{n-1}. \tag{4}$$

The density of Λ_n is $\lambda_n^{-1/2}$. It is clear from the definition that all Λ_n have the same determinant, but only those Λ_n with the maximal subcovering radius (h_n) can be extended to Λ_{n+1}'s.

Our main result is the following.

Theorem 1. *The density λ_n of any laminated lattice Λ_n for $n \leqslant 48$ is as shown in Table 6.1 (see also Fig. 1.5 of Chap. 1). The greatest subcovering radius h_n of any Λ_n for $n \leqslant 47$ is also shown in Table 6.1. When $h_n \leqslant \sqrt{3}$, h_n is also the greatest covering radius of any Λ_n. All Λ_n for $n \leqslant 25$ are shown in Fig. 6.1. At least one Λ_n is known for $26 \leqslant n \leqslant 48$.*

In dimensions 26 and above the number of inequivalent Λ_n appears to be very large (see §7). We see from Fig. 6.1 that for $n \leqslant 24$ there is only one Λ_n, $\Lambda_{13}^{\text{mid}}$, that is not contained in a Λ_{n+1}. However, Plesken (personal communication) has shown that $\Lambda_{13}^{\text{mid}}$ is a sublattice of Λ_{24}.

On the scale at which the minimal norm is 4, the Λ_n are integral lattices if and only if $n \leqslant 24$. (For $n \geqslant 25$, Λ_n contains Λ_{24}, a self-dual lattice, but does not contain Λ_{24} as a direct summand, and so is not integral.) The quadratic form associated with any laminated lattice is perfect. (This follows immediately from [Bar9, Part II, Th. 2.1].)

The proof of Theorem 1 depends heavily on the next two theorems, the first of which collects known results.

Theorem 2. *For $n = 0, 1, \ldots, 8$ the densest lattice packing in $\mathbf{R}^n$ is isomorphic to Λ_0, $A_1 \cong \mathbf{Z}$, A_2, $A_3 \cong D_3$, D_4, D_5, E_6, E_7, E_8 respectively. The laminated lattices $\Lambda_0, \Lambda_1, \ldots, \Lambda_8$ are unique and are isomorphic to these lattices. Their determinants and covering radii are shown in Table 6.1.*

Proof. For the first assertion see [Bli4], [Vet2]; for the second see [Lee7] and the previous chapter; and for the covering radii see for instance Chap. 21, [Con28], [Cox20].

Theorem 2 certainly establishes the results of Theorem 1 for $n \leqslant 8$, although of course it is much stronger than that.

Theorem 3. *Let L_{8m+r} $(m \geqslant 0, 0 \leqslant r \leqslant 8)$ be any lattice in $\mathbf{R}^{8m+r}$ of minimal norm $\geqslant 4$ with a section Λ_{8m} having covering radius r_{8m}. Then*

$$\det L_{8m+r} \geqslant \lambda_{8m}\, \lambda_r \left(\frac{4 - r_{8m}^2}{4} \right)^r . \tag{5}$$

If equality holds in (5) there is an additive map θ, a "gluing" map, from Λ_r into $\mathbf{R}\Lambda_{8m}$, such that the vectors of L_{8m+r} have the form

$$u + v^\theta + cv, \quad \text{where} \quad u \in \Lambda_{8m},\ u + v^\theta \in \mathbf{R}\Lambda_{8m},\ v \in \Lambda_r, \tag{6}$$

and

$$c = \left(\frac{4 - r_{8m}^2}{4} \right)^{1/2}, \tag{7}$$

and such that

$$N(u+v^{\theta}) \geqslant 4 - c^2 N(v) \tag{8}$$

for all $u \in \Lambda_{8m}$, $v \in \Lambda_r$, $(u,v) \neq (0,0)$. *In other words* L_{8m+r} *is obtained by gluing a scaled copy of* Λ_r *to* Λ_{8m}. *Furthermore* L_{8m+r} *is a* Λ_{8m+r}, *and*

$$\lambda_{8m+r} = \lambda_{8m}\,\lambda_r \left(\frac{4-r_{8m}^2}{4}\right)^r . \tag{9}$$

Proof. A typical vector of L_{8m+r} may be written as $v_1 + v_2$ where $v_1 \in \mathbf{R}\Lambda_{8m}$, $v_2 \in (\mathbf{R}\Lambda_{8m})^{\perp}$. The vectors v_2 span an r-dimensional lattice L_r, which is isomorphic to L_{8m+r}/Λ_{8m}. Thus $\det L_{8m+r} = \det \Lambda_{8m} \cdot \det L_r$.

By adding a vector of Λ_{8m} we can reduce the norm of v_1 to at most r_{8m}^2. But $4 \leqslant N(v_1+v_2) = N(v_1) + N(v_2)$, so the minimal norm of L_r is at least $4 - r_{8m}^2$. Therefore by Theorem 2

$$\det L_r \geqslant \left(\frac{4-r_{8m}^2}{4}\right)^r \lambda_r ,$$

which proves (5).

If equality holds in (5) then we must have $L_r = c\,\Lambda_r$, with c as in (7), and the gluing map θ described in the theorem must exist. If $0 \leqslant s \leqslant r$, Λ_r has a section Λ_s (by Theorem 2), and restricting v to Λ_s in (6) we obtain a section L_{8m+s} of L_{8m+r}. This shows that L_{8m+r} is a laminated lattice and completes the proof.

The K_n sequence. We shall not say as much about the rival sequence of sections K_n of the Leech lattice. Up to $n = 6$, $K_n \cong \Lambda_n$; $K_7 - K_{10}$ have a lower density than $\Lambda_7 - \Lambda_{10}$; but then K_{11}, K_{12} and K_{13} have higher densities than $\Lambda_{11} - \Lambda_{13}$. For $n = 18$ to 24 again $K_n \cong \Lambda_n$.

There is a similar sequence $\{K_n\}$ of sections of Λ_{48}, obtained by gluing K_n (for $0 \leqslant n \leqslant 24$) to the deep holes in Λ_{24}. The determinant κ_n of K_n is shown in Table 6.2 and Fig. 1.5 of Chap. 1, and MOG coordinates for $n \leqslant 24$ are given in Fig. 6.3.

Relations between densities. The sequences of determinants $\{\lambda_n\}$ and $\{\kappa_n\}$ possess a number of curious palindromic properties (see Tables 6.1, 6.2), namely

$$\frac{\lambda_{48-n}}{\lambda_n} = \frac{\kappa_{48-n}}{\kappa_n} = \left(\frac{1}{2}\right)^{24-n} \quad (0 \leqslant n \leqslant 48), \tag{10}$$

$$\frac{\lambda_{24-n}}{\lambda_n} = \frac{\kappa_{24-n}}{\kappa_n} = 1 \quad (0 \leqslant n \leqslant 24), \tag{11}$$

$$\frac{\lambda_{16-n}}{\lambda_n} = 2^{8-n} \quad (0 \leqslant n \leqslant 16), \quad \frac{\kappa_{12-n}}{\kappa_n} = 3^{6-n} \quad (0 \leqslant n \leqslant 12), \quad (12),(13)$$

$$\frac{\lambda_{8-n}}{\lambda_n} = 4^{4-n} \quad (0 \leqslant n \leqslant 8), \quad \frac{\lambda_{4-n}}{\lambda_n} = 8^{2-n} \quad (0 \leqslant n \leqslant 4). \tag{14),(15}$$

These imply the useful relations

$$\lambda_{24+n} = \frac{\lambda_n}{2^n}, \quad \kappa_{24+n} = \frac{\kappa_n}{2^n} \quad (0 \leqslant n \leqslant 24). \tag{16}$$

These identities are partially explained by the following theorem.

Theorem 4. *Let L_n be a (not necessarily integral) lattice in $\mathbf{R}^n$, and let $E = L_n \cap S$ be a k-dimensional section of L_n, where S is a k-dimensional subspace of $\mathbf{R}^n$. Let $F = L_n^* \cap S^\perp$, an $(n-k)$-dimensional section of L_n^*. Then*

$$\det F = \frac{\det E}{\det L_n}. \tag{17}$$

In particular, if $\det L_n = 1$,

$$\det F = \det E\ . \tag{18}$$

Proof. Let $e_1, \ldots, e_n$ be an orthonormal basis for $\mathbf{R}^n$, chosen so that $e_1, \ldots, e_k$ is a basis for S and $e_{k+1}, \ldots, e_n$ is a basis for $S^\perp$. Since E is a section of L_n, we can find an integral basis $v_1, \ldots, v_n$ for L_n so that $v_1, \ldots, v_k$ is an integral basis for E. Let $w_1, \ldots, w_n$ be the dual basis to $v_1, ..., v_n$, with $v_i \cdot w_j = \delta_{ij}$. Then $w_1, \ldots, w_n$ is an integral basis for L_n^* and $w_{k+1}, \ldots, w_n$ is an integral basis for F. There is a matrix $M = \begin{pmatrix} A & 0 \\ B & C \end{pmatrix}$, with inverse $\begin{pmatrix} A^{-1} & 0 \\ * & C^{-1} \end{pmatrix}$, such that

$$\begin{pmatrix} v_1 \\ \vdots \\ v_n \end{pmatrix} = M \begin{pmatrix} e_1 \\ \vdots \\ e_n \end{pmatrix}, \quad \begin{pmatrix} w_1 \\ \vdots \\ w_n \end{pmatrix} = (M^{-1})^{tr} \begin{pmatrix} e_1 \\ \vdots \\ e_n \end{pmatrix}.$$

We now compute determinants (see Eq. (5) of Chap. 1): $\det L_n = (\det A)^2 (\det C)^2$, $\det E = (\det A)^2$, $\det F = (\det C)^{-2}$, and (17) follows.

Corollary 5. *Let L_n be either* (i) *an integral unimodular lattice or* (ii) *a lattice which is equivalent to its dual and has been scaled so to have determinant* 1. *Then the smallest determinant of any k-dimensional section of L_n is equal to the smallest determinant of any $(n-k)$-dimensional section, for $0 \leqslant k \leqslant n$.*

Proof. In case (i) $L_n = L_n^*$, and in case (ii) the scaled version of L_n is congruent to L_n^* (cf. §1.4 of Chap. 1). In both cases $\det L_n = 1$, so $\det E = \det F$ in (17).

Corollary 6 [Lee10, Th. 4.5.1]. *Let L_n be one of the lattices D_4, E_8, K_{12}, Λ_{16}, Λ_{24}, P_{48p} or P_{48q}, scaled so as to have determinant* 1. *To every k-*

dimensional section of L_n there is a corresponding $(n-k)$-dimensional section with the same determinant. The same result applies to the particular Λ_{48} constructed in §7.

Proof. The lattices mentioned all satisfy the hypotheses of Corollary 5.

Corollary 7 (cf. [Lee10, Th. 4.5.2]). (a) *Each of $\Lambda_0, \ldots, \Lambda_8$, $\Lambda_{16}, \ldots, \Lambda_{23}$ has the smallest determinant of any section of the Leech lattice in its dimension.* (b) *For $n \leqslant 23$, Λ_n has the smallest determinant of any n-dimensional section of Λ_{n+1}.* (c) *For $n \leqslant 23$, K_n has the smallest determinant of any n-dimensional section of K_{n+1} containing or contained in K_{12}.*

Proof (a) $\Lambda_0, \ldots, \Lambda_8$ are the densest possible lattices in their dimensions (Theorem 2), which proves the first assertion, and the second follows from Corollary 6. (b) Corollary 6 also implies that $\Lambda_9, \ldots, \Lambda_{15}$ have the smallest determinant of any section of Λ_{16}. (c) Similarly $K_7, \ldots, K_{11}$ have the smallest determinant of any section of K_{12}, and $K_{13}, \ldots, K_{17}$ have the smallest determinant of any section of Λ_{24} containing K_{12}. For $0 \leqslant n \leqslant 6$ and $18 \leqslant n \leqslant 24$, $K_n = \Lambda_n$.

Remark. We are unable to exclude the possibility that there are sections of Λ_{24} in dimensions 9 to 15 with smaller determinant (and hence higher density) than either the Λ_n or K_n sequences.

Mordell's inequality is also a consequence of Theorem 4. This states that if δ_n denotes the highest possible center density of any lattice in $\mathbf{R}^n$, then

$$\delta_{n-1} \geqslant \frac{1}{2} \delta_n^{(n-2)/n} \tag{19}$$

([Cas2, Chap. X, Th. IV], [Mor4], [Opp1]).

Proof. Let $\delta(L)$ and $\mu(L)$ denote the center density and minimal norm of L, respectively. Let L_n be a densest lattice, with $\det L_n = 1$. Then $\det L_n^* = 1$, so $\delta(L_n^*) \leqslant \delta(L_n)$, $\mu(L_n^*) \leqslant \mu(L_n)$. Then $\mu(L_n^*) =$ determinant of densest 1-dimensional section of $L_n^* =$ determinant of densest $(n-1)$-dimensional section of $L_n \leqslant \mu(L_n)$, and now use (2).

Equality holds in (19) for $n = 1, 2, 4, 8$. If, as we suspect, K_{12}, Λ_{16}, Λ_{24} and P_{48p} are the densest lattices in these dimensions, equality also holds for $n = 12, 16, 24$ and 48. Some sections of P_{48p} are described in the following corollary.

Corollary 8. *In dimension $48-n$ the lattice P_{48p} has a section of density $\delta = (3^{48-2n}/2^{48-n}D)^{1/2}$ where D is given by the following table:*

$n = 0$	1	2	3	4	5	6	7	8	9	10
$D = 1$	2	3	4	4	4	4	4	$\frac{32}{9}$	$\frac{28}{9}$	$\frac{24}{9}$

Moreover in dimensions 43 and above these lattices have the smallest possible determinant of any section of any even unimodular 48-dimensional lattice with minimal norm 6.

Proof. It follows from Table 5.3 that coordinates for P_{48p} can be chosen so that the minimal vectors include all the vectors

$$v_{\pm i, \pm j} = (\pm 3^2, 0^{46}) \tag{20}$$

and the particular vector

$$v^* = (-2, 1^{14}, 0^{33}). \tag{21}$$

It therefore contains scaled copies of D_n (all the $v_{\pm i, \pm j}$ supported in the first n coordinates), A_n (the part of the above D_{n+1} with zero coordinate sum), and the lattices we shall call S_n (generated by v^* together with the above D_{n-1}). On the scale at which P_{48p} is unimodular, its sections orthogonal to the lattices

$$\Lambda_0, A_1, A_2, A_3 \cong D_3, D_4, D_5, D_6, D_7 \text{ or } S_7, S_8, S_9, S_{10}$$

have the same densities as those lattices by Corollary 5; the first assertion now follows by rescaling. The second assertion of the corollary follows from the optimality of $\Lambda_0, \ldots, D_5$ (Theorem 2).

3. Properties of Λ_0 to Λ_8

The lattices Λ_0 to Λ_8 (which were identified in Theorem 2) may be specified as the sections of the Leech lattice shown in Fig. 6.2. For instance the diagram for Λ_2 indicates that Λ_2 may be defined as the set of all points of the Leech lattice in which the first three coordinates add to zero and the rest are equal to zero. This lattice is spanned by $(4, -4, 0^{22})$ and $(4, 0, -4, 0^{21})$, and clearly is a scaled version of A_2.

For use in the next section we remark that, for any r, $\Lambda_r/2\Lambda_r$ is an elementary abelian group of order 2^r. The usual coordinates for D_n and E_n are more convenient for describing the congruence classes of $\Lambda_r/2\Lambda_r$. First, the nonzero classes of $D_3/2D_3$ have the structure of a projective plane of order 2, as shown in Fig. 6.4.

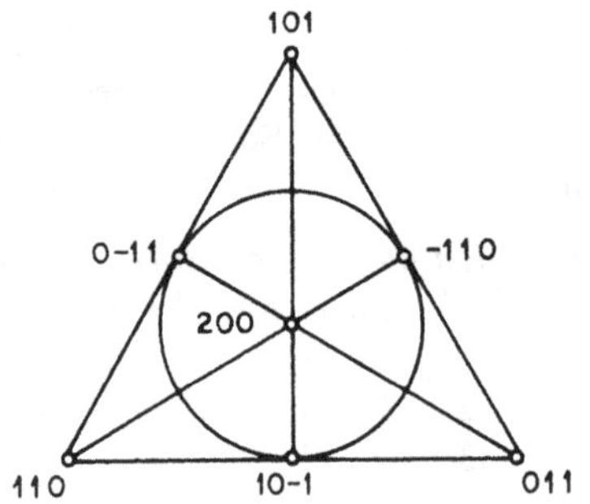

Figure 6.4. Representatives for classes of $D_3/2D_3$. Three classes on a line add to zero.

Representatives for the sixteen classes of $D_4/2D_4$ may be taken to be

$$\begin{array}{cccc} 0000 & 2000 & \bar{1}111 & \overline{1111} \\ 00\bar{1}1 & 0011 & \bar{1}100 & 1100 \\ 010\bar{1} & 0101 & \bar{1}010 & 1010 \\ 0\bar{1}10 & 0110 & \bar{1}001 & 1001 \end{array}$$

where the bars indicate minus signs, and there is an automorphism of D_4 rotating the last three columns.

The classes of $D_5/2D_5$ and typical representatives are:

(i)	1 class	(00000)
(ii)	20 classes	$(\pm 1^2, 0^3)$, (minimal vectors)
(iii)	1 class	$(\pm 2, 0^4)$
(iv)	5 classes	$(\pm 1^4, 0)_{\text{even}}$
(v)	5 classes	$(\pm 1^4, 0)_{\text{odd}}$.

The elements of E_6 are described by elements of E_8 in which the last three coordinates are equal. The classes and representatives for $E_6/2E_6$ are:

1 class	(0^8)
16 classes	$((\pm {}^1\!/_2)^5; (\pm {}^1\!/_2)^3)$, (minimal vectors)
20 classes	$(\pm 1^2, 0^3; 0^3)$, (minimal vectors)
1 class	$(\pm 2, 0^4; 0^3)$
16 classes	$(\mp {}^3\!/_2, (\pm {}^1\!/_2)^4; (\pm {}^1\!/_2)^3)$
10 classes	$(\pm 1^4, 0; 0^3)$.

The last 27 classes are each represented by ten norm 4 vectors forming a 5-dimensional coordinate frame (in these coordinates the minimal vectors have norm 2). Similarly the classes of $E_7/2E_7$ are:

1 class	represented by (0^8)
63 classes	represented by a pair of minimal (norm 2) vectors
63 classes	represented by 12 norm 4 vectors forming a 6-dimensional coordinate frame
1 class	represented by 56 vectors of norm 6, namely $(\pm 1^6; 0^2)$ with an odd number of minuses, and $\pm(\pm 2^1, 0^5; 1^2)$.

Finally $E_8/2E_8$ contains:

1 class	represented by (0^8)
120 classes	represented by a pair of minimal vectors
135 classes	represented by 16 vectors of twice the minimal norm forming an 8-dimensional coordinate frame.

Each deep hole in E_8 is an element of $\frac{1}{2} E_8$ that is congruent (modulo E_8) to 16 vectors which are halves of one of these coordinate frames (see

Chap. 21). In MOG notation (in which the minimal norm of E_8 is 32), these 135 types of deep holes are the following:
1 frame f_0 consisting of 16 vectors of the form

$$\begin{array}{|rr|}\hline 0 & 0 \\ \pm 4 & 0 \\ 0 & 0 \\ 0 & 0 \\ \hline \end{array}$$

35 positive tetrad frames $f^{+}_{abcd|efgh}$ consisting of all vectors like

$$\begin{array}{|rr|}\hline \pm 2 & 0 \\ \pm 2 & 0 \\ \pm 2 & 0 \\ 0 & \pm 2 \\ \hline \end{array}$$

where the support is a,b,c,d or e,f,g,h and the product of the nonzero entries is positive,
35 negative tetrad frames $f^{-}_{abcd|efgh}$ defined similarly, and
64 odd frames consisting of vectors like

$$\begin{array}{|rr|}\hline \pm 1 & \pm 1 \\ \pm 1 & \pm 1 \\ \mp 3 & \pm 1 \\ \pm 1 & \pm 1 \\ \hline \end{array} \; .$$

In the following section these 135 types will be referred to as the *frames* in E_8 (or Λ_8).

4. Dimensions 9 to 16

In this section we establish the assertions of Theorem 1 for $9 \leqslant n \leqslant 16$. Let Λ_{8+r} $(1 \leqslant r \leqslant 8)$ be any laminated lattice in $\mathbf{R}^{8+r}$. By definition Λ_{8+r} contains $\Lambda_8 \cong E_8$. From Theorem 3 there is a gluing map $\theta \colon \Lambda_r \to \mathbf{R}\Lambda_8$ such that Λ_{8+r} consists of the vectors

$$u + v^{\theta} + cv, \qquad u \in \Lambda_8,\ v \in \Lambda_r,\ c^2 = \tfrac{1}{2}\,.$$

If $v \in \Lambda_r$ has the minimal nonzero norm, 4, then from (6), $N(v^{\theta}) \geqslant 2$. But the biggest hole in Λ_8 has norm 2, so we conclude that *minimal vectors in Λ_r must be glued to deep holes in Λ_8.*

We saw in the previous section that the deep holes in Λ_8 are vectors in $\tfrac{1}{2}\Lambda_8$; and the value of v^{θ} is only important modulo Λ_8. Also the minimal vectors of Λ_r generate Λ_r, so $2\Lambda_r$ is mapped into Λ_8. Therefore θ induces a well-defined map

$$\Theta : \Lambda_r/2\Lambda_r \to \tfrac{1}{2}\Lambda_8/\Lambda_8\,. \tag{22}$$

The classes of $\Lambda_r/2\Lambda_r$ that are represented by minimal vectors of Λ_r must be mapped by Θ to frames in Λ_8 (see the previous section).

Proposition 9. *For $1 \leqslant r \leqslant 8$, any class of $\Lambda_r/2\Lambda_r$ not represented by a minimal vector of Λ_r is mapped by Θ either to 0 or to a frame (but not to a class represented by half a minimal vector of Λ_8).*

Proof. For $r \leqslant 2$ the hypothesis is vacuous. For $r \geqslant 3$, $r \neq 7$, we reduce to the case $r = 3$ by observing that any class of $\Lambda_r/2\Lambda_r$ is contained in a subgroup $\Lambda_3/2\Lambda_3$.

The case $r = 3$. Suppose the nonzero classes of $D_3/2D_3$, shown in Fig. 6.4, are mapped by Θ into the elements $A, \ldots, G$ of $\frac{1}{2}\Lambda_8/\Lambda_8$ shown in Fig. 6.5, where $A, \ldots, F$ are frames. Without loss of generality A is the frame f_0. B cannot be an odd frame, represented say by $(-3, 1^7)$, for then $C \equiv B + A$ would be represented by (1^8), which is not a deep hole. Thus B, and similarly C, D, E, are tetrad frames, and therefore G is represented by a vector with even coordinates. So if G is represented by half a minimal vector we can assume it is $(2^2, 0^6)$. But then $F \equiv A - G$ is represented by $(2, -2, 0^6)$, which is not a deep hole.

The case $r = 7$. The previous argument must be modified because there is one class of $\Lambda_7/2\Lambda_7$ which is represented by norm 6 vectors. However we may choose a basis $v_1, \ldots, v_7$ for $\Lambda_7/2\Lambda_7$ such that all the v_i and $v_i \pm v_j$ are represented by vectors of norm 2 or 4. By the previous argument no v_i^Θ or $(v_i \pm v_j)^\Theta$ may be represented by half a minimal vector. There is a quadratic form Q defined on $\frac{1}{2}\Lambda_8/\Lambda_8$ which takes the value 0 on the 0 class and on the frames, and the value 1 on the other 120 classes. Then the quadratic form $\Theta \circ Q$ maps the v_i and $v_i \pm v_j$ to zero and so is identically zero on $\Lambda_7/2\Lambda_7$. Thus the class of $\Lambda_7/2\Lambda_7$ represented by norm 6 vectors is also mapped to 0 or to a frame by Θ.

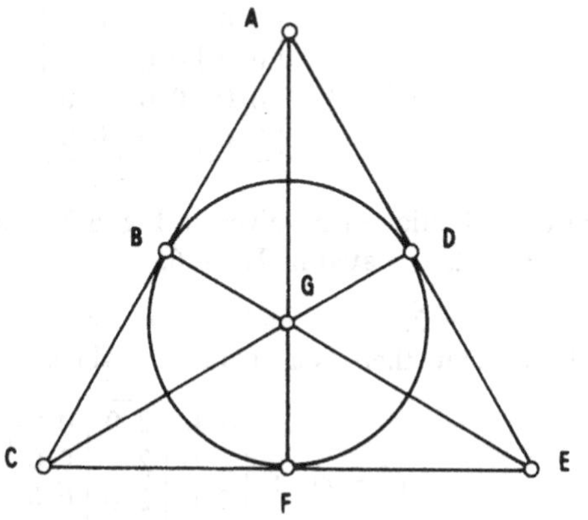

Figure 6.5. Image of Fig. 6.4 under Θ.

Proposition 10. *If* $v_1, \ldots, v_t$ *are minimal vectors of* Λ_r *that are independent modulo* (*ker* Θ, $2\Lambda_r$) *then we may assume that* $v_1^\Theta, \ldots, v_t^\Theta$ *are represented by an initial segment of*

$$\begin{array}{|cc|}\hline 4 & 0 \\ 0 & 0 \\ 0 & 0 \\ 0 & 0 \\ \hline\end{array}\,,\quad \begin{array}{|cc|}\hline 2 & 0 \\ 2 & 0 \\ 2 & 0 \\ 2 & 0 \\ \hline\end{array}\,,\quad \begin{array}{|cc|}\hline 2 & 2 \\ 2 & 2 \\ 0 & 0 \\ 0 & 0 \\ \hline\end{array}\,,\quad \begin{array}{|cc|}\hline 2 & 2 \\ 0 & 0 \\ 2 & 2 \\ 0 & 0 \\ \hline\end{array} \tag{23}$$

and so $t \leqslant 4$. *Thus the possible gluing maps* Θ *are completely determined by* t *and ker* Θ.

Proof. Without loss of generality v_1 is mapped to the frame f_0. As in the proof of Prop. 9, no minimal vector can now be mapped to an odd frame, so without loss of generality v_2^Θ is represented by the second element of (23), and so on. This completes the proof.

We can now determine all the possible laminated lattices Λ_9 to Λ_{16}. We choose the kernel of Θ in $\Lambda_r/2\Lambda_r$ to have dimension $k \geqslant r-4$, and so that no class in the kernel is represented by a minimal vector of Λ_r. Let $v_1, \ldots, v_r$ be a basis for $\Lambda_r/2\Lambda_r$ such that $v_1, \ldots, v_k$ is a basis for ker Θ and $v_{k+1}, \ldots, v_r$ are represented by minimal vectors. Then we map the $t = r-k$ vectors $v_{k+1}, \ldots, v_r$ onto an initial segment of (23).

(Λ_9). Here $r = t = 1$, $k = 0$ and Θ maps v_1 to the frame f_0. Thus there is a unique Λ_9, which may be described in MOG coordinates as being spanned by

$$\Lambda_8 : \begin{array}{|cc|cc|cc|}\hline * & * & 0 & 0 & 0 & 0 \\ * & * & 0 & 0 & 0 & 0 \\ * & * & 0 & 0 & 0 & 0 \\ * & * & 0 & 0 & 0 & 0 \\ \hline\end{array}$$

and

$$v_1^\theta + cv_1 : \begin{array}{|cc|cc|cc|}\hline 4 & 0 & 4 & 0 & 0 & 0 \\ 0 & 0 & 0 & 0 & 0 & 0 \\ 0 & 0 & 0 & 0 & 0 & 0 \\ 0 & 0 & 0 & 0 & 0 & 0 \\ \hline\end{array}\,.$$

This leads to the definition of Λ_9 given in Fig. 6.2. Also $N(cv_1) = 2 = \pi_8$, and $\lambda_9 = \pi_8\lambda_8 = 512$, as stated in Table 6.1.

(Λ_{10}). Here $r = t = 2$, $k = 0$ and Θ maps v_1 and v_2 to the first two elements of (23). Thus there is a unique Λ_{10} which is spanned by Λ_9 and

$$v = v_2^\theta + cv_2 : \begin{array}{|cc|cc|cc|}\hline 2 & 0 & 2 & 0 & 0 & 0 \\ 2 & 0 & 2 & 0 & 0 & 0 \\ 2 & 0 & 2 & 0 & 0 & 0 \\ 2 & 0 & 2 & 0 & 0 & 0 \\ \hline\end{array}$$

(see Fig. 6.2). The projection of v onto $\mathbf{R}\Lambda_9$ is

$$\begin{array}{|cc|cc|cc|} \hline 2&0&2&0&0&0\\ 2&0&0&0&0&0\\ 2&0&0&0&0&0\\ 2&0&0&0&0&0\\ \hline \end{array},$$

which is therefore a typical deep hole in Λ_9, of norm $h_9^2 = 5/2$, and the projection onto $(\mathbf{R}\Lambda_9)^{\perp}$ is

$$\begin{array}{|cc|cc|cc|} \hline 0&0&0&0&0&0\\ 0&0&2&0&0&0\\ 0&0&2&0&0&0\\ 0&0&2&0&0&0\\ \hline \end{array}$$

which has norm $\pi_9 = 3/2$. Also from either (4) or (9) we have $\lambda_{10} = 768$.

(Λ_{11}). There are two possibilities, depending on whether the central class in Fig. 6.4 is or is not in the kernel of Θ: (a) $t = 2$, $k = 1$, ker $\Theta = <(200)>$, leading to $\Lambda_{11}^{\max}$ (see Fig. 6.2); (b) $t = 3$, $k = 0$, Θ maps v_1, v_2, v_3 to the first three elements of (23), leading to $\Lambda_{11}^{\min}$. We shall see at the end of this section that these two lattices *are* distinct, as are the Λ_{12}'s and Λ_{13}'s constructed below.

(Λ_{12}). The kernel of Θ is a subgroup of the first row of the table of classes of $D_4/2D_4$ (see §3), and there is a unique subgroup of each of the orders 1, 2 and 4. Thus there are (at most) three possibilities for Λ_{12} : $\Lambda_{12}^{\max}$ (when $k=2$), shown in Fig. 6.2, $\Lambda_{12}^{\text{mid}}$ (when $k=1$), and $\Lambda_{12}^{\min}$ (when $k=0$).

(Λ_{13}). The kernel must be a subgroup of classes (i), (iii), (iv), (v) of $D_5/2D_5$, and have dimension $\geqslant 1$. However two classes from (iv) or (v) with distinct support have a difference of type (ii), and cannot both be in the kernel. Thus the kernel contains at most three nonzero classes. There are (at most) three possibilities: (a) $k=2$, ker $\Theta = \{(00000), (20000), (11110), (-11110)\}$, leading to $\Lambda_{13}^{\max}$ as shown in Fig. 6.2; (b) $k=1$, ker $\Theta = <(20000)>$, giving $\Lambda_{13}^{\text{mid}}$; (c) $k=1$, ker $\Theta = <(11110)>$, giving $\Lambda_{13}^{\min}$.

(Λ_{14}). Similar reasoning shows that there is a unique Λ_{14}, with ker $\Theta = <(20000;000), (11110;000)>$.

(Λ_{15}). There is a unique Λ_{15}, with ker $\Theta = <(200000;00), (111100;00), (110011;00)>$.

(Λ_{16}). There is a unique Λ_{16}, with ker $\Theta = <(20000000), (11110000), (11001100), (10101010)>$. The covering radius of Λ_{16} will be determined in the following section.

To see that these lattices are distinct we compute their kissing numbers τ. The minimal vectors fall into three classes: those in Λ_8 (240 in number), those gluing Λ_8 to $c\Lambda_r$ ($16\tau(\Lambda_r)$ of them), and those in $c\Lambda_r$ (the number of representations of elements of ker Θ by vectors of norm 4). For example

$$\tau(\Lambda_{13}^{max}) = 240 + 16 \cdot 40 + (10+8+8) = 906. \qquad (24)$$

The results appear in Table 6.3, and the inclusions between lattices of adjacent dimensions are displayed in Fig. 6.1. When $n \leqslant 15$ there are very few n-dimensional sections of Λ_{n+1} that contain Λ_8, so all such inclusions are easily found.

Table 6.3. Kissing numbers for laminated lattices.

Λ_0	Λ_1	Λ_2	Λ_3	Λ_4	Λ_5	Λ_6	Λ_7
0	2	6	12	24	40	72	126
Λ_8	Λ_9	Λ_{10}	Λ_{11}^{min}	Λ_{11}^{max}	Λ_{12}^{min}	Λ_{12}^{mid}	Λ_{12}^{max}
240	272	336	432	438	624	632	648
Λ_{13}^{min}	Λ_{13}^{mid}	Λ_{13}^{max}	Λ_{14}	Λ_{15}	Λ_{16}	Λ_{17}	Λ_{18}
888	890	906	1422	2340	4320	5346	7398
Λ_{19}	Λ_{20}	Λ_{21}	Λ_{22}	Λ_{23}	Λ_{24}	Λ_{25}	
10668	17400	27720	49896	93150	196560	196610-196656	

5. The deep holes in Λ_{16}

Having proved that Λ_{16} is unique, our next step is to find the deep holes in it, which we do by Simon Norton's method (see Chap. 22). First we enumerate the congruence classes of $\frac{1}{2}\Lambda_{16}/\Lambda_{16}$.

Proposition 11. *The* 2^{16} *classes of* $\frac{1}{2}\Lambda_{16}/\Lambda_{16}$ *consist of:*

(i) *the zero class,*

(ii) 2160 *classes each represented by a pair of vectors* $\pm \frac{1}{2} v$, *where* $v \in \Lambda_{16}$ *has norm* 4,

(iii) 30720 *classes each represented by a pair of vectors* $\pm \frac{1}{2} v$, *where* $v \in \Lambda_{16}$ *has norm* 6,

(iv) 135 *classes each represented by* 32 *vectors* $\pm \frac{1}{2} v_1, \ldots, \pm \frac{1}{2} v_{16}$, *where the* v_i *are mutually orthogonal norm* 8 *vectors (this accounts for* 4320 *of the norm* 8 *vectors),*

(v) 32400 *classes each represented by* 16 *vectors* $\pm \frac{1}{2} v_1, \ldots, \pm \frac{1}{2} v_8$, *where the* v_i *are norm* 8 *vectors (this accounts for the remaining* 518400 *norm* 8 *vectors — see Table* 4.12 *of Chap.* 4), *and finally*

(vi) 120 *classes each represented by the halves of a certain set of* 512 *norm* 12 *vectors in* Λ_{16}. *These representatives have the property that they* *can be extended to minimal vectors of the Leech lattice by supplementing them with (scalar multiples of) minimal vectors of Λ_8. Typical representatives are shown in Fig. 6.6, where the supplementing vectors are enclosed in broken lines. In Fig. 6.6a the* 512 *representatives are formed from the symmetric difference of any row and any column, with an even number of minus signs, while in Fig. 6.6b the* -3 *may be in any of the* 16 *positions and there are* 2^5 *possible sign combinations.*

The proof of this proposition is straightforward and is omitted.

0	2	2	2	2	2
2	0	0	0	0	0
2	0	0	0	0	0
2	0	0	0	0	0

(a)

-3	1	1	1	1	1
1	1	1	1	1	1
1	1	1	1	1	1
1	1	1	1	1	1

(b)

Figure 6.6. Typical deep holes in Λ_{16}. The broken line encloses (a scalar multiple of) the associated minimal vector of Λ_8.

Theorem 12. *The covering radius of* Λ_{16} *is* $\sqrt{3}$, *and the deep holes in* Λ_{16} *are those vectors of* $\frac{1}{2}\Lambda_{16}$ *which are congruent modulo* Λ_{16} *to one of the classes of type* (vi) *in Prop.* 11. *Thus apart from permutations and sign changes the deep holes are congruent modulo* Λ_{16} *to either Fig. 6.6a or Fig. 6.6b. There is a* 1-1 *correspondence between the congruence classes of deep holes and the pairs* $\pm u$ *of minimal vectors of* Λ_8.

Proof. Let the covering radius of Λ_{16} be $\sqrt{d}$. By compactness there is a point x which is at distance $\sqrt{d}$ from 0 and at distance $\geqslant \sqrt{d}$ from all other points of Λ_{16}. The existence of the vectors shown in Fig. 6.6 proves that $d \geqslant 3$. The covering radius of $\frac{1}{2}\Lambda_{16}$ is $\frac{1}{2}\sqrt{d}$, so we can write $x = u + \bar{x}$ where $u \in \frac{1}{2}\Lambda_{16}$ and $N(\bar{x}) \leqslant d/4$. From Prop. 11, u belongs to one of the classes (i) to (vi). In fact u must be in class (vi), for if $x = \frac{1}{2}v_r + \bar{x}$ where $v_r \in \Lambda_{16}$ has norm r then $N(\bar{x} \pm \frac{1}{2}v_r) \geqslant d$, so

$$d = N(x) \leqslant N(\tfrac{1}{2}v_r) + N(\bar{x}) \leqslant \frac{r+d}{4}, \tag{25}$$

which implies $r \geqslant 9$ and hence $r \geqslant 12$ by Prop. 11. Thus $x = \frac{1}{2}v_{12} + \bar{x}$ where $v_{12} \in \Lambda_{16}$, $N(v_{12}) = 12$, $N(\bar{x}) \leqslant d/4$.

Let us move the origin of coordinates to the point $\frac{1}{2}v_{12}$. From Prop. 11 there are 512 points $\phi_1, \ldots, \phi_{512}$ in $\frac{1}{2}\Lambda_{16}$ surrounding the new origin, with $N(\phi_i) = 3$, which without loss of generality we may assume consist of the point shown in Fig. 6.6a and the 511 other points obtained from it by permutations and sign changes. We define the map $T : \mathbf{R}^{16} \to \mathbf{R}^{16}$ by

$$T(y) = \sum_{i=1}^{512} (\phi_i \cdot y)\, \phi_i . \tag{26}$$

Then

$$T(y) \cdot y = \sum_{i=1}^{512} (\phi_i \cdot y)^2 \tag{27}$$

is a quadratic invariant of the orthogonal group which fixes the origin and sends Λ_{16} to itself. Since this is an irreducible group, $T(y) \cdot y$ must be a scalar multiple of the quadratic form $y \cdot y$, say

$$T(y) \cdot y = k\, y \cdot y . \tag{28}$$

(This result is an instance of a theorem of Hadwiger, and expresses the fact that the ϕ_i form a eutactic star [Cox20, §13.7], [Had1].) By taking y to have a single nonzero coordinate we find $k = 96$. Then with $y = \bar{x}$ we obtain

$$\sum_{i=1}^{512} (\phi_i \cdot \bar{x})^2 = 96\, N(\bar{x}) \leqslant 24d . \tag{29}$$

From (29) there is some ϕ_i, say ϕ, with

$$\phi \cdot \bar{x} \geqslant \left\{ \frac{96\, N(\bar{x})}{512} \right\}^{1/2} . \tag{30}$$

Let us write $d = 3+\delta$, $\delta \geqslant 0$, and $N(\bar{x}) = a^2$, $a \geqslant 0$. Since x is a deep hole, $N(\phi - \bar{x}) \geqslant 3+\delta$. From this and (30) we obtain $a^2 - a\sqrt{3}/2 - \delta \geqslant 0$, and so either

$$a \leqslant \frac{\sqrt{3}}{4} \left\{ 1 - \left(1 + \frac{16\delta}{3} \right)^{1/2} \right\} \tag{31}$$

or

$$a \geqslant \frac{\sqrt{3}}{4} \left\{ 1 + \left(1 + \frac{16\delta}{3} \right)^{1/2} \right\} . \tag{32}$$

If (31) holds then from $a \geqslant 0$ we have $a = 0$, $\delta = 0$, $x = \frac{1}{2} v_{12}$ and $d = 3$, as required. We complete the proof of the theorem by showing that (32) leads to a contradiction. From (32) and $a^2 \leqslant (3+\delta)/4$ it follows that $\delta = 0$, $d = 3$, and $a = N(\bar{x}) = \sqrt{3}/2$. Then, for all i, $|\phi_i \cdot \bar{x}| \leqslant 3/8$ and, from (29), $\phi_i \cdot \bar{x} = \pm 3/8$. It is now straightforward to use the explicit coordinates for the ϕ_i to show that this is impossible.

6. Dimensions 17 to 24

We next establish the assertions of Theorem 1 for $17 \leqslant n \leqslant 24$. If Λ_{16+r} $(1 \leqslant r \leqslant 8)$ is any laminated lattice then, arguing as in §4, we see that there is a gluing map

$$\Theta : \Lambda_r / 2\Lambda_r \rightarrow \tfrac{1}{2} \Lambda_{16} / \Lambda_{16} . \tag{33}$$

The images of the minimal vectors in Λ_r under Θ are congruence classes of deep holes modulo Λ_{16}. From §5 there is a monomorphism

$$\Phi : \Lambda_8/2\Lambda_8 \rightarrow \tfrac{1}{2}\,\Lambda_{16}/\Lambda_{16} \tag{34}$$

associating pairs of minimal vectors in Λ_8 with classes of deep holes in Λ_{16}.

Lemma 13. *If α, β, γ are congruence classes of $\Lambda_8/2\Lambda_8$ represented by $\pm a, \pm b, \pm c$, where a, b, c are minimal vectors of Λ_8, and $\alpha + \beta = \gamma$, then the labeling can be chosen so that $a + b = c$.*

Proof. Without loss of generality a and b can be chosen so that $a \cdot b \leqslant 0$. Therefore $N(a+b) \leqslant 2N(a)$, so $N(a+b) = N(a)$ or $2N(a)$. If the latter, then from §3 the minimal representatives of $a + b$ are 16 vectors of norm $2N(a)$, contradicting the existence of c. Thus $a + b$ is a minimal vector which we denote by c. This completes the proof.

We observe that Λ_r $(1 \leqslant r \leqslant 8)$ has a *strong generating set*. By this we mean a set S of minimal vectors $v_1, \ldots, v_r$ with the property that if we take the closure of S under the operations of (i) adjoining $-v$ if $v \in S$, and (ii) adjoining $v + v'$ if $v, v' \in S$ and $v + v'$ is a minimal vector, then we obtain all the minimal vectors of Λ_r.

Now consider the images $[w_1], \ldots, [w_r]$ of $[v_1], \ldots, [v_r]$ under Θ, where $v_1, \ldots, v_r$ is a strong generating set for Λ_r, and the square brackets denote "congruence class containing ...", and let $[x_i]$ be the image of $[w_i]$ under Φ^{-1}. Whenever we adjoin a vector $v + v'$ to S to reach another minimal vector of Λ_r, by Lemma 13 one of the corresponding vectors $\pm x \pm x'$ in Λ_8 is a minimal vector. Proceeding in this way, when we have found the closure of the strong generating set in Λ_r, we have obtained an equal number of minimal vectors in Λ_8. Since those vectors belong to the $\mathbf{Z}$-span of $x_1, \ldots, x_r$, the latter must be a copy of Λ_r.

Thus we have proved that what Λ_r is glued to in Λ_{16} by Θ is the Φ-image of a sublattice $\langle x_1, \ldots, x_r \rangle$ of Λ_8 which is a copy of Λ_r. Since Aut (Λ_8) is transitive on sublattices Λ_r, we conclude that $\Lambda_{17}, \ldots, \Lambda_{24}$ are unique (and in particular that Λ_{24} is the Leech lattice). Their determinants are given by (9), and the subcovering radii (except that of Λ_{24}) by (3) and (4). Finally, the covering radius of Λ_{24} is known from Chap. 23 ($\cong$ [Con20]).

7. Dimensions 25 to 48

In this section we establish the remaining assertions of Theorem 1. First we construct a sequence of lattices $L_{25} \subseteq L_{26} \subseteq \cdots \subseteq L_{48}$ containing Λ_{24}, and then prove that these are laminated lattices.

Let i denote the element of Aut (Λ_{24}) shown in Fig. 6.7, satisfying $i^2 = -1$. Then $(1 + i)/2$ is a similarity of $\mathbf{R}^{24}$ which halves the norms of vectors. The lattice L_{24+s} $(1 \leqslant s \leqslant 24)$ consists of the vectors $u + v^\theta + cv$, where $u \in \Lambda_{24}$, $v \in \Lambda_s$, $\theta = (1 + i)/2$ and $c = 1/\sqrt{2}$. The determinant of L_{24+s} agrees with the value of λ_{24+s} given in Table 6.1.

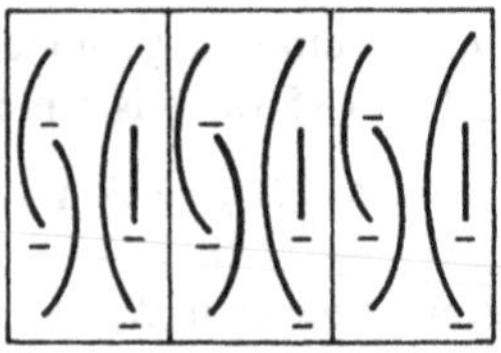

Figure 6.7. The element i of $\mathrm{Aut}(\Lambda_{24})$. In the first column the 1st and 3rd components are to be exchanged, and the 2nd and 4th, and then the 2nd and 3rd components are negated. Similarly for the other columns.

We claim that L_{24+s} is a Λ_{24+s}. For $1 \leqslant s \leqslant 8$ this follows immediately from Theorem 3, since all L_{24+s} contain Λ_{24}.

For $9 \leqslant s \leqslant 16$, we first consider *any* lattice M_{32+r} $(1 \leqslant r \leqslant 8)$ containing *any* Λ_{32}. Thus $\Lambda_{24} \subseteq \Lambda_{32} \subseteq M_{32+r}$. Projecting everything onto $(\mathbf{R}\Lambda_{24})^{\perp}$ we obtain $0 \subseteq c\Lambda_8 \subseteq M_{8+r}$, or $0 \subseteq \Lambda_8 \subseteq c^{-1} M_{8+r}$, where $c = 1/\sqrt{2}$ and $M_{8+r} = M_{32+r}/\Lambda_{24}$. We apply Theorem 3 to $c^{-1} M_{8+r}$, deducing $\det (c^{-1} M_{8+r}) \geqslant \lambda_8 \lambda_r \cdot 2^{-r}$, $\det M_{32+r} = \det \Lambda_{24} \cdot \det M_{8+r} \geqslant \lambda_r \cdot 2^{-2r}$. From Table 6.1 the lattices L_{32+r} meet this bound and therefore are laminated lattices, since no lattice containing Λ_{32} can have a smaller determinant.

For $17 \leqslant s \leqslant 24$ a similar argument shows that any lattice M_{40+r} containing any Λ_{40} has determinant $\det M_{40+r} \geqslant \lambda_r \cdot 2^{-8-3r}$, and since L_{40+r} meets this bound it is a Λ_{40+r}.

The Λ_{25}*'s*. We claim that there are exactly twenty-three Λ_{25}'s, one for each type of deep hole in Λ_{24} (cf. Chap. 23). From the definition of a laminated lattice it follows that there is at most one type of Λ_{25} for each type of deep hole in Λ_{24}. The kissing number of a Λ_{25} is $196560 + 2V$, where V is the number of vertices of the hole (see Table 16.1 in Chap. 16). This already shows that there are at least nine distinct Λ_{25}'s, the highest kissing number being $196560 + 96$ for the Λ_{25} constructed from the hole of type A_1^{24}.

To show that all the 23 holes produce distinct Λ_{25}'s, it is enough to prove that there is a unique Λ_{24} inside a Λ_{25}. In fact it is not difficult to show that in each Λ_{25} the original Λ_{24} is characterized by the following property: it is spanned by all minimal vectors of Λ_{25} that are orthogonal to at least 93150 other minimal vectors. We remark that the automorphism group of the Λ_{25} corresponding to a deep hole of type D_{24} has order 4 (whereas $|\mathrm{Aut}\,(\Lambda_n)| > 1000$ for $4 \leqslant n \leqslant 24$).

The Λ_{26}*'s*. Finally we sketch a heuristic argument suggesting that the number of distinct Λ_{26}'s is very large. Any Λ_{26} is obtained by gluing a Λ_2 to Λ_{24}. Let a, b, c be minimal vectors of Λ_2 with $a+b+c = 0$, and let us just consider the special class of Λ_{26}'s in which all of a^θ, b^θ, c^θ are deep holes in Λ_{24} of type A_{12}^2, with $a^\theta + b^\theta + c^\theta = 0$.

The subgroup of $\text{Aut}(\Lambda_{24})$ fixing an A_{12}^2 hole has order 52 (Table 16.1), so the number of triples α, β, γ of A_{12}^2 holes (modulo Λ_{24}) is $(g/52)^3$, where $g = |\text{Aut}(\Lambda_{24})|$. The sum $\alpha + \beta + \gamma$ is in $(1/13)\,\Lambda_{24}$, a lattice in which Λ_{24} is subgroup of index 13^{24}, so the sum will be zero for about $(g/52)^3/13^{24}$ triples, falling into at least $g^2/(52^3 \cdot 13^{24})$ orbits under Aut (Λ_{24}). Taking account of the fact that $-\alpha, -\beta, -\gamma$ and any permutation of α, β, γ lead to the same Λ_{26}, we can expect about $g^2/(12 \cdot 52^3 \cdot 13^{24}) > 75000$ distinct Λ_{26}'s of this restricted type.

Some calculations of Simon Norton concerning norm-doubling endomorphisms of Λ_{24} suggest that the number of distinct Λ_{48}'s may also be large, and we believe that there are many distinct Λ_n for all $n = 26, \ldots, 48$.

Acknowledgements. We are grateful to Simon Norton for informing us of these calculations and for determining the covering radius of Λ_{16}, and to E. S. Barnes and John Leech for their comments.

Appendix: The best integral lattices known.

Table 6.4, which is taken from [Con36] and also incorporates results of Plesken and Pohst [Ple6], [Poh2], shows the densest presently known integral lattices with minimal norms $\mu = 2$, 3 and 4.

For minimal norm $\mu = 2$ the smallest possible determinant is known for all n (see the second column of the table). It is equal to 1 in all dimensions $n \geqslant 14$. The third column gives examples of lattices having these determinants, taken from Chap. 16. The lattices are specified by their components (see §3 of Chap. 4) with a + to indicate the presence of additional glue vectors of norm greater than 2.

For minimal norms 3 and 4 the entries in general only give upper bounds on the smallest possible determinant. $\Lambda_n\{\mu\}$ indicates an integral laminated lattice of minimal norm μ, $\Lambda_n\{3\}^{\perp}$ denotes the orthogonal lattice to $\Lambda_n\{3\}$ in $\Lambda_{23}\{3\}$ for $0 \leqslant n \leqslant 12$, and $K_n\{3\} = K_{n+1}/K_1$.

The unique $\Lambda_{23}\{3\}$ is the *shorter Leech lattice* O_{23}. This is the unique 23-dimensional unimodular lattice of minimal norm 3 (see Chaps. 16, 19), and consists of those vectors of Λ_{24} that have even inner product with a fixed minimal vector $v \in \Lambda_{24}$, projected onto $v^{\perp}$. Its theta series is given in Eq. (7) of Chap. 19.

We have now explained all the entries in Table 6.4 except O_{24}, which is the *odd Leech lattice*. This is the unique 24-dimensional unimodular lattice of minimal norm 3 (Chap. 17), and may be defined as follows. Let $h\Lambda_{24}$ be the lattice constructed in §3.4 of Chap. 5. Then

$$O_{24} = h\Lambda_{24} \cup ((\tfrac{1}{2})^{24}) + h\Lambda_{24}, \tag{35}$$

whereas the Leech lattice itself (§4.4 of Chap. 5) is

$$\Lambda_{24} = h\Lambda_{24} \cup (-1\tfrac{1}{2}, (\tfrac{1}{2})^{23}) + h\Lambda_{24}. \tag{36}$$

O_{24} was first found by O'Connor and Pall in 1944 [O'Co1].

Table 6.4. Smallest determinant of any integral lattice presently known with minimal norm $\mu = 2$, 3 or 4, and examples of typical lattices having these determinants.

n	$\mu = 2$		$\mu = 3$		$\mu = 4$	
	Det	Lattice	Det	Lattice	Det	Lattice
0	1	Λ_0	1	$\Lambda_0\{3\}$	1	Λ_0
1	2	A_1	3	$\Lambda_1\{3\}$	4	Λ_1
2	3	A_2	8	$\Lambda_2\{3\}$	12	Λ_2
3	4	A_3	16	$\Lambda_3\{3\}$	32	Λ_3
4	4	D_4	32	$\Lambda_4\{3\}$	64	Λ_4
5	4	D_5	48	$\Lambda_5\{3\}$	128	Λ_5
6	3	E_6	64	$\Lambda_6\{3\}$	192	Λ_6
7	2	E_7	64	$\Lambda_7\{3\}$	256	Λ_7
8	1	E_8	128	$\Lambda_8\{3\}$	256	Λ_8
9	2	E_8A_1	192	$\Lambda_9\{3\}$	512	Λ_9
10	3	E_8A_2	243	$K_{10}\{3\}$	768	Λ_{10}
11	2	$(D_{10}A_1)^+$	256	$\Lambda_{11}\{3\}$	972	K_{11}
12	1	D_{12}^+	256	$\Lambda_{12}\{3\}$	729	K_{12}
13	2	$(D_6E_7)^+$	243	$K_{10}\{3\}^\perp$	972	K_{13}
14	1	E_7^{2+}	192	$\Lambda_9\{3\}^\perp$	768	Λ_{14}
15	1	A_{15}^+	128	$\Lambda_8\{3\}^\perp$	512	Λ_{15}
16	1	E_8^2	64	$\Lambda_7\{3\}^\perp$	256	Λ_{16}
17	1	$(A_{11}E_6)^+$	64	$\Lambda_6\{3\}^\perp$	256	Λ_{17}
18	1	A_9^{2+}	48	$\Lambda_5\{3\}^\perp$	192	Λ_{18}
19	1	$(A_7^2D_5)^+$	32	$\Lambda_4\{3\}^\perp$	128	Λ_{19}
20	1	$E_8D_{12}^+$	16	$\Lambda_3\{3\}^\perp$	64	Λ_{20}
21	1	A_3^{7+}	8	$\Lambda_2\{3\}^\perp$	32	Λ_{21}
22	1	$E_8E_7^{2+}$	3	$\Lambda_1\{3\}^\perp$	12	Λ_{22}
23	1	$E_8A_{15}^+$	1	O_{23}	4	Λ_{23}
24	1	E_8^3	1	O_{24}	1	Λ_{24}

7

Further Connections Between Codes and Lattices

N. J. A. Sloane

This chapter contains further investigations of the connections between codes and sphere packings. Constructions A and B of Chapter 5 are analyzed in greater detail and are generalized to complex lattices. We also study self-dual codes and lattices and their weight enumerators and theta series.

1. Introduction

We already saw in Chap. 5 that codes and sphere packings are related. In this chapter we analyze Constructions A and B of Chap. 5 in greater detail and generalize them to complex lattices. (More about Constructions A and C will be found in the following chapter.) We also study the theta series of the packings obtained from these constructions.

We shall see that the connections between codes and packings become stronger when we consider self-dual linear codes and self-dual lattice packings (§§3, 4), and strongest of all when we compare self-dual codes in which the weight of every codeword is a multiple of 4 with self-dual lattices in which the norm of every vector is even (§§6, 7). In the latter case there are parallel theorems giving upper and lower bounds on the best codes and lattices (Cor. 21 and Theorems 24, 25), and parallel theorems characterizing the weight enumerator of the code and the theta series of the lattice (Theorems 16, 17).

The weight enumerator of a self-dual code and the theta series of a self-dual lattice are very strongly constrained, for the former must be invariant under a fairly large group of transformations, and the latter must be a modular form for a fairly large subgroup of $SL_2(\mathbf{R})$. As a consequence these weight enumerators and theta series must lie in certain simply described rings, sometimes free rings with a small number of generators (see Theorems 6, 7, 16, 17, 28-33).

Extremal codes and lattices have the highest minimal distance or norm permitted by the preceding theorems (the precise definition is given in §4). We describe what is known about the classification of self-dual codes and lattices, and in particular of extremal codes and lattices. In general only finitely many extremal codes or lattices exist, and in certain cases they are all known (Theorems 11-13, 20, 34, 36). The extremal weight enumerators and theta series can (at least in principle) be determined explicitly. This makes it possible to show that certain coefficients do not vanish (Theorems 9, 20, 34), leading to upper bounds on the minimal distance or norm (Corollaries 10, 21 35), and that other coefficients are negative, showing that, for all n sufficiently large, extremal codes and lattices do not exist. (Extremal lattices are unrelated to *extreme* forms (§2.2 of Chap. 2).)

Sections 2-7 deal with binary codes and real lattices, and §§8-10 with nonbinary codes and complex lattices. This chapter is based on [Slo6]-[Slo10]. Broué and Enguehard [Bro5]-[Bro7] have also observed the many parallels between codes and lattices (see also [Mah1]-[Mah3], [Tas1]).

Notation. We use the notation for codes established in §2 of Chap. 3, and the theta functions defined in §4.1 of Chap. 4. In particular we recall from Chap. 3 that Type I, II, III, IV codes are respectively self-dual codes over $\mathbf{F}_2$, $\mathbf{F}_2$, $\mathbf{F}_3$, $\mathbf{F}_4$ in which the weight of every codeword is a multiple of 2, 4, 3, 2. As in §2.4 of Chap. 2, Type I or II lattices are real self-dual (i.e. integral unimodular) lattices in which the norm of every vector is a multiple of 1 or 2 respectively. If $f_1, f_2, \ldots$ are algebraically independent polynomials or power series then $\mathbf{C}[f_1, f_2, \ldots]$ denotes the ring of polynomials in $f_1, f_2, \ldots$ with complex coefficients (a free ring).

2. Construction A

We repeat the construction from §2 of Chap. 5, changing it very slightly in order to clarify the parallels between codes and packings.

Construction A. Let C be an (n, M, d) binary code. We assume the zero codeword 0 is in C. A sphere-packing $\Lambda(C)$ in $\mathbf{R}^n$ is obtained by taking as centers all $x = (x_1, \ldots, x_n)$ in $\mathbf{R}^n$ with

$$\sqrt{2}\, x \pmod 2 \in C. \tag{1}$$

Thus the centers consist of all vectors which can be obtained from the codewords of C by adding arbitrary even numbers to the components and then dividing by $\sqrt{2}$. The origin is always a center.

Example 1. *The face-centered cubic lattice* D_3. Let $C = \{000, 011, 101, 110\}$ with $n = 3$, $d = 2$. Some of the centers closest to the origin are $2^{-1/2}(\pm 1, \pm 1, 0)$, $2^{-1/2}(\pm 2, 0, 0), \ldots$. This is the face-centered cubic lattice D_3, the densest lattice sphere-packing in $\mathbf{R}^3$ (§6.3 of Chap. 4).

In general, if $d < 4$, the centers closest to the origin are the $2^d A_d(0)$ vectors of shape $2^{-1/2}(\pm 1)^d 0^{n-d}$ obtained from the codewords of weight d. Such centers are at squared distance $d/2$ from the origin. If $d > 4$, the centers closest to the origin are the $2n$ centers of type $(\pm\sqrt{2})^1 0^{n-1}$, at squared distance 2. Finally, if $d = 4$, both sets of centers are at the same distance from the origin. Therefore the radius of the spheres should be taken to be

$$\rho = 2^{-3/2} d^{1/2} \text{ if } d \leqslant 4, \text{ or } 2^{-1/2} \text{ if } d \geqslant 4, \tag{2}$$

and the number of spheres touching the sphere at the origin is

$$\tau = 2^d A_d(0) \text{ if } d < 4,\ 2n + 16A_4(0) \text{ if } d = 4,\ 2n \text{ if } d > 4. \tag{3}$$

((3), (4) of Chap. 5). The center density of $\Lambda(C)$ ((2) of Chap. 5) is

$$\delta = M \rho^n 2^{-n/2}. \tag{4}$$

(These formulae differ superficially from those in Chap. 5 because of the $\sqrt{2}$ in Eq. (1).)

Theorem 1. *$\Lambda(C)$ is a lattice packing if and only if C is a linear code.*

Proof. Elementary.

Theorem 2. *Suppose C is a linear code of dimension k. Then*

(i) $\det \Lambda(C) = 2^{n-2k}$,

(ii) $\Lambda(C^*) = \Lambda(C)^*$,

(iii) *$\Lambda(C)$ is integral if and only if $C \subseteq C^*$,*

(iv) *$\Lambda(C)$ is of Type* I *if and only if C is of Type* I,

(v) *$\Lambda(C)$ is of Type* II *if and only if C is of Type* II,

(vi) *$\mathrm{Aut}(\Lambda(C))$ contains a subgroup $2^n.\mathrm{Aut}(C)$.*

Proof. Without loss of generality assume C has generator matrix $[I\,B]$. Then

$$\frac{1}{\sqrt{2}}\begin{bmatrix} I & B \\ 0 & 2I \end{bmatrix} \tag{5}$$

is a generator matrix for $\Lambda(C)$. (i)-(v) now follow easily. (vi) $\mathrm{Aut}(\Lambda(C))$ certainly contains all the coordinate permutations in $\mathrm{Aut}(C)$, all sign changes $x_i \to -x_i$, and possibly other symmetries.

Remarks. Compare Theorem 1 of Chap. 8. If the $\sqrt{2}$ is omitted from (1), Construction A always produces an integral lattice.

Theorem 3 ([Ber7], [Bro6]). *Suppose C is linear, with weight enumerator $W_C(x,y)$. Then the theta series of $\Lambda(C)$ is given by*

$$\Theta_{\Lambda(C)}(z) = W_C(\theta_3(2z), \theta_2(2z)), \tag{6}$$

where $\theta_2(z)$, $\theta_3(z)$ are Jacobi theta functions defined in §4.1 *of Chap.* 4.

Proof. Consider a codeword $u = (u_1, \ldots, u_n)$ in C. The corresponding centers in $\Lambda(C)$ consist of the set

$$\Lambda(u) = \{(y_1, \ldots, y_n) : y_r \in \frac{1}{\sqrt{2}} u_r + \sqrt{2}\,\mathbf{Z},\ 1 \leqslant r \leqslant n\}.$$

From Eq. (44) of Chap. 4 we have

$$\Theta_{\sqrt{2}\mathbf{Z}}(z) = \Theta_{\mathbf{Z}}(2z) = \theta_3(2z),\quad \Theta_{\frac{1}{\sqrt{2}}+\sqrt{2}\mathbf{Z}}(z) = \theta_2(2z).$$

Therefore

$$\Theta_{\Lambda(u)}(z) = \theta_3(2z)^{n-wt(u)}\,\theta_2(2z)^{wt(u)},$$

$$\Theta_{\Lambda(C)}(z) = \sum_{u \in C} \Theta_{\Lambda(u)}(z) = W_C(\theta_3(2z), \theta_2(2z)).$$

Example 2. *The cubic lattice* $\mathbf{Z}^n$. The universe code $\mathbf{F}_2^n$ has $d = 1$, $W(x,y) = (x+y)^n$. This produces the lattice $\frac{1}{\sqrt{2}}\mathbf{Z}^n$, which has theta-function $\theta_3(\frac{1}{2}z)^n$, and establishes the identity ((22) of Chap. 4)

$$\theta_3(2z) + \theta_2(2z) = \theta_3(\tfrac{1}{2}z). \tag{7}$$

Example 3. *The square lattice* $\mathbf{Z}^2$. The code $\mathscr{C}_2 = \{00, 11\}$ has weight enumerator

$$x^2 + y^2 = \psi_2 \ \text{(say)}, \tag{8}$$

and $\Lambda(\mathscr{C}_2) \cong \mathbf{Z}^2$, giving the identity ((25) of Chap. 4)

$$\theta_3(2z)^2 + \theta_2(2z)^2 = \theta_3(z)^2. \tag{9}$$

Example 4. *The checkerboard lattice* D_n. As already observed in §2.4 of Chap. 4, if C is the $[n, n-1, 2]$ even-weight code we obtain the lattice D_n. In the standard version of D_n the minimal norm is 2, so in this case we omit the $\sqrt{2}$ in (1). The weight enumerator of C is

$$W_C(x,y) = \tfrac{1}{2}\{(x+y)^n + (x-y)^n\}, \tag{10}$$

so from (6) the theta series of D_n is (allowing for the new scale)

$$\Theta_{D_n}(z) = W_C(\theta_3(4z), \theta_2(4z)) \tag{11}$$

$$= \tfrac{1}{2}\{(\theta_3(4z) + \theta_2(4z))^n + (\theta_3(4z) - \theta_2(4z))^n\}$$

$$= \tfrac{1}{2}\{\theta_3(z)^n + \theta_4(z)^n\}, \tag{12}$$

using (22) of Chap. 4. On this scale Theorem 2(i) reads

$$\det \Lambda(C) = 2^{2n-2k}, \tag{13}$$

so det $D_n = 4$ (cf. §7.1 of Chap. 4).

Example 5. *The Gosset lattice* E_8. Let $\mathcal{H}_8$ be the [8, 4, 4] Hamming code and ψ_8 its weight enumerator (§2.4.2 of Chap. 3). As observed in §2.5 of Chap. 5, applying Construction A to $\mathcal{H}_8$ produces the lattice E_8. From (2)-(4) and Theorem 2 we find that E_8 has $\rho = 1/\sqrt{2}$, $\tau = 240$, $\delta = 2^{-4}$, det = 1, and is a Type II unimodular lattice. The theta series of E_8 is (from (6))

$$\Theta_{E_8}(z) = \theta_3(2z)^8 + 14\,\theta_3(2z)^4\,\theta_2(2z)^4 + \theta_2(2z)^8 \tag{14}$$

$$= \tfrac{1}{2}\,[\{\theta_3(2z)^2 + \theta_2(2z)^2\}^4 + \{\theta_3(2z)^2 - \theta_2(2z)^2\}^4 \tag{15}$$

$$+ 16\,\theta_3(2z)^4\,\theta_2(2z)^4]$$

$$= \tfrac{1}{2}\,\{\theta_2(z)^8 + \theta_3(z)^8 + \theta_4(z)^8\}, \tag{16}$$

in agreement with Eq. (40) of Chap. 2. (To go from (15) to (16) we use Eqs. (24), (25) of Chap. 4.) The lattice E_7 can be obtained in the same way from the Hamming code $\mathcal{H}_7$ (§2.5 of Chap. 5).

Example 6. *Best's* 10-*dimensional nonlattice* P_{10c}. Best's (10, 40, 4) code leads to a nonlattice 10-dimensional packing P_{10c} with $\delta = 2^{-7} \cdot 5$ (§2.6 of Chap. 5). Using the weight distribution given in Chap. 5 we find that this packing is distance invariant and has theta series

$$\theta_3(2z)^{10} + 22\theta_3(2z)^6\theta_2(2z)^4 + 12\theta_3(2z)^4\theta_2(2z)^6 + 5\theta_3(2z)^2\theta_2(2z)^8$$

$$= 1 + 372q^2 + 768q^3 + 5684q^4 + 6144q^5 + \cdots, \tag{17}$$

verifying that the kissing number is 372. The generalized Jacobi formula for periodic packings given in [Odl6] (which replaces z by $-1/z$ and multiplies by an appropriate constant) transforms (17) into

$$1 + \frac{4}{5}q + \frac{516}{5}q^2 + \frac{2048}{5}q^{5/2} + \frac{1728}{5}q^3 + \cdots. \tag{18}$$

Question. Is there a formal "dual" to Best's packing, a nonlattice packing with average theta series equal to (18)?

We do not even know if the hexagonal close-packing (§6.5 of Chap. 4) has a "dual" in this sense. If so, its average theta series would be, from (73) of Chap. 4 (cf. [Odl6])

$$1 + \frac{3}{2}q^{4/3} + 2q^{3/2} + 9q^{41/24} + 3q^{17/6} + 6q^4 + 9q^{113/24} + \cdots \tag{19}$$

3. Self-dual (or Type I) codes and lattices

Throughout this section C will denote a code of Type I and Λ a lattice of Type I. We shall see that the weight enumerator $W_C(x,y)$ and the theta series $\Theta_\Lambda(\tau)$ are strongly constrained, and in similar ways.

In fact from the MacWilliams identity for codes (Eq. (50) of Chap. 3) and Jacobi's formula for lattices (Eq. (19) of Chap. 4) we deduce:

Theorem 4.

$$W_C((x+y)/\sqrt{2}\,,\,(x-y)/\sqrt{2}) = W_C(x,y)\,, \tag{20}$$

$$W_C(x,-y) = W_C(x,y)\,. \tag{21}$$

Theorem 5.

$$\Theta_\Lambda(-1/z) = (z/i)^{n/2}\,\Theta_\Lambda(z)\,, \tag{22}$$

$$\Theta_\Lambda(z+2) = \Theta_\Lambda(z)\,. \tag{23}$$

Let G be an arbitrary finite (multiplicative) group of $m \times m$ complex matrices. Then a polynomial $f(x) = f(x_1, \ldots, x_m)$ is said to be *invariant* under G if $f(Bx^{tr}) = f(x)$ for all $B \in G$. (see §4.2 of Chap. 3). Theorem 4 implies that $W_C(x,y)$ is invariant under the group G_1 generated by the matrices

$$\frac{1}{\sqrt{2}}\begin{pmatrix}1 & 1\\ 1 & -1\end{pmatrix},\ \begin{pmatrix}1 & 0\\ 0 & -1\end{pmatrix}. \tag{24}$$

It is easy to see that G_1 is isomorphic to the dihedral group of order 16 (a reflection group — see §2 of Chap. 4). There are now several ways of proceeding; the one we like best uses invariant theory (cf. [Blil], [Fla2]-[Fla6], [Huf2], [Mil2], [Sta3]) and is described in some detail in [Slo7] and [Mac6, Chap. 19]. (See also [Ber7], [Bro6], [Gle1], [Mac3], [Tol1].) Whichever method is followed, the final result is:

Theorem 6 (Gleason). *If C is a binary self-dual code then*

$$W_C(x,y) \in \mathbf{C}[\psi_2, \xi_8], \tag{25}$$

where ψ_2 *is given by* (8) *and*

$$\xi_8 = x^2y^2(x^2-y^2)^2. \tag{26}$$

In words, $W_C(x,y)$ *can be written uniquely as a polynomial in* ψ_2 *and* ξ_8, *or equivalently as a polynomial in the weight enumerators of* $\mathcal{C}_2$ *and* $\mathcal{H}_8$.

Generalizations are given later in this chapter and in [Ber7], [Huf1], [Leo7], [Mac3], [Mac4], [Mal2], [Mal4], [Mal5], [Que6].

On the other hand, let G be an arbitrary subgroup of finite index in $SL_2(\mathbb{Z})$, and let χ be a character of G. A complex valued function $f(z)$ is said to be a *modular form of weight w for G with respect to* χ if

(i) $f(z)$ is holomorphic for $\mathrm{Im}(z) > 0$,

(ii) $$f\left(\frac{az+b}{cz+d}\right) = \left(\frac{cz+d}{\chi(\sigma)}\right)^w f(z) \tag{27}$$

for every $\sigma = \begin{pmatrix}a & b\\ c & d\end{pmatrix} \in G$, and

(iii) $f(z)$ is "holomorphic" at every cusp of G.

(References on modular forms are [Cas2], [Gun1], [Har4], [Hec1], [Hec2], [Kit3], [Kno1], [Lan9], [Lew1], [Mah1]-[Mah3], [Mod1], [Ogg1], [Pet4], [Rad1], [Ran6], [Sch6], [Sch7], [Ser1], [Shi1], [Sie1]-[Sie3].) Since many different definitions of weight appear in the literature, it is worth saying that in our notation the cusp form $\Delta_{24}(z)$ which generates the Ramanujan numbers is

$$\Delta_{24}(z) = q^2 \prod_{m=1}^{\infty} (1-q^{2m})^{24}, \text{ where } q = e^{\pi i z}, \tag{28}$$

(Eq. (37) of Chap. 4); this is a modular form of weight 12 for $SL_2(\mathbf{Z})$ with respect to the character $\chi = 1$.

Theorem 5 implies that $\Theta_\Lambda(z)$ is a modular form of weight $\frac{1}{2}n$ for the group generated by

$$U = \begin{pmatrix} 1 & 2 \\ 0 & 1 \end{pmatrix} : z \to z+2, \tag{29}$$

$$S = \begin{pmatrix} 0 & -1 \\ 1 & 0 \end{pmatrix} : z \to -\frac{1}{z}, \tag{30}$$

with respect to the character $\chi(U) = 1$, $\chi(S) = i$. Let $\mathcal{M}_n$ be the complex vector space spanned by all modular forms of weight $\frac{1}{2}n$ with respect to χ, and let $\mathcal{M} = \bigoplus_{n=0}^{\infty} \mathcal{M}_n$. Then $\mathcal{M}$ is a graded ring.

Theorem 7 (Hecke). (a) $\dim_{\mathbf{C}} \mathcal{M}_n = 1 + [n/8]$. (b) *If* Λ *is a self-dual lattice then*

$$\Theta_\Lambda(z) \in \mathbf{C}\,[\theta_3(z), \Delta_8(z)], \tag{31}$$

where $\Delta_8(z)$ *is the cusp form*

$$\Delta_8(z) = \frac{1}{16}\,\theta_2(z)^4\theta_4(z)^4 \tag{32}$$

$$= 2^{-8}\theta_2\left(\frac{z+1}{2}\right)^8$$

$$= q \prod_{m=1}^{\infty} \{(1-q^{2m-1})\,(1-q^{4m})\}^8 \tag{33}$$

$$= q - 8q^2 + 28q^3 - 64q^4 + 126q^5 - 224q^6 + \cdots \tag{34}$$

Proof. (a) follows from [Ogg1, Th. 4, p. I-41], since the hypotheses imply $\Theta_\Lambda(z) \in \mathcal{M}\,(2, \frac{1}{2}n, 1)$ in the notation of [Ogg1, p. xiv]. (b) Since the Poincaré series for $\mathcal{M}$ is

$$\sum_{n=0}^{\infty} (\dim_{\mathbf{C}} \mathcal{M}_n)\lambda^n = \frac{1}{(1-\lambda)(1-\lambda^8)}, \tag{35}$$

we expect to find a basis for $\mathcal{M}$ consisting of two algebraically independent modular forms, one in $\mathcal{M}_1$ and one in $\mathcal{M}_8$. We already know from Eq. (17) of Chap. 4 that $\theta_3(z) \in \mathcal{M}_1$. For the form in $\mathcal{M}_8$ we can either use $\Theta_{E_8}(z)$ from Eq. (16), or the simpler form

$$16\,\Delta_8(z) = \theta_3(z)^8 - \Theta_{E_8}(z)$$
$$= \{\theta_3(2z)^2+\theta_2(2z)^2\}^4 - \{\theta_3(2z)^8+14\theta_3(2z)^4\theta_2(2z)^4+\theta_2(2z)^8\}$$

from (9), (14),

$$= 4\theta_3(2z)^2\theta_2(2z)^2\{\theta_3(2z)^2-\theta_2(2z)^2\}^2 = \theta_2(z)^4\theta_4(z)^4 = \frac{1}{16}\,\theta_2\left(\frac{z+1}{2}\right)^8$$

by (24), (25) of Chap. 4,

$$= 16q \prod_{m=1}^{\infty} \{(1-q^{2m-1})\,(1-q^{4m})\}^8,$$

by (34) of Chap. 4. It remains to show that the Poincaré series for $\mathbf{C}[\theta_3(z), \Delta_8(z)]$ is equal to (35). For this it is enough to prove that for any dimension $n = 8\alpha + \nu$, $0 \leqslant \nu \leqslant 7$, all $\alpha + 1$ products

$$\theta_3(z)^{n-8r}\,\Delta_8(z)^r \quad (0 \leqslant r \leqslant \alpha)$$

are *linearly* independent. But this is immediate, for the rth product begins $q^r + \cdots$.

Remark. From the proof of Theorem 7, in particular Eq. (35), we see that the theta series of an n-dimensional Type I lattice can be written as

$$\Theta_\Lambda(z) = \sum_{r=0}^{\alpha} a_r\, \theta_3(z)^{n-8r}\,\Delta_8(z)^r, \tag{36}$$

if $n = 8\alpha + \nu$, $0 \leqslant \nu \leqslant 7$, where the a_i are integers. Thus $\Theta_\Lambda(z)$ is a polynomial of a particular form (an *isobaric* polynomial) in θ_3 and Δ_8, or equivalently in the theta series of $\mathbf{Z}$ and E_8.

Example 5 (cont.). It is well known (see e.g. [Oggl]) that the Eisenstein series

$$E_4(z) = 1 + 240 \sum_{m=1}^{\infty} \sigma_3(m) q^{2m} \tag{37}$$

belongs to $\mathcal{M}_8$. Since both series begin $1 + 240q^2 + \cdots$, it follows that $E_4(z)$ is equal to the theta series of the lattice E_8.

Theorem 8. [Slo10]. *If we expand $\Delta_8(z) = \Sigma a_r q^r$, the coefficients a_r alternate in sign and are multiplicative in the sense that*

$$|a_r| \cdot |a_s| = \sum_{\substack{d|(r,s)\\ d \text{ odd}}} d^3 |a_{rs/d^2}|. \tag{38}$$

(Compare Th. 21 below.)

Classification. Type I codes have been classified for length $n \leqslant 30$ [Ple10], [Ple12], [Ple17], and Type I lattices for dimension $n \leqslant 25$, as we discuss in §2.4 of Chap. 2, and Chaps. 16, 17.

4. Extremal Type I codes and lattices

It follows from Th. 6 that if $n = 2j = 8\alpha + 2\nu$, where $0 \leqslant \nu \leqslant 3$, then

$$W_C(x,y) = \sum_{r=0}^{\alpha} a_r \psi_2{}^{j-4r} \xi_8^r, \tag{39}$$

for uniquely determined integers $a_0, \ldots, a_\alpha$. Suppose we begin by choosing the a_i so that the right-hand side of (39) is

$$x^n + A_{2\alpha+2}^* x^{n-2\alpha-2} y^{2\alpha+2} + \cdots, \tag{40}$$

containing no powers of y between 0 and $2\alpha + 2$. We call (40) the *extremal weight enumerator* $W^*(x,y)$ of length n, and a code having this weight enumerator (if there is one) an *extremal code*(1) *of Type* I.

If such a code exists its minimal distance is $2\alpha + 2$, unless $A_{2\alpha+2}^* = 0$, in which case d may be greater than $2\alpha + 2$. But Theorem 9 shows that such "accidents" do not happen. Thus an extremal code (when such a code exists) has the greatest minimal distance of any Type I code.

Similarly for lattices: let $n = 8\alpha + \nu$, where $0 \leqslant \nu \leqslant 7$, write $\Theta_\Lambda(z)$ in the form (36), and choose $a_0, \ldots, a_\alpha$ so that the right-hand side of (36) becomes the *extremal theta series* in dimension n:

$$1 + A_{\alpha+1}^* q^{\alpha+1} + \cdots = \Theta^*(z) \quad \text{(say)}. \tag{41}$$

A lattice with this theta series is called an *extremal lattice*(1) *of Type* I.

Theorem 9 ([Mal1], [Mal3], [Sie3]). *In the extremal weight enumerator (resp. extremal theta series) the leading coefficient $A_{2\alpha+2}^*$ (resp. $A_{\alpha+1}^*$) is positive.*

The proof (of Theorem 9 and also Theorems 19 and 34) gives explicit formulae for the extremal weight enumerators and theta series — see [Mal1], [Mal3].

Corollary 10. *The minimal distance of a self-dual binary code satisfies*

$$d \leqslant 2[n/8] + 2. \tag{42}$$

The minimal norm of a self-dual lattice satisfies

$$\mu \leqslant [n/8] + 1. \tag{43}$$

---

(1) A code or lattice which is "extremal of Type I" might accidentally have Type II!

In an extremal code or lattice equality holds in (42), (43). Eq. (43) implies an upper bound on the density (see for example Eq. (2) of Chap. 6). For large values of n the JPL bound (see [McE4], and Eq. (2) of Chap. 9) implies the stronger result that any code with dimension $k = n/2$ has

$$\frac{d}{n} \le 0.182490\ldots, \tag{44}$$

and the Kabatiansky-Levenshtein bound (Eq. (45) of Chap. 1) implies

$$\frac{\mu}{n} \leqslant 0.102\ldots. \tag{45}$$

In view of (44), (45) it is not surprising that we have:

Theorem 11 [Mal1]. *The next coefficient $A^*_{2\alpha+4}$ (resp. $A^*_{\alpha+2}$) in the extremal weight enumerator (resp. theta series) is negative for all sufficiently large n. Therefore the corresponding extremal codes or lattices do not exist.*

In both cases the first negative coefficient appears when $n = 32$. E.g. the extremal theta-function for $n = 32$ is

$$\Theta^*(z) = 1 + 4700160q^5 - 8094720q^6 + \cdots .$$

Theorem 12 ([Mal3], [Ple10], [Ple17], [War2]). *Extremal self-dual codes of Type* I *exist if and only if n =* 2, 4, 6, 8, 12, 14, 22 *or* 24.

Theorem 13 (see Chap. 19). *Extremal self-dual lattices of Type* I *exist if and only if n =* 1–8, 12, 14, 15, 23 *or* 24.

The Nordstrom-Robinson code. The extremal weight enumerator of length 16 is

$$x^{16} + 112\, x^{10}y^6 + 30\, x^8y^8 + 112\, x^6y^{10} + y^{16}, \tag{46}$$

although no linear code with this weight enumerator exists. But there is a *nonlinear* code with this weight enumerator, the (16, 256, 6) Nordstrom-Robinson code (see §2.12 of Chap. 3). Similarly the extremal theta series in dimension 16 is

$$\Theta^*(z) = 1 + 7680\, q^3 + 4320\, q^4 + 276480\, q^5 + 61440\, q^6 + \cdots \tag{47}$$

If a corresponding packing were to exist, it would set new records for the density and kissing number in 16 dimensions. No such *lattice* exists (Chap. 19), and we conjecture that there is also no such nonlattice packing. Inspection of (47) suggests that such a packing might be a union of Λ_{16} and 15 of its translates. However, this is impossible, for we can prove:

Theorem 14. *Suppose P is a packing in $\mathbf{R}^{16}$ with minimal norm* 3, *consisting of a union of t translates of* Λ_{16}. *Then* $t \leqslant 9$, *and* 9 *can be attained.*

The proof is omitted. *Question.* Are there packings analogous to the Kerdock or Preparata nonlinear codes (§2.12 of Chap. 3)? (The analogous codes or lattices of Type II do not exist, as we shall see in §7.)

5. Construction B

We repeat the construction from §3 of Chap. 5, again with slight changes.

Construction B. Let C be an $[n, k, d=8]$ binary linear code in which the weight of every codeword is a multiple of 4. A sphere packing $\mathcal{L}(C)$ is obtained in $\mathbf{R}^n$ by taking as centers all $x = (x_1, \ldots, x_n)$ in $\mathbf{R}^n$ with

(i) $\sqrt{2}\, x \pmod 2 \in C$, and (48)

(ii) $4 \mid \sqrt{2} \sum_{i=1}^{n} x_i$. (49)

(The construction may be applied to other codes, but this version covers the most important cases.)

Theorem 15. *$\mathcal{L}(C)$ is an integral lattice with minimal norm, kissing number, center density, determinant and theta series given by*

$$\mu = 4, \tag{50}$$

$$\tau = 2n(n-1) + 128\, A_8, \tag{51}$$

$$\delta = 2^{k-1-n/2}, \tag{52}$$

$$\det \mathcal{L}(C) = 2^{n+2-2k}, \tag{53}$$

$$\Theta(z) = \tfrac{1}{2}\, W_C(\theta_3(2z), \theta_2(2z)) + \tfrac{1}{2}\, \theta_4(2z)^n. \tag{54}$$

Example 5 (cont.) The [8, 1, 8] repetition code produces the lattice E_8 again. In this case we obtain the theta series directly in the form (16).

Example 7. *The Barnes-Wall lattice* Λ_{16}. The [16, 5, 8] first-order Reed-Muller code produces the lattice Λ_{16} (§10 of Chap. 4, §3.4 of Chap. 5), with $\mu = 4$, $\tau = 2 \cdot 16 \cdot 15 + 128 \cdot 30 = 4320$, $\delta = 2^{-4}$, det $= 256$, and we obtain its theta series (Eq. (132) of Chap. 4) from Eq. (54).

Example 8. The [24, 12, 8] Golay code $\mathscr{C}_{24}$ produces the integral lattice $h\Lambda_{24}$ (§3.4 of Chap. 5), for which $\mu = 4$, $\tau = 98256$, $\delta = 2^{-1}$, det $= 4$.

6. Type II codes and lattices

We continue the study of self-dual codes and lattices from §4. In this section C will denote a Type II code and Λ a Type II lattice. It is here that we find the strongest analogies between codes and lattices.

Equations (21) and (23) can now be replaced by

$$W_C(x,iy) = W_C(x,y), \tag{55}$$

$$\Theta_\Lambda(z+1) = \Theta_\Lambda(z). \tag{56}$$

Therefore $W_C(x,y)$ is invariant under the larger group G_2 generated by $\frac{1}{\sqrt{2}}\begin{bmatrix} 1 & 1 \\ 1 & -1 \end{bmatrix}$ and $\begin{bmatrix} 1 & 0 \\ 0 & i \end{bmatrix}$. This is the complex reflection group 4[6]2 of order 192 (see [She2]), and we obtain (cf. [Slo7]):

Theorem 16 (Gleason). *If C is a Type* II *code then*

$$W_C(x,y) \in \mathbf{C}[\psi_8, \xi_{24}], \tag{57}$$

where ψ_8 *is the weight enumerator of* $\mathcal{H}_8$ (§2.4.2 *of Chap.* 3) *and*

$$\xi_{24} = x^4y^4(x^4-y^4)^4. \tag{58}$$

Equivalently, $W_C(x,y)$ *can be written as a polynomial in the weight enumerators of* $\mathcal{H}_8$ *and* $\mathcal{C}_{24}$.

Similarly $\Theta_\Lambda(z)$ is a modular form of weight $\frac{1}{2}n$ for the full modular group $SL_2(\mathbf{Z})$ generated by S (Eq. (30)) and T: $z \rightarrow z+1$, with respect to the character $\chi(T) = 1$, $\chi(S) = i$. Let $\mathcal{M}_n$ and $\mathcal{M}$ be as in §3.

Theorem 17 (Hecke). *If* Λ *is a Type* II *lattice then*

(a) $\dim_{\mathbf{C}} \mathcal{M}_n = 1 + [\frac{n}{24}]$ *if* $8|n$, $= 0$ *if* $8 \nmid n$.

(b) $$\Theta_\Lambda(z) \in \mathbf{C}[E_4(z), \Delta_{24}(z)], \tag{59}$$

where $E_4(z)$ *is the theta series of the lattice* E_8, *given by* (14), (16) *or* (37), *and* $\Delta_{24}(z)$ *is given by* (28) (*see also* (37)–(39) *of Chap.* 4). *Equivalently,* $\Theta_\Lambda(z)$ *can be written as an isobaric polynomial in the theta series of* E_8 *and* Λ_{24}.

Proof. (a) follows from [Ogg1, Th. 3, p. I-23]. (b) The Poincaré series is

$$\sum_{n=0}^{\infty} (\dim_{\mathbf{C}} \mathcal{M}_n)\lambda^n = \frac{1}{(1-\lambda^8)\,(1-\lambda^{24})}, \tag{60}$$

and we expect to find a basis for $\mathcal{M}$ consisting of two algebraically independent modular forms, one in $\mathcal{M}_8$ and one in $\mathcal{M}_{24}$. We already have $E_4(z) \in \mathcal{M}_8$ from (16), and $\Delta_{24}(z) = 2^{-8}\{\theta_2(z)\theta_3(z)\theta_4(z)\}^8$ (Eq. (38) of Chap. 4) $\in \mathcal{M}_{24}$. The remainder of the proof follows that of Theorem 7.

Remark. The proof of Theorem 16 shows that the theta series of a Type II lattice is (i) an isobaric polynomial in E_4 and Δ_{24}, and (ii) a symmetric function of $\theta_2(z)^8$, $\theta_3(z)^8$ and $\theta_4(z)^8$.

Corollary 18. *A Type* II *code or lattice exists if and only if n is a multiple of* 8.

Example 8 (cont.) *The Leech lattice* Λ_{24}. The Leech lattice Λ_{24} is obtained by taking the union of $h\Lambda_{24}$ and its translation by $(-1\frac{1}{2}, (\frac{1}{2})^{23})$ (§4.4 of Chap. 5). This doubles the density (without changing the minimal norm), so Λ_{24} has $\mu = 4$, $\delta = 1$, det $= 1$, and therefore is a Type II lattice. We can write down its theta series immediately from Theorem 16. Since there are no vectors of norm 2, it is $E_4(z)^3 - 720\,\Delta_{24}(z)$ (see Eqs. (138)-(141) of Chap. 4).

Theorem 19 (Ramanujan [Ram1]; see also [Mor1], [Ogg1].) *If we expand* $\Delta_{24}(z) = \sum_{m=1}^{m=\infty} \tau(m) q^{2m}$, *the function* $\tau(m)$, *the Ramanujan function, is multiplicative in the sense that*

$$\tau(r)\tau(s) = \sum_{d\,|\,(r,s)} d^{11}\tau(rs/d^2). \tag{61}$$

We have already mentioned estimates for the magnitude of $\tau(m)$ in §2.5 of Chap. 2.

Comparing (57) and (59) we observe that $\mathbf{C}[\psi_8, \xi_{24}]$ and $\mathbf{C}[E_4(z), \Delta_{24}(z)]$ are isomorphic graded rings — both have Poincaré series (6). In fact an isomorphism between them is given by Eq. (6), sending

$$f(x,y) \to f(\theta_3(2z), \theta_2(2z)). \tag{62}$$

Unfortunately this does not tell us how to find a lattice with given theta series, even when an analogous code is known. In fact (62) does not even map the weight enumerator of the Golay code onto the theta series of the Leech lattice. Broué and Enguehard [Bro6] give an interesting discussion of the isomorphism between these rings.

Classification. Type II codes have been classified for length $n \leqslant 32$ [Con21], [Ple10], [Ple12], and Type II lattices for dimension $n \leqslant 24$, as we have discussed in §2.4 of Chap. 2 (see Chap. 16).

7. Extremal Type II codes and lattices

The results are quite similar to those in §4, so we omit the details. Let $n = 8j = 24\alpha + 8\nu$, $0 \leqslant \nu \leqslant 2$. For codes, let us choose $a_0, \ldots, a_\alpha$ so that

$$\sum_{r=0}^{\alpha} a_r \psi_8^{j-3r} \xi_{24}^r = x^n + A_{4\alpha+4}^* \, x^{n-4\alpha-4}\, y^{4\alpha+4} + \cdots . \tag{63}$$

This is called an *extremal weight enumerator* and a code having this weight enumerator is an *extremal code of Type* II and length n. Similarly for lattices: choose $a_0, \ldots, a_\alpha$ so that

$$\sum_{r=0}^{\alpha} a_r E_4(z)^{j-3r}\, \Delta_{24}(z)^r = 1 + A_{2\alpha+2}^* \, q^{2\alpha+2} + \cdots . \tag{64}$$

This is an *extremal theta series* and a corresponding lattice is an *extremal lattice of Type* II in dimension n. The 48-dimensional extremal theta series was given in Eq. (57) of Chap. 2 (see Example 9 below).

Theorem 20 ([Mal1], [Mal3], [Sie3]). *In the extremal weight enumerator* (63), $A^*_{4\alpha+4} > 0$ *for all n, but* $A^*_{4\alpha+8} < 0$ *for all sufficiently large n. In the extremal theta series* (64), $A^*_{2\alpha+2} > 0$ *for all n, but* $A^*_{2\alpha+4} < 0$ *for all sufficiently large n.*

This coefficient first goes negative at around $n = 3720$ for codes and 41000 for lattices. For example there is no extremal code of length 3720, since the extremal weight enumerator of that length is

$$W^*(x,y) = x^{3720} + A^*_{624}\, x^{3096}y^{624} + A^*_{628}\, x^{3092}y^{628} + \cdots ,$$

where $A^*_{624} = 1.16... \cdot 10^{170}$, $A^*_{628} = -5.84... \cdot 10^{170}$ [Mal3].

Corollary 21. *The minimal distance of a Type* II *code satisfies*

$$d \leqslant 4[n/24] + 4 . \qquad (65)$$

The minimal norm of a Type II *lattice satisfies*

$$\mu \leqslant 2[n/24] + 2 . \qquad (66)$$

In [Mal1] it is shown that for any constant b

$$d(C) \leqslant \frac{n}{6} - b, \quad d(\Lambda) \leqslant \frac{n}{12} - b , \qquad (67)$$

for all sufficiently large n. Note that (65), (66) are stronger than (44), (45).

In an extremal code or lattice equality holds in (65) and (66). Extremal Type II codes are known of length 8 (one code, $\mathcal{H}_8$ [Ple10]), 16 (two codes, for example $\mathcal{H}_8 \oplus \mathcal{H}_8$ [Ple10]), 24 (one code $\mathcal{C}_{24}$ [Del13], [Mac6, Chap. 20], [Ple8], [Ple17]), 32 (five codes [Con21], [Koc3]), 40 (at least nine codes [Ior1], [Oze3], [Oze4], [Ton2]), and 48, 56, 64, 80, 88, 104, 136 (at least one each – see [Mac6, Fig. 19.2], [Moo3], [Huf3]). Most of the codes of length $\geqslant 48$ are quadratic residue or double circulant codes. The first open case is $n = 72$ (see [Con22], [Huf4], [Ple13], [Ple14], [Ple19], [Slo3]). The extremal weight enumerators for $n \leqslant 200$ are given in [Mal3].

Extremal Type II lattices are known in dimensions 8 (E_8 [Wit4]), 16 (two lattices, $E_8 \oplus E_8$ and D_{16}^+ [Wit4]), 24 (one lattice, Λ_{24} [Con4] ≅ Chap. 12, [Lee5], [Nie2]), 32 (BW_{32} [Bar18] and §8.2f of Chap. 8, and at least one other lattice [Bay1], [Bro7], [Che4], [Oze1], [Ven3], [Ven6] (although Λ_{32} of the previous chapter has determinant 1 on the standard scale, it is not integral), 40 (M_{40} [McK2] and §5 of Chap. 8, and at least two others [Bay1], [Oze3], [Oze4]), 48 (at least two lattices, P_{48p} and P_{48q}, see Example 9 below, and also [Oze2], [Pet1]), 56 (at least one lattice [Hsi6a]), and 64 (at least one lattice Q_{64}, see [Que5] and §2c of Chap. 8). See also the Postscript to Chap. 5. The extremal theta series for $n = 48$, 56, 64 and 72 (the first open case) are respectively

$$1 + 52416000q^6 + 39007332000q^8 + \cdots,$$
$$1 + 15590400q^6 + 36957286800q^8 + \cdots,$$
$$1 + 2611200\,q^6 + 19524758400q^8 + \cdots,$$
$$1 + 6218175600q^8 + 15281788354560q^{10} + \cdots. \tag{68}$$

Example 9. The lattices P_{48p} and P_{48q}. In this section we summarize the properties of these two very similar lattices. Both are even unimodular (Type II) extremal lattices and were defined in §5.7 of Chap. 5. P_{48q} is obtained by applying Construction B_3 (§5.4 of Chap. 5) to the [48, 24, 15] ternary quadratic residue code, and then doubling the density by adjoining a translate by $(-5/2, {}^{1}\!/\!{}_{2}{}^{47})$. P_{48p} is obtained in the same way from the [48, 24, 15] Pless code. Since the two codes have the same c.w.e. (Table 3.2), the vectors in P_{48q} and P_{48p} have the same shapes. The two lattices are not equivalent, however, since they have different automorphism groups. For P_{48q} we label the coordinate positions ∞, 0, 1, ..., 46 and let Q denote the set of nonzero quadratic residues modulo 47. Then P_{48q} is spanned by the vectors

$c(2^{24}, 0^{24})$,	47 vectors, supported on a translate $\{\{0\} \cup Q\} + i$, $0 \leqslant i \leqslant 46$,
$c(-5, 1^{47})$,	a single vector, and (69)
$c(\pm 6^2, 0^{46})$,	$2 \cdot 48 \cdot 47$ vectors,

where $c = 1/\sqrt{12}$. For P_{48p} we label the coordinate positions by ∞, 0, ..., 22, ∞', $0'$, ..., $22'$, and let Q denote the set of nonzero quadratic residues modulo 23 and N the nonresidues. Then P_{48p} is spanned by:

$c(2^{12}, (-2)^{12}, 0^{24})$,	23 vectors, with 2's at i and $\{Q+i\}'$ and -2's at ∞' and $\{N+i\}'$, $0 \leqslant i \leqslant 22$,
$c(-5, 1^{47})$,	a single vector, and (70)
$c(\pm 6^2, 0^{46})$,	$2 \cdot 48 \cdot 47$ vectors.

P_{48q} and P_{48p} have det $= 1$, minimal norm $= 6$, $\tau = 52416000$, their minimal vectors are described in Table 5.3, $\rho = \sqrt{3}/2$, $\Delta = 0.00000002318\ldots$ ($\delta = (3/2)^{24} = 16834.112\ldots$), and the covering radii R are not known, but $R \geq 2$, corresponding to putatively deep holes $c(2^{12}, 0^{36})$. We have $\mathrm{Aut}(P_{48q}) = SL_2(47)$, and $\mathrm{Aut}(P_{48p}) = SL_2(23) \times S_3$ [Tho7]. (It is because there is no suitable group containing both $L_2(23)$ and $L_2(47)$ that we know these lattices are inequivalent.) Theta series: see Eq. (68) and Eq. (57) of Chap. 2, Table 7.1. For cross sections see Corollary 8 of Chap. 6.

It is natural to ask if there are any *nonlinear* extremal Type II codes or *nonlattice* extremal Type II packings (analogous to the Nordstrom-Robinson code). In fact such codes or lattices do not exist, as J.-M. Goethals (personal communication) has remarked. For the code or lattice would be distance invariant, and then the argument of [Del13] (or

Table 7.1. Theta series of 48-dimensional lattices P_{48q}, P_{48q}.

m	$N(m)$	Factors of $N(m)$
0	1	1
6	52416000	$2^9 3^2 5^3 7 \cdot 13$
8	39007332000	$2^5 3^7 5^3 7^3 13$
10	6609020221440	$2^{11} 3^8 5 \cdot 7 \cdot 13 \cdot 23 \cdot 47$
12	437824977408000	$2^{12} 3^3 5^3 7 \cdot 13 \cdot 31 \cdot 103 \cdot 109$

[Mac6, p. 646]) shows that it must be linear.

Extremal codes whose length is a multiple of 24 are especially important in view of:

Theorem 22 (Assmus and Mattson [Ass3], [Mac6, Chap. 6]). *If C is an extremal Type* II *code and n is a multiple of* 24 *then the codewords of any nonzero weight form a* 5*-design* (*cf.* §3.1 *of Chap.* 3).

The parameters of these designs can be determined from [Mal3]. For example the codewords of minimal nonzero weight form a 5-design with $v = n = 24m$, $k = 4m + 4$ and $\lambda = \binom{5m-2}{m-1}$.

There is a similar result for lattices.

Theorem 23 (Venkov [Ven7]). *If Λ is an extremal Type* II *lattice and n is a multiple of* 24 *then the vectors of any nonzero norm form a spherical* 11*-design* (*cf.* §3.2 *of Chap.* 3).

The number of minimal vectors in an extremal Type II lattice in dimension $n = 24\alpha$ can be found from Eq. (7) of [Mal1]; it is

$$A^*_{2\alpha+2} = \frac{65520}{691} \frac{\alpha}{\alpha+1} \sum_{r=0}^{\alpha} (r+1)\sigma_{11}(r+1)\, \rho_{\alpha-r}^{(\alpha+1)} , \tag{71}$$

where $\rho_m^{(\alpha+1)}$ is the coefficient of q^m in the q-expansion of

$$\prod_{s=1}^{\infty} (1-q^s)^{-24(\alpha+1)} .$$

The numbers $A^*_{2\alpha+2}$ for $n = 24, 48, \ldots, 216$ and their prime factorizations are given in Table I of [Slo8].

Corollaries 10 and 21 give upper bounds on self-dual codes and lattices. There are corresponding lower bounds, which state that for large n self-dual codes meet the Gilbert-Varshamov bound of [Mac6, Chap. 17], and self-dual lattices meet the Minkowski bound of Chap. 1, Eq. (29). The following theorems are proved in [Mac6, Chap. 19], [Mac8], [Mil7, pp. 46-47], [Tho1]. We state the result only for Type II codes.

Theorem 24. *Let $\delta(n)$ be the largest multiple of 4 such that*

$$\binom{n}{4} + \binom{n}{8} + \cdots + \binom{n}{\delta(n)-4} < 2^{n/2-2} + 1 .$$

Then there exists a Type II *code of length n and minimal distance $d \geqslant \delta(n)$. For large n there exist Type* II *codes with*

$$\frac{d}{n} \geqslant H_2^{-1}(\tfrac{1}{2}) = 0.110...,$$

where

$$H_2(x) = -x \log_2 x - (1-x)\log_2(1-x) \tag{72}$$

is the binary entropy function.

Theorem 25. *Let $m(n)$ denote the nearest integer to*

$$\frac{1}{\pi}\left\{\frac{5}{3}\Gamma\left(\frac{n}{2}+1\right)\right\}^{2/n} . \tag{73}$$

Then there exists an n-dimensional Type I *lattice with minimal norm $\mu \geqslant m(n)$. For large n there exist Type* I *lattices with*

$$\log_2 \frac{\mu}{n} \geqslant -\log_2(2\pi e) = -4.09... \tag{74}$$

or equivalently

$$\frac{1}{n}\log_2 \Delta \geqslant -1 . \tag{75}$$

If $m(n)$ is defined as the nearest even integer to (73), *the same assertions hold for Type* II *lattices.*

The proofs use generalizations of the Minkowski-Siegel mass formulae for lattices and their analogs for codes (see Chap. 16).

8. Constructions A and B for complex lattices

We now consider nonbinary codes and complex lattices, using the notation of §2.6 of Chap. 2. In §§8-10 J denotes either the Eisenstein integers $\mathcal{E}$ or the Gaussian integers $\mathcal{G}$; §4 of Chap. 8 discusses the case $J = \mathbf{Z}[\xi]$, $\xi = (1+i)/\sqrt{2}$. The constructions may also be applied to other rings.

Let π be a prime in $J (= \mathcal{E}$ or $\mathcal{G})$ of norm $\pi\bar{\pi} = q$ (say), so that $J/\pi J \cong \mathbf{F}_q$. Then there is a map $\sigma : J \rightarrow J/\pi J \rightarrow \mathbf{F}_q$ (a ring homomorphism) with $\sigma(a) \in \mathbf{F}_q$ for all $a \in J$. We define σ on n-tuples by $\sigma(a) = \sigma(a_1, \ldots, a_n) = (\sigma(a_1), \ldots, \sigma(a_n))$.

Construction $\mathrm{A_c}$. Let C be an $[n,k]$ code of length n over $\mathbf{F}_q$. Then the J-lattice $\Lambda(C)$ consists of the points $\pi^{-1/2}\,\sigma^{-1}(C)$ in $\mathbf{C}^n$.

Stated another way, if we abuse notation and regard $\mathbf{F}_q$ as a subset of J, then $\Lambda(C)$ consists of the points $\pi^{-1/2}(c+\pi x)$ for all $c \in C$ and all $x \in J^n$. If C has generator matrix $[I_k\ B]$ then $\Lambda(C)$ has generator matrix

$$G = \frac{1}{\sqrt{\pi}} \begin{bmatrix} I_k & B \\ 0 & \pi I_{n-k} \end{bmatrix}, \tag{76}$$

and

$$\det \Lambda(C) = |\det G|^2 = |\pi|^{n-2k} = q^{(n-2k)/2}. \tag{77}$$

The corresponding $2n$-dimensional real lattice $\Lambda(C)_{\text{real}}$ (§2.6 of Chap. 2) has determinant

$$\det \Lambda(C)_{\text{real}} = 4^{-n}\, \partial^n\, q^{n-2k}. \tag{78}$$

The minimal squared norm μ of $\Lambda(C)$ or $\Lambda(C)_{\text{real}}$ and the packing radius ρ and kissing number τ of $\Lambda(C)_{\text{real}}$ are determined by the codewords of minimal Euclidean norm in C. The center density of $\Lambda(C)_{\text{real}}$ is

$$\delta = 2^n\, \rho^{2n}\, \partial^{-n/2}\, q^{k-n/2}. \tag{79}$$

The theta series of $\Lambda(C)$ (or equivalently of $\Lambda(C)_{\text{real}}$) is determined by the c.w.e. of C, and in some cases just by the Hamming weight enumerator.

Construction $\mathbf{B}_c$. $\Lambda'(C)$ consists of the points $\pi^{-1/2}(c+\pi x)$, for all $c \in C$ and all $x \in J^n$ such that $\Sigma x_i \equiv 0 \pmod{\pi}$. (In order for $\Lambda'(C)$ to be a lattice C must satisfy certain obvious conditions.) Then $\det \Lambda'(C) = q^{n/2-k+1}$, and $\Lambda'(C)_{\text{real}}$ has center density

$$\delta = 2^n\, \rho^{2n}\, \partial^{-n/2}\, q^{k-1-n/2}. \tag{80}$$

Example 10. *The case J = Eisenstein integers, $\pi = 2$.* Let $J = \mathcal{E}$, $\pi = 2$, so that $\pi\bar{\pi} = 4$, $\mathcal{E}/2\mathcal{E} \cong \mathbf{F}_4$. We partition $\mathcal{E}$ into $2\mathcal{E}$, $1+2\mathcal{E}$, $\omega+2\mathcal{E}$, $\bar{\omega}+2\mathcal{E}$, and identify $\mathbf{F}_4$ with $\{0,1,\omega,\bar{\omega}\} \subset \mathcal{E}$ (Fig. 7.1). Construction A_c produces an $\mathcal{E}$-lattice from a code C over $\mathbf{F}_4$.

Theorem 26 [Slo9]. *If C is an $[n,k,d]$ code then $\Lambda(C)$ has minimal norm $\mu = \min\{2, \tfrac{1}{2}d\}$ and $\det\Lambda = 2^{n-2k}$. Furthermore $\Lambda(C)_{\text{real}}$ has $\rho = \tfrac{1}{2}\sqrt{\mu}$, $\delta = 3^{-n/2}\, 4^k\, \rho^{2n}$, and theta series*

$$W_C(\phi_0(z), \phi_1(z)), \tag{81}$$

where $W_C(x,y)$ is the Hamming weight enumerator of C and $\phi_0(z)$, $\phi_1(z)$ are given in (12), (13) *of Chap.* 4. *If C is a Type* IV *self-dual code then Λ is a self-dual $\mathcal{E}$-lattice.*

Example 10a. *The Coxeter-Todd lattice K_{12}.* For example, when C is the [6,3,4] hexacode $\mathcal{C}_6$ (§2.5.2 of Chap. 3), $\Lambda(\mathcal{C}_6)$ is a 6-dimensional self-dual $\mathcal{E}$-lattice, for which $\Lambda(\mathcal{C}_6)_{\text{real}} \cong K_{12}$ (§9 of Chap. 4). The theta series given in (131) of Chap. 4 follows from Theorem 26. (This is the

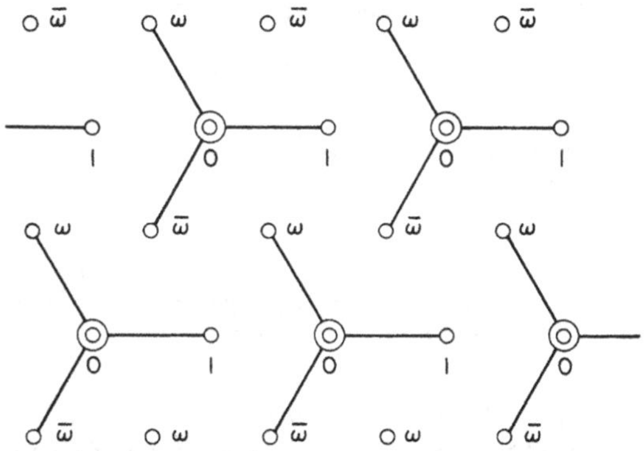

Figure 7.1. The Eisenstein integers $\mathscr{E}$ are represented by small circles, and $2\mathscr{E}$ by double circles, showing how $\mathscr{E}/2\mathscr{E} \cong \mathbf{F}_4 = \{0, 1, \omega, \bar{\omega}\}$.

"2-base" for K_{12} in the notation of [Con37]. The "4-base" is given in (128) of Chap. 4.)

Example 11. *The case* J = *Eisenstein integers,* $\pi = \sqrt{-3}$. Instead of 2 we may use the prime $\theta = \sqrt{-3} = \omega - \bar{\omega}$ in $\mathscr{E}$, obtaining $\mathscr{E}/\theta\mathscr{E} \cong \mathbf{F}_3$. We partition $\mathscr{E}$ into $\theta\mathscr{E}$, $1 + \theta\mathscr{E}$, $-1 + \theta\mathscr{E}$, and identify $\mathbf{F}_3$ with $\{0, 1, -1\} \subset \mathscr{E}$ (Fig. 7.2). Construction A_c produces an $\mathscr{E}$-lattice $\Lambda(C)$ from a code C over $\mathbf{F}_3$.

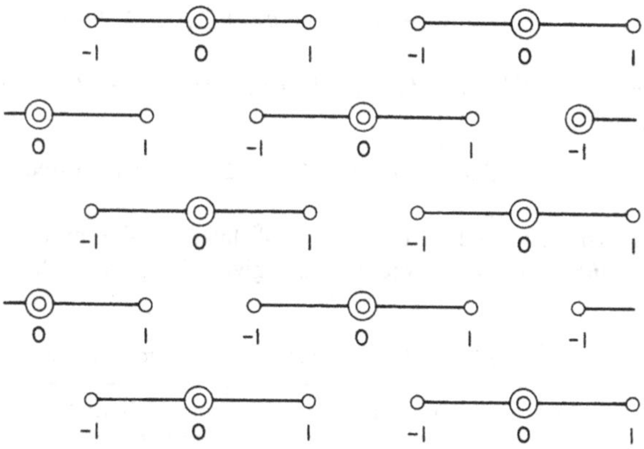

Figure 7.2. The double circles represent $\theta\mathscr{E}$, where $\theta = \sqrt{-3}$, showing how $\mathscr{E}/\theta\mathscr{E} \cong \mathbf{F}_3 = \{0, 1, -1\}$.

Theorem 27 [Slo10]. *If* C *is an* $[n, k, d]$ *code,* $\Lambda(C)$ *has minimal norm* $\mu = \min\{\sqrt{3}, d/\sqrt{3}\}$ *and* $\det \Lambda(C) = 3^{n/2-k}$. *Then* $\Lambda(C)_{\text{real}}$ *has*

$$\det = 4^{-n} 3^{2n-2k} , \tag{82}$$

$$\delta = 2^n 3^{k-n} \rho^{2n} , \tag{83}$$

$$\tau = 3^d A_d \text{ if } d \leqslant 2, \ 6n + 27A_3 \text{ if } d = 3, \ 6n \text{ if } d \geqslant 4, \tag{84}$$

and theta series

$$W_C \left\{ \phi_0\left(\frac{3z}{2}\right), \frac{1}{2}\phi_0\left(\frac{z}{2}\right) - \frac{1}{2}\phi_0\left(\frac{3z}{2}\right) \right\}. \tag{85}$$

Example 11a. *The Gosset lattice* E_6. If C is the repetition code $\{(000), (111), (222)\}$, $\Lambda(C)_{\text{real}}$ is E_6 (see (120) of Chap. 4).

Example 11b. E_8 *again*. The [4, 2, 3] tetracode $\mathscr{C}_4$ (§2.5.1 of Chap. 3) produces E_8.

Example 11c. K_{12} *again*. By applying Construction B_c to the repetition code of length 6 we obtain K_{12}. (This is the "3-base" for K_{12}, in the notation of [Con37].)

Example 12. *The complex Leech lattice*. The Leech lattice Λ_{24} may be constructed as a complex 12-dimensional $\mathscr{E}$-lattice from the [12, 6, 6] ternary Golay code $\mathscr{C}_{12}$ (§2.8.5 of Chap. 3). We start by applying Construction B_c to $\mathscr{C}_{12}$. The lattice $\sqrt{\theta}\,\Lambda'(\mathscr{C}_{12})$ consists of the points $c + \theta x$, where $c \in \mathscr{C}_{12}$, $x \in \mathscr{E}^{12}$ and $\Sigma\, x_i \equiv 0 \pmod{\theta}$. The complex Leech lattice consists of the union of this lattice and two translates. Thus (after multiplying by θ) we define the *complex Leech lattice* to consist of the vectors

$$0 + \theta c + 3x, \ 1 + \theta c + 3y, \ -1 + \theta c + 3z , \tag{86}$$

where $0 = (0^{12})$, $1 = (1^{12})$, $-1 = ((-1)^{12})$, $c \in \mathscr{C}_{12}$, and $x, y, z \in \mathscr{E}^{12}$ satisfy

$$\Sigma\, x_i \equiv 0, \ \Sigma\, y_i \equiv 1, \ \Sigma\, z_i \equiv -1 \pmod{\theta}.$$

It is easily checked that this *is* an $\mathscr{E}$-lattice. The minimal norm is 18, and the 196560 minimal vectors are given in Table 7.2. In this table the

Table 7.2. Minimal vectors in the complex Leech lattice.

Coordinates	Number
$\theta(\alpha^6, 0^6) = \theta c + 3x$	$264 \cdot 3^5 = 64152$
$(\beta, \alpha^{11}) = 1 + \theta c + 3y$ or $-1 + \theta c + 3z$	$2 \cdot 3^6 \cdot 12 \cdot 2 = 34992$
$((2\alpha)^2, \alpha^{10}) = 1 + \theta c + 3y$ or $-1 + \theta c + 3z$	$3^6 \cdot 66 \cdot 2 = 96228$
$((3\alpha)^2, 0^{10}) = 3x$	$66 \cdot 2 \cdot 3^2 = 1188$
	Total = 196560

symbols α stand for (possibly distinct) numbers of norm 1, i.e. ± 1, $\pm\omega$, $\pm\omega^2$, and β is a number of norm 7, i.e. one of $\pm(1+3\omega)$, $\pm(1+3\omega^2)$,.... The centers described in the first row of the table are obtained by taking one of the 264 codewords of weight 6 in $\mathscr{C}_{12}$, say $c = (0,1,0,0,0,0,-1,0,1,-1,-1,1)$, and choosing x so that

$$\theta c + 3x = (0,\theta\omega^a,0,0,0,0,-\theta\omega^b, 0,\theta\omega^c,-\theta\omega^d,-\theta\omega^e,\theta\omega^f)$$

with $a,b,c,\cdots \in \{0,1,2\}$ and $\Sigma x_i \equiv 0 \pmod{\theta\mathscr{E}}$. There are three choices for each of the first five nonzero components of x, after which x is uniquely determined, and so there are $264\cdot 3^5$ such centers. A typical center $1 + \theta c + 3y$ in the second row is

$$\begin{array}{rcccccccccccc}
(1 & 1 & 1 & 1 & 1 & 1 & 1 & 1 & 1 & 1 & 1 & 1) \\
+\theta\ (0 & 1 & 0 & 0 & 0 & 0 & -1 & 0 & 1 & -1 & -1 & 1) \\
+3\ (0 & \omega^2 & 0 & 0 & 0 & 0 & \omega & 0 & \omega^2 & \omega & \omega & -1) \\
=(1 & \omega^2 & 1 & 1 & 1 & 1 & \omega & 1 & \omega^2 & \omega & \omega & 2\omega-1)
\end{array}$$

The first vector may be 1 or -1, c may be any codeword of $\mathscr{C}_{12}$, and then there are $12\cdot 2$ choices for y: the element of norm 7 may appear in any of the 12 positions and can be chosen in two ways. Thus the number of centers of this type is $2\cdot 3^6\cdot 12\cdot 2$. The other centers may be counted in a similar manner.

The corresponding real lattice, when multiplied by $\sqrt{2}/3$, is the Leech lattice Λ_{24} — compare Eqs. (135), (136) of Chap. 4. Table 7.3 shows some of the parallels between the two lattices. The automorphism group of the complex Leech lattice will be discussed in Chap. 10. See also [Wil3], [Yos1].

It is natural to ask if there is an analogous construction of Λ_{24} as a 6-dimensional quaternionic lattice, using the hexacode $\mathscr{C}_6$ over $\mathbf{F}_4$. The MOG construction of Λ_{24}, described in detail in Chap. 11, is essentially this analog, although the construction is not quite what one would expect just from a comparison of the real and complex constructions. A construction using the icosian ring is described in §2 of Chap. 8.

Table 7.3. Analogies between the real and complex Leech lattices.

Real Leech lattice	Complex Leech lattice
Binary Golay code $\mathscr{C}_{24}$	Ternary Golay code $\mathscr{C}_{12}$
2	$\theta = \sqrt{-3}$
$\{+1,-1\}$	$\{1,\omega,\omega^2\}$
Steiner system $S(5,8,24)$	Steiner system $S(5,6,12)$
Mathieu group M_{24}	Mathieu group M_{12}
Automorphism group contains $\mathrm{diag}((-1)^{c_1},\ldots,(-1)^{c_{24}})$ for any $c \in \mathscr{C}_{24}$	Automorphism group contains $\mathrm{diag}(\omega^{c_1},\ldots,\omega^{c_{12}})$ for any $c \in \mathscr{C}_{12}$

Example 13. *The case* J = *Gaussian integers,* $\pi = 1 + i$. Let $J = \mathcal{G}$, $\pi = 1 + i$, so $\pi\bar{\pi} = 2$, $\mathcal{G}/\pi\mathcal{G} \cong \mathbf{F}_2$. Constructions $\mathrm{A_c}$ and $\mathrm{B_c}$ produce Gaussian lattices from binary codes. It is best to omit the factor $\pi^{-1/2}$ for these lattices. Then Construction $\mathrm{A_c}$ produces a lattice with theta series $W(\theta_3(2z)^2, \theta_2(2z)^2)$, where $W(x,y)$ is the weight enumerator of the code. The determinant of the real lattice Λ_{real} corresponding to a Gaussian lattice Λ is

$$\det \Lambda_{\mathrm{real}} = (\det \Lambda)^2 . \tag{87}$$

Example 13a. D_4 *again.* Construction $\mathrm{A_c}$ applied to $\mathcal{C}_2 = \{00, 11\}$ produces a two-dimensional Gaussian lattice $\mathcal{D}_2$ with generator and Gram matrices

$$M = \begin{pmatrix} 1 & 1 \\ 0 & 1+i \end{pmatrix}, \quad A = M\overline{M}^{tr} = \begin{pmatrix} 2 & 1-i \\ 1+i & 2 \end{pmatrix} . \tag{88}$$

The 24 minimal vectors have the form $(\pm 1, \pm 1)$, $(\pm 1, \pm i)$, $(\pm i, \pm i)$, $(0, \pm 1 \pm i)$. The corresponding real lattice is D_4; in this form the theta series is (cf. (92) of Chap. 4)

$$\Theta_{D_4}(z) = \theta_2(2z)^4 + \theta_3(2z)^4 = 1 + 24q^2 + 24q^4 + 96q^6 + \cdots . \tag{89}$$

Example 13b. E_8 *again.* Construction $\mathrm{B_c}$ applied to $\{0000, 1111\}$ produces a Gaussian version of E_8, consisting of all vectors $x = (x_1, x_2, x_3, x_4) \in \mathcal{G}^4$ satisfying $x \pmod{1+i} \equiv 0000$ or 1111, and $\Sigma x_i \equiv 0 \pmod 2$. The corresponding real lattice is the Hamming code version of E_8 (§8.1 of Chap. 4).

9. Self-dual nonbinary codes and complex lattices

There are analogous results to Theorems 6 and 7.

Theorem 28 (Gleason). *If C is a Type* III *code then $W_C(x,y)$ is invariant under a group G_3 of order* 48, *and belongs to* $\mathbf{C}[\psi_4, \xi_{12}]$, *where ψ_4 is the weight enumerator of $\mathcal{C}_4$* (Eq. (61) *of Chap.* 3) *and*

$$\xi_{12} = y^3(x^3 - y^3)^3 . \tag{90}$$

Equivalently, $W_C(x,y)$ can be written as a polynomial in the weight enumerators of $\mathcal{C}_4$ and $\mathcal{C}_{12}$.

To describe the c.w.e. we first define some further polynomials. Let

$$a = x^3 + y^3 + z^3, \quad p = 3xyz , \tag{91}$$

$$b = x^3y^3 + y^3z^3 + z^3x^3 , \tag{92}$$

$$\beta_6 = a^2 - 12b = x^6 + y^6 + z^6 - 10(x^3y^3 + y^3z^3 + z^3x^3) , \tag{93}$$

$$\pi_9 = (x^3 - y^3)(y^3 - z^3)(z^3 - x^3) , \tag{94}$$

$$\alpha_{12} = a(a^3 + 8p^3) = \overset{(3)}{\Sigma} x^{12} + 4 \overset{(6)}{\Sigma} x^9y^3 + 6 \overset{(3)}{\Sigma} x^6y^6 + 228 \overset{(3)}{\Sigma} x^6y^3z^3 . \tag{95}$$

Theorem 29 [Mal5]. *If C is a Type* III *code which contains the all-ones vector then the c.w.e. is invariant under a group G_4 of order* 2592, *and belongs to*

$$R \oplus \beta_6 \pi_9^2 R, \tag{96}$$

where

$$R = \mathbf{C}[\beta_6^2, \alpha_{12}, \pi_9^4]. \tag{97}$$

In other words the c.w.e. can be written uniquely as a polynomial in β_6^2, α_{12} and π_9^4, plus $\beta_6 \pi_9^2$ times another such polynomial.

Theorem 30. *If C is a Type* IV *code then $W_C(x,y)$ is invariant under a group G_5 of order* 12 *and belongs to* $\mathbf{C}[\psi_2', \xi_6]$, *where $\psi_2' = x^2 + 3y^2$ is the weight enumerator of the code $\mathscr{C}_2' = \{00, 11, \omega\omega, \bar{\omega}\bar{\omega}\}$ and*

$$\xi_6 = y^2(x^2 - y^2)^2. \tag{98}$$

Equivalently, $W_C(x,y)$ can be written as a polynomial in the weight enumerators of $\mathscr{C}_2'$ and $\mathscr{C}_6$.

Theorem 31 [Mac4]. *If C is a Type* IV *code which contains the all-ones vector then the c.w.e. is invariant under a group G_6 of order* 1152, *and belongs to* $\mathbf{C}[f_2, f_6, f_8, f_{12}]$ *where (using standard notation for symmetric functions of w, x, y, z)*

$$f_2 = (2) = w^2 + x^2 + y^2 + z^2, \tag{99}$$

$$f_6 = (6) + 15(222) = w^6 + \cdots + 15(w^2x^2y^2 + \cdots), \tag{100}$$

$$f_8 = (8) + 14(44) + 168(2222), \tag{101}$$

$$f_{12} = (12) + 22(66) + 330(6222) + 165(444) + 330(4422). \tag{102}$$

Here f_2 is the c.w.e. of $\mathscr{C}_2^1$, f_6 is the c.w.e. of the version of the hexacode defined by Eq. (64) of Chap. 3, f_8 is the c.w.e. of the $\mathbf{F}_4$ span of $\mathscr{H}_8$, and f_{12} is obtained from the c.w.e. of a certain code of length 12 (see [Mac4]). For proofs and generalizations see [Ber7], [Gle1], [Leo7], [Mac3], [Mac4], [Mac6, Chap. 19], [Mal2], [Mal4], [Mal5], [Slo6], [Slo7]. G_3 is generated by $\frac{1}{\sqrt{3}}\begin{bmatrix} 1 & 2 \\ 1 & -1 \end{bmatrix}$, $\begin{bmatrix} 1 & 0 \\ 0 & \omega \end{bmatrix}$ and is the reflection group 3[6]2. G_4 is generated by all permutations, diag$\{1, 1, \omega\}$ and

$$\frac{1}{\sqrt{3}} \begin{bmatrix} 1 & 1 & 1 \\ 1 & \omega & \bar{\omega} \\ 1 & \bar{\omega} & \omega \end{bmatrix},$$

has a center $Z(G_4)$ of order 12, and $G_4/Z(G_4)$ is the Hessian group of order 216 [Cox28]. G_5 is generated by ½$\begin{bmatrix} 1 & 3 \\ 1 & -1 \end{bmatrix}$, $\begin{bmatrix} 1 & 0 \\ 0 & -1 \end{bmatrix}$ and is a dihedral group. G_6 is generated by all 4×4 monomial matrices and the matrix H_4 of Chap. 4, §7.1, and is the reflection group [3, 4, 3].

Theorem 32 [Slo9]. *If* Λ *is a self-dual* $\mathcal{E}$*-lattice then* $\Theta_\Lambda(z) \in \mathbf{C}[\phi_0(z/2), \Delta_6(z)]$, *where*

$$\Delta_6(z) = \frac{1}{16}\,\theta'_1\left(\frac{z}{2}\right)^2 \theta'_1\left(\frac{3z}{2}\right)^2 \tag{103}$$

$$= q \prod_{m=1}^{\infty} (1-q^m)^6\,(1-q^{3m})^6 \tag{104}$$

$$= q - 6q^2 + 9q^3 + 4q^4 + 6q^5 - 54q^6 - 40q^7 + \cdots .$$

Equivalently, $\Theta_\Lambda(z)$ *can be written as an isobaric polynomial in the theta series of* $\mathcal{E}$ *and of the complex* ($\mathcal{E}$*-lattice) version of* K_{12}. *If we write* $\Delta_6(z) = \Sigma\, a_m q^m$, *the coefficients* a_m *are multiplicative, and in fact satisfy*

$$a_m\, a_n = \prod_{3 \nmid d,\ d \mid (m,n)} d^5 a_{mn/d^2}. \tag{105}$$

Theorem 33. *If* Λ *is a self-dual* $\mathcal{G}$*-lattice then* $\Theta_\Lambda(z) \in \mathbf{C}[\theta_3(z)^2, \Delta_8(z)]$. *Equivalently,* $\Theta_\Lambda(z)$ *can be written as an isobaric polynomial in the theta series of* $\mathcal{G}$ *and of the complex* ($\mathcal{G}$*-lattice) version of* E_8.

Proof. From (67) of Chap. 2, det $\Lambda_{real} = (\det \Lambda)^2 = 1$, so Λ_{real} is a Type I lattice of even dimension. Λ and Λ_{real} have the same theta series, and the result follows from Theorem 7.

Classification. Type III codes have been classified for length $n \leqslant 24$ [Con24], [Leo8], [Mal2], [Ple18], Type IV codes for length $n \leqslant 16$ [Con24], [Mac4], and self-dual codes over $\mathbf{F}_5$ for length $n \leqslant 12$ [Leo7]. Self-dual $\mathcal{E}$-lattices have been classified for dimension $n \leqslant 12$ [Fei2], and self-dual $\mathcal{G}$-lattices for $n \leqslant 7$ [Iya1] (it would now be possible to extend the latter classification using the results of Chap. 16). See also [Hsi6a], [Ple20].

To illustrate how rapidly the numbers of these codes and lattices grow as n increases, Table 7.4 shows $N_t(n)$, the total number of distinct Type IV codes of length n, as well as $N_e(n)$, the number of inequivalent codes and $N_i(n)$, the number of inequivalent indecomposable codes.

Table 7.4. The numbers of Type IV codes.

n	$N_t(n)$	$N_e(n)$	$N_i(n)$
2	3	1	1
4	27	1	0
6	891	2	1
8	114939	3	1
10	58963707	5	2
12	120816635643	10	4
14	989850695823099	21	10
16	32436417451427131131	55	31
24	$\sim 4 \cdot 10^{43}$	$> 10^8$	

10. Extremal nonbinary codes and complex lattices

Extremal weight enumerators, codes, theta series and lattices can now be defined from Theorems 28, 30 (just using the Hamming weight enumerators), 32 and 33, exactly as in §§4 and 7. There are analogs of Theorem 9 and Corollary 10.

Theorem 34. *In an extremal Type* III *or* IV *weight enumerator, or an extremal* $\mathcal{E}$- *or* $\mathcal{G}$*-lattice theta series, the first coefficient that is not zero by definition is always strictly positive, and for all sufficiently large n the next coefficient is negative.*

Corollary 35. *The minimal distance of a Type* III *code satisfies*

$$d \leqslant 3[n/12] + 3 , \tag{106}$$

and for a Type IV *code*

$$d \leqslant 2[n/6] + 2 . \tag{107}$$

The minimal norm of a self-dual $\mathcal{E}$*-lattice satisfies*

$$\mu \leqslant [n/6] + 1 , \tag{108}$$

and for a $\mathcal{G}$*-lattice*

$$\mu \leq [n/4] + 1 . \tag{109}$$

An extremal code or lattice satisfies (106)-(109) with equality. Extremal Type III codes exist for $n = 4$ ($\mathcal{C}_4$), 8 ($\mathcal{C}_4 \oplus \mathcal{C}_4$), 12 ($\mathcal{C}_{12}$), 16, 20, 24, 32, 36, 40, 44, 48, 56, 60 and 64, do not exist for $n = 72$, 96, 120, 144,..., and and other values are undecided (see [Bee1], [Daw1], [Mac6], [Mal2], [Mal3], [Ple18] and the Postscript to Chap. 5). Extremal Type IV codes exist for $n = 2$ ($\mathcal{C}_2'$), 4 ($\mathcal{C}_2' \oplus \mathcal{C}_2'$), 6 ($\mathcal{C}_6$), 8, 10, 14, 16, 18, 20, 22, 28, 30, and do not exist for $n = 12$, 102, 108, 114, 120, 122, 126, 128, ... The other values of n are undecided [Con24], [Mac4].

Since there are [24, 12, 8] Type II and [24, 12, 9] Type III codes, it would be nice to know whether an extremal [24, 12, 10] Type IV code exists ([Con22], [Con23], [Mac4]).

Extremal $\mathcal{E}$-lattices exist for $n = 1$ to 5 ($\mathcal{E}^n$), 6 (K_{12}), 8 to 11, but do not exist for $n = 7$, 12 and sufficiently large n [Fei2].

Theorem 36. *Extremal* $\mathcal{G}$*-lattices exist if and only if* $n = 1$ *to* 4, 6, 7 *or* 12.

Proof. This follows from Theorem 12.

Acknowledgements. We thank E. Bannai (who independently discovered Theorem 23) for drawing our attention to Venkov's paper [Ven7], and J.-M. Goethals and C. L. Mallows for their helpful comments.

8

Algebraic Constructions for Lattices

J. H. Conway and N. J. A. Sloane

In this chapter we construct dense lattice packings in dimensions 32, 36, 40, 48, 64, 96, ..., 65536, ... using a variety of techniques. The main constructions used are Constructions A_f (a more abstract version of Construction A), D (which uses a nested family of codes) and E (a powerful general construction which can be applied recursively). Other methods construct dense lattices from ideals in algebraic number fields. We also construct E_8 and the Leech lattice using icosians.

1. Introduction

This chapter makes use of a variety of constructions, generally more technical than in previous chapters. One common theme is that they use more complicated rings than **Z**, so that more knowledge of algebra is required. The lattices constructed include E_8 (§2), the Leech lattice (§§2, 3, 5, 7.3, 7.5), Quebbemann's lattices Q_{32} and Q_{64} (§§3, 4), the lattices C_{32} (§8.2h) and B_{36} (§8.2d), McKay's lattice M_{40} (§5), the lattices P_{48q} (§7.5) and P_{64c} (§8.2e), the Barnes-Wall lattices BW_n, $n = 2^m$ (§8.2f), the lattices B_n, $n = 2^m$, obtained from BCH codes (§8.2g), Craig's lattices $A_n^{(m)}$ (§§6, 7.3), and the infinite trees of lattices $\eta(\Lambda)$ obtained from Construction E (§10). The latter give reasonably dense lattices in dimensions up to at least 10^{10000} (see §10h, Table 8.7, and Tables 1.3, 1.4 of Chap. 1), although *eventually* their density is not very good.

The constructions used include two general versions of Construction A (§§3, 4), a version of Construction C that always produces lattices (Construction D, §8), and a powerful generalization of most of the previous constructions (Construction E, §§9, 10) that may be applied recursively. Section 7 gives a number of constructions using ideals in algebraic number fields. One of the most interesting of these is the result, already mentioned in §1.5 of Chap. 1, that towers of algebraic number fields of the type exhibited by Golod and Shafarevich correspond to infinite sequences of extremely dense lattices (§7.4).

All the packings mentioned in this chapter are lattices. We shall sometimes take the more general point of view mentioned in §2.2 of Chap. 2, and regard a lattice Λ as a discrete subgroup of a real vector space V in which norms and inner products are defined by means of a symmetric bilinear form $f : V \times V \to \mathbf{R}$. The norm of x is given by $f(x) = f(x,x)$, and we have

$$f(x,y) = \tfrac{1}{2}(f(x+y) - f(x) - f(y)). \qquad (1)$$

$f(x)$ is assumed to be a positive definite quadratic form.

We begin (in §2) by introducing certain quaternions called *icosians*, which are equivalent to the E_8 lattice equipped with a multiplicative structure. This leads to a simple construction of the Leech lattice as a three-dimensional icosian lattice (§2.2), a direct analog of the Turyn construction of the Golay code.

2. The icosians and the Leech lattice

2.1 The icosian group. The *icosian group* is a multiplicative group of order 120 consisting of the quaternions

$$\tfrac{1}{2}(\pm 2,0,0,0)^{\mathsf{A}},\ \tfrac{1}{2}(\pm 1,\pm 1,\pm 1,\pm 1)^{\mathsf{A}},\ \tfrac{1}{2}(0,\pm 1,\pm\sigma,\pm\tau)^{\mathsf{A}},$$

where $(\alpha,\beta,\gamma,\delta)$ means $\alpha + \beta i + \gamma j + \delta k$,

$$\sigma = \tfrac{1}{2}(1-\sqrt{5}), \quad \tau = \tfrac{1}{2}(1+\sqrt{5}),$$

and the superscript A means that all even permutations of the coordinates are permitted. We use the particular names

$$\omega = \tfrac{1}{2}(-1,1,1,1), \quad i_H = \tfrac{1}{2}(0,1,\sigma,\tau).$$

There is a homomorphism from this group to the alternating group A_5 on five letters $\{G,H,I,J,K\}$, defined by

$$\begin{aligned}
i &= \tfrac{1}{2}(0,2,0,0) &&\to (H,I)(J,K),\\
j &= \tfrac{1}{2}(0,0,2,0) &&\to (H,J)(K,I),\\
k &= \tfrac{1}{2}(0,0,0,2) &&\to (H,K)(I,J),\\
\omega &= \tfrac{1}{2}(-1,1,1,1) &&\to (I,J,K),\\
i_H &= \tfrac{1}{2}(0,1,\sigma,\tau) &&\to (G,I)(J,K),
\end{aligned}$$

in which the kernel is $\{\pm 1\}$. Table 8.1 below gives much more information about this homomorphism. Abstractly the icosian group is the perfect double cover $2.\mathsf{A}_5$ of A_5, and is sometimes called the *binary icosahedral group*.

The *icosian ring* $\mathcal{I}$ is the set of all finite sums $q_1 + \cdots + q_n$, where each q_t is in the icosian group. Elements of the icosian ring are simply called *icosians* ([Con16], [Du1], [Wil9]).

The typical icosian q has the form $q = \alpha + \beta i + \gamma j + \delta k$, where the coordinates α, β, γ, δ belong to the *golden field* $\mathbf{Q}(\tau)$, and so have the form

$a + b\sqrt{5}$ where $a,\ b \in \mathbf{Q}$. The *conjugate* icosian (cf. §2.6 of Chap. 2) is $\bar{q} = \alpha - \beta i - \gamma j - \delta k$, and $q\bar{q} = \alpha^2 + \beta^2 + \gamma^2 + \delta^2$.

We shall work with vectors $v = (q_1, q_2, \ldots)$ whose entries are icosians. The typical scalar multiple of v is $\lambda v = (\lambda q_1, \lambda q_2, \ldots)$, with scalars (which are also icosians) *on the left*. Correspondingly the congruence $q \equiv r \pmod{\lambda}$, where $q,\ r,\ \lambda$ are quaternions, means that $q - r = \lambda s$ for some icosian s. Two vectors v and $w = (r_1, r_2, \ldots)$ have a quaternionic inner product

$$(v, w) = q_1 \bar{r}_1 + q_2 \bar{r}_2 + \cdots .$$

We shall use two different norms for such vectors, the quaternionic norm

$$QN(v) = (v, v), \tag{2}$$

which is a number of the form $a + b\sqrt{5}$, with $a,\ b \in \mathbf{Q}$, and the Euclidean norm

$$EN(v) = a + b. \tag{3}$$

(It is easy to show that $EN(v) \geqslant 0$.) The icosians of quaternionic norm 1 are the elements of the icosian group.

With respect to the quaternionic norm the icosians belong to a four-dimensional space over $\mathbb{Q}(\tau)$; with the Euclidean norm they lie in an eight-dimensional space. In fact under the Euclidean norm the icosian ring $\mathcal{I}$ is isomorphic to an E_8 lattice in this space. Table 8.1 (taken from [Wil9]) displays the particular isomorphism we shall use.

Table 8.1 has 60 entries, one for each pair of elements $\pm q$ of the icosian group. The typical entry in this table:

$$\begin{array}{rrrrr} \multicolumn{5}{c}{\bar{\omega}^i = \omega_{IG} \rightarrow (HJK)} \\ -1 & - & - & 0 & 0 \\ -1 & - & - & 0 & 0 \\ 1 & + & + & -2 & 2 \\ 1 & + & + & 0 & 0 \end{array}$$

should be read as follows. The top line gives name(s) for q (in this case $\bar{\omega}^i = \omega_{IG} = ½(-1-i+j+k)$) and indicates the corresponding even permutation of $\{G,H,I,J,K\}$. The four quaternionic coordinates of $2q$ appear in the first column, followed by two columns giving the E_8 vectors representing $2q$ and $2\sigma q$. As usual $-$ stands for -1 and $+$ for $+1$. The formulae given in Eq. (74) of Chap. 2 are helpful for manipulating these quaternions.

The naming system for $q = ½(\alpha, \beta, \gamma, \delta)$ is as follows. The letters i, j, k indicate that $\alpha = 0$, ω indicates $\alpha = -1$, s indicates $\alpha = -\sigma$, t indicates $\alpha = -\tau$, and the subscripts indicate the signs of α, β, γ, δ:

Table 8.1. Icosians and corresponding elements of A_5 and E_8.

$1 \to$ identity	$i = i_G \to (HI)(JK)$	$j = j_G \to (HJ)(KI)$	$k = k_G \to (HK)(IJ)$
2 2 2 − +	0 0 0 − +	0 0 0 − +	0 0 0 − +
0 0 0 + −	2 2 2 − +	0 0 0 − +	0 0 0 + −
0 0 0 + −	0 0 0 + −	2 2 2 − +	0 0 0 − +
0 0 0 + −	0 0 0 − +	0 0 0 + −	2 2 2 − +

$\omega = \omega_{GH} \to (IJK)$	$\omega^i = \omega_{GI} \to (HKJ)$	$\omega^j = \omega_{GJ} \to (HIK)$	$\omega^k = \omega_{GK} \to (HJI)$
−1 − − − +	−1 − − + −	−1 − − + −	−1 − − + −
1 + + − +	1 + + − +	−1 − − − +	−1 − − + −
1 + + − +	−1 − − + −	1 + + − +	−1 − − − +
1 + + − +	−1 − − − +	−1 − − + −	1 + + − +

$\bar{\omega} = \omega_{HG} \to (IKJ)$	$\bar{\omega}^i = \omega_{IG} \to (HJK)$	$\bar{\omega}^j = \omega_{JG} \to (HKI)$	$\bar{\omega}^k = \omega_{KG} \to (HIJ)$
−1 − − 2 −2	−1 − − 0 0	−1 − − 0 0	−1 − − 0 0
−1 − − 0 0	−1 − − 0 0	1 + + 0 0	1 + + −2 2
−1 − − 0 0	1 + + −2 2	−1 − − 0 0	1 + + 0 0
−1 − − 0 0	1 + + 0 0	1 + + −2 2	−1 − − 0 0

$i_H \to (GI)(JK)$	$j_H \to (GJ)(KI)$	$k_H \to (GK)(IJ)$	$i_I \to (HG)(JK)$	$j_J \to (HG)(KI)$	$k_K \to (HG)(IJ)$
0 0 0 − +	0 0 0 − +	0 0 0 − +	0 0 0 0 0	0 0 0 0 0	0 0 0 0 0
1 0 2 − +	τ 2 0 − −	σ 0 0 + +	1 2 0 0 0	−τ −2 0 0 2	−σ 0 0 0 −2
σ 0 0 + +	1 0 2 − +	τ 2 0 − −	−σ 0 0 0 −2	1 2 0 0 0	−τ −2 0 0 2
τ 2 0 − −	σ 0 0 + +	1 0 2 − +	−τ −2 0 0 2	−σ 0 0 0 −2	1 2 0 0 0

$i_J \to (HI)(GK)$	$j_K \to (HJ)(GI)$	$k_I \to (HK)(GJ)$	$i_K \to (HI)(JG)$	$j_I \to (HJ)(KG)$	$k_J \to (HK)(IG)$
0 + − 0 0	0 + − 0 0	0 + − 0 0	0 − + − +	0 − + − +	0 − + − +
1 + + 0 0	τ + + −2 0	−σ + − 0 −2	1 + + − +	−τ − − + +	σ − + + +
−σ + − 0 −2	1 + + 0 0	τ + + −2 0	σ − + + +	1 + + − +	−τ − − + +
τ + + −2 0	−σ + − 0 −2	1 + + 0 0	−τ − − + +	σ − + + +	1 + + − +

$\omega_{HI} \to (GJK)$	$\omega_{HJ} \to (GKI)$	$\omega_{HK} \to (GIJ)$	$\omega_{JK} \to (GHI)$	$\omega_{KI} \to (GHJ)$	$\omega_{IJ} \to (GHK)$
−1 − − 0 0	−1 − − 0 0	−1 − − 0 0	−1 0 −2 + −	−1 0 −2 + −	−1 0 −2 + −
0 + − 0 0	σ − + 0 2	τ + + −2 0	0 0 0 − +	−σ 0 0 − −	τ 2 0 − −
τ + + −2 0	0 + − 0 0	σ − + 0 2	τ 2 0 − −	0 0 0 − +	−σ 0 0 − −
σ − + 0 2	τ + + −2 0	0 + − 0 0	−σ 0 0 − −	τ 2 0 − −	0 0 0 − +

$\omega_{IH} \to (GKJ)$	$\omega_{JH} \to (GIK)$	$\omega_{KH} \to (GJI)$	$\omega_{KJ} \to (GIH)$	$\omega_{IK} \to (GJH)$	$\omega_{JI} \to (GKH)$
−1 − − + −	−1 − − + −	−1 − − + −	−1 −2 0 0 0	−1 −2 0 0 0	−1 −2 0 0 0
0 − + − +	−σ + − − −	−τ − − + +	0 0 0 0 0	σ −2 0 0 2	−τ 0 0 0 2
−τ − − + +	0 − + − +	−σ + − − −	−τ −2 0 0 2	0 0 0 0 2	σ 0 0 0 0
−σ + − − −	−τ − − + +	0 − + − +	σ 0 0 0 2	−τ 0 0 0 0	0 −2 0 0 2

$t_{HI} \to (HGIJK)$	$t_{HJ} \to (HGJKI)$	$t_{HK} \to (HGKIJ)$	$t_{JK} \to (JGKHI)$	$t_{KI} \to (KGIHJ)$	$t_{IJ} \to (IGJHK)$
−τ −2 0 0 2	−τ −2 0 0 2	−τ −2 0 0 2	−τ −2 0 + +	τ −2 0 + +	−τ −2 0 + +
0 0 0 0 0	1 2 0 0 0	σ 0 0 0 2	0 0 0 − +	−1 0 −2 + −	σ 0 0 + +
σ 0 0 0 2	0 0 0 0 0	1 2 0 0 0	σ 0 0 + +	0 0 0 − +	−1 0 −2 + −
1 2 0 0 0	σ 0 0 0 2	0 0 0 0 0	−1 0 −2 + −	σ 0 0 + +	0 0 0 − +

$t_{IH} \to (IGHKJ)$	$t_{JH} \to (JGHIK)$	$t_{KH} \to (KGHJI)$	$t_{KJ} \to (KGJIH)$	$t_{IK} \to (IGKJH)$	$t_{JI} \to (JGIKH)$
−τ − − 2 0	−τ − − 2 0	−τ − − 2 0	−τ − − + +	−τ − − + +	−τ − − + +
0 + − 0 0	−1 − − 0 0	−σ + − 0 −2	0 + + −	1 + + − +	−σ + − − −
−σ + − 0 −2	0 + − 0 0	−1 − − 0 0	−σ + − −	0 + − + −	1 + + − +
−1 − − 0 0	−σ + − 0 −2	0 + − 0 0	1 + + − +	−σ + − − −	0 + − + −

$s_{JK} \to (KGJHI)$	$s_{KI} \to (IGKHJ)$	$s_{IJ} \to (JGIHK)$	$s_{HI} \to (IGHJK)$	$s_{HJ} \to (JGHKI)$	$s_{HK} \to (KGHIJ)$
−σ 0 0 − −	−σ 0 0 − −	−σ 0 0 − −	−σ + − − −	−σ + − − −	−σ + − − −
0 0 0 − +	−τ −2 0 + +	1 0 2 − +	0 − + − +	τ + + − −	1 + + − +
1 0 2 − +	0 0 0 − +	−τ −2 0 + +	1 + + − +	0 − + − +	τ + + − −
−τ −2 0 + +	1 0 2 − +	0 0 0 − +	τ + + − −	1 + + − +	0 − + − +

$s_{KJ} \to (JGKIH)$	$s_{IK} \to (KGIHJ)$	$s_{JI} \to (IGJKH)$	$s_{IH} \to (HGIKJ)$	$s_{JH} \to (HGJIK)$	$s_{KH} \to (HGKJI)$
−σ + − 0 −2	−σ + − 0 −2	−σ + − 0 −2	−σ 0 0 0 −2	−σ 0 0 0 −2	−σ 0 0 0 −2
0 − + 0 0	τ + + −2 0	−1 − − 0 0	0 0 0 0 0	−τ −2 0 0 2	−1 −2 0 0 0
−1 − − 0 0	0 − + 0 0	τ + + −2 0	−1 −2 0 0 0	0 0 0 0 0	−τ −2 0 0 2
τ + + −2 0	−1 − − 0 0	0 − + 0 0	−τ −2 0 0 2	−1 −2 0 0 0	0 0 0 0 0

(There are four mistakes in this table — see Preface to Second Edition.)

signs	subscript	signs	subscript
0(+00)	G	0+−−	I
−+++	GH	−0++	HI
−−−−	HG	−0−−	IH
−+−−	GI	−0+−	JK
−−++	IG	−0−+	KJ
0+++	H	...	...

For example ω_{XY} corresponds to the permutation (Z,T,U), where X, Y, Z, T, U is an even permutation of G, H, I, J, K. The triple $\{i_X, j_X, k_X\}$, where X is any of G, H, I, J, K, forms a system of unit quaternions (with $i_X^2 = -1$, $i_X j_X = k_X$, etc.).

The 240 minimal vectors of this version of E_8 have Euclidean norm 1, and quaternionic norm either 1 or σ^2. They consist of the elements q and σq, where q is any element of the icosian group.

Using the map in Table 8.1, a vector $v = (q_1, \ldots, q_n)$ with n icosian coordinates is identified with a vector having $8n$ rational coordinates.

2.2 The icosian and Turyn-type constructions for the Leech lattice. Let L be the three-dimensional lattice over the icosians consisting of all vectors (x,y,z), where x, y, $z \in \mathcal{I}$ satisfy

$$x \equiv y \equiv z \pmod{h}, \tag{4}$$

$$x + y + z \equiv 0 \pmod{\bar{h}}, \tag{5}$$

and h is the quaternion

$$h = \omega + \sigma = \tfrac{1}{2}(-\sqrt{5}, 1, 1, 1),$$

with $h\bar{h} = 2$. Then the Euclidean norm (Eq. (3)) converts L into a copy of the Leech lattice. This will be proved in §3, Example (c). (It is also established in [Wil1], [Wil9].)

Let G be the automorphism group of L (regarded as a three-dimensional icosian lattice). G is a double cover $2J_2$ of the Hall-Janko group $J_2 = HJ$ [Con16, p. 42], [Wil9]. In fact it follows from the definition of L that G contains a subgroup S (of order $2^7.3.3!$) generated by all permutations of x, y, z and by right multiplication of (x,y,z) by the diagonal matrices $\{1,i,i\}$, $\{i,j,k\}$, $\{\omega,\omega,\omega\}$. The minimal Euclidean norm of L is 4. Modulo scalar multiplication (on the left) by elements of the icosian group, there are four orbits of vectors in L of quaternionic norm 4, namely

$$(2,0,0),\ (0,h,h),\ (\bar{h},1,1),\ (1,\tau\omega,\sigma\bar{\omega}), \tag{6}$$

with 3, 24, 192 and 96 images, respectively, a total of 315. If r is one of these 315 "root" vectors and $v \in L$, then $(v,r) \in 2\mathcal{I}$ [Wil9, p. 163], and so the reflection in r,

$$v \to v - \frac{2(v,r)r}{(r,r)},$$

preserves L (so is in G), and preserves quaternionic norms and inner products. These 315 quaternionic reflections generate $2J_2$.

The minimal vectors in this version of the Leech lattice are the elements of L having quaternionic norm 4, $4\sigma^2$ or $2 + 2\sigma^2$. There are $120 \cdot 315 = 37800$ vectors of shape qr (where q is in the icosian group and r is a root) and quaternionic norm 4, an equal number of shape σqr and norm $4\sigma^2$, and $120 \cdot 1008 = 120960$ of shape qt and norm $2 + 2\sigma^2$, where t is one of the images of $(h,\sigma\omega h,0)$, $(1,1,\sigma\bar{\omega}h)$, $(\sigma\bar{\omega},\sigma\bar{\omega},\bar{h})$ or $(1,\sigma\bar{\omega},\sigma\omega^j+\omega^k)$ under S. (These have 48, 192, 192 and 576 images respectively.) For further information about this lattice and its group see [Wil9].

We remark that there is an easy and well-known construction of the Leech lattice from E_8, analogous to Turyn's construction of the Golay code from the Hamming code given in §12 of Chap. 11. In the usual MOG notation, the vectors v for which

(1)

v	v	0

, (2)

v	v	v

is in Λ_{24} form two E_8 lattices in the same space, say $E_8^{(1)}$ and $E_8^{(2)}$. We can now *define* Λ_{24} to be the lattice generated by the vectors

$$(a,b,c),\ a+b+c=0,\ a,\,b,\,c \in E_8^{(1)},$$
$$(x,x,x),\ x \in E_8^{(2)}.$$

The icosian construction of Λ_{24} is merely this Turyn-type construction equipped with extra structure. (See also [Coh2], [Cos1], [Lep2], [Tit6], [Tit7].)

3. A general setting for Construction A, and Quebbemann's 64-dimensional lattice

The following fairly general version of Construction A will be used in several examples. (It does not completely replace the versions of Chaps. 5 and 7, however, since it applies to linear codes and real lattices. Construction E of §9 is an even more general version.)

Let L be a lattice in $\mathbf{R}^n$ with an associated symmetric bilinear form f. Let p be a prime number which does not divide $\det L$, and set

$$\bar{L} = L/pL.$$

We denote the image of $\ell \in L$ in $\bar{L}$ by $\bar{\ell} = \ell + pL$. $\bar{L}$ has induced quadratic and bilinear forms given by $\bar{f}(\bar{\ell}) = f(\ell) \bmod p$, $\bar{f}(\bar{\ell},\bar{m}) = f(\ell,m) \bmod p$. (Note that if $p = 2$, there is no analog of Eq. (1) for $\bar{f}$.)

Let C be a subspace of $\bar{L}$. (C plays the role of the "code" in this construction.)

Construction A_f. We define a new lattice $\Lambda(C)$ in $\mathbf{R}^n$ by

$$\Lambda(C) = \{\ell \in L : \bar{\ell} \in C\}, \tag{7}$$

equipped with the bilinear form $f'(\ell,m) = p^{-1} f(\ell,m)$.

Clearly $pL \subseteq \Lambda(C) \subseteq L$. The dual lattice $\Lambda(C)^*$ is given by

$$\Lambda(C)^* = \left\{ x \in \mathbf{R}^n : \frac{1}{p} f(x, \Lambda(C)) \subseteq \mathbf{Z} \right\}. \tag{8}$$

The dual of C in $\bar{L}$, C^*, is defined with respect to $\bar{f}$:

$$C^* = \{x \in \bar{L} : \bar{f}(x,C) = 0\}.$$

Then it is immediate that

$$\Lambda(C)^* = \Lambda(C^*). \tag{9}$$

As usual the dual of L is defined with respect to the original form f:

$$L^* = \{x \in \mathbf{R}^n : f(x,L) \subseteq Z\}. \tag{10}$$

By abuse of notation we can regard C^*/C as a subgroup of $\Lambda(C^*)/\Lambda(C) = \Lambda(C)^*/\Lambda(C)$.

Theorem 1. *If* $C \subseteq C^*$ *then*

$$(\Lambda(C)^*/\Lambda(C))/(C^*/C) \cong L^*/L. \tag{11}$$

This implies

$$\det \Lambda(C) = \det \Lambda \cdot |C^*/C|. \tag{12}$$

If L is unimodular and $C = C^*$, $\Lambda(C)$ is unimodular.

Proof. $\bar{f}(c,c') = 0$ for c, $c' \in C$, so $f(\ell,\ell') \in p\mathbf{Z}$ for $\ell,\ell' \in \Lambda(C)$. Therefore $\Lambda(C) \subseteq \Lambda(C)^*$. Similarly $\Lambda(C)^* \subseteq L^*$, so

$$\Lambda(C) \subseteq \Lambda(C)^* \subseteq L^*.$$

We define a homomorphism

$$\theta : \Lambda(C)^*/\Lambda(C) \to L^*/L$$

by $\theta(v + \Lambda(C)) = v + L$. It is easy to check that θ is well-defined and (using $p \nmid \det L$) that θ is onto. Also $\ker \theta = \{v + \Lambda(C) : v + pL \in C^*\}$. Since $(\Lambda(C)/pL) \cap C^* = C$, $\ker \theta \cong C^*/C$, which completes the proof.

Examples (a). Taking $L = \mathbf{Z}^n$, $p = 2$ we obtain the linear version of Construction A (Chap. 7, §2).

(b). *Orthogonal geometries.* Let $M \subseteq \mathbf{R}^k$ be a lattice with a symmetric bilinear form f, let p be a prime not dividing det M, and let $\bar{M} = M/pM$. Suppose $\bar{M}$ has the structure of an orthogonal geometry, in particular that

$$\bar{M} = V \oplus V',$$

where V and V' are subspaces of $\bar{M}$ such that $\bar{f}(V) = \bar{f}(V') = 0$, and $V = V^*$, $V' = V'^*$ (with respect to $\bar{f}$).

Let $L = M^n \subseteq \mathbf{R}^{kn}$ with the bilinear form

$$f(\ell,m) = f((\ell_1,\ldots,\ell_n),(m_1,\ldots,m_n)) = \Sigma f(\ell_i,m_i).$$

We construct a self-dual "code" C as follows. Let B be an arbitrary subspace of V^n and set

$$B' = V'^n \cap B^{\perp} = \{v \in (V')^n : \bar{f}(v,B) = 0\},$$

$$C = B \oplus B'.$$

Then $C = C^*$ (with respect to $\bar{f}$). From Theorem 1, $\Lambda(C)$ (with the form $p^{-1} f$) is integral and

$$\Lambda(C)^*/\Lambda(C) \cong (M^*/M)^n. \tag{13}$$

(c) *The Leech lattice.* The Turyn-type construction of the Leech lattice given in §2.2 follows by taking $M = E_8$, f to be usual Euclidean inner product on E_8, and $p = 2$. Then $\bar{M} = M/2M$ is an 8-dimensional space over $\mathbf{F}_2$ which may be written $V \oplus V'$, where (in the icosian notation)

$$V = \langle h, ih, jh, \omega h\rangle, \quad V' = \langle \bar{h}, i\bar{h}, j\bar{h}, \omega\bar{h}\rangle.$$

We apply Construction A_f with $n = 3$, taking B to be the repetition code consisting of all (x,x,x), $x \in V$, and B' to consist of all $(x,y,z) \in (V')^3$ with $x + y + z = 0$. Then $\Lambda(C)$ is the lattice L defined in Eqs. (4), (5). It is now straightforward to check that this corresponds to an even unimodular 24-dimensional lattice of minimal norm 4 (compare the proof of Theorem 1 below), and so by Chap. 12 is the Leech lattice.

(d) *The lattice Q_{64}.* Quebbemann [Que5] used Construction A_f to obtain two important lattices in 32 and 64 dimensions. We postpone consideration of the 32-dimensional lattice to §4. The 64-dimensional lattice is obtained by taking M (in Example (b)) to be E_8, $k = 8$, $p = 3$ and $n = 8$. Then $\bar{M} = E_8/3E_8$ can be decomposed as $V \oplus V'$, where $V = \langle e_1,e_2,e_3,e_4\rangle$, $V' = \langle f_1,f_2,f_3,f_4\rangle$ and the e_i, f_i are given in Table 8.2. The coordinates in Table 8.2 are labeled to show the identification with the Hamming code version of E_8 (§8.1 of Chap. 4). The cosets $e_i + 3E_8$ and $f_i + 3E_8$ $(i = 1,\ldots,4)$ have minimal norm 6 (on the scale at which the minimal norm of E_8 is 2). Each pair $\{e_1,e_2\},\ldots,\{f_3,f_4\}$ generates a copy of the tetracode $\mathscr{C}_4$ (Chap. 3, §2.5.1). Aut$(\mathscr{C}_4)$ contains elements of order 8, and so there are automorphisms of $E_8/3E_8$ of order 8, such as

$$\pi = \{(12\infty4)(3506) \text{ then negate } 2,5\}$$

(with the coordinates labeled as in Table 8.2).

Let B be the subspace of V^8 spanned by the vectors (x, x, x, x, x, x, x, x), $x \in V$ and $(y, \pi y, \pi^2 y, \ldots, \pi^7 y)$, $y \in V$. From (11), B' consists of the vectors $(z_0, z_1, \ldots, z_7) \in (V')^8$ satisfying

$$z_0 + z_1 + \cdots + z_7 = 0, \tag{14}$$

$$z_0 + \pi^{-1} z_1 + \cdots + \pi^{-7} z_7 = 0. \tag{15}$$

We take $C = B \oplus B'$, and let $Q_{64} = \Lambda(C)$. Thus Q_{64} is spanned by the vectors

$$(3\ell, 0^7), \text{ where } \ell \text{ runs through a basis for } E_8, \tag{16}$$

$$(x^8),\ x \in \{e_1, e_2, f_1, f_2\}, \tag{17}$$

$$(y, \pi y, \ldots, \pi^7 y),\ y \in \{e_1, e_2, f_1, f_2\}, \tag{18}$$

$$(z_0, z_1, \ldots, z_7),\ z_i \in \{e_3, e_4, f_3, f_4\}, \text{ satisfying (14), (15)}. \tag{19}$$

Table 8.2. A basis for $E_8/3E_8$, an 8-dimensional vector space over the field $\mathbf{F}_3$ ($\bar{1}$ denotes -1).

		1	2	4	∞	3	5	6	0
V	e_1	1	1	1	0	0	0	0	0
	e_2	0	1	$\bar{1}$	1	0	0	0	0
	e_3	0	0	0	0	1	1	1	0
	e_4	0	0	0	0	0	1	$\bar{1}$	1
V'	f_1	$\bar{1}$	1	1	0	0	0	0	0
	f_2	0	1	$\bar{1}$	$\bar{1}$	0	0	0	0
	f_3	0	0	0	0	$\bar{1}$	1	1	0
	f_4	0	0	0	0	0	1	$\bar{1}$	$\bar{1}$

Theorem 2 [Que5]. *Q_{64} is an extremal even unimodular 64-dimensional lattice with minimal norm 6, kissing number 2611200 and center density $\delta = (3/2)^{24}$.*

Proof. (13) implies $\det Q_{64} = 1$, and the lattice is even by construction. Suppose there is a vector $v = (v_0, \ldots, v_7) \in Q_{64}$ with $\frac{1}{3} f(v,v) \leqslant 4$, i.e. $f(v,v) \leqslant 12$. Let $\bar{v} = (\bar{v}_0, \ldots, \bar{v}_7)$ where $\bar{v}_r = x + \rho^r y + z_r$, $x, y \in V$, $z_r \in V'$. Since $f(v_i, v_j)$ is even, there are at least two coordinates i, j for which $v_i = v_j = 0$ (otherwise $f(v,v) \geqslant 14$). Therefore $\pi^i y = \pi^j y$, so $y = 0$, $x = 0$ and $\bar{v}_r = z_r$. Since $z_r \in V'$, the norm of any element of $z_r + 3E_8$ is divisible by 6. Therefore at most two of the z_r are nonzero. (14), (15) now imply that all $z_r = 0$, and $v \in 3E_8^8$. Since $3E_8^8$ has minimal norm 18, we conclude that $v = 0$ and the minimal norm of Q_{64} is 6. The kissing number is then determined by Theorem 17 of Chap. 7.

Quebbemann's construction is slightly different from the one given here, and it is likely that there are several inequivalent lattices with these parameters. The existence of at least one such lattice was established in [Que3].

4. Lattices over $\mathbf{Z}[e^{\pi i/4}]$, and Quebbemann's 32-dimensional lattice

Sections 8-10 of Chap. 7 describe lattices over the Eisenstein and Gaussian integers and their connections with codes over the fields $\mathbf{F}_2$, $\mathbf{F}_3$ and $\mathbf{F}_4$. Quebbemann [Que6] has developed a similar theory relating lattices over the ring of cyclotomic integers $\mathbf{Z}[\zeta]$, where $\zeta = e^{\pi i/4} = (1+i)/\sqrt{2}$, to codes over $\mathbf{F}_9$. The results are not exactly parallel to those in Chap. 7, however, and it seems best to treat this case separately. The most interesting example is an 8-dimensional $\mathbf{Z}[\zeta]$-lattice which produces the 32-dimensional lattice Q_{32}. This is obtained from a complex analog of Construction A_f, replacing the symmetric bilinear form f by a Hermitian form.

The lattice $\mathbf{Z}[\zeta]$ *itself.* We begin by showing that $\mathbf{Z}[\zeta]$ can be regarded as a version of the two-dimensional Gaussian lattice $\mathcal{D}_2$ (Example 13a of Chap. 7) or of the four-dimensional real lattice D_4. It is plausible that $\mathbf{Z}[\zeta]$ corresponds to a four-dimensional real lattice, since the corresponding field $\mathbf{Q}(\zeta)$ is an extension of the rationals of degree 4, with basis

$$1,\ i,\ \sqrt{2},\ i\sqrt{2}. \tag{20}$$

The cyclotomic number $\zeta = e^{\pi i/4}$ satisfies $\zeta^2 = i$, $\zeta^4 = -1$, $\zeta^8 = 1$, and has minimal polynomial $X^4 + 1$ (over $\mathbf{Z}$). Elements of $\mathbf{Z}[\zeta]$ may be written either as

$$\alpha = \beta + \gamma\zeta, \quad \beta, \gamma \in \mathbf{Z}[i], \tag{21}$$

or as

$$\alpha = \alpha_0 + \alpha_1\zeta + \alpha_2\zeta^2 + \alpha_3\zeta^3, \quad \alpha_0, \alpha_1, \alpha_2, \alpha_3 \in \mathbf{Z}. \tag{22}$$

To identify $\mathbf{Z}[\zeta]$ with $\mathcal{D}_2$ we regard $\mathbf{Z}[\zeta]$ as a two-dimensional lattice over $\mathbf{Z}[i]$ with basis $1, \zeta$, and simply define a Hermitian inner product ϕ on $\mathbf{Z}[\zeta]$ by specifying that the Gram matrix of the basis is

$$\begin{bmatrix} 2 & 1-i \\ 1+i & 2 \end{bmatrix}. \tag{23}$$

Thus if $\alpha = \beta + \gamma\zeta$, $\alpha' = \beta' + \gamma'\zeta$, where $\beta, \beta', \gamma, \gamma' \in \mathbf{Z}[i]$,

$$\phi(\alpha,\alpha') = \beta\overline{\beta'}\phi(1,1) + \beta\overline{\gamma'}\phi(1,\zeta) + \gamma\overline{\beta'}\phi(\zeta,1) + \gamma\overline{\gamma'}\phi(\zeta,\zeta), \tag{24}$$

where $\phi(1,1) = \phi(\zeta,\zeta) = 2$, $\phi(1,\zeta) = \overline{\phi(\zeta,1)} = 1 - i$. We define norms in $\mathbf{Z}[\zeta]$ by $N(\alpha) = \phi(\alpha,\alpha)$.

Since (23) is the same as Eq. (88) of Chap. 7, $\mathbf{Z}[\zeta]$ with this inner product *is* $\mathcal{D}_2$. The minimal norm is 2, and the 24 elements of norm 2 are

ζ^ν, $\zeta^\nu(1-\zeta)$ and $\zeta^\nu(1-\sqrt{2}) = \zeta^\nu(1-\zeta+\zeta^3)$ for $\nu = 0, 1, \ldots, 7$ (forming three orbits of size 8 under multiplication by ζ). Although not as simple as the familiar representation of the minimal vectors of D_4 (the permutations of $(\pm 1, \pm 1, 0, 0)$), this set of 24 points forms a pleasantly symmetric configuration in the complex plane (Fig. 8.1). To recover the familiar coordinates we map

in $\mathbf{Z}[\zeta]$		in $\mathbf{Z}[i]^2$			in $\mathbf{Z}^4$
1	to	1	1	or	1 1 0 0
ζ		1	i		0 1 1 0
ζ^2		i	i		0 0 1 1
ζ^3		i	−1		−1 0 0 1

The inner product in $\mathbf{Z}^4$ is the real part of the inner product in $\mathbf{Z}[\zeta]$ or $\mathbf{Z}[i]^2$. Multiplication by ζ in $\mathbf{Z}[\zeta]$ corresponds to interchanging the two coordinates of $\mathbf{Z}[i]^2$ and multiplying the second coordinate by i, and to the operation "(0123) then negate 3" on $\mathbf{Z}^4$. The vectors of norm 4 in $\mathbf{Z}[\zeta]$ are obtained by multiplying the vectors of norm 2 by $\zeta + \zeta^{-1} = \sqrt{2}$. The theta series of $Z[\zeta]$ is given by Eq. (89) of Chap. 7.

We can write ϕ in another way. The trace map Tr from $\mathbf{Q}(\zeta)$ to $\mathbf{Q}(i)$ sends $a + bi + c\sqrt{2} + di\sqrt{2}$ (with a, b, c, $d \in \mathbf{Q}$) to $2a + 2bi$. For $\alpha = \beta + \gamma\zeta \in \mathbf{Q}(\zeta)$, $\beta, \gamma \in Q(i)$, we define

$$T(\alpha) = Tr(\alpha(1+2^{-1/2})) = Tr((\beta + \gamma\zeta)(1+2^{-1/2}))$$
$$= 2\beta + (1+i)\gamma, \tag{25}$$

a map from $\mathbf{Q}(\zeta)$ to $\mathbf{Q}(i)$. Then

$$\phi(\alpha, \alpha') = T(\alpha\overline{\alpha'}), \quad N(\alpha) = T(\alpha\bar{\alpha}). \tag{26}$$

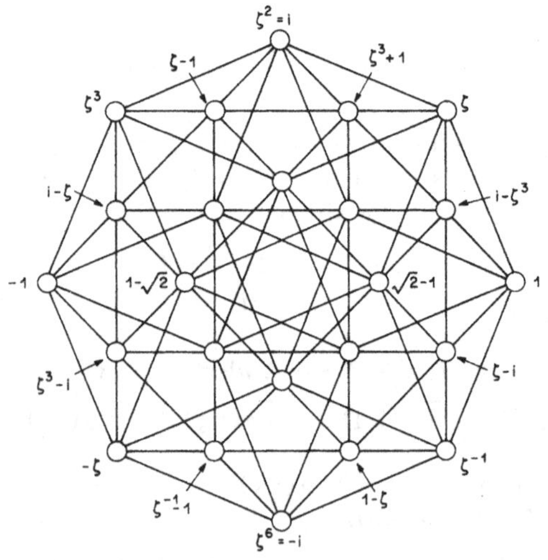

Figure 8.1. A projection of the polytope $\{3,4,3\}$ whose vertices are the 24 minimal vectors of D_4, here represented as elements of $\mathbf{Z}[\zeta]$.

The field $\mathbf{F}_9$. By a slight abuse of notation we can also regard the 8-th roots of unity ζ^ν as the nonzero elements of the field of order 9. To be more precise we let $\mathbf{F}_9$ consist of the elements 0, 1, ζ, $\zeta^2 = 1 - \zeta$, $\zeta^3 = -1 - \zeta$, $\zeta^4 = -1$, $\zeta^5 = -\zeta$, $\zeta^6 = -1 + \zeta$, $\zeta^7 = 1 + \zeta$, where $\zeta^2 + \zeta - 1 = 0$, $\zeta^8 = 1$ [Mac6, p. 95]. Now $\mathbf{Z}[i]/3\mathbf{Z}[i]$ is identified with $\mathbf{F}_9$ and $\mathbf{Z}[\zeta]/3\mathbf{Z}[\zeta]$ with $\mathbf{F}_9 \times \mathbf{F}_9$. The homomorphism $\sigma : Z[\zeta] \to Z[\zeta]/3\mathbf{Z}[\zeta] \to \mathbf{F}_9 \times \mathbf{F}_9$ is given by

$$\sigma(\zeta) = (\zeta, -\zeta) \tag{27}$$

or, using the representation (22), by

$$\sigma(\alpha) = (\alpha_0 + \alpha_1\zeta + \alpha_2\zeta^2 + \alpha_3\zeta^3,\ \alpha_0 - \alpha_1\zeta + \alpha_2\zeta^2 - \alpha_3\zeta^3)\,. \tag{28}$$

The following construction is the appropriate analog of Construction A_f.

Construction A_ϕ. If B and C are codes of length n over $\mathbf{F}_9$ with $C \subseteq B^*$, we define

$$\Lambda(B,C) = \{v \in \mathbf{Z}[\zeta]^n : \sigma(v) \in B \times C\} \tag{29}$$

with the hermitian inner product

$$\frac{1}{3}(\phi(v_1,w_1) + \cdots + \phi(v_n,w_n))\,. \tag{30}$$

The corresponding real lattice $\Lambda(B,C)_{\mathrm{real}}$ is obtained by regarding $\Lambda(B,C)$ as a sublattice of $\mathcal{D}_2^n \subseteq \mathbf{C}^{2n}$, and writing down real and imaginary parts (cf. Eq. (66) of Chap. 2). Then

$$\det \Lambda(B,C)_{\mathrm{real}} = 4^n (\det \Lambda(B,C))^4\,. \tag{31}$$

Next we shall relate the norms in $\Lambda(B,C)$ to the weights of the codewords in B and C. The map T of (25) induces a map $t : \mathbf{F}_9 \times \mathbf{F}_9 \to \mathbf{F}_9$; it is given by

$$t(b,c) = \zeta b + \zeta^3 c\,, \quad \text{for } b, c \in \mathbf{F}_9$$

(see Table 8.3). There is a corresponding "norm" $N : \mathbf{F}_9 \times \mathbf{F}_9 \to \mathbf{F}_3$ defined analogously to (26). For $a = (b,c) \in \mathbf{F}_9 \times \mathbf{F}_9$, let $\hat{a} = (c^3, b^3)$. The component-wise product $a\,\hat{a}$ is (bc^3, cb^3), and we define the norm of a by

$$N(a) = t(a\,\hat{a}) = \zeta bc^3 + (\zeta bc^3)^3\,, \tag{32}$$

which takes values in $\mathbf{F}_3 = \{0, 1, -1\}$.

Theorem 3 [Que6]. *The elements in* $\mathbf{Z}[\zeta]$ *of norm* 2 *map* 1-*to*-1 *under* σ *onto the elements of norm* -1 *in* $\mathbf{F}_9 \times \mathbf{F}_9$; *the elements of norm* 4 *map* 1-*to*-1 *onto the elements of norm* 1; *and the elements of norm* 6 *map* 3-*to*-1 *onto the nonzero elements of norm* 0.

Sketch of proof. This comes from the fact that $D_4/3D_4$ consists of 81 congruence classes: the zero class, 24 classes represented by single vectors

Table 8.3. $N((b,c))$ for $b, c \in \mathbf{F}_9$.

$b\backslash c$	0	1	ζ	ζ^2	ζ^3	ζ^4	ζ^5	ζ^6	ζ^7
0	0	0	0	0	0	0	0	0	0
1	0	−1	1	1	0	1	−1	−1	0
ζ	0	0	1	−1	−1	0	−1	1	1
ζ^2	0	−1	0	−1	1	1	0	1	−1
ζ^3	0	1	1	0	1	−1	−1	0	−1
ζ^4	0	1	−1	−1	0	−1	1	1	0
ζ^5	0	0	−1	1	1	0	1	−1	−1
ζ^6	0	1	0	1	−1	−1	0	−1	1
ζ^7	0	−1	−1	0	−1	1	1	0	1

of norm 2, 24 classes represented by single vectors of norm 4, and 32 classes each containing precisely three vectors of norm 6. The vectors in $\mathbf{Z}[\zeta] \cong D_4$ of nonzero norm (mod 3) map into vectors of nonzero norm in $\mathbf{F}_9 \times \mathbf{F}_9$.

Theorem 3 makes it possible to find the minimal norm in $\Lambda(B,C)$. We see that for each pair of codewords $b = (b_1, \ldots, b_n) \in B$, $c = (c_1, \ldots, c_n) \in C$, we must determine the number of coordinates i with $(b_i,c_i) \neq (0,0)$ for which $N(b_i,c_i)$ takes the values -1, 1 and 0: let these numbers be $p_1(b,c)$, $p_2(b,c)$ and $p_3(b,c)$ respectively. Let $p_0(b,c)$ be the number of i for which $(b_i,c_i) = (0,0)$, so that

$$n = \sum_{\nu=0}^{3} p_\nu(b,c). \tag{33}$$

If $v \in \mathbf{Z}[\zeta]^n$ is such that $\sigma(v) = (b,c) \neq (0,0)$, then, from Theorem 3,

$$N(v) \geqslant 2p_1(b,c) + 4p_2(b,c) + 6p_3(b,c), \tag{34}$$

and the number of v for which equality holds in (34) is $3^{p_3(b,c)}$. We call the quadruple $(p_0(b,c), p_1(b,c), p_2(b,c), p_3(b,c))$ the *type* of (b,c), and set

$$p(b,c) = 2p_1(b,c) + 4p_2(b,c) + 6p_3(b,c). \tag{35}$$

Furthermore one can show (as in the proof of Theorem 1) that if $b_1\bar{c}_1 + \cdots + b_n\bar{c}_n = 0$ then $p(b,c)$ is a multiple of 3, which justifies the division by 3 in (30), and proves that $\Lambda(B,C)$ is an integral $\mathbf{Z}[\zeta]$-lattice. Since $p(b,c)$ is obviously even, $\Lambda(B,C)$ is in fact an *even* integral $Z[\zeta]$-lattice. Also $\Lambda(B,C)^* = \Lambda(C^*,B^*)$, so if $C = B^*$ then Λ is an even unimodular $\mathbf{Z}[\zeta]$-lattice.

The elements $v \in \mathbf{Z}[\zeta]^n$ such that $\sigma(v) = (0,0)$ belong to $3\mathbf{Z}[\zeta]^n$, a sublattice of $\Lambda(B,C)$ with minimal norm 6 (using (30)). Therefore (from

(34)) the minimal norm of $\Lambda(B,C)$ is

$$\min\{d/3, 6\}, \tag{36}$$

where

$$d = \min\{p(b,c) : b \in B,\ c \in C,\ (b,c) \neq (0,0)\} \tag{37}$$

is a kind of "minimal distance" for the pair of codes (B,C). We summarize the preceding results:

Theorem 4 [Que6]. *$\Lambda(B,C)$ is an even integral $Z[\zeta]$- lattice and is unimodular if $C = B^*$. The minimal norm is given by* (36), (37). *The kissing number is given by*

$$\tau = \sum 3^{p_3(b,c)}, \quad \text{if } d < 18, \tag{38}$$

the sum being over all $b \in B$, $c \in C$ with $p(b,c) = d$, or by

$$\tau = 24n, \quad \text{if } d > 18, \tag{39}$$

or by the sum of these two expressions if $d = 18$.

Quebbemann also gives a result analogous to Theorem 3 of Chap. 7 for the theta series of $\Lambda(B,C)$.

Example. The Barnes-Wall lattice Λ_{16} again. Let B be the [4, 1, 4] repetition code over $\mathbf{F}_9$ and let $C = B^*$ be the [4, 3, 2] zero-sum code. If $b \neq 0$, b has four non-zero components and (from Table 8.3) $p(b, c) \geq 8$. If $b = 0$, $c \neq 0$, $p(b, c) \geq 12$. Since $6 \mid p(b, c)$ we conclude that d, the minimal value of $p(b, c)$ for $(b, c) \neq (0, 0)$, is 12. Therefore $\Lambda(B, C)_{\text{real}}$ is a 16-dimensional lattice with minimal norm $12/3 = 4$ and determinant 4^4 (from (31)).

To determine the kissing number we must find all $(b,c) \in B \times C$ with $p(b,c) = 12$. The possible types (p_0,p_1,p_2,p_3) are (from (33), (35)) $(0,2,2,0)$, $(0,3,0,1)$ and $(2,0,0,2)$. Consider type $(0,3,0,1)$, and assume $b = (1,1,1,1)$. Then to have $p_1(b,c) = p_2(b,c) = 2$ we see from Table 8.3 that c must contain two elements from $\{1,\zeta^5=-\zeta,\zeta^6=-1+\zeta\}$ and two from $\{\zeta^4=-1,\zeta,\zeta^2=1-\zeta\}$, with the sum of all four equal to 0. There are 162 possibilities, and the total number of pairs (b,c) of this type is $8 \cdot 162 = 1296$. A typical pair is $b = (1\,1\,1\,1)$, $c = (1,-\zeta,1-\zeta,1-\zeta)$, and the unique $v \in Z[\zeta]^n$ with $\sigma(v) = (b,c)$ and $N(v) = 12$ is

$$(1,\zeta^3-\zeta^2,-1+2\zeta-\zeta^2,-1+2\zeta-\zeta^2)\ .$$

The corresponding vector in $\mathbf{R}^{16}$ is

$$(1\ 0\ 1\ 0,\ 0\ 0\ -1\ -1,\ 1\ -1\ -1\ 1,\ 1\ -1\ -1\ 1)\,.$$

The number of (b,c) of type $(0,3,0,1)$ is 864, and the number of type $(2,0,0,2)$ is 48. From (38) the kissing number is $1296 + 864 \cdot 3 + 48 \cdot 3^2 = 4320$. $\Lambda(B,C)_{\text{real}}$ is the Barnes-Wall lattice Λ_{16} (§10 of Chap. 4).

There is also an analog of Theorems 17 and 32 of Chap. 7, with a similar proof.

Theorem 5 [Que6]. *If* Λ *is an even unimodular* $\mathbf{Z}[\zeta]$*-lattice then*

$$\Theta_\Lambda(z) \in \mathbf{C}\left[\Theta_{D_4}(z),\, \Delta_{16}(z)\right] \tag{40}$$

where

$$\Delta_{16}(z) = \frac{1}{96}\left\{\Theta_{\Lambda_{16}}(z) - \Theta_{D_4}(z)^4\right\} = q^2 \prod_{m=1}^{\infty} \{(1-q^{2m})(1-q^{4m})\}^8$$

$$= q^2 - 8q^4 + 12q^6 + 64q^8 - 210q^{10} - 96q^{12} + \cdots . \tag{41}$$

Remark. The theta series in Theorem 17 of Chap. 7, in the preceding theorem, and in Theorem 32 of Chap. 7 are respectively modular forms for the groups $H(1) \cong SL_2(\mathbf{Z})$, $H(2)$ and $H(3)$, where the Hecke group $H(\lambda)$ is generated by $z \to z + \lambda$ and $z \to -1/z$ ([Hec3, p. 609], [Ogg1, p. xiii], [Que6]).

As in Chap. 7 we may now define extremal theta series and lattices. An n-dimensional extremal $\mathbf{Z}[\zeta]$-lattice has minimal norm $2[n/4] + 2$, and [Que6] gives examples for $n \leqslant 11$. The initial examples are $\mathbf{Z}[\zeta]^n$ for $n = 1, 2, 3$, and then $\Lambda(B, B^*)$ where B is the repetition code of length n, for $4 \leqslant n \leqslant 7$. The next example is the most interesting.

Example. The 32*-dimensional real lattice* Q_{32}. For $n = 8$ Quebbemann [Que6] takes B to be the [8,2,7] Reed-Solomon code over $\mathbf{F}_9$ (Chap. 3, §2.5) generated by $(111...1)$ and $(1\omega\omega^2 ... \omega^7)$, so that $C = B^*$ is an [8,6,3] Reed-Solomon code. If $b \neq 0$ then $p(b,c) \geqslant 2 \cdot 7 = 14$, and if $b = 0$, $c \neq 0$ then $p(b,c) \geqslant 2 \cdot 3 \cdot 6 = 36$. Thus $d = 18$, so $\Lambda(B,C)$ has minimal norm 6, and is extremal. Its theta series is

$$\Theta_{D_4}(z)^8 - 192\Theta_{D_4}(z)^4\Delta_{16}(z) + 576\Delta_{16}(z)^2$$

$$= 1 + 261120q^6 + 18947520q^8 + 535818240q^{10}$$

$$+ 8320327680q^{12} + 83347937280q^{14} + \cdots , \tag{42}$$

Quebbemann found that there are 448 pairs $(b,c) \in B \times C$ of type $(5,0,0,3)$, 36288 of type $(1,6,0,1)$, 108864 of type $(1,5,2,0)$ and 31104 of type $(0,7,1,0)$, and so from (38) + (39) the kissing number is 261120, in agreement with (42).

The corresponding real lattice is a 32-dimensional lattice Q_{32} with determinant $= 2^{16}$, minimal norm $= 6$, kissing number $= 261120$, center density $\delta = 2^{-24}\, 3^{16} = 2.5658...$ and theta series (42). This is the densest packing known in 32 dimensions. Q_{32} was originally constructed in [Que5] by the method described in §3. That method also yields lattices Q_{30} in $\mathbf{R}^{30}$ with $\delta = 2^{-22}\, 3^{13.5} = 0.6584...$ and Q_{31} in $\mathbf{R}^{31}$ with $\delta = 2^{-23.5}\, 3^{15} = 1.2095...$ (see Fig. 1.5). A slightly less dense 32-dimensional lattice is constructed in §8.2(h) below.

5. McKay's 40-dimensional extremal lattice

McKay [McK2] describes a construction for the Leech lattice that also produces a 40-dimensional even unimodular lattice of minimal norm 4. Let $p \equiv -1 \pmod 4$ be prime and let H be a Hadamard matrix (§2.13 of Chap. 3) of order $p+1$ of the form $H = S - I$, where S is skew-symmetric. (This implies $S^2 = -pI$.) Let $n = 2(p+1) = 8k$. Then an n-dimensional lattice M_n is defined by the generator matrix

$$\frac{1}{\sqrt{k+1}} \begin{bmatrix} (k+1)I & 0 \\ S-2I & I \end{bmatrix}. \tag{43}$$

The Gram matrix is

$$\begin{bmatrix} (k+1)I & -S-2I \\ S-2I & 4I \end{bmatrix}, \tag{44}$$

so M_n is an integral even unimodular lattice if k is odd. It is a straightforward but tedious calculation (which we omit) to verify that if $k = 3$ or 5 then M_n has minimal norm 4. For $k = 1$, 3 and 5 there is essentially only one way to choose S [Lon1]. The resulting lattices are $M_8 \cong E_8$, $M_{24} \cong \Lambda_{24}$ and M_{40}, an extremal Type II lattice.

A slightly nicer description of M_{40} is obtained if the first column of (43) is negated. Thus as generator matrix for M_{40} we take

$$\frac{1}{\sqrt{6}} \begin{bmatrix} 6I & O \\ B & I \end{bmatrix}, \tag{45}$$

where B is constructed as follows. Let the rows and columns of B be labeled $\{\infty, 0, 1, \ldots, 18\}$. Then $B_{\infty\infty} = 2$, $B_{ii} = -2$ and $B_{\infty i} = B_{i\infty} = 1$ for $i \geqslant 0$, $B_{ij} = 1$ if $j - i$ is a quadratic residue modulo 19 and $B_{ij} = -1$ if $j - i$ is a nonresidue, for $i \neq j$, $i, j \geqslant 0$. M_{40} has det $= 1$, minimal norm $= 4$, kissing number $= 39600$ (the minimal vectors being given in Table 8.4), packing radius $= 1$, center density $\delta = 1$ (to be compared with $\delta = 4$ for the lattice Λ_{40} of Chaps. 5 and 6), and theta series

$$\begin{aligned} &E_4(z)^5 - 1200\, E_4(z)^2 \Delta_{24}(z) \\ &= 1 + 39600 q^4 + 87859200 q^6 + 20779902000 q^8 + \cdots . \end{aligned} \tag{46}$$

To describe the minimal vectors and the automorphism group we represent M_{40} as a 20-dimensional complex lattice, a typical vector being written $x = (x_\infty x_0 x_1 \ldots x_{18})$. Then the 20 real parts give the left-hand half of (45), and the 20 imaginary parts the right-hand half.

Let $Q = \{1,4,5,6,7,9,11,16,17\}$, $N = \{2,3,8,10,12,13,14,15,18\}$, and set $\pi(t) = +1$ if $t \in \{0\} \cup Q$, $\pi(t) = -1$ if $t \in \{\infty\} \cup N$. Then $\mathrm{Aut}(M_{40})$ has a subgroup isomorphic to $2.PGL_2(19)$ of order 13680 (this may in fact be all of $\mathrm{Aut}(M_{40})$) consisting of the transformations

$$\alpha_A : x_t \to \sqrt{\pi(ad-bc)}\, \pi(ct+d)\, x_{\frac{at+b}{ct+d}}, \tag{47}$$

for $A = \begin{pmatrix} a & b \\ c & d \end{pmatrix}$, $a, b, c, d \in \mathbf{F}_{19}$, $ad - bc = \pm 1$, where either square root may be taken. For example, if $A = \begin{pmatrix} 1 & 1 \\ 0 & 1 \end{pmatrix}$, $\alpha_A : x_t \to x_{t+1}$; if $A = \begin{pmatrix} -1 & 0 \\ 0 & 1 \end{pmatrix}$, $\alpha_A : x_t \to ix_{-t}$, which swaps the two halves of (45). In this complex notation M_{40} is generated (over $\mathbf{Z}$) by the vectors of type $(6, 0^{19})$, $(6i, 0^{19})$, and the images of $(2+i, 1^{19})$ under the transformations (47).

In Table 8.4 we write the components of the minimal vectors in the order $x_\infty x_0 [x_1 x_2 x_4 x_8 \cdots x_{10}]$, where the brackets enclose the cycle of length 18 under $x_t \to x_{2t}$. It seems that there are seven orbits of minimal vectors, and for each orbit we give a representative vector, the order of the subgroup of $2 \,.\, PGL_2(19)$ fixing it, and the size of the orbit. However, we have not been able to exclude the possibility that the group is slightly larger than indicated, and that some of the orbits fuse. Note that this is not a Gaussian lattice in the sense of Chap. 2, §2.6, since it is not closed under multiplication by i.

Table 8.4. The minimal vectors of M_{40} ($a = 1+i$, $b = 1-i$, $c = -1+i$, $d = -1-i$). The coordinates are in the order $x_\infty x_0 \, [x_1 x_2 x_4 \ldots x_5 x_{10}]$.

Coordinates	Stabilizer	Number
$2+i,\ 1\,[1^{18}]$	171	80
$a,\ 2i\,[(a,0)^9]$	9	1520
$i,\ 2+i\,[(i,1)^9]$	9	1520
$2+i,\ 1\,[(1,i,i,i,-i,-1)^3]$	3	4560
$2i,\ b\,[(c,0,a,0,0,b)^3]$	3	4560
$2i-1,\ -1\,[1,-1,1^2,-i,i^5,1^2,-1,-i,1,-1,-i,i]$	1	13680
$2,\ -2i\,[c,0,0,d,d,b,a,0,b,b,a,0^7]$	1	13680

6. Repeated differences and Craig's lattices

Let Λ be an n-dimensional lattice in $\mathbf{R}^m$, with $m \geqslant n$. A *simplicial basis* for Λ is a set of $n+1$ vectors $v_0, v_1, \ldots, v_n$ which span Λ and satisfy $v_0 + v_1 + \cdots + v_n = 0$. Then the *difference lattice* $\Delta\Lambda$ (with respect to this basis) is spanned by the vectors $v_i - v_j$. Clearly $\Delta\Lambda \subseteq \Lambda$.

Let $\tilde{M}$ be the $(n+1) \times m$ matrix with rows $v_0, \ldots, v_n$. By omitting the last row we obtain a generator matrix M for Λ. If $\tilde{\Delta}$ and Δ denote the matrices

$$\tilde{\Delta} = \begin{bmatrix} 1 & -1 & 0 & \cdots & 0 & 0 \\ 0 & 1 & -1 & \cdots & 0 & 0 \\ \cdot & \cdot & \cdot & \cdots & \cdot & \cdot \\ 0 & 0 & 0 & \cdots & 1 & -1 \\ -1 & 0 & 0 & \cdots & 0 & 1 \end{bmatrix}, \quad \Delta = \begin{bmatrix} 1 & -1 & 0 & \cdots & 0 & 0 \\ 0 & 1 & -1 & \cdots & 0 & 0 \\ \cdot & \cdot & \cdot & \cdots & \cdot & \cdot \\ 0 & 0 & 0 & \cdots & 1 & -1 \\ 1 & 1 & 1 & \cdots & 1 & 2 \end{bmatrix}$$

of sizes $(n+1) \times (n+1)$ and $n \times n$ respectively, then the rows of $\tilde{\Delta}\tilde{M}$

form a simplicial basis for $\Delta\Lambda$, and ΔM is a generator matrix for $\Delta\Lambda$. Although $\Delta\Lambda$ depends on the particular simplicial basis used, its determinant does not.

Theorem 6.

$$\det \Delta\Lambda = (n+1)^2 \det \Lambda . \tag{48}$$

Proof. This follows since $\det \Delta = n + 1$.

Examples (a). For A_n^* we may take

$$\tilde{M} = (n+1)^{-1/2}\,((n+1)I_{n+1} - J)\,, \tag{49}$$

where J is the all-ones matrix. Then $\Delta A_n^* = \sqrt{n+1}\,A_n$. The minimal norm has increased from n to $2n + 2$.

(b) *Craig's lattices.* The following lattices were first described by Craig [Cra5] using the construction of §7.3c. As a simplicial basis for A_n we take the rows of $\tilde{\Delta}$ itself, and define $A_n^{(m)} = \Delta^{m-1} A_n$ for $m = 1, 2, \ldots$. Thus the rows of $\tilde{\Delta}^m$ form a simplicial basis for $A_n^{(m)}$. For example $A_6^{(3)}$ is spanned by the rows of

$$\begin{bmatrix} 1 & -3 & 3 & -1 & 0 & 0 & 0 \\ 0 & 1 & -3 & 3 & -1 & 0 & 0 \\ \cdot & \cdot & \cdot & \cdot & \cdot & \cdot & \cdot \\ 3 & -1 & 0 & 0 & 0 & +1 & -3 \\ -3 & 3 & -1 & 0 & 0 & 0 & 1 \end{bmatrix}.$$

(These lattices should not be confused with Coxeter's lattices A_n^r [Cox10]; Coxeter's A_n^r is our $A_n[s]$, where $rs = n + 1$.)

Theorem 7 [Cra5]. *$A_n^{(m)}$ has determinant $(n+1)^{2m-1}$. If $n + 1 = p$ is a prime and $m < \frac{1}{2}n$, $A_n^{(m)}$ has minimal norm at least $2m$.*

Proof. The first assertion follows from Theorem 6. Since $A_n^{(m)} \subseteq A_n$, the minimal norm is even. To find the minimal norm we represent vectors $u = (u_0 u_1 \ldots u_n) \in A_n^{(m)}$ by polynomials $u(X) = u_0 + u_1 X + \cdots + u_n X^n$ in the ring $R = \mathbf{Z}[X]/(X^p - 1)$. Then $A_n^{(m)}$ is represented by the principal ideal $((X-1)^m)$ in R. Suppose $u \in A_n^{(m)}$ has norm $\leqslant 2m - 2$. Then there are two subsets $S, T \subseteq \{0, 1, \ldots, n\}$ of size $m - 1$, possibly containing repeated elements, such that

$$u(X) = \sum_{s \in S} X^s - \sum_{t \in T} X^t .$$

Now $u(X) = a(X)(X-1)^m$ in R, for some $a(X)$, so

$$u(X) = a(X)\,(X-1)^m + b(x)\,(X^p - 1) \text{ in } \mathbf{Z}[X].$$

By repeatedly differentiating this expression we find that

$$u^{(i)}(1) \equiv 0 \pmod{p} \text{ for } i = 0, \ldots, m-1 .$$

Hence

$$\sum_{s \in S} s^i \equiv \sum_{t \in T} t^i \pmod p, \ i = 0, \ldots, m-1 .$$

Therefore by Newton's identities [Mac6, p. 245], [Wae2, I, p. 81], the elementary symmetric functions of S and T of degree $< m$ agree, and so the polynomials

$$\prod_{s \in S} (X-s), \ \prod_{t \in T} (X-t)$$

coincide over $\mathbf{F}_p$. Since a polynomial of degree $m-1$ over a field has at most $m-1$ zeros, we conclude that $S = T$, and $u = 0$. This completes the proof.

It can be shown that the Gram matrix for $A_n^{(m)}$ is obtained from $\tilde{\Delta}^m \, (\tilde{\Delta}^m)^{tr}$ by omitting the last row and column, and has (i, j) entry equal to

$$\pm \begin{bmatrix} 2m \\ k \end{bmatrix} , \tag{50}$$

where $k = m + |i-j| \pmod{n+1}$, and the sign is $(-1)^{i+j}$ if $|i-j| \le n/2$, $(-1)^{n+1+i+j}$ if $|i-j| > n/2$.

When $n+1$ is a prime, Theorem 7 gives a lower bound on the density of $A_n^{(m)}$. The lower bound is maximized by taking $m = m_0$, the nearest integer to $\frac{1}{2} n / \log_e(n+1)$. Examples are given in Table 1.3, where it can be seen that these are the densest packings known for $148 \leqslant n \leqslant 3000$. For large n the density Δ satisfies

$$\frac{1}{n} \log_2 \Delta \ge -\frac{1}{2} \log_2 \log_2 n + o(1) ,$$

which (assuming that the lower bound is close to the actual value) is inferior to the lattices constructed in §9 (compare Table 1.4).

7. Lattices from algebraic number theory

7.1 Introduction. The connections between lattices and algebraic number theory have been studied by many authors from Minkowski onwards. In this section we describe some of the principal ways in which lattices can be obtained from ideals in algebraic number fields. This account has been influenced by the work of Lekkerkerker [Lek1, §4], Craig [Cra3]-[Cra5], Thompson [Tho7] and Litsin and Tsfasman [Lit3] (see also [Bay1], [Feil], [Que1], [Tau1]). For background information in algebraic number theory see for example [Cas5], [Coh6], [Has1], [Hec0], [Jan7], [Lan7], [LeV1], [Nar1], [Rib1], [Wae2], [Wei2].

7.2 Lattices from the trace norm. Let $K = \mathbf{Q}(\theta)$, of degree n, be an algebraic number field which is Galois with primitive element θ. Let the conjugates of a typical element $\alpha \in K$ be $\alpha = \alpha^{(1)}, \ldots, \alpha^{(n)}$, so that $(1), \ldots, (n)$ are names for the elements of the Galois group $G = Gal(K/\mathbf{Q})$. The *trace* and *norm* of $\alpha \in K$ are

$$Tr_{K/\mathbf{Q}}(\alpha) = \alpha^{(1)} + \cdots + \alpha^{(n)}, \tag{51}$$

$$\text{Norm}(\alpha) = \alpha^{(1)} \cdots \alpha^{(n)} . \tag{52}$$

We apologize for the confusion caused by the multiple uses of the word "norm". We use N for the norm of a lattice vector (§2.1 of Chap. 2) and Norm when the word is used in the sense of algebraic number theory.

The two most interesting cases for our purpose occur when K is either *totally real* (all the conjugates $\theta^{(j)}$ are real) or *totally complex* (none of the $\theta^{(j)}$ are real). If K is totally complex then $n = 2s$, say, and we label G so that

$$\alpha^{(s+1)} = \overline{\alpha^{(1)}}, \ldots, \alpha^{(2s)} = \overline{\alpha^{(s)}}$$

and "$(s+1)$" is complex conjugation. Elements $\alpha \in K$ may be represented as vectors in $\mathbf{R}^n$ by the mappings

$$\alpha \rightarrow v(\alpha) = (\alpha^{(1)}, \alpha^{(2)}, \ldots, \alpha^{(n)}) \tag{53}$$

if K is totally real, or

$$\alpha \rightarrow v(\alpha) = (\text{Re}\,\alpha^{(1)}, \text{Im}\,\alpha^{(1)}, \ldots, \text{Re}\,\alpha^{(s)}, \text{Im}\,\alpha^{(s)}) \tag{54}$$

if K is totally complex. The (lattice) norm of $v(\alpha)$ is

$$N(v(\alpha)) = \sum_{j=1}^{n} \alpha^{(j)2} = Tr_{K/Q}(\alpha^2) \tag{55a}$$

if K is totally real, or

$$N(v(\alpha)) = \sum_{j=1}^{s} |\alpha^{(j)}|^2 = \frac{1}{2} Tr_{K/Q}(\alpha\bar{\alpha}) \tag{55b}$$

if K is totally complex. The *trace norm* defined by (55a), (55b) is just one of many possible ways of defining a norm on K; another is given in §7.5. If we define $c = 1$ if K is totally real and $c = ½$ if K is totally complex, (55a) and (55b) may be combined into

$$N(v(\alpha)) = c\; Tr_{K/Q}(|\alpha|^2) . \tag{56}$$

Let $\mathcal{O}$ denote the ring of algebraic integers in K, and let $\omega_1, \ldots, \omega_n$ be an integral basis for $\mathcal{O}$. The matrix $\Omega = (\omega_j^{(k)})\ (1 \le j, k \le n)$ has the property that the (j, k)th entry of $\Omega\,\Omega^{tr}$ is

$$Tr_{K/Q}(\omega_j \omega_k) ,$$

and $(\det \Omega)^2 = |\text{disc}(K)|$ is the absolute value of the discriminant of K.

Finally, let $\mathcal{A} \subseteq \mathcal{O}$ be an integral ideal. A real lattice $\Lambda(\mathcal{A})$ is defined by

$$\Lambda(\mathcal{A}) = \{v(\alpha) : \alpha \in \mathcal{A}\} . \tag{57}$$

We first discuss the case when $\mathcal{A} = \mathcal{O}$.

Theorem 8. *If K is totally real or totally complex, $\Lambda(\mathcal{O})$ is an n-dimensional real lattice with* $\det \Lambda(\mathcal{O}) = c^n \, |\mathrm{disc}(K)|$ *and minimal norm at least cn.*

Proof. If K is totally real then Ω is a generator matrix for the lattice, so $\det \Lambda(\mathcal{O}) = (\det \Omega)^2 = |\mathrm{disc}(K)|$. For $\alpha \in \mathcal{O}$, $\alpha \neq 0$, we have

$$N(\nu(\alpha)) = Tr_{K/Q}(\alpha^2) \geq n\{\mathrm{Norm}(\alpha^2)\}^{1/n},$$

by the arithmetic-mean geometric-mean inequality. This is at least n since $\mathrm{Norm}(\alpha^2)$ is a positive integer. The proof in the totally complex case is similar.

Remark. If K is totally real, or if K is totally complex and complex conjugation is in the center of G, then $c^{-1}N(\nu(\alpha)) \in \mathbf{Z}$ and $c^{-1/2}\Lambda(\mathcal{O})$ is an integral lattice.

Theorem 9. *For any ideal $\mathcal{A} \subseteq \mathcal{O}$, $\Lambda(\mathcal{A})$ is an n-dimensional sublattice of $\Lambda(\mathcal{O})$, and*

$$\det \Lambda(\mathcal{A}) = c^n \, \mathrm{Norm}(\mathcal{A})^2 \, |\mathrm{disc}(K)|. \tag{58}$$

Proof. (Cf. [LeV1, II, p. 68].) We may choose an integral basis for $\mathcal{A}$ of the form

$$\begin{aligned} \eta_1 &= a_{11}\,\omega_1, \\ \eta_2 &= a_{21}\,\omega_1 + a_{22}\,\omega_2, \\ &\cdots \\ \eta_n &= a_{n1}\,\omega_1 + a_{n2}\,\omega_2 + \cdots + a_{nn}\,\omega_n, \end{aligned} \tag{59}$$

where $a_{ij} \in \mathbf{Z}$. Then $\det \Lambda(\mathcal{A}) = (a_{11}a_{22} \cdots a_{nn})^2 \det \Lambda(\mathcal{O})$, and one can show that $a_{11} \cdots a_{nn} = \mathrm{Norm}(\mathcal{A})$ = number of residue classes of $\mathcal{O}$ modulo $\mathcal{A}$.

For $\alpha \in \mathcal{O}$ let $M(\alpha)$ be the $n \times n$ integral matrix defined by

$$\alpha \begin{pmatrix} \omega_1 \\ \omega_2 \\ \cdots \\ \omega_n \end{pmatrix} = M(\alpha) \begin{pmatrix} \omega_1 \\ \omega_2 \\ \cdots \\ \omega_n \end{pmatrix}. \tag{60}$$

Then

$$\Omega \, \mathrm{diag}\{\alpha^{(1)}, \ldots, \alpha^{(n)}\} = M(\alpha)\,\Omega, \tag{61}$$

so $\det M(\alpha) = \mathrm{Norm}(\alpha)$. This immediately implies the following theorem.

Theorem 10 [Cra3]. *If $\mathcal{A} = (\alpha)$ is a principal ideal, $\Lambda(\mathcal{A})$ has generator matrix $M(\alpha)\,\Omega$. If $\mathcal{A} = (\alpha, \alpha', \alpha'', \ldots)$, $\Lambda(\mathcal{A})$ has generator matrix $M\,\Omega$, where M is a generator matrix for the sublattice of Z^n spanned by the rows of $M(\alpha)$, $M(\alpha')$, ...*

7.3 Examples from cyclotomic fields. Craig [Cra3]-[Cra5] has obtained a number of interesting examples by applying this construction to ideals in cyclotomic fields.

For any integer $m \geq 3$ let $\zeta = \zeta_m = e^{2\pi i/m}$. The *cyclotomic field* $\mathbf{Q}(\zeta_m)$ is an abelian extension of $\mathbf{Q}$ of degree $n = \phi(m)$, and is totally complex when m is odd. The ring of integers $\mathcal{O} = \mathbf{Z}[\zeta_m]$, with integral basis $\zeta, \zeta^2, \ldots, \zeta^n$, and

$$\operatorname{disc} \mathbf{Q}(\zeta_m) = (-1)^{n/2} m^n \prod_{p \mid m} p^{-n/(p-1)}. \tag{62}$$

The corresponding lattice $\Lambda(\mathbf{Z}[\zeta_m])$ has Gram matrix $A = (A_{jk})$, where ([Cra3])

$$A_{jk} = \frac{\mu(d)\phi(m)}{\phi(d)}, \quad d = \frac{m}{(m, k-j)}. \tag{63}$$

If m is an odd prime, $\Lambda(\mathbf{Z}[\zeta_m]) \cong A_{m-1}^*$.

(a) *The lattice E_6.* When $m = 9$ and $\mathcal{A} = ((1-\zeta_9)^2)$, Craig [Cra3] shows that $\Lambda(\mathcal{A}) \cong E_6$. He is able to construct five of the six 6-dimensional extreme forms [Bar6], [Bar7] in this manner.

(b) *The Leech lattice.* Suppose $m = 39$, $n = \phi(39) = 24$. In $\mathbf{Q}(\zeta_{39})$ the ideals (3) and (13) have prime factorizations

$$(3) = (\mathcal{P}_1 \overline{\mathcal{P}}_1 \mathcal{P}_2 \overline{\mathcal{P}}_2)^2, \quad (13) = (\mathcal{R} \overline{\mathcal{R}})^{12},$$

where $\mathcal{P}_1$, $\mathcal{P}_2$ have Norm 3^3 and $\mathcal{R}$ has Norm 13. Let $\mathcal{A} = \mathcal{P}_1 \mathcal{P}_2 \mathcal{R}$. Then Craig [Cra4] shows that $\Lambda(\mathcal{A}) = \Lambda_{24}$. However, this is not a very simple way to construct Λ_{24}, and the proof that $\Lambda(\mathcal{A})$ *is* Λ_{24} is fairly complicated. A more natural cyclotomic construction of Λ_{24} is given in §7.5.

(c) *The lattices $A_n^{(m)}$.* Craig's lattices $A_n^{(m)}$ described in §6 were originally constructed in [Cra5] as the lattices $\Lambda(\mathcal{A})$, where $\mathcal{A}$ is the ideal $((1-\zeta_p)^m)$ in the cyclotomic field $\mathbf{Q}(\zeta_p)$ and $p = n+1$ is a prime.

7.4 Lattices from class field towers. As pointed out by Litsyn and Tsfasman [Lit3], the infinite class field towers found by Golod and Shafarevich [Gol9], Martinet [Mar2], [Mar3] and others [Roq1] produce infinite sequences of very dense lattices. For example, these authors show that an infinite sequence or *tower* of totally complex fields

$$\mathbf{Q} \subset K_1 \subset K_2 \subset K_3 \subset \cdots \tag{64}$$

may be obtained by starting with a suitable K_1, and defining K_{i+1} to be the maximal unramified extension of K_i whose Galois group $\mathrm{Gal}(K_{i+1}/K_i)$ is an abelian 2-group. Martinet [Mar2] proves that if k is totally real field of degree $n \geq 10$, and q is a prime number that totally decomposes in k, then the totally complex field $K_1 = k(\sqrt{-q})$ may be used in (64). For example, we may take $k = \mathbf{Q}(\cos 2\pi/11, \sqrt{2})$ and $q = 23$, obtaining

$$K_1 = \mathbf{Q}(\cos \frac{2\pi}{11}, \sqrt{2}, \sqrt{-23}), \tag{65}$$

a field of degree 20 over Q with discriminant $2^{30}\, 11^{16}\, 23^{10}$. The subfield

$$\mathbf{Q}(\cos \frac{2\pi}{11}, \sqrt{-46}), \tag{66}$$

of (65), having degree 10 over Q and discriminant $2^{15}\, 11^{8}\, 23^{5}$, may also be used for K_1.

In the tower (64) let $[K_i : K_{i-1}] = h_i$, $[K_i : \mathbf{Q}] = n_i$ and $\operatorname{disc}(K_i) = D_i$. Then

$$n_{i+1} = h_{i+1}\, n_i, \tag{67}$$

and, since the extensions are unramified,

$$|D_{i+1}| = |D_i|^{h_{i+1}} \tag{68}$$

[Has1, Chap. 25, §5]. Applying the construction (57) to the ring $\mathcal{O}_i$ of algebraic integers in K_i, we obtain a lattice $\Lambda(\mathcal{O}_i)$ of dimension n_i. From Theorem 8 the packing radius and center density satisfy $\rho \geqslant \sqrt{n_i}/2\sqrt{2}$,

$$\delta \geqslant n_i^{n_i/2}\, 2^{-n_i}\, |D_i|^{-1/2}, \tag{69}$$

and so the density of $\Lambda(\mathcal{O}_i)$ satisfies

$$\frac{1}{n_i} \log_2 \Delta \geqslant - \frac{1}{2n_i} \log |D_i| - \log_2 \sqrt{\frac{2}{\pi e}} + O\left(\frac{\log_2 n_i}{n}\right).$$

However, from (67), (68),

$$\frac{\log_2 |D_{i+1}|}{2\, n_{i+1}} = \frac{\log_2 |D_i|}{2 n_i} = \cdots = \frac{\log_2 |D_1|}{2 n_1}.$$

So the sequence of lattices $\Lambda(\mathcal{O}_i)$ satisfies

$$\frac{1}{n_i} \log_2 \Delta_i \geqslant -\frac{1}{2} \log_2 \frac{2|D_1|^{1/n_1}}{\pi e} + O\left(\frac{\log_2 n_i}{n_i}\right). \tag{70}$$

Choosing the field (66) for K_1, we obtain

$$\frac{1}{n_i} \log_2 \Delta_i \geqslant -2.218.... \tag{71}$$

Unfortunately at present there does not seem to be any practicable method known for explicitly finding these lattices (or even their dimensions). In view of Odlyzko's bounds on the discriminants of number fields [Odl1]-[Odl3], lattices constructed in this way always satisfy

$$\frac{1}{n} \log_2 \Delta < -1.193, \tag{72}$$

and, assuming the generalized Riemann hypothesis,

$$\frac{1}{n} \log_2 \Delta < -1.694, \tag{73}$$

so they can never reach Minkowski's bound (Chap. 1, Eq. (29)).

7.5 Unimodular lattices with an automorphism of prime order. The lattices E_8, D_{24}^{+}, Λ_{24}, P_{48q} are even unimodular lattices in dimension $n = p + 1$ where p is a prime, having an automorphism of order p. The following analysis, due to Thompson [Tho7], makes it possible (at least in principle) to find all such lattices. (A somewhat similar approach has been used in [Bay1], [Cra5], [Fei1], [Que1].) Such a lattice arises naturally as a 2-dimensional sublattice Γ glued to a $(p-1)$-dimensional sublattice Δ, where Δ is defined by an ideal in a cyclotomic field, using a trace form that generalizes the one used in (56).

Let Λ be an n-dimensional even unimodular lattice, where $n = p + 1 \equiv 0 \pmod 8$ and p is prime, containing an element σ of order p in its automorphism group G. Let $\Gamma = \ker(1-\sigma)$, the sublattice fixed by σ, and let

$$\Delta = (1-\sigma)\Lambda = \{v \in \Lambda : v \cdot u = 0 \text{ for all } u \in \Gamma\}.$$

Thus Γ has dimension 2 and Δ has dimension $p - 1$.

We show that Δ has a natural representation as an integral ideal $\mathscr{A}$ in the cyclotomic field $K = \mathbf{Q}(\zeta)$, where $\zeta = e^{2\pi i/p}$. The ring of integers in K is $\mathbf{Z}[\zeta]$. Now ζ satisfies

$$1 + \zeta + \zeta^2 + \cdots + \zeta^{p-1} = 0,$$

and the element $\nu = 1 + \sigma + \cdots + \sigma^{p-1}$ in the group ring $\mathbf{Z}[\sigma]$ satisfies

$$\nu\Delta = (1-\sigma^p)\Lambda = 0.$$

Therefore we may regard Δ as a $\mathbf{Z}[\zeta]$-module, or more precisely there is an isomorphism

$$\phi : \Delta \xrightarrow{\cong} \mathscr{A}, \tag{74}$$

where $\mathscr{A}$ is an ideal of $\mathbf{Z}[\zeta]$.

Theorem 11 ([*Tho*7], *cf.* [*Fei*1, *Theorem* 6.1]). (a) *With* Λ, Δ *as above, there is a totally positive element* $\gamma \in K$ (*i.e.* γ *and all its conjugates are positive*) *such that*

$$u \cdot v = Tr_{K/\mathbf{Q}} (\phi(u)\gamma\overline{\phi(v)}) \tag{75}$$

for all $u, v \in \Delta$, *and*

(b)
$$\gamma \in \mathscr{A}^{-1}\,\overline{\mathscr{A}}^{-1}\,\mathscr{D}^{-1}, \tag{76}$$

where $\mathscr{D}$ is the different of $K/\mathbf{Q}$. (c) *Conversely, given any ideal $\mathscr{A}$ of $\mathbf{Z}[\zeta]$ and a totally positive element $\gamma \in \mathbf{Z}[\zeta]$ satisfying* (76), *$\mathscr{A}$ becomes a positive definite integral lattice if we define*

$$(\alpha,\beta) = Tr_{K/\mathbf{Q}}(\alpha\gamma\bar{\beta}), \quad \alpha,\beta \in \mathscr{A}. \tag{77}$$

Sketch of proof. (a) The inner product $u \cdot v$ on Δ extends in the obvious way to $\Delta \otimes \mathbf{Q}$; but from (74) $\Delta \otimes \mathbf{Q} \cong \mathscr{A} \otimes \mathbf{Q} = K$. So we have an inner product (,) on K satisfying $u \cdot v = (\phi(u), \phi(v))$ for u, $v \in \Delta$. Furthermore $\phi(\sigma u) = \zeta\phi(u)$ for $u \in \Delta$, which implies $(\alpha,\beta) = (\zeta\alpha,\zeta\beta)$ for α, $\beta \in \mathscr{A}$. Therefore the inner product on K is determined by the $p-1$ numbers

$$(1,\zeta^i) = a_i \in \mathbf{Q}, \quad 0 \leqslant i \leqslant p-2. \tag{78}$$

Now $\zeta^{-1}, \ldots, \zeta^{-(p-1)}$ is a basis for K over $\mathbf{Q}$, and there exists a dual basis $\theta_1, \ldots, \theta_{p-1}$ satisfying

$$Tr_{K/\mathbf{Q}}(\zeta^{-i}\theta_j) = \delta_{ij}$$

[Lan8, p. 212]. If we set

$$\gamma = \sum_{i=1}^{p-1} a_i\theta_i,$$

then

$$Tr_{K/\mathbf{Q}}(\gamma\zeta^{-i}) = a_i, \quad 0 \leqslant i \leqslant p-2, \tag{79}$$

and (75) follows from (78), (79). Furthermore, since Λ is positive definite,

$$Tr_{K/Q}(\alpha\gamma\bar{\alpha}) = (\alpha,\alpha) > 0$$

for all $\alpha \in K$, $\alpha \neq 0$. This implies that γ is totally positive. (b) We also know that $u \cdot v \in Z$ for u, $v \in \Delta$, so

$$Tr_{K/\mathbf{Q}}(\gamma\alpha\bar{\beta}) \in \mathbf{Z} \text{ for } \alpha,\beta \in \mathscr{A},$$

which means that γ is in the *complementary ideal* to $\mathscr{B} = \mathscr{A}\bar{\mathscr{A}}$, denoted by $\tilde{\mathscr{B}}$ (cf. [Rib1, p. 204], [Wei2, p. 107]). Now $\tilde{\mathscr{B}}$ is the fractional ideal $\tilde{\mathscr{B}} = \mathscr{B}^{-1}\mathscr{D}^{-1}$, where $\mathscr{D}$ is the different of $K/\mathbf{Q}$, and this implies (76). We omit the proof of the converse.

The next step is to consider how the sublattices Δ and Γ are glued together to form Λ (cf. §3 of Chap. 4). From Theorem 1 of Chap. 4 we have

$$\Delta^*/\Delta \cong \Gamma^*/\Gamma. \tag{80}$$

(For, by construction, $\Delta = \Delta \cap (\Lambda \otimes \mathbf{R})$, $\Gamma = \Gamma \cap (\Lambda \otimes \mathbf{R})$, so there is no self-glue.)

Eq. (80) is an isomorphism of $\mathbf{Z}[\sigma]$-modules. But σ acts trivially on Γ^*/Γ, and therefore acts trivially on Δ^*/Δ. If

$$\mathscr{A}^* = \{x \in K : Tr_{K/Q}(\alpha\gamma\bar{x}) \in \mathbf{Z} \text{ for all } \alpha \in \mathscr{A}\} \tag{81}$$

is the ideal corresponding to Δ^*, ζ acts trivially on $\mathscr{A}^*/\mathscr{A}$, i.e.

$$(1-\zeta)\mathscr{A}^* \subseteq \mathscr{A}. \tag{82}$$

Also $\mathscr{A}^* \neq \mathscr{A}$ (otherwise Δ would be a direct summand of Λ, and Λ would not be even, since dim Δ is not a multiple of 8).

Theorem 12 [Tho7]. (a) *With Λ, Δ, Γ as above, the ideal $\mathscr{A}$ representing Δ is such that*

$$(1-\zeta)^{3-p}\, \mathscr{A}^{-1}\, \overline{\mathscr{A}}^{-1} = (\gamma) \tag{83}$$

is a principal ideal with a totally positive generator γ, and

(b)
$$|\Delta^*/\Delta| = |\Gamma^*/\Gamma| = p\,. \tag{84}$$

(c) *Conversely, if $\mathscr{A}$ is an ideal of $\mathbf{Z}[\zeta]$ satisfying* (83) *where γ is totally positive then the inner product* (57) *makes $\mathscr{A}$ into an even integral lattice with determinant p.*

Sketch of proof. (a), (b). From (81), $\mathscr{A}^*$ is the complementary ideal to $\mathscr{A}\bar{\gamma}$, so

$$\mathscr{A}^* = \overline{\mathscr{A}}^{-1}\gamma^{-1}\mathscr{D}^{-1}. \tag{85}$$

From (82), (85),

$$(1-\zeta)\,\mathscr{D}^{-1} \subseteq \mathscr{A}\overline{\mathscr{A}}\gamma. \tag{86}$$

Also $\mathscr{A}\overline{\mathscr{A}}\gamma \subset \mathscr{D}^{-1}$, where the containment is proper since $\mathscr{A}^* \neq \mathscr{A}$. Thus

$$(1-\zeta)\,\mathscr{D}^{-1} \subseteq \mathscr{A}\overline{\mathscr{A}}\gamma \subset \mathscr{D}^{-1}.$$

But $(1-\zeta)\,\mathscr{D}^{-1}$ has index p in $\mathscr{D}^{-1}$, so finally we may conclude that

$$(1-\zeta)\,\mathscr{D}^{-1} = \mathscr{A}\overline{\mathscr{A}}\gamma\,, \tag{87}$$

$$(1-\zeta)\mathscr{A}^* = \mathscr{A}. \tag{88}$$

Therefore $\mathscr{A}$ has index p in $\mathscr{A}^*$, which proves (84). In this cyclotomic field we know that $\mathscr{D}$ is the principal ideal $((1-\zeta)^{p-2})$, and $\mathscr{D}^{-1} = ((1-\zeta)^{2-p})$. Then (83) follows from (87).

Examples. The Leech lattice arises by taking $n = 24$, $p = 23$, and letting $\mathscr{A}$ be a prime ideal dividing 2. Now $x^{23}+1$ factors modulo 2 into $(x+1)f_1(x)f_2(x)$, where

$$f_1(x) = x^{11} + x^{10} + x^6 + x^5 + x^4 + x^2 + 1,$$

$$f_2(x) = x^{11} + x^9 + x^7 + x^6 + x^5 + x + 1. \tag{89}$$

Therefore in $\mathbf{Z}[\zeta_{23}]$, (2) factors into $(2) = \mathcal{P}_1 \mathcal{P}_2$, where

$$\mathcal{P}_1 = (2, f_1(\zeta)),\ \mathcal{P}_2 = (2, f_2(\zeta)).$$

We take $\mathcal{A} = \mathcal{P}_1$, $\overline{\mathcal{A}} = \mathcal{P}_2$, and

$$\gamma = \frac{1}{2(1-\zeta)^{20}}, \tag{90}$$

which defines Δ. We must take Γ to be a 2-dimensional even lattice of determinant 23 and minimal norm 4, and so Γ may be chosen to have Gram matrix $\begin{bmatrix} 4 & 1 \\ 1 & 6 \end{bmatrix}$. One can now show that there is a unique way to glue Δ to Γ to produce an even unimodular lattice, which is Λ_{24}.

The lattice P_{48q} (Chap. 7, Example 9) arises in an exactly similar way. $x^{47} + 1$ factors modulo 2 into $(x+1)\, f_1(x)\, f_2(x)$, where f_1 and f_2 have degree 23. We take $\mathcal{A} = \mathcal{P}_1 = (3, f_1(\zeta))$, a prime ideal dividing 3, and $\Gamma = \begin{bmatrix} 6 & 1 \\ 1 & 8 \end{bmatrix}$. Again there is a unique way to glue Δ to Γ to produce P_{48q}.

8. Constructions D and D′

8.1 Construction D. This construction, first given in [Bar15], uses a nested family of binary codes to produce a lattice packing L in $\mathbf{R}^n$. It generalizes a construction of Barnes and Wall [Bar18], and has some features in common with Construction C of Chap. 5, §6.1, although differing from it in always producing lattice packings. (On the other hand Construction C can be applied to codes that are not nested and to nonlinear codes.)

Let $\gamma = 1$ or 2, and let $C_0 \supseteq C_1 \supseteq \cdots \supseteq C_a$ be binary linear codes, where C_i has parameters $[n, k_i, d_i]$, with $d_i \geqslant 4^i/\gamma$ for $i = 1, \ldots, a$, and C_0 is the $[n, n, 1]$ universe code $\mathbf{F}_2^n$. Choose a basis $c_1, \ldots, c_n$ for $\mathbf{F}_2^n$ such that (i) $c_1, \ldots, c_{k_i}$ span C_i for $i = 0, \ldots, a$, and (ii) if M denotes the matrix with rows $c_1, \ldots, c_n$, some permutation of the rows of M forms an upper triangular matrix. Define the map $\sigma_i : \mathbf{F}_2 \rightarrow \mathbf{R}$ by $\sigma_i(x) = x/2^{i-1}$ ($x = 0$ or 1), for $i = 1, \ldots, a$, and let the same symbol σ_i denote the map $\mathbf{F}_2^n \rightarrow \mathbf{R}^n$ given by

$$\sigma_i(x_1, \ldots, x_n) = (\sigma_i(x_1), \ldots, \sigma_i(x_n)).$$

Also let $k_{a+1} = 0$. Then the new lattice L in $\mathbf{R}^n$ consists of all vectors of the form

$$l + \sum_{i=1}^{a} \sum_{j=1}^{k_i} \alpha_j^{(i)} \sigma_i(c_j), \tag{91}$$

where $l \in (2\mathbf{Z})^n$ and $\alpha_j^{(i)} = 0$ or 1.

Theorem 13 [Bar15]. *L is a lattice, with minimal norm at least $4/\gamma$, determinant*

$$\det L = 4^{n - \sum_{i=1}^{a} k_i} \tag{92}$$

and hence center density

$$\delta \geqslant \gamma^{-n/2}\, 2^{\sum_{i=1}^{a} k_i - n} \tag{93}$$

This is a consequence of Theorems 15 and 16 below.

An integral basis for L is given by the vectors $\sigma_i(c_j)$ for $i = 1, \ldots, a$, $j = k_{i+1} + 1, \ldots, k_i$, plus $n - k_1$ vectors of the shape $(0, \ldots, 0, 2, 0, \ldots, 0)$.

8.2 Examples. (a) *Construction A.* When $a = \gamma = 1$, Construction D reduces to the linear case of Construction A (Chap. 5, §2.1). A typical example is obtained when C_1 is the [8,4,4] Hamming code $\mathcal{H}_8$. We may take

$$\begin{aligned}
c_1 &= (1,1,1,1,1,1,1,1),\\
c_2 &= (0,1,0,1,0,1,0,1),\\
c_3 &= (0,0,1,1,0,0,1,1),\\
c_4 &= (0,0,0,0,1,1,1,1),\\
c_5 &= (0,0,0,1,0,0,0,0),\\
c_6 &= (0,0,0,0,0,1,0,0),\\
c_7 &= (0,0,0,0,0,0,1,0),\\
c_8 &= (0,0,0,0,0,0,0,1).
\end{aligned}$$

Then L is a version of the E_8 lattice (see Example 5 of the previous chapter), is spanned by the rows of the matrix

$$\begin{bmatrix}
1 & 1 & 1 & 1 & 1 & 1 & 1 & 1\\
0 & 1 & 0 & 1 & 0 & 1 & 0 & 1\\
0 & 0 & 1 & 1 & 0 & 0 & 1 & 1\\
0 & 0 & 0 & 0 & 1 & 1 & 1 & 1\\
0 & 0 & 0 & 2 & 0 & 0 & 0 & 0\\
0 & 0 & 0 & 0 & 0 & 2 & 0 & 0\\
0 & 0 & 0 & 0 & 0 & 0 & 2 & 0\\
0 & 0 & 0 & 0 & 0 & 0 & 0 & 2
\end{bmatrix},$$

and has minimal norm 4, determinant 2^8, and $\delta = 2^{-4}$.

(b) *Construction B.* If $a = \gamma = 2$, C_1 is the $[n, n-1, 2]$ even weight code, C_2 has minimal distance 8 and the weight of every codeword of C_2 is a multiple of 4, then Construction D reduces to the version of

Construction B given in §5 of Chap. 7. For example, as in Example 7 of Chap. 7, if C_2 is the [16,5,8] Reed-Muller code we obtain $\Lambda_{16} \cong BW_{16}$.

(c) *Construction C.* Whenever Construction C of Chap. 5 is applied to codes that are linear and nested, Construction D produces a lattice packing with the same density.

(d) *Rao and Reddy's code and a lattice in* $\mathbf{R}^{36}$. Figure 8.2 shows a [48, 31, 8] code C found by Rao and Reddy [Rao1]. The dual code C^* has minimal distance 12; let $a \in C^*$ have weight 12. The codewords $c \in C$ which vanish on the support of a (with these twelve 0's deleted) form a [36, 20, 8] code C_2. The maximal weight occurring in C_2 is 32; let $b \in C_2$ have weight 32. By applying Construction D to the codes $C_1 = [36,35,2]$, C_2, $C_3 = \{0,b\}$ we obtain a 36-dimensional lattice B_{36} with $\delta = 4$ (see Table 1.2 and Fig. 1.5). Similarly one can obtain $\delta = \sqrt{2}$, 2 and $2\sqrt{2}$ in dimensions 33, 34 and 35. This is an improvement of Bos' earlier use of Construction C to obtain nonlattice packings with the same densities (personal communication).

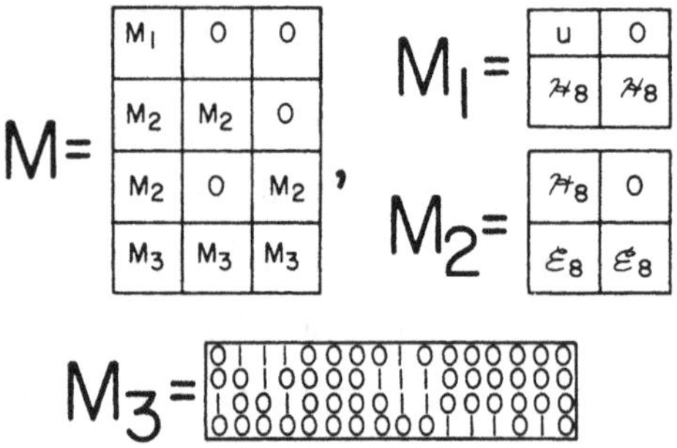

Figure 8.2. Generator matrix M for the [48, 31, 8] Rao-Reddy code. Here $u = (11111111)$, and $\mathscr{H}_8$ (Eq. (57) of Chap. 3), $\mathscr{E}_8$, M_1, M_2 are generator matrices for [8, 4, 4], [8, 7, 2], [16, 5, 8], [16, 11, 4] codes respectively.

(e) *A 64-dimensional lattice.* For a second example we use the codes in Eq. (13) of Chap. 5 to obtain a lattice P_{64c} in $\mathbf{R}^{64}$ with $\delta = 2^{22}$.

(f) *The Barnes-Wall lattices.* Generalizing example (b), we apply Construction D to the Reed-Muller codes used in §6.5 of Chap. 5. This yields the ***Barnes-Wall lattices*** BW_n for $n = 2^m$, $m \geqslant 2$. BW_n has center density

$$\delta = 2^{-5n/4}\, n^{n/4} \tag{94}$$

(as in §6.5 of Chap. 5), density

$$\frac{1}{n}\log_2 \Delta \sim -\frac{1}{4}\log_2 n \quad \text{as } n \to \infty, \tag{95}$$

and kissing number given by Eq. (17) of Chap. 5. Also $BW_4 \cong D_4 \cong \Lambda_4$, $BW_8 \cong E_8 = \Lambda_8$ and $BW_{16} \cong \Lambda_{16}$. After multiplication by $1/\sqrt{2}$, BW_{32} is an extremal Type II lattice with det = 1, $\delta = 1$, $\tau = 146880$. See Table 1.3.

(g) *Lattices from BCH codes.* Similarly the nonlattice packings obtained from BCH codes in §6.6 of Chap. 5 may now be converted to equally dense lattice packings B_n, for $n = 2^m$. The dimensions of these codes may be estimated from the BCH bound [Mac6, Chap. 7], and the densities of the corresponding lattices are shown in Tables 8.5, 1.3. Except for small n the actual densities of these lattices may be slightly greater than is shown. (Tables of BCH codes of modest length are given in [Mac6], [Pet3].) Asymptotically the density satisfies

$$\frac{1}{n} \log_2 \Delta \sim -\frac{1}{2} \log_2 \log_2 n \,, \tag{96}$$

as shown in §6.7 of Chap. 5. The kissing numbers are not known.

(h) Let $C_1 = [32,31,2]$, $C_3 = [32,1,32]$, and let C_2 be the [32, 17, 8] code given in [Che3]. Then Construction D produces a 32-dimensional lattice C_{32} for which $\delta = 2$ and $\tau = 249280$.

Table 8.5. Center density δ of n-dimensional lattices B_n obtained from BCH codes.

n	$\log_2 \delta$	n	$\log_2\delta$	n	$\log_2\delta$
4	−3	256	250	16384	57819
8	−4	512	698	32768	130510
16	−4	1024	1817	65536	290998
32	0	2048	4502	131072	642300
64	19	4096	10794	...	...
128	85	8192	25224		

8.3 Construction D′. This construction generalizes another of the constructions in [Bar18], and converts a set of parity-checks defining a family of codes into congruences for a lattice, in the same way that Construction D converts a set of generators for a family of codes into generators for a lattice.

Let $\gamma = 1$ or 2, and let $C_0 \supseteq C_1 \supseteq \cdots \supseteq C_a$ be binary linear codes, where C_i has parameters $[n, k_i, d_i]$ and $d_i \geq \gamma \cdot 4^i$ for $i = 0, \ldots, a$. Let $h_1, \ldots, h_n$ be linearly independent vectors in $\mathbf{F}_2^n$ such that (i) for $i = 0, \ldots, a$, the code C_i is defined by the $r_i = n - k_i$ parity-check vectors $h_1, \ldots, h_{r_i}$, (ii) some rearrangement of $h_1, \ldots, h_n$ forms the rows of an upper triangular matrix, and let $r_{-1} = 0$. Considering the vectors h_j as integral vectors in $\mathbf{R}^n$, with components 0 or 1, we define the new lattice L' to consist of those $x \in \mathbf{Z}^n$ that satisfy the congruences

$$h_j \cdot x \equiv 0 \,(\text{mod } 2^{i+1}) \tag{97}$$

for all $i = 0, \ldots, a$ and $r_{a-i-1} + 1 \leqslant j \leqslant r_{a-i}$.

Theorem 14 [Bar15]. *The minimal norm of L' is at least $\gamma \cdot 4^a$, and*

$$\log_4 \det L' = \sum_{i=0}^{a} r_i . \tag{98}$$

Proof. (98) holds because for each $i = 0, \ldots, a$ the lattice satisfies $r_{a-i} - r_{a-i-1}$ independent congruences modulo 2^{i+1}. The evaluation of the minimal norm is straightforward and is omitted.

By changing the scale and relabeling the codes, Construction D may be restated in such a way that the norms and determinants of L and L' agree. Then if the C_i are Reed-Muller codes, as in Example 8.2f, the two lattices coincide. In general however the two constructions produce inequivalent lattices with the same density.

9. Construction E.

The following rather general construction includes many of the earlier constructions as special cases. It was first given in [Bar15] and [Bos3].

The ingredients for the construction are a lattice Λ in $\mathbf{R}^m$, an endomorphism D of Λ that satisfies certain conditions, one of which is that $\Lambda / D\Lambda$ is an elementary abelian group E of order p^b, say, and a family of codes $C_0 \supseteq C_1 \supset \cdots \supseteq C_a$ of length n over E; the result is a family $L_0 \subseteq L_1 \subseteq \cdots \subseteq L_a$ of lattice packings in $\mathbf{R}^{mn}$.

Hypotheses.

(i) Let Λ be a lattice in $\mathbf{R}^m$ with minimal non-zero norm M.

(ii) Let D be an endomorphism of Λ which is also a similarity (i.e. a constant times an orthogonal transformation) and which satisfies

$$pD^{-1} = \sum_{i=0}^{r} a_i D^i \tag{99}$$

for integers $p \geqslant 1$, $r \geqslant 0$, $a_0, \ldots, a_r$. Let $T = D^{-1}$, so that, from (99), pT is also an endomorphism of Λ.

It follows that $\Lambda / D\Lambda$ has the structure of an elementary abelian group E of some order p^b $(b \geqslant 1)$. This implies that there is a b-dimensional sublattice $K \subseteq \Lambda$, spanned say by vectors $v_1, \ldots, v_b \in \Lambda$, such that $K \,/\, (D\Lambda \cap K) \cong E$. The assumptions also imply that

$$pK \subseteq K \subseteq \Lambda, \tag{100}$$

$$pK \subseteq p\Lambda \subseteq D\Lambda \subseteq \Lambda, \tag{101}$$

and therefore that

$$K \,/\, pK = K \,/\, (D\Lambda \cap K) \cong E . \tag{102}$$

Note that (100) and (101) imply $pK \subseteq D\Lambda \cap K$, so $K \,/\, pK \supseteq K \,/\, (D\Lambda \cap K)$. But both sides have order p^b and so must be equal, and (102) follows. Furthermore

$$|\det T| = p^{-b} = t^m, \text{ say}, \tag{103}$$

and T multiplies norms by t^2.

(iii) Assume that all $p^b - 1$ non-zero congruence classes of $T\Lambda/\Lambda$ have minimal norm at least $t^2 M$.

(iv) Let ϕ denote the obvious map from $\mathbf{Z}/p\mathbf{Z}$ to $\mathbf{Z}$ which takes the congruence class $\bar{x}$ to x, for $x \in \{0,1,\ldots,p-1\}$. The elements of E may be identified with the b-tuples $\bar{X} = (\bar{x}_1, \ldots, \bar{x}_b)$, where all $\bar{x}_i \in \mathbf{Z}/p\mathbf{Z}$. Then

$$\begin{aligned}\bar{X} \to V(\bar{X}) &= \phi(\bar{x}_1)v_1 + \cdots + \phi(\bar{x}_b)v_b \\ &= x_1 v_1 + \cdots + x_b v_b\end{aligned}$$

is a map from E into Λ, and

$$\bar{X} = (\bar{X}_1, \ldots, \bar{X}_n) \to V(\bar{X}) = (V(\bar{X}_1), \ldots, V(\bar{X}_n))$$

maps E^n into Λ^n.

(v) Let $C_0 \supseteq C_1 \supseteq \cdots \supseteq C_a$ be additive codes over E of length n (i.e. abelian subgroups of E^n), where C_i contains p^{bk_i} codewords and has minimal distance d_i (we indicate this by saying that C_i has parameters $[n,k_i,d_i]$), and suppose that C_0 is the trivial $[n,n,1]$ code. Let $c_1, \ldots, c_{bk_1} \in E^n$ be chosen so that (1) a typical codeword of C_i can be written as

$$\sum_{j=1}^{bk_i} \bar{x}_j c_j, \ \bar{x}_j \in \mathbf{Z}/p\mathbf{Z},$$

for $i = 1, \ldots, a$, and (2) some rearrangement of $c_1, \ldots, c_{bk_1}$ forms the rows of an upper triangular matrix.

Construction E. Let $L_0 = \Lambda^n$ and, for $i = 1,2,\ldots,a$, define

$$L_i = \bigcup \left\{ L_{i-1} + \sum_{j=1}^{bk_i} x_j T^i V(c_j) \right\}, \tag{104}$$

where the union is taken over all $x_1, \ldots, x_{bk_i} \in \{0,1,\ldots,p-1\}$. By abuse of notation we shall also use D and T to denote the maps $(D,D,\ldots,D)$ and $(T,T,\ldots,T)$ acting on $\mathbf{R}^{mn}$. It is clear that L_0 is a lattice and that D and pT are endomorphisms of L_0.

Theorem 15 [Bos3]. *For* $i = 1, \ldots, a$,

(a) *L_i is a lattice in $\mathbf{R}^{mn}$, and in fact*

$$L_i = \mathbf{Z}\langle L_{i-1}, T^i V(c_1), \ldots, T^i V(c_{bk_i})\rangle; \tag{105}$$

(b) *D maps L_i into L_{i-1}; and*

(c) *pT is an endomorphism of L_i.*

Proof. The proof is by induction on i, the results for $i = 0$ having already been mentioned. (a) To show that L_i is a lattice, we write any integer combination of the elements on the right-hand side of (104) as

$$l + p \sum_{j=1}^{bk_i} y_j T^i V(c_j) + \sum_{j=1}^{bk_i} z_j T^i V(c_j), \tag{106}$$

where $l \in L_{i-1}$ and $0 \leqslant z_j \leqslant p-1$. But

$$pT^i V(c_j) = pT \cdot T^{i-1} V(c_j) \in L_{i-1}$$

by the induction hypothesis, so (106) becomes

$$l' + \sum_{j=1}^{bk_i} z_j T^i V(c_j),$$

where $l' \in L_{i-1}$, which by (104) is in L_i. Thus L_i is a lattice, and therefore can be defined by the right-hand side of (105). Now (b) follows from

$$DT^i V(c_j) = T^{i-1} V(c_j) \in L_{i-1}.$$

From (105) and (b) we have

$$pT(L_i) = pD^{-1}(L_i) = \sum_{j=0}^{r} a_j D^j (L_i) \subseteq L_i,$$

which is (c).

Theorem 16 [Bos3]. *For $i = 0, 1, \ldots, a$, the determinant, minimal norm and center density of L_i are given by*

$$\log_p \det L_i = n \log_p \det \Lambda - 2b \sum_{j=1}^{i} k_j, \tag{107}$$

$$\overline{M} = \min \{ M, d_j t^{2j} M \text{ for } j = 1, \ldots, i \}, \tag{108}$$

and

$$\delta = \overline{M}^{mn/2} / 2^{mn} (\det L_i)^{1/2} \tag{109}$$

respectively.

Proof. Equation (107) is immediate from (104), and (108) follows from the definition of L_i and the fact that the minimal distance of C_i is d_i.

Usually we are only interested in the finest lattice L_a. The construction may now be applied to L_a, since it inherits D and T from Λ, and (99) still holds. The values of p and t are unchanged, while b becomes nb. A lattice obtained by applying Construction E to Λ is denoted by $\eta(\Lambda)$ (in Table 1.3 for example).

10. Examples of Construction E

In most of these examples T is a linear map from $\mathbf{R}^m$ to $\mathbf{R}^m$ that sends minimal vectors of Λ onto deep holes for that lattice. It is helpful to think of T as a "screw" map. For example the map of $\mathbf{R}^2$ defined by (113) is a clockwise rotation through 45° followed by a contraction by a factor of $\sqrt{2}$. The minimal vectors $(\pm 1,0)$ and $(0,\pm 1)$ are mapped onto the deep holes $2^{-1/2}(\pm 1,\pm 1)$.

(a) *Construction* D. If we take $\Lambda = 2\mathbf{Z}$ and $D = 2I$, where I denotes the identity map, then we find $T = \frac{1}{2}I$, $p = 2$, $b = 1$, $t = \frac{1}{2}$, and Construction E reduces to Construction D. Theorem 13 now follows from Theorems 15 and 16.

(b) *If D is a norm-doubler.* In most of the following examples D is a *norm-doubler*, so that $t = 1/\sqrt{2}$, $p = 2$ and $b = m/2$. We shall take the codes C_i, $i \geqslant 1$, to be maximal distance separable codes (§2.5 of Chap. 3) over the field $\mathbf{F}_{2^b}$, with parameters $[n, k_i = n - 2^i + 1, d_i = 2^i]$. In general, the largest n for which such codes are presently known to exist is $2^b + 1$ (see [Mac6, Chap. 11]). When $n = 2^b + 1$, C_i for $i \geqslant 1$ can be taken to be the cyclic code with generator polynomial

$$g(x) = \prod_{j=-s}^{s} (x + \xi^j),$$

where ξ is a primitive nth root of unity in the field of order 2^{2b} and $s = 2^{i-1} - 1$ (see the proof of Theorem 9, Chap. 11 of [Mac6][(1)]). Besides being additive, these codes are also closed under multiplication by elements of $\mathbf{F}_{2^b}$. However we make no use of this multiplicative structure in our construction. Clearly $C_0 \supseteq C_1 \supseteq C_2 \supseteq \cdots$. For n less than $2^b + 1$ the codes are shortened by setting the appropriate number of information symbols equal to zero. If n lies in the range $2 \leqslant n \leqslant 2^b + 1$, we use the codes $C_1, C_2, \ldots, C_a$, where a is determined by $2^a \leqslant n < 2^{a+1}$. Then the value of $\Sigma_{j=1}^{a} k_j$ for use in (107) is

$$an - 2^{a+1} + a + 2. \tag{110}$$

Since $d_j t^{2j} = 1$, the minimal norm is unchanged (see (108)). From (107), (109), (110) and Eq. (17) of Chap. 1 we obtain:

Theorem 17. *If an m-dimensional lattice Λ with center density $\delta = 2^\gamma$ has a norm-doubling map D satisfying the hypotheses stated at the beginning of §9, then Construction* E *produces an N-dimensional lattice Λ' with center density $\delta' = 2^{\gamma'}$, where $N = mn$, $1 \leqslant n \leqslant 2^{m/2} + 1$,*

$$\gamma' = n\gamma + \frac{m}{2}(an - 2^{a+1} + a + 2), \tag{111}$$

and $a = [\log_2 n]$. The density Δ of Λ' satisfies

$$\frac{1}{N}\log_2 \Delta = \frac{\gamma}{m} - \frac{1}{2}\log_2 \frac{2m}{\pi e} + \epsilon, \tag{112}$$

where $0 < \epsilon < (1/2n)\log_2(4n)$.

---

(1) That theorem is incorrect if the field size q is odd (see [Mac6, p. xii]) but that does not concern us since here q is even.

The examples that follow are descendants of the lattices $\mathbf{Z}^2$, K_{12}, Λ_{24} and P_{48q}.

(c) *Packings constructed from the lattice* $\mathbf{Z}^2$. Let Λ be the two-dimensional square lattice $\mathbf{Z}^2$ with minimal norm $M = 1$ (see Fig. 8.3). Although this is not a particularly dense packing, it has surprisingly good progeny. We let D map (1,0) to (1,1), and (0,1) to (1,−1), or in matrix notation, with D mapping v to vD,

$$D = \begin{pmatrix} 1 & 1 \\ 1 & -1 \end{pmatrix}, \quad T = D^{-1} = \frac{1}{2}\begin{pmatrix} 1 & 1 \\ 1 & -1 \end{pmatrix}, \qquad (113)$$

so that Eq. (99) reads $2D^{-1} = D$. Then $\Lambda/D\Lambda \cong \mathbf{Z}/2\mathbf{Z}$, $p^b = 2^1$, and we may take K to be the 1-dimensional lattice spanned by $v_1 = (1,0)$ (see Fig. 8.3).

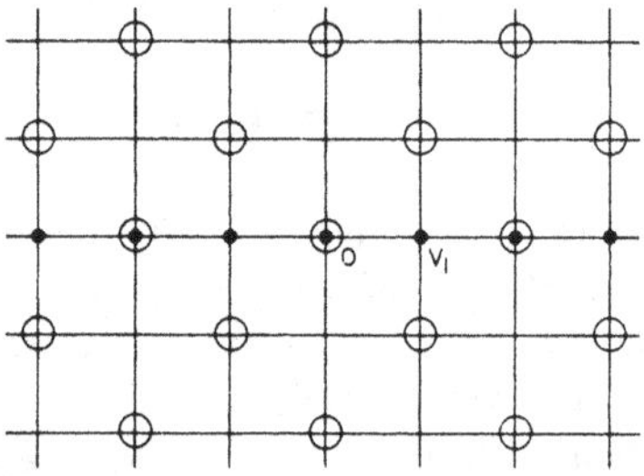

Figure 8.3. The lattice $\Lambda = \mathbf{Z}^2$ (represented by the underlying grid of points), and the sublattices $D\Lambda$ (small circles) and K (solid circles).

We apply Construction E to $\mathbf{Z}^2$, with $n = 2$ and C_1 equal to the code {00, 11} over $\mathbf{F}_2$, and obtain the lattice D_4, with $\gamma = -3$ (from (111)).

From now on we need not specify D, T or p^b, since these are inherited (see the remarks at the end of §9). For concreteness, however, we note that if D_4 is defined by Eq. (15) of Chap. 1 or Eq. (86) of Chap. 4, then $T = D^{-1}$ is given by (91) of Chap. 4 and is essentially unique.

Applying the construction to D_4 with $n = 1, \ldots, 5$ $(=2^b+1)$ we obtain the laminated lattices

$$\Lambda_4 = D_4, \ \Lambda_8 = E_8, \ \Lambda_{12}, \ \Lambda_{16}, \ \Lambda_{20}$$

(see Chap. 6) for which γ is

$$-3, -4, -5, -4, -3$$

respectively.

E_8 (the first grandchild of $\mathbf{Z}^2$) also has an essentially unique norm-doubler. If E_8 is defined by Eq. (99) of Chap. 4, then we may take

$$D = \begin{bmatrix} h & 0 & 0 & 0 \\ 0 & h & 0 & 0 \\ 0 & 0 & h & 0 \\ 0 & 0 & 0 & h \end{bmatrix}, \quad h = \begin{bmatrix} 1 & 1 \\ 1 & -1 \end{bmatrix}. \tag{114}$$

We apply Construction E to E_8, using MDS codes over $\mathbf{F}_{16}$ of lengths $n = 1,2,\ldots,17$. The lattices $\eta(E_8)$ obtained in this way are shown in Tables 8.6 and 1.3. These include Λ_{16} again, packings $\bar{\Lambda}_{32}$ and $\bar{\Lambda}_{40}$ with the same density as the laminated lattices Λ_{32} and Λ_{40}, excellent packings in dimensions 80 to 136, but indifferent packings in dimensions 24 and 48 to 72. (Nonlattice packings with these densities were already found by Bos.) The children of E_8 extend only to dimension 136. Beyond this point the lattices obtained by applying the construction to Λ_{12} and Λ_{16} are roughly comparable, up to dimension $16(2^8+1)$. The children of Λ_{20}, $\bar{\Lambda}_{32}$ and $\bar{\Lambda}_{40}$ are not quite as good as other known packings, although all of these families contain some very dense members. Above dimension 140 these packings are inferior to Craig's lattices (§6) and the lattices constructed in the following paragraphs.

Table 8.6. Center density δ of N-dimensional lattices $\eta(E_8)$ obtained from E_8 ($N = 8n$).

n	N	$\log_2 \delta$	n	N	$\log_2 \delta$	n	N	$\log_2 \delta$
1	8	−4	7	56	12	13	104	60
2	16	−4	8	64	20	14	112	68
3	24	−4	9	72	28	15	120	76
4	32	0	10	80	36	16	128	88
5	40	4	11	88	44	17	136	100
6	48	8	12	96	52			

(d) *Packings constructed from the Coxeter-Todd lattice K_{12}.* We use the complex 6-dimensional version of K_{12} defined in Example 10a of the previous chapter (the "2-base"). This has a norm-doubling map

$$D:(a,b,c,d,e,f) \to (\omega a+\omega b, \bar{\omega} a-\bar{\omega} b, \ldots, \omega e+\omega f, \bar{\omega} e-\bar{\omega} f),$$

for $a,\ldots,f \in \mathcal{E}$, satisfying $D^2 - \theta D - 2 = 0$, which we may regard as acting on K_{12}. Construction E produces lattices $\eta(K_{12})$ in $\mathbf{R}^{12n}$ for $1 \le n \le 65$ with (from (111))

$$\log_2 \delta = -3n \log_2 3 + 6(an - 2^{a+1} + a + 2), \tag{115}$$

where $a = \lceil \log_2 n \rceil$. In dimensions 228 to 780 these are denser than any other lattices constructed in §4, but are inferior to Craig's lattices.

(e) *Lattices constructed from a norm-doubling map for* Λ_{24}. The Leech lattice has (at least) two distinct norm doubling maps D. One is $D = I - i$, where i is the element of $\mathrm{Aut}(\Lambda_{24})$ shown in Fig. 6.7, with $i^2 = -1$ and $T = D^{-1} = \frac{1}{2}(I+i)$. If Λ_{24} is defined by Eq. (134) of Chap. 4 then the second norm-doubler D' is represented by the symmetric matrix

$$\frac{1}{4}\left|\begin{array}{c|ccccccc} 3 & -1 & -1 & -1 & -1 & -1 & \ldots & -1 \\ \hline -1 & -3 & 1 & 1 & 1 & 1 & \ldots & -1 \\ -1 & 1 & 1 & 1 & 1 & -1 & \ldots & -3 \\ -1 & 1 & 1 & 1 & -1 & 1 & \ldots & 1 \\ & & \ldots & & & & \ldots & \\ -1 & -1 & -3 & 1 & 1 & 1 & \ldots & -1 \end{array}\right| \qquad (116)$$

where in the second row there is a -3 in position 0, 1's in the positions that are quadratic residues modulo 23, and -1's elsewhere. Also we have $T' = (D')^{-1} = \frac{1}{2} D'$. Both T and T' map the minimal vectors of Λ_{24} onto deep holes of type A_1^{24} (cf. Chap. 23). MDS codes over $\mathbf{F}_{4096}$ exist for lengths n satisfying $1 \leqslant n \leqslant 4097$. Then Construction E (using either of the maps D) produces lattices $\eta(\Lambda_{24})$ in dimension $N = 24n$, $1 \leqslant n \leqslant 4097$, for which the center density satisfies (from (111))

$$\log_2 \delta = 12((m-1)n - 2^m + m + 1) \qquad (117)$$

if $2^{m-1} \leqslant n \leqslant 2^m - 1$. Some examples are shown in Tables 8.7, 1.3 and 1.4. Up to dimension 4096 these lattices run neck and neck with $\eta(\Lambda_{12})$ and $\eta(\Lambda_{16})$. For $N = 24\, n \leqslant 24 \cdot 4097 = 98328$, $\eta(\Lambda_{24})$ is a lattice in $\mathbf{R}^N$ with density Δ satisfying

$$\frac{1}{N} \log_2 \Delta = -1.2454... + \epsilon \qquad (118)$$

from (112), where $|\epsilon| < 12N^{-1}\log_2(N/6)$. The first term on the right side of (118) is $-\frac{1}{2} \log_2 (48/e\pi)$.

Table 8.7. Center density δ of N-dimensional lattices $\eta(\Lambda_{24})$ obtained from the Leech lattice ($N = 24n$).

n	N	$\log_2 \delta$	n	N	$\log_2 \delta$	n	N	$\log_2 \delta$
1	24	0	10	240	228	341	8184	26712
2	48	12	21	504	696	342	8208	26808
3	72	24	43	1032	1896	683	16392	61608
4	96	48	85	2040	4680	1366	32784	139488
5	120	72	170	4080	11316	2731	65544	311496
6	144	96	171	4104	11400	4097	98328	491832

(f) *Lattices constructed from a norm-trebling map for* Λ_{24}. Suppose the MOG coordinates for a Leech lattice vector

$$\begin{array}{|cc|cc|cc|} \hline w_1 & w_2 & w_3 & w_4 & w_5 & w_6 \\ x_1 & x_2 & x_3 & x_4 & x_5 & x_6 \\ \hline y_1 & y_2 & y_3 & y_4 & y_5 & y_6 \\ z_1 & z_2 & z_3 & z_4 & z_5 & z_6 \\ \hline \end{array} \qquad (119)$$

are represented by the quaternionic vector

$$(w_1+x_1i+y_1j+z_1k, \ldots, w_6+x_6i+y_6j+z_6k) \qquad (120)$$

in $\mathbf{H}^6$. Then multiplication by the quaternion $\theta = i+j+k$ is a norm-trebling endomorphism D of Λ_{24}. Using MDS codes over $\mathbf{F}_{3^{12}}$ and Construction E we obtain lattices η_3 (Λ_{24}) in dimension $N = 24n \leqslant 24(3^{12}+1) = 12754608$ with

$$\log_2 \delta = 12(\log_2 3)\left\{an - \frac{3^{a+1}-1}{2} + a + 1\right\}, \qquad (121)$$

where $a = [\log_3 n]$, and

$$\frac{1}{N}\log_2 \Delta = -1.434... + \epsilon. \qquad (122)$$

These appear to be the densest packings known in dimensions 10^5 to 10^8 (see Table 1.3).

(g) *Packings constructed from* P_{48q}. Our fourth starting point is P_{48q} (Example 9 of the previous chapter), which has a norm-doubling map D very similar to (116). If P_{48q} is defined by (69) of Chap. 7 then we may take D to be the symmetric matrix

$$\frac{1}{6}\left|\begin{array}{c|cccccccc} 5 & -1 & -1 & -1 & -1 & -1 & \ldots & -1 \\ \hline -1 & -5 & 1 & 1 & 1 & 1 & \ldots & -1 \\ -1 & 1 & 1 & 1 & 1 & -1 & \ldots & -5 \\ -1 & 1 & 1 & 1 & -1 & 1 & \ldots & 1 \\ & & & \ldots & & & \ldots & \\ -1 & -1 & -5 & 1 & 1 & 1 & \ldots & -1 \end{array}\right| \qquad (123)$$

where in the second row there is a -5 in position 0, 1's at the quadratic residues modulo 47, and -1's elsewhere. Also $T = D^{-1} = \frac{1}{2}D$. This produces lattices $\eta(P_{48q})$ in dimensions $N = 48n$ for $n \leqslant 8 \cdot 10^8$ with

$$\frac{1}{N}\log_2 \Delta = -1.4529... + \varepsilon. \qquad (124)$$

The 96-dimensional lattice is particularly good (see Table 1.3).

(h) *Higher dimensions.* The construction may be applied repeatedly to any of these packings, producing an infinite tree of lattices. We may derive a lower bound on the density obtained in this way in high dimensions by the following argument. Let Λ be any of the lattices described in this section, with say dimension $m = 2\mu$, center density $\delta = 2^{\gamma}$, and norm-doubling map D. We apply our construction using codes of length $n = 2^{\mu}$, obtaining a lattice Λ' $(= L_a)$ in dimension $m' = 2\mu'$ with center density $\delta' = 2^{\gamma'}$, where, from (111)

$$\mu' = \mu 2^{\mu}, \quad \gamma' = 2^{\mu}\gamma + \mu(\mu-2)2^{\mu} + \mu(\mu+2). \tag{125}$$

It is simpler to work with the density Δ rather than with δ, so let us define

$$\eta = -\frac{\log_2 \Delta}{m}.$$

(We know from Chap. 1 that for large dimensions m the best packings satisfy $0\cdot 599 \leqslant \eta \leqslant 1$.) Using (18) of Chap. 1, (125) becomes

$$\eta' = \eta + 1 - \frac{1}{4\mu}\log_2(2\pi\mu) + o(1/\mu). \tag{126}$$

The solution of (125), (126) is $\eta(m) = \log_2^* m$, the smallest value of k for which $\log_2 \log_2 \cdots \log_2 m$ (with k $\log_2$'s) is less than 1. In other words as the dimension $N \to \infty$, these lattices have density Δ satisfying

$$\frac{1}{N}\log_2 \Delta \geqslant -\log_2^* N. \tag{127}$$

(127) holds for any choice of the initial lattice. The actual rate of growth is quite slow. For example the 324-dimensional lattice $\eta(K_{12})$ has children with

$$\frac{1}{N}\log_2 \Delta = -2.006... + \epsilon \tag{128}$$

for $N \leq 10^{51}$, the 65544-dimensional lattice $\eta\ (\Lambda_{24})$ has children with

$$\frac{1}{N}\log_2 \Delta = -2.2005... + \epsilon \tag{129}$$

for $N \leqslant 10^{9870}$, where ϵ in (129) satisfies

$$|\epsilon| < \frac{32772}{N}\log_2\frac{N}{16386}, \tag{130}$$

and so on.

Acknowledgements. We thank E. S. Barnes, M. Craig, W. M. Kantor, H.-G. Quebbemann and J. G. Thompson for helpful correspondence and discussions. §§8-11 are based on [Bar15], [Bos3], [Con37].

9

Bounds for Codes and Sphere Packings

N. J. A. Sloane

This chapter surveys recent work on finding bounds for error-correcting codes, constant weight codes, spherical codes, sphere packings, and other packing problems in 2-point-homogeneous spaces. A simplified account is given of the general machinery developed by Kabatiansky and Levenshtein for setting up such problems as linear programs. Many other recent bounds are also described.

1. Introduction

The introduction of the techniques of harmonic analysis and linear programming into sphere packing and coding theory has resulted in a considerable strengthening of the classical bounds. Our aim in this chapter is to describe a derivation of the new bounds from the analytical theory, which we quote without proof in §2.

The first step towards these bounds was taken by MacWilliams, who proved the surprising result that if C is a linear code then the weight distribution of the dual code C^* is a certain linear transformation of the weight distribution of C. (Eq. (50) of Chap. 3, [Mac2], [Mac6, Chap. 5, Th. 1]. See §2 of Chap. 3 and [Mac6] for terminology from coding theory.) Of course this implies in particular that this linear transformation (it is actually a *Krawtchouk* transform, as we shall see in §3.1) of the weight distribution of C has nonnegative entries.

The next step was taken by Delsarte ([Del8], [Mac6, Chap. 5, Th. 6]) who showed that the same, Krawtchouk, transform of the weight distribution of *any* code, linear or nonlinear, has nonnegative entries (see Theorem 1 below). This result has had far-reaching consequences. One of the main problems in coding theory is to determine the function $A(n,d)$, which is the maximal number of binary vectors of length n that can be found with the property that any two of the vectors differ in at least d places. Using his result, Delsarte showed that the value of $A(n,d)$ is bounded by the solution to a certain linear programming problem. Although this reformulation has not led to a complete solution of the problem, it has been very fruitful. One of its merits is its generality. In a long paper in 1973 Delsarte [Del9] showed that a number of other combinatorial problems can also be bounded by linear programs, especially problems which ask for the largest subset of an *association scheme* satisfying certain constraints (see also [Ban11], [Del11], [Gab1], [Goe4], [Sei3], [Slo5]). One such question, closely related to the coding problem already described, is to determine the function $A(n,d,w)$, which gives the maximum number of vectors containing w 1's and $n-w$ 0's, and having the property that any two of the vectors differ in at least d places. Such a collection of vectors is called a *constant weight code.* The linear programming bound for $A(n,d,w)$ is, as we shall see in §3, quite similar to that for $A(n,d)$.

The most spectacular application of the linear programming method in coding theory came in 1977 when McEliece, Rodemich, Rumsey and Welch ([McE4]; see also [McE1], [McE2], [Mac6, Chap. 17]) obtained upper bounds on $A(n,d)$ and $A(n,d,w)$ for large n that were a significant improvement on the existing bounds, and are still the best bounds known. In order to state their results it is appropriate to define the *rate* of the best code to be

$$R(n,d) = \frac{1}{n} \log_2 A(n,d),$$

and to examine how this rate behaves as a function of the ratio d/n when d and n both increase, for $0 \leqslant d/n \leqslant 1$. In fact it is enough to consider the range $0 \leqslant d/n \leqslant 1/2$, for it is not difficult to show (using for example the Plotkin bound (45)) that, in the range $1/2 \leqslant d/n \leqslant 1$, $R(n,d) \to 0$ as $n \to \infty$. The McEliece et al. (or JPL) bound on $A(n,d)$ states that, for $0 \leqslant d/n \leqslant 1/2$,

$$R(n,d) \lesssim \min_{0 \leqslant u \leqslant 1-2d/n} \{1 + h(u^2) - h(u^2 + \frac{2du}{n} + \frac{2d}{n})\} \tag{1}$$

as $n \to \infty$, where $h(x) = H_2((1 - \sqrt{1-x})/2)$, and $H_2(x)$ is the binary entropy function defined in (72) of Chap. 7. In the range $0.273 < d/n \leqslant 0.5$ the minimum in (1) is attained at $u = 1 - 2d/n$, and the bound simplifies to

$$R(n,d) \lesssim H_2\left(\frac{1}{2} - \left\{\frac{d}{n}\left(1 - \frac{d}{n}\right)\right\}^{1/2}\right), \quad \text{as } n \to \infty. \tag{2}$$

In the other direction Gilbert proved in 1952 ([Gil1], [Mac6, Chap. 17, Th. 30]) that, for $0 \leqslant \delta \leqslant 1/2$, there exists an infinite sequence of codes with $d/n \geqslant \delta$ and rate

$$R \gtrsim 1 - H_2\left(\frac{d}{n}\right), \quad \text{as } n \to \infty. \tag{3}$$

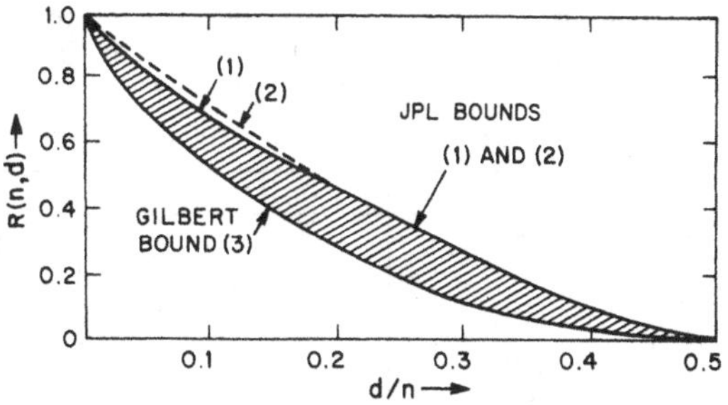

Figure 9.1. Asymptotic bounds on the best codes, showing $R(n,d) = n^{-1} \log_2 A(n,d)$ as a function of d/n as $n \to \infty$.

The bounds (1), (2) and (3) are plotted in Fig. 9.1. For large n the best codes lie somewhere in the shaded region.

The linear programming method has proved equally effective in studying codes of shorter length. [Bes3] gives a large number of bounds on $A(n,d)$ and $A(n,d,w)$ that were obtained in this way for codes of length $n \leqslant 24$. Some examples will be given in §3.3. Table 9.1 shows the present state of knowledge about small values of $A(n,d)$. Note that it is enough to consider even values of $d \geqslant 4$, since $A(n,2t-1) = A(n+1,2t)$, and $A(n,2) = 2^{n-1}$ [Mac6, Chap. 2]. The unmarked entries in the table are taken from [Bes3] or [Mac6]. References to other tables will be found in §2.1 of Chap. 3. See also [Bes2], [Roo1].

The next major advance was the paper by Kabatiansky and Levenshtein [Kab1], which succeeded in establishing bounds for sphere packings analogous to the JPL bounds for codes. In order to obtain their bounds Kabatiansky and Levenshtein extended the linear programming approach still further to include a number of packing problems on Riemannian manifolds. Their paper contains many new bounds, the most important being an upper bound on the highest density Δ_n of any packing of equal spheres in n-dimensional Euclidean space $\mathbf{R}^n$. The classical results of Minkowski and Blichfeldt (see §1.5 of Chap. 1, [Rog7]) state that

$$-n \leq \log_2 \Delta_n \lesssim -\frac{n}{2} + \log_2 \frac{n+2}{2}, \quad \text{for all } n. \tag{4}$$

Earlier improvements to the upper bound had been obtained by Rankin [Ran1], Rogers (Eq. (39), (40) of Chap. 1, [Rog2], [Rog7]) Sidel'nikov [Sid3], and Levenshtein (Eq. (42) of Chap. 1, [Lev4], [Lev7]). (Sidel'nikov [Sid1], [Sid2] and Levenshtein [Lev3] had also obtained new upper bounds to $A(n,d)$, but their results were superseded by the JPL bound. See also [Lev2], [Lev5], [Lev6].) Kabatiansky and Levenshtein showed that the upper bound in (4) can be replaced by

$$\log_2 \Delta_n \lesssim -0.599n, \quad \text{as } n \to \infty . \tag{5}$$

Table 9.1. The best codes: bounds for $A(n,d)$ (a: [Bes1], b: [Pul82], c: [Rom1], d: [Ham88], e: [Kaik98], f: P.R.J. Östergård, T. Baicheva, and E. Koler (1988), g: Etzion [CHLL], p. 58).

n	$d=4$	$d=6$	$d=8$	$d=10$
6	4	2	1	1
7	8	2	1	1
8	16	2	2	1
9	20	4	2	1
10	40^a	6	2	2
11	72^f	12	2	2
12	144^f	24	4	2
13	256	32	4	2
14	512	64	8	2
15	1024	128	16	4
16	2048	256	32	4
17	2720^c-3276	256-340	36-37	6
18	5312^g-6552	512-680	64-72^b	10
19	10496^d-13104	1024-1288	128-144	20
20	20480^a-26208	2048-2372	256-279	40
21	36864-43690	2560-4096	512	42^e-48^b
22	73728-87380	4096-6942	1024	50^e-88^b
23	147456-173784	8192-13774	2048	76^e-150
24	294912-344636	16384-24106	4096	128-280

Their technique is to consider, instead of Δ_n itself, the related quantity $A(n,\theta)$. We repeat the definition from §2.3 of Chap. 1. Let Ω_n denote the [$(n-1)$-dimensional] unit sphere in $\mathbf{R}^n$:

$$\Omega_n = \{x = (x_1, \ldots, x_n) \in \mathbf{R}^n : x_1^2 + \cdots + x_n^2 = 1\} .$$

A *spherical code C* of dimension n, size M and minimal angle θ is a set of M points of Ω_n with the property that

$$x \cdot y \leqslant \cos\theta \quad \text{for } x, y \in C, x \neq y .$$

$A(n,\theta)$ is the maximal size of such a code. Kabatiansky and Levenshtein prove that, for $0 < \theta < \pi/2$, and large n,

$$\frac{1}{n} \log_2 A(n,\theta) \le \frac{1+\sin\theta}{2\sin\theta} \log_2 \frac{1+\sin\theta}{2\sin\theta} - \frac{1-\sin\theta}{2\sin\theta} \log_2 \frac{1-\sin\theta}{2\sin\theta}, \qquad (6)$$

and use this to derive (5). The main goal of the present chapter is to give a simplified account of Kabatiansky and Levenshtein's general theory and to sketch how their bounds are obtained.

The linear programming method has also proved successful in low dimensions. As we shall see in Chap. 13, it can be used to solve the kissing number problem in 8 and 24 dimensions.

The general theme of this chapter is the problem of finding nice arrangements of points in various spaces. The four examples we are primarily concerned with are the following.

Example E1. The space is "Hamming space", i.e. the set $\mathbf{F}_2^n = \{0,1\}^n$ of binary vectors of length n, and a binary code is a subset of $\mathbf{F}_2^n$ (see §2.1 of Chap. 3).

Example E2. The space is "Johnson space", i.e. the subset $\mathbf{F}_2^{n,w} \subseteq \mathbf{F}_2^n$ consisting of all vectors containing w 1's and $n-w$ 0's. A subset C of $\mathbf{F}_2^{n,w}$ is a *constant weight code*, and $A(n,d,w)$ is the greatest cardinality of a constant weight code in which distinct codewords have Hamming distance at least d apart. (The name Johnson space was chosen because S. M. Johnson was the first to extensively study $A(n,d,w)$ – see [Mac6].)

Example E3. The space is the "spherical space" Ω_n, and a finite subset of Ω_n is a spherical code. In this example we always assume $n \geqslant 3$, the cases $n \leqslant 2$ being special (and trivial).

Example E4. Finally there is Euclidean space $\mathbf{R}^n$. A subset $C \subset \mathbf{R}^n$ such that $N(x-y) \geqslant 4\rho^2$ for $x, y \in C$, $x \neq y$, is a sphere packing of radius ρ.

In §2 we quote the machinery needed from harmonic analysis, the main purpose being to construct certain functions, the zonal spherical functions, associated with each of the packing problems to be considered. §3 shows how the linear programming bounds (Theorems 2-4) are simply derived from the results in §2. A short proof is also given of the "mean inequality" for codes (Theorem 5). §§3.3-3.5 given many applications of these bounds to Examples E1, E2 and E3. Table 9.2 for instance gives bounds on the number of spheres that can touch a pair of touching spheres of the same size. The crucial link between spherical codes (Example E3) and sphere packings (E4) is provided by Theorem 6. Finally §4 gives a brief account of other recent bounds. As background references [Dym1], [Mac1], [Mil3], [Ter1] and [Vil1] are particularly recommended.

2. Zonal spherical functions

In this section we prepare the way for the linear programming bounds by constructing a set of functions, called *zonal spherical functions*, for each of the problems to be considered.

Table 9.2. How many spheres can touch two spheres? Bounds on $(n, \cos^{-1} 1/3)$, from [Ass1]. The lower bound for $n = 4$ is taken from [Mac0].

n	$A(n, \cos^{-1} 1/3)$	n	$A(n, \cos^{-1} 1/3)$
1	2	6	32-37
2	5	7	56
3	9	8	64-78
4	14-15	9	96-107
5	20-24	10	≤146
		23	4600

2.1 The 2-point-homogeneous spaces. To start with we choose a group G, whose properties will determine everything that follows. Although Kabatiansky and Levenshtein's theory is somewhat more general, we shall restrict ourselves here to two classes of compact groups: finite groups and connected compact groups. We shall use Examples E1, E2 and E3 of the previous section as illustrations (these are the settings for error-correcting codes, constant weight codes and spherical codes respectively). The appropriate groups G are as follows. In Example E1 we take G to be the automorphism group of the n-dimensional cube, called the hyperoctahedral group, or $[3^{n-2}, 4]$ in Coxeter's notation (see [Cox28]). It has order $2^n \cdot n!$. (E2) G is the symmetric group S_n, of order $n!$. (E3) G is the special orthogonal group $SO(n)$, i.e. the group of isometries of Ω_n with determinant 1.

Next we need a *2-point-homogeneous G-space* $\mathbf{M}$ [Hel6, p. 289]. This implies three things. (i) $\mathbf{M}$ is a set on which G acts, so that there is an element $g(x) \in \mathbf{M}$ for every $g \in G$ and $x \in \mathbf{M}$, and furthermore $g_2(g_1(x)) = (g_2 g_1)(x)$ and $1(x) = x$ for all x, where g_1, g_2 are arbitrary elements of G and $1 \in G$ is the identity element. (ii) $\mathbf{M}$ is a metric space with a distance function τ defined on it. (iii) τ is strongly invariant under G: for any $x, x', y, y' \in \mathbf{M}$, $\tau(x,y) = \tau(x',y')$ if and only if there is an element $g \in G$ such that $g(x) = x'$ and $g(y) = y'$. We let $T \subseteq \mathbf{C}$ denote the range of τ.

Assumption (iii) certainly implies that G acts transitively on $\mathbf{M}$. In fact G is as close to being doubly transitive as it can be without changing the distances $\tau(x,y)$ between the points of $\mathbf{M}$.

Let H be the subgroup of G that fixes a particular element $x_0 \in \mathbf{M}$ (any x_0 will do, since G is transitive). Then $\mathbf{M}$ can be identified with the space G/H of left cosets gH (with the point x_0 corresponding to H itself).

Examples (continued). (E1) For $\mathbf{M}$ we take the set of vertices of the n-dimensional cube, $\{(-1)^x := ((-1)^{x_1}, \ldots, (-1)^{x_n})$ for $x = (x_1, \ldots, x_n) \in \mathbf{F}_2^n\}$, which of course we will often identify with $\mathbf{F}_2^n$. We let $\tau((-1)^x, (-1)^y)$ be the Hamming distance between x and y, so that

$T = \{0,1,\ldots,n\}$. A typical element of G has the form $g = \sigma\pi$, and consists of a permutation $\pi \in S_n$ followed by a sign change σ = diag $\{(-1)^{a_1},\ldots,(-1)^{a_n}\}$, $a_i = 0$ or 1. The subgroup H fixing the vertex $x_0 = (1,1,\ldots,1)$ is S_n itself, and the natural map $G \to G/H \cong \mathbf{M}$ sends g to $(-1)^a$.

(E2) $\mathbf{M}$ = Johnson space $\mathbf{F}_2^{n,w}$ with $w \leqslant n/2$, $\tau(x,y) = ½$ Hamming distance between x and y, $T = \{0,1,\ldots,w\}$, $H \cong S_{n-w} \times S_w$. (E3) $\mathbf{M} = \Omega_n$, $\tau(x,y) = x \cdot y$, $T = [-1, 1]$, $H = SO(n-1)$. (In this case $\cos^{-1}(x \cdot y)$ is the metric on $\mathbf{M}$, but it is simpler to work with $x \cdot y$.)

The final assumption we shall make about G is that (a) if G is infinite, then $\mathbf{M}$ is a connected Riemannian manifold and τ is a constant times the natural distance on the manifold, and (b) if G is finite, and $d_0 = \min \tau(x,y)$ for $x,y \in \mathbf{M}$, $x \neq y$, then $\mathbf{M}$ has the structure of a graph in which x is adjacent to y if and only if $\tau(x,y) = d_0$, and furthermore τ is a constant times the natural distance in the graph.

These assumptions are quite restrictive. In fact if G is infinite then Wang ([Wan2]; see also [Hel6, p. 535], [Tit1], [Wol1, Th. 8.12.2]) has proved that the following are the only possibilities for the space $\mathbf{M}$:

(a) the sphere Ω_n,

(b) real projective space $\mathbf{P}^n(\mathbf{R}) = SO(n+1)/O(n)$,

(c) complex projective space $\mathbf{P}^n(\mathbf{C}) = SU(n+1)/U(n)$,

(d) quaternionic projective space $\mathbf{P}^n(\mathbf{H}) = Sp(n+1)/(Sp(n) \times Sp(1))$,

(e) the Cayley projective plane $\mathbf{P}^2(\mathbf{Cay})$.

These are the compact Riemannian symmetric spaces of rank one, and (G,H) is an example of a Gelfand pair ([Bou0], [Gel1], [Kra2], [Let1], [Let2]).

The finite 2-point-homogeneous spaces $\mathbf{M}$ have not yet been completely classified, although they have been the subject of an enormous amount of research in connection with distance-transitive graphs, strongly regular graphs, two-weight codes and association schemes (see [Ban11], [Ban12], [Big2], [Big3a], [Big4], [Big5], [Bum1], [Cal1], [Cam2], [Coi1], [Coi2], [Cur1], [Del9], [Del11], [Del12], [Del14], [Dun1]-[Dun9], [Goe4], [Gor1], [Hub1], [Kan5], [Leo10], [Leo11], [Lin12], [McK1], [Sei1]-[Sei3], [Slo5], [Smi1]-[Smi3], [Sta4], [Sta8]). Some of the most important examples are:

(f) Hamming space $\mathbf{F}_2^n$, or more generally the space $\mathbf{F}_q^n$ of n-tuples from $\mathbf{F}_q$, with $\tau(x,y)$ = Hamming distance,

(g) Johnson space $\mathbf{F}_2^{n,w}$ as in Example E2,

(h) the space whose elements $x, y, \ldots$ are k-dimensional subspaces of $\mathbf{F}_q^n$, where $k \leqslant n/2$, with $\tau(x,y) = \mathrm{rank}(x \cap y)$,

(i) the space of maximal totally isotropic subspaces of an orthogonal, unitary, or symplectic geometry over a finite field, with $\tau(x,y)$ = rank $(x \cap y)$, and

(j) the spaces of bilinear, alternating bilinear, or hermitian forms over a finite field, with $\tau(x,y) = \text{rank}\,(x-y)$,

but this list is by no means complete. It is remarkable that the bounds obtained in §3 apply to subsets of any one of these spaces.

2.2 Representations of G. Since G is compact, it has a unique normalized measure μ, Haar measure, which is invariant under G (see for example [Hal6]). This induces a unique invariant measure on $\mathbf{M}$, which will also be denoted by μ. We assume that μ is normalized so that $\mu(\mathbf{M}) = 1$. In Example (E1), $\mu(x) = 2^{-n}$ for each $x \in \mathbf{F}_2^n$, and in (E2) $\mu(x) = 1/\binom{n}{w}$ for each $x \in \mathbf{F}_2^{n,w}$. In (E3) μ is normalized Lebesgue measure on the sphere, with $\mu(\Omega_n) = 1$.

Let $L^2(G)$ denote the vector space of complex-valued functions u on G satisfying

$$\int_G |u(g)|^2 \, d\mu(g) < \infty,$$

with the inner product

$$(u_1, u_2) = \int_G u_1(g) \, \overline{u_2(g)} \, d\mu(g).$$

Those $u \in L^2(G)$ that are constant on left cosets of H can be regarded as belonging to $L^2(\mathbf{M})$, which is defined similarly and has the inner product

$$(u_1, u_2) = \int_{\mathbf{M}} u_1(x) \, \overline{u_2(x)} \, d\mu(x).$$

We shall study how G acts on the functions in $L^2(G)$ and $L^2(\mathbf{M})$.

The left regular representation of G on $L^2(G)$ is given by

$$g \to R(g), \quad g \in G,$$

where

$$R(g)u(f) = u(g^{-1}f), \text{ for } u \in L^2(G), \; f \in G.$$

Then $R(g)$ is a unitary representation with respect to our scalar product, i.e. satisfies

$$(R(g)u_1, \; R(g)u_2) = (u_1, u_2)$$

for all $u_1, u_2 \in L^2(G)$ (see [Vil1, p. 28]).

By the Peter-Weyl theorem ([Coi1], [Coi2], [Gro1], [Vil1], [War3]) the space $L^2(G)$ decomposes into a countable direct sum of mutually orthogonal subspaces $\{V^{(m,i)}, m = 0, 1, 2, \ldots; i = 1, 2, \ldots, d'_m\}$ with the following properties. $V^{(m,i)}$ is a vector space of *continuous* functions, of dimension d'_m, and affords an irreducible unitary representation of G. The spaces $V^{(m,1)}, V^{(m,2)}, \ldots, V^{(m,d'_m)}$ are isomorphic and afford equivalent representations of G. Finally every irreducible unitary representation of G is obtained in this way.

To obtain the decomposition of $L^2(\mathbf{M})$ we must consider only those irreducible representations of G which are of "class 1 with respect to H", i.e. those representations $\rho(g)$ on a space $V^{(m,i)}$ which have the property that there is a nonzero vector $a \in V^{(m,i)}$ with $\rho(h)a = a$ for all $h \in H$ (see [Vil1], pp. 29, 52]). For the groups we are considering it is known (cf. [Coi, Th. 3.5], [Coi2, Chap. 2, §4], [Gan1, Prop. 3.4], [Hig2], [Tra1]) that if such an invariant vector a exists then it is unique apart from a scale factor (a subgroup H with this property is called "massive").

Then (see [Vil1, p. 54]) $L^2(\mathbf{M})$ decomposes into a countable direct sum of mutually orthogonal subspaces $\{V^{(k)}, k=0, 1, ...\}$. Each $V^{(k)}$ is one of the spaces $V^{(m,i)}$ mentioned above, and is a space of continuous functions of dimension $d_k = d'_m$. $V^{(k)}$ affords an irreducible representation $\rho^{(k)}(g)$ of G which is of class 1 with respect to H, and each such representation arises just once in the decomposition, the $V^{(k)}$ being mutually nonisomorphic. We take $V^{(0)}$ to be the space of constant functions, and $d_0 = 1$.

Let $\{e_i^k: i=1, \ldots, d_k\}$ be an orthonormal basis for $V^{(k)}$, chosen so that e_1^k is an invariant vector a. Let $\rho^{(k)}(g)$ be represented with respect to this basis by the matrix $\left(\rho_{ij}^{(k)}(g)\right)$, where

$$\rho^{(k)}(g)\, e_i^k = \sum_{j=1}^{d_k} \rho_{ji}^{(k)}(g) e_j^k .$$

It is clear that $\rho_{11}^{(k)}(h) = 1$ for $h \in H$. The functions $\rho_{i1}^{(k)}(g)$ are constant on left cosets of H, and can be written as $\rho_{i1}^{(k)}(x)$, $x \in \mathbf{M}$ [Vil1, p. 53]. Then

$$\sqrt{d_k}\, \rho_{11}^{(k)}(x), \ldots, \sqrt{d_k}\, \rho_{d_k 1}^{(k)}(x)$$

is an orthonormal basis for $V^{(k)}$ [Coi2, Th. 1.10]. Any element of $L^2(\mathbf{M})$ has a "Fourier expansion"

$$u(x) = \sum_k \sum_{i=1}^{d_k} c_{i1}^{(k)}\, \rho_{i1}^{(k)}(x), \quad x \in \mathbf{M}, \tag{7}$$

where the series converges in the norm of $L^2(\mathbf{M})$, and

$$c_{i1}^{(k)} = d_k \int_{\mathbf{M}} u(x)\, \overline{\rho_{i1}^{(k)}(x)} d\mu(x) \tag{8}$$

(see [Vil1, p. 54]).

2.3 Zonal spherical functions. To each space $V^{(k)}$ we associate the function

$$J_k(x,y) = d_k \sum_{i=1}^{d_k} \rho_{i1}^{(k)}(x)\, \overline{\rho_{i1}^{(k)}(y)}, \quad x, y \in \mathbf{M}, \tag{9}$$

which has a number of interesting properties.

(i) It is immediate from the definition that $J_k(x,y)$ is *positive definite*, in the sense that, for any $u \in L^2(\mathbf{M})$,

$$\int_{\mathbf{M}}\int_{\mathbf{M}} J_k(x,y)\, u(x)\overline{u(y)}\, d\mu(x)\, d\mu(y) \geqslant 0, \tag{10}$$

or alternatively for any integer n, any points $x_1, \ldots, x_n \in \mathbf{M}$, and any complex numbers $a_1, \ldots, a_n$ the relation

$$\sum_{i=1}^{n} \sum_{j=1}^{n} J_k(x_i, x_j)\, a_i \bar{a}_j \geqslant 0 \tag{11}$$

holds. (Conditions (10) and (11) are equivalent — see for example Bochner [Boc1].)

(ii) Since the representation $\rho(g)$ afforded by $V^{(k)}$ is unitary, it also follows from (9) that $J_k(x,y)$ is independent of the particular choice of basis for $V^{(k)}$, and furthermore that

$$J_k(gx, gy) = J_k(x,y) \tag{12}$$

for all $x,y \in \mathbf{M}$, $g \in G$, i.e. $J_k(x,y)$ is invariant under G, and so depends only on $\tau(x,y)$. Thus we can write

$$J_k(x,y) = d_k\, \Phi_k(\tau(x,y)), \tag{13}$$

where

$$\Phi_k(\tau(x,y)) = \sum_{i=1}^{d_k} \rho_{i1}^{(k)}(x)\, \overline{\rho_{i1}^{(k)}(y)}. \tag{14}$$

The function $\Phi_k(t)$, which is a continuous function of t defined on the set T (the range of τ), is called the *zonal spherical function* associated with $V^{(k)}$ (see [Coi2], [Dun7], [Erd1], [Gan1], [Müll], [Sta11], [Vil1]). If we let $\tau(x,x) = \tau_0$, then

$$\Phi_k(\tau_0) = \sum_{i=1}^{d_k} \left| \rho_{i1}^{(k)} \right|^2 = 1, \tag{15}$$

since $\{\rho_{i1}^{(k)}\}_i$ is the first column of a unitary matrix. Also $\Phi_0(t) = 1$.

(iii) From (9), $J_k(x, y) = \overline{J_k(y, x)}$, and (13) and $\tau(x,y) = \tau(y,x)$ then imply that $J_k(x,y)$ and $\Phi_k(t)$ are real.

(iv) The orthogonality of the spaces $V^{(k)}$ implies that the $\Phi_k(t)$ are orthogonal and satisfy

$$\int_T \Phi_k(t)\, \Phi_l(t)\, d\tilde{\mu}(t) = \frac{\delta_{kl}}{d_k}, \tag{16}$$

where $\tilde{\mu}$ is the measure on T given by

$$\tilde{\mu}(A) = \mu\, \{x \in \mathbf{M}: \tau(x_0,x) \in A\}.$$

(v) An alternative expression for $\Phi_k(t)$ may be obtained as follows ([Coi2, p. 38], [Vil1, p. 30]). By definition,

$$\rho_{11}^{(k)}(g) = (\rho^{(k)}(g)a, a), \tag{17}$$

where a is the unit vector in $V^{(k)}$ fixed by H. Furthermore it is easy to see that ([Vil1, p. 30])

$$\rho_{11}^{(k)}(h_1 \, g \, h_2) = \rho_{11}^{(k)}(g) \tag{18}$$

for any $h_1, h_2 \in H$, $g \in G$. Thus $\rho_{11}^{(k)}$ is constant on the double cosets HgH. If g_x and g_y are any elements of G such that

$$g_x(x_0) = x, \ \ g_y(x_0) = y,$$

then

$$\Phi_k(\tau(x,y)) = \rho_{11}^{(k)}(g_y^{-1} g_x). \tag{19}$$

In the literature $\rho_{11}^{(k)}(g)$, $F_k(x,y)$ and $\Phi_k(t)$ are all referred to as zonal spherical functions (or kernels). The name comes from the identification of points $x \in \mathbf{M}$ with left cosets gH. Then Hx represents a "sphere" in $\mathbf{M}$ centered at x_0 and passing through x, and is identified with the double coset HgH. From (18) $\rho_{11}^{(k)}(x)$ is constant on these spheres.

Example E1 (see [Dun1], [Dun8]). $L^2(\mathbf{M})$ consists of the complex-valued functions on the vertices of the n-cube, and has a basis consisting of the monomials $\phi_1^{a_1} \dots \phi_n^{a_n}$, where $(a_1, \dots, a_n) \in \mathbf{F}_2^n$ and ϕ_i is the ith coordinate function (with $\phi_i^2 = 1$). G permutes and/or negates the ϕ_i while H just permutes them. Also $V^{(k)}$ consists of the polynomials of degree k in $\phi_1, \dots, \phi_n$, with $d_k = \binom{n}{k}$, for $k = 0, 1, \dots, n$. The fixed vector $a = e_1^k$ in $V^{(k)}$ is $\Sigma \, \phi_1^{a_1} \dots \phi_n^{a_n}$, the sum being over all $(a_1, \dots, a_n)$ with weight k (i.e. which contain k ones). Then $\Phi_k(t)$ is a special case of a Krawtchouk polynomial (see [Mac6], [Sze1]):

$$\Phi_k(t) = K_k(t,n) = \binom{n}{k}^{-1} \sum_{j=0}^{k} (-1)^j \binom{t}{j} \binom{n-t}{k-j}, \tag{20}$$

for $k = 0, 1, \dots, n$.

Example E2 (see [Del12], [Dun2]). $L^2(\mathbf{M})$ consists of the complex-valued functions on the w-subsets of an n-set, and decomposes into orthogonal subspaces $\{V^{(k)}, 0 \leqslant k \leqslant w\}$. V_k has dimension $d_k = \binom{n}{k} - \binom{n}{k-1}$, and is described explicitly in [Del12] (where it is denoted by Harm(k)). V_k affords the irreducible representation of S_n corresponding to the two-rowed Young tableau $[n-k,k]$ (cf. [Mil3, §4.2]). The corresponding zonal spherical function is

$$\Phi_k(t) = Q_k(w-t), \tag{21}$$

where

$$Q_k(t) = \sum_{i=0}^{k} (-1)^i \frac{\binom{k}{i}\binom{n+1-k}{i}\binom{t}{i}}{\binom{w}{i}\binom{n-w}{i}} \tag{22}$$

is a particular kind of Hahn polynomial [Kar2].

Example E3 (the most extensively studied case: see for example [Coi2], [Del16], [Dun7], [Erd1], [Hob1], [Koo1], [Müll], [Vil1, Chap. IX]). $L^2(\Omega_n)$ decomposes into an infinite direct sum of orthogonal subspaces $\{V^{(k)}, k = 0, 1, ...\}$. $V^{(k)}$ consists of all functions in $L^2(\Omega_n)$ which are represented by homogeneous polynomials $f(x_1, \ldots, x_n)$ of total degree k in variables $x_1, \ldots, x_n$ that satisfy Laplace's equation

$$\nabla^2 f = \frac{\partial^2 f}{\partial x_1^2} + \cdots + \frac{\partial^2 f}{\partial x_n^2} = 0.$$

$V^{(k)}$ is usually denoted by Harm(k) and its elements are called *spherical harmonics*. It has dimension

$$d_k = \binom{n+k-1}{n-1} - \binom{n+k-3}{n-1}.$$

The corresponding zonal spherical function is a Gegenbauer or ultraspherical polynomial, which is a special case of a Jacobi polynomial. To avoid confusion of notation we write it as a Jacobi polynomial, using the standard notation of [Abr1]:

$$\Phi_k(t) = \frac{P_k^{(\alpha,\beta)}(t)}{\binom{k+\alpha}{k}}, \quad k = 0, 1, \ldots, \tag{23}$$

where $\alpha = \beta = (n-3)/2$. (The normalization is determined by (15), remembering that $\tau_0 = \tau(x,x) = x \cdot x = 1$.)

Many of the other 2-point-homogeneous spaces that we listed in Section 2.1 have also been studied in detail — see [Coi2], [Del9], [Del11], [Del12], [Del14], [Dun2]-[Dun6], [Koo1]-[Koo3], [Sta4]-[Sta11], [Vil1]. For the projective spaces $\mathbf{P}^n(\mathbf{R})$, $\mathbf{P}^n(\mathbf{C})$, $\mathbf{P}^n(\mathbf{H})$ and $\mathbf{P}^2(\mathbf{Cay})$ the zonal spherical functions are Jacobi polynomials

$$\Phi_k(t) = \frac{P_k^{(\alpha,\beta)}(2t^2-1)}{\binom{k+\alpha}{k}}, \tag{24}$$

where $\alpha = (n-2)/2$ and $\beta = -\frac{1}{2}$, 0, 1 and 3 respectively. For the finite 2-point-homogeneous spaces many (but not all) of the zonal spherical functions belong to a general family found by Askey and Wilson (see [Ask4], [Leo10], [Leo11]).

2.4 Positive-definite degenerate kernels. We have seen that the zonal spherical function $J_k(x,y)$ is a continuous, positive-definite, degenerate (because of (9)) kernel or p.d.k. on $\mathbf{M}$. It is also invariant under G by

(12). It is easy to see that the product of two p.d.k.'s is a p.d.k., and a sum of p.d.k.'s with positive coefficients is a p.d.k. The p.d.k.'s that are given by the zonal spherical functions are called *elementary*, and clearly any sum

$$F(x, y) = \sum_k \lambda_k J_k(x, y) \tag{25}$$

with $\lambda_k \geqslant 0$ and $\Sigma_k \lambda_k d_k < \infty$ of elementary p.d.k.'s is also an invariant p.d.k. It is surprising that the converse holds: any invariant p.d.k. can be written in the form (25) as a sum of elementary p.d.k.'s with positive coefficients. This was first proved for Example E3 by Schoenberg [Sch5], and generalized to include all the groups considered here and many others by Bochner [Boc1] and later writers [Bin1], [Gel1], [Kre1] (see also [Kab1, Th. 2]).

It is a corollary that the product of any two zonal spherical functions $\Phi_i(t)$ and $\Phi_j(t)$ can be expanded as a sum

$$\Phi_i(t)\Phi_j(t) = \sum_k c_{ijk} \Phi_k(t) \tag{26}$$

with $c_{ijk} \geqslant 0$ ([Koo3]). Of course this applies to those Krawtchouk, Hahn and Jacobi polynomials given above. Such expansions with positive coefficients have been extensively studied [Ask1], [Ask2], [Gas1].

3. The linear programming bounds

3.1 Codes and their distance distributions. Having constructed the zonal spherical functions $\phi_k(t)$ we can now proceed to derive the linear programming bounds. We shall study finite subsets C of any of the 2-point-homogeneous spaces $\mathbf{M}$; such subsets will be called *codes*, since they generalize the notions of error-correcting codes and spherical codes. Of course we are not interested in arbitrary subsets, but only those in which the distances between vectors are restricted in some way. We assume that a subset $S \subseteq T$ has been specified, and that we wish to find codes C for which $\tau(x, y) \in S$ for all $x, y \in C$, $x \neq y$; such a code is called an S-*code*, and the largest cardinality $|C|$ of an S-code will be denoted by $A(\mathbf{M}, S)$.

Example E1. If we take $S = \{d, d+1, \ldots, n\}$ then an S-code is an ordinary error-correcting code of minimal distance d, and $A(\mathbf{M}, S) = A(n, d)$. (E2) If $S = \{\delta, \delta+1, \ldots, n\}$ an S-code is a constant weight code with minimal distance 2δ, and $A(\mathbf{F}_2^{n,2}, S) = A(n, e, w)$. (E3) If $S = [-1, \cos\theta]$, an S-code is a spherical code with minimal angle θ, and $A(\Omega_n, S) = A(n, \theta)$.

The *distance distribution* $\{\alpha_t\}$ of any code C is defined by

$$\alpha_t = \frac{1}{|C|} \cdot (\text{number of ordered pairs } x, y \in C \text{ with } \tau(x, y) = t).$$

It is immediate that

$$\alpha_{\tau_0} = 1, \tag{27}$$

$$\alpha_t \geqslant 0 \quad \text{for } t \in T, \tag{28}$$

$$\sum_{t \in T} \alpha_t = |C|, \tag{29}$$

but there are some additional inequalities.

Theorem 1. *Let* $\beta_k = |C|^{-1} \cdot \sum_t \alpha_t \, \Phi_k(t)$, $k = 0, 1, \ldots,$ *denote the "transform" of the distance distribution by the zonal spherical functions. Then* $\beta_k \geqslant 0$ *for all* k.

Proof.

$$\beta_k = \frac{1}{|C|} \sum_t \alpha_t \, \Phi_k(t) = \frac{1}{|C|^2} \sum_{x, y \in C} \Phi_k(\tau(x, y))$$

$$\geqslant 0 \quad \text{by (11).}$$

If G (and $\mathbf{M}$) are finite this result is due to Delsarte [Del8], [Del9]; in the infinite case it can be found (implicitly or explicitly) in [Del16], [Kab1], [Llo1], [Odl5]. See also [Dun5]. In Example E1 $\{\beta_k\}$ is the Krawtchouk transform of $\{\alpha_t\}$. If C is a linear code then $\{\alpha_t\}$ is its weight distribution and by the MacWilliams identity (Eq. (50) of Chap. 3) $\{\beta_t\}$ is the weight distribution of the dual code.

3.2 The linear programming bounds. Theorem 1 and (27)-(29) make it possible to regard the problem of bounding $A(\mathbf{M}, S)$ as a linear programming problem:

Primal problem: Choose a natural number s, a subset $\{\tau_1, \ldots, \tau_s\}$ of S, and real numbers $\alpha_{\tau_1}, \ldots, \alpha_{\tau_s}$ so as to

$$\text{maximize} \sum_{i=1}^{s} \alpha_{\tau_i} \tag{30}$$

subject to

$$\alpha_{\tau_i} \geqslant 0, \ i = 1, \ldots, s, \tag{31}$$

$$\sum_{i=1}^{s} \alpha_{\tau_i} \Phi_k(\tau_i) \geqslant -1, \ k = 0, 1, \ldots. \tag{32}$$

This is a linear programming problem with perhaps infinitely many unknowns α_t $(t \in S)$ and constraints (32). If C is an S-code then its distance distribution $\{\alpha_t\}$ certainly satisfies the inequalities (31), (32). So if the maximal value of the sum (30) that can be attained (for any choice of s and $\tau_1, \ldots, \tau_s$) is A^*, then $A(\mathbf{M}, S) \leqslant 1 + A^*$. (The extra 1 arises because the term $\alpha_{\tau_0} = 1$ does not occur in (30).)

The dual problem is as always more convenient, and reads as follows (cf. [Duf1], [Dun9], [Sim0]).

Dual problem. Choose a natural number N and real numbers $f_1, \ldots, f_N$ so as to

$$\text{minimize} \sum_{k=1}^{N} f_k \tag{33}$$

subject to

$$f_k \geqslant 0, \quad k = 1, \ldots, N, \tag{34}$$

$$\sum_{k=1}^{N} f_k \, \Phi_k(t) \leqslant -1, \quad \text{for } t \in S. \tag{35}$$

Thus we have proved:

Theorem 2. *If A^* is the optimal solution to either of the primal or dual problems then $A(\mathbf{M}, S) \leqslant 1 + A^*$.*

The reason for preferring the dual problem is that by the duality theorem *any* feasible solution to the dual problem is an upper bound to the optimal solution of the primal problem, or in other words:

Theorem 3. *If $f_1, \ldots, f_N$ satisfy* (34) *and* (35) *then $A(\mathbf{M}, S) \leqslant 1 + f_1 + \cdots + f_N$.*

This can be restated in an equivalent form which is sometimes easier to use.

Theorem 4. *Suppose the space $\mathbf{M}$, the zonal spherical functions $\Phi_k(t)$, and a subset $S \subseteq T$, are given, and we wish to bound the cardinality $A(\mathbf{M}, S)$ of the largest subset C of $\mathbf{M}$ such that $\tau(x,y) \in S$ for all $x, y \in C$, $x \neq y$. If we can find a linear combination*

$$f(t) = \sum_{k=0}^{N} f_k \Phi_k(t), \tag{36}$$

for some N, which satisfies $f_0 > 0$, $f_k \geqslant 0$ for $k = 1, \ldots, N$, and $f(t) \leqslant 0$ for $t \in S$, then

$$A(\mathbf{M}, S) \leqslant f(\tau_0)/f_0. \tag{37}$$

Theorem 2, 3 and 4 are the general linear programming bounds we were seeking.

As a consequence of Theorem 1 any code C must satisfy further inequalities, the so-called mean inequalities. Let $F(x,y)$ be any invariant p.d.k. on $\mathbf{M}$. By §2.4 we can write $F(x,y) = f(\tau(x,y))$, where

$$f(t) = \sum_k f_k \Phi_k(t), \quad f_k \geqslant 0,$$

and the sum converges for all $t \in T$. The mean of F over C and over $\mathbf{M}$ are defined by

$$F(C) = \frac{1}{|C|^2} \sum_{x,y \in C} F(x,y), \tag{38}$$

$$F(\mathbf{M}) = \int_{\mathbf{M}}\int_{\mathbf{M}} F(x,y)\, d\mu(x)\, d\mu(y) \tag{39}$$

$$= \int_{\mathbf{M}} F(x,y)\, d\mu(x), \quad \text{for any } y \in \mathbf{M},$$

since $F(x,y)$ only depends on $\tau(x,y)$. Thus $F(\mathbf{M}) = f_0$, using (16) and $\Phi_0(t) = 1$.

Theorem 5 (*The mean inequality* [*Kab*1], [*Lev*5], [*Sid*3]). *For any invariant p.d.k. F we have*

$$F(C) \geqslant F(\mathbf{M}) \tag{40}$$

or equivalently

$$\frac{1}{|C|^2} \sum_{x,y \in C} f(\tau(x,y)) \geqslant f_0. \tag{41}$$

Proof. From Theorem 1, $\beta_k \geqslant \delta_{k,0}$. Multiplying by f_k and summing on k we obtain (41).

This gives an immediate proof of Theorem 4, avoiding the use of linear programming.

Second proof of Theorem 4. Let C be any S-code. From Theorem 5

$$f_0 = F(\mathbf{M}) \le F(C) = \frac{1}{|C|^2} \sum_{x,y \in C} F(x,y)$$

$$\leqslant \frac{1}{|C|^2} \sum_{x \in C} F(x,x) = \frac{1}{|C|} f(\tau_0).$$

Note that the only properties of zonal spherical functions used to prove Theorems 2-4 are that they are positive-definite kernels invariant under G. The theorem quoted in §2.4, that the most general continuous functions with these properties are positive sums of zonal spherical functions, (a) justifies our restriction to functions of the form (36), and (b) is used to derive the mean inequality and therefore in the second proof of Theorem 4.

In general the value of the linear programming bound $1 + A^*$ (see Theorem 2) is unknown. Nor is it known in general how close the bound comes to the true value of $A(\mathbf{M}, S)$. For small problems it is usually very close and occasionally does give the exact value of $A(\mathbf{M}, S)$

3.3 Bounds for error-correcting codes. In Example E1 there are just $n+1$ distinct Φ_k's, and so we may take $N = n$ in (33) and (36) (and similarly in all the cases where G and $\mathbf{M}$ are finite). Theorem 4 now states that if

$$f(t) = \sum_{k=0}^{n} f_k\, K_k(t\,;n) \tag{42}$$

is a polynomial satisfying $f_0 > 0$, $f_k \geqslant 0$ $(k=1,\ldots,n)$, and $f(t) \leqslant 0$ for $t = d, d+1, \ldots, n$, then $A(n,d) \leqslant f(0)/f_0$. We shall give a number of illustrations.

(i) For $n = 8$, $d = 4$ we can take

$$f(t) = \tfrac{1}{2}(t-4)(t-8) = K_0(t;8) + 3K_1(t;8) + 7K_2(t;8) \quad (43)$$

and deduce that $A(8,4) \leqslant 16$. On the other hand the Hamming code $\mathcal{H}_8$ (§2.4.2 of Chap. 3) shows that $A(8,4) \geqslant 16$. Thus $A(8,4) = 16$. The code is self-dual and the nonzero codewords have weights 4 and 8, the zeros of $f(t)$.

(ii) The technique suggested by the first example also works in some other cases. For example the Golay code $\mathcal{C}_{24}$ implies that $A(24,8) \geqslant 4096$. The lowest degree polynomial that satisfies the constraints and vanishes at the nonzero weights of the dual code is

$$\begin{aligned} f(t) &= \frac{1}{1728}(t-8)(t-12)^2(t-16)^2(t-24) \quad (44)\\ &= K_0(t;24) + \cdots + \frac{168245}{192}K_6(t;24), \end{aligned}$$

and proves that $A(24,8) = 4096$.

(iii) We shall prove that the Nadler code is optimal [Bes3]. This is a nonlinear code containing 32 codewords of length 13 and minimal distance 6; thus $A(13,6) \geqslant 32$. In this case there is no polynomial $f(t)$ that we can use in Theorem 4 to prove that $A(13,6) = 32$, so we shall apply Theorem 2 instead. Suppose C is an optimal code, containing $A(13,6)$ codewords. There is no loss of generality in assuming that the distances between the codewords of C are even (if not, delete one coordinate and add an overall parity check). Let $\alpha_t(u)$ be the number of codewords in C at distance t from a codeword $u \in C$; then

$$\alpha_t = \frac{1}{|C|}\sum_{u \in C}\alpha_t(u).$$

Also $\alpha_0 = 1$ and the remaining α_t's are zero except perhaps for α_6, α_8, α_{10} and α_{12}. The primal problem is to maximize $\alpha_6 + \alpha_8 + \alpha_{10} + \alpha_{12}$ subject to $\alpha_i \geqslant 0$ and

$$\begin{aligned} \alpha_6 - 3\alpha_8 - 7\alpha_{10} - 11\alpha_{12} &\geqslant +\binom{13}{1},\\ -6\alpha_6 - 2\alpha_8 + 18\alpha_{10} + 54\alpha_{12} &\geqslant -\binom{13}{2},\\ -6\alpha_6 + 14\alpha_8 - 14\alpha_{10} - 154\alpha_{12} &\geqslant -\binom{13}{3},\\ 15\alpha_6 - 5\alpha_8 - 25\alpha_{10} + 275\alpha_{12} &\geqslant -\binom{13}{4}, \end{aligned}$$

$$15\alpha_6 - 25\alpha_8 + 63\alpha_{10} - 297\alpha_{12} \geqslant - \binom{13}{5},$$

$$-20\alpha_6 + 20\alpha_8 - 36\alpha_{10} + 132\alpha_{12} \geqslant - \binom{13}{6},$$

(the remaining inequalities (32) being redundant). Unfortunately the solution only gives $A(13, 6) \leqslant 40$. However there is an additional inequality that we can impose. We must certainly have $\alpha_{12}(u) \leqslant A(13,6,12) = A(13,6,1) = 1$, for all $u \in C$, and $\alpha_{10}(u) \leqslant A(13,6,10) = A(13,6,3) = 4$. Furthermore these can be combined. For if $\alpha_{12}(u) = 1$ then $\alpha_{10}(u) = 0$, so $\alpha_{10}(u) + 4\,\alpha_{12}(u) \leqslant 4$, and averaging over $u \in C$ gives

$$\alpha_{10} + 4\,\alpha_{12} \leqslant 4.$$

Solving the primal problem by the simplex method with this additional inequality gives $A(13,6) \leqslant 32$. Therefore $A(13,6) = 32$ and the Nadler code is optimal. By analyzing the solution to this linear program in more detail, Goethals [Goe3] has shown that the code is unique.

(iv) The best available bounds on $A(n,d)$ for $n \leqslant 24$ and $d \leqslant 10$ are given in Table 9.1; many of the upper bounds were obtained from Theorem 2 and the simplex algorithm, with the addition of extra inequalities (as in the previous example) whenever possible.

(v) Returning to Theorem 4, the best choice for a *linear* polynomial is $f(t) = K_0(t; n) + n\,K_1(t; n)/(2d - n) = 2(d - t)/(2d - n)$, for $2d > n$. This satisfies the constraints of the theorem and proves that

$$A(n,d) \leqslant \frac{2d}{2d-n}, \text{ for } 2d > n.$$

This is the Plotkin bound (see [Mac6, Chap. 2, Th. 1]). By a combinatorial argument (see [Mac6, p. 41]) it may be improved to

$$A(n,d) \leqslant 2\left\lfloor \frac{d}{2d-n} \right\rfloor \tag{45}$$

for d even and $2d > n$. Similarly one can show that

$$A(2d,d) \leqslant 4d, \quad \text{for } d \text{ even.} \tag{46}$$

Provided Hadamard matrices of the appropriate orders exist, Levenshtein has shown that equality holds in (45) and (46) ([Lev1], [Mac6, Chap. 2, Th. 8]). Thus $A(n,d)$ is essentially known for $n \leqslant 2d$.

(vi) For n slightly greater than $2d$ McEliece (see [Mac6, Chap. 17, Th. 38]) has obtained bounds on $A(n,d)$ using Theorem 4. Tietäväinen [Tie1] has found considerably stronger bounds using analytic arguments and some of the inequalities of Theorem 1. Other bounds on $A(n,d)$ obtained by linear programming can be found in [Bes1]-[Bes3].

(vii) As we have already discussed in Section 1, for large n and d the best upper bound on $A(n,d)$ presently known is the JPL bound. The simpler part, given in (2), is obtained from Theorem 4 by using a polynomial of the form

$$f(t) = \frac{1}{a-t}\{K_{k+1}(t;n)\,K_k(a;n) - K_k(t;n)\,K_{k+1}(a;n)\}^2 \quad (47)$$

for suitably chosen values of a and k. The details can be found in [McE2], [McE4] or [Mac6, Chap. 17]. Equation (26) plays a crucial role. The more complicated part of their bound will be mentioned in the next section. The JPL bounds have been generalized to tree codes in [Aal1].

3.4 Bounds for constant-weight codes. *Example* E2, *continued.* The function $A(n,d,w)$ is important in its own right, and because it can be used to derive additional inequalities satisfied by the weight distribution of a code, as we have seen in example (iii) of the previous section. Tables of $A(n,d,w)$ for $n \leqslant 24$ and $d \leqslant 10$ are given in [Bes3], and many of the upper bounds there were obtained using Theorem 2 and the simplex method (again with the addition of extra inequalities whenever possible). The best tables of $A(n,d,w)$ presently available may be found in [Bes3] or [Gra2], but see also [Bes1], [Bro8], [Bro9], [Col1], [Con41], [Hon2]-[Hon5], [Kib1], [Klϕ1], [Zin1].

The more complicated part of the JPL bound, Eq. (1), is obtained by using Theorem 4 to bound $A(n,d,w)$ and converting this to a bound for $A(n,d)$ via Elias's inequality

$$A(n,2\delta) \leqslant \frac{2^n A(n,2\delta,w)}{\binom{n}{w}}, \quad \text{for } 0 \leqslant w \leqslant n \quad (48)$$

(see [Bas1], [Bas2], [McE2], [Mac6, Chap. 17, Th. 33]). For the details of the proof of (1) see [McE2].

3.5 Bounds for spherical codes and sphere packings. *Example* E3, *continued.* Theorem 4 now takes the following form: if

$$f(t) = \sum_{k=0}^{N} f_k\, P_k^{(\alpha,\alpha)}(t), \quad \alpha = \frac{n-3}{2}, \quad (49)$$

is a polynomial satisfying $f_0 > 0$, $f_k \geqslant 0$ $(k=1,\ldots,N)$, and $f(t) \leqslant 0$ for $-1 \leqslant t \leqslant \cos\theta$, then

$$A(n,\theta) \leqslant \frac{f(1)}{f_0}. \quad (50)$$

(i) The kissing number problem. This result will be used in Chap. 13 to obtain bounds on the kissing number problem, the case $A(n,\pi/3)$, and in Chap. 14 to prove that certain arrangements of spheres achieving these bounds are unique.

(ii) Tables of bounds on $A(n, \theta)$ for some other values of θ and for certain packing problems in the projective spaces $\mathbf{P}^n(\mathbf{R})$, $\mathbf{P}^n(\mathbf{C})$ and $\mathbf{P}^n(\mathbf{H})$ have been calculated in [Ass1]. For example Table 9.2 gives some bounds on $A(n, \cos^{-1} 1/3)$, which is the maximum number of nonoverlapping spheres that can touch a pair of touching spheres in $\mathbf{R}^{n+1}$ (Chap. 14, Th. 1). E.g. five billiard balls can be arranged so that they are all in contact with two touching billiard balls, implying $A(2, \cos^{-1} 1/3) = 5$.

The values $A(7, \cos^{-1} 1/3) = 56$ and $A(23, \cos^{-1} 1/3) = 4600$ correspond to arrangements of spheres in the E_8 and Leech lattices, and were independently found by Levenshtein [Lev7]. In Chap. 14 it is proved that the corresponding spherical codes are unique.

(iii) Kabatiansky and Levenshtein [Kab1] show that the best linear and quadratic polynomials are $f(t) = t - s$ and $f(t) = (t - s)(t + 1)$, where $s = \cos \theta$. The corresponding bounds are

$$A(n, \theta) \leqslant \left[\frac{\cos \theta - 1}{\cos \theta} \right], \text{ if } \cos \theta < 0, \tag{51}$$

$$A(n, \theta) \leqslant \left[\frac{2n(1 - \cos \theta)}{1 - n\cos \theta} \right], \text{ if } \cos \theta < 1/n. \tag{52}$$

Actually the results in [Kab1] apply to any of the examples in which $\mathbf{M}$ is a continuous manifold (see §2.1), but in this and the following subsection we shall just state their bounds for the case when $\mathbf{M}$ is the unit sphere Ω_n. Just as in the coding case (see §3.3v) the bound given by the linear polynomials was already known. It is Rankin's bound (Eq. (59) of Chap. 1, [Ran1]). Several other bounds on $A(n,\theta)$ that can be obtained by the linear programming approach have already been mentioned in §2.6 of Chap. 1 (including Astola's analogs ((68), (69) of Chap. 1, [Ast1]) of Tietäväinen's coding bounds).

(iv) The principal achievement of [Kab1], as we have discussed in §1, lies in the bounds obtained when the dimension n is large. As in the previous subsection we state the results only in the case of the sphere. Working by analogy with (47) the authors use the polynomial

$$f(t) = \frac{1}{t-s} \{ P_{k+1}^{(\alpha,\alpha)}(t) \ P_k^{(\alpha,\alpha)}(s) - P_k^{(\alpha,\alpha)}(t) \ P_{k+1}^{(\alpha,\alpha)}(s) \}^2 \tag{53}$$

of degree $2k + 1$, where $\alpha = (n-3)/2$, $s = \cos \theta$ and $k = 1, 2, \ldots$. This implies

$$A(n,\theta) \leqslant 4 \binom{n+k-2}{k} (1 - t_{1,k+1}^{(\alpha)}), \tag{54}$$

for $\cos \theta \leqslant t_{1,k}^{(\alpha)}$, where the Jacobi polynomial $P_k^{(\alpha,\alpha)}(x)$ has k simple zeros

$$1 > t_{1,k}^{(\alpha)} > \cdots > t_{k,k}^{(\alpha)}.$$

For large n Kabatiansky and Levenshtein show by a detailed investigation of $t_{1,k}^{(\alpha)}$ that (54) implies the bound (6). For $\theta < 63^\circ$ (6) implies the simpler result that, as $n \to \infty$,

$$\frac{1}{n} \log_2 A(n, \theta) \lesssim -\frac{1}{2} \log_2 (1-\cos\theta) - 0.099 \tag{55}$$

(Eq. (66) of Chap. 1), which for $\theta = \pi/3$ leads to the asymptotic bound on the kissing number given in (49) of Chap. 1.

To deduce their bound (5) on the density of an n-dimensional sphere packing, Kabatiansky and Levenshtein make use of the following analog of Elias's inequality (48).

Theorem 6 [Yag1]. *The highest density* Δ_n *of any sphere packing in* $\mathbf{R}^n$ *satisfies*

$$\Delta_n \leqslant (\sin \tfrac{1}{2}\theta)^n A(n+1, \theta), \quad \text{for } 0 < \theta \leqslant \pi. \tag{56}$$

Proof. Since the reference [Yag1] cited by [Kab1] is fairly inaccessible we sketch a proof. Let S be a large sphere of radius ρ in $\mathbf{R}^{n+1}$ and let Π be a hyperplane through the center of S. In Π we construct an n-dimensional packing of unit spheres with density Δ_n. By shifting the packing slightly we may assume that the portion of Π inside S contains at least $\Delta_n \rho^n$ centers. We project these centers, perpendicularly to Π, "upwards" onto S. The Euclidean distance between the new points is at least 2 and their angular separation is at least θ, where $\sin\theta/2 = 1/\rho$. The result is therefore a spherical code of minimum angle θ, containing $\Delta_n(\sin\theta/2)^{-n} \leq A(n+1, \theta)$ points.

The bound (5) follows from (55) and (56) by taking $\theta = 1.0995$.

For finite n a numerical bound on the density may be obtained by combining (54), (56) with information about $t_{1,k}^{(\alpha)}$ given in [Kab1, p. 12]. The bounds on $\log_2 \delta$ for $n \geqslant 48$ given in Table 1.3 were obtained in this way.

In [Lev6], [Lev7] Levenshtein has found other polynomials $f(t)$ that lead to the bound (6) and give good bounds for small values of n (see also [Sid4]).

4. Other bounds

(i) Lovász [Lov1] has solved a coding problem of long standing by determining the "capacity" (see [Sha4]) of the pentagon. His method, which gives a new way to bound the capacity of any graph, turns out to be closely related to Delsarte's linear programming bound for subsets of an association scheme (see [McE2], [McE3], [Sch9]).

(ii) An unusual method for obtaining lower bounds on $A(n,d,w)$ was given in [Gra2] (see also [Gra1], [Gra3], [Hon2], [Hon3], [Klϕ1]). This yields

$$A(n, 4, w) \geqslant \frac{1}{n} \binom{n}{w},$$

which implies

$$A(n,4,w) \sim n^{w-1}/w! \text{ as } n \to \infty;$$

and

$$A(n,2\delta,w) \gtrsim n^{w-\delta+1}/w! \text{ as } n \to \infty.$$

(iii) For codes in some of the other 2-point-homogeneous spaces mentioned in this paper see [Big1], [Big3], [Cam3], [Del15], [Ham3], [Ham4], [Kan4], [Sta6], [Tha1], [Tha2].

(iv) Lloyd [Llo1] has presented another method for deriving linear programming bounds for spherical codes, by defining an association scheme on Ω_n. Yet another way of approaching these problems has been given by Neumaier [Neu1]. Urakawa [Ura1] derived a bound on Δ_n that is equivalent to Levenshtein's 1979 bound [Lev7] by considering the spectrum of Laplace-Beltrami operators.

(v) Recently Tsfasman, Vladut and Zink [Tsf2] and others have shown that codes over $\mathbf{F}_q$, for $q = p^2$, $p \geqslant 7$, constructed from algebraic curves give an asymptotic improvement over the Gilbert bound (see §2.11 of Chap. 3).

Acknowledgements. R. Askey, E. Bannai, A. R. Calderbank, P. Diaconis, C. F. Dunkl, W. M. Kantor, T. Koornwinder, R. Lidl, A. M. Odlyzko, J. J. Seidel and D. Stanton provided many helpful suggestions.

10

Three Lectures on Exceptional Groups

J. H. Conway

The first lecture records certain exceptional properties of the groups $L_2(p)$ and gives a description of the Mathieu group M_{12} and some of its subgroups, followed by a digression on the Janko group J_1 of order 175560. With the exception of the Janko group material, all the structure described appears within the Mathieu group M_{24}, which is the subject of the second lecture, where M_{24} is constructed and its subgroups described in some detail. The information on M_{24} is then found useful in the third lecture, on the group $Co_0 = \cdot 0$ and its subgroups. An appendix describes the exceptional simple groups.

This chapter is based on lectures given in Oxford, England in 1970. The groups of the title have been called the sporadic groups since [Con2], but we have preferred not to update the language in this chapter.

1. First lecture

1.1 Some exceptional behavior of the groups $L_n(q)$. The general linear group $GL_n(q)$ is the group of all linear automorphisms of an n-dimensional vector space over the field $\mathbf{F}_q$, q being any prime power. The special linear group is the normal subgroup consisting of the automorphisms of determinant 1. The center of either of these groups consists of operations of the form $x \rightarrow kx$ (for $k \in \mathbf{F}_q$), and we obtain the corresponding projective groups $PGL_n(q)$ and $PSL_n(q)$ by factoring out these centers. $PSL_n(q)$ is a simple group (for $n \geqslant 2$) except in the two cases $n = 2$ and $q = 2$ or 3. It was called by Dickson [Dic1] the linear fractional group $LF(n,q)$, but we shall use Artin's abbreviation $L_n(q)$ [Art1].

When $n = 2$, we take a basis y, z for the space, so that the operations of the group $SL_n(q)$ have the form $y \rightarrow ay + bz$, $z \rightarrow cy + dz$ (with $ad - bc = 1$). The projective line $PL(q)$ consists of the $q + 1$ values of the formal ratio $x = y/z$ (which are conveniently thought of as the q field elements together with the formal ratio ∞) and, on the projective line,

$L_2(q)$ becomes the group of all operations of the form

$$x \rightarrow \frac{ax+b}{cx+d}$$

with $ad - bc = 1$, or equivalently with $ad - bc$ any non-zero square in $\mathbf{F}_q$. We use the following names for subsets of $PL(q)$:

$$\Omega = PL(q), \quad \Omega' = \mathbf{F}_q = \Omega\backslash\{\infty\}, \quad Q = \{x^2 : x \in \mathbf{F}_q\},$$
$$N = \Omega\backslash Q, \quad Q' = Q\backslash\{0\}, \quad N' = N\backslash\{\infty\}.$$

In fact $L_2(q)$ is generated by three operations

$$\alpha : x \rightarrow x+1, \quad \beta : x \rightarrow kx, \quad \gamma : x \rightarrow -x^{-1},$$

provided that Q' is the set of powers of the field element k; for $\beta^b\alpha^a$ takes x to $k^bx + a$, while $\beta^b\alpha^a\gamma\alpha^c$ takes x to $c - (k^bx+a)^{-1}$, and plainly every operation of $L_2(q)$ can be expressed in one of these forms.

The necessary set of defining relations varies slightly with the structure of q: when q is a prime congruent to 3 modulo 4 we have

$$L_2(q) = <\alpha,\beta,\gamma : \alpha^q = \beta^{\frac{1}{2}(q-1)} = \gamma^2 = \alpha^\beta \cdot \alpha^{-k} = (\beta\gamma)^2 = (\alpha\gamma)^3 = 1>,$$

since it is easy to see that these relations enable us to put every function of α, β, γ into one of the two forms above. (α^β denotes $\beta^{-1}\alpha\beta$.) In the cases $q = 3,5,7,11$, we shall take $k = 1,4,2,3$; for $p = 5$, the relation $(\alpha\beta\gamma)^5 = 1$ completes the above set.

It was proved by Galois (in a letter to Chevalier written on the eve of his fatal duel) that $L_2(p)$ cannot have a non-trivial permutation representation on fewer than $p+1$ symbols if $p > 11$ [Hup1, p. 214]. However, for $p = 3,5,7,11$ there exist transitive representations on exactly p symbols, and by studying these we shall illuminate some of the "unexpected" isomorphisms

$$L_2(3) \cong \mathsf{A}_4, \quad L_2(4) \cong L_2(5) \cong \mathsf{A}_5,$$
$$L_2(7) \cong L_3(2), \quad L_2(9) \cong \mathsf{A}_6, \quad L_4(2) \cong \mathsf{A}_8,$$

where A_n is the alternating group of degree n. These isomorphisms are visible inside the Mathieu groups, and in the second lecture we shall find the pieces fitting together. This study is also interesting as an example of a common but puzzling phenomenon: the four cases have much in common but in each particular case there is something peculiar to that case, so that there is no completely general pattern.

It happens that the p objects permuted by $L_2(p)$ ($p=3,5,7,11$) can in each case be taken as p involutory permutations of the set Ω. For $p = 3,5,7,11$, respectively, define π to be

$$(\infty 0)(12), \quad (\infty 0)(14)(23), \quad (\infty 0)(13)(26)(45),$$
$$(\infty 0)(16)(37)(9X)(58)(42) \quad \text{(where } X \text{ denotes 10)}$$

and define π_i as $\alpha^{-i}\pi\alpha^i$. (Mnemonic: π interchanges ∞ with 0, and takes x to nx or x/n according as $x \in Q'$ or $x \in N'$, where $n = 2,4,3,6$ in the respective cases.) It is convenient to define π_∞ to be the identity permutation of Ω.

Theorem 1. *The group $L_2(p)$ $(p=3,5,7,11)$ leaves invariant the set Π consisting of the p involutions $\pi_i (i \in \Omega')$.*

Proof. Π is obviously invariant under α and the mnemonic shows it to be invariant under β also. So we need only check invariance under γ. Here the miraculous enters — we find that $(\pi_i)^\gamma = \pi_{i\delta}$, where δ is

$$(\infty)(0)(1)(2), \qquad (\infty)(0)(12)(34), \qquad (\infty)(0)(12)(36)(4)(5),$$
$$(\infty)(0)(1)(2X)(34)(59)(67)(8)$$

in the four cases.

We now discuss the cases separately.

1.2 The case $p=3$. Here the π_i are the elements of a Klein 4-group which is contained in $L_2(3)$, and being invariant, is a normal subgroup of $L_2(3)$, which is therefore not simple. In this case the representation on p letters is not faithful, since we observe that δ is the identity. Since $L_2(3)$ is of order 12 and contains only even permutations of Ω, we have the isomorphism $L_2(3) \cong A_4$.

1.3 The case $p=5$. The group $L_2(5)$ induces only even permutations of Π, and being of order 60, can only be the alternating group on Π, and we have the isomorphism $L_2(5) \cong A_5$. Transforming Π by the operations of the symmetric group S_6 on Ω, we obtain just 6 such sets of five involutions. Each permutation of S_6 defines a permutation of these six sets, and so S_6 can be regarded as a permutation group on two distinct systems of six objects. The S_5 fixing Π is generated by the permutations $\pi_j^{-1}\pi_i\pi_j$ and contains the original $L_2(5)$, and so does not fix any element of Ω. It follows that the two permutation representations of S_6 on six objects are essentially distinct, in the sense that they are related by an *outer* automorphism of S_6. It is known that only for $n=6$ does S_n possess an outer automorphism.

The details are as follows. Let $\Pi_i = \pi_i^{-1}\Pi\pi_i$ $(i \in \Omega)$. The sets Π_i are the six sets of five involutions, and the permutation π_i of Ω induces the permutation (Π_∞,Π_i) of these six sets, and symmetrically the permutation (∞,i) induces the permutation taking Π_x to $\Pi_{x\pi_i}$. So there is an outer automorphism θ interchanging (∞,i) with π_i, and hence all the 15 involutions of shape $(ab)(c)(d)(e)(f)$ with those of shape $(uv)(wx)(yz)$. The former class corresponds to Sylvester's *duads* [Syl1], the latter to his *synthemes*, and the six sets Π_i are the *synthematic totals*. We have $\theta^2 = 1$, and $(i,j)^\theta = \pi_{i,j}$, where $\pi_{i,j} = \pi_j^{-1}\pi_i\pi_j = \pi_i^{-1}\pi_j\pi_i$ except that $\pi_{\infty,i} = \pi_{i,\infty} = \pi_i$.

1.4 The case $p = 7$. In this case the Π_i are the elements of an elementary abelian group E of order 8. $L_2(7)$ acts as a subgroup of the automorphism group $L_3(2)$ of E ($L_3(2)$ because we can regard E as a 3-dimensional vector space over $\mathbf{F}_2$), and since $|L_3(2)|=|L_2(7)|$ we have the isomorphism $L_2(7)\cong L_3(2)$.

The group E generates together with $L_2(7)$ a subgroup F of A_8 of order $8\cdot168$, and so of index 15 in A_8. Since $L_2(7)$ is generated by α, β, γ and is transitive on Π, the group F is generated by α, β, γ, π_0. But (a miracle!) we observe that $\gamma\delta=\pi_0$, so that F is equally generated by α, β, δ, π_0. The group F therefore contains two subgroups $L_3(2)$ complementary to E, namely the "original" $L_2(7) = <\alpha,\beta,\gamma>$, and the "exceptional" $L_2(7) = L_3(2) = <\alpha,\beta,\delta>$. These are not conjugate in F, for one of them is transitive on Ω, while the other fixes ∞. It is exceptional for the holomorph of an elementary abelian group to exhibit this behavior.

We can describe this situation in another way by saying that F has an outer automorphism θ which fixes E pointwise and also fixes the quotient group F/E pointwise, without of course fixing F. In fact we have $\theta^2 = 1$, $\alpha^\theta = \alpha$, $\beta^\theta = \beta$, $\gamma^\theta = \delta$, $\delta^\theta = \gamma$ and $\pi_i^\theta = \pi_i$. It is not hard to see that every subgroup of F isomorphic to $L_2(7)$ is conjugate to either $<\alpha,\beta,\gamma>$ or $<\alpha,\beta,\delta>$, so that θ is essentially the only outer automorphism of F. Another consequence of this situation is that F has two essentially distinct faithful representations on 8 letters (compare the behavior of S_6).

Since F has index 15 in A_8 we obtain 15 sets of 7 involutions like Π on transforming Π by the elements of A_8. These 15 sets we call the *even* sets – if we transform instead by the elements of $S_8\backslash A_8$ we obtain 15 further sets, the *odd* sets. Each of the 105 regular involutions of A_8 (i.e. those of shape $(ab)(cd)(ef)(gh)$) belongs to just one even set and just one odd set. There is therefore a natural graph with vertices the 30 sets and edges the 105 regular involutions of A_8, with each involution joining the two sets containing it. Each vertex of either set is joined to just 7 vertices of the other set.

We can derive a similar graph from an elementary abelian group of order 16 by taking as even and odd vertices its 15 involutions and 15 subgroups of order 8, joining each subgroup to the 7 involutions it contains. To show that these two graphs are isomorphic we must turn our 15 even sets Π_i into an elementary abelian group of order 16. We do this by defining the product $\Pi_i\Pi_j$ of two distinct even sets to be the unique third set joined in the graph to all the vertices joined to both Π_i and Π_j. There is a simple combinatorial proof that this does indeed define a group, and since each element of A_8 acts non-trivially on this group, A_8 is a subgroup of its automorphism group $L_4(2)$. Since $|L_4(2)| = \frac{1}{2}\cdot 8!$ we have the isomorphism $L_4(2) \cong A_8$. We do not go into further details here, since we shall give another proof of this isomorphism in the second lecture.

In this sort of work it is useful to have a simple way of specifying at a glance the structure of the groups that appear. We shall say that G is a group of type $A.B$ (or AB, when no confusion can arise) when we mean that G has a normal subgroup A whose quotient is isomorphic to B. In this notation, the cyclic group of order n is written simply as n, and the elementary abelian group of order p^n just as p^n. Thus the group F we have just discussed is a group of type $2^3L_3(2)$.

1.5 The case $p = 11$. In the cases $p = 3,5,7$ the group $<\alpha,\beta,\gamma,\delta>$ was A_4, S_6, $2^3L_3(2)$ respectively, and in the cases $p = 5,7$ this group had an outer automorphism θ fixing α and β and interchanging γ with δ. We assert that for $p = 11$ the group $<\alpha,\beta,\gamma,\delta>$ is the Mathieu group on 12 letters, M_{12}, and that $<\alpha,\beta,\gamma>$ and $<\alpha,\beta,\delta>$ are two subgroups of type $L_2(11)$ not interchanged by any automorphism of M_{12}.

We shall show here that $<\alpha,\beta,\gamma,\delta>$ is a proper subgroup of A_{12}, with a 6-dimensional projective representation over F_3. Take a 12-dimensional space $\mathscr{X}$ over $\mathbf{F}_3$, with basis vectors $x_i\,(i \in \Omega)$, and for $S \subseteq \Omega$ let x_S denote $\Sigma x_i\,(i \in S)$, with a similar notation in other cases. We consider the space $\mathscr{W}$ spanned by vectors $w_i\,(i \in \Omega)$, where $w_\infty = x_\Omega$, and $w_i = x_{N-i} - x_{Q-i}\,(i \in \Omega')$, where for instance $N - i$ means $\{n-i : n \in N\}$. The *ternary Golay code* $\mathscr{C}_{12}$ (§2.8.5 of Chap. 3) is the set of all 12-tuples $(c_\infty, c_0, \ldots, c_X)$ with $\Sigma c_i x_i \in \mathscr{W}$.

Theorem 2. *$\mathscr{C}_{12}$ is 6-dimensional, and $\Sigma c_i w_i = 0$ if and only if $(c_i) \in \mathscr{C}_{12}$.*

Proof. w_∞, w_1, w_3, w_4 w_5, w_9 are linearly independent in $\mathscr{W}$, so $\mathscr{C}_{12}$ is at least 6-dimensional. But $w_N = w_Q = 0$, so that $w_\Omega = 0$ and $w_{N-i} = w_{Q-i} = 0$ for $i \in \Omega'$. Thus $(c_i) \in \mathscr{C}_{12}$ implies that $\Sigma c_i w_i = 0$, and the w_i satisfy at least six linearly independent relations, from which the statements follow.

We define linear maps A, B, C, D on $\mathscr{X}$ by:

$$A : x_i \to x_{i+1}, \qquad B : x_i \to x_{3i}, \qquad C : x_i \to \pm\, x_{-1/i}, \qquad D : x_i \to x_{i\delta},$$

where $\delta = (\infty)(0)(1)(2X)(34)(59)(67)(8)$, and the sign $\pm$ is $+$ for $i \in Q$, $-$ for $i \in N$.

Theorem 3. *A,B,C,D preserve $\mathscr{W}$.*

Proof. We check the effects of these operations on the w_i, finding miraculously that

$$A : w_i \to w_{i-1},\; B : w_i \to w_{3i},\; C : w_i \to \mp\, w_{-1/i},\; D : w_i \to w_{i\delta},$$

where the sign $\mp$ is the opposite of the previous sign $\pm$.

It follows at once that α, β, γ, δ generate a proper subgroup M_{12} of A_{12}. The following portmanteau theorem, which we shall not prove, gives us considerable information about M_{12}. Most of it is easily proved by mimicking the methods of the next lecture for M_{24}.

Theorem 4. *M_{12} is a quintuply transitive group of order $12\cdot 11\cdot 10\cdot 9\cdot 8$. The group $<A,B,C,D>$ is a non-splitting extension $2M_{12}$, consisting precisely of those automorphisms of $\mathscr{X}$ which preserve $\mathscr{W}$. It has an outer automorphism θ that satisfies*

$$\theta^2 = 1, \quad A^\theta = A^{-1}, \quad B^\theta = B, \quad C^\theta = C^{-1} = -C, \quad D^\theta = D$$

and whose adjunction completes it to a group $2M_{12}2$.

The symmetry corresponding to θ is illuminated by defining $\mathcal{V} = \mathcal{X}/\mathcal{W}$, and vectors v_i to be the canonical images of the x_i in $\mathcal{V}$. The group $2M_{12}2$ is then a group of linear automorphisms of the space $\mathcal{V} \oplus \mathcal{W}$, and as such permutes the 48 vectors $\pm v_i$, $\pm w_i$, the automorphism θ interchanging each v_i with the corresponding w_i. The quotient group $M_{12}2$ permutes the 24 subgroups $V_i = \{0, v_i, -v_i\}$ and $W_i = \{0, w_i, -w_i\}$ of $\mathcal{V} \oplus \mathcal{W}$. We describe some of the subgroups of $2M_{12}2$ in these terms.

Fixing V_∞ we have a subgroup $2M_{11}$ with orbits of sizes 2, 22, 24 on the 48 vectors, namely $\{\pm v_\infty\}$ $\{\pm v_i : i \in \Omega'\}$ and $\{\pm w_i : i \in \Omega\}$. This must be a direct product $C_2 \times M_{11}$, (where C_2 is a cyclic group), since it has a subgroup M_{11} of index 2 fixing v_∞ and $-v_\infty$ separately, with orbits of sizes 1, 1, 22, 12, 12, namely $\{v_\infty\}$, $\{-v_\infty\}$, $\{\pm v_i : i \in \Omega'\}$ and $\{w_i : i \in \Omega\}$, $\{-w_i : i \in \Omega\}$. Note that the permutation representation of $2M_{11}$ on 22 objects here is *not* the direct product of permutation representations of C_2 on 2 objects and M_{11} on 11 objects, since the subgroup M_{11} is still transitive on all 22 objects. There is a second conjugacy class of subgroups M_{11} in M_{12}, obtained by interchanging the roles of the v_i and w_i.

There is a subgroup $SL_2(11) = \langle A, B, C \rangle$ with two orbits of size 24, and a subgroup $L_2(11)$ with orbits of sizes 1, 1, 11, 11, 1, 1, 11, 11, namely $\{v_\infty\}$, $\{v_i : i \in \Omega'\}$, $\{w_\infty\}$, $\{w_i : i \in \Omega'\}$ and their negatives. The group $L_2(11)$ is completed to a direct product $2 \times L_2(11)$ by adjoining -1. In the quotient group $M_{12}2$ these both become $L_2(11)$'s but one is maximal in M_{12}, and the other is not.

The subgroup $2 \times M_{10}$ fixing two subgroups V_∞, V_0 has orbits of sizes 2, 2, 20, 24. It has a subgroup M_{10} fixing v_∞ and $-v_\infty$ separately, and this has a further subgroup M_{10}' fixing also v_0, $-v_0$ separately. M_{10}' has orbits of sizes 1, 1, 1, 1, 20, 6, 6, 6, 6, the fixed points being obvious, and the orbits of size 6 being $\{w_i : i \in N\}$, $\{w_i : i \in Q\}$ and their negatives. Coincidence of orders implies the isomorphism $M_{10}' \cong \mathsf{A}_6$, but we can also see $M_{10}' \cong L_2(9)$ by translating the 10 subgroups $V_1, V_2, \ldots, V_X$ into the 10 points

$$0, 1, i, -i, 1-i, -1-i, i+1, \infty, i-1, -1$$

of the projective line $PL(9)$, when the permutations of M_{10}' become linear fractional transformations. We therefore have $\mathsf{A}_6 \cong M_{10}' \cong PSL_2(9)$. The three groups S_6, M_{10}, $PGL_2(9)$ are all distinct, being the three subgroups of index 2 in $\mathrm{Aut}(\mathsf{A}_6)$ ($\mathrm{Aut}(\mathsf{A}_6)/\mathsf{A}_6$ is a 4-group). S_6 is completed to $\mathrm{Aut}(\mathsf{A}_6)$ by our outer automorphism θ, interchanging duads with synthemes, M_{10} by an outer automorphism interchanging V_∞ with V_0, and $PGL_2(9)$ by its field automorphism, interchanging i with $-i$. The groups S_6, M_{10}, $PGL_2(9)$ may be distinguished as abstract groups by the numbers of classes of elements of orders 3 and 5. S_6 has two classes of order 3, one of order 5, M_{10} has one of each, and $PGL_2(9)$ has one of order 3 and two of order 5.

1.6 A presentation for M_{12}. Each of the triples α, β, γ and α, β, δ satisfies the defining relations for $L_2(11)$. It is remarkable that the conjunction of the relations so obtained with the miraculous relation $(\gamma\delta)^2 = \beta^2$, or equivalently $(\delta\gamma\beta)^2 = 1$, gives a presentation for M_{12}, namely

$$<\alpha,\beta,\gamma,\delta : \alpha^{11} = \beta^5 = \gamma^2 = \delta^2 = \alpha^\beta\cdot\alpha^{-3} = (\alpha\gamma)^3 = (\alpha\delta)^3 = (\beta\gamma)^2 = (\beta\delta)^2 = (\delta\gamma\beta)^2 = 1> .$$

This can be supplemented by the relations $\theta^2 = 1$, $\alpha^\theta = \alpha^{-1}$, $\beta^\theta = \beta$, $\gamma^\theta = \gamma^{-1}$, $\delta^\theta = \delta$ to yield a presentation for $M_{12}2$. We can get a presentation for $2M_{12}2$ by making products involving γ into -1 instead of 1, where -1 is an involutory central element.

If in the presentation for M_{12} we eliminate β by means of the miraculous relation, we get the considerably simpler presentation

$$<\alpha,\gamma,\delta : \alpha^{11} = \gamma^2 = \delta^2 = (\alpha\gamma)^3 = (\alpha\delta)^3 = (\gamma\delta)^{10} = \alpha^{\gamma\delta\gamma\delta}\cdot\alpha^2 = 1>,$$

which we can further transform into

$$<\alpha,\gamma,\eta : \alpha^{11} = \gamma^2 = (\gamma\eta)^2 = (\alpha\gamma)^3 = (\eta\gamma\alpha)^3 = \eta^{10} = \eta^{-2}\alpha\eta^2\cdot\alpha^2 = 1>$$

by replacing $\gamma\delta$ by η, eliminating δ. We get yet another by eliminating γ instead. (These are quite essentially distinct, though in appearance very similar, since we know that the subgroups $<\alpha,\beta,\gamma>$ and $<\alpha,\beta,\delta>$ are thoroughly distinct — indeed one is maximal and the other is not.)

1.7 Janko's group of order 175560. The way in which the exceptional representations of the small groups $L_2(p)$ can arise in other situations is well illustrated by considering the permutation representation of Janko's group J_1 of order 175560 on 266 letters ([Jan1], [Jan2]; [Cam1], [Eva1], [Gag1], [Mar1], [Per1], [Whi2]). The stabilizer of a point is an $L_2(11)$, with orbits of sizes 1, 11, 110, 132, 12. Knowing this much, it is easy to construct J_1 and so verify its existence. The centralizer $<i> \times A_5$ of an involution i of J_1 is such that the A_5 is contained in an $L_2(11)$. We take a particular subgroup $L_2(11) = L$. Associated with L is a set S of 11 involutions centralizing A_5's in L, and permuted like our 11 permutations π_i of the projective line $PL(11)$ — we let π_i be the permutation of $PL(11)$ corresponding to the involution i of S. The 12 cosets of L which form an orbit under L are permuted like the points x of $PL(11)$ — we let L_x be the coset corresponding to $x \in PL(11)$. Any element of L can be expressed as a product $f(\alpha,\beta,\gamma)$ of the generating permutations of L — the permutation it induces on S is the corresponding product $f(\alpha,\beta,\delta)$.

Now the effect of an involution i of S on the 266 cosets must be capable of being described in a manner invariant under L. This limits the possibilities, and in fact forces the following unique situation. The 266 cosets are L, Li, Lij, L_xj, L_x, for $x \in PL(11)$, i, $j \in S$, $i \neq j$.

Postmultiplication by the typical element of L permutes these by permuting x by $f(\alpha,\beta,\gamma)$ and i and j by $f(\alpha,\beta,\delta)$; so we need only describe the effect of postmultiplying by a typical $k \in S$.

Since k^2 is the identity and $Liji = Lij$, we need only consider expressions $Lijk$ and $L_x jk$ in which i,j,k are distinct. Now the permutation $\pi_i \pi_j \pi_k$ has two fixed points u and v in $PL(11)$ if it has any, and u and v may be distinguished by the condition that the permutation π_h which interchanges u and v also interchanges $u\pi_i$ and $v\pi_k$, but not $u\pi_k$ and $v\pi_i$. We define $u = [i,j,k]$, and then $[k,j,i]$ will be v. We then have

$$\begin{aligned} Lijk &= Lji && \text{if } \pi_i\pi_j\pi_k \text{ has no fixed point,} \\ Lijk &= L_x h && \text{if } [i,j,k] = x = [h,i,j], \\ L_x jk &= L_y h && \text{if } x = [h,k,j] \text{ and } y = [j,k,h], \\ L_x jk &= Lhi && \text{if } x\pi_j = [i,j,k] = [h,i,j]. \end{aligned}$$

These equations are quite easy to work with when we remember that the equation $[i,j,k] = u$ inverts to determine π_i, π_j, π_k as interchanging the pairs $(u, u\pi_k\pi_j)$, $(u\pi_i, u\pi_k)$, and $(u, u\pi_i\pi_j)$ respectively. (Any pair of distinct points of $PL(11)$ is interchanged by just one of the permutations π_i.)

That J_1 is a proper subgroup of A_{266} with the right order is also easy to verify. If we join each of the 266 points to the orbit of size 11 in its stabilizer we get a graph in which the joins are (L, Lh), (Li, Lhi), $(Lij, Lhij)$, $(L_x, L_y h)$, $(L_x j, L_y hj)$, where in each case h varies arbitrarily and $y = x\pi_h$. It is not hard to verify that the operations defined above preserve this graph. If instead we join each point to the corresponding set of 12 points, we get the joins (in which y and h are arbitrary)

$$(L, L_y), \quad (Li, L_y i), \quad (Lij, L_y ij),$$
$$(L_x, L), \quad (L_x, L_x h), \quad (L_x j, Lj), \quad (L_x j, L_x hj).$$

2. Second lecture

2.1 The Mathieu group M_{24}. We define M_{24} to be the group obtained by adjoining the permutation $\delta : x \to x^3/9\ (x \in Q)$ or $x \to 9x^3\ (x \in N)$ to the group $L_2(23)$ acting on the projective line $\Omega = PL(23)$. We list the generators in full:

$\alpha = (\infty)(0\ 1\ 2\ 3\ 4\ 5\ 6\ 7\ 8\ 9\ 10\ 11\ 12\ 13\ 14\ 15\ 16\ 17\ 18\ 19\ 20\ 21\ 22)$,
$\beta = (\infty)(15\ 7\ 14\ 5\ 10\ 20\ 17\ 11\ 22\ 21\ 19)\ (0)(3\ 6\ 12\ 1\ 2\ 4\ 8\ 16\ 9\ 18\ 13)$,
$\gamma = (\infty\ 0)(15\ 3)(7\ 13)(14\ 18)(5\ 9)(10\ 16)$
$(20\ 8)(17\ 4)(11\ 2)(22\ 1)(21\ 12)(19\ 6)$,
$\delta = (\infty)(14\ 17\ 11\ 19\ 22)(15)(20\ 10\ 7\ 5\ 21)$
$(0)(18\ 4\ 2\ 6\ 1)(3)(8\ 16\ 13\ 9\ 12)$.

It is sometimes convenient to replace the pair γ, δ by the product $\gamma\delta^2$, which generates the same group, since plainly γ and δ commute. We have

$\gamma = (\gamma\delta^2)^5$, $\delta = (\gamma\delta^2)^{-2}$, and

$\gamma\delta^2 = (\infty\ 0)(15\ 3)(14\ 2\ 22\ 4\ 19\ 18\ 11\ 1\ 17\ 6)(20\ 13\ 21\ 16\ 5\ 8\ 7\ 12\ 10\ 9)$, or, algebraically, $\gamma\delta^2$ interchanges ∞ with 0, and otherwise takes x to $-(x/2)^2$ (if $x \in Q$), or $(2x)^2$ (if $x \in N$). The operations γ, δ plainly normalize the group $<\beta>$, and in fact we have $\beta^\gamma = \beta^{-1}$, $\beta^\delta = \beta^3$, and $(\gamma\delta^2)^{-1}\beta(\gamma\delta^2) = \beta^2$.

Theorem 5. *M_{24} is quintuply transitive on Ω.*

Proof. By randomly multiplying the generators, we find permutations of cycle-shapes 1 23, $1^2 11^2$, $1^3 7^3$, $1^4 5^4$, 2^{12}, $1^8 2^8$ and 4^6 (for instance α, β, $\delta\alpha^2$, δ, γ, $(\alpha\delta)^3$ and $(\alpha^{13}\gamma\delta^2)^3$). The shapes 1 23 and 2^{12} show that M_{24} is transitive, and then that the stabilizer of any point must have a permutation of shape 1 23, and so be transitive on the remaining 23 points. In a similar way we see that the stabilizer of 2 points is transitive on the remaining 22 (using shapes $1^2 11^2$ and $1^3 7^3$), and the stabilizer of 3 points is transitive on the remaining 21 (using $1^3 7^3$ and $1^4 5^4$), so that M_{24} is quadruply transitive. Now the subgroup fixing a set of 4 points as a whole has permutations of shapes $1^4 5^4$ and 4^6, so is transitive on the remaining 20 points, whence M_{24} must act transitively on the set of all 5-element subsets of Ω. But the subgroup fixing any 5-element set has permutations of shapes $1^4 5^4$ and $1^8 2^8$ which induce permutations of shapes 5 and $1^3 2$ on the 5-elements, and the quintuple transitivity follows from this, since any two permutations of shapes 5 and $1^3 2$ generate the full symmetric group on 5 letters.

We define M_{24-k} to be the pointwise stabilizer of a k-element subset of Ω in $M_{24} (k \leqslant 5)$.

To show that M_{24} is a proper subgroup of the alternating group A_{24}, we turn the set $P(\Omega)$ of all subsets of Ω into a 24-dimensional vector space over $\mathbf{F}_2$ (by defining the sum $A+B$ of two sets to be their symmetric difference $(A \backslash B) \cup (B \backslash A)$) and show that M_{24} preserves a 12-dimensional subspace. The *binary Golay code* $\mathscr{C}_{12}$ (§2.8.2 of Chap. 3) is the space spanned by the 24 sets $N_i (i \in \Omega)$, where $N_\infty = \Omega$ and otherwise $N_i = N - i = \{n - i : n \in N\}$. If $S \subseteq \Omega$, we write N_S for ΣN_i $(i \in S)$.

Theorem 6. *$\mathscr{C}_{24}$ is at most 12-dimensional.*

Proof. We can verify directly that $N_\Omega = N_N = 0$, whence $N_{N_i} = 0$ for each $i \in \Omega$, and $N_C = 0$ for each $C \in \mathscr{C}_{24}$. If $\mathscr{C}_{24}$ is k-dimensional, we therefore have at least k independent linear relations between its generating sets N_i, so that $k \leqslant 24 - k$. (The relations above are consequences of well-known number-theoretical facts, but in any case it is obvious *a priori* that each of N_Ω, N_N must have the form $aN + bQ + c\{\infty\} + d\{0\}$, so there is not much to check.)

A similar calculation gives the exact dimension of $\mathscr{C}_{24}$, and a test for membership in $\mathscr{C}_{24}$.

Theorem 7. *$\mathscr{C}_{24}$ is exactly 12-dimensional, and we have $C \in \mathscr{C}_{24}$ if and only if $N_C = 0$.*

Proof. We find $N_{\{-2,0,2,3\}} = \{0,1,2,3,4,7,10,12\}$, a $\mathscr{C}$-set (i.e. member of $\mathscr{C}_{24}$) with least element 0. Adding $i (i \leqslant 10)$ to the elements of this we get $\mathscr{C}$-sets with least element i $(0 \leqslant i \leqslant 10)$. These obviously span an 11-dimensional space, and we get the extra dimension by adjoining any $\mathscr{C}$-set containing ∞.

Theorem 8. *M_{24} preserves $\mathscr{C}_{24}$.*

Proof. We have obviously $N_i\alpha = N_{i-1}$, $N_i\beta = N_{2i}$, and we verify the equations

$$N_0\,\gamma\delta^2 = N_\infty - N_0, \quad N_\infty\gamma\delta^2 = N_\infty, \quad N_1\,\gamma\delta^2 = N_{\{-1,0,1,3\}},$$
$$N_{-1}\,\gamma\delta^2 = N_{\{3,12,21\}},$$

from which we can deduce $N_i\gamma\delta^2 \in \mathscr{C}_{24}$ for all i, using $N_{2i}\gamma\delta^2 = N_i\beta\gamma\delta^2 = N_i\gamma\delta^2\beta^2$. (There is really only one non-trivial relation to be verified, since $\mathscr{C}_{24}$ is spanned by the N_i (for $i \in Q'$) together with $N_\infty = \Omega$.)

Theorem 9. *There exist 8-element $\mathscr{C}$-sets, called* (special) octads, *and each non-empty $\mathscr{C}$-set is the symmetric difference of a strictly smaller $\mathscr{C}$-set with an octad. Each 5-element set is contained in just one octad.*

Proof. We have already found one octad, $\{0,1,2,3,4,7,10,12\}$, and so each 5-element set is contained in at least one octad, by Theorem 5. Any $\mathscr{C}$-set with at least 5 points is therefore the symmetric difference of an octad containing those 5 points and a strictly smaller $\mathscr{C}$-set. If some non-empty $\mathscr{C}$-set had fewer than 5 elements, every set of the same cardinal would be a $\mathscr{C}$-set, by Theorem 5, and by taking symmetric differences we should obtain every two-element set, and so every set of even cardinal, as a $\mathscr{C}$-set, which cannot be. It follows that the octads are the smallest non-empty $\mathscr{C}$-sets, for from any smaller non-empty one we could obtain a still smaller one, and so on. No 5-element set can be contained in two distinct octads, for then their symmetric difference would be a $\mathscr{C}$-set with at most 6 members.

The last sentence of Theorem 9 asserts that the octads form a *Steiner system* $S(5, 8, 24)$ (§3.1 of Chap. 3). We can deduce from it that there are just

$$\binom{24}{5} \Big/ \binom{8}{5} = 759 \text{ octads,}$$

permuted transitively by M_{24}. Witt [Wit2], [Wit3] proved that the system $S(5, 8, 24)$ — likewise the systems $S(4, 7, 23)$, $S(3, 6, 22)$, $S(5, 6, 12)$, $S(4, 5, 11)$ which are involved with the other Mathieu groups — is essentially unique, and defined M_{24} to be its automorphism group.

2.2 The stabilizer of an octad. M_{24} contains permutations of shapes $1\ 3\ 5\ 15$ and $1^2 2\ 4\ 8^2$, for example $\delta\alpha^{11}$ and $\delta\alpha^5$. The octad containing the 5-cycle in the first case (being fixed) can only be the union of the 5- and 3-cycles, and in the second case the octad containing the 4-cycle and a 1-cycle must be the union of the 4-, 2-, and 1-cycles. The fourth power of the second operation is therefore a permutation of shape $1^8 2^8$ whose fixed points form an octad. We can therefore suppose that M_{24} contains the permutation $\lambda = (abcde)(fgh)(i)(jkl...x)$ and a permutation μ of shape $1^8 2^8$ interchanging i and j and fixing a,b,c,d,e,f,g,h. The permutations $\mu^{\lambda}, \mu^{\lambda^2}, \ldots$ have the same fixed points but now interchange i with k, i with l, etc., and so the pointwise stabilizer of $\{a,b,c,d,e,f,g,h\}$ has at least 15 involutions and order at least 16. The pointwise stabilizer of $\{a,b,c,d,e\}$ has at least three times this order, since it contains also the permutation λ^5, of order 3. It follows that M_{24} has order at least $24 \cdot 23 \cdot 22 \cdot 21 \cdot 20 \cdot 16 \cdot 3 = 244823040$.

Theorem 10. *The subgroup of* M_{24} *fixing an octad setwise is a group* $2^4 \mathsf{A}_8$, *an extension of an elementary abelian group of order* 16 *(fixing the octad pointwise) by the alternating group on* 8 *letters, which is isomorphic to* $L_4(2)$. M_{24} *has order exactly* 244823040, *and contains all permutations of* Ω *that preserve the Golay code* $\mathscr{C}_{24}$.

Proof. We consider the subgroup H of index 16 in the stabilizer of an octad that in addition to fixing the octad $\{a,b,c,d,e,f,g,h\}$ as a whole, fixes the point i. Since $\{a,b,c,d,e,f,g,h,i\}$ contains just one non-empty $\mathscr{C}$-set, the $\mathscr{C}$-sets disjoint from it form a space of codimension 8, and so dimension 4, and it is easy to see that every non-trivial element of H acts non-trivially on this space, so that H is a subgroup of its automorphism group $L_4(2)$, whose order is

$$(16-1)\ (16-2)\ (16-4)\ (16-8) = 20160.$$

The group of permutations of a,b,c,d,e,f,g,h induced by H is transitive and contains the 3-cycle (hgf) (induced by λ^5), and so must be the alternating group A_8 of order $\frac{1}{2} \cdot 8! = 20160$; the coincidence of orders establishes the isomorphism $H \cong \mathsf{A}_8 \cong L_4(2)$. Since $20160 = 244823040/759 \cdot 16$, the order of M_{24} is exactly 244823040, and the subgroup fixing a,b,c,d,e,f,g,h individually has order exactly 16, being transitive on the remaining 16 letters. It consists of the identity and the 15 involutions we found earlier, and so is an elementary abelian group. (It is the dual of the above 4-dimensional space of $\mathscr{C}$-sets, and is also permuted by the group H.) The upper bound by this argument for the order of M_{24} applies also to the group of all permutations of Ω preserving $\mathscr{C}_{24}$, and establishes the identity of the two groups.

Theorem 11. *Let* $\{a_1, a_2, \ldots, a_8\}$ *be an octad. Then the number of octads intersecting* $\{a_1, \ldots, a_i\}$ *in* $\{a_1, \ldots, a_j\}$ *(exactly) is the* $(j+1)th$ *entry in the* $(i+1)th$ *line of Table* 10.1.

Proof. Since M_{24} is transitive on those sets of any given cardinal that are contained in octads, the number of octads containing $\{a_1, \ldots, a_i\}$ is exactly

$$759 \cdot \binom{8}{i} \Big/ \binom{24}{i} \quad (\text{if } i \leqslant 5), \quad \text{or } 1 \quad (\text{if } i \geqslant 5),$$

and we have the rightmost entries. The others are deduced from these by repeated use of the rule that the sum of two adjacent entries in any line is

Table 10.1 How many octads?

759								
506	253							
330	176	77						
210	120	56	21					
130	80	40	16	5				
78	52	28	12	4	1			
46	32	20	8	4	0	1		
30	16	16	4	4	0	0	1	
30	0	16	0	4	0	0	0	1

Table 10.2 How many dodecads?

2576								
1288	1288							
616	672	616						
280	336	336	280					
120	160	176	160	120				
48	72	88	88	72	48			
16	32	40	48	40	32	16		
0	16	16	24	24	16	16	0	
0	0	16	0	24	0	16	0	0

the entry just above them in the previous line. Table 10.2 gives the number of umbral dodecads meeting $\{a_1, \ldots, a_i\}$ in $\{a_1, \ldots, a_j\}$ (see below).

2.3 The structure of the Golay code $\mathscr{C}_{24}$

Theorem 12. *$\mathscr{C}_{24}$ has weight distribution* $0^1\ 8^{759}\ 12^{2576}\ 16^{759}\ 24^1$.

Proof. Since by Theorem 11 two distinct octads intersect in 0 or 2 or 4 points their symmetric difference has cardinal 16 or 12 or 8. The theorem follows using Theorem 9 and the fact that $\mathscr{C}$-sets come in complementary pairs.

2.4 The structure of $\mathrm{P}(\Omega)/\mathscr{C}_{24}$

Theorem 13. *Every subset of Ω is congruent modulo $\mathscr{C}_{24}$ either to a unique set of cardinal at most* 3 *or to each of* 6 *distinct sets of cardinal* 4.

Proof. If a subset S has 5 or more members we obtain a congruent set with fewer members by taking the symmetric difference with an octad containing 5 members of S. If S has 4 members we obtain 5 more

4-element sets congruent to S by taking the symmetric differences of S with the 5 octads that contain it (Theorem 11). If two distinct sets of 4 or fewer elements are congruent modulo $\mathscr{C}_{24}$, their symmetric difference is an octad, and so they must be disjoint and have 4 elements each, so the theorem is best possible.

Thus $4096 = 1 + 24 + \binom{24}{2} + \binom{24}{3} + \frac{1}{6}\binom{24}{4}$. In other words $\mathscr{C}_{24}$ is a triple-error-correcting code with covering radius 4.

2.5 The maximal subgroups of M_{24}. The complete list of maximal subgroups of M_{24} is known. J. A. Todd [Tod3] observed eight distinct types in his paper on M_{24}, and a further type, the *octern group* of order 168 described below, has since been discovered. Chang Choi [Cho1] and Curtis [Cur2]-[Cur5] have given proofs that these nine groups are indeed the only conjugacy classes of maximal subgroups of M_{24}. With two exceptions these groups are easily described in terms of the Golay code $\mathscr{C}_{24}$, and we describe them in some detail. We shall give a further description of these groups in terms of the MOG in Chap. 11.

A 12-element $\mathscr{C}$-set is called an *(umbral) dodecad*, and a complementary pair of umbral dodecads is a *duum*. A triplet of mutually disjoint octads is a *trio*, and a system of 6 tetrads with the property that the union of any two is an octad is a *sextet*. (Mnemonics: **du**um = 2 **um**brals, **tri**o = 3 **o**ctads, **sext**et = 6 **t**etrads.) In general, we use *n-ad* for an n-element subset of Ω, except that the terms *octad* and *dodecad* usually presuppose the corresponding adjectives *special* and *umbral*.

The maximal subgroups M_{24} can now be described as the stabilizers of the concepts *monad, duad, triad, octad, sextet, trio, duum,* along with two further groups $L_2(23)$, $L_2(7)$ (compare Table 11.1 of the following chapter). The $L_2(23)$ is the group we started from, and the $L_2(7)$ is the last maximal subgroup. We give theorems asserting the exact amount of transitivity of each of these groups on various configurations in Ω. We say that a group is $a+b$ transitive on two sets A and B if it contains elements taking any a elements of A, and simultaneously any b elements of B, to preassigned positions.

Theorem 14. *The group $2^4\mathrm{A}_8$ fixing an octad is $6+1$, $3+2$, and $1+3$ transitive on the octad and its complement.*

Proof. The stabilizer of an octad *and* one point in its complement is the group $H = \mathrm{A}_8$, sextuply transitive on the octad, and, in its role as $L_4(2)$, doubly transitive on the remaining 15 points of the complement, which can be regarded as the involutions of a group 2^4 acted upon by $L_4(2)$. The stabilizer of two points of the complement is therefore a group $2^3L_3(2)$, which can only be the group we met in the first lecture, triply transitive on the octad. The stabilizer of one point in the octad together with two of the complement is $L_3(2)$, and is easily seen to be transitive on the remaining 14 points of the complement.

Theorem 15. *M_{24} is transitive on dodecads. The group fixing a dodecad setwise is the Mathieu group M_{12}, which is $5+0$, $3+1$, $1+3$, $0+5$ transitive on the dodecad and its complement. This group M_{12} has index 2 in a group $M_{12}2$, the stabilizer of a duum.*

Proof. The group stabilizing the sets $\{a,b\}$ and $\{c,d,e,f,g,h\}$ is plainly a subgroup 2^4S_6 inside our group 2^4A_8, and it is easy to see that its subgroup 2^4 is transitive on the 16 octads which by Theorem 11 intersect $\{a,b,c,d,e,f,g,h\}$ in $\{a,b\}$. If one of them is $\{a,b,i,j,k,l,m,n\}$, it follows that the stabilizer of the two sets $\{c,d,e,f,g,h\}$ and $\{i,j,k,l,m,n\}$ is a group S_6, being by symmetry sextuply transitive on each of the two sets. (The permutation representations on these sets cannot be permutation identical, for then — the notation being suitably chosen — there would be an element of order 3 fixing each of a,b,c,d,e,i,j,k, and our knowledge of the pointwise stabilizer of $\{a,b,c,d,e\}$ contradicts this.)

Since every dodecad is expressible in this way as the symmetric difference of two octads with two points in common (by Theorem 9), it follows that M_{24} is transitive on dodecads. Again, any 5 points of a dodecad belong to one of a pair of octads whose symmetric difference is the dodecad, and we deduce that the stabilizer of a dodecad is quintuply transitive on the dodecad (and by symmetry on the complementary dodecad). (The permutation representations are not permutation identical, for the subgroup S_6 has two orbits of size 6 in one dodecad, but an orbit of size 2 in the other.)

When we have identified this group with M_{12} we can say that the stabilizer of a point in one of the dodecads is a group M_{11}, triply transitive on the other. In fact, identification with our previous M_{12} is unnecessary, since we could easily *define* M_{12} to be the stabilizer of a dodecad in M_{24}. But it is quite easy — the groups $V_\infty, V_0, \ldots, V_X, W_\infty, W_0, \ldots, W_X$ of the first lecture become the points

$$\infty, 15, 7, 14, 5, 10, 20, 17, 11, 22, 21, 19, 0, 3, 13, 18, 9, 16, 8, 4, 2, 1, 12, 6$$

of Ω, and we easily check that the permutations α, β, γ, δ of that lecture do indeed correspond to permutations of M_{24} as defined here. (α and β become β and δ rather plainly, so we need only check the effects of γ, δ on the N_i, using the test of Theorem 7 for $\mathscr{C}$-sets. Only two of the calculations are really required.) The outer automorphism θ of the first lecture translates into the operation γ of this lecture, interchanging the two dodecads of the duum, here taken as N and Q.

Theorem 16. *M_{24} is transitive on monads, dyads, and triads, whose stabilizers are respectively groups M_{23}, $M_{22}2$, and $M_{21}S_3$. These groups are $a+b$ transitive on the two appropriate sets if $a+b \leqslant 5$ and $a \leqslant 1, 2, 3$ respectively.*

Proof. These statements follow at once from the quintuple transitivity of M_{24}.

If we give the points 2, 3, . . . , 22 the coordinates

$$100, 010, 001, \omega\bar\omega 0, 111, \bar\omega 11, 1\omega 1, 1\bar\omega 1, 0\bar\omega\omega, 11\bar\omega,$$
$$101, \bar\omega\omega 0, \bar\omega 1\omega, \omega 0\bar\omega, 110, \omega 1\bar\omega, 011, 0\omega\bar\omega, \omega 11, \bar\omega 0\omega$$

then the 21 octads containing $\infty, 0, 1$ yield the 21 lines of the the projective plane over $\mathbf{F}_4 = \{0, 1, \omega, \bar\omega\}$. Using standard techniques we can then identify M_{21} with $L_3(4)$ and deduce the simplicity of M_{21}, M_{22}, M_{23}, M_{24}.

Theorem 17. *M_{24} is transitive on sextets. The stabilizer of a sextet is a group $2^6 . 3 . S_6$, and the subgroup $2^6 . 3$ fixing the tetrads individually is $2 + 1 + 1 + 0 + 0 + 0$ transitive on the tetrads (in any order).*

Proof. The transitivity follows at once from the quadruple transitivity of M_{24} and the fact that a sextet is determined by any one of its tetrads (as, say, the family of all tetrads congruent to the given one modulo $\mathscr{C}_{24}$). Now in the stabilizer of the sextet containing $\{a, b, c, d\}$ and $\{e, f, g, h\}$, consider the subgroup fixing the four points a, b, e, i. This is contained within the group $H = A_8$ considered earlier, and so we can see what it is — a group S_3 permuting f, g, h in the obvious way. It permutes the three tetrads disjoint from $\{a, b, e, i\}$ in the same way, and so the group fixing the tetrads individually as well as the 4 points a, b, e, i has order 1. Since this group has index at most $4 \cdot 3 \cdot 4 \cdot 4 = 192$ in the group of all permutations fixing the tetrads individually, the latter has order at most 192. But the number of sextets is

$$\frac{1}{6}\binom{24}{4} = 1771 ,$$

and so the stabilizer of a sextet has order $244823040 / 1771 = 192 \cdot 6!$. We conclude that there are 192 permutations fixing the tetrads individually, and that all 6! permutations of the tetrads are induced by permutations fixing the sextet. The transitivity $2 + 1 + 1 + 0 + 0 + 0$ is also established. We leave to the reader the exact discussion of the group of order 192 — Todd remarks that as well as the identity it contains 45 permutations of shape $1^8 2^8$, 18 of shape 2^{12}, and 128 of shape $1^6 3^6$.

Theorem 18. *M_{24} is transitive on trios. The stabilizer of a trio is a group $2^6(S_3 \times L_3(2))$, and the subgroup $2^6 L_3(2)$ fixing the octads separately is $2 + 1 + 1$ and $3 + 1 + 0$ transitive on the three octads (in any order).*

Proof. The element $(abcde)(fgh)(i)(jkl...x)$ of shape 1 3 5 15 permutes the 30 octads disjoint from $\{a, b, c, d, e, f, g, h\}$ (Theorem 11) in two orbits of size 15, consisting of those containing, and those not containing, the point i. Each octad therefore belongs to 15 trios, and M_{24} is transitive on the trios, of which there must be $759 \cdot 15/3 = 3795$. Since the tetrads of a sextet can be grouped to form 15 distinct trios, and since $1771 \cdot 15 = 3795 \cdot 7$, each trio can be refined into 7 distinct sextets. Now each sextet corresponds naturally to an element of $P(\Omega)/\mathscr{C}_{24}$

(Theorem 13), and the 7 sextets refining a trio must therefore, together with the empty set, form a 3-dimensional vector space over $\mathbf{F}_2$ (a subspace of $P(\Omega)/\mathscr{C}_{24}$). The stabilizer of a trio has therefore two interesting normal subgroups — one fixing the octads individually, and another fixing each of the 7 sextets refining the trio. The quotient by the first is S_3, since all permutations of the three octads are induced, and the quotient by the second is $L_3(2)$ or a subgroup thereof, since the permutations of the 7 sextets must preserve the vector space structure. The index of their intersection therefore divides $6 \cdot 168$, and so the intersection, fixing both the octads and the sextets individually, has order of form $64n$ (since $64 \cdot 6 \cdot 168 \cdot 3795 = 244823040$). But the intersection is a subgroup of the group described in the previous theorem, and since an element $1^6 3^6$ cannot fix 7 sextets, it has order at most 64. This gives the structure of the group.

We make only the following remarks about the transitivity. If we fix the three octads individually, and then also a point of one of them, we obtain a subgroup $2^3 L_3(2)$ of index 8 in $2^6 L_3(2)$, which is represented on 7 points of the first octad (with kernel 2^3) and in two distinct ways on the 8 points of the other two octads (faithfully in each case, related by the outer automorphism of the first lecture). Fixing a point of the second octad in addition, we come down to a group $L_3(2)$, represented on 7 points in the first and second octads, on 8 points in the third, and doubly transitively in each case.

Theorem 19. *M_{24} contains subgroups of type $L_2(23)$, doubly transitive on Ω and transitive on the set of* 759 *octads.*

Proof. Such is the subgroup $<\alpha,\beta,\gamma>$. The double transitivity on Ω is obvious, and the transitivity on octads is proved as follows. An easy if rather tedious calculation shows that at most 8 linear fractional transformations preserve some given octad. This octad therefore has at least $|L_2(23)|/8 = 759$ images under $L_2(23)$.

Theorem 20. *M_{24} contains subgroups $L_2(7)$, transitive on Ω but with imprimitivity sets of size* 3, *not contained in any of the above subgroups of M_{24}.*

Proof. Consider the permutation Π =

$$(12\ 13\ 14)(21\ 7\ 18)(17\ 1\ 20)(2\ 19\ 15)(6\ 3\ 11)(\infty\ 5\ 10)(16\ 0\ 9)(4\ 22\ 8),$$

not in M_{24}. Direct calculation reveals that it commutes with the permutation γ and also with the permutation

$$(12)(21\ 17\ 2\ 6\ \infty\ 16\ 4)(13)(7\ 1\ 19\ 3\ 5\ 0\ 22)(14)(18\ 20\ 15\ 11\ 10\ 9\ 8)$$

of M_{24}. These generate a subgroup of type $L_2(7)$ whose imprimitivity blocks are the eight 3-cycles of Π. It is also easy to show that this $L_2(7)$ preserves no monad, dyad, triad, octad, sextet, trio, or duum, and consideration of orders shows that it cannot be contained in the group $L_2(23)$.

Thus the new maximal subgroup of M_{24} is easily described as the centralizer *in* M_{24} of the permutation Π, itself *not* in M_{24}. So it is the *octern stabilizer*, the cycles of Π being ordered triads, or *terns*.

In Chang Choi's analysis, subgroups of M_{24} are divided into three classes — intransitive, transitive but imprimitive, and primitive — and his arguments for the second and third classes are very short. Here we present a quick way to deal with the intransitive groups.

Theorem 21. *Any intransitive subgroup of* M_{24} *fixes a monad, dyad, triad, octad, dodecad, or sextet, and is so contained in one of the groups of Todd's list.*

Proof. Let S be a non-empty proper subset of Ω invariant under the group. If $S \in \mathscr{C}_{24}$, then S or its complement is an octad or dodecad fixed by the group. Otherwise S is congruent modulo $\mathscr{C}_{24}$ to a unique monad, dyad or triad, or to the six tetrads of a sextet, fixed by the group, as we see from Theorem 13.

Similar ideas enable us to deal with some of the imprimitive groups, as we shall see in a moment. They also enable us to give a complete classification of the subsets of Ω under the action of M_{24}.

2.6 The structure of P(Ω)

Theorem 22. *The subsets of* Ω *fall into* 49 *orbits under* M_{24}, *related as in* Fig. 10.1. *Each node corresponds to one orbit, and the nodes are joined by lines indicating the number of ways a set of one type can be converted to one of another type by the addition or removal of a single point. Thus in an umbral heptad* (U_7), *there is one point whose removal leaves a special hexad* (S_6), *while the removal of any of the six others leaves an umbral hexad* (U_6). *In the complement of an umbral heptad there are two points whose addition gives a transverse octad* (T_8), *and* 15 *that yield umbral octads* (U_8).

Proof. We discuss 8-element sets as an example. Each 8-element set S is congruent modulo $\mathscr{C}_{24}$ to one of

(i) the empty set,
(ii) a unique 2-element set T,
(iii) each of the tetrads $T_0, \ldots, T_5$ of a sextet.

In case (i), S is a *special octad* S_8. M_{24} is transitive on these.

In case (ii), S is obtained from the special octad $S+T$ by adding one point and subtracting another. Since the stabilizer of a special octad is $1+1$ transitive on it and its complement, M_{24} is transitive on sets of this type, which we call *transverse octads* T_8.

In case (iii) $S + T_i$ is a special octad if T_i contains two points of S and two points of its complement, and an umbral dodecad if T_i is disjoint from S. Counting points of S shows that there are four tetrads of the first kind and two of the second, so that S can be obtained (in two ways) by removing four points from an umbral dodecad. Since the stabilizer of an umbral dodecad is quadruply transitive on the dodecad, M_{24} is transitive on sets like S (which we call *umbral octads* U_8).

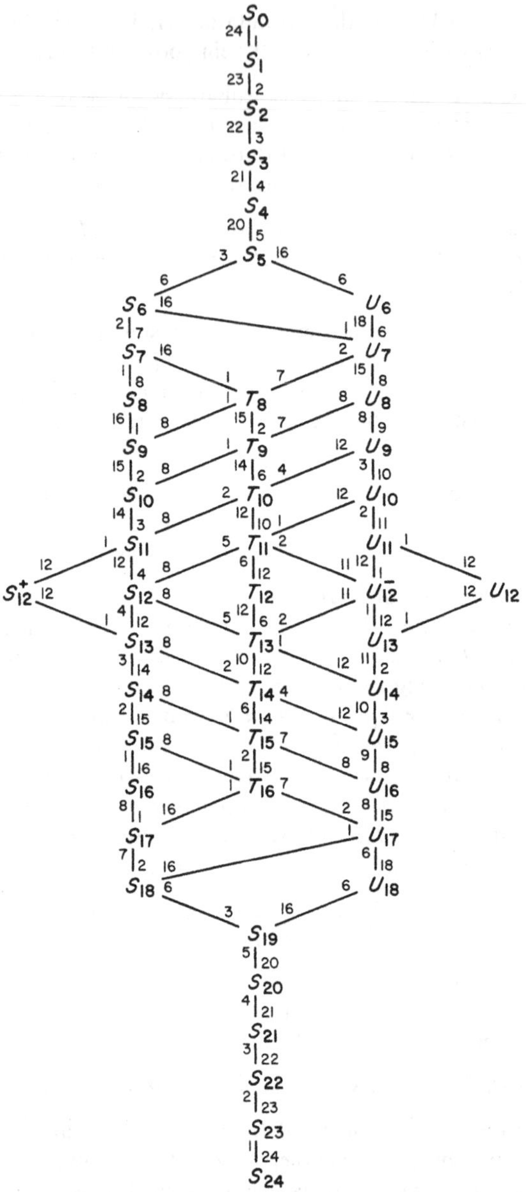

Figure 10.1. The action of M_{24} on $P(\Omega)$ (Theorem 22).

In general, a set of cardinal $n < 12$ is called *special* (S_n) if it contains or is contained in a special octad, otherwise *umbral* (U_n) if it is contained in an umbral dodecad, and *transverse* (T_n) if not. A non-umbral dodecad is *extraspecial* (S_{12}^{+}) if it contains three special octads, *special* (S_{12}) if it contains just one, *penumbral* $(U_{\overline{12}})$ if it contains all but one of the points of an umbral dodecad, and *transverse* (T_{12}) in all other cases. Sets of more than 12 points are described by the same adjectives as their complements.

Figure 10.1 enables us to read off many interesting relationships at a glance. Thus on removing any point from a U_8 we find a U_7, in which there is a unique further point whose removal leaves an S_6. The 8 points of a U_8 therefore split into 4 sets of 2 in a natural manner, so that the stabilizer of a U_8 acts imprimitively on the U_8. We can also see from the diagram that the stabilizer of any 8-element set also stabilizes a special octad. For S_8, this is the S_8 itself; for T_8, the unique S_8 sharing 7 points with the T_8; and for U_8, the union of the two tetrads whose addition separately would convert the U_8 to a U_{12}.

Theorem 23. *If a subgroup of M_{24} has two imprimitivity sets of size 12, or three of size 8, it is contained in one of the groups of Todd's list.*

Proof. If the 12-ads are umbral, this is obvious. If not, they are congruent modulo $\mathscr{C}_{24}$ to (the same) one of the sets of size 1, 2, or 3, or to the six tetrads of a sextet. In the other case, we observe that the group stabilizes the three corresponding special octads. If the union of these is Ω, they form a fixed trio, and if not, the group is intransitive.

In a similar way, with more complicated arguments, one can handle the case of four sets of size 6, or six sets of size 4, but other ideas are needed beyond this point.

All maximal subgroups of the Mathieu groups M_{23}, M_{22}, M_{12}, M_{11} are also known. Table 10.3 gives descriptions. The left hand column gives the action on the set Ω, with semicolons separating the orbits of the Mathieu group under discussion. The symbol [1; 1; 7, 15] for example indicates a subgroup of M_{22} (fixing the two 1's) that has orbits of size 7 and 15 on the remaining 22 points. In this case there are two conjugacy classes according to which of the two fixed points completes the orbit of size 7 to an octad. Again, the symbol [8, 16] indicates a group with two orbits of sizes 8 and 16, while [4^6] means that the group is transitive, with six

Table 10.3 Maximal subgroups of the Mathieu groups.

M_{24}		M_{23}		M_{22}		M_{12}		M_{11}	
[24]	$L_2(23)$	[1; 23]	23.11	[1; 1; 1, 21]	M_{21}	[12; 12]	$L_2(11)$	[1; 11; 1, 11]	$L_2(11)$
[1, 23]	M_{23}	[1; 1, 22]	M_{22}	[1; 1; 2, 4^5]	$2^4.S_5$	[1, 11; 12]	M_{11}	[1; 1, 10; 6^2]	M_{10}
[2, 22]	$M_{22}.2$	[1; 2, 21]	$M_{21}.2$	[1; 1; 6, 16]	$2^4.A_6$	[12; 1, 11]	M_{11}	[1; 2, 9; 3^4]	$M_9.2$
[3, 21]	$M_{21}.S_3$	[1; 3, 4^5]	$2^4(3{\times}S_5)$	[1; 1; 7, 15]	A_7	[2, 10; 6^2]	$M_{10}.2$	[1; 3, 8; 4, 8]	$M_8.S_3$
[4^6]	$2^6.3S_6$	[1; 7, 16]	$2^4.A_7$	[1; 1; 7, 15]	A_7	[6^2; 2, 10]	$M_{10}.2$	[1; 5, 6; 2, 10]	S_5
[8^3]	$2^6(L_3(2){\times}S_3)$	[1; 8, 15]	A_8	[1; 1; 8, 14]	$2^3.L_3(2)$	[3, 9; 3^4]	$M_9.S_3$		
[12^2]	$M_{12}.2$	[1; 11, 12]	M_{11}	[1; 1; 11, 11]	$L_2(11)$	[3^4; 3, 9]	$M_9.S_3$		
[3^8]	$L_2(7)$			[1; 1; 10, 6^2]	M_{10}	[4, 8; 4, 8]	$M_8.S_4$		
[8, 16]	$2^4.A_8$					[6×2; 6×2]	$2{\times}S_5$		
						[4×3; 4×3]	$A_4{\times}S_3$		
						[4^3; 4^3]	$4^2.D_{12}$		

imprimitivity sets of size 4. (Thus $[4^6]$ is *not* the same as $[4,4,4,4,4,4]$.) The symbol 6×2 denotes a set of 12 points with at the same time six imprimitivity sets of size 2 and two of size 6, so forming a 6×2 table in a natural way. It is very easy to recognize the groups from this information, so we refrain from further comment.

3. Third lecture

3.1 The group $Co_0 = \cdot 0$ and some of its subgroups. This group is the group of automorphisms of the Leech lattice in 24-dimensional space $\mathbf{R}^{24}$. It contains as subquotients 12 exceptional simple groups, and it seems fair to say that these groups are easiest studied inside $\cdot 0$.

3.2 The geometry of the Leech lattice. Let $\mathbf{R}^{24}$ be spanned by the orthonormal basis $v_i (i \in \Omega = PL(23))$, and define v_S to be $\Sigma v_i (i \in S)$ whenever $S \subseteq \Omega$. We use $\mathscr{C}_{24}$ for the binary Golay code, $\mathscr{C}(8)$ for the set of special octads, $\mathscr{C}(12)$ for the set of umbral dodecads, and finally $\Omega(n)$ for the set of all n-ads. Let Λ_0 be the lattice spanned by the vectors $2v_C$ for $C \in \mathscr{C}(8)$.

Theorem 24. *Λ_0 contains all vectors $4v_T (T \in \Omega(4))$, and $4v_i - 4v_j (i,j \in \Omega)$. A vector belongs to Λ_0 if and only if its coordinate-sum is a multiple of 16 and the coordinates not divisible by 4 fall in the places of a $\mathscr{C}$-set, the coordinates being all even.*

Proof. If T, U, V are three tetrads of a sextet, Λ_0 contains

$$2v_{T+U} + 2v_{T+V} - 2v_{U+V} = 4v_T .$$

We illustrate this by the addition sum

$$\begin{array}{rlll} & 2222 & 2222 & 0000\ldots \\ + & 2222 & 0000 & 2222\ldots \\ - & 0000 & 2222 & 2222\ldots \\ \hline = & 4444 & 0000 & 0000\ldots \end{array}$$

and the sum

$$\begin{array}{rrrrrr} & 4 & 4 & 4 & 4 & 0\ldots \\ - & 0 & 4 & 4 & 4 & 4\ldots \\ \hline = & 4 & 0 & 0 & 0 & -4\ldots \end{array}$$

proves the second statement similarly. The final statement then follows since $\mathscr{C}(8)$ spans $\mathscr{C}_{24}$ and the set of vectors $4v_T$ and $4v_i - 4v_j$ spans the lattice of all points with coordinate-sum a multiple of 16 and each coordinate a multiple of 4.

The *Leech lattice* Λ (called Λ_{24} in the rest of the book) is the lattice spanned by $v_\Omega - 4v_\infty$ together with the vectors of Λ_0. This is the definition given in §11 (ii) of Chap. 4. The addition sum

$$\begin{array}{rrrr} -3 & 1 & 1 & 1\ldots \\ 4 & -4 & 0 & 0\ldots \\ \hline 1 & -3 & 1 & 1\ldots \end{array}$$

shows that Λ contains each vector $v_\Omega - 4v_i$.

Theorem 25. *The vector* $(x_\infty, x_0, \ldots, x_{22})$ *is in* Λ *if and only if*

(i) *the coordinates* x_i *are all congruent modulo* 2, *to* m, *say;*
(ii) *the set of* i *for which* x_i *takes any given value modulo* 4 *is a* $\mathscr{C}$*-set;*
(iii) *the coordinate-sum is congruent to* $4m$ *modulo* 8.
For $x, y \in \Lambda$ *the scalar product* $x \cdot y$ *is a multiple of* 8, *while* $x \cdot x$ *is a multiple of* 16.

Proof. (i), (ii), (iii) and the statement about scalar products are satisfied by the generating vectors of Λ, and so for all vectors of Λ, by linearity. If x satisfies (i), (ii), (iii), we can subtract a suitable multiple of $v_\Omega - 4v_\infty$ to make all coordinates even and their sum a multiple of 16, and then $x \in \Lambda$ by Theorem 24.

We can use these conditions to enumerate the vectors in Λ of any given small length. We write $\Lambda(n)$ for the set of $x \in \Lambda$ with $x \cdot x = 16n$, and we find, for instance, that $\Lambda(1)$ is empty, while $\Lambda(2)$ contains 196560 vectors, namely $2^7 \cdot 759$ vectors of shape $((\pm 2)^8 0^{16})$ (the non-zero coordinates having positive product and being in the places of an octad), $2^{12} \cdot 24$ vectors of shape $(\mp 3(\pm 1)^{23})$ (the lower sign taken on a $\mathscr{C}$-set), and all the $2 \cdot 24 \cdot 23$ possible vectors of the shape $((\pm 4)^2 0^{22})$, in an obvious notation. The corresponding decompositions of $\Lambda(3)$ and $\Lambda(4)$ are also included in Table 4.13 of Chap. 4, these sets being orbits under the group N below, and the signs of the coordinates being suppressed.

3.3 The group $\cdot 0$, and its subgroup N. We define the group Co_0 or $\cdot 0$ (pronounced "dotto") to be the group of all Euclidean congruences of $\mathbf{R}^{24}$ that fix the origin and preserve the Leech lattice Λ as a whole. Since each vector $8v_i$ is in Λ, the elements of $\cdot 0$ have rational orthogonal matrices in which all the denominators divide 8. If π is a permutation of Ω, we extend π to a congruence of $\mathbf{R}^{24}$ by defining $v_i \pi = v_{i\pi}$, and if $S \subseteq \Omega$, we define a congruence ϵ_S by $v_i \epsilon_S = v_i \, (i \notin S)$ or $-v_i \, (i \in S)$.

Theorem 26. *The following conditions on an element* λ *of* $\cdot 0$ *are equivalent:*

(i) $v_i \lambda = \pm v_j$ *for some* $i, j \in \Omega$, *and some sign* $\pm$;
(ii) $\lambda = \pi \epsilon_C$ *for some* $\pi \in M_{24}$ *and some* $C \in \mathscr{C}_{24}$.
These operations form a subgroup $N = 2^{12} M_{24}$.

Proof. Plainly (ii) implies (i). We show (i) implies (ii). Now (i) implies that the ith row of the matrix of λ contains a single non-zero entry ± 1 in its jth place. Since λ is orthogonal, the jth place of every other row must be zero. Now the image of $4v_i + 4v_k$ under λ is 4 times the sum of the ith and kth rows, and so has a coordinate ± 4. Since this vector must be in $\Lambda(2)$, it has shape $((\pm 4)^2 0^{22})$, and the kth row has also a single non-zero coordinate ± 1. It follows that $\lambda = \pi \epsilon_S$ for some permutation π and some set S. But if $C \in \mathscr{C}_{24}$, the non-zero coordinates of $2v_c \lambda$ are in the places of $C\pi$, so π preserves $\mathscr{C}_{24}$, and is in M_{24}. Again, the coordinates congruent to 3 modulo 4 in $(v_\Omega - 4v_\infty)\lambda$ are in the places of S, so that $S \in \mathscr{C}_{24}$, and λ satisfies (ii). The final statement is obvious.

Suspecting that N is a proper subgroup of $\cdot 0$ we seek an additional operation. If $\cdot 0$ is transitive on $\Lambda(4)$, it will contain an element λ taking $8v_\infty$ to some vector $4v_T$, and a row in the matrix of λ will then have ½ in every position of T and zeros elsewhere. But 4 times the sum or difference of this row and any other must be a vector of $\Lambda(2)$, and this limits the possibilities — if this vector of $\Lambda(2)$ has shape $((\pm 2)^8 0^{16})$, the other row has four entries $\pm$½ in the places of some tetrad from the same sextet as T, and if the shape is $((\pm 4)^2 0^{22})$, we get entries $\pm$½ in the places of T. This suggests that we try a matrix which is the direct sum of six 4×4 matrices of $\pm$½'s in the places of a sextet. The signs must be chosen carefully — if Ξ is a sextet we let $\eta = \eta_\Xi$ be the map taking v_i to $v_i - \frac{1}{2} v_T$ $(i \in T \in \Xi)$, and then define $\xi = \xi_T$ to be $\eta \epsilon_T$, Ξ being the sextet of T.

Now since two octads intersect in 0, 2, 4, or 8 points, any octad is either the union of two tetrads of Ξ or contains two points from each of four tetrads, or has three points from one tetrad and one from each of the others. Supposing the coordinates properly ordered, we apply η:

$$
\begin{array}{rllllll}
x = & 2222 & 2222 & 0000 & 0000 & 0000 & 0000 \\
x\eta = & \overline{2222} & \overline{2222} & 0000 & 0000 & 0000 & 0000 \\[6pt]
x = & 2220 & 2000 & 2000 & 2000 & 2000 & 2000 \\
x\eta = & \overline{1113} & 1\overline{111} & 1\overline{111} & 1\overline{111} & 1\overline{111} & 1\overline{111} \\[6pt]
x = & 2200 & 2200 & 2200 & 2200 & 0000 & 0000 \\
x\eta = & 00\overline{22} & 00\overline{22} & 00\overline{22} & 00\overline{22} & 0000 & 0000 \\[6pt]
x = & \bar{3}111 & 1111 & 1111 & 1111 & 1111 & 1111 \\
x\eta = & \bar{3}111 & \overline{1111} & \overline{1111} & \overline{1111} & \overline{1111} & \overline{1111}
\end{array}
$$

($\bar{n}$ denotes $-n$). Since the vectors $x\eta$ are not all in Λ, η is *not* in $\cdot 0$, but on changing the sign of any tetrad we get a vector of Λ, and so $\xi = \eta \epsilon_T$ *is* in $\cdot 0$.

In [Con3] we proved that N is a maximal subgroup of $\cdot 0$, and at the same time computed the order of $\cdot 0$. The following method involves explicit calculations with the element ξ_T, but is perhaps as elegant.

Theorem 27. *$\cdot 0$ is transitive on each of the three sets $\Lambda(2)$, $\Lambda(3)$, $\Lambda(4)$, and N has index $|\Lambda(4)|/48$ in $\cdot 0$, which is generated by N together with any element ξ_T.*

Proof. We have already shown that ξ_T takes certain elements of $\Lambda(2)_2$ (see Table 4.13) to elements of $\Lambda(2)_3$, and in a similar way we see that if i and j are in distinct tetrads of Ξ, ξ_T takes $4v_i + 4v_j$ to a member of $\Lambda(2)_2$, so that the subgroup $\langle N, \xi_T \rangle$ of $\cdot 0$ is transitive on $\Lambda(2)$. Similar calculations (they are all easy) establish the transitivity on $\Lambda(3)$ and $\Lambda(4)$. Now the particular vector $8v_\infty$ has just 48 images under N (the vectors $\pm 8v_i$), and every operation of $\cdot 0$ taking $8v_\infty$ to one of these is in N. From these remarks the rest of the theorem follows. (If $\lambda \in \cdot 0$ we can find $\mu \in \langle N, \xi_T \rangle$ with $8v_\infty \mu = 8v_\infty \lambda$, so $\lambda \in N\mu$.)

Thus $Co_0 = \cdot 0$ has order

$$2^{22} \cdot 3^9 \cdot 5^4 \cdot 7^2 \cdot 11 \cdot 13 \cdot 23 = 8315553613086720000.$$

The permutation representation of $\cdot 0$ on $\Lambda(4)$ is imprimitive, since the vectors of $\Lambda(4)$ come in naturally defined *coordinate-frames*, each consisting of 24 mutually orthogonal pairs of opposite vectors. We can distinguish these frames geometrically — the vectors of the frame containing x are all those vectors of $\Lambda(4)$ that are congruent to x modulo 2Λ. To prove this, it suffices by transitivity to consider the particular case $x = 8v_\infty$.

Theorem 28. *Every vector of* Λ *is congruent modulo* 2Λ *to one of*:

(i) *the zero vector*;
(ii) *each vector of a unique pair* $x, -x$ $(x \in \Lambda(2))$;
(iii) *each vector of a unique pair* $x, -x$ $(x \in \Lambda(3))$;
(iv) *each of the* 48 *vectors of a coordinate-frame in* $\Lambda(4)$.

Proof. Let x,y be two vectors of $\Lambda(0) \cup \Lambda(2) \cup \Lambda(3) \cup \Lambda(4)$ congruent modulo 2Λ, with $y \neq \pm x$. Then since $x \pm y \in 2\Lambda$ we have $(x \pm y)\cdot(x \pm y) \geqslant 128$, whence $x \cdot y = 0$ and $x \cdot y = y \cdot y = 64$, since we know that $x \cdot x$ and $y \cdot y$ are both at most 64. It follows that x and y are members of $\Lambda(4)$ both in the same coordinate-frame. We therefore have at least $|\Lambda(0)| + |\Lambda(2)|/2 + |\Lambda(3)|/2 + |\Lambda(4)|/48$ distinct classes of $\Lambda/2\Lambda$, and since this number comes to exactly 2^{24} we have accounted for every class.

Let us say that a vector $x \in \Lambda(n)$ (of norm $16n$) has *type n*, and that x has type n_{ab} if it is also the sum of two vectors of types a and b.

Theorem 29. *Every vector* x *of type* n *has some type* n_{ab} *in which* $a + b = \frac{1}{2}(n+k)$, *where* $k = 0,2,3$ *or* 4 (*corresponding to the cases of Theorem* 28), *and these possibilities are exclusive.* $\cdot 0$ *is transitive on vectors of each of the types*

$$2,\ 3,\ 4,\ 5,\ 6_{22},\ 6_{32},\ 7,\ 8_{22},\ 8_{32},\ 8_{42},\ 9_{33},\ 9_{42},\ 10_{33},\ 10_{42},\ 10_{52},\ 11_{43},\ 11_{52}$$

(*which include all vectors of type* $n < 12$).

Proof. For the first part, we suppose x congruent to $y \in \Lambda(0) \cup \Lambda(2) \cup \Lambda(3) \cup \Lambda(4)$ modulo 2Λ, and let $\frac{1}{2}(x+y)$ and $\frac{1}{2}(x-y)$ have types a and b. For the second part, we give only a few typical samples.

Type 6_{32}. Each such vector is $x-y$, where $x \in \Lambda(3)$, $y \in \Lambda(2)$, $x \cdot y = -8$, and so $x+y \in \Lambda(4)$. (This follows from the first part by a sign-change.) To within transformation by some element of $\cdot 0$, we can take $x+y$ as $(8,0,0,\ldots)$, from which we get $x = (5,1,1,\ldots)\epsilon_C$, $y = (3,\bar{1},\bar{1},\ldots)\epsilon_C$ for some $C \in \mathscr{C}_{24}$, $\infty \notin C$ (this is the only way to express $(8,0,0,\ldots)$ as $x+y$, $x \in \Lambda(2)$, $y \in \Lambda(3)$). To within a further transformation we can suppose C empty, so that $x-y$ is equivalent under $\cdot 0$ to $(2,2,2,\ldots)$.

Type 5. In this case the typical vector is $x - y$, where $x,y \in \Lambda(2)$, $x \cdot y = -8$, so that $x + y \in \Lambda(3)$, and can be taken as $(5,1,1,\ldots)$. From this, without loss of generality, we get $x = (3,(-1)^7, 1^{16})$, $y = (2^8,0^{16})$ or vice versa, or $x = (4^2,0^{22})$, $y = (1,-3,1^{22})$ or vice versa, so that $x - y$ is one of $\pm(1,(-3)^7,1^{16})$ or $\pm(-3,-7,1^{22})$, the coordinates -3 in the first case being in a special heptad S_7. Applying a suitable ξ_T we see that

$$\begin{matrix} 1\overline{333} & \overline{3333} & 1111 & 1111 & 1111 & 1111, \\ \bar{3}111 & \bar{7}111 & 1111 & 1111 & 1111 & 1111 \end{matrix}$$

become

$$\begin{matrix} \overline{5111} & 3333 & \overline{1111} & \overline{1111} & \overline{1111} & \overline{1111}, \\ 3\overline{111} & \bar{5}333 & \overline{1111} & \overline{1111} & \overline{1111} & \overline{1111}, \end{matrix}$$

which are plainly equivalent under M_{24}.

These proofs get easier as we go further, since we then have more transitivities already known. We can avoid explicit calculations with ξ_T by using instead the counting methods of [Con3].

3.4 Subgroups of ·0. Many subgroups of $\cdot 0$ are easiest considered in relation to the infinite group Co_∞ or $\cdot\infty$ of all euclidean congruences of Λ, including translations. Any finite subgroup G of $\cdot\infty$ fixes a point (not necessarily a lattice point), and so there is a translation t of $\mathbf{R}^{24}$ (not necessarily in $\cdot\infty$) such that $G^t \subseteq \cdot 0$. If S is the name of a simplex, we use $\cdot S$ for the subgroup of those elements of $\cdot\infty$ which fix every vertex of S, $*S$ for the subgroup of elements fixing S as a whole, and $!S$ for the subgroup fixing the centroid of S. Plainly $\cdot S \subseteq *S \subseteq !S$. A simplex is then named by the types of its edges. Thus the stabilizer of two points whose difference is a vector of type n is written $\cdot n$, the stabilizer of the vertices of a triangle whose sides have types a,b,c, is written $\cdot abc$, and so on — see Fig. 10.2. The particular groups $\cdot 1$, $\cdot 2$, $\cdot 3$ are also called Co_1, Co_2, Co_3, respectively.

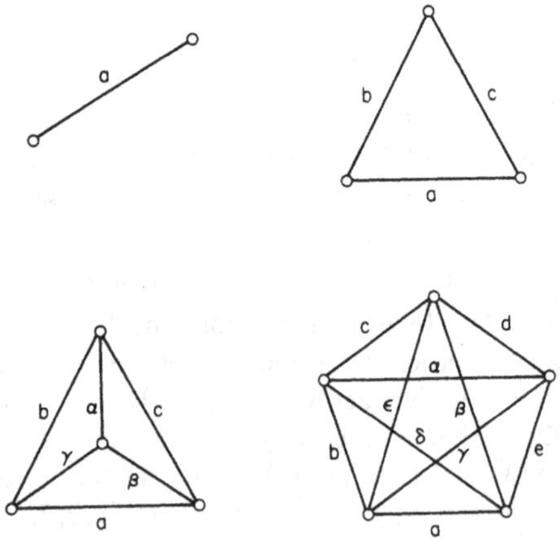

Figure 10.2.

These groups are identified in Table 10.4. Many of the identifications are easy. Thus $\cdot 6_{32}$ can be taken as the stabilizer in $\cdot 0$ of the vector $(2,2,2,...)$. The argument of Theorem 29 shows that this is expressible in just 24 ways as a sum $x + y$ ($x \in \Lambda(3), y \in \Lambda(2)$), and we can see all of them: $x = (5,1^{23})$, $y = (-3,1^{23})$ and their images under coordinate permutations. Since the vectors $x - y$ are the $8v_i$, we have $\cdot 6_{32} \subseteq N$, and so $\cdot 6_{32} = M_{24}$. At the same time, we get $\cdot 632 = M_{23}$, stabilizing the same two points together with $(5,1^{23})$. The groups $\cdot 632$ and $\cdot 432$ are identical, since a parallelogram with sides $\sqrt{2}$ and $\sqrt{3}$ and one diagonal $\sqrt{6}$ has $\sqrt{4}$ for the other diagonal. Since $\cdot 432 = \cdot 632 = M_{23}$, the group M_{23} must be contained in each of the groups $\cdot 2$, $\cdot 3$, $\cdot 4$, $\cdot 6_{32}$ — we can see this also by taking suitable coordinates for the vectors involved, namely $(-3,1^{23})$, $(5,1^{23})$, $(8,0^{23})$, and (2^{24}).

Table 10.4 Various groups associated with the Leech lattice. HS, McL, p^n, p^{1+2n} denote respectively the Higman-Sims group, the McLaughlin group, an elementary abelian group of order p^n, and an extraspecial group of order p^{2n+1} respectively.

Name	*Order*	*Structure*	*Name*	*Order*	*Structure*
$\cdot 0$	$2^{22}3^9 5^4 7^2 11.13.23$	$2.Co_1$	$\cdot 222$	$2^{15}3^6 5.7.11$	$PSU_6(2)$
$\cdot 1$	$2^{21}3^9 5^4 7^2 11.13.23$	Co_1	$\cdot 322$	$2^7 3^6 5^3 7.11$	McL
$\cdot 2$	$2^{18}3^6 5^3 7.11.23$	Co_2	$\cdot 332$	$2^9 3^2 5^3 7.11$	HS
$\cdot 3$	$2^{10}3^7 5^3 7.11.23$	Co_3	$\cdot 333$	$2^4 3^7 5.11$	$3^5.M_{11}$
$\cdot 4$	$2^{18}3^2 5.7.11.23$	$2^{11}M_{23}$	$\cdot 422$	$2^{17}3^2 5.7.11$	$2^{10}.M_{22}$
$\cdot 5$	$2^8 3^6 5^3 7.11$	$McL.2$	$\cdot 432$	$2^7 3^2 5.7.11.23$	M_{23}
$\cdot 6_{22}$	$2^{16}3^6 5.7.11$	$PSU_6(2).2$	$\cdot 433$	$2^{10}3^2 5.7$	$2^4.A_8$
$\cdot 6_{32}$	$2^{10}3^3 5.7.11.23$	M_{24}	$\cdot 442$	$2^{12}3^2 5.7$	$2^{1+8}.A_7$
$\cdot 7$	$2^9 3^2 5^3 7.11$	HS	$\cdot 443$	$2^7 3^2 5.7$	$M_{21}.2$
$\cdot 8_{22}$	$2^{18}3^6 5^3 7.11.23$	Co_2	$\cdot 522$	$2^7 3^6 5^3 7.11$	McL
$\cdot 8_{32}$	$2^7 3^6 5^3 7.11$	McL	$\cdot 532$	$2^8 3^6 5.7$	$PSU_4(3).2$
$\cdot 8_{42}$	$2^{15}3^2 5.7$	$2^{1+8}.A_8$	$\cdot 533$	$2^4 3^2 5^3 7$	$PSU_3(5)$
$\cdot 9_{33}$	$2^5 3^7 5.11$	$3^5.M_{11}.2$	$\cdot 542$	$2^7 3^2 5.7.11$	M_{22}
$\cdot 9_{42}$	$2^7 3^2 5.7.11.23$	M_{23}	$\cdot 633$	$2^6 3^3 5.11$	M_{12}
$\cdot 10_{33}$	$2^{10}3^2 5^3 7.11$	$HS.2$	$*2 = !2$	$2^{19}3^6 5^3 7.11.23$	$(\cdot 2) \times 2$
$\cdot 10_{42}$	$2^{17}3^2 5.7.11$	$2^{10}.M_{22}$	$*3 = !3$	$2^{11}3^7 5^3 7.11.23$	$(\cdot 3) \times 2$
$\cdot 11_{43}$	$2^{10}3^2 5.7$	$2^4.A_8$	$*4$	$2^{19}3^2 5.7.11.23$	$(\cdot 4) \times 2$
$\cdot 11_{52}$	$2^8 3^6 5.7$	$PSU_4(3).2$	$!4$	$2^{22}3^3 5.7.11.23$	$2^{12}.M_{24}$
			$!333$	$2^7 3^9 5.11$	$3^6.2.M_{12}$
			$!442$	$2^{15}3^4 5.7$	$2^{1+8}.A_9$

3.5 The Higman-Sims and McLaughlin groups. In a similar way we see that M_{22} is contained in each of the groups $\cdot 222$, $\cdot 322$, $\cdot 332$. Since $\cdot 222$ is the group $PSU_6(2)$ it follows that M_{22} has a 6-dimensional projective representation over $\mathbf{F}_4$ (and from this it follows fairly easily that the multiplier of M_{22} has order divisible by 3).

D. G. Higman and C. C. Sims [Hig3] have described a simple group of order $44352000 = 100|M_{22}|$ as the group of even permutations of a certain graph on 100 vertices, and J. McLaughlin [McL1] has discovered a simple group of order 898128000 which is a group of automorphisms of a graph on 275 vertices. We here identify the Higman-Sims group HS with our $\cdot 332$, and sketch the identification of McLaughlin's group McL with $\cdot 322$.

Let $X = 4v_i + v_\Omega$, $Y = 4v_j + v_\Omega$, $Z = 0$, where i and j are distinct monads, so that XYZ is a triangle of type 332. Then there are exactly 100 points T for which $XYZT$ has type 332222, namely the point $P = 4v_i + 4v_j$, 22 points $Q_k = v_\Omega - 4v_k$ ($k \in \Omega \setminus \{i, j\}$), and 77 points $R_K = 2v_K$ ($\{i, j\} \subseteq K \in \mathscr{C}(8)$). If we say that two of these points are *incident* when their difference has type 3, then the incidences are (P, Q_k), (Q_k, R_K) ($k \in K$), and $(R_K, R_{K'})$ ($K \cap K' = \{i, j\}$), and the incidence graph is visibly identical with the Higman-Sims graph.

Now $X - Y + P - Z = 8v_i$, so that the stabilizer in $\cdot\infty$ of X,Y,Z,P is a subgroup of N, and in fact this stabilizer is the group M_{22} of permutations of Ω that fix i and j. We identify $\cdot 332$ with the Higman-Sims group, and at the same time provide an easy proof of the latter's existence, by showing that $\cdot\infty$ has operations fixing X, Y, Z but disturbing P. For let λ be an operation of $\cdot\infty$ such that $X\lambda = 2v_C$, $Y\lambda = 2v_D$, $Z\lambda = 0$, where $C \in \mathscr{C}(12)$, $D \in \mathscr{C}(12)$, $C + D \in \mathscr{C}(8)$. (Such λ exist by the transitivity of $\cdot\infty$ on triangles of type $\cdot 332$.) Then the subgroup H of those operations of N that fix each of $X\lambda$, $Y\lambda$, $Z\lambda$ fixes no one of the 100 points $T\lambda$ that differ by type 2 vectors from each of $X\lambda$, $Y\lambda$, $Z\lambda$, and so the group $\lambda H \lambda^{-1}$ has operations that fix each of X,Y,Z, but not P. (The argument has proved transitivity of $\cdot\infty$ on tetrahedra of type 332222 given transitivity on triangles of type 332.)

Sims [Sim1] has shown that a doubly transitive group on 176 letters described by G. Higman [Hig4] is isomorphic to the Higman-Sims group. (See also [Smi8]-[Smi10].) G. Higman's group is the automorphism group of a "geometry" of 176 "points" and 176 "quadrics," there being 50 points on each quadric and 50 quadrics through each point. Now there are 352 points G such that GX has type 3 while GY and GZ have type 2, and these naturally form 176 pairs $P_K = \{A_K, B_K\}$ ($K \in \mathscr{C}(8)$, $K \cap \{i, j\} = \{i\}$), $A_K = 2v_K$, and $B_K = X - A_K$. There are similarly 176 pairs $Q_K = \{C_K, D_K\}$ obtained by interchanging the roles of i and j. If we regard "points" as the pairs P_K, and "quadrics" as pairs Q_K, and say that P_K is on $Q_{K'}$ just if $|K \cap K'| = 2$, then we obtain G. Higman's geometry and another proof of the identity of his group with the Higman-Sims group.

If we consider instead a triangle of type 322, we find that there are just 275 points which complete it to a tetrahedron of type 322222, and that the incidence graph (where x and y are incident if and only if $x - y$ has type 3) is now McLaughlin's graph. The details of the identification are rather complicated, but the result is a fairly simple definition of McLaughlin's graph. If we take the triangle XYZ to be $X = 0$, $Y = 4v_i + v_\Omega$, $Z = -4v_j + v_\Omega$ $(i \neq j)$, then the 275 points fall naturally into the following three sets: 22 points U_k $(k \in \Omega \setminus \{i, j\})$, 77 points V_K ($\{i, j\} \subseteq K \in \mathscr{C}(8)$), and 176 points $G_{K'}$ ($K' \in \mathscr{C}(8)$, $\{i, j\} \cap K' = \{i\}$), and the incidences can be simply described by combinatorial conditions on k, K, K'.

3.6 The group $Co_3 = \cdot 3$. We shall also show that $Co_3 = \cdot 3$ has a doubly transitive permutation representation on 276 letters, the stabilizer of a point being a group $McL.2$. To this end we consider the stabilizer of the vector $x = (5, 1^{23})$. There are 276 unordered pairs $\{y,z\}$ with $x = y + z$, where $y,z \in \Lambda(2)$, namely 23 with y (say) of shape $(4^2, 0^{22})$ and 253 with y of shape $(2^8, 0^{16})$, the coordinate y_∞ being non-zero in each case.

Letting $\{y_0, z_0\}$ be one of these pairs, we have for each other pair $\{y,z\}$ that either $z - y_0$ or $y - y_0$ is of type 2. Assuming transitivity of $\cdot 0$ on triangles 322 and tetrahedra 322222 then suffices for the result: $\cdot 3$ is doubly transitive on the 276 unordered pairs $\{y,z\}$. The stabilizer in $\cdot 3$ of the point y_0 has index 2 in the stabilizer of the pair $\{y_0, z_0\}$, and is visibly $\cdot 322$. If we take $y_0 = 4v_\infty + 4v_0$ the situation is invariant under the group M_{22} fixing ∞ and 0, and the 276 pairs fall into orbits of sizes 1, 22, 77, 176 under this group, which is contained in three distinct subgroups of $\cdot 3$ — an M_{23} with orbits 1 + 22 and 77 + 176, the McLaughlin group $\cdot$ 322 with orbits 1 and 22 + 77 + 176, and a Higman-Sims group $\cdot$ 332 with orbits 1 + 22 + 77 and 176, the latter neatly displaying both the 100 letter representation of Higman and Sims and the doubly transitive 176 letter representation of Higman.

A pretty series of subgroups of $\cdot 0$ arises in the following way [Tho7]. The centralizer of a certain element x of order 3 in $\cdot 0$ has the form $\langle x \rangle \times 2\mathrm{A}_9$, the $2\mathrm{A}_9$ being the Schur double cover of A_9, and containing a natural sequence of subgroups $2\mathrm{A}_n (2 \leq n \leq 9)$. The centralizers B_n of $2\mathrm{A}_n$ are, for $n = 2, 3, \dots, 9$, groups $\cdot 0$, $6Suz$, $2G_2(4)$, $2HJ$, $2U_3(3)$, $2L_3(2)$, $2\mathrm{A}_4$, C_6, where HJ (also called J_2) is the Hall-Janko simple group [Hal4] and Suz is Suzuki's sporadic simple group [Suz1]. It follows that HJ has a multiplier of order divisible by 2, and Suz a multiplier of order divisible by 6, and also that HJ has a 6-dimensional projective representation that can be written over $\mathbf{Q}(\sqrt{-3}, \sqrt{-5})$, while Suz has a 12-dimensional projective representation over $\mathbf{Q}(\sqrt{-3})$. The latter can be obtained as follows. Take an element ω of order 3 with no fixed point, and so satisfying (as a matrix) the equation $\omega^2 + \omega + 1 = 0$. Then, in the ring of 24×24 matrices, ω generates a copy of the complex numbers in which it is identified with $e^{2\pi i/3}$. If we define, for $x \in \Lambda$, $x(a + be^{2\pi i/3}) = ax + b(x\omega)$, the Leech lattice becomes the *complex Leech lattice* Λ_C (see Chap. 7, Example 12), a 12-dimensional lattice (or module) over the ring $\mathbf{Z}[e^{2\pi i/3}]$ of Eisenstein integers, and the automorphism group of Λ_C is the group $6Suz$. The complex Leech lattice has a natural coordinate system that displays a remarkable analogy between it and the real Leech lattice, with 2 everywhere becoming $\theta = \sqrt{-3} = \omega - \omega^2$ (see Table 7.3).

The details are as follows. Define vectors x_i, y_i, z_i ($i \in PL(11)$) by

$$x_i + y_i + z_i = 0, \quad y_i = z_i \epsilon_N, \quad z_\infty = \nu_\Omega + 4\nu_\infty ,$$

$$z_i = (\nu_\Omega + 4\nu_{15}) \epsilon_{\{\infty,15,1,2,3,4,6,18\}} \beta^{-i} \quad (i \in GF(11)).$$

Let ω be the operation of $\cdot 0$ that takes $x_i \to y_i \to z_i \to x_i$ for each i. In terms of the coordinates x_i, and the *ternary* Golay code $\mathscr{C}_{12}$, Λ_C is now spanned (up to a scale factor) by the vectors θx_C ($C \in \mathscr{C}_{12}$), $3x_i - 3\omega^s x_j$, $x_\Omega + 3x_i$ ($\Omega = PL(11)$). These are obtained from $2\nu_C$, $4\nu_i \mp 4\nu_j$, $\nu_\Omega - 4\nu_i$ in the real case by replacing 2 by θ, 4 by $\theta^2 = -3$, and ± 1 by ω^s. It is interesting that the order of the binary Golay code is 2^{12}, and that of the ternary Golay code is $\theta^{12} = 3^6$, and that $6\,Suz$ is generated by its monomial part together with a matrix of 3×3 blocks whose entries have the form $\pm\omega^i/\theta$. (Compare the real case, when we have 4×4 blocks with entries $\pm 1/2$.)

3.7 Involutions in $\cdot 0$. Each involution in $\cdot 0$ is conjugate to an involution ϵ_C ($C \in \mathscr{C}_{24}$). If C is an octad, the centralizer preserves the corresponding 8-dimensional subspace, which intersects Λ in a copy of the 8-dimensional lattice E_8. The subgroup of the centralizer that fixes every point of this space is therefore in N (since the space contains vectors ν_i), and is easily seen to be an extraspecial group 2^{1+8} of order 2^9. The quotient by this extraspecial group is a subgroup of the Weyl group of E_8, and in fact it is the derived group $W(E_8)'$, so that the whole centralizer is a group $2^{1+8}(W(E_8))'$. If C is a 16-ad, we get the same centralizer.

If C is a dodecad, the fixed space can contain no vector of shape $((\pm 2)^8 0^{16})$ or $((\mp 3)(\pm 1)^{23})$, and so contains in $\Lambda(2)$ only the $2^2 \cdot 66$ vectors of shape $((\pm 4)^2 0^{22})$ whose non-zero coordinates are in the fixed space. We call two such vectors *skew* unless they are equal, opposite, or orthogonal. Then the only vectors of the set which are skew to all the vectors skew to $x = 4\nu_i + 4\nu_j$ say, are just the four vectors $\pm 4\nu_i \pm 4\nu_j$ having the same non-zero coordinates as x. It follows (since $(4\nu_i + 4\nu_j) + (4\nu_i - 4\nu_j) = 8\nu_i$) that the centralizer of ϵ_C is in N, and so is the group $2^{12}M_{12}$ centralizing ϵ_C in N. (The centralizer of this type of involution ϵ_C in $\cdot 3$ is $<\epsilon_C> \times M_{12}$, as is seen by taking the fixed vector of $\cdot 3$ to be $(2^{12}\, 0^{12})$ — compare the situation in Janko's group J_1.)

The only remaining involution in $\cdot 0$ is $\epsilon_\Omega = -1$, which is central in the whole group.

3.8 Congruences for theta series. Let

$$\Theta_\Lambda(z) = \Sigma\, N(m) q^{2m} = 1 + 196560 q^4 + \cdots$$

be the theta series of Λ (Eq. (138)-(141) of Chap. 4). From §6 of Chap. 7 we know that $\Theta_\Lambda(z)$ is a modular form of weight 12 for $SL_2(\mathbf{Z})$. Since the Eisenstein series $1 + c\, \Sigma\, \sigma_{11}(m) q^{2m}$ ($c = 65520/691$) and the cusp form

$\Delta(z) = \Sigma\tau(m)q^{2m}$ are modular forms of weight 12, it follows from Theorem 17 of Chap. 7 that

$$N(m) = \frac{65520}{691}(\sigma_{11}(m) - \tau(m)),$$

where the $\tau(m)$ are the Ramanujan numbers (Eq. (37)-(39) of Chap. 4).

Ramanujan's remarkable congruence $\tau(m) \equiv \sigma_{11}(m) \pmod{691}$ is particularly evident, and indeed we can use the formula to find congruences for $\tau(m)$ modulo any prime power dividing $g/c = 2^{18}3^7 5^3 7 \cdot 11 \cdot 23 \cdot 691$ (where g is the order of $\cdot 0$). Thus to modulus 23 we have $N(m) \equiv N(m)^{(\alpha)}$, where $N(m)^{(\alpha)}$ is the number of vectors of $\Lambda(m)$ that are fixed by the element α of order 23. But the fixed vectors of α are those of the shape $av_\infty + bv_{\Omega\setminus\infty}$, and since this has type m if and only if $16m = a^2 + 23b^2$ we obtain $\tau(m) \equiv \sigma_{11}(m) \equiv 0 \pmod{23}$ whenever m is a non-residue modulo 23.

3.9 A connection between $\cdot 0$ and Fischer's group Fi_{24}. $\cdot 0$ can be regarded as a permutation group on the 196560 vectors of $\Lambda(2)$. It has a subgroup $N = 2^{12}M_{24}$ that permutes these in three orbits of sizes $2^7 \cdot 759$, $2^{12} \cdot 24$, and $2^2\binom{24}{2}$. The elementary part 2^{12} of N is permuted by M_{24} like the binary Golay code $\mathscr{C}_{24}$. The extension $2^{12}M_{24}$ is split.

The group Fi_{24} of B. Fischer [Fis1] is a permutation group on 306936 objects (the involutions in a certain conjugacy class). It has a subgroup $N^* = 2^{12}M_{24}$ that permutes these in three orbits of sizes $2^5 \cdot 759$, $2^0 \cdot 24$, and $2^{10}\binom{24}{2}$. The elementary part 2^{12} of N^* is permuted by M_{24} like the quotient $P(\Omega)/\mathscr{C}_{24}$. The extension $2^{12}M_{24}$ is non-split.

These facts suggest that there is a relation between the two groups, and this is borne out by the fact that the subgroups of M_{24} corresponding to the three orbits are identical, while the elementary subgroups are dual. The appropriate name for this relationship — whatever it is — is twinning, as we see from the following parable. "Once upon a time there was an egg (the elementary group $P(\Omega)$ of order 2^{24}), which after fertilization (by the action of M_{24}) split into two (the group $\mathscr{C}_{24}$ and its quotient $P(\Omega)/\mathscr{C}_{24}$), which grew into two healthy twins ($\cdot 0$ and Fi_{24})."

That the parable is meaningful is suggested by the existence of similar structure relating $\cdot 0$ to itself via its subgroup $6\,Suz$, with elements of order 3 like ω taking the place of Fischer's involutions. The elementary 2^{12} in the subgroup N^* of Fi_{24} has 24 of Fischer's 3-transpositions (the special involutions of his group), permuted in the natural way by M_{24}, the centralizer of any one being a group $2Fi_{23}$. In a similar way $\cdot 0$ has a group $3^6.2M_{12}$ in which the 3^6 has 12 special subgroups of order 3 (isomorphic to the groups V_i of the first lecture) permuted naturally by M_{12}, the centralizer of any one being the group $6\,Suz$.

Acknowledgements. I should like to thank D. Livingstone and his colleagues for some remarks which I have made use of in the second lecture, and in particular, for the material of Table 10.3, and J. G. Thompson for his sustained interest in the subject of the third lecture.

Appendix on the exceptional simple groups

We discuss under six headings the 26 non-abelian simple groups not occurring among the infinite families of Chevalley and twisted Chevalley groups and alternating groups. With each group we give its order, preceded by its Schur multiplier and followed by its outer automorphism group (see [Con16], [Gri1], [Gri3], [Gri9]). Asterisks indicate groups that belong to one of the infinite families but are included here by analogy.

The Mathieu groups

M_{24}	$1(2^{10}3^3 5\cdot 7\cdot 11\cdot 23)1$	M_{12}	$2(2^6 3^3 5\cdot 11)2$
M_{23}	$1(2^7 3^2 5\cdot 7\cdot 11\cdot 23)1$	M_{11}	$1(2^4 3^2 5\cdot 11)1$
M_{22}	$12(2^7 3^2 5\cdot 7\cdot 11)2$	$*A_6$	$6(2^3 3^2 5)2\times 2$
$*M_{21}$	$4\times 4\times 3(2^6 3^2 5\cdot 7)2\times S_3$	$*A_7$	$6(2^3 3^2 5\cdot 7)2$

We have already discussed the isomorphisms $M_{21}\cong L_3(4)$ and $M_{10}'\cong A_6 \cong L_2(9)$, but the unusual multipliers show that the usual definitions of these two groups are unrevealing. For alternating groups other than A_6, A_7 our symbol reads $2(n!/2)2$. The groups under the next two headings are in some sense the Mathieu groups "writ large".

The Fischer groups

Fi_{24}	$3(2^{21}3^{16}5^2 7^3 11\cdot 13\cdot 17\cdot 23\cdot 29)2$	Fi_{22}	$6(2^{17}3^9 5^2 7\cdot 11\cdot 13)2$
Fi_{23}	$1(2^{18}3^{13}5^2 7\cdot 11\cdot 13\cdot 17\cdot 23)1$	$*Fi_{21}$	$2\times 2\times 3(2^{15}3^6 5\cdot 7\cdot 11)S_3$

Here Fi_n has a conjugacy class of involutions any one of which has centralizer a group $2Fi_{n-1}$ and any two of which have product of order at most 3. A maximal commuting set of such involutions has exactly n of them, generating a group 2^{n-12} whose normalizer is a group $2^{n-12}M_n$. We have the isomorphism $Fi_{21}\cong U_6(2)$. (See also [Con7], [Enr1], [Enr2], [Hun1], [Par3].)

Lattice stabilizers in $\cdot 0$

$\cdot 1 = Co_1$	$2(2^{21}3^9 5^4 7^2 11\cdot 13\cdot 23)1$	$*\cdot 222$	$2\times 2\times 3(2^{15}3^6 5\cdot 7\cdot 11)S_3$
$\cdot 2 = Co_2$	$1(2^{18}3^6 5^3 7\cdot 11\cdot 23)1$	$\cdot 322\cong McL$	$3(2^7 3^6 5^3 7\cdot 11)2$
$\cdot 3 = Co_3$	$1(2^{10}3^7 5^3 7\cdot 11\cdot 23)1$	$\cdot 332\cong HS$	$2(2^9 3^2 5^3 7\cdot 11)2$

The first group contains M_{24}, the next two contain M_{23}, and the remainder contain M_{22}. We have the isomorphism $\cdot 222\cong U_6(2)$. (See also [Fin2].)

The Suzuki chain

Suz	$6(2^{13}3^{7}5^{2}7\cdot 11\cdot 13)2$	$HJ = J_2$	$2(2^{7}3^{3}5^{2}7)2$
$*G_2(4)$	$2(2^{12}3^{3}5^{2}7\cdot 13)2$	$*U_3(3)$	$1(2^{5}3^{3}7)2$

These are the central quotients of the groups B_n described in §3.6. We have the isomorphism $U_3(3) \cong G_2(2)'$.

It is perhaps worthy of note that the group $U_4(3)$ has an unusual multiplier $(3 \times 3 \times 4)$. The group $3^2U_4(3)$ is the centralizer of two commuting elements of order 3 in $\cdot 0$, and $U_4(3)$ is closely connected with a number of the groups above.

Centralizers in the Monster

$M = F_1$	$1(2^{46}3^{20}5^{9}7^{6}11^{2}13^{3}17\cdot 19\cdot 23\cdot 29\cdot 31\cdot 41\cdot 47\cdot 59\cdot 71)1$		
$B = F_{2+}$	$2(2^{41}3^{13}5^{6}7^{2}11\cdot 13\cdot 17\cdot 19\cdot 23\cdot 31\cdot 47)1$		
$Co_1 = F_{2-}$	$2(2^{21}3^{9}5^{4}7^{2}11\cdot 13\cdot 23)1$		
$Fi_{24} = F_{3+}$	$3(2^{21}3^{16}5^{2}7^{3}11\cdot 13\cdot 17\cdot 23\cdot 29)2$		
$Suz = F_{3-}$	$6(2^{13}3^{7}5^{2}7\cdot 11\cdot 13)2$	$He = F_{7+}$	$1(2^{10}3^{3}5^{2}7^{3}17)2$
$Th = F_{3\|3}$	$1(2^{15}3^{10}5^{3}7^{2}13\cdot 19\cdot 31)1$	$*A_7 = F_{7-}$	$6(2^{3}3^{2}5\cdot 7)2$
$HN = F_{5+}$	$1(2^{14}3^{6}5^{6}7\cdot 11\cdot 19)2$	$M_{12} = F_{11+}$	$2(2^{6}3^{3}5\cdot 11)2$
$HJ = F_{5-}$	$2(2^{7}3^{3}5^{2}7)2$	$*L_3(3) = F_{13+}$	$1(2^{4}3^{3}13)2$

The Monster group M, which was discovered a few years after these lectures were given, led to the discovery of several additional simple groups and casts further light on some of those already known. These groups usually arise from studying the structure of centralizers of suitable elements of the Monster. Here we consider elements of prime order, and write F_{p+}, F_{p-} etc. for the nonabelian composition factor in the centralizer of an element in the conjugacy class of M called $p+$, $p-$ etc. in [Con16], [Con17]. Some of the groups that are obtained in this way are shown above.

$M = F_1$, the Monster, or Fischer-Griess group, or Friendly Giant, was discovered independently by Fischer and Griess in 1973, and constructed by Griess in 1980 [Gri4]-[Gri8]. A simplified construction is described in Chap. 29. The baby Monster $B = F_{2+}$ was discovered by Fischer slightly earlier and was constructed by Leon and Sims [Leo9], [Sim3].

The existence of the Thompson group $Th = F_{3|3}$ (constructed by J. G. Thompson and P. E. Smith [Tho2]) and of the Harada-Norton group $HN = F_{5+}$ (constructed by Norton [Har0], [Har1], [Nor2], [Nor8]) was suggested by that of the Monster, although the entirely analogous groups of Fischer ($Fi'_{24} = F_{3+}$) and Held ($He = F_{7+}$) were already known [Hel1], [Hel2]. All of these groups are best understood in terms of their relationship with the Monster. We remark that $Co_1 = F_{2-}$, $Suz = F_{3-}$, $HJ = J_2 = F_{5-}$, $M_{12} = F_{11+}$.

The remaining groups

J_1	$1(2^3 3\cdot 5\cdot 7\cdot 11\cdot 19)1$	Ru	$2(2^{14}3^3\cdot 5^3\cdot 7\cdot 13\cdot 29)1$
J_3	$3(2^7 3^5 5\cdot 17\cdot 19)2$	$O'N$	$3(2^9 3^3 5\cdot 7^3\cdot 11\cdot 19\cdot 31)2$
J_4	$1(2^{21}3^3 5\cdot 7\cdot 11^3\cdot 23\cdot 29\cdot 31\cdot 37\cdot 43)1$	Ly	$1(2^8 3^7 5^6 7\cdot 11\cdot 31\cdot 37\cdot 67)1$

Most of these were discovered via their centralizers of involutions. $J_2 = HJ$ (the Hall-Janko group) and J_3 (the Higman-Janko-McKay group ([Hig5], [Jan4], [Con45], [Gor6], [Lin4], [Lin5], [Smi4], [Tit4], [Wal1], [Wei3]-[Wei5]), have involutions with the same centralizer, as do He, M_{24}, and $L_5(2)$. The centralizers of suitable elements of order 3 are $3A_7$ in the Held group He, and $3McL$ in the Lyons group Ly ([Lyo1], [Sim2]), exhibiting the 3-parts of the multipliers of these groups. J_1 is a subgroup of $G_2(11)$, and $G_2(5)$ is a subgroup of Lyons' group.

The three remaining groups are Janko's group J_4 [Jan6], [Con12], [Nor3], Rudvalis's group Ru [Con10], [Con44], [Rud3], and O'Nan's group $O'N$ [And2], [O'Na1]. It seems that the simplest way of computing with all six groups is to use matrices over various finite fields. The groups

$$J_1 \qquad 3.J_3 \qquad J_4 \qquad Ru \qquad 3O'N \qquad Ly$$

have representations of degrees

$$7 \qquad 9 \qquad 112 \qquad 28 \qquad 45 \qquad 111$$

over the fields of orders

$$11 \qquad 4 \qquad 2 \qquad 2 \qquad 7 \qquad 5$$

respectively (cf. [Con16], [Eva1], [Mey3], [Ryb1]).

Postscript. The maximal subgroups of $\cdot 1 \cong Co_1$ have been completely enumerated by Wilson [Wil4], following earlier work of Curtis [Cur2], [Cur5]. Maximal subgroups of other sporadic simple groups have been classified in [But1], [Fin2]-[Fin4], [Kle1], [Lem1], [Nor8], [Wil1]-[Wil17], [Wol0], [Yos2]. Character tables and much other information about sporadic groups will be found in [Con16]. General references for simple groups are [Car2], [Car3], [Cox28], [Gor2]-[Gor5], [Tit3]-[Tit9]. See also [Soi1], [Soi3].

11

The Golay Codes and The Mathieu Groups

J. H. Conway

This chapter contains a detailed description of the binary Golay code of length 24, the Steiner system $S(5,8,24)$, and the Mathieu group M_{24}. The MOG (Miracle Octad Generator) and the hexacode are computational tools that make it easy to perform calculations with these objects. The MINIMOG and the tetracode perform similar services for the ternary Golay code of length 12, the Steiner system $S(5,6,12)$, and the Mathieu group M_{12}.

1. Introduction

The Golay code $\mathscr{C}_{24}$, the Steiner system $S(5,8,24)$, and the Mathieu group M_{24} are beautiful combinatorial objects with a great wealth of structure and applications. The MOG, and its companion-at-arms the hexacode, are computational tools that enable one to perform mental calculations on these objects with great ease. In particular, one can check at sight that a word is in $\mathscr{C}_{24}$, complete an octad of $S(5,8,24)$ from any five of its points, or write down many permutations of M_{24} in a few moments. There is also a MINIMOG, with companion the tetracode, which together perform similar services for M_{12}. Some discoveries about the "lexicography" of M_{24} and M_{12} indicate that these devices may have some theoretical as well as practical value.

Many calculations in the larger sporadic simple groups reduce to calculations inside M_{24}, which has consequently been used in a number of constructions. In particular, we mention the construction of the Conway groups via the Leech lattice (see Chap. 10), the Fischer groups Fi_{24}, Fi_{23}, Fi_{22} (§3.9 of Chap. 10), the Janko group J_4 [Con12], [Nor3], and most recently Griess's construction of the Monster group (see [Gri4]-[Gri8] and Chap. 29).

The MOG array was invented by R. T. Curtis [Cur2], [Cur3] in the course of his work on the Leech lattice, and he and others have made valuable use of it for many other investigations involving M_{24}. The special virtues of the approach via the hexacode emerged much later, during the construction by Norton, Parker, Benson, Conway and Thackray of the largest Janko group J_4 [Con12], [Nor3]. Although that company included several practiced MOG users, it was found that the hexacode effected still greater simplifications in the enormous number of M_{24}-related calculations that were required in the course of the construction. More recently, it has performed a similar role for the many calculations with the Leech lattice that were carried out in the determination of its covering radius and the enumeration of its deep holes (see Chap. 23).

There are almost as many different constructions of M_{24} as there have been mathematicians interested in that most remarkable of all finite groups. Usually, when one constructs a large group from some "simpler" object, the construction itself will have a smaller group of symmetries, and one finds that elements inside this "visible subgroup" are "easy" to understand, and the others "hard." It is a remarkable fact that the MOG construction manages to have several "visible groups," each of them a maximal subgroup of M_{24}! We shall construct it here in a way that is specially related to the *sextet group* $2^6{:}3{\cdot}S_6$, since that offers maximal opportunities for the use of the hexacode, which is the main topic of this chapter. However, we feel that the reader should be aware that the MOG was first used by Curtis in connection with the *octad group* $2^4{:}A_8$, that when he wrote it up it seemed easiest to use the *trio group* $2^6{:}(S_3 \times L_2(7))$, and that still later the *triad group* $P\Gamma L_3(4)$ was found to be just as simply related to the MOG. In fact much of its power stems from the ability to switch groups while seeking the element one desires.

In the first part of the chapter we shall describe the construction of the hexacode and MOG, and show how to use them for the basic calculations that arise in every investigation with M_{24}. The later part has a section for each maximal subgroup, described where possible in close relation to the hexacode or MOG.

2. Definitions of the hexacode

As usual $\mathbf{F}_4 = \{0, 1, \omega, \bar{\omega}\}$, with the relations

$$1 + \omega = \bar{\omega}\,,\ \ 1 + \bar{\omega} = \omega\,,\ \ \omega + \bar{\omega} = \omega\bar{\omega} = 1\,,\ \ \omega^2 = \bar{\omega}\,,\ \ \bar{\omega}^2 = \omega\,,\ \ \omega^3 = 1\,.$$

The hexacode $\mathscr{C}_6$ (§2.5.2 of Chap. 3) is a 3-dimensional code of length 6 over $\mathbf{F}_4$. In order to display certain symmetries, we usually group the 6 digits of a hexacodeword into 3 *couples*. The code may be defined in several different ways:

Definition 1. $\mathscr{C}_6$ is generated by the words

$$\omega\bar{\omega}\ \omega\bar{\omega}\ \omega\bar{\omega},\quad \omega\bar{\omega}\ \bar{\omega}\omega\ \bar{\omega}\omega,\quad \bar{\omega}\omega\ \omega\bar{\omega}\ \bar{\omega}\omega,\quad \bar{\omega}\omega\ \bar{\omega}\omega\ \omega\bar{\omega}\,.$$

No No Yes Yes No Yes !

Definition 2. $\mathscr{C}_6$ has a word $W(\phi) = ab\ cd\ ef$ for each quadratic function $\phi(x) = ax^2 + bx + c$ defined over $\mathbf{F}_4$. The first three digits (a,b,c) specify the function ϕ, and the last four (c,d,e,f) give its values at $0, 1, \omega, \bar{\omega}$:

$$c = \phi(0),\ d = \phi(1),\ e = \phi(\omega),\ f = \phi(\bar{\omega}).$$

Definition 3. Each word $ab\ cd\ ef$ of $\mathscr{C}_6$ has a *slope, s*. Then $ab\ cd\ ef$ is a word of $\mathscr{C}_6$ with slope s if and only if it satisfies the following relations:

The 1*-rule*: $a + b = c + d = e + f = s,$

The ω*-rule*: $a + c + e = a + d + f = b + c + f = b + d + e = \omega s,$

The $\bar{\omega}$*-rule*: $b + d + f = b + c + e = a + d + e = a + c + f = \bar{\omega} s.$

(Thus the digits of each couple add to s, while three digits from distinct couples add to ωs or $\bar{\omega} s$ according as it is the left or right possibility that is chosen an odd number of times.)

The code under any of these definitions is 3-dimensional; in the first case because the four given words add to zero, but obey no further linear relation; in the second case obviously; and in the third case since the relations

$$a + b = c + d = e + f = s\ ,\quad a + c + e = \omega s$$

between the seven quantities a,b,c,d,e,f,s imply all the rest. (The first three of these imply that the replacement of any digit by its mate will alter the sum by s.)

Given the 3-dimensionality, the identity of the three codes follows readily by checking that some 3 independent words can be found in all three codes. We suggest that the reader checks this for the generators of the second version:

$$W(x^2) = 10\ 01\ \bar{\omega}\omega\ ,\quad W(x) = 01\ 01\ \omega\bar{\omega}\ ,\quad W(1) = 00\ 11\ 11\ .$$

From either the first or the third definition, it is obvious that $\mathscr{C}_6$ has the following symmetries:

1) scalar multiplication by any power of ω,
2) flipping of the two digits in any two of the couples,
3) bodily permutation of the three couples in any way.

A little calculation now reveals that under such symmetries every hexacodeword is an image of one of the following five:

$01\ 01\ \omega\bar{\omega}$	$\omega\bar{\omega}\ \omega\bar{\omega}\ \omega\bar{\omega}$	$00\ 11\ 11$	$11\ \omega\omega\ \bar{\omega}\bar{\omega}$	$00\ 00\ 00$
(36 images)	(12 images)	(9 images)	(6 images)	(1 image)

Since its images amount to more than half the code, the reader is advised to memorize the first of these — the others are easily reconstructed from this one and the known symmetries.

3. Justification of a hexacodeword

It is important to be able to recognize at sight whether a word is or is not a hexacodeword. Here's how to do this.

First check that it obeys the *shape rule:* up to permutations and flips of its three couples the word should have one of the forms

$$0a\ 0a\ bc, \quad bc\ bc\ bc, \quad 00\ aa\ aa, \quad aa\ bb\ cc, \quad 00\ 00\ 00$$

(where a,b,c are $1,\omega,\bar{\omega}$ in some order), and then check that it obeys the *sign rule*: if we say that a couple of the form

$$\begin{array}{llll} 0\,x & \text{or} & y\ \omega y & \text{has sign } + \\ x\,0 & \text{or} & y\ \bar{\omega} y & \text{has sign } - \\ 0\,0 & \text{or} & y\ y & \text{has sign } 0 \end{array}$$

with $x,y \neq 0$, then the three couples have signs $+++$, $+--$, $-+-$, $--+$ or 000 (i.e. are either all 0 or have product $+$).

Thus $0\bar{\omega}\ 1\omega\ \bar{\omega}0$ obeys the shape rule, with $a = \bar{\omega}$, $b = 1$, $c = \omega$, but the signs of the three couples are $++-$, so that it disobeys the sign rule, and is not a hexacodeword. It would become a hexacodeword on flipping any couple, for example $0\bar{\omega}\ \omega 1\ \bar{\omega}0$, with signs $+--$.

We refer to the process of checking the shape and sign rules as the *justification* of a hexacodeword. Which of the words at the foot of this page are hexacodewords ? (Answers are on the page behind.)

4. Completing a hexacodeword

In the application, one is often faced with the problem of determining a hexacodeword from some partial information about its digits. In many cases this reduces to one of two standard problems:

The 3-problem: determine a hexacodeword from any three of its digits.

The 5-problem: determine a hexacodeword from any five of its digits, *one of which may be mistaken.*

In either case, the answer is unique, and so we earnestly recommend that you use

The Best Method: guess the correct answer, and then justify it.

This is both easier and quicker than the methods that follow. But if you're stuck, you can always solve such problems like this:

The 3-problem: If the given digits are from distinct couples (such as a,c,e), use the ω- or $\bar{\omega}$-rule to find s, and then the 1-rule for the remaining digits (here b,d,f).

$00\ 11\ \omega\omega \quad 10\ 10\ 10 \quad \omega 0\ \omega 0\ \bar{\omega}1 \quad \omega\omega\ \omega\omega\ 00 \quad 1\bar{\omega}\ 1\bar{\omega}\ 1\bar{\omega} \quad 11\ \bar{\omega}\bar{\omega}\ \omega\omega$ **?**

Example. ω? 1? 0? $\Rightarrow$ $\omega s = \omega+1+0 = \bar{\omega}$, so $s = \omega$, $W = \omega 0\ 1\bar{\omega}\ 0\omega$, with signs $--+$. (Although the answer must be right if we've calculated correctly, the reader is recommended always to check the answers by the shape and sign rules.)

If instead the three given digits include a couple (as do a,b,c), use the 1-rule to find s and a fourth digit (d), and the shape and sign rules (or the ω- and $\bar{\omega}$-rules) for the other two (e,f).

Example. $\omega 1$ 0? ?? $\Rightarrow$ $s = \omega+1 = \bar{\omega}$, so $d = \bar{\omega}$, $W = \omega 1\ 0\bar{\omega}\ \bar{\omega}0$, with signs $-+-$. (Here the shape rule gives $\{e,f\} = \{0,\bar{\omega}\}$, and the sign rule then gives their order, $\bar{\omega}0$.)

The 5-problem. Any 5-problem reduces to several 3-problems, as follows. The incorrect digit, if any, can affect at most one couple. But on deleting the information for any couple, we still have at least 3 given digits, enough to determine a hexacodeword. The required word must be one of these three. Notice that in one case we actually have *four* digits – the calculation can be speeded up slightly by noticing that these will be part of a hexacodeword if and only if they are *consistent*, i.e. add to 0 in $\mathbf{F}_4$.

Example. $\omega 1$ 0ω 1? The three 3-problems are:

$\omega 1$ 0ω ?? , inconsistent, since $\omega+1+0+\omega \neq 0$,
$\omega 1$?? 1? , so $s = \bar{\omega}$, $W = \omega 1\ \omega 1\ 1\omega$ (signs $--+$, but not an answer) .
?? 0ω 1?, so $s = \omega$, $W = \omega 0\ 0\omega\ 1\bar{\omega}$ (signs $-+-$, which is) .

Another example. $\omega 1$ $0\bar{\omega}$ 1? This time the two couples are consistent, and so we need only try $\omega 1$ $0\bar{\omega}$?? , which gives $\omega 1\ 0\bar{\omega}\ \bar{\omega}0$ (with signs $-+-$).

The reader, who is naturally keen to acquire the ability to make lightning calculations involving the Golay code, should now practice by computing the hexacodewords defined by the 3- and 5-problems at the foot of the page. Answers will again be found on the page behind. We strongly advise that as many of these problems as possible be solved by the Best Method, rather than by the slower methods outlined above.

5. The Golay code $\mathscr{C}_{24}$ and the MOG

The Golay code $\mathscr{C}_{24}$ has already been mentioned several times (see in particular §2.8.2 of Chap. 3 and §2 of Chap. 10). In this chapter we shall often use blank and non-blank symbols instead of the digits 0 and 1. Robert Curtis [Cur2] found that the easiest way to compute with words of $\mathscr{C}_{24}$ is to arrange their digits in a 6×4 array, called by him the Miracle Octad Generator, or MOG.

Words of $\mathscr{C}_{24}$ in this notation are readily obtained from hexacodewords by

---

?ω 0? ω?	11 ?0 ??	$\omega 1$?ω $\bar{\omega}1$	$\bar{\omega}0$?ω 11	?
?ω ?1 ω?	$\omega 0$ $\omega 0$ 0?	?1 ?ω ?ω	0? $\omega 1$??	

either i) replacing each digit by an *odd* interpretation, in any way such that the top row becomes *odd;*

or ii) replacing each digit by an *even* interpretation, in any way such that the top row becomes *even.*

A row or column is called *odd* or *even* according as it contains an odd or even number of non-zero digits. The odd and even interpretations of the digits are shown in Fig. 11.1. Note that the odd interpretations have either just one non-zero digit, or just one zero digit, in each column, while the even interpretations differ from the odd ones only in their treatment of the top row.

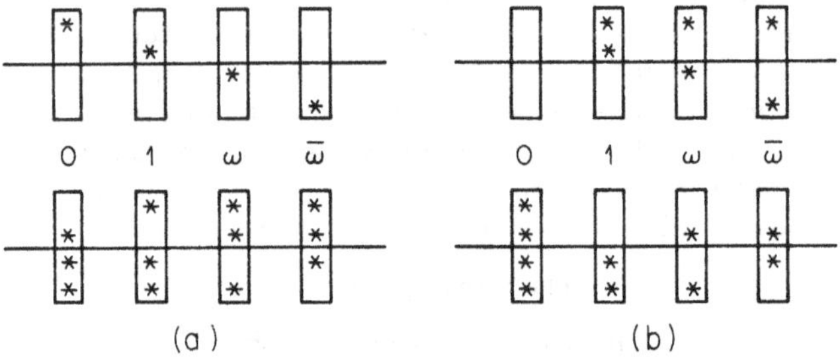

Figure 11.1. Interpreting hexacode digits.
(a) Odd interpretations. (b) Even interpretations.

Justifying $\mathscr{C}$-sets. The following procedure can therefore be used to verify that a given set is a $\mathscr{C}$*-set* (i.e. the set of places where a codeword of $\mathscr{C}_{24}$ is non-zero).
1) For each column, and for the top row, compute its *count* (the number of non-zero digits it contains).
2) For each column, compute also its *score*, obtained by adding $0, 1, \omega, \bar{\omega}$ for any non-zero digit in the first, second, third, fourth row, respectively. (Alternatively, a column has a given score digit if and only if it is one of the four interpretations of that digit).
Then the set is a $\mathscr{C}$-set if and only if the *counts* have all the same parity, while the *scores* form a hexacodeword. We shall call this process *justifying* the $\mathscr{C}$-set.

In Fig. 11.2 we have given several interpretations of the standard hexacodeword 01 01 $\omega\bar{\omega}$. We recommend to the beginner our practice of writing counts above, and scores below, the MOG diagram. Of course in Fig. 11.2 the score words are all 01 01 $\omega\bar{\omega}$.

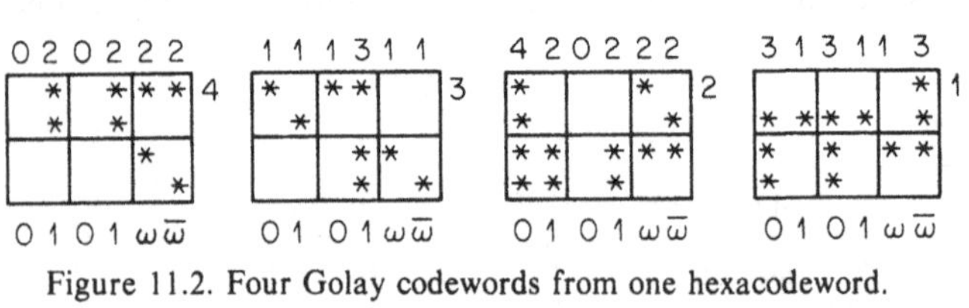

Figure 11.2. Four Golay codewords from one hexacodeword.

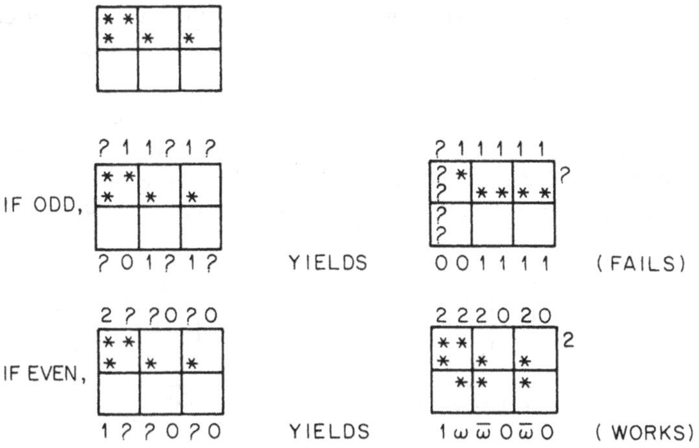

Figure 11.3. An example.

Which of the figures at the foot of the page correspond to Golay codewords? You should know where to find the answers.

6. Completing octads from 5 of their points

An *octad* is (the $\mathscr{C}$-set corresponding to) a Golay codeword of weight 8. The octads of $\mathscr{C}_{24}$ constitute a Steiner system $S(5,8,24)$ (§3.1 of Chap. 3, §2.1 of Chap. 10); which is to say that any 5 points of the 24 belong to just one octad. It is of considerable value to be able to locate the octad containing 5 given points. This is easily reduced to 3-problems and 5-problems when we note that to change the score of a given column we must change at least one point in that column, and at least two points if we are to preserve the parity of that column.

If the given points are located so as to make three columns odd and three even, then we must correct the three columns having whichever is the wrong parity, and so can afford to make no change to the others. We therefore suppose first that the right parity is *odd*, second that it is *even*, and then in either case we know three correct digits of the corresponding

---

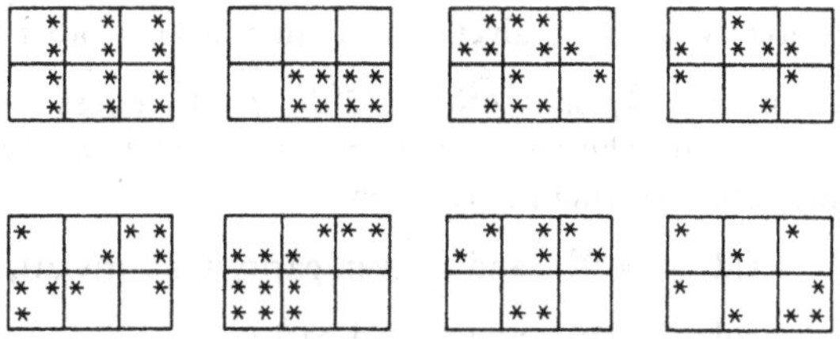

?

$\omega\omega$ 00 $\omega\omega$ 11 00 11 $\omega 1$ 1ω $\omega 1$ $\bar\omega\bar\omega$ $\omega\omega$ 11

1ω $\omega 1$ $\omega 1$ $\omega 0$ $\omega 0$ $\bar\omega 1$ $\bar\omega 1$ 0ω 0ω $0\bar\omega$ $\omega 1$ $\bar\omega 0$

!

hexacodeword, from which we can complete that word. Only one of the two choices will yield an octad compatible with the given five points. Figure 11.3 shows an example.

If instead the given five points are distributed so as to make five of the columns have the same parity, then that parity must be the correct one, and we can find the desired hexacodeword (whence the octad) by solving the corresponding 5-problem (see the example in Fig. 11.4).

Figure 11.4. Another example.

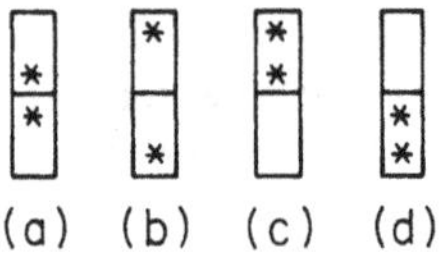

Figure 11.5. The first column is (a) rather than (b); the fifth is (c) rather than (d).

It is perhaps worthwhile to spell out in more detail the way in which the octad is reconstructed from the hexacodeword in this example. Which of the two even interpretations should we choose for the two digits that have to be altered? For the first column there is no problem. We already have a star in that column, and so are forced to choose Fig. 11.5a rather than Fig. 11.5b. But now we have determined all of the top row except its fifth entry, and so must choose Fig. 11.5c rather than Fig. 11.5d for the fifth column, so as to make the top row even.

Experience shows that when we have three odd and three even columns, and are in doubt about the correct parity, it is usually best to consider the odd case first (partly because odd interpretations are easier to see, and partly because for octads the parity condition on the top row is automatically satisfied in the odd case). However, the correct parity can often be determined directly from the given information in some other way.

Thus octads must necessarily split $4^2 0^4$ or $2^4 0^2$ or $3 1^5$ across the six columns, and so from the distribution of the five given points across the columns we can deduce the following information:

$4 1 0^4$ or $3 2 0^4$	yield	$4^2 0^4$	, and so even parity (and trivial),
$3 1^2 0^3$ or $1^5 0$	yield	$3 1^5$	, and so odd parity,
$2^2 1 0^3$	yields	$2^4 0^2$	, and so even parity,
but $2 1^3 0^2$	may yield	$3 1^5$	(odd), or $2^4 0^2$ (even).

This time the foot of the page contains a number of sets of five points for you to complete to octads.

Locating other $\mathcal{C}$-sets. Any set consisting of an odd number of the 24 points can be converted to a $\mathcal{C}$-set by changing the status (member/non-member) of 1 or 3 points. Our discussion of the octad-completion problem generalizes almost trivially to this one — once again we must solve a 5-problem or one of two 3-problems according as the given points are located so as to make the columns be five of one parity and one of the other, or three of each.

Sets consisting of an even number of points are *either* convertible to $\mathcal{C}$-sets in a unique way by changing the status of 0 or 2 points, *or* in six distinct ways by changing the status of 4 points. In the first case, the columns all have the same parity, or will be four of one parity and two of the other, and we solve a 5-problem (but with 6 given digits!) or two 3-problems (with 4), as before. In the second case, we can change the status of an arbitrary point, and will then be able to change three more to achieve a $\mathcal{C}$-set.

7. The maximal subgroups of M_{24}

As we have already discussed in §2.5 of Chap. 10, there are just nine conjugacy classes of maximal subgroups of M_{24}. With the exception of $L_2(23)$, they can be characterized by the non-trivial partition of the 24 elements they fix, as shown in Table 11.1.

Several of these have been used as starting points for constructions of M_{24}, notably $L_2(23)$ by Carmichael, $P\Gamma L_3(4)$ by Witt and Tits, $2^6{:}(S_3 \times L_2(7))$ by Turyn, $2^4{:}A_8$ by Curtis, and of course $2^6{:}3^{\cdot}S_6$ for our own construction.

In the rest of the chapter we shall devote a section to each maximal subgroup in turn, relating it to the MOG as far as is possible. The sections are in a rather haphazard order — first the projective group, since it

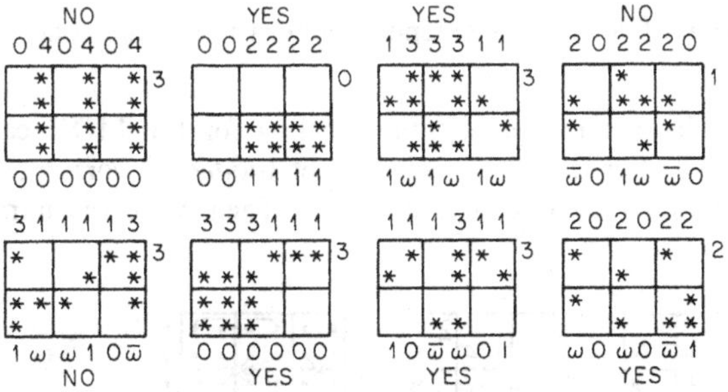

!

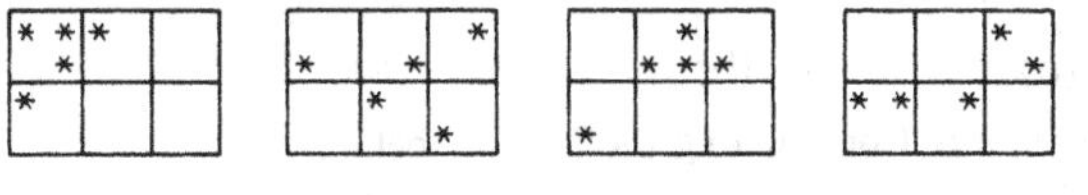

?

Table 11.1. Maximal subgroups of M_{24}.

Group structure	Partition	Name
M_{23}	1, 23	monad group
$M_{22}:2$	2, 22	duad group
$P\Gamma L_3(4)$	3, 21	triad group
$2^6:3^{\cdot}S_6$	4^6	sextet group
$2^4:A_8$	8, 16	octad group
$M_{12}:2$	12^2	dodecad group
$2^6:S_3\times L_2(7)$	8^3	trio group
$L_2(23)$	24	projective group
$L_2(7)$	3^8	octern group

provides the most usual labeling for the 24 points, then the four groups most closely related to the MOG (the sextet, octad, triad and trio groups), a few short sections on the monad, duad and octern groups, and then a specially long section on $M_{12}:2$, which deserves an even longer one!

8. The projective subgroup $L_2(23)$

One of the maximal subgroups of M_{24} is the group $PSL_2(23)$, or $L_2(23)$ for short, which acts on the 24 points, when labeled $\infty, 0, 1, 2, \ldots, 21, 22$, by linear fractional transformations $z \longrightarrow (az+b)/(cz+d)$, with $ad-bc$ a non-zero square modulo 23. The Golay code $\mathscr{C} = \mathscr{C}_{24}$ can then be defined as the code spanned by the images of the set Q consisting of 0 and the quadratic residues modulo 23:

$$Q = \{0,1,2,3,4,6,8,9,12,13,16,18\} .$$

(This is the definition of $\mathscr{C}_{24}$ as an extended quadratic residue code — see §2.8 of Chap. 3.)

It is a remarkable fact that a suitable labeling for the MOG can be constructed as follows. For the 12-element $\mathscr{C}$-set (*dodecad*) corresponding to Q we shall take that indicated in Fig. 11.6a. Then write the numbers

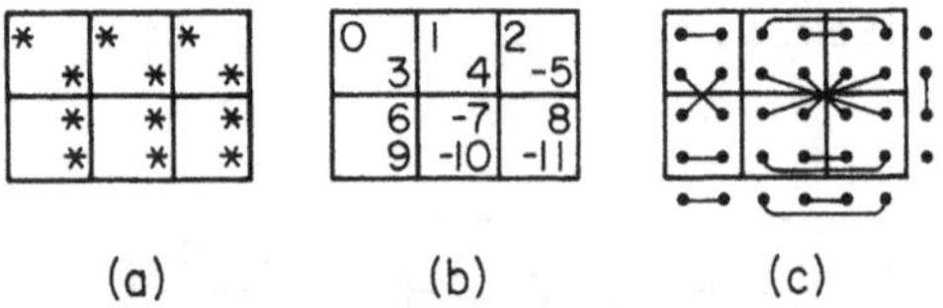

Figure 11.6. Constructing the MOG labeling.

0	∞	1	11	2	22
19	3	20	4	10	18
15	6	14	16	17	8
5	9	21	13	7	12

Figure 11.7. The standard MOG labeling.

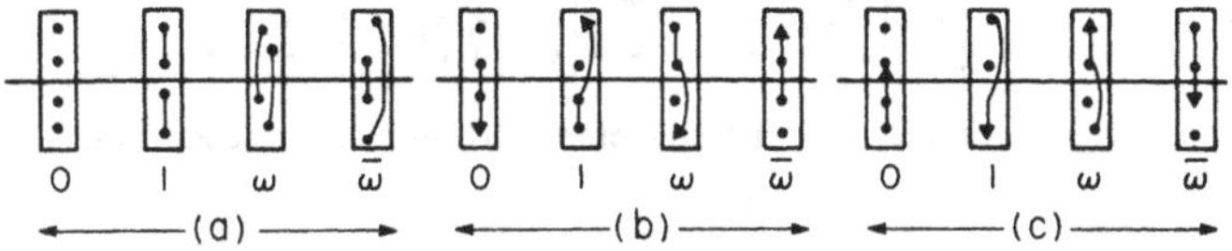
0
1
ω
ω̄
0
1
ω
ω̄
0
1
ω
ω̄
(a)
(b)
(c)

Figure 11.8. Interpretations giving elements of $2^6{:}3$.

0, 1, 2, 3, 4, 5, 6, 7, 8, 9, 10, 11 into these places in the natural order, and affix signs where necessary so as to produce elements of Q (Fig. 11.6b). The remaining numbers (∞ and the quadratic non-residues modulo 23) are found using the action of the permutation $\gamma : z \longrightarrow -1/z$ shown in Fig. 11.6c. The resulting labeling, displayed in Fig. 11.7, is called the *standard (modulo* 23) *MOG labeling.* Any element of $L_2(23)$ now becomes a permutation of M_{24}.

9. The sextet group $2^6{:}3^{\cdot}S_6$

A *sextet* (mnemonic: *six tet*rads) is a collection of six 4-element sets of which the union of any two is an octad. It is known that M_{24} is transitive on sextets, and that any tetrad belongs to a uniquely determined sextet. The *standard sextet* is the one whose six tetrads are the columns of the MOG array.

The stabilizer of a sextet is a group $2^6{:}3^{\cdot}S_6$, in which the subgroup $2^6{:}3$ consists of the permutations that preserve every tetrad, while the quotient group S_6 describes the action on the tetrads. We shall describe this group for the standard sextet in some detail.

The elementary abelian subgroup 2^6 is isomorphic to the additive group of the hexacode, and its elements can be found from hexacodewords by interpreting their digits as in Fig. 11.8a. This group has two further cosets in the subgroup $2^6{:}3$, and elements of these cosets can be found by interpreting digits in the alternative ways shown in Figs. 11.8b and 11.8c. So a knowledge of the hexacodewords gives us the ability to recognize all

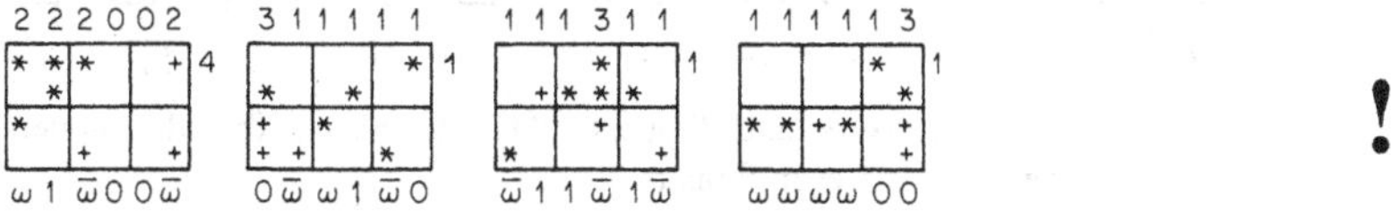
2 2 2 0 0 2
4
ω 1 ω̄ 0 0 ω̄
3 1 1 1 1 1
1
0 ω̄ ω 1 ω̄ 0
1 1 1 3 1 1
1
ω̄ 1 1 ω̄ 1 ω̄
1 1 1 1 1 3
1
ω ω ω ω 0 0
!

192 elements of the group $2^6{:}3$ at sight! Figure 11.9 shows the three elements obtained like this from the word 01 01 $\omega\bar{\omega}$.

But this is not all! By suitably interpreting the digits of suitable hexacodewords, we can also obtain all the involutions in the subgroup $3^{\cdot}S_6$ that preserves the top row of the MOG array. (Any permutation of the top row extends to just three elements of this group, but this is a non-split extension, and contains no subgroup S_6.)

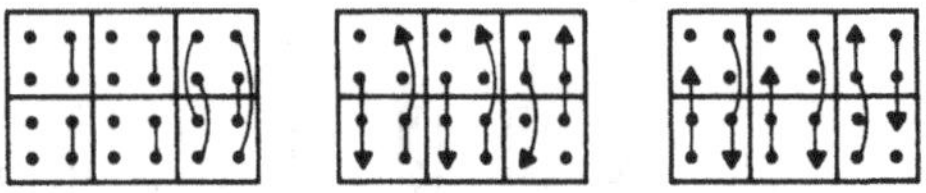

Figure 11.9. Three elements from one hexacodeword.

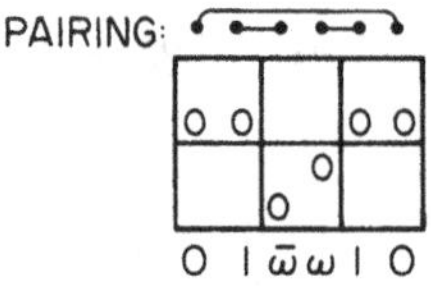

Figure 11.10. A word, a pairing, and blobs.

The blob trick. Here's how the trick is done. We take any hexacodeword *W* with two 0 digits, and any pairing of the digits in this word for which the 0 digits form a pair. For each non-zero digit of *W* we place a *blob* in the corresponding row and column of the MOG, but in the columns corresponding to 0 digits we place blobs instead in the row corresponding to the *non-zero* digit that appears at least twice in *W*. Figure 11.10 shows our pairing and the blobs for the word 01 $\bar{\omega}\omega$ 10.

Now, using the terms defined in Fig. 11.11, and according to the cycle-shape $1^4 2$ or $1^2 2^2$ or 2^3 of our permutation, we

($1^4 2$) : replace the 0-pair by an odd flipper, and other digits by odd fixers;

($1^2 2^2$) : replace the 0 digits by even fixers, and other pairs by even flippers;

(2^3) : replace the 0-pair *and* the other pairs by odd flippers.

These fixers or flippers must always take blobs to blobs. Figure 11.12 shows the results for our example.

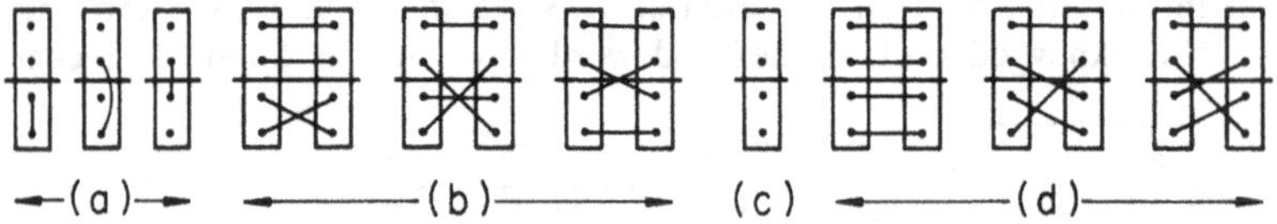

Figure 11.11. The fixers and flippers: (a) odd fixers, (b) odd flippers, (c) even fixer, and (d) even flippers.

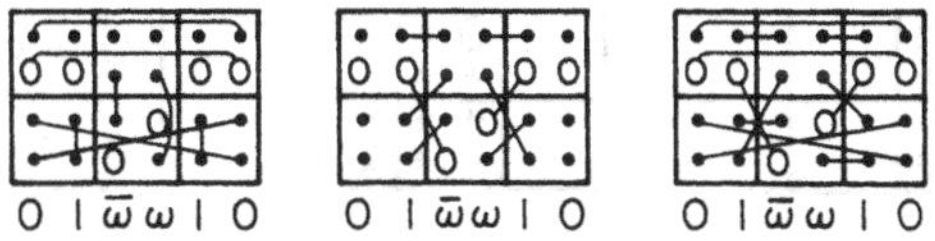

Figure 11.12. Three more permutations from a hexacodeword.

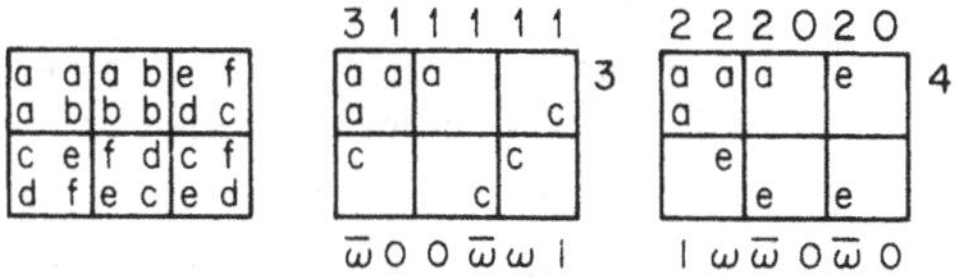

Figure 11.13. A sextet and two octads.

Figure 11.14. A labeling, and some more interpretations.

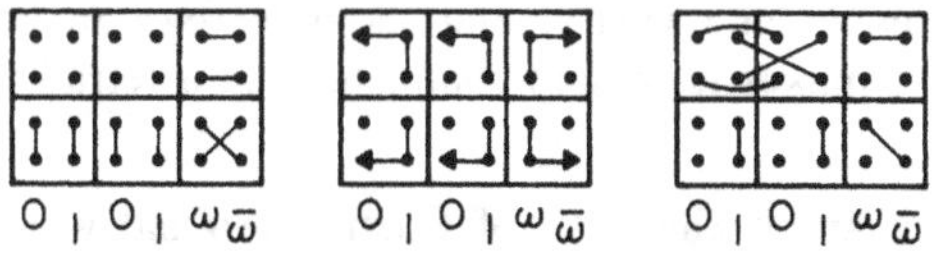

Figure 11.15. Three new interpretations of 01 01 $\omega\bar\omega$.

Other sextets. As we remarked, any tetrad belongs to a unique sextet, which can be found by completing octads from various sets of five points including that tetrad. An example is shown in Fig. 11.13, with some of the octads that justify it. (This sextet was in fact completed mentally at the typewriter from its tetrad *aaaa*. It is instructive to check that the octads formed by any two of its tetrads really are all octads.)

Some other sextets are particularly elegant, and the results for our standard sextet can be transferred to these with little trouble. For instance the six 2×2 blocks into which we regularly subdivide our MOG, and the elements of *its* group $2^6{:}3^{\cdot}S_6$ can be obtained from the labeling and interpretations shown in Fig. 11.14. Figure 11.15 shows some of these elements.

10. The octad group $2^4{:}A_8$

It is known that M_{24} is transitive on octads, and that the stabilizer of an octad is a group $2^4{:}A_8$. It is significant that the alternating group A_8 is isomorphic to the linear group $GL_4(2)$. In fact the group is best understood by saying that the complement of the fixed octad acquires the structure of an affine 4-space over the field $\mathbf{F}_2$, or a vector space if we fix also a particular point of it called the *origin*.

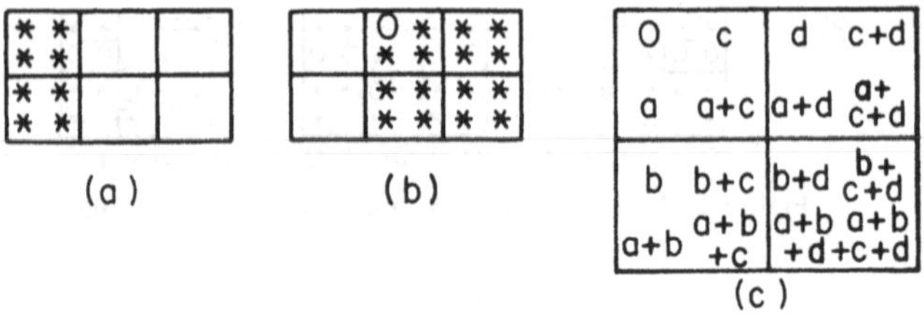

Figure 11.16. (a) The standard octad and (b) its complementary square, with origin marked 0. (c) The entries in the square as vectors.

Figure 11.16 shows (a) our *standard octad* and (b) its complement, the *standard square,* and the *standard origin* in that square. Figure 11.16c illustrates the vector space structure in the standard square.

The octad and its complementary square are linked by the fact that there is a 1-1 correspondence between the partitions of the octad into two tetrads, and of the square into four parallel 2-spaces, given by the requirement that the resulting family of six tetrads should be a sextet. In Fig. 11.17 we show all these 35 sextets. It is useful to note that the tetrads (that is, 2-spaces) appearing in the square are precisely those that meet its four columns in one parity, and its four rows in another (possibly the same). This makes it easy to complete such a tetrad from any three of its points.

(This remark is the basis of Curtis's original way of finding octads in the MOG [Cur2], which uses the fact that every octad meeting the standard one in four points is visible in one of the 35 sextets of Fig. 11.17.

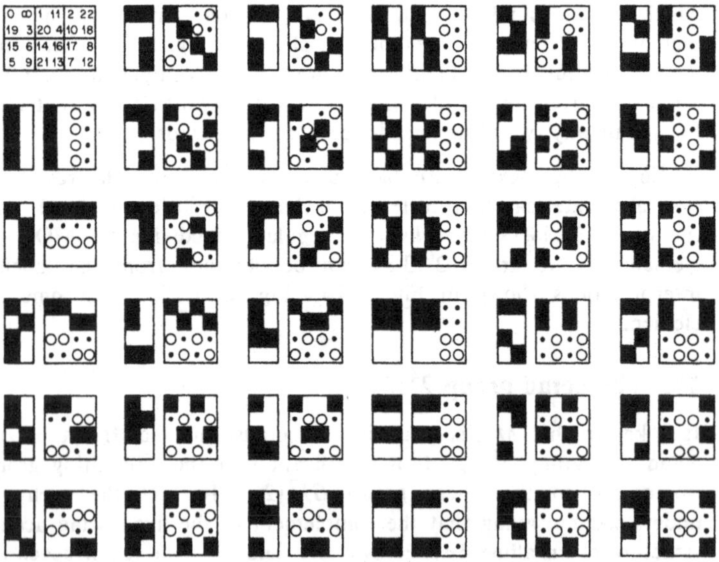

Figure 11.17. The 35 standard sextets of the MOG.

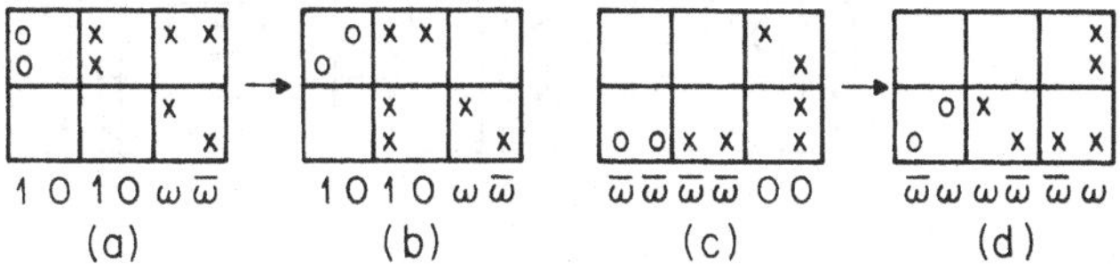

Figure 11.18. Transposing two octads.

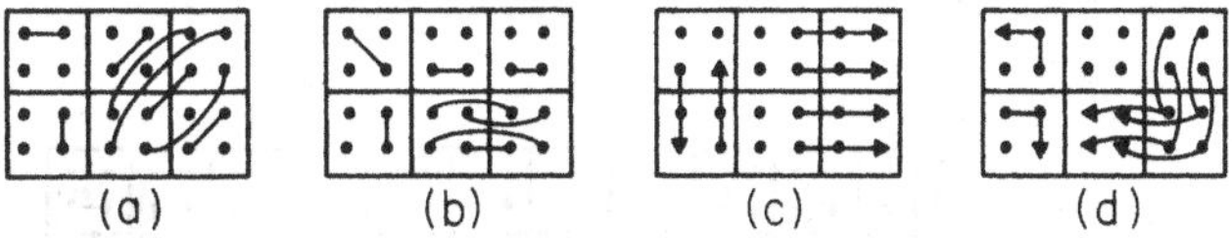

Figure 11.19. Some permutations of 2^4:A_8.

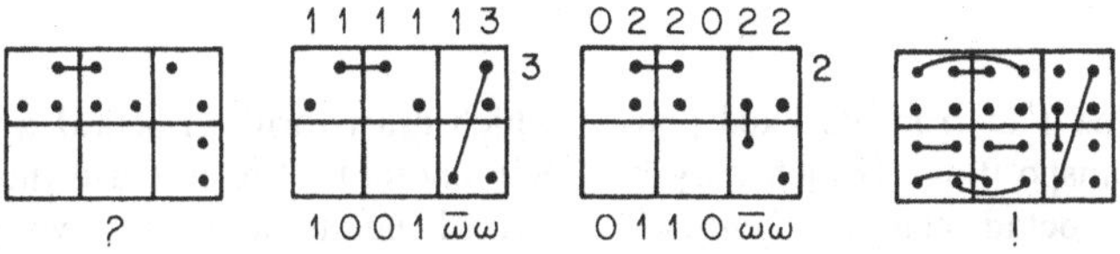

Figure 11.20. A problem, some calculations, and its solution.

Indeed, every octad except the three of the standard trio meets at least one octad of that trio in four points, and so the figure can actually be used to find all octads.)

The group $2^4 : A_8$. The normal subgroup 2^4 consists of all elements that fix the standard octad pointwise. In affine language, these are just the *translations* of the complementary square.

The general element of 2^4:A_8 acts as an element of the alternating group A_8 on the standard octad, and as an affine symmetry of the 4-space on the complementary square. Every affine transformation of the square extends to a unique element of M_{24} by adjoining the appropriate even permutation of the octad. Conversely, every even permutation of the octad extends to 16 elements of M_{24} — we can force uniqueness by specifying a destination for any one point outside the octad. By requiring that the origin be fixed, we restrict to a subgroup A_8 complementing the group 2^4.

We illustrate some of these ideas by an easy calculation. The operation of *transposing* the square is obviously an affine symmetry — let us find the element of A_8 that extends it to a permutation in M_{24}. We note that the two octads of Figs. 11.18a and c each have six points (×) in the square, and only two (o) in the octad. We obtain the octads of Figs. 11.18b and d by transposing the parts lying in the square, and then completing to octads. Now the desired permutation is easily completed (Fig. 11.19a), using the fact that it is an involution, and must be an even permutation. Some other elements of M_{24} obtained in this way are also shown in Fig. 11.19.

The involutions of M_{24} lie in two conjugacy classes, having cycle shapes $1^8 2^8$ and 2^{12}. There is an easy way to compute an arbitrary permutation of

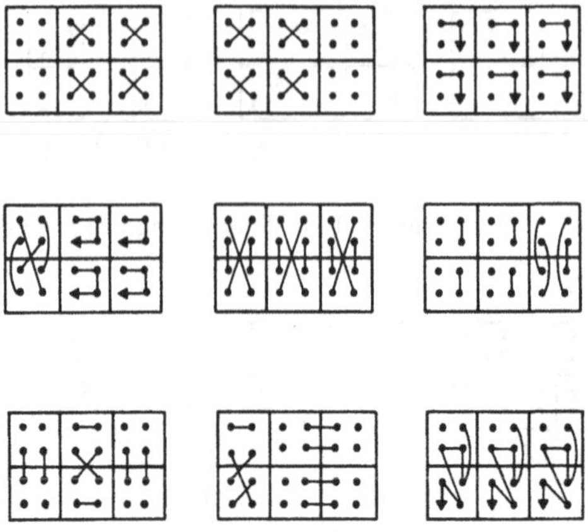

Figure 11.21. Elements in $2^4:A_8$ and other groups.

type $1^8 2^8$ from its fixed points (which must form an octad) and one of its transpositions (which may be freely chosen). We just use the fact that if an octad contains four of the fixed points and the two points of a transposition, then its remaining two points must also be a transposition of the permutation. Figure 11.20 illustrates the calculation.

The process of extending permutations by completing octads is one that soon palls, however, and it is handy to know that it can often be avoided. In particular, we can find many permutations in our group $2^4:A_8$ by using other groups for which the calculation is trivial. If the desired element preserves the sextet made from the six columns, or that made from the six 2×2 subsquares, we can use the sextet group. If it fixes the standard trio, we can use the group $2^6:(S_3 \times L_2(7))$. Finally, if it fixes the standard triad, or equivalently, the standard way of regarding the square as a 2-space over $\mathbf{F}_4$ (rather than merely a 4-space over $\mathbf{F}_2$) we can use the triad group $P\Gamma L_3(4)$, explained in §11. A miscellany of elements obtainable in these ways is displayed in Fig. 11.21.

11. The triad group and the projective plane of order 4

When 3 points of the 24 are distinguished, the remaining 21 points acquire the structure of a projective plane over $\mathbf{F}_4$. The points of this plane can be coordinatized with triples (x,y,z) of elements of $\mathbf{F}_4$ not all zero, with the usual understanding that $(x,y,z) = (kx,ky,kz)$. The five points with $z = 0$ are specified by the ratio y/x and form the *line at infinity,* while the other 16 points can be given normalized coordinates $(x,y,1)$, which we abbreviate to (x,y), and they form an *affine plane* over $\mathbf{F}_4$.

In the MOG, we take the three distinguished points to be the lower three points in the first column, the line at infinity to be the remainder of the standard octad, and the affine plane to be the standard square, the coordinatization being as in Fig. 11.22.

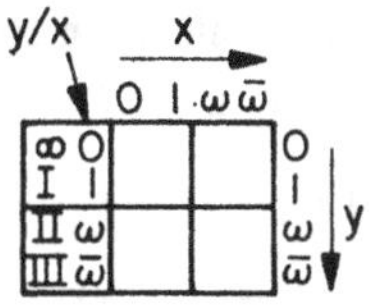

Figure 11.22. The MOG as a projective plane.

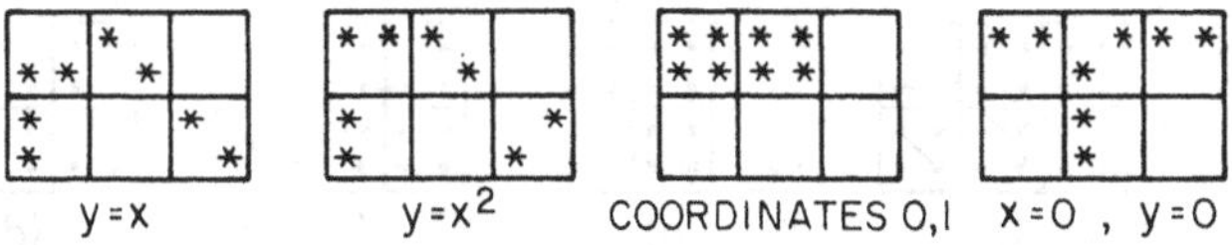

Figure 11.23. Four octads seen from a projective viewpoint.

We follow Witt [Wit2], [Wit3] in calling the three special points I, II, III, and we shall refer to them as the *Romans*. The octads of the Steiner system have the following four forms:

(3+5) All three Romans together with a line of the plane.

(2+6) Two Romans with a six-point oval (hyperconic) in the plane.

(1+7) One Roman, together with a subplane defined over $\mathbf{F}_2$.

(0+8) No Romans, and the sum of two lines of the plane.

(The sum is modulo 2, so that the intersection of the two lines does not appear.) A *hyperconic* consists of the 5 points of a conic together with the unique point on no chord of that conic. Figure 11.23 shows examples of all four types of octad. We shall call this the Witt-Tits construction, because Witt constructed M_{24} from $PSL_3(4)$, while Tits [Tit2] explored the implied geometry in some detail.

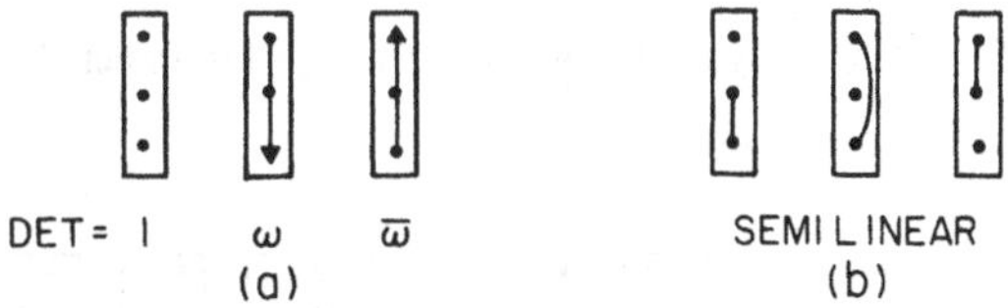

Figure 11.24. Appropriate ways to permute the Romans.

Any automorphism of the plane now extends uniquely to a permutation of M_{24} by adjoining the appropriate permutation of the Romans, as follows. If it results from a linear (over $\mathbf{F}_4$) transformation of the coordinates, the appropriate permutation depends on the determinant in the way shown in Fig. 11.24a. If instead it results from a semilinear map, we get one of the permutations shown in Fig. 11.24b, and the first of those permutations if this map is in $P\Sigma L_3(4)$. In summary

$PSL_3(4)$ fixes all three Romans (and is M_{21}),

$PGL_3(4)$ permutes them cyclically,

$P\Sigma L_3(4)$ fixes I, and permutes {II, III}.

You can now write down a permutation of M_{24} for any element of $P\Gamma L_3(4)$. Figure 11.25 shows some examples.

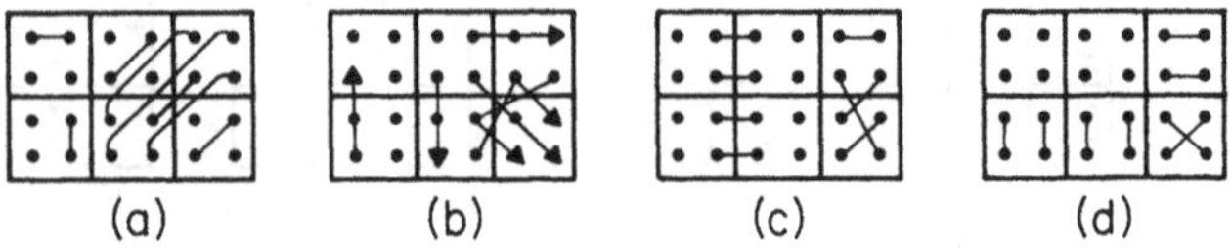

Figure 11.25. Four elements of $P\Gamma L_3(4)$:
(a) $x' = y, y' = x, z' = z$; (b) $x' = \omega x, y' = \omega y, z' = z$;
(c) $x' = z, y' = y, z' = x$; (d) $x' = \bar{x}, y' = \bar{y}, z' = \bar{z}$.

12. The trio group $2^6:(S_3 \times L_2(7))$

M_{24} is transitive on **trios** (mnemonic: **three octads**) consisting of three disjoint octads. The subgroup fixing a trio is a group $2^6:(S_3 \times L_2(7))$, it being significant that $L_2(7)$ is isomorphic to $L_3(2)$. This subgroup is best understood by saying that the three octads acquire the structure of three parallel affine 3-spaces over $\mathbf{F}_2$.

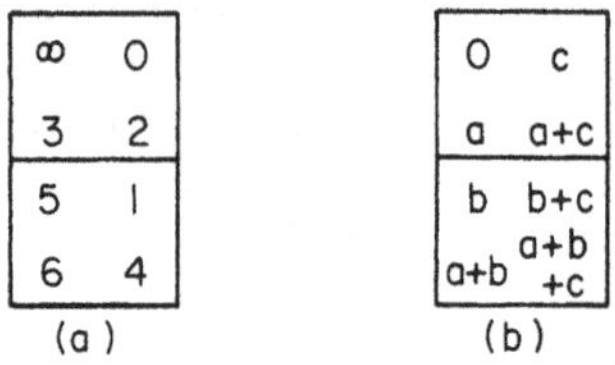

Figure 11.26. Two labelings for an octad.

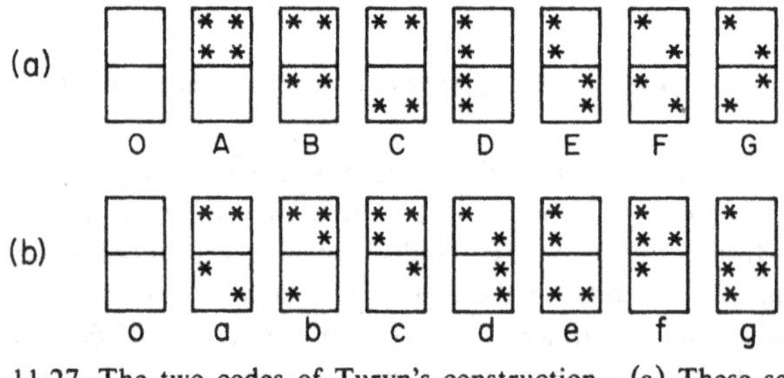

Figure 11.27. The two codes of Turyn's construction. (a) These sets and their complements form the line-code. (b) These sets and their complements form the point-code.

We take the standard trio to be that formed by the left, middle, and right couples of columns of the MOG, and label each of these octads as shown in Fig. 11.26a. The affine structure on any of these octads is that corresponding to Fig. 11.26b.

According to *Turyn's construction* [Mac6, Chap. 18, Th. 12], the Golay code $\mathscr{C}_{24}$ consists of the words

$$(X+t\ ,\quad Y+t\ ,\quad Z+t\ )$$

and their complements, where $X+Y+Z=0$, and X,Y,Z belong to the *line-code* of Fig. 11.27a, while t belongs to the *point-code* of Fig. 11.27b. The tetrads of the line-code are the 2-spaces of the affine 3-space and yield octads when placed in any two of the three standard octads. The tetrads of the point code are precisely those tetrads that yield dodecads of $\mathscr{C}_{24}$ when repeated in all three octads of the standard trio. (One of these dodecads is the one labeled with elements of Q in the standard labeling of the MOG – see Fig. 11.7.)

The group $2^6:(S_3\times L_2(7))$ contains the following subgroups:
1) The elementary abelian normal subgroup 2^6, whose elements are obtained by combining any three translations adding to 0.
2) The subgroup S_3 consisting of the bodily permutations of the three octads of the standard trio.
3) The group $L_2(7)$, whose elements are precisely those permutations of M_{24} that act in the same way on all three octads.
Figure 11.28 contains some examples. The first picture is made of three translations adding to 0 (namely, those by vectors b,c, and $b+c$), the second is the cyclic permutation of the three standard octads, and the remaining three pictures are the elements

$$\alpha:z \longrightarrow z+1\ ,\quad \beta:z \longrightarrow 4z\ ,\quad \gamma:z \longrightarrow -1/z$$

of $L_2(7)$ under the labeling of Fig. 11.26a.

We remark that the illustrated tetrads of either of the line- and point-codes are images of each other under the element α of Fig. 11.28. If we call two tetrads *incident* when they meet in two points, Fig. 11.29a shows how the two codes get their names.

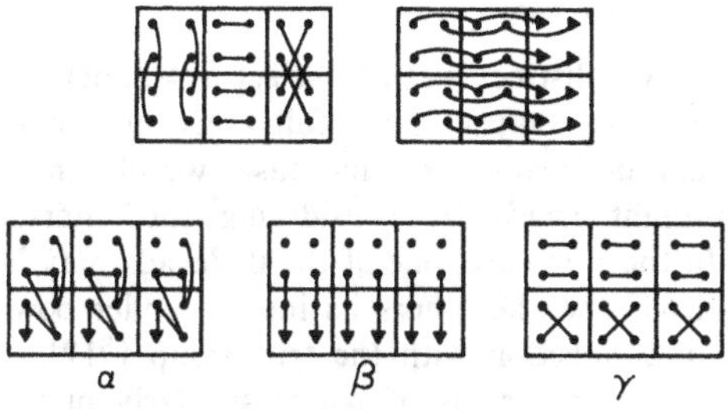

Figure 11.28. Five elements of the trio group.

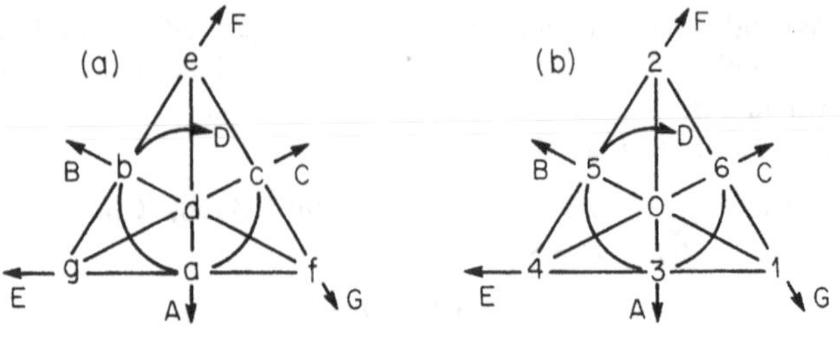

Figure 11.29. Incidences between points and lines.

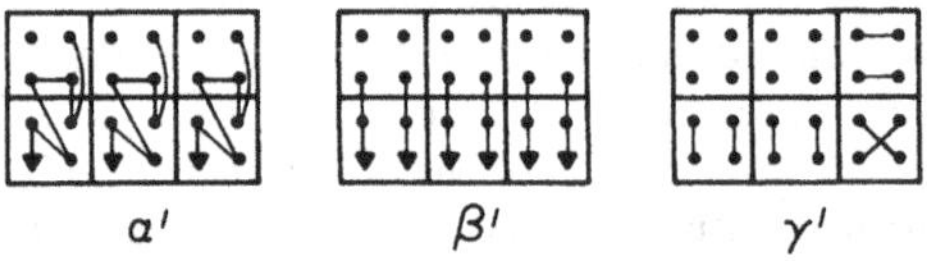

Figure 11.30. Generators for another $L_2(7)$.

If we consider the group $2^6{:}L_2(7)$ that preserves all three octads, we find that it acts on any one of the octads as a group $2^3{:}L_2(7)$, in which the normal subgroup 2^3 consists of our affine translations, while a complementary $L_2(7)$ is the one determined by the labeling of Fig. 11.26a. However, another complement to the normal 2^3 is the $L_3(2)$ that fixes the point ∞ and acts as automorphisms of the projective plane whose seven points are the other numbers $0,1,2,3,4,5,6$ and whose seven lines are obtained from the tetrads of the line-code by deleting ∞ (see Fig. 11.29b). So to obtain the isomorphism from $L_2(7)$ onto $L_3(2)$ we simply multiply any given permutation of $L_2(7)$ by the affine translation that restores ∞ to its rightful place.

Figure 11.30 shows three generators for the $L_2(7)$ $(= L_3(2))$ that preserves all three octads, and also fixes a point from each of the first two. These were obtained from the permutations α,β,γ of Fig. 11.28 by applying this process to the first two octads.

13. The octern group

This, the smallest maximal subgroup of M_{24}, is the stabilizer of a particular partition of the 24 points into eight cyclically ordered triads (mnemonic: *eight terns*).

There are two distinct ways to select a 4-dimensional subspace of $\mathscr{C}_{24}$ consisting of the empty and full sets together with 7 pairs of complementary dodecads. In each case, we obtain a partition of the 24 points into eight triads by considering the intersections of triples of dodecads. In the first case, one of the dodecads may be taken to be the set Q of Fig. 11.6a, and the others its images under powers of the element α we defined in connection with the trio group (§12). The resulting set of triads consists of the triads of points similarly numbered in Fig. 11.26a, and the stabilizer is the subgroup $S_3{\times}L_2(7)$ of the trio group.

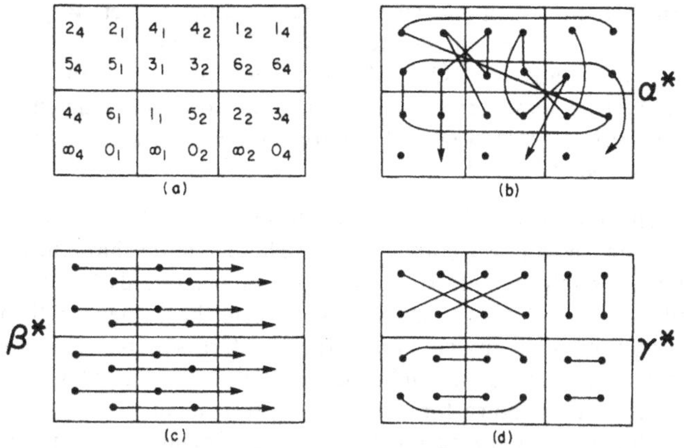

Figure 11.31. (a) Numbering for the octern group. (b)-(d) The elements $\alpha^*, \beta^*, \gamma^*$.

In the second case the triads are $\{n_1, n_2, n_4\}$ for $n = \infty, 0, 1, \ldots, 6$ (Fig. 11.31a). The stabilizing group (the *octern group*) is the $L_2(7)$ generated by the elements α^*, β^*, γ^* shown in Fig. 11.31b-1.31d, with effects

$$\alpha^* : n_t \longrightarrow (n+1)_t,$$
$$\beta^* : n_t \longrightarrow (2n)_{2t},$$
$$\gamma^* : n_t \longrightarrow \left(-\frac{1}{n}\right)_{t/n^2 \text{ or } t^2},$$

where n and t are to be read modulo 7, and the second alternative in the last line is used only for $n = \infty$ and $n = 0$. One of the dodecads consists of all the points 0_t, 1_t, 2_t or 4_t in Fig. 11.31a. This group can be defined as the centralizer in M_{24} of the *octern element* (not in M_{24}):

$$(\infty_4 \infty_2 \infty_1)\ (0_1 0_2 0_4)\ (1_1 1_2 1_4)\ \cdots\ (6_1 6_2 6_4)$$

14. The Mathieu group M_{23}

By fixing any one point of the 24 (say the point ∞), we obtain the maximal subgroup M_{23}, which could alternatively be defined as the stabilizer of the Steiner system $S(4,7,23)$ whose 253 heptads are obtained by deleting ∞ from all the octads of $S(5,8,24)$ that contain it. Although M_{23} is a smaller group than M_{24}, it cannot be too strongly stressed that all properties of M_{23} are best understood by embedding it in M_{24}, and we shall therefore say no more about it.

15. The group $M_{22}{:}2$

By fixing two points of the 24 (say ∞ and 0), we obtain the Mathieu group M_{22}, which of course is not maximal, since we can add to it the permutations which interchange those points, to get $M_{22}{:}2$. This is the stabilizer of the Steiner system $S(3,6,22)$ obtained by deleting ∞ and 0 from all octads of $S(5,8,24)$ that contain them both. It is of course equally true to say that these groups are also best studied by embedding them into M_{24}. However, there are some remarks about the hexads which may be found helpful.

Any one of the 77 hexads is disjoint from 16 others, and meets the remaining 60 in two points each. (These relationships turn the family of 77 hexads into a rank 3 graph.) One of the hexads for the M_{22} fixing ∞ and 0 is the rest of the standard octad, and the 16 hexads disjoint from this are any row plus any column in the standard square. The other 60 hexads come in 15 sets of 4, and can be read from Fig. 11.17. There are 15 sextets of that figure for which one tetrad includes $\{\infty, 0\}$, and we obtain the 60 hexads by taking the remaining two points of that tetrad with one of the four tetrads corresponding to it in the standard square.

16. The group M_{12}, the tetracode and the MINIMOG

The Mathieu group M_{12} is both a subgroup of M_{24} and an analog of M_{24}. We treat it in the latter role first.

M_{12} may be defined as the stabilizer of a Steiner system $S(5,6,12)$, but it is most easily understood in connection with the ternary Golay code $\mathscr{C}_{12}$ (§2.8.5 of Chap. 3, §1.5 of Chap. 10), which we shall define here in terms of the tetracode $\mathscr{C}_4$ (§2.5.1 of Chap. 3) and the MINIMOG 4×3 array.

The tetracode $\mathscr{C}_4$ consists of just 9 words over $\mathbf{F}_3$:

o ooo, o +++, o − − −,

+ o+−, + +−o, + −o+,

− o−+, − +o−, − −+o

(We usually prefer to use the symbols o, +, − rather than 0, 1, 2.) In the typical word *a bcd*, the first two digits (a, b) specify a linear function $\phi(x) = ax + b$, while the last three give the values $\phi(0)$, $\phi(1)$, $\phi(2)$:

$$a \quad b = \phi(\mathrm{o}) \quad c = \phi(+) \quad d = \phi(-).$$

In other words the leading digit defines the *slope* s of the word, and the other three digits cyclically increase by s. This makes it very easy to solve the two types of problem:

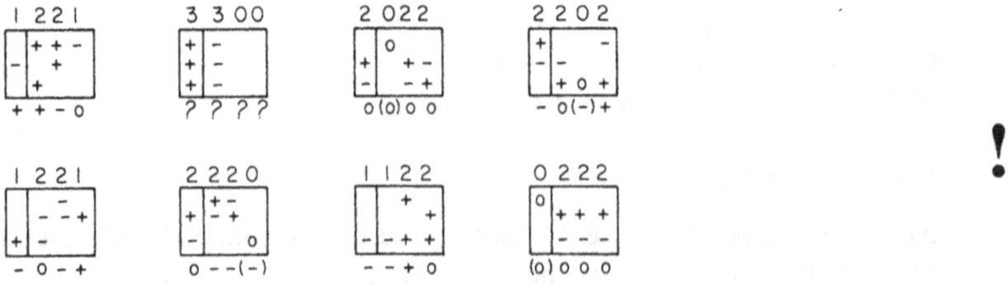

The 2-problem: complete a tetracodeword from any 2 of its digits.

The 4-problem: correct a tetracodeword given all 4 of its digits, *one of which may be mistaken.*

The MINIMOG is a 4×3 array whose rows are labeled o, +, −, and whose columns correspond to the four digits of the tetracode. We shall use the word *column* (abbreviated "col") also for the length 12 word with + digits in the entries of one column (and o digits elsewhere), and the word *tetrad* ("tet") similarly for the weight 4 word with + digits in the places corresponding to a tetracodeword and o digits elsewhere (see Fig. 11.32).

$$
\begin{array}{c|cccc}
o & 6 & 3 & 0 & 9 \\
+ & 5 & 2 & 7 & 10 \\
- & 4 & 1 & 8 & 11
\end{array}
$$

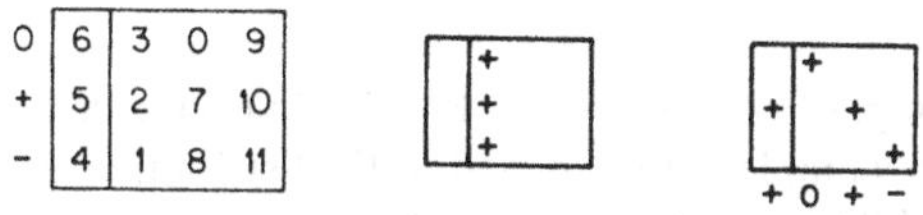

Figure 11.32. The MINIMOG array with shuffle numbering, a "col" and a "tet."

Then the ternary Golay code $\mathscr{C}_{12}$ is defined by the observation that

modulo $\mathscr{C}_{12}$, any column is congruent to the negative of any tetrad.

(Compare $\mathscr{C}_{24}$, modulo which any column is congruent to (the negative of) any hexad.) In particular, this tells us that the combinations

$$\text{col} - \text{col}, \quad \text{col} + \text{tet}, \quad \text{tet} - \text{tet}, \quad \text{col} + \text{col} - \text{tet}$$

all yield $\mathscr{C}_{12}$-words, and in fact they all yield *signed hexads* ($\mathscr{C}_{12}$-words of weight 6) — see the examples in Fig. 11.33. We have also entered in the figure the column-distribution of the 6 points, and the *odd-men-out*. (If a column contains just one non-zero digit, or alternatively just one zero digit, then the *odd-man-out* is the name of the corresponding row. Otherwise there is no odd-man-out for that column, and we write ? under it.)

If we ignore signs, then from these signed hexads we get the 132 hexads of the Steiner system $S(5, 6, 12)$. These turn out to be *all* possible hexads for which

    1)   the odd-men-out form part of a tetracodeword, and

    2)   the column-distribution is not 3 2 1 0 in any order.

The problem of completing such a hexad from any 5 of its points is easily solved in terms of 2-problems or 4-problems, and then the signs can be added, if desired, by using the expression in terms of columns and tetrads.

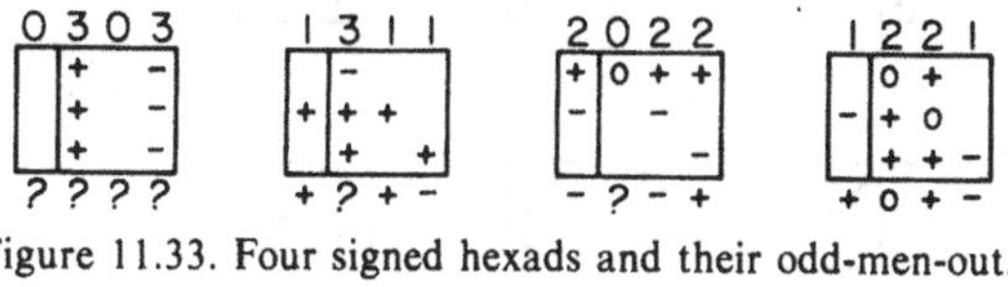

Figure 11.33. Four signed hexads and their odd-men-out.

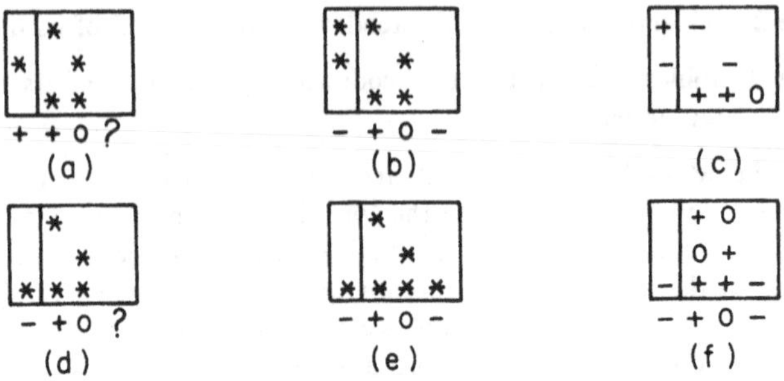

Figure 11.34. From pentads to signed hexads.

For example, the 5 points of Fig. 11.34a have column-distribution $2^2 1 0$ and so must complete to a hexad with distribution $2^3 0$ or $2^2 1^2$. In either case the middle two digits of the odd-man-out sequence are correct, and so we solve the 2-problem ? +o? ⇛ − +o−. This shows that it is the first column that must be amended, and so the answer is Fig. 11.34b. The answer to the problem in Fig. 11.34d is given in Fig. 11.34e.

Now to put in the signs! For the second of these two problems, this presents no difficulty, since the column-distribution $2^2 1^2$ tells us that the hexad has the form col + col − tet, and therefore the signs are as shown in Fig. 11.34f (we've inserted those o digits to show you where the subtracted tetrad was). But for the original problem, the column-distribution implies a form tet − tet, and we need *two* tetrads, neither of which is the odd-men-out tetrad − +o−, since in this case we can see that those odd men really are out! The answer is that we take the two tetrads o −−− and + o+− that agree with that tetrad in the digit corresponding to the empty column (where we might have written ? rather than −), and so derive the signs as shown in Fig. 11.34c ([o −−−] − [+ o+−]). This is one of many cases when the process is easier performed than described.

At the foot of the page we have set up a number of 5-element sets. On the page behind we give their completions to hexads of $\mathscr{C}_{12}$, together with a proper choice of signs. Note that there are two correct ways to sign any hexad of $S(5,6,12)$ — the one you first thought of, and its negative.

The relation between the MOG and the MINIMOG. The MINIMOG can be inserted into the MOG as shown in Fig. 11.35. With this inclusion, the hexads of $S(5,6,12)$ are just those 6-element subsets of the *right* region (marked by letters *r* in the figure) which can be completed by two points

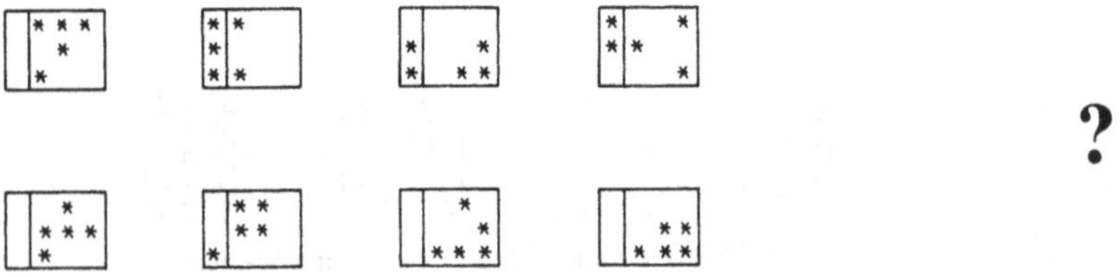

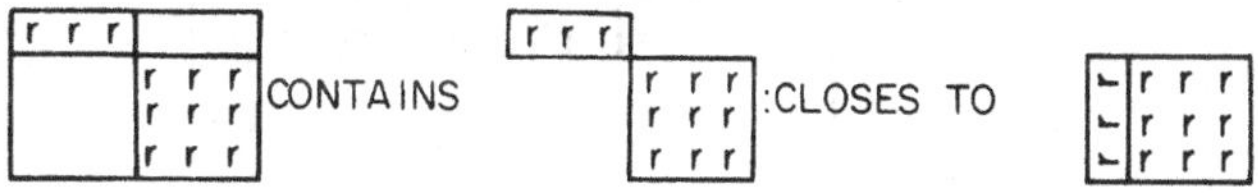

Figure 11.35. How the MOG contains the MINIMOG.

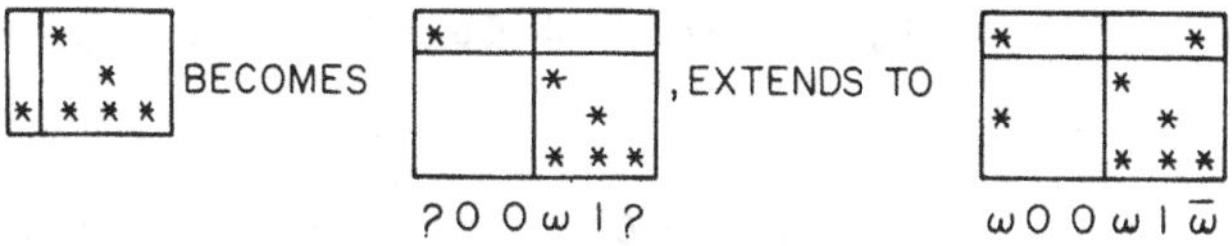

Figure 11.36. How a hexad becomes an octad.

of the *left* region to form an octad of $S(5,8,24)$. Figure 11.36 shows an example.

17. Playing cards and other games

The numbering of Fig. 11.32 has many remarkable properties. In the course of an investigation into the mathematics of shuffling playing cards, Diaconis *et al.* [Dia2] discovered a remarkable new way to generate M_{12}. More generally, for any divisor n of 12, let us take n cards, and number them $0, 1, 2, \ldots, n-1$. Let us consider the group M_n of permutations of these cards generated by the operations

$$r_n \ : \ t \longrightarrow n-1-t \ ,$$

$$s_n \ : \ t \longrightarrow \min\{2t \ , \ 2n-1-2t\} \ .$$

In playing-card language, r_n is the operation of reversing the order of the cards, while s_n is a special permutation called the Mongean shuffle. Then for $n = 1, 2, 3, 4, 6, 12$ we find these groups are respectively

$$S_1, S_2, S_3, A_4, PGL_2(5), M_{12},$$

where the $PGL_2(5)$ acts on the cards in the way it permutes $0, \infty, 1, 2, 3, 4$, while M_{12} acts upon them as on the numbers in **Fig. 11.32.**

Many remarkable facts about M_{12} appear when we regard it as acting on a set ordered in this particular way. We consider the M_{12} as embedded in M_{24} acting on two such populations, as in Fig. 11.37. This has a simple relation to the projective numbering of Fig. 11.7, given by

$$n^+ \ = \ \text{whichever of } n, -n \text{ is in } Q \ ,$$

$$n^- \ = \ \text{whichever of } 8/n, -8/n \text{ is in } \Omega \backslash Q.$$

The *duality* of this M_{12} is the permutation of M_{24} that interchanges u^+ with u^-. It is a particularly interesting outer automorphism of M_{12}.

Whenever $ab = 12$, there is a standard partition of the 12 cards into a sets of b, namely one of the partitions indicated by the bars in the

0^+	0^-	1^+	7^-	2^+	8^-
2^-	3^+	5^-	4^+	10^-	5^+
1^-	6^+	6^-	7^+	9^-	8^+
3^-	9^+	4^-	10^+	11^-	11^+

Figure 11.37. The shuffle numbering for M_{24}.

```
|0 1 2 3 4 5 6 7 8 9 10 11|        FOR a = 1, b = 12
|0 1 2 3 4 5|6 7 8 9 10 11|        FOR a = 2, b = 6
|0 1 2 3|4 5 6 7|8 9 10 11|        FOR a = 3, b = 4
|0 1 2|3 4 5|6 7 8|9 10 11|        FOR a = 4, b = 3
|0 1|2 3|4 5|6 7|8 9|10 11|        FOR a = 6, b = 2
|0|1|2|3|4|5|6|7|8|9|10|11|        FOR a = 12, b = 1
```

Figure 11.38. The standard partitions.

appropriate line of Fig. 11.38. Then the stabilizer of such a partition turns out to be a group $M_a \times M_b$, where M_a acts on the a rows and M_b on the b columns of the appropriate array of Fig. 11.39. Moreover, the standard duality takes the stabilizer of the standard partition into a sets of b into that of the standard partition into b sets of a, with the map s_a on the rows dualizing to s_a^{-1} on the columns! The map r_a on the rows dualizes to a map which reverses the order of the columns, although in the case $a = 6$ this is not r_a alone, but its product with the r_2 that interchanges the two rows.

```
(a)
0 1 2 3 4 5 6 7 8 9 10 11

(b)
0 1 2 3 4 5
6 7 8 9 10 11

(c)
0 1 2 3
7 6 5 4
8 9 10 11

(d)
0 1 2
3 4 5
6 7 8
9 10 11

(e)
0 1
3 2
4 5
7 6
8 9
11 10

(f)
0
1
2
3
4
5
6
7
8
9
10
11
```

Figure 11.39. Six ways to lay down the cards.

The group generated by the *turnaround permutations*

$$T_1, T_2, T_3, T_4, T_6, T_8, T_{12}, T_{24}$$

shown in Fig. 11.40 is $2 \times M_{12}$, the central group of order 2 being generated by T_{24}. This gives a strikingly simple definition of M_{12}.

Identifying points with the same number in Fig. 11.40 produces the same M_{12} that we have just defined.

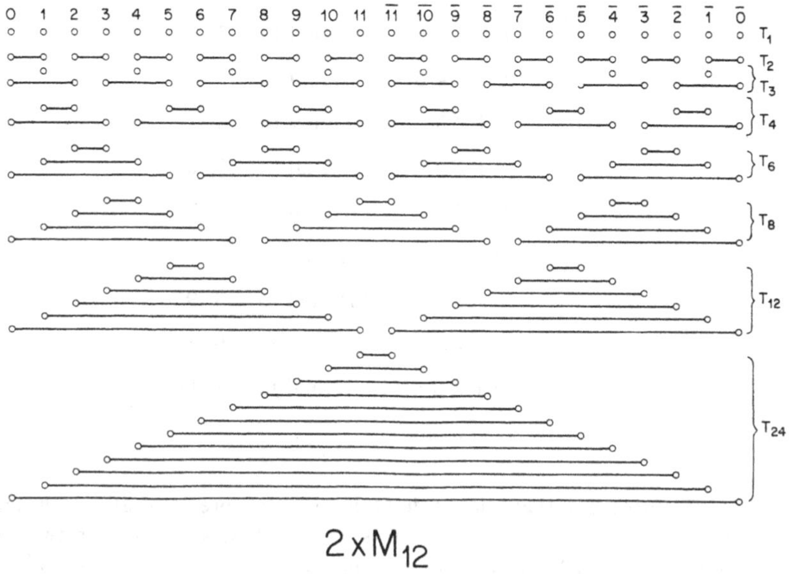

Figure 11.40. The turnaround permutations $T_1, T_2, \ldots, T_{24}$ that generate $2 \times M_{12}$.

Some more surprises happen when we plot the hexads of $S(5,6,12)$ in terms of the sums of their elements, in this shuffle numbering. The resulting histogram (Fig. 11.41) suggests that there should be something very special about the 11 hexads with sum 21, and indeed there is! We shall call these the *light* hexads, and their complements, which are the 11 hexads with sum 45, the *heavy* ones.

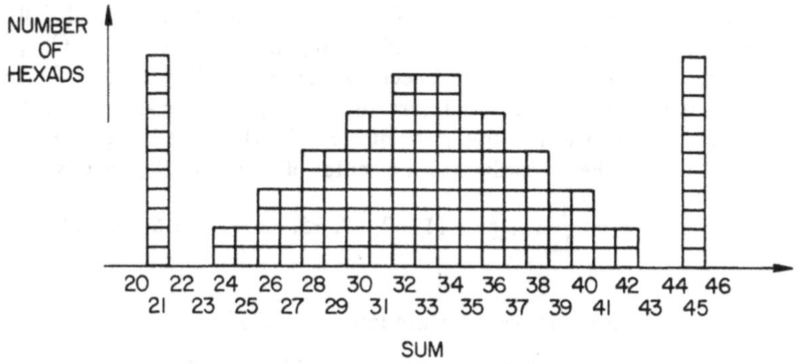

Figure 11.41. Pyramid between obelisks.

Table 11.2. Duality between hexads and duads.

1 2 3 4 5 6	is dual to	0 1
0 1 2 3 7 8	is dual to	1 2
0 1 2 4 5 9	is dual to	2 3
0 1 3 4 6 7	is dual to	3 4
0 1 2 3 5 10	is dual to	4 5
0 1 2 4 6 8	is dual to	5 6
0 2 3 4 5 7	is dual to	6 7
0 1 2 3 6 9	is dual to	7 8
0 1 3 4 5 8	is dual to	8 9
0 1 2 5 6 7	is dual to	9 10
0 1 2 3 4 11	is dual to	10 11

It turns out that the 11 light hexads are precisely all the hexads of non-negative integers whose sum is 21, and that when we join two hexads if they have just 3 points in common, then these 11 hexads naturally form a chain, in which each joins just those adjacent to it. Moreover, under our duality, these hexads correspond to the 11 duads $\{0,1\}$, $\{1,2\}$, ... , $\{10,11\}$ as in Table 11.2. [The hexad $\{a,b,c,d,e,f\}$ is dual to the duad $\{x,y\}$ if and only if $\{a^+,b^+,c^+,d^+,e^+,f^+,x^-,y^-\}$ is an octad of $S(5,8,24)$.] A last remark is that the number of heavy hexads that contain (respectively)

$$0,1,2,3,4,5,6,7,8,9,10,11 \quad \text{is} \quad 1,1,2,3,4,5,6,7,8,9,10,10\,.$$

The Steiner system $S(5,6,12)$ is obtained from the 11 light hexads by complementation in $\{0,1,2,3,4,5,6,7,8,9,10,11\}$ and by addition of hexads whenever the sum is again a hexad. The Golay code $\mathscr{C}_{24}$ is spanned by the corresponding 11 octads together with the universal set.

It will not be easy to explain all the above observations. They are certainly connected with hyperbolic geometry and with the "hole" structure of the Leech lattice. For example, using the terminology of Chap. 24, if we take coordinates for the Leech lattice centered at a hole of type A_1^{24}, then the vertices of that hole are the 48 f-points $(\pm 4,0^{23})$, and the next nearest points are the 4096 g-points $(\pm 1^{24})$, the signs being determined by a $\mathscr{C}$-set. In these coordinates the center of a certain D_{24}-hole is

$$c = \frac{1}{23}(11^{12}\,;\,-11,-9,-7,-5,-3,-1,1,3,5,7,9,11)$$

when the coordinate positions are labeled in the order

$$0^-,1^-,2^-,\ldots,11^-\,;\,0^+,1^+,2^+,\ldots,11^+\,.$$

The vertices of this hole are the f-points that have inner product 44/23 with c, and the g-points whose coordinates -1 are against the 11 octads mentioned above.

The lexicography of the hexacode and the Golay code $\mathscr{C}_{24}$. If we define a list of numbers in the scale of 4 by the requirement that each should be the smallest such number differing in at least 4 digits from all previous ones, then it turns out that the first 64 of them:

$$000000\ ,\quad 001111\ ,\quad 002222\ ,\quad 003333\ ,\quad 010123\ ,\ldots,\ 333300$$

are just the hexacodewords (with $0,1,2,3$ for $0,1,\omega,\bar{\omega}$). This observation turns out to be equivalent to the "ax^2+bx+c" definition of the hexacode, and R. A. Wilson [Wil6a] has generalized it in various ways. By using the scale of 2, and demanding that the numbers differ from previous ones in at least 8 digits, M. J. T. Guy found the Golay code $\mathscr{C}_{24}$, with the order obtained by reading down the MOG columns, from left to right. Such "lexicographic codes" are discussed in greater detail in [Con41].

18. Further constructions for M_{12}

The arithmetic progression loop. There is a loop multiplication, $*$, on the 12-element set $L = \{\infty,0,1,2,3,4,5,6,7,8,9,X = 10\}$ defined by the following two assertions.

1) ∞ is an identity, and is the square of every element (e.g. $\infty * 5 = 5 * \infty = 5$, $5*5 = \infty$).

2) If a,b,c are three distinct numbers in arithmetic progression, and $b-a$ is a quadratic residue modulo 11, then $a*b = b*a = c$. The quadratic residues are $1,-2,3,4,5$, so for example $7*8 = 9$, $6*8 = 4$.

This loop yields two constructions for M_{12}.

(I) M_{12} is generated by the permutations π_a $(a \in L)$, where $\pi_a : x \longrightarrow x*a$. E.g. $\pi_0 = (\infty 0)(1\,2\,9\,7\,4\,8\,3\,6\,5\,X)$, and the other π_a are obtained by translating the numbers in this by a, modulo 11.

(II) We consider the set of all permutations π of L for which there exists a π' such that

$$a*b = c \longrightarrow a^{\pi} * b^{\pi} = c^{\pi'} .$$

Then M_{12} is generated by these π and π'. In fact the permutations π constitute the standard $L_2(11)$ of maps of the form $x \longrightarrow (ax+b)/(cx+d)$ with $ad-bc$ a quadratic residue, and the π' form another subgroup $L_2(11)$ of M_{12}. This yields the generators for M_{12} used in Chap. 10, thus:

$$\alpha = \alpha' = (\infty)(0\,1\,2\,3\,4\,5\,6\,7\,8\,9\,X) \qquad (t \longrightarrow t+1)\ ,$$

$$\beta = \beta' = (\infty)(0)(1\,3\,9\,5\,4)(2\,6\,7\,X\,8) \qquad (t \longrightarrow 3t)\ ,$$

$$\gamma = (\infty 0)(1\,X)(2\,5)(3\,7)(4\,8)(6\,9) \qquad (t \longrightarrow -1/t)\ ,$$

$$\gamma' = (\infty)(0)(1)(2\,X)(3\,4)(5\,9)(6\,7)(8)\ .$$

The permutation γ' was called δ in Chap. 10.

Here is a dictionary between this modulo 11 numbering and the shuffle numbering:

shuffle:	0	1	2	3	4	5	6	7	8	9	10	11
modulo 11:	∞	1	9	3	4	5	0	8	6	2	X	7
mnemonic:	∞	+1	−2	+3	+4	+5	0	−3	+6	−9	−12	−15

The Rubicon or Rubik-icosahedron construction of $\mathbf{M_{12}}$. If we call permutations α and β of twelve letters *congruent* when $\alpha\beta^{-1} \in M_{12}$, there are $12!/|M_{12}| = 7! = 5040$ congruence classes, as shown in Table 11.3.

Table 11.3. Classes of permutations of S_{12} modulo M_{12}.

Even permutations:

(a)	1	class represented by the identity permutation
(b)	440	classes represented by the 3-cycles
(c)	495	classes each with 3 representatives of shape 2^2
(d)	1584	classes each with 12 representative of shape 5
	2520	total

Odd permutations:

(e)	66	classes represented by the 2-cycles
(f)	990	classes each with 3 representatives of shape 4
(g)	1320	classes each with 12 representatives of shape $2^1 3^1$
(h)	144	classes each with 110 representatives of shape 6
	2520	total

Examples of the classes containing more than one shortest representative are (in the shuffle numbering):

Class c:

$$(0\,9)(3\,6) \equiv (1\,2)(4\,5) \equiv (7\,8)(10\;11)\ .$$

These are the squares of:

Class f:

$$(0\,3\,9\,6) \equiv (1\,4\,2\,5) \equiv (7\;11\;8\;10)\ .$$

Class g:

$$\begin{aligned} & (1\,2)(9\,6\,3) \equiv (0\,2)(10\,7\,4) \equiv (0\,1)(11\,8\,5) \\ \equiv\ & (4\,5)(0\,6\,9) \equiv (3\,5)(1\,7\,10) \equiv (3\,4)(2\,8\,11) \\ \equiv\ & (7\,8)(9\,3\,0) \equiv (6\,8)(10\,4\,1) \equiv (6\,7)(11\,5\,2) \\ \equiv\ & (10\,11)(0\,3\,6) \equiv (9\,11)(1\,4\,7) \equiv (9\,10)(2\,5\,8). \end{aligned}$$

(To understand these, arrange the numbers in the standard 4×3 array as in Fig. 11.39d.)

For classes d and h we change to the modulo 11 numbering $\{\infty, 0, 1, \ldots, X\}$.

Class d:

$$\begin{array}{ccccccc} & (1\,9\,4\,3\,5) & \equiv & (2\,X\,6\,8\,7) & \equiv & (\infty\,5\,7\,8\,9) \\ \equiv & (0\,7\,5\,3\,X) & \equiv & (\infty\,4\,X\,2\,5) & \equiv & (\infty\,9\,6\,X\,3) \\ \equiv & (\infty\,3\,2\,7\,1) & \equiv & (0\,X\,4\,9\,8) & \equiv & (0\,8\,1\,5\,2) \\ \equiv & (0\,6\,9\,1\,7) & \equiv & (\infty\,1\,8\,6\,4) & \equiv & (0\,2\,3\,4\,6)\,. \end{array}$$

Class h: The map

$$(\infty)(0)(1\,X)(2\,9)(3\,8)(4\,7)(5\,6) \quad (t \longrightarrow -t)$$

belongs to a class of type h. This class is invariant under the group $L = L_2(11)$ of maps $t \rightarrow (at+b)/(ct+d)$ where $ad-bc$ is a quadratic residue modulo 11. The 110 shortest representatives of this class are the conjugates of $(\infty\,0\,2\,3\,4\,6)$ by elements of L. (Since the permutation $(\infty\,0\,2\,3\,4\,6)\ (1\,9\,7\,X\,8\,5)$ is in L, each of these representatives is one of the cycles of an element of order 6 in L.)

We note that the permutations in our example for class d are just the twelve clockwise 5-cycles τ_i around the vertices i of the icosahedron shown in Fig. 11.42.

It turns out that M_{12} is actually generated by the corresponding quotients $\tau_i \tau_j^{-1}$. In other words M_{12} is the set of all permutations of the form

twist untwist twist untwist ... twist untwist

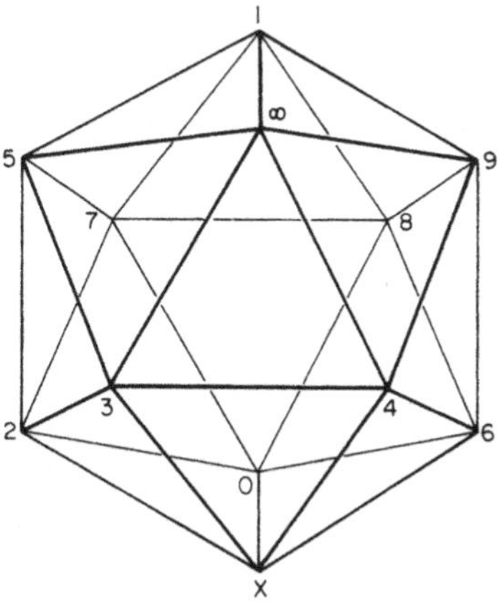

Figure 11.42. An icosahedron.

of the vertices of a kind of "Rubik" icosahedron in which the "twists" τ_i and "untwists" τ_i^{-1} are physically possible.

Remarks. (i) The twist about one vertex followed by the untwist about the opposite one is a generating rotation of the icosahedron. So the rotation group of the icosahedron is contained in this M_{12}. (ii) Mathematically it is simpler to regard M_{12} as generated by the twelve permutations $\tau_i \sigma$, where

$$\sigma = (\infty 0)(1X)(29)(38)(47)(56)$$

is the antipodal inversion of the icosahedron. It turns out that

$$\tau_0 \sigma = (\infty 0)(129748365X)$$

(Memo: alternately double and negate!) is the permutation π_0 mentioned earlier in this section.

The Mathieu groups have now been known for a century and a quarter, but are still capable of surprising us. Some references to them are [Ass2], [Bet1], [Car1], [Cur6], [Cur7], [Gib1], [Hug1], [Jon5], [Kra1], [Mas1], [Mat1], [Mat2], [Pee1], [Ple15], [Ple15a], [Sol1], [Sta12], [Ton3], [War1], [Whi1].

12

A Characterization of the Leech Lattice

J. H. Conway

We give a short proof that Leech's remarkable lattice is characterized by some of its simplest properties.

The main result of this chapter is Theorem 7, which characterizes the Leech lattice in four different ways. Although we must quote two theorems to open and close the proof, the reader can take these on trust if he wishes, and he will find that the proof is otherwise completely self-contained. The argument immediately gives the order of the automorphism group (Co_0) of the lattice, and can be used to give other information about the group and the lattice.

We proceed at once to the proof. We shall show that Leech's is the only lattice in fewer than 32 dimensions that has one point per unit volume and in which the square of every non-zero distance is an even integer greater than 2. A lattice with one point per unit volume is called *unimodular,* and then also *even* if every squared distance or norm is an even integer (§2.4 of Chap. 2). In an even unimodular lattice of dimension n we use N_m for the number of vectors of norm m. Our main theorem then asserts that Leech's is the only even unimodular lattice with $n < 32$ and $N_2 = 0$.

Theorem 1. *If* Λ *is an even unimodular lattice with* $n < 32$ *and* $N_2 = 0$, *then we have* $n = 24$, $N_4 = 196560$, $N_6 = 16773120$, $N_8 = 398034000$.

Proof. Hecke's theorem (Theorem 17 and Cor. 18 of Chap. 7) says that if Λ is an even unimodular lattice then n is a multiple of 8, and the theta series $\Theta_\Lambda(z)$ is a modular form of weight $n/2$ for the full modular group. For $n = 8$ or $n = 16$ the space of such forms has dimension 1 and so the condition $N_0 = 1$ determines N_{2m} for all m, and in fact we find $N_2 > 0$ in each case. For $n = 24$ the corresponding space has dimension 2 and now the values $N_0 = 1$, $N_2 = 0$ determine N_{2m}. In fact we find that the coefficients of the extremal theta series are

$$N_{2m} = \frac{65520}{691} \left(\sigma_{11}(m) - \tau(m) \right) ,$$

where $\sigma_{11}(m)$ is the sum of the 11-th powers of the divisors of m and τ is Ramanujan's function. The given values are readily computed from this.

From now on we shall suppose that Λ is a lattice satisfying the hypotheses of Theorem 1. We call two vectors of Λ *equivalent* if their difference is twice a lattice-vector — i.e. we consider $\Lambda/2\Lambda$. If $\{b_1, \ldots, b_{24}\}$ is a basis for Λ then $\{2b_1, \ldots, 2b_{24}\}$ is a basis for 2Λ, and so there are just 2^{24} equivalence classes. We call a vector x *short* if it has length at most $\sqrt{8}$, and enquire which equivalence classes contain short vectors. Since x and $-x$ are always equivalent, the non-zero short vectors in an equivalence class will occur in pairs (compare Theorem 2.8 of Chap. 10).

Theorem 2. *Each equivalence class contains a short vector. The equivalence classes that contain more than a single pair of short vectors are precisely those that contain vectors of length $\sqrt{8}$, and these classes all contain exactly* 24 *mutually orthogonal pairs of vectors of that length.*

Proof. Let $x = OX$ and $y = OY$ be equivalent short vectors, with $y \neq \pm x$. Replacing y by $-y$ if needs be, we can suppose that the angle $XOY \leqslant \pi/2$. Then since $OX^2 \leqslant 8$ and $OY^2 \leqslant 8$ we must have $XY^2 \leqslant 16$ (Fig. 12.1). But XY is twice a lattice vector, and so $XY \geqslant 2\sqrt{4}$, whence $XY^2 \geqslant 16$. It now follows that every inequality in the argument must in fact have been an equality, so that two distinct equivalent short vectors not forming a pair must be orthogonal and both of length $\sqrt{8}$, and so there can be at most 24 pairs of such vectors. The number of equivalence classes that contain short vectors is therefore at least

$$\frac{N_0}{1} + \frac{N_4}{2} + \frac{N_6}{2} + \frac{N_8}{48} .$$

Since the displayed number is exactly 2^{24} each equivalence class is accounted for, the other inequalities of our argument must also have been exact, and the conclusions of the theorem follow.

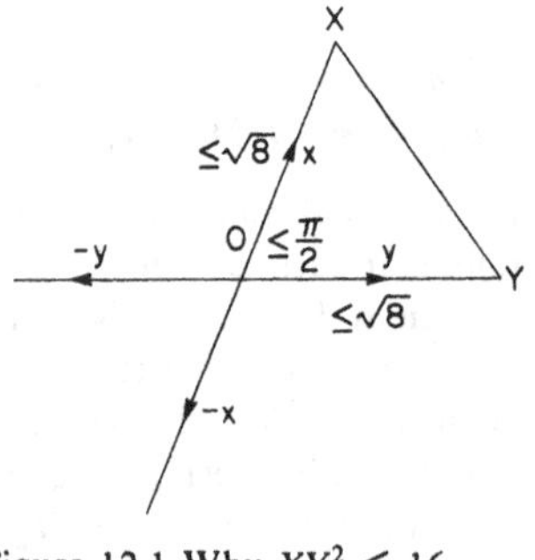

Figure 12.1 Why $XY^2 \leqslant 16$.

Now let $c = 1/\sqrt{8}$. Then we can suppose that Λ contains all vectors of the shapes $c(\pm 8, 0^{23})$ and $c((\pm 4)^2, 0^{22})$, these being 24 mutually orthogonal pairs of vectors of length $\sqrt{8}$ and their halved differences. The symbol $((\pm 4)^2, 0^{22})$ denotes the typical vector with two coordinates ± 4 and 22 coordinates zero. These vectors generate the sublattice Λ_0 of all vectors $c(x_1, \ldots, x_{24})$ in which each x_i is a multiple of 4 and their sum is a multiple of 8.

If x and y are vectors of Λ the scalar product $x \cdot y$ is necessarily an integer, since it equals $\frac{1}{2}((x+y)^2 - x^2 - y^2)$. Letting $x = c(x_1, \ldots, x_{24})$ be the typical vector of Λ, and taking scalar products with the generating vectors of Λ_0, we see that $c^2 \cdot 8x_i$ and $c^2(4x_i - 4x_j)$ are integral for all i, j, or in other words that the coordinates x_i are integers all of the same parity. (This explains our choice of the scale factor c.) If Λ has any vector x whose coordinates x_i are all odd, we can suppose (by changing the signs of suitable coordinate-vectors) that it has a vector x that satisfies $x_i \equiv 1 \pmod 4$ for each i.

We now investigate what congruence conditions modulo 4 are imposed on the x_i. To this end, we let $\Omega = \{1, 2, \ldots, 24\}$ and let $\mathscr{C}$ be the set of all subsets C of Ω for which there exists a lattice-point x with $x_i \equiv 2$ or $0 \pmod 4$ according as $i \in C$ or $i \notin C$ (x is said to *correspond* to C). In general, if A and B are subsets of Ω, we use $A + B$ for the symmetric difference $(A \backslash B) \cup (B \backslash A)$, and define their *distance* to be the cardinal of $A + B$. We call two sets *close* if their distance is at most 4. If x and y correspond respectively to C and D, then $x + y$ corresponds to $C + D$, so that the set $\mathscr{C}$ is closed under symmetric difference.

Theorem 3. *Each non-empty $\mathscr{C}$-set has at least 8 members. The number of $\mathscr{C}$-sets with exactly 8 members is at most 759, and the total number of $\mathscr{C}$-sets at most 2^{12}.*

Proof. Let $C \neq \varnothing$ be a $\mathscr{C}$-set with n members, and x a corresponding vector. By subtracting a suitable member of Λ_0 we can reduce all but one of the x_i to 0 or 2, and the remaining x_i to ± 2. But then $x \cdot x = c^2 \cdot 4n$, and since $x \cdot x \geqslant 4$ we must have $n \geqslant 8$. Now no 5-element subset of Ω can be contained in two distinct $\mathscr{C}$-sets of cardinal 8, for then their symmetric difference would be a $\mathscr{C}$-set of cardinal at most 6. The number of $\mathscr{C}$-sets of cardinal 8 is therefore at most $\binom{24}{5}/\binom{8}{5} = 759$.

Now suppose the set A is close to the distinct $\mathscr{C}$-sets $A + X$ and $A + Y$. Then $X + Y$ is a $\mathscr{C}$-set and so X and Y must be disjoint sets with 4 members each. It follows that if A is close to more than one $\mathscr{C}$-set it is close to at most 6, and distant 4 from each of these. If we define $f(A, C)$ to be 1, 1/6, or 0 according as the distance from A to C is less than, equal to, or greater than 4, we can express this result as

$$\sum_{C \in \mathscr{C}} f(A, C) \leqslant 1 \qquad \text{for each } A\,,$$

whence

$$\sum_{A \subseteq \Omega} \sum_{C \in \mathscr{C}} f(A, C) \leqslant 2^{24}.$$

On the other hand we have plainly

$$\sum_{A \subseteq \Omega} f(A, C) = 1 + 24 + \binom{24}{2} + \binom{24}{3} + \frac{1}{6}\binom{24}{4},$$

and since the displayed number is exactly 2^{12} there can be at most $2^{24}/2^{12} = 2^{12}$ $\mathscr{C}$-sets.

We now enumerate the vectors x with $x \cdot x = 4$. Each such x corresponds to a solution of the equation $\Sigma x_i^2 = 32$ in which the x_i are integers all of the same parity, the number of $x_i \equiv 2 \pmod 4$ being zero or at least 8. The only possibilities are therefore of the shapes $c((\pm 2)^8, 0^{16})$, $c(\mp 3, (\pm 1)^{23})$, and $c((\pm 4)^2, 0^{22})$. In the first case the non-zero coordinates must form a $\mathscr{C}$-set, and we obtain at most 2^7 such vectors per $\mathscr{C}$-set since otherwise we should have two lattice-vectors differing by a vector of shape $c(\pm 4, 0^{23})$, which is impossibly short. Hence we have at most $759 \cdot 2^7$ vectors of this shape. In the second case, the lower sign must be taken on a $\mathscr{C}$-set, as we see by subtracting a vector with every coordinate congruent to 1 modulo 4, and since there are 24 positions for the exceptional coordinate ∓ 3, we have at most $24 \cdot 2^{12}$ vectors of this shape. We have already observed that Λ contains all the $\binom{24}{2} \cdot 2^2$ possible vectors of the third shape.

Theorem 4. *Every 5-element subset of $\mathscr{C}$ is contained in just one 8-element $\mathscr{C}$-set. The set $\mathscr{C}$ is the closure under symmetric difference of the set of 8-element $\mathscr{C}$-sets.*

Proof. The preceding argument has shown that Λ contains at most

$$2^7 \cdot 759 + 2^{12} \cdot 24 + 2^2 \cdot \binom{24}{2}$$

vectors x with $x \cdot x = 4$, and since the displayed number is exactly 196560 every inequality of that argument must in fact be an equality. In particular each 5-element subset of Ω is contained in just one of the 759 8-element $\mathscr{C}$-sets. If now C is any non-empty $\mathscr{C}$-set we can find an 8-element $\mathscr{C}$-set with at least 5 members in common with C, and so C is the symmetric difference of this 8-element $\mathscr{C}$-set and a $\mathscr{C}$-set of smaller cardinal. Repeating the argument, we see that C is an iterated symmetric difference of 8-element $\mathscr{C}$-sets.

Theorem 5. *The vector $x = c(x_1, \ldots, x_{24})$ is contained in Λ just if*
(i) *the x_i are integers all of the same parity,*
(ii) *the set of i where x_i takes any given value* (*mod* 4) *is a $\mathscr{C}$-set, and*
(iii) $\Sigma x_i \equiv 0$ *or* 4 (*mod* 8) *according as each $x_i \equiv 0$ or* 1 (*mod* 2).

Proof. For each 8-element $\mathscr{C}$-set, Λ contains 2^7 corresponding vectors $c((\pm 2)^8, 0^{16})$, which are either all those with evenly many + signs or all the others. But we know that Λ contains the 24 vectors of the shape $c(-3, 1^{23})$, and it cannot therefore contain any vector of the shape

$c(-2, 2^7, 0^{16})$, for we obtain from this the impossibly short lattice vector $c(1^8, (-1)^{16})$ by subtraction of a suitable vector $c(-3, 1^{23})$. Λ must therefore contain the 759 vectors $c(2^8, 0^{16})$ that correspond to the 759 8-element $\mathcal{C}$-sets.

Let L be the set of all x satisfying (i)-(iii), and let $x = c(x_1, \ldots, x_{24})$ be any vector of either Λ or L. We shall show that x is a linear combination of the vectors $c(-3, 1^{23})$, the 759 vectors $c(2^8, 0^{16})$ just discussed, and the generators of Λ_0, all of which belong to both Λ and L. Subtracting $c(-3, 1^{23})$ if necessary, we may suppose that the x_i are even. Subtracting a number of vectors $c(2^8, 0^{16})$, we may suppose the x_i to be multiples of 4. Finally, subtracting a member of Λ_0, we can reduce all but one of the x_i to 0, and the remaining one to 0 or 4. But since $x \cdot x < 4$ every coordinate is now zero, our result is proved, and Λ has been identified with L.

Theorem 6. *Λ is unique to within isomorphism.*

Proof. The first sentence of Theorem 4 asserts that the 8-element $\mathcal{C}$-sets form a Steiner system $S(5,8,24)$. To complete the identification of Λ we appeal to a theorem of Witt ([Wit3]; [Lün1], [Mac6, Chap. 20, §5]), asserting that the system $S(5,8,24)$ is unique to within a permutation of Ω. The lattice is therefore completely determined.

Our argument gives slightly more. There are $N_8/48$ sets of 24 mutually orthogonal pairs of vectors of length $\sqrt{8}$, and any one of these sets could have been used to define our coordinate-system. In other words, the automorphism group of Λ is transitive on the $N_8/48$ natural coordinate-frames. On the other hand, the number of automorphisms fixing any coordinate-frame is exactly 2^{12} times the order of the Mathieu group M_{24}, since any such automorphism can be obtained by changing the signs of a $\mathcal{C}$-set of coordinates and then permuting Ω by a permutation fixing the system $S(5,8,24)$, and Witt defined M_{24} to be the group of all such permutations. Hence the order of the automorphism group is $N_8/48$ times 2^{12} times the order of M_{24}, namely

$$2^{22}3^9 5^4 7^2 11 \cdot 13 \cdot 23 = 8315553613086720000 .$$

Theorem 7. *Λ is the only even unimodular lattice L with $n < 32$ that satisfies any one of the following.*

(i) *L is not directly congruent to its mirror-image.*

(ii) *No reflection leaves L invariant.*

(iii) $N_2(L) = 0$.

(iv) $N_{2m}(L) = 0$ *for some* $m \geqslant 0$.

Proof. Plainly (i) implies (ii), and (iii) implies (iv). We show that (ii) implies (iii).

If L has a vector x that satisfies $x \cdot x = 2$, then the reflection in the hyperplane through the origin perpendicular to x takes y to

$$y - 2\frac{x \cdot y}{x \cdot x}x ,$$

that is, to $y - (x \cdot y)x$, which is a lattice-vector whenever y is. Hence this reflection leaves L invariant.

To complete the proof of the theorem we need only show that condition (iv) determines Λ. But $N_{2m}(L)$ has the form

$$\frac{65520}{691}\left(\sigma_{11}(m) - k\tau(m)\right)$$

where since $N_2(L) \geqslant 0$ we must have $k \leqslant 1$. But then

$$N_{2m}(L) \geq \frac{65520}{691}(\sigma_{11}(m) - |\tau(m)|) > 0$$

for all $m > 1$.

Since the discovery of the above proof, Niemeier [Nie2] has completed his enumeration of the even unimodular lattices in 24 dimensions. He finds 24 such lattices, which are characterized by their configurations of vectors of length $\sqrt{2}$, and confirms that Leech's lattice is the only one without such vectors (see Chaps. 16 and 17). Another characterization of the Leech lattice is given in §5 of Chap. 18.

13

Bounds on Kissing Numbers

A. M. Odlyzko and N. J. A. Sloane

Upper bounds are given on the maximal number, τ_n, of nonoverlapping unit spheres that can touch a unit sphere in n-dimensional Euclidean space, for $n \leqslant 24$. In particular it is shown that $\tau_8 = 240$ and $\tau_{24} = 196560$.

1. A general upper bound

The kissing number problem, as discussed in §2 of Chap. 1, asks for the maximal number τ_n of nonoverlapping equal n-dimensional spheres that can touch another sphere of the same size. The following theorem leads to upper bounds on τ_n. We assume $n \geqslant 3$.

Theorem. *If $f(t)$ is a real polynomial such that* (C1) $f(t) \leqslant 0$ *for* $-1 \leqslant t \leqslant 1/2$, *and* (C2) *the coefficients in the expansion of $f(t)$ in terms of Jacobi polynomials* [*Chap.* 22 *of Abr*1]

$$f(t) = \sum_{i=0}^{k} f_i P_i^{\alpha,\alpha}(t),$$

where $\alpha = (n-3)/2$, *satisfy* $f_0 > 0, f_1 \geqslant 0, \ldots, f_k \geqslant 0$, *then* τ_n *is bounded by* $\tau_n \leqslant f(1)/f_0$.

Although this follows from the general theory developed in Chap. 9 (see especially Eq. (49), (50)), for completeness we sketch a simplified proof. A *spherical code* $\mathcal{C}$ (§2.3 of Chap. 1) is any finite subset of the unit sphere in n dimensions. For $-1 \leqslant t \leqslant 1$ let

$$A_t = \frac{\delta_t}{|\mathcal{C}|} \text{ (number of ordered pairs } c, c' \in \mathcal{C} \text{ for which } (c, c') = t),$$

where δ_t is a Dirac delta-function, $|\mathcal{C}|$ is the cardinality of $\mathcal{C}$, and (,) is the usual inner product. Then

$$\int_{-1}^{1} A_t dt = |\mathscr{C}|.$$

For all $k \geqslant 0$ we have

$$\int_{-1}^{1} A_t P_k^{\alpha,\alpha}(t) dt = \frac{1}{|\mathscr{C}|} \sum_{c,c' \in \mathscr{C}} P_k^{\alpha,\alpha}(\ (c,c')\ ) \geqslant 0,$$

since the kernel $P_k^{\alpha,\alpha}(\ (x,y)\ )$ is positive definite ((10), (11) of Chap. 9).

If there is an arrangement of τ unit spheres $S_1, \ldots, S_\tau$ touching another unit sphere S_0, the points of contact of S_0 with $S_1, \ldots, S_\tau$ form a spherical code $\mathscr{C}$ with $A_t = 0$ for $\frac{1}{2} < t < 1$. It follows that an upper bound on τ_n is given by the optimal solution to the following linear programming problem: choose the $A_t (-1 \leqslant t \leqslant \frac{1}{2})$ so as to maximize $\int A_t dt$ subject to the constraints $A_t \geqslant 0$ for $-1 \leqslant t \leqslant \frac{1}{2}$, and

$$\int_{-1}^{1/2} A_t P_k^{\alpha,\alpha}(t) dt \geqslant -P_k^{\alpha,\alpha}(1), \text{ for } k = 0,1, \ldots .$$

The theorem now follows by passing to the dual problem, and using the fact that any feasible solution to the dual problem is an upper bound to the optimal solution of the original problem.

2. Numerical results

For $n = 8$ we apply the theorem with

$$f(t) = \frac{320}{3}(t+1)(t+\tfrac{1}{2})^2 t^2 (t-\tfrac{1}{2})$$

$$= P_0 + \frac{16}{7} P_1 + \frac{200}{63} P_2 + \frac{832}{231} P_3 + \frac{1216}{429} P_4 + \frac{5120}{3003} P_5 + \frac{2560}{4641} P_6 \quad (1)$$

where P_i stands for $P_i^{2.5,2.5}(t)$, and obtain $\tau_8 \leqslant 240$. For $n = 24$ we take

$$f(t) = \frac{1490944}{15}(t+1)(t+\tfrac{1}{2})^2(t+\tfrac{1}{4})^2 t^2 (t-\tfrac{1}{4})^2(t-\tfrac{1}{2})$$

$$= P_0 + \frac{48}{23} P_1 + \frac{1144}{425} P_2 + \frac{12992}{3825} P_3 + \frac{73888}{22185} P_4$$

$$+ \frac{2169856}{687735} P_5 + \frac{59062016}{25365285} P_6 + \frac{4472832}{2753575} P_7$$

$$+ \frac{23855104}{28956015} P_8 + \frac{7340032}{20376455} P_9 + \frac{7340032}{80848515} P_{10}, \quad (2)$$

where P_i stands for $P_i^{10.5,10.5}(t)$, and obtain $\tau_{24} \leqslant 196560$. Since each sphere in the E_8 lattice packing in 8 dimensions touches 240 others (§8.1 of Chap. 4), and each sphere in the Leech lattice packing in 24 dimensions touches 196560 others (§11 of Chap. 4), we have determined τ_8 and τ_{24}.

For other values of n below 24 we were unable to find such simple and effective polynomials. The results are summarized in Table 1.5 of Chap. 1, which also gives the degree of the polynomial used. The best polynomial we have found for $n = 4$, for example, is the 9th degree polynomial

$f(t) = P_0 + a_1P_1 + a_2P_2 + a_3P_3 + \cdots + a_9P_9$, where $a_1 = 2.412237$, $a_2 = 3.261973$, $a_3 = 3.217960$, $a_4 = 2.040011$, $a_5 = 0.853848$, $a_6 = a_7 = a_8 = 0$, $a_9 = 0.128520$ (shown to 6 decimal places, although we actually used 17 places), and P_i stands for $P_i^{0.5,0.5}(t)$. This implies $\tau_4 \leqslant 25.5585$. This polynomial was found by the following method. First replace (C1) by a finite set of inequalities at the points $t_j = -1 + 0.0015j$ $(0 \leqslant j \leqslant 1000)$. Second, choose a value of k, and use linear programming to find $f_1^*, \ldots, f_k^*$ so as to minimize

$$\sum_{i=1}^{k} f_i^* P_i^{\alpha,\alpha}(1)$$

subject to the constraints

$$f_i^* \geqslant 0 \quad (1 \leqslant i \leqslant k), \quad \sum_{i=1}^{k} f_i^* P_i^{\alpha,\alpha}(t_j) \leqslant -1 \quad (0 \leqslant j \leqslant 1000).$$

Let $f^*(t)$ denote the polynomial $1 + \sum_{i=1}^{k} f_i^* P_i^{\alpha,\alpha}(t)$. Of course this need not satisfy (C1) for *all* points t in the interval $[-1, \frac{1}{2}]$. Let ϵ be chosen to be greater than the maximal value of $f^*(t)$ on $[-1, \frac{1}{2}]$ (ϵ may be calculated by finding the zeros of the derivative of $f^*(t)$). Then $f(t) = f^*(t) - \epsilon$ satisfies (C1) and (C2), and so

$$\tau_n \leqslant \frac{f^*(1) - \epsilon}{1 - \epsilon}.$$

All the upper bounds shown in Table 1.5, except those for $n = 3$ and 17, were obtained in this way. The degree k was allowed to be as large as 30, but in all the cases considered the degree of the best polynomial (given in the third column of the table) did not exceed 14. For $n = 8$ and $n = 24$ the form of the polynomials obtained in this way led us to (1) and (2), but for the other values of n no such simple expression suggested itself.

For $n = 17$ we made use of the additional inequalities

$$\int_{-1}^{-\sqrt{3}/2} A_t dt \leqslant 1 \quad \text{and} \quad \int_{-1}^{-\sqrt{2/3}} A_t dt \leqslant 2$$

to obtain $\tau_{17} \leqslant 12215$. Other inequalities of this type could probably be used to obtain further improvements of these results. Unfortunately, for $n = 3$ our methods only gave $\tau_3 \leqslant 13$, whereas the actual value is 12 (§2.1 of Chap. 1).

For comparison the values of Coxeter's bounds on τ_n ([Cox16], [Bör2]) for $n = 4, 5, 6, 7$ and 8 as given in [Lee10] are 26, 48, 85, 146 and 244 respectively. The lower bounds in Table 1.5 are obtained from the packings in Table 1.2.

14

Uniqueness of Certain Spherical Codes

E. Bannai and N. J. A. Sloane

We show that there is essentially only one way of arranging 240 (resp. 196560) nonoverlapping unit spheres in $\mathbf{R}^8$ (resp. $\mathbf{R}^{24}$) so that they all touch another unit sphere Ω_n, and only one way of arranging 56 (resp. 4600) spheres in $\mathbf{R}^8$ (resp. $\mathbf{R}^{24}$) so that they all touch two further, touching spheres. The following tight spherical t-designs are also unique: the 5-design in Ω_7, the 7-designs in Ω_8 and Ω_{23}, and the 11-design in Ω_{24}.

1. Introduction

It was shown in the previous chapter that the maximal number of nonoverlapping unit spheres in $\mathbf{R}^8$ (resp. $\mathbf{R}^{24}$) that can touch another unit sphere is 240 (resp. 196560). Arrangements of spheres meeting these bounds can be obtained from the E_8 and Leech lattices, respectively. We now prove that these are the only arrangements meeting these bounds. In [Ban7], [Ban8] it was shown that there are no tight spherical t-designs for $t \geqslant 8$ except for the tight 11-design in Ω_{24} (cf. §3.2 of Chap. 3). The present chapter also shows that this and three other tight t-designs are unique. (Other results on the uniqueness of these lattices and their associated codes and groups may be found in [Con4] ≅ Chap. 12, [Cur3], [Del13], [Jon5], [Kne4], [Lün1], [Mac6], [Nie2], [Ple8], [Ple17], [Sta12], [Wit3].)

Our notation is that Ω_n denotes the unit sphere in $\mathbf{R}^n$ and (,) is the usual inner product. An (n, M, s) *spherical code* (§2.3 of Chap. 1) is a subset $\mathscr{C}$ of Ω_n of size M for which $(u, v) \leqslant s$ for all $u, v \in \mathscr{C}, u \neq v$.

Examples of spherical codes may be obtained from sphere packings via the following theorem, whose proof is trivial.

Theorem 1. *In a packing of unit spheres in* $\mathbf{R}^n$ *let* $S_1, \ldots, S_k$ *be a set of spheres such that* S_i *touches* S_j *for all* $i \neq j$. *Suppose there are further* *spheres* $T_1, \ldots, T_M$ *each of which touches all the* S_i. *Then, after rescaling, the centers of* $T_1, \ldots, T_M$ *form an* $(n - k + 1, M, 1/(k + 1))$ *spherical code.*

Example 2. In the E_8 lattice packing in $\mathbf{R}^8$ (§8.1 of Chap. 4) there are 240 spheres touching each sphere, 56 that touch each pair of touching spheres, 27 that touch each triple of mutually touching spheres, and so on. From Theorem 1 the centers of these spheres give rise to (8, 240, 1/2), (7, 56, 1/3), (6, 27, 1/4), (5, 16, 1/5), (4, 10, 1/6) and (3, 6, 1/7) spherical codes.

Example 3. Similarly the Leech lattice in $\mathbf{R}^{24}$ (§11 of Chap. 4) gives rise to (24, 196560, 1/2), (23, 4600, 1/3), (22, 891, 1/4), (21, 336, 1/5), (20, 170, 1/6), ... spherical codes.

If $\mathscr{C}$ is an (n, M, s) spherical code and $u \in \mathscr{C}$, the *distance distribution of* C *with respect to* u is the system of numbers $\{A_t(u): -1 \leqslant t \leqslant 1\}$, where

$$A_t(u) = |\{v \in \mathscr{C} : (u, v) = t\}|,$$

and the *distance distribution of* $\mathscr{C}$ is the system of numbers $\{A_t: -1 \leqslant t \leqslant 1\}$, where

$$A_t = \frac{1}{M} \sum_{u \in \mathscr{C}} A_t(u).$$

Then (see Chap. 13) the A_t satisfy $A_1 = 1$, $A_t = 0$ for $s < t < 1$,

$$\sum_{-1 \leqslant t \leqslant s} A_t = M - 1,$$

$$\sum_{-1 \leqslant t \leqslant s} A_t P_k(t) \geqslant -P_k(1), \text{ for } k = 1,2,3, \ldots,$$

where $P_k(x) = P_k^{(n-3)/2,(n-3)/2}(x)$ is a Jacobi polynomial [Chap. 22 of Abr1]. For a specified value of s an upper bound to M is therefore given by the following linear programming problem.

(P1) Choose $\{A_t: -1 \leqslant t \leqslant s\}$ so as to maximize

$$\sum_{-1 \leqslant t \leqslant s} A_t$$

subject to the inequalities

$$A_t \geqslant 0,$$

$$\sum_{-1 \leqslant t \leqslant s} A_t P_k(t) \geqslant -P_k(1), \text{ for } k = 1,2,3,\ldots. \tag{1}$$

Then, as we have seen in Eq. (49), (50) of Chap. 9 and Theorem 1 of the previous chapter, the dual problem can be stated as follows.

(P2) Choose an integer N and a polynomial $f(t)$ of degree N, say

$$f(t) = \sum_{k=0}^{N} f_k P_k(t),$$

so as to minimize $f(1)/f_0$ subject to the inequalities

$$f_0 > 0, f_k \geqslant 0 \text{ for } k = 1,2,\ldots,N, \tag{2}$$

$$f(t) \leqslant 0 \text{ for } -1 \leqslant t \leqslant s. \tag{3}$$

Since any feasible solution to the dual problem is an upper bound to the optimal solution of the primal problem, we have

$$M \leqslant f(1)/f_0 \tag{4}$$

for any polynomial $f(t)$ satisfying (2) and (3).

2. Uniqueness of the code of size 240 in Ω_8

Theorem 4 (see Chap. 13). *If $\mathscr{C}$ is an $(8, M, \frac{1}{2})$ code then $M \leqslant 240$.*

Proof. Use the polynomial

$$f(t) = \frac{320}{3}(t+1)(t+\tfrac{1}{2})^2 t^2 (t-\tfrac{1}{2}). \tag{5}$$

Theorem 5. *If (a) $\mathscr{C}$ is an $(8, 240, \frac{1}{2})$ code then (b) $\mathscr{C}$ is a tight spherical 7-design in Ω_8, (c) $\mathscr{C}$ carries a 4-class association scheme,*[(1)] *(d) the intersection numbers of this association scheme are uniquely determined, and (e) the distance distribution of $\mathscr{C}$ with respect to any $u \in \mathscr{C}$ is given by*

$$A_1(u) = A_{-1}(u) = 1,\ A_{1/2}(u) = A_{-1/2}(u) = 56,\ A_0(u) = 126. \tag{6}$$

Proof. Let $\{A_t\}$ be the distance distribution of $\mathscr{C}$. Then $\{A_t\}$ is an optimal solution to the primal problem (P1), and the polynomial $f(t)$ in (5) is an optimal solution to the dual problem (P2). The dual variables $f_1, \ldots, f_6$ are nonzero, so by the theorem of complementary slackness [Sim0] the primal constraints (1) must hold with equality for $k = 1, \ldots, 6$.

The dual constraints (3) do not hold with equality except for $t = -1, \pm \frac{1}{2}$ and 0. Therefore the primal variables must vanish everywhere except perhaps for $A_{-1}, A_{\pm 1/2}$ and A_0. From (1) these numbers satisfy the equations

$$A_{-1}P_k(-1) + A_{-1/2}P_k(-\tfrac{1}{2}) + A_0 P_k(0) + A_{1/2}P_k(\tfrac{1}{2}) = -P_k(1), \tag{7}$$

for $k = 1,2,\ldots,6$. Thus

---

(1) Cf. [Ban11], [Del9], [Goe4], [Hig2], [Sei3], [Slo5].

$$\begin{bmatrix} 1 & 1 & 1 & 1 \\ -\frac{7}{2} & -\frac{7}{4} & 0 & \frac{7}{4} \\ \frac{63}{8} & \frac{9}{8} & -\frac{9}{8} & \frac{9}{8} \\ -\frac{231}{16} & \frac{33}{64} & 0 & -\frac{33}{64} \\ \frac{3003}{128} & -\frac{429}{256} & \frac{143}{128} & -\frac{429}{256} \\ -\frac{9009}{256} & \frac{1287}{1024} & 0 & -\frac{1287}{1024} \\ \frac{51051}{1024} & \frac{663}{2048} & -\frac{1105}{1024} & \frac{663}{2048} \end{bmatrix} \begin{bmatrix} A_{-1} \\ A_{-1/2} \\ A_0 \\ A_{1/2} \end{bmatrix} = \begin{bmatrix} 239 \\ -\frac{7}{2} \\ -\frac{63}{8} \\ -\frac{231}{16} \\ -\frac{3003}{128} \\ -\frac{9009}{256} \\ -\frac{51051}{1024} \end{bmatrix} \tag{8}$$

The unique solution is

$$A_{-1} = 1,\ A_{-1/2} = A_{1/2} = 56,\ A_0 = 126. \tag{9}$$

Since $A_{-1}(u) \leqslant 1$ and $A_{-1} = 1$, we have $A_{-1}(u) = 1$ for all $u \in \mathscr{C}$, and so the code is antipodal [Del16, p. 373]. Therefore (7) also holds for $k = 7$ and by [Del16, Theorem 5.5] $\mathscr{C}$ is a spherical 7-design. By [Del16, Definition 5.13] the design is tight, since $|\mathscr{C}| = 2\binom{10}{3}$. By [Del16, Theorem 7.5] $\mathscr{C}$ carries a 4-class association scheme. Therefore $A_t(u) = A_t$ is independent of u for all t. This proves (b), (c) and (e). The numbers (9) are the valencies of the association scheme, and by [Del16, Theorem 7.4] determine all the intersection numbers. This proves (d).

Theorem 6. *If condition (b) of Theorem 5 holds then so do (a), (c), (d) and (e).*

Proof. By definition $|\mathscr{C}| = 2\binom{10}{3}$. From [Del16, Theorem 5.12] the inner products between the members of $\mathscr{C}$ are ± 1 and the zeros of

$$C_3(x) = 160(x + \tfrac{1}{2})x(x - \tfrac{1}{2}).$$

Thus all the A_t are zero except perhaps for $A_{\pm 1}, A_{\pm 1/2}$ and A_0. From [Del16, Theorem 5.5] Eq. (7) holds for $k = 1,2, \ldots, 7$. The rest of the proof is the same as for Theorem 5.

In Example 2 we saw that the minimal vectors in the E_8 lattice form an (8, 240, ½) code. Thus conditions (a)-(e) of Theorem 5 apply to this code. Conversely we have:

Theorem 7. *If $\mathscr{C}$ is a tight spherical 7-design in Ω_8 there is an orthogonal transformation mapping $\mathscr{C}$ onto the minimal vectors of the E_8 lattice.*

Proof. From Theorem 6 the possible inner products in $\mathscr{C}$ are 0, $\pm 1/2$, ± 1. Let $\mathscr{C} = \{u_1, \ldots, u_{240}\}$ and let L be the lattice in $\mathbf{R}^8$ consisting of the vectors

$$\sum_{i=1}^{240} \sqrt{2}\, a_i u_i, \quad a_i \in \mathbf{Z}.$$

Then L is an eight-dimensional even integral lattice generated by vectors of norm 2, and is therefore (see §3 of Chap. 4) a direct sum of various lattices $A_n (n \geqslant 1)$, $D_n (n \geqslant 4)$ and $E_n (n = 6, 7, 8)$. The only lattice of this type with at least 240 minimal vectors is E_8, so L is isometric to E_8 and $\mathscr{C}$ is isometric to the set of minimal vectors in E_8.

By combining Theorems 5 and 7 we obtain:

Theorem 8. *There is a unique way (up to isometry) to arrange* 240 *nonoverlapping unit spheres in* $\mathbf{R}^8$ *so that they all touch another unit sphere.*

3. Uniqueness of the code of size 56 in Ω_7

Theorem 9. *If* $\mathscr{C}$ *is a* $(7, M, 1/3)$ *code then* $M \leqslant 56$.

Proof. From (4), using the polynomial

$$f(t) = (t+1)(t+1/3)^2(t-1/3).$$

Theorem 10. *If (a)* $\mathscr{C}$ *is a* $(7, 56, 1/3)$ *code then (b)* $\mathscr{C}$ *is a tight spherical* 5*-design in* Ω_7, *(c)* $\mathscr{C}$ *carries a* 3*-class association scheme, (d) the intersection numbers of this association scheme are uniquely determined, and (e) the distance distribution of* $\mathscr{C}$ *with respect to any* $u \in \mathscr{C}$ *is given by*

$$\begin{aligned} A_1(u) &= A_{-1}(u) = 1, \\ A_{1/3}(u) &= A_{-1/3}(u) = 27. \end{aligned} \tag{10}$$

Conversely (b) implies (a), (c), (d) and (e).

Proof. The proof is parallel to the proofs of Theorems 5 and 6.

For example the $(7, 56, 1/3)$ code given in Example 2 has properties (a)-(e). Conversely we have:

Theorem 11. *If* $\mathscr{C}$ *is a tight spherical* 5*-design in* Ω_7 *there is an orthogonal transformation mapping* $\mathscr{C}$ *onto the* $(7, 56, 1/3)$ *code obtained from the* E_8 *lattice.*

Proof. Let $\mathscr{C}$ consist of the points $u_1, \ldots, u_{56}$ lying on a unit sphere in $\mathbf{R}^7$ centered at P. Choose a point 0 (in $\mathbf{R}^8$) so that $\measuredangle\, u_i 0 P = \pi/3$ for all i, and thus

$$\cos \measuredangle\, u_i 0 u_j = (1 + 3\cos \measuredangle\, u_i P u_j)/4$$

for all i,j. Let v be a unit vector along $0P$ (see Fig. 14.1). From

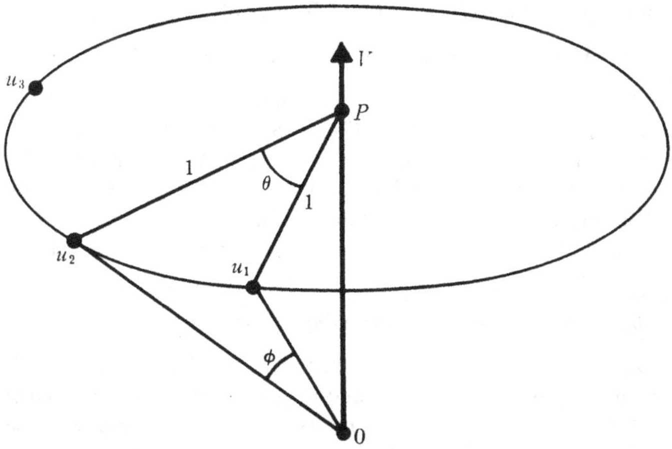

Figure 14.1. The construction used in the proof of Theorem 11: $\measuredangle\, u_i 0 P = \pi/3$ for all i, $|0P| = 1/\sqrt{3}$, $|0u_1| = |0u_2| = 2/\sqrt{3}$, and $\cos\phi = (1 + 3\cos\theta)/4$.

Theorem 10 $\cos \measuredangle\, u_i P u_j$ takes the values ± 1 and $\pm 1/3$, so $\cos \measuredangle\, u_i 0 u_j$ takes the values 0, $\pm \frac{1}{2}$ and 1. It follows that the vectors $\sqrt{3/2}\, 0u_i$ $(1 \leqslant i \leqslant 56)$ span an even integral lattice, containing at least $2(56 + 1) = 114$ minimal vectors (corresponding to $\pm \mathscr{C}, \pm v$). This lattice must therefore be either E_8 or $E_7 \oplus A_1$, and the latter is incompatible with (10).

By combining Theorems 10 and 11 we obtain:

Theorem 12. *There is a unique way (up to isometry) to arrange 56 nonoverlapping unit spheres in* $\mathbf{R}^8$ *so that they all touch two further, touching, unit spheres.*

4. Uniqueness of the code of size 196560 in Ω_{24}

Theorem 13 (see Chap. 13). *If $\mathscr{C}$ is a* (24, *M*, ½) *code then* $M \leqslant 196560$.

Proof. From (4), using the polynomial

$$f(t) = (t + 1)(t + \tfrac{1}{2})^2 (t + \tfrac{1}{4})^2 t^2 (t - \tfrac{1}{4})^2 (t - \tfrac{1}{2}).$$

Theorem 14. *If (a) $\mathscr{C}$ is a* (24, 196560, ½) *code then (b) $\mathscr{C}$ is a tight spherical* 11*-design in* Ω_{24}, *(c) $\mathscr{C}$ carries a* 6*-class association scheme, (d) the intersection numbers of this association scheme are uniquely determined, and (e) the distance distribution of $\mathscr{C}$ with respect to any $u \in \mathscr{C}$ is given by*

$$A_1(u) = A_{-1}(u) = 1,\ A_{1/2}(u) = A_{-1/2}(u) = 4600,$$

$$A_{1/4}(u) = A_{-1/4}(u) = 47104,\ A_0(u) = 93150. \qquad (11)$$

Conversely (b) implies (a), (c), (d) and (e).

Proof. The proof is parallel to those of Theorems 5 and 6.

In Example 3 we saw that the minimal vectors in the Leech lattice Λ when suitably scaled form a (24, 196560, ½) code. We shall require an explicit description of this code, and take Λ to consist of the vectors in (135), (136) of Chap. 4. Let Λ_4 denote the set of 196560 minimal vectors (of norm 4) — see §11 of Chap. 4. Then $\frac{1}{2}\Lambda_4$ is a (24, 196560, ½) code to which conditions (a)-(e) of Theorem 14 apply. Conversely we have:

Theorem 15. *If $\mathcal{C}$ is a tight spherical 11-design in Ω_{24} there is an orthogonal transformation mapping $\mathcal{C}$ onto $\frac{1}{2}\Lambda_4$.*

Proof. From Theorem 14 the distance distribution of $\mathcal{C}$ with respect to any $u \in \mathcal{C}$ is given by (11), and in particular the inner products in $\mathcal{C}$ are $0, \pm\frac{1}{4}, \pm\frac{1}{2}, \pm 1$. Let $\mathcal{C} = \{u_1, u_2, \ldots, u_{196560}\}$, and let L be the lattice in $\mathbf{R}^{24}$ consisting of the vectors

$$\sum_{i=1}^{196560} 2a_i u_i, \quad a_i \in \mathbf{Z}. \tag{12}$$

Then

$$(2u_i, 2u_j) \in \{0, \pm 1, \pm 2, \pm 4\} \tag{13}$$

and L is a 24-dimensional even integral lattice. We shall establish Theorem 15 by showing that there is an orthogonal transformation mapping L onto Λ and $\mathcal{C}$ onto $\frac{1}{2}\Lambda_4$.

Lemma 16. *The minimal norm (v,v) for $v \in L$, $v \neq 0$, is 4.*

Proof. The minimal norm is even, so suppose it is 2, with $(v,v) = 2$, $v \in L$. For $u \in 2\mathcal{C}$ we have

$$|(u,v)| = |u|\cdot|v|\cdot|\cos \measuredangle\, u0v| \leqslant 2\sqrt{2},$$

so $(u,v) \in \{0, \pm 1, \pm 2\}$ since L is integral. Suppose $(u,v) = 0$ for α choices of u, $(u,v) = 1$ for β choices, and $(u,v) = 2$ for γ choices, with $\alpha + 2\beta + 2\gamma = 196560$. Without loss of generality we may assume $v = (\sqrt{2}, 0, 0, \ldots, 0)$.

Since $\mathcal{C}$ is an 11-design,

$$\frac{1}{196560} \sum_{i=1}^{196560} f(u_i) = \frac{1}{\omega_{24}} \int_{\Omega_{24}} f(\xi)\, d\omega(\xi) \tag{14}$$

holds for any homogeneous polynomial $f(\xi_1, \xi_2, \ldots, \xi_{24})$ of total degree $\leqslant 11$, where ω_{24} is the surface area of Ω_{24} (Eq. (81) of Chap. 3). Let us choose $f = f_k = \xi_1^k$, for $k = 2$ and 4, so that

$$f_k(u_i) = 2^{-k/2}((u_i, v))^k.$$

The right hand side of (14) can be evaluated from

$$\frac{1}{\omega_{24}} \int_{\Omega_{24}} f_k(\xi) d\omega(\xi) = \frac{1}{196560} \sum_{u \in (1/2)\Lambda_4} f_k(u)$$

$$= \frac{8190}{196560} \text{ if } k = 2, \text{ or } \frac{945}{196560} \text{ if } k = 4 .$$

The equations (14) now read

$$2\beta \cdot \frac{1^2}{8} + 2\gamma \cdot \frac{2^2}{8} = 8190 , \quad 2\beta \cdot \frac{1^4}{64} + 2\gamma \cdot \frac{2^4}{64} = 945,$$

which imply $\beta = 33600$, $\gamma = -210$, an impossibility.

Lemma 17. *The set L_4 of vectors of norm 4 in L coincides with $2\mathscr{C}$.*

Proof. By construction L_4 contains $2\mathscr{C}$. Conversely take $u,v \in L_4$. Then $(u,v) \neq 3$, or else

$$(u - v, u - v) = (u, u) - 2(u, v) + (v, v) = 2,$$

contradicting Lemma 16. Similarly $(u,v) \neq -3$. Therefore $(u,v) \in \{0, \pm 1, \pm 2, \pm 4\}$ and $\measuredangle\, u0v \geqslant \pi/3$ for $u \neq v$. From Theorem 13, $|L_4| \leqslant 196560 = |2\mathscr{C}|$. Therefore $L_4 = 2\mathscr{C}$.

For $n \geqslant 3$, the vectors

$$g_1 = \sqrt{2}(e_1 + e_2), \; g_2 = \sqrt{2}(e_1 - e_2),$$
$$g_3 = \sqrt{2}(e_2 - e_3), \ldots, g_n = \sqrt{2}(e_{n-1} - e_n), \tag{15}$$

with respect to an orthonormal basis $\{e_1, \ldots, e_n\}$ for $\mathbf{R}^n$ span a lattice D_n (§7.1 of Chap. 4). There are $2n(n-1)$ minimal vectors $((\pm\sqrt{2})^2 0^{n-2})$ in D_n. These lattices are nested: $D_3 \subseteq D_4 \subseteq \cdots$. (It must be understood that all lattices D_n in this chapter are scaled to have minimal norm 4.)

Lemma 18. (i) *For any pair of vectors u,v in Λ_4 with $\measuredangle\, u0v = \pi/2$ there are 44 vectors w in Λ_4 with $\measuredangle\, u0w = \measuredangle\, v0w = \pi/3$.* (ii) *The same statement holds with Λ_4 replaced by $L_4 = 2\mathscr{C}$.* (iii) *There are $2n-4$ minimal vectors w in D_n such that $\measuredangle\, g_1 0w = \measuredangle\, g_2 0w = \pi/3$.*

Proof. (i) and (iii) are straightforward, and (ii) follows from (i) since Λ_4 and $2\mathscr{C}$ are association schemes with the same parameters (Theorem 14).

Lemma 19. *L contains a sublattice D_3.*

Proof. For the generators g_1, g_2, g_3 of D_3 we can take any triple $u, v, w \in L_4$ with $\measuredangle\, u0v = \pi/2$, $\measuredangle\, u0w = \measuredangle\, v0w = \pi/3$. Such a triple exists by Lemma 18(ii).

Lemma 20. *L contains a sublattice D_n, for $n = 3,4, \ldots, 24$.*

Proof. We proceed by induction on n. Suppose the assertion holds for $n \geqslant 3$. By choosing a suitable orthonormal basis $e_1, \ldots, e_n$, we see that L_4 contains vectors $g_1, \ldots, g_n$ given by (15) and spanning D_n. By Lemma 18 (ii) there are 44 vectors w in L_4 with

$\measuredangle g_1 0 w = \measuredangle g_2 0 w = \pi/3$. By Lemma 18 (iii) at least one of these is not a minimal vector of D_n. Then this vector w is not in $\mathbf{R}D_n$. (For suppose $w = c_1 e_1 + \cdots + c_n e_n$. Since $\measuredangle g_1 0 w = \measuredangle g_2 0 w = \pi/3$, $c_1 = \sqrt{2}$ and $c_2 = 0$. For $3 \leqslant i \leqslant n$,

$$\sqrt{2}(e_1 \pm e_i) \in L_4 \cap D_n \subseteq 2\mathscr{C},$$

and therefore

$$(w, \sqrt{2}(e_1 \pm e_i)) \in \{0, \pm 1, \pm 2\}$$

from (13). This implies $c_3 = c_4 = \cdots = c_n = 0$, and contradicts $(w,w) = 4$.) Choose e_{n+1} so that $\{e_1, \ldots, e_{n+1}\}$ is an orthonormal basis for $\mathbf{R}\langle D_n, w\rangle$, and suppose

$$w = c_1 e_1 + \cdots + c_n e_n + c_{n+1} e_{n+1}.$$

The above argument shows that $c_1 = \sqrt{2}, c_2 = \cdots = c_n = 0$, and $c_{n+1} = \pm\sqrt{2}$. Therefore $\langle D_n, w\rangle = D_{n+1} \subseteq L$.

Lemma 21. *L is isometric to Λ.*

Proof. From Lemma 20 we may choose an orthonormal basis $e_1, \ldots, e_{24}$ so that $2\mathscr{C}$ contains the vectors $(\pm\sqrt{2})^2 0^{22}$. Let $u = (u_1, \ldots, u_{24})/\sqrt{8}$ be any vector in $2\mathscr{C}$. From (13) the inner products of u with the vectors $(\pm\sqrt{2})^2 0^{22}$ are $0, \pm 1, \pm 2, \pm 4$. By considering the inner products with $(\sqrt{2}, \pm\sqrt{2}, 0, \ldots, 0)$ we obtain

$$u_1^2 + u_2^2 + \cdots + u_{24}^2 = 32,$$

$$\tfrac{1}{2}(u_1 \pm u_2) \in \{0, \pm 1, \pm 2, \pm 4\},$$

$$u_1, u_2, \ldots \in \{0, \pm 1, \pm 2, \pm 3, \pm 4, \pm 5\}.$$

Suppose $u_1 = \pm 5$. Then another u_i, say u_2, is zero. The inner product of $\mathbf{u}$ with $(\sqrt{2}, \sqrt{2}, 0, \ldots, 0)$ is $5/2$, a contradiction. Proceeding in this way it is not difficult to show that the only possibilities for the components of u are

$$((\pm 2)^8 0^{16})/\sqrt{8},\ ((\pm 4)^2 0^{22})/\sqrt{8}, \text{ and } ((\pm 1)^{23}(\pm 3)^1)/\sqrt{8}.$$

In particular $u_1, \ldots, u_{24}$ are integers with the same parity.

It remains to show that these vectors are the same as those in Λ_4. To see this we define a binary linear code $\mathscr{D}$ of length 24 by taking as codewords all binary vectors c for which there is a vector $u \in L$ with

$$u = (0 + 2c + 4x)/\sqrt{8}$$

for some $x \in \mathbf{Z}^{24}$. Then as in Theorem 3 of Chap. 12 it follows that $\text{wt}(c) \geqslant 8$ for $c \neq 0$, and that there are at most 759 codewords of weight 8. Therefore $|\mathscr{D}| \leqslant A(24,8) = 2^{12}$, from Table 9.1. The argument following Theorem 3 of Chap. 12 now shows that the only way that $2\mathscr{D}$ can contain

196560 vectors u is for these vectors to coincide with the minimal vectors in Λ_4.

This completes the proof of Theorem 15. By combining Theorems 14 and 15 we obtain:

Theorem 22. *There is a unique way (up to isometry) to arrange* 196560 *nonoverlapping unit spheres in* $\mathbf{R}^{24}$ *so that they all touch another unit sphere.*

Remark. Since the Leech lattice has two distinct mirror image forms, the isometry of Theorem 22 might necessarily involve a reflection.

5. Uniqueness of the code of size 4600 in Ω_{23}

Theorem 23. *If* $\mathscr{C}$ *is a* (23, M, 1/3) *code then* $M \leqslant 4600$.

Proof. From (4), using the polynomial

$$f(t) = (t+1)(t+1/3)^2t^2(t-1/3)\ .$$

Theorem 24. *If (a)* $\mathscr{C}$ *is a* (23, 4600, 1/3) *code then (b)* $\mathscr{C}$ *is a tight spherical* 7*-design in* Ω_{23}, *(c)* $\mathscr{C}$ *carries a* 4*-class association scheme, (d) the intersection numbers of this association scheme are uniquely determined, and (e) the distance distribution of* $\mathscr{C}$ *with respect to any* $u \in \mathscr{C}$ *is given by*

$$A_1(u) = A_{-1}(u) = 1,\ A_{1/3}(u) = A_{-1/3}(u) = 891,\ A_0(u) = 2816.$$

Conversely (b) implies (a), (c), (d) and (e).

For example the (23, 4600, 1/3) code given in Example 3 has properties (a)-(e). Conversely we have:

Theorem 25. *If* $\mathscr{C}$ *is a tight spherical* 7*-design in* Ω_{23} *there is an orthogonal transformation mapping* $\mathscr{C}$ *onto the* (23, 4600, 1/3) *code obtained from the Leech lattice.*

Proof. As in the proof of Theorem 11 we embed $\mathscr{C} = \{u_1, \ldots, u_{4600}\}$ in $\mathbf{R}^{24}$, choosing 0 so that $\measuredangle\, u_i 0P = \pi/3$ for all i (cf. Fig. 14.1). Then

$$\cos \measuredangle\, u_i 0 u_j \in \{-\tfrac{1}{2}, 0, \tfrac{1}{4}, \tfrac{1}{2}, 1\}.$$

Let L be the even integral lattice in $\mathbf{R}^{24}$ spanned by the vectors $\sqrt{3}\, 0u_i$. For convenience we set $U_i = \sqrt{3} 0u_i$.

Lemma 26. *The minimal norm* (v,v) *for* $v \in L$, $v \neq 0$, *is* 4.

Proof. Suppose $v \in L$ with $(v,v) = 2$, and write $v = v' + v''$ with $v' \parallel 0P$, $v'' \perp 0P$, $|v'| = y$, $|v''| = (2-y^2)^{1/2}$, and $U_i = U_i' + U_i''$ with $U_i' \parallel 0P$, $U_i'' \perp 0P$, $|U_i'| = 1$, $|U_i''| = 3^{1/2}$. Then

$$(U_i, v) = (U'_i, v') + (U''_i, v'') \in \{0, \pm 1, \pm 2\},$$

$$\cos \measuredangle U''_i 0 v'' \in \frac{\{0, \pm 1, \pm 2\} - y}{\sqrt{3}\sqrt{2 - y^2}}.$$

Since $\mathscr{C}$ is a tight 7-design, the set $\{\cos \measuredangle U''_i 0 v'' \colon 1 \leqslant i \leqslant 4600\}$ is symmetric about 0. Therefore $y \in \{0, \pm 1/2, \pm 1\}$. First suppose $y = 0$. Then

$$\cos \measuredangle U''_i 0 v'' \in \left\{-\frac{2}{\sqrt{6}}, -\frac{1}{\sqrt{6}}, 0, \frac{1}{\sqrt{6}}, \frac{2}{\sqrt{6}}\right\}.$$

Let these values occur $\gamma, \beta, \alpha, \beta, \gamma$ times respectively. Then by evaluating the 0th, 2nd and 4th moments of $\mathscr{C}$ with respect to v'', as in the proof of Lemma 16, we obtain the equations $\alpha + 2\beta + 2\gamma = 4600$, $\beta/3 + 4\gamma/3 = 200$, $\beta/8 + 8\gamma/9 = 24$, which imply $\gamma = -14$, an impossibility. Similarly for the other values of y.

Lemma 27. *L contains a sublattice isometric to* D_n, *for* $n = 3, 4, \ldots, 24$.

Proof. This is similar to the proof of Lemma 20, starting from the fact that if we take $u_1, u_2 \in \mathscr{C}$ with $\measuredangle u_1 0 u_2 = \pi/2$, there are 42 vectors $u_i \in \mathscr{C}$ with $\measuredangle u_1 0 u_i = \measuredangle u_2 0 u_i = \pi/3$. Furthermore the vector $v = 20P \in L$ also satisfies $\measuredangle u_1 0 v = \measuredangle u_2 0 v = \pi/3$.

Lemma 28. *L is isometric to* Λ, *and* $\mathscr{C}$ *is isometric to the* (23, 4600, 1/3) *code obtained from the Leech lattice.*

Proof. Let L_4 denote the set of minimal vectors in L. From Lemma 27 we may assume that L_4 contains all the vectors $((\pm 4^2 0^{22}))/\sqrt{8}$, and that $v = 2{\cdot}0P$ is $(440\ldots0)/\sqrt{8}$. As in Lemma 21 it follows that the vectors in L_4 have the form $((\pm 2)^8 0^{16})/\sqrt{8}$, $((\pm 4^2 0^{22})/\sqrt{8}$ and $((\pm 1)^{23}(\pm 3)^1)/\sqrt{8}$. Furthermore the vectors U_i begin $(2,2\ldots)/\sqrt{8}$, $(4,0\ldots)/\sqrt{8}$, $(0,4\ldots)/\sqrt{8}$, $(3,1\ldots)/\sqrt{8}$, or $(1,3\ldots)/\sqrt{8}$. The code $\mathscr{D}$ is defined as in Lemma 21: it is a linear code of minimum distance 8 containing at most 2^{12} codewords. The zero codeword corresponds to the vectors U_i that begin $(4,0\ldots)/\sqrt{8}$ or $(0,4\ldots)/\sqrt{8}$, and there are at most $2{\cdot}2{\cdot}22$ of these. The codewords of weight 8 that begin 11... correspond to the vectors U_i that begin $(2,2\ldots)/\sqrt{8}$. The number of such codewords is at most 77, since $A(22,8,6) = 77$ [Mac6, Fig. 3, p. 688], and there are at most $2^5{\cdot}77$ corresponding U_i. The remaining U_i come from codewords beginning 10... or 01..., and there are at most $2{\cdot}2^{10}$ of them (since $A(22,8) \leqslant 2^{10}$ by Table 9.1). But $2{\cdot}2{\cdot}22 + 2^5{\cdot}77 + 2{\cdot}2^{10} = 4600$, so all the inequalities in the argument must be exact. In particular the codewords of weight 8 that begin 11... must form the unique Steiner system S(3, 6, 22) ([Lün1], [Mac6, p. 645], [Wit3]), and hence L must be the Leech lattice.

This completes the proof of Theorem 25. By combining Theorem 24 and 25 we obtain:

Theorem 29. *There is a unique way (up to isometry) to arrange* 4600 *unit spheres in* $\mathbf{R}^{24}$ *so that they all touch two further, touching, unit spheres.*

(Once again, the isometry might necessarily involve a reflection.)

Acknowledgements. We should like to acknowledge helpful conversations with C. L. Mallows, A. M. Odlyzko and J. G. Thompson.

15

On the Classification of Integral Quadratic Forms

J. H. Conway and N. J. A. Sloane

This chapter gives an account of the classification of integral quadratic forms. It is particularly designed for readers who wish to be able to do explicit calculations. Novel features include an elementary system of rational invariants (defined without using the Hilbert norm residue symbol), an improved notation for the genus of a form, an efficient way to compute the number of spinor genera in a genus, and some conditions which imply that there is only one class in a genus. We give tables of the binary forms with $-100 \leqslant \det \leqslant 50$, the indecomposable ternary forms with $|\det| \leqslant 50$, the genera of forms with $|\det| \leqslant 11$, the genera of p-elementary forms for all p, and the positive definite forms with determinant 2 up to dimension 18 and determinant 3 up to dimension 17.

1. Introduction

The project of classifying integral quadratic forms has a long history, to which many mathematicians have contributed. The binary (or two-dimensional) forms were comprehensively discussed by Gauss. Gauss and later workers also made substantial inroads into the problem of ternary and higher-dimensional forms. The greatest advances since then have been the beautiful development of the theory of rational quadratic forms (Minkowski, Hasse, Witt), and Eichler's complete classification of indefinite forms in dimension 3 or higher in terms of the notion of spinor genus.

Definite forms correspond to lattices in Euclidean space. For small dimensions they can be classified using Minkowski's generalization of Gauss's notion of reduced form, but this method rapidly becomes impracticable when the dimension reaches 6 or 7. However there is a geometric method used by Witt and Kneser which (after the work of Niemeier) is effective roughly until the sum of the dimension and the (determinant)$^{1/2}$ exceeds 24, beyond which point it seems that the forms are inherently unclassifiable. The situation is summarized in Fig. 15.1.

There are several novel features of this chapter.

(1) We present (in §5) an elementary system of rational invariants for quadratic forms, defined without using the Hilbert norm residue symbol, and whose values are certain integers modulo 8. The modulo-8 version of the 2-adic invariant seems to have first arisen in topological investigations (see [Cas1], [Hir4]). Practitioners in the subject know that the effect of the "product formula" is to yield congruence conditions modulo 8 on the signature. With the invariants we use these congruences emerge immediately rather than at the end of a long calculation (see Eqs. (15) and (16)).

(2) We also give a simply described system of p-adic invariants for integral forms that yields a handy notation for the genus of a quadratic form (§7).

(3) In terms of this new notation we enumerate (in §8.1 and Table 15.4) all genera of quadratic forms having determinant of magnitude less than 12, and

(4) classify the genera of p-elementary forms for all p (§8.2).

In view of Eichler's theory of spinor genera (see Theorem 14), these results actually give the integral equivalence classes in the case of indefinite forms of dimension $\geqslant 3$.

(5) We also give a simple description of the spinor genus, including a notation for the various spinor genera in the genus of a given form, and an easy computational way of finding their number (§9).

Dimension	Definite	Indefinite
1	Trivial	Trivial
2	Gauss: reduced forms.	Gauss: cycles of reduced forms.
3	Minkowski:	
.	reduced	
.	forms	
.	...	
7		Eichler:
.	Kneser-	spinor
.	Niemeier	genus
.	gluing	
.	method.	
24	...	
.		
.	Impracticable.	

Figure 15.1 How quadratic forms are classified.

(6) We include some theorems giving conditions under which a genus contains just one class (Theorem 20 and its Corollaries). In particular we show that, if f is an indefinite form of dimension n and determinant d, and there is more than one class in the genus of f, then

$$4^{[\frac{n}{2}]} d \quad \text{is divisible by } k^{\binom{n}{2}} \tag{1}$$

for some nonsquare natural number $k \equiv 0$ or 1 (mod 4).
(7) We also give a number of tables:
– reduced binary quadratic forms whose determinant d satisfies $|d| \leqslant 50$ for definite forms, $|d| \leqslant 100$ for indefinite forms (Tables 15.1, 15.2),
– indecomposable ternary forms with $|d| \leqslant 50$ for definite forms and $|d| \leqslant 100$ for indefinite forms (Tables 15.6, 15.7),
– genera of quadratic forms with $|d| < 12$ (Table 15.4),
– definite quadratic forms of determinant 2 and dimension $\leqslant 18$ (15.8),
– definite quadratic forms of determinant 3 and dimension $\leqslant 17$ (15.9).

Our treatment is addressed to a reader who wishes to be able to do explicit calculations while gaining some understanding of the general theory. The chapter is arranged as follows. §§2 and 4 contain definitions and other mathematical preliminaries. §3 deals with binary forms. §§5,6 treat the classification of forms over the rational numbers, and §7 the classification over the p-adic integers and the associated notion of the genus of a form. §8 gives some applications of the results of §7. The spinor genus of a form and the classification of indefinite forms are dealt with in §9, and definite forms in §10. The final section discusses the computational complexity of the classification problem.

2. Definitions

2.1 Quadratic forms. Let x be the row vector (x_1, x_2) and A the symmetric matrix $\begin{pmatrix} a & b \\ b & c \end{pmatrix}$. Then the expression

$$f(x) = xAx^{tr} = ax_1^2 + 2bx_1x_2 + cx_2^2 \tag{2}$$

is called the *binary quadratic form* with matrix A. Calculations with this form often involve the associated *bilinear form*

$$f(x, y) = xAy^{tr} = ax_1y_1 + bx_1y_2 + bx_2y_1 + cx_2y_2 .$$

Replacing A by a symmetric matrix of arbitrary dimension n we obtain the notion of an *n-ary quadratic form* and of an *n-ary symmetric bilinear form* (see §2.2 of Chap. 2). From the very large number of references on quadratic forms let us mention in particular [Bor5], [Cas3], [Cas4], [Coh5], [Dav2a], [Dic2], [Dic3], [Eic1], [Gau1], [Hsi1]-[Hsi9], [Jon3], [Kne1]-[Kne9], [Lam1], [Mil7], [Min0]-[Min3], [O'Me1]-[O'Me4], [Orz1], [Ran3], [Ran5], [Ran5a], [Rie1], [Rys1]-[Rys14], [Sch0]-[Sch2], [Sch13], [Sch14], [Ser1], [Smi5], [Smi6], [Tau2], [Wae5], [Wat3]-[Wat22], [Wit1], [Zag1].

We are concerned with the classification of integral quadratic forms under integral equivalence. There are two definitions of integrality for a quadratic form. The binary form (2) is *integral as a quadratic form* if a, $2b$ and c belong to $\mathbf{Z}$, and *integral as a symmetric bilinear form* if its matrix entries a, b, c belong to $\mathbf{Z}$. The latter is the definition used by Gauss [Gau1]; Cassels [Cas3] calls such a form *classically integral.* Although some authors hold strong opinions about which definition of integrality should be used (Watson [Wat3] refers to "Gauss's mistake of introducing binomial coefficients into the notation"), it makes very little difference for the classification theory. The two definitions are not really rivals but collaborators. For if f is integral in either sense then $2f$ is integral in the other sense, and f is equivalent to g if and only if $2f$ is equivalent to $2g$. Since for algebraic purposes a form is usually most conveniently specified by its matrix entries, we prefer the second definition, and so in this book we call f an *integral form* if and only if its matrix entries are integers (i.e. if and only if it is classically integral, or integral as a symmetric bilinear form).

2.2 Forms and lattices; integral equivalence. Some geometric ideas are appropriate. We consider a rational vector space V with inner product $(\, , \,)$ and refer to (x, x) as the *norm* of x. If V is 2-dimensional, and spanned by vectors e_1 and e_2 with

$$(e_1, e_1) = a, \quad (e_1, e_2) = b, \quad (e_2, e_2) = c \ ,$$

then the norm of the vector $x = x_1 e_1 + x_2 e_2$ is just $f(x)$, and its inner product with the vector $y = y_1 e_1 + y_2 e_2$ is $f(x, y)$.

The vectors $x = x_1 e_1 + x_2 e_2$ for which x_1 and x_2 are integers form a *lattice* in V, and e_1, e_2 is an *integral basis* for this lattice. The other integral bases for this lattice have the form $\alpha e_1 + \beta e_2$, $\gamma e_1 + \delta e_2$, where $\alpha, \beta, \gamma, \delta$ are integers with $\alpha\delta - \beta\gamma = \pm 1$. We shall therefore call two binary forms f and g with matrices A and B *integrally equivalent,* or say they are in the same *class,* and write $f \sim g$, if there exists a matrix

$$M = \begin{bmatrix} \alpha & \beta \\ \gamma & \delta \end{bmatrix} \tag{3}$$

with integer entries and determinant ± 1 for which $B = M A M^{tr}$. Geometrically, f and g refer to the same lattice with different integral bases. The equivalence is *proper* if $\det M = +1$, *improper* if $\det M = -1$.

The general case is exactly similar (§2.2 of Chap. 2). A symmetric matrix $A = (a_{ij})$ with integer entries determines a (classically) integral form f, whose values are the norms of the members of a lattice Λ (in an n-dimensional vector space) that has an integral basis $e_1, \ldots, e_n$ with $(e_i, e_j) = a_{ij}$. If $M = (m_{ij})$ has integer entries then the vectors $e_1{}' = \Sigma m_{1j} e_j, \ldots, e_n{}' = \Sigma \, m_{nj} e_j$ will generate a sublattice of Λ, and this will be all of Λ if and only if M^{-1} has integer entries; this happens if and only if $\det M = \pm 1$.

We therefore say that two n-ary forms f and g with matrices A and B are (*properly* or *improperly*) *integrally equivalent* if their matrices are related by

$$B = M\,A\,M^{tr} \tag{4}$$

for some M with integer entries and determinant $+1$ or -1 respectively.

It is clear that the number $d = \det A$ is an invariant of f for integral equivalence (i.e. if f and g are integrally equivalent, $\det A = \det B$), and we shall call d the *determinant* of f. The reader should be aware that many authors use the term *discriminant* for this number multiplied by certain powers of 2 and -1 that depend on the dimension (and the author).

In practice, when transforming f into an equivalent form, we derive the matrix B from A by performing elementary row operations (multiplying a row by a unit or adding multiples of one row to another), followed by the exactly corresponding column operations.

These notions can be immediately generalized to arbitrary rings R (with 1). A form f is defined over R if it is represented by a matrix A with entries from R, and two forms f and g are equivalent over R if (4) holds for some M with entries from R and with a determinant which is a unit of R (i.e. an element of R with an inverse in R). For example, taking R to be the rational numbers $\mathbf{Q}$, we see that the forms $x^2 + y^2$, $2z^2 + 2t^2$, although not integrally equivalent, are rationally equivalent, as shown by the formulae

$$x = z + t,\quad z = \tfrac{1}{2}(x + y)\,,$$
$$y = z - t,\quad t = \tfrac{1}{2}(x - y)\,.$$

If f and g are rationally equivalent, the ratio of their determinants is the square of a nonzero rational number.

3. The classification of binary quadratic forms

Almost everything one can say about the classification of binary quadratic forms was already said by Gauss in his *Disquisitiones arithmeticae* [Gau1]. The complete classification in the indefinite binary case uses cycles of reduced forms, and is totally unlike Eichler's complete classification of indefinite forms in dimensions $\geqslant 3$. A very clear account has recently been given by Edwards [Edw1], who also discusses the connections with the ideal theory of quadratic number rings, and so here we shall only present the results, together with a brief indication of these connections. Other treatments of the binary case may be found in [Cas3], [Dav1], [Jon3], [LeV1], [Wat3], [Zag1], etc.

3.1 Cycles of reduced forms

Theorem 1 (Quoted with changes in notation from [Edw1, p. 325].)

Let $\begin{pmatrix} a_0 & b_0 \\ b_0 & a_1 \end{pmatrix}$ *(abbreviated* $a_0{}^{b_0}a_1$*) be an integral binary quadratic form of* *determinant* $d = a_0 a_1 - b_0^2$, *and suppose that* $-d$ *is not a square. We define a sequence*

$$\begin{pmatrix} a_0 & b_0 \\ b_0 & a_1 \end{pmatrix}, \begin{pmatrix} a_1 & b_1 \\ b_1 & a_2 \end{pmatrix}, \dots, \begin{pmatrix} a_i & b_i \\ b_i & a_{i+1} \end{pmatrix}, \dots \tag{5}$$

(*abbreviated*

$$a_0 \, {}^{b_0} \, a_1 \, {}^{b_1} \, a_2 \dots a_i \, {}^{b_i} \, a_{i+1} \dots)$$

of binary forms of determinant d *by the rules*:

a_i *and* b_i *determine* a_{i+1} *as* $\dfrac{b_i^2+d}{a_i}$,

b_i *and* a_{i+1} *determine* b_{i+1} *as the largest solution of*

$$b_i + b_{i+1} \equiv 0 \pmod{a_{i+1}} \tag{6}$$

for which

$$b_{i+1}^2 + d < 0 \tag{7}$$

if such solutions exist, and otherwise as the smallest solution of (6) *in absolute value, taking* b_{i+1} *positive in case of a tie.*

Then the sequence of forms derived in this way from the given form is ultimately periodic, the forms in the period being called a cycle of reduced forms. *Furthermore two binary forms are properly equivalent if and only if they lead to the same cycle of reduced forms.*

The condition that $-d$ be not a square implies that the a_i are nonzero, and so a_{i+1} and b_{i+1} are well-defined. Only minor modifications are required when $-d$ is a square — see §3.3.

As an example, the sequence of forms arising from $x^2 - 67y^2$ is

$$\begin{array}{cccccccccccccc} 0 & 0 & 8 & 7 & 5 & 2 & 7 & 7 & 2 & 5 & 7 & 8 & 8 & \\ +1 & -67 & +1 & -3 & +6 & -7 & +9 & -2 & +9 & -7 & +6 & -3 & +1 & -3\dots \end{array} \tag{8}$$

There are ten reduced forms in the cycle. On the other hand $-x^2 + 67y^2$ leads to the disjoint sequence obtained by changing the signs of the lower terms in (8) (the a_i's), and is therefore an inequivalent form.

3.2 Definite binary forms. In the case of a definite (say positive definite) binary form, it follows from Theorem 1 that the cycle contains either a single form

$$\begin{pmatrix} a & b \\ b & a \end{pmatrix}, \quad \text{with } b = 0 \text{ or } 2b = |a|, \tag{9}$$

or else two forms

$$\begin{pmatrix} a & b \\ b & c \end{pmatrix}, \begin{pmatrix} c & -b \\ -b & a \end{pmatrix}. \tag{10}$$

The term reduced is normally applied only to the single form (9) or to whichever of the pair (10) satisfies

$$(1,1) \text{ entry} < (2,2) \text{ entry}, \text{ or } a = c \text{ and } b > 0 .$$

It is not difficult to show that a positive definite form $\begin{pmatrix} a & b \\ b & c \end{pmatrix}$ is reduced in this sense if and only if it satisfies

$$-a < 2b \leqslant a \leqslant c, \text{ with } b \geqslant 0 \text{ if } a = c \tag{11}$$

[Dic3, Theorem 99], [Jon3, Theorem 76]. Furthermore every positive definite form is properly equivalent to a unique such reduced form. Since (11) implies $b^2 \leqslant d/3$, all reduced forms are easily enumerated: for each b with $|b| \leqslant \sqrt{d/3}$ we factor $d+b^2 = ac$ in all possible ways consistent with (11) (see Table 15.1).

The reduction condition (11) expresses the fact that

a is the absolutely smallest value taken (at a vector e_1 say) by the form,

and

c is the absolutely smallest value that can be taken by the form at a vector e_2 independent of e_1.

An appropriate generalization of these conditions to higher dimensions leads to the notion of a Minkowski reduced form (see §10.1).

(11) is also equivalent to the assertion that the root $z = x/y$ of $f(x, y) = 0$ in the upper half plane lies in the region $|z| \geqslant 1$, $-1/2 \leqslant \mathrm{Re}\, z \leqslant 1/2$ shaded in Fig. 15.2. As the form undergoes unimodular transformations

$$\begin{pmatrix} \alpha & \beta \\ \gamma & \delta \end{pmatrix} \in SL_2(\mathbf{Z}) ,$$

i.e. with $\alpha, \beta, \gamma, \delta \in \mathbf{Z}$ and $\alpha\beta - \gamma\delta = 1$, this root transforms into $\dfrac{\alpha z+\beta}{\gamma z+\delta}$,

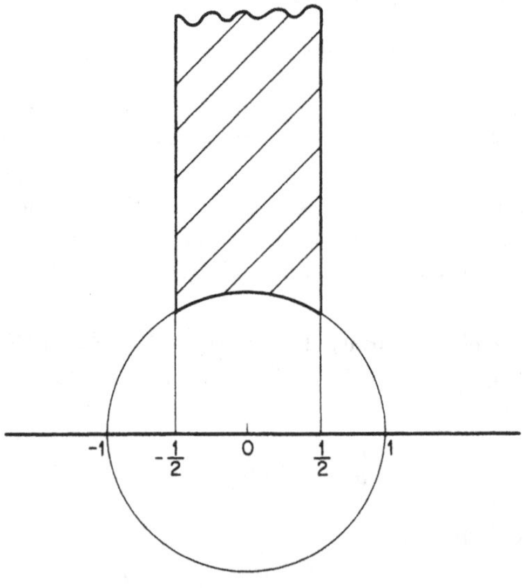

Figure 15.2 The fundamental region for $PSL_2(\mathbf{Z})$.

and the region mentioned is a fundamental region for the action of $SL_2(\mathbf{Z})$ on the upper half plane (see for example [LeV1, Vol. 1, Chap. 1]). This notion also generalizes to higher dimensions, and it is an important theorem that in the generalization the fundamental region has only finitely many walls (which correspond to inequalities on the matrix entries of the form).

3.3 Indefinite binary forms. In the indefinite case, supposing that $-d$ is not a square, it can be shown from Theorem 1 that the reduced forms (those in the cycle) are precisely the ones satisfying

$$0 < b < \sqrt{-d} < \min\{b+|a|,\ b+|c|\} \tag{12}$$

[Gau1, §183]. Again the reduced forms are easily found: for each positive integer $b < \sqrt{-d}$ we factor $d+b^2 = ac$ in all possible ways satisfying (12).

When $-d$ is a square, Theorem 1 must be modified slightly, and in particular the inequality in (7) must be replaced by $\leqslant$. Then the process of Theorem 1 terminates in a form $\begin{pmatrix} a & b \\ b & 0 \end{pmatrix}$ with $b = \sqrt{-d}$. If the process is extended *backwards* it terminates in a form $\begin{pmatrix} 0 & b \\ b & c \end{pmatrix}$. Gauss [Gau1, §§206-210] proved that two forms $\begin{pmatrix} a & b \\ b & 0 \end{pmatrix}$, $\begin{pmatrix} a' & b \\ b & 0 \end{pmatrix}$ are properly equivalent if and only if $a \equiv a' \pmod{2b}$, and are improperly equivalent if and only if

$$aa' \equiv \gcd(a,b)^2 \pmod{2b \gcd(a,b)} .$$

We therefore extend the notion of reduced form in this case to include, besides the forms satisfying (12), all forms of the shape

$$0\ {}^{b}\ a \quad \text{or} \quad a\ {}^{b}\ 0 \quad (-b < a \leqslant b) .$$

Then the "cycle" of reduced forms becomes a finite sequence

$$0\ {}^{b_0}\ a_1\ {}^{b_1}\ a_2\ {}^{b_2} \ldots a_k\ {}^{b_0}\ 0 .$$

Gauss declared that tables of binary quadratic forms should not be published, since they are so easily computed [Leh1, p. 69]. Nevertheless we feel the usefulness of our paper is enhanced by Tables 15.1 and 15.2, which enumerate *all* reduced binary quadratic forms with $-100 \leqslant d \leqslant 50$. The notation is that introduced in Theorem 1. Table 15.1 gives the positive definite forms with $d \leqslant 50$, and Table 15.2 the cycles of reduced indefinite forms. In Table 15.2 four related cycles

$$\ldots +a\ {}^{b} -c\ {}^{d} +e\ {}^{f} -g\ {}^{h} \ldots$$

$$\ldots -a\ {}^{b} +c\ {}^{d} -e\ {}^{f} +g\ {}^{h} \ldots$$

$$\ldots {}^{h} +g\ {}^{f} -e\ {}^{d} +c\ {}^{b} -a \ldots$$

$$\ldots {}^{h} -g\ {}^{f} +e\ {}^{d} -c\ {}^{b} +a \ldots$$

Table 15.1. **Reduced positive definite binary forms.**

d	Forms
1	$1^{0}1$
2	$1^{0}2$
3	$1^{0}3,\ 2^{1}2$
4	$1^{0}4,\ 2^{0}2$
5	$1^{0}5,\ 2^{1}3$
6	$1^{0}6,\ 2^{0}3$
7	$1^{0}7,\ 2^{1}4$
8	$1^{0}8,\ 2^{0}4,\ 3^{1}3$
9	$1^{0}9,\ 3^{0}3,\ 2^{1}5$
10	$1^{0}10,\ 2^{0}5$
11	$1^{0}11,\ 2^{1}6,\ 3^{\pm 1}4$
12	$1^{0}12,\ 2^{0}6,\ 3^{0}4,\ 4^{2}4$
13	$1^{0}13,\ 2^{1}7$
14	$1^{0}14,\ 2^{0}7,\ 3^{\pm 1}5$
15	$1^{0}15,\ 3^{0}5,\ 2^{1}8,\ 4^{1}4$
16	$1^{0}16,\ 2^{0}8,\ 4^{0}4,\ 4^{2}5$
17	$1^{0}17,\ 2^{1}9,\ 3^{\pm 1}6$
18	$1^{0}18,\ 2^{0}9,\ 3^{0}6$
19	$1^{0}19,\ 2^{1}10,\ 4^{\pm 1}5$
20	$1^{0}20,\ 2^{0}10,\ 4^{0}5,\ 3^{\pm 1}7,\ 4^{2}6$
21	$1^{0}21,\ 3^{0}7,\ 2^{1}11,\ 5^{2}5$
22	$1^{0}22,\ 2^{0}11$
23	$1^{0}23,\ 2^{1}12,\ 3^{\pm 1}8,\ 4^{\pm 1}6$
24	$1^{0}24,\ 2^{0}12,\ 3^{0}8,\ 4^{0}6,\ 5^{1}5,\ 4^{2}7$
25	$1^{0}25,\ 5^{0}5,\ 2^{1}13$
26	$1^{0}26,\ 2^{0}13,\ 3^{\pm 1}9,\ 5^{\pm 2}6$
27	$1^{0}27,\ 3^{0}9,\ 2^{1}14,\ 4^{\pm 1}7,\ 6^{3}6$
28	$1^{0}28,\ 2^{0}14,\ 4^{0}7,\ 4^{2}8$
29	$1^{0}29,\ 2^{1}15,\ 3^{\pm 1}10,\ 5^{\pm 1}6$
30	$1^{0}30,\ 2^{0}15,\ 3^{0}10,\ 5^{0}6$
31	$1^{0}31,\ 2^{1}16,\ 4^{\pm 1}8,\ 5^{\pm 2}7$
32	$1^{0}32,\ 2^{0}16,\ 4^{0}8,\ 3^{\pm 1}11,\ 4^{2}9,\ 6^{2}6$
33	$1^{0}33,\ 3^{0}11,\ 2^{1}17,\ 6^{3}7$
34	$1^{0}34,\ 2^{0}17,\ 5^{\pm 1}7$
35	$1^{0}35,\ 5^{0}7,\ 2^{1}18,\ 3^{\pm 1}12,\ 4^{\pm 1}9,\ 6^{1}6$
36	$1^{0}36,\ 2^{0}18,\ 3^{0}12, 4^{0}9,\ 6^{0}6,\ 4^{2}10,\ 5^{\pm 2}8$
37	$1^{0}37,\ 2^{1}19$
38	$1^{0}38,\ 2^{0}19,\ 3^{\pm 1}13,\ 6^{\pm 2}7$
39	$1^{0}39,\ 3^{0}13,\ 2^{1}20,\ 4^{\pm 1}10,\ 5^{\pm 1}8,\ 6^{3}8$
40	$1^{0}40,\ 2^{0}20,\ 4^{0}10,\ 5^{0}8,\ 4^{2}11,\ 7^{3}7$
41	$1^{0}41,\ 2^{1}21,\ 3^{\pm 1}14,\ 6^{\pm 1}7,\ 5^{\pm 2}9$
42	$1^{0}42,\ 2^{0}21,\ 3^{0}14,\ 6^{0}7$
43	$1^{0}43,\ 2^{1}22,\ 4^{\pm 1}11$
44	$1^{0}44,\ 2^{0}22,\ 4^{0}11,\ 3^{\pm 1}15,\ 5^{\pm 1}9,\ 4^{2}12,\ 6^{\pm 2}8$
45	$1^{0}45,\ 3^{0}15,\ 5^{0}9,\ 2^{1}23,\ 7^{2}7,\ 6^{3}9$
46	$1^{0}46,\ 2^{0}23,\ 5^{\pm 2}10$
47	$1^{0}47,\ 2^{1}24,\ 3^{\pm 1}16,\ 4^{\pm 1}12,\ 6^{\pm 1}8,\ 7^{\pm 3}8$
48	$1^{0}48,\ 2^{0}24,\ 3^{0}16,\ 4^{0}12,\ 6^{0}8,\ 7^{1}7,\ 4^{2}13,\ 8^{4}8$
49	$1^{0}49,\ 7^{0}7,\ 2^{1}25,\ 5^{\pm 1}10$
50	$1^{0}50,\ 2^{0}25,\ 5^{0}10,\ 3^{\pm 1}17,\ 6^{\pm 2}9$

are all represented in the table by a single entry

$$\dots a\,{}^{b}\,c\,{}^{d}\,e\,{}^{f}\,g\,{}^{h}\ \dots .$$

In restoring the signs it is helpful to note that the lower digits alternate in sign. Some care is needed in recovering the original cycles, because some of the four cycles just mentioned may coincide, and so an entry in the table may represent one, two or four cycles. We have used ordinary parentheses ( ) and curly brackets { } to further reduce the size of the table. Entries bounded by parentheses indicate whole or half periods. Thus for $d = -99$ the entry

$$(5\,{}^{8}\,7\,{}^{6}\,9\,{}^{3}\,10\,{}^{7})$$

represents the four distinct cycles

$$\dots 5\,{}^{8}\,-7\,{}^{6}\,9\,{}^{3}\,-10\,{}^{7}\,5\,{}^{8}\,-7 \dots$$

$$\dots -5\,{}^{8}\,7\,{}^{6}\,-9\,{}^{3}\,10\,{}^{7}\,-5\,{}^{8}\,7 \dots$$

$$\dots 7\,{}^{8}\,-5\,{}^{7}\,10\,{}^{3}\,-9\,{}^{6}\,7\,{}^{8}\,-5 \dots$$

$$\dots -7\,{}^{8}\,5\,{}^{7}\,-10\,{}^{3}\,9\,{}^{6}\,-7\,{}^{8}\,5 \dots$$

However, the entry

$$(3\,{}^{5}\,4\,{}^{3}\,7\,{}^{4})$$

for $d = -37$ yields only two inequivalent cycles, namely

$$\dots 3\,{}^{5}\,-4\,{}^{3}\,7\,{}^{4}\,-3\,{}^{5}\,4\,{}^{3}\,-7\,{}^{4}\,3\,{}^{5}\,-4 \dots$$

and its reversal. Most entries in the table are contained within curly brackets, which indicate reflections about the outermost digits. Thus for $d = -13$ the entry

$$\{1\,{}^{3}\,4\,{}^{1}\,3\,{}^{2}\}$$

represents the single cycle

$$\dots 1\,{}^{3}\,-4\,{}^{1}\,3\,{}^{2}\,-3\,{}^{1}\,4\,{}^{3}\,-1\,{}^{3}\,4\,{}^{1}\,-3\,{}^{2}\,3\,{}^{1}\,-4\,{}^{1}\,1\,{}^{3}\,-4 \dots$$

while for $d = -14$,

$$\{1\,{}^{3}\,5\,{}^{2}\,2\}$$

represents the two cycles

$$\dots 1\,{}^{3}\,-5\,{}^{2}\,2\,{}^{2}\,-5\,{}^{3}\,1\,{}^{3}\,-5 \dots$$

and

$$\dots -1\,{}^{3}\,5\,{}^{2}\,-2\,{}^{2}\,5\,{}^{3}\,-1\,{}^{3}\,5 \dots .$$

Table 15.2a. Reduced indefinite binary forms.

d	Forms
−1	$0^{1}\}$, $0^{1}1\}$
−2	$\{1^{1}\}$
−3	$\{1^{1}2\}$
−4	$0^{2}\}$, $0^{2}1\}$, $0^{2}2\}$
−5	$\{1^{2}\}$, $\{2^{1}\}$
−6	$\{1^{2}2\}$
−7	$\{1^{2}3^{1}2\}$
−8	$\{1^{2}4\}$, $\{2^{2}\}$
−9	$0^{3}\}$, $0^{3}1\}$, $0^{3}2\}$, $0^{3}3\}$
−10	$\{1^{3}\}$, $\{2^{2}3^{1}\}$
−11	$\{1^{3}2\}$
−12	$\{1^{3}3\}$, $\{2^{2}4\}$
−13	$\{1^{3}4^{1}3^{2}\}$, $\{2^{3}\}$
−14	$\{1^{3}5^{2}2\}$
−15	$\{1^{3}6\}$, $\{2^{3}3\}$
−16	$0^{4}\}$, $0^{4}1\}$, $0^{4}2\}$, $0^{4}3^{2}4\}$, $0^{4}4\}$
−17	$\{1^{4}\}$, $\{2^{3}4\}$
−18	$\{1^{4}2\}$, $\{3^{3}\}$
−19	$\{1^{4}3^{2}5^{3}2\}$
−20	$\{1^{4}4\}$, $\{2^{4}\}$, $\{4^{2}\}$
−21	$\{1^{4}5^{1}4^{3}3\}$, $\{2^{3}6\}$
−22	$\{1^{4}6^{2}3^{4}2\}$
−23	$\{1^{4}7^{3}2\}$
−24	$\{1^{4}8\}$, $\{2^{4}4\}$, $\{3^{3}5^{2}4\}$
−25	$0^{5}\}$, $0^{5}1\}$, $0^{5}2\}$, $0^{5}3^{4}\}$, $0^{5}4^{3}\}$, $0^{5}5\}$
−26	$\{1^{5}\}$, $\{2^{4}5^{1}\}$
−27	$\{1^{5}2\}$, $\{3^{3}6\}$
−28	$\{1^{5}3^{4}4\}$, $\{2^{4}6^{2}4\}$
−29	$\{1^{5}4^{3}5^{2}\}$, $\{2^{5}\}$
−30	$\{1^{5}5\}$, $\{2^{4}7^{3}3\}$
−31	$\{1^{5}6^{1}5^{4}3^{5}2\}$
−32	$\{1^{5}7^{2}4\}$, $\{2^{4}8\}$, $\{4^{4}\}$
−33	$\{1^{5}8^{3}3\}$, $\{2^{5}4^{3}6\}$
−34	$\{1^{5}9^{4}2\}$, $\{{}^{5}3^{4}6^{2}5^{3}\}$
−35	$\{1^{5}10\}$, $\{2^{5}5\}$
−36	$0^{6}\}$, $0^{6}1\}$, $0^{6}2\}$, $0^{6}3\}$, $0^{6}4\}$, $0^{6}5^{4}4\}$, $0^{6}6\}$
−37	$\{1^{6}\}$, $\{2^{5}6^{1}\}$, $(3^{5}4^{3}7^{4})$
−38	$\{1^{6}2\}$
−39	$\{1^{6}3\}$, $\{2^{5}7^{2}5^{3}6\}$
−40	$\{1^{6}4\}$, $\{2^{6}\}$, $\{5^{5}3^{4}8\}$, $\{4^{4}6^{2}\}$
−41	$\{1^{6}5^{4}\}$, $\{2^{5}8^{3}4^{5}\}$
−42	$\{1^{6}6\}$, $\{2^{6}3\}$
−43	$\{1^{6}7^{1}6^{5}3^{4}9^{5}2\}$
−44	$\{1^{6}8^{2}5^{3}7^{4}4\}$, $\{2^{6}4\}$
−45	$\{1^{6}9^{3}4^{5}5\}$, $\{2^{5}10\}$, $\{3^{6}\}$, $\{6^{3}\}$
−46	$\{1^{6}10^{4}3^{5}7^{2}6^{4}5^{6}2\}$
−47	$\{1^{6}11^{5}2\}$
−48	$\{1^{6}12\}$, $\{2^{6}6\}$, $\{3^{6}4\}$, $\{4^{4}8\}$
−49	$0^{7}\}$, $0^{7}1\}$, $0^{7}2\}$, $0^{7}3^{5}8^{3}5^{7}0$, $0^{7}4^{5}6^{7}0$, $0^{7}7\}$
−50	$\{1^{7}\}$, $\{2^{6}7^{1}\}$, $\{5^{5}\}$

Table 15.2b. Reduced indefinite binary forms.

d	Forms
-51	$\{1^{7}2\}, \{3^{6}5^{4}7^{3}6\}$
-52	$\{1^{7}3^{5}9^{4}4\}, \{2^{6}8^{2}6^{4}\}, \{4^{6}\}$
-53	$\{1^{7}4^{5}7^{2}\}, \{2^{7}\}$
-54	$\{1^{7}5^{3}9^{6}2\}, \{3^{6}6\}$
-55	$\{1^{7}6^{5}5\}, \{2^{7}3^{5}10\}$
-56	$\{1^{7}7\}, \{2^{6}10^{4}4\}, \{4^{6}5^{4}8\}$
-57	$\{1^{7}8^{1}7^{6}3\}, \{2^{7}4^{5}8^{3}6\}$
-58	$\{1^{7}9^{2}6^{4}7^{3}\}, \{2^{6}11^{5}3^{7}\}$
-59	$\{1^{7}10^{3}5^{7}2\}$
-60	$\{1^{7}11^{4}4\}, \{2^{6}12\}, \{3^{6}8^{2}7^{5}5\}, \{4^{6}6\}$
-61	$\{1^{7}12^{5}3^{7}4^{5}9^{4}5^{6}\}, \{2^{7}6^{5}\}$
-62	$\{1^{7}13^{6}2\}$
-63	$\{1^{7}14\}, \{2^{7}7\}, \{3^{6}9^{3}6\}$
-64	$0^{8}\}, 0^{8}1\}, 0^{8}2\}, 0^{8}3^{7}5^{8}0, 0^{8}4\}, 0^{8}6^{4}8\}, 0^{8}7^{6}4\}, 0^{8}8\}$
-65	$\{1^{8}\}, \{2^{7}8^{1}\}, \{{}^{7}4^{5}10\}, \{5^{5}8^{3}7^{4}\}$
-66	$\{1^{8}2\}, \{3^{6}10^{4}5^{6}6\}$
-67	$\{1^{8}3^{7}6^{5}7^{2}9^{7}2\}$
-68	$\{1^{8}4\}, \{2^{8}\}, \{4^{6}8^{2}\}$
-69	$\{1^{8}5^{7}4^{5}11^{6}3\}, \{2^{7}10^{3}6\}$
-70	$\{1^{8}6^{4}9^{5}5\}, \{2^{8}3^{7}7\}$
-71	$\{1^{8}7^{6}5^{4}11^{7}2\}$
-72	$\{1^{8}8\}, \{2^{8}4\}, \{3^{6}12\}, \{4^{6}9^{3}7^{4}8\}, \{6^{6}\}$
-73	$\{1^{8}9^{1}8^{7}3^{8}\}, \{2^{7}12^{5}4^{7}6^{5}8^{3}\}$
-74	$\{1^{8}10^{2}7^{5}\}, \{2^{8}5^{7}\}$
-75	$\{1^{8}11^{3}6\}, \{2^{7}13^{6}3\}, \{5^{5}10\}$
-76	$\{1^{8}12^{4}5^{6}8^{2}9^{7}3^{8}4\}, \{2^{8}6^{4}10^{6}4\}$
-77	$\{1^{8}13^{5}4^{7}7\}, \{2^{7}14\}$
-78	$\{1^{8}14^{6}3\}, \{2^{8}7^{6}6\}$
-79	$\{1^{8}15^{7}2\}, (3^{8}5^{7}6^{5}9^{4}7^{3}10^{7})$
-80	$\{1^{8}16\}, \{2^{8}8\}, \{4^{6}11^{5}5\}, \{4^{8}\}, \{8^{4}\}$
-81	$0^{9}\}, 0^{9}1\}, 0^{9}2\}, 0^{9}3\}. 0^{9}4^{7}8^{9}0. 0^{9}5^{6}9^{3}8^{5}7^{9}0, 0^{9}6\}, 0^{9}9\}$
-82	$\{1^{9}\}, \{2^{8}9^{1}\}, (3^{7}11^{4}6^{8})$
-83	$\{1^{9}2\}$
-84	$\{1^{9}3\}, \{2^{8}10^{2}8^{6}\}, \{4^{6}12\}, \{4^{8}7^{5}\}$
-85	$\{1^{9}4^{7}9^{2}\}, \{2^{9}\}, \{5^{5}12^{7}3^{8}7^{6}\}, \{{}^{''''7}6^{5}10\}$
-86	$\{1^{9}5^{6}10^{4}7^{3}11^{8}2\}$
-87	$\{1^{9}6\}, \{2^{9}3\}$
-88	$\{1^{9}7^{5}9^{4}8\}, \{2^{8}12^{4}6^{8}4\}, \{4^{6}13^{7}3^{8}8\}, \{4^{8}6\}$
-89	$\{1^{9}8^{7}5^{8}\}, \{2^{9}4^{7}10^{3}8^{5}\}$
-90	$\{1^{9}9\}, \{3^{9}\}, \{2^{8}13^{5}5\}, \{6^{9}9^{3}\}$
-91	$\{1^{9}10^{1}9^{8}3^{7}14\}, \{2^{9}5^{6}11^{5}6^{7}7\}$
-92	$\{1^{9}11^{2}8^{6}7^{8}4\}, \{2^{8}14^{6}4\}$
-93	$\{1^{9}12^{3}7^{4}11^{7}4^{9}3\}, \{2^{9}6\}$
-94	$\{1^{9}13^{4}6^{8}5^{7}9^{2}10^{8}3^{7}15^{8}2\}$
-95	$\{1^{9}14^{5}5\}, \{2^{9}7^{5}10\}$
-96	$\{1^{9}15^{6}4\}, \{3^{9}5^{6}12\}, \{2^{8}16\}, \{4^{8}8\}, \{6^{6}10^{4}8\}$
-97	$\{1^{9}16^{7}3^{8}11^{3}8^{5}9^{4}\}, \{2^{9}8^{7}6^{5}12^{7}4^{9}\}$
-98	$\{1^{9}17^{8}2\}, \{7^{7}\}$
-99	$\{1^{9}18\}, \{2^{9}9\}, \{3^{9}6\}, (5^{8}7^{6}9^{3}10^{7})$
-100	$0^{10}\}, 0^{10}1\}, 0^{10}2\}, 0^{10}3^{8}12^{4}7^{10}0, 0^{10}4\}, 0^{10}5\}, 0^{10}6^{8}\},$

When $-d$ is a perfect square, the cycles become chains terminated by 0's. Thus for $d = -16$,

$$0 \; {}^{4} \; 3 \; {}^{2} \; 4\}$$

represents

$$0 \; {}^{4} \; 3 \; {}^{2} \; -4 \; {}^{2} \; 3 \; {}^{4} \; 0$$

and

$$0 \; {}^{4} \; -3 \; {}^{2} \; 4 \; {}^{2} \; -3 \; {}^{4} \; 0 \, .$$

Numerous other tables exist in the literature (see [Bra1], [Cas3, p. 357], [Edw1, p. 333], [Inc1], [Jon1], [Leg1], [Leh1, pp. 68-72], [Som2]), but most of these omit some classes of forms (for example the imprimitive forms, or those with $-d$ a square, or indefinite forms).

3.4 Composition of binary forms. For binary forms, under suitable restrictions, there is a notion of *composition* found by Gauss, that gives the forms a group structure ([Gau1, §234], [Cas3, Chap. 4], [Dic2, Chap. 3], [Edw1, §8.6], [Jon3, §4.4]). There is no generalization to dimensions $\geqslant 3$. Composition is best understood in terms of the multiplication of ideal classes in the corresponding quadratic number rings.

For simplicity we shall suppose that $-d$ is a square-free number not congruent to 1 (mod 4), since then the set $\mathbf{Z}[\sqrt{-d}\,]$ of algebraic integers in $\mathbf{Q}\,(\sqrt{-d})$ is precisely the set of numbers of the form $r+s\sqrt{-d}$ for $r, s \in \mathbf{Z}$. It is also natural to restrict attention to *properly primitive* quadratic forms $f(x,y) = ax^2 + 2bxy + cy^2$ (those for which the numbers a, $2b$, c have no common factor) of determinant $ac - b^2 = d$.

Any ideal $\mathcal{I}$ in $\mathbf{Z}\,[\sqrt{-d}\,]$ has a two-member basis (over $\mathbf{Z}$), so that

$$\mathcal{I} = \langle r + s\sqrt{-d} \; , \; t + u\sqrt{-d} \rangle \, ,$$

and its *norm* is usually taken to be the positive integer $|ru - st|$ [Rei1, pp. 293, 330]. A *principal* ideal has a single generator $r + s\sqrt{-d}$ (say) over $\mathbf{Z}\,[\sqrt{-d}\,]$, while over $\mathbf{Z}$ it is generated by $r + s\sqrt{-d}$ and $\sqrt{-d}\,(r + s\sqrt{-d})$. This ideal has norm $r^2 + s^2 d$, and will be denoted by

$$\langle r + s\sqrt{-d} \rangle \, .$$

To get the proper correspondence between Gauss's group of forms under composition and the group of ideal classes we must introduce the notion of a *normed* or *oriented* ideal, an ideal $\mathcal{I}$ of norm N corresponding to two normed ideals $\mathcal{I}_N$ and $\mathcal{I}_{-N}$. A *principal normed ideal* is

$$\langle r + s\sqrt{-d} \rangle_{r^2+s^2 d} \, ,$$

rather than

$$\langle r + s\sqrt{-d} \rangle_{-(r^2+s^2 d)} \, .$$

Multiplication of normed ideals is defined by

$$\mathscr{I}_N \cdot \mathscr{I}_M = \mathscr{I}\mathscr{J}_{NM} \,. \tag{13}$$

Two (normed) ideals $\mathscr{I}$ and $\mathscr{J}$ are said to be in the same (normed) *ideal class* if and only if there exist principal (normed) ideals $\mathscr{P}$ and $\mathscr{Q}$ for which $\mathscr{I}\mathscr{P} = \mathscr{J}\mathscr{Q}$, and under composition these ideal classes form a group, the (normed) *ideal class group*.

The proper (i.e. determinant $+1$) equivalence classes of properly primitive forms of determinant d correspond to classes of normed ideals as follows. (We remind the reader that we are supposing $-d$ to be square-free and not congruent to 1 mod 4). One solution of the equation $f(x, y) = ax^2 + 2bxy + cy^2 = 0$ is

$$\frac{x}{y} = \frac{-b+\sqrt{-d}}{a},$$

and we shall say that the normed ideal

$$< a, -b+\sqrt{-d} >_a$$

corresponds to f.

Composition may now be defined as follows. To compose two quadratic forms we first pass to the corresponding normed ideals, multiply them using (13), and then, working modulo principal normed ideals, convert the product to a normed ideal of shape

$$< a, \ -b+\sqrt{-d} >_a ,$$

and reduce the corresponding quadratic form $\begin{pmatrix} a & b \\ b & c \end{pmatrix}$ of determinant d by the cycle method of Theorem 1.

For example, the quadratic form $\begin{pmatrix} 3 & 2 \\ 2 & -1 \end{pmatrix}$, which belongs to the reduced cycle

$$\begin{array}{ccccccccc} & 2 & & 2 & & 1 & & 1 & \\ 3 & & -1 & & 3 & & -2 & & 3 \end{array}$$

corresponds to the normed ideal

$$< 3, \ -2+\sqrt{7} >_3 .$$

The square of this is the normed ideal of norm 9 whose ideal part is generated by

$$3^2 = 9, \ 3(-2+\sqrt{7}) = -6+3\sqrt{7}, \ (-2+\sqrt{7})^2 = 11-4\sqrt{7} ,$$

and this is easily seen to be the same as the ideal $<9, \ -5+\sqrt{7}>_9$ corresponding to the form $\begin{pmatrix} 9 & 5 \\ 5 & 2 \end{pmatrix}$, which gives the reduction sequence

$$\begin{array}{ccccccccccccc} & 5 & & 1 & & 2 & & 2 & & 1 & & 1 & \\ 9 & & 2 & & -3 & & 1 & & -3 & & 2 & & -3 \cdots \end{array}$$

Thus the composition of the equivalence class containing $\begin{pmatrix} 3 & 2 \\ 2 & -1 \end{pmatrix}$ with itself yields the class containing $\begin{pmatrix} 2 & 1 \\ 1 & -3 \end{pmatrix}$. In fact every form of determinant -7 is in one of these two classes, and so the group is cyclic of order 2.

It is perhaps worthwhile to mention that the forms improperly equivalent to f belong to the inverse class in the group (see [Cas3, Chap. 14, Theorem 2.1]). Thus the inverse class is represented by either $\begin{pmatrix} c & b \\ b & a \end{pmatrix}$ or $\begin{pmatrix} a & -b \\ -b & c \end{pmatrix}$. A form improperly equivalent to itself is usually called an *ambiguous* form. Edwards [Edw1, Chaps. 7, 8] gives a clear account of the group structure in the general case, with many examples. He works, however, with an axiomatically defined notion of *divisor*, in place of our normed ideals.

3.5 Genera and spinor genera for binary forms. The theory of genera (§7) was originally developed by Gauss only for the binary case, where it presents special features and is intimately connected with the group of forms under composition. In fact two forms are in the same genus if and only if their quotient is a square in this group. The genus of a binary quadratic form is usually indicated by certain "characters" which can be computed from the numbers represented by the form. We shall not describe the easy conversion from this notation into the one used in §7.

The theory of spinor genera described in §9 assumes throughout that the dimension is at least three, and in the binary case there are several differences. Our treatment is no longer appropriate since there do not exist spinor operators corresponding to all sequences $(r_{-1}, r_2, r_3, \ldots)$ of p-adic unit square classes. Estes and Pall [Est1] have given a full investigation of spinor genera in the binary case and have shown in particular that two forms are in the same spinor genus if and only if their quotient is a fourth power in the group.

4. The p-adic numbers

We now return to the general case of a quadratic form of arbitrary dimension. The notion of integral equivalence (defined in §2) is a very subtle one, which is best approached by studying the weaker notions of equivalence over the larger rings

Q, the rational numbers,

$\mathbf{Q}_p$, the p-adic rational numbers, and

$\mathbf{Z}_p$, the p-adic integral numbers,

where p ranges over the "primes" $-1, 2, 3, 5, \ldots$. The multiplicative group of nonzero rational numbers is a direct product of its cyclic subgroups generated by -1 and the positive prime numbers $2, 3, 5, \ldots$. In this chapter we shall refer to -1 as a prime number, although it presents several special features which arise from the fact that the subgroup it generates is cyclic of order 2 rather than of infinite order. By convention

$\mathbf{Q}_{-1}$ and $\mathbf{Z}_{-1}$ are both equal to the ring $\mathbf{R}$ of real numbers. The number theory of $\mathbf{Q}$ or $\mathbf{Z}$ involves an infinity of prime numbers; by passing to $\mathbf{Q}_p$ or $\mathbf{Z}_p$ we concentrate on just one prime at a time.

4.1 The *p*-adic numbers. We give a brief description of the main properties of the p-adic numbers that will be needed later. Further information is readily available in a number of books [Bac1], [Bor5], [Cas3], [Kob1], [Mah4], [O'Me1].

Any real number is the limit of a Cauchy sequence of rationals in the ordinary metric

$$d_{-1}(x, y) = |x-y| .$$

If p is any positive prime, we can also equip the rationals with the *p-adic metric*

$$d_p(x, y) = \frac{1}{p^k} ,$$

where p^k is the exact power of p in the factorization of $x-y$. For instance

$$d_5(2, 52) = \frac{1}{5^2} = \frac{1}{25} ,$$

$$d_3(\frac{1}{27}, 0) = \frac{1}{3^{-3}} = 27 .$$

We note that for $x \neq y$ the product of $d_p(x,y)$ over all primes p (including -1) is $+1$.

The *p-adic rational numbers* $\mathbf{Q}_p$ are the limit points of Cauchy sequences of ordinary rational numbers with respect to the p-adic metric. The *p-adic integers* (or *rational p-adic integers*) $\mathbf{Z}_p$ are obtained by requiring the terms in the sequence to be ordinary integers.

For example, the difference between the nth and any later term in the sequence 4, 34, 334, 3334, ... is divisible by 5^n (in fact by 10^n), and so this sequence is 5-adically convergent to a 5-adic integer a say. In this case a is actually a rational number. For $3a$ is the limit of the sequence 12, 102, 1002, 10002, ... which 5-adically converges to 2, since its nth term differs from 2 by a multiple of 5^n. Thus $a = 2/3$.

In a similar way we see that any rational number whose denominator is not divisible by p is a p-adic integer.

4.2 *p*-adic square classes. As another example, we see that 6 has a square root among the 5-adic integers, since $1^2 \equiv 6 \pmod 5$, $(-9)^2 \equiv 6 \pmod{25}$, $(16)^2 \equiv 6 \pmod{125}$, $(-109)^2 \equiv 6 \pmod{625}$, ..., the sequence $1, -9, 16, -109$, ... being capable of being continued indefinitely in such a way that the n-th term is congruent to all later terms modulo 5^n. It is therefore a 5-adic Cauchy sequence whose limit is a square root of 6. Similarly, for any positive odd prime p, an integer not divisible by p that is a quadratic residue modulo p is a p-adic square [Bac1, p. 59].

Two p-adic rational numbers are said to be in the same *p-adic square class* if their ratio is a p-adic rational square. The square classes for the various values of p may be described as follows (cf. [Bac1, pp. 59-60], [Cas3, p. 40], [Wat3, p. 33]):

$$+u, -u \quad \text{for } p = -1 ,$$

$$u_1, u_3, u_5, u_7, 2u_1, 2u_3, 2u_5, 2u_7 \quad \text{for } p = 2 ,$$

$$u_+, u_-, pu_+, pu_- \quad \text{for } p \geqslant 3 ,$$

where

for $p = -1$, u is any positive real number (a (-1)-adic unit),

for $p = 2$, u_i represents any 2-adic unit congruent to i (mod 8),

for $p \geqslant 3$, u_+ (resp. u_-) is any p-adic unit that is a quadratic residue (nonresidue) mod p.

4.3 An extended Jacobi-Legendre symbol. In agreement with our policy of regarding -1 as a prime, it is natural to define the greatest common divisor of two integers

$$n = (-1)^a 2^b 3^c \ldots , \quad \nu = (-1)^\alpha 2^\beta 3^\gamma \ldots \tag{14}$$

with $a = 0$ or 1, $\alpha = 0$ or 1, and $b, \beta, c, \gamma, \ldots = 0, 1, 2, \ldots$ to be

$$(n, \nu) = (-1)^{\min(a, \alpha)} 2^{\min(b, \beta)} 3^{\min(c, \gamma)} \ldots .$$

Note that this implies that two negative numbers cannot have $(n, \nu) = 1$ and so will not be counted as coprime. We can define the Jacobi-Legendre symbol $\left(\frac{\nu}{n}\right)$ in all cases where $(n, \nu) = 1$ as follows:

$$\left(\frac{\nu}{p}\right) = \pm 1 \text{ for a prime } p \geqslant 3, \text{ according as } \nu \text{ is or is not congruent to a square mod } p .$$

$$\left(\frac{\nu}{-1}\right) = 1 \qquad (\text{since in this case } \nu > 0) , \text{ and}$$

$$\left(\frac{\nu}{2}\right) = 1 \text{ if } \nu \equiv \pm 1 \pmod 8 , \ -1 \text{ if } \nu \equiv \pm 3 \pmod 8 .$$

(Note that $\left(\frac{\nu}{p}\right)$ is equally well defined if ν is a p-adic integer not divisible by p, since every p-adic integer is congruent modulo p to a rational integer.) The symbol $\left(\frac{\nu}{n}\right)$ for n given by (14) is then defined to be

$$\left(\frac{\nu}{-1}\right)^a \left(\frac{\nu}{2}\right)^b \left(\frac{\nu}{3}\right)^c \ldots .$$

With this definition the law of quadratic reciprocity asserts that

$$\left(\frac{\nu}{n}\right) = \left(\frac{n}{\nu}\right),$$

whenever either symbol is defined, unless n and ν are both congruent to -1 (mod 4), when

$$\left(\frac{\nu}{n}\right) = -\left(\frac{n}{\nu}\right).$$

4.4 Diagonalization of quadratic forms. It is well-known that any quadratic form over a field of characteristic $\neq 2$ (for example $\mathbf{Q}$ or $\mathbf{Q}_p$) may be diagonalized, and in fact any vector at which the form takes a nonzero value may be taken as the first term of a diagonal basis (see for example [Jon3, Theorem 2]).

Furthermore for $p \neq 2$ any form can be diagonalized over $\mathbf{Z}_p$. For $p = -1$ this is covered by the previous assertion. Otherwise we proceed as follows. We first find a matrix entry that is divisible by the lowest power of p. If this is a diagonal entry, say a_{11}, then we can start the diagonalization by subtracting multiples of the first row from the others so as to clear the rest of the first column, following this by the corresponding column operations to clear the rest of the first row. On the other hand if a nondiagonal entry, say a_{12}, is divisible by the least power of p, and all diagonal entries are divisible by a higher power, we can reduce to the first case by adding the second row to the first and then the second column to the first. This replaces a_{11} by $a_{11}+2a_{12}+a_{22}$, which, since $p \neq 2$, is now divisible by the lowest power of p to occur in any entry.

The same method works if $p = 2$ unless we arrive at a stage when some off-diagonal entry a_{12} say is divisible by the least possible power $q = 2^k$, while all diagonal entries are divisible by 2^{k+1}. In this case the leading 2×2 submatrix has the form

$$\begin{pmatrix} qa & qb \\ qb & qc \end{pmatrix},$$

where a and c are divisible by 2 but b is not, so that $d = ac-b^2$ is not divisible by 2. This implies that any pair of integers (x, y) is a 2-adically integral linear combination of (a, b) and (b, c), so that (since all entries are divisible by q) we can subtract suitable multiples of the first two rows from the others (followed by the corresponding column operations), so as to remove $\begin{pmatrix} qa & qb \\ qb & qc \end{pmatrix}$ as a direct summand. Thus we have proved the following result.

Theorem 2. *For $p \neq 2$ any p-adically integral form can be diagonalized by a p-adically integral transformation. For $p = 2$ there is a p-adically* *integral transformation expressing the form as a direct sum of forms with matrices*

$$(qx)\ ,\quad \begin{bmatrix} qa & qb \\ qb & qc \end{bmatrix}\ ,$$

where q is a power of 2, a and c are divisible by 2, but x, b and $d = ac - b^2$ are not.

Note. The reader who consults other works on quadratic forms will notice that what we call the prime -1 is usually given the name ∞. Over more general algebraic number rings there will be several such "primes", corresponding to different archimedean valuations, and in Hasse's original publications they were given symbols such as $1', 1'', \ldots$. The most appropriate name for them is "unit primes", since they arise from properties of the group of units of the underlying ring. Unfortunately the pernicious habit has grown up of calling them "infinite primes" instead. When we started to write this chapter we hesitated between the notations ∞ and -1 for the archimedean prime in our case, but eventually found that the unconventional name -1 made things so much more simple that its omission would be indefensible.

5. Rational invariants of quadratic forms

In §§5 and 6 we shall investigate when two integral quadratic forms are equivalent over the rational numbers; the main results are stated in Theorems 3, 4 and 5. Rational invariants for quadratic forms are usually defined via the Hilbert norm residue symbol ([Cas3, Chap. 6], [Jon3, Chap. 3], [Wat3, Chap. 3]). Our treatment avoids the use of this symbol and furthermore transforms the standard "product formula" into the readily usable sum formula (15) or "oddity formula" (16). Since we are working over $\mathbf{Q}$ we may assume that the form has already been diagonalized (cf. §4.4).

5.1 The invariants and the oddity formula. Any rational or p-adic integer A can be written uniquely in the form $A = p^\alpha a$ where a is prime to p (meaning that a is positive if $p = -1$, cf. §4.3). Then $p(A) = p^\alpha$ is called the *p-part* of A, and $p'(A) = a$ is the *p'-part*. We shall introduce the term *p-adic antisquare* to mean a number of the form $p^{\text{odd}} \cdot u_-$ when $p \geqslant 3$, and $2^{\text{odd}} \cdot u_{\pm 3}$ for $p = 2$, since both the p- and p'-parts of such a number are non-squares. (There are no (-1)-adic antisquares.) In terms of our extended Jacobi-Legendre symbol, $p^\alpha a$ is a p-adic antisquare if and only if

$$p^\alpha \text{ is not a square and } \left(\frac{a}{p}\right) = -1\ .$$

The *p-signature* of the integral quadratic form $f = \operatorname{diag}\{p^\alpha a, p^\beta b, p^\gamma c, \ldots\}$ is defined to be

$$p^\alpha + p^\beta + p^\gamma + \cdots + 4m \quad (p \neq 2)\,,$$
$$a + b + c + \cdots + 4m \quad (p = 2)\,,$$

where m is the number of p-adic antisquares among $p^\alpha a$, $p^\beta b$, $p^\gamma c$, ... Thus the (-1)-signature of f is just its ordinary signature, which (by Sylvester's law of inertia, §6.2) is an invariant for real equivalence. For $p \geqslant 2$ the p-signature is only to be regarded as defined modulo 8, and we shall see that the p-signature is an invariant for rational equivalence. The 2-signature, which often behaves specially, is also called the *oddity* of f.

What is usually termed the product formula relating the different p-adic invariants [Cas3, p. 76], [Jon3, Th. 29] becomes in this notation the sum formula

$$\text{2-signature} - \text{dimension} \equiv \sum_{\text{odd } p} p\text{-signature} - \text{dimension} \pmod 8 \quad (15)$$

or

$$\sum_{\text{all } p} p\text{-excess }(f) \equiv 0 \pmod 8\,,$$

where we define the *p-excess* to be

$$p\text{-signature} - \text{dimension}\,, \quad (p \neq 2)\,,$$
$$\text{dimension} - p\text{-signature}\,, \quad (p = 2)\,.$$

For practical calculations it is better to treat the contributions from $p = -1$ and 2 separately, leading to the *oddity formula*:

$$\text{signature }(f) + \sum_{p \geqslant 3} p\text{-excess }(f) \equiv \text{oddity }(f) \pmod 8\,. \quad (16)$$

Example. For the form $f = \text{diag}\,\{1, 3, -3\}$ we compute:

for $p = -1$, the signature $= 1{+}1{-}1 = 1$,
for $p = 3$, the 3-excess $= 0{+}2{+}2{+}4 = 8$

(since -3 is a 3-adic antisquare), and

for $p \geqslant 5$, the p-excess $= 0{+}0{+}0 = 0$,

while

for $p = 2$, the oddity $= 1{+}3{-}3 = 1$,

so that (16) reads

$$1{+}8{+}0{+}0{+}\cdots \equiv 1 \pmod 8\,.$$

Then our first main theorem, the proof of which will be given in §6, is the following.

Theorem 3. *Two nonsingular forms of the same dimension are equivalent over the rationals if and only if*
(i) *the quotient of their determinants is a rational square, and*
(ii) *for each p they have the same p-excess.*
Condition (ii) *may be replaced by the equivalent condition*:
(ii)′ *they have the same signature, the same oddity, and, for all $p \geqslant 3$, the same p-excesses modulo* 8.

It is obvious that if two forms are equivalent over the rationals then they must be equivalent over $\mathbf{Q}_p$ for all p. Since on the other hand the invariants used in Theorem 3 are p-adic invariants, this theorem can be restated as follows.

Theorem 4 (The weak Hasse principle). *A necessary and sufficient condition for two rational forms to be equivalent over $\mathbf{Q}$ is that they be equivalent over $\mathbf{Q}_p$ for all p.*

5.2 Existence of rational forms with prescribed invariants. The next theorem states that the oddity formula is essentially the only relation between the p-adic invariants for all p.

Theorem 5 (The strong Hasse principle). *If for each p we are given a p-adic form $f^{(p)}$ of determinant d, satisfying*

$$\textit{signature}\ (f^{(-1)}) + \sum_{p \geqslant 3} p\text{-excess}\ (f^{(p)}) \equiv \textit{oddity}\ (f^{(2)}) \pmod 8,$$

then there exists a rational form f which is equivalent to $f^{(p)}$ over $\mathbf{Q}_p$ for each p.
By Theorem 4, if such a form exists it is unique up to equivalence.

Sketch of proof. (For further details see for example [Cas3, Chap. 6, Theorem 1.3] or [Jon3, Theorem 29].) The theorem is first reduced to the case when f is a binary form. Then we wish to find a rational form $f = \text{diag}\ \{A, B\}$ of prescribed determinant d and with a given signature and non-trivial p-excesses for finitely many primes p. The idea is to choose a large prime q and to take $f = \text{diag}\ \{p_1 p_2 \cdots q, p_1 p_2 \cdots qd\}$, where the p_i are chosen from these primes and the divisors of $2d$. The value of the p-excess for each prime p dividing d is controlled by the residue class of q modulo p, or if $p = 2$ by the residue class of q modulo 8. All these values may therefore be simultaneously adjusted by requiring q to be in a suitable arithmetic progression modulo $4d$. The existence of such primes q is guaranteed by Dirichlet's theorem (1837) on primes in arithmetic progressions (see for example [Apo1]). For primes $p \neq q$ that do not divide d, the p-excess is trivial, while for q itself the q-excess must be correct by the oddity formula.

Remarks. (1) This proof essentially dates back to Legendre, at a time when Dirichlet's theorem was an unproved conjecture. Gauss later eliminated the dependence on that conjecture by finding a proof using the genera of integral binary quadratic forms (see [Cas3, Chap. 14, §5).

(2) The possible values of a given p-adic invariant for an n-dimensional form can be computed from its possible values for 1-dimensional forms,

which are easily listed. For instance, p-excesses are always even, and are divisible by 4 if $p \equiv 1 \pmod 4$.

5.3 The conventional form of the Hasse-Minkowski invariant. In the literature on quadratic forms the subtle part of the p-adic invariant for a form is usually expressed as a number equal to ± 1, called the Hasse-Minkowski invariant (rather than our p-excess). There are several different conventions, but a common one [Cas3, p. 55] is that the Hasse-Minkowski invariant for $f = \text{diag}\,\{A_1, A_2, \ldots, A_n\}$ is

$$(f)_p = (A_1, A_2, \ldots, A_n)_p = \prod_{1 \leqslant i < j \leqslant n} (A_i, A_j)_p \tag{17}$$

where $(x, y)_p$ is the so-called Hilbert norm residue symbol. This invariant can be recovered from the p-excess as follows. $(f)_p$ is equal to

$$+1 \qquad \text{or} \qquad -1$$

according as the p-excess of f is or is not congruent modulo 8 to that of

$$f_0 = \text{diag}\,\{A_1 A_2 \ldots A_n, 1, 1, \ldots, 1\},$$

the "standard" form having the same determinant and dimension as f.

6. The invariance and completeness of the rational invariants

This section is devoted to proving Theorem 3.

6.1 The p-adic invariants for binary forms. The equivalence class of a *binary* form over a field is completely determined by its determinant d (modulo squares) and by any nonzero number a that it represents. For if $f(e_1) = a$, we can take e_1 to be the first vector of a diagonal basis, with respect to which we must have $f = \text{diag}\,\{a, d/a\}$.

For the p-adic rationals there are only finitely many square classes, and we can enumerate all possible forms (see Table 15.3), and thereby check that two forms have the same determinant and p-adic invariants (signature, p-excess, and oddity) if and only if they are equivalent. As an example we consider the 2-adic forms of determinant $2u_3$. Each such form is equivalent to one of

$$\text{diag}\,\{u_1, 2u_3\}, \qquad \text{diag}\,\{u_3, 2u_1\}, \qquad \text{diag}\,\{u_5, 2u_7\}, \qquad \text{diag}\,\{u_7, 2u_5\}$$

for which the oddities are respectively

$$1+3+4 \equiv 0, \qquad 3+1 \equiv 4, \qquad 5+7 \equiv 4, \qquad 7+5+4 \equiv 0$$

Table 15.3. Correspondence between the p-adic invariants and the numbers represented by binary forms.

p	Det	Square classes represented	Signature
-1	$+u$	$+u$	2
		$-u$	-2
	$-u$	$+u, -u$	0
p	Det	Square classes represented	Oddity
2	u_1	$u_1, u_5, 2u_1, 2u_5$	2
		$u_3, u_7, 2u_3, 2u_7$	6
	u_3	u_1, u_3, u_5, u_7	4
		$2u_1, 2u_3, 2u_5, 2u_7$	0
	u_5	$u_1, u_5, 2u_3, 2u_7$	6
		$u_3, u_7, 2u_1, 2u_5$	2
	u_7	all	0
	$2u_1$	$u_1, u_3, 2u_1, 2u_3$	2
		$u_5, u_7, 2u_5, 2u_7$	6
	$2u_3$	$u_1, u_7, 2u_3, 2u_5$	0
		$u_3, u_5, 2u_1, 2u_7$	4
	$2u_5$	$u_1, u_3, 2u_5, 2u_7$	2
		$u_5, u_7, 2u_1, 2u_3$	6
	$2u_7$	$u_1, u_7, 2u_1, 2u_7$	0
		$u_3, u_5, 2u_3, 2u_5$	4

p	Det	Square classes represented	p-excess, for p (mod 8) $\equiv$ 1	3	5	7
$\geqslant 3$	u_+	all	0	—	0	—
		u_+, u_-	—	0	—	0
		pu_+, pu_-	—	4	—	4
	u_-	all	—	0	—	0
		u_+, u_-	0	—	0	—
		pu_+, pu_-	4	—	4	—
	pu_+	u_+, pu_+	0	2	4	6
		u_-, pu_-	4	6	0	2
	pu_-	u_+, pu_-	4	6	0	2
		u_-, pu_+	0	2	4	6

(mod 8). Since x^2+6y^2 (the first form) represents 7, it is equivalent to the fourth form, and similarly the second and third forms are equivalent. But an easy calculation reveals that x^2+6y^2 does not represent any number of the form $4^m(8k+3)$, and so these two pairs of forms are not equivalent. Thus there are two distinct 2-adic binary forms of determinant $2u_3$. The first has oddity 0 and represents the numbers u_1, u_7, $2u_3$ and $2u_5$, while the second has oddity 4 and represents the numbers u_3, u_5, $2u_1$ and $2u_7$. This explains the entries for $p = 2$, det $= 2u_3$ in Table 15.3.

After a similar discussion for all the cases with $p = 2$ in Table 15.3 we conclude that the oddity is a 2-adic invariant, and together with the determinant is a complete invariant for 2-adic rational equivalence. Similarly for the other primes. This completes the proof of the "only if" part of Theorem 3 for binary forms.

The reader will notice that there is, for each p, one form that represents all nonzero numbers. This is the *isotropic* form diag $\{A, -A\}$, of determinant -1, so called because it also represents zero nontrivially.

6.2 The p-adic invariants for n-ary forms. It is easy to extend the above remarks to show that our invariants really are invariants for forms of all dimensions. The proof reduces to showing that any equivalence between diagonal forms can be broken down into a chain of *binary* equivalences (i.e. ones effecting just two diagonal terms).

To see this, note that for

$$f = \text{diag}\ \{a,b,c, \dots\}, \quad f' = \text{diag}\ \{a',b', c', \dots\}$$

to be equivalent, a' must be representable by f. We choose a representation involving the smallest number of terms, say

$$a' = ax^2 + by^2 + cz^2 + dt^2 ,$$

and then we have the binary equivalences

$$\text{diag}\ \{a,b,c,d,e, \dots\} \sim \text{diag}\ \{a_2,b^*,c,d,e, \dots\}$$

$$\sim\text{diag}\ \{a_3,b^*,c^*,d,e, \dots\} \sim\text{diag}\ \{a_4,b^*,c^*,d^*,e, \dots\} ,$$

where $a_2 = ax^2 + by^2$, $a_3 = ax^2 + by^2 + cz^2$, $a_4 = \cdots$ are nonzero. These show that f and f' are equivalent to diag $\{a',b^*,c^*, \dots\}$, and so by Witt's cancellation theorem (which we shall prove in a moment) the forms diag $\{b,c,d, \dots\}$ and diag $\{b^*,c^*,d^*,\dots\}$ are equivalent. Since by induction the latter equivalence reduces to a chain of binary ones, we deduce the same for the equivalence of f and f'.

The remainder of this section will be devoted to the proof that two forms having the same p-adic invariants for all p (including -1) are equivalent over the rationals.

Let f and g be two rational quadratic forms with the same nonzero determinant (modulo a square factor), and the same signature, oddity and

p-excess for all $p \geqslant 3$. Then we shall prove that for a suitable nonsingular form h, $f \oplus h$ is equivalent to $g \oplus h$ and will deduce that f is equivalent to g by repeated application of Witt's cancellation theorem.

Theorem 6 (Witt's cancellation theorem [Wit1], [Cas3, p. 21], [Sch0, p. 22], [Sch2, Chap. 1]). *Over any field of characteristic* $\neq 2$, *if* diag $\{a, b, c, \ldots\} \sim$ diag $\{a, b', c', \ldots\}$ *and* $a \neq 0$, *then* diag $\{b, c, \ldots\} \sim$ diag $\{b', c', \ldots\}$.

Proof. This is equivalent to the following geometrical assertion. Let V be a vector space over the field, equipped with the bilinear form $f(x, y) = ax_1y_1 + bx_2y_2 + cx_3y_3 + \cdots$. Then if vectors v and w in V have the same nonzero norm a, there is an automorphism of V fixing f and taking v to w. Now for any $r \in V$ of nonzero norm, the reflection

$$x \rightarrow x - 2\frac{f(x, r)}{f(r, r)}r \tag{18}$$

is an automorphism of V preserving f. Not both the vectors $r = v \pm w$ can have zero norm, and so one of the corresponding reflections exists and takes v to $\pm w$, and can be followed by negation if necessary. This establishes the theorem.

Remark. Sylvester's law of inertia (the invariance of signature under real equivalence [Jon3, Theorem 2]) follows immediately from Theorem 6, since if

$$\text{diag}\,\{(+1)^{r+k}, (-1)^s\} \sim \text{diag}\,\{(+1)^r, (-1)^{s+k}\}$$

over the reals, then we can deduce that

$$\text{diag}\,\{(+1)^k\} \sim \text{diag}\,\{(-1)^k\},$$

implying $k = 0$ (since one form is positive definite, the other negative definite).

We shall say that a form has *trivial invariants* if all its p-excesses are congruent to zero modulo 8 and its determinant is a perfect square. Then the desired result will follow from:

Theorem 7. *If F has trivial invariants then F is equivalent over the rationals to a form of the shape*

$$\text{diag}\,\{\pm 1, \pm 1, \ldots, \pm 1\}.$$

To see that the result we want is a consequence of Theorem 7 we argue as follows. Since p-excesses are always even, if f and g have the same invariants then

$$f \oplus f \oplus f \oplus f \quad \text{and} \quad g \oplus f \oplus f \oplus f \tag{19}$$

will both have (the same) trivial invariants, and by Theorem 7 will be equivalent to forms of the shape diag $\{\pm 1, \pm 1, \ldots\}$. Since signature (f)

= signature (g), the numbers of positive and negative terms agree, and the two forms (19) are equivalent. By Witt cancellation (Theorem 6) we deduce that $f \sim g$.

6.3 The proof of Theorem 7. We shall suppose throughout that F is a diagonal form with square-free integer entries and with trivial invariants, and will actually show that, for a sufficiently large N,

$$F \oplus \text{diag}\,\{(+1)^N, (-1)^N\} \sim \text{diag}\,\{\pm 1, \pm 1, ...\}$$

over the rationals. We shall do this by gradually reducing the primes appearing in the entries of F. Let p be the largest such prime, and call the entries of F divisible by p the *p-terms*. Note that any p-term has the form $pq_1q_2...q_k$ where $-1 \leqslant q_i < p$ for all i. We suppose first that $p \geqslant 3$.

Theorem 8. (The replacement lemma). *Assume* $p \geqslant 3$. *If* $-1 \leqslant a, b < p$, *and* ab *is congruent to a square (mod* p*), then we may replace any* p*-term "pat" by "pbt" without introducing any prime larger than* p.

Note. The replacement process involves adjoining more direct summands ± 1 to F.

Proof. (Based on [Con8, p. 401].) We can write $ab = x^2 - py$, with $|x| < \frac{1}{2}p$, and so $|y| < p$. Then the identity

$$pat\,(b/p)^2 - pbt\,(x/p)^2 = -ybt$$

shows that the form diag $\{pat, -pbt\}$ represents $-ybt$, and therefore, by the observation at the beginning of §6.1, is equivalent to diag $\{yat, -ybt\}$. Also $x^2 - y^2 =$ diag $\{1, -1\}$ represents all numbers, in particular $-pbt$, and so

$$\text{diag}\,\{1, -1\} \sim \text{diag}\,\{-pbt, pbt\}\ .$$

Then we have

$$\text{diag}\,\{pat, 1, -1\} \sim \text{diag}\,\{pat, -pbt, pbt\} \sim \text{diag}\,\{yat, -ybt, pbt\}\ ,$$

which is the desired replacement. This completes the proof of Theorem 8.

We now suppose $p > 2$. By repeated use of the replacement lemma we can replace each p-term by pu^k, and so by p or pu, where $u = r + 1$ is the least positive non-residue modulo p. But also diag$\{p, pr\}$ represents pu, and so is equivalent to diag$\{pu, pur\}$. We may therefore replace

$$p, p \text{ by } p, pr, \text{ then } pu, pur, \text{ then } pu, pu\ , \qquad (20)$$

or vice versa, so that the first of two or more p-terms may be chosen arbitrarily.

The determinant condition tells us there exist evenly many p-terms. We may replace the first one by $-p$ and the second by p or pu (and definitely by p if there are more than two p-terms). If the second is p we can now eliminate the first two p-terms using

$$\operatorname{diag}\{-p,p\} \sim \operatorname{diag}\{-1,1\}. \tag{21}$$

If not, the *only* p-terms are $-p,pu$, and the p-excess differs by 4 from that of $\operatorname{diag}\{-p,p\}$ (which is zero), and so is non-trivial, contradicting the supposition.

So when $p > 2$ we have been able to reduce the size of p by repeated application of the replacement lemma. If $p = 2$ all p-terms are ± 2, and there are an even number of them since the determinant is a square. They can then be eliminated using the equivalences $\operatorname{diag}\{2,2\} \sim \operatorname{diag}\{1,1\}$, $\operatorname{diag}\{2,-2\} \sim \operatorname{diag}\{1,-1\}$, $\operatorname{diag}\{-2,-2\} \sim \operatorname{diag}\{-1,-1\}$. This completes the proofs of Theorems 7 and 3.

7. The genus and its invariants

Two integral quadratic forms are said to be in the same *genus* if they are equivalent over the p-adic integers for all primes p (including -1). As we shall see in §9, for indefinite forms of dimension $n \geqslant 3$, there is usually only one equivalence class of forms in a genus. In fact this holds whenever $|\det f| < 128$, and when it fails $4^{[n/2]}\det f$ must be divisible by $k^{\binom{n}{2}}$ for some nonsquare natural number $k \equiv 0$ or $1 \pmod 4$.

In this section we give a complete system of invariants for p-adic integral equivalence for each p, and then show how to combine them to obtain a handy characterization of the genus.

No proofs will be offered. For $p \neq 2$ several accounts are readily available (e.g. Cassels [Cas3]), and all cases are handled by O'Meara [O'Me1], who gives the invariants for forms over arbitrary number fields. The correctness of the simple system of invariants and transformation rules for $p = 2$ given here was originally verified (by J.H.C.) by showing that they suffice to put every form into B. W. Jones' canonical form [Jon2], yet are consistent with G. Pall's complete system of invariants [Pal1]. A direct verification has now been given by K. Bartels [Bar19].

We remark that much of the importance of the genus — for instance in topological investigations — arises from the fact that two forms f and g are in the same genus if and only if $f \oplus \begin{bmatrix} 0 & 1 \\ 1 & 0 \end{bmatrix}$ and $g \oplus \begin{bmatrix} 0 & 1 \\ 1 & 0 \end{bmatrix}$ are integrally equivalent. This follows from properties of the spinor genus.

7.1 p-adic invariants. As we saw in Theorem 2, any form can be decomposed over the p-adic integers as a direct sum

$$f = f_1 \oplus pf_p \oplus p^2 f_{p^2} \oplus \cdots \oplus q f_q \oplus \cdots \tag{22}$$

in which each f_q is a p-adic *unit form,* meaning a p-adic integral form whose determinant is prime to p (if $p \geqslant 2$) or a positive definite form (if $p = -1$). The summands qf_q in (22) are called *Jordan constituents* of f, and (22) itself is a *Jordan decomposition* of f. We call q the *scale* of the constituent qf_q. Note that, when $p = -1$, (22) is the familiar result that any form can be written as a sum of definite forms:

$$f = 1 \cdot f_1 \oplus (-1) f_{-1},$$

where f_1 and f_{-1} are positive definite.

For $p \neq 2$ the set of values of q occurring in (22), together with the *dimensions* $n_q = \dim f_q$ and *signs*

$$\epsilon_q = \left(\frac{\det f_q}{p}\right),$$

form a complete set of invariants for f (see Theorem 9). The case $p = 2$ presents additional complexities and will be discussed in §§7.3-7.6. In the case $p = -1$, $\epsilon_q = +1$ since $\det f_q$ is positive, and so n_{+1} and n_{-1} are the only invariants (this is Sylvester's law of inertia, §6.2). By convention, $+1$ and -1 will usually be abbreviated to $+$ and $-$, and so we write n_+ for n_{+1}, n_- for n_{-1}.

7.2 The *p*-adic symbol for a form. For $p = -1$ we express the fact that $n_+ = a$ and $n_- = b$ by the (-1)*-adic symbol*

$$+^a \; -^b .$$

For other odd p we shall use a *p-adic symbol* which is a formal product of the "factors"

$$q^{\epsilon_q n_q} .$$

For example if $p = 3$, the symbol

$$1^{-2}\, 3^{+5}\, 9^{+1}\, 27^{-3} \tag{23}$$

represents a form

$$f = f_1 \oplus 3f_3 \oplus 9f_9 \oplus 27f_{27}$$

with $\dim f_1 = 2$, $\dim f_3 = 5$, $\dim f_9 = 1$, $\dim f_{27} = 3$, where the determinants of f_3, f_9 are quadratic residues modulo 3 and those of f_1, f_{27} are nonresidues.

In these symbols we may adopt certain obvious abbreviations, such as replacing (23) by $1^{-2}\, 3^5\, 9\, 27^{-3}$.

Theorem 9. *For $p \neq 2$, two quadratic forms f and g are equivalent over the p-adic integers if and only if they have the same invariants n_q, ϵ_q for each power q of p, or equivalently if and only if they have the same p-adic symbol.*

The proofs of Theorems 9 and 10 are omitted (see the remarks at the beginning of this section). Theorem 9 makes two assertions. First, q, ϵ_q, n_q are a complete set of invariants for the Jordan constituents qf_q, and second, the invariants of the Jordan constituents are invariants of f. Both assertions must be modified when $p = 2$.

7.3 2-adic invariants. Let the 2-adic Jordan decomposition of f be

$$f = f_1 \oplus 2f_2 \oplus 4f_4 \oplus \cdots \oplus qf_q \oplus \cdots . \qquad (24)$$

Then the invariants of qf_q are the quantities
q, the *scale* of qf_q,
$S_q = \mathrm{I}$ or II (see below), the *type* of f_q, which is the *scaled type* of qf_q,
$n_q = \dim f_q$ (the *dimension* of f_q or qf_q),
$\epsilon_q = \left(\frac{\det f_q}{2}\right)$ (the *sign* of f_q or qf_q), and
t_q, the *oddity* of f_q (see §5.1).

We define S_q to be I if qf_q represents an odd multiple of q, and otherwise II (compare §2.4 of Chap. 2). Equivalently, S_q is I if and only if there is an odd entry on the main diagonal of the matrix representing f_q, and otherwise II. If f_q (of type I) has been diagonalized, then t_q is its trace, read modulo 8. If f_q has type II, $t_q = 0$.

7.4 The 2-adic symbol. The 2-adic symbol representing a given Jordan decomposition (24) of f is a formal product of factors

$$q_{t_q}^{\epsilon_q n_q} \quad \text{or} \quad q^{\epsilon_q n_q} ,$$

where the former indicates a constituent qf_q for which f_q has type I and

$$\left(\frac{\det f_q}{2}\right) = \epsilon_q , \quad \dim f_q = n_q , \quad \text{oddity } (f_q) = t_q ,$$

while the latter indicates a constituent qf_q for which f_q has type II and

$$\left(\frac{\det f_q}{2}\right) = \epsilon_q , \quad \dim\ f_q = n_q , \quad \text{oddity } (f_q) = 0 .$$

We shall sometimes write

$$q_{\mathrm{I}}^{\epsilon_q n_q} \text{ for } q_{t_q}^{\epsilon_q n_q} , \ q_{\mathrm{II}}{}^{\epsilon_q n_q} \text{ for } q^{\epsilon_q n_q}$$

(The value of t_q is often unimportant).

For example $1^{-2}2_5^{+3}4_3^{-1}8^{+4}$ (or $1_{\mathrm{II}}^{-2}\ 2_5^{+3}\ 4_3^{-1}\ 8_{\mathrm{II}}^{+4}$) represents a form having a Jordan decomposition

$$f_1 \oplus 2f_2 \oplus 4f_4 \oplus 8f_8 ,$$

in which f_1, f_2, f_4, f_8 have dimensions 2, 3, 1, 4 and determinants congruent to ± 3, ± 1, ± 3, ± 1 (mod 8) respectively, f_1 and f_8 have type II, while f_2 and f_4 are of type I and can be put into diagonal form with traces congruent to 5 and 3 (mod 8) respectively.

7.5 Equivalences between Jordan decompositions. So far we have described the invariants for a *Jordan constituent* qf_q. Unfortunately a form may have several essentially different Jordan decompositions, and so can have several different 2-adic symbols. The precise rule giving all such equivalences is as follows.

Theorem 10. *Two forms f and f' with respective invariants*

$$n_q, S_q, \epsilon_q, t_q \quad \textit{and} \quad n'_q, S'_q, \epsilon'_q, t'_q$$

are 2-adically equivalent just if
(i) $n_q = n'_q$, $S_q = S'_q$ *for all* q, *and*
(ii) *for each integer* m *(including negative integers) for which* f_{2^m} *has type* II, *we have*

$$\sum_{q < 2^m} (t_q - t'_q) \equiv 4(\min(a,m) + \min(b,m) + \cdots) \pmod 8$$

where $2^a, 2^b, \ldots$ *are the values of* q *for which* $\epsilon_q \neq \epsilon'_q$.

Although Theorem 10 completely describes all the 2-adic equivalences, it is often simpler to use the following ideas.

Compartments and trains. Suppose f has a Jordan decomposition (24). By an *interval* of forms we mean all the forms qf_q, even those of zero dimension, for which $q_1 \leqslant q \leqslant q_2$, where q_1, q_2 are powers of 2. A *compartment* is a maximal interval in which all forms are of scaled type I, and a *train* is a maximal interval having the property that for each pair of adjacent forms at least one is of scaled type I. Thus in

$$1^{+2}[2_6^{-2}\, 4_5^{+3}]8^{+0}[16_1^{+1}]32^{+2}\!:\; 64^{-2}\!:\; 128^{-4}[256_3^{-1}]512^{+0} \tag{25}$$

the square brackets enclose the compartments and the trains are separated by colons. Notice that here one train has two compartments, while another train has none.

There are two ways in which such symbols may be altered and yet still represent 2-adically equivalent forms.

(i) **Oddity fusion.** Two 2-adic symbols represent the same form if the only change is that the oddities have been altered in a way that does not affect their total sum over any compartment.

So we may replace the individual oddity markers in a compartment by their total (modulo 8), written as a subscript to the entire compartment. For example we may replace $[2_6^{-2}\, 4_5^{+3}]$ in (25) by $[2^{-2}\, 4^{+3}]_3$.

(ii) **Sign walking.** A form is unaltered if the signs ϵ_q of any two terms in a train are simultaneously changed, provided certain oddities are altered by 4. Let $\epsilon_{q_1}, \epsilon_{q_2}$, $q_1 < q_2$, be the signs we wish to change. We imagine walking along the train from the term for q_1 to that for q_2. Our walk consists of a number of steps between adjacent forms f_q and f_{2q}, and each such step involves just one compartment, since at least one of f_q, f_{2q} is of type I, and if they are both of type I they are in the same compartment. Then the rule is that the total oddity of a compartment must be changed by 4 modulo 8, precisely when the number of steps that involve that compartment is odd.

Suppose for example we wish to change the signs corresponding to f_2 and f_{16} in the train

$$1^2\,[2^{-2}\,4^3]_3\,8^0\,[16^1]_1\,32^2\,. \tag{26}$$

The walk has three steps, from f_2 to f_4, f_4 to f_8, and f_8 to f_{16}. The first two steps affect the first compartment and the third step affects the second compartment. The resulting symbol is therefore

$$1^2\,[2^2\,4^3]_3\,8^0[16^{-1}]_5\,32^2 \tag{27}$$

Alternatively, a walk of just one step from f_2 to f_4 in (26) would lead to

$$1^2[2^2\,4^{-3}]_7\,8^0\,[16^1]_1\,32^2\,. \tag{28}$$

All of (26), (27), (28) represent equivalent forms.

The effect of the three-step walk could also be achieved by three separate one-step walks, except that at an intermediate stage the symbol would contain a factor 8^{-0}, corresponding to an impossible Jordan constituent. Transformations involving such impossible constituents are quite legal provided the end results are meaningful.

7.6 A canonical 2-adic symbol. Using these rules we can arrange that there is at most one minus sign per train, which can be attached to any form of nonzero dimension. One convenient rule is to put this sign on the earliest nonzero dimensional form of a train. If this convention is adopted and only the total oddities of the compartments are given, the resulting symbol is absolutely unique and may be taken as a canonical symbol for the form. Thus for (25) the canonical symbol is

$$1^{-2}[2^{+2}4^{+3}]_7\,8^{+0}[16^{+1}]_1\,32^{+2}\colon\; 64^{-2}\colon\; 128^{+4}[256^{+1}]_7\,512^{+0}\,,$$

which would be further abbreviated in practice to

$$1^{-2}[2^2\,4^3]_7\,[16]_1\,32^2\colon\; 64^{-2}\colon\; 128^4\,[256]_7\,.$$

Then two forms are 2-adically equivalent if and only if their canonical symbols are identical.

7.7 Existence of forms with prescribed invariants. It is important to specify exactly which conceivable systems of invariants actually correspond to quadratic forms. There are three sets of conditions.

The determinant conditions for each p. The p-adic square classes of the determinant as computed from the p-adic symbols must agree with its known value. In other words the product of all the signs ϵ_q in the p-adic symbol must be $\left(\frac{a}{p}\right)$ (29)

where $\det(f) = p^\alpha a$ and $(a, p) = 1$.

The oddity condition, relating all p. From the p-adic symbols we can compute the invariants appearing in the oddity formula. Thus for $p = -1$,

$$\text{signature}(f) = r - s\ ,$$

if the (-1)-adic symbol is $+^r -^s$; and for $p \geqslant 3$,

$$p\text{-excess}(f) \equiv \sum_q n_q(q-1) + 4k_p \pmod 8\ ,$$

where the n_q are the dimensions of the Jordan constituents and k_p is the number of *antisquare terms* (i.e. q not a square and $\epsilon_q = -1$) in the p-adic symbol; and for $p = 2$,

$$\text{oddity}(f) \equiv \sum_q t_q + 4k_2 \pmod 8\ ,$$

from the 2-adic symbol. Then these quantities must be related by the oddity formula

$$\text{signature}(f) + \sum_{p \geqslant 3} p\text{-excess}(f) \equiv \text{oddity}(f) \pmod 8\ . \tag{30}$$

The existence condition for each Jordan constituent. Each term in the p-adic symbol must correspond to an existing form. If $p \neq 2$, for each Jordan constituent qf_q of dimension n and sign ϵ we must have

$$\text{if } n = 0 \text{ or } p = -1 \text{ then } \epsilon = +\ . \tag{31}$$

For $p = 2$ the following must hold:

$$\text{for } n = 0,\quad \text{type} = \text{II and } \epsilon = +\ , \tag{32}$$

$$\text{for } n = 1,\ \begin{cases} \epsilon = + \ \Rightarrow\ t \equiv \pm 1 \pmod 8\ , \\ \epsilon = - \ \Rightarrow\ t \equiv \pm 3 \pmod 8\ , \end{cases} \tag{33}$$

$$\begin{matrix}\text{for } n = 2 \\ \text{and type I}\end{matrix} \begin{cases} \epsilon = + \ \Rightarrow\ t \equiv 0 \text{ or } \pm 2 \pmod 8\ , \\ \epsilon = - \ \Rightarrow\ t \equiv 4 \text{ or } \pm 2 \pmod 8\ , \end{cases} \tag{34}$$

while for general n we have $t \equiv n \pmod 2$, and

$$t \equiv 0 \pmod 8 \text{ for type II, so that } n \text{ odd} \Rightarrow \text{type I}. \tag{35}$$

Theorem 11. *If a system of putative p-adic symbols for each p satisfies the determinant, oddity and p-adic existence conditions* (29)–(35), *then there exists an integral quadratic form with these p-adic symbols.*

When working with the abbreviated form of a 2-adic symbol, it is helpful to know that the only existence condition on a compartment containing two or more Jordan constituents is that its total oddity must have the same parity as its total dimension.

7.8 A symbol for the genus. We can combine the significant portions of our p-adic symbols to give a handy notation for the entire genus (that happily generalizes some notation we have used elsewhere [Con13], [Con33], [Con34]). This symbol has the form

$$\mathrm{I}_{r,s}(\cdots) \quad \text{or} \quad \mathrm{II}_{r,s}(\cdots) ,$$

where the Roman numeral is the type of the entire form, i.e. the type of its 2-adic Jordan constituent f_1; the subscripts indicate the (-1)-adic symbol $+^r -^s$; and the parenthesis contains the usual symbols

$$q^{\pm m},\ q_t^{\pm m},\ q_{\mathrm{I}}^{\pm m},\ q_{\mathrm{II}}^{\pm m}$$

for the powers $q > 1$ of all primes 2, 3, ... The subscripts t, I or II may be omitted when their values can be deduced from the oddness of m or the oddity formula (30).

The symbols

$$1^{\pm m},\ 1_t^{\pm m}\ (\text{or}\ \ 1_{\mathrm{I}}^{\pm m})$$

corresponding to the constituents f_1 in each p-adic Jordan decomposition have been omitted. However

- the sign $\pm$ can be recovered from $\det f$,
- the number m can be recovered from $\dim f = r+s$,
- the type I or II (for $p = 2$) is displayed, and
- the oddity t (when relevant) can be computed from the oddity formula (30).

For example $\mathrm{I}_{r,s}(2)$ has determinant $(-1)^s 2$, and so its p-excess is 0 for $p \geqslant 3$. From (30) the oddity is $r-s$, and therefore the 2-adic symbol is

$$1_{\mathrm{I}}^{\pm(r+s-1)}\, 2_{\mathrm{I}}^{1} = [1^{+(r+s-1)}\, 2^1]_{r-s} ,$$

using (29). Similarly $\mathrm{II}_{r,s}(3)$ (whose determinant is $(-1)^s 3 = \pm 3$) has 3-adic symbol $1^{\pm(r+s-1)} 3^1$ (with the same $\pm$), yielding a 3-excess of 2, and 2-adic symbol $1_{\mathrm{II}}^{-(r+s)}$, since $\det f \equiv \pm 3 \pmod 8$. The 2-adic symbol for $\mathrm{I}_{r,s}(3)$ would be $1_{r-s+2}^{-(r+s)}$.

A variant of this notation indicates the total dimension $r+s$ explicitly and treats -1 like any other prime, thus writing

$$\mathrm{I}_{r+s}(-^s X) \quad \text{or} \quad \mathrm{II}_{r+s}(-^s X)$$

for

$$\mathrm{I}_{r,s}(X) \quad \text{or} \quad \mathrm{II}_{r,s}(X)$$

respectively.

8. Classification of forms of small determinant and of p-elementary forms

8.1 Forms of small determinant. Using the notation introduced in the previous section, the distinct genera of forms of any given determinant can be classified in a systematic way. One first writes down all possible p-adic symbols for $p = -1$, 2 and all p dividing the determinant (it is best to handle $p = 2$ last). The determinant condition (29) and the rules for manipulating trains (§7.5) are used to control the signs. Then the oddity formula (30) and the existence conditions (§7.7) lead to congruences (modulo 8) relating the signature to the oddity parameters. We illustrate this process by classifying the genera of forms with determinants ± 1 and ± 3.

Determinant ± 1. Let the (-1)-adic symbol be $+^r -^s$, with $r+s = n$. Since the determinant is $\equiv \pm 1 \pmod 8$, the possible 2-adic symbols are

$$1_t^{+(r+s)} \quad \text{and} \quad 1^{+(r+s)},$$

for which the respective oddity conditions read

$$r-s \equiv t \quad \text{and} \quad r-s \equiv 0 \quad (\text{mod } 8).$$

The former determines t and corresponds to the genus symbol $I_{r,s} = I_{r,s}(1)$, which exists for all possible signatures except $r = s = 0$. The latter gives $II_{r,s} = II_{r,s}(1)$, which exists only for signatures divisible by 8. (The existence conditions are trivially satisfied.) This explains lines 1 and 2 of Table 15.4 below.

Determinant ± 3. The possible p-adic symbols for

$p =$	-1	3	2

are

$I_{r,s}(3)$:	$+^r -^s$	$1^{\pm(r+s-1)}3^1$	$1_t^{-(r+s)}$
$I_{r,s}(3^{-1})$:	$+^r -^s$	$1^{\mp(r+s-1)}3^{-1}$	$1_t^{-(r+s)}$
$II_{r,s}(3)$:	$+^r -^s$	$1^{\pm(r+s-1)}3^1$	$1^{-(r+s)}$
$II_{r,s}(3^{-1})$:	$+^r -^s$	$1^{\mp(r+s-1)}3^{-1}$	$1^{-(r+s)}$,

where the ambiguous sign $\pm$ is $-^s$, and the 2-adic sign is $-$ since the determinant is $\equiv \pm 3 \pmod 8$. These lead to the respective oddity conditions

$$r-s+2 \equiv t,$$
$$r-s+6 \equiv t,$$
$$r-s+2 \equiv 0,$$
$$r-s+6 \equiv 0,$$

the first two of which determine t, while the last two give conditions on the signature. The existence conditions on the Jordan constituents are automatically satisfied except for small n, when they are best handled by consideration of all pairs (r, s) with $r+s = n$. Thus in the case $I_{r,s}(3)$ and for (r,s) equal to

$$(0,0), \qquad (1,0), \qquad (0,1), \qquad (2,0), \qquad (1,1), \qquad (0,2)$$

we find the 2-adic symbol to be

$$1_2^{-0}, \qquad 1_3^{-1}, \qquad 1_1^{-1}, \qquad 1_4^{-2}, \qquad 1_2^{-2}, \qquad 1_0^{-2}$$

respectively, for which the existence condition holds in only three cases:

$$\times \qquad \surd \qquad \times \qquad \surd \qquad \surd \qquad \times$$

The three failures are classified as $n_{r-s} = 0,\ 1_{-1},\ 2_{-2}$ in Table 15.4.

Similar arguments lead to the following theorem.

Theorem 12. *All genera of forms with* $|determinant| \leqslant 11$ *are as indicated in Table* 15.4.

The theory of spinor genera (see Corollary 21 below) shows that for indefinite cases of dimension $\geqslant 3$ these are also the integral equivalence classes.

In Table 15.4 we have used the convention that all ambiguous signs in any row of the table are linked. The last column lists the exceptions, if any, in the form n_σ, where n is the dimension and σ the forbidden signature. If all signatures of dimension n are excluded the subscript is omitted. The asterisk * towards the end of the table indicates that when $n = 1$, this genus must be interpreted as $II_{r,s}(2^{-1} \times 5^{-1})$.

8.2 p-elementary forms. The same methods enable as to classify the so-called *p-elementary* forms, that is, forms for which L^*/L is a nontrivial elementary abelian p-group, where L is the lattice of the form and L^* is the dual lattice. These forms were partially classified by Rudakov and Shafarevich [Rud1], [Rud2], and the results used by Nikulin in [Nik1]-[Nik6] (see also [Dol1]).

Theorem 13. (a) *For* $p \geqslant 3$ *the distinct genera of p-elementary forms are*

$$I_{r,s}(p^{\pm k}) \quad \text{for all signatures} \quad r-s\ ,$$

and

$$II_{r,s}(p^{\pm k}) \quad \text{for} \quad r-s \equiv \pm 2-2-(p-1)k \pmod 8\ ,$$

except that in either case when $k = n$ $(= r+s)$ *the sign must be* $\left(\frac{-1}{p}\right)^s$.

Table 15.4. All genera of forms with |det| ≤ 11.

|Det|	Genus	Signature (mod 8)	Exceptions n_{r-s}
1	$\mathrm{I}_{r,s}$	all	0
1	$\mathrm{II}_{r,s}$	0	none
2	$\mathrm{I}_{r,s}(2)$	all	0,1
2	$\mathrm{II}_{r,s}(2)$	± 1	none
3	$\mathrm{I}_{r,s}(3^{\pm 1})$	all	$0, 1_{\mp 1}, 2_{\mp 2}$
3	$\mathrm{II}_{r,s}(3^{\pm 1})$	∓2	none
4	$\mathrm{I}_{r,s}(4_{\pm 1})$	all	$0, 1, 2_{\mp 2}, 3_{\mp 3}$
4	$\mathrm{II}_{r,s}(4^{1})$	± 1	none
4	$\mathrm{II}_{r,s}(4^{-1})$	± 3	none
4	$\mathrm{I}_{r,s}(2_{\mathrm{I}}^{2})$	all	0,1,2
4	$\mathrm{II}_{r,s}(2_{\mathrm{I}}^{2})$	0, ± 2	0
4	$\mathrm{I}_{r,s}(2_{\mathrm{II}}^{2})$	all	$0, 1, 2, 3_{\pm 3}, 4_{\pm 4}$
4	$\mathrm{II}_{r,s}(2_{\mathrm{II}}^{2})$	0	0
4	$\mathrm{II}_{r,s}(2_{\mathrm{II}}^{-2})$	4	none
5	$\mathrm{I}_{r,s}(5)$	all	0
5	$\mathrm{I}_{r,s}(5^{-1})$	all	$0, 1, 2_0$
5	$\mathrm{II}_{r,s}(5)$	4	none
5	$\mathrm{II}_{r,s}(5^{-1})$	0	0
6	$\mathrm{I}_{r,s}(2 \times 3^{\pm 1})$	all	0,1
6	$\mathrm{II}_{r,s}(2 \times 3^{\pm 1})$	odd	$1_{\pm 1}$
7	$\mathrm{I}_{r,s}(7^{\pm 1})$	all	$0, 1_{\mp 1}, 2_{\mp 2}$
7	$\mathrm{II}_{r,s}(7^{\pm 1})$	± 2	none
8	$\mathrm{I}_{r,s}(8_{\pm 1})$	all	$0, 1, 2_{\mp 2}, 3_{\mp 3}$
8	$\mathrm{I}_{r,s}(8_{\pm 3}^{-1})$	all	$0, 1, 2_0, 2_{\mp 2}, 3_{\mp 1}$
8	$\mathrm{II}_{r,s}(8)$	± 1	none
8	$\mathrm{II}_{r,s}(8^{-1})$	± 1	1
8	$\mathrm{I}_{r,s}(2 \times 4)$	all	0,1,2
8	$\mathrm{II}_{r,s}(2 \times 4)$	even	0
8	$\mathrm{I}_{r,s}(2^{3})$	all	0,1,2,3
8	$\mathrm{II}_{r,s}(2^{3})$	odd	1
9	$\mathrm{I}_{r,s}(9^{\pm 1})$	all	$0, 1_{\mp 1}$
9	$\mathrm{II}_{r,s}(9^{\pm 1})$	0	0
9	$\mathrm{I}_{r,s}(3^{2})$	all	$0, 1, 2_0$
9	$\mathrm{I}_{r,s}(3^{-2})$	all	$0, 1, 2_{\pm 2}$
9	$\mathrm{II}_{r,s}(3^{2})$	4	none
9	$\mathrm{II}_{r,s}(3^{-2})$	0	0
10	$\mathrm{I}_{r,s}(2 \times 5^{\pm 1})$	all	0,1
10	$\mathrm{II}_{r,s}(2 \times 5)$	± 3	none
10	$\mathrm{II}_{r,s}(2 \times 5^{-1})$	± 1	none*
11	$\mathrm{I}_{r,s}(11^{\pm 1})$	all	$0, 1_{\mp 1}, 2_{\mp 2}$
11	$\mathrm{II}_{r,s}(11^{\pm 1})$	∓ 2	none

Table 15.5. The 2-elementary forms.

$\mathrm{I}_{r,s}(2_{\mathrm{I}}^{k})$,	$0 < k < n$;
$\mathrm{I}_{r,s}(2_{\mathrm{II}}^{k})$,	k even $< n$, and if $k = n-1$ then $r-s \equiv \pm 1 \pmod 8$, if $k = n-2$ then $r-s \not\equiv 4 \pmod 8$;
$\mathrm{II}_{r,s}(2_{\mathrm{I}}^{k})$,	$0 < k \equiv n \pmod 2$, and if $k = 1$ then $r-s \equiv \pm 1 \pmod 8$, if $k = 2$ then $r-s \not\equiv 4 \pmod 8$;
$\mathrm{II}_{r,s}(2_{\mathrm{II}}^{k})$,	n and k even, $r-s \equiv 0 \pmod 8$;
$\mathrm{II}_{r,s}(2_{\mathrm{II}}^{-k})$,	n and k even, $0 < k < n$, $r-s \equiv 4 \pmod 8$.

When these forms are indefinite of dimension $\geqslant 3$, *each genus contains just one spinor genus, and so, by Theorem* 14, *just one class.*

(b) *The distinct genera of* 2*-elementary forms are as shown in Table* 15.5. *In all indefinite cases these genera again contain just one class.*

9. The spinor genus

9.1 Introduction. The spinor genus, introduced by Eichler ([Eic1]; see also [Ear1], [Ear3]-[Ear5], [Hsi2]), is a refinement of the notion of genus, and its importance stems from the following remarkable result.

Theorem 14. (Eichler [Eic1]; see also [Cas3, Chap. 11, Theorem 1.4], [Wat3, Theorem 63]). *For indefinite forms of dimension at least* 3, *a spinor genus contains exactly one integral equivalence class of forms.*

The current use of the term differs slightly from Eichler's (by replacing the orthogonal group by the special orthogonal group in some places). The description given here is in essence due to Watson [Wat3], although since Watson's description is rather complicated our treatment is based on the version given by Cassels [Cas3], which we recommend to the reader who would like to see the proofs. Cassels does not however give quite enough information about spinor norms to justify the claim that the spinor genus is a practicably computable invariant. We have therefore quoted from the relevant theorem of Watson, and have taken some pains to produce a mechanical rule for computing the spinor kernel.

The invariants we described in the previous section are invariants of the genus. If two forms are in the same genus we can find a rational transformation relating them whose denominator can be made relatively prime to any given integer. If this transformation is integral (i.e. if the denominator is 1) the forms are in the same class. The concept of spinor genus arises when we apply local arguments to this rational transformation to see what obstruction there is to making it an integral one.

Each genus is partitioned into a number of spinor genera (the number is always a power of 2). There is a group of *spinor operators* acting transitively on the spinor genera, so that we can obtain any spinor genus in the genus from a fixed one by applying a suitable spinor operator. Thus the division into spinor genera is determined once we name the spinor operators and are able to decide when a given spinor operator acts trivially, i.e. is in the *spinor kernel.* Our Theorems 15-17 serve as operational definitions for the notions of spinor operator and spinor kernel.

The computations depend on the calculation of the spinor norms of operations in certain orthogonal groups. Excellent references are Kneser [Kne3], Hsia [Hsi1], Earnest and Hsia [Ear2]. These references have the additional merit of discussing the problem over more general rings and of giving the best possible conditions on the indices of prime divisors of the discriminant to ensure that an indefinite genus contains just one class.

9.2 The spinor genus. If f and g are forms of determinant d in the same genus, then they are rationally equivalent by some transformation whose denominator is prime to $2d$ [Cas3, Chap. 9], [Wat3, p. 78]. Hence we can find corresponding lattices L and M for which

$$[L:\ L\cap M] = [M:\ L\cap M] = r \quad \text{(say)}\ , \tag{36}$$

for some number r which is prime to $2d$. We may paraphrase Watson's redefinition of spinor genus by saying that f and g are in the same spinor genus if and only if r is an automorphous number (as defined below).

Theorem 15. (Obtained from Theorem 70 of [Wat3].) (a) *The spinor genus of g (or M) is determined by that of f (or L) together with the number r. Using SG to denote spinor genus we shall write*

$$SG(g) = SG(f) * \Delta(r)\ ,$$

$$SG(M) = SG(L) * \Delta(r)\ .$$

and call "$\Delta(r)$" a spinor operator.
(b) *Furthermore, if the dimension is at least 3, then $SG(f) * \Delta(r)$ is defined for every natural number r prime to $2d$.*

To complete the definition we need to know how to find the r's for which $\Delta(r)$ is in the spinor kernel, i.e. $\Delta(r)$ fixes every spinor genus. This is done in the next section.

Remarks. (i) Part (b) of Theorem 15 fails for dimension 2. (ii) The theorem is usually applied to indefinite forms of dimension $\geqslant 3$, in which case by Theorem 14 we need not distinguish between the spinor genus of a form and the form itself.

9.3 Identifying the spinor kernel. Any element of the orthogonal group of f, over a field not of characteristic 2, can be written as a product of reflections in certain vectors $v_1, v_2, \ldots, v_k$. The *spinor norm* (defined only up to multiplication by square factors) of this operation is $f(v_1) \cdots f(v_k)$. This operation is *proper* (i.e. of determinant 1) or *improper* (determinant -1) according as k is even or odd. The proper operations form the *special orthogonal group* of the form. Only some elements of the orthogonal group of f have integral matrix entries: these are the *integral automorphisms* of f.

We call $r \in \mathbf{Q}$ *automorphous* if it is the spinor norm of a proper integral automorphism of f. Similarly a p-adic number $A = p^\alpha a \in \mathbf{Q}_p$ is *p-adically automorphous* if it is the spinor norm of a proper p-adic integral automorphism of f.

The following theorems are due to Eichler [Eic1] and Watson [Wat3]. We have obtained them by translating Cassels's versions into our notation – see in particular Theorem 3.1 of Cassels [Cas3, Chap. 11] and its corollary.

Theorem 16. *The spinor kernel consists of the spinor operators $\Delta(r)$ for which the positive integer r is an automorphous number not divisible by any prime divisor of $2d$.*

It is important that we can compute the spinor kernel "locally", in fact by performing a simple calculation for each prime in a certain finite set Π. In the terminology of the next section we have:

Theorem 17. *The spinor kernel is generated by the spinor operators $\Delta_p(A)$ for which $p \in \Pi$ and A is a p-adically automorphous number.*

9.4 Naming the spinor operators for the genus of f. Let Π be any finite set of primes that contains -1, 2 and all primes dividing $d = \det f$, where f is some form in the genus. (We shall see later that often some primes can be removed from Π with no loss of information).

Since the spinor operators depend only on the square class of r, we can name them by sequences $(\ldots, r_p, \ldots)_{p \in \Pi}$, in which each r_p is a p-adic unit square class. In this notation the group operation is componentwise multiplication. Rational or p-adic integers can be regarded as spinor operators in the following way.

(i) To any rational integer r not divisible by any $p \in \Pi$ there corresponds the spinor operator whose p-coordinate is the p-adic square class of r for each p. We shall write this as

$$\Delta(r) = (r, r, \ldots) \tag{37}$$

(the square class being understood).

(ii) To each p-adic integer $A = p^\alpha a$ there corresponds the spinor operator $\Delta_p(A)$ whose p_1-coordinate (for $p_1 \neq p$) is the p_1-adic square class of p^α, and whose p-coordinate is the p-adic square class of a:

$$\Delta_p(A) = (p^\alpha, p^\alpha, \ldots, p^\alpha, a, p^\alpha, \ldots) . \tag{38}$$

9.5 Computing the spinor kernel from the p-adic symbols. In view of Theorem 17 the spinor kernel is determined once we know the p-adically automorphous numbers. A vector v is called a *p-adic root vector* for f if and only if the reflection in v is a p-adically integral automorphism of f. Of course this reflection has determinant -1.

Theorem 18. (Based on Theorem 81 of [Wat3]). *A p-adic number is p-adically automorphous for f if and only if it is the product of an even number of norms of p-adic root vectors for f.*

Remark. For odd p it follows from [Cas3, p. 115, Corollary 1] that every p-adically integral automorphism is the product of p-adically integral reflections. For $p = 2$ this is usually true but not always [O'Me4].

An algorithm for finding the p-adically automorphous numbers for a form.

In the rest of this section we describe a mechanical rule for finding the p-adically automorphous numbers, that can be justified using Theorem 18. §9.6 contains some examples and simplifications.

The algorithm is rather complicated to state, but is completely mechanical and very easy to apply. If $p \neq 2$ we first diagonalize f, or if $p = 2$ we express f as a direct sum of forms of the shapes

$$(qx) \quad \text{and} \quad \begin{pmatrix} qa & qb \\ qb & qc \end{pmatrix},$$

where q is a power of 2, a and c are divisible by 2, but x, b and $ac-b^2$ are not (see Theorem 2). We now prepare a list in two parts:

(I) If $p \neq 2$, all the diagonal entries, or if $p = 2$ the diagonal entries qx. (These numbers are not yet to be interpreted modulo squares, and of course the list may contain repeated entries.)

(II) Only if $p = 2$: the numbers $2qu_1$, $2qu_3$, $2qu_5$, $2qu_7$ for every direct summand $\begin{pmatrix} qa & qb \\ qb & qc \end{pmatrix}$.

Then the group of p-adically automorphous numbers is generated by the p-adic square classes of the ratios (or products) of all pairs of numbers from the total list, supplemented by:

(i) all p-adic units if

either $p \geqslant 3$ and f_q has dimension $\geqslant 2$ for any q,
or $p = 2$ and $f_q \oplus f_{2q} \oplus f_{4q} \oplus f_{8q}$ has dimension $\geqslant 3$ for any q,

and

(ii) the square classes

$2u_1,\quad 2u_3,\quad u_5,\quad u_3,\quad u_7,$

whenever $p = 2$ and part (I) of the list contains two entries whose ratio has the form

$u_1,\quad u_5,\quad (1 \text{ or } 4 \text{ or } 16)u_{\text{odd}},\quad (2 \text{ or } 8)u_{1 \text{ or } 5},\quad (2 \text{ or } 8)u_{3 \text{ or } 7}$

respectively.

Example. For the form $f = \text{diag}\{3, 16\}$ and the prime $p = 2$, part (I) of the list consists of $\{3 = u_3, 16 = 16u_1\}$ and part (II) is empty. Since the ratio of 16 to 3 has the form $16u_3$, we supplement the total list by u_5, according to (ii). Thus the 2-adically automorphous numbers are generated by the square classes of $\{16u_1, u_3, u_5\}$, i.e. are $\{u_1, u_3, u_5, u_7\}$.

The supplementation rules correspond to the possibilities for root vectors not necessarily in the basis. Thus e_1+e_2, of norm $2 = 2u_1$, is a 2-adic root vector for $\begin{pmatrix}1 & 0\\ 0 & 1\end{pmatrix}$, illustrating the first rule in (ii).

9.6 Tractable and irrelevant primes

The following considerations may be used to simplify the calculations. If there is a prime p such that, for each p-adic unit u, the spinor operator

$$\Delta_p(u) = (1, 1, ..., 1, u, 1, ...)_{...,p,...} \tag{39}$$

lies in the spinor kernel, then plainly the p-coordinate can be deleted, since it conveys no information modulo the spinor kernel. Such p are called *tractable*. For example, -1 is always tractable, as is the prime 2 in the above example.

However, a tractable prime may still have some effect on spinor genus calculations, since if p itself is automorphous, the spinor kernel will contain $\Delta_p(p)$, which has nontrivial values in coordinates other than the p-th. For example, if f is indefinite then -1 is (-1)-adically automorphous, so that $\Delta_{-1}(-1) = (+1, -1, -1, ...)_{-1,2,3,...}$ is in the spinor kernel.

If the p-adically automorphous numbers are *precisely* the square classes of the p-adic units (as happens when $p \neq -1$ or 2 and $p \nmid \det(f)$), then p is not only tractable but *irrelevant*. Irrelevant primes do not affect the computation of the spinor genus in any way.

Example. We consider the form

$$f = \begin{bmatrix} 2 & 1 & 0 \\ 1 & 2 & 0 \\ 0 & 0 & 18 \end{bmatrix}$$

(discussed in [Wat3, p. 115]). For $p \neq 2$ we can diagonalize f, obtaining diag $\{2, 3/2, 18\}$. To find the spinor kernel we proceed as follows.

$p = -1$: list (I) $= \{2 = u, 3/2 = u, 18 = u\}$, so u is the only (-1)-adically automorphous number.
$p = 2$: list (I) $= \{18 = 2u\}$, list (II) $= \{2u_1, 2u_3, 2u_5, 2u_7\}$, so the 2-adically automorphous numbers are $\{u_1, u_3, u_5, u_7\}$, and 2 is tractable.
$p = 3$: list (I) $= \{2 = u_-, 3/2 = 3u_-, 18 = 9u_-\}$, so the 3-adically automorphous numbers are $\{u_+, 3u_+\}$.

We need retain only the 3-coordinate since 3 is the only intractable prime, and the spinor kernel is generated by

$$\Delta_2(u_1, u_3, u_5 \text{ or } u_7) = (u_+)_3 ,$$

$$\Delta_3(u_+ \text{ or } 3u_+) = (u_+)_3$$

So $(u_-)_3$ is not in the spinor kernel, and therefore the genus of f contains two distinct spinor genera, one containing f, the other containing $f \bullet (u_-)_3$. A representative for the second spinor genus is ([Wat3, p. 115])

$$\begin{bmatrix} 6 & 3 & 0 \\ 3 & 6 & 0 \\ 0 & 0 & 2 \end{bmatrix} .$$

9.7 When is there only one class in the genus? In practice one usually finds that all primes are tractable and so the spinor genus coincides with the genus (and therefore, in the case of indefinite forms of dimension at least 3, the genus contains only one class). This section gives some conditions which guarantee that this will happen.

Theorem 19. *If f is indefinite and the genus of f contains more than one class, then for some p (possibly -1), f can be p-adically diagonalized and the diagonal entries all involve distinct powers of p.*

Proof. Suppose the contrary. Then $\dim f \geqslant 3$, since otherwise in the (-1)-adic (real) diagonalization the terms involve distinct powers of -1. We now quote Eichler's theorem (Theorem 14) to see that the class coincides with the spinor genus and there must therefore be an intractable prime p. If $p \geqslant 3$ we know that none of the p-adic Jordan constituents f_q can have dimension $\geqslant 2$ and the result follows. So we may conclude that 2 is the only intractable prime. No nontrivial 2-adic Jordan constituent may have scaled type II, since then all 2-adic units are automorphous. Thus f is 2-adically diagonalizable. If any two of the diagonal terms involve the same power of 2, the algorithm implies that 5 is 2-adically automorphous, and so $(u_5)_2$ is in the spinor kernel. But since -1 is (-1)-adically automorphous, $(u_7)_2$ is also in the spinor kernel, and these generate all the possibilities. Therefore the diagonal entries may only involve distinct powers of 2. This completes the proof.

With a little more care the argument can be refined to give the following result.

Theorem 20. *Suppose f is an indefinite form of dimension n and determinant d.*
(a) *If $p \geqslant 3$ is an intractable prime then*

$$d \text{ is divisible by } p^{\binom{n}{2}} . \tag{40}$$

(b) *A prime $p \equiv 3 \pmod 4$ cannot be the only intractable prime.*

(c) *If 2 is an intractable prime then*

$$4^{[\frac{n}{2}]} d \text{ is divisible by } 4^{\binom{n}{2}}. \tag{41}$$

(d) *If 2 is the only intractable prime then*

$$4^{[\frac{n}{2}]} d \text{ is divisible by } 8^{\binom{n}{2}}. \tag{42}$$

(We recall that a prime p is tractable if $\Delta_p(u)$ is in the spinor kernel for every p-adic unit u.)

Proof. (a) If an odd prime p is intractable then from part (i) of the algorithm the powers of p in the diagonal terms of f are all distinct, and so, when arranged in increasing order, must be at least $p^0, p^1, p^2, \ldots$, yielding a product of at least $p^{\binom{n}{2}}$.

(b) If $p \equiv 3 \pmod 4$ is the only intractable prime then every element of the spinor kernel has a name $(u_\pm)_p$. But -1 is (-1)-adically automorphous and is a non-residue modulo p, so $(-1)_p$ is in the spinor kernel. Therefore the spinor kernel has order 2 and p is tractable, a contradiction.

(c) If 2 is intractable then from part (II) of the algorithm there is no type II summand, i.e. the form is diagonalizable. Moreover no three powers of 2 in the diagonal terms can lie in the range 2^t to 2^{t+3} (inclusive), for any t. Therefore, when arranged in increasing order, the powers of 2 must be at least

$$2^0, 2^0, 2^4, 2^4, 2^8, 2^8, \ldots .$$

When multiplied by 1, 4, 1, 4, ... this sequence becomes

$$4^0, 4^1, 4^2, 4^3, \ldots,$$

which implies (41).

(d) If 2 is the only intractable prime then every element of the spinor kernel has a name $(u_1)_2$, $(u_3)_2$, $(u_5)_2$ or $(u_7)_2$, and since -1 is (-1)-adically automorphous, $(u_7)_2$ is in the spinor kernel. Moreover, if any two powers of 2 in the diagonal terms have ratio 1 or 4 or 16, then by part (ii) of the algorithm $(u_5)_2$ is in the spinor kernel and all possibilities are generated. So the even powers of 2 are at least

$$2^0, 2^6, 2^{12}, \ldots,$$

and the odd powers of 2 are at least

$$2^1, 2^7, 2^{13}, \ldots .$$

The least possible values are therefore

$$2^0, 2^1, 2^6, 2^7, 2^{12}, 2^{13}, \dots ,$$

which on multiplication by 1, 4, 1, 4, ... become

$$8^0, 8^1, 8^2, 8^3, \dots ,$$

establishing (42).

Theorem 21. *If f is an indefinite form of dimension n and determinant d, with more than one class in its genus, then*

$$4^{[\frac{n}{2}]} d \text{ is divisible by } k^{\binom{n}{2}} \tag{43}$$

for some nonsquare natural number $k \equiv 0$ or $1 \pmod 4$.

This is a strengthening of [Wat3, Corollaries 1 and 2 to Theorem 69]. Compare [Kne3], [Hsi1], [Ear2]. Without going into too much detail it is worth mentioning that this result is nearly best possible. If $4^{[n/2]}d$ is divisible by $k^{\binom{n}{2}}$ for such a k, then there is a genus of forms with determinant d or some small multiple of d that contains more than one class.

Proof. If the dimension is 2 the assertion is trivial. For certainly $d \neq \pm 1$, or else there is only one class in the genus. Therefore some prime $p \geqslant 2$ divides d, and (43) holds with $k = 4p$.

For dimensions $n \geq 3$, as in the proof of Theorem 19, there must be at least one intractable prime $p \geq 2$. If some $p \equiv 1 \pmod 4$ is intractable then, from (40), (43) holds with $k = p$. If two primes p and q congruent to 3 (mod 4) are intractable then (43) holds with $k = pq$. If both 2 and $p \equiv 3 \pmod 4$ are intractable then $k = 4p$ will do, from (40) and (41); if 2 is the only intractable prime then $k = 8$ will do, from (42); and by part (b) of Theorem 20 this has exhausted all the possibilities.

Corollary 22. *Suppose f is an indefinite form of dimension n and determinant d, with more than one class in its genus. Then $|d| \geqslant d_0$, where d_0 is given by the following table.*

n	2	3	4, 6, 8, ...	5, 7, 9, ...
d_0	17	128	$5^{\binom{n}{2}}$	$2 \cdot 5^{\binom{n}{2}}$

Proof. For large n, (42) implies that $\det f$ is divisible roughly by $8^{\binom{n}{2}}$, whereas if 5 is intractable then from (40) we need only $5^{\binom{n}{2}} \mid \det f$. The cutoff point turns out to be at $n = 4$, and for $n \geqslant 4$ the smallest determinant (for an indefinite form with more than one class in its genus) is that of the form

$$\begin{pmatrix}2&3\\3&2\end{pmatrix} \oplus 5^2\begin{pmatrix}2&3\\3&2\end{pmatrix} \oplus 5^4\begin{pmatrix}2&3\\3&2\end{pmatrix} \oplus \cdots \oplus 5^{2m}\begin{pmatrix}2&3\\3&2\end{pmatrix} \oplus 5^{2m+2}(2)$$

omitting the final term if n is even. This determinant is

$$5^{\binom{n}{2}} \text{ if } n \text{ is even}, \quad 2\cdot 5^{\binom{n}{2}} \text{ if } n \text{ is odd}. \tag{44}$$

But (42) is better than (44) when $n = 3$. For $n = 2$, we find from Table 15.2 that the binary forms

$$\begin{Bmatrix}2&3\\3&-4\end{Bmatrix} \text{ and } \begin{Bmatrix}-2&3\\3&4\end{Bmatrix}$$

are in distinct classes although both belong to the genus $\mathrm{II}_{1,1}(17)$. The pair of ternary forms with determinant -128 is given at the end of this chapter.

For definite forms the question of when there is only one class in the genus behaves completely differently, and is the subject of a series of papers by Watson [Wat5]-[Wat7], [Wat14]-[Wat22].

10. The classification of positive definite forms

10.1 Minkowski reduction. We shall not say very much about this important notion, since our main interest is in forms of large dimension where it is impracticable. (For further information see the references on reduction algorithms listed in §1.4 of Chap. 2.) Let f be a positive definite n-dimensional form. f is said to be *Minkowski reduced* if it has been expressed in terms of an integral basis $e_1, \ldots, e_n$ such that for each t, $1 \leqslant t \leqslant n$,

$$f(e_t) \leqslant f(v) \quad \text{for all integral vectors } v \text{ for which } e_1, \ldots, e_{t-1}, v \text{ can be continued to an integral basis.} \tag{45}$$

In other words each successive e_t is chosen so that $f(e_t)$ is as small as is possible. By letting v range over all integral vectors, the condition (45) implies inequalities on the matrix entries a_{ij}. It turns out [Cas3, p. 256, Theorem 1.3] that only finitely many of these inequalities are necessary ("the fundamental region has finitely many walls"), but unfortunately their number tends to infinity very rapidly with the dimension.

Some of these inequalities may be easily written down. (i) It is immediate from (45) that

$$0 < a_{11} \leqslant a_{22} \leqslant \cdots \leqslant a_{nn}. \tag{46}$$

(ii) If we let $v = e_t - \sum_{s \in S} \epsilon_s e_s$ (for some set S of subscripts $s < t$ and coefficients $\epsilon_s = \pm 1$) the inequality $f(e_t) \leqslant f(v)$ becomes

$$2\Big| \sum_{s \in S} \epsilon_s a_{st} - \sum_{\substack{r, s \in S \\ r < s}} \epsilon_r \epsilon_s a_{rs} \Big| \leqslant \sum_{s \in S} a_{ss}.$$

The cases $S = \{s\}, \{r, s\}, \{q, r, s\}, \ldots$ lead to

$$2|a_{st}| \leqslant a_{ss} \quad (s < t)\,, \tag{47}$$

$$2|a_{rs} \pm a_{rt} \pm a_{st}| \leqslant a_{rr}+a_{ss} \quad (r < s < t)\,, \tag{48}$$

$$\begin{aligned} &2|\alpha a_{qt}+\beta a_{rt}+\gamma a_{st}-\alpha\beta a_{qr}-\alpha\gamma a_{qs}-\beta\gamma a_{rs}| \\ &\quad \leqslant a_{qq}+a_{rr}+a_{ss} \quad (q < r < s < t)\,, \end{aligned} \tag{49}$$

with $\alpha, \beta, \gamma = \pm 1$, etc.

It is a theorem of Minkowski (see for example [Cas3, p. 257, Lemma 1.2]) that for dimension $\leqslant 4$ a reduced form may be defined using vectors v with coefficients equal to 0 or ± 1. In fact the inequalities

(46), (46)-(47), (46)-(48), (46)-(49)

(in which $q, r, s, t \leqslant n$) define a Minkowski reduced form for

$n = 1$, 2, 3, 4

respectively.

For $n = 5$, 6, 7 and 8, defining systems of inequalities for Minkowski reduced forms have been given by Minkowski, Ryskov, Tammela and Novikova – see [Aff1], [Aff2], [Gru1], [Nov1], [Rys2], [Rys3], [Rys8], [Rys14], [Tam1]-[Tam4]. But the coefficients of v can no longer be restricted to 0 and ± 1.

Many theorems in the geometry of numbers are consequences of the inequalities (45) (see the references mentioned at the beginning of this section). In particular they can be used to show that there are only finitely many classes of forms of any given determinant [Cas1, p. 256, Theorem 1.1] and in some case to enumerate these forms [Wae5]. However, even in dimension 3, the process is tedious for moderately large determinants, and for much higher dimensions it is out of the question.

Tables 15.6 and 15.7 give all indecomposable, reduced, definite ternary forms with determinant $|d| \leqslant 50$ and the indefinite forms with $|d| \leqslant 100$. An entry $a\,{}_b\,c\,{}_f\,g\,{}_h$ represents the form having matrix

$$\begin{bmatrix} a & b & h \\ b & c & f \\ h & f & g \end{bmatrix},$$

and h is omitted when it is zero. Table 15.6 gives the positive definite forms with $d \leqslant 50$, and was computed using the inequalities (46)-(48). Table 15.7 gives the indefinite forms with $|d| \leqslant 100$, and was computed using the theory of spinor genera.

Table 15.6. **Indecomposable positive definite ternary forms.**

d	Forms
4	$2_1 2_1 2$
7	$2_1 2_1 3$
8	$2_1 3_1 2$
10	$2_1 2_1 4$
12	$2_1 4_1 2$, $3_1 2_1 3$
13	$2_1 2_1 5$, $2_1 3_1 3$
16	$2_1 5_1 2$, $2_1 2_1 6$, $3_1 3_1 3_{-1}$
17	$3_1 2_1 4$
18	$2_1 3_1 4$
19	$2_1 2_1 7$, $2_1 4_1 3$
20	$2_1 4_2 4$, $3_1 3_1 3_1$
21	$3_1 3_1 3$
22	$2_1 2_1 8$, $3_1 2_1 5$
23	$2_1 3_1 5$
24	$2_1 7_1 2$, $4_1 2_1 4$, $3_1 3_1 4_{-1}$
25	$2_1 2_1 9$, $2_1 5_1 3$
26	$2_1 4_1 4$
27	$3_1 2_1 6$, $2_1 4_2 5$
28	$2_1 8_1 2$, $2_1 2_1 10$, $2_1 3_1 6$, $2_1 5_2 4$, $3_1 3_1 4_1$
29	$3_1 3_1 4$
30	$3_1 4_1 3$
31	$2_1 2_1 11$, $4_1 2_1 5$
32	$2_1 9_1 2$, $3_1 2_1 7$, $3_1 3_1 5_{-1}$, $3_1 4_2 4$, $4_2 4_2 4$
33	$2_1 3_1 7$, $2_1 4_1 5$
34	$2_1 2_1 12$, $2_1 5_1 4$, $2_1 4_2 6$
35	$3_1 4_1 4_{-1}$
36	$2_1 10_1 2$, $2_1 6_2 4$, $2_1 5_1 5_{-1}$, $3_1 3_1 5_1$, $4_1 4_1 4_{-2}$
37	$2_1 2_1 13$, $2_1 7_1 3$, $3_1 2_1 8$, $2_1 5_2 5$, $3_1 3_1 5$
38	$2_1 3_1 8$, $4_1 2_1 6$
39	$3_1 5_1 3$, $3_1 4_1 4_1$
40	$2_1 11_1 2$, $2_1 2_1 14$, $2_1 4_1 6$, $5_1 2_1 5$, $3_1 3_1 6_{-1}$, $4_1 3_1 4$
41	$2_1 4_2 7$, $3_1 4_1 4$
42	$3_1 2_1 9$
43	$2_1 2_1 15$, $2_1 8_1 3$, $2_1 3_1 9$, $2_1 5_1 5$, $3_1 4_2 5$
44	$2_1 12_1 2$, $2_1 7_2 4$, $3_1 3_1 6_1$, $3_1 5_2 4$, $4_1 4_2 4$, $4_2 4_2 5$
45	$4_1 2_1 7$, $2_1 5_1 6_{-1}$, $3_1 3_1 6$
46	$2_1 2_1 16$, $2_1 5_2 6$, $3_1 4_1 5_{-1}$
47	$3_1 2_1 10$, $2_1 4_1 7$, $2_1 6_2 5$
48	$2_1 13_1 2$, $2_1 3_1 10$, $2_1 4_2 8$, $2_1 6_3 6$, $3_1 6_1 3$, $3_1 3_1 7_{-1}$, $4_1 5_1 4_{-2}$, $4_2 5_2 4$
49	$2_1 2_1 17$, $2_1 9_1 3$, $5_1 2_1 6$, $5_1 3_1 5_{-2}$
50	$2_1 7_1 4$, $3_1 5_1 4_1$, $4_1 4_1 4_{-1}$

Three forms should be added to this table:

$2_1 6_1 2$ at determinant 20, $2_1 6_1 3$ at determinant 31, and $2_1 6_1 4$ at determinant 42.

Table 15.7. Indecomposable indefinite ternary forms.

d	Forms
∓ 8	$\pm 2\ _1\ \pm 2\ _1\ \mp 2$
∓ 28	$\pm 2\ _1\ \mp 6\ _1\ \pm 2$
∓ 32	$\pm 2\ _1\ \mp 2\ _1\ \pm 6$
∓ 56	$\pm 2\ _1\ \pm 14\ _1\ \mp 2$
∓ 64	$\pm 4\ _2\ \pm 4\ _2\ \mp 4$
∓ 68	$\pm 2\ _1\ \pm 6\ _1\ \mp 6$
∓ 72	$\pm 6\ _1\ \pm 2\ _1\ \mp 6$
∓ 72	$\pm 2\ _1\ \mp 2\ _1\ \pm 14$
∓ 92	$\pm 2\ _1\ \mp 22\ _1\ \pm 2$

10.2 The Kneser gluing method. The integral lattices generated by vectors of norm 1 and 2 are completely classified. Such a lattice can be written as a direct sum of the particular lattices

$$\mathrm{I}_n\,(n \geqslant 1),\quad A_n\,(n \geqslant 1),\quad D_n\,(n \geqslant 4),\quad E_6,\ E_7,\ E_8.$$

Certain other lattices can be found by *gluing* these (and possibly other) components together. This technique, due to Witt and Kneser [Kne4], is described in §3 of Chap. 4. In Chaps. 16 and 17 we shall describe how, by a combination of gluing and other methods, the unimodular lattices of dimension $n \leqslant 25$ have been enumerated. However it is worth pointing out that the enumerations of unimodular lattices can be used to find lattices of other determinants in a fairly simple way, as the following section will illustrate (cf. [Kne4], [Ple12]).

10.3 Positive definite forms of determinant 2 and 3. By following the method used by Kneser [Kne4], and making use of the results of Chap. 16, in this section we classify the forms of determinant 2 up to dimension 18 and determinant 3 up to dimension 17. This is enough to demonstrate the techniques used, and since beyond this point the tables become unwieldy, is a good place to stop. The results for determinant + dimension $\leqslant 17$ agree with Kneser's.

Theorem 23. *All positive definite forms of determinant* 2 *and dimension* $\leqslant 18$, *or determinant* 3 *and dimension* $\leqslant 17$ *are as shown in Tables* 15.8 *and* 15.9.

Note. The tables explain these lattices in terms of unimodular lattices taken from Chap. 16. As in that chapter a unimodular lattice is specified by its component root lattices.

Outline of proof. Determinant 2. If L_n has determinant 2, then L_n^* / L_n has order 2 and there is a vector $v \in L_n^* \setminus L_n$ with $2v \in L_n$ and

$v \cdot v \equiv 1/2 \pmod 1$. In the particular case $L_1 = A_1$, write w instead of v, with $w \cdot w = 1/2$. Then $L_n \oplus A_1$ can be extended by the glue vector $v+w$ to give a lattice L_{n+1} (say) of determinant 1. Conversely, $L_n = w^{\perp}$ (in L_{n+1}) $= \{x \in L_{n+1} : \ x \cdot w = 0\}$. Thus all n-dimensional lattices L_n of determinant 2 are uniquely obtained as the orthogonal lattices to norm 2 vectors w in $(n+1)$-dimensional lattices L_{n+1} of determinant 1.

All such L_{n+1} can be found (for $n+1 \leqslant 23$) in Chap. 16. Suppose $L_{n+1} = M_{n+1-k} \oplus \mathrm{I}_k$, where M_{n+1-k} has minimal norm $\geqslant 2$. There are now two possibilities for w.

(a) $w \in \mathrm{I}_k$ (if $k \geqslant 2$), so that $L_n = w^{\perp} = M_{n+1-k} \oplus A_1 \oplus \mathrm{I}_{k-2}$. If L_n has minimal norm 2 then we must have $L_n = M_{n-1} \oplus A_1$. These lattices are shown in column (a) of Table 15.8.

We know from Table 15.4 that there are just two genera with determinant 2, namely $\mathrm{I}_n(2)$ and $\mathrm{II}_n(2)$. $L_n = M_{n-1} \oplus A_1$ is even (i.e. in $\mathrm{II}_n(2)$) exactly when M_{n-1} is. For $n = 17$ there are three lattices in

Table 15.8. Positive definite lattices of determinant 2, minimal norm $\geqslant 2$.

Dim.	(a)	(b) $<2>^{\perp}$ in	n_{I}	n_{II}	n_{tot}
0	—	—	0	0	0
1	A_1	—	0	1	1
2	—	—	0	0	1
3	—	—	0	0	1
4	—	—	0	0	1
5	—	—	0	0	1
6	—	—	0	0	1
7	—	E_8	0	1	2
8	—	—	0	0	2
9	$E_8 \oplus A_1$	—	0	1	3
10	—	—	0	0	3
11	—	D_{12}	1	0	4
12	—	—	0	0	4
13	$D_{12} \oplus A_1$	E_7^2	2	0	6
14	—	A_{15}	1	0	7
15	$E_7^2 \oplus A_1$	D_{16}, E_8^2, D_8^2	2	2	11
16	$A_{15} \oplus A_1$	$A_{11}E_6$	3	0	14
17	$E_8^2 \oplus A_1$	$A_{17}A_1$			
	$D_{16} \oplus A_1$	$D_{10}E_7A_1$			
	$D_8^2 \oplus A_1$	D_6^3			
		A_9^2	6	4	24
18	$A_{11}E_6 \oplus A_1$	$E_6^3O_1$			
		$A_{11}D_7O_1$			
		$A_7^2D_5$	6	0	30

column (a), two of which are even, although the table only indicates their number. The actual components can be found from Chap. 16.

(b) Alternatively we can take w to be any norm 2 vector in a lattice M_{n+1} of determinant 1 and minimal norm 2. For example there are exactly two inequivalent choices for w in the 18-dimensional lattice $A_{11}E_6$. These lattices are shown in column (b) of Table 15.8. In Tables 15.8, 15.9 the symbol $<c>$ denotes a one-dimensional lattice $u\mathbf{Z}$ with $u \cdot u = c$.

The last three columns of Table 15.8 give n_{I} (resp. n_{II}), the number of odd (resp. even) lattices with determinant 2 and minimal norm $\geqslant 2$ in each dimension, and n_t, the number of lattices with determinant 2 and minimal norm $\geqslant 1$.

Determinant 3. If L_n has determinant 3 there is a vector $v \in L_n^* \setminus L_n$ with $3v \in L_n$ and $v \cdot v \equiv \pm 1/3 \pmod 1$. It is easy to see that if $v \cdot v \equiv 1/3 \pmod 1$ then the 3-adic symbol for L_n is $1^{n-1}\,3^1$, and otherwise it is $1^{n-1}\,3^{-1}$.

First we consider the case $v \cdot v \equiv 1/3 \pmod 1$. We take A_2 with $w \in A_2^* \setminus A_2$, $w \cdot w = 2/3$ (Chap. 4, Eq. (55)), and extend $L_n \oplus A_2$ by the glue vector $v+w$ to obtain a lattice L_{n+2} of determinant 1. Conversely, $L_n = A_2^{\perp}$ in L_{n+2}. If $L_{n+2} = M_{n+2-k} \oplus \mathrm{I}_k$ there are two possibilities. (a) $A_2 \subset \mathrm{I}_k$, so (if L_n has minimal norm $\geqslant 2$), $L_n = M_{n-1} \oplus <3>$, where M_{n-1} has minimal norm 2 and determinant 1. (b) $A_2 \subset M_{n+2-k}$. For example $A_2^{\perp}$ in E_8 gives $L_6 = E_6$.

Second, consider $v \cdot v \equiv -1/3 \pmod 1$. We take a 1-dimensional lattice M_1 (say) $= <3>$, with generator w of norm 3, and extend $L_n \oplus M_1$ by the glue vector $v + 1/3\,w$ to get a lattice L_{n+1} of determinant 1. Conversely, L_n is the orthogonal lattice to a norm 3 vector w in an $(n+1)$-dimensional lattice L_{n+1} of determinant 1. If $L_{n+1} = M_{n+1-k} \oplus \mathrm{I}_k$ there are now three possibilities to consider. (c) $w \in \mathrm{I}_k$, (d) the projection of w onto M_{n+1-k} has norm 2, while the projection onto I_k has norm 1, and (e) $w \in M_{n+1-k}$.

In case (d), suppose $L_{n+1} = M_n \oplus \mathrm{I}_1$, where M_n has minimal norm 2. Let $w = v+e$ where $v \in M_n$, $v \cdot v = 2$, $e \in \mathrm{I}_1$, $e \cdot e = 1$, and let $K_{n-1} = v^{\perp}$ in M_n, so that $\det K_{n-1} = 2$ (from the first part of the proof). Then $L_n = w^{\perp}$ contains K_{n-1}, and also the vector $u = v - 2e$, which generates a 1-dimensional lattice $<6>$ orthogonal to K_{n-1}. In fact L_n is $K_{n-1} \oplus <6>$ extended by a glue vector $t + 1/2\,u$, where t is a nonzero glue vector for K_{n-1}^* / K_{n-1}.

Table 15.9 shows the lattices of minimal norm $\geqslant 2$ in the five cases, and the number of even (n_{II}) and odd (n_{I}) lattices of minimal norm 2 in each dimension. There are four genera of forms of determinant 3 (see Table 15.4), namely $\mathrm{I}_n(3^{\pm 1})$ and $\mathrm{II}_n(3^{\pm 1})$. Columns (a) and (b) belong to either $\mathrm{I}_n(3^1)$ or $\mathrm{II}_n(3^1)$, and columns (c), (d) and (e) to either $\mathrm{I}_n(3^{-1})$ or $\mathrm{II}_n(3^{-1})$. The final column, n_{tot}, gives the number of lattices with determinant 3 and minimal norm $\geqslant 1$.

Table 15.9. Positive definite lattices of determinant 3, minimal norm $\geqslant 2$.

Dim	(*a*)	(*b*) $A_2^\perp$ in	(*c*)	(*d*) $<3>^\perp$ in	(*e*) $<3>^\perp$ in	n_{I}	n_{II}	n_{tot}
0	–	–	–	–	–	0	0	0
1	$<3>$	–	–	–	–	1	0	1
2	–	–	A_2	–	–	0	1	2
3	–	–	–	–	–	0	0	2
4	–	–	–	–	–	0	0	2
5	–	–	–	–	–	0	0	2
6	–	E_8	–	–	–	0	1	3
7	–	–	–	–	–	0	0	3
8	–	–	–	$E_8 \oplus I_1$	–	1	0	4
9	$E_8 \oplus <3>$	–	–	–	–	1	0	5
10	–	D_{12}	$E_8 \oplus A_2$	–	–	1	1	7
11	–	–	–	–	D_{12}	1	0	8
12	–	E_7^2	–	$D_{12} \oplus I_1$	–	2	0	10
13	$D_{12} \oplus <3>$	A_{15}	–	–	E_7^2	3	0	13
14	–	D_{16}, E_8^2, D_8^2	$D_{12} \oplus A_2$	$E_7^2 \oplus I_1$	A_{15}	4	2	19
15	$E_7^2 \oplus <3>$	$A_{11}E_6$	–	$A_{15} \oplus I_1$	D_8^2	5	0	24
16	$A_{15} \oplus <3>$	$A_{17}A_1$ $D_{10}E_7A_1$ D_6^3 A_9^2	$E_7^2 \oplus A_2$	$E_8^2 \oplus I_1$ $D_{16} \oplus I_1$ $D_8^2 \oplus I_1$	$A_{11}E_6$	12	0	36
17	$E_8^2 \oplus <3>$ $D_{16} \oplus <3>$ $D_8^2 \oplus <3>$	$E_6^3O_1$ $A_{11}D_7O_1$ $A_7^2D_5$	$A_{15} \oplus A_2$	$A_{11}E_6 \oplus I_1$	$A_{17}A_1$ $D_{10}E_7A_1$ D_6^3 A_9^2	17	0	53

11. Computational complexity

Finally we give a brief discussion of the complexity of the classification problem. (The complexities of other questions connected with lattices are discussed in §1.4 of Chap. 2.) The following are some of the principal questions that we have encountered.

(C1) Find the number of classes of integral quadratic forms of dimension n and determinant d. (C2) Exhibit one form in each class. (C3) Given two forms, determine if they are in the same class. (C4) If they are, find an explicit equivalence. (G1) Find the number of genera of forms of dimension n and determinant d. (G2) Exhibit one form in each genus. (G3) Given two forms, determine if they are in the same genus. (G4) If they are, find an explicit rational equivalence whose denominator is prime to any given number. A closely related problem is to find which numbers are represented by a form. (S) Given a form f of dimension n and determinant d, and an integer k, is there an integral solution to $f(x) = k$? If so, find all solutions.

We shall not say much about (S). For $n = 2$ it was solved by Gauss [Gau1, §§180, 205, 212] (see also [Bor5, p. 142], [Coh5, p. 1], [Edw1, pp. 317, 330], [Lag1], [Lag2], [Mor6]). The complexity of Gauss's solution appears to be dominated by the complexity of factoring k, which is at most $\exp(c\sqrt{\log k \log\log k})$ for some constant c [Mor8], [Pom1]. (Certainly

Gauss's solution is more efficient than the $O(\sqrt{k})$ solutions proposed in [Dij1] and [Bac3] in the case $f = x^2+y^2$.) For general n not much is known. This problem includes the determination of the minimal nonzero norm of a lattice — see §1.4 of Chap. 2.

For problem (C1), in the case $n = 2$ there are explicit formulae for the class number, which can be evaluated in a number of steps that is a polynomial in d [Cas3, p. 371], [Dic2, Chap. VI].

For positive definite forms of fixed dimension n, all of (C1)-(C4) and (G1)-(G4) can be solved by algorithms whose running time is a polynomial in d. (By using Minkowski-reduced forms (§10.1), all entries in the matrices and vectors involved can be bounded by simple functions of d (compare §3.2).) However these polynomials typically behave like d^{n^2}, so the growth as a function of n is likely to be worse than exponential. Furthermore the mass formula (see Chap. 16) shows that the class number for definite forms grows at least as fast as n^{n^2} [Mil7, p. 50]. Since it seems unlikely that one can find the class number for $n > 2$ without determining all the classes, the complexity of (C1) and (C2) for definite forms as a function of n seems to be worse than exponential.

In the remainder of this section we consider indefinite forms. If the determinant is given in factored form, problem (C1) is easy. Using the method of §§7,8, the number of steps required is a polynomial in the number of factors of d. If d is a prime, for example, the answer is given in Theorem 13.

Our invariants for the genus and spinor genus also provide quick (and polynomial-time) solutions to (G1), (G3) and usually (C3). We illustrate with a notorious example. Dickson and Ross in 1930 were unable to decide whether the ternary forms $x^2-3y^2-2yz-23z^2$ and $x^2-7y^2-6yz-11z^2$ were equivalent [Dic3, p. 147]. It is now known that they are equivalent [Ben2], [Cas3, pp. 132, 251], but we shall establish this using our invariants. Replacing the forms by their negatives for convenience, the problem is to decide the integral equivalence of

$$\begin{pmatrix} 3 & 1 & 0 \\ 1 & 23 & 0 \\ 0 & 0 & -1 \end{pmatrix} \quad \text{and} \quad \begin{pmatrix} 7 & 3 & 0 \\ 3 & 11 & 0 \\ 0 & 0 & -1 \end{pmatrix}$$

We first compute the genus in each case. By rational transformations of denominators 3 and 7, respectively, the forms may be diagonalized to

$$\text{diag}\,\{3, \frac{68}{3}, -1\} \text{ and } \text{diag}\,\{7, \frac{68}{7}, -1\}\,.$$

Therefore, since 3 and 7 are odd and prime to the determinant -68, we can read off the relevant p-adic symbols

$$\begin{array}{lll} p = -1: & +^2 - & +^2 - \\ p = 17: & 1^{-2}\, 17^{-1} & 1^{-2}\, 17^{-1} \\ p = 2: & 1_2^{-2}\, 4_3^{-1} & 1_6^{+2}\, 4_7^{+1} \end{array}$$

and since the first 2-adic symbol is converted to the second by a 2-step walk, the forms are indeed in the same genus.

Let us compute the spinor kernel for the first form. Plainly 2 is the only intractable prime, although -1 is still relevant since it tells us that -1 is (-1)-adically automorphous and so $(u_7)_2$ is in the spinor kernel. Since two of the terms in the 2-adic diagonalization have ratio $-3 = u_5$, $(u_5)_2$ is also in the spinor kernel, which therefore includes everything, and so there is only one class in the genus. Thus the two forms are integrally equivalent. In fact the unimodular matrix

$$M = \begin{pmatrix} -3 & 2 & 10 \\ -2 & 3 & 14 \\ -1 & 1 & 5 \end{pmatrix} \tag{50}$$

transforms the first form into the second.

There do not seem to be good algorithms for the remaining problems (C2)-(C4), (G2) and (G4). *Logically* there is no difficulty: the proofs of the theorems on genus and spinor genus (Theorems 9, 10, 14 etc.) are at bottom computationally effective. For example, for problems (C3), (C4) and (G4), if two forms are known to be in the same genus, we can in principle search through all rational matrices until a rational equivalence of denominator r prime to $2d$ is found. The forms are then in the same class if and only if $\Delta(r)$ is in the spinor kernel. If they are in the same class we can continue the search until an integral equivalence is found. (The matrix (50) was essentially found by this procedure, after using the diophantine equations resulting from Eq. (4) to restrict the search.) For problems (C2) and (G2) we search through all integral matrices in turn, applying these techniques, until we have found the correct number of distinct classes or genera of forms of determinant d. But these are only logicians' solutions, and it is not clear to us that in general they can be converted into practical algorithms.

Often there is some convenient artifice. Sometimes one can make use of Gauss's complete theory of binary forms. We illustrate by finding two inequivalent indefinite ternary forms of determinant -128 of the same genus (see Corollary 22). It is easy to see that the genus of $f = \mathrm{diag}\,\{-1, 64, 2\}$, namely $I_{2,1}(2\times 64)$, contains two spinor genera and hence two classes, namely f and $f * \Delta(3)$. We must find a representative for the second class. Thus we wish to find two lattices L and M in this genus whose intersection has index 3 in each of them. Among the binary forms of determinant -64 we find the form $\begin{pmatrix} -9 & 1 \\ 1 & 7 \end{pmatrix}$, which remains integral when its first row and column are divided by 3 and simultaneously its second row and column are multiplied by 3, yielding $\begin{pmatrix} -1 & 1 \\ 1 & 63 \end{pmatrix}$ So the ternary forms

$$(a) \quad \begin{pmatrix} -1 & 1 & 0 \\ 1 & 63 & 0 \\ 0 & 0 & 2 \end{pmatrix} \quad \text{and} \quad (b) \quad \begin{pmatrix} -9 & 1 & 0 \\ 1 & 7 & 0 \\ 0 & 0 & 2 \end{pmatrix} \tag{51}$$

represent lattices L and M for which generators can be chosen in the form e_1, e_2, e_3 for L and $3e_1, e_2/3, e_3$ for M. Now L also possesses the diagonal basis e_1, e_1+e_2, e_3, showing that it corresponds to the original form f. The matrix (51b) is therefore a representative for the second class in this genus.

16

Enumeration of Unimodular Lattices

J. H. Conway and N. J. A. Sloane

In this chapter we state explicit formulae for the Minkowski-Siegel mass constants for unimodular lattices. We give Niemeier's list of 24-dimensional even unimodular lattices, use the mass constant to verify that it is correct, and then enumerate all unimodular lattices of dimension $n \leqslant 23$.

1. The Niemeier lattices and the Leech lattice

The even unimodular (or Type II) lattices in 24 dimensions were enumerated by Niemeier in 1968 [Nie2]. There are 23 such lattices with minimal norm 2, and one, the Leech lattice Λ_{24}, with minimal norm 4. (Λ_{24} was already known to be the unique 24-dimensional Type II lattice with minimal norm 4—see Chap. 12.)

The Niemeier lattices are shown in Table 16.1. The first column labels each lattice with a single Greek letter, for use in Chap. 17. Each Niemeier lattice Λ is obtained by gluing certain component lattices together by means of glue vectors, as described in §3 of Chap. 4. The second column gives the component lattices, which belong to the list A_n $(n \geqslant 1)$, D_n $(n \geqslant 4)$, E_6, E_7 and E_8. In this chapter we usually refer to lattices merely by their components, for example $A_{11}D_7E_6$ represents one of the Niemeier lattices. Elsewhere we should write $(A_{11}D_7E_6)^+$ to indicate the presence of glue. If Λ has components $L_1, \ldots, L_k$, the glue vectors for Λ have the form $y = (y_1, \ldots, y_k)$, where each y_i can be regarded as a coset representative (or glue vector) for L_i^* modulo L_i. These coset representatives, labeled [0], [1], ..., $[d-1]$ for a component of determinant d, are listed in Chap. 4.

The set of glue vectors $(y_1, \ldots, y_k)$ for Λ forms an additive group called the *glue code*. The third column of Table 16.1 gives generators for the glue code. If a glue vector contains parentheses, this indicates that all vectors obtained by cyclically shifting the part of the vector inside the

Table 16.1. The 24-dimensional even unimodular lattices L. The order of the glue code is $|G_\infty|$. Also Aut (L) has order $|G_0|\ |G_1|\ |G_2|$, where $|G_0|$ can be found from (2). h is the Coxeter number, $24h$ is the number of vectors of norm 2, V is the number of Leech lattice points around the corresponding deep hole (see Chap. 23).

Name	Components	Generators for glue code	$\lvert G_\infty\rvert$	$\lvert G_1\rvert$	$\lvert G_2\rvert$	h	V
α	D_{24}	[1]	2	1	1	46	25
β	$D_{16}E_8$	[10]	2	1	1	30	26
γ	E_8^3	[000]	1	1	6	30	27
δ	A_{24}	[5]	5	2	1	25	25
ϵ	D_{12}^2	[12], [21]	4	1	2	22	26
ζ	$A_{17}E_7$	[31]	6	2	1	18	26
η	$D_{10}E_7^2$	[110], [301]	4	1	2	18	27
θ	$A_{15}D_9$	[21]	8	2	1	16	26
ι	D_8^3	[(122)]	8	1	6	14	27
κ	A_{12}^2	[15]	13	2	2	13	26
λ	$A_{11}D_7E_6$	[111]	12	2	1	12	27
μ	E_6^4	[1(012)]	9	2	24	12	28
ν	$A_9^2D_6$	[240], [501], [053]	20	2	2	10	27
ξ	D_6^4	[even perms of {0123}]	16	1	24	10	28
o	A_8^3	[(114)]	27	2	6	9	27
π	$A_7^2D_5^2$	[1112], [1721]	32	2	4	8	28
ρ	A_6^4	[1(216)]	49	2	12	7	28
σ	$A_5^4D_4$	[2(024)0], [33001], [30302], [30033]	72	2	24	6	29
τ	D_4^6	[111111], [0(02332)]	64	3	720	6	30
υ	A_4^6	[1(01441)]	125	2	120	5	30
ϕ	A_3^8	[3(2001011)]	256	2	1344	4	32
χ	A_2^{12}	[2(11211122212)]	729	2	$\lvert M_{12}\rvert$	3	36
ψ	A_1^{24}	[1(00000101001100110101111)]	4096	1	$\lvert M_{24}\rvert$	2	48
ω	Leech	–	–	1	1	0	–

parentheses are also glue vectors. For example, the glue vectors for the Niemeier lattice D_8^3 in Table 16.1 are described by [(122)], indicating that the glue words are spanned by

$$[122] = ((\tfrac{1}{2})^8, (0^7 1), (0^7 1))$$

$$[212] = ((0^7 1), (\tfrac{1}{2})^8, (0^7 1))$$

$$[221] = ((0^7 1), (0^7 1), (\tfrac{1}{2})^8)$$

– the glue vectors for D_n are given in §7.1 of Chap. 4. The full glue code for this example contains the eight vectors [000], [122], [212], [221], [033], [303], [330], [111]. The number of glue vectors, $|G_\infty|$ (this notation is explained at the end of Chap. 24), is given in the fourth column.

The automorphism group Aut(Λ) is compounded of groups G_0, G_1, G_2 and has order

$$|\mathrm{Aut}(\Lambda)| = g_0 g_1 g_2 , \tag{1}$$

as described in Eq. (4) of Chap. 4. The group G_0 is the direct product of subgroups of order

$$(n+1)!,\ 2^{n-1}n!,\ 2^7 3^4 5,\ 2^{10}3^4 5\cdot 7,\ 2^{14}3^5 5^2 7\ , \tag{2}$$

for components A_n, D_n, E_6, E_7, E_8 respectively (these groups are the Weyl groups for the corresponding root lattices). The orders of G_1 and G_2 are given in columns 5 and 6 of the table. Column 7 gives the Coxeter number h for each lattice (which is the Coxeter number of any of its components). Then the lattice contains $24h$ vectors of norm 2.

The glue codes for the Niemeier lattices E_6^4, D_4^6, A_2^{12}, A_1^{24} are respectively the tetracode C_4, the hexacode C_6, and the Golay codes C_{12}, C_{24} (§§2.5.1, 2.5.2, 2.8.5, 2.8.2 of Chap. 3). The glue codes for A_{12}^2 and $A_{11}D_7E_6$ are given in full in Chap. 24. For A_4^6 the group $G_2(A_4^6)$ is isomorphic to $PGL_2(5)$ acting on $\{\infty, 0, 1, 2, 3, 4\}$. For A_3^8 the group $G_2(A_3^8)$ is isomorphic to $2^3.PSL_2(7)$ acting on the extended Hamming code of length 8 over the integers modulo 4. The glue codes are also described (often using different coordinates) by Venkov in Chap. 18. Other references dealing with the Niemeier lattices are [Ero1]-[Ero3].

The Niemeier lattice A_1^{24} is equivalent to the lattice obtained by applying Construction A (§2 of Chap. 7) to $\mathscr{C}_{24}$, has $\rho = 1/\sqrt{2}$, $\delta = 2^{-12}$, $\tau = 48$ and is of course an even integral lattice.

The lattices in Table 16.1 are arranged in decreasing order of h, and in tied cases by increasing size of G_2. This ordering has the property that occurrences from any one of the families A_n, D_n, E_n are in descending order of n.

In §3 we give two different verifications that Niemeier's list is correct (and another is given in Chap. 18). First we state the mass formulae.

2. The mass formulae for lattices

As mentioned in §2.4 of Chap. 2, unimodular lattices have the property that there are explicit formulae, the *mass formulae*,[1] which give appropriately weighted sums of the theta-series of all the inequivalent lattices of a given dimension. Here we just give the formulae for the number of inequivalent lattices, the *Minkowski-Siegel mass constants*.

Theorem 1. *Let* Ω *be the set of all inequivalent odd unimodular lattices of dimension* n. *Then*

$$\sum_{\Lambda \in \Omega} \frac{1}{|\mathrm{Aut}(\Lambda)|}$$

is equal to ½ *for* $n = 1$, *and otherwise to*

---

[1] The German term "Massformel" actually means "measure formula". The standard mistranslation is a happy one.

$$\frac{(1-2^{-k})(1+2^{1-k})}{2\cdot k!}\,|B_k\cdot B_2B_4\cdots B_{2k-2}|,\ \text{if}\ \ n=2k\equiv 0\ (mod\ 8),$$

$$\frac{2^k+1}{k!2^{2k+1}}\,|B_2B_4\cdots B_{2k}|,\ \text{if}\ \ n=2k+1\equiv \pm 1\ (mod\ 8),$$

$$\frac{1}{(k-1)!2^{2k+1}}\,|E_{k-1}\cdot B_2B_4\cdots B_{2k-2}|,\ \text{if}\ \ n=2k\equiv \pm 2\ (mod\ 8),$$

$$\frac{2^k-1}{k!2^{2k+1}}\,|B_2B_4\cdots B_{2k}|,\ \text{if}\ \ n=2k+1\equiv \pm 3\ (mod\ 8),$$

$$\frac{(1-2^{-k})(1-2^{1-k})}{2\cdot k!}\,|B_k\cdot B_2B_4\cdots B_{2k-2}|,\ \text{if}\ \ n=2k\equiv 4\ (mod\ 8).$$

Here B_k and E_k are the kth Bernoulli and Euler numbers, respectively: $B_0=1$, $B_1=-1/2$, $B_2=1/6,\ldots$, $E_0=1$, $E_2=-1$, $E_4=5,\ldots$ [Abr1, p. 810].

Theorem 2. *Let Ω be the set of all inequivalent even unimodular lattices of dimension n. Then*

$$\sum_{\Lambda\in\Omega}\frac{1}{|\mathrm{Aut}(\Lambda)|}=\frac{|B_k|}{2k}\prod_{j=1}^{k-1}\frac{|B_{2j}|}{4j},$$

if $n=2k\equiv 0\ (mod\ 8)$, the right-hand side being interpreted as 1 when $n=0$.

Theorem 3. *Let Ω be the set of all inequivalent unimodular Eisenstein lattices of dimension n, let*

$$M_n=\sum_{\Lambda\in\Omega}\frac{1}{|\mathrm{Aut}(\Lambda)|},$$

and let $a_n=3^{n-1}|B_n(1/3)|/n$, where $B_n(x)$ is the nth Bernoulli polynomial [Abr1, Chap. 23]. Then $M_1=1/6$,

$$M_n=M_{n-1}B_n\frac{3^{n/2}+(-1)^{n/2}}{2n}\quad \text{if } n \text{ even},$$

$$M_n=\frac{M_{n-1}a_n}{3^{(n-1)/2}+(-1)^{(n-1)/2}}\quad \text{if } n \text{ odd}.$$

Remarks. (i) These theorems have a long history. The versions given here are taken from [Con34], [Fei2], [Ser1], [Slo10]. For further information and generalizations see [Bra3], [Bro5], [Bro6], [Cas3], [Cos1], [Eic1], [Hsi6a], [Ko2], [Mag3], [Mil7], [O'Me1], [Pal1], [Pal2], [Pfe1]-[Pfe3], [Que3], [Sie1]-[Sie3]. The reader is warned that there are mistakes in many of the published formulae for the mass constants. (ii) Numerical values of the mass constants in Theorems 1 and 2 for $n \leqslant 32$ are given in Tables 16.2-16.5. (iii) Examples illustrating the use of these theorems were given in §2.4 of Chap. 2. Other examples appear in §§3,4 below. (iv) Analogous theorems for codes may be found in [Bro5], [Mac4], [Mac6, Chap. 19, §6], [Mac8], [Mal2], [Ple7], [Ple8], [Ple16], [Ple17], [Tho1].

Table 16.2. Exact values of the Minkowski-Siegel mass constants for Type I lattices.

n	mass
1	$\frac{1}{2}$
2	$\frac{1}{8}$
3	$\frac{1}{48}$
4	$\frac{1}{384}$
5	$\frac{1}{3840}$
6	$\frac{1}{46080}$
7	$\frac{1}{645120}$
8	$\frac{1}{10321920}$
9	$\frac{17}{2786918400}$
10	$\frac{1}{2229534720}$
11	$\frac{31}{735746457600}$
12	$\frac{31}{5885971660800}$
13	$\frac{691}{765176315904000}$
14	$\frac{42151}{192824431607808000}$
15	$\frac{29713}{385648863215616000}$
16	$\frac{505121}{12340763622899712000}$

3. Verifications of Niemeier's list

As an application of Theorem 2 we prove:

Theorem 4 [Nie2]. *The even unimodular lattices of dimension* 24 *are as shown in Table* 16.1.

Proof ([Con34]; see also [Ero3]). We shall verify that

$$\sum_{\Lambda} \frac{1}{|\mathrm{Aut}(\Lambda)|} = \frac{1027637932586061520960267}{129477933340026851560636148613120000000}, \quad (3)$$

Table 16.2 (cont.)

n	mass
17	$\frac{642332179}{18881368343036559360000}$
18	$\frac{692319119}{1510509467442924748800 0}$
19	$\frac{8003636403977}{7748913567982203961344 0000}$
20	$\frac{248112728523287}{6199130854385763169075 20000}$
21	$\frac{593468652605200909}{2169695799035017109176 32000000}$
22	$\frac{50904295073459007001}{1507367607750643465322 496000000}$
23	$\frac{1015740532498234470066371}{1317439289174062388691861504000000}$
24	$\frac{7018767079562800188158623 61}{2107902862678499821906978406 4000000}$
25	$\frac{847150594803046516236122 72842147}{304653960800063180142673290854 40000000}$
26	$\frac{146163356358943888761884726 84851927}{318714912836989173072335135047 68000000}$
27	$\frac{189435275177214686743048699 5462923265007}{124298816006425777498210702668 59520000000}$
28	$\frac{103450603774276940430378894 8222302395020 3227}{994390528051406219985685621348 76160000000}$
29	$\frac{428500982395959068211562873935 6169586687220752159}{288373253134907803795848830191 14086400000000}$
30	$\frac{156429914319579070270102710292957201465 7258504511950 39}{346047903761889364555018596229369036800000000}$
31	$\frac{447543572700878404232772149927275573042725639059585587715489}{150184790232659984216878070763546161971200000000}$
32	$\frac{4160226313319828373442537746357476271577534295749799151870993 94241}{96118265748902389898801965288669543661568 00000000}$

where the sum is taken over all lattices Λ in Table 16.1. $|\mathrm{Aut}(\Lambda)|$ is given by (1), where $|G_1|$ and $|G_2|$ can be found in columns 5 and 6 of the table, and $|G_0|$ is the product of the $|G_0|$'s for the components of Λ, which can be obtained from Eq. (2) (except for the Leech lattice, for which $|\mathrm{Aut}(\Lambda)| = |Co_0|$ is given in §3.3 of Chap. 10). The second column of Table 16.6 gives $|\mathrm{Aut}(\Lambda)|^{-1}$ times the denominator of the right-hand side of (3). Since the sum is equal to the numerator of the right-hand side of (3), the list is complete.

Table 16.3. Decimal expansions of the mass constants for Type I lattices.

n	Mass	n	Mass	n	Mass	n	Mass
1	0.5	9	6.100×10^{-9}	17	3.402×10^{-14}	25	2.781×10^{-6}
2	0.125	10	4.485×10^{-10}	18	4.583×10^{-14}	26	4.586×10^{-4}
3	2.083×10^{-2}	11	4.213×10^{-11}	19	1.033×10^{-13}	27	1.524×10^{-1}
4	2.604×10^{-3}	12	5.267×10^{-12}	20	4.002×10^{-13}	28	1.040×10^{2}
5	2.604×10^{-4}	13	9.031×10^{-13}	21	2.735×10^{-12}	29	1.486×10^{5}
6	2.170×10^{-5}	14	2.186×10^{-13}	22	3.377×10^{-11}	30	4.520×10^{8}
7	1.551×10^{-6}	15	7.705×10^{-14}	23	7.710×10^{-10}	31	2.980×10^{12}
8	9.688×10^{-8}	16	4.093×10^{-14}	24	3.330×10^{-8}	32	4.328×10^{16}

Table 16.4. The Minkowski-Siegel mass constants for lattices of Type II.

n	mass
0	1
8	$\frac{1}{696729600}$
16	$\frac{691}{277667181515243520000}$
24	$\frac{1027637932586061520960267}{129477933340026851560636148613120000000}$
32	$\frac{4890529010450384254108570593011950899382291953107314413193123}{121325280941552041649762780685623131486814208000000000}$

Table 16.5. Decimal expansions of the mass constants for Type II lattices.

n	0	8	16	24	32
Mass	1	1.435×10^{-9}	2.489×10^{-18}	7.937×10^{-15}	4.031×10^{7}

A more illuminating proof appears in Chap. 18. We sketch yet another proof, which uses some results from later chapters. Let G denote the group of all autochronous automorphisms of the even unimodular 26-dimensional Lorentzian lattice $\mathrm{II}_{25,1}$, and let H be the reflection subgroup of G. The groups G and H are given in Chap. 27, where in particular it is shown that H is a Coxeter group whose graph is isomorphic to the Leech lattice. From Vinberg's work [Vin5] it follows that the even unimodular 24-dimensional Euclidean lattices of minimal norm 2 are described by those maximal subdiagrams of the Leech lattice that are unions of the extended Coxeter-Dynkin diagrams $A_n\,(n \geqslant 1)$, $D_n\,(n \geqslant 4)$, E_6, E_7 and E_8. But in Chap. 23 it is shown that there are precisely 23 such subdiagrams, which are exactly those corresponding to the Niemeier lattices of minimal norm 2. This result, plus the fact that the Leech lattice is the unique even unimodular lattice of minimal norm $\geqslant 4$ (Chap. 12), provides another

Table 16.6. Verification that Niemeier's list is complete.

D_{24}	24877125
E_8^3	63804560820
$D_{16}E_8$	271057837050
A_{24}	4173688995840
D_{12}^2	67271626831500
$A_{17}E_7$	3483146354688000
$D_{10}E_7^2$	4134535541136000
$A_{15}D_9$	33307587016704000
D_8^3	156983146327507500
A_{12}^2	834785957117952000
E_6^4	373503391765504000
$A_{11}D_7E_6$	8082641116053504000
D_6^4	19144966823230248000
$A_9^2D_6$	106690862731906252800
A_8^3	225800767686574080000
$A_7^2D_5^2$	2700612462901377024000
A_6^4	8361079854908571648000
D_4^6	1196560426451890500000
$A_5^4D_4$	52278522738634063872000
A_4^6	180674574584719324741632
A_3^8	437599241673834240000000
A_2^{12}	3129279325918986240000000
A_1^{24}	315227121719590080000000
Leech	1557057285233049600 0
Total	10276379325860615209602 67

proof of Niemeier's result. It also clarifies the one-to-one correspondence between the deep holes in the Leech lattice (Chap. 23) and the Niemeier lattices. We must emphasize, however, that in no sense is this a short-cut to Niemeier's result, for the proofs in Chap. 23 require extensive computations.

4. The enumeration of unimodular lattices in dimensions $n \leqslant 23$

References to earlier work on the classification of unimodular lattices are given in §2.4 of Chap. 2, and the known results are summarized in Table 2.2 of that chapter. The odd lattices in dimensions 24 and 25 are discussed in Chap. 17. In the rest of this chapter we shall enumerate the unimodular lattices of dimension $n \leqslant 23$.

Theorem 5 [Con34]. *The unimodular lattices of dimension not exceeding* 23 *and containing no vectors of norm* 1 *are as shown in Table* 16.7.

Remark. The unimodular lattices of minimal norm 1 are now easily determined, since they have the form $\mathbf{Z}^k \oplus L$, where the minimal norm of L is at least 2 (see Eq. (4)).

Proof. We refer to the lattices in Table 16.1 as the Niemeier lattices. The proof of Theorem 5 is based on the remark (justified below) that any lattice of the desired type is associated with a certain Niemeier lattice, and consists in finding all lattices that can be obtained from the Niemeier lattices.

The lattices in Table 16.7 have components V_1, V_2, ... (shown in the column headed V). The V_i are taken from the list $O_n (n \geqslant 1)$, an *empty component* (that is, one containing no vector of norm $\leqslant 2$); $\mathbf{Z}$, the one-dimensional lattice of integers; and $A_n (n \geqslant 1)$, $D_n (n \geqslant 1)$, E_6, E_7 and E_8. As usual the subscript on a component indicates its dimension. We sometimes use I_m to denote the m-dimensional integer lattice $\mathbf{Z}^m$.

Remark. The notation used in this chapter to specify the components of a lattice differs slightly from that used in [Con27]. For example the lattice $E_6^3\, O_1$ was there called $E_6^3[3]$. The latter notation is more informative but less general (it fails when the empty component has dimension greater than one), and somewhat confusing, since [3] is also the name of a glue digit. We therefore recommend the $\ldots O_n$ notation for general use.

Chains of lattices. Every unimodular lattice Λ appears in a uniquely determined chain of the form

$$\Lambda_n,\ \Lambda_n \oplus \mathrm{I}_1,\ \Lambda_n \oplus \mathrm{I}_2, \ldots,\ \Lambda = \Lambda_n \oplus \mathrm{I}_m\ , \ldots\ , \tag{4}$$

where the initial lattice Λ_n does not represent 1. We call Λ_n the *reduced* version of Λ. (Beware: elsewhere in this book Λ_n usually denotes a laminated lattice.) The summand I_m is the sublattice of Λ generated by vectors of norm 1, and Λ_n is its orthogonal complement. In these circumstances we have

$$|\mathrm{Aut}(\Lambda_n \oplus \mathrm{I}_m)| = |\mathrm{Aut}(\Lambda_n)| \cdot 2^m m!\ .$$

Our enumeration process considers all lattices of a chain simultaneously. For some purposes the most appropriate lattice to consider is the initial lattice Λ_n, but for other purposes it is the 23-dimensional lattice $\Lambda_n \oplus \mathrm{I}_{23-n}$.

The associated Niemeier lattice. We now describe how an odd unimodular lattice Λ_n of dimension $n \leqslant 23$ is associated with a uniquely determined Niemeier lattice N. For brevity we omit the easy justifications of some statements, since in any case the mass formula provides an independent verification. Let $m = 24-n$. Both Λ_n and I_m have sublattices of index 2 containing only vectors of even norm, denoted by Λ_n^0 and I_m^0, respectively. The dual of Λ_n^0 consists of Λ_n^0 and three cosets, say Λ_n^1, Λ_n^2, Λ_n^3, and similarly the dual of I_m^0 consists of the cosets I_m^0, I_m^1, I_m^2, I_m^3. The notation can be chosen so that

$$\Lambda_n = \Lambda_n^0 \cup \Lambda_n^2\ ,\quad \mathrm{I}_m = \mathrm{I}_m^0 \cup \mathrm{I}_m^2\ ,$$

and the set of all vectors

$$\{x+y : x \in \Lambda_n^k,\ y \in \mathrm{I}_m^k,\ k = 0,1,2,3\}$$

is an even unimodular 24-dimensional lattice. The latter is the Niemeier lattice N associated with Λ_n.

Constructing Λ_n *from* N. The lattice I_m^0 is better known as D_m, and seen from the point of view of N the process we have just described appears as follows. We look for a copy of the lattice D_m in N and observe that Λ_n is the set of vectors x in the orthogonal complement $D_m^\perp$ for which either $x + [0]$ or $x + [2]$ is in N, where [0], [1], [2], [3] are the glue vectors for D_m in $D_m^\perp$, [2] being a glue vector of norm 1 (see §7.1 of Chap. 4). Λ_n will be the initial member of its chain if D_m is not contained in a $D_{m'}$ in N with $m' > m$ using the same glue vector [2].

Alternatively we can refer to the 23-dimensional member of the chain, for which the appropriate D_m is a D_1 generated by a single vector $v = 2e$ of norm 4 and $[2] = e$. This 23-dimensional lattice is the set of vectors $x \in e^\perp$ for which $x + ne \in N$ for some integer n. The other members of the chain can be obtained by removing summands I_k.

Thus our process can be viewed from either end of the chain containing Λ_n. On the one hand we search for a maximal sublattice D_m (and glue vector [2]) in the Niemeier lattice N, and locate the initial member of the chain in $D_m^\perp$. Alternatively we find each possible vector $v = 2e$ of norm 4 in N and locate the 23-dimensional member of the chain in the subspace $v^\perp$.

The Niemeier lattices of minimal norm 2. Let N be one of the 23 Niemeier lattices with minimal norm 2, and let W_1, W_2,... be its components. The norm 4 vector v used in the construction may then be written as

$$v = v_1 + v_2 + \cdots ,$$

where v_1, v_2, v_3,... are either (1) minimal representatives of glue digits corresponding to a glue word of norm 4; (2) minimal vectors r, s of norm 2 in two distinct components W_i, W_j, and 0 in the other components; or ($\geqslant$3) 0 except for one v_i which is a vector of norm 4 in its component W_i. (Again we omit the easy proof of this statement.) The cases have been numbered so as to correspond with the value of m mentioned above.

In case (1) the D_1 generated by v is maximal and we obtain a 23-dimensional lattice not representing 1. Our symbol for this case is the glue word mentioned.

In case (2) the D_2 generated by r and s is maximal with $[2] = \frac{1}{2}(r+s)$, and so the reduced lattice is 22-dimensional. The symbol for this case is a word with *'s in positions i and j and 0's elsewhere.

The cases ($\geqslant$3) can be further subdivided as follows. Case (3): $v = v_i$ is the vector $(1, 1, -1, -1, 0^{k-3})$ in a component $W_i = A_k$ for some $k \geqslant 3$. The maximal D_m is the sublattice A_3 of A_k supported in the four nonzero coordinates of v_i. The reduced lattice is 21-dimensional, and the symbol for this case has a + in position i and 0's elsewhere.

Table 16.7. The unimodular lattices of dimension $\leqslant 23$ that contain no vectors of norm 1. All are odd (or of type I) except for four lattices: the empty lattice, E_8, E_8^2, and D_{16}, which are even (or of type II). The notation is explained in §4.

dim	N	w	V	$c(w)$	g_1g_2	t_2
0	D_{24}	–	$\varnothing$	1	1	0
8	$D_{16}E_8$	-0	E_8	1	1	240
12	D_{12}^2	-0	D_{12}	2	1	264
14	$D_{10}E_7^2$	-00	E_7^2	1	2	252
15	$A_{15}D_9$	0-	A_{15}	1	2	240
16	E_8^3	-00	E_8^2	3	2	480
16	$D_{16}E_8$	0-	D_{16}	1	1	480
16	D_8^3	-00	D_8^2	3	2	224
17	$A_{11}D_7E_6$	0-0	$A_{11}E_6$	1	2	204
18	$A_{17}E_7$	0-	$A_{17}A_1$	1	2	308
18	$D_{10}E_7^2$	0-0	$D_{10}E_7A_1$	2	1	308
18	D_6^4	-000	D_6^3	4	6	180
18	$A_9^2D_6$	00-	A_9^2	1	4	180
19	E_6^4	-000	$E_6^3O_1$	4	12	216
19	$A_{11}D_7E_6$	00-	$A_{11}D_7O_1$	1	2	216
19	$A_7^2D_5^2$	00-0	$A_7^2D_5$	2	4	152
20	D_{24}	+	D_{20}	1	1	760
20	$D_{16}E_8$	+0	$D_{12}E_8$	1	1	504
20	D_{12}^2	+0	$D_{12}D_8$	2	1	376
20	$D_{10}E_7^2$	+00	$E_7^2D_6$	1	2	312
20	$A_{15}D_9$	0+	$A_{15}D_5$	1	2	280
20	D_8^3	+00	$D_8^2D_4$	3	2	248
20	$A_{11}D_7E_6$	0+0	$A_{11}E_6A_3$	1	2	216
20	D_6^4	+000	$D_6^3A_1^2$	4	6	184
20	$A_9^2D_6$	00+	$A_9^2A_1^2$	1	4	184
20	$A_7^2D_5^2$	00+0	$A_7^2D_5O_1$	2	4	152
20	D_4^6	$+0^5$	D_4^5	6	120	120
20	$A_5^4D_4$	0^4+	A_5^4	1	16	120
21	A_{24}	+	$A_{20}O_1$	1	2	420
21	$A_{17}E_7$	+0	$A_{13}E_7O_1$	1	2	308
21	$A_{15}D_9$	+0	$A_{11}D_9O_1$	1	2	276
21	A_{12}^2	+0	$A_{12}A_8O_1$	2	2	228
21	$A_{11}D_7E_6$	+00	$D_7A_7E_6O_1$	1	2	212
21	$A_9^2D_6$	+00	$A_9D_6A_5O_1$	2	2	180
21	A_8^3	+00	$A_8^3A_4O_1$	3	4	164
21	$A_7^2D_5^2$	+000	$A_7D_5^2A_3O_1$	2	4	148
21	A_6^4	+000	$A_6^3A_2O_1$	4	6	132
21	$A_5^4D_4$	+0000	$A_5^3D_4A_1O_1$	4	12	116
21	A_4^6	$+0^5$	$A_4^5O_1$	6	40	100
21	A_3^8	$+0^7$	A_3^7	8	336	84
22	$D_{16}E_8$	**	$D_{14}E_7A_1$	1	1	492
22	E_8^3	**0	$E_8E_7^2$	3	2	492
22	D_{12}^2	**	$D_{10}^2A_1^2$	1	2	364
22	$A_{17}E_7$	**	$A_{15}D_6O_1$	1	2	300
22	$D_{10}E_7^2$	0**	$D_{10}D_6^2$	1	2	300
22	$D_{10}E_7^2$	**0	$D_8E_7D_6A_1$	2	1	300
22	$A_{15}D_9$	**	$A_{13}D_7A_1O_1$	1	2	268
22	D_8^3	**0	$D_8D_6^2A_1^2$	3	2	236
22	A_{12}^2	**	$A_{10}^2O_2$	1	4	220
22	E_6^4	**00	$E_6^2A_5^2$	6	8	204
22	$A_{11}D_7E_6$	0**	$A_{11}D_5A_5A_1$	1	2	204
22	$A_{11}D_7E_6$	*0*	$A_9D_7A_5O_1$	1	2	204

Table 16.7 (concluded)

dim	N	w	V	$c(w)$	g_1g_2	t_2
22	$A_{11}D_7E_6$	**0	$A_9E_6D_5A_1O_1$	1	2	204
22	D_6^4	**00	$D_6^2D_4^2A_1^2$	6	4	172
22	$A_9^2D_6$	**0	$A_7^2D_6O_2$	1	4	172
22	$A_9^2D_6$	*0*	$A_9A_7D_4A_1O_2$	2	2	172
22	A_8^3	**0	$A_8A_6^2O_2$	3	4	156
22	$A_7^2D_5^2$	00**	$A_7^2A_3^2A_1^2$	1	8	140
22	$A_7^2D_5^2$	**00	$D_5^2A_5^2O_2$	1	8	140
22	$A_7^2D_5^2$	*0*0	$A_7D_5A_5A_3A_1O_1$	4	2	140
22	A_6^4	**00	$A_6^2A_4^2O_2$	6	4	124
22	D_4^6	$**0^4$	$D_4^4A_1^6$	15	144	108
22	$A_5^4D_4$	*000*	$A_5^3A_3A_1^3O_1$	4	12	108
22	$A_5^4D_4$	**000	$A_5^2D_4A_3^2O_2$	6	8	108
22	A_4^6	$**0^4$	$A_4^4A_2^2O_2$	15	16	92
22	A_3^8	$**0^6$	$A_3^6A_1^2O_2$	28	96	76
22	A_2^{12}	$**0^{10}$	$A_2^{10}O_2$	66	2880	60
22	A_1^{24}	$**0^{22}$	A_1^{22}	276	887040	44
23	$D_{16}E_8$	10	$A_{15}E_8$	1	2	480
23	A_{24}	5	$A_{19}A_4$	2	2	400
23	D_{12}^2	12	$D_{11}A_{11}O_1$	2	2	352
23	$A_{17}E_7$	60	$A_{11}E_7A_5$	2	2	288
23	$D_{10}E_7^2$	110	$A_9E_7E_6O_1$	2	2	288
23	$D_{10}E_7^2$	211	$D_9E_6^2O_2$	1	4	288
23	$A_{17}E_7$	31	$A_{14}E_6A_2O_1$	2	2	288
23	$A_{15}D_9$	80	$D_9A_7^2$	1	4	256
23	$A_{15}D_9$	21	$A_{13}A_8A_1O_1$	2	2	256
23	$A_{15}D_9$	42	$A_{11}D_8A_3O_1$	2	2	256
23	D_8^3	033	$D_8A_7^2O_1$	3	4	224
23	D_8^3	122	$D_7^2A_7O_2$	3	4	224
23	A_{12}^2	15	$A_{11}A_7A_4O_1$	4	2	208
23	A_{12}^2	32	$A_{10}A_9A_2A_1O_1$	4	2	208
23	E_6^4	0111	$E_6D_5^3O_2$	8	12	192
23	$A_{11}D_7E_6$	620	$E_6D_6A_5^2O_1$	1	4	192
23	$A_{11}D_7E_6$	401	$D_7A_7D_5A_3O_1$	2	2	192
23	$A_{11}D_7E_6$	330	$A_8E_6A_6A_2O_1$	2	2	192
23	$A_{11}D_7E_6$	111	$A_{10}A_6D_5O_2$	2	2	192
23	$A_{11}D_7E_6$	222	$A_9D_6D_5A_1O_2$	2	2	192
23	D_6^4	2222	$D_5^4O_3$	1	48	160
23	$A_9^2D_6$	501	$A_9A_5A_4^2O_1$	2	4	160
23	D_6^4	0123	$D_6D_5A_5^2O_2$	12	4	160
23	$A_9^2D_6$	240	$A_7D_6A_5A_3A_1O_1$	4	2	160
23	$A_9^2D_6$	312	$A_8A_6D_5A_2O_2$	4	2	160
23	$A_9^2D_6$	121	$A_8A_7A_5A_1O_2$	4	2	160
23	A_8^3	036	$A_8A_5^2A_2^2O_1$	6	4	144
23	A_8^3	411	$A_7^2A_4A_3O_2$	6	4	144
23	A_8^3	177	$A_7A_6^2A_1^2O_2$	6	4	144
23	$A_7^2D_5^2$	4400	$D_5^2A_3^4O_1$	1	16	128
23	$A_7^2D_5^2$	4022	$A_7D_4^2A_3^2O_2$	2	8	128
23	$A_7^2D_5^2$	2031	$A_7A_5A_4^2A_1O_2$	4	4	128
23	$A_7^2D_5^2$	2220	$D_5A_5^2D_4A_1^2O_2$	4	4	128
23	$A_7^2D_5^2$	1112	$A_6^2D_4A_4O_3$	4	4	128
23	$A_7^2D_5^2$	1303	$A_6D_5A_4^2A_2O_2$	8	2	128
23	A_6^4	5111	$A_5^3A_4A_1O_3$	8	6	112
23	A_6^4	0124	$A_6A_5A_4A_3A_2A_1O_2$	24	2	112
23	D_4^6	002332	$D_4^2A_3^4O_3$	45	96	96
23	$A_5^4D_4$	00331	$A_5^2A_3A_2^4O_2$	6	16	96
23	$A_5^4D_4$	02220	$A_5D_4A_3^3A_1^3O_2$	8	12	96
23	$A_5^4D_4$	31110	$D_4A_4^3A_2^2O_3$	8	12	96
23	$A_5^4D_4$	04111	$A_5A_4^2A_3^2A_1O_3$	24	4	96
23	A_4^6	011111	$A_4A_3^5O_4$	12	40	80
23	A_4^6	001234	$A_4^2A_3^2A_2^2A_1^2O_3$	60	8	80
23	A_3^8	0^42^4	$A_3^4A_1^8O_3$	14	384	64
23	A_3^8	0^321^33	$A_3^3A_2^4A_1^2O_4$	112	48	64
23	A_2^{12}	0^61^6	$A_2^6A_1^6O_5$	264	1440	48
23	A_1^{24}	$0^{16}1^8$	$A_1^{16}O_7$	759	645120	32
23	Λ_{24}	min	O_{23}	196560	84610842624000	0

Case (4): $v = v_i$ is a vector of the form $(\pm 1^4, 0^{k-4})$ in a component D_k. In this case the maximal D_m is the D_4 supported in the four nonzero coordinates of v, and the reduced lattice is 20-dimensional. Again the symbol contains 0's except for a + in position i. We remark that the case $k = 4$, when there is a component D_4, is rather special because all three nontrivial cosets of D_4 in its dual contain vectors of norm 1 which can serve as the glue vector [2]. Fortunately, in both the cases $N = D_4^6$ and $N = A_5^4 D_4$, Aut(N) contains elements permuting all three cosets of any D_4 component, so we may suppose $v = (\pm 1^4, 0^{k-4})$.

Cases ($\geqslant$5): $v = v_i$ is the vector $(2, 0^{k-1})$ in a D_k $(k \geqslant 5)$ or any norm 4 vector in E_6, E_7 or E_8. The maximal D_m in these four cases is D_k D_5, D_6 or D_8 respectively, and the symbol has a $-$ in position i and 0's elsewhere.

Inequivalent lattices. The above remarks make it easy to determine when two lattices constructed in this way are equivalent. Each lattice Λ_n is specified by a Niemeier lattice

$$N = W_1 W_2 \cdots W_a$$

and a symbol

$$w = d_1 d_2 \cdots d_a$$

which is either a glue word or a permutation of $**0^{a-2}$, $+0^{a-1}$ or -0^{a-1}. Two lattices are equivalent if and only if they are specified by the same N and their symbols are equivalent under Aut(N). The components V_1, $V_2, \ldots$ of Λ_n are determined by the W_i and d_i as shown in Table 16.8.

The automorphism group of Λ_n. To compute the order of the automorphism group of Λ_n we argue as follows. The lattices Λ_n, $\Lambda_{n+1}, \ldots$ of a chain whose reduced lattice is Λ_n have

$$|\mathrm{Aut}(\Lambda_{n+m})| = |\mathrm{Aut}(\Lambda_n)| \cdot 2^m m! \ ,$$

so that it suffices to compute $|\mathrm{Aut}(\Lambda_{23})|$. Now a 23-dimensional lattice Λ_{23} is completely determined by the associated Niemeier lattice N and either of the norm 4 vectors $\pm v$ of $\Lambda_{23}^{\perp} \cap N$. Hence

$$|\mathrm{Aut}(\Lambda_{23})| = \frac{2|\mathrm{Aut}(N)|}{c(v)} \ , \tag{5}$$

where Aut(N) is described in §3 and $c(v)$ is the number of images of v under Aut(N). This number is easily computed as the product

$$c(v) = c(d_1)\, c(d_2) \cdots c(d_a)\, c(w) \ , \tag{6}$$

where $w = d_1 d_2 \cdots d_a$ is the symbol for Λ_N, $c(w)$ is the number of images of w under Aut(N), and $c(d_i)$ is the number of choices for the component v_i of the vector v which would lead to the digit d_i. For example $c(*)$ is the number of minimal vectors in W_i. The numbers $c(d_i)$ can be found by

Table 16.8. Each lattice Λ_n is specified by the components $W_1, W_2, \ldots$ of a Niemeier lattice N and a certain symbol $w = d_1 d_2 \cdots$. This table gives the components $V_1, V_2, \ldots$ of Λ_n corresponding to each component W of N and each digit d of w, as well as the numbers $c(d)$ defined in §4. The six parts of the table correspond to the cases $W = A_n$, D_4, $D_n (n \geqslant 5)$, E_6, E_7 and E_8.

A_n		
d	$V_1 V_2 \cdots$	$c(d)$
0	A_n	1
i	$A_{i-1} A_{n-i}$	$\binom{n+1}{i}$
*	A_{n-2}	$2\binom{n+1}{2}$
+	A_{n-4}	$6\binom{n+1}{4}$

D_4		
d	$V_1 V_2 \cdots$	$c(d)$
0	D_4	1
1	A_3	8
2	A_3	8
3	A_3	8
*	A_1^3	24
+	–	24

$D_n (n \geqslant 5)$		
d	$V_1 V_2 \cdots$	$c(d)$
0	D_n	1
1	A_{n-1}	2^{n-1}
2	D_{n-1}	$2n$
3	A_{n-1}	2^{n-1}
*	$D_{n-2} A_1$	$4\binom{n}{2}$
+	D_{n-4}	$16\binom{n}{4}$
–	–	$2n$

	E_6		E_7		E_8	
d	$V_1 V_2 \cdots$	$c(d)$	$V_1 V_2 \cdots$	$c(d)$	$V_1 V_2 \cdots$	$c(d)$
0	E_6	1	E_7	1	E_8	1
1	D_5	27	E_6	56		
2	D_5	27				
*	A_5	72	D_6	126	E_7	240
–	–	270	A_1	756	–	2160

elementary counting arguments and are given in Table 16.8. The values of $c(w)$, which are usually small, were found by careful consideration of the group action, and are given in Table 16.7.

The orders of the automorphism groups of the lattices Λ_n in Table 16.7 were calculated from (5) and (6). The lattices with minimal norm 1 were then found by forming direct sums $\Lambda_n \oplus I_m$, and the mass constants were checked against the formulae given in Theorem 1. (This completes the formal proof of Theorem 5.) The numbers of lattices found in each dimension are shown in Table 2.2.

The fifth column of Table 16.7 gives the number $g_1(\Lambda_n) g_2(\Lambda_n) = |G_1(\Lambda_n)| \, |G_2(\Lambda_n)|$, and provides the reader with an alternative and easier method of computing $|\mathrm{Aut}(\Lambda_n)|$ via the formula (cf. Eq. (1))

$$|\mathrm{Aut}(\Lambda_n)| = \prod_{i=1}^{k} |G_0(V_i)| \cdot g_1(\Lambda_n) g_2(\Lambda_n). \tag{7}$$

Here $V_1, \ldots, V_k$ are the components of Λ_n (given in the column headed V), and the values of $|G_0(V_i)|$ are given in Eq. (2). For example the penultimate line of Table 16.7 refers to the lattice $A_1^{16} O_7$, for which

$$|\mathrm{Aut}(\Lambda_n)| = (2!)^{16} \cdot 1^7 \cdot 645120 .$$

The values of $g_1 g_2$ were actually "back-computed" from (5), (6) and (7) after the mass formula check had been applied. We note that $(V_1, \ldots, V_k)$ identifies Λ_n uniquely. The final column of Table 16.7 gives the number (t_2) of norm 2 vectors in Λ_n.

The Leech lattice. It remains to consider the unique Niemeier lattice with minimal norm greater than 2, the Leech lattice. In this lattice there is just one orbit of norm 4 vectors (Theorem 27 of Chap. 10). Our construction produces a certain 23-dimensional lattice with minimal norm 3, the *shorter Leech lattice* O_{23}, which we have already encountered in the Appendix to Chap. 6. It appears in the last line of Table 16.7, and is the only lattice in the table with minimal norm 3.

Acknowledgements. The extensive arithmetical calculations needed to prove Theorems 4 and 5 were performed using ALTRAN [Bro14] and MACSYMA [Mat3].

17

The 24-Dimensional Odd Unimodular Lattices

R. E. Borcherds

This chapter completes the classification of the 24-dimensional unimodular lattices by enumerating the odd lattices. These are (essentially) in one-to-one correspondence with neighboring pairs of Niemeier lattices.

1. Introduction

The even unimodular lattices in 24 dimensions were classified by Niemeier [Nie2] and the results are given in the previous chapter, together with the enumeration of the even and odd unimodular lattices in dimensions less than 24. There are twenty-four Niemeier lattices, and in the present chapter they will be referred to by their components D_{24}, $D_{16}E_8, \ldots$ (with the Leech lattice being denoted by Λ_{24}), and also by the Greek letters $\alpha, \beta, \ldots$ (see Table 16.1).

The *odd* unimodular lattices in 24 and 25 dimensions were classified in [Bor1]. In this chapter we list the odd 24-dimensional lattices. Only those with minimal norm at least 2 are given, i.e., those that are strictly 24-dimensional, since the others can easily be obtained from lower dimensional lattices (see the summary in Table 2.2 of Chapter 2).

Tables of all the 665 25-dimensional unimodular lattices and the 121 even 25-dimensional lattices of determinant 2 are available electronically from [BorchHP]. The 665 25-dimensional unimodular lattices are also available from the electronic *Catalogue of Lattices* [NeSl].

Two lattices are called *neighbors* if their intersection has index 2 in each of them [Kne4], [Ven2] (see Introduction to Third Edition for a discussion of this concept).

We now give a brief description of the algorithm used in [Bor1] to enumerate the 25-dimensional unimodular lattices.

The first step is to observe that there is a one-to-one correspondence between 25-dimensional unimodular lattices (up to isomorphism) and orbits of norm -4 vectors in the even Lorentzian lattice $II_{25,1}$: the lattice Λ corresponds to the norm -4 vector v if and only if the sublattice of even vectors of Λ is isomorphic to the lattice $v^{\perp}$. So we can classify 25-dimensional unimodular lattices if we can classify negative norm vectors in $II_{25,1}$.

We classify orbits of vectors of norm $-2n \leq 0$ in $II_{25,1}$ by induction on n as follows. First of all the primitive norm 0 vectors correspond to the Niemeier lattices as in Section 1 of Chapter 26. So there are exactly 24 orbits of primitive norm 0 vectors, and any norm 0 vector can be obtained from a primitive one by multiplying it by some constant.

Suppose we have classified all orbits of vectors of norms $-2m$ with $0 \geq -2m > -2n$, and that we have a vector v of norm $-2n$. We fix a fundamental Weyl chamber for the reflection group of $II_{25,1}$ containing v, as in Chapter 26. We look at the root system of the lattice $v^{\perp}$, and find that one of the following three things can happen:

1. There is a norm 0 vector z with $(z, v) = 1$. It turns out to be trivial to classify such norm $-2n$ vectors v: there is one orbit corresponding to each orbit of norm 0 vectors. They correspond to lattices $v^{\perp}$ which are the sum of a Niemeier lattice and a 1-dimensional lattice generated by a vector of norm $2n$.
2. There is no norm 0 vector z with $(z, v) = 1$ and the root system of $v^{\perp}$ is nonempty. In this case we choose a component of the root system of $v^{\perp}$ and let r be its highest root. Then the vector $u = v + r$ has norm $-2(n-1)$, and the assumption about no norm 0 vectors z with $(z, v) = 1$ easily implies that u is still in the Weyl chamber of $II_{25,1}$. Hence we have reduced v to some known vector u of norm $-2(n-1)$, and with a little effort it is possible to reverse this process and construct v from u.
3. Finally suppose that there are no roots in $v^{\perp}$. As v is in the Weyl chamber this implies that $(v, r) \leq -1$ for all simple roots r. By Theorem 1 of Chapter 27 there is a norm 0 (Weyl) vector w_{25} with the property that $(w_{25}, r) = -1$ for all simple roots r. Therefore the vector $u = v - w_{25}$ has the property that $(u, r) \leq 0$ for all simple roots r. So u is in the Weyl chamber, and has norm $-2n-(u, w_{25})$ which is larger than $-2n$ unless v is a multiple of w_{25}. So we can reconstruct v from the known vector u as $v = u + w_{25}$.

In every case we can reconstruct v from known vectors, so we get an algorithm for classifying the norm $-2n$ vectors in $II_{25,1}$. (This algorithm breaks down in higher-dimensional Lorentzian lattices for two reasons: it is too difficult to classify the norm 0 vectors, and there is usually no analogue of the Weyl vector w_{25}.)

We now apply the algorithm above to find the 121 orbits of norm -2 vectors from the (known) norm 0 vectors, and then apply it again to find the 665 orbits of norm -4 vectors from the vectors of norm 0 and -2.

The neighbors of a strictly 24 dimensional odd unimodular lattice can be found as follows. If a norm -4 vector $v \in II_{25,1}$ corresponds to the sum of a strictly 24 dimensional odd unimodular lattice Λ and a 1-dimensional lattice, then there are exactly two norm-0 vectors of $II_{25,1}$ having inner product -2 with v, and these norm 0 vectors correspond to the two even neighbors of Λ.

The enumeration of the odd 24-dimensional lattices. Figure 17.1 shows the neighborhood graph for the Niemeier lattices, which has a node for each Niemeier lattice. If A and B are neighboring Niemeier lattices, there are three integral lattices containing $A \cap B$, namely A, B, and an odd unimodular lattice C (cf. [Kne4]). An edge is drawn between nodes A and B in Fig. 17.1 for each strictly 24-dimensional unimodular lattice arising in this way. Thus there is a one-to-one correspondence between the strictly 24-dimensional odd unimodular lattices and the edges of our neighborhood graph. The 156 lattices are shown in Table 17.1. Figure 17.1 also shows the corresponding graphs for dimensions 8 and 16.

For each lattice Λ in the table we give its components (in the notation of the previous chapter) and its even neighbors (represented by 2 Greek letters as in Table 16.1). The final column gives the orders $g_1 \cdot g_2$ of the groups $G_1(\Lambda)$, $G_2(\Lambda)$ defined as follows. We may write $Aut(\Lambda) = G_0(\Lambda).G_1(\Lambda).G_2(\Lambda)$ where G_0 is the reflection group. The group G_1 is the subgroup of $Aut(\Lambda)$ of elements fixing a fundamental chamber of the Weyl group and not interchanging the two neighbors. The group $G_2(\Lambda)$ has order 1 or 2 and interchanges the two neighbors of Λ if it has order 2. (It turns out that $G_2(\Lambda)$ has order 2 if and only if the two components of Λ are isomorphic.) The components are written as a union of orbits under $G_1(\Lambda)$, with parentheses around two orbits if they fuse under $G_2(\Lambda)$.

The first lattice in the table is the odd Leech lattice O_{24}, which is the only one with no norm 2 vectors. The number of norm 2 vectors is given by the formula

$$8h(A) + 8h(B) - 16$$

where $h(A)$ and $h(B)$ are the Coxeter numbers of the even neighbors of the lattice. These Coxeter numbers satisfy the inequality $h(B) \leq 2h(A) - 2$ and the lattices for which equality holds are indicated by a thick line in Figure 17.1. The Weyl vector $\rho(\Lambda)$ of the lattice Λ has norm given by the formula $\rho(\Lambda)^2 = h(A)h(B)$.

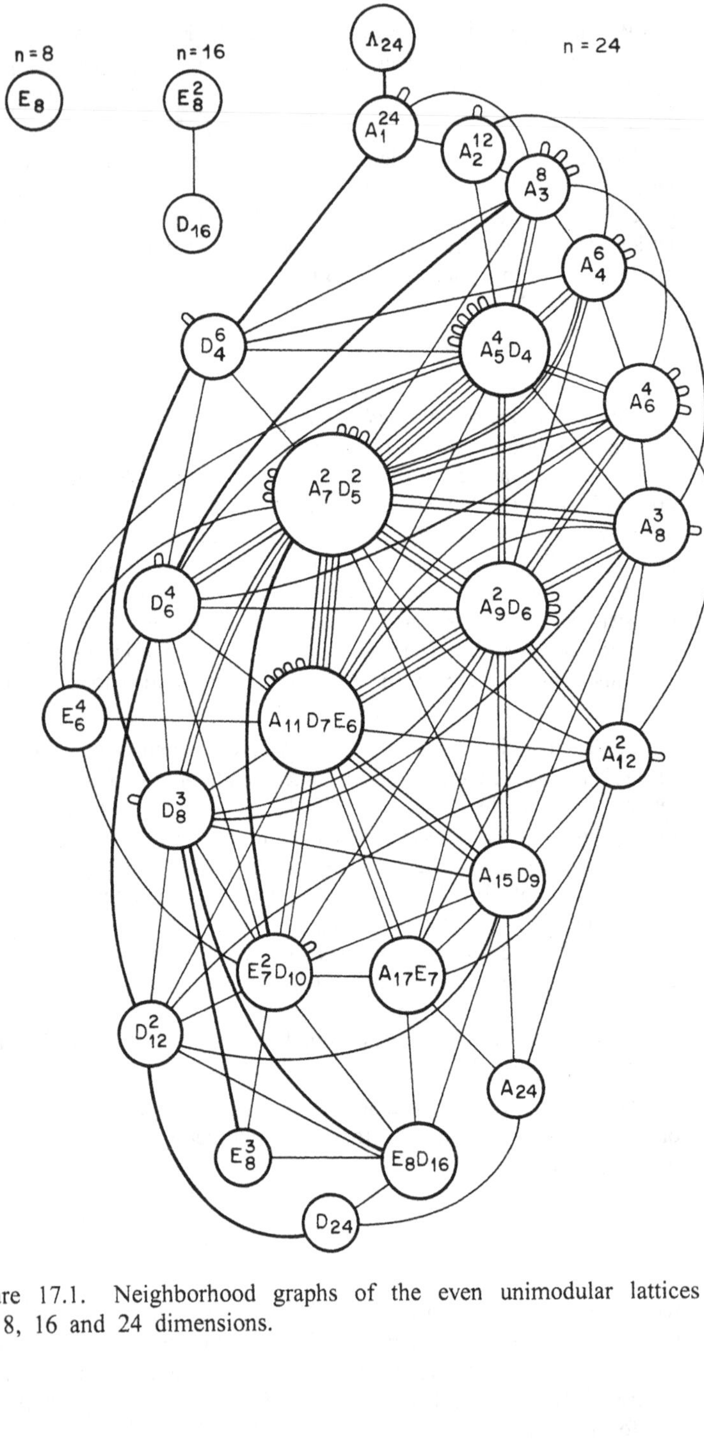

Figure 17.1. Neighborhood graphs of the even unimodular lattices in $n = 8$, 16 and 24 dimensions.

Table 17.1a

	Components		$g_1 g_2$
1	O_{24}	$\omega\ \psi$	$2^{12} M_{24}$
2	$A_1^8 O_{16}$	$\psi\ \psi$	$10321920 \cdot 2$
3	$A_1^{12} O_{12}$	$\psi\ \chi$	190080
4	$A_1^{16} O_8$	$\psi \phi$	43008
5	$A_2^2 A_1^{10} O_{10}$	$\chi\ \chi$	$2880 \cdot 2$
6	A_1^{24}	$\psi\ \tau$	138240
7	$A_2^4 A_1^8 O_8$	$\chi\ \phi$	384
8	$A_2^6 A_1^6 O_6$	$\chi\ \upsilon$	240
9	$A_3^2 A_1^{12} O_6$	$\phi\ \phi$	$384 \cdot 2$
10	$A_3 A_2^4 A_1^6 O_7$	$\phi\ \phi$	$48 \cdot 2$
11	$A_2^8 O_8$	$\phi\ \phi$	$336 \cdot 2$
12	$A_2^8 A_1^4 O_4$	$\chi\ \sigma$	384
13	$A_3^2 A_2^4 A_1^4 O_6$	$\phi\ \upsilon$	16
14	$A_3^4 A_1^8 O_4$	$\phi\ \tau$	384
15	$A_3^4 A_1^4 A_1^4 O_4$	$\phi\ \sigma$	48
16	$A_3^2 A_3 A_2^4 A_1^2 O_5$	$\phi\ \sigma$	16
17	$A_4 A_3 A_2^4 A_1^4 O_5$	$\upsilon\ \upsilon$	$8 \cdot 2$
18	$A_3^4 A_2^2 A_1^2 O_6$	$\upsilon\ \upsilon$	$16 \cdot 2$
19	$A_3^4 A_2^4 O_4$	$\phi\ \rho$	24
20	$A_3^6 O_6$	$\upsilon\ \tau$	240
21	$A_4^2 A_2^4 A_1^4 O_4$	$\upsilon\ \sigma$	16
22	$A_4 A_3^2 A_3 A_2^2 A_1^2 O_5$	$\upsilon\ \sigma$	4
23	$A_3^4 A_3^2 A_1^4 O_2$	$\phi\ \pi$	32
24	$A_4^2 A_3^2 A_2^2 A_1^2 O_4$	$\upsilon\ \rho$	4
25	$D_4^2 A_1^{16}$	$\tau\ \tau$	$576 \cdot 2$
26	$D_4 A_3^4 A_1^4 O_4$	$\tau\ \sigma$	48
27	$A_5 (A_3 A_3) A_2^4 A_1 O_4$	$\sigma\ \sigma$	$8 \cdot 2$
28	$A_4^2 (A_3 A_3) A_3 (A_1 A_1) O_5$	$\sigma\ \sigma$	$4 \cdot 2$
29	$A_5 A_3^3 (A_1^3 A_1^3) A_1 O_3$	$\sigma\ \sigma$	$12 \cdot 2$
30	$4 A_3^4 A_1^4 O_4$	$\sigma\ \sigma$	$16 \cdot 2$
31	$D_4 A_4 A_2^6 O_4$	$\sigma\ \sigma$	$12 \cdot 2$
32	A_3^8	$\phi\ \xi$	384
33	$A_4^2 A_4 A_3 A_2^2 A_1^2 O_3$	$\upsilon\ \pi$	4
34	$A_4^2 A_3^4 O_4$	$\upsilon\ \pi$	16
35	$A_5 A_4 A_3 A_3 A_2 A_2 A_1 O_4$	$\sigma\ \rho$	2
36	$A_4^4 A_1^4 O_4$	$\sigma\ \rho$	24
37	$A_4^3 A_3^3 O_3$	$\upsilon\ o$	12
38	$D_4^2 A_3^4 O_4$	$\tau\ \pi$	32
39	$A_5 A_4^2 A_3 A_3 A_1 O_4$	$\sigma\ \pi$	4
40	$A_5^2 A_3^2 A_1^2 A_1^2 A_1^2 O_2$	$\sigma\ \pi$	8
41	$A_5 D_4 A_3^2 A_3 A_1^2 A_1 O_3$	$\sigma\ \pi$	4
42	$D_4 A_4^2 A_4 A_2^2 O_4$	$\sigma\ \pi$	4
43	$A_6 A_3^3 A_2^3 O_3$	$\rho\ \rho$	$6 \cdot 2$
44	$A_5^2 A_3^2 A_2^2 O_4$	$\rho\ \rho$	$4 \cdot 2$
45	$A_5 A_4^3 A_1^3 O_4$	$\rho\ \rho$	$6 \cdot 2$
46	$A_4^4 A_3^2 O_2$	$\upsilon\ \nu$	16

Table 17.1b

	Components		g_1g_2
47	$A_5^2A_4A_3A_2^2O_3$	σ o	4
48	$A_6A_4A_4A_3A_2A_1A_1O_3$	ρ π	2
49	$A_5^2A_4^2A_1^2O_4$	ρ π	4
50	$D_4^4A_1^8$	τ ξ	48
51	$A_5^2D_4A_3^2A_1^2O_2$	σ ξ	8
52	$A_5A_5A_4^2A_3O_3$	σ ν	4
53	$A_5^2D_4A_3^2A_1^2O_2$	σ ν	4
54	$A_6A_5A_4A_3A_2A_1O_3$	ρ o	2
55	$A_6D_4(A_4A_4)A_2O_4$	π π	$2 \cdot 2$
56	$D_5A_4(A_4A_4)(A_2A_2)O_3$	π π	$2 \cdot 2$
57	$A_7(A_3^2A_3^2)A_1^4O_1$	π π	$8 \cdot 2$
58	$A_5^2D_4^2A_1^2O_4$	π π	$8 \cdot 2$
59	$D_5A_5(A_3A_3)A_3(A_1A_1)A_1O_2$	π π	$2 \cdot 2$
60	$D_5D_4A_3^4O_3$	π π	$8 \cdot 2$
61	$A_5^4O_4$	ρ ξ	24
62	$A_6^2A_3^2A_2^2O_2$	ρ ν	4
63	$A_6A_5A_5A_3A_2O_3$	ρ ν	2
64	$A_5^4A_1^4$	σ μ	48
65	$A_5^2A_5D_4A_3A_1O_1$	σ λ	4
66	$A_6^2A_4A_3A_1^2O_3$	π o	4
67	$A_7A_4^2A_3^2O_3$	π o	4
68	$D_5A_5^2D_4A_1^2O_3$	π ξ	4
69	$D_5^2A_3^4O_2$	π ξ	16
70	$A_6D_5A_4A_4A_2O_3$	π ν	2
71	$A_7A_5D_4A_3A_1A_1A_1O_2$	π ν	2
72	$A_7A_5A_4^2A_1O_3$	π ν	4
73	$A_7A_5^2A_2^2O_3$	o o	$4 \cdot 2$
74	D_4^6	τ ι	48
75	$A_6A_6A_5A_4A_1O_2$	ρ λ	2
76	$A_7A_6A_5A_2A_1O_3$	o ν	2
77	$A_8A_4^2A_3^2O_2$	o ν	4
78	$A_6^2A_5^2O_2$	ρ κ	4
79	$D_5^2A_5^2A_1^2O_2$	π μ	8
80	$A_6A_6D_5A_4O_3$	π λ	2
81	$A_7D_5D_4A_3A_3O_2$	π λ	2
82	$D_5D_5A_5^2A_1^2O_2$	π λ	4
83	$A_7D_5A_5A_3A_1A_1A_1O_1$	π λ	2
84	$D_6D_4^3A_1^6$	ξ ξ	$6 \cdot 2$
85	$D_6A_5^2A_3^2O_2$	ξ ν	4
86	$A_8(A_5A_5)A_3O_3$	ν ν	$2 \cdot 2$
87	$D_6A_5^2A_3^2O_2$	ν ν	$4 \cdot 2$
88	$A_7^2D_4(A_1^2A_1^2)O_2$	ν ν	$4 \cdot 2$
89	$A_7^2A_4^2O_2$	π κ	4
90	$A_8A_6A_5A_2A_1O_2$	o λ	2
91	$A_7^2D_4^2O_2$	π ι	8
92	$A_7D_5^2A_3^2O_1$	π ι	4

Table 17.1c

	Components		$g_1 g_2$
93	$A_8 A_7 A_4 A_3 O_2$	$o\ \kappa$	2
94	$D_5^4 O_4$	$\xi\ \mu$	48
95	$D_6 D_5 A_5^2 O_3$	$\xi\ \lambda$	4
96	$A_8 A_6 D_5 A_2 O_3$	$\nu\ \lambda$	2
97	$A_7 D_6 A_5 A_3 A_1 O_2$	$\nu\ \lambda$	2
98	$A_9 A_5 D_4 A_3 A_1 A_1 O_1$	$\nu\ \lambda$	2
99	$A_7^3 O_3$	$o\ \iota$	12
100	$A_7^2 D_5 D_4 O_1$	$\pi\ \theta$	4
101	$A_9 A_6 A_5 A_2 O_2$	$\nu\ \kappa$	2
102	$A_8^2 A_3^2 O_2$	$\nu\ \kappa$	4
103	$A_7^2 D_6 A_1^2 O_2$	$\nu\ \iota$	4
104	$D_6^2 D_4^2 A_1^4$	$\xi\ \iota$	4
105	$A_7^2 D_5^2$	$\pi\ \eta$	8
106	$A_8 A_7^2 O_2$	$o\ \theta$	4
107	$E_6 D_5 A_5^2 A_1^2 O_1$	$\mu\ \lambda$	4
108	$E_6(A_6 A_6) A_4 O_2$	$\lambda\ \lambda$	$2 \cdot 2$
109	$A_7 E_6 D_4 (A_3 A_3) O_1$	$\lambda\ \lambda$	$2 \cdot 2$
110	$A_9 (D_5 D_5)(A_1 A_1) A_1 O_2$	$\lambda\ \lambda$	$2 \cdot 2$
111	$D_7 (A_5 A_5) A_5 A_1 O_1$	$\lambda\ \lambda$	$2 \cdot 2$
112	$A_{10} A_6 A_5 A_1 O_2$	$\lambda\ \kappa$	2
113	$A_9 A_8 A_5 O_2$	$\nu\ \theta$	2
114	$A_9 D_6 A_5 A_3 O_1$	$\nu\ \theta$	2
115	$A_8^2 A_7 O_1$	$o\ \zeta$	4
116	$D_7 A_7 D_5 A_3 O_2$	$\lambda\ \iota$	2
117	$A_9^2 A_2^2 O_2$	$\kappa\ \kappa$	$4 \cdot 2$
118	$D_6^2 D_6 D_4 A_1^2$	$\xi\ \eta$	2
119	$A_9 A_7 D_6 A_1 O_1$	$\nu\ \eta$	2
120	$A_9 A_7 D_6 A_1 O_1$	$\nu\ \zeta$	2
121	$A_{11} D_5 D_4 A_3 O_1$	$\lambda\ \theta$	2
122	$A_9 D_7 A_5 A_1 A_1 O_1$	$\lambda\ \theta$	2
123	$D_8 D_4^4$	$\iota\ \iota$	$4 \cdot 2$
124	$A_{11} A_8 A_3 O_2$	$\kappa\ \theta$	2
125	$E_6^2 D_5^2 O_2$	$\mu\ \eta$	8
126	$A_9 E_6 D_6 A_1 O_2$	$\lambda\ \eta$	2
127	$D_7 A_7 E_6 A_3 O_1$	$\lambda\ \eta$	2
128	$A_{10} E_6 A_6 O_2$	$\lambda\ \zeta$	2
129	$A_{11} D_6 A_5 A_1 O_1$	$\lambda\ \zeta$	2
130	D_6^4	$\xi\ \epsilon$	8
131	$D_8 A_7^2 O_2$	$\iota\ \theta$	4
132	$A_{12} A_7 A_4 O_1$	$\kappa\ \zeta$	2
133	$D_8 D_6^2 A_1^2 A_1^2$	$\iota\ \eta$	2
134	$A_{11} D_7 D_5 O_1$	$\lambda\ \epsilon$	2
135	$A_{11}^2 O_2$	$\kappa\ \epsilon$	4
136	$D_9 A_7^2 O_1$	$\theta\ \eta$	4
137	$A_{13} D_6 A_3 A_1 O_1$	$\theta\ \zeta$	2
138	$D_8^2 D_4^2$	$\iota\ \epsilon$	2

Table 17.1d

	Components		$g_1 g_2$
139	$E_7 D_6 D_6 D_4 A_1$	η η	$1 \cdot 2$
140	$A_9 E_7 A_7 O_1$	η ζ	2
141	$A_{12} A_{11} O_1$	κ δ	2
142	$A_{11} D_9 A_3 O_1$	θ ϵ	2
143	$D_{10} D_6^2 A_1^2$	η ϵ	2
144	$A_{15} A_8 O_1$	θ δ	2
145	D_8^3	ι γ	6
146	$D_8^2 D_8$	ι β	2
147	$A_{16} A_7 O_1$	ζ δ	2
148	$A_{15} D_8 O_1$	θ β	2
149	$D_8 E_7^2 A_1^2$	η γ	2
150	$D_{10} E_7 D_6 A_1$	η β	1
151	$A_{15} E_7 A_1 O_1$	ζ β	2
152	$D_{12} D_8 D_4$	ϵ β	1
153	$E_8 D_8^2$	γ β	2
154	D_{12}^2	ϵ α	2
155	$A_{23} O_1$	δ α	2
156	$D_{16} D_8$	β α	1

18

Even Unimodular 24-Dimensional Lattices

B. B. Venkov

Niemeier's classification of even unimodular 24-dimensional lattices is simplified. The methods involve the theory of modular forms, algebraic coding, and root systems.

1. Introduction

We consider even unimodular Euclidean lattices Λ in $\mathbf{R}^n$ or, in more classical terminology, classes of integral, positive definite, even quadratic forms of determinant 1 in n variables. Such lattices exist only if $n = 8k$. A complete classification of such lattices is known only for $k = 1, 2, 3$. When $k = 4$ there are already more than 80 million lattices (see Table 16.5.). When $k = 1$ or 2 the classification is very simple: there are precisely one 8-dimensional and two 16-dimensional lattices of the above type. A complete classification of the 24-dimensional lattices was obtained in 1968 by Niemeier [Nie2]. According to Niemeier, there are precisely 24 classes of even unimodular lattices. Each lattice Λ is uniquely determined by its set of minimal vectors $\Lambda(2) = \{\lambda \in \Lambda : (\lambda,\lambda) = 2\}$, which form root systems of the following types:

$\varnothing$,

$24A_1,\ 12A_2,\ 8A_3,\ 6A_4,\ 4A_6,\ 3A_8,\ 2A_{12},\ A_{24},$

$6D_4,\ 4D_6,\ 3D_8,\ 2D_{12},\ D_{24},$

$4E_6,\ 3E_8,$

$4A_5+D_4,\ 2A_7+2D_5,\ 2A_9+D_6,\ A_{15}+D_9,\ E_8+D_{16},\ 2E_7+D_{10},\ E_7+A_{17},$

$$E_6+D_7+A_{11}. \tag{1}$$

The lattice corresponding to the empty root system is the Leech lattice. Niemeier obtained his classification by Kneser's method [Kne4], which involves studying lattices that are "neighbors" of a given one (see the previous chapter), requires extensive calculations, and in no way explains the appearance of the strange list (1) of root systems. Here we offer another approach to the classification of the even unimodular 24-dimensional lattices, based on an *a priori* proof of the fact that the root system of a 24-dimensional lattice is one of the systems in (1) and that each root system in (1) can be realized in one and only one way. In §2 we prove that the root system $\Lambda(2)$ of an even unimodular lattice Λ in $\mathbf{R}^{24}$ possesses the following properties: (i) either $\Lambda(2) = \varnothing$ or rank $\Lambda(2) = 24$; (ii) all irreducible components of $\Lambda(2)$ have the same Coxeter number h; (iii) $|\Lambda(2)| = 24h$, where $|\Lambda(2)|$ is the number of elements in $\Lambda(2)$. This is proved by means of modular forms (using theta series with spherical coefficients). It is easy to show that the only root systems possessing properties (i)-(iii) are those in the list (1).

We then show that each root system in (1) can be realized as $\Lambda(2)$ for a unique lattice. This is actually a problem in algebraic coding theory (see §3), and the existing literature on algebraic coding quickly enables us to obtain the final answer (§4). The last section is devoted to the Leech lattice.

2. Possible configurations of minimal vectors

Suppose $\Lambda \subset \mathbf{R}^{24}$ is an even unimodular lattice, $\Lambda(2n) = \{\lambda \in \Lambda : (\lambda,\lambda) = 2n\}$, and $\Lambda(2)$ is the root system of Λ.

Proposition 1. *If* $\alpha \in \mathbf{R}^{24}$, *then*

$$\sum_{y \in \Lambda(2)} (y, \alpha)^2 = \frac{1}{12}(\alpha, \alpha)|\Lambda(2)|. \tag{2}$$

Proof. We use the classical result (see [Hec2] or [Oggl, Chap. VI]) which says that, for an even unimodular lattice Λ in $V = \mathbf{R}^f$, the theta series with spherical coefficients

$$f_{\nu,p} = \sum_n \left(\sum_{x \in \Lambda(2n)} P_\nu(x) \right) q^n, \quad q = e^{2\pi i z}, \ \mathrm{Im}\, z > 0,$$

is the Fourier expansion of a modular form for $PSL(2,\mathbf{Z})$ of weight $\frac{1}{2}f + \nu$. This form is a cusp form if $\nu > 0$. Here P_ν is a homogeneous polynomial of degree ν which is a Λ-spherical function, i.e. $\Delta_\Lambda P_\nu = 0$, where Δ_Λ is the Laplace operator in a coordinate system in V in which the inner product $(\ ,\ )$ reduces to a sum of squares. The 2-dimensional Λ-spherical polynomials have a very simple form: they are linear combinations of polynomials

$$P_{2,\alpha}(x) = (x, \alpha)^2 - \frac{1}{f}(\alpha, \alpha)(x, x),$$

where $\alpha \in V$ (see [Hec2]). Thus, for any $\alpha \in V$,

$$\sum_n \left[\sum_{x \in \Lambda(2n)} \left[(x, \alpha)^2 - \frac{1}{f}(\alpha, \alpha)(x, x) \right] \right] q^n \tag{3}$$

is a cusp form of weight $\frac{1}{2}f + 2$ for the full modular group. In particular, for $f = 24$ the series (3) is a cusp form of weight 14; but such a form is equal to zero, since any modular form of weight k is a polynomial in two modular forms, E_2 and E_3, of weights 4 and 6, and the cusp forms in this polynomial ring constitute the principal ideal generated by the unique parabolic form Δ of weight 12 (see [Ser1]). Thus

$$\sum_{x \in \Lambda(2n)} (x, \alpha)^2 = \frac{1}{24} 2n(\alpha, \alpha)|\Lambda(2n)|,$$

which for $n = 1$ yields (2).

Corollary 1. *Either* $\Lambda(2) = \varnothing$ *or rank* $\Lambda(2) = 24$.

Proof. Indeed, suppose rank $\Lambda(2) < 24$, i.e. the vector subspace generated by $\Lambda(2)$ has dimension < 24. Then $\mathbf{R}^{24}$ contains a vector $v \neq 0$ orthogonal to all of $\Lambda(2)$. Putting $\alpha = v$ in (2), we see that $|\Lambda(2)| = 0$.

Corollary 2. *All irreducible components of the root system* $\Lambda(2)$ *have the same Coxeter number* h, *and* $|\Lambda(2)| = 24h$.

We now make use of the following property of the Coxeter number [Bou1, Chap. VI, §1.11, Proposition 32]: if in an irreducible root system R all roots have the same length, then the number of elements of R not orthogonal to a fixed $\alpha \in R$ is equal to $4h - 6$. We take the usual normalization for an invariant inner product, where $(\alpha,\alpha) = 2$ for $\alpha \in R$. Now if $\alpha, \alpha' \in R$, then (α, α') can assume only the values $0, \pm 1, \pm 2$, and $(\alpha, \alpha') = \pm 2$ only if $\alpha' = \pm\alpha$. Therefore $4h - 6 = 2 + 2\beta(R)$, where $\beta(R)$ is the number of roots $\alpha' \in R$ such that $(\alpha', \alpha) = 1$ for a fixed α. This does not depend on the choice of α, since the Weyl group $W(R)$ acts transitively on the roots. Thus $\beta(R) = 2h(R) - 4$.

Proof of Corollary 2. If $\alpha \in \Lambda(2)$, then, according to (2).

$$\sum_{y \in \Lambda(2)} (y, \alpha)^2 = \frac{1}{6}|\Lambda(2)|,$$

It is clear that only the elements of that irreducible component R_α containing α make a contribution to the left-hand side of the equality, and that

$$\sum_{y \in \Lambda(2)} (y, \alpha)^2 = 2 \cdot 2^2 + 2\beta(R_\alpha).$$

Since $\beta(R_\alpha) = 2h(R_\alpha) - 4$, it follows that $|\Lambda(2)| = 24h(R_\alpha)$, as required.

Thus we have obtained the following result.

Proposition 2. *If* $\Lambda(2) \neq \varnothing$, *then the root system* $\Lambda(2)$ *in* $\mathbf{R}^{24}$ *possesses the following properties*:

(i) *rank* $\Lambda(2) = 24$,
(ii) *all irreducible components of* $\Lambda(2)$ *have the same Coxeter number* h,
(iii) $|\Lambda(2)| = 24h$.

Conditions (i)-(iii) are very strong, and it is not difficult to find all root systems in $\mathbf{R}^{24}$ satisfying them.

Proposition 3. *Suppose R is a (possibly reducible) root system in* $\mathbf{R}^{24}$ *whose roots all have the same length. Suppose also that R satisfies conditions* (i)–(iii) *of Proposition* 2. *Then R is isomorphic to one of the* 23 *nontrivial root systems in* (1).

Proof. The components of R can be A_i, D_j or E_k. Let

$$R = \sum_{i=1}^{24} \alpha_i A_i + \sum_{j=1}^{24} \beta_j D_j + \sum_{k=6}^{8} \gamma_k E_k.$$

We take the values of the Coxeter numbers from Chap. 4:

$$h(A_l) = l + 1, \ \ h(D_l) = 2l - 2, \ \ h(E_6) = 12, \ \ h(E_7) = 18, \ \ h(E_8) = 30.$$

Since all components of R must have the same Coxeter number, it follows that at most one α, at most one β, and at most one γ can be different from zero. Moreover, if R contains components from the A and D series, i.e. $\alpha_i \neq 0$ and $\beta_j \neq 0$, then $i = 2j - 3$. Similarly, if $\alpha_i \neq 0$ and $\gamma_k \neq 0$, then $i = 11$ if $k = 6$, $i = 17$ if $k = 7$, and $i = 29$ if $k = 8$. Finally, if $\beta_j \neq 0$ and $\gamma_k \neq 0$, then $j = 7$ if $k = 6$, $j = 10$ if $k = 7$, and $j = 16$ if $k = 8$.

Thus R has the following form:

$$\text{either } R = \alpha_i A_i \text{ for some } i, \tag{4}$$

$$\text{or } R = \beta_j D_j \text{ for some } j, \tag{5}$$

$$\text{or } R = \gamma_k E_k \text{ for some } k, \tag{6}$$

$$\text{or } R = \alpha_{2j-3} A_{2j-3} + \beta_j D_j \text{ for certain } \beta_j \neq 0, \alpha_{2j-3} \neq 0, \tag{7}$$

$$\text{or } R = \alpha_{11} A_{11} + \beta_7 D_7 + \gamma_6 E_6, \gamma_6 \neq 0, \alpha_{11} \text{ or } \beta_7 \text{ positive}, \tag{8}$$

$$\text{or } R = \alpha_{17} A_{17} + \beta_{10} D_{10} + \gamma_7 E_7, \gamma_7 \neq 0, \alpha_{17} \text{ or } \beta_{10} \text{ positive}, \tag{9}$$

$$\text{or } R = \alpha_{29} A_{29} + \beta_{16} D_{16} + \gamma_8 E_8, \gamma_8 \neq 0, \beta_{16} \text{ positive}. \tag{10}$$

Condition (i) of Proposition 2 means that

$$\sum_{i,j,k} (i\alpha_i + j\beta_j + k\gamma_k) = 24.$$

This yields the following possibilities for R. In case (4):

$$24A_1, 12A_2, \ 8A_3, 6A_4, 4A_6, 3A_8, 2A_{12}, A_{24}.$$

In case (5):

$$6D_4, 4D_6, 3D_8, 2D_{12}, D_{24}.$$

In case (6):

$$4E_6,\ 3E_8.$$

In case (7), condition (i) becomes $j\beta_j + (2j-3)\alpha_{2j-3} = 24$, $j \geqslant 4$, β_j and $\alpha_{2j-3} \geqslant 1$ integers. It is easy to see that the indeterminate equation $j\beta + (2j-3)\alpha = 24$, $j \geqslant 4, \alpha \geqslant 1, \beta \geqslant 1$, has only the following solutions: $(j, \alpha, \beta) = (4, 4, 1), (5, 2, 2), (6, 2, 1), (9, 1, 1)$. This leads to the following possibilities for R in case (7):

$$4A_5 + D_4,\ 2A_7 + 2D_5,\ 2A_9 + D_6,\ A_{15} + D_9.$$

In case (8), condition (i) becomes $11\alpha_{11} + 7\beta_7 + 6\gamma_6 = 24$, which admits a unique possibility for R:

$$E_6 + D_7 + A_{11}.$$

In case (9) we have $17\alpha_{17} + 10\beta_{10} + 7\gamma_7 = 24$, which admits two possibilities:

$$2E_7 + D_{10},\ E_7 + A_{17}.$$

Finally, in case (10) we obtain $16\beta_{16} + 8\gamma_8 = 24$, which yields

$$E_8 + D_{16}.$$

This proves Proposition 3.

Note that we did not use property (iii) in the proof of Proposition 3. Since all of the root systems obtained above are realized as $\Lambda(2)$ for even unimodular lattices Λ, a fact which will be proved in §4, it follows that (iii) is a consequence of (i) and (ii), i.e. if a 24-dimensional root system possesses properties (i) and (ii) of Proposition 2, then it also possesses property (iii).

Corollary. *Suppose Λ is an even unimodular lattice in $\mathbf{R}^{24}$. Then either $\Lambda(2) = \varnothing$ or $\Lambda(2)$ is one of the 23 nontrivial Niemeier root systems listed in* (1).

3. On lattices with root systems of maximal rank

The aim of this section is to make several general observations about even unimodular lattices having a root system of maximal rank and about the connection between such lattices and algebraic codes. In what follows we will consider only those root systems in which all roots have the same length, i.e. those root systems whose irreducible components have types A, D, or E.

Let us consider the following construction. To each irreducible root system $R \neq E_8$ we associate a triple $(T(R), G(R), l_R)$ consisting of a finite Abelian group $T(R)$, a finite group $G(R)$ acting on $T(R)$, and an

R-valued $G(R)$-invariant function l_R on $T(R)$. We define this triple as follows.

1. If $R = A_i$, then $T(A_i) = Z/(i+1)Z = \{0, 1, ..., i\}$, and $G(A_1) = 1$, $G(A_i) = \mathbf{Z}/2\mathbf{Z}$ for $i > 1$. A nontrivial element $\sigma \in G(A_i)$, $i > 1$, acts on $T(A_i)$ like multiplication by -1, i.e. $\sigma(k) = i + 1 - k$. Also $l_{A_i}(k) = k(i + 1 - k)/(i + 1)$.

2. If $R = D_j$, $j \geqslant 4$, then

$$T(D_j) = \{d_0, d_1, d_2, d_3\} = \begin{cases} \mathbf{Z}/2\mathbf{Z} \oplus \mathbf{Z}/2\mathbf{Z} & \text{if } j \equiv 0 \pmod 2, \\ \mathbf{Z}/4\mathbf{Z} & \text{if } j \equiv 1 \pmod 2. \end{cases}$$

The notation is chosen so that d_0 is the zero element of the group $T(D_j)$, and $d_i \equiv i \pmod 4$ if $j \equiv 1 \pmod 2$. Next, $G(D_4) = S_3$, $G(D_j) = \mathbf{Z}/2\mathbf{Z}$ for $j \geqslant 5$. The action is the following: a nontrivial element $\sigma \in \mathbf{Z}/2\mathbf{Z}$ fixes d_0 and d_2, and $\sigma d_1 = d_3$. When $j = 4$, $G(D_4) = S_3 = GL(2,2)$ coincides with the full automorphism group of $T(D_4)$. The function l_{D_j} is defined as follows: $l(d_0) = 0$, $l(d_2) = 1$, $l(d_1) = l(d_3) = j/4$. In the case where $j \equiv 0 \pmod 2$, it is sometimes convenient to identify $T(D_j)$ with the additive group of the field $\mathbf{F}_4$; $G(D_j)$, $j > 4$, can then be identified with the Galois group $\mathrm{Gal}(\mathbf{F}_4 / \mathbf{F}_2)$.

3. If $R = E_6$, then $T(E_6) = \mathbf{Z}/3\mathbf{Z} = \{0, 1, 2\}$, $G(E_6) = \mathbf{Z}/2\mathbf{Z} = \{\sigma\}$, $\sigma(1) = 2$, and $l(0) = 0$, $l(1) = l(2) = 4/3$.

4. If $R = E_7$, then $T(E_7) = \mathbf{Z}/2\mathbf{Z} = \{0, 1\}$, $G(E_7) = 1$, and $l(0) = 0$, $l(1) = 3/2$.

The invariant meaning of this triple is the following: $T(R) = P/Q$, where Q is the root lattice and P the weight lattice for R; $G(R)$ is isomorphic to the quotient group of the automorphism group $A(R)$ of the root system R by the Weyl group $W(R)$ with the natural faithful action of $A(R)/W(R)$ on P/Q; l is the norm of the microweight lying in a given coset with respect to Q. Recall that an element $p \in P$ is called a *microweight* if $(p,r) = 0, \pm 1$ for all $r \in R$. In each coset in P/Q there is precisely one orbit under the Weyl group $W(R)$ consisting of microweights, and l on a coset with respect to Q is the norm of any microweight lying in this coset.

If a root system consists of n isomorphic components $\neq E_8$, i.e. has the form nR, then we put $T(nR) = T(R)^n$, we define $G(nR)$ to be the natural semidirect product

$$G(nR) = G(R)^n \cdot S_n$$

(the wreath product of $G(R)$ by S_n) with the natural monomial action on $T(nR)$, and we define l by additivity:

$$l(x_1, ..., x_n) = \sum l(x_i).$$

Clearly l is $G(nR)$-invariant.

Finally, for an arbitrary root system $R = \Sigma\, n_i R_i$, where R_i are the irreducible components, we define our triple as follows. We discard the components E_8 and put

$$T(R) = \bigoplus_i T(n_i R_i), \quad G(R) = \prod_i G(n_i R_i), \quad l_R = \sum_i l_{n_i R_i},$$

where the summation extends over all types of irreducible root systems other than E_8. We call a subgroup $A \subset T(R)$ even and self-dual if $|A|^2 = |T(R)|$ and the function l_R assumes even integral values >2 on $A \setminus \{0\}$. There may be no such subgroups, but if they exist, the set of them is $G(R)$-invariant, since the function l_R is $G(R)$-invariant.

Proposition 4. *Suppose R is a root system of rank n. There exists a natural one-to-one correspondence between the classes of even unimodular n-dimensional lattices (to within isomorphism) with root system isomorphic to R and the orbits of even self-dual subgroups $A \subset T(R)$ relative to the group $G(R)$.*

The proof amounts to a simple verification and is omitted. The correspondence assigns to a lattice Λ the quotient $A = \Lambda / Q(\Lambda(2))$ with respect to the sublattice generated by the roots, which is viewed as a subgroup of $Q(\Lambda(2))^0 / Q(\Lambda(2)) = P/Q$.

Example 1. Suppose $R = nA_1$ is the simplest root system of rank n. In this case the group $T(R)$ can be identified with the space of sequences of length n over the field $\mathbf{F}_2$, and $T(R) = (\mathbf{Z}/2\mathbf{Z})^n$. The group $G(R)$ is isomorphic to S_n, and the function l is given by $l(\alpha) = l(\alpha_1, \ldots, \alpha_n) = \frac{1}{2}\, wt(\alpha)$, where $wt(\alpha)$ is the Hamming weight (§2.1 of Chap. 3). An even self-dual subgroup is a binary self-dual Type II code (§2.2 of Chap. 3) with minimal distance $\geqslant 8$. The classification of such codes under the action of S_n is a famous problem of binary coding theory (see §6 of Chap. 7). Thus this same problem is equivalent to that of classifying the even unimodular n-dimensional lattices with root system nA_1.

Example 2. Suppose $R = nA_2$. In this case $T(R) = (\mathbf{Z}/3\mathbf{Z})^n$, $G(R) = (\mathbf{Z}/2\mathbf{Z})^n \cdot S_n$, $l(\alpha) = l(\alpha_1, \ldots, \alpha_n) = (2/3) \cdot$ (the number of nonzero α_i), and, exactly as above, the problem of classifying the $2n$-dimensional lattices with root system nA_2 is equivalent to that of classifying the ternary self-dual Type III codes with minimal distance > 3 (§2.2 of Chap. 3, §9 of Chap. 7).

In the general case, i.e. for any root system R, the problem of classifying the even self-dual subgroups of $T(R)$ relative to $G(R)$ can also be viewed as a problem of coding theory. This problem includes and differs insignificantly from that of classifying the self-dual codes over all rings $\mathbf{Z}/n\mathbf{Z}$ and the field $\mathbf{F}_4$. An equivalent problem is to classify the even unimodular lattices having a root system of maximal rank. In the sequel, self-dual subgroups of $T(R)$ will often be called codes.

Remarks (i) The usual concept of isomorphism in considering codes over commutative rings is defined by monomial matrices (§2.2 of Chap. 3). Proposition 4 shows that to classify lattices with root system of maximal rank of type nR (where R is an irreducible root system) in most cases it is necessary to classify codes relative to monomial groups in which the nonzero elements are equal to ± 1. Only lattices of type nD_4 lead to complete monomial groups.

(ii) The classification of lattices with root systems of maximal rank of type nD_{2k} depends very weakly on k: in the corresponding code only the function l depends on k. This enables us to reduce the problem of classifying such lattices of type nD_{2k} to the case of small k. An analogous remark holds for other cases of isomorphisms between the $T(R_i)$. For example $T(D_{2i+1}) = T(D_5) = T(A_3)$, $T(E_6) = T(A_2)$, and $T(E_7) = T(A_1)$.

(iii) Codes corresponding to lattices of maximal rank and of type nD_{2k} are not necessarily linear codes over $\mathbf{F}_4$; that is, self-dual subgroups of $T(nD_{2k}) = T(D_{2k})^n = \mathbf{F}_4^n$ need not be invariant under multiplication by scalars of $\mathbf{F}_4$. For example, this occurs in the 24-dimensional lattice with root system $3D_8$ (see case XI in §4); if the corresponding self-dual subgroup were a linear subspace, it would have dimension 3/2.

4. Construction of the Niemeier lattices

We are now in a position to prove the main theorem.

Theorem (Niemeier [Nie2]). *To within isomorphism there exist precisely 24 even unimodular lattices in* $\mathbf{R}^{24}$. *Each lattice is uniquely determined by its root system. The possible root systems are the* 24 *listed in* (1).

Proof. We already know from §2 that the root system of a given even unimodular lattice $\Lambda \subset \mathbf{R}^{24}$ belongs to the list (1). Therefore it suffices to prove the existence and uniqueness of a lattice with each of the 24 root systems in (1). The case of the empty root system is considered in §5. All of the other root systems have maximal rank 24; hence they come within the scope of the theory of §3. Thus for each of the root systems in (1) it is necessary to calculate the triple $(T(R), G(R), l_R)$ and verify the existence and uniqueness of the corresponding code. We will number the root systems and lattices in the order in which they appear in (1).

I. $R = 24A_1$. Then $T(R) = (\mathbf{Z}/2\mathbf{Z})^{24}$, $G(R) = S_{24}$, $l(\alpha) = \frac{1}{2} wt(\alpha)$, and $A \subset T(R)$ is a binary self-dual Type II code with minimal distance 8. The Golay code $\mathscr{C}_{24}$ (§2.8.2 of Chap. 3) is the unique such code (Pless [Ple8]; see also [Ras1] or §7 of Chap. 7). Its automorphism group is the Mathieu group M_{24} (Chaps. 10, 11), which is 5-transitive on the components of R. Thus there exists a unique lattice Γ_1 with root system $24A_1$, and the automorphism group of Γ_1 is the extension

$$1 \to W(24A_1) \to \operatorname{Aut} \Gamma_1 \to M_{24} \to 1.$$

II. $R = 12A_2$. Then $T(R) = (\mathbf{Z}/3\mathbf{Z})^{12}$, $G(R) = (\mathbf{Z}/2\mathbf{Z})^{12} \cdot S_{12}$, and A must be a ternary self-dual Type III code with minimal distance > 3. The Golay code $\mathscr{C}_{12}$ (§2.8.5 of Chap. 3) is the unique such code (Pless [Ple8];

see also §10 of Chap. 7). Its automorphism group is isomorphic to $2 \cdot M_{12}$, and acts 5-transitively on the components of R.

III. $R = 8A_3$. Then $T(R) = (\mathbf{Z}/4\mathbf{Z})^8$ and $G(R) = (\mathbf{Z}/2\mathbf{Z})^8 \cdot S_8$. If we identify $\mathbf{Z}/4\mathbf{Z}$ with the numbers 0, ± 1, 2, then l has the form $l(0) = 0, l(\pm 1) = 3/4$, $l(2) = 1$. Thus in the space of sequences $\alpha = (\alpha_1, \ldots, \alpha_8)$, $\alpha_i = 0, \pm 1, 2$, we must look for subgroups A of order 4^4 such that $l(\alpha) = \Sigma\, l(\alpha_i) \in 2\mathbf{Z}$ and $l(\alpha) > 2$ for all $\alpha \in A \setminus \{0\}$. A direct calculation shows that for such an A we can take the subgroup spanned by the vectors

$$x_1 = (0, 0, 0, 1, 1, 1, 1, 2),$$

$$x_2 = (0, 0, 1, 0, 2, 1, -1, 1),$$

$$x_3 = (0, 1, 0, 0, 1, -1, 2, 1),$$

$$x_2 = (1, 0, 0, 0, 1, 2, -1, -1),$$

and such a subgroup is unique (to within $G(R)$, of course). It can be shown that the permutation part of the group Aut A is isomorphic to $GA(3,2)$ (the full affine group of order 3 over the field $\mathbf{F}_2$, having order 1344) and that the elements of order 2 of A form a binary Hamming code. The automorphism group of the lattice Γ acts 3-transitively on the components of R.

IV. $R = 6A_4$. Then $T(R) = (\mathbf{Z}/5\mathbf{Z})^6$ and $G(R) = (\mathbf{Z}/2\mathbf{Z})^6 \cdot S_6$. If we identify $\mathbf{Z}/5\mathbf{Z}$ with the numbers 0, ± 1, ± 2, then the function l has the form $l(0) = 0, l(\pm 1) = 4/5$, $l(\pm 2) = 6/5$. In the space of sequences $\alpha = (\alpha_1, \ldots, \alpha_6)$, $\alpha_i = 0, \pm 1, \pm 2$, we must find 3-dimensional subspaces $A \subset T(R)$ such that $l(\alpha) = \Sigma l(\alpha_i)$ for $\alpha \in A \setminus \{0\}$ is even and >2. Elementary calculations show that for A we can take the subspace spanned by $x_1 = (0, 0, 1, 1, -2, -2)$, $x_2 = (2, -2, 0, 0, 1, 1)$, $x_3 = (1, 1, 2, -2, 0, 0)$, and that A is unique to within the action of $G(R)$.

V. $R = 4A_6$. Then $T(R) = (\mathbf{Z}/7\mathbf{Z})^4$ and $G(R) = (\mathbf{Z}/2\mathbf{Z})^4 \cdot S_4$. If we identify $\mathbf{Z}/7\mathbf{Z}$ with ± 1, ± 2, ± 3, then $l(0) = l(\pm 1) = 6/7$, $l(\pm 2) = 10/7$ and $l(\pm 3) = 12/7$. We must look for 2-dimensional subspaces of $T(R)$ on which l assumes even values >2. Again such a subspace exists and is unique. As a representative we can take the subspace spanned by $x_1 = (1, 2, 3, 0)$, $x_2 = (0, 3, -2, 1)$.

VI. $R = 3A_8$. Then $T(R) = (\mathbf{Z}/9\mathbf{Z})^3$ and $G(R) = (\mathbf{Z}/2\mathbf{Z})^3 \cdot S_3$; $l(0) = 0$, $l(\pm 1) = 8/9$, $l(\pm 2) = 14/9$, $l(\pm 3) = 2$ and $l(\pm 4) = 20/9$. The desired subspace of $T(R)$ exists and is unique (relative to $G(R)$): it is generated by $x_1 = (3, 3, 3)$, $x_2 = (1, -2, -2)$, $x_3 = (-2, -2, 1)$.

VII. $R = 2A_{12}$. Then $T(R) = (\mathbf{Z}/13\mathbf{Z})^2$ and $G(R) = (\mathbf{Z}/2\mathbf{Z})^2 \cdot S_2$; $l(0) = 0, l(\pm 1) = 12/13$, $l(\pm 2) = 22/13$, $l(\pm 3) = 30/13$, $l(\pm 4) = 36/13$, $l(\pm 5) = 40/13$ and $l(\pm 6) = 42/13$. The desired code is the 1-dimensional subspace generated by the vector (1, 5).

VIII. $R = A_{24}$. Then $T(R) = \mathbf{Z}/25\mathbf{Z}$, $G(R) = \mathbf{Z}/2\mathbf{Z}$, and the desired code consists of the elements divisible by 5.

IX. $R = 6D_4$. Then $T(R) = \mathbf{F}_4^6$ and $G(R) = (S_3)^6 \cdot S_3$. If $\alpha = (\alpha_1, \ldots, \alpha_6) \in T(R)$, then $l(\alpha)$ is the number of nonzero α_i. The desired code is a self-dual Type 4 code over $\mathbf{F}_4$ with minimal distance > 2. The hexacode (§2.5.2 of Chap. 3) is the unique such code ([Mac4], §10 of Chap. 7). Its automorphism group is $3 \cdot A_6$. It is interesting to observe that this is one of the three known cases (the others are the two Golay codes) where the automorphism group of an extended quadratic residue code over $\mathbf{F}_p$ is larger than $SL(2, \mathbf{F}_p)$ (see [Ras1] and Chap. 10).

X. $R = 4D_6$. Then $T(R) = \mathbf{F}_4^4$ and $G(R) = (\mathbf{Z}/2\mathbf{Z})^4 \cdot S_4$; $l(0) = 0$, $l(1) = 1$ and $l(\omega) = l(\omega^2) = 3/2$, where ω is a primitive element of $\mathbf{F}_4$. The desired code exists and is unique: as a representative we may take the subgroup spanned by $(1, 1, 1, 1)$, $(\omega, \omega, \omega, \omega)$, $(0, 1, \omega^2, \omega)$, $(0, \omega^2, \omega, 1)$.

XI. $R = 3D_8$. Then $T(R) = \mathbf{F}_4^3$ and $G(R) = (\mathbf{Z}/2\mathbf{Z})^3 \cdot S_3$; $l(0) = 0$, $l(1) = 1$ and $l(\omega) = l(\omega^2) = 2$. As the desired subgroup of $T(R)$ we can take the subgroup generated by $(1, 1, \omega)$, $(1, \omega, 1)$, $(\omega, 1, 1)$. The uniqueness is very easy.

XII. $R = 2D_{12}$. Then $T(R) = \mathbf{F}_4^2$ and $G(R) = (\mathbf{Z}/2\mathbf{Z})^2 \cdot S_2$; $l(0) = 0$, $l(1) = 1$ and $l(\omega) = l(\omega^2) = 3$. The code is generated by $(1, \omega)$, $(\omega, 1)$. The uniqueness is obvious.

XIII. $R = D_{24}$. Then $T(R) = \mathbf{F}_4$ and $G(R) = \mathbf{Z}/2\mathbf{Z}$; $l(0) = 0$, $l(1) = 1$ and $l(x) = l(x^2) = 6$. The code consists of 0 and x. The lattice obtained is, of course D_{24}^+ (§7.3 of Chap. 4), or Γ_{24} in the notation of [Ser1, Chap. V, Example 1.4.3].

XIV. $R = 4E_6$. Then $T(R) = (\mathbf{Z}/3\mathbf{Z})^4$ and $G(R) = (\mathbf{Z}/2\mathbf{Z})^4 \cdot S_4$; $l(0) = 0$ and $l(\pm 1) = 4/3$. In the space of sequences $\alpha = (\alpha_1, \alpha_2, \alpha_3, \alpha_4)$, $\alpha_i = 0, \pm 1$, we must look for 2-dimensional even self-dual subspaces. The problem has a unique solution: the subspace spanned by $(1, 1, 1, 0)$, $(0, -1, 1, 1)$ (the tetracode $\mathscr{C}_4$, §2.5.1 of Chap. 3).

XV. $R = 3E_8$. Then $T(R) = 0$ and the code is trivial. The lattice is isomorphic to $E_8 \oplus E_8 \oplus E_8$ (or $\Gamma_8 \oplus \Gamma_8 \oplus \Gamma_8$ in Serre's notation).

XVI. $R = 4A_5 + D_4$. Then $T(R) = (\mathbf{Z}/6\mathbf{Z})^4 \times \mathbf{F}_4$ and $G(R) = (\mathbf{Z}/2\mathbf{Z})^4 \cdot S_4 \times S_3$; $l(0) = 0$, $l(\pm 1) = 5/6$, $l(\pm 2) = 4/3$, $l(3) = 3/2$ and $l(b) = 1$, $b \in \mathbf{F}_4$, $b \neq 0$. In the space of sequences $\alpha = (\alpha_1, \alpha_2, \alpha_3, \alpha_4, b)$, $\alpha_i = 0, \pm 1, \pm 2, 3$, $b \in \mathbf{F}_4$, we must find a subgroup A of order 72 such that for all $\alpha \in A, \alpha \neq 0$, we have $l(\alpha) = \Sigma l(\alpha_i) + l(b) \in 2Z$ and $l(\alpha) > 2$. For such a subgroup we may take that generated by $(0, 1, 2, -1, \omega)$, $(1, 1, 1, 3, 0)$, $(3, 3, 0, 0, 1)$, where $\omega \in \mathbf{F}_4$ is a primitive element. The uniqueness of such a subgroup to within the action of $G(R)$ is easily verified. (Details are given in [Nie2, p. 163].)

XVII. $R = 2A_7 + 2D_5$. Then $T(R) = (\mathbf{Z}/8\mathbf{Z})^2 \times (\mathbf{Z}/4\mathbf{Z})^2$ and $G(R) = (\mathbf{Z}/2\mathbf{Z})^2 \cdot S_2 \times (\mathbf{Z}/2\mathbf{Z})^2 \cdot S_2$; the function l on the first two summands has the form $l(0) = 0$, $l(\pm 1) = 7/8$, $l(\pm 2) = 3/2$, $l(\pm 3) = 15/8$, $l(4) = 2$, and on the last two it has the form $l(0) = 0$, $l(\pm 1) = 5/4$, $l(2) = 1$. For the desired subgroup of order 32 we may take that generated by (3, 1, 1, 0), (2, 0, −1, 1). Uniqueness is very easy.

XVIII. $R = 2A_9 + D_6$. Then $T(R) = (\mathbf{Z}/10\mathbf{Z})^2 \times \mathbf{F}_4$ and $G(R) = (\mathbf{Z}/2\mathbf{Z})^2 \cdot S_2 \times \mathbf{Z}/2\mathbf{Z}$; the function l on the first two summands has the form $l(0) = 0$, $l(\pm 1) = 9/10$, $l(\pm 2) = 8/5$, $l(\pm 3) = 21/10$, $l(\pm 4) = 12/5$, $l(5) = 5/2$, and on $\mathbf{F}_4$ it has the form $l(0) = 0$, $l(1) = 1$, $l(\omega) = l(\omega^2) = 3/2$. The desired subspace of order 20 exists and is unique: as generators we may take $(5, 5, 1)$ $(1, 2, \omega)$.

XIX. $R = A_{15} + D_9$. Then $T(R) = \mathbf{Z}/16\mathbf{Z} \times \mathbf{Z}/4\mathbf{Z}$ and $G(R) = \mathbf{Z}/2\mathbf{Z} \times \mathbf{Z}/2\mathbf{Z}$; the function l on $\mathbf{Z}/16\mathbf{Z}$ has the form $l(0) = 0$, $l(\pm 1) = 15/16$, $l(\pm 2) = 7/4$, $l(\pm 3) = 39/16$, $l(\pm 4) = 3$, $l(\pm 5) = 55/16$, $l(\pm 6) = 15/4$, $l(\pm 7) = 63/16$, $l(8) = 4$, and on $\mathbf{Z}/4\mathbf{Z}$ it has the form $l(0) = 0$, $l(\pm 1) = 9/4$, $l(2) = 1$. The desired subgroup of order 8 is generated by the element (2, 1). The uniqueness is obvious.

XX. $R = E_8 + D_{16}$. Since the sublattice generated by E_8 is unimodular, the whole lattice is decomposable and is isomorphic to $E_8 \oplus D_{16}^+$.

XXI. $R = 2E_7 + D_{10}$. Then $T(R) = (\mathbf{Z}/2\mathbf{Z})^2 \times \mathbf{F}_4$ and $G(R) = S_2 \times \mathbf{Z}/2\mathbf{Z}$; the function l on the first two summands has the form $l(0) = 0$, $l(1) = 3/2$, and on the last it has the form $l(0) = 0$, $l(1) = 1$, $l(\omega) = l(\omega^2) = 5/2$. For the desired subgroup of order 4 we may take that generated by $(1, 0, \omega)$, $(0, 1, \omega^2)$. The uniqueness is obvious.

XXII. $R = E_7 + A_{17}$. Then $T(R) = \mathbf{Z}/2\mathbf{Z} \times \mathbf{Z}/18\mathbf{Z}$ and $G(R) = \mathbf{Z}/2\mathbf{Z}$; the function l on the first summand has the form $l(0) = 0$, $l(1) = 3/2$, and on the second it has the form $l(0) = 0$, $l(\pm 1) = 17/18$, $l(\pm 2) = 16/9$, $l(\pm 3) = 5/2$, $l(\pm 4) = 28/9$, $l(\pm 5) = 65/18$, $l(\pm 6) = 4$, $l(\pm 7) = 77/18$, $l(\pm 8) = 40/9$, $l(9) = 2$. The desired subgroup of order 6 is generated by the element (1, 3). The uniqueness is obvious.

XXIII. $R = E_6 + D_7 + A_{11}$. Then $T(R) = \mathbf{Z}/3\mathbf{Z} \times \mathbf{Z}/4\mathbf{Z} \times \mathbf{Z}/12\mathbf{Z}$ and $G(R) = \mathbf{Z}/2\mathbf{Z} \times \mathbf{Z}/2\mathbf{Z} \times \mathbf{Z}/2\mathbf{Z}$; the function l on the first summand has the form $l(0) = 0$, $l(\pm 1) = 4/3$, on the second the form $l(0) = 0$, $l(\pm 1) = 7/4$, $l(2) = 1$, and on the third the form $l(0) = 0$, $l(\pm 1) = 11/12$, $l(\pm 2) = 5/3$, $l(\pm 3) = 9/4$, $l(\pm 4) = 8/3$, $l(\pm 5) = 35/12$, $l(6) = 3$. The desired subgroup of order 12 is generated by (1, 1), and the existence and uniqueness are obvious.

Thus we have considered all possible nontrivial root systems, and in each case we have obtained precisely one lattice. This proves the main theorem in the case of lattices with a nonempty root system and provides a complete description of the lattices. (An alternative description is given in Table 16.1.)

5. A characterization of the Leech lattice

To complete the proof of the main theorem it remains to consider the case of an empty root system, i.e. we must consider a lattice Λ is in which $(x, x) \neq 2$ for $x \in \Lambda$. In this case there also exists a lattice and it is unique: it is the remarkable Leech lattice. A characterization of Leech's lattice as the unique even unimodular lattice in $\mathbf{R}^{24}$ with $(x, x) \neq 2$ was given by Conway (see Chap. 12). Conway's proof is very pretty and can hardly be improved. However, if we assume (as is natural in our approach) that the classification of lattices with nontrivial root systems is already known, then the characterization of Leech's lattice can be obtained more simply as follows.

Suppose $\Lambda \subset \mathbf{R}^{24}$ is an even unimodular lattice with $\Lambda(2) = \emptyset$. We want to prove that Λ is isomorphic to Leech's lattice. From Theorem 17 of Chap. 7, the condition $\Lambda(2) = \emptyset$ implies that the theta series of Λ is the same as the theta series of Leech's lattice, in particular, $\Lambda(8) \neq \emptyset$. Let $u \in \Lambda(8)$ and consider the lattice $\Gamma = \Lambda^u = \Lambda' \cup (u/2 + \Lambda')$, where $\Lambda' = \{\lambda \in \Lambda : (\lambda, u) \equiv 0 \pmod 2\}$. Then Γ is an even unimodular lattice in $\mathbf{R}^{24}$ (it is a neighbor of Λ in the sense of the previous chapter). The lattice Γ has a nonempty root system, since $u/2 \in \Gamma(2)$. We shall prove that the root system for Γ is kA_1. It suffices to show that nonproportional roots of $\Gamma(2)$ are orthogonal. This is very easy. Suppose $x,y \in \Gamma(2)$, $x \neq \pm y$, so that $x,y \in u/2 + \Lambda'$, and suppose $(x,y) \neq 0$. Changing the sign of y if necessary, we may assume that $(x,y) = -1$, but then $x - y$ is a root, i.e. $(x - y, x - y) = 2$. But $x - y \in \Lambda' \subset \Lambda$ and $\Lambda(2) = \emptyset$, which is impossible. Thus $\Gamma(2) = kA_1$; but then, scanning the list (1), we see that $k = 24$ and Γ is described in paragraph I of §4, i.e. in $\mathbf{R}^{24}$ we can choose a basis $e_1, \ldots, e_{24}$ such that $(e_i, e_j) = 0$ if $i \neq j$, $(e_i, e_i) = 1/2$, and $\Sigma n_i e_i \in \Gamma$ if and only if $n_i \in Z$ and $\{i : n_i \equiv 1 \pmod 2\}$ is a codeword in the binary Golay code. Since the property of being adjacent lattices is symmetric, there is a $\nu \in \Gamma$ such that $\Lambda = \Gamma^\nu$. Suppose $\nu = \Sigma m_i e_i$, $m_i \in \mathbf{Z}$; we will prove that all of the m_i are odd. Indeed, if some m_α is even, then $(\nu, 2e_\alpha) = m_\alpha \equiv 0 \pmod 2$; hence $2e_\alpha \in \Gamma^\nu = \Lambda$, which is impossible, since $\Lambda(2) = \emptyset$. Thus each $m_i \equiv 1 \pmod 2$. Since ν is defined mod 2Γ, we may assume that $m_i = \pm 1$, and the action of the Weyl group $W(\Gamma(2))$ allows us to assume that $m_1 = 1$, i.e. as ν we can choose $e_1 + \cdots + e_{24}$. For such a ν, Γ^ν is the Leech lattice, as required. This completes the proof of the main theorem.

[Translated by G. A. Kendall]

19

Enumeration of Extremal Self-Dual Lattices

J. H. Conway, A. M. Odlyzko and N. J. A. Sloane

We saw in Chapter 7 that the minimal norm of a unimodular lattice in $\mathbf{R}^n$ does not exceed $[n/8] + 1$. If the minimal norm is equal to this quantity the lattice is called extremal. In this chapter we show that there are unique extremal lattices in dimensions 1, 2, 3, 4, 5, 6, 7, 8, 12, 14, 15, 23, 24, and no other such lattices.

1. Dimensions 1-16

The theorem stated in the title is Theorem 13 of Chap. 7. We proceed at once to the proof. The following result was proved by Kneser [Kne4], and can also be obtained from Table 16.7.

Theorem 1. *The only extremal unimodular lattices in* $\mathbf{R}^n$ *for* $n \leqslant 16$ *are* $\mathbf{Z}^n$ $(1 \leqslant n \leqslant 7)$, E_8, D_{12}^+, $(E_7+E_7)^+$, A_{15}^+, *in* $\mathbf{R}^1$ *to* $\mathbf{R}^7$, $\mathbf{R}^8$, $\mathbf{R}^{12}$, $\mathbf{R}^{14}$, $\mathbf{R}^{15}$ *respectively.*

Here $\Lambda = (L_1 + \cdots + L_k)^+$ indicates that Λ is obtained by gluing component lattices $L_1, \ldots, L_k$ together (cf. §3 of Chap. 4).

2. Dimensions 17-47

Theorem 2. *The only extremal unimodular lattices in* $\mathbf{R}^n$ *for* $17 \leqslant n \leqslant 47$ *are the shorter Leech lattice* O_{23} *in* $\mathbf{R}^{23}$ *and the Leech lattice* Λ_{24} *in* $\mathbf{R}^{24}$.

Proof. We could read the results for dimensions 17 to 24 straight from the tables in Chaps. 16 and 17, but we prefer to give independent arguments. Suppose first that Λ is a Type II (or even) lattice in $\mathbf{R}^n$, which is also extremal as a unimodular lattice. From Cor. 21, Chap. 7, the minimal norm μ of Λ satisfies $\mu \leqslant 2[n/24] + 2$. Also $\mu = [n/8] + 1$ and $8 | n$, which imply $n = 8$, 16 or 24, in which dimensions the even unimodular lattices were enumerated by Witt and Niemeier — see §2.4 of Chap. 2 and Table 16.1. Only E_8 and Λ_{24} are extremal.

From now on we suppose that Λ is an odd unimodular lattice which is extremal. The theta-series of Λ is therefore the extremal theta-series in dimension n (§4 of Chap. 7). Let $n = 8\alpha + \nu$, $0 \leqslant \nu \leqslant 7$. This means that

$$\Theta_\Lambda(z) = \sum_{r=0}^{\alpha} a_r \, \theta_3(z)^{n-8r} \, \Delta_8(z)^r \tag{1}$$

(see Eq. (36) of Chap. 7), where the integer coefficients $a_0, \ldots, a_\alpha$ are chosen so that

$$\Theta_\Lambda(z) = 1 + A_{\alpha+1}\, q^{\alpha+1} + A_{\alpha+2}\, q^{\alpha+2} + \cdots \tag{2}$$

(containing no power of q with exponent between 0 and $\alpha+1$).

Let $\Lambda_0 = \{u \in \Lambda \,|\, u \cdot u \in 2\mathbf{Z}\}$ be the even sublattice of Λ. Then $\det \Lambda = 1$, $\det \Lambda_0 = 4$, and Λ_0 has theta-series

$$\Theta_{\Lambda_0}(z) = \tfrac{1}{2}\, \{\Theta_\Lambda(z) + \Theta_\Lambda(z+1)\}. \tag{3}$$

To eliminate almost all of the remaining cases we make use of an argument due to Ward [War2] (who applied it to the weight enumerators of codes). It will turn out that the condition that the theta-series of the dual lattice Λ_0^* has integer coefficients implies that the coefficients a_r in (1) must be divisible by high powers of two. From the Jacobi identity (Chap. 4, Eq. (19))

$$\Theta_\Lambda(z) = \left(\frac{i}{z}\right)^{n/2} \Theta_\Lambda\left(-\frac{1}{z}\right),$$

$$\Theta_{\Lambda_0^*}(z) = \Theta_\Lambda(z) + \left(\frac{i}{z}\right)^{n/2} \Theta_\Lambda\left(-\frac{1}{z}+1\right). \tag{4}$$

Hence

$$\left(\frac{i}{z}\right)^{n/2} \Theta_\Lambda\left(-\frac{1}{z}+1\right)$$

has integer coefficients in its q-expansion. Then

$$\left(\frac{i}{z}\right)^{n/2} \Theta_\Lambda\left(-\frac{1}{z}+1\right)$$

$$= \left(\frac{i}{z}\right)^{n/2} \sum_{r=0}^{\alpha} (-1)^r 2^{-8r} a_r \theta_4\left(-\frac{1}{z}\right)^{n-8r} \theta_2\left(-\frac{1}{2z}\right)^{8r}$$

$$= \sum_{r=0}^{\alpha} (-1)^r 2^{-4r} a_r \, \theta_2(z)^{n-8r} \theta_4(2z)^{8r}, \tag{5}$$

which implies

$$2^{12r-n} \,|\, a_r \quad \text{for } n < 12r\,. \tag{6}$$

The a_r are easily computed (by equating (1) and (2)) and condition (6) eliminates all n in the range $17 \leqslant n \leqslant 47$ except 22, 23 and 24. For example, when $n = 17$, $a_0 = 1$, $a_1 = -34$, $a_2 = -204$, and $2^7 \nmid a_2$.

Finally, $n = 22$ is eliminated by calculating $\Theta_{\Lambda_0^*}(z)$ explicitly. It is $1 - 11 q_{3/2} + \cdots$, and the negative coefficient shows that Λ does not exist.

When $n = 24$ the extremal theta series is equal to the theta series of the Leech lattice, and the uniqueness of the Leech lattice (Chap. 12) shows that $\Lambda = \Lambda_{24}$.

In dimension 23 an extremal lattice Λ must have theta series

$$\theta_3(z)^{23} - 46\theta_3(z)^{15}\,\Delta_8(z) = 1 + 4600q^3 + 93150q^4 + \cdots. \tag{7}$$

The even sublattice Λ_0 (with theta series $1 + 93150q^4 + \cdots$) has dual lattice Λ_0^* with theta series (from (4), (5))

$$\begin{aligned}
&1 + 4600q^3 + 93150q^4 + \cdots + 2^{23}q^{23/4}(1 + 23q^2 + \cdots)\\
&\quad + 23{\cdot}2^{12}q^{15/4}(1 + 15q^2 + \cdots)(1 - 16q^2 + \cdots)\\
&\quad = 1 + 4600q^3 + 94208q^{15/4} + 93150q^4 + \cdots.
\end{aligned}$$

From this we can see that the four cosets of Λ_0 in Λ_0^* contain nonzero vectors of norms

$$4 + 2m,\ \frac{15}{4} + 2m,\ 3 + 2m,\ \frac{15}{4} + 2m,$$

and have coset representatives $y_0 = 0$, y_1, y_2, y_3 (say), for various integers $m \geqslant 0$. The union of the cosets Λ_0 and $\Lambda_0 + y_2$ is Λ. Now the lattice M of even integers also has four cosets in its dual M^*, whose nonzero vectors have norms

$$4 + 2m,\ \tfrac{1}{4} + 2m,\ 1 + 2m,\ \tfrac{1}{4} + 2m\,,$$

with coset representatives $z_0 = 0$, $z_1 = \tfrac{1}{2}$, $z_2 = 1$, $z_3 = -\tfrac{1}{2}$. It follows that there is a 24-dimensional lattice $(\Lambda_0 + M)^+$ obtained by extending $\Lambda_0 + M$ by the glue vectors $y_i + z_i$ ($i = 0, 1, 2, 3$), which is extremal and so must be the Leech lattice. Λ is therefore uniquely characterized as the projection onto the 23-space orthogonal to v of the vectors of Λ_{24} that have even inner product with a minimal vector v in Λ_{24}. This is the *shorter Leech lattice* O_{23} (see also the Appendix to Chap. 6 and Table 16.7). This completes the proof of Theorem 2.

3. Dimensions $n \geqslant 48$

Theorem 3. *No extremal lattice exists in* $\mathbf{R}^n$ *for* $n \geqslant 48$.

Proof. With the same notation as before we show that

$$\Theta_{\Lambda_0^*}(z) = \Theta_\Lambda(z) + \sum_{r=0}^{\alpha} (-16)^{-r} a_r \theta_2(z)^{n-8r} \theta_4(2z)^{8r}$$

contains a negative coefficient for $n \geqslant 48$, which proves the theorem. The q-expansion of the right-hand side is

$$1 + (-1)^\alpha 2^{\nu-4\alpha} a_\alpha q^{\nu/4}$$
$$+ (-1)^{\alpha+1} 2^{\nu-4\alpha} q^{\nu/4+2} \{(16\alpha-\nu) a_\alpha + 4096 a_{\alpha-1}\} + \cdots .$$

We will show that $a_\alpha < 0$ and $a_{\alpha-1} < 0$ for $\alpha \geqslant 6$, from which the negative coefficient is apparent. Applying the Bürmann-Lagrange theorem to (1) and (2) we obtain, exactly as in [Mal1, Eq. (6)],

$$a_s = -\frac{n}{s!} \frac{d^{s-1}}{dq^{s-1}} \left\{ \frac{d\theta_3}{dq} \cdot \theta_3^{8s-n-1} . h^s \right\}_{q=0}, \tag{8}$$

for $0 \leqslant s \leqslant \alpha$, where, from Eq. (33) of Chap. 7,

$$h = \prod_{m=1}^{\infty} \{(1-q^{2m-1})(1-q^{4m})\}^{-8} .$$

Now $d\theta_3/dq$ has non-negative coefficients in its q-expansion. When $s = \alpha$ or $\alpha-1$ we have

$$-1 \geqslant 8s - n - 1 \geqslant -16 .$$

Let $1 \leqslant k \leqslant 16$. Then, from Eq. (35) of Chap. 4,

$$\theta_3^{-k} = \prod_{m=1}^{\infty} (1-q^{2m})^{-k} \prod_{m=1}^{\infty} (1+q^{2m-1})^{-2k} .$$

The first product has non-negative coefficients. On the other hand,

$$h^s \prod_{m=1}^{\infty} (1+q^{2m-1})^{-2k} = \prod_{m=1}^{\infty} (1-q^{2m-1})^{-(8s-2k)}$$
$$\times \prod_{m=1}^{\infty} (1-q^{4m-2})^{-2k} \prod_{m=1}^{\infty} (1-q^{4m})^{-8s} .$$

If $n \geqslant 48$, $\alpha \geqslant 6$, $s \geqslant 5$, then every product on the right-hand side has non-negative coefficients and the first has strictly positive coefficients. Therefore the expression in braces in (8) has strictly positive coefficients, showing that $a_\alpha < 0$ and $a_{\alpha-1} < 0$. This completes the proof of Theorem 3.

Acknowledgement. Some of the theta series were calculated on the MACSYMA system [Mat3].

20

Finding the Closest Lattice Point

J. H. Conway and N. J. A. Sloane

This chapter describes algorithms which, given an arbitrary point of $\mathbf{R}^n$, find the closest point of some given lattice. The lattices discussed include the root lattices A_n, D_n, E_6, E_7, E_8 and their duals. These algorithms can be used for vector quantizing or for decoding lattice codes for a bandlimited channel.

1. Introduction

As discussed in §3 of Chap. 2 and §1 of Chap. 3, lattices are practically important because they can be used as vector quantizers and as codes for a bandlimited channel. For these applications it is essential that there exist fast "decoding" algorithms which, given an arbitrary point of the space, find the closest lattice point.

In this chapter we describe such algorithms for the lattices $A_n (n \geqslant 1)$, $D_n (n \geqslant 3)$, E_6, E_7, E_8 and their duals. This chapter is based on [Con29] and [Con40]. The latter reference also gives an algorithm for decoding the Leech lattice (omitted here because of its length), algorithms for finding the closest codeword to various binary codes including the Golay code ("soft" decoding algorithms — see §4 below), and an extensive bibliography. See also [Ado1], [Bud1], [Che2], [Con26], [Con38], [For2], [Gor0], [Wol2].

Practical applications of these lattices also require algorithms for labeling lattice points. Suppose our code or quantizer uses a set $C = \{c_1, \ldots, c_m\}$ of lattice points. Efficient "encoding" algorithms are needed which map i $(1 \leqslant i \leqslant M)$ into c_i, and vice versa. The first problem can be solved using the algorithms described in this chapter, and the second by using a dual basis for the lattice — see [Con35].

The problem discussed in the following sections is this: for a lattice Λ in $\mathbf{R}^n$, find a *decoding algorithm* which, given an arbitrary point $x \in \mathbf{R}^n$, finds a lattice point $u \in \Lambda$ that minimizes the distance from x to u. We shall use the descriptions of the lattices and their duals given in Chap. 4. As usual it easier to work with the squared distance or norm $N(x-u) = (x-u) \cdot (x-u)$ than with the distance itself.

2. The lattices $\mathbf{Z}^n$, D_n and A_n

The algorithm for finding the closest point of $\mathbf{Z}^n$ to an arbitrary point $\mathbf{x} \in \mathbf{R}^n$ is particularly simple. For a real number x, let

$$f(x) = \text{closest integer to } x\,.$$

In case of a tie, choose the integer with the smallest absolute value. For $x = (x_1, \ldots, x_n) \in \mathbf{R}^n$, let

$$f(x) = (f(x_1), \ldots, f(x_n))\,.$$

For future use we also define $g(x)$, which is the same as $f(x)$ except that the *worst* coordinate of x — that furthest from an integer — is rounded the *wrong way*. In case of a tie, the coordinate with the lowest subscript is rounded the wrong way.

More formally, for $x \in \mathbf{R}$ we define $f(x)$ and the function $w(x)$ which rounds the wrong way as follows. (Here m is an integer.)

$$\begin{array}{rlll} \text{If } x = 0 & \text{then } f(x) = 0, & w(x) = 1. & \\ \text{If } 0 < m \leqslant x \leqslant m + \tfrac{1}{2} & \text{then } f(x) = m, & w(x) = m+1. & \\ \text{If } 0 < m + \tfrac{1}{2} < x < m+1 & \text{then } f(x) = m+1, & w(x) = m. & (1) \\ \text{If } -m - \tfrac{1}{2} \leqslant x \leqslant -m < 0 & \text{then } f(x) = -m, & w(x) = -m-1. & \\ \text{If } -m-1 < x < -m - \tfrac{1}{2} & \text{then } f(x) = -m-1, & w(x) = -m. & \end{array}$$

(Ties are handled so as to give preference to points of smaller norm.) We also write

$$x = f(x) + \delta(x)\,, \tag{2}$$

so that $|\delta(x)| \leqslant \tfrac{1}{2}$ is the distance from x to the nearest integer.

Given $x = (x_1, \ldots, x_n) \in \mathbf{R}^n$, let k $(1 \leqslant k \leqslant n)$ be such that

$$|\delta(x_k)| \leqslant |\delta(x_i)| \text{ for all } 1 \leqslant i \leqslant n$$

and

$$|\delta(x_k)| = |\delta(x_i)| \Rightarrow k \leq i\,.$$

Then $g(x)$ is defined by

$$g(x) = (f(x_1), f(x_2), \ldots, w(x_k), \ldots, f(x_n))\,.$$

Algorithm 1. To find the closest point of $\mathbf{Z}^n$ to x.

Given $x \in \mathbf{R}^n$, the closest point of $\mathbf{Z}^n$ is $f(x)$. (If x is equidistant from two or more points of $\mathbf{Z}^n$, this procedure finds the one with the smallest norm.)

To see that the procedure works, let $u = (u_1, \ldots, u_n)$ be any point of $\mathbf{Z}^n$. Then

$$N(u-x) = \sum_{i=1}^{n} (u_i - x_i)^2,$$

which is minimized by choosing $u_i = f(x_i)$ for $i = 1, \ldots, n$. Because of (1) ties are broken correctly, favoring the point with the smallest norm.

Next we consider D_n (§7.1 of Chap. 4).

Algorithm 2. To find the closest point of D_n to x

Given $x \in \mathbf{R}^n$, the closest point of D_n is whichever of $f(x)$ and $g(x)$ has an even sum of coordinates (one will have an even sum, the other an odd sum). If x is equidistant from two or more points of D_n this procedure produces a nearest point having the smallest norm.

This procedure works because $f(x)$ is the closest point of $\mathbf{Z}^n$ to x and $g(x)$ is the next closest. $f(x)$ and $g(x)$ differ by 1 in exactly one coordinate, and so precisely one of $\Sigma f(x_i)$ and $\Sigma g(x_i)$ is even and the other is odd.

Example: Find the closest point of D_4 to

$$x = (0.6, -1.1, 1.7, 0.1).$$

We compute $f(x) = (1,-1,2,0)$ and $g(x) = (0,-1,2,0)$, since the first coordinate of x is the furthest from an integer. The sum of the coordinates of $f(x)$ is $1 - 1 + 2 + 0 = 2$, which is even, while that of $g(x)$ is $0 - 1 + 2 + 0 = 1$, odd. Therefore $f(x)$ is the point of D_4 closest to x. To illustrate how ties are handled, suppose

$$x = (\tfrac{1}{2}, \tfrac{1}{2}, \tfrac{1}{2}, \tfrac{1}{2}).$$

In fact x is now equidistant from eight points of D_4, namely $(0,0,0,0)$, any permutation of $(1,1,0,0)$, and $(1,1,1,1)$. The algorithm computes

$$f(x) = (0,0,0,0), \ \text{sum} = 0, \ \text{even},$$
$$g(x) = (1,0,0,0), \ \text{sum} = 1, \ \text{odd},$$

and selects $f(x)$. Indeed $f(x)$ does have the smallest norm of the eight neighboring points. The algorithm takes about $4n$ steps to decode D_n.

The lattice A_n (§6.1 of Chap. 4) consists of the points $u = (u_0, u_1, \ldots, u_n) \in \mathbf{Z}^{n+1}$ with $\Sigma u_i = 0$.

Algorithm 3. To find the closest point of A_n to x

Step 1. Given $x \in \mathbf{R}^{n+1}$, compute $s = \Sigma x_i$ and replace x by

$$x' = x - \frac{s}{n+1}(1, 1, \ldots, 1). \tag{3}$$

Step 2. Calculate $f(x') = (f(x'_0), \ldots, f(x'_n))$ and the *deficiency* $\Delta = \Sigma f(x'_i)$.

Step 3. Sort the x'_i in order of increasing value of $\delta(x'_i)$ (defined in (2)). We obtain a rearrangement of the numbers $0, 1, \ldots, n$, say i_0, $i_1, \ldots, i_n$, such that

$$-\tfrac{1}{2} \leqslant \delta(x'_{i_0}) \leqslant \ldots \leqslant \delta(x'_{i_n}) \leqslant \tfrac{1}{2}.$$

Step 4. If $\Delta = 0$, $f(x')$ is the closest point of A_n to x.

If $\Delta > 0$, the closest point is obtained by subtracting 1 from the coordinates $f(x'_{i_0}), \ldots, f(x'_{i_{\Delta-1}})$.

If $\Delta < 0$, the closest point is obtained by adding 1 to the coordinates $f(x'_{i_n}), f(x'_{i_{n-1}}), \ldots, f(x'_{i_{n-\Delta+1}})$.

Remarks. Step 1 projects x onto a point x' in the hyperplane $\Sigma x_i = 0$ containing A_n. Then $f(x')$ is the closest point of $\mathbf{Z}^{n+1}$ to x', and Steps 3 and 4 make the smallest changes to the norm of $f(x')$ needed to make $\Sigma f(x'_i)$ vanish.

The most time-consuming step is the sorting operation, which takes $O(n \log n)$ steps [Knu1], [Rei5]. As pointed out to us by A. M. Odlyzko, this can be reduced to a constant times n, by using the Rivest-Tarjan algorithm [Knu1, p. 216] to find the $|\Delta|$ largest (if $\Delta > 0$) or smallest (if $\Delta < 0$) of the numbers $\delta(x'_i)$.

For $n = 2$ and 3 there are better algorithms. A_2 is the hexagonal lattice (Fig. 1.3a), and is best decoded using the fact that it is the union of a rectangular lattice and a translate, as suggested by Gersho ([Ger3, p. 165], [Con38, p. 299]). A_3 is equivalent to D_3, and is best decoded by the D_3 algorithm.

3. Decoding unions of cosets.

A procedure Φ for finding the closest point of a lattice Λ to a given point x can be easily converted to a procedure for finding the closest point of a coset $r + \Lambda$ to x. For if $\Phi(x)$ is the closest point of Λ to x, then $\Phi(x - r) + r$ is the closest point of $r + \Lambda$ to x.

Suppose further that L is a lattice (or in fact any set of points) which is a union of cosets of Λ:

$$L = \bigcup_{i=0}^{t-1} (r_i + \Lambda). \tag{4}$$

(If L is a lattice, $t = [L : \Lambda] = (\det \Lambda / \det L)^{1/2}$ is the index of Λ in L.) Then Φ can be used as the basis for the following procedure for finding the closest point of L.

Algorithm 4. To find the closest point of L (a union of cosets of a lattice) to a given point x.

Given x, compute

$$y_i = \Phi(x - r_i) + r_i \tag{5}$$

for $i = 0, 1, \ldots, t-1$. Compare each of $y_0, \ldots, y_{t-1}$ with x and choose the closest.

Examples (a) E_6 and E_6^* (in the complex versions defined by (120), (126) of Chap. 4) both contain a sublattice equivalent to A_2^3, spanned by the vectors $\omega^\nu(\theta,0,0)$, $\omega^\nu(0,\theta,0)$, $\omega^\nu(0,0,\theta)$. This has index 3 in E_6 and index 9 in E_6^*. Explicit coset representatives r_i are given in [Con38].

(b) The versions of E_7 and E_7^* defined in §2.5 of Chap. 5 contain a sublattice $2\mathbf{Z}^7$, to index 8 and 16 respectively. The coset representatives form the [7, 3, 4] and [7, 4, 3] Hamming codes respectively [Con38].

(c) E_8 contains D_8 as a sublattice of index 2, leading to the fast decoding algorithm for E_8 described in [Con26], [Con29]. This decodes E_8 in about 104 steps. An even faster algorithm is given in §6.

(d) K_{12} has several possible sublattices, but the best seems to be the one equivalent to A_2^6, of index 64, with coset representatives forming the hexacode [Con38].

(e) Λ_{24} has a sublattice of index 8192 that is equivalent to D_{24}, which leads to a relatively slow decoding algorithm. Faster algorithms are described in [Con40], [For2].

(f) D_n has index 4 in D_n^*, and A_n has index $n+1$ in A_n^*, so Algorithm 4 can be applied to D_n^* and A_n^*. Coset representatives (the glue vectors for D_n and A_n) are given in Chap. 4.

4. "Soft decision" decoding for binary codes

Let C be an $[n,k]$ binary code. By replacing the 0's and 1's of the codewords by the real numbers $+1$ and -1 we may regard C as embedded in $\mathbf{R}^n$. A "soft decision" (or "maximum likelihood") decoder for C is an algorithm which, when presented with an arbitrary point $x \in \mathbf{R}^n$, finds a codeword $c \in C$ that minimizes the Euclidean distance from x to c. Note that

$$N(x-c) = (x-c)\cdot(x-c) = x\cdot x - 2x\cdot c + n, \tag{6}$$

since the components of c are ± 1, so it is enough to find a $c \in C$ that maximizes the inner product $x \cdot c$. Many examples (including the Golay codes $\mathscr{C}_{23}$ and $\mathscr{C}_{24}$) and an extensive bibliography are given in [Con40].

Here we just give one example, for use in §6. A first-order Reed-Muller code (§2.7 of Chap. 3) has parameters $[n,k] = [2^m, m+1]$, and may be decoded in about $m\,2^m$ steps using the fast Hadamard transform. (The so-called "Green machine" described in [Gre2], [Gre3], [Har8], [Mac6, Chap. 14], [Pos1]. Litsin and Shekhovtsov [Lit2] have recently shown that even this algorithm can be speeded up.)

5. Decoding lattices obtained from Construction A

If C is an $[n,k]$ binary code, written in ± 1 notation, Construction A (§2.1 of Chap. 5, §2 of Chap. 7) produces a lattice $\Lambda(C)$ consisting of the vectors of the form

$$c + 4z, \quad c \in C, \ z \in \mathbf{Z}^n. \tag{7}$$

(The set of points (7) is strictly speaking not a lattice, but the translate of a lattice by the vector $(1,1,\ldots,1)$.)

The following lemma makes it possible to use a decoding algorithm for C to decode $\Lambda(C)$.

Lemma. *Suppose $x = (x_1,\ldots,x_n)$ lies in the cube $-1 \leqslant x_i \leqslant 1$ $(i=1,\ldots,n)$. Then no point of $\Lambda(C)$ is closer to x than the closest codeword of C.*

Proof. Suppose the contrary, and let $u = (u_1,\ldots,u_n)$ be a closest lattice point to x. By hypothesis some x_i's are neither $+1$ nor -1. By subtracting a suitable vector $4z$, we may change these coordinates to $+1$ or -1 (depending on their parity) to produce a point of $\Lambda(C)$ that is in C, and is at least as close to x as u is, a contradiction.

Algorithm 5. To find the closest point of $\Lambda(C)$ to a given point x.

(i) Given $x = (x_1,\ldots,x_n)$, we first reduce all x_i to the range $-1 \leqslant x_i < 3$ by subtracting a vector $4z$.

(ii) Let S denote the set of i for which $1 < x_i < 3$. For $i \in S$, replace x_i by $2 - x_i$.

(iii) Since x is now in the cube $-1 \leqslant x_i \leqslant 1$ $(i=1,\ldots,n)$, by the Lemma we are justified in applying the decoder for C to x, obtaining an output $c = (c_1,\ldots,c_n)$ say.

(iv) For $i \in S$ change c_i to $2 - c_i$. Then $c + 4z$ is a closest point of $\Lambda(C)$ to the original vector x.

The total number of steps required is roughly $5n$ plus the number to decode C.

6. Decoding E_8

Since $E_8 = \Lambda(\mathcal{H}_8)$, and the Hamming code $\mathcal{H}_8$ (§2.4.2 of Chap. 3) is a first order Reed-Muller code, E_8 can also be decoded by this algorithm. $\mathcal{H}_8$ can be decoded in about $3\cdot 8 + 8 = 32$ steps by the fast Hadamard transform (§4), and so E_8 can be decoded in about 72 steps. It is worth noting that this is faster than the more obvious algorithm given in Example (c) of the previous section.

21

Voronoi Cells of Lattices and Quantization Errors

J. H. Conway and N. J. A. Sloane

In n-dimensional space, what is the average squared distance of a random point from the closest point of the lattice A_n (or D_n, E_n, A_n^* or D_n^*)? If a point is picked at random inside a regular simplex, octahedron, 600-cell or other polytope, what is its average squared distance from the centroid? The answers are given here, together with a description of the Voronoi cells of the above lattices. The results have applications to quantization and to the design of codes for a bandlimited channel. For example, a quantizer based on the eight-dimensional lattice E_8 has a mean squared error per symbol of 0.0717... when applied to uniformly distributed data, compared with 0.08333... for the best one-dimensional quantizer.

1. Introduction

Let Λ be an n-dimensional lattice in $\mathbf{R}^n$. The Voronoi cells (§1.2 of Chap. 2) around the lattice points are congruent polytopes, and we denote by $V(0)$ the Voronoi cell around the origin. Then, as we saw in Eq. (87) of Chap. 2, the normalized, dimensionless, *second moment of inertia* of $V(0)$,

$$G(\Lambda) = \frac{1}{n} (\det \Lambda)^{-\frac{n+2}{2n}} \int_{V(0)} x \cdot x \, dx \,, \qquad (1)$$

is equal to the mean squared error per symbol when the lattice is used as a vector quantizer for uniformly distributed inputs. If instead the lattice is used as a code for a bandlimited channel, the integral of $\exp(-x \cdot x / 2\sigma^2)$ over $V(0)$ gives the probability of correct decoding (§1.3 of Chap. 3). Furthermore the vertices of $V(0)$ furthest from Λ are the deep holes in Λ, and their distance from Λ is the covering radius R of Λ (§1.2 of Chap. 2). Spheres of radius R around the lattice points just cover $\mathbf{R}^n$, so R determines the *thickness* of Λ (§1.1 of Chap. 2).

For all these applications it is important to understand the structure of the polytope $V(0)$. In this chapter we give fairly explicit descriptions of $V(0)$ for the root lattices $A_n (n \geqslant 1)$, $D_n (n \geqslant 3)$, E_6, E_7, E_8 (§2 of Chap. 4) and their duals (except for E_6^* and E_7^*). Furthermore it will turn out that for these lattices it is possible to evaluate (1) exactly.

In §2 we study the second moment of an arbitrary polytope P, not necessarily the Voronoi cell of a lattice. Of course Voronoi cells $V(0)$ have the special property of being *space-filling* polytopes: the whole space $\mathbf{R}^n$ can be tiled by translated copies of $V(0)$. In three dimensions polytopes with the latter property are called parallelohedra. (See [McM1], [Ven0].)

If P is a polytope with centroid $\hat{x}$, then its volume, *unnormalized second moment* and *normalized second moment* are

$$Vol\,(P) = \int_P dx, \quad U(P) = \int_P \|x - \hat{x}\|^2\, dx\,, \tag{2}$$

$$I(P) = \frac{U(P)}{Vol\,(P)}, \tag{3}$$

respectively, and its *dimensionless second moment* is

$$G(P) = \frac{1}{n}\,\frac{U(P)}{Vol\,(P)^{1+2/n}} = \frac{1}{n}\,\frac{I(P)}{Vol\,(P)^{2/n}}\,. \tag{4}$$

For applications to quantizing we are interested in the minimal value of $G(P)$ taken over all polytopes P that tile by translation (cf. [Bar16]).

Summary of results. In §§2 and 3 we compute $G(P)$ for a number of important polytopes, including all the regular polytopes (see Theorem 4).

Table 21.1. Comparison of dimensionless second moments $G(P)$ for the sphere and various 3-dimensional polyhedra P (* indicates a space-filling polyhedron).

P	$G(P)$
tetrahedron	.1040042 ...
cube*	.0833333 ...
octahedron	.0825482 ...
hexagonal prism*	.0812227 ...
rhombic dodecahedron*	.0787451 ...
truncated octahedron*	.0785433 ...
dodecahedron	.0781285 ...
icosahedron	.0778185 ...
sphere	.0769670 ...

Table 21.2. Comparison of dimensionless second moments $G(P)$ for the sphere and various 4-dimensional polytopes P (* indicates a space-filling polytope).

P	$G(P)$
simplex {3,3,3}	.1092048 ...
hypercube* {4,3,3}	.0833333 ...
16-cell* {3,3,4}	.0816497 ...
24-cell* {3,4,3}	.0766032 ...
120-cell {5,3,3}	.0751470 ...
600-cell {3,3,5}	.0750839 ...
sphere	.0750264 ...

The three- and four-dimensional polytopes are compared in Tables 21.1 and 21.2. The chief tools used are Dirichlet's integral (Theorem 1), an explicit formula for the second moment of an n-simplex (Theorem 2), and a recursion formula giving the second moment of a polytope in terms of its cells (Theorem 3).

In §3 we consider the lattices A_n, D_n, E_n and their duals, determine their Voronoi cells $V(0)$, the second moments $G(\Lambda) = G(V(0))$, and their covering radii R. The second moments of these and other lattices are summarized in Table 2.3 and Fig. 2.9 of Chap. 2. It is worth emphasizing the remarkably low value of the mean squared error for the lattice E_8 (see (28) and Fig. 2.9). Fast algorithms for quantizing with this lattice were described in the preceding chapter.

2. Second moments of polytopes

A *polytope* in this chapter means a closed and bounded convex region of $\mathbf{R}^n$ that is the intersection of a finite number of half-spaces (cf. [Cox20, §7.4]). If we use a minimal collection of such half-spaces then the part of the polytope that lies in one of the bounding hyperplanes is called a *face*. If all edge-lengths are equal this length will usually be denoted by 2ℓ. The main source for information about polytopes is Coxeter [Cox20], but there is an extensive literature, particularly for low-dimensional figures. See for example [Ale1], [Cox1], [Cox5]-[Cox7], [Cox10], [Cox14], [Cox18], [Cox22], [Cox23], [Cox25], [Cox27], [Cox28], [Cun1], [Elt1], [Fej9], [Fej10], [Grü1], [Hil2], [Hol1], [Loe1], [McM2], [Pug1], [Wen1], [Wen2]. Second moments *about an axis* are tabulated for many simple polyhedra in standard engineering handbooks (see also [Sat1]), but these references do not give the values for $G(P)$.

2.A. Dirichlet's integral. A few special figures can be handled using Dirichlet's integral.

Theorem 1 [Wht1, §12.5]. *Let f be continuous and $\alpha_1,...,\alpha_n > 0$. Then*

$$\int f(x_1+\cdots+x_n)\, x_1^{\alpha_1}\cdots x_n^{\alpha_n}\, \frac{dx_1}{x_1}\cdots\frac{dx_n}{x_n}$$

$$= \frac{\Gamma(\alpha_1)\cdots\Gamma(\alpha_n)}{\Gamma(\alpha_1+\cdots+\alpha_n)} \int_0^1 f(\tau)\, \tau^{\Sigma\alpha_i-1} d\tau\,, \tag{5}$$

where the integral on the left is taken over the region bounded by $x_1 \geqslant 0,...,x_n \geqslant 0$ and $x_1+\cdots+x_n \leqslant 1$.

2.B. Generalized octahedron or crosspolytope. Consider for example the n-dimensional generalized octahedron or crosspolytope β_n [Cox20, §7.2] of edge-length 2ℓ. Taking $f=1$, $\alpha_1=1$ (to get the volume) or $\alpha_1=3$ (to get the second moment) and $\alpha_i=1$ for $i\geqslant 2$ in Theorem 1 we find

$$\frac{Vol(\beta_n)}{(2\ell)^n} = \frac{2^{n/2}}{n!}\,,\quad \frac{I(\beta_n)}{(2\ell)^2} = \frac{n}{(n+1)\,(n+2)}\,,$$

$$G(\beta_n) = \frac{(n!)^{2/n}}{2(n+1)\,(n+2)} \tag{6}$$

$$\to \frac{1}{2e^2} = 0.0676676... \text{ as } n \to \infty\,.$$

2.C. The n-sphere. As a second application, for the n-dimensional solid sphere (or ball) S_n of radius ρ it is easy to deduce from Theorem 1 that

$$\frac{Vol(S_n)}{\rho^n} = \frac{\pi^{n/2}}{\Gamma(\frac{1}{2}n+1)} \tag{7}$$

(which establishes (16), (17) of Chap. 1), $I(S_n) = n\rho^2/(n+2)$, and

$$G(S_n) = \frac{\Gamma(\frac{1}{2}n+1)^{2/n}}{(n+2)\pi} \tag{8}$$

$$\to \frac{1}{2\pi e} = 0.0585498... \text{ as } n \to \infty$$

It is clear from the definition that the sphere has the smallest value of $G(P)$ of any figure, so (8) establishes Zador's lower bound, the *sphere bound* (the left side of Eq. (82), Chap. 2). Unfortunately quantizers cannot be built with either spheres (unless $n = 1$) or generalized octahedra (unless $n = 1$, 2 or 4) as Voronoi cells, since these objects do not fill space.

2.D. n-Dimensional simplices. The next result makes it possible to find the second moment of any figure, provided it can be decomposed into simplices (cf. [Hoh1]).

Theorem 2. *Let P be an arbitrary simplex in $\mathbf{R}^n$ with vertices $v_i = (v_{i1},...,v_{in})$ for $0 \leqslant i \leqslant n$. Then (a) the centroid of P is the barycenter*

$$\hat{v} = \frac{1}{n+1}\left(v_0+\cdots+v_n\right) \tag{9}$$

of the vertices,

$$(b) \qquad Vol(P) = \frac{1}{n!} \det \begin{bmatrix} 1 & v_{01} \dots v_{0n} \\ 1 & v_{11} \dots v_{1n} \\ & \dots \quad \dots \\ 1 & v_{n1} \dots v_{nn} \end{bmatrix}, \text{ and} \qquad (10)$$

(*c*) *the normalized second moment about the origin* 0 *is*

$$I_0 = \frac{n+1}{n+2} \|\hat{v}\|^2 + \frac{1}{(n+1)\,(n+2)} \sum_{i=0}^{n} \|v_i\|^2 . \qquad (11)$$

In other words I_0 *is equal to the second moment of a system of* $n+1$ *particles each of mass* $1/((n+1)(n+2))$ *placed at the vertices and one particle of mass* $(n+1)/(n+2)$ *placed at the barycenter.*

Proof. (a) is elementary, (b) is well-known (cf. [All1], [Goo1, p. 349]), and (c) follows from [Goo1, Eq. (24), the case $n=2$].

2.E. Regular simplex. For example if P is a regular n-simplex of edge-length 2ℓ then

$$\frac{Vol(P)}{(\sqrt{2}\ell)^n} = \frac{\sqrt{n+1}}{n!}, \quad \frac{I(P)}{(\sqrt{2}\ell)^2} = \frac{n}{(n+1)\,(n+2)},$$

$$G(P) = \frac{(n!)^{2/n}}{(n+1)^{1+\frac{1}{n}}(n+2)} \qquad (12)$$

$$\to e^{-2} = 0.135335\ldots \text{ as } n \to \infty .$$

For $n=1,2$ and 3 the values of $G(P)$ are $1/12$, $1/(6\sqrt{3}) = 0.0962250\ldots$, and $3^{2/3}/20 = 0.104004\ldots$, and $G(P)$ *increases* monotonically with n.

2.F. Volume and second moment of a polytope in terms of its faces. Instead of decomposing a figure into simplices one may proceed by induction, expressing the volume and second moment of a polytope in terms of the volume and second moment of its faces, then in terms of its $(n-2)$-dimensional faces, and so on. Theorem 3 is the basis for this procedure.

Suppose P is an n-dimensional polytope with N_1 congruent faces F_1, F_1', F_1'', . . . , N_2 congruent faces F_2, F'_2, F_2'', . . . , and so on. Suppose also that P contains a point 0 such that all of the generalized pyramids $0F_1$, $0F_1'$,... are congruent, all of $0F_2$, $0F_2'$,... are congruent, ... Let $a_i \in F_i$ be the foot of the perpendicular from 0 to F_i, let $h_i = \|0a_i\|$, and let $V_{n-1}(i)$ be the volume of F_i and $U_{n-1}(i)$ the unnormalized second moment of F_i about a_i.

Theorem 3. *The volume and unnormalized second moment about* 0 *of* P *are given by*

$$Vol(P) = \sum_i \frac{N_i h_i}{n} V_{n-1}(i) ,$$

$$U(P) = \sum_i \frac{N_i h_i}{n+2} \left[h_i^2 V_{n-1}(i) + U_{n-1}(i) \right] .$$

Proof. Follows from elementary calculus by dividing each generalized pyramid OF_1 into slabs parallel to the face F_i.

2.G. Truncated octahedron. For example let P be the truncated octahedron with vertices consisting of all permutations of $\sqrt{2}\ell(0, \pm 1, \pm 2)$. P has $N_1 = 6$ square faces and $N_2 = 8$ faces which are regular hexagons, all with edge-length 2ℓ. The second moments of these faces can be calculated directly, or else found in §2.I. Then from the theorem we find that

$$Vol(P) = \frac{6 \cdot \ell\sqrt{8}}{3} 4\ell^2 + \frac{8 \cdot \ell\sqrt{6}}{3} 6\sqrt{3}\ell^2 = 64\sqrt{2}\ell^3 ,$$

$$U(P) = \frac{6 \cdot \ell\sqrt{8}}{5}\left[8\ell^2 \cdot 4\ell^2 + \frac{8\ell^4}{3}\right] + \frac{8 \cdot \ell\sqrt{6}}{5}\left[6\ell^2 \cdot 6\sqrt{3}\ell^2 + 10\sqrt{3}\ell^4\right]$$

$$= 304\sqrt{2}\ell^5 ,$$

hence

$$G(P) = \frac{1}{3}\frac{I(P)}{Vol(P)^{2/3}} = \frac{19}{192\sqrt[3]{2}} = 0.0785433\ldots . \tag{13}$$

2.H. Second moment of regular polytopes. The next theorem gives an explicit formula for the second moment of any regular polytope. Suppose P is an n-dimensional regular polytope [Cox20]. For $0 \leqslant j \leqslant n$ choose a j-dimensional face F_j of P so that $F_0 \subseteq F_1 \subseteq \ldots \subseteq F_n = P$, and let 0_j be the center of F_j, $R_j = \|0_n 0_j\|$, and for $j \geqslant 1$ let $r_j = \|0_{j-1}0_j\|$. Thus r_j is the inradius of F_j measured from 0_j, and $r_j^2 = R_{j-1}^2 - R_j^2$. Let $N_{j,j-1}$ be the number of $(j-1)$-dimensional faces of F_j. Then it is known that the symmetry group of P has order

$$g = N_{n,n-1}\, N_{n-1,n-2} \cdots N_{2,1} N_{1,0}$$

[Cox20, §7.6], and the volume of P [Cox20, §7.9] is

$$Vol(P) = N_{n,n-1} \cdots N_{2,1} N_{1,0} \cdot \frac{r_1 r_2 \cdots r_n}{n!} . \tag{14}$$

Theorem 4. *The second moment of any n-dimensional regular polytope P about its center 0_n is given by*

$$I(P) = \frac{2}{(n+1)\,(n+2)} (R_0^2 + 2R_1^2 + 3R_2^2 + \cdots + nR_{n-1}^2) , \tag{15}$$

or equivalently

$$I(P) = \frac{2}{(n+1)\,(n+2)} (r_1^2 + 3r_2^2 + 6r_3^2 + \cdots + \frac{n(n+1)}{2} r_n^2) . \tag{16}$$

Proof. The proof is by induction, the 1-dimensional case being immediate. From Theorem 3 we have

$$U(P) = \frac{N_{n,n-1}\, r_n}{n+2} (r_n^2 V_{n-1}(P) + U_{n-1}(P)) ,$$

where (from (14) and the induction hypothesis)

$$V_{n-1}(P) = N_{n-1,n-2} \cdots N_{2,1} N_{1,0} \cdot \frac{r_1 r_2 \cdots r_{n-1}}{(n-1)!} ,$$

$$U_{n-1}(P) = V_{n-1}(P) \cdot \frac{2}{n(n+1)} (r_1^2 + \cdots + \frac{n(n-1)}{2} r_{n-1}^2) .$$

Then

$$I(P) = \frac{U(P)}{Vol(P)} = \frac{n}{n+2} \{ r_n^2 + \frac{2}{n(n+1)} (r_1^2 + \cdots + \frac{n(n-1)}{2} r_{n-1}^2) \} ,$$

which simplifies to (16).

The values of g, $Vol(P)$ and R_j are tabulated for all regular polytopes in [Cox20, Table I, pp. 292-295]. We have already dealt with the simplex and generalized octahedron. For an n-dimensional cube $G(P) = 1/12$ for all n (since the cube is a direct product of line segments). We now treat the remaining regular polytopes.

2.I. Regular polygons. If P is a regular p-gon of edge-length 2ℓ, then from Theorem 4 we find

$$Vol(P) = p\ell^2 \cot \frac{\pi}{p} , \quad I(P) = \frac{\ell^2}{6}(1 + 3 \cot^2 \frac{\pi}{p}) ,$$

$$G(P) = \frac{1}{6p}(\operatorname{cosec} \frac{2\pi}{p} + \cot \frac{\pi}{p}) . \tag{17}$$

For $p = 3,4$ and 6, $G(P) = 1/(6\sqrt{3})$, $1/12$ and $5/(36\sqrt{3})$.

2.J. Icosahedron and dodecahedron. For the icosahedron

$$Vol = \frac{20\ell^3\tau^2}{3} , \quad G = \frac{1}{20} \left[\frac{6\tau}{5} \right]^{2/3} = 0.0778185... , \tag{18}$$

where $\tau = (\sqrt{5} + 1)/2$, and for the dodecahedron

$$Vol = 4\sqrt{5}\ell^3\tau^4 , \quad G = \frac{11\tau + 17}{300} \left[\frac{2}{\tau\sqrt{5}} \right]^{2/3} = 0.0781285... . \tag{19}$$

2.K. The exceptional 4-dimensional polytopes. There are three "exceptional" regular polytopes in four dimensions, the 24-cell, the 120-cell and the 600-cell. For the 24-cell

$$Vol = 32\ell^4, \quad G = \frac{13}{120\sqrt{2}} = 0.0766032... ; \tag{20}$$

for the 120-cell

$$Vol = 120\sqrt{5}\ell^4\tau^8 , \quad G = \frac{43\tau + 13}{300\sqrt{6}\, 5^{1/4}} = 0.0751470... ; \tag{21}$$

and for the 600-cell

$$Vol = 100\ell^4\tau^3 , \quad G = \frac{(3\tau + 4)\,\tau^{1/2}}{150} = 0.0750839... . \tag{22}$$

The three- and four-dimensional polytopes that have been considered are compared in Tables 21.1 and 21.2.

3. Voronoi cells and the mean squared error of lattice quantizers

3.A. The Voronoi cell of a root lattice. The Voronoi cells of the root lattices $A_n (n \geqslant 1)$, $D_n (n \geqslant 3)$, E_6, E_7 and E_8 may be obtained in a uniform manner. The method is based on finding a fundamental simplex for the affine Weyl group of the lattice (§2 of Chap. 4). We denote the (finite) Weyl group of Λ by $W(\Lambda)$, and the (infinite) affine Weyl group by $W_a(\Lambda)$.

The affine Weyl group $W_a(\Lambda)$ is described by the extended Coxeter-Dynkin diagram shown in Figs. 21.1-21.3 (and in the last column of Table 4.1). This diagram can be read in several different ways ([Bou1], [Cox20], [Gro3], [Haz1]). First, it provides a presentation for $W_a(\Lambda)$, defining the group in terms of generators and relations (see §2 of Chap. 4); however we shall not make use of this interpretation here. Second, it can be used to specify a *fundamental simplex* S for $W_a(\Lambda)$. This is an n-dimensional closed simplex whose images under the action of $W_a(\Lambda)$ are distinct and tile $\mathbf{R}^n$. In other words we can write

$$\mathbf{R}^n = \bigcup_{g \in W_a(\Lambda)} g(S) , \tag{23}$$

where (except for the boundaries of $g(S)$, a set of measure zero) each point $\mathbf{x} \in \mathbf{R}^n$ belongs to a unique image $g(S)$. In this interpretation the nodes of the diagram represent the hyperplanes that are the walls of the fundamental simplex [Cox20, §11.3]. The angle between two walls or hyperplanes is indicated by the branch of the diagram joining the corresponding nodes. If the hyperplanes are at an angle of $\pi/3$ the nodes are joined by a single branch, if the angle is $\pi/4$ they are joined by a

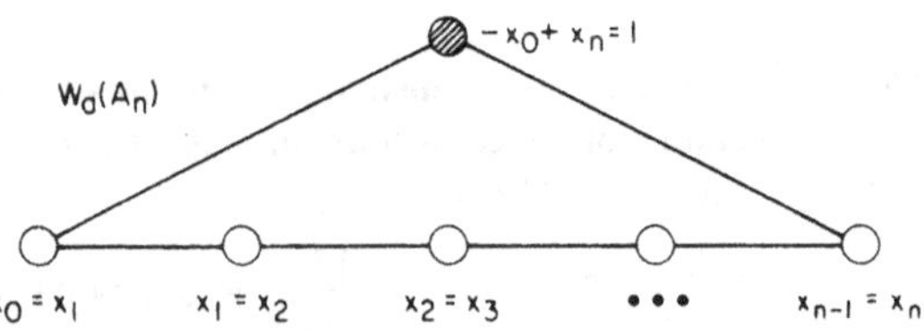

Figure 21.1. Extended Coxeter-Dynkin diagram for $W_a(A_n)$. The extending node is indicated by a solid circle. The $n+1$ nodes are labeled with the equations to the hyperplanes that are the walls of the fundamental simplex. The labelings in Figs. 21.1-21.3 are based on [Bou1, pp. 250-270] and [Con25].

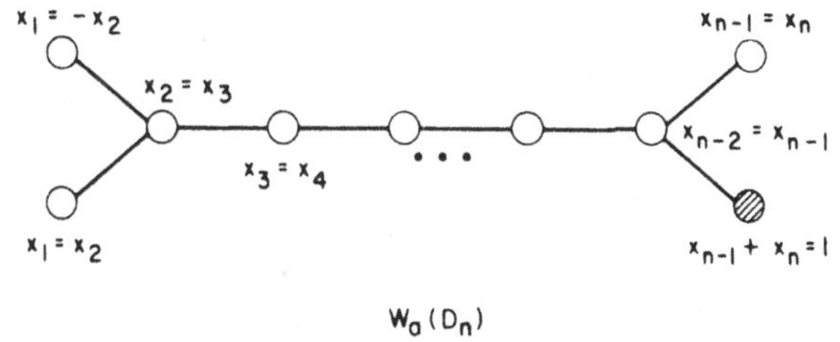

Figure 21.2. Extended Coxeter-Dynkin diagram for $W_a(D_n)$. There are $n+1$ nodes.

double branch (see Fig. 21.4), if the angle is π/p with $p>4$ they are joined by a branch labeled p, and finally if the hyperplanes are perpendicular the nodes are not joined by a branch. The nodes in Figs. 21.1-21.3 have been labeled with the equations to the corresponding hyperplanes.

In the third interpretation the nodes in the extended Coxeter-Dynkin diagram are taken to represent the *vertices* of a fundamental simplex, rather than the bounding hyperplanes. Each node represents the vertex opposite to the corresponding hyperplane (some examples are shown in Figs. 21.6-21.8) — cf. [Cox20, §11.6].

One of the nodes in the diagram is indicated by a solid circle. This is an *extending node*; removing it leaves an (ordinary) Coxeter-Dynkin diagram for the Weyl group $W(\Lambda)$. Of the $n+1$ hyperplanes represented by the extended diagram, all except that corresponding to the extending node pass through the origin. It is helpful to think of the latter hyperplane as forming the *roof* of the fundamental simplex. The vertex of the fundamental simplex opposite the roof is the origin.

For later use we remark that the finite Weyl group $W(\Lambda)$ also has a fundamental domain, consisting of an infinite cone centered at the origin. A fundamental simplex for $W_a(\Lambda)$ is obtained by taking the finite part of the cone beneath the roof. The intersection of this cone with the roof, or more precisely with a unit sphere centered at the origin, is a *spherical simplex*. The ordinary Coxeter-Dynkin diagram for $W(\Lambda)$ describes this spherical simplex in the same way as the extended diagram describes the fundamental simplex for $W_a(\Lambda)$. These spherical simplices and (ordinary) Coxeter-Dynkin diagrams can be used to define the Weyl groups of all the root *systems* (and not just the root *lattices* A_n, D_n and E_n). In §§3.E, 3.F we shall require the spherical simplices corresponding to $W(A_n)$ and $W(C_n)$, shown in Fig. 21.4. The Weyl group $W(A_n)$ is usually written as $[3^{n-1}]$ and is isomorphic to the symmetric group on $n+1$ letters. $W(C_n)$ is written $[3^{n-2},4]$ and has order $2^n n!$. (See §2 of Chap. 4.)

Lemma. *The origin is the closest lattice point to any interior point of the fundamental simplex.*

Proof. Let u be the closest lattice point to $x \in S$. Suppose $u \neq 0$. Then $u \notin S$, and u and x are on opposite sides of a reflecting hyperplane of

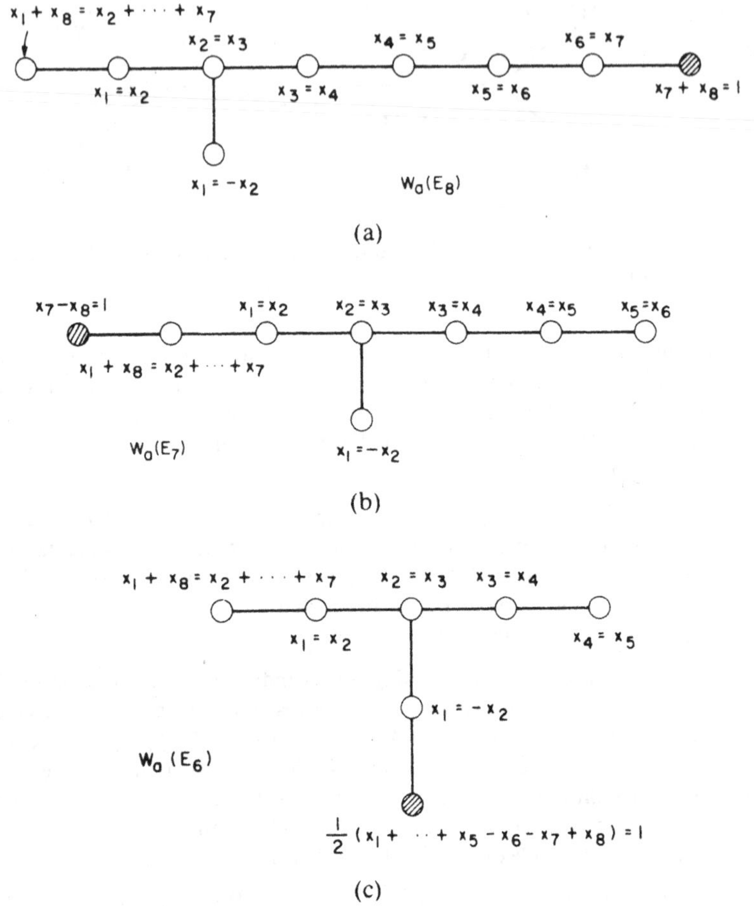

Figure 21.3. Extended Coxeter-Dynkin diagrams for (a) $W_a(E_8)$, (b) $W_a(E_7)$ and (c) $W_a(E_6)$.

$W_a(\Lambda)$. Let $u' \in \Lambda$ be the image of u in this hyperplane, and let y be the foot of the perpendicular from x to the line uu' (see Fig. 21.5). Then

$$\|xu'\|^2 = \|xy\|^2 + \|yu'\|^2 < \|xy\|^2 + \|yu\|^2 = \|xu\|^2,$$

and x is closer to u' than to u, a contradiction. Therefore $u = 0$.

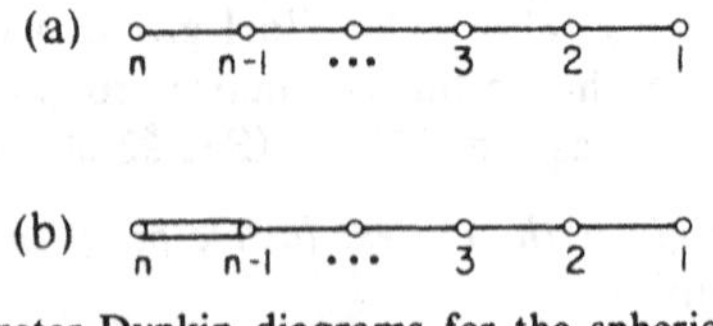

Figure 21.4. Coxeter-Dynkin diagrams for the spherical simplices of (a) $W(A_n)$ and (b) $W(C_n)$. (The labeling of the nodes is for convenience only and has no geometrical significance.)

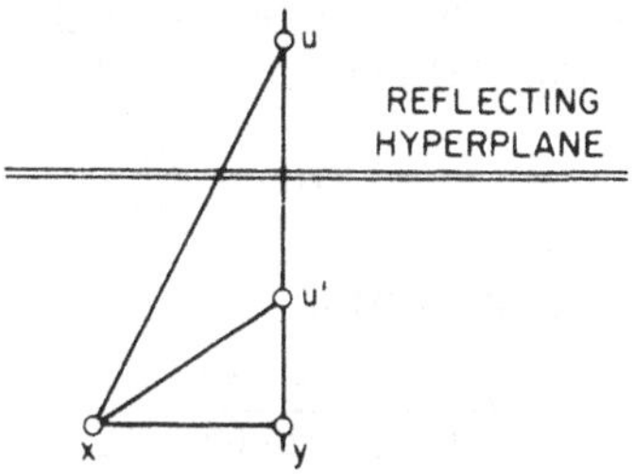

Figure 21.5. Why 0 is closer to x than u is.

The connection between the fundamental simplex and the Voronoi cell is given by the following theorem.

Theorem 5. *For any root lattice Λ, the Voronoi cell around the origin is the union of the images of the fundamental simplex under the Weyl group $W(\Lambda)$.*

Proof. Let x be any point of the Voronoi cell around the origin. From (23), $x \in g(S)$ for some $g \in W_a(\Lambda)$. Suppose x is an interior point of $g(S)$. By the lemma, the closest lattice point to x is $g(0)$. Therefore $g(0) = 0$, $g \in W(\Lambda)$, and

$$x \in \bigcup_{g \in W(\Lambda)} g(S) .$$

We omit the discussion of the case when x is a boundary point of $g(S)$. The converse statement, that $x \in \bigcup g(S)$ implies x is in the Voronoi cell, follows by reversing the steps.

It follows from Theorem 5 that the Voronoi cell is the union of $|W(\Lambda)|$ copies of the fundamental simplex S. Furthermore the faces of the Voronoi region are the images of the roof of the fundamental simplex under $W(\Lambda)$. Thus the Voronoi cell is bounded by hyperplanes which are the perpendicular bisectors of the lines joining 0 to its nearest neighbors in the lattice.

Corollary. *The number of $(n-1)$-dimensional faces of the Voronoi cell of a root lattice is equal to the kissing number of the lattice.*

This is not true for all lattices, as we shall see in §3.G. The second moment of the Voronoi region can now be obtained from that of the fundamental simplex.

3.B. Voronoi cell for A_n. We first find the vertices $v_0, v_1, \ldots, v_n$ of the fundamental simplex S. These are found by omitting each of the hyperplanes of Fig. 21.1 in turn and calculating the point of intersection of the remaining n hyperplanes. The results are shown in Fig. 21.6, where each node is labeled with the coordinates of the vertex opposite the corresponding hyperplane. The ith vertex is

$$v_i = \left(\left(\frac{-j}{n+1}\right)^i, \left(\frac{i}{n+1}\right)^j \right)$$

where $i+j = n+1$, for $0 \leqslant i \leqslant n$, and is the same as the glue vector $[i]$ for A_n in A_n^* (see Eq. (55) of Chap. 4). Also

$$\|v_i\|^2 = \frac{ij}{n+1} . \tag{24}$$

The barycenter of S (Eq. (9)) is

$$\hat{v} = (\frac{-n}{2n+2}, \frac{-n+2}{2n+2}, \ldots, \frac{n-2}{2n+2}, \frac{n}{2n+2})$$

and from (11) the normalized second moment about the origin is

$$I(S) = \frac{n+1}{n+2} \|\hat{v}\|^2 + \frac{1}{(n+1)\ (n+2)} \sum_{i=0}^{n} \|v_i\|^2 = \frac{n}{12} + \frac{n}{6(n+1)} .$$

Now $|W(A_n)| = (n+1)!$ and $\det A_n = n+1$ (§6.1 of Chap. 4). By Theorem 5 the Voronoi cell around the origin, $V(0)$, is the union of $|W(A_n)|$ copies of S, so

$$I(V(0)) = \frac{U(V(0))}{Vol(V(0))} = \frac{U(S)}{Vol(S)} = I(S) , \tag{25}$$

$$G(A_n) = \frac{1}{n} \frac{I(V(0))}{Vol(V(0))^{2/n}} = \frac{1}{(n+1)^{1/n}} \left[\frac{1}{12} + \frac{1}{6(n+1)} \right] \tag{26}$$

$$\sim \frac{1}{12} \text{ as } n \to \infty$$

(in agreement with [Dav2]). Once the Voronoi cell has been found we can also determine the points in $\mathbf{R}^n$ at maximal distance from the lattice, since these are necessarily vertices of the Voronoi cells. From (24) it follows that the covering radius of A_n is $R = (ab/(n+1))^{1/2} = \rho(2ab/(n+1))^{1/2}$, where ρ is the packing radius and $a = [(n+1)/2]$ and $b = n+1-a$. Typical points at this distance from A_n are the vertex v_a of $V(0)$ and its images under $W(A_n)$.

The lattice A_1 consists of equally spaced points on the real line, and $G(A_1) = 1/12$. The lattice A_2 is the hexagonal lattice, the fundamental

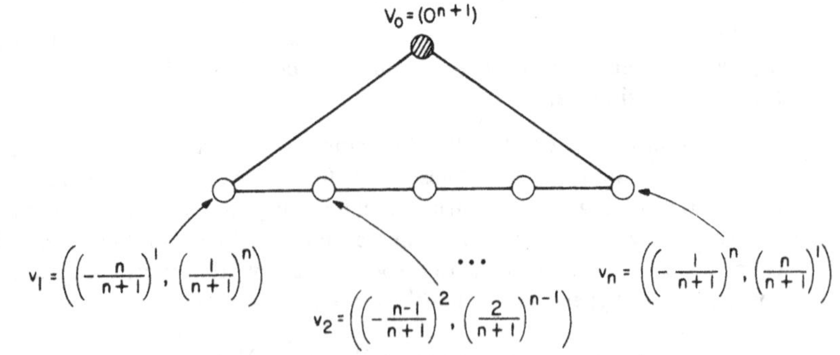

Figure 21.6. Vertices of fundamental simplex for A_n.

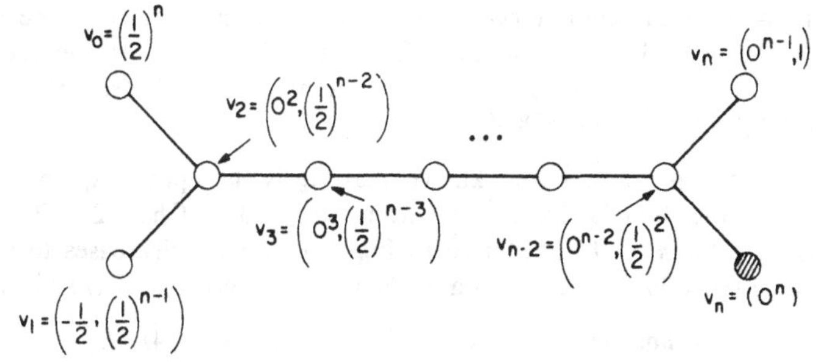

Figure 21.7. Vertices of fundamental simplex for D_n.

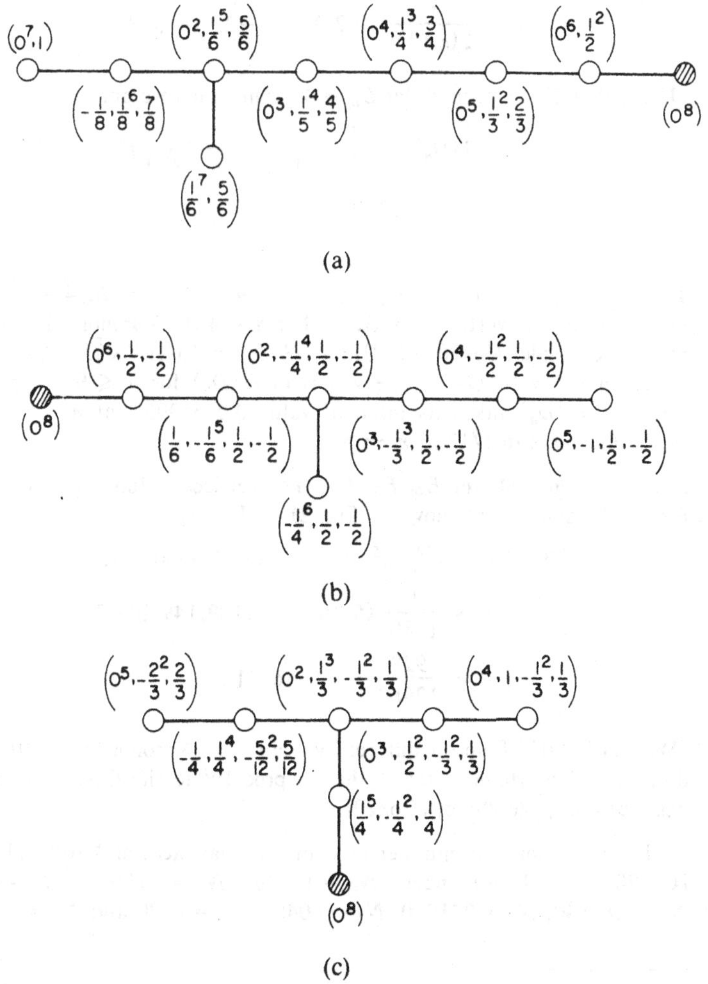

Figure 21.8. Vertices of fundamental simplices for (a) E_8, (b) E_7 and (c) E_6.

simplex is an equilateral triangle, the Voronoi cell is a hexagon, and $G(A_2) = 5/(36\sqrt{3})$ (compare §2.I). The lattice A_3 is the face-centered cubic lattice, the Voronoi cell is a rhombic dodecahedron (Fig. 2.2a) and $G(A_3) = 2^{-11/3} = 0.0787451 \ldots$.

For $n = 1$ and 2 it is known that A_n is the optimal quantizer, but for $n = 3$ the dual lattice A_3^* is better (see §3.3 of Chap. 2). The values of $G(A_n)$ for $n \leqslant 9$ are plotted in Fig. 2.9. $G(A_n)$ decreases to its minimal value 0.0773907... at $n = 8$ and then slowly increases to 1/12 as $n \to \infty$.

3.C. Voronoi cell for $D_n (n \geqslant 4)$. D_n (for $n \geqslant 4$), E_6, E_7 and E_8 are handled in the same way as A_n and our treatment will be brief. The vertices $v_0, \ldots, v_n$ of a fundamental simplex for D_n, $n \geqslant 4$, are shown in Fig. 21.7. Their barycenter is

$$\hat{v} = \frac{1}{2(n+1)} (0,2,3,\ldots,n-2,n-1,n+1),$$

$|W(D_n)| = 2^{n-1} \cdot n!$ and $\det D_n = 4$, from which we obtain

$$G(D_n) = \frac{1}{2^{2/n}} \left(\frac{1}{12} + \frac{1}{2n(n+1)}\right) \tag{27}$$

$$\to \frac{1}{12} \text{ as } n \to \infty .$$

The covering radius of D_n (for $n \geqslant 4$) is $R = \sqrt{n/4} = \rho\sqrt{n/2}$, as illustrated by the vertex $v_0 = (\frac{1}{2}^n)$. For $n = 4$ the Voronoi cell is a 24-cell (see for example [Cox20, §8.7]), and $G(D_4) = 13/(120\sqrt{2}) = 0.0766032\ldots$, in agreement with (20). The values of $G(D_n)$ for $n \leqslant 9$ are plotted in Fig. 2.9. $G(D_n)$ takes its minimal value 0.0755905... at $n = 6$ and then slowly increases to 1/12 as $n \to \infty$.

3.D. Voronoi cells for E_6, E_7, E_8. The vertices of fundamental simplices for E_6, E_7 and E_8 are shown in Fig. 21.8. For E_8

$$|W(E_8)| = 2^{14} \cdot 3^5 \cdot 5^2 \cdot 7 = 696729600, \quad \det = 1,$$

$$\hat{v} = \frac{1}{1080} (5,35,55,79,109,149,209,751),$$

$$G(E_8) = \frac{929}{12960} = 0.0716821 \ldots \tag{28}$$

(Watson [Wat2] found a different value.) The Voronoi cell $V(0)$ is an 8-dimensional polytope which is the reciprocal(1) to the Gosset polytope 4_{21} described on page 204 of [Cox20].

Let N_i denote the number of i-dimensional faces of $V(0)$. Then from [Cox20, §11.8] we have $N_0 = 19440$, $N_1 = 207360$, $N_2 = 483840$, $N_3 = 483840$, $N_4 = 241920$, $N_5 = 60480$, $N_6 = 6720$ and $N_7 = 240$. The

(1) It is not difficult to give a direct proof of this statement; it also follows from Theorem 8 below.

19440 vertices consist of 2160 at distance 1 from 0 and 17280 at distance $2\sqrt{2}/3$. The former are the images of the vertex $(0^7,1)$ of the fundamental simplex under $W(E_8)$, and are at the maximal possible distance from E_8, while the latter are the images of the vertex $((1/6)^7,5/6)$ under $W(E_8)$. Thus $R = 1 = \rho\sqrt{2}$. The other seven vertices of the fundamental simplex are not vertices of the Voronoi cell.

For $E_7, |W(E_7)| = 2^{10} \cdot 3^4 \cdot 5 \cdot 7 = 2903040$, det = 2,

$$\hat{v} = -\frac{1}{96}(1,5,8,12,18,30,-42,42),$$

$$G(E_7) = \frac{163}{2016} \cdot 2^{-1/7} = 0.0732306\ldots . \tag{29}$$

The covering radius of E_7 is $R = \sqrt{3/2} = \rho\sqrt{3}$, as illustrated by the vertex $(0^5, -1, ½, -½)$.

For $E_6, |W(E_6)| = 2^7 \cdot 3^4 \cdot 5 = 51840$, det = 3,

$$\hat{v} = \frac{1}{42}(0,3,5,8,14,-14,-14,14),$$

$$G(E_6) = \frac{5}{56 \cdot 3^{1/6}} = 0.0743467\ldots . \tag{30}$$

The Voronoi cells for E_7 and E_6 are the reciprocals of the Gosset polytopes 2_{31} and 1_{22} described in [Cox20, §11.8]. The covering radius of E_6 is $R = 2/\sqrt{3} = \rho\sqrt{8/3}$, as illustrated by the vertices $(0^5, -2/3, -2/3, 2/3)$ and $(0^4, 1, -1/3, -1/3, 1/3)$.

For E_6^* and E_7^* see [Con38], [Wor1], [Wor2].(2)

3.E. Voronoi cell for D_n^*. In order to determine the Voronoi cells for the dual lattices A_n^* and D_n^* we shall use *Wythoff's construction*, as described in [Cox5] and [Cox20, §11.6]. The idea is to construct new polytopes out of the spherical simplices described in §3.A, the vertices of the new polytope being indicated by drawing rings around certain nodes in the ordinary Coxeter-Dynkin diagram. More precisely, let $v_1,\ldots,v_n$ be the vertices of a spherical simplex for a Weyl group $W(\Lambda)$. If a single node of the diagram is ringed, say that corresponding to v_i, the vertices of the new polytope are the images of v_i under the Weyl group. If two or more nodes are ringed, say those corresponding to v_i, $v_j,\ldots$, the symbol represents a polytope whose vertices are the images under $W(\Lambda)$ of some interior point of the spherical subsimplex with vertices v_i, $v_j,\ldots$. We can adjust the metrical properties of the polytope (for example, equalize its edge lengths) by choosing this interior point suitably. Some 1-, 2- and 3-dimensional examples are shown in Fig. 21.9; others may be found in [Cox5], [Cox20].

---

(2) Worley [Wor1], [Wor2] finds the following exact values

$$G(E_6^*) = 12619 \cdot 3^{1/6}/204120 = 0.0742437,$$

$$G(E_7^*) = 21361 \cdot 2^{1/7}/322560 = 0.0731165.$$

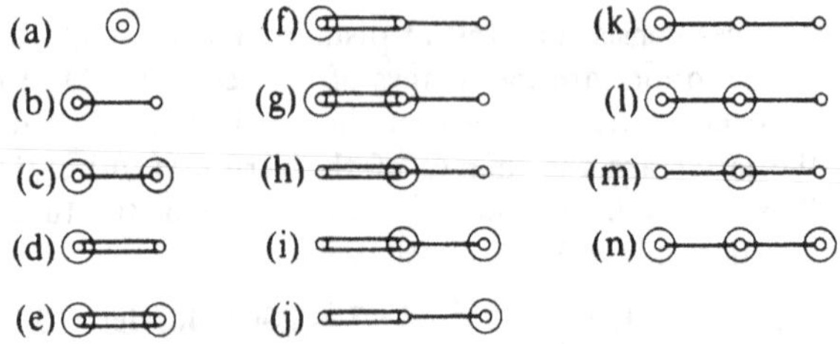

Figure 21.9. Examples of polytopes obtained by Wythoff's construction: (a) edge, (b) triangle, (c) hexagon, (d) square, (e) octagon, (f) cube, (g) truncated cube, (h) cuboctahedron, (i) truncated octahedron, (j) octahedron, (k) tetrahedron, (l) truncated tetrahedron, (m) octahedron (again), (n) truncated octahedron (again).

We now use this construction to find the Voronoi cell for D_n^*, $n \geqslant 3$. We change the scale so that D_n^* becomes the union of $(2\mathbf{Z})^n$ and $(1^n) + (2\mathbf{Z})^n$. The closest points to the origin in the first set consist of $2n$ points of the form $(\pm 2, 0^{n-1})$, and the closest in the second set consist of 2^n points of the form $(\pm 1^n)$. The Voronoi cell $V(0)$ is the intersection of the Voronoi cells determined by these two sets. The first of these, P say, is a cube centered at 0 with vertices $(\pm 1^n)$. The second, Q say, is a generalized octahedron with vertices $(\pm n/2, 0^{n-1})$. Furthermore Q can be obtained by reciprocating P in a sphere of radius $\rho = \sqrt{n/2}$ centered at the origin.

Thus the Voronoi cell $V(0)$ is the intersection of P and a reciprocal polytope $Q = P^*$. In other words $V(0)$ is obtained by *truncating* P in the manner described on page 147 of [Cox20], and is therefore specified by ringing one or two nodes of the Coxeter-Dynkin diagram (Fig. 21.4b) for the spherical simplex of P ([Cox5], [Cox20, §§8.1, 11.7]).

The radii R_j (defined in §2.H) for the cube P are given by $R_j = \sqrt{n-j}$ [Cox20, Table I, p. 295]. If n is even the radius ρ of the sphere of reciprocation is equal to $R_{n/2}$, and we must ring the node labeled $n/2$ in

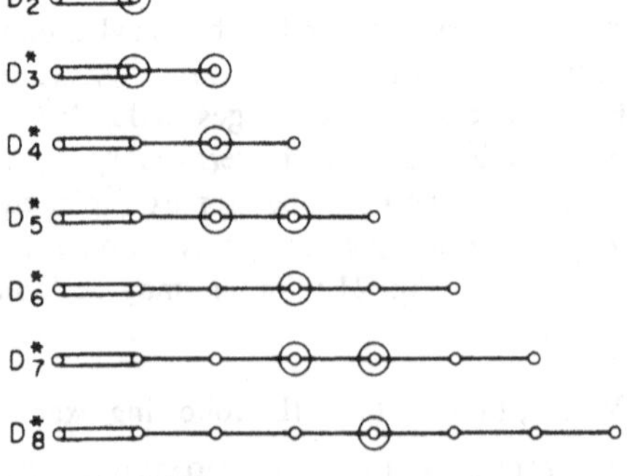

Figure 21.10. Voronoi cells for the lattices D_n^*.

Fig. 21.4b. If n is odd ρ lies between $R_{(n-1)/2}$ and $R_{(n+1)/2}$ and both nodes $(n-1)/2$ and $(n+1)/2$ must be ringed. We have therefore established the following theorem.

Theorem 6. *The Voronoi cell around the origin of the lattice* D_n^* *is the polytope defined by the diagrams in Fig.* 21.10.

The coordinates for $\beta(n,k)$ and $\delta(n,k)$ given below show that the edge-lengths of the Voronoi regions are all equal. In Coxeter's notation [Cox20, §8.1] the Voronoi cell for D_{2t}^* is

$$\begin{Bmatrix} 3\ 3\ \ldots\ 3 \\ 3\ 3\ \ldots\ 3\ 4 \end{Bmatrix}$$

with $t-1$ 3's in each row, and for D_{2t+1}^* it is

$$\left\{ 3 \begin{matrix} 3\ 3\ \ldots\ 3 \\ 3\ 3\ \ldots\ 3\ 4 \end{matrix} \right\}$$

with $t-1$ 3's in each row.

We shall determine the second moments of these Voronoi cells recursively, using Theorem 3. In order to do this it will be necessary to find the second moments of all the polytopes $\alpha(n,k)$, $\beta(n,k)$, $\gamma(n,k)$ and $\delta(n,k)$ defined as in Fig. 21.11. In this notation the Voronoi cell of D_n^* is (up to a scale factor) equal to $\beta(n,n/2)$ if n is even and to $\delta(n,(n-1)/2)$ if n is odd. Let $R_\alpha(n,k)$, $V_\alpha(n,k)$ and $U_\alpha(n,k)$ denote respectively the circumradius, volume and unnormalized second moment about the center of $\alpha(n,k)$, with a similar notation for $\beta(n,k)$, $\gamma(n,k)$ and $\delta(n,k)$.

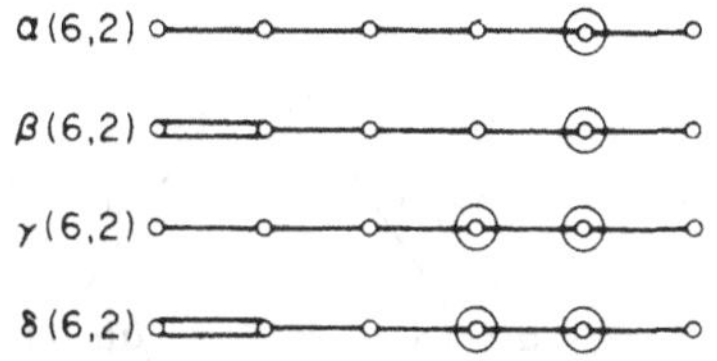

Figure 21.11. The polytopes $\alpha(n,k)$, $\beta(n,k)$, $\gamma(n,k)$ and $\delta(n,k)$. In general $\alpha(n,k)$ has n nodes with the k-th node from the right ringed, and $\gamma(n,k)$ has n nodes with the k-th and $(k+1)$st ringed (except that $\gamma(n,0) = \alpha(n,1)$ and $\gamma(n,n) = \alpha(n,n)$). $\beta(n,k)$ and $\delta(n,k)$ are the same as $\alpha(n,k)$ and $\gamma(n,k)$ respectively, except that the left branch is a double bond. By convention $\alpha(0,0)$ and $\gamma(0,0)$ represent a point.

For the vertices of the polytope $\alpha(n,k)$ it is convenient to take the points in $\mathbf{R}^{n+1}$ whose coordinates are all the permutations of $(0^{n-k+1}, 1^k)$ [Cox20, §8.7]. The centroid of $\alpha(n,k)$ is the point

$$\frac{k}{n+1}\,(1^{n+1}),$$

and so the circumradius is

$$R_\alpha(n,k) = \sqrt{\frac{k(n-k+1)}{n+1}} .$$

Similarly

$\beta(n, k)$ has vertices $(0^{n-k}, \pm 1^k)$, $R_\beta(n, k) = \sqrt{k}$,

$\gamma(n, k)$ has vertices $(0^{n-k}, 1, 2^k)$, $R_\gamma(n, k) = \sqrt{\frac{4k(n-k)+n}{n+1}}$,

$\delta(n, k)$ has vertices $(0^{n-k-1}, \pm 1, \pm 2^k)$, $R_\delta(n, k) = \sqrt{4k+1}$.

(These polytopes appear with different names in [Cox22].) Each of these polytopes has two kinds of faces, obtained by deleting either the left or the right node of its diagram [Cox20, §§ 7.6, 11.6, 11.7]. For example deleting the left node of the $\alpha(n,k)$ diagram produces an $\alpha(n-1, k-1)$, while deleting the right node produces an $\alpha(n-1,k)$. Thus in general $\alpha(n,k)$ has faces of type $\alpha(n-1,k-1)$ and $\alpha(n-1,k)$. (If $k=0$ the first type is absent, while if $k=n$ the second type is absent.) The number of faces of each type is given by the ratio of the orders of the underlying Weyl groups (obtained by ignoring the rings on the diagram). Thus the number of $\alpha(n-1,k-1)$-type cells of an $\alpha(n,k)$ is

$$\frac{|[3^{n-1}]|}{|[3^{n-2}]|} = \frac{(n+1)!}{n!} = n+1 .$$

This is also the number of $\alpha(n-1,k)$-type faces. We represent this process of finding the faces by the graph shown in Fig. 21.12.

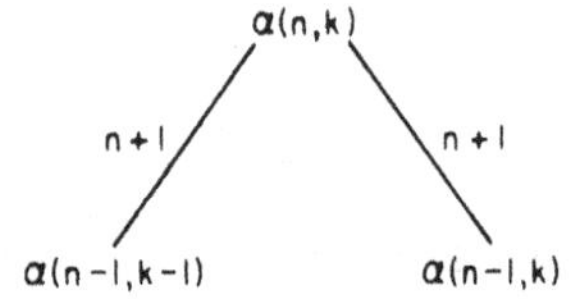

Figure 21.12. The polytope $\alpha(n,k)$ has $n+1$ faces which are $\alpha(n-1,k-1)$'s and $n+1$ faces which are $\alpha(n-1,k)$'s.

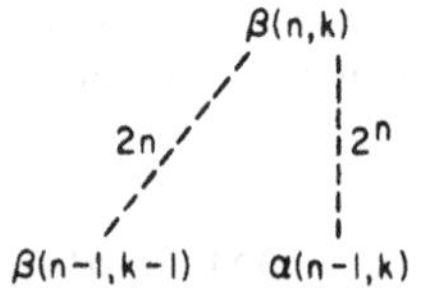

Figure 21.13. $\beta(n,k)$ has $2n$ faces which are $\beta(n-1,k-1)$'s and 2^n faces which are $\alpha(n-1,k)$'s. (We use broken lines here to make the structure of Figs. 21.14, 21.15 more visible.)

We can now apply Theorem 3 to $\alpha(n,k)$, obtaining

$$V_\alpha(n,k) = \frac{(n+1)h_L}{n} V_\alpha(n-1,k-1) + \frac{(n+1)h_R}{n} V_\alpha(n-1,k),$$

$$U_\alpha(n,k) = \frac{(n+1)h_L}{n+2} (h_L^2 V_\alpha(n-1,k-1) + U_\alpha(n-1,k-1)) + \frac{(n+1)h_R}{n+2} (h_R^2 V_\alpha(n-1,k) + U_\alpha(n-1,k)),$$

where

$$h_L^2 = R_\alpha(n,k)^2 - R_\alpha(n-1,k-1)^2 = \frac{(n-k+1)^2}{n(n+1)},$$

$$h_R^2 = R_\alpha(n,k)^2 - R_\alpha(n-1,k)^2 = \frac{k^2}{n(n+1)},$$

the subscripts on h standing for left and right. If we write

$$V_\alpha(n,k) = v_\alpha(n,k) \frac{\sqrt{n+1}}{n!}, \tag{31}$$

$$U_\alpha(n,k) = u_\alpha(n,k) \frac{\sqrt{n+1}}{(n+2)!} \tag{32}$$

then v_α and u_α are integers satisfying the recurrences

$$v_\alpha(n,k) = (n-k+1)\, v_\alpha(n-1,k-1) + k\, v_\alpha(n-1,k), \tag{33}$$

for $n \geqslant 2$ and $1 \leqslant k \leqslant n$, with $v_\alpha(n,0) = v_\alpha(n,n+1) = 0$ for $n \geqslant 1$, $v_\alpha(1,1) = 1$, and

$$u_\alpha(n,k) = (n-k+1)^3\, v_\alpha(n-1,k-1) + k^3\, v_\alpha(n-1,k) + (n-k+1)\, u_\alpha(n-1,k-1) + k\, u_\alpha(n-1,k), \tag{34}$$

for $n \geqslant 2$ and $1 \leqslant k \leqslant n$, with $u_\alpha(n,0) = u_\alpha(n,n+1) = 0$ for $n \geqslant 1$, $u_\alpha(1,1) = 1$. The first few values of v_α and u_α are shown in Table 21.3. With the help of [Slo2] the v_α may be identified as Eulerian numbers [Rio1, p. 215], and are given by

$$v_\alpha(n,k) = \sum_{j=0}^{k} (-1)^j \binom{n+1}{j} (k-j)^n \,. \tag{35}$$

There is a more complicated formula for $u_\alpha(n,k)$ which we omit.

Similarly for the polytope $\beta(n,k)$ we have the graph shown in Fig. 21.13, and writing

$$V_\beta(n,k) = v_\beta(n,k) \frac{2^n}{n!}, \tag{36}$$

$$U_\beta(n,k) = u_\beta(n,k) \frac{2^n}{(n+2)!}, \tag{37}$$

we obtain the recurrences

$$v_\beta(n,k) = n\, v_\beta(n-1,k-1) + k\, v_\alpha(n-1,k), \tag{38}$$

for $n \geqslant 2$ and $1 \leqslant k \leqslant n$, with $v_\beta(n,0) = 0$ for $n \geqslant 1$, $v_\beta(1,1) = 1$, and

$$\begin{aligned} u_\beta(n,k) &= k^3(n+1)v_\alpha(n-1,k) + ku_\alpha(n-1,k) \\ &\quad + n^2(n+1)v_\beta(n-1,k-1) + nu_\beta(n-1,k-1), \end{aligned} \tag{39}$$

for $n \geqslant 2$ and $1 \leqslant k \leqslant n$, with $u_\beta(n,0) = 0$ for $n \geqslant 1$, $u_\beta(1,1) = 2$ (see Table 21.3). Furthermore one can show by induction that

$$v_\beta(n,k) = \sum_{i=1}^{k} v_\alpha(n,i), \tag{40}$$

which implies $v_\beta(n,n) = n!$. Since the $v_\alpha(n,k)$ satisfy $v_\alpha(n,k) = v_\alpha(n,n-k+1)$ it follows that

$$v_\beta(2t,t) = \frac{1}{2}(2t)! \tag{41}$$

For $\gamma(n,k)$ and $\delta(n,k)$ we have a pair of graphs similar to Figs. 21.12, 21.13 (simply replace α by γ and β by δ). As before we set

$$V_\gamma(n,k) = v_\gamma(n,k)\,\frac{\sqrt{n+1}}{n!}\,,\; U_\gamma(n,k) = u_\gamma(n,k)\,\frac{\sqrt{n+1}}{(n+2)!}\,,$$
$$V_\delta(n,k) = v_\delta(n,k)\,\frac{2^n}{n!}\,,\; U_\delta(n,k) = u_\delta(n,k)\,\frac{2^n}{(n+2)!}\,,$$

and obtain the recurrences

$$v_\gamma(n,k) = (2n-2k+1)v_\gamma(n-1,k-1) + (2k+1)v_\gamma(n-1,k), \tag{42}$$

for $n \geqslant 1$ and $0 \leqslant k \leqslant n$, with $v_\gamma(n,-1) = v_\gamma(n,n+1) = 0$ for $n \geqslant 0$, $v_\gamma(0,0) = 1$;

$$\begin{aligned} u_\gamma(n,k) &= (2n-2k+1)^3\, v_\gamma(n-1,k-1) + (2k+1)^3\, v_\gamma(n-1,k) \\ &\quad + (2n-2k+1)\, u_\gamma(n-1,k-1) + (2k+1)\, u_\gamma(n-1,k), \end{aligned} \tag{43}$$

for $n \geqslant 1$ and $0 \leqslant k \leqslant n$, with $u_\gamma(n,-1) = u_\gamma(n,n+1) = 0$ for $n \geqslant 0$, $u_\gamma(0,0) = 0$;

$$v_\delta(n,k) = 2nv_\delta(n-1,k-1) + (2k+1)v_\gamma(n-1,k), \tag{44}$$

for $n \geqslant 1$ and $0 \leqslant k \leqslant n$, with $v_\delta(n,-1) = 0$ for $n \geqslant 0$, $v_\delta(0,0) = 1$;

$$\begin{aligned} u_\delta(n,k) &= (2k+1)^3(n+1)v_\gamma(n-1,k) + (2k+1)u_\gamma(n-1,k) \\ &\quad + 8n^2(n+1)v_\delta(n-1,k-1) + 2n\, u_\delta(n-1,k-1), \end{aligned} \tag{45}$$

Table 21.3. The first few values of $v_\alpha(n,k)$, $u_\alpha(n,k)$, $v_\beta(n,k)$ and $u_\beta(n,k)$. The diagonals correspond to $k = 1,2,\ldots$.

$v_\alpha(n,k)$

n						
6	1	57	302	302	57	1
5	1	26	66	26	1	
4	1	11	11	1		
3	1	4	1			
2	1	1				
1	1					

$u_\alpha(n,k)$

n						
6	6	1158	8916	8916	1158	6
5	5	400	1290	400	5	
4	4	116	116	4		
3	3	24	3			
2	2	2				
1	1					

$v_\beta(n,k)$

n						
6	1	58	360	662	719	720
5	1	27	93	119	120	
4	1	12	23	24		
3	1	5	6			
2	1	2				
1	1					

$u_\beta(n,k)$

n						
6	12	2568	28848	69624	80388	80640
5	10	950	5490	8250	8400	
4	8	312	880	960		
3	6	84	120			
2	4	16				
1	2					

for $n \geqslant 1$ and $0 \leqslant k \leqslant n$, with $u_\delta(n,-1) = 0$ for $n \geqslant 0$, $u_\delta(0,0) = 0$ (see Table 21.4). Also

$$v_\gamma(n,k) = \sum_{j=0}^{k} (-1)^j \binom{n+1}{j} (2k+1-2j)^n , \; v_\delta(n,k) = \sum_{i=0}^{k} v_\gamma(n,i),$$

$$v_\delta(n,n) = 2^n n!, \quad v_\delta(2t+1,t) = 2^{2t}(2t+1)! \,. \tag{46}$$

The j-dimensional faces of $\alpha(n,k),\ldots,\delta(n,k)$ for any j can be found from Figs. 21.14, 21.15.

Table 21.4. $v_\gamma(n,k)$, $u_\gamma(n,k)$, $v_\delta(n,k)$ and $u_\delta(n,k)$. The diagonals correspond to $k = 0,1,\ldots$.

$v_\gamma(n,k)$

n											
5	1		237		1682		1682		237		1
4		1		76		230		76		1	
3			1		23		23		1		
2				1		6		1			
1					1		1				
0						1					

$u_\gamma(n,k)$

n											
5	5		10065		124330		124330		10065		5
4		4		2416		10520		2416		4	
3			3		477		477		3		
2				2		60		2			
1					1		1				
0						0					

$v_\delta(n,k)$

n											
5	1		238		1920		3602		3839		3840
4		1		77		307		383		384	
3			1		24		47		48		
2				1		7		8			
1					1		2				
0						1					

$u_\delta(n,k)$

n											
5	10		20840		369740		954120		1074490		1075200
4		8		5224		41240		61048		61440	
3			6		1140		3654		3840		
2				4		188		256			
1					2		16				
0						0					

The special cases we are most interested in are $\beta(2t,t)$ and $\delta(2t+1,t)$, the Voronoi regions for D_{2t}^* and D_{2t+1}^* respectively. For D_{2t}^* we have

$$Vol(V(0)) = v_\beta(2t,t)\,\frac{2^{2t}}{(2t)!} = 2^{2t-1},$$

$$U(V(0)) = u_\beta(2t,t)\,\frac{2^{2t}}{(2t+2)!},$$

$$G(D_{2t}^*) = \frac{u_\beta(2t,t)}{2^{2-1/t}\,t\,(2t+2)!},\ \rho = 1\ , \tag{47}$$

and covering radius

$$R = R_\rho(2t,t) = \sqrt{t} = \rho\sqrt{t}\ .$$

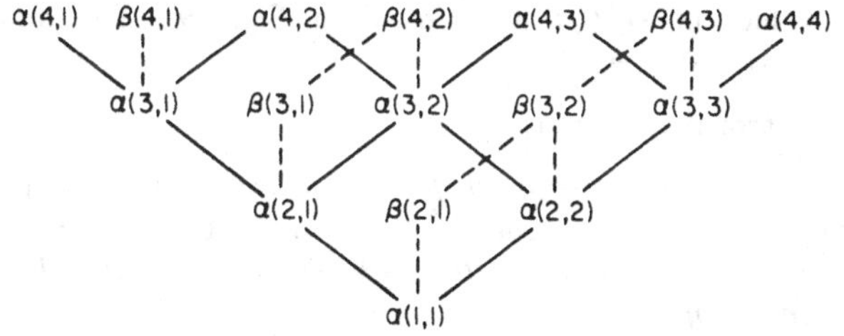

Figure 21.14. Interconnections between the $\alpha(n,k)$ and $\beta(n,k)$.

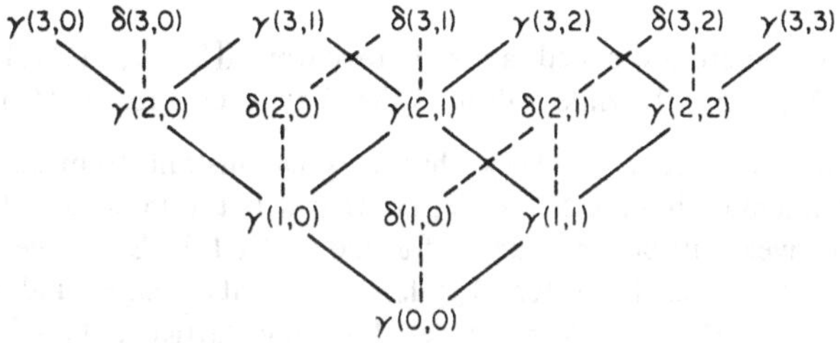

Figure 21.15. Interconnections between the $\gamma(n,k)$ and $\delta(n,k)$.

For this version of D_{2t+1}^* (which differs by a scale factor from the definitions given in §7.4 of Chap. 4) we have

$$Vol(V(0)) = v_\delta(2t+1,t)\frac{2^{2t+1}}{(2t+1)!} = 2^{4t+1},$$

$$U(V(0)) = u_\delta(2t+1,t)\ \frac{2^{2t+1}}{(2t+3)!},$$

$$G(D_{2t+1}^*) = \frac{u_\delta(2t+1,t)}{(2t+1)\ (2t+3)!\ 2^{f(t)}}, \tag{48}$$

where

$$f(t) = \frac{2(2t^2+5t+1)}{2t+1},$$

$$\rho = \sqrt{3}\ (\text{if } t = 1),\ \ \rho = 2\ (\text{if } t > 1),$$

$$R = R_\delta(2t+1,t) = \sqrt{4t+1}$$

$$= \rho\sqrt{5/3}\ \ (\text{if } t = 1),\ \ \rho\sqrt{t+1/4}\ (\text{if } t > 1)\ .$$

For example $D_3^* \cong A_3^*$ is the body-centered cubic lattice, the Voronoi cell is a truncated octahedron (Fig. 2.2b), and $G(D_3^*) = 19/(192\ {}^3\sqrt{2})$ (in agreement with (13)). Also

$$G(D_4^*) = 13/(120\sqrt{2}) = G(D_4),$$

$$G(D_5^*) = \frac{2641}{23040 \cdot 2^{3/5}} = 0.0756254..., \tag{49}$$

$$G(D_6^*) = \frac{601 \cdot 2^{1/3}}{10080} = 0.0751203.... \tag{50}$$

The values of $G(D_n^*)$ are plotted in Fig. 2.9. The minimal value is 0.0746931... at $n = 9$.

3.F. Voronoi cell for A_n^*.

Theorem 7. *The Voronoi cell for the lattice A_n^* is the polytope P_n defined in Fig. 21.16. If we rescale A_n^* by multiplying it by $n + 1$, the vertices of the Voronoi cell consist of the $(n+1)!$ points obtained by permuting the coordinates of*

$$\sigma = \left[\frac{-n}{2}, \frac{-n+2}{2}, \frac{-n+4}{2}, \ldots, \frac{n-2}{2}, \frac{n}{2} \right] .$$

P_n is sometimes called a permutohedron ([Bow1], [Cox18, pp. 72-73], [Cox23, p. 574]), while σ itself is the Weyl vector for A_n [Cox25].

Proof. It is easy to check that σ is equidistant from the walls of the fundamental simplex S for A_n; i.e. that σ is the in center of S. Let P be the convex hull of the images of σ under $W(A_n)$. Since the walls of S are reflecting hyperplanes for $W_a(A_n)$, P and its images under $W_a(A_n)$ tile $\mathbf{R}^n$. Thus P is the Voronoi cell for some lattice $\Lambda \subseteq A_n^*$. But Λ must contain all the glue vectors $[i]$ for A_n in A_n^* (§6.1 of Chap. 4), since these are the images of 0 in the walls of P. Since these points span A_n^*, $\Lambda = A_n^*$.

The second moment of P_n may be found as follows. First, the covering radius of A_n^*, $R(n)$ say, is the circumradius of P_n, which is

$$R(n) = \sqrt{\sigma \cdot \sigma} = \left\{ \frac{1}{2} \binom{n+2}{3} \right\}^{1/2} = R_n \text{ (say)},$$

$$= \rho\{(n+2)/3\}^{1/2} , \tag{51}$$

since now $\rho = \sqrt{n(n+1)/4}$, and the volume of P_n is $V_n = (n+1)^{n-1/2}$. Let $I_n = I(P_n)$ be the normalized second moment. A typical face of P_n is obtained by deleting say the rth node from the left in Fig. 21.16, and is a prism $P_r \times P_s$ with $r + s = n-1$ (see Fig. 21.17). The number of such faces is

$$\frac{|W(A_n)|}{|W(A_r)|\,|W(A_s)|} = \binom{n+1}{r+1} . \tag{52}$$

Furthermore $I(P_r \times P_s) = I_r + I_s$.

Let h_{rs} be the height of the perpendicular from the center of P_n to a typical face $P_r \times P_s$. Then (see Fig. 21.18)

$$h_{rs}^2 = R(r+s+1)^2 - R(r)^2 - R(s)^2$$
$$= \frac{(r+1)\,(s+1)\,(n+1)}{4}, \text{ using (51)} .$$

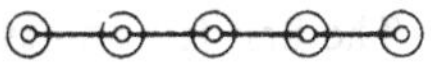

Figure 21.16. Voronoi cell $P_n = V(0)$ for A_n^*. There are n nodes, all ringed.

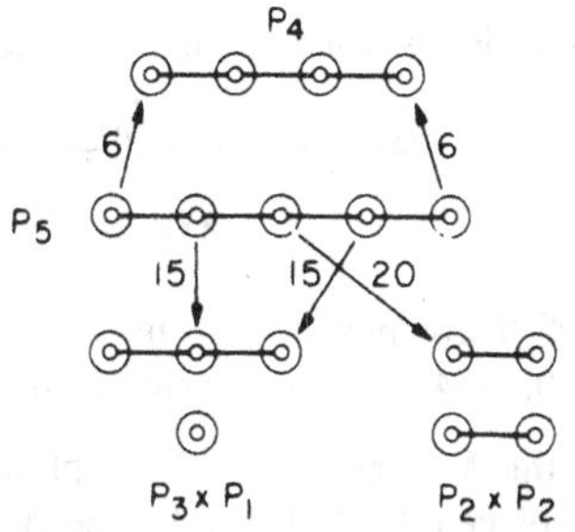

Figure 21.17. The three types of faces of P_5.

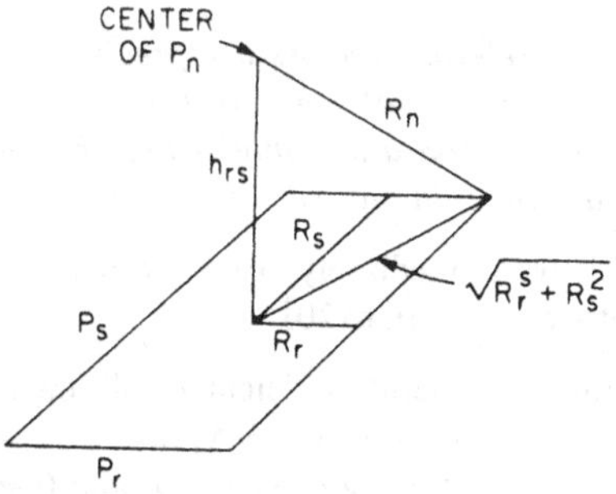

Figure 21.18. Calculation of height h_{rs} of perpendicular from center of P_n to face $P_r \times P_s$.

We now apply Theorem 3, obtaining

$$I_n V_n = \frac{1}{n+2} \sum_{r=0}^{n-1} \binom{n+1}{r+1} h_{rs} V_r V_s \, (h_{rs}^2 + I_r + I_s)$$

(with $r+s = n-1$), which, if we write $J_n = I_{n-1}/n$, becomes

$$J_n = \frac{1}{2(n+1)n^{n-1}} \sum_{r=1}^{n-1} \binom{n}{r} r^r s^s \left(\frac{n}{4} + \frac{J_r}{s} + \frac{J_s}{r}\right).$$

where $r+s = n-2$. Using Abel's identity [Rio2, §1.5] to simplify the first term, this becomes

$$J_n = \frac{n!}{8(n+1)n^{n-2}} \sum_{k=0}^{n} \frac{n^k}{k!} - \frac{n^2}{4(n+1)}$$

$$+ \frac{1}{n+1} \sum_{r=1}^{n-1} \binom{n}{r} \left(\frac{r}{n}\right)^r \left(\frac{n-r}{n}\right)^{n-r-1} J_r \tag{53}$$

for $n \geqslant 2$, with $J_1 = 0$. The first few values are as follows:

n:	1	2	3	4	5	6	7
J_n:	0	$\frac{1}{12}$	$\frac{5}{18}$	$\frac{19}{32}$	$\frac{389}{375}$	$\frac{1045}{648}$	$\frac{78077}{33614}$

Finally, the dimensionless second moment of the Voronoi cell of A_n^* is

$$G(A_n^*) = \frac{J_{n+1}}{n(n+1)^{1-\frac{1}{n}}} . \tag{54}$$

The values for $n \leqslant 9$ are plotted in Fig. 2.9. The curve is extremely flat, the minimal value 0.0754913... occurring at $n = 16$.

G. The walls of the Voronoi cell. As explained in §1.2 of Chap. 2, the walls of the Voronoi cell $V(0)$ for a lattice Λ are defined by the *relevant* vectors of Λ. For a root lattice we saw in the Corollary to Theorem 5 that the relevant vectors in a root lattice are precisely the minimal vectors. In this case there is a simple description of $V(0)$.

Theorem 8. *If the relevant vectors are precisely the minimal vectors, then $V(0)$ is the reciprocal of the vertex figure of Λ at the origin. Equivalently, $V(0)$ is (on a suitable scale) the reciprocal of the polytope whose vertices are the minimal vectors of Λ.*

Proof. This follows immediately from the definitions of vertex figure and reciprocal polytope — see [Cox20].

The following is a useful sufficient condition for a lattice to have this property. Let us write $\Lambda = \Lambda_0 \cup \Lambda_1 \cup \cdots$, where the ith shell Λ_i consists of all $v \in \Lambda$ with $v \cdot v = \lambda_i$ (say), and $0 = \lambda_0 < \lambda_1 < \cdots$.

Theorem 9. *Suppose that* (i) $\Lambda_r \subset \Lambda_1 + \Lambda_1 + \cdots + \Lambda_1$ *(r times), and* (ii) $r\lambda_1 \leqslant \lambda_r$, *for all* $r = 1, 2, \ldots$ *Then the relevant vectors are precisely the minimal vectors.*

Condition (i) states that Λ_1 spans Λ, and moreover does it economically in the sense that any vector in Λ_r is the sum of not more than r vectors of Λ_1. In practice this condition is very easily checked by induction. An important class of lattices satisfying (i) are those obtained by applying Construction A to a linear binary code with minimal distance $\leqslant 4$ that is spanned by the codewords of minimal weight, and if $d < 4$ has the additional property that no coordinate of the code is always zero.

Proof of Theorem 9. Suppose the contrary, so that there is a point $u \in \Lambda_r$ with $r > 1$ and a point $x \in \mathbf{R}^n$ such that $x \cdot u > \frac{1}{2}\lambda_r$, but $x \cdot v \leqslant \frac{1}{2}\lambda_1$ for all $v \in \Lambda_1$. From (i), $u = \sum n_i v_i$ with $v_i \in \Lambda_1$, $n_i > 0$ and $\sum n_i \leqslant r$. Then $x \cdot u = \sum n_i (x \cdot v_i) \leqslant \frac{1}{2}\lambda_1 \sum n_i \leqslant \frac{1}{2} r\lambda_1 \leqslant \frac{1}{2}\lambda_r$, a contradiction.

Voronoi has given a simple characterization of the relevant vectors in any lattice (see [Vor1, Vol. **134** (1908), p. 277] and [Ven5]).

Theorem 10. *A nonzero vector* $v \in \Lambda$ *is relevant if and only if* $\pm v$ *are the only shortest vectors in the coset* $v + 2\Lambda$.

Proof. Note that each nonzero vector $v \in \Lambda$ determines a half-space $H_v = \{x \in \mathbf{R}^n : x \cdot v \leqslant \frac{1}{2} v \cdot v\}$, and $V(0)$ is the intersection of these half-spaces. (Only if) Suppose v, w satisfy $v \equiv w \pmod{2\Lambda}$, $v \neq \pm w$, $N(w) \leqslant N(v)$. Then $t = \frac{1}{2}(v+w)$, $u = \frac{1}{2}(v-w)$ are nonzero vectors in Λ. If $x \in H_t \cap H_u$ then $x \cdot t \leqslant \frac{1}{2} t \cdot t$, $x \cdot u \leqslant \frac{1}{2} u \cdot u$, implying $x \cdot v \leqslant \frac{1}{2} v \cdot v$, and so H_v is not needed to define $V(0)$, i.e. v is not relevant. (If) Suppose v is not relevant. In particular, the point $\frac{1}{2} v$ must lie on or outside some H_w for $w \neq 0$, $w \neq v$, i.e. $\frac{1}{2} v \cdot w \geqslant \frac{1}{2} w \cdot w$. Therefore $N(v-2w) \leqslant N(v)$. Furthermore $v - 2w \neq \pm v$ and is also in the coset $v + 2\Lambda$.

For example, for the Leech lattice we see from Theorem 2 of Chap. 12 that the Voronoi cell is bounded by the 196560 + 16773120 = 16969680 hyperplanes corresponding to the vectors of norm 4 and 6.

Acknowledgements. During the early stages of this work we were greatly helped by several discussions with Allen Gersho. Some of the calculations were performed on the MACSYMA system [Mat3]. We should also like to thank E. S. Barnes and H. S. M. Coxeter for their comments.

22

A Bound for the Covering Radius of the Leech Lattice

S. P. Norton

This chapter describes a method for bounding the covering radius of a lattice. When applied to the Leech lattice it gives an answer very close to the true value.

The problem of determining the covering radius of the Leech lattice is solved in the following chapter by Conway, Parker and Sloane. Before the discovery of their line of approach, the following close approximation to the true answer was obtained. This is considered worth recording not only for historical reasons, and because the proof is very short, but also because the method is not dependent on the properties of the Leech lattice and is therefore applicable to other lattices.

The only property of the 24-dimensional Leech lattice that we use is that the midpoint of the line joining any pair of lattice points satisfies one of the following conditions.

(a) It is itself a lattice point.

(b) It lies halfway between two lattice points of type 2 or 3 relative to each other (i.e. distance 2 or $\sqrt{6}$, assuming, as we do from now on, that the lattice is unimodular), with all other lattice points being further away.

(c) It is the midpoint of 24 mutually orthogonal lines joining pairs of lattice points of type 4 (distance $2\sqrt{2}$), so that we can choose a Cartesian coordinate system with it at the origin, where all points whose coordinates are a permutation of $(\pm\sqrt{2}, 0^{23})$ are lattice points.

From this property (Theorem 28 of Chap. 10), we derive the following result.

Theorem. *Every point in 24-space is within a distance* $(2k)^{1/2} = 1.4518442...$ *of some lattice point, where* $k = (37 - \sqrt{73})/27 = 48/(37 + \sqrt{73}) = 1.0539258...$.

The distance $(2k)^{1/2}$ is therefore a bound for the covering radius, since spheres with this radius, centered on lattice points, will cover the entire space. (In the following chapter it is proved that the actual value of the covering radius is $\sqrt{2}$.) The packing radius of this lattice is 1.

Proof. We start by putting a normed space structure on our 24-space, with the zero point as one of the lattice points, and the norm of a vector as the distance of the corresponding point from the zero point.

Lemma. *If (the point corresponding to) $2v$ satisfies the result of the theorem, then so does v.*

Proof. By hypothesis there is a vector u, with $2u$ in the lattice, such that $\|2v-2u\| \leqslant 2k$, so that $\|v-u\| \leqslant \frac{1}{2}k$. Now since u is the midpoint of the line joining 0 to $2u$, it satisfies one of (a)-(c) above. In case (a), u is a lattice point, and v is obviously within the required distance of it. In case (b), let x and y be the two lattice points closest to u, so that $2u = x+y$. The last equation, together with $\|u-x\| = \frac{1}{4}\|x-y\| = 1$ or $\frac{3}{2}$, implies that $\|v-x\| + \|v-y\| = 2(\|v-u\| + \|u-x\|) \leqslant k+3 < 4k$, from which the result of the theorem follows, since one of $\|v-x\|$, $\|v-y\|$ will be less than $2k$. In case (c) we apply a translation that takes u to the origin (and v to say $v' = v-u$), and use the coordinate system defined in the condition. Let x' be one of the 48 lattice points named in this condition with the greatest inner product with v', and let this inner product be a. Then $\|v'\| \leqslant 12a^2$, and $\|v'-x'\| = \|v'\| + 2-2a$, as $v' \cdot x' = a$ and $\|x'\| = 2$. From this we can show by computation that the hypothesis $\|v'\| \leqslant \frac{1}{2}k$ implies $\|v'-x'\| \leqslant 2k$, so that v is within the required distance of a lattice point. This proves the lemma. It turns out that the peculiar-looking value of k is the least that allows the final deduction.

To complete the proof of the theorem, we note that the lemma implies immediately that every dyadic rational point satisfies the result of the theorem, which is automatically true for lattice points. But it is obvious that the distance from the nearest lattice point is continuous on 24-space, and the inverse image of $[0, 2k]$, being the set of points satisfying the result of the theorem, will be both closed and dense (as the dyadic rationals are dense in $\mathbf{R}$). The theorem now follows directly.

Postscript. The method described here has proved even more successful when applied to other lattices. Illustrations may be found in §5 of Chap. 6 and in [Con36], [Con37], [Con43].

23

The Covering Radius of the Leech Lattice

J. H. Conway, R. A. Parker and N. J. A. Sloane

We investigate the points in 24-dimensional space that are at maximal distance from the Leech lattice, i.e. the "deep holes" in that lattice. The distance of such a point from the Leech lattice is $1/\sqrt{2}$ times the minimal distance between the lattice points. Furthermore there are 23 inequivalent types of deep hole, one for each of the 23 even unimodular 24-dimensional lattices found by Niemeier.

1. Introduction

The main result of this chapter is the following, which confirms a conjecture made by Leech soon after his discovery of the lattice.

Theorem 1. *The covering radius of the Leech lattice is* $\sqrt{2}$ *times the packing radius.*

The covering radius R is defined in Eq. (3) of Chap. 2. An upper bound of $R \leqslant 1.452...\rho$ (where ρ is the packing radius) was obtained by Norton in [Nor4] (≅ the previous chapter). In the course of proving Theorem 1 we shall classify all the "deep holes" (defined in §1.2 of Chap. 2) in the Leech lattice. There is a remarkable connection between these holes and the Niemeier lattices. The Leech lattice is the unique 24-dimensional even unimodular lattice with minimal norm 4 (Chap. 12), and Niemeier ([Nie2]; §1 of Chap. 16) found that there are 23 other 24-dimensional even unimodular lattices, all with minimal norm 2.

Theorem 2. *There are* 23 *inequivalent deep holes in the Leech lattice under congruences of that lattice, and they are in one-one correspondence with the* 23 *Niemeier lattices* (*see Table* 23.1).

An outline of the remainder of this chapter is as follows. In §2 we associate a Coxeter-Dynkin diagram (§2 of Chap. 4) to any hole in the Leech lattice. Two kinds of diagrams occur; those for which all components are *ordinary* Coxeter-Dynkin diagrams, and those that contain

Table 23.1. The 23 types of "deep hole" in the Leech lattice. Here h = Coxeter number, V = number of vertices of hole. Note that the component of highest dimension identifies the hole uniquely.

Components	h	V	Fig.	Components	h	V	Fig.
D_{24}	46	25	23.22	$A_9^2 D_6$	10	27	23.15
$D_{16}E_8$	30	26	23.29	D_6^4	10	28	23.25
E_8^3	30	27	23.30	A_8^3	9	27	23.14
A_{24}	25	25	23.11	$A_7^2 D_5^2$	8	28	23.13
D_{12}^2	22	26	23.28	A_6^4	7	28	23.12
$A_{17}E_7$	18	26	23.19	$A_5^4 D_4$	6	29	23.24
$D_{10}E_7^2$	18	27	23.27	D_4^6	6	30	23.23
$A_{15}D_9$	16	26	23.18	A_4^6	5	30	23.7
D_8^3	14	27	23.26	A_3^8	4	32	Th. 11
A_{12}^2	13	26	23.17	A_2^{12}	3	36	Th. 10
$A_{11}D_7E_6$	12	27	23.16	A_1^{24}	2	48	Th. 8
E_6^4	12	28	23.31				

an *extended* diagram as a subgraph. The holes of the first kind have radius less than $\sqrt{2}$ times the packing radius, and can therefore be ignored (but see Chap. 25), while the second kind have radius greater than or equal to $\sqrt{2}$ times the packing radius. In §§3-5 we shall determine all holes of the second kind, without any further assumption about the radius of the hole. We do this by systematically classifying all holes whose diagram contains a subgraph of type $A_1, A_2, ..., A_{24}, D_4, D_5, ..., D_{24}, E_6, E_7$ or E_8. At the conclusion of this process we find that all these holes have radius equal to $\sqrt{2}$ times the packing radius, and so the proof of Theorem 1 is complete.

Furthermore, precisely 23 inequivalent holes of this kind occur, and their Coxeter-Dynkin diagrams exactly describe the Witt lattice components of the 23 Niemeier lattices, completing the proof of Theorem 2.

Certain sublattices of the Leech lattice (the $\mathcal{S}$-lattices, defined in §3) were classified by Curtis [Cur2]. In the course of proving Theorems 1 and 2 we also find all sublattices of the Leech lattice that contain a set of points whose distances correspond to an extended Coxeter-Dynkin diagram. The sublattices corresponding to A_5, A_6, ... have an especially simple structure that is described by the A_n-tree of Fig. 23.9, while those corresponding to D_5, D_6, ... are described by the D_n-tree of Fig. 23.20. Further information about these sublattices is given in §§3-5.

We have tried to keep this chapter as short as possible. But for future applications it has seemed desirable to give explicit coordinates for the vertices and center of a hole of each type.

2. The Coxeter-Dynkin diagram of a hole

The Leech lattice will be denoted by Λ. The *norm* $N(\mathbf{x}) = \mathbf{x} \cdot \mathbf{x}$ of a vector $\mathbf{x}$ is its squared length, while its *type* is $\frac{1}{2}N(\mathbf{x})$. We choose the scale so that the first four shells in Λ, the so-called *short* vectors (see Chap. 12), consist of 1 vector of type 0, 196560 vectors of type 2, 16773120 vectors of type 3, and 398034000 vectors of type 4. The type 2 vectors can be taken to have the shapes

$\delta((\pm 4)^2, 0^{22})$,

$\delta((\pm 2)^8, 0^{16})$, where the ± 2's occupy an octad and have positive product,

$\delta(\mp 3, (\pm 1)^{23})$, where the entries congruent to 1 modulo 4 occupy a $\mathscr{C}$-set,

and $\delta = 1/\sqrt{8}$; while the type 3 vectors are

$\delta(\pm 4, (\pm 2)^8, 0^{15})$, where the ± 2's occupy an octad and have negative product,

$\delta((\pm 2)^{12}, 0^{12})$, where the ± 2's occupy a dodecad and have positive product,

$\delta(\pm 5, (\pm 1)^{23})$, where the entries congruent to 1 modulo 4 occupy a $\mathscr{C}$-set,

$\delta((\mp 3)^3, (\pm 1)^{21})$, where the entries congruent to 1 modulo 4 occupy a $\mathscr{C}$-set.

The packing radius of Λ is now 1. As in the previous chapter, Theorem 28 of Chapter 10 plays a central role. We restate it as follows.

Theorem 3. *Every vector in Λ is congruent modulo 2Λ to a short vector. Furthermore the only congruences among the short vectors are that each vector of type 2 or 3 is congruent to its negative, while the vectors of type 4 fall into congruence classes of size 48, each class consisting of a coordinate frame (i.e. a set of 24 mutually orthogonal pairs).*

Let $(\mathbf{c}, P)$ be a hole in Λ of radius R, and let the vertices of P be $\mathbf{v}_1, \ldots, \mathbf{v}_\nu$. The $\mathbf{v}_i$ are in Λ, so $N(\mathbf{v}_i - \mathbf{v}_j) = 4, 6, 8, \ldots$ for $i \neq j$. On the other hand $N(\mathbf{v}_i - \mathbf{v}_j) \leqslant 8$, for if $N(\mathbf{v}_i - \mathbf{v}_j) \geqslant 10$ then Theorem 3 implies that the midpoint $\frac{1}{2}(\mathbf{v}_i + \mathbf{v}_j)$ is also the midpoint of a pair $\mathbf{x}, \mathbf{x}' \in \Lambda$ with $\mathbf{x} - \mathbf{x}'$ a short vector, and then at least one of $\mathbf{x}$ or $\mathbf{x}'$ is closer to $\mathbf{c}$ than $\mathbf{v}_i$ and $\mathbf{v}_j$ are. Thus $N(\mathbf{v}_i - \mathbf{v}_j) = 4$, 6 or 8.

The diagram for this hole is constructed as follows. There is one node (also labeled $\mathbf{v}_i$) corresponding to each vector $\mathbf{v}_i$. Two nodes $\mathbf{v}_i$, $\mathbf{v}_j$ are

(i) not joined (or *disjoined*) if $N(\mathbf{v}_i - \mathbf{v}_j) = 4$,

(ii) joined by an edge if $N(\mathbf{v}_i - \mathbf{v}_j) = 6$, or

(iii) joined by two edges if $N(\mathbf{v}_i - \mathbf{v}_j) = 8$.

The resulting graph with these nodes and edges is the *hole diagram*. By the end of the chapter we shall have proved that this diagram is actually a Coxeter-Dynkin diagram, or more precisely that the following theorem holds.

Theorem 4. *The diagram for a hole in the Leech lattice is a graph whose connected components are taken from the list* $a_n (n \geqslant 1)$, $d_n (n \geqslant 4)$, e_6, e_7, e_8, $A_n (n \geqslant 1)$, $D_n (n \geqslant 4)$, E_6, E_7, E_8 *(see Fig.* 23.1).

The components denoted by lower case letters are called *ordinary* Coxeter-Dynkin diagrams, the others, *extended*. (In this chapter the terms ordinary and extended diagram will always refer to the connected graphs shown in Fig. 23.1.)

The following construction is helpful when studying a hole ($\mathbf{c}$, P). We embed the space in which Λ lies in $\mathbf{R}^{25}$, with Λ contained in a hyperplane H, and choose a point $\mathbf{c}'$ on the line through $\mathbf{c}$ perpendicular to H as follows. If the radius $R \geqslant \sqrt{2}$ we take $\mathbf{c}' = \mathbf{c}$, but if $R < \sqrt{2}$ we choose $\mathbf{c}'$ so that its distance from the vertices of P is $\sqrt{2}$ (Fig. 23.2). If $R < \sqrt{2}$ two nodes $\mathbf{v}_i$, $\mathbf{v}_j$ in the hole diagram are not joined, or are joined by one or

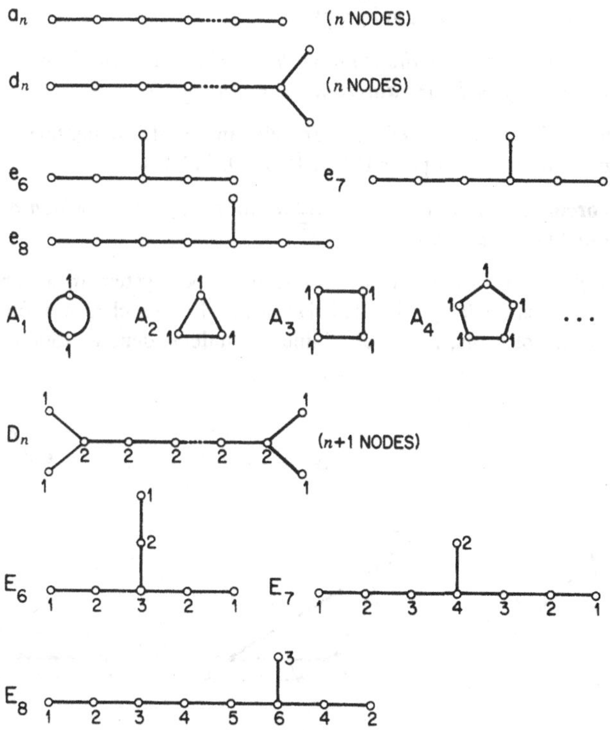

Figure 23.1. Ordinary Coxeter-Dynkin diagrams $a_n (n \geqslant 1)$, $d_n (n \geqslant 4)$, e_6, e_7, e_8 and extended Coxeter-Dynkin diagrams $A_n (n \geqslant 1)$, $D_n (n \geqslant 4)$, E_6, E_7, E_8. The latter are labeled with the integers c_i.

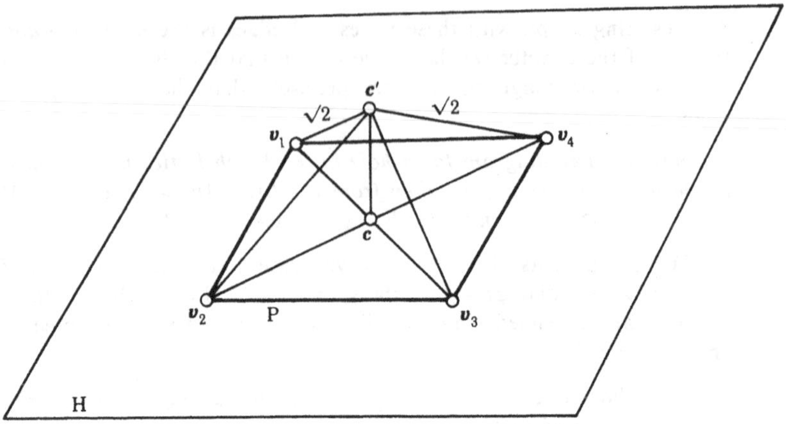

Figure 23.2. Construction of the point $\mathbf{c}'$.

two edges according to whether the inner product $(\mathbf{v}_i-\mathbf{c}', \mathbf{v}_j-\mathbf{c}')$ is 0, -1 or -2 respectively (see Fig. 23.3).

The first step is to dispose of the case when the diagram contains no extended diagram as a subgraph.

Theorem 5. *A hole diagram with no extended diagram embedded in it contains only ordinary diagrams as components.*

Proof. This is proved by an elegant and elementary combinatorial argument (given on page 195 of [Cox20, §11.5]).

Theorem 6. *The hole corresponding to a diagram in which all components are ordinary has radius* $R < \sqrt{2}$.

Proof. Such a diagram also describes a root system in which all the roots have the same length [Boul], [Huml]. Let us choose a fundamental set (or base) of roots, i.e. a set of linearly independent vectors $\mathbf{V}_1, \ldots, \mathbf{V}_\nu$ in an

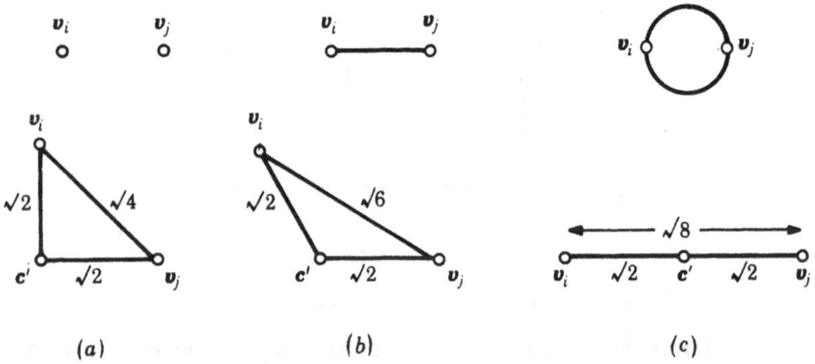

Figure 23.3. The conditions for a pair of nodes in the diagram for a hole of radius $R < \sqrt{2}$ to be (a) disjoined, (b) joined by an edge, or (c) joined by two edges.

appropriate Euclidean space $\mathbf{S}$, of constant length and having the same mutual distances as $\mathbf{v}_1, \ldots, \mathbf{v}_\nu$. Therefore the set of points $\{\mathbf{V}_i\}$ is isometric to the set $\{\mathbf{v}_i\}$. Since the origin in $\mathbf{S}$ is not a linear combination of the $\mathbf{V}_i$, it follows that the corresponding point $\mathbf{c}'$ is not in the hyperplane containing the $\mathbf{v}_i$. Thus $R < \sqrt{2}$, by definition of $\mathbf{c}'$. This completes the proof.

On the other hand, if $\mathbf{V}_1, \ldots, \mathbf{V}_\mu$ are a set of fundamental roots corresponding to an extended diagram, there are positive integers $c_1, \ldots, c_\mu$ such that $\Sigma\, c_i \mathbf{V}_i = 0$ ([Boul], [Cox20, p. 194], [Hum1, p. 58]). These integers are shown in Fig. 23.1. If this diagram occurs as a subgraph of a hole diagram, $\mathbf{v}_1, \ldots, \mathbf{v}_\mu$ are the corresponding vertices and $\mathbf{c}$ is the center of the hole, then $\Sigma\, c_i(\mathbf{v}_i - \mathbf{c}) = 0$. Thus the center can be found from

$$\mathbf{c} = \sum_{i=1}^{\mu} c_i \mathbf{v}_i \Big/ \sum_{i=1}^{\mu} c_i . \tag{1}$$

Suppose we can find points $\mathbf{v}_1^{(1)}, \ldots, \mathbf{v}_{\mu_1}^{(1)}$ in the Leech lattice Λ whose mutual distances define an extended diagram Δ_1, and such that their circumcenter $\mathbf{c}$ is distant $\sqrt{2}$ from all of them. The next theorem, which is essential in all that follows, asserts that under certain conditions $\mathbf{c}$ is the center of a unique hole in Λ of radius $\sqrt{2}$.

Theorem 7. *Suppose* $\mathbf{c} \in \mathbf{R}^{24}$ *is distant* $\sqrt{2}$ *from Leech vectors* $\mathbf{v}_1^{(1)}, \ldots, \mathbf{v}_{\mu_1}^{(1)}; \mathbf{v}_1^{(2)}, \ldots, \mathbf{v}_{\mu_2}^{(2)}; \ldots; \mathbf{v}_1^{(s)}, \ldots, \mathbf{v}_{\mu_s}^{(s)}$ *such that the vectors* $\mathbf{v}_j^{(i)} - \mathbf{c}$ *span* $\mathbf{R}^{24}$, *and, for* $i = 1, \ldots, s$, *the mutual distances of* $\mathbf{v}_1^{(i)}, \ldots, \mathbf{v}_{\mu_i}^{(i)}$ *define an extended diagram* Δ_i. *Then there is a unique hole in* Λ *with center* $\mathbf{c}$ *and radius* $\sqrt{2}$, *whose vertices are precisely the* $\mathbf{v}_j^{(i)}$ *and whose hole diagram has components* $\Delta_1, \Delta_2, \ldots, \Delta_s$. *In this case we say that the hole is of type* $\Delta_1 \Delta_2 \cdots \Delta_s$.

Proof. We first show that there is a hole of radius $\sqrt{2}$ centered at $\mathbf{c}$, and the $\mathbf{v}_j^{(i)}$ are all its vertices. Suppose $\mathbf{z} \in \Lambda$ is distinct from the $\mathbf{v}_j^{(i)}$ and satisfies $N(\mathbf{z}-\mathbf{c}) \leqslant 2$. For any $\mathbf{v}_j^{(i)}$ we have

$$\begin{aligned} 4 &\leqslant N(\mathbf{z} - \mathbf{v}_j^{(i)}) \text{ (since both are in } \Lambda) \\ &= N(\mathbf{z} - \mathbf{c} - (\mathbf{v}_j^{(i)} - \mathbf{c})) \\ &\leqslant 2 - 2(\mathbf{z} - \mathbf{c}) \cdot (\mathbf{v}_j^{(i)} - \mathbf{c}) + 2 , \end{aligned}$$

so

$$(\mathbf{z} - \mathbf{c}) \cdot (\mathbf{v}_j^{(i)} - \mathbf{c}) \leqslant 0 . \tag{2}$$

Let $\mathbf{z}_i$ be the projection of $\mathbf{z}$ onto the subspace of $\mathbf{R}^{24}$ spanned by $\{\mathbf{v}_j^{(i)} - \mathbf{c}: j = 1, \ldots, \mu_i\}$. Since the $\mathbf{v}_j^{(i)} - \mathbf{c}$ span $\mathbf{R}^{24}$ we can write $\mathbf{z} - \mathbf{c} = (\mathbf{z}_1, \ldots, \mathbf{z}_s)$, and from (2) $\mathbf{z}_i \cdot (\mathbf{v}_j^{(i)} - \mathbf{c}) = (\mathbf{z} - \mathbf{c}) \cdot (\mathbf{v}_j^{(i)} - \mathbf{c}) \leqslant 0$. As mentioned at the end of §2 there are positive integers $c_j^{(i)}$ satisfying $\sum_j c_j^{(i)} (\mathbf{v}_j^{(i)} - \mathbf{c}) = 0$. It follows that $\mathbf{z}_i = 0$ for all i, and $\mathbf{z} = \mathbf{c}$, which is a contradiction since $\mathbf{c} \notin \Lambda$.

A similar argument shows that if $\mathbf{b}$ is any point of $\mathbf{R}^{24}$ at a constant distance of $\sqrt{R}$ from all the $\mathbf{v}_j^{(i)}$ then $\mathbf{b} = \mathbf{c}$. Thus the center of the hole, and therefore the hole itself, is unique.

3. Holes whose diagram contains an A_n subgraph

In this section we use Theorem 7 to classify all holes whose diagram contains an A_n diagram as a subgraph, without any further assumption about the radius of the hole.

Theorem 8. *Any hole that has diagram containing a subgraph A_1 is of type A_1^{24}. Furthermore there is a unique hole of this type. We express these two statements by the formula*

$$A_1 \twoheadrightarrow A_1^{24} .$$

Proof. Since Co_∞, the automorphism group of Λ, is transitive on pairs of vectors $\mathbf{u}, \mathbf{v} \in \Lambda$ with $N(\mathbf{u}-\mathbf{v}) = 8$, there is essentially only one way to choose the vertices of the first A_1 diagram. Taking the origin at the center of the diagram we may assume these vertices to be $\mathbf{u} = (\sqrt{2}, 0, ..., 0)$, $\mathbf{v} = (-\sqrt{2}, 0, ..., 0)$. The origin is also at the center of 23 other A_1 diagrams, with vertices $(0, ..., 0, \pm\sqrt{2}, 0, ..., 0)$. Since these 48 vectors span $\mathbf{R}^{24}$, the desired conclusions follow from Theorem 7. When represented in the usual coordinates a center of a hole of this type has the form $\mathbf{c} = \delta(4, 0, ..., 0)$, where $\delta = 1/\sqrt{8}$.

Following Curtis [Cur2] we define an *$\mathcal{S}$-lattice* to be a sublattice L of Λ such that every vector of L is congruent modulo 2Λ to a vector in L of type 0, 2 or 3. An $\mathcal{S}$-lattice of type $2^i 3^j$ contains $2i$ vectors of type 2 and $2j$ of type 3.

Theorem 9. ([Cur2]). *There are* 12 *types of $\mathcal{S}$-lattice, and each occurs uniquely (up to isomorphism) inside Λ.*

Theorem 10.

$$A_2 \twoheadrightarrow A_2^{12} .$$

Proof. Consider a hole diagram containing an A_2 subgraph with vertices $\mathbf{u}, \mathbf{v}, \mathbf{w}$. Then the vectors $\mathbf{v}-\mathbf{u}$, $\mathbf{w}-\mathbf{u}$ span an $\mathcal{S}$-lattice of type 3^3. By the previous theorem we may take $\mathbf{u} = \mathbf{0}$, $\mathbf{v} = \delta(5, 1^{23})$, $\mathbf{w} = \delta(5, 1^{11}, -1^{12})$, with center $\mathbf{c} = \delta(10/3, (2/3)^{11}, 0^{12})$. Then $\mathbf{c}$ is the center of 11 other A_2's having vertices of the form

$$\mathbf{u}' = \delta(4, 4, 0^{10}, 0^6, 0^6) ,$$

$$\mathbf{v}' = \delta(3, -1, 1^{10}, 1^6, -1^6) ,$$

$$\mathbf{w}' = \delta(3, -1, 1^{10}, -1^6, 1^6) ,$$

etc. These are permuted by an element of order 11 fixing $\mathbf{u}$, $\mathbf{v}$ and $\mathbf{w}$. Again Theorem 7 completes the proof.

Theorem 11.

$$A_3 \twoheadrightarrow A_3^8 .$$

Proof. We take one vertex of A_3 as the origin and label the others **u**, **v**, **w**, and calculate the following inner products:

	u	**v**	**w**	**u+v−w**
u	6	2	4	4
v	2	4	2	4
w	4	2	6	0
u+v−w	4	4	0	8

Since $\mathbf{u}+\mathbf{v}-\mathbf{w}$ is of type 4, by Theorem 3 it belongs to a coordinate frame, and we may assume $\mathbf{u}+\mathbf{v}-\mathbf{w} = \delta(8, 0^{23})$ and $\mathbf{u} = \delta(4, \ldots)$. From the list of type 3 vectors given at the beginning of §2, $\mathbf{u} = \delta(4, (\pm 2)^8, 0^{15})$, where the 2's occupy an octad and there are an odd number of minus signs. It will be convenient from here on to write vectors in $\mathbf{R}^{24}$ in MOG format (see Chap. 11), and to omit the factor δ.

We may now assume without loss of generality that

$$\mathbf{u} = \begin{array}{|cc|cc|cc|}\hline -2 & 2 & 4 & 0 & 0 & 0 \\ 2 & 2 & 0 & 0 & 0 & 0 \\ 2 & 2 & 0 & 0 & 0 & 0 \\ 2 & 2 & 0 & 0 & 0 & 0 \\ \hline \end{array}$$

The subgroup of Co_∞ fixing the origin is the group Co_0, and the subgroup of Co_0 fixing **u** is $2^4 \mathsf{A}_7$ (A_n denotes an alternating group and S_n a symmetric group). To find **v** we use the facts that **v** is a type 2 vector, its inner product with **u** is 2, and that the group $2^4 \cdot \mathsf{A}_7$ is transitive on the last 15 coordinates of **u** to write

$$\mathbf{v} = \begin{array}{|cc|cc|cc|}\hline 0 & 0 & 4 & 0 & 4 & 0 \\ 0 & 0 & 0 & 0 & 0 & 0 \\ 0 & 0 & 0 & 0 & 0 & 0 \\ 0 & 0 & 0 & 0 & 0 & 0 \\ \hline \end{array},$$

and then

$$\mathbf{w} = \begin{array}{|cc|cc|cc|}\hline -2 & 2 & 0 & 0 & 4 & 0 \\ 2 & 2 & 0 & 0 & 0 & 0 \\ 2 & 2 & 0 & 0 & 0 & 0 \\ 2 & 2 & 0 & 0 & 0 & 0 \\ \hline \end{array}.$$

We conclude that there is essentially a unique A_3 diagram, which we can assume to have the vertices **0**, **u**, **v**, **w** given above. The center is

$$\mathbf{c} = \begin{array}{|cc|cc|cc|}\hline -1 & 1 & 2 & 0 & 2 & 0 \\ 1 & 1 & 0 & 0 & 0 & 0 \\ 1 & 1 & 0 & 0 & 0 & 0 \\ 1 & 1 & 0 & 0 & 0 & 0 \\ \hline \end{array}.$$

For the second part of the proof we observe that **c** is also the center of seven other A_3's such as that shown in Fig. 23.4, and which are based on octads meeting the left-hand octad of the MOG in four places. (The others are the images of the one shown in Fig. 23.4 under powers of the permutation α given in Fig. 23.8 below.) Again Theorem 7 completes the proof.

Theorem 12.

$$A_4 \Rightarrow A_4^6 .$$

Proof. We label the vertices **0**, **u**, **v**, **w**, **x**. By examining inner products (as in the previous proof) we find that **u**, **v**, **w**, **x** generate an $\mathcal{S}$-lattice of type $2^5 3^{10}$, which is unique by Theorem 9. Therefore we may assume that the vertices are labeled as in Fig. 23.5a, with center

$$c = (1/5) \left[\begin{array}{cc|cc|cc} 4 & 4 & 2 & 6 & 6 & 6 \\ 8 & 4 & 2 & 2 & 2 & 2 \\ 4 & 4 & 6 & 2 & 2 & 2 \\ 4 & 4 & 6 & 2 & 2 & 2 \end{array}\right] .$$

This is also the center of five other A_4's, namely the one shown in Fig. 23.5b and its images under the following element of order 5 in Co_0

$$\left[\begin{array}{cc|cc|cc} \cdot & \cdot & b_0 & c_0 & c_2 & c_3 \\ \cdot & a_0 & \cdot & d_0 & b_4 & b_1 \\ a_3 & a_4 & c_4 & d_3 & d_4 & b_2 \\ a_2 & a_1 & c_1 & d_2 & b_3 & d_1 \end{array}\right] .$$

(This has four fixed points indicated by dots and four cycles $(a_0, \ldots, a_4)$, $(b_0, \ldots, b_4), \ldots, (d_0, \ldots, d_4)$.)

The remaining A_n's can be handled uniformly. Consider a hole diagram containing an A_n, where n is large, labeled as in Fig. 23.6 and

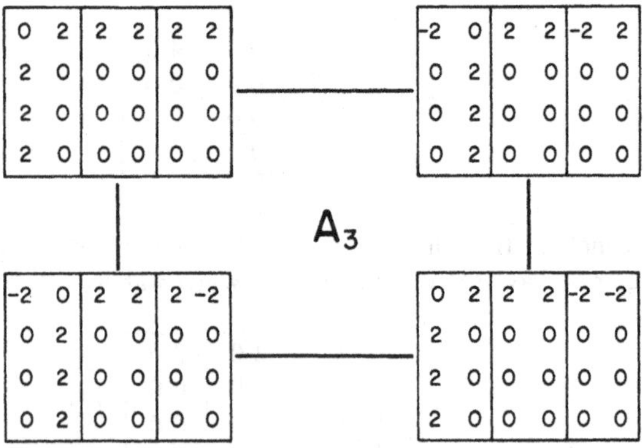

Figure 23.4. One of seven other A_3's used in the proof of Theorem 11.

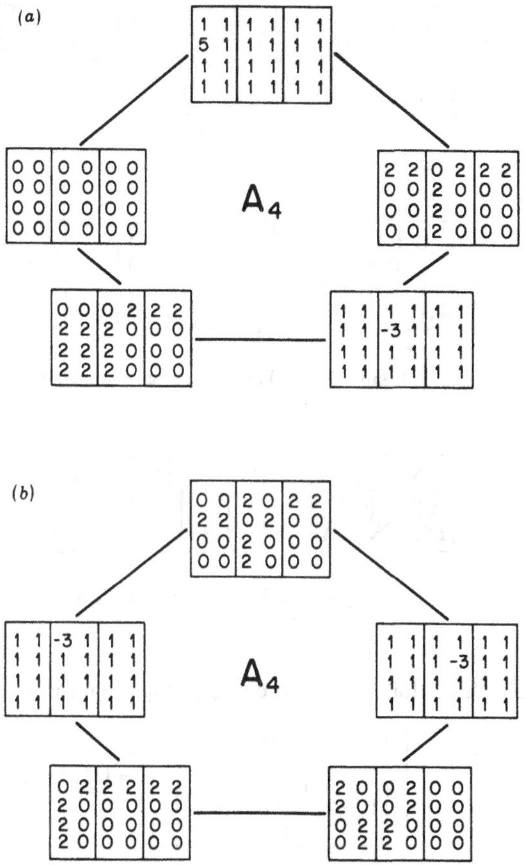

Figure 23.5. Some vertices of the A_4^6 hole.

with $\mathbf{v}_0 = \mathbf{0}$. As in Theorem 11 we examine the inner products and discover that $\mathbf{v}_1 + \mathbf{v}_2 - \mathbf{v}_4$ is of type 4, and so can be taken to be $\delta(8, 0^{23})$. Proceeding as before we find that $\mathbf{v}_1, \ldots, \mathbf{v}_5$ can be taken to be those shown in Fig. 23.7. The subgroup of Co_0 fixing $\mathbf{v}_1, \ldots, \mathbf{v}_5$ is a $PSL_2(7)$ generated by the permutations α, β, γ shown in Fig. 23.8.

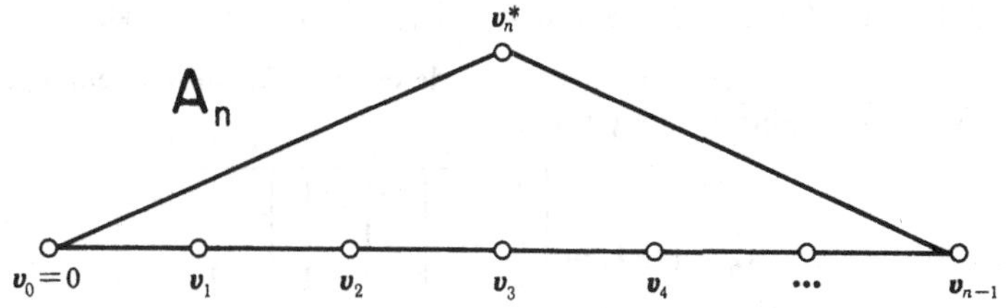

Figure 23.6. Labels for the nodes of an A_n diagram.

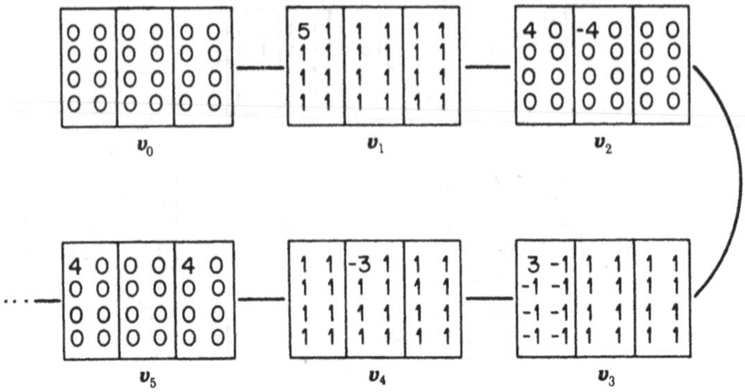

Figure 23.7. The beginning of the A_n-tree.

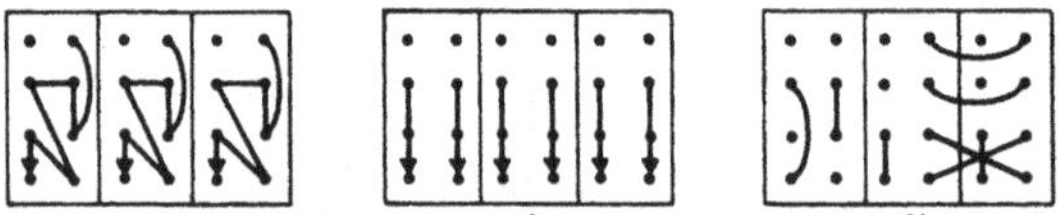

Figure 23.8. Generators for $PSL_2(7)$.

From this it follows that each $\mathbf{v}_m$ in Fig. 23.6 for $m \geqslant 6$ is of the form

3	−1
$(1^4, -1^3)$	$(1^{12}, -1^3)$

$= [ijk]$ (say) ,

where i, j, k are the locations of the -1's in the left octad with the following labeling:

∞ 0		
3 2		
5 1		
6 4		

Furthermore $N(\mathbf{v}_5 - [ijk]) = 6$ if and only if $[ijk]$ is one of [124], [235], [346], [450], [561], [602] or [013] (corresponding to the *lines* in the projective plane of order 2 with the usual labeling). Also we have $N([ijk] - [rst]) = 6$ if and only if $\{i,j,k\} \cap \{r,s,t\} = \varnothing$.

These remarks make it easy to determine the subsequent $\mathbf{v}_m$. Without loss of generality we may take

$\mathbf{v}_6 = [124] =$

3	1	−1	−1	−1	−1
1	−1	1	1	1	1
1	−1	1	1	1	1
1	−1	1	1	1	1

,

$\mathbf{v}_7 = [356] =$

3 1	−1 1	1 1
−1 1	−1 1	1 1
−1 1	−1 1	1 1
−1 1	−1 1	1 1

,

$\mathbf{v}_8 = [012]$ and $\mathbf{v}_9 = [345]$. Now only the identity element of Co_0 fixes $\mathbf{v}_1, \ldots, \mathbf{v}_9$, and the remaining $\mathbf{v}_m$ are not unique (see Fig. 23.9). We summarize these results in a theorem.

Theorem 13. *All possibilities for the vertices* $\mathbf{v}_0, \ldots, \mathbf{v}_{n-1}$ *of an* A_n *diagram are displayed in the* A_n*-tree of Fig.* 23.9.

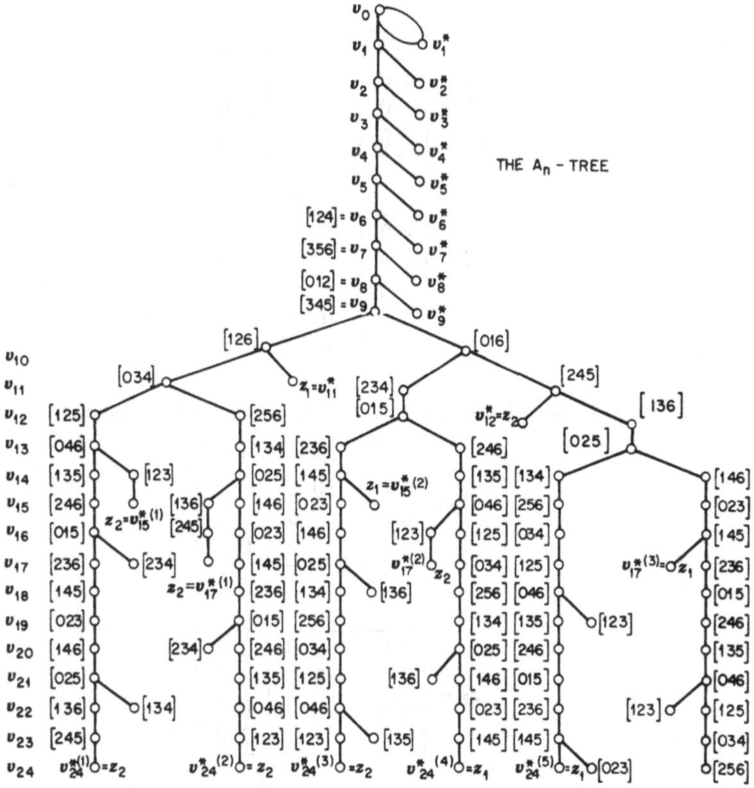

Figure 23.9. The A_n-tree.

So far we have mentioned only n of the $n+1$ nodes in the A_n diagram. We must now consider how to choose the last node $\mathbf{v}_n^*$. For $A_5, \ldots, A_9$ there is a unique choice for $\mathbf{v}_n^*$, as follows:

$\mathbf{v}_5^* =$

5 −1	−1 1	1 1
1 −1	−1 1	1 1
1 −1	−1 1	1 1
1 −1	−1 1	1 1

,

$$\mathbf{v}_6^* = \begin{array}{|cc|cc|cc|} \hline 4 & 0 & -2 & 2 & 0 & 0 \\ 0 & 0 & 2 & 2 & 0 & 0 \\ 0 & 0 & 2 & 2 & 0 & 0 \\ 0 & 0 & 2 & 2 & 0 & 0 \\ \hline \end{array},$$

$$\mathbf{v}_7^* = \begin{array}{|cc|cc|cc|} \hline 4 & 0 & -2 & 2 & 2 & 2 \\ 0 & 0 & 2 & 2 & 2 & 2 \\ 0 & 0 & 0 & 0 & 0 & 0 \\ 0 & 0 & 0 & 0 & 0 & 0 \\ \hline \end{array},$$

$$\mathbf{v}_8^* = \begin{array}{|cc|cc|cc|} \hline 4 & 0 & -2 & 0 & 2 & 0 \\ 0 & 0 & 2 & 0 & 2 & 0 \\ 0 & 0 & 2 & 0 & 2 & 0 \\ 0 & 0 & 2 & 0 & 2 & 0 \\ \hline \end{array},$$

$$\mathbf{v}_9^* = \begin{array}{|cc|cc|cc|} \hline 4 & 0 & -2 & 0 & 2 & 0 \\ 0 & 0 & 2 & 0 & 2 & 0 \\ 0 & 0 & 0 & 2 & 0 & 2 \\ 0 & 0 & 0 & 2 & 0 & 2 \\ \hline \end{array}.$$

For A_n ($n \geqslant 10$) there are at most two choices for $\mathbf{v}_n^*$, namely

$$\mathbf{z}_1 = \begin{array}{|cc|cc|cc|} \hline 4 & 0 & -2 & 0 & 2 & 0 \\ 0 & 0 & 0 & 2 & 0 & 2 \\ 0 & 0 & 2 & 0 & 2 & 0 \\ 0 & 0 & 0 & 2 & 0 & 2 \\ \hline \end{array},$$

$$\mathbf{z}_2 = \begin{array}{|cc|cc|cc|} \hline 4 & 0 & -2 & 0 & 2 & 0 \\ 0 & 0 & 0 & 2 & 0 & 2 \\ 0 & 0 & 0 & 2 & 0 & 2 \\ 0 & 0 & 2 & 0 & 2 & 0 \\ \hline \end{array},$$

and these are also shown in Fig. 23.9. Inspection of the tree reveals that we have proved the following theorem (the assertions about the $A_1, \ldots, A_4$ holes having been established earlier).

Theorem 14. *There is a unique A_n diagram in Λ for $n = 1, \ldots, 9, 11, 12$. There are at most two A_{15}'s, three A_{17}'s and five A_{24}'s, and none of type A_n for $n = 10, 13, 14, 16, 18, \ldots, 23$.*

It remains to consider the holes containing these diagrams. We postpone consideration of A_5 until Theorem 18.

Theorem 15.

$$
\begin{aligned}
A_6 &\twoheadrightarrow A_6^4 && (Fig.\ 23.10)\ ,\\
A_7 &\twoheadrightarrow A_7^2 D_5^2 && (Fig.\ 23.11)\ ,\\
A_8 &\twoheadrightarrow A_8^3 && (Fig.\ 23.12)\ ,\\
A_9 &\twoheadrightarrow A_9^2 D_6 && (Fig.\ 23.13)\ ,\\
A_{11} &\twoheadrightarrow A_{11} D_7 E_6 && (Fig.\ 23.14)\ ,\\
A_{12} &\twoheadrightarrow A_{12}^2 && (Fig.\ 23.15)\ .
\end{aligned}
$$

Proof ($A_{12} \twoheadrightarrow A_{12}^2$). The unique A_{12} diagram obtained above is shown in Fig. 23.15a and has center

$$
\mathbf{c} = (1/13) \left|\begin{array}{cc|cc|cc}
39 & 3 & -13 & 5 & 13 & 5\\
3 & 1 & 5 & 9 & 7 & 9\\
1 & 1 & 7 & 9 & 11 & 7\\
3 & 1 & 7 & 7 & 7 & 9
\end{array}\right| .
$$

To find an A_{12} disjoined from this diagram we need only consider vectors $[ijk]$ that are not lines and that intersect the triples $[i'j'k']$ occurring in the first A_{12} in one or two places. The result is shown in Fig. 23.15b. Theorem 6 then completes the proof. A similar argument applies to the other cases, the corresponding centers being

$$
A_6^4 : (1/7) \left|\begin{array}{cc|cc|cc}
21 & 1 & -7 & 5 & 7 & 3\\
1 & 1 & 5 & 5 & 3 & 3\\
1 & 1 & 5 & 5 & 3 & 3\\
1 & 1 & 5 & 5 & 3 & 3
\end{array}\right| ,
$$

$$
A_7^2 D_5^2 : (1/4) \left|\begin{array}{cc|cc|cc}
12 & 1 & -4 & 2 & 4 & 2\\
1 & 0 & 3 & 3 & 3 & 3\\
1 & 0 & 2 & 2 & 2 & 2\\
1 & 0 & 2 & 2 & 2 & 2
\end{array}\right| ,
$$

$$
A_8^3 : (1/9) \left|\begin{array}{cc|cc|cc}
27 & 3 & -9 & 3 & 9 & 3\\
1 & 1 & 5 & 5 & 7 & 5\\
1 & 1 & 5 & 5 & 7 & 5\\
1 & 1 & 5 & 5 & 7 & 5
\end{array}\right| ,
$$

$$
A_9^2 D_6 : (1/5) \left|\begin{array}{cc|cc|cc}
15 & 1 & -5 & 2 & 5 & 2\\
1 & 0 & 3 & 3 & 3 & 3\\
1 & 0 & 2 & 4 & 3 & 3\\
1 & 1 & 2 & 3 & 3 & 4
\end{array}\right| ,
$$

$$A_6^4$$

(a) $v_0 - v_1 - v_2 - v_3 - v_4 - v_5 - v_6^*$

(b) $[012]-[345]-[016]-[234]-[056]-[123]-[456]$

$[014]-[236]-[145]-[036]-[125]-[034]-[256]$

$[025]-[136]-[024]-[135]-[246]-[035]-[146]$

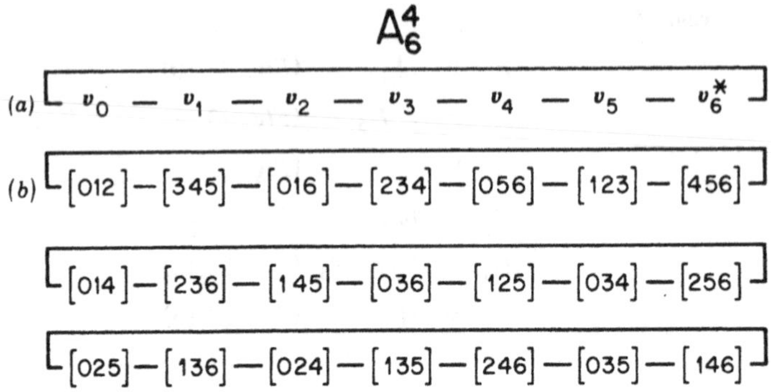

Figure 23.10. Vertices of the A_6^4 hole.

$$A_7^2 \, D_5^2$$

(a) $v_0 - v_1 - v_2 - v_3 - v_4 - v_5 -[124]- v_7^*$

(b) $[012]-[456]-[123]-[046]-[125]-[034]-[126]-[345]$

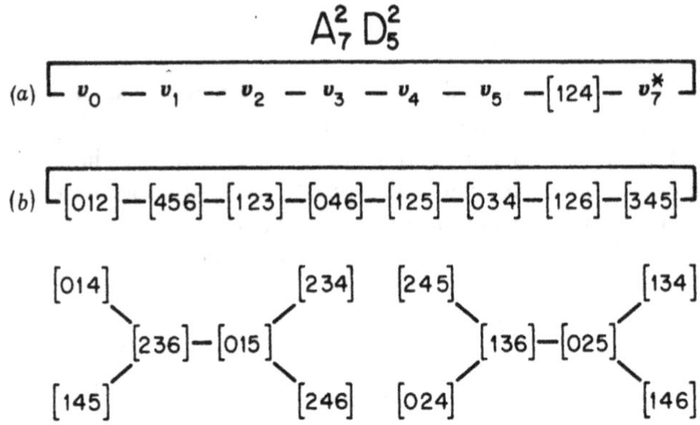

Figure 23.11. The $A_7^2 D_5^2$ hole.

(a)
$v_0 - v_1 - v_2 - v_3 - v_4 - v_5 -[124]-[356] - v_8^*$

(b)
$[015]\quad [236]\quad [145]\quad [023]\quad [456]\quad [123]\quad [046]\quad [135]\quad [246]$

$[016]\quad [245]\quad [136]\quad [025]\quad [134]\quad [256]\quad [034]\quad [126]\quad [345]$

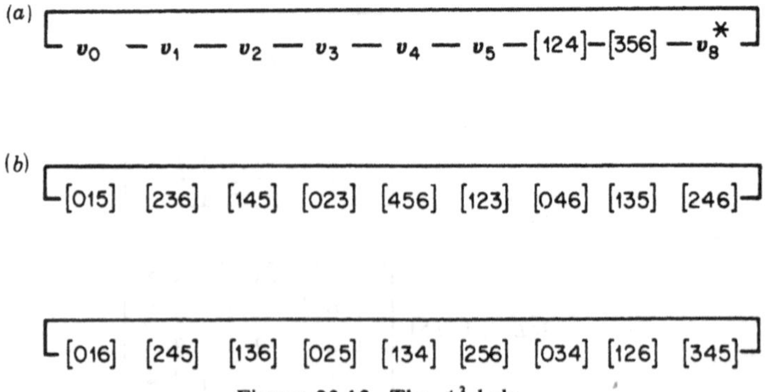

Figure 23.12. The A_8^3 hole.

(a) $v_0 - v_1 - v_2 - v_3 - v_4 - v_5 - [124] - [356] - [012] - v_9^*$

(b) $[015]\ [234]\ [016]\ [245]\ [136]\ [025]\ [146]\ [023]\ [145]\ [236]$

$$[126] \qquad\qquad [123]$$
$$[034] - [125] - [046]$$
$$[256] \qquad\qquad [135]$$

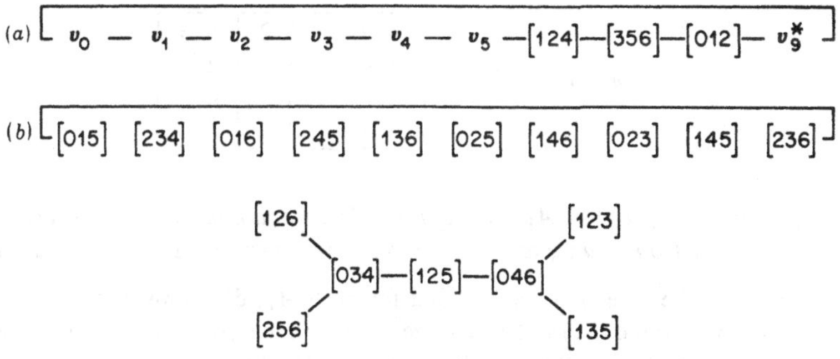

Figure 23.13. The $A_9^2 D_6$ hole, the unique hole containing D_6^g.

$$A_{11}$$

(a) $v_0\ v_1\ v_2\ v_3\ v_4\ v_5\ [124]\ [356]\ [012]\ [345]\ [126]\ z_1$

(b)

$$[023]$$
$$|$$
$$[146] \qquad E_6$$
$$|$$
$$[245] - [136] - [025] - [134] - [256]$$

$$[234] \qquad\qquad\qquad [123]$$
$$[015] - [246] - [135] - [046] \qquad D_7$$
$$[236] \qquad\qquad\qquad [125]$$

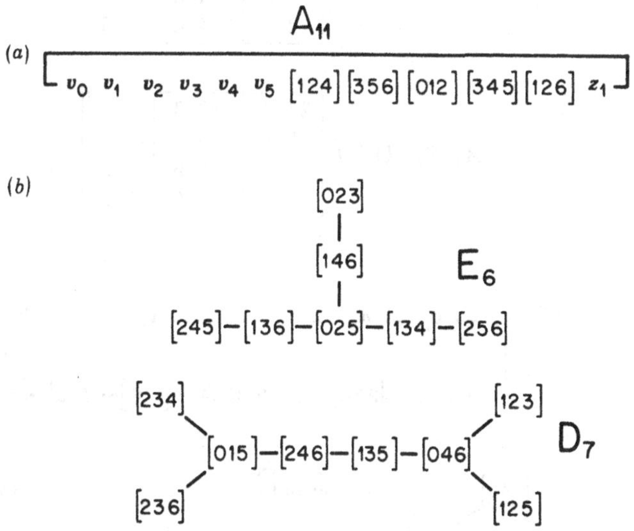

Figure 23.14. The $A_{11} D_7 E_6$ hole.

$$A_{12}^2$$

(a) $v_0\ v_1\ v_2\ v_3\ v_4\ v_5\ [124]\ [356]\ [012]\ [345]\ [016]\ [245]\ z_2$

(b) $[015]\ [236]\ [145]\ [023]\ [146]\ [025]\ [134]\ [256]\ [034]\ [125]\ [046]\ [135]\ [246]$

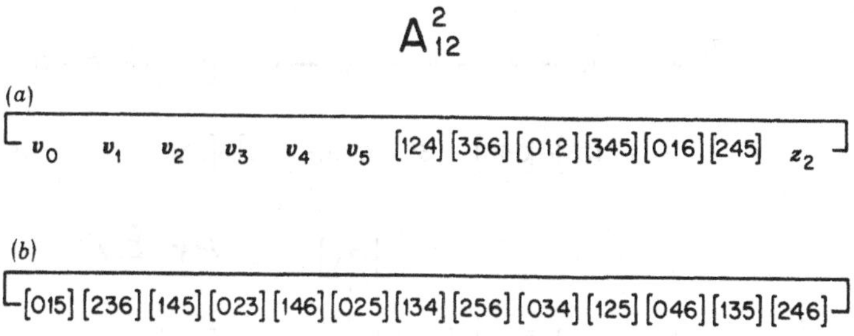

Figure 23.15. The A_{12}^2 hole.

$$A_{11}D_7E_6 : (1/6) \quad \begin{array}{cc|cc|cc} 18 & 2 & -6 & 2 & 6 & 3 \\ 1 & 0 & 3 & 4 & 3 & 5 \\ 1 & 0 & 3 & 4 & 4 & 4 \\ 1 & 1 & 3 & 3 & 4 & 3 \end{array} \;.$$

Theorem 16. *Every A_{15} is disjoined from a unique D_9, and every A_{17} is disjoined from a unique E_7 (examples are given in Figs. 23.16 and 23.17).*

Proof. The assertion was verified for both A_{15} diagrams and all three A_{17} diagrams found from the A_n-tree. (It will be seen later that there is a unique hole of each type.) The centers of the two diagrams illustrated are

$$A_{15}D_9 : (1/8) \quad \begin{array}{cc|cc|cc} 24 & 2 & -8 & 3 & 8 & 3 \\ 1 & 1 & 4 & 7 & 4 & 5 \\ 1 & 0 & 4 & 6 & 5 & 5 \\ 2 & 1 & 4 & 4 & 5 & 5 \end{array} \;,$$

$$A_{17}E_7 : (1/9) \quad \begin{array}{cc|cc|cc} 27 & 2 & -9 & 3 & 9 & 4 \\ 1 & 1 & 4 & 7 & 4 & 6 \\ 2 & 0 & 5 & 6 & 7 & 5 \\ 2 & 1 & 4 & 6 & 6 & 5 \end{array} \;.$$

(a)
$$v_0 - v_1 - v_2 - v_3 - v_4 - v_5 - [124]-[356]$$
$$z_1 - [145]-[236]-[015]-[234]-[016]-[345]-[012]$$

(b)
$$[135] \qquad A_{15}D_9 \qquad [136]$$
$$[046]-[125]-[034]-[256]-[134]-[025]$$
$$[123] \qquad\qquad\qquad\qquad [146]$$

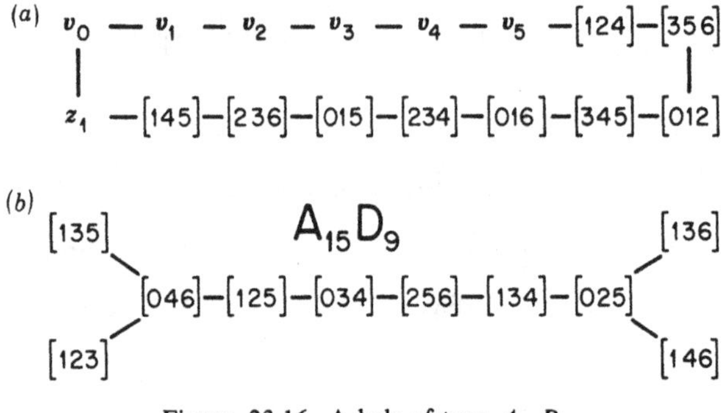

Figure 23.16. A hole of type $A_{15}D_9$.

(a)
$$v_0 - v_1 - v_2 - v_3 - v_4 - v_5 - [124]-[356]-[012]$$
$$z_2 - [123]-[046]-[135]-[246]-[015]-[234]-[016]-[345]$$

(b)
$$[136] \qquad A_{17}E_7$$
$$[145]-[023]-[146]-[025]-[134]-[256]-[034]$$

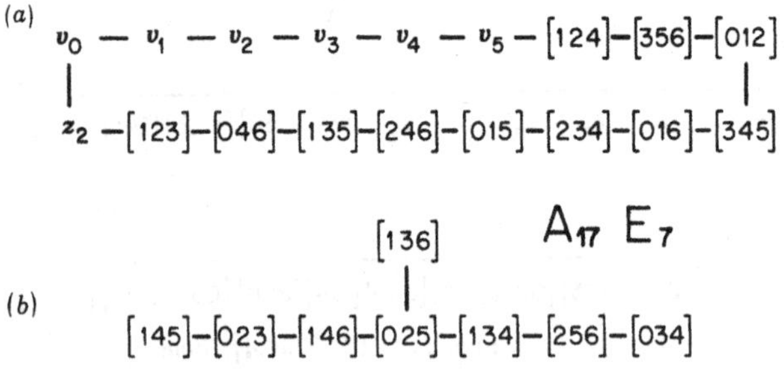

Figure 23.17. A hole of type $A_{17}E_7$.

4. Holes whose diagram contain a D_n subgraph

The diagrams containing a D_n subgraph can be handled uniformly, just as the A_n's were. For this purpose we label the positions in the MOG as shown in Fig. 23.18. There are three distinguished positions labeled I, II, III, the remaining 21 being identified with the points of a projective plane of order 4, as in §11 of Chap. 11. We take an oval C in that plane consisting of the points $\{\infty, 0, 1, 2, 3, 4\}$, and label the remaining fifteen points with *synthemes* (to use Sylvester's term [Syl1]) such as $\infty 0.14.23$. Of the 21 lines in the plane, 15 are *secants* which meet the oval in two points, and any such line is labeled with the *duad* of those two points. For example the duad $\infty 0$ meets the oval in ∞ and 0, and also contains the points

$$\infty 0.12.34, \quad \infty 0.13.24, \quad \infty 0.14.23$$

The remaining six lines are the *axes* of C, and do not meet the oval (see Fig. 23.19). They are labeled with *totals*, a total $a|bcdef$ being an abbreviation for the set of five synthemes $ad.ce.bf$, $ae.bc.df$, $af.be.cd$, $ab.cf.de$, $ac.bd.ef$, which are the points on the axis. The vertices in the D_n diagrams will involve the following vectors:

- $\emptyset$, the zero vector;
- $[P]$, where P is a point in the plane or I, II or III, has -3 at position P and 1's elsewhere;
- $[\hat{P}]$ has 5 at position P and 1's elsewhere;
- $[L]$, where L is a line in the plane, has 2's at the five points of the line and at I, II, III, and 0's elsewhere;

∞ 0 1 4 2 3	∞ 0 1 3 2 4	∞ 4 0 3 1 2	∞ 2 0 1 3 4	∞ 3 0 2 1 4	∞ 1 0 4 2 3
I	∞ 0 1 2 3 4	∞ 1 0 2 3 4	∞ 3 0 4 1 2	∞ 2 0 3 1 4	∞ 4 0 1 2 3
II	∞	∞ 2 0 4 1 3	∞ 4 0 2 1 3	1	3
III	0	∞ 3 0 1 2 4	∞ 1 0 3 2 4	4	2

Figure 23.18. Labels for positions in the MOG, based on an oval in the plane of order 4.

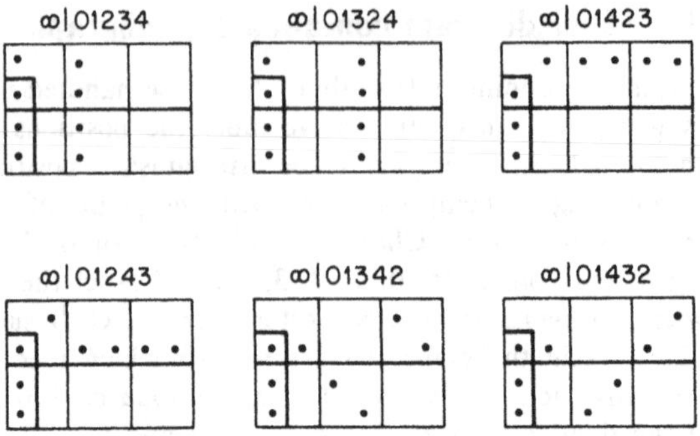

Figure 23.19. The six axes of C.

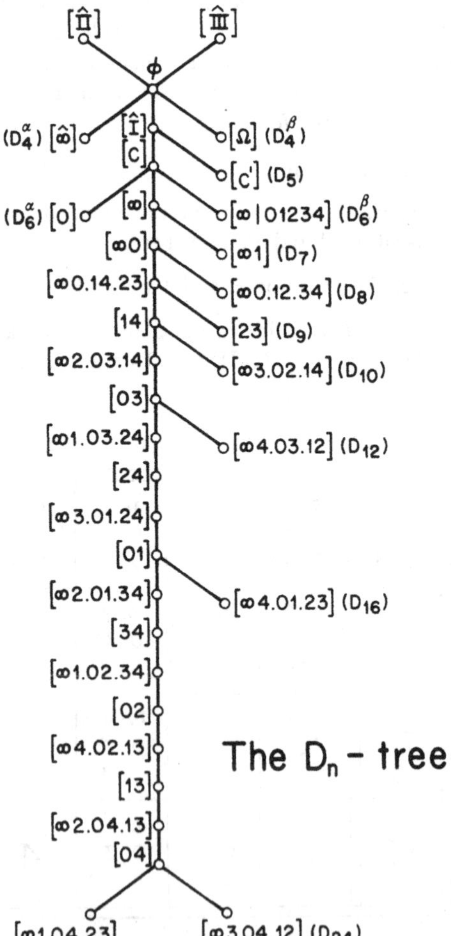

Figure 23.20. The D_n-tree.

$$[C] = \begin{array}{|cc|cc|cc|}\hline 0 & 0 & 0 & 0 & 0 & 0 \\ 0 & 0 & 0 & 0 & 0 & 0 \\ 2 & 2 & 0 & 0 & 2 & 2 \\ 2 & 2 & 0 & 0 & 2 & 2 \\ \hline \end{array} \; ;$$

$$[C'] = \begin{array}{|cc|cc|cc|}\hline 0 & 0 & 0 & 0 & 2 & 2 \\ 0 & 0 & 0 & 0 & 2 & 2 \\ 2 & 2 & 0 & 0 & 0 & 0 \\ 2 & 2 & 0 & 0 & 0 & 0 \\ \hline \end{array} \; ;$$

$$[\Omega] = \begin{array}{|cc|cc|cc|}\hline 2 & 0 & 2 & 0 & 0 & 0 \\ 2 & 0 & 2 & 0 & 0 & 0 \\ 2 & 0 & 0 & 2 & 2 & 2 \\ 2 & 0 & 0 & 2 & 2 & 2 \\ \hline \end{array} \; .$$

The edges in the diagrams can be found easily from the incidence relations in the plane. For example

$$\begin{aligned} N([P],[L]) &= 6 \quad \text{if and only if } P \text{ is on } L, \\ N([\hat{P}],[L]) &= 6 \quad \text{if and only if } P \text{ is not on } L, \\ N([P],[C]) &= 6 \quad \text{if and only if } P \text{ is on } C, \\ N([L],[C]) &= 6 \quad \text{if and only if } L \text{ is an axis of } C, \end{aligned}$$

Arguments similar to those used in the previous section now lead to the following theorems.

Theorem 17. *All possibilities for a D_n diagram are displayed in the D_n-tree of Fig. 23.20. There are at most two distinct D_4's and D_6's, the diagrams of type D_5, D_7, D_8, D_9, D_{10}, D_{12}, D_{16} and D_{24} are unique, and there is no diagram of type D_n for $n = 11, 13, 14, 15, 17, \ldots, 23$.*

Theorem 18.

$D_4 \Rightarrow D_4^6$ (*Fig.* 23.21) *or* $A_5^4 D_4$ (*Fig.* 23.22) ,

$D_5 \Rightarrow A_7^2 D_5^2$,

$D_6 \Rightarrow D_6^4$ (*Fig.* 23.23) *or* $A_9^2 D_6$,

$D_7 \Rightarrow A_{11} D_7 E_6$,

$D_8 \Rightarrow D_8^3$ (*Fig.* 23.24) ,

$D_9 \Rightarrow A_{15} D_9$,

$D_{10} \Rightarrow D_{10} E_7^2$ (*Fig.* 23.25) ,

$D_{12} \Rightarrow D_{12}^2$ (*Fig.* 23.26) ,

$D_{16} \Rightarrow D_{16} E_8$ (*Fig.* 23.27) ,

$D_{24} \Rightarrow D_{24}$.

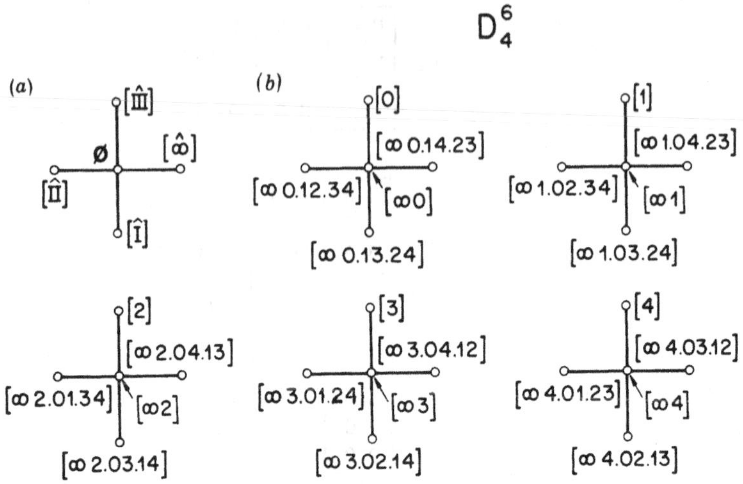

Figure 23.21. The D_4^6 hole, the unique hole containing D_4^α.

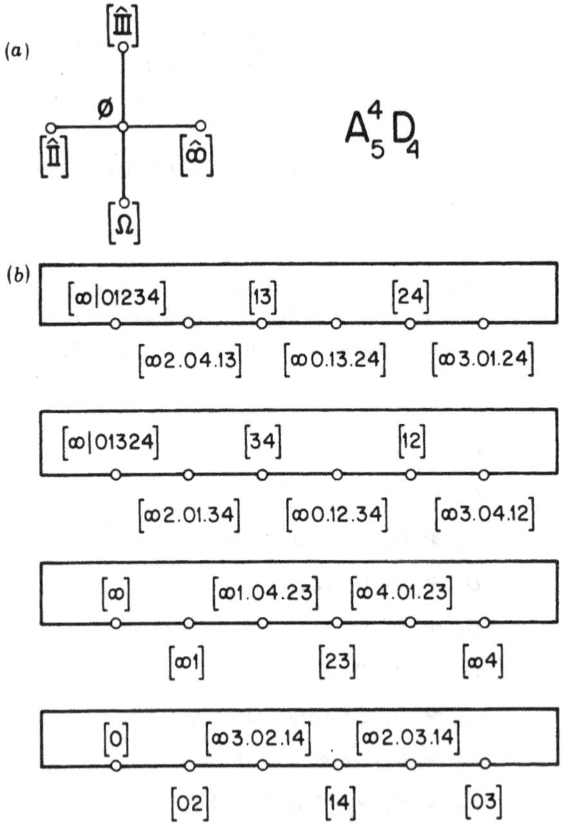

Figure 23.22. The $A_5^4 D_4$ hole, the unique hole containing D_4^β.

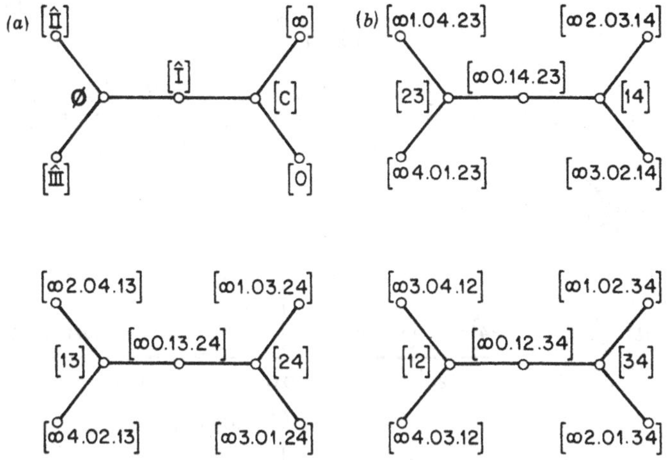

Figure 23.23. The D_6^4 hole, the unique hole containing D_6^{α}.

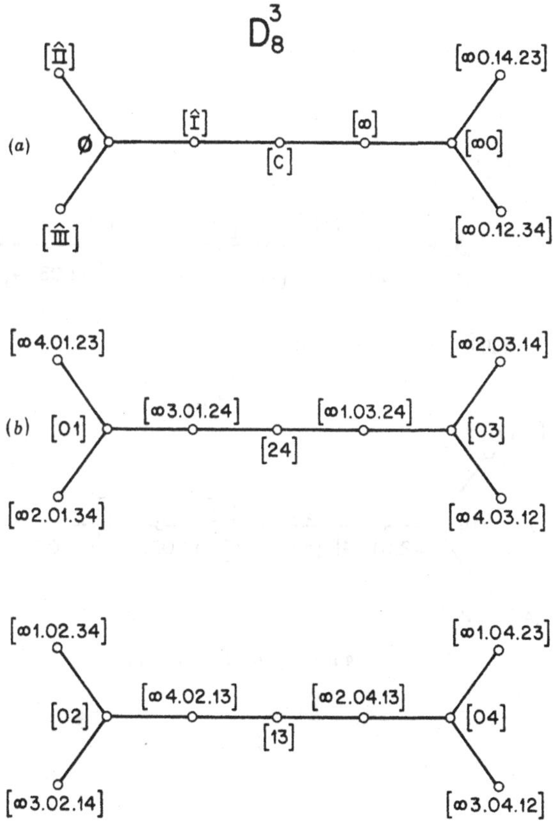

Figure 23.24. The D_8^3 hole.

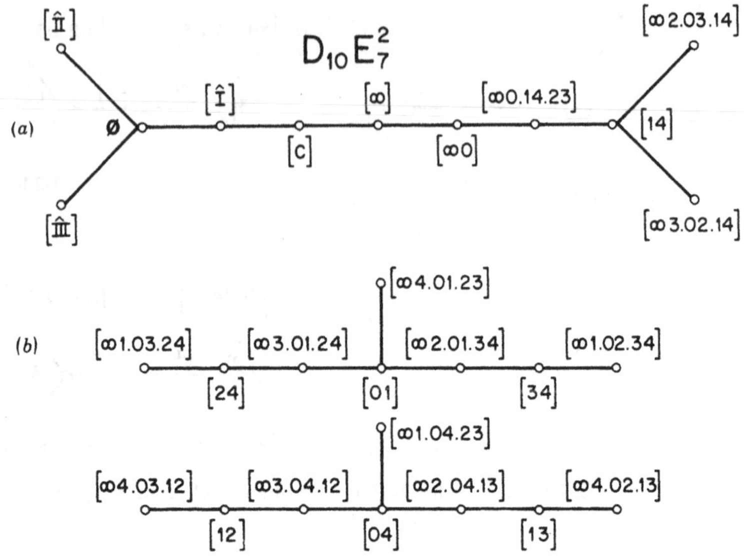

Figure 23.25. The $D_{10}E_7^2$ hole.

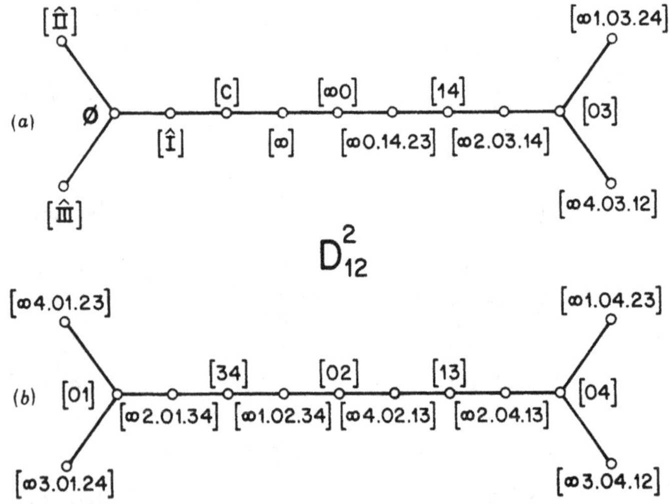

Figure 23.26. The D_{12}^2 hole.

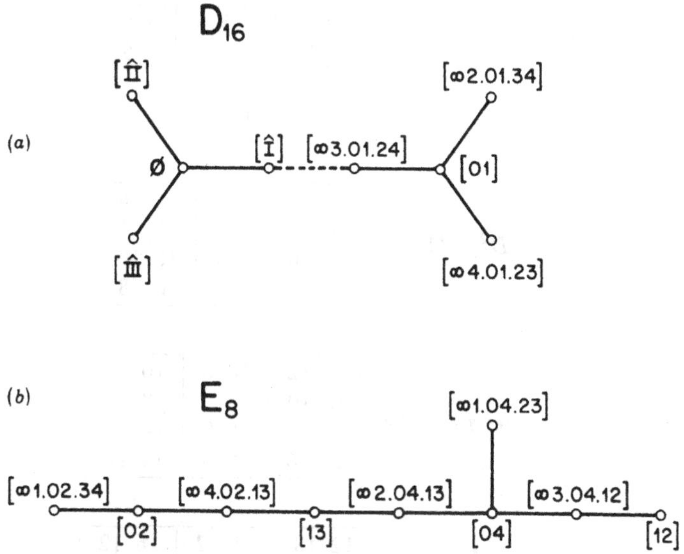

Figure 23.27. The $D_{16}E_8$ hole. (a) D_{16} (from the D_n-tree), (b) E_8.

There is no need to give diagrams for the holes of types $A_7^2 D_5^2$, $A_9^2 D_6$, $A_{11}D_7E_6$ or $A_{15}D_9$ since we have already seen these holes from a different point of view in §3. The D_{24} diagram is the main trunk of the D_n-tree in Fig. 23.20. The centers of these holes are:

D_4^6 : (1/3)

2 2	2 2	2 2
4 2	2 2	2 2
4 4	2 2	2 2
4 2	2 2	2 2

,

$A_5^4 D_4$: (1/6)

5 3	5 3	3 3
9 3	5 3	3 3
9 3	3 5	5 5
9 3	3 5	5 5

,

D_6^4 : (1/5)

3 3	3 3	3 3
7 3	3 3	3 3
7 3	3 3	5 5
7 3	3 3	5 5

,

D_8^3 : (1/7)

4 6	4 4	4 4
10 4	4 4	4 4
10 4	4 4	6 6
10 8	4 4	6 6

,

$$D_{10}E_7^2 : (1/9) \quad \left|\begin{array}{cc|cc|cc} 5 & 7 & 5 & 5 & 5 & 5 \\ 13 & 7 & 5 & 5 & 5 & 5 \\ 13 & 5 & 5 & 5 & 9 & 7 \\ 13 & 9 & 5 & 5 & 9 & 7 \end{array}\right| ,$$

$$D_{12}^2 : (1/11) \quad \left|\begin{array}{cc|cc|cc} 6 & 8 & 6 & 6 & 8 & 6 \\ 16 & 8 & 6 & 6 & 6 & 6 \\ 16 & 6 & 6 & 6 & 10 & 10 \\ 16 & 12 & 6 & 6 & 10 & 8 \end{array}\right| ,$$

$$D_{16}E_8 : (1/15) \quad \left|\begin{array}{cc|cc|cc} 8 & 12 & 10 & 8 & 10 & 8 \\ 22 & 10 & 8 & 8 & 8 & 8 \\ 22 & 8 & 8 & 8 & 14 & 12 \\ 22 & 16 & 8 & 8 & 14 & 12 \end{array}\right| ,$$

$$D_{24} : (1/23) \quad \left|\begin{array}{cc|cc|cc} 12 & 18 & 14 & 12 & 16 & 12 \\ 34 & 16 & 12 & 12 & 12 & 14 \\ 34 & 12 & 12 & 12 & 20 & 20 \\ 34 & 24 & 12 & 12 & 22 & 18 \end{array}\right| .$$

Theorem 19. *There is a unique A_{15} diagram and a unique hole of type $A_{15}D_9$.*

Proof. D_9 is unique (Theorem 17) and is disjoined from A_{15} (Theorem 18), so by Theorem 7 the hole is unique. From Theorem 16 each A_{15} is disjoined from a D_9, so A_{15} is unique.

5. Holes whose diagram contains an E_n subgraph

Theorem 20. *There are two possible E_8 diagrams in Λ.*

Proof. From the A_n-tree the backbone of an E_8 diagram can be assumed to be $\mathbf{v}_0, \mathbf{v}_1, \ldots, \mathbf{v}_5$, [124], [356]. The remaining node is an $[ijk]$ which since it is joined to $\mathbf{v}_5$ must be a line. The subgroup of Co_0 fixing $\mathbf{v}_0, \ldots,$ [356] is an S_3 generated by β and γ (see Fig. 23.8), and has three orbits on lines, with representatives [045], [346] and [124]. Since [124] has already appeared in the diagram only two possibilities remain.

The next theorem, whose proof follows the standard pattern, shows that both possibilities occur and are distinct.

Theorem 21.

$$E_8 \Rightarrow E_8^3 \text{ (Fig. 23.28) or } D_{16}E_8 .$$

We have already encountered the latter hole in §4. The center of the E_8^3 hole is

$$\mathbf{c} = (1/15) \left[\begin{array}{cc|cc|cc} 45 & 6 & -15 & 6 & 15 & 6 \\ 1 & 2 & 8 & 7 & 10 & 7 \\ 4 & 2 & 8 & 10 & 10 & 10 \\ 1 & -1 & 8 & 10 & 10 & 10 \end{array}\right].$$

Similar arguments lead to Theorems 22 and 24.

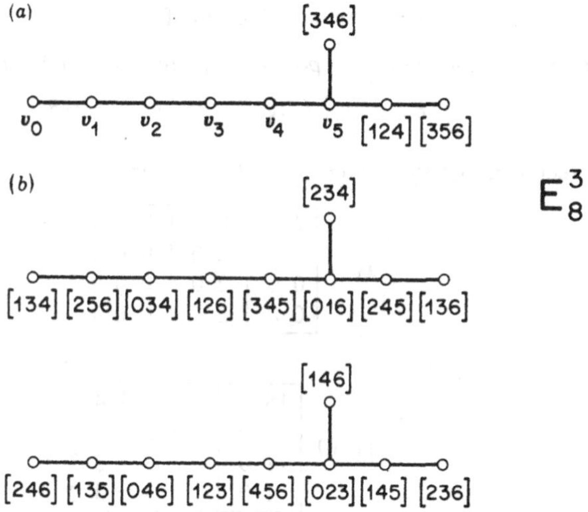

Figure 23.28. The E_8^3 hole.

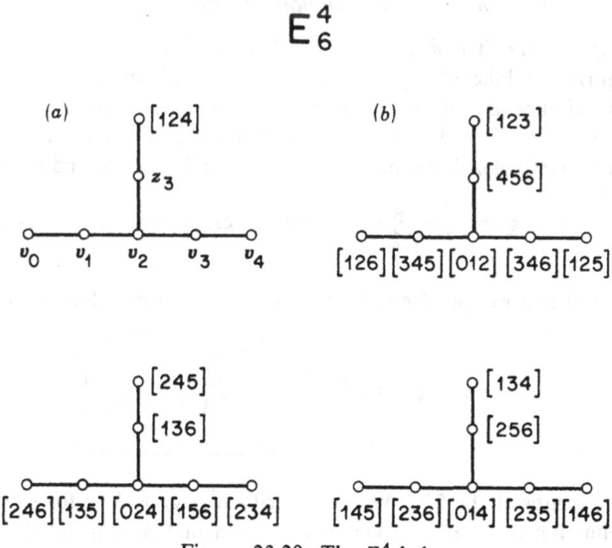

Figure 23.29. The E_6^4 hole.

Theorem 22. *There are two possible E_7 diagrams in Λ, and*

$$E_7 \Rightarrow D_{10}E_7^2 \text{ or } A_{17}E_7 .$$

Theorem 23. *There is a unique A_{17} diagram and a unique hole of type $A_{17}E_7$.*

Proof. Every A_{17} is disjoined from an E_7 (Theorem 16) and every D_{10} is disjoined from two E_7's (Theorem 18). From Theorem 22 this accounts for all E_7's, so there is a unique diagram of type A_{17}.

Theorem 24. *There are two possible E_6 diagrams in Λ, and*

$$E_6 \Rightarrow E_6^4 \text{ (Fig. 23.29) or } A_{11}D_7E_6 .$$

The vector $\mathbf{z}_3$ and the center of the E_6^4 hole are

$$\mathbf{z}_3 = \begin{array}{|cc|cc|cc|} 2 & 2 & 0 & 2 & 2 & 2 \\ 0 & 0 & 2 & 0 & 0 & 0 \\ 0 & 0 & 2 & 0 & 0 & 0 \\ 0 & 0 & 2 & 0 & 0 & 0 \end{array} ,$$

$$\mathbf{c} = (1/6) \begin{array}{|cc|cc|cc|} 18 & 3 & -6 & 4 & 4 & 4 \\ 1 & 0 & 5 & 3 & 3 & 3 \\ 1 & 0 & 5 & 3 & 3 & 3 \\ 1 & 0 & 5 & 3 & 3 & 3 \end{array} .$$

Now only the A_{24} holes are left.

Theorem 25. *There is a unique hole of type A_{24}.*

Proof. After Theorem 14, it is known that Λ contains at most five inequivalent labeled A_{24} diagrams. We check in the following way that these diagrams all arise from the same type of hole. Let $\mathbf{f}_0 = \mathbf{v}_0$, $\mathbf{f}_1 = \mathbf{v}_1, \ldots, \mathbf{f}_{21} = [146]$, $\mathbf{f}_{22} = [023]$, $\mathbf{f}_{23} = [145]$, $\mathbf{f}_{24} = \mathbf{z}_1$ be the vertices of the fourth A_{24} diagram in the A_n-tree in Fig. 23.9, and define

$$\mathbf{g}_r = \frac{1}{25} \sum_{i=0}^{24} i(25-i)(\mathbf{f}_{r+c} - \mathbf{c}), \quad r = 0, 1, \ldots, 24 ,$$

where the subscripts are to be read modulo 25 throughout this proof, and

$$\mathbf{c} = (1/25) \begin{array}{|cc|cc|cc|} 75 & 7 & -25 & 9 & 25 & 11 \\ 3 & 1 & 13 & 17 & 13 & 17 \\ 5 & 1 & 13 & 15 & 23 & 13 \\ 5 & 3 & 11 & 15 & 15 & 15 \end{array}$$

is the center of the $\mathbf{f}_i$'s. We then verify that $\mathbf{g}_r \in \Lambda - \mathbf{c}$ for $r = 0, 5, 10, 15, 20$, but not for the other values of r. (In fact Λ is spanned by the vectors $\{\mathbf{f}_i - \mathbf{f}_0, \mathbf{g}_{5r} - \mathbf{f}_0\}$. This is one of the 23 "holy constructions" of the Leech lattice — see the next chapter.) Therefore no element of Co_∞ can send $\mathbf{f}_0$ to $\mathbf{f}_t$, $\mathbf{f}_1$ to $\mathbf{f}_{t+1}, \ldots, \mathbf{f}_{24}$ to $\mathbf{f}_{24+t}$ for $t = 1, 2, 3$, or 4, and so there are five labeled A_{24} diagrams which are inequivalent under Co_∞. This accounts for all the diagrams, and completes the proof.

We have found a unique hole for every one of the Niemeier lattices (see Table 23.1), proving Theorems 2 and 4. Each of these holes has radius $\sqrt{2}$, so the proof of Theorem 1 is also complete.

24

Twenty-Three Constructions for the Leech Lattice

J. H. Conway and N. J. A. Sloane

In the previous chapter we classified the points at maximal distance from the Leech lattice (the "deep holes" in that lattice), and showed that there are 23 classes of such holes, the classes being in one-to-one correspondence with the 23 Niemeier lattices in 24 dimensions. We now present 23 constructions for the Leech lattice, one for each class of hole or Niemeier lattice. Two of these are the usual constructions of the Leech lattice from the Golay codes over $\mathbf{F}_2$ and $\mathbf{F}_3$.

1. The "holy constructions"

For each of the 23 Niemeier lattices or classes of deep hole there is a "holy construction" of the Leech lattice. In each case we shall define a set of *fundamental vectors* (f_i) and a set of *glue vectors* (g_w). It then turns out that the Niemeier lattice is the set of all integer combinations

$$\sum m_i f_i + \sum n_w g_w \text{ with } \sum n_w = 0 \ , \tag{1}$$

while the set of all integer combinations

$$\sum m_i f_i + \sum n_w g_w \text{ with } \sum m_i + \sum n_w = 0 \tag{2}$$

is a copy of the Leech lattice.

We begin by defining certain fundamental vectors f_i and glue vectors g_x for each root lattice, as shown in Fig. 24.1. The f_i form an extended fundamental set of roots (or basis) for the lattice (compare Figs. 4.5-4.9 of Chap. 4), and are described abstractly by specifying that their inner products should satisfy

$(f_i, f_i) = 2,$
$(f_i, f_j) = -$ (the number of lines joining f_i to f_j in the Coxeter-Dynkin diagram), for $i \neq j$.

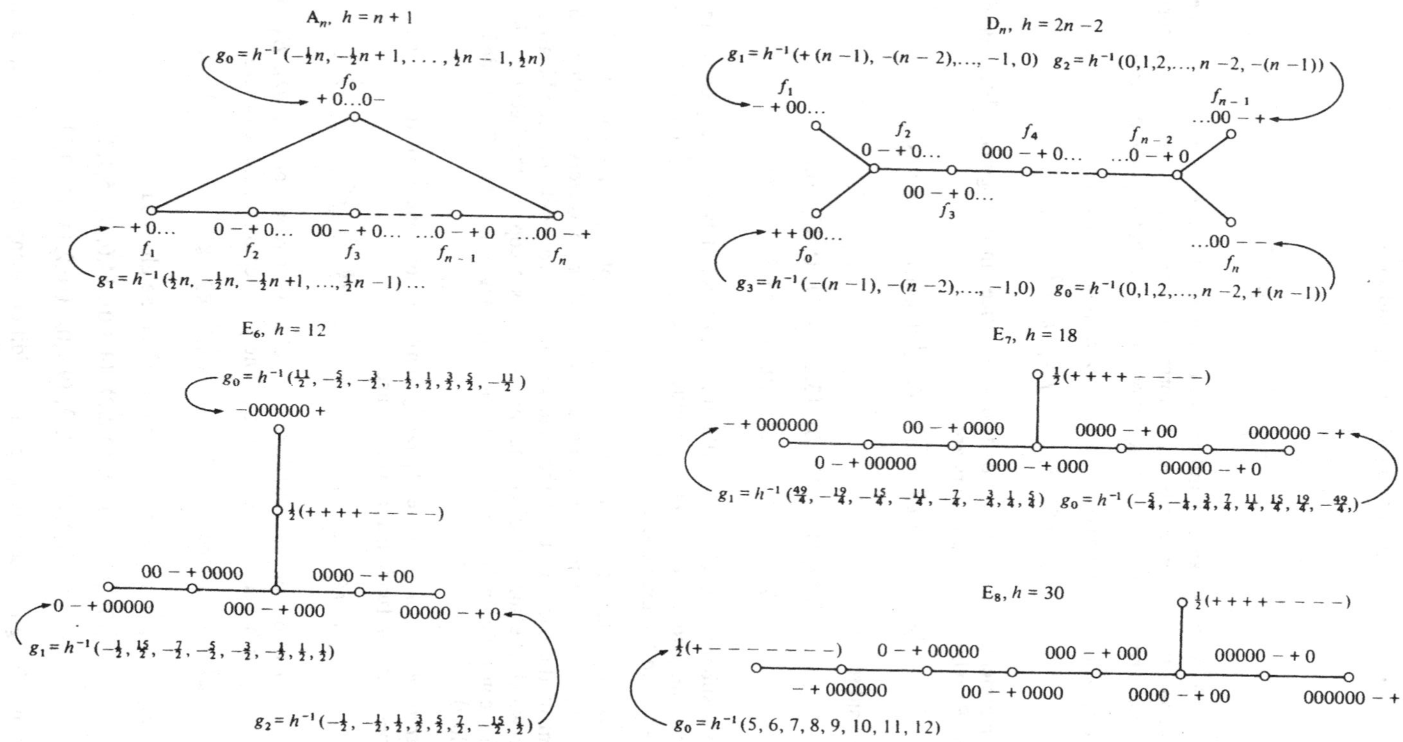

Figure 24.1. Fundamental vectors f_i, glue vectors g_x, and Coxeter number h for the root lattices $A_n (n \geqslant 1)$, $D_n (n \geqslant 4)$, E_6, E_7 and E_8 (plus and minus signs stand for +1 and −1 respectively).

Correction. In the E_8 diagram, the coordinates of the top node should be ½(+++++−−−).

Each glue vector g_x points to a certain f_i, and satisfies

$$(g_x, f_i) = \frac{1}{h} - 1 \quad \text{for this } f_i \ ,$$

$$(g_x, f_j) = \frac{1}{h} \quad \text{for the other } f_j \ ,$$

where h is the Coxeter number (§2 of Chap. 4). The f_i and g_x for the components are then combined to form the fundamental and glue vectors for the corresponding holy construction, in a way that we now illustrate.

Example 1. For the case A_{12}^2 the fundamental vectors are

$$(f_i) \quad f_0^A, f_1^A, \ldots, f_{12}^A, f_0^B, f_1^B, \ldots, f_{12}^B$$

(fundamental sets of roots for two A_{12} lattices in orthogonal 12-spaces A and B) while the glue vectors are

$$(g_w) \quad g_0^A + g_0^B, g_1^A + g_5^B, g_2^A + g_{10}^B, \ldots, g_{12}^A + g_8^B \ .$$

The subscripts

$$\{(0,0), (1,5), (2,10), (3,2), (4,7), (5,12), (6,4),$$
$$(7,9), (8,1), (9,6), (10,11), (11,3), (12,8)\}$$

are a subgroup of $C_{13} \times C_{13}$ (the direct product of two cyclic groups) and form the *glue code*. There is a glue vector $g_w = g_x^A + g_y^B$ for each word $w = xy$ in the glue code.

We now assert that the Niemeier lattice of type A_{12}^2 consists of all integer combinations of the fundamental and glue vectors for which the sum of the coefficients of the glue vectors is zero, while the combinations for which the sum of *all* coefficients is zero is a copy of the Leech lattice. This construction is described by the hole diagram in Fig. 24.2a (defined below).

Example 2. For the case $A_{11}D_7E_6$ there are components of three distinct types, and the fundamental vectors are

$$(f_i) \quad f_0^A, f_1^A, \ldots, f_{12}^A, f_0^B, f_1^B, \ldots, f_7^B, f_0^C, f_1^C, \ldots, f_6^C \ ,$$

the superscripts picking out fundamental sets of roots for A_{11}, D_7, E_6 in mutually orthogonal spaces A, B, C. The glue vectors in this case are

$$(g_w) \quad g_0^A + g_0^B + g_0^C, g_1^A + g_1^B + g_1^C, g_2^A + g_2^B + g_2^C, g_3^A + g_3^B + g_0^C, \ldots,$$

corresponding to the words $w = (x, y, z)$ of the glue code

$$\{(0,0,0), (1,1,1), (2,2,2), (3,3,0), (4,0,1), (5,1,2),$$
$$(6,2,0), (7,3,1), (8,0,2), (9,1,0), (10,2,1), (11,3,2)\}.$$

Again the glue code is a group, addition being modulo 12, 4, 3 in coordinates x, y, z respectively. The hole diagram for this construction is shown in Fig. 24.2b.

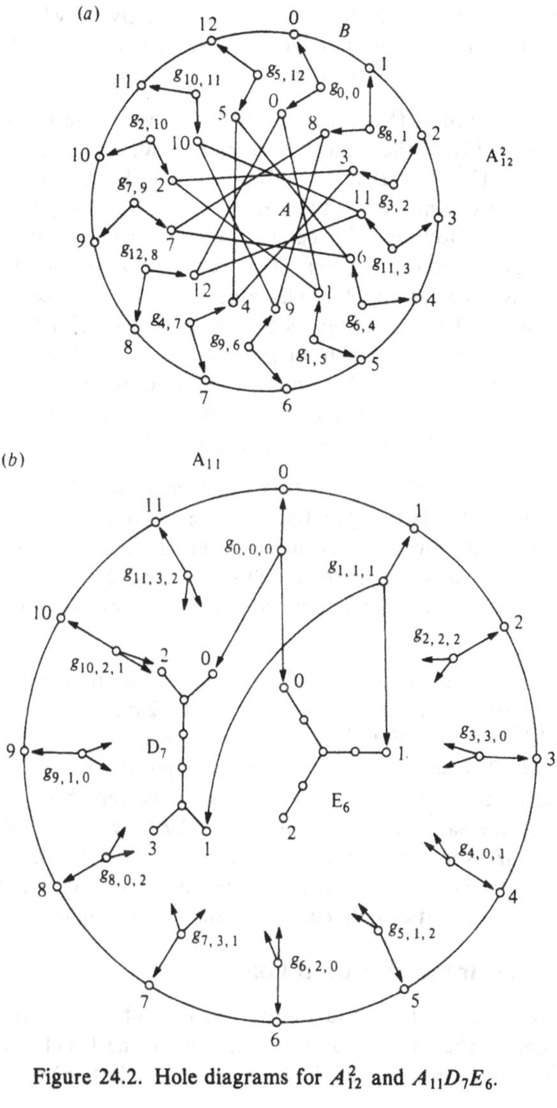

Figure 24.2. Hole diagrams for A_{12}^2 and $A_{11}D_7E_6$.

The other constructions are entirely similar. In each case our fundamental vectors f_i are the union of extended fundamental sets of roots $f_i^A, f_i^B, \ldots$ for the component root lattices, in orthogonal spaces, and our glue vectors are the vectors

$$g_w = g_x^A + g_y^B + g_z^C + \cdots$$

for which the word $w = xyz\ldots$ lies in a certain glue code. In all cases the glue code is abstractly an additive group whose generators are given in Table 16.1. The symbols of the glue code corresponding to an A_n component are read modulo $n+1$, those for D_{2n+1}, E_6, E_7 or E_8

components are read modulo 4,3,2 or 1 respectively, while those for a D_{2n} component belong to the four-group with elements {0,1,2,3} satisfying $1+1 = 2+2 = 3+3 = 0$, $1+2 = 3$, $2+3 = 1$, $3+1 = 2$.

As to the proofs that these constructions work, we remark first that our construction (1) of the typical Niemeier lattice is identical to that given in Chap. 16. The fact that (2) always gives the Leech lattice still quite astonishes us, and we have only been able to give a case-by-case verification, as follows. For each of the 23 Niemeier lattices, the previous chapter gives an explicit set of Leech lattice vectors having the same mutual distances as the f_i. The glue vectors g_x for each component are easily found. The vector hg_0 is the Weyl vector (ρ in Bourbaki's notation [Bou1]), and is half the sum of the positive roots. We were able to verify computationally that, with a suitable labeling, the vectors corresponding to our g_w were indeed Leech lattice vectors, establishing the desired result. We would like to see a more uniform proof.(1)

There is a "holy construction" analogous to (2) associated with any even unimodular lattice (1) that can be obtained by gluing root lattices, but it will only give an even unimodular lattice when the component root lattices have the same Coxeter number h, and $24h$ divides $(h+1)(n-24)$. It is remarkable that the Niemeier lattices are precisely the 24-dimensional combinations permitted by this rule (see Chap. 18).

The Coxeter number h has the following property. For each root lattice there are positive integers c_i with $\Sigma c_i f_i = 0$ (see Fig. 23.1). In Fig. 24.1 each g_x points to an f_i with $c_i = 1$ (a *special* node — see §2 of Chap. 4). The sum of all the c_i for a component is the Coxeter number h. Furthermore for any integer k the set of vectors $\Sigma n_i f_i$ with $\Sigma n_i = k$ is obviously the same as the set of vectors $\Sigma n_i f_i$ with $\Sigma n_i \equiv k$ (modulo h). This remark shows that the equalities in (1) and (2) may be replaced by congruences modulo h, and hence the intersection of the Niemeier lattice with the Leech lattice defined by (1) and (2) has index h in each of them.

2. The environs of a deep hole

The results given here and in the previous chapter enable us to form a clear idea of the environs of any deep hole in the Leech lattice. If we take the origin of coordinates at the hole, the Leech lattice vectors become

$$\sum m_i f_i + \sum n_w g_w \quad \text{with} \quad \sum m_i + \sum n_w = 1$$

(or equivalently with $\Sigma m_i + \Sigma n_w \equiv 1$ modulo h), and the nearest lattice points to the hole are just the f_i. The radius of the hole is $(f_i, f_i)^{1/2} = \sqrt{2}$. The g_w are lattice points only slightly more distant, having $(g_w, g_w) = 2 + 2h^{-1}$.

It is easy to check from our description that f_i is distant $\sqrt{4}$ or $\sqrt{6}$ from f_i according to whether $(f_i, f_j) = 0$ or -1, i.e. according to whether

(1)R. E. Borcherds has recently found such a proof [Bor2].

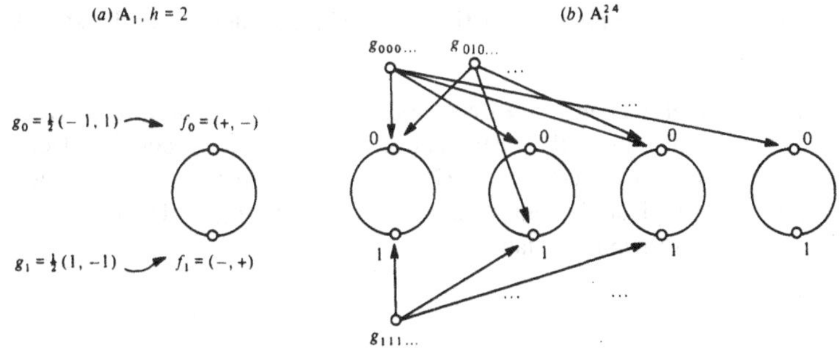

Figure 24.3. (a) Fundamental and glue vectors for A_1. (b) Hole diagram for A_1^{24}. There is a glue vector $g_{xyz...}$ for each codeword in the binary Golay code.

f_i and f_j are unjoined or singly joined in the diagram. Again $g_{xyz...}$ is distant $\sqrt{6}$ from $f_x^A, f_y^B, f_z^C, ...$ and distant $\sqrt{4}$ from all other f_i. We therefore indicate distances of $\sqrt{6}$ between f_i and f_j or between f_i and g_w by joins in the *hole diagram*, which has a node for each f_i and g_w. Some examples appear in Fig. 24.2. (Distances between one g_w and another are *not* indicated in the diagram.)

The case A_1^{24} is slightly exceptional. The diagram for A_1 is given in Fig. 24.3a, where the pair of joins between f_0 and f_1 indicates that their inner product (f_0, f_1) is -2, showing that $f_1 = -f_0$, and the distance between f_0 and f_1 is $\sqrt{8}$. For the holy construction of type A_1^{24} we change to a different coordinate system and rescale, obtaining

$$f_0^A = (-4, 0^{23}), \quad f_0^B = (0, -4, 0^{22}), ...,$$
$$f_1^A = (4, 0^{23}), \quad f_1^B = (0, 4, 0^{22}), ...,$$

and there is a glue vector

$$g_{xyz...} = ((-1)^x, (-1)^y, (-1)^z, ...)$$

for each word $xyz...$ in the binary Golay [24,12] code $\mathscr{C}_{24}$ (§2.8.2 of Chap. 3). After differencing we obtain Leech's original construction for his lattice (see §4.4 of Chap. 5). The hole diagram is sketched in Fig. 24.3b.

For A_2^{12} it is best to use 12 complex coordinates, and to take

$$f_k^A = (-3\omega^k, 0^{11}), \quad f_k^B = (0, -3\omega^k, 0^{10}), ...$$

for $k = 0, 1, 2$, where $\omega = e^{2\pi i/3}$. There is a glue vector

$$g_{xyz...} = (\omega^x, \omega^y, \omega^z, ...)$$

for each word $xyz...$ in the ternary Golay [12,6] code $\mathscr{C}_{12}$ (§2.8.5 of Chap. 3). This is the complex version of the Leech lattice (Example 12 of Chap. 7, §3.6 of Chap. 10). It seems likely that some of our other constructions correspond to the various quaternionic Leech lattices [Tit6], [Tit7].

The symmetry group of a deep hole. For each entry in Table 16.1, the group of automorphisms of the Leech lattice that fix the corresponding deep hole has structure $G_\infty \cdot G_1 \cdot G_2$, where G_∞ is a group isomorphic to the glue code (or group). A typical element $\pi_t \in G_\infty$ takes g_w to g_{w+t}. The orders of G_∞, G_1 and G_2 are shown in the table.

25

The Cellular Structure of the Leech Lattice

R. E. Borcherds, J. H. Conway and L. Queen

We complete the classification of the holes in the Leech lattice, and of the associated Delaunay cells, by showing that there are precisely 284 types of shallow hole.

1. Introduction

In Chap. 23 it was shown that there are 23 types of deep hole in the Leech lattice Λ_{24}, and that these holes are in one-to-one correspondence with the Niemeier lattices. The existence of this correspondence, and the recently discovered correspondence between the conjugacy classes in the Monster group and certain modular functions [Con17], suggested that it might be worth enumerating the shallow holes in the Leech lattice and completing the classification of its Delaunay cells, in case any "deep structure" became apparent. Although this has not yet happened, the complete list of deep and shallow holes has already found several uses, and it seems worth while to put it on record. The main result is the following.

Theorem 1. *There are* 307 *types of hole in the Leech lattice, consisting of* 23 *types of deep hole and* 284 *types of shallow hole. They are listed in Table* 25.1.

2. Names for the holes

We use the notation of Chap. 23, and describe sets of Leech lattice points by graphs, with a node for each lattice point, and where two nodes x and y are

not joined	if $N(x-y) = 4$,
joined by a single edge	if $N(x-y) = 6$, or
joined by two edges	if $N(x-y) = 8$.

Larger numbers of joins will not arise here.

It was shown in Chap. 23 that the vertices of a deep hole in Λ_{24} are described by a graph that is a disjoint union of extended Coxeter-Dynkin diagrams which have total subscript (or dimension) 24 and constant Coxeter number h. There are just 23 possible combinations, which can be seen in the first 23 lines of Table 25.1. Using the same graphical notation we prove the following result.

Lemma 2. *The vertices of a shallow hole in the Leech lattice are sets of 25 points of Λ_{24} for which the corresponding graph is a union of ordinary Coxeter-Dynkin diagrams.*

Proof. Theorems 5 and 6 of Chap. 23 show that the graph is a union of ordinary Coxeter-Dynkin diagrams, and by dimension considerations the hole must contain at least 25 vertices. But the fundamental roots corresponding to such a diagram (whether or not it is connected) are linearly independent in the vector space they lie in, and so are affinely independent. This shows that the graph cannot contain more than 25 nodes. It is just as easy to verify that any such set of 25 points is the vertex set of a shallow hole. The argument is similar to Theorem 7 of Chap. 23, which is the corresponding result for deep holes. Thus all the shallow holes in the Leech lattice are simplices.

3. The volume formula

Let $P_1, P_2, \ldots, P_N$ be a system of representatives for all the holes in Λ_{24} under the full automorphism group Co_∞ of Λ_{24}. Let $\mathrm{vol}(P_i)$ denote the volume of P_i and $g(P_i)$ the order of its automorphism group (i.e. the subgroup of Co_∞ fixing P_i). Then we have the *volume formula*:

$$1 = \text{volume of a fundamental domain of } \Lambda_{24}$$

$$= \sum_{i=1}^{N} \mathrm{vol}(P_i) \times \text{no. of images of } P_i \text{ under } Co_0$$

$$= \sum_{i=1}^{N} \frac{|Co_0|}{g(P_i)} \mathrm{vol}(P_i) ,$$

where $|Co_0|$ denotes the order of Co_0.

Table 25.1. A list of all 307 holes in the Leech lattice. The first 23 entries are the deep holes. The entries give the name of a hole P_i, the order $g(P_i)$ of its automorphism group, its scaled volume

$$\mathrm{svol}(P_i) = \mathrm{vol}(P_i) \cdot 24! ,$$

the norm $s(P_i)$ of its Weyl vector, and the determinant $d(P_i)$ of the Cartan matrix. The volume formula then becomes

$$\sum_i \mathrm{svol}(P_i)/g(P_i) = 24!/|Co_0| = 74613 .$$

The name of a hole indicates the orbits of its automorphism group on the components of the diagram. Thus the hole $a_7^2 a_3^2 a_3 a_1^2$ has two components of type a_7 that are equivalent under the automorphism group, also two equivalent components of type a_1, and three components of type a_3, only two of which are equivalent.

Table 25.1a

Name	g	svol	s	d
D_{24}	2	92	—	4
A_{24}	10	125	—	25
$A_{17}E_7$	12	1944	—	36
$D_{16}E_8$	2	1800	—	4
$A_{15}D_9$	16	2048	—	64
D_{12}^2	8	1936	—	16
A_{12}^2	52	2197	—	169
$A_{11}D_7E_6$	24	20736	—	144
$D_{10}E_7^2$	8	23328	—	16
$A_9^2D_6$	80	20000	—	400
E_8^3	6	27000	—	1
D_8^3	48	21952	—	64
A_8^3	324	19683	—	729
$A_7^2D_5^2$	256	131072	—	1024
E_6^4	432	186624	—	81
D_6^4	384	160000	—	256
A_6^4	1176	117649	—	2401
$A_5^4D_4$	3456	559872	—	5184
D_4^6	138240	2985984	—	4096
A_4^6	30000	1953125	—	15625
A_3^8	688128	16777216	—	4^8
A_2^{12}	138568320	387420489	—	3^{12}
A_1^{24}	1002795171840	68719476736	—	2^{24}

Table 25.1b

Name	g	svol	s	d
d_{25}	1	140	9800/2	4
a_{25}	1	195	2925/2	26
$d_{24}a_1$	1	186	8649/2	8
$a_{24}a_1$	2	255	2601/2	50
$a_{23}a_2$	2	288	2304/2	72
$d_{22}a_2a_1$	1	282	6627/2	24
$d_{21}a_4$	1	240	5760/2	20
$a_{21}a_3a_1$	1	396	1782/2	176
$d_{20}d_5$	1	200	5000/2	16
$a_{20}a_5$	1	315	1575/2	126
$d_{19}e_6$	1	162	4374/2	12
$a_{19}d_6$	1	240	1440/2	80
$a_{19}a_4a_1^2$	2	520	1352/2	400
$d_{18}e_7$	1	126	3969/2	8
$a_{18}e_7$	1	171	1539/2	38
$a_{18}a_6a_1$	1	399	1197/2	266
$d_{17}e_8$	1	92	4232/2	4
$a_{17}e_8$	2	141	2209/2	18
$a_{17}a_8$	2	297	1089/2	162
$a_{17}a_8$	2	297	1089/2	162
$a_{17}e_7a_1$	2	222	1369/2	72
$a_{17}d_7a_1$	1	288	1152/2	144
$a_{17}d_7a_1$	2	288	1152/2	144
$a_{17}d_6a_2$	2	342	1083/2	216
$a_{17}a_6a_2$	2	441	1029/2	378
$a_{17}a_5a_3$	2	468	1014/2	432
$a_{17}a_4a_3a_1$	2	600	1000/2	720
$d_{16}d_9$	1	152	2888/2	16
$d_{16}a_9$	1	230	2645/2	40
$d_{16}e_8a_1$	1	122	3721/2	8

Table 25.1c

Name	g	svol	s	d
$d_{16}a_8a_1$	1	306	2601/2	72
$d_{16}e_7a_2$	1	186	2883/2	24
$d_{16}e_6a_3$	1	252	2646/2	48
$d_{16}a_6a_2a_1$	1	462	2541/2	168
$d_{16}d_5a_4$	1	320	2560/2	80
$d_{16}a_5a_4$	1	390	2535/2	120
$a_{16}d_9$	1	204	1224/2	68
$a_{16}e_8a_1$	1	187	2057/2	34
$a_{15}d_{10}$	2	200	1250/2	64
$a_{15}d_9a_1$	2	264	1089/2	128
$a_{15}e_8a_1^2$	2	248	1922/2	64
$a_{15}e_7a_3$	2	264	1089/2	128
$a_{15}d_7a_2a_1$	2	408	867/2	384
$a_{15}e_6d_4$	2	288	864/2	192
$a_{15}d_6a_4$	2	360	810/2	320
$a_{15}d_5d_5$	2	320	800/2	256
$d_{14}d_{10}a_1$	1	188	2209/2	32
$d_{14}a_{10}a_1$	1	286	1859/2	88
$d_{14}a_9a_1a_1$	1	380	1805/2	160
$d_{14}e_8a_2a_1$	1	186	2883/2	24
$d_{14}e_7a_3a_1$	1	256	2048/2	64
$d_{14}a_7a_2a_1a_1$	1	576	1728/2	384
$d_{14}e_6a_4a_1$	1	330	1815/2	120
$d_{14}a_6a_4a_1$	1	490	1715/2	280
$d_{14}d_5a_5a_1$	1	408	1734/2	192
$a_{14}a_9a_2$	2	405	729/2	450
$a_{14}e_8a_2a_1$	1	285	1805/2	90
$a_{14}d_7a_2a_2$	1	450	750/2	540
$a_{14}a_6a_3a_2$	1	630	630/2	1260
$a_{14}a_4^2a_2a_1$	2	825	605/2	2250

Table 25.1d

Name	g	svol	s	d
$d_{13}d_{12}$	1	136	2312/2	16
$d_{13}a_{12}$	1	208	1664/2	52
$d_{13}a_{11}a_1$	1	276	1587/2	96
$d_{13}a_9a_2a_1$	1	420	1470/2	240
$d_{13}e_8a_4$	1	160	2560/2	20
$d_{13}a_8a_4$	1	360	1440/2	180
$d_{13}e_7a_5$	1	204	1734/2	48
$d_{13}a_7d_5$	1	304	1444/2	128
$d_{13}e_6a_6$	1	252	1512/2	84
$a_{13}a_{12}$	1	273	819/2	182
$a_{13}d_{10}a_2$	1	294	1029/2	168
$a_{13}a_9a_3$	1	420	630/2	560
$a_{13}e_8a_4$	1	245	1715/2	70
$a_{13}e_8a_2a_1a_1$	1	378	1701/2	168
$a_{13}e_7a_4a_1$	1	350	875/2	280
$a_{13}d_7a_5$	1	336	672/2	336
$a_{13}a_7a_4a_1$	1	560	560/2	1120
$a_{13}e_6d_5a_1$	1	336	672/2	336
$a_{13}a_6a_6$	1	441	567/2	686
$d_{12}^2a_1$	2	180	2025/2	32
$d_{12}d_{10}a_2a_1$	1	276	1587/2	96
$d_{12}d_9a_4$	1	240	1440/2	80
$d_{12}e_8d_5$	1	136	2312/2	16
$d_{12}d_8d_5$	1	208	1352/2	64
$d_{12}e_7d_6$	1	156	1521/2	32
$d_{12}d_7e_6$	1	180	1350/2	48
$a_{12}^2a_1$	4	351	729/2	338
$a_{12}e_8d_5$	1	208	1664/2	52
$a_{12}e_8a_4a_1$	1	325	1625/2	130
$a_{12}e_7a_6$	1	273	819/2	182

Table 25.1e

Name	g	svol	s	d
$a_{12}d_7e_6$	1	234	702/2	156
$d_{11}e_8e_6$	1	114	2166/2	12
$d_{11}e_7^2$	2	112	1568/2	16
$a_{11}d_{10}a_3a_1$	1	408	867/2	384
$a_{11}d_{10}a_2^2$	2	432	864/2	432
$a_{11}d_9a_5$	2	324	729/2	288
$a_{11}a_9d_4a_1$	1	480	480/2	960
$a_{11}e_8e_6$	1	174	1682/2	36
$a_{11}e_8d_5a_1$	1	276	1587/2	96
$a_{11}e_8a_2^2a_1^2$	2	576	1536/2	432
$a_{11}d_8e_6$	2	228	722/2	144
$a_{11}e_7d_7$	1	204	867/2	96
$a_{11}e_7a_5a_1^2$	2	456	722/2	576
$a_{11}d_7e_6a_1$	2	300	625/2	288
$a_{11}d_7a_6a_1$	1	420	525/2	672
$a_{11}d_7a_5a_2$	2	468	507/2	864
$a_{11}d_7a_3a_2^2$	2	648	486/2	1728
$a_{11}d_7a_3a_2^2$	2	648	486/2	1728
$o_{11}a_7d_5a_1a_1$	1	576	432/2	1536
$a_{11}e_6e_6a_1^2$	2	360	600/2	432
$a_{11}e_6d_5a_3$	2	384	512/2	576
$a_{11}e_6d_5a_2a_1$	2	468	507/2	864
$a_{11}e_6d_4a_4$	2	420	490/2	720
$a_{11}d_6a_5a_3$	2	504	441/2	1152
$a_{11}a_6d_4a_2^2$	2	756	378/2	3024
$a_{11}d_5a_5a_3a_1$	2	672	392/2	2304
$a_{11}d_5a_4a_2^2a_1$	2	900	375/2	4320
$a_{11}a_5a_5d_4$	2	576	384/2	1728
$d_{10}^2a_3a_1^2$	2	384	1152/2	256
$d_{10}a_{10}a_5$	1	330	825/2	264

Table 25.1f

Name	g	svol	s	d
$d_{10}d_9a_5a_1$	1	312	1014/2	192
$d_{10}a_9d_6$	1	260	845/2	160
$d_{10}a_9a_3a_2a_1$	1	600	750/2	960
$d_{10}e_8e_7$	1	94	2209/2	8
$d_{10}d_8d_6a_1$	1	248	961/2	128
$d_{10}a_8e_7$	1	198	1089/2	72
$d_{10}a_8a_5a_2$	1	486	729/2	648
$d_{10}e_7^2a_1$	2	148	1369/2	32
$d_{10}e_7d_7a_1$	1	192	1152/2	64
$d_{10}e_7d_6a_2$	1	228	1083/2	96
$d_{10}e_7a_6a_2$	1	294	1029/2	168
$d_{10}e_7a_5a_3$	1	312	1014/2	192
$d_{10}e_7a_4a_3a_1$	1	400	1000/2	320
$d_{10}a_7d_6a_2$	1	384	768/2	384
$d_{10}a_7a_3^2a_1^2$	2	832	676/2	2048
$d_{10}d_6^2a_3$	2	320	800/2	256
$d_{10}d_6a_5a_4$	1	420	735/2	480
$d_{10}d_6a_5a_3a_1$	1	528	726/2	768
$d_{10}a_6a_5a_3a_1$	1	672	672/2	1344
$d_{10}a_5^3$	2	540	675/2	864
$a_{10}a_9d_6$	1	330	495/2	440
$a_{10}e_8e_7$	1	143	1859/2	22
$a_{10}e_8e_6a_1$	1	231	1617/2	66
$a_{10}e_8a_4a_2a_1$	1	495	1485/2	330
$a_{10}e_7a_7a_1$	1	352	704/2	352
$a_{10}d_7a_6a_2$	1	462	462/2	924
$d_9^2a_7$	2	240	900/2	128
$d_9a_9a_6a_1$	1	420	630/2	560
$d_9e_8^2$	2	76	2888/2	4
$d_9d_8^2$	2	176	968/2	64

Table 25.1g

Name	g	svol	s	d
$d_9a_8^2$	2	324	648/2	324
$d_9a_7^2a_1^2$	4	544	578/2	1024
$a_9a_9e_7$	2	270	729/2	200
$a_9^2d_7$	4	320	512/2	400
$a_9^2d_6a_1$	4	420	441/2	800
$a_9^2d_5a_1^2$	4	560	392/2	1600
$a_9^2d_4a_2a_1$	4	660	363/2	2400
$a_9^2a_4a_3$	4	600	360/2	2000
$a_9^2a_4a_3$	2	600	360/2	2000
$a_9e_8^2$	2	115	2645/2	10
$a_9e_8e_7a_1$	1	190	1805/2	40
$a_9e_8d_5a_2a_1$	1	420	1470/2	240
$a_9e_8a_4^2$	2	425	1445/2	250
$a_9d_8e_7a_1$	1	260	845/2	160
$a_9a_8^2$	2	405	405/2	810
$a_9e_7d_7a_2$	1	300	750/2	240
$a_9e_7a_6a_3$	1	420	630/2	560
$a_9e_7a_4^2a_1$	2	550	605/2	1000
$a_9a_7d_5a_3a_1$	1	640	320/2	2560
$a_9d_5a_4^2a_2a_1$	2	900	270/2	6000
$a_9a_4^2a_4^2$	4	875	245/2	6250
$e_8^3a_1$	6	61	3721/2	2
$e_8^2a_8a_1$	2	153	2601/2	18
$e_8^2e_7a_2$	2	93	2883/2	6
$e_8^2e_6a_3$	2	126	2646/2	12
$e_8^2a_6a_2a_1$	2	231	2541/2	42
$e_8^2d_5a_4$	2	160	2560/2	20
$e_8^2a_5a_4$	2	195	2535/2	30
$e_8a_8e_6a_2a_1$	1	351	1521/2	162
$e_8a_8d_5a_4$	1	360	1440/2	180

Table 25.1h

Name	g	svol	s	d
$e_8e_7^2a_3$	2	128	2048/2	16
$e_8e_7a_7a_2a_1$	1	288	1728/2	96
$e_8e_7e_6a_4$	1	165	1815/2	30
$e_8e_7a_6a_4$	1	245	1715/2	70
$e_8e_7d_5a_5$	1	204	1734/2	48
$e_8a_7e_6a_4$	1	300	1500/2	120
$e_8a_7d_5^2$	2	304	1444/2	128
$e_8e_6^2a_5$	2	207	1587/2	54
$e_8e_6a_6d_5$	1	252	1512/2	84
$d_8^3a_1$	6	232	841/2	128
$d_8^2e_7a_1a_1$	2	248	961/2	128
$d_8^2e_6a_3$	2	264	726/2	192
$d_8^2d_6a_2a_1$	2	360	675/2	384
$d_8^2d_5d_4$	2	288	648/2	256
$d_8^2d_5a_4$	2	320	640/2	320
$d_8e_7^2a_2a_1$	2	228	1083/2	96
$d_8e_7d_6a_3a_1$	1	320	800/2	256
$d_8e_7a_7a_2a_1$	1	384	768/2	384
$d_8e_7a_5a_4a_1$	1	420	735/2	480
$d_8e_7a_5a_3a_1a_1$	1	528	726/2	768
$d_8a_7e_6a_4$	1	360	540/2	480
$d_8a_7d_5^2$	2	352	484/2	512
$d_8e_6a_5^2a_1$	2	468	507/2	864
$d_8d_6^2d_4a_1$	2	368	529/2	512
$d_8d_6^2a_3a_1^2$	2	512	512/2	1024
$d_8d_6d_5a_5a_1$	1	432	486/2	768
$a_8^3a_1$	12	513	361/2	1458
$a_8a_7d_5^2$	2	432	324/2	1152
$a_8e_6^2a_2^2a_1$	2	567	441/2	1458
$a_8e_6d_5a_4a_2$	1	540	360/2	1620

Table 25.1i

Name	g	svol	s	d
$e_7^3d_4$	6	140	1225/2	32
$e_7^2d_7a_4$	2	200	1000/2	80
$e_7^2e_6d_5$	2	156	1014/2	48
$e_7^2d_6d_5$	2	176	968/2	64
$e_7d_7a_7a_3a_1$	1	416	676/2	512
$e_7d_7d_6a_5$	1	264	726/2	192
$e_7d_7a_6a_5$	1	336	672/2	336
$e_7a_7^2a_3a_1$	2	544	578/2	1024
$e_7a_7a_6a_4a_1$	1	560	560/2	1120
$e_7a_7a_5a_4a_1a_1$	1	720	540/2	1920
$e_7e_6^3$	6	153	867/2	54
$e_7d_6^3$	3	216	729/2	128
$e_7a_6^3$	3	441	567/2	686
$e_7a_5^3a_1^3$	6	936	507/2	3456
$d_7a_7^2a_3a_1$	4	608	361/2	2048
$d_7a_7a_6a_3a_2$	1	672	336/2	2688
$d_7a_7a_5^2a_3a_2$	2	960	300/2	6144
$d_7d_6^3$	6	256	512/2	256
$d_7a_6^3$	3	490	350/2	1372
$d_7a_3^6$	24	1408	242/2	16384
$a_7^3a_1^4$	24	1024	256/2	8192
$a_7^2e_6d_5$	2	384	384/2	768
$a_7^2d_6d_5$	4	416	338/2	1024
$a_7^2d_5^2a_1$	8	544	289/2	2048
$a_7^2d_5a_4a_1^2$	4	800	250/2	5120
$a_7^2d_5a_3a_2a_1$	4	864	243/2	6144
$a_7^2a_5a_3a_1^2a_1$	4	1152	216/2	12288
$a_7^2a_3^2a_3a_1^2$	8	1280	200/2	16384
$a_7e_6^2a_5a_1$	2	432	432/2	864
$a_7d_6^2a_3^2$	4	576	324/2	2048

Table 25.1j

Name	g	svol	s	d
$a_7a_6^3$	3	588	252/2	2744
$a_7a_5^3a_1^3$	6	1152	192/2	13824
$a_7a_3^6$	24	1536	144/2	32768
$e_6^4a_1$	48	225	625/2	162
$e_6^3a_6a_1$	6	315	525/2	378
$e_6^3a_5a_2$	12	351	507/2	486
$e_6^3a_3a_2^2$	12	486	486/2	972
$e_6^2a_5^2a_3$	8	504	392/2	1296
$e_6^2a_5a_4a_2^2$	4	675	375/2	2430
$e_6^2a_5a_2^4$	8	891	363/2	4374
$e_6d_5a_5^2a_2^2$	4	756	294/2	3888
$e_6a_5^3d_4$	12	612	289/2	2592
$d_6^4a_1$	24	336	441/2	512
$d_6^3d_5a_1^2$	6	448	392/2	1024
$d_6^3d_4a_2a_1$	6	528	363/2	1536
$d_6^3a_4a_3$	6	480	360/2	1280
$d_6^3d_4a_1^3$	6	608	361/2	2048
$d_6^2a_5d_4a_3a_1$	2	672	294/2	3072
$d_6^2d_4^2a_3a_1^2$	4	768	288/2	4096
$d_6a_5^3d_4$	6	648	243/2	3456
$d_6d_4^4a_1^3$	24	960	225/2	8192
$a_6^4a_1$	24	735	225/2	4802
$a_6a_5^3d_4$	6	756	189/2	6048
$d_5a_5^4$	16	720	200/2	5184
$d_5d_4^5$	120	640	200/2	4096
$d_5a_4^5$	20	1000	160/2	12500
$a_5^4d_4a_1$	48	936	169/2	10368
$a_5^4a_2a_1^3$	48	1512	147/2	31104
$a_5a_4^5$	20	1125	135/2	18750
$a_5a_2^{10}$	720	3645	75/2	354294

Table 25.1k

Name	g	svol	s	d
$d_4^6a_1$	2160	832	169/2	8192
$d_4^5a_2a_1^3$	360	1344	147/2	24576
$d_4^4a_3a_1^6$	144	2048	128/2	65536
$d_4^4a_1^9$	432	2816	121/2	131072
$d_4a_3^7$	336	1792	98/2	65536
$d_4a_1^{21}$	120960	14336	49/2	8388608
$a_4^6a_1$	240	1375	121/2	31250
$a_4a_3^7$	168	1920	90/2	81920
$a_3^8a_1$	2688	2304	81/2	131072
$a_3a_2^{11}$	7920	4374	54/2	708588
$a_3a_1^{22}$	887040	16384	32/2	16777216
$a_2^{12}a_1$	190080	5103	49/2	1062882
$a_2a_1^{23}$	10200960	18432	27/2	25165824
$a_1^{24}a_1$	244823040	20480	25/2	33554432

The volume of a hole P can be expressed in terms of familiar concepts in the Lie theory (cf. [Bou1]). For a deep hole,

$$\mathrm{vol}(P) = \frac{1}{24!} h\sqrt{d} ,$$

and for a shallow hole

$$\mathrm{vol}(P) = \frac{1}{24!} \sqrt{sd} ,$$

where h is the Coxeter number, d is the determinant of the Cartan matrix of the ordinary Coxeter-Dynkin diagram, and s is the norm of the Weyl vector. For a connected component the values of these quantities are shown in Table 25.2 (note that $s = h(h+1)n/12$). For a disconnected graph, h and d are the products of the values for the components, while s is their sum. One can also show that the radius of a shallow hole is $(2-1/s)^{1/2}$.

Table 25.2. The parameters h, d, s for connected diagrams.

	a_n (or A_n)	d_n (or D_n)	e_6(or E_6)	e_7(or E_7)	e_8(or E_8)
h:	$n+1$	$2n-2$	12	18	30
d:	$n+1$	4	3	2	1
s:	$\frac{n(n+1)(n+2)}{12}$	$\frac{(n-1)n(2n-1)}{6}$	78	$\frac{399}{2}$	620

4. The enumeration of the shallow holes

We shall only give a brief description of how the shallow holes were enumerated and Theorem 1 proved. Two methods of classification were used.

Method 1. This method was used to find all the shallow holes that contain a particular ordinary Coxeter-Dynkin diagram X as a component, by finding all occurrences of X as a set of points in the Leech lattice. This method was used only when X has at least seven points, since for others the classification becomes too complicated. We drew the graph of all points in Λ_{24} not joined to any point of X, and then deleted enough points from this graph to make the union of X with the remaining graph be a collection of ordinary diagrams with a total of 25 points. The group orders of such holes, found by inspection, are usually 1 or 2.

As an example we consider the diagram $X = a_{15}$. We find from Fig. 23.9 that there are just nine types of ordered a_{15} diagrams in the Leech lattice. When we identify reversals these reduce to five distinct types,

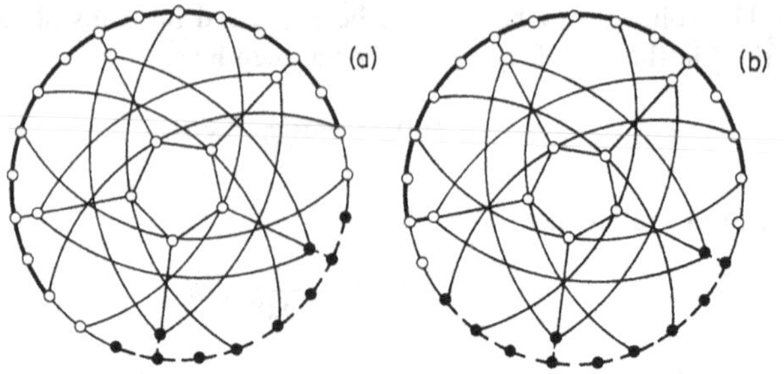

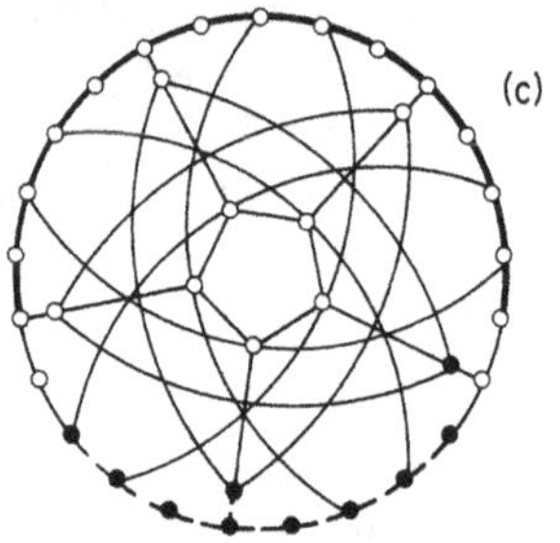

Figure 25.1. The enumeration of shallow holes whose diagram contains a component a_{15} (indicated by heavy lines). For types a-d the set of points disjoined from a_{15} (indicated by shaded nodes and broken lines) has no 10-point subgraph of mutually disjoined ordinary diagrams, so only case e leads to holes.

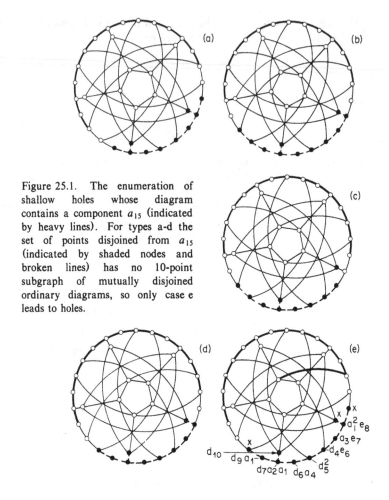

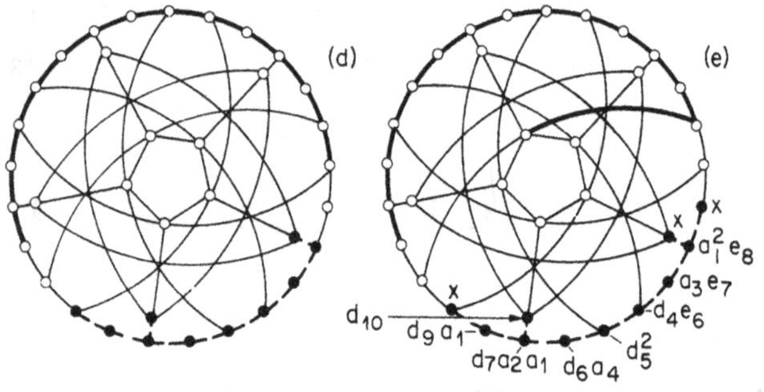

which are drawn in Figs. 25.1a-e as heavy lines. (The background of these figures is taken from Fig. 27.3, and shows all edges between pairs of the 35 points nearest to the center of a deep hole of type A_{24}. This is a convenient portion of the Leech lattice to work with, and in particular contains all the points of Λ_{24} not joined to the a_{15} diagrams.)

In four of figures a-e one finds that in the subgraph not joined to the a_{15} diagram (indicated by shaded nodes and broken lines) it is impossible to find a union of disjoint ordinary Coxeter-Dynkin diagrams with a total of $10 (= 25-15)$ nodes, and so the corresponding type of a_{15} diagram cannot be part of a shallow hole. However, for the fifth type (figure e), the disjoined subgraph contains 11 points, and by omitting each point in turn we can break it into Coxeter-Dynkin diagrams, as shown in the figure. Three of the eleven possibilities fail (those marked with an x), since they leave a graph containing an *extended* diagram. The remaining eight succeed.

Thus there are eight types of shallow hole with a component of type a_{15} in their diagram: $a_{15}e_8a_1^2$, $a_{15}e_7a_3$, $a_{15}e_6d_4$, $a_{15}d_5^2$, $a_{15}d_6a_4$, $a_{15}d_7a_2a_1$, $a_{15}d_9a_1$, and $a_{15}d_{10}$.

Method 2. We take a given deep hole and find all shallow holes having a face in common with it. The environs of a deep hole are described in sufficient detail in the previous chapter to make this quite easy. The group orders of these holes can be found by considering the subgroup of those automorphisms of the corresponding deep hole that fix the face in question and then making due allowance for the fact that the shallow hole may touch several deep holes in the same manner.

The complete enumeration was carried out by combining the two methods and using the volume formula as a check. A second check was supplied by the fact that the quantity *sd* for each hole must be a perfect square (since the volume of a hole is a rational number). Most of the enumeration was performed twice: first by L.Q. and J.H.C., then independently by R.E.B who completed the list. The group orders were computed by both teams.

Remarks. (i) It turns out that (in contrast to the deep holes) the diagram of a shallow hole does not determine it uniquely. There are two holes of each of the types $a_{17}a_8$, $a_{17}d_7a_1$, $a_{11}d_7a_3a_2^2$ and $a_9^2a_4a_3$.

(ii) There is a unique hole of type $a_1^{24}a_1$. This has the shape of a regular simplex with 25 vertices, although its group is not transitive on the vertices. The group is the Mathieu group M_{24} acting with orbits of size 24 and 1. This hole has the smallest radius, namely $(2-2/25)^{1/2}$, while the shallow hole of largest radius is d_{25}, with radius $(2-1/4900)^{1/2}$. The deep holes have radius $\sqrt{2}$.

(iii) If the closed polytopes corresponding to the deep holes of type D_4^6 are deleted from $\mathbf{R}^{24}$, the resulting space is disconnected.

Acknowledgement. We should like to thank R. A. Parker for several helpful discussions.

26

Lorentzian Forms for the Leech Lattice

J. H. Conway and N. J. A. Sloane

The "holy constructions" of Chapter 24 lead to some very simple definitions of the Leech lattice using Lorentzian coordinates.

1. The unimodular Lorentzian lattices

We begin by defining the lattices $I_{n,1}$ and $II_{n,1}$. Let $\mathbf{R}^{n,1}$ denote the real $(n+1)$-dimensional vector space of vectors $x = (x_0,\ldots,x_{n-1} | x_n)$ with inner product and norm defined by

$$x \cdot y = x_0 y_0 + \cdots + x_{n-1} y_{n-1} - x_n y_n , \qquad N(x) = x \cdot x .$$

$x_0, \ldots, x_{n-1}$ are sometimes called the space-like coordinates and x_n the time-like coordinate. If $x \in \mathbf{R}^{n,1}$, then $x^{\perp}$ is the set of $y \in \mathbf{R}^{n,1}$ with $x \cdot y = 0$.

Just as for Euclidean lattices (i.e. lattices in $\mathbf{R}^n$, see §2.4 of Chap. 2), a lattice Λ in $\mathbf{R}^{n,1}$ is called *integral* if $x \cdot y \in \mathbf{Z}$ for all $x,y \in \Lambda$, and *unimodular* if it has a basis $v^{(0)},\ldots,v^{(n)}$ such that the determinant of the Gram matrix $(v^{(i)} \cdot v^{(j)})$ is ± 1. An integral lattice is *even*, or of type II, if $N(x) \in 2\mathbf{Z}$ for all $x \in \Lambda$, and otherwise is *odd* or of type I.

In contrast to the Euclidean case, the classification of integral unimodular lattices in $\mathbf{R}^{n,1}$ is easy: there is a unique odd unimodular lattice in $\mathbf{R}^{n,1}$ for all n, and a unique even unimodular lattice when $n \equiv 1$ (modulo 8) (see [Mil7], [Neu5], [Ser1] and Chap. 15). These lattices will be denoted by $I_{n,1}$ and $II_{n,1}$ respectively. (In Chap. 15 we used $I_{n,1}$ and $II_{n,1}$ to denote genera of forms, but there is no confusion in using the same symbols for the unique forms in the genera.)

$I_{n,1}$ can be taken to be the set of vectors $x = (x_0,\ldots,x_{n-1} | x_n)$ with all $x_i \in \mathbf{Z}$. For $n \equiv 1$ (modulo 8), $II_{n,1}$ can be taken to be the set of x for which all x_i are all in $\mathbf{Z}$ or all in $\mathbf{Z} + ½$ and which satisfy

$$x_0 + \cdots + x_{n-1} - x_n \in 2\mathbf{Z} .$$

As we shall see here and in the following chapters, the lattice $II_{25,1}$ is especially interesting.

Background references on Lorentzian or hyperbolic space are [Cox21a], [Mil6], [Neu5].

If Λ is an integral unimodular lattice in $\mathbf{R}^{n,1}$, and $t \in \Lambda$ has norm -1, then $t^{\perp} \cap \Lambda$ is an integral unimodular n-dimensional Euclidean lattice. (It is Euclidean by Sylvester's law of inertia (§6.2 of Chap. 15), because there can not exist two mutually orthogonal vectors x,y in $\mathbf{R}^{n,1}$ with $x \cdot x \leqslant 0$ and $y \cdot y \leqslant 0$.) For example it has been known for a long time ([Cox10, p. 419], [Neu3], [Neu5]) that for

$$t = (1,1,1,1,1,1,1,1,1 \mid 3) ,$$

a vector of norm -1 in $I_{8,1}$, the lattice $t^{\perp} \cap I_{8,1}$ is a copy of E_8.

Similarly, let $u \in \Lambda$ be an isotropic vector, i.e. have norm 0. Then $u \in \Lambda^{\perp}$, and $(u^{\perp} \cap \Lambda)/\langle u \rangle$ — which we abbreviate to $(u^{\perp} \cap \Lambda)/u$ — may be regarded as an integral unimodular $(n-1)$-dimensional Euclidean lattice. (Since $(x+\lambda u) \cdot (y + \mu u) = x \cdot y$ for $x,y \in u^{\perp}$ and $\lambda, \mu \in \mathbf{Z}$, the inner product is well-defined on $u^{\perp}/u$.) For example if

$$u = (1,1,1,1,1,1,1,1,1,1 \mid 3) ,$$

an isotropic vector in $I_{9,1}$, then $(u^{\perp} \cap I_{9,1})/u$ is again a copy of the E_8 lattice ([Cox10], [Neu3], [Neu5]). The 240 minimal vectors of E_8 are represented by $9 \cdot 8$ vectors of shape $(1,-1,0^7 \mid 0)$ and $2\binom{9}{3}$ vectors of shape $\pm(1^3, 0^6 \mid 1)$.

2. Lorentzian constructions for the Leech lattice

Some years ago J. H. Conway and R. T. Curtis found the following construction for the Leech lattice Λ_{24}. Their proof was long (and unpublished). Using one of the "holy constructions" of Chap. 24 we can now give a short proof.

Theorem 1. *If* $t = (3,5,7,\ldots,45,47,51 \mid 145)$, *a vector of norm* -1 *in* $I_{24,1}$, *then* $t^{\perp} \cap I_{24,1}$ *is a copy of the Leech lattice.*

Proof. From the above remarks we know that $t^{\perp} \cap I_{24,1}$ is an integral unimodular lattice. To show that it is Λ_{24} we work instead in the affine hyperplane

$$H = \{v \in \mathbf{R}^{24,1} : v \cdot t = -2\} ,$$

and we shall show that $H \cap I_{24,1}$ is a translate of Λ_{24}. More precisely, if the points of $H \cap I_{24,1}$ are arbitrarily labeled $P_0, P_1, \ldots$, we shall show that $\{P_i - P_0 : P_i \in H \cap I_{24,1}\}$ is a copy of Λ_{24}.

Straightforward calculation shows that $H \cap I_{24,1}$ contains all the points mentioned in Fig. 26.1, consisting of 25 "fundamental vectors" $f_0, \ldots, f_{24}$ (in the notation of Chap. 24) forming a D_{24} diagram, plus two "glue

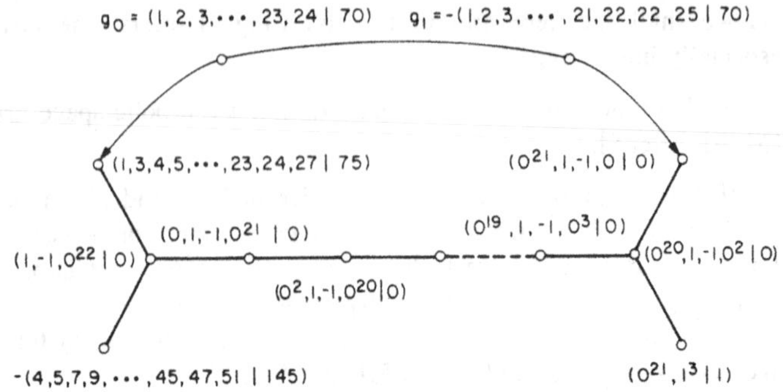

Figure 26.1. Hole diagram of type D_{24} for Leech lattice.

vectors" g_0 and g_1. Two points in the figure are joined by an edge (or a directed edge) if they are distance $\sqrt{6}$, all other pairs being distant $\sqrt{4}$ apart.

Since Fig. 26.1 is a copy of the D_{24} hole diagram (see Chap. 24), it follows that the set $\{P_i - P_0\}$ contains Λ_{24}. But Λ_{24} is unimodular, so this lattice *is* the Leech lattice.

A similar argument establishes the next result.

Theorem 2. *If* $u = (1,3,5,...,45,47,51 \mid 145)$, *an isotropic vector in* $\mathrm{I}_{25,1}$, *then* $(u^{\perp} \cap \mathrm{I}_{25,1})/u$ *is a copy of the Leech lattice.*

Proof. Prefix all the points of Fig. 26.1 with an initial zero.

However still more elegant coordinates for Λ_{24} may be obtained as follows. It is well known that the only solutions of

$$0^2 + 1^2 + 2^2 + \cdots + m^2 = n^2$$

in positive integers are $(m,n) = (1,1)$ and $(24,70)$ ([Lju1], [Mor6, p. 258], [Wat23]). In particular

$$w = (0,1,2,3,...,23,24 \mid 70)$$

is an isotropic vector in $\mathrm{II}_{25,1}$. Let us define the *Leech roots* to be the vectors $r \in \mathrm{II}_{25,1}$ that satisfy

$$r \cdot r = 2\,, \quad r \cdot w = -1\,. \tag{1}$$

(The reason for this name will appear in the next chapter.) Our main result is the following.

Theorem 3. (*a*) *The Leech roots are in one-to-one correspondence with the points of the Leech lattice. This correspondence is an isometry for the metric defined by*

$$d(r,s)^2 = N(r-s)\,. \tag{2}$$

(*b*) *Furthermore* $(w^{\perp} \cap \mathrm{II}_{25,1})/w$ *is also a copy of the Leech lattice.*

Proof. We know from the remarks in §1 that $(w^\perp \cap II_{25,1})/w$ is an integral unimodular lattice. The hyperplane

$$H = \{s \in \mathbf{R}^{25,1} : s \cdot w = -1\} \tag{3}$$

is a translate of $w^\perp$, and for any $s \in H$, $(s + \lambda w) \cdot (s + \lambda w) = s \cdot s - 2\lambda$. For $s \in H \cap II_{25,1}$, $s \cdot s$ is even and $\lambda \in \mathbf{Z}$ can be chosen uniquely so that $s + \lambda w$ is a Leech root. Thus the Leech roots are a set of coset representatives for $(w^\perp \cap II_{25,1})/w$.

The set of Leech roots we have found can be embedded in a set of Leech roots that contains the A_{24} hole diagram (see Fig. 27.3), and so by Chap. 24 $(w^\perp \cap II_{25,1})/w$ is a copy of the Leech lattice.

Remarks. Although distances in H are correctly computed using $d(x,y) = N(x-y)^{1/2}$, the reader should note that norms and inner products of individual vectors in H are not invariant under the addition of w. If r,s are distinct Leech roots then, since $N(r-s) \geqslant 4$, we have $r \cdot s \leqslant 0$.

We may also work with the isotropic representatives for the points of H.

Theorem 4. *If u is any isotropic vector of $H \cap II_{25,1}$, then $(u^\perp \cap II_{25,1})/u$ is a copy of the Leech lattice.*

Proof. We see that $r = u-w$ is a Leech root, and that the reflection in r,

$$R_r : x \to x - (x \cdot r)r ,$$

is a symmetry of $II_{25,1}$ that sends u to w, and $(u^\perp \cap II_{25,1})/u$ to $(w^\perp \cap II_{25,1})/w$.

Theorem 5. *After identifying the Leech lattice with the set of Leech roots in the hyperplane H (see (3)), let u denote the isotropic representative of the center of a deep hole in H. Then $(u^\perp \cap II_{25,1})/u$ is a copy of the Niemeier lattice corresponding to that hole.*

Proof. We already know that $(u^\perp \cap II_{25,1})/u$ is a 24-dimensional even unimodular lattice, and so is either Λ_{24} or one of the Niemeier lattices. We take a particular case, and suppose that U is the center of a deep hole of type $A_{11}D_7E_6$. The Leech lattice vectors closest to U are the fundamental vectors

$$(f_i) \qquad f_0^A, f_1^A, \ldots, f_{12}^A, f_0^B, f_1^B, \ldots, f_7^B, f_0^C, f_1^C, \ldots, f_6^C ,$$

the superscripts picking out fundamental sets of roots for A_{11}, D_7, E_6 in mutually orthogonal spaces A, B, C (see Example 2 of Chap. 24). From Eq. (1) of Chap. 23 we have

$$U = \frac{1}{h} \sum_i c_i^A f_i^A = \frac{1}{h} \sum_i c_i^B f_i^B = \frac{1}{h} \sum_i c_i^C f_i^C , \tag{4}$$

where h is the Coxeter number of the hole and the c_i are defined in Fig. 23.1. Since the covering radius of Λ_{24} is $\sqrt{2}$,

$$N(f_i^A - U) = 2 , \quad N(f_i^A) = 2 ,$$

so

$$f_i^A \cdot U = \tfrac{1}{2} N(U) \tag{5}$$

But from (4),

$$h\, N(U) = \sum c_i^A f_i^A \cdot U = \tfrac{1}{2} N(U) \sum c_i^A = \tfrac{1}{2} h\, N(U) .$$

Therefore $N(U) = 0$, so we can take $u = U$, and from (5) $f_i^A \in u^\perp$. Thus all the f_i are in $u^\perp \cap \mathrm{II}_{25,1}$, and the lattice is the Niemeier lattice whose components agree with the components of the hole diagram.

Again, the sum of the vectors on the 25-gon visible in Fig. 27.3 is $((1/2)^{25} | 5/2)$, which is isotropic and proportional to the center of the corresponding hole of type A_{24} in the Leech lattice in H. Thus for $u = (1^{25} | 5)$, we see that $(u^\perp \cap \mathrm{II}_{25,1})/u$ is a Niemeier lattice of type A_{24}.

In a similar way the vectors

$$u_1 = (1^8, 3^9, 5^8 | 17) , \quad u_2 = (1^{13}, 3^{12} | 11) ,$$
$$u_3 = (1^{18}, 3^7 | 9) , \quad u_4 = (1^{15}, 3^9, 5 | 11) ,$$

have $(u_i^\perp \cap \mathrm{II}_{25,1})/u_i$ equal to the Niemeier lattices of types $A_8^3, A_{12}^2, A_{17}E_7, A_{15}D_9$ respectively, and other examples can be produced at will.

The final lemma shows that in these examples it does not matter whether we work in $\mathrm{I}_{n,1}$ or $\mathrm{II}_{n,1}$.

Lemma 6. *If* $u = \tfrac{1}{2}(u_0, u_1, \ldots, u_{n-1} | u_n) \in \mathrm{II}_{n,1}$ *is isotropic and all* u_i *are odd integers then*

$$u^\perp \cap \mathrm{I}_{n,1} = u^\perp \cap \mathrm{II}_{n,1} . \tag{6}$$

Proof. If $x = (x_0, \ldots, x_{n-1} | x_n) \in u^\perp \cap \mathrm{I}_{n,1}$ then

$$x_0 u_0 + \cdots + x_{n-1} u_{n-1} - x_n u_n = 0 ,$$
$$x_0 + \cdots + x_{n-1} - x_n \equiv 0 \pmod 2 ,$$

and $x \in \mathrm{II}_{n,1}$. Therefore, in (6), $LHS \subseteq RHS$. But $\det \mathrm{I}_{n,1} = \det \mathrm{II}_{n,1}$ $(=1)$, so $LHS = RHS$.

27

The Automorphism Group of the 26-Dimensional Lorentzian Lattice

J. H. Conway

The automorphism group of the lattice $II_{25,1}$ is shown to be a certain infinitely generated Coxeter group extended by the negative of the identity operation and the group of all automorphisms of the Leech lattice.

1. Introduction

The automorphism groups of the unimodular Lorentzian lattices $I_{n,1}$ and $II_{n,1}$ (defined in the previous chapter) are of considerable interest. In the present chapter we investigate $II_{25,1}$; the odd lattices $I_{n,1}$ will be considered in the succeeding chapter.

The study of these automorphism groups seems to have originated with Coxeter and Whitrow [Cox30] (see also [Cox28]), who identified the group of $I_{3,1}$ as a hyperbolic reflection group. More recently, Vinberg [Vin1]-[Vin4], [Vin7], Meyer [Mey2], and Vinberg and Kaplinskaja [Vin15] have shown that the reflection subgroup of $\mathrm{Aut}(I_{n,1})$ has finite index only if $n \leqslant 19$, and have given Coxeter diagrams for the reflection subgroups in these cases. In [Vin7] Vinberg has also shown that the reflection subgroups of $\mathrm{Aut}(II_{9,1})$ and $\mathrm{Aut}(II_{17,1})$ have finite index and are described by the Coxeter diagrams shown in Figs. 27.1 and 27.2.

The vectors with positive time coordinate and negative or zero norm in a Lorentzian space, taken modulo positive scalar factors, become the ordinary or ideal points, respectively, of the associated hyperbolic space. The symmetry group of the hyperbolic space can be identified with the *autochronous* symmetries of the Lorentzian space (those symmetries not interchanging the positive and negative time cones). The full group of symmetries of the Lorentzian space is the direct product of the autochronous subgroup with the group of order 2 generated by negation.

If L is a Euclidean or Lorentzian lattice, the reflecting hyperplanes that correspond to all the roots of L partition the corresponding Euclidean or hyperbolic space into fundamental regions (§2 of Chap. 4). The roots corresponding to the walls of any one fundamental region are a set of *fundamental roots*, for which the corresponding reflections generate the smallest group that contains all reflections in L. The latter group we call the reflection group of L. Since $II_{n,1}$ is unimodular, the roots all have norm 2.

We shall describe reflection groups by Coxeter diagrams using Vinberg's conventions [Vin7, p. 325]. These diagrams contain a node for each fundamental root R_i, and the reflection group is defined by the relations

$$R_i^2 = 1,$$
$$(R_i R_j)^2 = 1, \quad \text{if nodes } i \text{ and } j \text{ are unjoined,}$$
$$(R_i R_j)^3 = 1, \quad \text{if nodes } i \text{ and } j \text{ are joined.}$$

There are no other defining relations. These relations correspond to reflecting hyperplanes that intersect. Two nodes corresponding to non-intersecting hyperplanes are joined by a heavy line if the reflecting hyperplanes are parallel, or by a broken line if they are divergent.

2. The main theorem

Theorem 1. *For* $n = 9,17,25$ *there are respectively* $10,19,\infty$ *fundamental roots for* $II_{n,1}$, *which can be taken to be the vectors* $r \in II_{n,1}$, *with*

$$r \cdot r = 2 \text{ and } r \cdot w_n = -1 \ , \tag{1}$$

where the Weyl vectors[(1)] w_n *are given by*

$$w_9 = (0,1,\ldots,8\,|\,38), \qquad N(w_9) = -1240 \ ,$$
$$w_{17} = (0,1,\ldots,16\,|\,46), \qquad N(w_{17}) = -620 \ ,$$
$$w_{25} = (0,1,\ldots,24\,|\,70), \qquad N(w_{25}) = 0 \ .$$

The group of all autochronous automorphisms of $II_{n,1}$ *is the split extension of the Coxeter group whose generators are the corresponding reflections, by*

for $n = 9$, *the trivial group,*
for $n = 17$, *a group of order 2, or*
for $n = 25$, *an infinite group abstractly isomorphic to the group* Co_∞ *of all automorphisms of the Leech lattice (including translations).*

The corresponding Coxeter diagrams are displayed for $n = 9$ *in Fig. 27.1 and for* $n = 17$ *in Fig. 27.2. For* $n = 25$, *a portion of the diagram is*

---

(1) It would perhaps be more conventional to call $-w$ the Weyl vector (compare [Bou1]), but our choice is better in a hyperbolic space.

shown in Fig. 27.3. The full diagram has one node for each Leech lattice vector and (using Vinberg's conventions) two nodes r,s are joined by

no line	*if*	$N(r-s) = 4$,
an ordinary line	*if*	$N(r-s) = 6$,
a heavy line	*if*	$N(r-s) = 8$, *or*
a broken line	*if*	$N(r-s) \geqslant 10$.

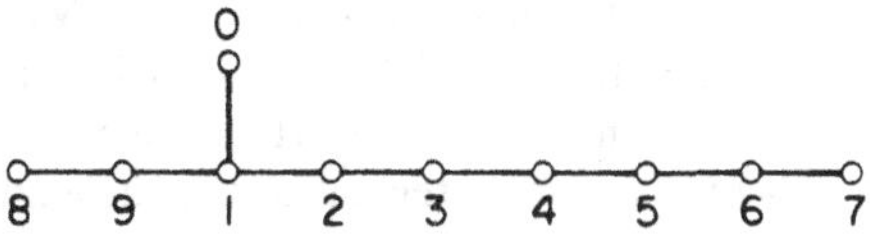

Figure 27.1. The Coxeter diagram for $II_{9,1}$. The points are: i : $(0^i, +1, -1, 0^{7-i} \mid 0)$, for $0 \leqslant i \leqslant 7$; 8 : $((½)^9 \mid ½)$; 9 : $(-1^2, 0^7 \mid 0)$.

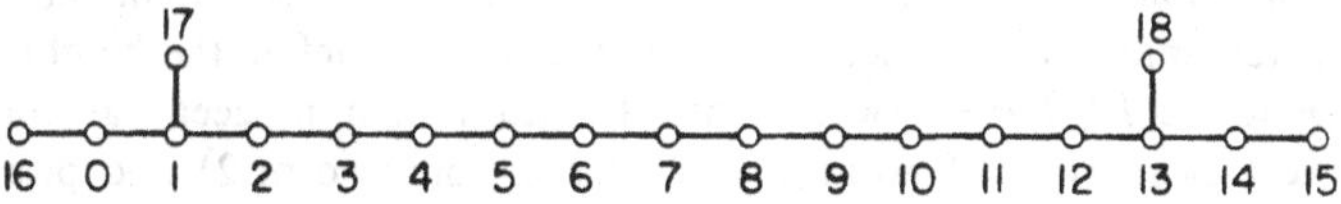

Figure 27.2. The Coxeter diagram for $II_{17,1}$. The points are: i : $(0^i, +1, -1, 0^{15-i} \mid 0)$, for $0 \leqslant i \leqslant 15$; 16 : $(-½, (½)^{16} \mid 3/2)$; 17 : $(-1^2, 0^{15} \mid 0)$; 18 : $(0^{14}, 1^3 \mid 1)$.

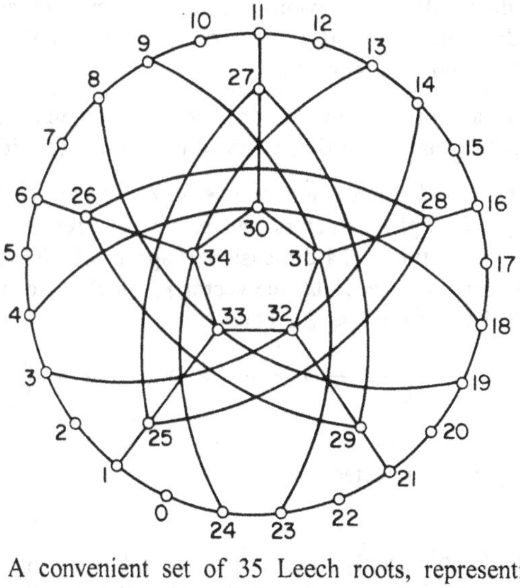

Figure 27.3. A convenient set of 35 Leech roots, representing the Leech lattice points closest to a deep hole of type A_{24}. The coordinates of the points are as follows: i : $(0^i, +1, -1, 0^{23-i}|0)$ for $0 \le i \le 23$; 24 : $(-1/2, (1/2)^{23}, 3/2|5/2)$; 25 : $(-1^2, 0^{23}|0)$; 26 : $(0^7, 1^{18}|4)$; 27 : $((1/2)^{12}, (3/2)^{13}|11/2)$; 28 : $((1/2)^{17}, (3/2)^8|9/2)$; 29 : $(0^{22}, 1^3|1)$; 30 : $(0^5, 1^{14}, 2^6|6)$; 31 : $(0^{10}, 1^{14}, 2|4)$; 32 : $(0^4, 1^{11}, 2^{10}|7)$; 33 : $((1/2)^9, (3/2)^{11}, (5/2)^5|15/2)$; 34 : $(0^{14}, 1^{11}|3)$.

Proof. It is easy to check that for $n = 9$ and 17 all r that satisfy the conditions (1) are displayed in Figs. 27.1 and 27.2. Since the diagrams of these figures agree with those given by Vinberg [Vin7, p. 347], the assertion of the theorem holds for $n = 9$ and 17.

When $n = 25$, Theorem 3 of the previous chapter shows that there is a one-to-one correspondence between the points of the Leech lattice and the vectors r that satisfy (1), namely the Leech roots. This correspondence is an isometry for the metric defined by $d(r,s)^2 = N(r-s)$. The assertion of the theorem will therefore follow if we can show that the Leech roots are precisely the fundamental roots for $II_{25,1}$, since they have the joins and symmetries indicated in the statement of the theorem.

We show this using the algorithm described by Vinberg for finding the fundamental roots for any discrete hyperbolic reflection group ([Vin7], §4 of Chap. 28). We take Vinberg's vector x_0 to be w_{25} (which satisfies his conditions, except of course those entailing that the algorithm terminates after finitely many vectors are produced), and define the height of a vector r to be $h(r) = -r \cdot w_{25}$. The algorithm then proceeds as follows. The vectors in $II_{25,1}$ of norm 1 or 2 (for us, only norm 2) and positive height are enumerated in increasing order of height$\cdot$norm$^{-1/2}$ (for us, in order of height). Vectors of height zero have norm at least 4, by the result of the previous chapter, and so need not be considered in our case. (In more general cases such vectors require special treatment when starting the algorithm.) A vector is rejected by the algorithm if it has strictly positive inner product with any previously accepted vector; otherwise it is accepted as a fundamental root. It is not necessary to test the condition between vectors of the same height.

It is clear that in our case all vectors of height 1 and norm 2 are accepted. We shall show that every other norm 2 candidate x is rejected.

Let $h(x) = h$, and define $v = x/h$. Then v lies in the affine hyperplane of height 1 vectors and so by the previous chapter specifies a point in the rational 24-dimensional space of the Leech lattice. By Chap. 23 there is a Leech lattice vector r, which we can regard as a Leech root, such that $N(v-r) \leqslant 2$. Thus

$$\frac{2}{h^2} - \frac{2}{h} x \cdot r + 2 \leqslant 2 \ , \quad x \cdot r \geqslant \frac{1}{h} \ ,$$

showing that x is rejected.

Remarks. (i) Abstractly, the reflection subgroup of $\mathrm{Aut}(II_{25,1})$ is the Coxeter group with a generator R_r for each Leech lattice vector r and having the presentation

$$\begin{aligned} (R_r)^2 &= 1 \ , \\ (R_r R_s)^2 &= 1 \quad \text{if } N(r-s) = 4 \ , \\ (R_r R_s)^3 &= 1 \quad \text{if } N(r-s) = 6 \end{aligned}$$

(and no other relations). The group of all autochronous automorphisms of $II_{25,1}$ is this Coxeter group extended by the group Co_∞ of its diagram automorphisms.

(ii) Several alternative proofs are known, all of which, however, depend on the main theorem of Chap. 23. It would be desirable to find a more direct proof.

(iii) We can show that for $n = 33, 41, \ldots$ there is no vector w_n having constant inner product with the fundamental roots for $II_{n,1}$.

(iv) The group Co_∞ of diagram automorphisms acts transitively on the fundamental roots! Compare $II_{9,1}$, where the diagram has no nontrivial automorphism, and $II_{17,1}$, where the automorphism group has order 2. (See also [Vin8]-[Vin12], [Bor3b].)

(v) Is there any connection with the Fischer-Griess Monster group? (Compare Chap. 30.)

28

Leech Roots and Vinberg Groups

J. H. Conway and N. J. A. Sloane

This chapter investigates the remarkable properties of the Leech roots, which are the fundamental roots for the even unimodular lattice in Lorentzian space $\mathbf{R}^{25,1}$, and correspond one for one with the points of the Leech lattice. We give an extensive table of the Leech roots in both Euclidean and hyperbolic coordinates. We also provide the first of what promise to be many applications by showing that the Leech roots simplify and explain the remarkable results of Vinberg, Kaplinskaja and Meyer on the reflection groups of unimodular Lorentzian lattices in dimensions below 20. They also enable us to make some progress on the study of these groups in the next few dimensions.

1. The Leech roots

The Leech roots are the vectors r in the even unimodular lattice $\mathrm{II}_{25,1}$ for which

$$r \cdot r = 2\,, \quad r \cdot w = -1\,,$$

where w is the Weyl vector

$$w = (0,1,2,...,23,24\,|\,70)\,.$$

Let T_1 be the affine hyperplane

$$T_1 = \{x \in \mathbf{R}^{25,1} : x \cdot w = -1\}\,.$$

Table 28.1. Leech roots r_i in hyperbolic coordinates $(x_0, x_1, \ldots, x_{24}\,|\,x_{25})$. (The 27 columns of the table give the index i and then $x_0, \ldots, x_{25}$, except that the roots with $i < 0$ are fractional and have been multiplied by 2. For example $r_{-1} = (-1/2, 1/2, \ldots, 1/2, 3/2\,|\,5/2)$. The table contains all roots with $x_{25} < 25$.)

i	x_0	x_1	x_2	x_3	x_4	x_5	x_6	x_7	x_8	x_9	x_{10}	x_{11}	x_{12}	x_{13}	x_{14}	x_{15}	x_{16}	x_{17}	x_{18}	x_{19}	x_{20}	x_{21}	x_{22}	x_{23}	x_{24}	x_{25}
0	−1	−1	0	0	0	0	0	0	0	0	0	0	0	0	0	0	0	0	0	0	0	0	0	0	0	0
1	1	−1	0	0	0	0	0	0	0	0	0	0	0	0	0	0	0	0	0	0	0	0	0	0	0	0
2	0	1	−1	0	0	0	0	0	0	0	0	0	0	0	0	0	0	0	0	0	0	0	0	0	0	0
3	0	0	1	−1	0	0	0	0	0	0	0	0	0	0	0	0	0	0	0	0	0	0	0	0	0	0
4	0	0	0	1	−1	0	0	0	0	0	0	0	0	0	0	0	0	0	0	0	0	0	0	0	0	0
5	0	0	0	0	1	−1	0	0	0	0	0	0	0	0	0	0	0	0	0	0	0	0	0	0	0	0
6	0	0	0	0	0	1	−1	0	0	0	0	0	0	0	0	0	0	0	0	0	0	0	0	0	0	0
7	0	0	0	0	0	0	1	−1	0	0	0	0	0	0	0	0	0	0	0	0	0	0	0	0	0	0
8	0	0	0	0	0	0	0	1	−1	0	0	0	0	0	0	0	0	0	0	0	0	0	0	0	0	0
9	0	0	0	0	0	0	0	0	1	−1	0	0	0	0	0	0	0	0	0	0	0	0	0	0	0	0
10	0	0	0	0	0	0	0	0	0	1	−1	0	0	0	0	0	0	0	0	0	0	0	0	0	0	0
11	0	0	0	0	0	0	0	0	0	0	1	−1	0	0	0	0	0	0	0	0	0	0	0	0	0	0
12	0	0	0	0	0	0	0	0	0	0	0	1	−1	0	0	0	0	0	0	0	0	0	0	0	0	0
13	0	0	0	0	0	0	0	0	0	0	0	0	1	−1	0	0	0	0	0	0	0	0	0	0	0	0
14	0	0	0	0	0	0	0	0	0	0	0	0	0	1	−1	0	0	0	0	0	0	0	0	0	0	0
15	0	0	0	0	0	0	0	0	0	0	0	0	0	0	1	−1	0	0	0	0	0	0	0	0	0	0
16	0	0	0	0	0	0	0	0	0	0	0	0	0	0	0	1	−1	0	0	0	0	0	0	0	0	0
17	0	0	0	0	0	0	0	0	0	0	0	0	0	0	0	0	1	−1	0	0	0	0	0	0	0	0
18	0	0	0	0	0	0	0	0	0	0	0	0	0	0	0	0	0	1	−1	0	0	0	0	0	0	0
19	0	0	0	0	0	0	0	0	0	0	0	0	0	0	0	0	0	0	1	−1	0	0	0	0	0	0
20	0	0	0	0	0	0	0	0	0	0	0	0	0	0	0	0	0	0	0	1	−1	0	0	0	0	0
21	0	0	0	0	0	0	0	0	0	0	0	0	0	0	0	0	0	0	0	0	1	−1	0	0	0	0
22	0	0	0	0	0	0	0	0	0	0	0	0	0	0	0	0	0	0	0	0	0	1	−1	0	0	0
23	0	0	0	0	0	0	0	0	0	0	0	0	0	0	0	0	0	0	0	0	0	0	1	−1	0	0
24	0	0	0	0	0	0	0	0	0	0	0	0	0	0	0	0	0	0	0	0	0	0	0	1	−1	0
25	0	0	0	0	0	0	0	0	0	0	0	0	0	0	0	0	0	0	0	0	0	0	1	1	1	1
26	0	0	0	0	0	0	0	0	0	0	0	0	0	0	1	1	1	1	1	1	1	1	1	1	1	3
27	0	0	0	0	0	0	0	1	1	1	1	1	1	1	1	1	1	1	1	1	1	1	1	1	1	4
28	0	0	0	0	0	0	0	0	0	0	1	1	1	1	1	1	1	1	1	1	1	1	1	1	2	4
29	0	0	0	0	0	1	1	1	1	1	1	1	1	1	1	1	1	1	1	2	2	2	2	2	2	6
30	0	0	0	0	0	0	0	0	1	1	1	1	1	1	1	1	1	1	2	2	2	2	2	2	2	6
31	0	0	0	0	1	1	1	1	1	1	1	1	1	1	1	2	2	2	2	2	2	2	2	2	2	7
32	0	0	0	0	0	0	1	1	1	1	1	1	1	1	1	1	2	2	2	2	2	2	2	2	3	7
33	0	0	0	0	0	1	1	1	1	1	1	1	1	2	2	2	2	2	2	2	2	2	2	3	3	8
34	0	0	0	0	1	1	1	1	1	1	1	2	2	2	2	2	2	2	2	2	2	3	3	3	3	9
35	0	0	0	1	1	1	1	1	1	1	1	1	1	2	2	2	2	2	2	2	3	3	3	3	3	9

Table 28.1 (cont.)

i	x_0	x_1	x_2	x_3	x_4	x_5	x_6	x_7	x_8	x_9	x_{10}	x_{11}	x_{12}	x_{13}	x_{14}	x_{15}	x_{16}	x_{17}	x_{18}	x_{19}	x_{20}	x_{21}	x_{22}	x_{23}	x_{24}	x_{25}
36	0	0	0	0	0	0	1	1	1	1	1	1	2	2	2	2	2	2	2	2	3	3	3	3	3	9
37	0	0	0	1	1	1	1	1	1	1	2	2	2	2	2	2	2	2	3	3	3	3	3	3	3	10
38	0	0	0	0	0	1	1	1	1	1	1	2	2	2	2	2	2	3	3	3	3	3	3	3	3	10
39	0	0	0	0	1	1	1	1	1	1	1	1	2	2	2	2	2	2	3	3	3	3	3	3	4	10
40	0	0	0	0	1	1	1	1	1	2	2	2	2	2	2	2	2	3	3	3	3	3	3	4	4	11
41	0	0	0	1	1	1	1	1	1	1	1	2	2	2	2	2	3	3	3	3	3	3	3	4	4	11
42	0	0	0	1	1	1	1	1	1	2	2	2	2	2	3	3	3	3	3	3	3	3	4	4	4	12
43	0	0	0	1	1	1	1	1	2	2	2	2	2	2	2	2	3	3	3	3	3	4	4	4	4	12
44	0	0	1	1	1	1	1	1	1	1	2	2	2	2	2	3	3	3	3	3	3	4	4	4	4	12
45	0	0	0	0	0	1	1	1	1	2	2	2	2	2	2	3	3	3	3	3	3	4	4	4	4	12
46	0	0	0	0	1	1	1	1	1	1	2	2	2	2	2	2	3	3	3	3	4	4	4	4	4	12
47	0	0	1	1	1	1	1	1	2	2	2	2	2	2	3	3	3	3	3	4	4	4	4	4	4	13
48	0	0	0	0	1	1	1	1	1	2	2	2	2	3	3	3	3	3	3	4	4	4	4	4	4	13
49	0	0	0	0	1	1	1	1	2	2	2	2	2	2	3	3	3	3	3	3	4	4	4	4	5	13
50	0	0	0	1	1	1	1	1	1	2	2	2	2	2	2	3	3	3	3	4	4	4	4	4	5	13
51	0	0	0	1	1	1	1	1	2	2	2	2	2	3	3	3	3	4	4	4	4	4	4	4	5	14
52	0	0	0	1	1	1	1	1	2	2	2	2	3	3	3	3	3	3	3	4	4	4	4	5	5	14
53	0	0	0	1	1	1	1	2	2	2	2	2	2	2	3	3	3	3	4	4	4	4	4	5	5	14
54	0	0	1	1	1	1	1	1	1	2	2	2	2	3	3	3	3	3	4	4	4	4	4	5	5	14
55	0	0	0	1	1	1	2	2	2	2	2	2	3	3	3	3	3	4	4	4	4	4	5	5	5	15
56	0	0	1	1	1	1	1	1	2	2	2	3	3	3	3	3	3	4	4	4	4	4	5	5	5	15
57	0	0	1	1	1	1	1	2	2	2	2	2	2	3	3	3	4	4	4	4	4	4	5	5	5	15
58	0	0	0	0	1	1	1	1	2	2	2	2	3	3	3	3	4	4	4	4	4	4	5	5	5	15
59	0	0	1	1	1	1	1	2	2	2	2	2	3	3	3	3	3	3	4	4	4	5	5	5	5	15
60	0	0	0	0	1	1	1	2	2	2	2	2	2	3	3	3	3	4	4	4	4	5	5	5	5	15
61	0	0	0	1	1	1	1	1	1	2	2	2	3	3	3	3	3	4	4	4	4	5	5	5	5	15
62	0	0	1	1	1	1	2	2	2	2	2	3	3	3	3	3	4	4	4	4	5	5	5	5	5	16
63	0	0	0	1	1	1	1	1	2	2	2	3	3	3	3	4	4	4	4	4	5	5	5	5	5	16
64	0	0	0	1	1	1	1	2	2	2	2	2	3	3	3	3	4	4	4	5	5	5	5	5	5	16
65	0	0	0	1	1	1	1	2	2	2	2	2	3	3	3	4	4	4	4	4	4	5	5	5	6	16
66	0	0	0	1	1	1	1	2	2	2	2	3	3	3	3	3	3	4	4	4	5	5	5	5	6	16
67	0	0	1	1	1	1	1	1	2	2	2	2	3	3	3	3	4	4	4	4	5	5	5	5	6	16
68	0	0	0	1	1	1	2	2	2	2	2	3	3	3	4	4	4	4	4	5	5	5	5	5	6	17
69	0	0	1	1	1	1	1	2	2	2	2	3	3	3	3	4	4	4	5	5	5	5	5	5	6	17
70	0	0	0	1	1	1	2	2	2	2	3	3	3	3	3	4	4	4	4	4	5	5	5	6	6	17
71	0	0	1	1	1	1	1	2	2	2	2	3	3	3	4	4	4	4	4	4	5	5	5	6	6	17

Table 28.1 (cont.)

i	x_0	x_1	x_2	x_3	x_4	x_5	x_6	x_7	x_8	x_9	x_{10}	x_{11}	x_{12}	x_{13}	x_{14}	x_{15}	x_{16}	x_{17}	x_{18}	x_{19}	x_{20}	x_{21}	x_{22}	x_{23}	x_{24}	x_{25}
72	0	0	1	1	1	1	1	2	2	2	3	3	3	3	3	3	4	4	4	5	5	5	5	6	6	17
73	0	0	1	1	1	1	2	2	2	2	2	2	3	3	3	4	4	4	4	5	5	5	5	6	6	17
74	0	0	0	0	1	1	1	2	2	2	2	3	3	3	3	4	4	4	4	5	5	5	5	6	6	17
75	0	0	1	1	1	1	2	2	2	2	3	3	3	3	4	4	4	5	5	5	5	5	5	6	6	18
76	0	0	1	1	1	1	2	2	2	2	3	3	3	4	4	4	4	4	4	5	5	5	6	6	6	18
77	0	0	1	1	1	1	2	2	2	3	3	3	3	3	3	4	4	4	5	5	5	5	6	6	6	18
78	0	0	1	1	1	2	2	2	2	2	2	3	3	3	4	4	4	4	5	5	5	5	6	6	6	18
79	0	1	1	1	1	1	1	2	2	2	3	3	3	3	4	4	4	4	5	5	5	5	6	6	6	18
80	0	0	0	0	1	1	2	2	2	2	3	3	3	3	4	4	4	4	5	5	5	5	6	6	6	18
81	0	0	0	1	1	1	1	2	2	2	2	3	3	4	4	4	4	4	5	5	5	5	6	6	6	18
82	0	0	0	1	1	1	1	2	2	2	3	3	3	3	3	4	4	5	5	5	5	5	6	6	6	18
83	0	0	0	1	1	1	1	2	2	2	3	3	3	3	4	4	4	4	4	5	5	6	6	6	6	18
84	0	0	0	1	1	1	2	2	2	2	2	3	3	3	3	4	4	4	5	5	5	6	6	6	6	18
85	0	0	1	1	1	2	2	2	2	3	3	3	3	4	4	4	4	5	5	5	5	6	6	6	6	19
86	0	0	0	1	1	1	2	2	2	2	3	3	3	4	4	4	5	5	5	5	5	6	6	6	6	19
87	0	0	0	1	1	1	2	2	2	3	3	3	3	3	4	4	4	5	5	5	6	6	6	6	6	19
88	0	0	1	1	1	1	1	2	2	2	3	3	3	4	4	4	4	5	5	5	6	6	6	6	6	19
89	0	0	0	1	1	1	2	2	2	3	3	3	3	4	4	4	4	4	5	5	5	6	6	6	7	19
90	0	0	0	1	1	2	2	2	2	2	3	3	3	3	4	4	4	5	5	5	5	6	6	6	7	19
91	0	0	1	1	1	1	1	2	2	3	3	3	3	3	4	4	4	5	5	5	5	6	6	6	7	19
92	0	0	1	1	1	1	2	2	2	2	2	3	3	4	4	4	4	5	5	5	5	6	6	6	7	19
93	0	0	1	1	1	1	2	2	2	2	3	3	3	3	4	4	4	4	5	5	6	6	6	6	7	19
94	0	0	0	1	1	2	2	2	2	3	3	3	4	4	4	4	5	5	5	5	6	6	6	6	7	20
95	0	0	1	1	1	1	2	2	2	3	3	3	3	4	4	5	5	5	5	5	6	6	6	6	7	20
96	0	0	1	1	1	1	2	2	2	3	3	3	4	4	4	4	4	5	5	6	6	6	6	6	7	20
97	0	0	1	1	1	2	2	2	2	2	3	3	3	4	4	4	5	5	5	6	6	6	6	6	7	20
98	0	0	1	1	1	1	2	2	2	3	3	3	4	4	4	4	5	5	5	5	5	6	6	7	7	20
99	0	0	1	1	1	1	2	2	3	3	3	3	3	4	4	4	4	5	5	5	6	6	6	7	7	20
100	0	0	1	1	1	2	2	2	2	2	3	3	4	4	4	4	4	5	5	5	6	6	6	7	7	20
101	0	0	1	1	1	2	2	2	2	3	3	3	3	3	4	4	5	5	5	5	6	6	6	7	7	20
102	0	1	1	1	1	1	2	2	2	2	3	3	3	4	4	4	5	5	5	5	6	6	6	7	7	20
103	0	0	0	1	1	1	1	2	2	3	3	3	3	4	4	4	5	5	5	5	6	6	6	7	7	20
104	0	0	0	1	1	1	2	2	2	2	3	3	3	4	4	4	4	5	5	6	6	6	6	7	7	20
105	0	0	1	1	1	2	2	2	2	3	3	4	4	4	4	5	5	5	5	6	6	6	6	7	7	21
106	0	0	1	1	1	2	2	2	3	3	3	3	4	4	4	4	5	5	6	6	6	6	6	7	7	21
107	0	0	0	1	1	1	2	2	2	3	3	3	4	4	4	5	5	5	6	6	6	6	6	7	7	21

Table 28.1 (cont.)

i	x_0	x_1	x_2	x_3	x_4	x_5	x_6	x_7	x_8	x_9	x_{10}	x_{11}	x_{12}	x_{13}	x_{14}	x_{15}	x_{16}	x_{17}	x_{18}	x_{19}	x_{20}	x_{21}	x_{22}	x_{23}	x_{24}	x_{25}
108	0	0	1	1	1	2	2	2	3	3	3	3	4	4	4	5	5	5	5	5	6	6	7	7	7	21
109	0	0	1	1	2	2	2	2	2	3	3	3	4	4	4	4	5	5	5	6	6	6	7	7	7	21
110	0	1	1	1	1	1	2	2	3	3	3	3	4	4	4	4	5	5	5	6	6	6	7	7	7	21
111	0	0	0	1	1	1	2	2	2	3	3	4	4	4	4	4	5	5	5	6	6	6	7	7	7	21
112	0	1	1	1	1	2	2	2	2	3	3	3	3	4	4	5	5	5	5	6	6	6	7	7	7	21
113	0	0	0	1	1	1	2	2	3	3	3	3	3	4	4	5	5	5	5	6	6	6	7	7	7	21
114	0	0	0	1	1	2	2	2	2	2	3	3	4	4	4	5	5	5	5	6	6	6	7	7	7	21
115	0	0	1	1	1	1	1	2	2	3	3	3	4	4	4	5	5	5	5	6	6	6	7	7	7	21
116	0	0	0	1	1	2	2	2	2	3	3	3	3	4	4	4	5	5	6	6	6	6	7	7	7	21
117	0	0	1	1	1	1	2	2	2	2	3	3	4	4	4	4	5	5	6	6	6	6	7	7	7	21
118	0	0	1	1	1	1	2	2	2	3	3	3	3	4	4	4	5	5	5	6	6	7	7	7	7	21
119	0	1	1	1	1	2	2	2	3	3	3	4	4	4	4	5	5	5	6	6	6	7	7	7	7	22
120	0	0	0	1	1	2	2	2	3	3	3	3	4	4	5	5	5	5	6	6	6	7	7	7	7	22
121	0	0	1	1	1	1	2	2	2	3	3	4	4	4	5	5	5	5	6	6	6	7	7	7	7	22
122	0	0	1	1	1	1	2	2	3	3	3	3	4	4	4	5	5	6	6	6	6	7	7	7	7	22
123	0	0	1	1	1	2	2	2	2	3	3	3	4	4	4	5	5	5	6	6	7	7	7	7	7	22
124	0	0	0	1	1	2	2	2	3	3	3	4	4	4	4	5	5	5	6	6	6	6	7	7	8	22
125	0	0	1	1	1	1	2	2	3	3	3	3	4	4	5	5	5	5	6	6	6	6	7	7	8	22
126	0	0	1	1	1	2	2	2	2	3	3	3	4	4	4	5	5	6	6	6	6	6	7	7	8	22
127	0	0	1	1	1	1	2	2	3	3	3	4	4	4	4	5	5	5	5	6	6	7	7	7	8	22
128	0	0	1	1	1	2	2	2	2	3	3	3	4	4	5	5	5	5	5	6	6	7	7	7	8	22
129	0	0	1	1	1	2	2	2	2	3	3	4	4	4	4	4	5	5	6	6	6	7	7	7	8	22
130	0	0	1	1	1	2	2	2	3	3	3	3	3	4	4	5	5	5	6	6	6	7	7	7	8	22
131	0	1	1	1	1	1	2	2	2	3	3	3	4	4	4	5	5	5	6	6	6	7	7	7	8	22
132	0	0	1	1	1	2	2	2	3	3	3	4	4	4	5	5	6	6	6	6	6	7	7	7	8	23
133	0	0	1	1	1	2	2	2	3	3	3	4	4	5	5	5	5	5	6	6	7	7	7	7	8	23
134	0	0	1	1	1	2	2	2	3	3	4	4	4	4	4	5	5	6	6	6	7	7	7	7	8	23
135	0	0	1	1	1	2	2	3	3	3	3	3	4	4	5	5	5	6	6	6	7	7	7	7	8	23
136	0	0	1	1	2	2	2	2	2	3	3	4	4	4	5	5	5	6	6	6	7	7	7	7	8	23
137	0	1	1	1	1	1	2	2	3	3	3	4	4	4	5	5	5	6	6	6	7	7	7	7	8	23
138	0	0	1	1	1	2	2	2	3	3	4	4	4	4	5	5	5	5	6	6	6	7	7	8	8	23
139	0	0	1	1	1	2	2	3	3	3	3	4	4	4	4	5	5	6	6	6	6	7	7	8	8	23
140	0	0	1	1	2	2	2	2	3	3	3	3	4	4	5	5	5	6	6	6	6	7	7	8	8	23
141	0	1	1	1	1	2	2	2	2	3	3	4	4	4	5	5	5	6	6	6	6	7	7	8	8	23
142	0	0	0	1	1	1	2	2	3	3	3	4	4	4	5	5	5	6	6	6	6	7	7	8	8	23
143	0	0	1	1	2	2	2	2	3	3	3	4	4	4	4	5	5	5	6	6	7	7	7	8	8	23

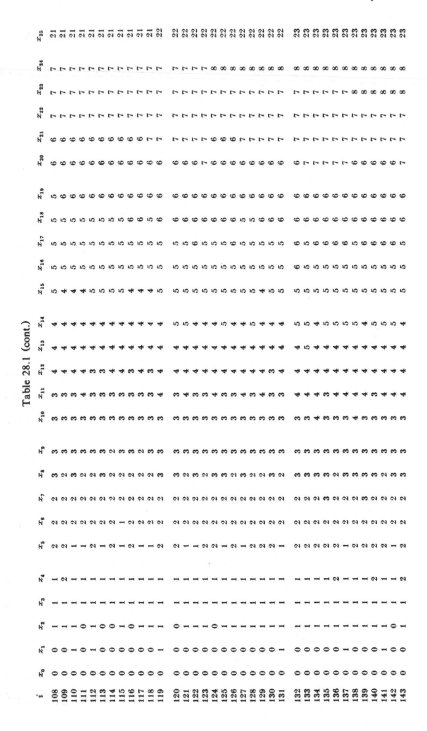

Table 28.1 (cont.)

i	x_0	x_1	x_2	x_3	x_4	x_5	x_6	x_7	x_8	x_9	x_{10}	x_{11}	x_{12}	x_{13}	x_{14}	x_{15}	x_{16}	x_{17}	x_{18}	x_{19}	x_{20}	x_{21}	x_{22}	x_{23}	x_{24}	x_{25}
144	0	1	1	1	1	2	2	2	3	3	3	3	4	4	5	5	5	5	6	6	7	7	7	8	8	23
145	0	0	0	1	1	2	2	2	2	3	3	4	4	4	5	5	5	5	6	6	7	7	7	8	8	23
146	0	0	0	1	1	2	2	2	3	3	3	3	4	4	4	5	5	6	6	6	7	7	7	8	8	23
147	0	0	1	1	1	1	2	2	2	3	3	4	4	4	4	5	5	6	6	6	7	7	7	8	8	23
148	0	0	1	1	2	2	2	2	3	3	4	4	4	5	5	5	6	6	6	6	7	7	7	8	8	24
149	0	0	1	1	2	2	2	3	3	3	3	4	4	5	5	5	5	6	6	7	7	7	7	8	8	24
150	0	1	1	1	1	2	2	2	3	3	4	4	4	5	5	5	5	6	6	7	7	7	7	8	8	24
151	0	1	1	1	1	2	2	3	3	3	3	4	4	4	5	5	6	6	6	7	7	7	7	8	8	24
152	0	0	0	1	1	2	2	2	3	3	4	4	4	4	5	5	6	6	6	7	7	7	7	8	8	24
153	0	0	1	1	1	1	2	2	3	3	3	4	4	5	5	5	6	6	6	7	7	7	7	8	8	24
154	0	0	1	1	2	2	2	3	3	3	4	4	4	4	5	5	5	6	6	6	7	7	8	8	8	24
155	0	1	1	1	1	2	2	3	3	3	3	4	4	5	5	5	5	6	6	6	7	7	8	8	8	24
156	0	0	0	1	1	2	2	2	3	3	4	4	4	5	5	5	5	6	6	6	7	7	8	8	8	24
157	0	1	1	1	2	2	2	2	3	3	3	4	4	4	5	5	6	6	6	6	7	7	8	8	8	24
158	0	0	0	1	1	2	2	3	3	3	3	4	4	4	5	5	6	6	6	6	7	7	8	8	8	24
159	0	0	1	1	1	1	2	2	3	3	4	4	4	4	5	5	6	6	6	6	7	7	8	8	8	24
160	0	0	1	1	1	2	2	2	2	3	3	4	4	5	5	5	6	6	6	6	7	7	8	8	8	24
161	0	0	0	1	2	2	2	2	3	3	3	4	4	4	5	5	5	6	6	7	7	7	8	8	8	24
162	0	0	1	1	1	1	2	3	3	3	3	4	4	4	5	5	5	6	6	7	7	7	8	8	8	24
163	0	0	1	1	1	2	2	2	2	3	4	4	4	4	5	5	5	6	6	7	7	7	8	8	8	24
164	0	0	1	1	1	2	2	2	3	3	3	3	4	5	5	5	5	6	6	7	7	7	8	8	8	24
165	0	0	1	1	1	2	2	2	3	3	3	4	4	4	4	5	6	6	6	7	7	7	8	8	8	24
166	0	0	1	1	1	2	2	2	3	3	3	4	4	4	5	5	5	6	6	6	7	8	8	8	8	24
167	0	0	1	1	1	2	2	2	3	3	3	4	4	4	5	5	5	6	6	7	7	7	7	8	9	24
−1	−1	1	1	1	1	1	1	1	1	1	1	1	1	1	1	1	1	1	1	1	1	1	1	1	3	5
−2	1	1	1	1	1	1	1	1	1	1	1	1	1	1	1	1	1	3	3	3	3	3	3	3	3	9
−3	1	1	1	1	1	1	1	1	1	1	1	1	3	3	3	3	3	3	3	3	3	3	3	3	3	11
−4	1	1	1	1	1	1	1	1	1	3	3	3	3	3	3	3	3	3	3	3	5	5	5	5	5	15
−5	−1	1	1	1	1	1	1	1	1	1	1	3	3	3	3	3	3	3	3	5	5	5	5	5	5	15
−6	−1	1	1	1	1	1	1	1	1	3	3	3	3	3	3	3	5	5	5	5	5	5	5	5	5	17
−7	1	1	1	1	1	1	1	1	3	3	3	3	3	3	3	5	5	5	5	5	5	5	5	7	7	19
−8	1	1	1	1	1	1	3	3	3	3	3	3	3	3	5	5	5	5	5	5	5	7	7	7	7	21
−9	−1	1	1	1	1	1	1	1	3	3	3	3	3	5	5	5	5	5	5	5	5	7	7	7	7	21
−10	−1	1	1	1	1	1	1	3	3	3	3	3	3	3	3	5	5	5	5	5	7	7	7	7	7	21
−11	−1	1	1	1	1	1	3	3	3	3	3	3	3	5	5	5	5	5	7	7	7	7	7	7	7	23
−12	1	1	1	1	1	1	1	3	3	3	3	3	3	5	5	5	5	5	5	7	7	7	7	7	9	23

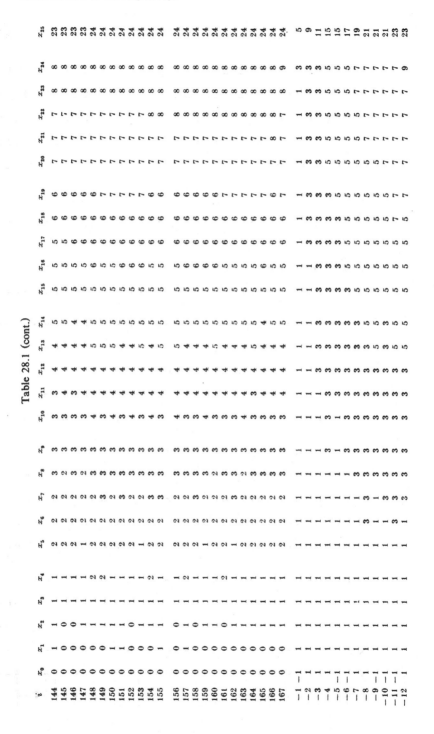

Table 28.1 (cont.)

i	x_0	x_1	x_2	x_3	x_4	x_5	x_6	x_7	x_8	x_9	x_{10}	x_{11}	x_{12}	x_{13}	x_{14}	x_{15}	x_{16}	x_{17}	x_{18}	x_{19}	x_{20}	x_{21}	x_{22}	x_{23}	x_{24}	x_{25}
−13	1	1	1	1	1	1	3	3	3	3	3	5	5	5	5	5	5	5	7	7	7	7	7	9	9	25
−14	−1	1	1	1	1	1	1	3	3	3	3	3	5	5	5	5	5	7	7	7	7	7	7	9	9	25
−15	1	1	1	1	1	3	3	3	3	3	5	5	5	5	5	5	7	7	7	7	7	7	9	9	9	27
−16	−1	1	1	1	1	1	3	3	3	3	3	5	5	5	5	7	7	7	7	7	7	7	9	9	9	27
−17	−1	1	1	1	1	1	3	3	3	3	5	5	5	5	5	5	5	7	7	7	7	9	9	9	9	27
−18	−1	1	1	1	1	3	3	3	3	3	3	3	5	5	5	5	7	7	7	7	7	9	9	9	9	27
−19	1	1	1	1	1	1	1	3	3	3	3	5	5	5	5	5	7	7	7	7	7	9	9	9	9	27
−20	−1	1	1	1	1	3	3	3	3	3	5	5	5	5	7	7	7	7	7	7	9	9	9	9	9	29
−21	1	1	1	1	1	1	3	3	3	3	5	5	5	5	5	7	7	7	7	9	9	9	9	9	9	29
−22	1	1	1	1	1	3	3	3	3	3	3	5	5	5	5	7	7	7	7	7	9	9	9	9	11	29
−23	1	1	1	1	1	3	3	3	3	5	5	5	5	5	7	7	7	7	9	9	9	9	9	9	11	31
−24	1	1	1	1	3	3	3	3	3	3	5	5	5	5	7	7	7	7	7	9	9	9	9	11	11	31
−25	−1	1	1	1	1	1	3	3	3	5	5	5	5	5	7	7	7	7	7	9	9	9	9	11	11	31
−26	−1	1	1	1	1	3	3	3	3	3	5	5	5	5	5	7	7	7	9	9	9	9	9	11	11	31
−27	1	1	1	1	3	3	3	3	5	5	5	5	5	7	7	7	7	7	9	9	9	9	11	11	11	33
−28	−1	1	1	1	1	3	3	3	3	5	5	5	7	7	7	7	7	7	9	9	9	9	11	11	11	33
−29	−1	1	1	1	1	3	3	3	5	5	5	5	5	5	7	7	7	9	9	9	9	9	11	11	11	33
−30	−1	1	1	1	3	3	3	3	3	3	5	5	5	7	7	7	7	9	9	9	9	9	11	11	11	33
−31	1	1	1	1	1	1	3	3	3	5	5	5	5	7	7	7	7	9	9	9	9	9	11	11	11	33
−32	−1	1	1	1	3	3	3	3	3	5	5	5	5	5	7	7	7	7	9	9	9	11	11	11	11	33
−33	1	1	1	1	1	3	3	3	3	3	5	5	5	7	7	7	7	7	9	9	9	11	11	11	11	33
−34	−1	1	1	1	3	3	3	3	5	5	5	5	7	7	7	7	7	9	9	9	11	11	11	11	11	35
−35	1	1	1	1	1	3	3	3	5	5	5	5	5	7	7	7	9	9	9	9	11	11	11	11	11	35
−36	1	1	1	1	1	3	3	3	5	5	5	5	7	7	7	7	7	9	9	9	9	11	11	11	13	35
−37	1	1	1	1	3	3	3	3	3	5	5	5	5	7	7	7	9	9	9	9	9	11	11	11	13	35
−38	−1	1	1	1	1	3	3	3	3	5	5	5	5	7	7	7	7	9	9	9	11	11	11	11	13	35
−39	1	1	1	1	3	3	3	3	5	5	5	7	7	7	7	7	9	9	9	11	11	11	11	11	13	37
−40	−1	1	1	1	1	3	3	3	5	5	5	5	7	7	7	9	9	9	9	11	11	11	11	11	13	37
−41	1	1	1	1	3	3	3	5	5	5	5	5	7	7	7	7	9	9	9	9	11	11	11	13	13	37
−42	−1	1	1	1	1	3	3	3	5	5	5	7	7	7	7	7	9	9	9	9	11	11	11	13	13	37
−43	−1	1	1	1	3	3	3	3	3	5	5	5	7	7	7	9	9	9	9	9	11	11	11	13	13	37
−44	−1	1	1	1	3	3	3	3	5	5	5	5	5	7	7	7	9	9	9	11	11	11	11	13	13	37
−45	1	1	1	1	1	3	3	3	3	5	5	5	7	7	7	7	9	9	9	11	11	11	11	13	13	37
−46	−1	1	1	1	3	3	3	3	5	5	5	7	7	7	9	9	9	9	11	11	11	11	11	13	13	39
−47	1	1	1	3	3	3	3	5	5	5	5	7	7	7	7	9	9	9	9	11	11	11	13	13	13	39
−48	−1	1	1	1	3	3	3	3	5	5	7	7	7	7	7	9	9	9	9	11	11	11	13	13	13	39

Table 28.1 (cont.)

i	x_0	x_1	x_2	x_3	x_4	x_5	x_6	x_7	x_8	x_9	x_{10}	x_{11}	x_{12}	x_{13}	x_{14}	x_{15}	x_{16}	x_{17}	x_{18}	x_{19}	x_{20}	x_{21}	x_{22}	x_{23}	x_{24}	x_{25}
−49	−1	1	1	1	3	3	3	5	5	5	5	5	7	7	9	9	9	9	9	11	11	11	13	13	13	39
−50	1	1	1	1	1	3	3	3	5	5	5	7	7	7	9	9	9	9	9	11	11	11	13	13	13	39
−51	−1	1	1	1	3	3	3	5	5	5	5	7	7	7	7	7	9	9	11	11	11	11	13	13	13	39
−52	−1	1	1	3	3	3	3	3	5	5	5	5	7	7	7	9	9	9	11	11	11	11	13	13	13	39
−53	1	1	1	1	1	3	3	5	5	5	5	5	7	7	7	9	9	9	11	11	11	11	13	13	13	39
−54	1	1	1	1	3	3	3	3	3	5	5	7	7	7	7	9	9	9	11	11	11	11	13	13	13	39
−55	1	1	1	1	3	3	3	3	5	5	5	5	7	7	7	9	9	9	9	11	11	13	13	13	13	39
−56	−1	1	1	3	3	3	3	5	5	5	5	7	7	7	9	9	9	11	11	11	11	13	13	13	13	41
−57	1	1	1	1	3	3	3	3	5	5	7	7	7	7	9	9	9	11	11	11	11	13	13	13	13	41
−58	1	1	1	1	3	3	3	5	5	5	5	7	7	7	9	9	9	9	11	11	13	13	13	13	13	41
−59	1	1	1	1	3	3	3	5	5	5	5	7	7	7	9	9	9	11	11	11	11	11	13	13	15	41
−60	1	1	1	1	3	3	3	5	5	5	7	7	7	7	7	9	9	9	11	11	11	13	13	13	15	41
−61	1	1	1	3	3	3	3	3	5	5	5	7	7	7	9	9	9	9	11	11	11	13	13	13	15	41
−62	−1	1	1	1	1	3	3	5	5	5	5	7	7	7	9	9	9	9	11	11	11	13	13	13	15	41
−63	−1	1	1	1	3	3	3	3	5	5	5	7	7	7	7	9	9	11	11	11	11	13	13	13	15	41
−64	1	1	1	1	3	3	5	5	5	5	7	7	7	9	9	9	9	11	11	11	13	13	13	13	15	43
−65	1	1	1	3	3	3	3	5	5	5	7	7	7	7	9	9	11	11	11	11	13	13	13	13	15	43
−66	−1	1	1	1	3	3	3	5	5	5	5	7	7	9	9	9	11	11	11	11	13	13	13	13	15	43
−67	−1	1	1	1	3	3	3	5	5	5	7	7	7	7	9	9	9	11	11	13	13	13	13	13	15	43
−68	1	1	1	3	3	3	3	5	5	5	7	7	7	9	9	9	9	11	11	11	11	13	13	15	15	43
−69	−1	1	1	1	3	3	3	5	5	5	7	7	7	7	9	9	11	11	11	11	11	13	13	15	15	43
−70	−1	1	1	1	3	3	3	5	5	5	7	7	7	9	9	9	9	9	11	11	13	13	13	15	15	43
−71	−1	1	1	1	3	3	5	5	5	5	5	7	7	7	9	9	9	11	11	11	13	13	13	15	15	43
−72	−1	1	1	3	3	3	3	3	5	5	7	7	7	7	9	9	9	11	11	11	13	13	13	15	15	43
−73	1	1	1	1	1	3	3	5	5	5	7	7	7	7	9	9	9	11	11	11	13	13	13	15	15	43
−74	1	1	1	1	3	3	3	3	5	5	5	7	7	9	9	9	9	11	11	11	13	13	13	15	15	43
−75	1	1	1	3	3	3	5	5	5	7	7	7	7	9	9	9	11	11	11	13	13	13	13	15	15	45
−76	−1	1	1	1	3	3	5	5	5	5	7	7	9	9	9	9	11	11	11	13	13	13	13	15	15	45
−77	−1	1	1	3	3	3	3	5	5	5	7	7	7	9	9	11	11	11	11	13	13	13	13	15	15	45
−78	1	1	1	1	3	3	3	5	5	5	7	7	7	9	9	9	11	11	13	13	13	13	13	15	15	45
−79	−1	1	1	1	3	3	5	5	5	7	7	7	7	9	9	9	11	11	11	11	13	13	15	15	15	45
−80	−1	1	1	3	3	3	3	5	5	5	7	7	9	9	9	9	11	11	11	11	13	13	15	15	15	45
−81	1	1	1	1	3	3	3	5	5	5	7	7	7	9	9	11	11	11	11	11	13	13	15	15	15	45
−82	−1	1	1	3	3	3	3	5	5	7	7	7	7	9	9	9	9	11	11	13	13	13	15	15	15	45
−83	1	1	1	1	3	3	3	5	5	5	7	7	9	9	9	9	9	11	11	13	13	13	15	15	15	45
−84	−1	1	1	3	3	3	5	5	5	5	7	7	7	7	9	9	11	11	11	13	13	13	15	15	15	45

Table 28.1. (cont.)

i	x_0	x_1	x_2	x_3	x_4	x_5	x_6	x_7	x_8	x_9	x_{10}	x_{11}	x_{12}	x_{13}	x_{14}	x_{15}	x_{16}	x_{17}	x_{18}	x_{19}	x_{20}	x_{21}	x_{22}	x_{23}	x_{24}	x_{25}
−85	1	1	1	1	3	3	3	5	5	7	7	7	7	7	9	9	11	11	11	13	13	13	15	15	15	45
−86	1	1	1	1	3	3	5	5	5	5	5	7	7	9	9	9	11	11	11	13	13	13	15	15	15	45
−87	1	1	1	3	3	3	3	3	5	5	7	7	7	9	9	9	11	11	11	13	13	13	15	15	15	45
−88	−1	1	1	1	1	3	3	5	5	5	7	7	7	9	9	9	11	11	11	13	13	13	15	15	15	45
−89	−1	1	1	3	3	3	5	5	5	7	7	7	9	9	9	11	11	11	11	13	13	15	15	15	15	47
−90	1	1	1	1	3	3	5	5	5	7	7	7	9	9	9	9	11	11	13	13	13	15	15	15	15	47
−91	1	1	1	3	3	3	3	5	5	7	7	7	7	9	9	11	11	11	13	13	13	15	15	15	15	47
−92	−1	1	1	1	3	3	3	5	5	5	7	7	9	9	9	11	11	11	13	13	13	15	15	15	15	47
−93	1	1	1	1	3	3	5	5	5	7	7	7	9	9	9	11	11	11	11	13	13	13	15	15	17	47
−94	1	1	1	3	3	3	3	5	5	7	7	7	9	9	9	9	11	11	13	13	13	13	15	15	17	47
−95	1	1	1	3	3	3	5	5	5	5	7	7	7	9	9	11	11	11	13	13	13	13	15	15	17	47
−96	−1	1	1	1	3	3	3	5	5	7	7	7	7	9	9	11	11	11	13	13	13	13	15	15	17	47
−97	1	1	1	3	3	3	5	5	5	5	7	7	9	9	9	9	11	11	11	13	13	15	15	15	17	47
−98	−1	1	1	1	3	3	3	5	5	7	7	7	9	9	9	9	11	11	11	13	13	15	15	15	17	47
−99	−1	1	1	1	3	3	5	5	5	5	7	7	7	9	9	11	11	11	11	13	13	15	15	15	17	47
−100	−1	1	1	3	3	3	3	5	5	5	7	7	7	9	9	9	11	11	13	13	13	15	15	15	17	47
−101	1	1	1	3	3	3	5	5	5	7	7	9	9	9	9	11	11	13	13	13	13	15	15	15	17	49
−102	−1	1	1	1	3	3	5	5	5	7	7	7	9	9	11	11	11	13	13	13	13	15	15	15	17	49
−103	1	1	1	3	3	3	5	5	7	7	7	7	9	9	9	11	11	11	13	13	15	15	15	15	17	49
−104	−1	1	1	1	3	3	5	5	5	7	7	9	9	9	9	11	11	11	13	13	15	15	15	15	17	49
−105	−1	1	1	3	3	3	3	5	5	7	7	7	9	9	11	11	11	11	13	13	15	15	15	15	17	49
−106	−1	1	1	3	3	3	5	5	5	5	7	7	9	9	9	11	11	13	13	13	15	15	15	15	17	49
−107	1	1	1	1	3	3	3	5	5	7	7	7	9	9	9	11	11	13	13	13	15	15	15	15	17	49
−108	1	1	1	3	3	5	5	5	5	7	7	7	9	9	9	11	11	11	13	13	13	15	15	17	17	49
−109	−1	1	1	1	3	3	5	5	7	7	7	7	9	9	9	11	11	11	13	13	13	15	15	17	17	49
−110	−1	1	1	3	3	3	3	5	5	7	7	9	9	9	9	11	11	11	13	13	13	15	15	17	17	49
−111	−1	1	1	3	3	3	5	5	5	5	7	7	9	9	11	11	11	11	13	13	13	15	15	17	17	49
−112	1	1	1	1	3	3	3	5	5	7	7	7	9	9	11	11	11	11	13	13	13	15	15	17	17	49
−113	−1	1	1	3	3	3	5	5	5	7	7	7	7	9	9	11	11	13	13	13	13	15	15	17	17	49
−114	1	1	1	1	3	3	5	5	5	5	7	7	9	9	9	11	11	13	13	13	13	15	15	17	17	49
−115	−1	1	1	3	3	3	5	5	5	7	7	7	9	9	9	9	11	11	13	13	15	15	15	17	17	49
−116	1	1	1	1	3	3	5	5	5	7	7	7	7	9	9	11	11	11	13	13	15	15	15	17	17	49
−117	1	1	1	3	3	3	3	5	5	5	7	7	9	9	9	11	11	11	13	13	15	15	15	17	17	49

We saw in Chap. 26 that, modulo w, T_1 may be regarded as a 24-dimensional Euclidean space, in which the Leech roots form a (nonlinear) set isometric to the Leech lattice. The main result of Chap. 27 is that the reflection subgroup of $\mathrm{Aut}(\mathrm{II}_{25,1})$ is generated by the reflections in the Leech roots.

In the present chapter we enumerate some of the Leech roots: see Tables 28.1 and 28.2. We show in §§3-5 how the fundamental roots for $\mathrm{I}_{n,1}$ ($n \leqslant 19$) may be easily derived from the Leech roots (see especially Theorems 1 and 2 in §6). The description in terms of Leech roots provides a natural explanation for the striking symmetries observed in the fundamental regions of $\mathrm{Aut}(\mathrm{I}_{14,1}),\ldots,\mathrm{Aut}(\mathrm{I}_{19,1})$ by Vinberg [Vin1]-[Vin4], [Vin7], Meyer [Mey2] and Vinberg and Kaplinskaja [Vin15]. We also give a partial description of the fundamental roots for $\mathrm{I}_{20,1},\ldots,\mathrm{I}_{24,1}$ and their relationship to the Leech roots (see Table 28.3). Figures 28.1a-h show (part of) the Coxeter-Vinberg diagrams for $\mathrm{Aut}(\mathrm{I}_{n,1})$, $13 \leqslant n \leqslant 20$.

2. Enumeration of the Leech roots

We discovered the relationship between the groups $\mathrm{Aut}(\mathrm{I}_{n,1})$ and the Leech lattice after examining an extensive list of Leech roots. Since these roots promise to have many other applications (see for example the final chapter), we have thought it desirable to provide a substantial tabulation. In Table 28.1 the roots are labeled (in the left-hand column) by nonnegative or negative integers according to whether their coordinates are integral or fractional. The ordering is otherwise lexicographic, first by the value of the timelike coordinate $t = x_{25}$, then by $|x_{24}|$, $|x_{23}|$, ...

It is in fact true that every Leech root has $t \geqslant 0$, and since this is not entirely obvious we provide a proof. Let z denote the vector

$$z = \frac{1}{70}\,(0^{25}\,|\,1) \ ,$$

where the notation indicates that the first 25 coordinates are zero. Since $z \cdot w = -1$ (i.e. $z \in T_1$), z can be regarded as a point in the real 24-space that contains the Leech lattice. If $x = (x_0,\ldots,x_{24}\,|\,t)$ is another point of T_1 of norm 2 we have

$$N(x-z) = 2 + \frac{t}{35} - \frac{1}{4900} \ ,$$

showing that t increases linearly with the squared distance from x to z. If x is one of the particular Leech roots $r_0,\ldots,r_{24}$ (see Table 28.1), we have $t = 0$, and so

$$N(x-z) = 2 - \frac{1}{4900} \ .$$

The coordinates of $r_0,\ldots,r_{24}$ show that they do not lie in a 23-space, and so they form a simplex whose circumcenter is z. In the language of Chap. 23 these points form an ordinary Dynkin diagram d_{25}, and they are all the vertices of a shallow hole, of radius $(2-1/4900)^{1/2}$, centered at z. Other

Leech roots r must therefore give strictly larger values of $N(r-z)$, and so have strictly positive values of t.

The Leech roots (r_i) in Table 28.1 include all those with $t \leqslant 25$, and were found by a backtracking program. We shall refer to the entries in this table as the hyperbolic coordinates for the roots. In Table 28.2 we give Euclidean coordinates for these roots. (The blocks of four numbers correspond to columns in the MOG arrangement of coordinates — see Chap. 11.)

Table 28.2 Euclidean coordinates $(y_0, \ldots, y_{23})$ for the same Leech roots r_i as in Table 28.1. (The columns give $i, y_0, \ldots, y_{23}$. The roots r_{-i} have not been multiplied by 2. Here + = +1, − = −1, b = −2, d = −4.)

0	4 0 0 0	0 0 0 0	0 0 0 0	0 0 0 0	0 0 0 0	0 0 0 0
1	0 0 0 0	*d* 0 0 0	0 0 0 0	0 0 0 0	0 0 0 0	0 0 0 0
2	− − − −	+ + + +	− − − −	+ + + +	+ + + +	+ + + +
3	0 0 0 0	0 0 0 0	4 0 0 0	0 0 0 0	0 0 0 0	0 0 0 0
4	+ − − −	− + + +	− + + +	− + + +	− + + +	− + + +
5	0 0 0 0	0 0 0 0	0 0 0 0	0 0 0 0	4 0 0 0	0 0 0 0
6	+ − − +	− − + +	+ − + −	+ + + +	− + − +	+ + + +
7	0 0 0 0	0 4 0 0	0 0 0 0	0 0 0 0	0 0 0 0	0 0 0 0
8	+ + − −	− − + +	+ + − −	− − + +	+ + + +	+ + + +
9	0 0 0 0	0 0 0 0	0 0 0 0	0 4 0 0	0 0 0 0	0 0 0 0
10	+ − − −	− + + +	+ − + +	+ − + +	+ − + +	+ − + +
11	0 0 0 0	0 0 0 0	0 0 0 0	0 0 0 0	0 0 0 0	0 4 0 0
12	+ − − +	− + − +	+ + − −	+ + + +	+ + + +	− − + +
13	0 0 0 0	0 0 4 0	0 0 0 0	0 0 0 0	0 0 0 0	0 0 0 0
14	+ − + −	− + − +	+ − + −	− + − +	+ + + +	+ + + +
15	0 0 0 0	0 0 0 0	0 0 0 0	0 0 4 0	0 0 0 0	0 0 0 0
16	+ − − −	− + + +	+ + − +	+ + − +	+ + − +	+ + − +
17	0 0 0 0	0 0 0 0	0 0 0 0	0 0 0 0	0 0 0 0	0 0 4 0
18	+ + − −	− + + −	+ − + −	+ + + +	+ + + +	− + − +
19	0 0 0 0	0 0 0 4	0 0 0 0	0 0 0 0	0 0 0 0	0 0 0 0
20	+ − − +	− + + −	+ − − +	− + + −	+ + + +	+ + + +
21	0 0 0 0	0 0 0 0	0 0 0 0	0 0 0 4	0 0 0 0	0 0 0 0
22	+ − − −	− + + +	+ + + −	+ + + −	+ + + −	+ + + −
23	0 0 0 0	0 0 0 0	0 0 0 0	0 0 0 0	0 0 0 4	0 0 0 0
24	+ − − −	− + + +	+ − − −	− + + +	+ − − −	− + + +
25	0 0 0 0	0 0 0 0	0 0 0 0	0 0 0 0	0 0 0 0	0 0 0 4
26	0 0 *d* 0	0 0 0 0	0 0 0 0	0 0 0 0	0 0 0 0	0 0 0 0
27	+ − + −	− − + +	+ − − +	+ + + +	+ + + +	− + + −
28	0 0 0 0	0 0 0 0	0 0 0 0	0 0 0 0	0 4 0 0	0 0 0 0
29	+ − + −	− + + −	+ + − −	+ + + +	− − + +	+ + + +
30	0 *d* 0 0	0 0 0 0	0 0 0 0	0 0 0 0	0 0 0 0	0 0 0 0
31	+ + − +	− + + +	+ − − −	+ + − +	+ − + +	+ + + −
32	0 0 0 0	0 0 0 0	0 0 0 0	0 0 0 0	0 0 4 0	0 0 0 0
33	+ + − −	− + − +	+ − − +	+ + + +	− + + −	+ + + +
34	+ + + −	− + + +	+ − − −	+ + + −	+ + − +	+ − + +
35	+ + − −	− + − +	− + + −	+ + + +	+ − − +	+ + + +
36	0 0 0 *d*	0 0 0 0	0 0 0 0	0 0 0 0	0 0 0 0	0 0 0 0
37	+ − + +	− + + +	− + − −	− + + +	+ + − +	+ + + −
38	+ − − −	− + + +	+ − − −	− + + +	− + + +	+ − − −
39	0 0 0 0	0 0 0 0	0 0 0 0	0 0 0 0	0 0 0 0	4 0 0 0
40	+ − + +	− + + +	+ − − −	+ − + +	+ + + −	+ + − +
41	+ + − +	− + + +	− − + −	− + + +	+ + + −	+ − + +
42	+ + − +	− + + +	− − − +	+ − + +	+ + − +	− + + +
43	+ − + +	− + + +	− − + −	+ + + −	+ − + +	− + + +
44	+ + − +	− + + +	+ + + −	− + − −	+ + − +	− + + +
45	+ − − −	− + + +	+ − − −	+ − − −	− + + +	− + + +

Table 28.2 (cont.)

46	0 0 0 0	0 0 0 0	0 0 0 *d*	0 0 0 0	0 0 0 0	0 0 0 0
47	+ − − +	− + + −	+ + + +	+ + + +	+ − − +	− + + −
48	+ − − −	− + − −	+ − − −	+ − + +	+ + − +	+ + + −
49	0 0 0 0	0 0 0 0	0 0 0 0	4 0 0 0	0 0 0 0	0 0 0 0
50	+ − − +	− + + −	− + + −	+ − − +	+ + + +	+ + + +
51	+ − − +	− + − +	− − + +	+ + + +	+ + + +	+ + − −
52	+ − + −	− + + −	− − + +	+ + + +	+ + − −	+ + + +
53	+ − − +	− − + +	− + − +	+ + + +	+ − + −	+ + + +
54	+ − − +	− + − +	+ + + +	− − + +	+ + − −	+ + + +
55	+ + + −	− + + +	− − − +	− + + +	+ − + +	+ + − +
56	+ − + +	− + + +	+ − + +	− + + +	+ + − +	− − − +
57	+ + − +	− − − +	+ − + +	− + + +	+ − + +	− + + +
58	0 0 0 0	0 0 0 0	0 *d* 0 0	0 0 0 0	0 0 0 0	0 0 0 0
59	+ − + −	− − + +	+ + + +	− + + −	+ − − +	+ + + +
60	+ − − −	− − − +	+ − − −	+ + + −	+ − + +	+ + − +
61	+ − − −	− + + +	− − + −	− − + −	+ + − +	+ + − +
62	+ + + −	− + + +	+ + + −	− + + +	+ − + +	− − + −
63	+ − − −	− + + +	− − + −	+ + − +	+ + − +	− − + −
64	+ − − −	− − + −	− − + −	− + + +	+ − + +	+ + + −
65	+ + − −	− − + +	− − + +	+ + − −	+ + + +	+ + + +
66	+ − + −	− − + +	− + + −	+ + + +	+ + + +	+ − − +
67	0 0 0 0	0 0 0 0	0 0 4 0	0 0 0 0	0 0 0 0	0 0 0 0
68	+ + − −	− + + −	− + − +	+ + + +	+ + + +	+ − + −
69	+ − − +	− − + +	+ + + +	− + − +	+ + + +	+ − + −
70	+ + + −	− + + +	− + − −	+ + − +	+ + + −	− + + +
71	+ + − −	− − + +	+ + + +	+ + + +	+ + − −	− − + +
72	+ − + +	− − + −	+ + + −	− + + +	+ + + −	− + + +
73	+ + − −	− + + −	+ + + +	− + − +	+ − + −	+ + + +
74	+ − − −	− − + −	+ − − −	+ + − +	+ + + −	+ − + +
75	+ + − +	− + + +	+ + − +	− + + +	+ + + −	− + − −
76	+ + + −	− + − −	+ + − +	− + + +	+ + − +	− + + +
77	+ − + +	− + + +	+ + − +	− − − +	+ − + +	− + + +
78	+ + − +	− + + +	+ + − +	− + + +	− − − +	+ − + +
79	− − − +	+ − + +	+ + − +	− + + +	+ + − +	− + + +
80	0 0 0 0	0 0 0 0	0 0 *d* 0	0 0 0 0	0 0 0 0	0 0 0 0
81	+ − − −	− − − +	− − − +	− + + +	+ + − +	+ − + +
82	+ − − +	− − + +	− − − −	− + − +	+ + + +	− + − +
83	+ − − −	− − + −	− + − −	+ + + −	+ + − +	− + + +
84	+ − − −	− + + +	− + − −	− + − −	+ − + +	+ − + +
85	2 0 0 0	*b* 2 0 2	2 0 0 0	0 0 2 0	0 0 0 2	0 2 0 0
86	+ + − −	− + − +	− − − −	− + + −	+ + + +	− + + −
87	+ − − −	− + + +	− + − −	+ − + +	+ − + +	− + − −
88	+ − − −	− − − +	+ + + −	− + + +	+ + − +	− + − −
89	+ − + −	− + − +	− + − +	+ − + −	+ + + +	+ + + +
90	+ + − +	− + + +	− + − −	+ + + −	− + + +	+ + − +
91	+ − − +	− − + +	+ + + +	+ − + −	+ + + +	− + − +
92	+ + − −	− + − +	+ + + +	− + + −	+ + + +	+ − − +
93	0 0 0 0	0 0 0 0	0 4 0 0	0 0 0 0	0 0 0 0	0 0 0 0
94	+ + + −	− + + +	− − + −	+ − + +	− + + +	+ + + −
95	+ + − −	− + − +	+ + + +	+ − − +	+ + + +	− + + −
96	+ − + −	− + + −	+ + + +	− − + +	+ + + +	+ + − −
97	+ + − +	− + − −	+ + + −	− + + +	− + + +	+ + + −
98	+ + + −	− + + +	+ − + +	− − + −	+ + + −	− + + +
99	+ − + −	− + − +	+ + + +	+ + + +	+ − + −	− + − +
100	+ + + −	− + + +	+ + + −	− + + +	− + − −	+ + − +
101	+ + − +	− + + +	+ + + −	+ − + +	− − + −	− + + +
102	− + − −	+ + − +	+ + + −	− + + +	+ + + −	− + + +
103	+ − − −	− − − +	− − + −	+ − + +	+ + + −	− + + +
104	+ − − −	− + − −	− + − −	− + + +	+ + + −	+ + − +
105	2 0 0 0	*b* 2 2 0	2 0 0 0	0 0 0 2	0 2 0 0	0 0 2 0
106	+ − + +	− + + +	+ − + +	− + + +	− − + −	+ + + −

Table 28.2 (cont.)

107	+ − − −	− + + +	− − − +	− − − +	+ + + −	+ + + −
108	+ + + −	− + + +	+ − + +	+ + − +	− − − +	− + + +
109	+ + + +	− + + −	+ − + −	− − + +	+ − − +	+ + + +
110	− − + −	+ + + −	+ − + +	− + + +	+ − + +	− + + +
111	+ − + −	− + + −	− − − −	− − + +	+ + + +	− − + +
112	0 0 *b* 0	0 2 0 0	2 0 0 0	0 0 0 2	0 0 0 2	0 2 2 2
113	+ − − −	− + − −	− − − +	+ + − +	+ − + +	− + + +
114	+ + − −	− + + −	− − − −	− + − +	− + − +	+ + + +
115	+ − − −	− − + −	+ − + +	− − − +	+ + − +	− + + +
116	+ − − +	− + − +	− − − −	− − + +	− − + +	+ + + +
117	0 0 0 0	0 0 0 0	0 0 0 0	*d* 0 0 0	0 0 0 0	0 0 0 0
118	+ − − −	− + − −	+ + + −	− − + −	+ − + +	− + + +
119	0 *b* 0 0	0 2 2 2	2 0 0 0	0 2 0 0	0 0 0 2	0 0 2 0
120	+ − − −	− + + +	− − − +	+ + + −	− − − +	+ + + −
121	+ − − −	− + + +	+ + − +	− − + −	+ + − +	− − + −
122	+ − − −	− + + +	+ − + +	− + − −	+ − + +	− + − −
123	+ − − −	− + + +	+ + + −	− − − +	− − − +	+ + + −
124	+ − + +	− + + +	− − − +	+ + − +	− + + +	+ − + +
125	0 0 0 0	0 0 0 0	0 0 0 4	0 0 0 0	0 0 0 0	0 0 0 0
126	+ + − +	− + + +	+ − + +	− − − +	− + + +	+ + − +
127	+ − + −	− + + −	+ + + +	+ + − −	+ + + +	− − + +
128	+ + − −	− + + −	+ + + +	+ − + −	− + − +	+ + + +
129	+ − + +	− + + +	+ + + −	− − + −	− + + +	+ − + +
130	+ − − +	− + − +	+ + + +	+ + − −	− − + +	+ + + +
131	− − − −	+ + + +	+ + + +	− − − −	+ + + +	+ + + +
132	+ + − +	− + + +	+ − + +	+ + + −	− + + +	− − + −
133	+ − + −	− + − +	+ + + +	+ + + +	− + − +	+ − + −
134	+ − + +	− + + +	+ + + −	+ + − +	− + + +	− + − −
135	+ + − −	− − + +	+ + + +	+ + + +	− − + +	+ + − −
136	+ + − +	− + + +	+ + + −	+ − + +	+ + − +	+ − − −
137	− − − −	+ + + +	+ + + +	+ + + +	+ + + +	− − − −
138	+ − + +	− + + +	+ + − +	+ + + −	− + − −	− + + +
139	2 0 0 0	*b* 0 2 2	2 0 0 0	0 2 0 0	0 0 2 0	0 0 0 2
140	+ + − +	− + + +	+ − + +	+ + + −	+ − − −	+ + − +
141	0 0 *b* 0	0 2 2 2	2 0 0 0	0 0 2 0	0 2 0 0	0 0 0 2
142	+ − − −	− + + +	− − − +	+ + + −	+ + + −	− − − +
143	+ − + +	− + + +	+ + + −	+ + − +	+ − − −	+ − + +
144	− − − −	+ + + +	+ + + +	+ + + +	− − − −	+ + + +
145	+ − − −	− + + +	− + − −	+ − + +	− + − −	+ − + +
146	+ − − −	− + + +	− − + −	+ + − +	− − + −	+ + − +
147	+ − − −	− + + +	+ + + −	− − − +	+ + + −	− − − +
148	+ + + +	− + − +	+ − + −	+ + + +	+ + − −	− + + −
149	2 0 0 0	*b* 0 0 0	2 0 0 0	0 2 2 2	2 0 0 0	2 0 0 0
150	0 *b* 0 0	0 2 0 0	2 0 0 0	0 2 2 2	0 2 0 0	0 2 0 0
151	0 0 *b* 0	0 0 2 0	2 0 0 0	0 2 2 2	0 0 2 0	0 0 2 0
152	+ − − +	− + + −	− − − −	+ + + +	− + + −	− + + −
153	+ − − −	− + − −	+ − + +	− + + +	+ + + −	− − + −
154	+ + + +	− − + +	+ + − −	+ + + +	+ − − +	− + − +
155	0 0 0 *b*	0 0 0 2	2 0 0 0	0 2 2 2	0 0 0 2	0 0 0 2
156	+ − + −	− + − +	− − − −	+ + + +	− + − +	− + − +
157	− + − +	+ + + +	+ − + −	+ + + +	+ − − +	− − + +
158	+ + − −	− − + +	− − − −	+ + + +	− − + +	− − + +
159	0 0 0 0	0 0 0 0	0 0 0 0	0 0 0 0	0 0 0 0	*d* 0 0 0
160	+ + − −	− + − +	+ − + −	− − + +	− + − +	− − + +
161	+ − − +	− + + −	− − − −	+ + + +	+ − − +	+ − − +
162	+ − − −	− − + −	+ + − +	− + + +	+ − + +	− − − +
163	+ − − +	− + + −	+ + − −	− − + +	− + − +	− + − +
164	+ − − −	− + − −	+ − + +	− + + +	− − − +	+ + − +
165	+ − − +	− + + −	+ − + −	− + − +	− − + +	− − + +
166	+ − − −	− + + +	+ + + −	+ + + −	− − − +	− − − +
167	+ − − +	− + + −	+ + + +	+ + + +	− + + −	+ − − +

Table 28.2 (cont.)

−1	+ + + +	+ + + +	+ + + +	+ + + +	+ + + +	+ + + +
−2	− − − +	− + + +	+ + + −	− + + +	+ − + +	+ + − +
−3	− + − −	− + + +	+ − + +	− + + +	+ + − +	+ + + −
−4	− − + −	− + + +	+ + + −	+ − + +	+ + − +	− + + +
−5	+ − − −	+ + + −	+ + + −	− + + +	+ + − +	+ − + +
−6	+ − − +	+ + + +	+ − + −	− − + +	+ + + +	− + + −
−7	− − − +	− + + +	+ − + +	+ + − +	+ + + −	− + + +
−8	− + − −	− + + +	+ + − +	+ + + −	+ − + +	− + + +
−9	+ − − −	+ + − +	+ − + +	+ + + −	+ + − +	− + + +
−10	+ − − −	+ − + +	+ + + −	+ + − +	+ − + +	− + + +
−11	+ − − −	+ + − +	+ + − +	− + + +	+ − + +	+ + + −
−12	− − − −	− − − −	+ + + +	+ + + +	+ + + +	+ + + +
−13	− − + −	− + + +	+ + − +	− + + +	+ + + −	+ − + +
−14	+ − − −	+ − + +	+ − + +	− + + +	+ + + −	+ + − +
−15	− + + +	− + + +	+ − − −	− + + +	− + + +	− + + +
−16	+ + − −	+ + + +	+ − − +	− + − +	+ + + +	− − + +
−17	+ − + −	+ + + +	+ + − −	− + + −	+ + + +	− + − +
−18	+ + − −	+ + + +	+ − + −	− + + −	− − + +	+ + + +
−19	− − − −	− − + +	+ − + −	− + + −	+ + + +	− − + +
−20	+ + − −	+ + + +	+ + − −	+ + + +	− + − +	− + + −
−21	− − − −	− + + −	+ + − −	− + − +	+ + + +	− + + −
−22	− + − −	− + + +	+ + + −	+ + − +	− + + +	+ − + +
−23	− − − +	− + + +	+ + − +	+ − + +	− + + +	+ + + −
−24	− + − +	− + + −	+ + − −	+ + + +	+ + − −	+ + + +
−25	+ − − −	+ + + −	+ + − +	+ − + +	+ + + −	− + + +
−26	+ − − +	+ + + +	+ + − −	− + − +	− + + −	+ + + +
−27	− − + +	− + − +	+ − − +	+ + + +	+ − − +	+ + + +
−28	+ − + −	+ + + +	+ − − +	− − + +	− + − +	+ + + +
−29	+ − − +	+ + + +	+ − − +	+ + + +	− − + +	− + − +
−30	+ + − +	+ + − +	+ − − −	− + + +	+ + − +	+ + − +
−31	− − − −	− + − +	+ − − +	− − + +	+ + + +	− + − +
−32	+ − − +	+ + + +	+ + − −	+ − + −	+ − − +	+ + + +
−33	− − − −	− + − +	+ + − −	− + + −	− + − +	+ + + +
−34	+ − + −	+ + + +	+ − + −	+ + + +	+ − − +	+ + − −
−35	− − − −	− + − +	+ − + −	+ + + +	− − + +	− + + −
−36	− − + −	− + + +	+ − + +	+ + + −	− + + +	+ + − +
−37	− + − +	− + − +	+ − + −	+ − + −	+ + + +	+ + + +
−38	+ − − −	+ + − +	+ + + −	+ − + +	− + + +	+ + − +
−39	− − + +	− + + −	+ − + −	+ + + +	+ + + +	+ − + −
−40	+ − − −	+ + + −	+ − + +	+ + − +	− + + +	+ + + −
−41	− + + −	− − + +	+ − + −	+ + + +	+ − + −	+ + + +
−42	+ − + −	+ + + +	+ − + −	+ + + +	− + + −	− − + +
−43	+ + − −	+ + + +	+ − + −	+ − − +	+ + − −	+ + + +
−44	+ − − +	+ + − −	+ − + −	+ + + +	+ − + −	+ + + +
−45	− − − −	− + + −	+ − + −	− − + +	− + + −	+ + + +
−46	+ − − +	+ + + +	+ − − +	+ + + +	+ + − −	+ − + −
−47	0 0 0 0	*b* 0 2 0	0 0 0 0	0 2 0 2	2 0 0 2	0 0 2 2
−48	+ − + +	+ + + −	+ − − −	+ + − +	+ + − +	− + + +
−49	+ + − −	+ − + −	+ − − +	+ + + +	+ − − +	+ + + +
−50	− − − −	− + + −	+ − − +	+ + + +	− + − +	− − + +
−51	+ − + +	+ − + +	+ − − −	− + + +	+ − + +	+ − + +
−52	+ − − +	+ + + +	− − + +	− + − +	+ − − +	+ + + +
−53	− − − −	− − + +	+ − − +	− + − +	− − + +	+ + + +
−54	− − − +	− + + +	+ − − −	− − − +	+ + − +	+ − + +
−55	− − − −	− + + −	+ − + −	+ + − −	+ − − +	+ + + +
−56	2 0 *b* 0	0 0 2 2	0 0 0 0	0 2 2 0	2 0 0 2	0 0 0 0
−57	− − − +	− + + +	+ − − −	+ + + −	+ + − +	− + − −
−58	− − − −	− − + +	+ + − −	+ + + +	+ − − +	+ − + −
−59	− + − +	− − + +	+ − − +	+ + + +	+ + + +	+ − − +
−60	− − + +	− − + +	+ + − −	+ + − −	+ + + +	+ + + +

Table 28.2 (cont.)

−61	− − − +	− + + +	− + + +	+ + + −	+ + − +	+ − + +
−62	+ − − −	+ − + +	+ + − +	+ + + −	− + + +	+ − + +
−63	+ − − +	+ + + +	+ − + −	+ + − −	+ + + +	+ − − +
−64	− + + −	− + − +	+ + − −	+ + + +	+ + + +	+ + − −
−65	− + − +	− − + +	− + + −	+ + + +	+ + + +	− + + −
−66	+ + − −	+ − − +	+ − + −	+ + + +	+ + + +	+ − + −
−67	+ − − +	+ − + −	+ + − −	+ + + +	+ + + +	+ + − −
−68	0 0 0 0	*b* 0 0 2	0 0 0 0	0 2 2 0	2 2 0 0	0 2 0 2
−69	+ + − +	+ − + +	+ − − −	+ + + −	+ + + −	− + + +
−70	+ − + −	+ − − +	+ + − −	+ + + +	+ + − −	+ + + +
−71	+ + − −	+ + + +	+ + − −	+ + + +	+ − + −	+ − − +
−72	+ − − +	+ + + +	− + + −	+ + + +	+ + − −	− + − +
−73	− − − −	− − + +	+ + − −	+ + + +	− + + −	− + − +
−74	− − − −	− + − +	+ − + −	+ + + +	+ + − −	+ − − +
−75	0 0 0 0	*b* 2 0 0	0 0 0 0	0 0 2 2	2 0 2 0	0 2 2 0
−76	+ + + −	+ + + −	+ − − −	− + + +	+ + + −	+ + + −
−77	2 0 *b* 0	0 0 0 0	0 0 0 0	0 2 0 2	2 2 0 0	0 2 2 0
−78	− − − +	− − − +	+ − − −	− + + +	+ + + −	+ + + −
−79	+ + + −	+ + − +	+ − − −	+ − + +	+ − + +	− + + +
−80	+ + + −	+ − + +	− − + −	− + + +	+ + − +	− + + +
−81	− + − −	− − − +	+ − − −	+ + − +	+ + − +	− + + +
−82	2 *b* 0 0	0 0 0 0	0 0 0 0	0 0 2 2	2 0 0 2	0 2 0 2
−83	− − + −	− − + −	+ − − −	− + + +	+ + − +	+ + − +
−84	+ + − +	+ + + −	− + − −	− + + +	+ − + +	− + + +
−85	− − − +	− − + −	+ − − −	+ − + +	+ − + +	− + + +
−86	− + − −	− + − −	+ − − −	− + + +	+ − + +	+ − + +
−87	− − − +	− + − −	− − + −	− + + +	+ + − +	− + + +
−88	+ − − −	+ − − −	+ − − −	− + + +	− + + +	− + + +
−89	2 0 0 *b*	0 2 2 0	0 0 0 0	0 0 0 0	2 0 0 2	0 2 2 0
−90	− − + −	− + + +	+ − − −	− − + −	+ − + +	+ + + −
−91	0 *b* *b* 0	*b* 0 0 2	0 0 0 0	0 0 0 0	2 0 0 2	0 2 2 0
−92	+ − − −	+ − + +	+ − − −	− + − −	+ + − +	+ + + −
−93	− + + −	− + + −	+ − − +	+ − − +	+ + + +	+ + + +
−94	− − + +	− − + +	− − + +	− − + +	+ + + +	+ + + +
−95	− + − +	− + − +	− + − +	− + − +	+ + + +	+ + + +
−96	+ − − +	+ − − +	+ − − +	+ − − +	+ + + +	+ + + +
−97	− + + −	− + + −	− + + −	− + + −	+ + + +	+ + + +
−98	+ − + −	+ − + −	+ − + −	+ − + −	+ + + +	+ + + +
−99	+ + − −	+ + − −	+ + − −	+ + − −	+ + + +	+ + + +
−100	+ − − +	+ − − +	− + + −	− + + −	+ + + +	+ + + +
−101	0 0 0 0	*b* 2 2 2	0 0 0 0	0 0 0 0	2 2 2 2	0 0 0 0
−102	+ + − −	+ + + +	+ − − +	+ − + −	+ + + +	+ + − −
−103	− − + −	− + + +	− + + +	+ + − +	+ − + +	+ + + −
−104	+ − + −	+ + + +	+ + − −	+ − − +	+ + + +	+ − + −
−105	+ − − −	+ − + +	− + + +	+ − + +	+ + − +	+ + + −
−106	+ + − −	+ + + +	− + + −	− + − +	+ + + +	+ + − −
−107	− − − −	− − + +	+ − + −	+ − − +	+ + + +	+ + − −
−108	0 0 0 0	*b* 2 2 2	0 0 0 0	0 0 0 0	0 0 0 0	2 2 2 2
−109	+ − + −	+ + + +	+ − − +	+ + − −	+ − + −	+ + + +
−110	2 *b* 0 0	0 0 2 2	0 0 0 0	0 0 0 0	2 2 0 0	0 0 2 2
−111	+ + − −	+ + + +	− + − +	− + + −	+ + − −	+ + + +
−112	− − − −	− − + +	+ − − +	+ − + −	+ + − −	+ + + +
−113	2 0 *b* 0	0 2 0 2	0 0 0 0	0 0 0 0	2 0 2 0	0 2 0 2
−114	− + − −	− + + +	+ − − −	− + − −	+ + + −	+ + − +
−115	+ − + −	+ + + +	− + + −	− − + +	+ − + −	+ + + +
−116	− − − −	− + − +	+ + − −	+ − − +	+ − + −	+ + + +
−117	− − − −	− − + +	− + + −	− + − +	+ + − −	+ + + +

3. The lattices $I_{n,1}$ for n ⩽ 19

Examination of the coordinates given by Vinberg [Vin4] reveals that the fundamental roots $v = (v_0,\ldots,v_{n-1} | v_n)$ for $I_{n,1}$, $n \leqslant 17$, can be arranged in the following two "batches": (a) the vectors v (of norm 2) for which

$$v^+ = (0,\ldots,0,0,v_0,\ldots,v_{n-1} | v_n)$$

is a Leech root, and (b) the vectors v (of norm 1) for which

$$v^+ = (0,\ldots,0,1,v_0,\ldots,v_{n-1} | v_n)$$

is a Leech root. Similarly the partial lists of coordinates given by Vinberg and Kaplinskaja [Vin15] suggest that for $I_{18,1}$ the fundamental roots are (a) and (b), and (c) the vectors v (of norm 1) for which

$$v^{++} = (\pm 1,1,1,1,1,1,1,1,v_0,\ldots,v_{17} | v_{18})$$

is a vector obtained by doubling a Leech root with fractional coordinates; and also that for $I_{19,1}$ the fundamental roots are (a) and (b), and (c′) the vectors v (of norm 2) for which

$$v^{++} = (\pm 1,1,1,1,1,1,v_0,\ldots,v_{18} | v_{19})$$

is obtained by doubling a Leech root with fractional coordinates.

These observations, proved below, provide an immediate way to decide if a specified vector v is a fundamental root for $I_{n,1}$ $(n \leqslant 19)$: does the supplemented vector $r = v^+$ or $\frac{1}{2}v^{++}$ satisfy $r \cdot r = 2$ and $r \cdot w = -1$?

4. Vinberg's algorithm and the initial batches of fundamental roots

Vinberg [Vin7] describes an algorithm for finding a set of fundamental roots for the reflection subgroup of any discrete hyperbolic group. For $\mathrm{Aut}(I_{n,1})$ the algorithm proceeds as follows. For a suitable vector x_0 (which may be called the controlling vector), we first determine the subgroup H generated by the reflections that fix x_0. We then choose a set of fundamental roots $v_1,\ldots,v_n$ for H. Further vectors $v_{n+1}, v_{n+2},\ldots,v_k,\ldots$ are determined inductively by the following conditions:

(i) $v_k \in I_{n,1}$, $N(v_k) = 1$ or 2,

(ii) the v_k are enumerated in increasing order of $-v_k \cdot x_0/N(v_k)^{1/2}$,

(iii) $v_k \cdot v_j \leqslant 0$ for all $1 \leqslant j < k$.

We shall regard $I_{n,1}$ $(1 \leqslant n \leqslant 25)$ as embedded in $I_{25,1}$ by preceding its vectors by an initial string of $25-n$ zeros. Our observations stem from the fact that w is then a suitable choice for x_0. This is equivalent to using the projection

$$w_n = (25-n,\ldots,23,24 | 70)$$

of w into $I_{n,1}$, and this does not affect inner products with elements of $I_{n,1}$. It is easy to check that w_n does satisfy the conditions on x_0 given in Vinberg [Vin7].

It is convenient to group the candidate vectors $v \in I_{n,1}$ that are to be considered by the algorithm into batches according to their norm and *height* $h(v) = -v \cdot w$. We shall use $n_{L,h}$ to denote the set of all $v \in I_{n,1}$ that have height h and norm 2 (i.e. are long roots), and $n_{S,h}$ to denote the height h roots of norm 1 (short roots).

A candidate v is accepted by the algorithm (with w as the controlling vector) if and only if v has non-positive inner products with the vectors of all strictly earlier batches taken in the order $n_{L,1}, n_{S,1}, n_{L,2}, n_{S,2}, n_{L,3}, n_{L,4}, n_{S,3}, \ldots$ specified by condition (ii). Theorems 1 and 2 determine the first batches of long and short roots, respectively, that are non-empty.

Theorem 1. *For $n \geqslant 2$, the first batch of long roots is $n_{L,1}$, and consists of the vectors $v = (v_0, \ldots, v_{n-1} | v_n)$ for which $v^+ = (0, \ldots, 0, v_0, \ldots, v_{n-1} | v_n)$ is a Leech root. All candidates from this batch are accepted.*

Proof. The smallest possible value for h is 1, and the vectors of $n_{L,1}$ are easily checked to be Leech roots (when extended by 0's). Since $n_{L,1}$ is the earliest possible batch, no candidate is rejected. The Leech root r_{24} is in $I_{n,1}$ for $n \geqslant 2$ and shows that the batch exists (i.e. is nonempty). This completes the proof.

Vinberg takes for his controlling vector x_0 the vector $z = (0^n | 1)$. It is easy to see that the algorithm with his controlling vector and initial vectors accepts sufficiently many more vectors from our first batch to span the space $\mathbf{R}^{n,1}$. It follows that the two forms of the algorithm determine the same fundamental region.

Theorem 2. *For $1 \leqslant n \leqslant 24$ the first non-empty batch of short roots is $n_{S,25-n}$, and consists of the vectors $v = (v_0, \ldots, v_{n-1} | v_n)$ for which $v^+ = (0, \ldots, 0, 1, v_0, \ldots, v_{n-1} | v_n)$ is a Leech root. Again all candidates from this batch are accepted.*

Proof. If v is in $n_{S,h}$ where $1 \leqslant h \leqslant 25-n$ then the vector

$$v^+ = (0^{h-1}, 1, 0^{25-n-h}, v_0, \ldots, v_{n-1} | v_n)$$

has norm 2 and height 1 and so is a Leech root. This implies that $h = 25-n$, since otherwise the inner product of v^+ and the particular Leech root $(0^{h-1}, 1, -1, 0^{24-h} | 0)$ would be positive. Thus v has the required form.

The batch $n_{S,25-n}$ exists since it contains the vector $(-1,0,\ldots,0|0)$ of $I_{n,1}$. We have not been able to find a uniform proof that all candidates from this batch are accepted. For $n \leqslant 19$ this can be read from the results of Vinberg and Kaplinskaja. For $20 \leqslant n \leqslant 24$ we observe that the corresponding set of vectors v^+ forms a single orbit under the stabilizer of $I_{n,1}^{\perp}$ in the automorphism group of the Leech lattice. So if one member of this batch is accepted, all must be. But $(-1,0,\ldots,0|0)$ is accepted by

Table 28.3. The first two batches of roots

Dimension n	$\lvert n_{L,1} \rvert$	$\lvert n_{S,25-n} \rvert$	Known values of h for which $n_{L,h} \neq \emptyset$	$n_{S,h} \neq \emptyset$
1	0	1	none	24 only
2	1	2	1 only	23 only
3	3	1	1 only	22 only
4	4	1	1 only	21 only
5	5	1	1 only	20 only
6	6	1	1 only	19 only
7	7	1	1 only	18 only
8	8	1	1 only	17 only
9	9	1	1 only	16 only
10	10	2	1 only	15 only
11	12	1	1 only	14 only
12	13	1	1 only	13 only
13	14	1	1 only	12 only
14	15	2	1 only	11 only
15	17	1	1 only	10 only
16	18	2	1 only	9 only
17	20	2	1 only	8 only
18	22	3	1 only	7, 23 only
19	25	5	1, 17 only	6 only
20	30	12	1,17,...	5,13,...
21	42	56	1,9,17,...	4,8,...
22	100	1100	1,5,9,...	3,5,7,...
23	4600	953856	1,3,5,...	2,3,4,...
24	∞	∞	1,4,5,6,...	1,2,3,...

Vinberg's form of the algorithm, and therefore will be accepted by ours. This completes the proof.

The sizes $\lvert n_{L,1} \rvert$ and $\lvert n_{S,25-h} \rvert$ of the first nonempty batches of long and short roots can be computed from knowledge of the stabilizer of $I_{n,1}^{\perp}$ in the automorphism group of the Leech lattice, and are shown in Table 28.3.

The first batches of long roots can be described as follows. For $n = 18$ and 19, $n_{L,1}$ is the subset of roots called Σ_n^* by Vinberg and Kaplinskaja. In particular, the Coxeter diagram for the roots in $19_{L,1}$ is the incidence graph of the Petersen graph. The subset $20_{L,1}$ consists of 30 roots forming a copy of Tutte's 8-cage graph (see [Har2, p. 174] and Figs. 28.1g, 28.1h below). The 42 roots in $21_{L,1}$ correspond to the points and lines of a projective plane of order 4, joined by incidence, while the 100 points of $22_{L,1}$ are a copy of the Higman-Sims graph (see §3.5 of Chap. 10). The last finite diagram in this sequence is $23_{L,1}$, whose 4600 points may be regarded as the points of the Leech lattice at the minimal distance from two points which are themselves at the minimal distance. They are therefore invariant under the group Co_2 (see Chap. 10). The first batches of short roots could be described similarly.

5. The later batches of fundamental roots

We have much less understanding of the later batches of fundamental roots. A computer program was prepared to run Vinberg's algorithm (with controlling vector $x_0 = z$ rather than w) on the Cray-1 computer at Bell Laboratories. For $n = 18$ and 19 the algorithm ran to completion and confirmed the results (obtained partly by calculation and partly by analysis) of Vinberg and Kaplinskaja [Vin15]. For $n = 18$ there are 12 more fundamental roots, belonging to the batch $18_{S,23}$, as described in §4. For $n = 19$ there are 20 more, belonging to $19_{L,17}$. Since the algorithm is very time-consuming, and since Vinberg and Kaplinskaja do not list all the roots for $n = 18$ and 19, we have made Tables 28.1 and 28.2 extensive enough to include the corresponding Leech roots v^{++}.

For $n \geqslant 20$, Vinberg [Vin7] has shown that the reflection subgroup of $\mathrm{Aut}(\mathrm{I}_{n,1})$ has infinite index[(1)] and so there are infinitely many fundamental roots. Our extensive but still incomplete computations show that some vectors are accepted from at least the batches shown in Table 28.3. In $20_{L,17}$ and $20_{S,13}$, for example, the algorithm appears to accept the vectors $v = (v_0, \ldots, v_{19} | v_{20})$ for which

$$v^{+++} = (\pm 1, 1, 1, 1, 2, v_0, \ldots, v_{19} | v_{20})$$

can be expressed as $2r + s$, where r and s are Leech roots.

We have also shown that there are symmetries of the fundamental region for $\mathrm{II}_{20,1}$ that do not preserve the vector w. One such symmetry takes w_{20} to

$$w'_{20} = (13, 14, 15, \ldots, 30, 31, 48 | 110)$$

and transforms the initial batches $20_{L,1}$ and $20_{S,5}$ to image batches $20'_{L,1}$ and $20'_{S,5}$. The two 8-cage graphs $20_{L,1}$ and $20'_{L,1}$ intersect in 24 roots, and it seems likely that the full set of long fundamental roots is a union of 8-cages intersecting in this way, while the set of short roots is the union of infinitely many disjoint images of $20_{S,5}$. It would be nice to know more![(2)]

The 100 fundamental roots in $22_{L,1}$ are obtained from the Leech roots that begin 000... . The latter consist of

42	beginning	0000...	(producing $21_{L,1}$),
56	beginning	0001...	(producing $21_{S,4}$), and
2	beginning	0002...	

---

(1) More extensive information about unimodular lattices in the Euclidean space $\mathbf{R}^{19}$ upon which this theorem is based can be found in Chap. 16.

(2) R. E. Borcherds (personal communication) has found more, and in particular has confirmed this guess about the case $n = 20$.

Table 28.4. Further Leech roots in hyperbolic coordinates. (This together with Table 28.1 includes the 100 Leech roots beginning 000..., and therefore includes the first two batches of fundamental roots for $I_{21,1}$ and the first batch of fundamental roots for $I_{22,1}$.)

x_0	x_1	x_2	x_3	x_4	x_5	x_6	x_7	x_8	x_9	x_{10}	x_{11}	x_{12}	x_{13}	x_{14}	x_{15}	x_{16}	x_{17}	x_{18}	x_{19}	x_{20}	x_{21}	x_{22}	x_{23}	x_{24}	x_{25}
0	0	0	1	2	2	2	3	3	3	4	4	4	5	5	5	6	6	6	7	7	7	8	8	9	25
0	0	0	1	1	2	2	2	3	3	3	4	4	5	5	5	6	6	6	7	7	8	8	8	9	25
0	0	0	1	1	2	2	3	3	3	4	4	4	5	5	6	6	6	7	7	8	8	8	8	9	26
0	0	0	1	1	2	2	3	3	3	4	4	5	5	5	5	6	6	7	7	7	8	8	9	9	26
0	0	0	1	2	2	2	2	3	3	4	4	4	5	5	6	6	6	7	7	7	8	8	9	9	26
0	0	0	1	2	2	2	3	3	3	4	4	5	5	6	6	6	7	7	7	8	8	8	9	9	27
0	0	0	1	2	2	2	3	3	4	4	4	5	5	5	6	6	6	7	7	8	8	9	9	9	27
0	0	0	1	1	2	2	3	3	4	4	4	5	5	6	6	6	7	7	8	8	8	9	9	10	28
0	0	0	1	2	2	2	3	3	4	4	5	5	5	6	6	7	7	8	8	8	9	9	9	10	29
0	0	0	1	2	2	3	3	3	4	4	4	5	5	6	6	7	7	7	8	8	9	9	10	10	29
0	0	0	1	2	2	3	3	4	4	4	5	5	6	6	6	7	7	8	8	9	9	9	10	10	30
0	0	0	1	2	2	3	3	3	4	4	5	5	6	6	7	7	8	8	8	9	9	10	10	11	31
0	0	0	1	2	2	3	3	4	4	5	5	6	6	6	7	7	8	8	9	9	10	10	10	11	32
0	0	0	2	2	2	3	3	4	4	5	5	6	6	7	7	8	8	9	9	10	10	11	11	12	34
0	0	0	2	2	3	3	4	4	5	5	6	6	7	7	8	8	9	9	10	10	11	11	12	12	36

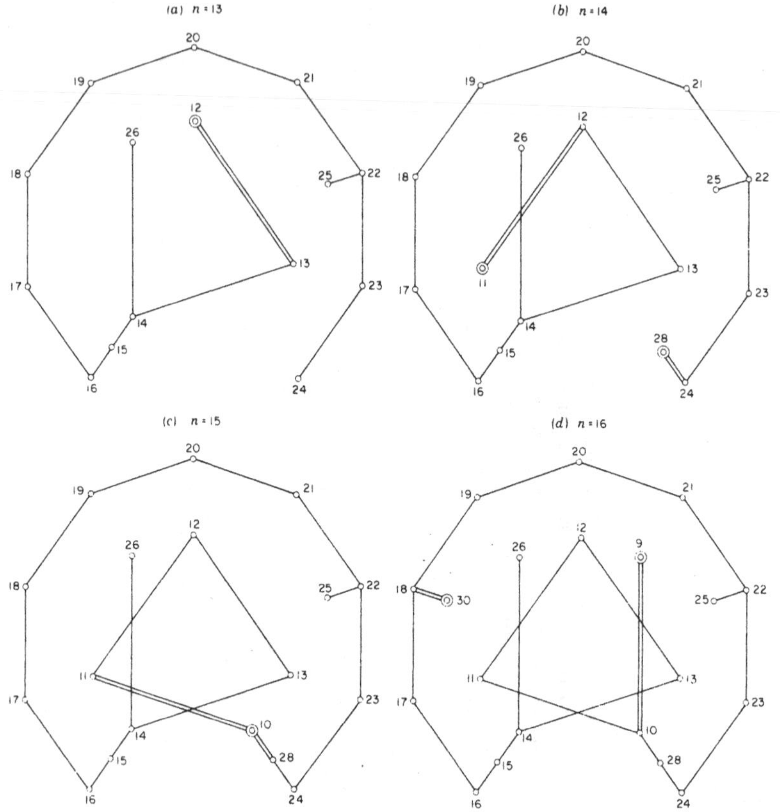

Figure 28.1. Coxeter-Vinberg diagrams for $\mathrm{Aut}(\mathrm{I}_{13,1})$ — $\mathrm{Aut}(\mathrm{I}_{20,1})$. The numbers attached to the nodes indicate the corresponding Leech roots.

All except fifteen of those roots can be found in Table 28.1. The last fifteen occur much further along in our ordering of the Leech roots, and are given separately in Table 28.4.

Figures 28.1a-h show the Coxeter-Vinberg diagrams for the fundamental roots of $\mathrm{I}_{n,1}$, $13 \leqslant n \leqslant 20$. The fundamental roots in $n_{L,1}$ are represented by single circles and those in $n_{S,25-n}$ by double circles. The remaining roots (for $n = 18-20$) are not shown. The numbers attached to the roots in $n_{L,1}$ and $n_{S,25-n}$ give the subscripts i of the corresponding Leech roots $v^+ = r_i$, which must then be projected onto $\mathbf{R}^{n,1}$ as described in §5. For example, node 9 in the diagram for $\mathrm{I}_{16,1}$ (Fig. 28.1d) refers to the Leech root

$$r_9 = (0^8, +1, -1, 0^{15} \,|\, 0) \ .$$

Since this node has a double circle it is a short root, so r_9 must be projected onto $(-1, 0^{15} \,|\, 0)$, which is the fundamental root for $\mathrm{I}_{16,1}$ corresponding to node 9. In the next figure node 9 carries a single circle, and we project r_9 onto $(+1, -1, 0^{15} \,|\, 0)$, a fundamental root for $\mathrm{I}_{17,1}$.

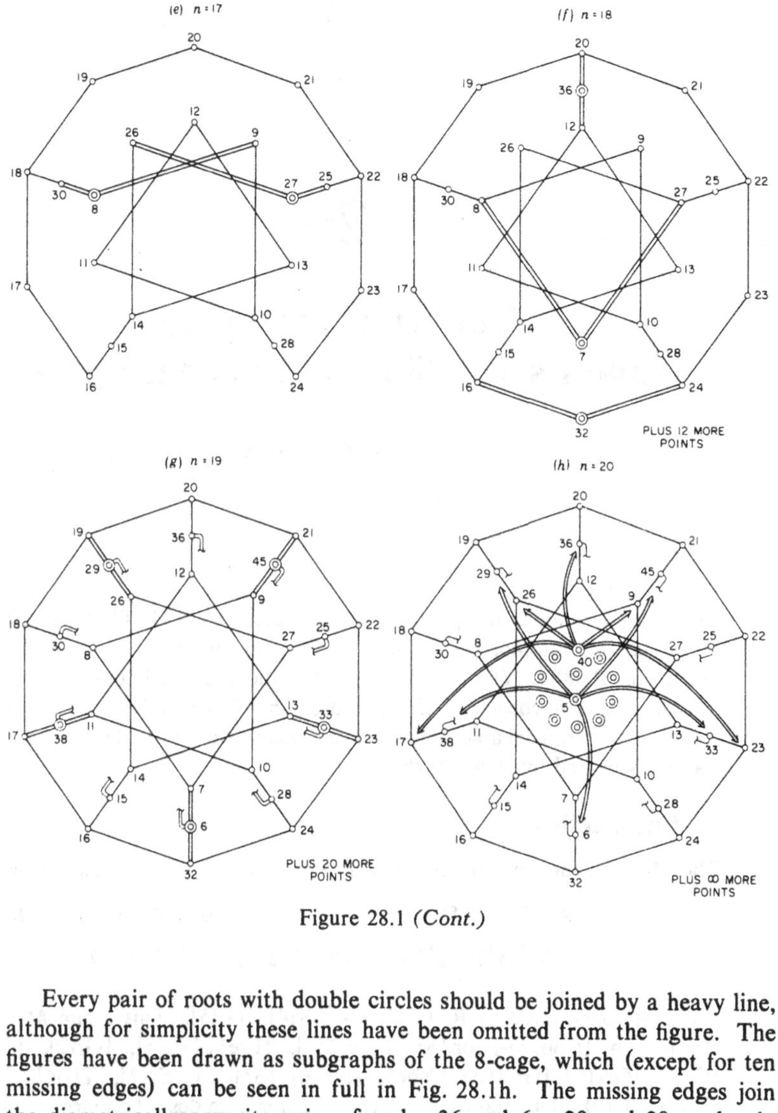

Figure 28.1 *(Cont.)*

Every pair of roots with double circles should be joined by a heavy line, although for simplicity these lines have been omitted from the figure. The figures have been drawn as subgraphs of the 8-cage, which (except for ten missing edges) can be seen in full in Fig. 28.1h. The missing edges join the diametrically opposite pairs of nodes 36 and 6,...,29 and 28, and only their ends have been shown. The reader should note that for each dimension greater than 13 the diagram has a nontrivial symmetry.

In the center of Fig. 28.1h there is a circle of ten short roots enclosing two more. The ten nodes around this circle should be labeled 34, 39, 40, 46, 48, 49, 58, 60, 74 and 80, and two in the center should be labeled 5 and 31. The double lines from node 40 to 9, 17, 23, 26 and 36, and from 5 to 6, 29, 33, 38 and 45 have been indicated by long double arrows. Node 31 should be joined to 15, 18, 25, 28 and 36 by double lines, while the remaining joins from the nodes 34, 39, ... can be found by rotating the figure.

29

The Monster Group and its 196884-Dimensional Space

J. H. Conway

The Monster simple group was first constructed by R. L. Griess in 1981 as the group of automorphisms of a certain algebra in Euclidean space of dimension 196884. This chapter describes the simplified construction given in [Con15]. The construction is based on the binary Golay code, the Leech lattice, and a remarkable nonassociative loop (closely related to the Golay code) discovered by R. A. Parker.

1. Introduction

The Monster (or Friendly Giant or Fischer-Griess) simple group of order

$$808017424794512875886459904961710757005754368000000000$$
$$= 2^{46}3^{20}5^{9}7^{6}11^{2}13^{3}17\cdot 19\cdot 23\cdot 29\cdot 31\cdot 41\cdot 47\cdot 59\cdot 71 \qquad (1)$$

was first constructed by R. L. Griess [Gri4], [Gri5]. This group M is also studied in [Con16]-[Con18], [Con24a], [Fre1]-[Fre5], [Gri2], [Gri6]-[Gri8], [Gri10], [Gri10a], [Nor5], [Nor6], [Smi13], [Soi2], [Tho4]-[Tho6], [Tit8], [Tit9]. The main goal of this chapter is to describe more briefly the simplified construction given in [Con15]. The verification that the construction works reduces to elementary but tedious calculations. The often repeated phrase "explicit calculation" is an implicit reference to the Appendices of [Con15], where these calculations are performed in detail.

Figure 29.1 shows various subgroups of the Monster that arise in our construction, with some information about their structures. The group $\overline{N}$ is the normalizer of a certain fourgroup in the Monster, whose non-trivial elements are $\overline{x}_{-1}$, $\overline{y}_{-1}$, $\overline{z}_{-1}$. Since $\overline{N}$ permutes these, it has an obvious homomorphism onto S_3, and $\overline{N}_x$, $\overline{N}_y$, $\overline{N}_z$ are the preimages of the three subgroups S_2 of this S_3 — in other words they consist of the elements of $\overline{N}$

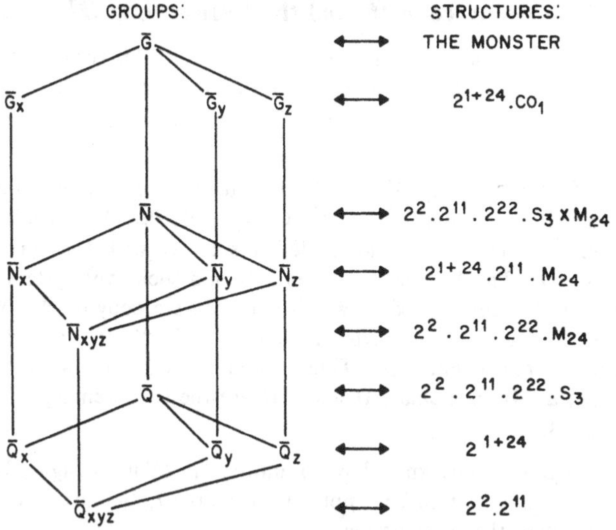

Figure 29.1. Some subgroups of the Monster.

that commute with $\bar{x}_{-1}$, $\bar{y}_{-1}$, $\bar{z}_{-1}$ respectively. The intersection of these three groups, which is also the intersection of any two of them, is the group $\overline{N}_{xyz}$ that centralizes the fourgroup $\bar{x}_{-1}$, $\bar{y}_{-1}$, $\bar{z}_{-1}$. The groups $\overline{Q}$, $\overline{Q}_x$, $\overline{Q}_y$, $\overline{Q}_z$, $\overline{Q}_{xyz}$ are normal 2-subgroups of the above groups, while the groups $\overline{G}_x$, $\overline{G}_y$, $\overline{G}_z$ are the full centralizers in the Monster of the respective involutions $\bar{x}_{-1}$, $\bar{y}_{-1}$, $\bar{z}_{-1}$.

The time has come to explain all those bars, and quickly describe our construction. The main idea is that we give an explicit construction for a certain group N of order

$$201889334420601569280 = 2^{48} \cdot 3^4 \cdot 5 \cdot 7 \cdot 11 \cdot 23,$$

as a group of permutations of triples of elements in a non-associative loop $\mathscr{P}$ discovered by R. A. Parker [Par2]. This group N turns out to be a fourfold cover of the group $\overline{N}$ described above. The definition of N makes it very easy to define certain representations of its subgroups N_x, N_y, N_z (the preimages in N of $\overline{N}_x$, $\overline{N}_y$, $\overline{N}_z$) on some quite large-dimensional Euclidean spaces, and we can put these representations together to build a representation for $\overline{N}$ on a space of dimension 196884, on which we also define an algebra structure invariant under $\overline{N}$.

In fact we give three different definitions of this algebra, which make it clear that it is actually invariant under each of three rather large groups $\overline{G}_x$, $\overline{G}_y$, $\overline{G}_z$ that are not subgroups of $\overline{N}$, and we can *define* the Monster to be the group $\overline{G}$ that these generate. We give a surprisingly easy proof that $\overline{G}$ is finite, but the computation of its order, and its identification with the abstract group already known as the Monster, depend on a theorem of Stephen D. Smith.

2. The Golay code $\mathscr{C}$ and the Parker loop $\mathscr{P}$

As usual, we use $\mathscr{C}$ for the binary Golay code, consisting of 2^{12} binary codewords of length 24, with weight distribution

$$0^1 \quad 8^{759} \quad 12^{2576} \quad 16^{759} \quad 24^1$$

(see (2.8.2) of Chap. 3). In this chapter we shall use multiplicative rather than additive notation for the Golay code $\mathscr{C}$, but will continue as in previous chapters to identify elements of $\mathscr{C}$ with the corresponding sets ($\mathscr{C}$-sets). The Parker loop $\mathscr{P}$ is a (non-associative) system with a center $\{1,-1\}$ of order 2, modulo which it becomes a copy of $\mathscr{C}$. We shall write $a, b, c, d, e, f, \ldots$ for typical elements of $\mathscr{P}$, and use $d \to \tilde{d}$ for the homomorphism onto $\mathscr{C}$. Thus d and $-d$ are the two preimages of the $\mathscr{C}$-set $\tilde{d}$. In particular, Ω and $-\Omega$ are the two preimages of the universal $\mathscr{C}$-set $\tilde{\Omega}$.

So you already know how to multiply in $\mathscr{P}$ up to sign. The exact rules for the signs are rather subtle (Appendix 1), but are not needed for an overview of the construction.

3. The Mathieu Group M_{24}; the Standard Automorphisms of $\mathscr{P}$

The automorphisms of the Golay code $\mathscr{C}$ form the Mathieu group M_{24} of permutations of $\tilde{\Omega}$. (Regarded as an abstract group, of course, $\mathscr{C}$ has other automorphisms.) The group M_{24} is 5-transitive on $\tilde{\Omega}$, of order $24 \cdot 23 \cdot 22 \cdot 21 \cdot 20 \cdot 16 \cdot 3 = 244823040$.

The *standard automorphisms* of $\mathscr{P}$ are those that become elements of M_{24} when we ignore the signs of Parker loop elements. They form a group we call $\mathscr{S}$. (Regarded merely as an abstract group, $\mathscr{P}$ has some other automorphisms.) A standard automorphism is called *even* if it fixes both Ω and $-\Omega$, and *odd* if it interchanges them.

We shall use π for the typical standard automorphism of $\mathscr{P}$, and $\tilde{\pi}$ for the corresponding element of M_{24}. The homomorphism $\pi \to \tilde{\pi}$ is 2^{12}-to-1 (Appendix 1), and $\mathscr{S}$ has structure $2^{12} \cdot M_{24}$.

4. The Golay Cocode $\mathscr{C}^*$ and the Diagonal Automorphisms

The π for which $\tilde{\pi}$ is trivial are the *diagonal automorphisms* — they merely change the signs of certain elements of P. The diagonal automorphisms correspond one-for-one with the elements of the Golay cocode $\mathscr{C}^*$ of weight distribution

$$0^1 \quad 1^{24} \quad 2^{276} \quad 3^{2024} \quad 4^{1771},$$

and we shall give them the same names $\delta, \epsilon, \zeta, \ldots$. As usual we take the elements of $\mathscr{C}^*$ to be subsets of $\tilde{\Omega}$ modulo addition of $\mathscr{C}$-sets. The rule is that δ changes the sign of the preimages of just those $\mathscr{C}$-sets it intersects oddly — in other words

$$\delta : d \to d^{\delta} = (-1)^{|\tilde{d} \cap \delta|} d.$$

As a standard automorphism δ is even or odd according as $|\delta|$ is even or odd.

The diagonal automorphisms form an elementary abelian normal subgroup of order 2^{12} in the group $\mathcal{S}$ of all standard automorphisms of $\mathcal{P}$. In fact $\mathcal{S}$ has no subgroup isomorphic to M_{24} — its structure is a non-split extension $2^{12} \cdot M_{24}$ of the diagonal automorphisms by M_{24}.

5. The Group N of Triple Maps

We define N to be the group of permutations of triples of elements of $\mathcal{P}$ generated by the following maps (which we specify by giving the image of a typical triple (a,b,c)):

x_d :	(dad, db, cd)	x_π :	(a^π, b^π, c^π)	or	$((a^\pi)^{-1}, (c^\pi)^{-1}, (b^\pi)^{-1})$
y_d :	(ad, dbd, dc)	y_π :	(a^π, b^π, c^π)	or	$((c^\pi)^{-1}, (b^\pi)^{-1}, (a^\pi)^{-1})$
z_d :	(da, bd, dcd)	z_π :	(a^π, b^π, c^π)	or	$((b^\pi)^{-1}, (a^\pi)^{-1}, (c^\pi)^{-1})$
(according as π is			even	or	odd),

where d ranges over the Parker loop $\mathcal{P}$, and π over its standard automorphism group S.

There is an obvious "triality" automorphism that cyclically permutes x, y, z in these and later notations. In future we shall often make only one of three statements or definitions related by triality, and write "(&c)" to indicate that the reader should derive the others for himself.

6. The Kernel K and the Homomorphism $g \to \bar{g}$

The success of our construction depends vitally on the fact that when taken modulo a certain kernel K of order 4, N becomes the normalizer of a certain fourgroup in the Monster. The kernel K consists of the identity and the three elements

$$k_x = y_\Omega z_{-\Omega}, \quad k_y = z_\Omega x_{-\Omega}, \quad k_z = x_\Omega y_{-\Omega}.$$

We shall write $g \to \bar{g}$ for the canonical homomorphism $N \to N/K$, and $\bar{\Gamma} = \Gamma/(K \cap \Gamma)$ for the image of a subgroup Γ of N under this homomorphism, and in particular, $\bar{N}$ for N/K.

On a first reading, it is wise to ignore elements of the kernel, and so identify every element g with the corresponding $\bar{g}$, even though this is not consistently possible.

7. The Structures of Various Subgroups of $\bar{N}$

The structures we give here can be verified by explicit calculations, but most of them become obvious in terms of the representations defined in later sections.

The group N itself has structure

$$\begin{array}{cccccccccc} & (2^2 & \times & 2^2) & \cdot & 2^{11} & \cdot & 2^{22} & \cdot & (S_3 \times M_{24}) \\ & & & & & & & & & \\ & k_x & & x_{-1} & & x_\delta & & x_d & & x_\delta \quad x_\pi \\ & k_y & & y_{-1} & & y_\delta & & y_d & & y_\delta \quad y_\pi \\ & k_z & & z_{-1} & & z_\delta & & z_d & & z_\delta \quad z_\pi \\ & & & & & (\delta \text{ even}) & & & & (\delta \text{ odd}) \end{array}$$

(Below the various composition factors we have given generators modulo earlier factors. We note that $x_\delta = y_\delta = z_\delta$ (δ even), and $x_d y_d z_d = 1$.) Of course the structure of $\bar{N}$ is obtained by omitting the first factor 2^2.

The homomorphism onto S_3 is obvious — any of our triple maps takes a typical triple (a,b,c) to another triple (A,B,C) in which each of A, B, C depends on just one of a, b, c. The elements for which A depends only on a form a subgroup N_x (&c), and the intersection of N_x, N_y, N_z (which is also the intersection of any two of them) is called N_{xyz}. The group N_x has a normal subgroup $Q_x = <x_d, x_\delta>$ (&c), and the intersection of all (or any two) of Q_x, Q_y, Q_z is called Q_{xyz}.

8. The Leech Lattice Λ_{24} and the Group $\bar{Q}_x$

The reader should already be familiar with the Leech lattice Λ_{24}. For our purposes, it is convenient to note that it is generated by the vectors (where the usual factor of $8^{-1/2}$ is to be understood)

$$\lambda_d = (2_{\text{on } \tilde{d}}, 0_{\text{elsewhere}}),$$

$$\lambda_i = (3_{\text{on } i}, -1_{\text{elsewhere}}),$$

as d ranges through $\mathcal{P}$ and i through $\tilde{\Omega}$ (of course, λ_d depends only on $\tilde{d}$). If $i, j, k, \ldots$ are elements of $\tilde{\Omega}$, we write

$$\lambda_{ijk\ldots} = \lambda_i + \lambda_j + \lambda_k + \cdots .$$

It can be checked that if $\{i,j,k,\ldots\}$ is a $\mathcal{C}$-set, then $\lambda_{ijk\ldots} \in 2\Lambda_{24}$. So when we work modulo $2\Lambda_{24}$, we can regard $\lambda_{ijk\ldots}$ as determined even when $\delta = \{i,j,k,\ldots\}$ is specified only modulo addition of $\mathcal{C}$-sets, and so we call it λ_δ ($\delta \in \mathcal{C}^*$).

There is a wonderful homomorphism, with kernel $\{x_1, x_{-1}\}$, from $\bar{Q}_x = <\bar{x}_d, \bar{x}_\delta>$ onto $\Lambda_{24}/2\Lambda_{24}$. It is defined by

$$\overline{x_d} \to \tilde{\lambda}_d, \quad \overline{x_\delta} \to \tilde{\lambda}_\delta,$$

where $\tilde{\lambda}_r$ denotes the image of λ_r in $\Lambda_{24}/2\Lambda_{24}$. We note that every element of $\Lambda_{24}/2\Lambda_{24}$ can be expressed in the form

$$\tilde{\lambda}_d + \tilde{\lambda}_\delta \quad (d \in \mathcal{P}, \delta \in \mathcal{C}^*),$$

the expression being unique apart from the identity $\tilde{\lambda}_d = \tilde{\lambda}_{-d}$.

9. Short Elements

We shall write $\tilde{\lambda}_{d\cdot\delta} = \tilde{\lambda}_d + \tilde{\lambda}_\delta$, $x_{d\cdot\delta} = x_d x_\delta$ (&c), and will use r for the generic subscript $d \cdot \delta$ that occurs here. A vector λ or λ_r that is a minimal vector in the Leech lattice will be called *short*, as will the corresponding objects $\tilde{\lambda}$, $\tilde{\lambda}_r$, x_r, $\overline{x_r}$, X_r (&c).

Here are all the short vectors up to sign, with the names of the corresponding elements of $\Lambda_{24}/2\Lambda_{24}$:

$$\begin{array}{rcllll}
\tilde{\lambda}_{ij} & : & (4_{\text{on } i}, & -4_{\text{on } j}, & 0_{\text{elsewhere}}), & \\
\tilde{\lambda}_{\Omega\cdot ij} & : & (4_{\text{on } i}, & 4_{\text{on } j}, & 0_{\text{elsewhere}}), & \\
\tilde{\lambda}_{d\cdot i} & : & (3_{\text{on } i}, & 1_{\text{on } \tilde{d}}, & -1_{\text{elsewhere}}), & (i \notin d), \\
\tilde{\lambda}_{(\Omega)d\cdot\delta} & : & (-2_{\text{on } \delta}, & 2_{\text{on rest of } \tilde{d}}, & 0_{\text{elsewhere}}), &
\end{array}$$

where in the last line $\tilde{d}$ is an octad, δ is an even subset of $\tilde{d}$, and the factor Ω is present just when $\frac{1}{2}|\delta|$ is odd. (The actual vectors of Λ_{24} displayed above are $\lambda_i - \lambda_j$ for $\tilde{\lambda}_{ij}$, $\lambda_d - \lambda_{\Omega\cdot ij}$ for $\tilde{\lambda}_{\Omega d\cdot ij}$, and $\lambda_d - \lambda_{\Omega\cdot ij} - \lambda_{\Omega\cdot kl}$ for $\tilde{\lambda}_{d\cdot ijkl}$.) So a subscript r indicates shortness just if it appears in the above list.

10. The Basic Representations of N_x

The group N_x has the following representations (for proof see Appendix 4).

A representation on a 24-space 24_x spanned by vectors i_x ($i \in \tilde{\Omega}$). In this, x_d and x_δ act trivially, y_d and z_d both act by changing the sign of those i_x for which $i \in \tilde{d}$, and x_π acts as the permutation $\tilde{\pi}$.

A representation on a 4096-space 4096_x spanned by vectors $(d)_x^-$ and $(d)_x^+$ that satisfy the relations, in which σ denotes a sign $-$ or $+$:

$$(-d)_x^\sigma = -(d)_x^\sigma, \quad (\sigma\,\Omega\, d)_x^\sigma = (d)_x^\sigma.$$

The actions of the generators on 4096_x are given in Table 29.1.

A representation on a 98280-space 98280_x spanned by vectors X_r that are permuted exactly like the short elements $\overline{x_r}$ of $\overline{Q_x}$ under conjugation, and satisfy $X_{-r} = -X_r$.

The entries in Table 29.1 are easily checked from the exact definitions in Appendix 4, and show in particular that the elements k_x, k_y, k_z, x_{-1} all act as plus or minus the identity in these representations, according to the display:

	k_x	k_y	k_z	x_{-1}
24_x	+	−	−	+
4096_x	+	−	−	−
98280_x	+	+	+	+

We can now build the desired representation 196884_x of N_x as

$$300_x + 98304_x + 98280_x,$$

where 300_x is the symmetric tensor square of 24_x, and 98304_x the tensor product $4096_x \otimes 24_x$. The above display shows that k_x, k_y, k_z all act trivially in 196884_x, which we can therefore regard as a representation of the quotient group $\overline{N}_x = N_x/K$.

Table 29.1. Action of the generators of N on basis vectors.
Notes: $D = d^{-1}$, $\delta' = \tilde{d} \cap \tilde{e}$, $u = |\{i\} \cap \tilde{e}|$, $m = |\delta \cap \tilde{e}|$, $n = ½|\delta'| + m$.

g		Action of g on the vectors $X_{d.\delta}$ (δ even)	$X_{d.\delta}$ (δ odd)	i_x	d_x^-	d_x^+
x_e		$X_{(-1)^n d.\delta}$	$X_{(-1)^m ede.\delta}$	i_x	$(ed)_x^-$	$(de)_x^+$
y_e		$X_{(-\Omega)^n d.\delta\delta'}$	$X_{(-\Omega)^m ed.\delta}$	$(-1)^u i_x$	$(ede)_x^-$	$(ed)_x^+$
z_e		$X_{\Omega^n d.\delta\delta'}$	$X_{\Omega^m de.\delta}$	$(-1)^u i_x$	$(de)_x^-$	$(ede)_x^+$
$x_\epsilon = y_\epsilon = z_\epsilon$	(ϵ even)	$X_{d^\epsilon.\delta}$		i_x	$(d^\epsilon)_x^-$	$(d^\epsilon)_x^+$
x_ϵ		$X_{D^\epsilon.\delta}$		i_x	$(D^\epsilon)_x^+$	$(D^\epsilon)_x^-$
y_ϵ	(ϵ odd)	$Z_{D^\epsilon.\delta}$		i_z	$(D^\epsilon)_z^+$	$(D^\epsilon)_z^-$
z_ϵ		$Y_{D^\epsilon.\delta}$		i_y	$(D^\epsilon)_y^+$	$(D^\epsilon)_y^-$
$x_\pi = y_\pi = z_\pi$	(π even)	$X_{d^\pi.\delta^\pi}$		$(i^\pi)_x$	$(d^\pi)_x^-$	$(d^\pi)_x^+$

11. The Dictionary

Here is a dictionary that identifies the three spaces 196884_x, 196884_y, 196884_z:

$$\begin{aligned}
(ii)_x &= (ii)_y &&= (ii)_z = (ii), \text{ say,} \\
(ij)_x &= Y_{ij} + Y_{\Omega\cdot ij} &&= Z_{ij} - Z_{\Omega\cdot ij} && (\&c), \\
X_{d\cdot i} &= d^+ \otimes_y i &&= d^- \otimes_z i && (\&c), \\
X_{(\Omega)d\cdot\delta} &= \frac{1}{8}\sum_\epsilon -_{\delta,\epsilon} Y_{(\Omega)d\cdot\epsilon} &&= \frac{1}{8}\sum_\epsilon -_{\epsilon,\delta} Z_{(\Omega)d\cdot\epsilon} && (\&c).
\end{aligned}$$

Here in line 1, $(ii)_x$ means $(i)_x \otimes (i)_x$,
in line 2, $(ij)_x$ means $(i)_x \otimes (j)_x + (j)_x \otimes (i)_x$,
in line 3, $d^\sigma \otimes_x i$ means $(d)_x^\sigma \otimes (i)_x$; and,
in line 4, $|\tilde{d}| = 8$, δ and ϵ range over the 64 $\mathscr{C}^*$-sets that are representable by even subsets of d, and $-_{\delta,\epsilon} = (-1)^{|\delta\epsilon|/2+|\epsilon|/2}$

There are fairly obvious inner products on these spaces, under which basis vectors are orthogonal except when equal or opposite, and the basis vectors have norm 1 except that $(ij)_x = (i)_x \otimes (j)_x + (j)_x \otimes (i)_x$ has norm 2 (&c).

The identifications given by the dictionary equate these inner products. One can also check by explicit calculation that the actions of elements of N (as given in Table 29.1), are preserved by the dictionary.

12. The Algebra

From now on, we can regard ourselves as having defined a *single* space of dimension 196884, a single inner product on that space, and a well-defined action of N upon it. In the same spirit, we next define an algebra multiplication ($*$) on the space, by making three definitions for $*$ on 196884_x, 196884_y, 196884_z, that are checked to be equivalent by explicit calculations. Here is the definition for 196884_x:

$$M_x * N_x = 2(MN+NM)_x,$$

$$M_x * X_r = (M_x, \lambda_r \otimes_x \lambda_r)\, X_r,$$

$$X_r * X_s = \begin{cases} \pm\lambda_r \otimes_x \lambda_r & (x_{r\cdot s} = x_{\pm 1}), \\ X_{r\cdot s} & (x_{r\cdot s} \text{ short}), \\ 0 & (\text{otherwise}), \end{cases}$$

$$d^\sigma \otimes_x \lambda * M_x = d^\sigma \otimes_x \lambda M + \frac{1}{8} \operatorname{tr} M.\ d^\sigma \otimes_x \lambda,$$

$$d^\sigma \otimes_x \lambda * X_r = \frac{1}{8}\, e^\tau \otimes_x [\lambda - 2(\lambda,\lambda_r)\lambda_r], \text{ where (\$) holds},$$

$$d^\sigma \otimes_x \lambda * e^\tau \otimes_x \mu = \sum_{(\$)} [(\lambda,\mu) - 2(\lambda,\lambda_r)(\mu,\lambda_r)] X_r$$

$$\pm\ [(\lambda,\mu) I_x + 4\lambda \otimes_x \mu + 4\mu \otimes_x \lambda] \text{ perhaps},$$

the last term being present just if $(d)_x^\sigma = \pm(e)_x^\tau$.

Here λ and μ denote vectors of 24_x, and if M is a 24×24 symmetric matrix, then M_x is the corresponding element of 300_x. The vectors $(d)_x^\sigma$ and $(e)_x^\tau$ are typical basis elements of 4096_x, and (\$) is the condition that x_r is a short element taking $(d)_x^\sigma$ to $(e)_x^\tau$. We only enter one of two equal products $u * v$ and $v * u$.

13. The Definition of the Monster Group $\overline{G}$, and its Finiteness

The group $\overline{N}_x$ has structure $2^{1+24} \cdot 2^{11} M_{24}$, in which the quotient group $2^{11} M_{24}$ can be regarded as the group of *monomial* automorphisms of Λ_{24} taken modulo $2\Lambda_{24}$. It can be enlarged to a group $\overline{G}_x$ of structure $2^{1+24} \cdot Co_1$ by replacing the group of monomial automorphisms of Λ_{24} by the group Co_1 of all automorphisms of Λ_{24} taken modulo $2\Lambda_{24}$.

When this is done (see Appendix 5), it is obvious that the group $\overline{G}_x$ still has a representation on the space 196884_x, and that our definition of the algebra on 196884_x is actually invariant under $\overline{G}_x$. (The main difference is that $\overline{N}_x$ permutes the vectors X_r in three orbits according to the shapes $(\pm 4^2 0^{22})$, $(\pm 2^8 0^{16})$, $(\mp 3 \pm 1^{23})$ of the corresponding Leech lattice vectors, whereas $\overline{G}_x$ fuses these three types.)

Of course the algebra is equally invariant under the analogous groups $\overline{G}_y$, $\overline{G}_z$. We define the Monster group M to be the group $\overline{G}$ generated by $\overline{G}_x$, $\overline{G}_y$, $\overline{G}_z$ (in fact it is generated by any two of them). This group $\overline{G}$ plainly preserves the algebra multiplication and the inner product defined on our 196884-space.

The fact that this group is finite is almost a triviality. Explicit calculations show that if t is the vector

$$\frac{1}{16}[(ii)+(jj)-(ij)_x-(ij)_y-(ij)_z],$$

then the map $v \to v * t$ is diagonalizable with a single eigenvalue of 1 (at t), and that all the other eigenvalues are at most $1/4$ and correspond to eigenvectors orthogonal to t.

It follows that t can have only finitely many images under $\overline{G}$. To see this, observe that if t had infinitely many images, then one of them, t' say, would be very close to t and have the same norm. So we could write

$$t' = t + \theta w + O(\theta^2), \tag{2}$$

where w is a nonzero vector orthogonal to t and θ is a small real number. But then

$$t' * t' = t * t + 2\theta w * t + O(\theta^2), \tag{3}$$

and since $\|w * t\| \leqslant 1/4\|w\|$, (2) and (3) prohibit the necessary equation $t' * t' = t'$. Since the images of v span the space, $\overline{G}$ is isomorphic to the group of permutations it induces upon them, and so is a finite group.

14. Identifying the Monster

In order to identify $\overline{G}$ with the Monster group, we first note that the centralizer of $\overline{x}_{-1}$ in $\overline{G}$ plainly contains G_x, for which the irreducible constituents of 196884_x have dimensions 1, 299, 98280, and 98304. In fact these spaces are invariant under the full centralizer of $\overline{x}_{-1}$ in $\overline{G}$. This is because the 98304-space is just the space on which $\overline{x}_{-1}$ has eigenvalue -1, and explicit calculations show that on this space the mean squared eigenvalue of the map $v \to v * u$ is respectively

$$\frac{2}{3}(u,u),\ \frac{1}{24}(u,u),\ \frac{3}{64}(u,u),$$

if u is a vector in one of the constituents 1, 299, 98280. The 2×98280 vectors X_r are now determined as the unit vectors that are eigenvectors for all the maps $v \to v * M$, M in 300_x. The only non-trivial element fixing all the X_r is $\overline{x}_{-1}$ itself. It follows that the centralizer of $\overline{x}_{-1}$ in $\overline{G}$ is exactly the group $2^{1+24} \cdot Co_1$ we can see. Stephen D. Smith [Smi12] has shown that a finite group that has an involution with this centralizer has order given by (1), and has another involution whose centralizer is $2.F_2$, where F_2 is the Baby Monster — in other words $\overline{G}$ is the group usually known as the Fischer-Griess Monster.

Appendix 1. Computing in $\mathcal{P}$.

In $\mathcal{P}$ we have the relations (the *sign rules*):

$$a^2 = (-1)^{|\bar{a}|/4},$$

$$ba = (-1)^{|\bar{a} \cap \bar{b}|/2} \cdot ab,$$

$$a(bc) = (-1)^{|\bar{a} \cap \bar{b} \cap \bar{c}|} \cdot (ab)c,$$

and it is easy to see that in fact these provide an abstract definition for $\mathcal{P}$ (if it exists). For let us suppose that $e_1, e_2, \ldots, e_{12}$ are preimages of the 12 elements of some basis of $\mathcal{C}$, so that any element of $\mathcal{P}$ can be written as $\pm F(e_1, \ldots, e_{12})$, where F is some bracketed product of some sequence of its arguments, possibly with repetitions or omissions.

Then an obvious condition that an equation between two such products should hold in $\mathcal{P}$ is that it hold in $\mathcal{C}$. On the other hand, $\mathcal{C}$ is an elementary abelian 2-group, and the category of such groups is defined by the laws $a^2 = 1$, $ba = ab$, $a(bc) = (ab)c$. This means that if the two products are equal in $\mathcal{C}$ we can verify this fact by repeated applications of these laws. Using the sign rules above, we can therefore check whether the two products have the same or opposite sign in $\mathcal{P}$.

The argument also shows that any element $\bar{\pi}$ of M_{24} lifts to 2^{12} distinct standard automorphisms. For if $\bar{\pi}$ takes $\bar{e}_1, \ldots, \bar{e}_{12}$ to $\bar{f}_1, \ldots, \bar{f}_{12}$ we can pick (in any of 2^{12} ways), preimages $f_1, \ldots, f_{12}$ for $\bar{f}_1, \ldots, \bar{f}_{12}$. No matter how this is done, $f_1, \ldots, f_{12}$ will satisfy the same relations as $e_1, \ldots, e_{12}$, so there is an automorphism $\pi: e_i \to f_i$.

Appendix 2. A Construction for $\mathcal{P}$.

Here is the shortest explicit construction we know for $\mathcal{P}$. The elements of $\mathcal{P}$ will be just the elements of $\mathcal{C}$, equipped with formal signs. The loop multiplication $A \cdot B$ is defined in terms of the addition $A + B$ in $\mathcal{C}$ by the formula

$$A \cdot B = (-1)^{|\theta(A) \cap B|} \; (A+B)$$

where θ is a certain function from $\mathcal{C}$ to $\mathcal{C}^*$. The sign rules for $\mathcal{P}$ then reduce to the following requirements on θ (the congruences being modulo 2):-

$$p(A):\quad |\theta(A) \cap A| \equiv \tfrac{1}{4}|A|,$$

$$q(A,B):\quad |\theta(A) \cap B| + |\theta(B) \cap A| \equiv \tfrac{1}{2}|A \cap B|,$$

$$r(A,B,C):\quad \left.\begin{array}{l} |\theta(A) \cap B| + |\theta(A+B) \cap C| + \\ + |\theta(B) \cap C| + |\theta(A) \cap (B+C)| \end{array}\right\} \equiv |A \cap B \cap C|.$$

We satisfy these as follows. We first require that θ have the property

$$\theta(A+B) = \theta(A) + \theta(B) + (A \cap B)$$

(the addition here being in $\mathscr{C}^*$), which says that it is a quadratic function from $\mathscr{C}$ to $\mathscr{C}^*$ whose associated symmetric bilinear function is $A \cap B$. In characteristic 2 such a requirement still leaves us free to specify θ at a basis $E_1, E_2, \ldots, E_{12}$ for $\mathscr{C}$, so we can demand further that for $n = 1, 2, \ldots, 12$ in turn we have:

$$|\theta(E_n) \cap E_i| \equiv \begin{cases} \tfrac{1}{2}|E_i \cap E_n| - |\theta(E_i) \cap E_n| & (i < n), \\ \tfrac{1}{4}|E_n| & (i = n), \\ 0, \text{ say}, & (i > n). \end{cases}$$

(Recall that a $\mathscr{C}^*$-set is exactly specified by giving the parities of its intersections with the E_i, since $\mathscr{C}$ and $\mathscr{C}^*$ are dual groups.)

These equations ensure that $p(A)$ and $q(A,B)$ hold when A and B are among the E_i. But then, using the quadratic property and the well known formulae

$$|X+Y| = |X| + |Y| - 2|X \cap Y|, \quad (X+Y) \cap Z = X \cap Z + Y \cap Z,$$

we find that $r(A,B,C)$ holds universally, that $r(A,B,C)$ enables us to deduce $q(A,B+C)$ from $q(A,B)$ and $q(A,C)$, and finally that $q(A,B)$ enables us to deduce $p(A+B)$ from $p(A)$ and $p(B)$. Inductively, this shows that all of $p(A)$, $q(A,B)$, $r(A,B,C)$ hold universally.

Appendix 3. Some Relations in Q_x.

By direct computation with triples, one can verify that the elements x_d and x_δ satisfy the relations:

$$[x_\delta, x_\epsilon] = 1,$$

$$[x_d, x_e] = x_{-1}^{|\tilde{d} \cap \tilde{e}| + |\tilde{d}| \cdot |\tilde{e}|/4},$$

$$[x_d, x_\epsilon] = x_{-1}^{|\tilde{d} \cap \tilde{\epsilon}|/2},$$

$$x_\delta^2 = 1, \quad x_d^2 = x_{-1}^{|\tilde{d}|/4},$$

$$x_\delta x_\epsilon = x_{\delta\epsilon}, \quad x_d x_e = z_{-1}^{|\tilde{d} \cap \tilde{e}|/2} \cdot x_{de} \cdot x_{\tilde{d} \cap \tilde{e}},$$

where in the last relation $\tilde{d} \cap \tilde{e}$ is to be regarded as a $\mathscr{C}^*$-set. We shall take the last (and hardest) of these as an example. We have

$$x_d x_e : (a,b,c) \to ({*}{*}{*}, e.db, cd.e), \tag{4}$$

$$x_{de} : (a,b,c) \to ({*}{*}{*}, de.b, c.de), \tag{5}$$

where the symbols *** indicate some more complicated expressions whose exact values won't be needed. Now since $z_{-1} : (a,b,c) \to (-a,-b,c)$ and $[d,e] = (-1)^{|\tilde{d} \cap \tilde{e}|/2}$ we have

$$z_{-1}^{|\tilde{d} \cap \tilde{e}|/2}.x_{de} : (a,b,c) \to (***, ed.b, c.de),$$

which needs only some reassociation to become (4). But the diagonal automorphism $\tilde{d} \cap \tilde{e}$ actually reassociates any three-term product two of whose terms are d and e. For example, since $|\tilde{d} \cap \tilde{e}|$ is even, we have

$$(c.de)^{\tilde{d} \cap \tilde{e}} = c^{\tilde{d} \cap \tilde{e}}.de = (-1)^{|\tilde{c} \cap \tilde{d} \cap \tilde{e}|}.c.de = cd.e$$

from the definitions in Appendix 1. Accordingly

$$z_{-1}^{|\tilde{d} \cap \tilde{e}|/2}.x_{de}.x_{\tilde{d} \cap \tilde{e}} : (a,b,c) \to (***, e.db, cd.e)$$

agreeing with (4). (We need not check the first coordinate, since it is easily checked that triples with $abc = 1$ are taken to other such triples.)

This relation shows that z_{-1} and $k_x = x_\Omega z_{-1}$ are in Q_x, and from the previous relations we see that modulo z_{-1} and k_x the group Q_x becomes abelian, after which it is easy to check that its order is at most 2^{26}, and that of $\overline{Q_x}$ at most 2^{25}.

But the relations show much more; that indeed there is a 2:1 homomorphism from $\overline{Q_x}$ onto $\Lambda_{24}/2\Lambda_{24}$, taking $\bar{x}_r$ to λ_r. To see this, one only has to check the images of the above relations. Once again we consider only the last one. We have:

$$\lambda_d = (2_{\text{on } \tilde{d}}, 0_{\text{elsewhere}}),$$

$$\lambda_e = (2_{\text{on } \tilde{e}}, 0_{\text{elsewhere}}),$$

and so

$$\lambda_d + \lambda_e = (2_{\text{on } \widetilde{de}}, 4_{\text{on } \tilde{d} \cap \tilde{e}}, 0_{\text{elsewhere}}).$$

Now $\tilde{d} \cap \tilde{e}$ is an even subset of $\tilde{\Omega}$, and so can be expressed as the union of m 2-element sets ij, where $m = |\tilde{d} \cap \tilde{e}|/2$. From the definitions,

$$(4_{\text{on } ij}, 0_{\text{elsewhere}}) = \lambda_\Omega + \lambda_{ij},$$

$$(4_{\text{on } \tilde{d} \cap \tilde{e}}, 0_{\text{elsewhere}}) = m\lambda_\Omega + \lambda_{\tilde{d} \cap \tilde{e}},$$

which imply the desired relation

$$\lambda_d + \lambda_e = \lambda_{de} + m\lambda_\Omega + \lambda_{\tilde{d} \cap \tilde{e}}.$$

(We remark that $z_{-1} \equiv x_\Omega$ modulo K, since $z_{-1}x_\Omega = k_x$, so that the last of the relations with which we introduced this Appendix reads

$$\bar{x}_d \cdot \bar{x}_e = \bar{x}_{\Omega^m.de} \cdot \bar{x}_{d \cap e}$$

in $\overline{Q_x}$.)

The relations we have given therefore show that the group $\overline{Q_x}$ is at most a double cover of $\Lambda_{24}/2\Lambda_{24}$, and so has structure (at most) 2^{1+24}. But it is easy to see that no non-trivial element of this double cover fixes every triple (a,b,c), so this is the exact structure. We shall see in Appendix 4 that Q_x is the kernel of the representation 24_x of N_x, the image being a group of structure $2^{12}:M_{24}$. This gives the structure of N_x, and the other structures in §7 are easily deduced.

Appendix 4. Constructing Representations for N_x.

The representation 24_x. We extend $\mathscr{P}$ by adjoining a formal zero satisfying

$$d0 = 0d = 00 = 0 \quad (d \in \mathscr{P}).$$

Obviously we can regard N as acting on triples of the larger system $\mathscr{P}_0 = \mathscr{P} \cup \{0\}$. We now define formal objects i_x, $-i_x$

$$i_x = \{(d,0,0) : i \notin \bar{d}\}, \qquad -i_x = \{(d,0,0) : i \in d\} \quad (\&c).$$

It is easy to see that N_x acts on these as follows

$$y_d,\, z_d : i_x \to (-1)^{|\bar{d} \cap i|} i_x, \quad x_\pi : i_x \to (i^{\pi})_x,$$

so demonstrating that a quotient group of structure $2^{12}:M_{24}$ acts faithfully, and that the kernel is exactly Q_x.

The formal objects $\pm i_x$ are permuted exactly like the coordinate vectors

$$(\pm 1_{\text{on } i}, 0_{\text{elsewhere}})$$

for the space of the Leech lattice, and so we can abstractly identify them with these vectors, so defining a 24-dimensional representation 24_x of N_x.

The representation 4096_x. We define

$$d_x^- = \{(0,d,0),\, (0,-\Omega d,0)\}, \qquad d_x^+ = \{(0,0,d),\, (0,0,\Omega d)\} \quad (\&c),$$

for each $d \in \mathscr{P}$. Then there are $2 \cdot 4096$ distinct objects d_x^σ, which are permuted by N_x, and there is an invariant pairing on these, given by changing the sign of d. We can therefore identify these objects with $2 \cdot 4096$ vectors d_x^σ spanning a 4096-dimensional space 4096_x by adding the relations $(-d)_x^\sigma = -(d)_x^\sigma$. The definitions have been arranged so that k_x (which takes (a,b,c) to $(a,-\Omega b,\Omega c)$) acts trivially, and in fact the kernel of this representation is the fourgroup $\{1, y_\Omega, z_{-\Omega}, k_x\}$. (The elements k_y and k_z both act as -1.)

The representation 98280_x. The group $\overline{Q_x}$ has $2 \cdot 98280$ short elements $\bar{x}_r$ corresponding to the 98280 short elements λ_r of $\Lambda_{24}/2\Lambda_{24}$. Under conjugation, $\overline{N_x}$ permutes these in such a way that the pairing x_r, x_{-r} is preserved.

We can think of this as an action of $\overline{N_x}$ on a 98280-dimensional space 98280_x spanned by vectors X_r that are permuted just like the x_r, and satisfy $X_{-r} = -X_r$. We usually prefer to regard it as a representation of N_x in which K acts trivially. One can check that the full kernel of this representation is the group of order 8 generated by K and x_{-1}.

Appendix 5. Building the Group $\bar{G}_x$.

Now in all our constructions we enlarge the group $2^{12}{:}M_{24}$ of monomial automorphisms of Λ_{24} to the group $2 \cdot Co_1$ of all automorphisms of Λ_{24}, and so obtain a group $\bar{G}_x$ strictly larger than $\bar{N}_x$. In fact, $\bar{G}_x$ is essentially defined just by making sure that this process works for 24_x and 4096_x — we shall see that 98280_x takes care of itself.

It is well-known that an extraspecial 2-group 2^{1+2n} has a unique 2^n-dimensional representation up to equivalence. In particular, $\bar{Q}_x$ has a unique 4096-dimensional representation.

It now follows that if β is an automorphism of $\bar{Q}_x$, then there is a matrix M (unique up to scalar factors, by Schur's lemma), such that

$$m_{4096}(g) = M^{-1} \cdot m_{4096}(g) \cdot M$$

for all g in $\bar{Q}_x$, where $m_{4096}(g)$ is the matrix representing g in 4096_x. We can make M unique up to sign by demanding that it have real entries and determinant 1. The two such matrices will be called the *representors* of β.

Every automorphism of this group 2^{1+24} yields an automorphism of the quotient group 2^{24} ($\cong \Lambda_{24}/2\Lambda_{24}$) when we factor by x_{-1}. The automorphisms of 2^{24} that happen in this way are precisely those that fix the quadratic form (modulo 2) defined on $\Lambda_{24}/2\Lambda_{24}$. In particular, there is such an automorphism α^*, say, for each automorphism α of the real Leech lattice Λ_{24}.

We define G_x to be the group of ordered pairs of matrices of the form

$$(24 \times 24 \text{ matrix of } \alpha, \quad 4096 \times 4096 \text{ matrix of a representor of } \alpha^*).$$

If we restrict the elements α to be the monomial automorphisms, we get a subgroup that is almost the same as N_x. More precisely, it is isomorphic to $N_x/\langle k_x\rangle$, since k_x is represented trivially in 24_x and 4096_x.

The group G_x obviously has 24 and 4096-dimensional representations extending those of $N_x/\langle k_x\rangle$. However, it also has a 98280 dimensional representation, since it still permutes the short $\bar{x}_r$. (To see this, observe that $\bar{G}_x = G_x/\langle k_y\rangle$ normalizes $\bar{Q}_x$, and therefore permutes by conjugation the $2 \cdot 98280$ short elements $\bar{x}_r$. In the usual way, we regard this as an action of G_x in which k_y acts trivially. So G_x has a representation of degree 98280 extending that of N_x.)

We note that G_x has elements that intermix the X_r corresponding to Leech lattice vectors λ_r of the three shapes $(\pm 4^2, 0^{22})$, $(\pm 2^8, 0^{16})$, $(\mp 3, \pm 1^{23})$.

30

A Monster Lie Algebra?

R. E. Borcherds, J. H. Conway,
L. Queen and N. J. A. Sloane

We define a remarkable Lie algebra of infinite dimension, and conjecture that it may be related to the Fischer-Griess Monster group.

The idea was mooted in *Monstrous Moonshine* [Con17] that there might be an infinite-dimensional Lie algebra (or superalgebra) L that in some sense "explains" the Fischer-Griess "Monster" group M. In this chapter we produce some candidates for L based on properties of the Leech lattice described in the preceding chapters. These candidates are described in terms of a particular Lie algebra L_∞ of infinite rank.

We first review some of our present knowledge about these matters. It was proved by character calculations in [Con17, p. 317] that the centralizer C of an involution of class 2A in the Monster group has a natural sequence of modules affording the head characters (restricted to C). In [Kac4], V. Kac has explicitly constructed these as C-modules. Now that Atkin, Fong and Smith ([Atk1], [Fon1], [Smi13]) have verified the relevant numerical conjectures of [Con17] for M, we know that these modules can be given the structure of M-modules. More recently, Frenkel, Lepowsky and Meurman [Fre1]-[Fre5] have given a simple construction for the Monster along these lines, but this sheds little light on the conjectures.

Some of the conjectures of [Con17] have analogs in which M is replaced by a compact simple Lie group, and in particular by the Lie group E_8. Most of the resulting statements have now been established by Kac and others. However, it seems that this analogy with Lie groups may not be as close as one would wish, since two of the four conjugacy classes of elements of order 3 in E_8 were shown in [Que7]-[Que9] to yield examples of modular functions neither of which are the Hauptmodul for any modular group. This disproves the conjecture made on p. 267 of [Kac3], and is particularly distressing since it was the Hauptmodul property that prompted the discovery of the conjectures in [Con17], and it is this property that gives those conjectures almost all their predictive power. (Other references to recent work on the Monstrous Moonshine conjectures will be found in §1.4 of Chap. 1.)

The properties of the Leech lattice that we shall use stem mostly from the facts about "deep holes" in that lattice reported in Chap. 23. Let $w = (0,1,2,3,...,24 \mid 70)$. The main result of Chap. 26 is that the subset of vectors r in $II_{25,1}$ for which $r \cdot r = 2$, $r \cdot w = -1$ ("Leech roots") is isometric to the Leech lattice, under the metric defined by $d(r,s)^2 = \text{norm}(r-s)$. The main result of Chap. 27 is that $\text{Aut}(II_{25,1})$ is obtained by extending the Coxeter subgroup generated by the reflections in these Leech roots by its group of graph automorphisms together with the central inversion -1. It is remarkable that the walls of the fundamental region for this Coxeter group (which correspond one-for-one with the Leech roots) are transitively permuted by the graph automorphisms, which form an infinite group abstractly isomorphic to the group Co_∞ of all automorphisms of the Leech lattice, including translations.

Vinberg [Vin7] shows that for the earlier analogs $II_{9,1}$ and $II_{17,1}$ of $II_{25,1}$ the fundamental regions for the reflection subgroups have respectively 10 and 19 walls, and the graph automorphism groups have orders 1 and 2. For the later analogs $II_{33,1},...$, there is no "Weyl vector" like w, so it appears that $II_{25,1}$ is very much a unique object.

We can use the vector w to define a root system in $II_{25,1}$. If $v \in II_{25,1}$ then we define the height of v by $-v \cdot w$, and we say that v is positive or negative according as its height is positive or negative. We now define a Kac-Moody Lie algebra L_∞, of infinite dimension and rank, as follows: L_∞ has three generators $e(r)$, $f(r)$, $h(r)$ for each Leech root r, and is presented by the following relations:

$$
\begin{aligned}
&[e(r),h(s)] = r \cdot s\, e(r)\ ,\\
&[f(r),h(s)] = -r \cdot s\, f(r),\\
&[e(r),f(r)] = h(r),\\
&[e(r),f(s)] = 0,\\
&[h(r),h(r)] = 0 = [h(r),h(s)],\\
&e(r)\,\{\text{ad}\, e(s)\}^{1-r \cdot s} = 0 = f(r)\,\{\text{ad}\, f(s)\}^{1-r \cdot s},
\end{aligned}
$$

where r and s are distinct Leech roots. (We have quoted these relations from Moody's excellent survey article [Moo2]. Moody supposes that the number of fundamental roots is finite, but since no argument ever refers to infinitely many fundamental roots at once, this clearly does not matter. See also [Kac5].)

Then we conjecture that L_∞ provides a natural setting for the Monster, and more specifically that the Monster can be regarded as a subquotient of the automorphism group of some naturally determined subquotient algebra of L_∞.

The main problem is to "cut L_∞ down to size". Here are some suggestions. A rather trivial remark is that we can replace the Cartan subalgebra H of L_∞ by the homomorphic image obtained by adding the relations

$$c_1\, h(r_1) + c_2\, h(r_2) + \cdots = 0$$

for Leech roots $r_1, r_2, \ldots$, whenever $c_1, c_2, \ldots$ are integers for which

$$c_1 r_1 + c_2 r_2 + \cdots = 0 .$$

A more significant idea is to replace L_∞ by some kind of completion allowing us to form infinite linear combinations of the generators, and then restrict to the subalgebra fixed by all the graph automorphisms. The resulting algebra, supposing it can be defined, would almost certainly not have any notion of root system.

Other subalgebras of L_∞ are associated with the holes in the Leech lattice, which are either "deep" holes or "shallow" holes (see Chap. 23).

(i) By Chap. 23, any deep hole corresponds to a Niemeier lattice N, which has a Witt part which is a direct sum of root lattices chosen from the list A_n ($n = 1,2,\ldots$), D_n ($n = 4,5,\ldots$), E_6, E_7 and E_8. Only 23 particular combinations arise, and we shall take $A_{11}D_7E_6$ as our standard example. The graph of Leech roots contains a finite subgraph which is the disjoint union of extended Dynkin diagrams corresponding to these Witt components W of N, and so our algebra L_∞ has a subalgebra $L[N]$ which is a direct sum of the Euclidean Lie algebras $E(W)$ corresponding to those components (see [Kac1], [Kac5], [Moo1]). For example, L_∞ has a subalgebra

$$E(A_{11}) + E(D_7) + E(E_6) .$$

Each such subalgebra of L_∞ can be extended to a larger subalgebra $L^\ddagger[N]$ having one more fundamental root, corresponding to a "glue vector" of the appropriate hole (see Chap. 25). In the corresponding graph, the new node is joined to a single special node (§2 of Chap. 4) in each component. The graph for $L^\ddagger[A_{11}D_7E_6]$ is shown in Fig. 30.1. These hyperbolic algebras $L^\ddagger[N]$, having finite rank, are certainly more manageable than L_∞ itself.

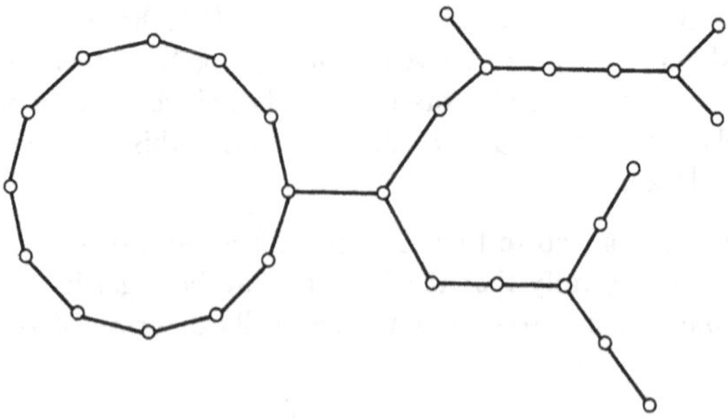

Figure 30.1 The fundamental root diagram for $L^\ddagger[A_{11}D_7E_6]$.

Since the 23 Niemeier lattices yield 23 constructions for the Leech lattice (Chap. 25), it is natural to ask if we can obtain 23 different constructions for the Monster using the Lie algebras $L[N]$ or $L^{\ddagger}[N]$.

(ii) Each shallow hole in the Leech lattice (see Chap. 24) corresponds to a maximal subalgebra of L_∞ of finite rank.

We have made various calculations concerning L_∞ (finding the multiplicities of certain roots via the Weyl-MacDonald-Kac formula, etc.). It is worth noting that these calculations are facilitated by the remarkable recent discovery that the Mathieu group M_{12} is generated by the two permutations

$$t \to |2t| , \qquad t \to 11-t \pmod{23} ,$$

of the set $\{0,1,2,3,4,5,6,7,8,9,10,11\}$, where $|x|$ denotes the unique y in this set for which $y \equiv \pm x \pmod{23}$ (see Chap. 11 for this result and its history). The simplest transformation (see Chap. 28) between the usual Euclidean coordinates for the Leech lattice and its Lorentzian coordinates uses this description of M_{12}.

Bibliography

We have tried to give an adequate bibliography for the major topics discussed in this book. The numbers in square brackets at the end of an entry give the chapters where it is cited, and thus also serve as an author index. [B] indicates that the reference is cited elsewhere in this bibliography. Certain items, not otherwise mentioned in the book, have been included for completeness. There are about 1550 references. Further references on number theory and many related subjects will be found in [LeV2] and [Guy1], and [Mac6] contains an extensive bibliography on coding theory up to 1977. The books [Gru1a], [Lek1] and the survey articles in [Töl1] and [Gru2] also contain relevant bibliographies.

Abbreviations

AJM = American Journal of Mathematics.
AMAH = Acta Mathematica Academiae Scientiarum Hungaricae.
AMM = American Mathematical Monthly.
ASUH = Abhandlungen Mathematischen Seminar der Universität Hamburg.
BAMS = Bulletin of the American Mathematical Society.
BLMS = Bulletin of the London Mathematical Society.
BSTJ = AT&T Technical Journal (formerly Bell System Technical Journal).
CJM = Canadian Journal of Mathematics.
CMB = Canadian Mathematical Bulletin.
COMM = Institute of Electrical and Electronics Engineers, Transactions on Communications (formerly Transactions on Communication Technology).
CR = Comptes Rendus Hebdomadaires des Séances de l'Académie des Sciences, Paris, Série A.
DAN = Doklady Akademii Nauk SSR.
Discr. Math. = Discrete Mathematics.
DMJ = Duke Mathematical Journal.
IANS = Izvestiya Akademii Nauk SSSR, Seriya Matematicheskaya.
IC = Information and Control.
IJM = Illinois Journal of Mathematics.
Izv. = Mathematics of the USSR — Izvestiya.
J. Alg. = Journal of Algebra.
JAMS = Journal of the Australian Mathematical Society.
JCT = Journal of Combinatorial Theory.
JLMS = Journal of the London Mathematical Society.
JNT = Journal of Number Theory.

JRAM = Journal für die Reine und Angewandte Mathematik.
JSM = Journal of Soviet Mathematics.
KNAW = Proceedings of the Koninklijke Nederlandse Akademie van Wetenschappen.
LNM = Lecture Notes in Mathematics (Springer-Verlag).
LOMI = Zapiski Nauchnykh Seminarov Leningradskogo Otdeleniya Matematicheskogo Instituta imeni V. A. Steklova Akademii Nauk SSSR.
Mat. Sb. = Matematicheskii Sbornik.
Math. = Mathematika.
MTAC = Mathematics of Computation (formerly Mathematical Tables and Aids to Computation).
PAMS = Proceedings of the American Mathematical Society.
PCPS = Mathematical Proceedings of the Cambridge Philosophical Society (formerly Proceedings of the Cambridge Philosophical Society).
PGIT = Institute of Electrical and Electronics Engineers, Transactions on Information Theory.
PIEEE = Proceedings of the Institute of Electrical and Electronics Engineers.
PIT = Problems of Information Transmission.
PJM = Pacific Journal of Mathematics.
PLMS = Proceedings of the London Mathematical Society.
PNAS = Proceedings of the National Academy of Sciences of the U.S.A.
PPI = Problemy Peredachi Informatsii.
PRS = Proceedings of the Royal Society, London.
PSIM = Proceedings of the Steklov Institute of Mathematics.
PSPM = Proceedings of Symposia in Pure Mathematics, American Mathematical Society, Providence, RI.
Sb. = Mathematics of the USSR — Sbornik.
SIAD = SIAM Journal on Algebraic and Discrete Methods.
SIAJ = SIAM Journal on Applied Mathematics.
SMD = Soviet Mathematics, Doklady.
Springer-Verlag = Springer-Verlag: New York, Berlin, Heidelberg, London, Paris and Tokyo.
TAMS = Transactions of the American Mathematical Society.
TMIS = Trudy Matematicheskogo Instituta imeni V. A. Steklova.

A

[Aal1] M. J. Aaltonen, *Linear programming bounds for tree codes*, PGIT **25** (1979), 85-90 [9].

[Abr1] M. Abramowitz and I. A. Stegun, *Handbook of Mathematical Functions*, National Bureau of Standards Appl. Math. Series **55**, U. S. Dept. Commerce, Washington DC, 1972 [3, 9, 13, 14, 16].

[Ado1] J.-P. Adoul, *La quantification vectorielle des signaux: approche algebrique*, Ann. Télécommun. **41** (1986), 158-177 [2, 20].

[Aff1] L. Afflerbach, *Minkowskische Reduktionsbedingungen für positiv definite quadratische Formen in 5 Variablen*, Monatsh. Math. **94** (1982), 1-8 [2, 15].

[Aff2] — and H. Grothe, *Calculation of Minkowski-reduced lattice bases*, Computing **35** (1985), 269-276 [2, 15].

[Ahm1] F. R. Ahmed, K. Huml and B. Sedlác ek, editors, *Crystallographic Computing Techniques*, Munksgaard, Copenhagen, 1976 [2].

[Air1] T. J. Aird and J. R. Rice, *Systematic search in high dimensional sets*, SIAM J.

Num. Anal. **14** (1977), 296-312 [1].

[Aki1] I. Y. Akimova, *The problem of optimal arrangement and generalization of a theorem of Fejes Tóth*, Tekh. Kibern. **20** (No. 2, 1982) = Eng. Cybernetics **22** (No. 2, 1982), 149 [2].

[Aki2] — *Application of Voronoi diagrams in combinatorial problems (a survey)*, Tekh. Kibern. **22** (No. 2, 1984), 102-109 = Eng. Cybernetics **22** (No. 4, 1984), 6-12 [2].

J. M. Alexander: see T. L. Saaty.

[Ale1] A. D. Alexandrov, *Convex Polyhedra*, Moscow, 1950; German translation, Akademie-Verlag, Berlin, 1958 [21].

[All1] E. L. Allgower and P. H. Schmidt, *Computing volumes of polyhedra*, MTAC **46** (1986), 171-174 [21].

[And1] J. B. Anderson and R. de Buda, *Better phase-modulation performance using trellis phase codes*, Elect. Lett. **12** (1976), 587-588 [3].

[And2] S. Andrilli, *Existence and uniqueness of O'Nan's simple group*, Ph.D. Dissertation, Rutgers University, 1979 [10].

[Apo1] T. M. Apostol, *Introduction to Analytic Number Theory*, Springer-Verlag, 1976 [15].

[Arn1] V. I. Arnold and A. L. Krylov, *Uniform distribution of points on a sphere and some ergodic properties of solutions of linear ordinary differential equations in a complex region*, DAN **148** (1963), 9-12 = SMD **4** (1963), 1-5 [1].

[Art1] E. Artin, *The orders of the classical simple groups*, Comm. Pure Appl. Math. **8** (1955), 455-472 = Coll. Papers pp. 398-415 [10].

M. Aschbacher: see R. L. Griess Jr., G. Mason.

[Ash1] A. Ash, *Eutactic forms*, CJM **29** (1977), 1040-1054.

[Ash2] — *On the existence of eutactic forms*, BLMS **12** (1980), 192-196.

[Ash3] P. F. Ash and E. D. Bolker, *Recognizing Dirichlet tessellations*, Geom. Dedic. **19** (1985), 175-206 [2].

T. Ashida: see S. R. Hall.

[Asi1] D. Asimov, *The grand tour-a tool for viewing multidimensional data*, SIAM J. Sci. Stat. Comput. **6** (1985), 128-143.

[Ask1] R. Askey, *Orthogonal polynomials and positivity*, Studies in Applied Math. **6**, Special Functions and Wave Propagation, Soc. Indust. Appl. Math., Phil. PA, 1970, pp. 64-85 [9].

[Ask2] — *Orthogonal polynomials and special functions*, Soc. Indust. Appl. Math., Phil. PA, 1975 [9].

[Ask2a] — editor, *Theory and Application of Special Functions*, Ac. Press, NY, 1975 [B].

[Ask3] — *Orthogonal polynomials old and new, and some combinatorial connections*, in [Jac1], 1984, pp. 67-84.

[Ask4] — and J. Wilson, *A set of orthogonal polynomials that generalize the Racah coefficients or* $6-j$ *symbols*, SIAM J. Math. Anal. **10** (1979), 1008-1016 [9].

[Ass1] S. F. Assmann and N. J. A. Sloane, in preparation [9].

[Ass2] E. F. Assmus, Jr., and H. F. Mattson, Jr., *Perfect codes and the Mathieu groups*, Archiv. Math. **17** (1966), 121-135 [3, 11].

[Ass3] — — *New 5-designs*, JCT **6** (1969), 122-151 [3, 7].

[Ass4] — — *Coding and combinatorics*, SIAM Rev. **16** (1974), 349-388 [3].

[Ast1] J. T. Astola, *The Tietäväinen bound for spherical codes*, Discr. Appl. Math. **7** (1984), 17-21 [1, 9].

B. S. Atal: see also J. L. Flanagan, M. R. Schroeder.

[Ata1] — *Predictive coding of speech at low bit rates*, COMM **30** (1982), 600-614

[2].
[Ata2] — and M. R. Schroeder, *Predictive coding of speech signals and subjective error criteria*, IEEE Trans. Acoust. Speech, Signal Proc. **27** (1979), 247-254 [2].
[Ata3] — — *Stochastic coding of speech signals at very low bit rates*, in *Links for the Future*, ed. P. Dewilde and C. A. May, IEEE Press, NY, 1984, pp. 1610-1613 [2].
A. O. L. Atkin: see also J. Larmouth.
[Atk1] — P. Fong and S. D. Smith, personal communication [30].
[Atk2] M. D. Atkinson, editor, *Computational Group Theory*, Ac. Press, NY, 1984 [B].
[Aus1] L. Auslander, *An account of the theory of crystallographic groups*, PAMS **16** (1965), 1230-1236 [4].
[Avi1] D. Avis and B. K. Bhattacharya, *Algorithms for computing d-dimensional Voronoi diagrams and their duals*, in [Pre1a], 1983, pp. 159-180.

B

[Bab1] L. Babai, *On Lovász' lattice reduction and the nearest lattice point problem*, Combinatorica **6** (1986), 1-13 [2].
[Bab2] V. F. Babenko, *On the optimal error bound for cubature formulas on certain classes of continuous functions*, Analysis Math. **3** (1977), 3-9 [1].
[Bac1] G. Bachman, *Introduction to p-Adic Numbers and Valuation Theory*, Ac. Press, NY, 1964 [15].
[Bac2] P. Bachmann, *Zahlentheorie* IV, *Die Arithmetik der Quadratische Formen* **2**, Teubner, Leipzig, 1923 [2].
[Bac3] R. Backhouse, *Writing a number as a sum of two squares: a new solution*, Info. Proc. Lett. **14** (1982), 15-17 [15].
R. Bajcsy: see R. Mohr.
[Bal1] W. W. R. Ball and H. S. M. Coxeter, *Mathematical Recreations and Essays*, Univ. Toronto Press, Toronto, 12th ed., 1974 [4].
[Bam1] R. P. Bambah, *On lattice coverings by spheres*, Proc. Nat. Inst. Sci. India **20** (1954), 25-52 [2, 4].
[Bam2] — *Lattice coverings with four-dimensional spheres*, PCPS **50** (1954), 203-208 [2, 4].
[Bam3] — and N. J. A. Sloane, *On a problem of Ryskov concerning lattice coverings*, Acta Arith. **42** (1982), 107-109 [2, 4].
[Ban1] E. Bannai, *On some spherical t-designs*, JCT **A26** (1979), 157-161 [3].
[Ban2] — *Orthogonal polynomials, algebraic combinatorics and spherical t-designs*, PSPM **37** (1980), 465-468 [3].
[Ban3] — *On the weight distribution of spherical t-designs*, Europ. J. Comb. **1** (1980), 19-26 [3].
[Ban4] — *Spherical t-designs which are orbits of finite groups*, J. Math. Soc. Japan **36** (1984), 341-354 [3].
[Ban5] — *Spherical designs and group representations*, Contemp. Math. **34** (1984), 95-107 [3].
[Ban6] — A. Blokhuis, J. J. Seidel and P. Delsarte, *An addition formula for hyperbolic space*, JCT **A36** (1984), 332-341 [3].
[Ban7] — and R. M. Damerell, *Tight spherical designs* I, J. Math. Soc. Japan **31** (1979), 199-207 [3, 14].
[Ban8] — — *Tight spherical designs* II, JLMS **21** (1980), 13-30 [3, 14].
[Ban9] — and S. G. Hoggar, *On tight t-designs in compact symmetric spaces of rank*

one, Proc. Jap. Acad. **A61** (1985), 78-82 [3].

[Ban10] — — *Tight t-designs and squarefree integers*, Europ. J. Combin. to appear [3].

[Ban11] — and T. Ito, *Algebraic Combinatorics* I: *Association Schemes*, Benjamin, Menlo Park CA, 1984 [3, 9, 14].

[Ban12] — — *Current research on algebraic combinatorics*, Graphs Combin., **2** (1984), 287-308 [3, 9].

[Ban13] — and N. J. A. Sloane, *Uniqueness of certain spherical codes*, CJM **33** (1981), 437-449 ≅ Chap. 14 of this book [1, 3, 14].

E. P. Baranovskii: see also S. S. Ryskov.

[Bar1] — *Packings, coverings, partitionings and certain other distributions in spaces of constant curvature*, Itogi Nauki-Ser. Mat. (Algebra, Topologiya, Geometriya) 1969, 185-225 = Progress in Math. **9** (1971), 209-253 [1].

[Bar2] — and S. S. Ryskov, *Primitive five-dimensional parallelohedra*, DAN **212** (1973), 532-535 = SMD **14** (1973) 1391-1395.

[Bar3] — — and S. S. Shushbaev, *A geometric estimate of the number of representations of a real number by a positive quadratic form and an estimate of the remainder of the multidimensional zeta-function*, TMIS **158** (1981) 3-8 = PSIM (1983, No. 4) 1-7.

[Bar4] E. S. Barnes, *Note on extreme forms*, CJM **7** (1955), 150-154 [2].

[Bar5] — *The covering of space by spheres*, CJM **8** (1956), 293-304 [2].

[Bar6] — *The perfect and extreme senary forms*, CJM **9** (1957), 235-242 [1, 2, 8].

[Bar7] — *The complete enumeration of extreme senary forms*, Phil. Trans. Royal Soc. London **A249** (1957), 461-506 [1, 2, 8].

[Bar8] — *On a theorem of Voronoi*, PCPS **53** (1957), 537-539 [2].

[Bar9] — *The construction of perfect and extreme forms*, Acta Arith. **5** (1959) 57-79 and 205-222 [2, 6].

[Bar10] — *Criteria for extreme forms*, JAMS **1** (1959), 17-20 [2].

[Bar11] — *Minkowski's fundamental inequality for reduced positive quadratic forms*, JAMS **A26** (1978), 46-52 [2].

[Bar12] — and M. J. Cohn, *On the reduction of positive quaternary quadratic forms*, JAMS **A22** (1976), 54-64 [2].

[Bar13] — — *On Minkowski reduction of positive quaternary quadratic forms*, Math. **23** (1976), 156-158 [2].

[Bar14] — and T. J. Dickson, *Extreme coverings of n-space by spheres*, JAMS **7** (1967), 115-127; **8** (1968), 638-640 [2].

[Bar15] — and N. J. A. Sloane, *New lattice packings of spheres*, CJM **35** (1983), 117-130 [1, 8].

[Bar16] — — *The optimal lattice quantizer in three dimensions*, SIAD **4** (1983), 30-41 [2, 21].

[Bar17] — and D. W. Trenerry, *A class of extreme lattice-coverings of n-space by spheres*, JAMS **14** (1972), 247-256 [2].

[Bar18] — and G. E. Wall, *Some extreme forms defined in terms of Abelian groups*, JAMS **1** (1959), 47-63 [1, 4, 5, 6, 7, 8].

[Bas1] L. A. Bassalygo, *New upper bounds for error-correcting codes*, PPI **1** (No. 4, 1965), 41-44 = PIT **1** (No. 4, 1965), 32-35 [9].

[Bas2] — and V. A. Zinoviev, *Some simple consequences of coding theory for combinatorial problems of packings and coverings*, PPI **34** (No. 2, 1983), 291-295 = PIT **34** (1983), 629-631 [9].

[Bat1] P. T. Bateman, *On the representations of a number as the sum of three squares*, TAMS **71** (1951), 70-101 [4].

[Bau1] J. R. Baumgardner and P. O. Frederickson, *Icosahedral discretization of the two-sphere*, SIAM J. Num. Anal. **22** (1985), 1107-1115 [1].

[Bay1] E. Bayer-Fluckiger, *Definite unimodular lattices having an automorphism of given characteristic polynomial*, Comment. Math. Helvet. **59** (1984), 509-538 [2, 4, 7, 8].

[Bec1] J. Beck, *Sums of distances between points on a sphere*, Math. **31** (1984), 33-41 [1].

[Bee1] G. F. M. Beenker, PGIT **30** (1984), 403-405 [7].

[Bel1] Bell Telephone Laboratories, *Transmission Systems for Communications*, Bell Telephone Labs, Murray Hill NJ, 1964 [3].

L. K. Bell: see C. E. Briant.

[Bel2] R. Bellman, *A Brief Introduction to Theta-Functions*, Holt, Rinehart and Winston, NY, 1961 [4].

[Ben0] M. Benard, *Characters and Schur indices of the unitary reflection group* $[3\ 2\ 1]^3$, PJM **58** (1975), 309-321 [4].

[Ben1] C. Bender, *Bestimmung der grössten Anzahl gleich Kugeln, welche sich auf eine Kugel von demselben Radius, wie die übrigen, auflegen lassen*, Archiv Math. Physik (Grunert) **56** (1874), 302-306 [1].

[Ben2] J. W. Benham and J. S. Hsia, *Spinor equivalence of quadratic forms*, JNT **17** (1983), 337-342 [15].

[Ben3] C. H. Bennett, *Serially deposited amorphous aggregates of hard spheres*, J. Appl. Phys. **43** (1972), 2727-2734 [1].

[Ben4] G. H. Bennett, *Pulse Code Modulation and Digital Transmission*, Marconi Instruments, St. Albans, England, 1978 [3].

C. T. Benson: see L. C. Grove.

[Ber1] T. Berger, *Rate Distortion Theory*, Prentice-Hall, Englewood Cliffs NJ, 1971 [2].

[Ber2] — *Optimum quantizers and permutation codes*, PGIT **18** (1972), 759-765 [2].

[Ber3] — F. Jelinek and J. K. Wolf, *Permutation codes for sources*, PGIT **18** (1972), 160-169 [2].

E. R. Berlekamp: see also C. E. Shannon.

[Ber4] — *Algebraic Coding Theory*, McGraw-Hill, NY, 1968 [3, 5].

[Ber5] — *Coding theory and the Mathieu groups*, IC **18** (1971), 40-64 [5].

[Ber6] — editor, *Key Papers in the Development of Coding Theory*, IEEE Press, NY, 1974 [3].

[Ber7] — F. J. MacWilliams and N. J. A. Sloane, *Gleason's theorem on self-dual codes*, PGIT **18** (1972), 409-414 [7].

[Ber8] — R. J. McEliece and H. C. A. van Tilborg, *On the inherent intractability of certain coding problems*, PGIT **24** (1978), 384-386 [2].

[Ber9] J. D. Berman and K. Hanes, *Volumes of polyhedra inscribed in the unit sphere in* E^3, Math. Ann. **188** (1970), 78-84 [1].

[Ber10] — — *Optimizing the arrangement of points on the unit sphere*, MTAC **31** (1977), 1006-1008 [1].

[Ber11] J. D. Bernal, *The structure of liquids*, PRS **A280** (1964), 299-322 [1].

[Bes1] M. R. Best, *Binary codes with a minimum distance of four*, PGIT **26** (1980), 738-742 [3, 5, 9].

[Bes2] — and A. E. Brouwer, *The triply shortened binary Hamming code is optimal*, Discr. Math. **17** (1977), 235-245 [9].

[Bes3] — — F. J. MacWilliams, A. M. Odlyzko and N. J. A. Sloane, *Bounds for binary codes of length less than* 25, PGIT **24** (1978), 81-93 [3, 5, 9].

[Bet1] T. Beth, W. Fumy and H. P. Reiss, *Der Wunderschöne Oktaden-Generator*, Mitt. Math. Sem. Giessen **163** (1984), 169-179 [11].

[Bez1] A. Bezdek, *Solid packing of circles in the hyperbolic plane*, Studia Sci. Math. Hungar. **14** (1979), 203-207 [1].

[Bez2] — *Uber Ionenpackungen*, Studia Sci. Math. Hungar. **18** (1983), 277-285.

[Bez3] — *Circle packings into convex domains of the Euclidean and hyperbolic plane and sphere*, Geom. Dedic. **21** (1986), 249-255 [1].

B. K. Bhattacharya: see D. Avis.

[Bie1] L. Bieberbach, *Uber die Bewegungsgruppen der Euklidischen Räume*, Math. Ann. **70** (1911), 297-336; **72** (1912), 400-412 [4].

[Big1] N. L. Biggs, *Perfect codes in graphs*, JCT **B15** (1973), 289-296 [9].

[Big2] — *Algebraic Graph Theory*, Camb. Univ. Press, 1974 [9].

[Big3] — *Designs, factors and codes in graphs*, Quart. J. Math. Oxford **26** (1975), 113-119 [9].

[Big3a] — A. G. Boshier and J. Shawe-Taylor, *Cubic distance-regular graphs*, JLMS **33** (1986), 385-394 [9].

[Big4] — and D. H. Smith, *On trivalent graphs*, BLMS **3** (1971), 155-158 [9].

[Big5] — and A. T. White, *Permutation Groups and Combinatorial Structures*, Camb. Univ. Press, 1979 [9].

[Bil1] E. Biglieri and M. Elia, *On the existence of group codes for the Gaussian channel*, PGIT **18** (1972), 399-402 [3].

[Bil2] — — *Optimum permutation modulation codes and their asymptotic performance*, PGIT **22** (1976), 751-753 [3].

[Bil3] — — *Cyclic-group codes for the Gaussian channel*, PGIT **22** (1976), 624-629 [3].

[Bin1] N. H. Bingham, *Positive definite functions on spheres*, PCPS **73** (1973), 145-156 [9].

B. J. Birch: see J. Larmouth.

[Bir1] G. Birkhoff and S. MacLane, *A Survey of Modern Algebra*, Macmillan, NY, 4th ed., 1977 [2].

[Bla1] N. M. Blachman, *The closest packing of equal spheres in a larger sphere*, AMM **70** (1963), 526-529 [1].

[Bla2] R. E. Blahut, *Theory and Practice of Error Control Codes*, Addison-Wesley, Reading MA, 1983 [3].

I. F. Blake: see also I. B. Gibson.

[Bla3] — *The Leech lattice as a code for the Gaussian channel*, IC **19** (1971), 66-74 [3].

[Bla4] — *Distance properties of group codes for the Gaussian channel*, SIAJ **23** (1972), 312-324 [3].

[Bla5] — editor, *Algebraic Coding Theory History and Development*, Dowden, Stroudsburg PA, 1973 [3].

[Bla6] — *Configuration matrices of group codes*, PGIT **20** (1974), 95-100 [3].

[Bla7] — *Properties of generalized Pless codes*, in *Proc. 12th Allerton Conf. Circ. Syst. Theory*, Univ. Ill., Urbana, 1974, pp. 787-789 [3].

[Bla8] — *On a generalization of the Pless symmetry codes*, IC **27** (1975), 369-373 [3].

[Bla9] — and R. C. Mullin, *The Mathematical Theory of Coding*, Ac. Press, NY, 1975 [3].

I. Blech: see D. Shechtman.

[Ble1] M. N. Bleicher, *Lattice coverings of n-space by spheres*, CJM **14** (1962), 632-650 [2].

H. F. Blichfeldt: see also G. A. Miller.

[Bli1] — *Finite Collineation Groups*, Univ. Chicago Press, 1917 [7].

[Bli2] — *On the minimum value of positive real quadratic forms in 6 variables*, BAMS **31** (1925), 386 [1].

[Bli3] — *The minimum value of quadratic forms, and the closest packing of spheres*, Math. Ann. **101** (1929), 605-608 [1].

[Bli4] — *The minimum values of positive quadratic forms in six, seven and eight variables*, Math. Zeit. **39** (1935), 1-15 [1, 4, 6].

[Bli5] F. van der Blij, *History of the octaves*, Simon Stevin **34** (1961), 106-125 [4].

[Bli6] — and T. A. Springer, *The arithmetics of octaves and of the group* G_2, KNAW **A62** (1959), 406-418 [4].

A. Blokhuis: see E. Bannai.

[Boc1] S. Bochner, *Hilbert distances and positive definite functions*, Ann. Math. **42** (1941), 647-656 [9].

[Boe1] A. H. Boerdijk, *Some remarks concerning close-packing of equal spheres*, Philips Res. Rep. **7** (1952), 303-313 [1].

E. D. Bolker: see P. F. Ash.

B. Bollobás: see J. J. Seidel.

[Boo1] W. M. Boothby and G. L. Weiss, editors, *Symmetric Spaces*, Dekker, NY, 1972 [9, B].

[Bor1] R. E. Borcherds, *The Leech lattice and other lattices*, Ph.D. Dissertation, Univ. of Cambridge, 1984 [2, 17].

[Bor2] — *The Leech lattice*, PRS **A398** (1985), 365-376 [24].

[Bor3] — *The 24-dimensional odd unimodular lattices*, Chap. 17 of this book [2, 17].

[Bor4] — J. H. Conway, L. Queen and N. J. A. Sloane, *A Monster Lie algebra*? Adv. in Math. **53** (1984), 75-79 ≅ Chap. 30 of this book [30].

[Bor5] Z. I. Borevich and I. R. Shafarevich, *Number theory*, Nauka, Moscow, 1964; English translation, Ac. Press, NY, 1966 [15].

[Bor6] D. Borwein, J. M. Borwein and K. F. Taylor, *Convergence of lattice sums and Madelung's constant*, J. Math. Phys. **26** (1985), 2999-3009 [2].

J. M. Borwein: see D. Borwein.

[Bör1] K. Böröczky, *Packing of spheres in spaces of constant curvature* (in Hungarian), Mat. Lapok **25** (1974), 265-306; **26** (1975), 67-90 [1].

[Bör2] — *Packing of spheres in spaces of constant curvature*, AMAH **32** (1978), 243-261 [1, 13].

[Bör3] — *The problem of Tammes for* $n = 11$, Studia Sci. Math. Hung. **18** (1983), 165-171 [1].

[Bör4] — *Closest packing and loosest covering*, in *Diskrete Geometrie, 3rd Kolloq.*, Inst. f. Math., Univ. Salzburg, 1985, pp. 329-334.

[Bos1] A. Bos, *Sphere-packings in Euclidean space*, in [Sch8], 1979, pp. 161-177.

[Bos2] — *Upper bounds for sphere packings in Euclidean space*, preprint [1].

[Bos3] — J. H. Conway and N. J. A. Sloane, *Further lattice packings in high dimensions*, Math. **29** (1982), 171-180 [1, 8].

A. G. Boshier: see N. L. Biggs.

[Bot1] F. V. Botya, *Bravais types of five-dimensional lattices*, TMIS **163** (1984) = PSIM **163** (No. 4, 1985), 23-27 [1, 3].

[Bou0] P. Bougerol, *Un Mini-Cours sur les Couples de Guelfand*, Pub. du Laboratoire de Statistique et Probabilités, Univ. Paul Sabatier, Toulouse, 1983 [9].

[Bou1] N. Bourbaki, *Groupes et Algèbres de Lie, Chapitres* 4, 5 *et* 6, Hermann, Paris,

1968 [4, 18, 21, 23, 24, 25, 27].

[Bou2] M. Bourdeau and A. Pitré, *Tables of good lattices in four and five dimensions*, Num. Math. **47** (1985), 39-43 [1].

[Bow1] V. J. Bowman, *Permutation polyhedra*, SIAJ **22** (1972), 580-589 [21].

[Bow2] A. Bowyer, *Computing Dirichlet tessellations*, Comp. J. **24** (1981), 162-166 [2].

H. Brändström: see L. H. Zetterberg.

[Bra1] H. Brandt and O. Intrau, *Tabellen reduzierter positiver ternärer quadratischen Formen*, Abh. Sächs. Akad. Wiss. Math.-Nat. Kl. **45** (No. 4, 1958), 261 pp. [15].

[Bra2] R. Brauer and C.-H. Sah, editors, *Theory of Finite Groups*, Benjamin, NY, 1969 [B].

[Bra3] H. Braun, *Zur Theorie der hermitischen Formen*, ASUH **14** (1941), 61-150 [16].

[Bri1] C. E. Briant, B. R. C. Theobald, J. W. White, L. K. Bell, D. M. P. Mingos and A. J. Welch, *Synthesis and X-ray structural characterization of the centered icosahedral gold cluster compound* $[Au_{13}(PMe_2Ph)_{10}Cl_2](PF_6)_3$; *the realization of a theoretical prediction*, J. Chem. Soc., Chem. Comm. (1981), 201-202 [1].

[Bri2] E. Brieskorn, *Milnor lattices and Dynkin diagrams*, PSPM **40** (1983), 153-165 [2].

J. Brillhart: see M. A. Morrison.

[Bro0] P. L. H. Brooke, *On matrix representations and codes associated with the simple group of order* 25920, J. Alg. **91** (1984), 536-566 [3].

[Bro1] — *Tables of Codes Associated with Certain Finite Simple Groups*, Ph.D. Dissertation, Univ. of Cambridge, June 1984 [3].

[Bro2] — *On the Steiner system* $S(2, 4, 28)$ *and codes associated with the simple group of order* 6048, J. Alg. **97** (1985), 376-406 [3].

[Bro3] W. Brostow, J.-P. Dussault and B. L. Fox, *Construction of Voronoi polyhedra*, J. Comp. Phys. **29** (1978), 81-92 [2].

[Bro4] M. Broué, *Le réseau de Leech et le groupe de Conway*, Thèse de 3e cycle, Paris, 1970 [4].

[Bro5] — *Codes correcteurs d'erreurs auto-orthogonaux sur le corps à deux eléments et formes quadratique entières definies positives à discriminant* +1, Discr. Math. **17** (1977), 247-269 [7, 16].

[Bro6] — and M. Enguehard, *Polynômes des poids de certains codes et fonctions thêta de certains réseaux*, Ann. Sci. Ecole Norm. Sup. **5** (1972), 157-181 [7, 16].

[Bro7] — — *Une famille infinie de formes quadratiques entières; leurs groupes d'automorphismes*, Ann. Sci. Ecole Norm. Sup. **6** (1973), 17-52 [7].

A. E. Brouwer: see also M. R. Best.

[Bro8] — *A few new constant weight codes*, PGIT **26** (1980), 366 [9].

[Bro9] — P. Delsarte and P. Piret, *On the* $(23, 14, 5)$ *Wagner code*, PGIT **26** (1980), 742-743 [9].

[Bro10] H. Brown, R. Bülow, J. Neubüser, H. Wondratschek and H. Zassenhaus, *Crystallographic Groups of Four-Dimensional Space*, Wiley, NY, 1978 [3].

[Bro11] — J. Neubüser and H. J. Zassenhaus, *On integral groups* I: *the reducible case*, Num. Math. **19** (1972), 386-399 [3].

[Bro12] — — — *On integral groups* II: *the irreducible case*, Num. Math. **20** (1972), 22-31 [3].

[Bro13] — — — *On integral groups* III: *normalizers*, MTAC **27** (1973) 167-182

[3].
[Bro14] W. S. Brown, *ALTRAN User's Manual*, Bell Laboratories, Murray Hill, 4th ed., 1977 [16].
[Bru1] V. Brun, *On regular packing of equal circles touching each other on the surface of a sphere*, Comm. Pure Appl. Math. **29** (1976), 583-590 [1].
J. A. Bucklew: see also N. C. Gallagher, Jr.
[Buc1] — *Compounding and random quantization in several dimensions*, PGIT **27** (1981), 207-211 [2].
[Buc2] — *Upper bounds to the asymptotic performance of block quantizers*, PGIT **27** (1981), 577-581 [2].
[Buc3] — and G. L. Wise, *Multidimensional asymptotic quantization theory with rth power distortion measures*, PGIT **28** (1982), 239-247 [2].
[Bud1] P. de Buda, *Encoding and decoding algorithms for an optimal lattice-based code*, in *Conference Record-Internat. Conf. Commun. ICC '81*, IEEE Press, NY, Vol. 3, 1981, pp. 65.3.1-65.3.5 [2, 20].
R. de Buda: see also J. B. Anderson.
[Bud2] — *The upper error bound of a new near-optimal code*, PGIT **21** (1975), 441-445 [3].
[Bud3] — and W. Kassem, *About lattices and the random coding theorem*, in *Abstracts of Papers, IEEE Inter. Symp. Info. Theory 1981*, IEEE Press, NY 1981, p. 145 [3].
R. Bülow: see also H. Brown.
[Bül1] — J. Neubüser and H. Wondratschek, *On crystallography in higher dimensions*, Acta Cryst. **A27** (1971), 517-535 [3].
[Bum1] C. Bumiller, *On rank three graphs with a large eigenvalue*, Discr. Math. **23** (1978), 183-187 [9].
H. Burzlaff: see also H. Zimmerman.
[Bur1] — and H. Zimmerman, *On the metrical properties of lattices*, Z. Krist. **170** (1985), 247-262 [2].
[Bur2] — — and P. M. de Wolff, *Crystal lattices*, in [Hah1], 1983, 734-744 [2].
[Bus1] P. Buser, *A geometric proof of Bieberbach's theorems on crystallographic groups*, L'Enseignement Math. **31** (1985), 137-145 [4].
[But1] G. Butler, *The maximal subgroups of the sporadic simple group of Held*, J. Alg. **69** (1981), 67-81 [10].
A. Buzo: see Y. Linde.

C

J. W. Cahn: see D. Shechtman.
[Cal1] A. R. Calderbank and W. M. Kantor, *The geometry of two-weight codes*, BLMS **18** (1986), 97-122 [9].
[Cal2] — T.-A. Lee and J. E. Mazo, *Baseband trellis codes with a spectral null at zero*, PGIT **34** (1986), 425-434 [3].
[Cal3] — and J. E. Mazo, *A new description of trellis codes*, PGIT **30** (1984), 784-792 [3].
[Cal4] — — and H. M. Shapiro, *Upper bounds on the minimum distance of trellis codes*, BSTJ **62** (1983), 2617-2646 [3].
[Cal5] — — and V. K. Wei, *Asymptotic upper bounds on the minimum distance of trellis codes*, COMM **33** (1985), 305-309 [3].
[Cal6] — and N. J. A. Sloane, *Four-dimensional modulation with an eight-state trellis code*, BSTJ **64** (1985), 1005-1018 [3].
[Cal7] — — *An eight-dimensional trellis code*, PIEEE **74** (1986), 757-759 [3].

[Cal8] — — *New trellis codes based on lattices and cosets*, PGIT **33** (1987), 177-195 [3].

[Cal9] — and D. B. Wales, *A global code invariant under the Higman-Sims group*, J. Alg. **75** (1982), 233-260 [3].

P. J. Cameron: see also N. J. A. Sloane.

[Cam1] — *Another characterization of the small Janko group*, J. Math. Soc. Japan **25** (1973), 591-595 [10].

[Cam2] — and J. H. van Lint, *Graphs, Codes and Designs*, Camb. Univ. Press, 1980 [3, 9].

[Cam3] — J. A. Thas and S. E. Payne, *Polarities of generalized hexagons and perfect codes*, Geom. Dedic. **5** (1976), 525-528 [9].

C. M. Campbell: see R. A. Wilson.

[Cam4] C. N. Campopiano and B. G. Glazer, *A coherent digital amplitude and phase modulation scheme*, COMM **10** (1962), 90-95 [3].

[Car1] R. D. Carmichael, *Introduction to the Theory of Groups of Finite Order*, Ginn, Boston, 1937; Dover, NY, 1956 [11].

[Car2] R. W. Carter, *Simple Groups of Lie Type*, Wiley, NY, 1972 [10].

[Car3] — *Finite Groups of Lie Type*, Wiley, NY, 1985 [10].

[Cas1] J. W. S. Cassels, *Uber die Aquivalenz 2-adischer quadratischer Formen*, Comment. Math. Helvet. **37** (1962), 61-64 [15].

[Cas2] — *An Introduction to the Geometry of Numbers*, Springer-Verlag, 1971 [1, 2, 6].

[Cas3] — *Rational Quadratic Forms*, Ac. Press, NY, 1978 [2, 3, 4, 7, 15, 16].

[Cas4] — *Rational quadratic forms*, in *Proc. Intern. Math. Conf.*, ed. L. H. Y. Chen et al., North-Holland, Amsterdam, 1982, pp. 9-26 [15].

[Cas5] — and A. Fröhlich, editors, *Algebraic Number Theory*, Ac. Press, NY, 1967 [8, B].

[Cha1] J. H. H. Chalk, *Algebraic lattices*, in [Gru2], 1983, pp. 97-110 [2].

[Cha2] R. C. Chang and R. C. T. Lee, *On the average length of Delaunay triangulations*, BIT **24** (1984), 269-273 [2].

[Cha3] G. Chapline, *Unification of gravity and elementary particle interactions in* 26 *dimensions*, Phys. Lett. **B158** (1985), 393-396 [1].

[Cha4] T. W. Chaundy, *The arithmetic minima of positive quadratic forms*, Quart. J. Math. **17** (1946), 166-192, but see also the review: H. S. M. Coxeter, Math. Rev. **8** (1947), 137-138 [6].

[Che1] C. L. Chen, *Computer results on the minimum distance of some binary cyclic codes*, PGIT **16** (1970), 359-360 [3, 5].

L. H. Y. Chen: see J. W. S. Cassels.

[Che2] D.-Y. Cheng, A. Gersho, B. Ramamurthi and Y. Shoham, *Fast search algorithms for vector quantization and pattern matching*, in *Proc. Inter. Conf. Acoust. Speech Sig. Proc. ICASSP '84*, IEEE Press, NY, 1984, Vol. 1, pp. 9.11.1-9.11.4 [20].

[Che3] Y. Cheng and N. J. A. Sloane, *Codes from symmetry groups and a* [32,17,8] *code*, SIAM J. Disc. Math. **2** (1989), 28-37 [8].

[Che4] A. G. Chernyakov, *An example of a* 32-*dimensional even unimodular lattice*, LOMI **86** (1979), 170-179 = JSM **17** (1979), 2068-2075 [7].

[Cho1] C. Choi, *On subgroups of* M_{24}, TAMS **167** (1972), 1-27 and 29-47 [10].

[Cla1] B. W. Clare and D. L. Kepert, *The closest packing of equal circles on a sphere*, PRS **A405** (1986), 329-344 [1].

[Coh1] A. M. Cohen, *Finite complex reflection groups*, Ann. Sci. Ecole Norm. Sup. **9** (1976), 379-436 [3,4].

[Coh2] — *Finite quaternionic reflection groups*, J. Alg. **64** (1980), 293-324 [3, 8].
[Coh3] G. D. Cohen, M. R. Karpovsky, H. F. Mattson, Jr., and J. R. Schatz, *Covering radius—survey and recent results*, PGIT **31** (1985), 328-343 [3].
[Coh4] — A. C. Lobstein and N. J. A. Sloane, *Further results on the covering radius of codes*, PGIT **32** (1986) 680-694 [3].
[Coh4a] J. Cohen and T. Hickey, *Two algorithms for determining volumes of convex polyhedra*, J. Assoc. Comput. Mach. **26** (1979), 401-414 [2].
[Coh5] H. Cohn, *A Second Course in Number Theory*, Wiley, NY, 1962 [15].
[Coh6] — *A Classical Invitation to Algebraic Numbers and Fields*, Springer-Verlag, 1978 [8, B].
M. J. Cohn: see E. S. Barnes.
S. Cohn-Vossen: see D. Hilbert.
[Coi1] R. R. Coifman and G. Weiss, *Representations of compact groups and spherical harmonics*, L'Enseignement Math. **14** (1968), 121-173 [9].
[Coi2] — — *Analyse harmonique non-commutative sur certains espaces homogenes*, LNM **242**, 1971 [9].
[Col1] M. J. Colbourn, *Some new upper bounds for constant weight codes*, PGIT **26** (1980), 478 [9].
[Col2] M. J. Collins, editor, *Finite Simple Groups* II, Ac. Press, NY, 1980 [B].
[Con1] R. Connelly, *The rigidity of circle and sphere packings*, J. Structural Topology **14** (1988), 43-60; **16** (1990), 57-76.
J. H. Conway: see also R. E. Borcherds, A. Bos, R. A. Parker.
[Con2] — *A perfect group of order* 8,315,553,613,086,720,000 *and the sporadic simple groups*, PNAS **61** (1968), 398-400 [3].
[Con3] — *A group of order* 8,315,553,613,086,720,000, BLMS **1** (1969), 79-88 [3, 10].
[Con4] — *A characterisation of Leech's lattice*, Invent. Math. **7** (1969), 137-142 ≅ Chap. 12 of this book [7, 12].
[Con5] — *Three lectures on exceptional groups*, in [Pow1], 1971, pp. 215-247 ≅ Chap. 10 of this book [10].
[Con6] — *Groups, lattices, and quadratic forms*, in *Computers in Algebra and Number Theory*, SIAM-AMS Proc. **IV**, Amer. Math. Soc., Providence RI, 1971, pp. 135-139.
[Con7] — *A construction for the smallest Fischer group*, in [Gag2], 1973, pp. 27-35 [10].
[Con8] — *Invariants for quadratic forms*, JNT **5** (1973), 390-404 [15].
[Con9] — *The miracle octad generator*, in [Cur0], 1977, pp. 62-68.
[Con10] — *A quaternionic construction for the Rudvalis group*, in [Cur0], 1977, pp. 69-81 [10].
[Con11] — *Monsters and moonshine*, Math. Intelligencer **2** (No. 4, 1980), 165-172 [1].
[Con12] — *The hunting of* J_4, Eureka **41** (1981), 46-54 [10, 11].
[Con13] — *The automorphism group of the 26-dimensional even unimodular Lorentzian lattice*, J. Alg. **80** (1983), 159-163 ≅ Chap. 27 of this book [15, 27].
[Con14] — *Hexacode and tetracode - MOG and MINIMOG*, in [Atk2], 1984, pp. 359-365.
[Con15] — *A simple construction for the Fischer-Griess Monster group*, Invent. Math. **79** (1985), 513-540 [29].
[Con16] — R. T. Curtis, S. P. Norton, R. A. Parker and R. A. Wilson, *ATLAS of Finite Groups*, Oxford Univ. Press, 1985 [3, 4, 8, 10, 29].

[Con16a] — and K. M. Knowles, *Quasiperiodic tiling in two and three dimensions*, J. Phys. **A 19** (1986), 3645-3653 [1].

[Con17] — and S. P. Norton, *Monstrous moonshine*, BLMS **11** (1979), 308-339 [1, 10, 25, 29, 30].

[Con18] — — and L. H. Soicher, *The Bimonster, the group* Y_{555}, *and the projective plane of order* 3, in *Computers in Algebra, Chicago 1985*, Dekker NY, 1988, pp. 27-50 [29].

[Con19] — A. M. Odlyzko and N. J. A. Sloane, *Extremal self-dual lattices exist only in dimensions* 1 *to* 8, 12, 14, 15, 23 *and* 24, Math. **25** (1978), 36-43 ≅ Chap. 19 of this book [7, 19].

[Con20] — R. A. Parker and N. J. A. Sloane, *The covering radius of the Leech lattice*, PRS **A380** (1982), 261-290 ≅ Chap. 23 of this book [2, 6, 23].

[Con21] — and V. Pless, *On the enumeration of self-dual codes*, JCT **A28** (1980), 26-53 [4, 7].

[Con22] — — *On primes dividing the group order of a doubly-even* (72, 36, 16) *code, and the group order of a quaternary* (24, 12, 10) *code*, Discr. Math. **38** (1982), 143-156 [7].

[Con23] — — *Monomials of order* 7 *and* 11 *cannot be in the group of a* (24, 12 10) *self-dual quaternary code*, PGIT **29** (1983), 137-140 [7].

[Con24] — — and N. J. A. Sloane, *Self-dual codes over GF*(3) *and GF*(4) *of length not exceeding* 16, PGIT **25** (1979), 312-322 [4, 5, 7].

[Con24a] — and A. D. Pritchard, *Hyperbolic reflections for the Bimonster and* $3Fi_{24}$, in *Groups and Combinatorics*, Camb. Univ. Press, 1992.

[Con25] — and L. Queen, *Computing the character table of a Lie group*, in [McK3], 1982, pp. 51-87 [21].

[Con26] — and N. J. A. Sloane, *Fast* 4- *and* 8-*dimensional quantizers and decoders*, in *Nat. Telecomm. Conf. Record-1981*, IEEE Press, NY, 1981, Vol. **3**, pp. F4.2.1 to F4.2.4 [20].

[Con27] — — *On the enumeration of lattices of determinant one*, JNT **15** (1982), 83-94 [2, 4, 7, 16].

[Con28] — — *Voronoi regions of lattices, second moments of polytopes, and quantization*, PGIT **28** (1982), 211-226 ≅ Chap. 21 of this book [2, 4, 6, 21].

[Con29] — — *Fast quantizing and decoding algorithms for lattice quantizers and codes*, PGIT **28** (1982), 227-232 [2, 20].

[Con30] — — *Twenty-three constructions for the Leech lattice*, PRS **A381** (1982), 275-283 ≅ Chap. 24 of this book [24].

[Con31] — — *Lorentzian forms for the Leech lattice*, BAMS **6** (1982), 215-217 ≅ Chap. 26 of this book [26].

[Con32] — — *Laminated lattices*, Ann. Math. **116** (1982), 593-620 ≅ Chap. 6 of this book [6].

[Con33] — — *Leech roots and Vinberg groups*, PRS **A384** (1982), 233-258 ≅ Chap. 28 of this book [15, 28].

[Con34] — — *The unimodular lattices of dimension up to* 23 *and the Minkowski-Siegel mass constants*, Europ. J. Combinatorics **3** (1982), 219-231 ≅ Chap. 16 of this book [2, 7, 15, 16].

[Con35] — — *A fast encoding method for lattice codes and quantizers*, PGIT **29** (1983), 820-824 [2, 20].

[Con36] — — *Complex and integral laminated lattices*, TAMS **280** (1983), 463-490 [2, 6, 22].

[Con37] — — *The Coxeter-Todd lattice, the Mitchell group, and related sphere packings*, PCPS **93** (1983), 421-440 [2, 4, 7, 8, 22].

[Con38] — — *On the Voronoi regions of certain lattices*, SIAD **5** (1984), 294-305 [2, 4, 20, 21].

[Con39] — — *A lower bound on the average error of vector quantizers*, PGIT **31** (1985), 106-109 [2].

[Con40] — — *Soft decoding techniques for codes and lattices, including the Golay code and the Leech lattice*, PGIT **32** (1986), 41-50 [2, 4, 20].

[Con41] — — *Lexicographic codes: error-correcting codes from game theory*, PGIT **32** (1986), 337-348 [6, 9, 11].

[Con42] — — *Low-dimensional lattices* — see Supplementary Bibliography.

[Con43] — — *Covering space with spheres*, in preparation [2, 22].

[Con44] — and D. B. Wales, *Construction of the Rudvalis group of order 145,926,144,000*, J. Alg. **27** (1973), 538-548 [10].

[Con45] — — *Matrix generators for* J_3, J. Alg. **29** (1974), 474-476 [10].

[Coo1] G. R. Cooper and R. W. Nettleton, *A spread-spectrum technique for high-capacity mobile communications*, IEEE Trans. Veh. Tech. **27** (1978), 264-275 [1].

D. J. Costello, Jr.: see Shu Lin.

[Cos1] P. J. Costello & J. S. Hsia, *Even unimodular 12-dimensional quadratic forms over* $\mathbf{Q}(\sqrt{5})$, Adv. in Math. **64** (1987), 241-278 [2, 8, 16].

H. S. M. Coxeter: see also W. W. R. Ball, T. W. Chaundy. A partial list of H. S. M. Coxeter's publications is given on pp. 5-13 of [Dav3].

[Cox1] — *The pure Archimedean polytopes in six and seven dimensions*, PCPS **24** (1928), 1-9 [4, 21].

[Cox2] — *Groups whose fundamental regions are simplexes*, JLMS **6** (1931), 132-136 [4].

[Cox3] — *Discrete groups generated by reflections*, Ann. Math. **35** (1934), 588-621 [4].

[Cox4] — *The complete enumeration of finite groups of the form* $R_i^2 = (R_i R_j)^{k_{ij}} = 1$, JLMS **10** (1935), 21-25 [4].

[Cox5] — *Wythoff's construction for uniform polytopes*, PLMS **38** (1935), 327-339 = Chap. 3 of [Cox18] [21].

[Cox6] — *Regular and semi-regular polytopes*, Math. Z. **46** (1940) 380-407 [21].

[Cox7] — *The polytope* 2_{21}*, whose twenty-seven vertices correspond to the lines on the general cubic surface*, AJM **62** (1940), 457-486 [4, 21].

[Cox8] — *Quaternions and reflections*, AMM **53** (1946), 136-146 [2, 4].

[Cox9] — *Integral Cayley numbers*, DMJ **13** (1946), 561-578 = Chap. 2 of [Cox18] [4].

[Cox10] — *Extreme forms*, CJM **3** (1951), 391-441 [2, 4, 8, 21, 26].

[Cox11] — *The product of the generators of a finite group generated by reflections*, DMJ **18** (1951), 765-782.

[Cox12] — *Twelve points in* $PG(5, 3)$ *with* 95040 *self-transformations*, PRS **A247** (1958), 279-293 = Chap. 7 of [Cox18].

[Cox13] — *Polytopes over* $GF(2)$ *and their relevance for the cubic surface group*, CJM **11** (1959), 646-650 [4].

[Cox14] — *Introduction to Geometry*, Wiley, NY, 1961 [1, 21].

[Cox15] — *The problem of packing a number of equal nonoverlapping circles on a sphere*, Trans. N.Y. Acad. Sci. **24** (No. 3, 1962), 320-331 [1].

[Cox16] — *An upper bound for the number of equal nonoverlapping spheres that can touch another of the same size*, PSPM **7** (1963), 53-71 = Chap. 9 of [Cox18] [1, 13].

[Cox17] — *Geometry*, in *Lectures on Modern Mathematics*, ed. T. L. Saaty, Wiley,

NY, Vol. **III**, 1965, pp. 58-94 = Chap. 12 of [Cox18] [2, 4].
[Cox18] — *Twelve Geometric Essays*, Southern Illinois Press, Carbondale IL, 1968 [1, 2, 4, 21, B].
[Cox19] — *Finite groups generated by unitary reflections*, ASUH **31** (1967), 125-135.
[Cox20] — *Regular Polytopes*, Dover, NY, 3rd ed., 1973 [3, 4, 5, 6, 21, 23].
[Cox21] — *Regular Complex Polytopes*, Camb. Univ. Press, 2nd ed., 1990 [1, 2, 3].
[Cox21a] — *Non-Euclidean Geometry*, Univ. Toronto Press, 5th ed., 1978 [26].
[Cox22] — *Polytopes in the Netherlands*, Nieuw Arch. Wisk. **26** (1978), 116-141 [21].
[Cox23] — *Regular and semi-regular polytopes* II, Math. Z. **188** (1985), 559-591 [21].
[Cox24] — *A packing of* 840 *balls of radius* 9° 0′ 19′′ *on the* 3*-sphere*, in *Intuitive Geometry*, ed. K. Böröczky and G. Fejes Tóth, North-Holland, Amsterdam, 1987, pp. 127-137 [1].
[Cox25] — *Regular and semi-regular polytopes* III, Math. Z., **200** (1988), 3-45 [21].
[Cox26] — L. Few and C. A. Rogers, *Covering space with equal spheres*, Math. **6** (1959), 147-157 [2].
[Cox27] — M. S. Longuet-Higgins and J. C. P. Miller, *Uniform polyhedra*, Phil Trans. Royal Soc. **246** (1954), 401-450 [21].
[Cox28] — and W. O. J. Moser, *Generators and Relations for Discrete Groups*, Springer-Verlag, 4th ed., 1980 [3, 4, 7, 9, 10, 21, 27].
[Cox29] — and J. A. Todd, *An extreme duodenary form*, CJM **5** (1953), 384-392 [1, 4, 6].
[Cox30] — and G. J. Whitrow, *World structure and non-Euclidean honeycombs*, PRS **A201** (1950), 417-437 [27].
[Cra1] A. P. Cracknell, *Irreducible representations of point groups and space groups—the last* 50 *years and the next* 10 *years*, Comm. Match in Math. Chem. **9** (1980), 227-241 [3].
[Cra2] M. Craig, *A characterization of certain extreme forms*, IJM **20** (1976), 706-717 [2, 3].
[Cra3] — *Extreme forms and cyclotomy*, Math. **25** (1978), 44-56 [2, 8].
[Cra4] — *A cyclotomic construction for Leech's lattice*, Math. **25** (1978), 236-241 [2, 8].
[Cra5] — *Automorphisms of prime cyclotomic lattices*, preprint [2, 8].
R. E. Crochiere: see J. L. Flanagan.
[Cun1] H. M. Cundy and A. P. Rollett, *Mathematical Models*, Oxford Univ. Press, 2nd ed., 1961 [2, 21].
[Cur0] M. P. J. Curran, *Topics in Group Theory and Computation*, Ac. Press, NY, 1977 [B].
[Cur1] C. W. Curtis, W. M. Kantor and G. M. Seitz, *The* 2*-transitive permutation representations of the finite Chevalley groups*, TAMS **218** (1976), 1-57 [9].
R. T. Curtis: see also J. H. Conway.
[Cur2] — *On subgroups of* ·0, I: *lattice stabilizers*, J. Alg. **27** (1973), 549-573 [10, 11, 23].
[Cur3] — *A new combinatorial approach to* M_{24}, PCPS **79** (1976), 25-42 [10, 11, 14].
[Cur4] — *The maximal subgroups of* M_{24}, PCPS **81** (1977), 185-192 [10].
[Cur5] — *On subgroups of* ·0, II: *local structure*, J. Alg. **63** (1980), 413-434 [10].
[Cur6] — *Eight octads suffice*, JCT **A36** (1984), 116-123 [11].
[Cur7] — *The Steiner system* $S(5, 6, 12)$, *the Mathieu group* M_{12} *and the "kitten"*, in [Atk2], 1984, pp. 352-358 [11].

[Cus1] E. L. Cusack, *Error control codes for QAM signalling*, Elect. Lett. **20** (1984), 62-63.

D

[Dad1] E. C. Dade, *The maximal finite groups of* 4×4 *integral matrices*, IJM **9** (1965), 99-122 [3].
R. M. Damerell: see E. Bannai.
[Dan1] L. Danzer, *Finite point-sets on* S^2 *with minimum distance as large as possible*, Discr. Math. **60** (1986), 3-66 [1].
[Dav1] H. Davenport, *The Higher Arithmetic*, Hutchinson, London, 1952 [15].
[Dav2] — *The covering of space by spheres*, Rend. Circ. Mat. Palermo, **1** (1952), 92-107 = Coll. Works **II** (1977), 609-624 [2, 21].
[Dav2a] — and G. L. Watson, *The minimal points of a positive definite quadratic form*, Math. **1** (1954), 14-17 = Coll. Works **II** (1977), 684-688 [15].
[Dav3] C. Davis, B. Grünbaum and F. A. Sherk, editors, *The Geometric Vein: The Coxeter Festschrift*, Springer-Verlag 1981 [B].
[Dav4] L. D. Davisson and R. M. Gray, editors, *Data Compression*, Dowden, Stroudsberg PA, 1976 [2].
[Daw1] E. Dawson, *Self-dual ternary codes and Hadamard matrices*, Ars Comb. **A19** (1985), 303-308 [7].
M. Dekesel: see P. A. Devijver.
B. N. Delaunay: see B. N. Delone.
[Deg1] P. Deligne, *Formes modulaires et représentations l-adiques*, Sém. Bourbaki **21** (No. 335, 1969), 139-172 [2].
[Deg2] — *La conjecture de Weil*, Inst. Hautes Etudes Sci. Publ. Math. **53** (1974), 273-307 [2].
[Del0] B. N. Delone, *Neue Darstellung der geometrischen Kristallographie*, Z. Krist **84** (1933), 109-149 [2].
[Del1] — *Sur la sphère vide*, Bull. Acad. Sci. USSR, Classe Sci. Mat. Nat. (1934), 793-800 [2].
[Del2] — N. P. Dolbilin, S. S. Ryskov and M. I. Stogrin, *A new construction in the theory of lattice coverings of an n-dimensional space by equal spheres*, IANS **34** (1970), 289-298 = Izv. **4** (1970), 293-302 [2].
[Del3] — R. V. Galiulin and M. I. Stogrin, *The types of Bravais lattices*, in *Current Problems in Math.* Vol. 2, Moscow 1973, pp. 119-257. Math. Rev. **54** #1068 [1].
[Del4] — and S. S. Ryskov, *Solution of the problem of least dense lattice covering of a four-dimensional space by equal spheres*, DAN **152** (1963), 523-524 = SMD **4** (1963), 1333-1334 [2, 4].
[Del5] — — *Extremal problems in the theory of positive quadratic forms*, TMIS **112** (1971) 203-229 = PSIM **112** (1971), 211-231.
[Del6] — N. N. Sandakova and S. S. Ryskov, *An optimal cubature lattice for bilaterally smooth functions of two variables*, DAN **162** (1965), 1230-1233 = SMD **6** (1965), 836-839 [1].
[Del7] — and M. I. Stogrin, *Simplified proof of the Schönflies theorem*, DAN **219** (1974), 95-98 = Sov. Phys. Dokl. **19** (1975), 727-729 [4].
P. Delsarte: see also E. Bannai, A. E. Brouwer.
[Del8] — *Bounds for unrestricted codes, by linear programming*, Philips Res. Reports **27** (1972), 272-289 [9].
[Del9] — *An algebraic approach to the association schemes of coding theory*, Philips Res. Reports Supplements, No. 10 (1973) [9, 14].

[Del10] — *Four fundamental parameters of a code and their combinatorial significance*, IC **23** (1973), 407-438 [9].

[Del11] — *Hahn polynomials, discrete harmonics, and t-designs*, SIAJ **34** (1978), 157-166 [3, 9].

[Del12] — *Bilinear forms over a finite field, with applications to coding theory*, JCT **A25** (1978), 226-241 [9].

[Del13] — and J.-M. Goethals, *Unrestricted codes with the Golay parameters are unique*, Discr. Math **12** (1975), 211-224 [7, 14].

[Del14] — — *Alternating bilinear forms over GF(q)*, JCT **A19** (1975), 26-50 [9].

[Del15] — — and J. J. Seidel, *Bounds for systems of lines, and Jacobi polynomials*, Philips Res. Reports **30** (1975), 91*-105* [3, 9].

[Del16] — — — *Spherical codes and designs*, Geom. Dedic. **6** (1977), 363-388 [1, 3, 9, 14].

[Dem1] A. P. Dempster, *The minimum of a definite ternary quadratic form*, CJM **9** (1957), 232-234 [1].

[Den1] R. H. F. Denniston, *Some new 5-designs*, BLMS **8** (1976), 263-267 [3].

N. Deo: see E. M. Reingold.

[Dev1] P. A. Devijver and M. Dekesel, *Insert and delete algorithms for maintaining dynamic Delaunay triangulations*, Patt. Recog. Lett. **1** (1982), 73-77 [2].

[Dev2] — — *Computing multidimensional Delaunay tessellations*, Patt. Recog. Lett. **1** (1983), 311-316 [2].

P. Dewilde: see B. S. Atal, S. G. Wilson.

[Dia1] P. Diaconis, *Lectures on the Use of Group Representations in Probability and Statistics*, Inst. Math. Statist. Lecture Note Series, to appear [9].

[Dia2] — R. L. Graham and W. M. Kantor, *The mathematics of perfect shuffles*, Adv. Appl. Math. **4** (1983), 175-196 [11].

[Dia3] W. J. Diamond, *Practical Experiment Designs*, Wadsworth, Belmont CA, 1981 [3].

J. C. Diaz: see P. Keast.

L. E. Dickson: see also G. A. Miller.

[Dic1] — *Linear Groups with an Exposition of the Galois Field Theory*, Dover, NY, 1958 [4, 10].

[Dic2] — *History of the Theory of Numbers*, Carnegie Institution of Washington Publications, No. 256, 1919, 1920, 1923, 3 vols; Chelsea, NY, 1966 [4, 15].

[Dic3] — *Studies in the Theory of Numbers*, Univ. of Chicago Press, 1930; Chelsea, NY (no date) [15].

T. J. Dickson: see also E. S. Barnes.

[Dic4] — *An extreme covering of 4-space by spheres*, JAMS **6** (1966), 179-192 [2].

[Dic5] — *The extreme coverings of 4-space by spheres*, JAMS **7** (1967), 490-496 [2].

[Dic6] — *A sufficient condition for an extreme covering of n-space by spheres*, JAMS **8** (1968), 56-62 [2].

[Dic7] — *On Voronoi reduction of positive definite quadratic forms*, JNT **4** (1972), 330-341 [2].

[Die1] U. Dieter, *How to calculate shortest vectors in a lattice*, MTAC **29** (1975), 827-833 [2].

[Dij1] E. W. Dijkstra, *A Discipline of Programming*, Prentice-Hall, Englewood Cliffs NJ, 1976, Chap. 19 [15].

[Dir1] P. G. L. Dirichlet, *Uber die Reduktion der positiven quadratische Formen mit drei unbestimmten ganzen Zahlen*, JRAM **40** (1850), 216-219 [2].

[Dob1] D. Dobkin and R. J. Lipton, *Multidimensional searching problems*, SIAM

J. Comput. **5** (1976), 181-186 [1].
N. P. Dolbilin: see B. N. Delone.
[Dol1] I. Dolgachev, *Integral quadratic forms: applications to algebraic geometry*, Sém. Bourbaki **611**, Astérisque **105** (1983), 251-278 [2, 15].
[Don1] J. L. Donaldson, *Minkowski reduction of integral matrices*, MTAC **33** (1978), 201-216 [2].
E. de Doncker: see D. Roose.
C. P. Downey: see also J. K. Karlof.
[Dow1] — and J. K. Karlof, *On the existence of* $[M, n]$ *group codes for the Gaussian channel with M and n odd*, PGIT **23** (1977), 500-503 [3].
[Dow2] — — *Optimum* $[M, 3]$ *group codes for the Gaussian channel*, PGIT **24** (1978), 760-761 [3].
[Dow3] — — *Computation methods for optimal* $[M, 3]$ *group codes for the Gaussian channel*, Util. Math. **18** (1980), 51-70 [3].
[Dow4] D. E. Downey and N. J. A. Sloane, *The covering radius of cyclic codes of length up to* 31, PGIT **31** (1985), 446-447 [3].
[Du1] P. Du Val, *Homographies, Quaternions and Rotations*, Oxford Univ. Press, 1964 [8].
[Duf1] R. J. Duffin, *Infinite programs*, in *Linear inequalities and related systems*, ed. H. W. Kuhn and A. W. Tucker, Princeton Univ. Press, 1956, pp. 157-170 [9].
[Dun1] C. F. Dunkl, *A Krawtchouk polynomial addition theorem and wreath products of symmetric groups*, Indiana Univ. Math. J. **25** (1976), 335-358 [9].
[Dun2] — *Spherical functions on compact groups and applications to special functions*, Symposia Mathematica **22** (1977), 145-161 [9].
[Dun3] — *An addition theorem for some q-Hahn polynomials*, Monatsh. Math. **85** (1977), 5-37 [9].
[Dun4] — *An addition theorem for Hahn polynomials: the spherical functions*, SIAM J. Math. Anal. **9** (1978), 627-637 [9].
[Dun5] — *Discrete quadrature and bounds on t-design*, Mich. Math. J. **26** (1979), 81-102 [3, 9].
[Dun6] — *Orthogonal functions on some permutation groups*, PSPM **34** (1979), 129-147 [9].
[Dun6a] — *Orthogonal polynomials on the sphere with octahedral symmetry*, TAMS **282** (1984), 555-575 [9].
[Dun6b] — *Orthogonal polynomials with symmetry of order three*, CJM **36** (1984), 685-717 [9].
[Dun6c] — *Reflection groups and orthogonal polynomials on the sphere*, Math. Z., **197** (1988), 33-60 [9].
[Dun7] — and D. E. Ramirez, *Topics in Harmonic Analysis*, Appleton-Century-Crofts, NY, 1971 [9].
[Dun8] — — *Krawtchouk polynomials and the symmetrization of hypergroups*, SIAM J. Math. Anal. **5** (1974), 351-366 [9].
[Dun9] — — *A linear programming problem in harmonic analysis*, Linear Alg. Applic. **14** (1976), 107-116 [9].
[Dun10] J. G. Dunn, *The performance of a class of n-dimensional quantizers for a Gaussian source*, in *Proc. Symp. Signal Trans. Process.*, IEEE Press, NY, 1965, pp. 76-81 [2].
J.-P. Dussault: see W. Brostow.
[Dym1] H. Dym and H. P. McKean, *Fourier Series and Integrals*, Ac. Press, NY, 1972 [3, 4, 9].

E

[Ear1] A. G. Earnest, *Spinor genera of unimodular* **Z***-lattices in quadratic fields*, PAMS **64** (1977), 189-195 [15].

[Ear2] — and J. S. Hsia, *Spinor norms of local integral rotations* II, PJM **61** (1975), 71-86; **115** (1984), 493-494 [15].

[Ear3] — — *Spinor genera under field extensions* I, Acta Arith. **32** (1977), 115-128 [15].

[Ear4] — — *Spinor genera under field extensions* II: 2 *unramified in the bottom field*, AJM **100** (1978), 523-538 [15].

[Ear5] — — *Spinor genera under field extensions* III: *quadratic extensions*, in *Number Theory and Algebra*, ed. H. Zassenhaus, Ac. Press, NY 1977, pp. 43-62 [15].

[Ede1] H. Edelsbrunner and R. Seidel, *Voronoi diagrams and arrangements*, Discr. Comput. Geom. **1** (1986), 25-44 [2].

[Edg1] W. L. Edge, *The geometry of an orthogonal group in six variables*, PLMS **8** (1958), 416-446 [4].

[Edg2] — *The partitioning of an orthogonal group in six variables*, PRS **A247** (1958), 539-549 [4].

[Edg3] — *An orthogonal group of order* $2^{13} \cdot 3^5 \cdot 5^2 \cdot 7$, Annali di Mat. **61** (1963), 1-95 [4].

[Edw1] H. M. Edwards, *Fermat's Last Theorem: A Genetic Introduction to Algebraic Number Theory*, Springer-Verlag, 1977 [15].

[Eic1] M. Eichler, *Quadratische Formen und Orthogonal Gruppen*, Springer-Verlag, 1952 [15, 16].

[Ein1] G. Einarsson, *Performance of polyphase signals on a Gaussian channel*, Ericsson Tech. **23** (No. 4, 1967), 411-438 [3].

[Ein2] — *Polyphase coding for a Gaussian channel*, Ericsson Tech. **24** (No. 2, 1968), 75-130 [3].

M. Elia: see E. Biglieri.

[Eli1] P. Elias, *Bounds and asymptotes for the performance of multivariate quantizers*, Ann. Math. Stat. **41** (1970), 1249-1259 [2].

[Els1] V. Elser, *The diffraction pattern of projected structures*, Acta Cryst. **A42** (1986), 36-43 [1].

[Els2] — and N. J. A. Sloane, *A highly symmetric four-dimensional quasicrystal*, J. Phys. **A 20**, (1987), 6161-6168 [1].

[Elt1] E. L. Elte, *The semiregular polytopes of the hyperspaces*, Doctor of Math. and Science Dissertation, State Univ. of Gronigen, 1912 [21].

[Emd1] P. van Emde Boas, *Another NP-complete partition problem and the complexity of computing short vectors in lattices*, Math. Dept. Report 81-04, Univ. Amsterdam, 1981 [2].

M. Enguehard: see M. Broué.

[Enr1] G. M. Enright, *A description of the Fischer group* F_{22}, J. Alg. **46** (1977), 334-343 [10].

[Enr2] — *Subgroups generated by transpositions in* F_{22} *and* F_{23}, Comm. Alg. **6** (1978), 823-837 [10].

[Erd1] A. Erdelyi, editor, *Higher Transcendental Functions*, McGraw-Hill, NY, 3 vols., 1953, especially Vol. II, Chap. XI [9].

[Erd2] P. Erdös and C. A. Rogers, *Covering space with convex bodies*, Acta Arith. **7** (1962), 281-285 [2].

[Ero1] V. A. Erokhin, *Theta-series of even unimodular 24-dimensional lattices*,

LOMI **86** (1979), 82-93, 190 = JSM **17** (1981), 1999-2008 [16].

[Ero2] — *On theta-series of even unimodular lattices*, LOMI **112** (1981), 59-70, 199 = JSM **25** (1984), 1012-1020 [16].

[Ero3] — *Automorphism groups of 24-dimensional even unimodular lattices*, LOMI **116** (1982), 68-73, 162 = JSM **26** (1984), 1876-1879 [16].

[Est1] D. Estes and G. Pall, *Spinor genera of binary quadratic forms*, JNT **5** (1973), 421-432 [15].

[Eva1] D. M. Evans, *The 7-modular representations of Janko's smallest simple group*, J. Alg. **96** (1985), 35-44 [10].

F

[Fal1] D. D. Falconer and R. D. Gitlin, editors, *Special Issue on Voiceband Telephone Data Transmission*, IEEE J. Select. Areas Commun. **2** (No. 5, 1984) [3, B].

H. M. Farkas: see H. E. Rauch.

[Fei1] W. Feit, *On integral representations of finite groups*, PLMS **29** (1974), 633-683 [3, 8].

[Fei2] — *Some lattices over* $\mathbf{Q}(\sqrt{-3})$, J. Alg. **52** (1978), 248-263 [2, 4, 7, 16].

[Fej1] G. Fejes Tóth, *New results in the theory of packing and covering*, in [Gru2], 1983, pp. 318-359 [1, 2].

[Fej2] — and L. Fejes Tóth, *Dictators on a planet*, Studia Sci. Math. Hung. **15** (1980), 313-316 [1].

[Fej2a] — P. Gritzmann and J. M. Wills, *Sausage-skin problems for finite covering*, Math. **31** (1984), 117-136 [1].

[Fej2b] — — — *Finite sphere packing and sphere covering*, Discrete Comput. Geom. **4** (1989), 19-40 [1].

L. Fejes Tóth: see also G. Fejes Tóth.

[Fej3] — *Uber einen geometrischen Satz*, Math. Z. **46** (1940) 79-83 [1].

[Fej4] — *Kreisausfüllungen der hyperbolischen Ebene*, AMAH **4** (1953), 103-110 [1].

[Fej5] — *Kreisüberdeckungen der hyperbolischen Ebene*, AMAH **4** (1953), 111-114 [1].

[Fej6] — *On close-packings of spheres in spaces of constant curvature*, Publ. Math. Debrecen **3** (1953), 158-167 [1].

[Fej7] — *Kugelunterdeckungen und Kugelüberdeckungen in Räumen konstanter Krümmung*, Archiv Math. **10** (1959), 307-313 [1].

[Fej8] — *Sur la représentation d'une population infinie par une nombre fini d'elements*, AMAH **10** (1959), 299-304 [2].

[Fej8a] — *What the bees know and what they do not know*, BAMS **70** (1964), 468-481 [4].

[Fej9] — *Regular Figures*, Pergamon, Oxford, 1964 [1, 2, 4, 21].

[Fej10] — *Lagerungen in der Ebene, auf der Kugel und in Raum*, Springer-Verlag, 2nd ed., 1972 [1, 2, 3, 4, 21, B].

[Fej11] — *Solid packing of circles in the hyperbolic plane*, Studia Sci. Math. Hungar. **15** (1980), 299-302 [1].

[Fej12] — *Stable packings of circles on the sphere*, Structural Topology **11** (1985), 9-14.

[Fer1] G. Ferraris and G. Ivaldi, *A simple method for Bravais lattice determination*, Acta Cryst. **A39** (1983), 595-596 [2].

L. Few: see also H. S. M. Coxeter.

[Few1] — *Covering space by spheres*, Math. **3** (1956), 136-139 [2, 4].

[Fie1] H. Fiedler, W. Jurkat and O. Körner, *Asymptotic expansions of finite theta series*, Acta Arith. **32** (1977), 723-745.

[Fie2] K. L. Fields, *Stable, fragile and absolutely symmetric quadratic forms*, Math. **26** (1979), 76-79.

[Fin1] U. Fincke and M. Pohst, *Improved methods for calculating vectors of short length in a lattice, including a complexity analysis*, MTAC **44** (1985), 463-471 [2].

[Fin2] L. Finkelstein, *The maximal subgroups of Conway's group* C_3 *and McLaughlin's group*, J. Alg. **25** (1973), 58-89 [10].

[Fin3] — and A. Rudvalis, *Maximal subgroups of the Hall-Janko-Wales group*, J. Alg. **24** (1973), 486-493 [10].

[Fin4] — — *The maximal subgroups of Janko's simple group of order* 50,232,960, J. Alg. **30** (1974), 122-143 [10].

[Fin5] J. L. Finney, *A procedure for the construction of Voronoi polyhedra*, J. Comp. Phys. **32** (1979), 137-143 [2].

[Fis1] B. Fischer, *Finite groups generated by 3-transpositions* I, Invent. Math. **13** (1971), 232-246 [10].

W. Fisher: see E. Koch.

[Fla1] J. L. Flanagan, M. R. Schroeder, B. S. Atal, R. E. Crochiere, N. S. Jayant and J. M. Tribolet, *Speech coding*, COMM **27** (1979), 710-737 and 932 [2].

[Fla2] L. Flatto, *Basic sets of invariants for finite reflection groups*, BAMS **74** (1968), 730-734 [7].

[Fla3] — *Invariants of finite reflection groups and mean value problems* II, AJM **92** (1970), 552-561 [7].

[Fla4] — *Invariants of finite reflection groups*, L'Enseignement Math. **24** (1978), 237-292 [7].

[Fla5] — and M. M. Wiener, *Invariants of finite reflection groups and mean value problems*, AJM **91** (1969), 591-598 [7].

[Fla6] — — *Regular polytopes and harmonic polynomials*, CJM **22** (1970), 7-21 [7].

[Fla7] — and D. J. Newman, *Random coverings*, Acta Math. **138** (1977), 241-264.

[Fol1] J. H. Folkman and R. L. Graham, *A packing inequality for compact convex subsets of the plane*, CMB **12** (1969), 745-752 [1].

P. Fong: see also A. O. L. Atkin.

[Fon1] — *Characters arising in the Monster-modular connection*, PSPM **37** (1980), 557-559 [1, 30].

[For1] G. D. Forney, Jr., personal communication [3].

[For2] — *Coset codes* — see Supplementary Bibliography [2, 3, 20].

[For3] — R. G. Gallager, G. R. Lang, F. M. Longstaff and S. U. Qureshi, *Efficient modulation for band-limited channels*, in [Fal1], 1984, pp. 632-646 [3].

[Fos1] G. J. Foschini, R. D. Gitlin and S. B. Weinstein, *Optimization of two-dimensional signal constellations in the presence of Gaussian noise*, COMM **22** (1971), 28-38 [3].

[Fos2] — — — *On the selection of a two-dimensional signal constellation in the presence of phase jitter and Gaussian noise*, BSTJ **52** (1973), 927-965 [3].

[Fow1] R. J. Fowler, M. S. Paterson and S. L. Tanimoto, *Optimal packing and covering in the plane are NP-complete*, Info. Proc. Lett. **12** (1981), 133-137 [2].

B. L. Fox: see W. Brostow.

[Fra1] W. Fraser and C. C. Gotlieb, *A calculation of the number of lattice points in*

the circle and sphere, MTAC **16** (1962), 282-290 [4].
P. O. Frederickson: see J. R. Baumgardner.
[Fre1] I. B. Frenkel, J. Lepowsky and A. Meurman, *A natural representation of the Fischer-Griess Monster with the modular function J as character*, PNAS **81** (1984), 3256-3260 [29, 30].
[Fre2] — — — *A moonshine module for the Monster*, in [Lep1], 1984, pp. 231-273 [29, 30].
[Fre3] — — — *An E_8-approach to* F_1, in [McK3], 1985, pp. 99-120 [29, 30].
[Fre4] — — — *An introduction to the Monster*, in *Unified String Theories*, M. G. Green and D. Gross, eds., World Scientific, Singapore 1986, pp. 533-546 [29, 30].
[Fre5] — — — *Vertex Operator Algebras and the Monster*, Ac. Press, NY 1988 [29, 30].
[Fri1] F. Fricker, *Einführung in die Gitterpunktlehre*, Birkhäuser, Boston, 1982 [4].
[Fri2] D. Fried and W. M. Goldman, *Three-dimensional affine crystallographic groups*, Adv. in Math. **47** (1983), 1-49 [4].
A. Fröhlich: see J. W. S. Cassels.
[Fro1] K. K. Frolov, *On the connection between quadrature formulas and sublattices of the lattice of integral vectors*, DAN **232** (No. 1, 1977) = SMD **18** (1977), 37-41 [1].
W. Fumy: see T. Beth.

G

[Gab1] E. M. Gabidulin and V. R. Sidorenko, *A general bound for code volume*, PPI **12** (No. 4, 1976), 31-35 = PIT **12** (1976), 226-269 [9].
[Gag1] T. M. Gagen, *A characterization of Janko's simple group*, PAMS **19** (1968), 1393-1395 [10].
[Gag2] — editor, *Finite Groups '72*, North-Holland, Amsterdam, 1973 [B].
R. V. Galiulin: see also B. N. Delone.
[Gal0] — *Delaunay systems*, Kristallografiya **25** (1980), 901-907 = Sov. Phys. Cryst. **25** (1980), 517-520.
R. G. Gallager: see also G. D. Forney, Jr., C. E. Shannon.
[Gal1] — *Information Theory and Reliable Communication*, Wiley, NY 1968 [3].
[Gal2] N. C. Gallagher, Jr. and J. A. Bucklew, *Some recent developments in quantization theory*, in *Proc. 12th Ann. Sympos. System Theory*, Virginia Beach VA, May 19-20, 1980 [2].
[Gam1] A. F. Gameckii, *On the theory of covering an n-dimensional Euclidean space with equal spheres*, DAN **146** (1962), 991-994 = SMD **3** (1962), 1410-1414 [2, 4].
[Gam2] — *The optimality of the principal lattice of Voronoi of the first type among the lattices of the first type of any number of dimensions*, DAN **151** (1963), 482-484 = SMD **4** (1963), 1014-1016 [2, 4].
[Gan1] R. Gangolli, *Spherical functions on semisimple Lie groups*, in [Boo1], 1972, pp. 41-92 [9].
[Gao1] J. Gao, L. D. Rudolph and C. R. P. Hartmann, *Iteratively maximum likelihood decodable spherical codes and a method for their construction*, PGIT **34** (1988), 480-485 [3].
R. A. Gaskins: see I. J. Good.
Z. Gáspár: see T. Tarnai.
[Gas1] G. Gasper, *Positivity and special functions*, in [Ask2a], 1975, pp. 375-433 [9].
[Gau1] C. F. Gauss, *Disquisitiones Arithmeticae*, Fleischer, Leipzig, 1801; English

translation, Yale University Press, 1966 [15].

[Gau2] — *Besprechung des Buchs von L. A. Seeber: Untersuchungen über die Eigenschaften der positiven ternären quadratischen Formen usw.*, Göttingsche Gelehrte Anzeigen (1831, July 9) = Werke, **II** (1876), 188-196 [1].

[Gel1] I. M. Gelfand, *Spherical functions on symmetric Riemannian spaces* DAN **70** (No. 1, 1970) 5-8 = Amer. Math. Soc. Transl. (2) **37** (1964), 39-43 [9].

[Gel2] M. Geller, *Comment on "New Jacobian theta functions and the evaluation of lattice sums" by I. J. Zucker*, J. Math. Phys. **18** (1977), 187 [2].

A. Gersho: see also D.-Y. Cheng, C.-S. Wang.

[Ger1] — *Principles of quantization*, IEEE Trans. Circuit Syst. **25** (1978), 427-436 [2].

[Ger2] — *Asymptotically optimal block quantization*, PGIT **25** (1979), 373-380 [2].

[Ger3] — *On the structure of vector quantizers*, PGIT **28** (1982), 157-166 [2, 20].

[Ger4] — and V. B. Lawrence, *Multidimensional signal constellations for voiceband data transmission*, in [Fal1], 1984, pp. 687-702 [3].

[Ger5] L. J. Gerstein, *Classes of definite hermitian forms*, AJM **100** (1978), 81-97 [2].

[Gib1] I. B. Gibson and I. F. Blake, *Decoding the binary Golay code with miracle octad generators*, PGIT **24** (1978), 261-264 [11].

[Gil1] E. N. Gilbert, *A comparison of signalling alphabets*, BSTJ **31** (1952), 504-522 [3, 9].

[Gin1] V. V. Ginzburg, *Multidimensional signals for a continuous channel*, PPI **20** (1984) 28-46 = PIT **20** (1984) 20-28 [3].

[Gis1] H. Gish and J. N. Pierce, *Asymptotically efficient quantizing*, PGIT **14** (1968), 676-683 [2].

R. D. Gitlin: see D. D. Falconer, G. J. Foschini.

[Gla1] J. W. L. Glaisher, *On the representations of a number as a sum of four squares and on some allied arithmetical functions*, Quart. J. Math. **36** (1905), 305-358 [4].

[Gla2] — *The arithmetical functions $P(m)$, $Q(m)$, $\Omega(m)$*, Quart. J. Math. **37** (1906), 36-48 [4].

[Gla3] — *On the representations of a number as the sum of two, four, six, eight, ten, and twelve squares*, Quart. J. Math. **38** (1907), 1-62 [4].

[Gla4] — *On the representations of a number as the sum of fourteen and sixteen squares*, Quart. J. Math. **38** (1907), 178-236 [4].

[Gla5] — *On the representations of a number as the sum of eighteen squares*, Quart. J. Math. **38** (1907), 289-351 [4].

[Gla6] — *On the numbers of representations of a number as a sum of $2r$ squares, where $2r$ does not exceed eighteen*, PLMS **5** (1907), 479-490 [4].

[Gla7] M. L. Glasser, *The evaluation of lattice sums*, J. Math. Phys. **14** (1973), 409-413, 701-703; **15** (1974), 188-189, 520; **16** (1975), 1237-1238 [2].

[Gla8] — and I. J. Zucker, *Lattice sums*, Theoret. Chem.: Advances and Perspectives **5** (1980), 67-139 [2].

B. G. Glazer: see C. N. Campopiano.

[Gle1] A. M. Gleason, *Weight polynomials of self-dual codes and the MacWilliams identities*, Actes, Congrès Intern. Math., Gauthier-Villars, Paris, 1971, **3**, pp. 211-215 [7].

[God1] P. Goddard and D. Olive, *Algebras, lattices and strings*, in [Lep1], 1985, pp. 51-96 [1].

J.-M. Goethals: see also P. Delsarte, F. J. MacWilliams.

[Goe1] — *Two dual families of nonlinear binary codes*, Elect. Lett. **10** (1974), 471-472 [5].
[Goe2] — *Nonlinear codes defined by quadratic forms over GF*(2), IC **31** (1976), 43-74 [5].
[Goe3] — *The extended Nadler code is unique*, PGIT **23** (1977), 132-135 [9].
[Goe4] — *Association schemes*, in *Algebraic Coding Theory and Applications*, ed. G. Longo, CISM Courses and Lectures No. 258, Springer-Verlag, 1979, pp. 243-283 [9, 14].
[Goe5] — and J. J. Seidel, *Cubature formulae, polytopes and spherical designs*, pp. 203-218 of [Dav3] [1, 3].
[Goe6] — — *Spherical designs*, PSPM **34** (1979), 255-272 [3].
[Gol1] M. J. E. Golay, *Notes on digital coding*, PIEEE **37** (1949), 637 [3].
[Gol2] — *Binary coding*, PGIT **4** (1954), 23-28 [3, 5].
[Gol3] M. Goldberg, *Packing of* 18 *equal circles on a sphere*, Elem. Math. **20** (1965), 59-61 [1].
[Gol4] — *Packing of* 19 *equal circles on a sphere*, Elem. Math. **22** (1967), 108-110 [1].
[Gol5] — *An improved packing of* 33 *equal circles on a sphere*, Elem. Math. **22** (1967), 110-111 [1].
[Gol6] — *Axially symmetric packing of equal circles on a sphere*, Ann. Univ. Sci. Budapest **10** (1967), 37-48; **12** (1969), 137-142 [1].
[Gol7] — *Stability configurations of electrons on a sphere*, MTAC **23** (1969), 785-786 [1].
[Gol8] — *On the densest packing of equal spheres in a cube*, Math. Mag. **44** (1971), 199-208 [1].
W. M. Goldman: see D. Fried.
D. Golke: see H. E. Ong.
[Gol9] E. S. Golod and I. R. Shafarevich, *On class field towers*, IANS **28** (1964), 261-272 = Amer. Math. Soc. Transl. (2) **48** (1965), 91-102 [1, 8].
[Goo1] I. J. Good and R. A. Gaskins, *The centroid method of numerical integration*, Num. Math. **16** (1971), 343-359 [21].
[Gop1] V. D. Goppa, *A new class of linear error-correcting codes*, PPI **6** (No. 3, 1970) = PIT **6** (1970), 207-212 [3].
[Gop2] — *Codes associated with divisors*, PPI **13** (1977), 33-39 = PIT **13** (1977) 22-27 [3].
[Gop3] — *Codes on algebraic curves*, DAN **259** (1981), 1289-1290 = SMD **24** (1981), 170-172 [3].
[Gop4] — *Algebraic-geometric codes*, IANS **46** (1982), 762-781 = Izv. **21** (1983), 75-91 [3].
[Gop5] — *Codes and information*, Usp. Mat. Nauk **39** (1984), 77-120 = Russ. Math. Surv. **39** (1984), 87-141 [3].
[Gor0] D. M. Gordon, *Minimal permutation sets for decoding the binary Golay codes*, PGIT **28** (1982), 541-543 [20].
[Gor1] L. M. Gordon and R. Levingston, *The construction of some automorphic graphs*, Geom. Dedic. **10** (1981), 261-267 [9].
[Gor2] D. Gorenstein, *The classification of finite simple groups* I. *Simple groups and local analysis*, BAMS **1** (1979), 43-199 [3, 10].
[Gor3] — *Finite Simple Groups*, Plenum, NY, 1982 [3, 10].
[Gor4] — *The Classification of Finite Simple Groups*, Plenum, NY, 1983, Vol. 1 [3, 10].
[Gor5] — *Classifying the finite simple groups*, BAMS **14** (1986), 1-98 [3, 10].

[Gor6] — and K. Harada, *A characterization of Janko's two new simple groups*, J. Fac. Sci. Univ. Tokyo **16** (1970), 331-406 [10].

[Gos1] T. Gosset, *On the regular and semi-regular figures in space of n dimensions*, Messenger Math. **29** (1900), 43-48 [4].

C. C. Gotlieb: see W. Fraser.

R. L. Graham: see also P. Diaconis, J. H. Folkman.

[Gra1] — and N. J. A. Sloane, *On constant weight codes and harmonious graphs*, in *Proc. West coast conference on combinatorics, graph theory and computing*, Utilitas Math. Publ. Co., Winnipeg, Canada, 1979, pp. 25-40 [9].

[Gra2] — — *Lower bounds for constant weight codes*, PGIT **26** (1980), 37-43 [3, 9].

[Gra3] — — *On additive bases and harmonic graphs*, SIAD **1** (1980), 382-404 [9].

[Gra4] — — *On the covering radius of codes*, PGIT **31** (1985), 385-401 [2, 3].

[Gra4a] — — *Penny-packing and two-dimensional codes*, Discrete Comput. Geom. to appear [1].

[Gra5] — H. S. Witsenhausen and H. J. Zassenhaus, *On tightest packings in the Minkowski plane*, PJM **41** (1972), 699-715 [1].

D. Gratias: see also D. Shechtman.

[Gra6] — and L. Michel, editors, *International Workshop on Aperiodic Crystals*, J. de Physique **47** (Colloq. C3, suppl. to No. 7, July 1986) [1].

R. M. Gray: see also L. D. Davisson, Y. Linde, Y. Yamada.

[Gra7] — editor, *Special Issue on Quantization*, PGIT **28** (1982), 127-262 [2].

M. B. Green: see also I. B. Frenkel.

[Gre1] — *Unification of forces and particles in superstring theories*, Nature **314** (1985), 409-414 [1].

[Gre2] R. R. Green, *A serial orthogonal decoder*, JPL Space Programs Summary, Vol. 37-39-IV (1966), 247-252 [20].

[Gre3] — *Analysis of a serial orthogonal decoder*, JPL Space Programs Summary, Vol. 37-53-III (1968), 185-187 [20].

[Gri1] R. L. Griess, Jr., *Schur multipliers of the known finite simple groups* I, BAMS **78** (1972), 68-71 [10].

[Gri2] — *The structure of the "Monster" simple group*, in [Sco3], 1976, pp. 113-118 [29].

[Gri3] — *Schur multipliers of the known finite simple groups* II, PSPM **37** (1980), 279-282 [10].

[Gri4] — *A construction of* F_1 *as automorphisms of a* 196,883-*dimensional algebra*, PNAS **78** (1981), 689-691 [3, 10, 29].

[Gri5] — *The friendly giant*, Invent. Math. **69** (1982), 1-102 [3, 10, 29].

[Gri6] — *The sporadic simple groups and construction of the Monster*, in *Proc. Intern. Congr. Math. Warsaw 1983*, North-Holland, Amsterdam, 1984, pp. 369-384 [10, 29].

[Gri7] — *Quotients of infinite reflection groups*, Math. Ann. **263** (1983), 267-278 [10, 29].

[Gri8] — *The Monster and its nonassociative algebra*, in [McK3], 1985, pp. 121-157 [10, 29].

[Gri9] — *Schur multipliers of the known finite simple groups* III, in Proc. Rutgers Group Theory Year 1983-1984, ed. M. Aschbacher et al., Camb. Univ. Press, 1984, pp. 69-80 [10].

[Gri10] — *Code loops*, J. Alg. **100** (1986), 224-234 [29].

[Gri10a] — *Sporadic groups, code loops and nonvanishing cohomology*, J. Pure Appl. Alg. **44** (1978), 191-214 [29].

P. Gritzmann: see also G. Fejes Tóth.
[Gri11] — *Lattice covering of space with symmetric convex bodies*, Math. **32** (1985), 311-315 [2].
[Gri12] — *Finite packing of equal balls*, JLMS **33** (1986), 543-553 [1].
[Gri13] — and J. M. Wills, *An upper estimate for the lattice point enumerator*, Math. **33** (1986), 196-202 [1].
[Gri14] — — *Finite packing and covering*, Studia Sci. Math. Hungar **21** (1986), 149-162 [1].
[Gro0] H. Groemer, *Some basic properties of packing and covering constants*, Discr. Comput. Geom **1** (1986), 183-193 [1].
D. Gross: see I. B. Frenkel.
F. Gross: see W. R. Scott.
H. Gross: see B. L. van der Waerden.
[Gro1] K. I. Gross, *On the evolution of noncommutative harmonic analysis*, AMM **85** (1978), 525-548 [9].
[Gro2] E. Grosswald, *Representations of Integers as Sums of Squares*, Springer-Verlag, 1985 [4].
H. Grothe: see L. Afflerbach.
[Gro3] L. C. Grove and C. T. Benson, *Finite Reflection Groups*, Springer-Verlag, 2nd ed., 1985 [4, 21].
[Gru1] P. M. Gruber, *Geometry of numbers*, in [Töl1], 1979, pp. 186-225 [1, 2, 15].
[Gru1a] — and C. G. Lekkerkerker, *Geometry of Numbers*, 2nd ed., North-Holland, Amsterdam, 1987 [1, B].
[Gru2] — and J. M. Wills, editors, *Convexity and Its Applications*, Birkhäuser, Boston, 1983 [B].
B. Grünbaum: see also C. Davis.
[Grü1] — *Convex Polytopes*, Wiley, NY, 1967 [21].
[Grü2] — and G. C. Shephard, *Patterns on the 2-sphere*, Math. **28** (1981), 1-35 [1].
[Grü3] — — *Tilings and Patterns*, Freeman, NY, 1987 [1].
[Gud1] P. M. Gudivok, A. A. Kirilyuk, V. P. Rud'ko and A. I. Tsitkin, *Finite subgroups of the group* $GL(n, \mathbf{Z})$, Kibernetika **18** (No. 6, 1982) 71-82 = Cybernetics **18** (1982), 788-803 [3].
[Gun1] R. C. Gunning, *Lectures on Modular Forms*, Princeton Univ. Press, 1962 [2, 7].
[Gün1] S. Günther, *Ein stereometrisches Problem*, Archiv Math. Physik (Grunert) **57** (1875), 209-215 [1].
[Gus1] W. H. Gustafson, *Remarks on the history and applications of integral representations*, in [Rog10], 1981, pp. 1-36 [3].
[Guy1] R. K. Guy, editor, *Reviews in Number Theory 1973-83*, Amer. Math. Soc., Providence RI, 6 vols., 1984 [B].

H

[Hab1] W. Habicht and B. L. van der Waerden, *Lagerung von Punkten auf der Kugel*, Math. Ann. **123** (1951), 223-234 [1].
[Had1] H. Hadwiger, *Uber ausgezeichnete Vektorsterne und reguläre Polytope*, Comment. Math. Helvet. **13** (1940), 90-107 [6].
[Hah1] T. Hahn, editor, *International Tables for Crystallography*, Reidel, Dordrecht, 1983 [1, B].
J. L. Hall: see M. R. Schroeder.
M. Hall, Jr.: see also N. J. A. Sloane.

[Hal1] — *Hadamard matrices of order* 16, Jet Propulsion Lab. Research Summary 36-10, Vol. 1 (Sept. 1, 1961), pp. 21-26 [3].

[Hal2] — *Hadamard matrices of order* 20, Jet Propulsion Lab. Technical Report 32-761 Jet Propulsion Lab., Pasadena CA, 1965 [3].

[Hal3] — *Combinatorial Theory*, Blaisdell, Waltham MA, 1967 [3].

[Hal4] — and D. B. Wales, *The simple group of order* 604,800 J. Alg. **9** (1968), 417-450 [10].

[Hal5] S. R. Hall and T. Ashida, editors, *Methods and Applications in Crystallographic Computing*, Oxford Univ. Press, 1984 [2].

[Hal6] P. R. Halmos, *Measure Theory*, Van Nostrand, Princeton NJ, 1965 [9].

B. I. Halperin: see D. R. Nelson.

H. Hämäläinen: see I. Honkala.

I. Hambleton: see C. R. Riehm.

[Ham1] C. M. Hamill, *On a finite group of order* 6,531,840, PLMS **52** (1951), 401-454 [4].

[Ham2] J. Hammer, *Unsolved Problems Concerning Lattice Points*, Pitman, San Francisco CA, 1977 [1].

[Ham3] P. Hammond, *q-Coverings and completely regular codes in distance-regular graphs*, Quart. J. Math. Oxford **29** (1978), 23-30 [9].

[Ham4] — and D. H. Smith, *Perfect codes in the graphs* O_k, JCT **B19** (1975), 239-255 [9].

[Han1] H. Hanani, *On quadruple systems*, CJM **12** (1960), 145-157 [3, 5].

[Han2] — *Balanced incomplete block designs and related designs*, Discr. Math. **11** (1975), 255-369 [3].

[Han3] H. Hancock, *Development of the Minkowski Geometry of Numbers*, Dover, NY, 1964 [1].

K. Hanes: see J. D. Berman.

W. Hanrath: see W. Plesken.

K. Harada: see also D. Gorenstein.

[Har0] — *On the simple group* F *of order* $2^{14} \cdot 3^6 \cdot 5^6 \cdot 7 \cdot 11 \cdot 19$, in [Sco3], 1976, 119-276 [10].

[Har1] — *The automorphism group and the Schur multiplier of the simple group of order* $2^{14} \cdot 3^6 \cdot 5^6 \cdot 7 \cdot 11 \cdot 19$, Osaka J. Math. **15** (1978), 633-635 [10].

[Har2] F. Harary, *Graph Theory*, Addison-Wesley, Reading MA, 1969 [28].

[Har3] G. H. Hardy, *On the representation of a number as the sum of any number of squares, and in particular of five*, TAMS **21** (1920), 255-284 = *Coll. Papers*, I, 345-374 [4].

[Har4] — *Ramanujan*, Camb. Univ. Press, 1940; Chelsea, NY, 1959 [2, 4, 7].

[Har5] — and E. M. Wright, *An Introduction to the Theory of Numbers*, Oxford Univ. Press, 5th ed., 1980 [2, 4].

[Har6] E. M. Hartley, *A sextic primal in five dimensions*, PCPS **46** (1950), 91-105 [4].

[Har7] — *Two maximal subgroups of a collineation group in five dimensions*, PCPS **46** (1950), 555-569 [4].

C. R. P. Hartmann: see J. Gao.

[Har8] M. Harwit and N. J. A. Sloane, *Hadamard Transform Optics*, Ac. Press, NY, 1979 [3, 20].

[Has0] H. Hasse, *Uber die Klassenzahl Abelscher Zahlkörper*, Akademie-Verlag, Berlin, 1952 [8].

[Has1] — *Number Theory*, Springer-Verlag, 1980 [8].

[Haz1] M. Hazewinkel, W. Hesselink, D. Siersma and F. D. Veldkamp, *The ubiquity* of *Coxeter-Dynkin diagrams* (*an introduction to the A-D-E problem*), Nieuw Arch. Wisk. **25** (1977), 257-307 [1, 2, 4, 21].

[Hec0] E. Hecke, *Vorlesungen über die Theorie der Algebraischen Zahlen*, Chelsea, NY 1970; English translation, Springer-Verlag, 1981 [8].

[Hec1] — *Uber Modulfunktionen und die Dirichletschen Reihen mit Eulerscher Produktentwickling*, Math. Ann. **114** (1937), 1-28 and 316-351 = [Hec3], pp. 644-707 [7].

[Hec2] — *Analytische Arithmetik der positiven quadratischen Formen*, Kgl. Danske Vid. Selskab. Mat.-fys. Medd. **13** (No. 12, 1940) = [Hec3], pp. 789-918 [2, 7, 18].

[Hec3] — *Mathematische Werke*, Vandenhoeck and Ruprecht, Göttingen, 1970 [2, 7, 8, B].

[Hed1] A. Hedayat and W. D. Wallis, *Hadamard matrices and their applications*, Ann. Stat. **6** (1978), 1184-1238 [3].

[Hel1] D. Held, *The simple groups related to* M_{24}, J. Alg. **13** (1969), 253-296 [10].

[Hel2] — *The simple groups related to* M_{24}, II, JAMS **16** (1973), 24-28 [10].

[Hel3] B. Helfrich, *An algorithm to construct Minkowski-reduced lattice-bases*, Lect. Notes Comp. Sci. **182** (1985), 173-179 [2].

[Hel4] — Algorithms to construct Minkowski reduced and Hermite reduced lattice bases, *Theoretical Computer Science* **41** (No. 2-3, 1986), 125-139 [2].

[Hel5] S. Helgason, *Differential Geometry and Symmetric Spaces*, Ac. Press, NY, 1962 [9].

[Hel6] — *Differential Geometry, Lie Groups, and Symmetric Spaces*, Ac. Press, NY, 1978 [9].

[Hel7] H. J. Helgert and R. D. Stinaff, *Minimum-distance bounds for binary linear codes*, PGIT **19** (1973), 344-356 [3].

[Hem1] C. Hermann, *Kristallographie in Räumen beliebiger Dimensionszahl* I, *Die Symmetrieoperationen*, Acta Cryst. **2** (1949), 139-145 [3].

[Her0] C. Hermite, *Second letter to Jacobi on number theory*, JRAM **40** (1850), 279-290 = Oeuvres **I** (1905), 122-130 [2].

[Her1] I. N. Herstein, *Topics in Algebra*, Ginn, Waltham MA, 1964 [2].

W. Hesselink: see M. Hazewinkel.

T. Hickey: see J. Cohen.

[Hig1] J. R. Higgins, *Five short stories about the cardinal series*, BAMS **12** (1985), 45-89 [3].

[Hig2] D. G. Higman, *Coherent configurations*, Geom. Dedicata **4** (1975), 1-32; **5** (1976), 413-424 [9, 14].

[Hig3] — and C. C. Sims, *A simple group of order* 44,352,000, Math. Z. **105** (1968), 110-113 [10].

G. Higman: see also M. B. Powell.

[Hig4] — *On the simple group of D. G. Higman and C. C. Sims*, IJM **13** (1969), 74-80 [10].

[Hig5] — and J. McKay, *On Janko's simple group of order* 50,232,960, BLMS **1** (1969), 89-94 [10].

[Hil1] D. Hilbert, *Mathematische Probleme*, Archiv. Math. Phys. **1** (1901), 44-63 and 213-237 = *Gesamm. Abh.*, **III**, 290-329. English translation in BAMS **8** (1902), 437-479 = PSPM **28** (1976), 1-34 [1].

[Hil2] — and S. Cohn-Vossen, *Anschauliche Geometrie*, Julius Springer, Berlin, 1932. English translation, *Geometry and the Imagination*, Chelsea, NY, 1952 [1, 21].

[Hil3] H. Hiller, *Crystallography and cohomology of groups*, AMM **93** (1986),

765-779 [4].

S. Hirasawa: see Y. Sugiyama.

[Hir1] J. W. P. Hirschfeld, *Projective Geometries over Finite Fields*, Oxford Univ. Press, 1979.

[Hir2] — *Linear codes and algebraic curves*, in *Geometrical Combinatorics*, Pitman, Boston, 1984, pp. 35-53 [3].

[Hir3] — *Finite Projective Spaces of Three Dimensions*, Oxford Univ. Press, 1986.

[Hir4] F. Hirzebruch, W. D. Neumann and S. S. Koh, *Differentiable Manifolds and Quadratic Forms*, Dekker, NY, 1971 [2, 15].

[Hla1] E. Hlawka, *Zur Geometrie der Zahlen*, Math. Z. **49** (1944), 285-312 [1].

[Hla2] — *Zur angenäherten Berechnung mehrfacher Integraler*, Monatsh. Math. **66** (1962) 140-151 [1].

[Hla3] — 90 *Jahre Geometrie der Zahlen*, Jahr. Uber. Math. (1980), 9-41 [1].

[Hoa1] M. R. Hoare, *Structure and dynamics of simple microclusters*, Adv. Chem. Phys. **40** (1979), 49-135 [1].

[Hoa2] — and J. A. McInnes, *Morphology and statistical statics of simple microclusters*, Adv. Phys. **32** (1983), 791-821 [1].

[Hob1] E. W. Hobson, *The Theory of Spherical and Ellipsoidal Harmonics*, Camb. Univ. Press, 1931; Chelsea, NY, 1955 [9].

[Hof1] N. Hofreiter, *Uber Extremformen*, Monatsh. Math. Phys. **40** (1933), 129-152 [2].

S. G. Hoggar: see also E. Bannai.

[Hog1] — *Bounds for quaternionic line systems and reflection groups*, Math. Scand. **43** (1978), 241-249 [3].

[Hog2] — *Parameters of t-designs in* $\mathbf{F}P^{d-1}$, Europ. J. Combin. **5** (1984), 29-36 [3].

[Hog3] — *t-Designs in projective spaces*, Europ. J. Combin. **3** (1982), 233-254 [3].

[Hoh1] B. von Hohenbalken, *Finding simplicial subdivisions of polytopes*, Math. Programming **21** (1981), 233-234 [21].

[Hol1] A. Holden, *Shapes, Space, and Symmetry*, Columbia Univ. Press, NY, 1971 [2, 21].

[Hon1] Y. Hong, *On spherical t-designs in* $\mathbf{R}^2$, Europ. J. Combin. **3** (1982), 255-258 [3].

[Hon2] I. Honkala, *Some optimal constant weight codes*, Ars Combin. **B20** (1985), 43-47 [9].

[Hon3] — *Some upper bounds for* $A(n, d)$ *and* $A(n, d, w)$, PGIT to appear [9].

[Hon4] — H. Hämäläinen and M. Kaikkonen, *A modification of the Zinoviev lower bound for constant weight codes*, Discr. Appl. Math. **11** (1985), 307-310 [9].

[Hon5] — — — *Some lower bounds for constant weight codes*, Disc. Appl. Math. **18** (1987), 95-98 [9].

[Hop1] R. Hoppe, *Bemerkung der Redaction*, Archiv Math. Physik (Grunert) **56** (1874), 307-312 [1].

[How1] D. J. Howarth and H. Jones, *The cellular method of determining electronic wave functions and eigenvalues in crystals, with applications to sodium*, PRS **65** (1952), 355-368 [2].

J. S. Hsia: see also J. W. Benham, P. J. Costello, A. G. Earnest.

[Hsi1] — *Spinor norms of local integral rotations* I, PJM **57** (1975), 199-206 [15].

[Hsi2] — *Representations by spinor genera*, PJM **63** (1976), 147-152 [15].

[Hsi3] — *Arithmetic theory of integral quadratic forms*, in *Proc. Queen's Number Theory Conf.* 1979, Queen's Papers Pure Appl. Math. **54**, Queen's Univ., Kingston, Ontario, 1980, pp. 173-204 [15].

[Hsi4] — *On the classification of unimodular quadratic forms*, JNT **12** (1980),

327-333 [15].

[Hsi5] — *Regular positive ternary quadratic forms*, Math. **28** (1981), 231-238 [2, 15].

[Hsi6] — *Algebraic methods in integral quadratic forms*, preprint [15].

[Hsi6a] — *Even positive definite unimodular quadratic forms over real quadratic fields*, Rocky Mtn. J. Math. **19** (1989), 725-733 [2, 7, 15].

[Hsi7] — and D. C. Hung, *Theta series of ternary and quaternary quadratic forms*, Invent. Math. **73** (1983), 151-156 [2, 15].

[Hsi8] — — *Theta series of quaternary quadratic forms over* **Z** *and* $\mathbf{Z}[(1+\sqrt{p})/2]$, Acta Arith. **45** (1985), 75-91 [2, 15].

[Hsi9] — Y. Kitaoka and M. Kneser, *Representations of positive definite quadratic forms*, JRAM **301** (1978), 132-141 [15].

[Hua1] L. K. Hua and Y. Wang, *Applications of Number Theory to Numerical Analysis*, Springer-Verlag, 1981 [1].

[Hub1] X. L. Hubaut, *Strongly regular graphs*, Discr. Math. **13** (1975), 357-381 [9].

V. Huber-Dyson: see K. W. Roggenkamp.

[Huf1] W. C. Huffman, *The biweight enumerator of self-orthogonal binary codes*, Discr. Math. **26** (1979), 129-143 [7].

[Huf2] — *Polynomial invariants of finite linear groups of degree two*, CJM **32** (1980), 317-330 [7].

[Huf3] — *Automorphisms of codes with applications to extremal doubly even codes of length* 48, PGIT **28** (1982), 511-521 [7].

[Huf4] — and V. I. Iorgov, *A* [72, 36, 16] *doubly even code does not have an automorphism of order* 11, PGIT **33** (1987), 749-752 [7].

[Huf5] — and N. J. A. Sloane, *Most primitive groups have messy invariants*, Adv. in Math. **32** (1979), 118-127 [3].

[Hug1] D. R. Hughes, *A combinatorial construction of the small Mathieu designs and groups*, Ann. Discr. Math. **15** (1982), 259-264 [11].

[Hug2] — and F. C. Piper, *Projective Planes*, Springer-Verlag, 1973 [3].

[Hug3] — — *Design Theory*, Camb. Univ. Press, 1985 [3].

K. Huml: see F. R. Ahmed.

[Hum1] J. E. Humphreys, *Introduction to Lie Algebras and Representation Theory*, Springer-Verlag, 1972 [4, 23].

D. C. Hung: see J. S. Hsia.

[Hun1] D. C. Hunt, *A characterization of the finite simple group* M(23), J. Alg. **26** (1972), 431-439 [10].

[Hup1] B. Huppert, *Endliche Gruppen* I, Springer-Verlag, 1967 [3,10].

[Hur1] A. Hurwitz, *Uber die Zahlentheorie der Quaternionen*, Nach. Gesellschaft Wiss. Göttingen, Math.-Phys. Klasse (1896), 313-340 = *Math. Werke*, Birkhäuser, Basel, **II**, 1933, pp. 303-330 [2, 4].

D. Husemoller: see J. Milnor.

B. G. Hyde: see M. O'Keeffe.

I

[Igu1] J. Igusa, *Theta Functions*, Springer-Verlag, 1972 [4].

[Iha1] Y. Ihara, *Some remarks on the number of rational points of algebraic curves over finite fields*, J. Fac. Sci. Univ. Tokyo, Sec. 1A, **28** (1982), 721-724 [3].

[Inc1] E. L. Ince, *Cycles of Reduced Ideals in Quadratic Fields*, Math. Tables **IV**, British Assoc. Adv. Sci., Camb. Univ. Press, 1966 [15].

[Ing1] I. Ingemarsson, *Commutative group codes for the Gaussian channel*, PGIT **19**

(1973), 215-219 [3].
[Int1] International Union of Crystallography, *International Tables for X-Ray Crystallography*, Kynoch Press, Birmingham, England, 3 vols., 1959 [4].
O. Intrau: see H. Brandt.
V. I. Iorgov: see also W. C. Huffman.
[Ior1] — *Binary self-dual codes with automorphisms of odd order*, PPI **19** (No. 4, 1983), 11-24 = PIT **19** (1983), 260-270 [7].
[Ior2] — *A method for constructing inequivalent self-dual codes with applications to length* 56, PGIT **33** (1987), 77-82 [7].
[Ior3] — *Doubly-even extremal codes of length* 64, PPI **22** (No. 4, 1986), 35-42 = PIT **22** (1986), 277-284 [7].
[Ito1] N. Ito, *Symmetry codes over* $GF(3)$, JCT **A29** (1980) 251-253 [3].
[Ito2] — J. S. Leon and J. Q. Longyear, *Classification of* 3-(24, 12, 5) *designs and 24-dimensional Hadamard matrices*, JCT **A31** (1981), 66-93 [3].
T. Ito: see E. Bannai.
G. Ivaldi: see G. Ferraris.
N. Iwahori: see J. G. Thompson.
[Iya1] K. Iyanaga, *Class numbers of definite Hermitian forms*, J. Math. Soc. Japan **21** (1969), 359-374 [7].

J

[Jac1] D. M. Jackson and S. A. Vanstone, editors, *Enumeration and Design*, Ac. Press, 1984 [3, B].
[Jac2] M. Jacob, editor, *Dual Theory*, North-Holland, Amsterdam, 1974 [1].
I. M. Jacobs: see J. M. Wozencraft.
[Jan1] Z. Janko, *A new finite simple group with Abelian Sylow* 2-*subgroups*, PNAS **53** (1965), 657-658 [10].
[Jan2] — *A new finite simple group with Abelian* 2-*Sylow subgroups and its characterization*, J. Alg. **3** (1966), 147-186 [10].
[Jan3] — *A characterization of a new simple group*, in *Proc. First Internat. Conf. Theory of Groups*, ed. L. Kovacs and B. Neumann, Gordon and Breach, NY, 1967, pp. 205-208 [10].
[Jan4] — *Some new simple groups of finite order* I, First Naz. Alta Math. Symposia Math. **1** (1968), 25-65 [10].
[Jan5] — *Some new simple groups of finite order*, in [Bra2], 1969, pp. 63-64 [10].
[Jan6] — *A new finite simple group of order* 86,775,571,046,077,562,880 *which possesses* M_{24} *and the full cover of* M_{22} *as subgroups*, J. Alg. **42** (1976), 564-596 [10].
[Jan7] G. J. Janusz, *Algebraic Number Fields*, Ac. Press, NY, 1973 [8].
N. S. Jayant: see also J. L. Flanagan.
[Jay1] — editor, *Waveform Quantization and Coding*, IEEE, NY, 1976 [2].
F. Jelinek: see T. Berger.
[Jer1] A. J. Jerri, *The Shannon sampling theorem—its various extensions and applications: a tutorial review*, PIEEE **65** (1977), 1565-1596 [3].
[Jod1] W. S. Jodrey and E. M. Tory, *Computer simulation of close random packing of equal spheres*, Phys. Rev. **A32** (1985), 2347-2351.
[Jon1] B. W. Jones, *A table of Eisenstein-reduced positive ternary quadratic forms of determinant* ≤ 200, Bull. Nat. Res. Council U.S.A. **97** (1935), 1-51 [15].
[Jon2] — *A canonical rational form for the ring of* 2-*adic integers*, DMJ **11** (1944), 715-727 [15].
[Jon3] — *The Arithmetic Theory of Quadratic Forms*, Math. Assoc. America and

Wiley, NY, 1950 [15].

[Jon4] G. A. Jones, *Geometric and asymptotic properties of Brillouin zones in lattices*, BLMS **16** (1984), 241-263 [2].

H. Jones: see D. J. Howarth.

[Jon5] W. Jónsson, *On the Mathieu groups M_{22}, M_{23}, M_{24} and the uniqueness of the associated Steiner systems*, Math. Z. **125** (1972), 193-214 [11, 14].

[Jul1] D. Julin, *Two improved block codes*, PGIT **11** (1965), 459 [3, 5].

[Jus1] J. Justesen, *A class of constructive asymptotically good algebraic codes*, PGIT **18** (1972), 652-656 [3].

K

[Kab1] G. A. Kabatiansky and V. I. Levenshtein, *Bounds for packings on a sphere and in space*, PPI **14** (No. 1, 1978), 3-25 = PIT **14** (No. 1, 1978), 1-17 [1, 9].

[Kac1] V. G. Kac, *Simple irreducible graded Lie algebras of finite growth*, IANS **32** (1968), 1323-1367 = Izv. **2** (1968), 1271-1311 [1, 30].

[Kac2] — *Infinite-dimensional algebras, Dedekind's η-function, classical Möbius function and the very strange formula*, Adv. in Math. **30** (1978), 85-136 [1].

[Kac3] — *An elucidation of "Infinite-dimensional algebras... and the very strange formula." $E_8^{(1)}$ and the cube root of the modular invariant j*, Adv. in Math. **35** (1980), 264-273 [1, 30].

[Kac4] — *A remark on the Conway-Norton conjecture about the "Monster" simple group*, PNAS **77** (1980) 5048-5049 [1, 30].

[Kac5] — *Infinite Dimensional Lie Algebras*, Birkhäuser, Boston, 2nd ed. 1985 [30].

[Kac6] — and D. H. Peterson, *Infinite-dimensional Lie algebras, theta functions and modular forms*, Adv. in Math. **53** (1984), 125-264 [1].

P. J. Kachoyan: see I. H. Sloan.

M. Kaikkonen: see I. Honkala.

[Kan1] T. Kaneko and H. Miyakawa, *A construction algorithm for linear codes for the Gaussian channel*, Elect. Commun. Japan **A62** (No. 4, 1979), 18-27 [3].

[Kan2] R. Kannan, *Improved algorithms for integer programming and related lattice problems*, Proc. 15th ACM Symp. Theory of Computing, 1983, pp. 193-206 [2].

W. M. Kantor: see also A. R. Calderbank, C. W. Curtis, P. Diaconis.

[Kan3] — *Generation of linear groups*, in [Dav3], 1981, pp. 497-509 [4].

[Kan4] — *Spreads, translation planes and Kerdock sets* I, SIAD **3** (1982), 151-165 [3, 9].

[Kan5] — and R. A. Liebler, *The rank* 3 *permutation representations of the finite classical groups*, TAMS **271** (1982), 1-71 [9].

I. M. Kaplinskaja: see E. B. Vinberg.

[Kar1] Z. A. Karian, *The covering problem for the fields $Q(-1)^{1/2}$ and $Q(-3)^{1/2}$*, JNT **8** (1976), 233-244 [2].

[Kar2] S. Karlin and J. L. McGregor, *The Hahn polynomials, formulas and an application*, Scripta Math. **26** (1961), 33-46 [9].

J. K. Karlof: see also C. P. Downey.

[Kar3] — and C. P. Downey, *Odd group codes for the Gaussian channel*, SIAJ **34** (1978), 715-716 [3].

M. R. Karpovsky: see G. D. Cohen.

M. Kasahara: see Y. Sugiyama.

[Kas1] T. Kasami and N. Tokura, *Some remarks on BCH bounds and minimum weights of binary primitive BCH codes*, PGIT **15** (1969), 408-413 [5].

W. Kassem: see also R. de Buda.

[Kas2] — *Optimal Lattice Codes for the Gaussian Channel*, Ph.D Dissertation, McMaster Univ., Hamilton, Ontario, 1981 [3].

G. L. Katsman: see also M. A. Tsfasman.

[Kat1] — *Packing density for spheres of radius* 1 *in Hamming space*, PPI **17** (No. 2, 1981), 52-56 = PIT **17** (1981), 114-117 [3].

[Kat2] — M. A. Tsfasman and S. G. Vladuts, *Modular curves and codes with a polynomial construction*, PGIT **30** (1984), 353-355 [3].

[Kat3] N. M. Katz, *An overview of Deligne's proof of the Riemann hypothesis for varieties over finite fields*, PSPM **28** (1976), 275-305 [2].

[Kau1] G. Kaur, *Extreme quadratic forms for coverings in four variables*, Proc. Nat. Inst. Sci. India **A32** (1966), 414-417 [2].

[Kea1] P. Keast and J. C. Diaz, *Fully symmetric integration formulas for the surface of the sphere in s dimensions*, SIAM J. Num. Anal. **20** (1983), 406-419 [1].

K. Keating: see B. K. Teo.

[Kel1] O. H. Keller, *Geometrie der Zahlen*, Springer-Verlag, 1954 [1, 2].

[Ken1] D. G. Kendall and R. A. Rankin, *On the number of points of a given lattice in a random hypersphere*, Quart. J. Math. Oxford **4** (1953), 178-189.

[Ken2] P. Kent, *An efficient new way to represent multi-dimensional data*, Computer J. **28** (1985), 184-190.

D. L. Kepert: see B. W. Clarke.

[Ker1] A. M. Kerdock, *A class of low-rate nonlinear codes*, IC **20** (1972), 182-187 [3].

[Ker2] B. W. Kernigham and Shen Lin, *Heuristic solution of a signal design optimization problem*, BSTJ **52** (1973), 1145-1159 [3].

[Ker3] R. Kershner, *The number of circles covering a set*, AJM **61** (1939), 665-671 [2].

[Kib1] R. E. Kibler, *Some new constant weight codes*, PGIT **26** (1980), 364-365 [9].

[Kil1] K. E. Kilby and N. J. A. Sloane, *On the covering radius problem for codes* I: *bounds on normalized covering radius*, SIAD **8** (1987), 604-618 [3].

[Kil2] — — *On the covering radius problem for codes* II: *codes of low dimension; normal and abnormal codes*, SIAD **8** (1987), 619-627 [3].

[Kin1] K. Kinoshita and M. Kobayashi, *Two-dimensionally arrayed optical-fiber splicing with a* CO_2 *laser*, Appl. Optics **21** (1982), 3419-3422 [1].

A. A. Kirilyuk: see P. M. Gudivok.

Y. Kitaoka: see also J. S. Hsia.

[Kit1] — *On the relation between the positive definite quadratic forms with the same representation numbers*, Proc. Jap. Acad. **47** (1971), 439-441 [2].

[Kit2] — *Positive definite quadratic forms with the same representation numbers*, Archiv Math. **28** (1977), 495-497 [2].

[Kit3] — *Representation of positive definite quadratic forms*, in [Rie1], 1983, pp. 203-208 [2, 7].

[Kit3a] C. J. Kitrick, *Geodesic domes*, Structural Topology **11** (1985), 15-20.

[Kit4] C. Kittel, *Introduction to Solid State Physics*, Wiley, NY, 5th ed. 1976 [1, 2, 4].

[Kle1] P. B. Kleidman and R. A. Wilson, *The maximal subgroups of* J_4, PLMS **41** (1988), 956-966 [10].

[Klϕ1] T. Klϕve, *A lower bound for* $A(n, 4, w)$, PGIT **27** (1981), 257-258 [9].

[Kna1] W. Knapp and P. Schmid, *Codes with prescribed permutation group*, J. Alg. **67** (1980), 415-435 [3].

[Kne1] M. Knebusch and W. Scharlau, *Algebraic Theory of Quadratic forms*, Birkhäuser, Boston, 1980 [15].

M. Kneser: see also J. S. Hsia.

[Kne2] — *Two remarks on extreme forms*, CJM **7** (1955), 145-149 [2, 15].

[Kne3] — *Klassenzahlen indefinite quadratischer Formen in drei order mehr Veränderlichen*, Archiv Math. **7** (1956), 323-332 [15].

[Kne4] — *Klassenzahlen definiter quadratischer Formen*, Archiv Math. **8** (1957), 241-250 [2, 7, 14, 15, 17, 18, 19].

[Kne5] — *Lineare Relationen zwischen Darstellungsanzahlen quadratischer Formen*, Math. Ann. **168** (1967), 31-39 [2, 4, 15].

[Kne6] — *Uber die Ausnahme-Isomorphismem zwischen endlichen klassischen Gruppen*, ASUH **31** (1967), 136-140 [15].

[Kne7] — *Witt's Satz über quadratische Formen und die Erzeugung orthogonaler Gruppen durch Spiegelungen*, Math.-Phys. Semesterber. **17** (1970), 33-45 [15].

[Kne8] — *Representations of integral quadratic forms*, in [Rie1], 1983, pp. 159-172 [15].

[Kne9] — *Erzeugung ganzzahliger orthogonaler Gruppen durch Spiegelungen*, Math. Ann. **255** (1981), 453-462 [15].

O. Knop: see T. W. Melnyk.

[Kno1] M. I. Knopp, *Modular Functions in Analytic Number Theory*, Markham, Chicago, 1970 [2, 7].

K. M. Knowles: see J. H. Conway.

[Knu1] D. E. Knuth, *The Art of Computer Programming*, Vol. 3, *Sorting and Searching*, Addison-Wesley, Reading MA, 1973 [20].

[Ko1] C. Ko, *Determination of the class number of positive quadratic forms in nine variables with determinant unity*, JLMS **13** (1938), 102-110 [2].

[Ko2] — *On the positive definite quadratic forms with determinant unity*, Acta Arith. **3** (1939), 79-85 [16].

M. Kobayashi: see K. Kinoshita.

[Kob1] N. Koblitz, *p-Adic Numbers, p-Adic Analysis, and Zeta Functions*, Springer-Verlag, 1977 [15].

[Koc1] E. Koch, *Wirkungsbereichspolyeder und Wirkungsbereichsteilungen zu kubischen Gitterkomplexen mit weniger also drei Freiheitsgraden*, Z. Krist. **138** (1973), 196-215 [2].

[Koc2] — and W. Fischer, *Types of sphere packings for crystallographic point groups, rod groups and layer groups*, Z. Krist. **148** (1978), 107-152 [2].

[Koc3] H. V. Koch, *On self-dual, doubly-even codes of length* 32, JCT **A 51** (1989), 63-76 [7].

S. S. Koh: see F. Hirzebruch.

[Koi1] M. Koike, *Mathieu group* M_{24} *and modular forms*, Nagoya Math. J. **99** (1985), 147-157 [1].

[Kon1] T. Kondo, *The automorphism group of the Leech lattice and elliptic modular functions*, J. Math. Soc. Japan **37** (1985), 337-361 [1].

[Kon2] — and T. Tasaka, *The theta functions of sublattices of the Leech lattice*, Nagoya Math. J. **101** (1986), 151-179 [1].

[Koo1] T. Koornwinder, *The addition formula for Jacobi polynomials and spherical harmonics*, SIAJ **25** (1973), 236-246 [9].

[Koo2] — *Explicit formulas for special functions related to symmetric spaces*, PSPM **26** (1973), 351-354 [9].

[Koo3] — *Positivity proofs for linearization and connection coefficients or orthogonal polynomials satisfying an addition formula*, JLMS **18** (1978), 101-114 [9].

[Koo4] — *Krawtchouk polynomials, a unification of two different group theoretic interpretations*, SIAM J. Math. Anal. **13** (1982), 1011-1023 [9].

[Kor1] A. Korkine and G. Zolotareff, *Sur les formes quadratique positive quaternaires*, Math. Ann. **5** (1872), 581-583 [1, 4].

[Kor2] — — *Sur les formes quadratiques*, Math. Ann. **6** (1873), 366-389 [4].

[Kor3] — — *Sur les formes quadratique positives*, Math. Ann. **11** (1877), 242-292 [1, 4].

L. Kovacs: see Z. Janko.

[Kra1] E. S. Kramer, S. S. Magliveras and D. M. Mesner, *t-Designs from the large Mathieu groups*, Discr. Math. **36** (1981), 171-189 [11].

[Kra2] M. Kramer, *Sphärische Untergruppen in Kompakten Zusammen-hängenden Liegruppen*, Compos. Math. **38** (1979), 129-153 [9].

[Kra3] A. Krazer, *Lehrbuch der Thetafunktionen*, Chelsea, NY, 1970 [4].

[Kre1] M. G. Krein, *Hermitian-positive kernels on homogeneous spaces*, Ukranian Math. J. (No. 4, 1949), 64-98; (No. 1, 1950), 10-59 = Amer. Math. Soc. Transl. (2) **34** (1963), 69-108, 109-164 [9].

[Kri1] H. A. Krieger and C. A. Schaffner, *The global optimization of incoherent phase signals*, SIAJ **21** (1971), 573-589 [3].

J. B. Kruskal: see L. A. Shepp.

A. L. Krylov: see V. I. Arnold.

H. W. Kuhn: see R. J. Duffin.

L

[Lac1] G. Lachaud, *Les codes geómetriques de Goppa*, Sém. Bourbaki, No. 641, 1985 [3].

[Lag1] J. C. Lagarias, *On the computational complexity of determining the solvability or unsolvability of the equation* $X^2 - DY^2 = -1$, TAMS **260** (1980), 485-508 [15].

[Lag2] — *Worst-case complexity bounds for algorithms in the theory of integral quadratic forms*, J. Algor. **1** (1980), 142-186 [15].

[Lag3] — H. W. Lenstra, Jr., and C. P. Schnorr, *Korkine-Zolotarev bases and successive minima of a lattice and its reciprocal lattice*, Combinatorica **10** (1990), 333-348 [2].

[Lag4] — and A. M. Odlyzko, *Solving low-density subset sum problems*, JACM **32** (1985), 229-246 [2].

[Lam1] T. Y. Lam, *The Algebraic Theory of Quadratic Forms*, Benjamin, Reading MA, 1973 [15].

[Lam2] P. J. C. Lamont, *The number of Cayley integers of given norm*, Proc. Edinburgh Math. Soc. **25** (1982), 101-103.

[Lan0] E. Landau, *Vorlesungen über Zahlentheorie*, Chelsea, NY, 1969 [2].

[Lan1] H. J. Landau, *Sampling, data transmission, and the Nyquist rate*, PIEEE **55** (1967), 1701-1706 [3].

[Lan2] — *How does a porcupine separate its quills*? PGIT **17** (1971), 157-161 [2].

[Lan3] — and H. O. Pollak, *Prolate spheroidal wave functions, Fourier analysis and uncertainty* II, BSTJ **40** (1961), 65-84 [3].

[Lan4] — — *Prolate spheroided wave functions, Fourier analysis and uncertainty* III: *the dimension of the space of essentially time- and band-limited signals*, BSTJ **41** (1962), 1295-1336 [3].

[Lan5] E. S. Lander, *Symmetric Designs: An Algebraic Approach*, Camb. Univ. Press, 1983 [3].

G. R. Lang: see G. D. Forney, Jr.

[Lan7] S. Lang, *Algebraic Number Theory*, Addison-Wesley, Reading MA, 1970 [8].

[Lan8] — *Algebra*, Addison-Wesley, Reading MA, revised 1971 [8].

[Lan9] — *Introduction to Modular Forms*, Springer-Verlag, 1976 [2,7].

[Lar1] J. Larmouth, *The enumeration of perfect forms*, in *Computers in Number Theory*, ed. A.O.L. Atkin and B. J. Birch, Ac. Press, NY, 1971, pp. 237-239 [1, 2].

V. B. Lawrence: see A. Gersho.

[Laz1] D. E. Lazic, *Class of block codes for the Gaussian channel*, Elect. Lett. **16** (1980), 185-186 [3].

D. W. Leavitt: see S. S. Magliveras.

A. Lebowitz: see H. E. Rauch.

[Lee0] A. J. Lee, *A note on the Campbell sampling theorem*, SIAJ **41** (1981), 553-557 [3].

[Lee1] D. T. Lee and F. P. Preparata, *Computational geometry - a survey*, IEEE Trans. Comp. **33** (1984), 1072-1101; **34** (1985), 584 [2].

R. C. T. Lee: see R. C. Chang.

S. L. Lee: see G. R. Welti.

T.-A. Lee: see A. R. Calderbank.

[Lee2] J. Leech, *The problem of the thirteen spheres*, Math. Gazette **40** (1956), 22-23 [1].

[Lee3] — *Equilibrium of sets of particles on a sphere*, Math. Gazette **41** (1957), 81-90 [1].

[Lee4] — *Some sphere packings in higher space*, CJM **16** (1964), 657-682 [1, 5, 6].

[Lee5] — *Notes on sphere packings*, CJM **19** (1967), 251-267 [1, 4, 5, 6, 7].

[Lee6] — *Five-dimensional nonlattice sphere packings*, CMB **10** (1967), 387-393 [5, 6].

[Lee7] — *Six and seven dimensional nonlattice sphere packings*, CMB **12** (1969), 151-155 [5, 6].

[Lee8] — and N. J. A. Sloane, *New sphere packings in dimensions* 9-15, BAMS **76** (1970), 1006-1010 [5].

[Lee9] — — *New sphere packings in more than* 32 *dimensions*, Proc. 2nd Chapel Hill Conf. on Comb. Math. Appl., Univ. North Carolina, Chapel Hill, 1970, pp. 345-355 [5].

[Lee10] — — *Sphere packing and error-correcting codes*, CJM **23** (1971), 718-745 ≅ Chap. 5 of this book [1, 5, 6, 13].

[Lee11] — and T. Tarnai, *Arrangements of* 22 *circles on a sphere*, Ann. Univ. Sci. Budapest Ser. Math. **31** (1988), 27-37 [1].

[Leg1] A.-M. Legendre, *Theórie des Nombres*, 4-th edition, Blanchard, Paris, 1955, 2 vols. [15].

[Leh1] D. H. Lehmer, *Guide to Tables in the Theory of Numbers*, Bull. Nat. Res. Council **105**, Nat. Acad. Sci., Washington DC, 1941 [15].

[Leh2] — *Ramanujan's function* $\tau(n)$, DMJ **10** (1943), 483-492 [4].

C. G. Lekkerkerker: see also P. M. Gruber.

[Lek1] — *Geometry of Numbers*, Wolters-Noordhoff, Groningen, and North-Holland, Amsterdam, 1969 (see also [Gru1a]) [1, 8, B].

A. Lempel: see J. Ziv.

[Lem1] W. Lempken, *Die Untergruppenstruktur der endlichen, einfachen Gruppe* J_4, Dissertation, Univ. Mainz, 1985 [10].

[Len1] A. K. Lenstra, H. W. Lenstra, Jr., and L. Lovász, *Factoring polynomials with rational coefficients*, Math. Ann. **261** (1982), 515-534 [2].

H. W. Lenstra, Jr.: see also J. C. Lagarias, A. K. Lenstra, C. Pomerance.

[Len2] — *Codes from algebraic number fields*, in *Fundamental Contributions in the Netherlands since 1945*, North-Holland, Amsterdam, **II** (1986), pp. 95-104 [3].

J. S. Leon: see also N. Ito.

[Leo1] — *An algorithm for computing the automorphism group of a Hadamard matrix*, JCT **A27** (1979), 289-306 [3].

[Leo2] — *On an algorithm for finding a base and a strong generating set for a group given by generating permutations*, MTAC **35** (1980), 941-974.

[Leo3] — *Computing automorphism groups of error-correcting codes*, PGIT **28** (1982), 496-511.

[Leo4] — *Computing automorphism groups of combinatorial objects*, in [Atk2], 1984, pp. 321-335.

[Leo5] — J. M. Masley and V. Pless, *Duadic codes*, PGIT **30** (1984), 709-714 [3].

[Leo6] — — — *On weights in duadic codes*, JCT **A44** (1987), 6-21 [3].

[Leo7] — V. Pless and N. J. A. Sloane, *Self-dual codes over* $GF(5)$, JCT **A32** (1982), 178-194 [4, 7].

[Leo8] — — — *On ternary self-dual codes of length* 24, PGIT **27** (1981), 176-180 [4,7].

[Leo9] — and C. C. Sims, *The existence and uniqueness of a simple group generated by* $\{3, 4\}$*-transpositions*, BAMS **83** (1977), 1039-1040 [10].

[Leo10] D. A. Leonard, *Parameters of association schemes that are both P- and Q-polynomial*, JCT **A36** (1984), 355-363 [9].

[Leo11] — *Discrete orthogonal polynomials, duality, and association schemes*, SIAM J. Math Anal. **13** (1982), 656-663 [9].

J. I. Lepowsky: see also I. B. Frenkel.

[Lep1] — S. Mandelstam and I. M. Singer, editors, *Vertex Operators in Mathematics and Physics*, Springer-Verlag, 1985 [B].

[Lep2] — and A. E. Meurman, *An* E_8*-approach to the Leech lattice and the Conway group*, J. Alg. **77** (1982), 484-504 [4, 8].

[Let1] G. Letac, *Problèmes classiques de probabilité sur un couple de Gelfand*, LNM **861** (1981), 93-120 [9].

[Let2] — *Les fonctions sphériques d'un couple de Gelfand symetrique et les chaines de Markov*, Adv. Appl. Prob. **14** (1982), 272-294 [9].

V. I. Levenshtein: see also G. A. Kabatiansky.

[Lev1] — *The application of Hadamard matrices to a problem in coding*, Problemy Kibernetiki **5** (1961), 123-136 = Problems of Cybernetics **5** (1964), 166-184 [3, 9].

[Lev2] — *On upper bounds for codes with codewords of a fixed weight*, PPI **7** (1971, No. 4), 3-12 = PIT **7** (1971), 281-287 [9].

[Lev3] — *On the minimal redundancy of binary error-correcting codes*, PPI **10** (1974, No. 2), 26-42 = IC **28** (1975), 268-291 [9].

[Lev4] — *Maximal packing density of equal spheres in n-dimensional Euclidean space*, Mat. Zametki **18** (1975, No. 2), 301-311 = Math. Notes Acad. Sci. USSR **18** (1975), 765-771 [1, 9].

[Lev5] — *Methods for obtaining bounds in metric problems of coding theory*, in *Proc. 1975 IEEE-USSR Joint Workshop on Information Theory*, IEEE, NY, 1976, pp. 126-144 [1, 9].

[Lev6] — *On choosing polynomials to obtain bounds in packing problems*, in *Proc. Seventh All-Union Conf. Coding Theory and Information Transmission*, Moscow, II, 1978, pp. 103-108 [9].

[Lev7] — *On bounds for packings in n-dimensional Euclidean space*, DAN **245**

(1979), 1299-1303 = SMD **20** (1979), 417-421 [1, 9].

[Lev8] — *Bounds on the maximal cardinality of a code with bounded modulus of the inner product*, DAN **263** (1982) 1303-1308 = SMD **25** (1982), 526-531 [9].

[Lev9] — *Bounds for packings in metric spaces and certain applications*, Problemy Kibernetiki, **40** (1983), 44-110 [1, 9].

[Lev10] D. Levine and P. J. Steinhardt, *Quasicrystals: a new class of ordered structures*, Phys. Rev. Lett. **53** (1984), 2477-2480 [1].

[Lev11] — — *Quasicrystals* I: *definition and structure*, Phys. Rev. **B34** (1986), 596-616 [1].

[LeV1] W. J. LeVeque, *Topics in Number Theory*, Addison-Wesley, Reading MA, 2 vols., 1956 [8, 15].

[LeV2] — editor, *Reviews in Number Theory*, Amer. Math. Soc., Providence RI, 6 vols., 1974 [B].

R. Levingston: see L. M. Gordon.

[Lew1] J. Lewittes, *Construction of modular forms*, Israel J. Math. **24** (1976), 145-183 [2, 7].

[Li1] D. Li, *Indefinite binary forms representing the same numbers*, PCPS **92** (1982), 29-33 [2].

R. A. Liebler: see W. M. Kantor.

Shen Lin: see B. W. Kernighan.

[Lin0] Shu Lin and D. J. Costello, Jr., *Error Control Coding*, Prentice-Hall, Englewood Cliffs NJ, 1983 [3].

[Lin1] Y. C. Lin and D. E. Williams, *Calculated geometry for ten-coordination*, Canad. J. Chem. **51** (1973), 312-316 [1].

[Lin2] Y. Linde, A. Buzo and R. M. Gray, *An algorithm for vector quantizer design*, COMM **28** (1980), 84-95 [2].

[Lin3] C. C. Lindner and A. Rosa, editors, *Topics on Steiner Systems*, North-Holland, Amsterdam, 1980 [3].

[Lin4] J. H. Lindsey II, *On a projective representation of the Hall-Janko group*, BAMS **74** (1968), 1094 [10].

[Lin5] — *Linear groups of degree* 6 *and the Hall-Janko group*, in [Bra2], 1969, pp. 97-100 [10].

[Lin6] — *A correlation between* $PSU_4(3)$, *the Suzuki group, and the Conway group*, TAMS **157** (1971), 189-204 [4].

[Lin7] — *Finite linear groups of degree* 6, CJM **5** (1971), 771-790 [4].

[Lin8] — *On the Suzuki and Conway groups*, PSPM **21** (1971), 107-109 [4].

[Lin9] — *Sphere-packing in* R^3, Math. **33** (1986), 137-147 [1].

[Lin9a] — *Sphere-packing* II, preprint [1].

[Lin9b] L. Lines, *Solid Geometry with Chapters on Space-Lattices, Sphere-Packs and Crystals*, Macmillan, London, 1935; Dover, NY 1965.

J. H. van Lint: see also P. J. Cameron, N. J. A. Sloane.

[Lin10] — *Combinational Theory Seminar, Eindhoven University of Technology*, LNM **382**, 1974 [3].

[Lin11] — *Introduction to Coding Theory*, Springer-Verlag, 1982 [3].

[Lin12] — and A. Schrijver, *Construction of strongly regular graphs, two-weight codes and partial geometries by finite fields*, Combinatorica **1** (1981), 63-73 [9].

[Lin13] — and R. M. Wilson, *On the minimum distance of cyclic codes*, PGIT **32** (1986), 23-40 [3].

R. J. Lipton: see D. Dobkin.

S. N. Litsyn: see also V. A. Zinoviev.

[Lit1] — *New non-linear codes with a minimum distance of* 3, Problems of Control and Inform. Theory **13** (1984), 13-15 [3, 5].

[Lit2] — and O. I. Shekhovtsov, *Fast decoding algorithms for first-order Reed-Muller codes*, PPI **19** (No. 2, 1983), 3-7 = PIT **19** (1983), 87-91 [20].

[Lit3] — and M. A. Tsfasman, *Algebraic-geometric and number-theoretic packings of spheres*, Uspekhi Mat. Nauk **40** (1985), 185-186 [1, 8].

[Lit4] — — *A note on lower bounds*, PGIT **32** (1986), 705-706 [3].

[Lit5] — — *Constructive high-dimensional sphere packings*, DMJ *54* (1987), 147-161 [1].

[Lju1] W. Ljunggren, *New solution of a problem proposed by E. Lucas*, Norsk Mat. Tid. **34** (1952), 65-72 [26].

[Llo1] S. P. Lloyd, *Hamming association schemes and codes on spheres*, SIAM J. Math. Anal. **11** (1980), 488-505 [9].

A. C. Lobstein: see G. D. Cohen.

[Loe1] A. L. Loeb, *Space Structures: Their Harmony and Counterpoint*, Addison-Wesley, Reading MA, 1976 [2, 21].

G. Longo: see J.-M. Goethals.

F. M. Longstaff: see G. D. Forney, Jr.

M. S. Longuet-Higgins: see H. S. M. Coxeter.

J. Q. Longyear: see also N. Ito.

[Lon1] — *Criteria for a Hadamard matrix to be skew-equivalent*, CJM **28** (1976), 1216-1223 [3, 8].

[Lon2] — *There are nine different skew-symmetric Hadamard matrices of order* 24, Ars Combin. **10** (1980), 115-122 [3].

[Loo1] L. H. Loomis, *An Introduction to Abstract Harmonic Analysis*, Van Nostrand, Princeton NJ, 1953 [4].

L. Lovász: see also A. K. Lenstra.

[Lov1] — *On the Shannon capacity of a graph*, PGIT **25** (1979), 1-7 [9].

W. E. Love: see A. L. Patterson.

[Lub1] A. Lubotzky, R. Phillips and P. Sarnak, *Hecke operators and distributing points on the sphere* I, Comm. Pure Appl. Math. **39** (1986), S149-S186 [1].

[Lub2] — — — II: Comm. Pure Appl. Math. **40** (1987), 401-420 [1].

[Lün1] H. Lüneburg, *Transitive Erweiterungen endlicher Permutationsgruppen* LNM **84**, 1969 [12, 14].

[Lyo1] R. Lyons, *Evidence for a new finite simple group*, J. Alg. **20** (1972), 540-569 [10].

M

[Mac0] A. L. Mackay, *The packing of three-dimensional spheres on the surface of a four-dimensional hypersphere*, J. Phys. **A13** (1980), 3373-3379 [1, 9].

[Mac1] G. W. Mackey, *Unitary Group Representations in Physics, Probability and Number Theory*, Benjamin, Reading MA 1978 [9].

S. MacLane: see G. Birkhoff.

F. J. MacWilliams: see also E. R. Berlekamp, M. R. Best.

[Mac2] — *A theorem on the distribution of weights in a systematic code*, BSTJ **42** (1963), 79-84 [3, 9].

[Mac3] — C. L. Mallows and N. J. A. Sloane, *Generalizations of Gleason's theorem on weight enumerators of self-dual codes*, PGIT **18** (1972), 794-805 [7].

[Mac4] — A. M. Odlyzko, N. J. A. Sloane and H. N. Ward, *Self-dual codes over GF*(4), JCT **A25** (1978), 288-318 [7, 16, 18].

[Mac5] — and N. J. A. Sloane, *Pseudo-random sequences and arrays*, PIEEE **64** (1976), 1715-1729.

[Mac6] — — *The Theory of Error-Correcting Codes*, North Holland, Amsterdam, 2nd printing, 1978; Russian translation, Svyaz, Moscow, 1979 [2-5, 7-9, 11, 12, 14, 16, 20, B].

[Mac7] — — and J.-M. Goethals, *The MacWilliams identities for nonlinear codes*, BSTJ **51** (1972), 803-819 [3].

[Mac8] — — and J. G. Thompson, *Good self-dual codes exist*, Discr. Math. **3** (1972), 153-162 [7, 16].

S. S. Magliveras: see also E. S. Kramer.

[Mag1] — and D. W. Leavitt, *Simple six-designs exist*, in *Proc. 14th Southeast. Conf., Boca Raton*, Congr. Num. **40** (1983), 195-205 [3].

[Mag2] — — *Simple* 6-(33, 8, 36) *designs from* $P\Gamma L_2$ (32), in [Atk2], 1984, pp. 337-352 [3].

[Mag3] W. Magnus, *Uber die Anzahl der in einem Geschlecht enthaltenen Klassen von positiv definiten quadratischen Formen*, Math. Ann. **114** (1937), 465-475; **115** (1938), 643-644 [16].

[Mah1] D. P. Maher, *Self-Orthogonal Codes and Modular Forms*, Ph.D. Dissertation, Lehigh Univ., Bethlehem PA, 1976 [7].

[Mah2] — *Lee polynomials of codes and theta functions of lattices*, CJM **30** (1978), 738-747 [7].

[Mah3] — *Modular forms from codes*, CJM **32** (1980), 40-58 [7].

K. Mahler: see also B. Segre.

[Mah4] — *p-Adic Numbers and their Functions*, Camb. Univ. Press, 2nd ed. 1981 [15].

C. L. Mallows: see also F. J. MacWilliams.

[Mal1] — A. M. Odlyzko and N. J. A. Sloane, *Upper bounds for modular forms, lattices, and codes*, J. Alg. 36 (1975), 68-76 [7, 19].

[Mal2] — V. Pless and N. J. A. Sloane, *Self-dual codes over GF*(3), SIAJ **31** (1976), 649-666 [3, 5, 7, 16].

[Mal3] — and N. J. A. Sloane, *An upper bound for self-dual codes*, IC **22** (1973), 188-200 [7].

[Mal4] — — *Weight enumerators of self-orthogonal codes*, Discr. Math. **9** (1974), 391-400 [7].

[Mal5] — — *Weight enumerators of self-orthogonal codes over GF*(3), SIAD **2** (1981), 452-480 [7].

[Mal6] A. V. Malysev, *On the representation of integers by positive quadratic forms with four or more variables* I, Izv. **23** (1959), 337-364 = Amer. Math. Soc. Transl. (2) **46** (1965), 17-47 [4].

S. Madelstam: see J. I. Lepowsky.

Y. I. Manin: see also S. G. Vladuts.

[Man1] — *What is the maximum number of points on a curve over* $\mathbf{F}_2$?, J. Fac. Sci. Univ. Tokyo, Sec. 1A, **28** (1982), 715-720 [3].

H. B. Mann: see also E. C. Posner.

[Man2] — *On the number of information symbols in Bose-Chaudhuri codes*, IC **5** (1962), 153-162 [5].

[Mar1] R. P. Martineau, *A characterization of Janko's simple group of order* 175,560, PLMS **19** (1969), 709-729 [10].

[Mar2] J. Martinet, *Tours de corps de classes et estimations de discriminants*, Invent. Math. **44** (1978), 65-73 [1, 8].

[Mar3] — *Petits discriminants*, Ann. Inst. Fourier **29** (1979), 159-170 [8].

J. M. Masley: see J. S. Leon.

[Mas1] D. R. Mason, *On the construction of the Steiner system* $S(5, 8, 24)$, J. Alg. **47** (1977), 77-79 [11].

[Mas2] G. Mason, *Modular forms and the theory of Thompson series*, in *Proc. Rutgers Group Theory Year 1983-1984*, ed. M. Aschbacher et al., Camb. Univ. Press, 1984, pp. 391-407 [1].

[Mas3] — M_{24} *and certain automorphic forms*, in [McK3], 1985, 223-244 [1].

[Mat1] E. Mathieu, *Mémoire sur l'étude des fonctions de plusieurs quantités*, J. Math. Pures Appl. **6** (1861), 241-243 [11].

[Mat2] — *Sur la fonction cinq fois transitive de* 24 *quantités*, J. Math. Pures Appl. **18** (1873), 25-46 [11].

[Mat3] Mathlab Group, *MACSYMA Reference Manual*, Version 10, Laboratory for Computer Science, MIT, Cambridge MA, 1983 [16, 19, 21].

H. F. Mattson, Jr.: see E. F. Assmus, Jr., G. D. Cohen.

[Mau1] A. Maus, *Delaunay triangulation and the convex hull of n points in expected linear time*, BIT **24** (1984), 151-163 [2].

V. C. Mavron: see D. H. Smith.

[Max1] J. Max, *Quantizing for minimum distortion*, PGIT **6** (1960), 7-12 [2].

C. A. May: see B. S. Atal, S. G. Wilson.

J. E. Mazo: see also A. R. Calderbank.

[Maz1] — *Some theoretical observations on spread-spectrum communications*, BSTJ **58** (1979), 2013-2023 [1].

T. P. McDonough: see D. H. Smith.

R. J. McEliece: see also E. R. Berlekamp.

[McE1] — *The Theory of Information and Coding*, Addison-Wesley, Reading MA, 1977 [3, 9].

[McE2] — *The bounds of Delsarte and Lovász, and their application to coding theory*, in *Algebraic Coding Theory and Applications*, ed. G. Longo, CISM Courses and Lectures No. 258, Springer-Verlag, 1979, pp. 107-178 [9].

[McE3] — E. R. Rodemich and H. C. Rumsey, Jr., *The Lovász bound and some generalizations*, J. Combinatorics, Inform. Syst. Sci. **3** (1978), 134-152 [9].

[McE4] — — — and L. R. Welch, *New upper bounds on the rate of a code via the Delsarte-MacWilliams inequalities*, PGIT **23** (1977), 157-166 [7, 9].

J. L. McGregor: see S. Karlin.

J. A. McInnes: see M. R. Hoare.

[McK1] B. D. McKay, *Transitive graphs with fewer than twenty vertices*, MTAC **33** (1979), 1101-1121 [9].

J. McKay: see also G. Higman.

[McK2] — *A setting for the Leech lattice*, in [Gag2], 1973, pp. 117-118 [7, 8].

[McK3] — editor, *Finite Groups - Coming of Age*, Contemp. Math. **45** (1985), 1-350 [B].

H. P. McKean: see H. Dym.

[McL0] A. D. McLaren, *Optimal numerical integration on a sphere*, MTAC **17** (1963), 361-383 [1].

[McL1] J. McLaughlin, *A simple group of order* 898,128,000, in [Bra2], 1969, pp. 109-111 [10].

[McL2] A. M. McLoughlin, *The complexity of computing the covering radius of a code*, PGIT **30** (1984), 800-804 [2].

[McM1] P. McMullen, *Convex bodies which tile space by translation*, Math. **27** (1980), 113-121; **28** (1981), 191 [21].

[McM2] — and G. C. Shephard, *Convex Polytopes and the Upper Bound Conjecture*,

Camb. Univ. Press, 1971 [21].

[Mel1] T. W. Melnyk, O. Knop and W. R. Smith, *Extremal arrangements of points and unit changes on a sphere: equilibrium configurations revisited*, Can. J. Chem. **55** (1977), 1745-1761 [1].

D. M. Mesner: see E. S. Kramer.

A. E. Meurman: see I. B. Frenkel, J. I. Lepowsky.

[Mey1] B. Meyer, *On the symmetries of spherical harmonics*, CJM **6** (1954), 135-157 [1].

[Mey2] J. Meyer, *Präsentation der Einheitengruppe der quadratischen Form* $F(X) = -X_0^2 + X_1^2 + \cdots + X_{18}^2$, Archiv Math. **29** (1977), 261-266 [27, 28].

[Mey3] W. Meyer, W. Neutsch and R. A. Parker, *The minimal 5-representation of Lyons' sporadic group*, Math. Ann. **272** (1985), 29-39 [10].

L. Michel: see D. Gratias.

D. Middleton: see D. P. Petersen.

A. D. Mighell: see A. Santoro.

[Mil1] E. Y. Miller and S. Washburn, *A sublattice of the Leech lattice*, JCT **A42** (1986), 9-14 [4].

[Mil2] G. A. Miller, H. F. Blichfeldt and L. E. Dickson, *Theory and Applications of Finite Groups*, Wiley, NY, 1916; Dover, NY, 1961 [7].

J. C. P. Miller: see H. S. M. Coxeter.

[Mil3] W. Miller, Jr., *Symmetry Groups and Their Applications*, Ac. Press, NY, 1972 [9].

[Mil4] W. H. Mills, *A new 5-design*, Ars Combin. **6** (1978), 193-195 [3].

[Mil5] J. Milnor, *Hilbert's Problem 18: on crystallographic groups, fundamental domains, and on sphere packing*, PSPM **28** (1976), 491-506 [1].

[Mil6] — *Hyperbolic geometry: the first 150 years*, BAMS **6** (1982), 9-24 = PSPM **39** (1983, Part I), 25-40 [26].

[Mil7] — and D. Husemoller, *Symmetric Bilinear Forms*, Springer-Verlag, 1973 [1, 7, 15, 16, 26].

[Mim1] Y. Mimura, *On 2-lattices over real quadratic integers*, Math. Sem. Notes (Kobe Univ.) **7** (1959), 327-342.

D. M. P. Mingos: see C. E. Briant.

[Min0] H. Minkowski, *Grundlagen für eine Theorie der quadratischen Formen mit ganzzahligen Koeffizienten*, Mém. prés. par divers savants à l'Académie des Sci. Inst. nat. de France, **29** (No. 2, 1884) = Ges. Abh. **I** (1911), 3-144 [4, 15].

[Min1] — *Sur la reduction des formes quadratiques positives quaternaires*, CR **96** (1883), 1205 = Ges. Abh. **I** (1911), 195 [2, 15].

[Min2] — *Uber positive quadratische Formen*, JRAM **99** (1886), 1-9 = Ges. Abh. **I** (1911), 149-156 [15].

[Min3] — *Zur Theorie der positiven quadratischen Formen*, JRAM **101** (1887), 196-202 = Ges. Abh. **I** (1911), 212-218 [15].

[Min4] — *Dichteste gitterförmige Lagerung kongruenter Körper*, Nachr. Ges. Wiss. Göttingen (1904), 311-355 = Ges. Abh. II (1911), 3-42 [1].

[Min5] — *Geometrie der Zahlen*, Teubner, Leipzig, 1910 [1].

[Min6] — *Diophantische Approximationen*, Teubner, Leipzig, 1927 [1].

[Mit1] H. H. Mitchell, *Determination of the ordinary and modular ternary linear groups*, TAMS **12** (1911), 207-242 [4].

[Mit2] — *Determination of the finite quaternary linear groups*, TAMS **14** (1913), 123-142 [4].

[Mit3] — *Determination of all primitive collineation groups in more than four variables which contain homologies*, AJM **36** (1914), 1-12 [4].

[Mit4] — *The subgroups of the quaternary abelian linear groups*, TAMS **15** (1914), 379-396 [4].

[Mit5] W. C. Mitchell, *The number of lattice points in a k-dimensional hypersphere*, MTAC **20** (1966), 300-310 [4].

[Mit6] D. S. Mitrinović, *Analytic Inequalities*, Springer-Verlag, 1970 [3].

H. Miyakawa: see T. Kaneko.

[Mod1] *Modular Functions of One Variable*, various editors, LNM **320** (1973), **349** (1973), **350** (1973), **476** (1975), **601** (1977), **627** (1977) [7].

[Moh1] R. Mohr and R. Bajcsy, *Packing volumes by spheres*, IEEE Trans. Patt. Anal. Mach. Int. **5** (1983), 111-116 [1].

J. Molk: see J. Tannery.

[Moo1] R. V. Moody, *Euclidean Lie algebras*, CJM **21** (1969), 1432-1454 [30].

[Moo2] — *Root systems of hyperbolic type*, Adv. in Math. **33** (1979), 144-160 [30].

[Moo3] E. H. Moore, *Using the group of a code to compute its minimum weight*, preprint [7].

[Mor1] L. J. Mordell, *On Mr. Ramanujan's empirical expansions of modular functions*, PCPS **19** (1919), 117-124 [7].

[Mor2] — *On the representations of numbers as a sum of $2r$ squares*, Quart. J. Math. **48** (1920), 93-104 [4].

[Mor3] — *The definite quadratic forms in eight variables with determinant unity*, J. Math. Pures Appl. **17** (1938), 41-46 [2].

[Mor4] — *Observation on the minimum of a positive definite quadratic form in eight variables*, JLMS **19** (1944), 3-6 [1, 6].

[Mor5] — *The minimum of a definite ternary quadratic form*, JLMS **23** (1948), 175-178 [1].

[Mor6] — *Diophantine Equations*, Ac. Press, NY, 1969 [2, 15, 26].

[Mor7] C. J. Moreno, *Algebraic Curves over Finite Fields with Applications to Coding Theory*, to appear [3].

[Mor8] M. A. Morrison and J. Brillhart, *A method of factoring and the factorization of F_7*, MTAC **29** (1975), 183-205 [15].

W. O. J. Moser: see H. S. M. Coxeter.

[Mud1] D. J. Muder, *Putting the best face on a Voronoi polyhedron*, PLMS **56** (1988), 329-348 [1].

[Mue1] E. L. Muetterties and C. M. Wright, *Molecular polyhedra of high coordination number*, Quart. Rev. Chem. Soc. **21** (1967), 109-194 [1].

[Mül1] C. Müller, *Spherical Harmonics*, LNM **17**, 1966 [9].

R. C. Mullin: see I. F. Blake.

[Mum1] D. Mumford, *Tata Lectures on Theta*, Birkhäuser, Boston, 2 vols., 1983, 1984 [4].

N

T. Namekawa: see Y. Sugiyama.

[Nar1] W. Narkiewicz, *Elementary and Analytical Theory of Algebraic Numbers*, Państ. Wyd. Nauk., Warsaw, 1973 [8].

[Nel1] D. R. Nelson and B. I. Halperin, *Pentagonal and icosahedral order in rapidly cooled metals*, Science **229** (1985), 233-238 [1].

R. W. Nettleton: see G. R. Cooper.

J. Neubüser: see H. Brown, R. Bülow.

[Neu1] A. Neumaier, *Distances, graphs and designs*, Europ. J. Combin. **1** (1980), 163-174 [9].

[Neu2] — *Combinational configurations in terms of distances*, Memorandum 81-07, Math. Dept., Eindhoven Univ. Technology, 1981 [3].

[Neu3] — *Lattices of simplex type*, SIAD **4** (1983), 145-160 [26].

[Neu4] — *On norm three vectors in integral Euclidean lattices* I, Math. Z. **183** (1983), 565-574.

[Neu5] — and J. J. Seidel, *Discrete hyperbolic geometry*, Combinatorica **3** (1983), 219-237 [26].

B. Neumann: see Z. Janko.

W. D. Neumann: see F. Hirzebruch.

W. Neutsch: see W. Meyer.

D. J. Newman: see also L. Flatto.

[New1] — *The hexagon theorem*, PGIT **28** (1982), 137-139 [2].

[New2] M. Newman, *A table of the coefficients of the power of* η (τ), KNAW **A59** (1956), 204-216 [4].

[New3] — *Integral Matrices*, Ac. Press, NY, 1972.

[Nie1] H. Niederreiter, *Quasi-Monte Carlo methods and pseudo-random numbers*, BAMS **84** (1978), 957-1041 [1].

[Nie2] H.-V. Niemeier, *Definite quadratische Formen der Dimension* 24 *und Diskriminante* 1, JNT **5** (1973), 142-178 [2, 4, 7, 12, 14-18, 23].

J. Nievergelt: see E. M. Reingold.

[Nik1] V. V. Nikulin, *On quotient groups of the automorphism groups of hyperbolic forms modulo subgroups generated by 2-reflections*, DAN **248** (1979), 1307-1309 = SMD **20** (1979), 1156-1158 [15].

[Nik2] — *Integral symmetric bilinear forms and some of their applications*, IANS **43** (1979) 111-177 = Izv. **14** (1980), 103-167 [2, 15].

[Nik3] — *On arithmetic groups generated by reflections in Lobachevsky spaces*, IANS **44** (1980), 637-669 = Izv. **16** (1981), 573-601 [15].

[Nik4] — *On classification of arithmetic groups generated by reflections in Lobachevsky spaces*, IANS **45** (1981) 113-142 = Izv. **18** (1982), 99-123 [15].

[Nik5] — *Quotient groups of groups of automorphisms of hyperbolic forms with respect to subgroups generated by 2-reflections. Algebraic-geometric applications*, Itogi Nauk. Tekhniki, Ser. Sov. Prob. Mat. **18** (1981), 3-114 = JSM **22** (1983), 1401-1475 [15].

[Nik6] — *Involutions of integral quadratic forms and their applications to real algebraic geometry*, IANS **47** (1983), 109-158 = Izv. **22** (1984), 99-172 [15].

[Nor1] A. W. Nordstrom and J. P. Robinson, *An optimum nonlinear code*, IC **11** (1967), 613-616 [3].

S. P. Norton: see also J. H. Conway.

[Nor2] — *Construction of Harada's groups*, Ph.D. Dissertation, Univ. of Cambridge, 1976 [10].

[Nor3] — *The construction of* J_4, PSPM **37** (1980), 271-277 [10,11].

[Nor4] — *A bound for the covering radius of the Leech lattice*, PRS **A380** (1982), 259-260 ≡ Chap. 22 of this book [2, 22].

[Nor5] — *More on moonshine*, in [Atk2], 1984, pp. 185-193 [1, 29].

[Nor6] — *The uniqueness of the Fischer-Griess monster*, in [McK3], 1985, pp. 271-185 [29].

[Nor7] — personal communication [29].

[Nor8] — and R. A. Wilson, *Maximal subgroups of the Harada-Norton group*, J. Alg. **103** (1986), 362-376 [10].

[Nov1] N. V. Novikova, *Korkine-Zolotarev reduction domains for positive quadratic forms in $n \leq 8$ variables and reduction algorithms for these domains,* DAN **270** (1983), 48-51 = SMD **27** (1983), 557-560 [2, 15].

[Now1] W. Nowacki, *Bibliography of Mathematical Crystallography,* Inter. Union Cryst., 1978.

[Nun1] R. de M. Nunes Mendes, *Symmetries of spherical harmonics,* TAMS **204** (1975), 161-178 [1].

O

[O'Co1] R. E. O'Connor and G. Pall, *The construction of integral quadratic forms of determinant* 1, DMJ **11** (1944), 319-331 [4, 6].

A. M. Odlyzko: see also M. R. Best, J. H. Conway, J. C. Lagarias, F. J. MacWilliams, C. L. Mallows.

[Odl1] — *Some analytic estimates of class numbers and discriminants,* Invent. Math. **29** (1975), 275-286 [8].

[Odl2] — *Lower bounds for discriminants of number fields,* Acta Arith. **29** (1976), 275-297 [8].

[Odl3] — *Lower bounds for discriminants of number fields* II, Tôhoku Math. J. **29** (1977), 209-216 [8].

[Odl4] — *Cryptanalytic attacks on the multiplicative knapsack cryptosystem and on Shamir's fast signature scheme,* PGIT **30** (1984), 594-601 [2].

[Odl5] — and N. J. A. Sloane, *New bounds on the number of unit spheres that can touch a unit sphere in n dimensions,* JCT **A26** (1979), 210-214 ≅ Chap. 13 of this book [1, 9, 13].

[Odl6] — — *A theta-function identity for nonlattice packings,* Studia Sci. Math. Hung. **15** (1980), 461-465 [2, 7].

[Ogg1] A. Ogg, *Modular Forms and Dirichlet Series,* Benjamin, NY 1969 [2, 7, 8, 18].

[O'Ke1] M. O'Keeffe and B. G. Hyde, *Plane nets in crystal chemistry,* Phil. Trans. Roy. Soc. London **A295** (1980), 553-623 [1].

D. Olive: see P. Goddard.

[Oli1] B. M. Oliver, J. R. Pierce and C. E. Shannon, *The philosophy of PCM,* PIEEE **36** (1948), 1324-1331 [3].

[Oli2] R. K. Oliver, *On Bieberbach's analysis of discrete Euclidean groups,* PAMS **80** (1980), 15-21 [4].

[Olv1] F. W. J. Olver, *Bessel Functions* III: *Zeros and Associated Values,* Royal Soc. Math. Tables **7**, Camb. Univ. Press, 1960 [1].

[O'Me1] O. T. O'Meara, *Introduction to Quadratic Forms,* Springer-Verlag, 1971 [2, 15, 16].

[O'Me2] — *The construction of indecomposable positive definite quadratic forms,* JRAM **276** (1975), 99-123 [15].

[O'Me3] — *Hilbert's eleventh problem: the arithmetic theory of quadratic forms,* PSPM **28** (1976), 379-400 [15].

[O'Me4] — and B. Pollak, *Generation of local integral orthogonal groups,* Math. Zeit. **87** (1965), 385-400; **93** (1966), 171-188 [15].

J. K. O'Mura: see A. J. Viterbi.

[O'Na1] M. O'Nan, *Some evidence for the existence of a new simple group,* PLMS **32** (1976), 421-479 [10].

[Ong1] H. E. Ong, *Perfect quadratic forms over real-quadratic number fields,* Geom. Dedic. **20** (1986), 51-77 [2].

[Ong2] — and D. Golke, *An algorithm for the computation of perfect polyhedral cones over real quadratic number fields*, Lect. Notes Comp. Sci. **204** (1985), 487-488 [2].

[Opp1] A. Oppenheim, *Remark on the minimum of quadratic forms*, JLMS **21** (1946), 251-252 [1, 6].

[Orz1] G. Orzech, editor, *Conference on Quadratic Forms—1976*, Queen's Papers Pure Appl. Math. **46**, Queen's Univ., Kingston, Ontario, 1977 [15, B].

[Otr1] G. Otremba, *Zur Theorie der hermiteschen Formen in imaginär-quadratischen Zahlkörpern*, JRAM **249** (1971), 1-19 [2].

[Oze1] M. Ozeki, *On even unimodular positive definite quadratic lattices of rank* 32, Math. Z. **191** (1986), 283-291 [7].

[Oze2] — *On the configurations of even unimodular lattices of rank* 48, Archiv Math. **46** (1986), 54-61 [7].

[Oze3] — *Hadamard matrices and doubly even self-dual error-correcting codes*, JCT **A 44** (1987), 274-287 [7].

[Oze4] — *Examples of even unimodular extremal lattices of rank* 40 *and their Siegel theta-series of degree* 2, JNT **28** (1988), 119-131 [7].

P

G. Pall: see also D. Estes, R. E. O'Connor.

[Pal1] — *The arithmetical invariants of quadratic forms*, BAMS **51** (1945), 185-197 [15, 16].

[Pal2] — *The weight of a genus of positive n-ary quadratic forms*, PSPM **8** (1965), 95-105 [16].

R. A. Parker: see also J. H. Conway, W. Meyer.

[Par1] — *The computer calculation of modular characters* (*the meat-axe*), in [Atk2], 1984, pp. 267-274 [3].

[Par2] — and J. H. Conway, *A remarkable Moufang loop, with an application to the Fischer group* Fi_{24}, in preparation [29].

[Par3] D. Parrott, *Characterization of the Fischer groups*, TAMS **269** (1981), 303-347 [10].

M. S. Paterson: see R. J. Fowler.

[Pat1] A. L. Patterson and W. E. Love, *Remarks on the Delaunay reduction*, Acta. Cryst. **10** (1957), 111-116 [2].

S. E. Payne: see P. J. Cameron.

[Pee1] D. E. Peek, *The icosahedral puzzle group*, Abstracts Amer. Math. Soc., #798-20-74, **3** (1982), 465 [11].

[Per1] M. Perkel, *A characterization of* J_1 *in terms of its geometry*, Geom. Dedic. **9** (1980), 291-298 [10].

[Pet1] M. Peters, *Definite unimodular* 48*-dimensional quadratic forms*, BLMS **15** (1983), 18-20 [7].

[Pet2] D. P. Petersen and D. Middleton, *Sampling and reconstruction of wave-number-limited functions in N-dimensional Euclidean space*, IC **5** (1962), 279-323 [3].

D. H. Peterson: see V. G. Kac.

[Pet3] W. W. Peterson and E. J. Weldon, Jr., *Error-Correcting Codes*, MIT Press, Cambridge, MA, 2nd ed., 1972 [3, 5, 8].

[Pet4] H. Petersson, *Modulfunktionen und Quadratische Formen*, Springer-Verlag, 1982 [2, 7].

[Pfe1] H. Pfeuffer, *Bemerkung zur Berechnung dyadischer Darstellungsdichten einer quadratischen Form über algebraischen Zahlkörpern*, JRAM **236**

(1969), 219-220 [16].
[Pfe2] — *Quadratsummen in totalreellen algebraischen Zahlkörpern*, JRAM **249** (1971), 208-216 [16].
[Pfe3] — *Einklassige Geschlechter totalpositiver quadratischer Formen in totalreellen algebraischen Zahlköpern*, JNT **3** (1971), 371-411 [16].
[Phe1] K. T. Phelps, *Every finite group is the automorphism group of some linear code*, Congr. Numer. **49** (1985), 139-141 [3].
R. Phillips: see A. Lubotzky.
J. N. Pierce: see H. Gish, V. Pless.
J. R. Pierce: see B. M. Oliver.
F. C. Piper: see D. R. Hughes.
P. Piret: see also A. E. Brouwer.
[Pir1] — *Good linear codes of length* 27 *and* 28, PGIT **26** (1980), 227 [9].
A. Pitre: see M. Bourdeau.
[Ple1] W. Plesken, *Bravais groups in low dimensions*, Comm. Match. in Math. Chem. **10** (1981) 97-119 [1, 3].
[Ple2] — *Applications of the theory of orders to crystallographic groups*, in [Rog10], 1981, pp. 37-92 [3].
[Ple3] — *Finite unimodular groups of prime degree and circulants*, J. Alg. **97** (1985), 286-312 [3].
[Ple4] — and W. Hanrath, *The lattices of six-dimensional Euclidean space*, MTAC **43** (1984), 573-587 [3].
[Ple5] — and M. Pohst, *On maximal finite irreducible subgroups of* $GL(n, \mathbf{Z})$, MTAC **31** (1977), 536-573; **34** (1980), 245-301 [3].
[Ple6] — — *Constructing integral lattices with prescribed minimum* I, MTAC **45** (1985), 209-221 [6].
V. Pless: see also J. H. Conway, J. S. Leon, C. L. Mallows.
[Ple7] — *The number of isotropic subspaces in a finite geometry*, Rend. Cl. Scienze fisiche, matematiche e naturali, Acc. Naz. Lincei **39** (1965), 418-421 [16].
[Ple8] — *On the uniqueness of the Golay codes*, JCT **5** (1968), 215-228 [7, 14, 16, 18].
[Ple9] — *On a new family of symmetry codes and related new five-designs*, BAMS **75** (1969), 1339-1342 [3].
[Ple10] — *A classification of self-orthogonal codes over* $GF(2)$, Discr. Math. **3** (1972), 209-246 [7].
[Ple11] — *Symmetry codes over* $GF(3)$ *and new five-designs*, JCT **A12** (1972), 119-142 [3].
[Ple12] — *The children of the* (32,16) *doubly even codes*, PGIT **24** (1982), 738-746 [7, 15].
[Ple13] — 23 *does not divide the order of the group of a* (72, 36, 16) *doubly-even code*, PGIT **28** (1982), 112-117 [7].
[Ple14] — *On the existence of some extremal self-dual codes*, in [Jac1], 1984, pp. 245-250 [7].
[Ple15] — *A decoding scheme for the ternary Golay code*, in Proc. 20th Allerton Conf. Comm. Control., Univ. of Ill., Urbana, 1982, pp. 682-687 [11].
[Ple15a] — *Decoding the Golay codes*, PGIT **32** (1986), 561-567 [11].
[Ple16] — and J. N. Pierce, *Self-dual codes over* $GF(q)$ *satisfy a modified Varshamov-Gilbert bound*, IC **23** (1973), 35-40 [16].
[Ple17] — and N. J. A. Sloane, *On the classification and enumeration of self-dual codes*, JCT **A18** (1975), 313-335 [7, 14, 16].
[Ple18] — — and H. N. Ward, *Ternary codes of minimum weight* 6, *and the*

classification of self-dual codes of length 20, PGIT 26 (1980), 305-316 [4, 7].

[Ple19] — and J. G. Thompson, 17 *does not divide the order of the group of a* (72, 36, 16) *doubly-even code*, PGIT **28** (1982), 537-541 [7].

[Ple20] — and V. D. Tonchev, *Self-dual codes over* $GF(7)$, PGIT **33** (1987), 723-727 [7].

M. Pohst: see also U. Fincke, W. Plesken.

[Poh1] — *On the computation of lattice vectors of minimal length, successive minima and reduced bases with applications*, ACM SIGSAM Bull. **15** (1981), 37-44 [2].

[Poh2] — *On integral lattice constructions*, Abstracts Amer. Math. Soc. **3** (1982), pp. 152, Abstract 793-12-14 [6].

B. Pollak: see O. T. O'Meara.

H. O. Pollak: see H. J. Landau, D. Slepian.

[Pom1] C. Pomerance, *Analysis and comparison of some integer factoring algorithms*, in *Computational Methods in Number Theory*, ed. H. W. Lenstra, Jr. and R. Tijdeman, Math. Centre Tracts **154**, Math. Centre, Amsterdam, 1982, pp. 89-139 [15].

[Pos1] E. C. Posner, *Combinatorial structures in planetary reconnaissance*, in *Error Correcting Codes*, ed. H. B. Mann, Wiley, NY, 1969, pp. 15-46 [20].

[Pow1] M. B. Powell and G. Higman, editors, *Finite Simple Groups*, Ac. Press, NY, 1971 [B].

F. P. Preparata: see also D. T. Lee.

[Pre1] — *A class of optimum nonlinear double-error-correcting codes*, IC **13** (1968), 378-400 [3].

[Pre1a] — editor, *Advances in Computing Research* I: *Computational Geometry*, Jai Press, Greenwich, CT 1983.

[Pre2] — and M. I. Shamos, *Computational Geometry*, Springer-Verlag, 1985 [2].

I. Prigogine: see G. Pry.

A. D. Pritchard: see J. H. Conway.

[Pro1] G. Promhouse and S. E. Tavares, *The minimum distance of all binary cyclic codes of odd lengths from* 69 *to* 99, PGIT **24** (1978), 438-442 [3, 5].

[Pry1] G. Pry and I. Prigogine, *Rayons et nombres de coordination de quelques réseaux simples*, Acad. Roy. Belgique, Bull. Cl. Sci. **28** (1942), 866-873 [4].

[Pug1] A. Pugh, *Polyhedra: A Visual Approach*, Univ. of Calif., Berkeley CA, 1976 [21].

Q

[Que1] H.-G. Quebbemann, *Zur Klassifikation unimodularer Gitter mit Isometrie von Primzahlordnung*, JRAM **326** (1981), 158-170 [8].

[Que2] — *Gitter ohne Automorphismen der Determinante* -1, Manuscripta Math. **40** (1982), 245-253.

[Que3] — *An application of Siegel's formula over quaternion orders to the existence of extremal lattices*, Math. **31** (1984), 12-16 [8, 16].

[Que4] — *Definite lattices over real algebraic function domains*, Math. Ann. **272** (1985), 461-476 [2].

[Que5] — *A construction of integral lattices*, Math. **31** (1984), 137-140 [1, 7, 8].

[Que6] — *Lattices with theta-functions for* $G(\sqrt{2})$ *and linear codes*, J. Alg. **105** (1987), 443-450 [1, 2, 7, 8].

L. Queen: see also R. E. Borcherds, J. H. Conway.

[Que7] — *Some relations between finite groups, Lie groups and modular functions*,

Ph.D. Dissertation, Univ. of Cambridge, 1980 [30].

[Que8] — *Modular functions and finite simple groups*, PSPM **37** (1980), 561-566 [30].

[Que9] — *Modular functions arising from some finite groups*, MTAC **37** (1981), 547-580 [30].

S. U. Qureshi: see G. D. Forney, Jr.

R

[Rad1] H. Rademacher, *Topics in Analytic Number Theory*, Springer-Verlag, 1973 [4, 7].

R. V. Raev: see V. D. Tonchev.

[Rag1] D. Raghavarao, *Constructions and Combinational Problems in Design of Experiments*, Wiley, NY, 1971 [3].

B. Ramamurthi: see D.-Y. Cheng.

[Ram1] S. Ramanujan, *On certain arithmetical functions*, Trans. Camb. Phil. Soc. **22** (1916), 159-184 = Paper 18 of [Ram2] [7].

[Ram2] — *Collected Papers*, Camb. Univ. Press, 1927; Chelsea, NY, 1962 [B].

D. E. Ramirez: see C. F. Dunkl.

R. A. Rankin: see also D. G. Kendall.

[Ran1] — *On the closest packing of spheres in n dimensions*, Ann. Math. **48** (1947), 1062-1081 [9].

[Ran2] — *A minimum problem for the Epstein zeta-function*, Proc. Glasgow Math. Assoc. **1** (1952-3), 149-158.

[Ran3] — *On positive definite quadratic forms*, JLMS **28** (1953), 309-314 [15].

[Ran4] — *The closest packing of spherical caps in n dimensions*, Proc. Glasgow Math. Assoc. **2** (1955), 139-144 [1].

[Ran4a] — *On packings of spheres in Hilbert space*, Proc. Glasgow Math. Assoc. **2** (1955), 145-146.

[Ran5] — *On the minimal points of positive definite quadratic forms*, Math. **3** (1956), 15-24 [15].

[Ran5a] — *On the minimal points of perfect quadratic forms*, Math. Z. **84** (1964), 228-232 [15].

[Ran6] — *Modular Forms and Functions*, Camb. Univ. Press, 1977 [2, 4, 7].

[Ran7] — *Fourier coefficients of cusp forms*, PCPS **100** (1986), 5-29 [2].

[Rao1] V. V. Rao and S. M. Reddy, *A* (48,31,8) *linear code*, PGIT **19** (1973), 709-711 [8].

[Ras1] R. Rasala, *Split codes and the Mathieu groups*, J. Alg. **42** (1976), 422-471 [18].

[Rau1] H. E. Rauch and H. M. Farkas, *Theta Functions with Applications to Riemann Surfaces*, Williams and Wilkins, Baltimore MD, 1974 [4].

[Rau2] — and A. Lebowitz, *Elliptic Functions, Theta Functions, and Riemann Surfaces*, Williams and Wilkins, Baltimore MD, 1973 [4].

S. M. Reddy: see V. V. Rao.

[Ree1] I. S. Reed and G. Solomon, *Polynomial codes over certain finite fields*, SIAJ **8** (1960), 300-304 [3].

[Rei1] L. W. Reid, *The Elements of the Theory of Algebraic Numbers*, Macmillan, NY, 1910 [15].

[Rei2] I. Reiner, *A survey of integral representation theory*, BAMS **76** (1970), 159-227 [3].

[Rei3] — *Maximal Orders*, Ac. Press, NY, 1975 [3].

[Rei4] — *Topics in Integral Representation Theory*, LNM **744**, 1979 [3].

[Rei5] E. M. Reingold, J. Nievergelt and N. Deo, *Combinatorial Algorithms: Theory and Practice*, Prentice-Hall, Englewood Cliffs NJ, 1977 [20].
H. P. Reiss: see T. Beth.
[Ren1] R. J. Renka, *Interpolation of data on the surface of a sphere*, ACM Trans. Math. Software **10** (1984), 417-436 and 437-439 [1].
[Rib1] P. Ribenboim, *Algebraic Numbers*, Wiley, NY, 1972 [8].
J. R. Rice: see T. J. Aird.
[Rie1] C. R. Riehm and I. Hambleton, editors, *Quadratic and Hermitian Forms*, Canad. Math. Soc. Conf. Proc. **4**, Amer. Math. Soc., Providence RI, 1983 [15, B].
[Rio1] J. Riordan, *An Introduction to Combinatorial Analysis*, Wiley, NY, 1958 [21].
[Rio2] J. Riordan, *Combinatorial Identities*, Wiley, NY, 1968 [21].
[Rit1] D. Ritchie and S. F. Slovin, *Protoachlya polysporus revisited*, Mycologia **66** (1974), 362-364 [1].
E. F. Robertson: see R. A. Wilson.
J. P. Robinson: see A. W. Nordstrom.
[Rob1] R. M. Robinson, *Arrangement of* 24 *points on a sphere*, Math. Ann. **144** (1961), 17-48 [1].
[Rob2] — *Finite sets of points on a sphere with each nearest to five others*, Math. Ann. **179** (1969), 296-318 [1].
E. R. Rodemich: see R. J. McEliece.
C. A. Rogers: see also H. S. M. Coxeter, P. Erdös.
[Rog1] — *A note on coverings*, Math. **4** (1957), 1-6 [2].
[Rog2] — *The packing of equal spheres*, PLMS **8** (1958), 609-620 [1, 9].
[Rog3] — *Lattice coverings of space*, Math. **6** (1959), 33-39 [2].
[Rog4] — *An asymptotic expansion for certain Schläfli functions*, JLMS **36** (1961), 78-80.
[Rog5] — *The chance that a point should be near the wrong lattice point*, JLMS **37** (1962), 161-163.
[Rog6] — *Covering a sphere with spheres*, Math. **10** (1963), 157-164 [2].
[Rog7] — *Packing and Covering*, Camb. Univ. Press, 1964 [1, 2, 9].
[Rog8] — *Probabilistic and combinatorial methods in the study of the geometry of Euclidean spaces*, in *Proc. Intern. Congr. Math.*, Vancouver, 1974, pp. 497-500.
[Rog9] K. W. Roggenkamp, *Lattices over Orders* II, LNM **142**, 1970 [2].
[Rog10] — editor, *Integral Representations and Applications*, LNM **882**, 1981 [3, B].
[Rog11] — and V. Huber-Dyson, *Lattices over Orders* I, LNM **115**, 1970 [2].
A. P. Rollett: see H. M. Cundy.
[Rom1] A. M. Romanov, *New binary codes of minimal distance* 3, PPI **19** (1983), 101-102 [3, 5, 9].
[Roo1] C. Roos and C. de Vroedt, *Upper bounds for $A(n, 4)$ and $A(n, 6)$ derived from Delsarte's linear programming bound*, Discr. Math. **40** (1982), 261-276 [9].
[Roo2] D. Roose and E. de Doncker, *Automatic integration over a sphere*, J. Comput. Appl. Math. **7** (1981), 203-224 [1].
[Roq1] P. Roquette, *On class field towers*, in [Cas5], 1967, pp. 231-249 [8].
A. Rosa: see C. C. Lindner.
[Rud1] A. N. Rudakov and I. R. Shafarevich, *Supersingular K*3 *surfaces over fields of characteristic* 2, IANS **42** (1978), 848-869 = Izv. **13** (1979), 147-165 [15].
[Rud2] — — *Surfaces of type K*3 *over fields of finite characteristic*, IANS **18** (1981), 115-207 = JSM 22 (1983), 1476-1533 [15].

V. P. Rud'ko: see P. M. Gudivok.
L. D. Rudolph: see J. Gao.
A. Rudvalis: see also L. Finkelstein.

[Rud3] — *A rank 3 simple group of order* $2^{14} \cdot 3^3 \cdot 5^3 \cdot 7 \cdot 13 \cdot 29$, J. Alg. **86** (1984), 181-258 [10].

H. C. Rumsey, Jr.: see R. J. McEliece.

[Rus1] J. A. Rush and N. J. A. Sloane, *An improvement to the Minkowski-Hlawka bound for packing superballs*, Math. **34** (1987), 8-18 [1].

[Ryb1] A. J. E. Ryba, *A new construction of the O'Nan simple group*, J. Alg. **112** (1988), 173-197 [10].

S. S. Ryskov: see also E. P. Baranovskii, B. N. Delone.

[Rys1] — *Effectuation of a method of Davenport in the theory of coverings*, DAN 175 (1967) 303-305 = SMD **8** (1967), 865-867 [2, 15].

[Rys2] — *On the reduction theory of positive quadratic forms*, DAN **198** (1971), 1028-1031 = SMD **12** (1971), 946-950 [2, 15].

[Rys3] — *On Hermite, Minkowski and Voronoi reduction of positive quadratic forms in n variables*, DAN **207** (1972) 1054-1056 = SMD **13** (1972), 1676-1679 [2, 15].

[Rys4] — *On maximal finite groups of integer* $(n \times n)$*-matrices*, DAN **204** (1972, No. 3) 561-564 = SMD **13** (1972), 720-724 [15].

[Rys5] — *Maximal finite groups of integral* $(n \times n)$*-matrices and full groups of integral automorphisms of positive quadratic forms (Bravais lattices)*, TMIS **128** (1972), 183-211, 261 = PSIM **128** (1972) 217-250 [1, 15].

[Rys6] — *On the question of the final* ζ *- optimality of lattices that yield the densest packing of n-dimensional balls*, Sibirsk. Mat. Z. **14** (1973), 1065-1075 and 1158 = Sib. Math. J. **14** (1973), 743-750 [15].

[Rys7] — *The perfect form* A_n^{κ} *: the existence of lattices with a nonfundamental division simplex, and the existence of perfect forms which are not Minkowski-reducible to forms having identical diagonal coefficients*, LOMI **33** (1973), 65-71 = JSM **6** (1976), 672-676 [4, 15].

[Rys8] — *The Hermite-Minkowski theory of reduction of positive definite quadratic forms*, LOMI **33** (1973), 37-64 = JSM **6** (1976), 651-671 [2, 15].

[Rys9] — *The geometry of positive quadratic forms*, Proc. Int. Congress Math., Vancouver, 1974, **I**, pp. 501-506 = Amer. Math. Soc. Transl. (2) **109** (1977), 27-32 [15].

[Rys10] — *On the problem of the determination of the perfect quadratic forms in many variables*, TMIS **142** (1976), 215-239 = PSIM **142** (1976), 233-259 [15].

[Rys11] — editor, *The Geometry of Positive Quadratic Forms*, TMIS **152** (1980) = PSIM (1982, No. 3) [15].

[Rys12] — and E. P. Baranovskii, *Solution of the problem of least dense lattice covering of five-dimensional space by equal spheres*, DAN **222** (1975), 39-42 = SMD **16** (1975) 586-590 [2, 4, 15].

[Rys13] — — *C-types of n-dimensional lattices and 5-dimensional primitive parallelohedra (with application to the theory of coverings)*, TMIS **137** (1976, No. 4) = PSIM (1978, No. 4) [2, 4, 15].

[Rys14] — — *Classical methods in the theory of lattice packings*, Uspekhi Mat. Nauk **34** (No. 4, 1979), 3-63 = Russian Math. Surveys **34** (No. 4, 1979), 1-68 [2, 15].

[Rys15] — — *Perfect lattices as admissable centerings*, Itogi Nauki Tekh., Ser. Probl. Geom. **17** (1985), 3-49 [2].

S

T. L. Saaty: see also H. S. M. Coxeter.

[Saa1] — and J. M. Alexander, *Optimization and the geometry of numbers: packing and covering*, SIAM Rev. **17** (1975), 475-519.

C.-H. Sah: see R. Brauer.

N. N. Sandakova: see B. N. Delone.

[San1] A. Santoró and A. D. Mighell, *Determination of reduced cells*, Acta Cryst. **A26** (1970), 124-127 [1].

P. Sarnak: see A. Lubotsky.

[Sat1] J. Satterly, *The moments of inertia of some polyhedra*, Math. Gazette, **42** (1958), 11-13 [21].

[Saw1] K. Sawade, *A Hadamard matrix of order* 268, Graphs Comb. **1** (1985), 185-187 [3].

[Say1] S. I. Sayegh, *A condition for optimality of two-dimensional signal constellations*, COMM **33** (1985), 1220-1222 [3].

[Say2] D. Sayre, editor, *Computational Crystallography*, Oxford Univ. Press, 1982 [2].

C. A. Schaffner: see H. A. Krieger.

W. Scharlau: see also M. Knebusch.

[Sch0] — *Quadratic Forms*, Queen's Papers on Pure and Applied Mathematics No. 22, Queen's Univ., Kingston Ontario, 1970 [15].

[Sch1] — *A historical introduction to the theory of integral quadratic forms*, in [Orz1], 1976, pp. 284-339 [15].

[Sch2] — *Quadratic and Hermitian Forms*, Springer-Verlag, 1985 [15].

J. R. Schatz: see G. D. Cohen.

[Sch3] J. Scherk, *An overview of supersymmetry and supergravity*, Lecture Notes Phys. **116** (1980), 343-357 [1].

P. Schmid: see W. Knapp.

P. H. Schmidt: see E. L. Allgower.

C. P. Schnorr: see also J. C. Lagarias, H. W. Lenstra, Jr.

[Sch4] — *A hierarchy of polynomial time basis reduction algorithms*, in *Theory of Algorithms*, North-Holland, 1985, 375-386 [2].

[Sch5] I. J. Schoenberg, *Positive definite functions on spheres*, DMJ **9** (1942), 96-107 [9].

[Sch6] B. Schoeneberg, *Das Verhalten von mehrfachen Thetareichen bei Modulsubstitutionen*, Math. Ann. **116** (1939), 511-523 [2, 7].

[Sch7] — *Elliptic Modular Functions*, Springer-Verlag, 1974 [2, 7].

A. Schrijver: see also J. H. van Lint.

[Sch8] — editor, *Packing and Covering in Combinatorics*, Math. Centre Tracts **106**, Math. Centre, Amsterdam, 1979 [B].

[Sch9] — *A comparison of the Delsarte and Lovaśz bounds*, PGIT **25** (1979), 425-429 [9].

M. R. Schroeder: see also B. S. Atal, J. L. Flanagan.

[Sch10] — B. S. Atal and J. L. Hall, *Optimizing digital speech encoders by exploiting masking properties of the human ear*, J. Acoust. Soc. Am. **66** (1979), 1647-1652 [2].

[Sch11] K. Schütte and B. L. van der Waerden, *Auf welcher Kugel haben* 5, 6, 7, 8 *oder* 9 *Punkte mit Mindesabstand Eins Platz*?, Math. Ann. **123** (1951), 96-124 [1].

[Sch12] — — *Das Problem der dreizehn Zugeln*, Math. Ann. **125** (1953), 325-334 [1].

[Sch13] R. Schulze-Pillot, *Thetareihen positiv definiter quadratischer Formen*, Invent. Math. **75** (1984), 283-299 [15].

[Sch14] — *Darstellungsmasse von Spinorgeschlechtern ternärer quadratischer Formen*, JRAM **352** (1984), 114-132 [15].

[Sch15] J. H. Schwarz, editor, *Superstrings*, World Scientific, Singapore, 1985 [1].

[Sch16] R. L. E. Schwarzenberger, *N-Dimensional Crystallography*, Pitman, San Francisco CA, 1980 [1].

[Sch17] — *Graphical representation of n-dimensional space groups*, PLMS **44** (1982), 244-266.

[Sco1] P. R. Scott, *On perfect and extreme forms*, JAMS **4** (1964), 56-77 [2].

[Sco2] — *The construction of perfect and extreme forms*, CJM **18** (1966), 145-158 [2].

[Sco3] W. R. Scott and F. Gross, *Proceedings of the Conference on Finite Groups*, Ac. Press, NY, 1976 [B].

B. Sedlác ek: see F. R. Ahmed.

[Seg1] B. Segre and K. Mahler, *On the densest packing of circles*, AMM **51** (1944), 261-270 [1].

J. J. Seidel: see also E. Bannai, P. Delsarte, J.-M. Goethals, A. Neumaier, N. J. A. Sloane.

[Sei1] — *Strongly regular graphs*, in *Progress in Combinatorics*, edited by W. T. Tutte, Ac. Press, NY, 1969, pp. 185-197 [9].

[Sei2] — *Strongly regular graphs*, in *Surveys in Combinatorics*, ed. B. Bollobás, Camb. Univ. Press, 1979, pp. 157-180 [9].

[Sei3] — *Delsarte's theory of association schemes*, in *Graphs and Other Combinatorial Topics*, Teubner, Leipzig, 1983, pp. 249-258 [9, 14].

R. Seidel: see H. Edelsbrunner.

[Sei4] — *Constructing higher-dimensional convex hulls at logarithmic cost per face*, in *Proc. 18th Annual ACM Sympos. Theory of Computing*, Assoc. Computing Machinery, N.Y. 1986, pp. 404-413 [2].

F. Seitz: see E. Wigner.

G. M. Seitz: see C. W. Curtis.

[Sem1] N. V. Semakov and V. A. Zinoviev, *Complete and quasi-complete balanced codes*, PPI **5** (No. 2, 1969), 28-36 = PIT **5** (No. 2, 1969), 11-13 [3].

[Ser1] J.-P. Serre, *Cours d'Arithmetique*, Presses Universitaires de France, Paris, 1970; English translation, Springer-Verlag, 1973 [2, 4, 7, 15, 16, 18, 26].

[Ser2] — *Corps Locaux*, Hermann, Paris, 2nd ed., 1968; English translation, Springer-Verlag, 1979 [2].

[Ser3] — *Sur le nombre des points rationnels d'une courbe algebrique sur un corps fini*, CR **296** (1983), 397-402 [3].

[Sey1] P. D. Seymour and T. Zaslavsky, *Averaging sets: a generalization of mean values and spherical designs*, Adv. in Math. **52** (1984), 213-240 [3].

I. R. Shafarevich: see Z. I. Borevich, E. S. Golod, A. N. Rudakov.

M. I. Shamos: see F. P. Preparata.

[Sha0] E. A. Shamsiev and S. S. Shushbaev, *Construction of cubature formulae that are invariant with respect to groups of integral automorphisms of perfect quadratic forms*, Dokl. Akad. Nauk UzSSR (No. 3, 1982), 6-8; Math. Rev. **83** #41034 [1].

C. E. Shannon: see also B. M. Oliver. All of the following except [Sha6] are reprinted in [Sle5].

[Sha1] — *A mathematical theory of communication*, BSTJ **27** (1948), 379-423 and 623-656 [3].

[Sha2] — *Communication in the presence of noise*, PIEEE **37** (1949), 10-21 [3].

[Sha3] — *Prediction and entropy of printed English*, BSTJ **30** (1951), 50-64 [3].

[Sha4] — *The zero error capacity of a noisy channel*, PGIT **2** (1956), 8-19 [3, 9].

[Sha5] — *Certain results in coding theory for noisy channels*, IC **1** (1957), 6-25 [3].

[Sha6] — *Probability of error for optimal codes in a Gaussian channel*, BSTJ **38** (1959), 611-656 [1, 3].

[Sha7] — *Coding theorems for a discrete source with a fidelity criterion*, IRE Nat. Conv. Record (Part 4, 1959), 142-163 [3].

[Sha8] — R. G. Gallager and E. R. Berlekamp, *Lower bounds to error probability for coding on discrete memoryless channels*, IC **10** (1967), 65-103 and 522-552 [3].

H. M. Shapiro: see A. R. Calderbank.

J. Shawe-Taylor: see N. L. Biggs.

[She0] D. Shechtman, I. Blech, D. Gratias and J. W. Cahn, *Metallic phase with long-range orientation order and no translational symmetry*, Phys. Rev. Lett. **53** (1984), 1951-1953 [1].

O. I. Shekhovtsov: see S. N. Litsyn.

G. C. Shephard: see also B. Grünbaum, P. McMullen.

[She1] — *Unitary groups generated by reflections*, CJM **5** (1953), 364-383 [4].

[She2] — and J. A. Todd, *Finite unitary reflection groups*, CJM **6** (1954), 274-304 [3, 4, 7].

[She3] L. A. Shepp, *Computerized tomography and nuclear magnetic resonance*, J. Computer Assisted Tomography **4** (1980), 94-107 [1].

[She4] — and J. B. Kruskal, *Computerized tomography: the new medical X-ray technology*, AMM **85** (1978), 420-439 [1].

F. A. Sherk: see C. Davis.

[Shi1] G. Shimura, *Introduction to the Arithmetic Theory of Automorphic Functions*, Princeton Univ. Press, 1971 [7].

Y. Shoham: see D.-Y. Cheng.

S. S. Shushbaev: see E. P. Baranovskii, E. A. Shamsiev.

[Shu1] — *An algorithm for the calculation of groups of integer automorphisms of perfect forms in $n \geq 2$ variables*, Voprosy Vychisl. Prikl. Mat. (Tashkent) **22** (1973), 3-13; Math. Rev. **58** #16516 [3].

[Shu2] — *The determination of the groups of integer automorphisms of perfect quadratic forms*, Dokl. Akad. Nauk UzSSR, (No. 1, 1974), 16-18 [3].

[Shu3] — *The Dirichlet-Voronoi domain for the second perfect form*, Voprosy Vychisl. Prikl. Mat. (Tashkent) **45** (1977), 3-11; Math. Rev. **58** #27811 [2].

[Shu4] — *Voronoi neighborhood of the perfect form ϕ_{15} $(x_1, x_2, \ldots, x_7)$*, Voprosy Vychisl. Prikl. Mat. (Tashkent), **77** (1985), 48-56 [2].

[Sid1] V. M. Sidel'nikov, *On the density of sphere packings on the surface of an n-dimensional Euclidean sphere and on the cardinality of a binary code with a given minimum distance*, DAN **213** (No. 5, 1973), 1029-1032 = SMD **14** (1973), 1851-1855 [1, 9].

[Sid2] — *Upper bounds on the cardinality of a binary code with a given minimum distance*, PPI **10** (No. 2, 1974), 43-51 = IC **28** (1975), 292-303 [9].

[Sid3] — *New bounds for densest packings of spheres in n-dimensional Euclidean space*, Mat. Sb. **95** (No. 137, 1974), 148-158 = Sb. **24** (1974), 147-157 [1, 9].

[Sid4] — *On extremal polynomials used to estimate the size of codes*, PPI **16** (No. 3, 1980), 17-30 = PIT **16** (1980), 174-186 [9].

V. R. Sidorenko: see E. M. Gabidulin.

[Sie1] C. L. Siegel, *Uber die analytische Theorie der quadratischen Formen*, Ann. Math. **36** (1935), 527-606 = Gesam. Abh. **I**, 326-405, Springer-Verlag, 1966 [7, 16].

[Sie2] — *Uber die Fourierschen Koeffizienten der Eisensteinschen Reihen*, Danske Vid. Selskab. Mat.-fys. Meddelelser **34** (1964), Nr. 6 = Gesam. Abh. **III**, 443-458 [7, 16].

[Sie3] — *Berechnung von Zetafunktionen an ganzzahligen Stellen*, Göttingen Nach. **10** (1969), 87-102 = Gesam. Abh. **IV**, 82-97 [7, 16].

D. Siersma: see M. Hazewinkel.

[Sig1] F. Sigrist, *Sphere packing*, Math. Intelligencer **5** (No. 3, 1983), 34-38 [1].

[Sim0] M. Simonnard, *Linear Programming*, Prentice-Hall, Englewood Cliffs, NJ, 1966 [9, 14].

C. C. Sims: see also D. G. Higman, J. S. Leon.

[Sim1] — *On the isomorphism of two groups of order* 44,352,000, in [Bra2], 1969, pp. 101-108 [10].

[Sim2] — *The existence and uniqueness of Lyons' group*, in [Gag2], 1973, 138-141 [10].

[Sim3] — *How to construct a Baby Monster*, in [Col2], 1980, 339-345 [10].

I. M. Singer: see J. I. Lepowsky.

[Sku1] B. F. Skubenko, *Dense lattice packings of spheres in Euclidean spaces of dimension* $n \leq 16$, LOMI **82** (1979), 144-146 = JSM **18** (1982), 958-960 [4, 6].

[Sku2] — *A remark on an upper bound on the Hermite constant for the densest lattice packing of spheres*, LOMI **82** (1979), 147-148 = JSM **18** (1982), 960-961.

H. A. Sleeper: see S. G. Wilson.

[Sle1] D. Slepian, *Bounds on communication*, BSTJ **42** (1963), 681-707 [3].

[Sle2] — *Permutation modulation*, PIEEE (1965), 228-236 [3].

[Sle3] — *Group codes for the Gaussian channel*, BSTJ **47** (1968), 575-602 [3].

[Sle4] — *On neighbor distances and symmetry in group codes*, PGIT **17** (1971), 630-632 [3].

[Sle5] — editor, *Key Papers in the Development of Information Theory*, IEEE Press, NY, 1974 [3, B].

[Sle6] — *Some comments on Fourier analysis, uncertainty and modeling*, SIAM Rev. **25** (1983), 379-393 [3].

[Sle7] — and H. O. Pollak, *Prolate spheroidal wave functions, Fourier analysis and uncertainty* I, BSTJ **40** (1961), 43-63 [3].

[Sln1] I. H. Sloan, *Lattice methods for multiple integration*, J. Comput. Appl. Math. **12** (1985), 131-143 [1].

[Sln2] — and P. J. Kachoyan, *Lattice methods for multiple integration: theory, error analysis and examples*, SIAM J. Num. Anal. **24** (1987), 116-128 [1].

N. J. A. Sloane: see also S. F. Assmann, R. P. Bambah, E. Bannai, E. S. Barnes, E. R. Berlekamp, M. R. Best, R. E. Borcherds, A. Bos, A. R. Calderbank, Y. Cheng, G. D. Cohen, J. H. Conway, D. E. Downey, V. Elser, R. L. Graham, M. Harwit, W. C. Huffman, K. E. Kilby, J. Leech, J. S. Leon, F. J. MacWilliams, C. L. Mallows, A. M. Odlyzko, V. Pless, J. A. Rush, B. K. Teo.

[Slo1] — *Sphere packings constructed from BCH and Justesen codes*, Math. **19** (1972), 183-190 [1, 5].

[Slo2] — *A Handbook of Integer Sequences*, Ac. Press, NY, 1973 [21].

[Slo3] — *Is there a* (72, 36) $d = 16$ *self-dual code?*, PGIT **19** (1973), 251 [7].

[Slo4] — *A Short Course on Error-Correcting Codes*, Springer-Verlag, 1975.

[Slo5] — *An introduction to association schemes and coding theory*, in [Ask2a], pp. 225-260 [9, 14].

[Slo6] — *Weight enumerators of codes*, in *Combinatorics*, ed. M. Hall Jr. and J. H. van Lint, Reidel, Dordrecht, Holland, 1975, pp. 115-142 [7].

[Slo7] — *Error-correcting codes and invariant theory: new applications of a nineteenth-century technique*, AMM **84** (1977), 82-107 [7].

[Slo8] — *Binary codes, lattices and sphere packings*, in *Combinatorial Surveys*, ed. P. J. Cameron, Ac. Press, NY, 1977, pp. 117-164 [4, 7].

[Slo9] — *Codes over* $GF(4)$ *and complex lattices*, J. Alg. **52** (1978), 168-181 [7].

[Slo10] — *Self-dual codes and lattices*, PSPM **34** (1979), 273-308 [3, 7, 16].

[Slo11] — *A note on the Leech lattice as a code for the Gaussian channel*, IC **46** (1980), 270-272 [3].

[Slo12] — *Tables of sphere packings and spherical codes*, PGIT **27** (1981), 327-338 [1, 4].

[Slo13] — *Recent bounds for codes, sphere packings and related problems obtained by linear programming and other methods*, Contemp. Math. **9** (1982), 153-185 ≅ Chap. 9 of this book [3, 9].

[Slo13a] — *Encrypting by random rotations*, Lecture Notes Computer Sci. **149** (1983), 71-128 [1].

[Slo14] — *The packing of spheres*, Scientific American **250** (No. 1, 1984), 116-125 [1].

[Slo15] — *A new approach to the covering radius of codes*, JCT **A42** (1986), 61-86 [3].

[Slo16] — *Unsolved problems related to the covering radius of codes*, in *Proc. Conf. Specific Prob. Commun. (SPOC-85)*, Springer-Verlag, 1987, to appear [3].

[Slo17] — *Theta-series and magic numbers for diamond and certain ionic crystal structures*, J. Math. Phys. **28** (1987), 1653-1657 [1, 2, 4].

[Slo18] — and J. J. Seidel, *A new family of nonlinear codes obtained from conference matrices*, Ann. New York Acad. Sci. **175** (1970), 363-365 [3, 5].

[Slo19] — and B. K. Teo, *Theta series and magic numbers for closed-packed spherical clusters*, J. Chem. Phys. **83** (1985), 6520-6534 [1, 4].

[Slo20] — and D. S. Whitehead, *A new family of single-error correcting codes*, PGIT **16** (1970), 717-719 [3, 5].

S. F. Slovin: see D. Ritchie.

[Sma1] I. Smalley, *Simple regular sphere packings in three dimensions*, Math. Mag. **36** (1963), 295-299 [4].

D. H. Smith: see also N. L. Biggs, P. Hammond.

[Smi1] — *On tetravalent graphs*, JLMS **6** (1973), 659-662 [9].

[Smi2] — *Distance-transitive graphs*, in *Combinatorics*, ed. T. P. McDonough and V. C. Mavron, Camb. Univ. Press, 1974, pp. 145-153 [9].

[Smi3] — *Distance-transitive graphs of valency four*, JLMS **8** (1974), 377-384 [9].

[Smi4] F. L. Smith, *A general characterization of the Janko simple group* J_2, Archiv Math. **25** (1974), 17-22 [10].

[Smi5] H. J. S. Smith, *Report on the theory of numbers*, Report of the British Association (1859) 228-267; (1860), 120-169; (1861), 292-340; (1862), 503-526; (1863), 768-786; (1865), 322-375 = Coll. Math. Papers I, 38-364 [4, 15].

[Smi6] — *On the orders and genera of quadratic forms containing more than three indeterminates*, PRS **16** (1867), 197-208 = Coll. Math. Papers **I**, 510-523 [2, 4, 15].

[Smi7] K. T. Smith, D. C. Solmon and S. L. Wagner, *Practical and mathematical aspects of the problem of reconstructing objects from radiographs*, BAMS **83** (1977), 1227-1270 [1].

[Smi8] M. S. Smith, *On rank* 3 *permutation groups*, J. Alg. **33** (1975), 22-42 [10].

[Smi9] — *On the isomorphism of two simple groups of order* 44,352,000, J. Alg. **41** (1976), 172-174 [10].

[Smi10] — *A combinatorial configuration associated with the Higman-Sims group*, J. Alg. **41** (1976), 175-195 [10].

N. K. Smith: see S. G. Wilson.

[Smi11] R. F. Smith, *The construction of definite indecomposable Hermitian forms*, AJM **100** (1978), 1021-1048 [2].

S. D. Smith: see also A. O. L. Atkin.

[Smi12] — *Large extraspecial subgroups of widths* 4 *and* 6, J. Alg. **58** (1979), 251-281 [29].

[Smi13] — *On the Head characters of the Monster simple group*, in [McK3], 1985, pp. 303-313 [1, 29, 30].

W. R. Smith: see T. W. Melnyk.

[Sob1] I. M. Sobol, *On systematic search in a hypercube*, SIAM J. Num. Anal. **16** (1979), 790-793 [1].

[Sob2] S. L. Sobolev, *Cubature formulae on the sphere invariant under finite groups of rotations*, DAN **146** (No. 2, 1962), 310-313 = SMD **3** (1962), 1307-1310 [1, 3].

[Sob3] — *Introduction to the Theory of Cubature Formulae*, Nauka, Moscow, 1974 [1, 3].

L. H. Soicher: see also J. H. Conway.

[Soi1] — *Presentations for some groups related to* Co_1, in *Computers in Algebra, Chicago 1985*, Dekker NY 1988, pp. 151-154 [10].

[Soi2] — *From the Monster to the Bimonster*, J. Alg. **121** (1989), 275-280 [29].

[Soi3] — *Presentations for Conway's group* Co_1, PCPS **102** (1987), 1-3 [10].

D. C. Solmon: see K. T. Smith.

G. Solomon: see also I. S. Reed.

[Sol1] — and M. M. Sweet, *A Golay puzzle*, PGIT **29** (1983), 174-175 [11].

[Som1] D. M. Y. Sommerville, *An Introduction to the Geometry of n Dimensions*, Dover, NY, 1958 [1].

[Som2] J. Sommer, *Vorlesungen über Zahlentheorie*, Teubner, Leipzig, 1907 [15].

[Spe1] T. Speevak, *An efficient algorithm for obtaining the volume of a special kind of pyramid and application to convex polyhedra*, MTAC **46** (1986), 531-536 [2].

T. A. Springer: see F. van der Blij.

[Sta1] K. C. Stacey, *The enumeration of perfect septenary forms*, JLMS **10** (1975), 97-104 [1, 2].

[Sta2] — *The perfect septenary forms with* $\Delta_4 = 2$, JAMS **A22** (1976), 144-164 [1, 2].

[Sta3] R. P. Stanley, *Invariants of finite groups and their application to combinatorics*, BAMS **1** (1979), 475-511 [7].

[Sta4] D. Stanton, *Some q-Krawtchouk polynomials on Chevalley groups*, AJM **102** (1980), 625-662 [9].

[Sta5] — *Product formulas for q-Hahn polynomials*, SIAM J. Math. Anal. **11** (1980), 100-107 [9].

[Sta6] — *Another infinite family of perfect codes*, preprint [9].

[Sta7] — *Three addition theorems for some q-Krawtchouk polynomials*, Geom.

Dedic. **10** (1981), 403-425 [9].
[Sta8] — *A partially ordered set and q-Krawtchouk polynomials*, JCT **A30** (1981), 276-284 [9].
[Sta9] — *Generalized n-gons and Chebychev polynomials*, JCT **A34** (1983), 15-27 [9].
[Sta10] — *Harmonics on posets*, JCT **A40** (1985), 136-149 [9].
[Sta11] — *Orthogonal polynomials and Chevalley groups*, in [Ask2a], 1984, pp. 87-128 [9].
[Sta11a] — *t-Designs in classical association schemes*, Graphs Comb. **2** (1986), 283-286.
[Sta12] R. G. Stanton, *The Mathieu groups*, CJM **3** (1951), 164-174 [11, 14].
I. A. Stegun: see M. Abramowitz.
P. J. Steinhardt: see D. Levine.
[Ste1] F. Stenger, *Numerical methods based on Whittaker cardinal or sinc functions*, SIAM Rev. **23** (1981), 165-224 [1, 3].
R. D. Stinaff: see H. J. Helgert.
M. I. Stogrin: see also B. N. Delone.
[Sto1] — *The Voronoi, Venkov and Minkowski reduction domains*, DAN **207** (1972), 1070-1073 = SMD **13** (1972), 1698-1702 [2].
[Sto2] — *Locally quasidensest lattice packings of spheres*, DAN **218** (1974), 62-65 = SMD **15** (1974), 1288-1292.
A. P. Street: see W. D. Wallis.
[Str1] R. L. Streit, *Optimization of discrete arrays of arbitrary geometry*, J. Acoust. Soc. Am. **69** (1981), 199-212 [1].
[Str2] J. Strohmajer, *Uber die Verteilung von Punkten auf der Kugel*, Ann. Univ. Sci. Budapest, Sect. Math. **6** (1963), 49-53 [1].
[Str3] A. H. Stroud, *Approximate Calculation of Multiple Integrals*, Prentice-Hall, Englewood Cliffs NJ, 1971 [1].
[Sug1] Y. Sugiyama, M. Kasahara, S. Hirasawa and T. Namekawa, *A modification of the constructive asympotically good codes of Justesen for low rates*, IC **25** (1974), 341-350 [3].
[Sug2] — — — — *A new class of asymptotically good codes beyond the Zyablov bound*, PGIT **24** (1978), 198-204 [3].
[Sug3] — — — — *Superimposed concatenated codes*, PGIT **26** (1980), 735-736 [3].
[Suz1] M. Suzuki, *A simple group of order* 448,345,497,600, in [Bra2], 1969, pp. 113-119 [10].
M. M. Sweet: see G. Solomon.
[Syl1] J. J. Sylvester, *Elementary researches in the analysis of combinatorial aggregation*, Phil. Mag. **24** (1844), 285-296 = Coll. Math. Papers **I** (1904), 91-102 [10, 23].
[Sze1] G. Szegö, *Orthogonal polynomials*, Amer. Math. Soc., Providence RI, 4th ed., 1975 [9].
[Sze2] E. Székely, *Sur le problème de Tammes*, Ann. Univ. Sci. Budapest. Eötvös, Sect. Math. **17** (1974), 157-175 [1].

T

[Tam1] P. Tammela, *The Hermite-Minkowski domain of reduction of positive definite quadratic forms in six variables*, LOMI **33** (1973), 72-89 = JSM **6** (1976), 677-688 [2, 15].
[Tam2] — *On the reduction theory of positive quadratic forms*, DAN **209** (1973),

1299-1302 = SMD **14** (1973), 651-655 [2, 15].

[Tam3] — *On the reduction theory of positive quadratic forms*, LOMI **50** (1975), 6-96 = JSM **11** (1979), 197-277 [2, 15].

[Tam4] — *Minkowski's fundamental reduction domain for positive quadratic forms of seven variables*, LOMI **67** (1977), 108-143, 226 = JSM **16** (1981), 836-857 [2, 15].

[Tam5] P. M. L. Tammes, *On the origin of number and arrangement of the places of exit on the surface of pollen-grains*, Recueil des travaux botaniques néerlandais **27** (1930), 1-84 [1].

S. L. Tanimoto: see R. J. Fowler.

[Tan1] J. Tannery and J. Molk, *Eléments de la théorie des Fonctions Elliptiques*, Chelsea, NY, 2nd ed., 4 vols., 1972 [4].

[Tar1] T. Tarnai, *Packing of* 180 *equal circles on a sphere*, Elem. Math. **38** (1983), 119-122; **39** (1984), 129 [1].

[Tar2] — *Note on packing of* 19 *equal circles on a sphere*, Elem. Math. **39** (1984), 25-27 [1].

[Tar3] — *Spherical circle-packing in nature, practice and theory*, Structural Topology **9** (1984), 39-58 [1].

[Tar4] — and Z. Gáspár, *Improved packing of equal circles on a sphere and rigidity of its graph*, PCPS **93** (1983), 191-218 [1].

T. Tasaka: see also T. Kondo.

[Tas1] — *On even lattices of* 2*-squares type and self-dual codes*, J. Fac. Sci. Tokyo Sect. 1A Math. **28** (1981), 701-714 [7].

[Tau1] O. Taussky, *Introduction into connections between algebraic number theory and integral matrices*, Appendix 2 to [Coh6], 1978 [8].

[Tau2] O. Taussky, editor, *Ternary Quadratic Forms and Norms*, Dekker NY, 1982 [15].

S. E. Tavares: see G. Promhouse.

K. F. Taylor: see D. Borwein.

S. Tazaki: see Y. Yamada.

[Tei1] L. Teirlinck, *Non-trivial t-designs without repeated blocks exist for all t*, Discr. Math. **65** (1987), 301-311 [3].

B. K. Teo: see also N. J. A. Sloane.

[Teo1] — and K. Keating, *Novel triicosahedral structure of the largest metal alloy cluster*: $[(Ph_3P)_{12}Au_{13}Ag_{12}Cl_6]^{m+}$, J. Am. Chem. Soc. **106** (1984), 2224-2226 [1].

[Teo2] — and N. J. A. Sloane, *Magic numbers in polygonal and polyhedral atom counting*, Inorg. Chem. **24** (1985), 4545-4558 [1, 4].

[Teo3] — — *Atomic arrangements and electronic requirements for close-packed circular and spherical clusters*, Inorg. Chem. **25** (1986), 2315-2322 [1, 4].

[Ter1] A. Terras, *Harmonic Analysis on Symmetric Spaces and Applications* I, Springer-Verlag, 1985 [9].

J. Thas: see also P. J. Cameron.

[Tha1] — *Two infinite families of perfect codes in metrically regular graphs*, JCT **B23** (1977), 236-238 [9].

[Tha2] — *Polar spaces, generalized hexagons, and perfect codes*, JCT **A29** (1980), 87-93 [9].

B. R. C. Theobald: see C. E. Briant.

[Thi1] J. Thierry-Mieg, *Remarks concerning the* $E_8 \times E_8$ *and* D_{16} *string theories*, Phys. Lett. **B156** (1985), 199-202 [1].

[Thi2] — *Anomaly cancellation and Fermionisation in* 10, 18 *and* 26 *dimensional*

superstrings, Phys. Lett. **B171** (1986), 163-169 [1].
J. G. Thompson: see also F. J. MacWilliams, V. Pless.
[Tho1] — *Weighted averages associated to some codes*, Scripta Math. **29** (1973), 449-452 [7, 16].
[Tho2] — *A simple subgroup of* $E_8(3)$, in *Finite Groups Symposium*, ed. N. Iwahori, Japan Soc. Promotion Science, Tokyo, 1976, pp. 113-116 [10].
[Tho3] — *Finite groups and even lattices*, J. Alg. **38** (1976), 523-524 [3].
[Tho4] — *Uniqueness of the Fischer-Griess Monster*, BLMS **11** (1979), 340-346 [29].
[Tho5] — *Finite groups and modular functions*, BLMS **11** (1979), 347-351 [1, 29].
[Tho6] — *Some numerology between the Fischer-Griess Monster and the elliptic modular function*, BLMS **11** (1979), 352-353 [1, 29].
[Tho7] — personal communication [4, 5, 7, 8, 10, 17].
[Tho8] T. M. Thompson, *From Error-Correcting Codes through Sphere Packings to Simple Groups*, Math. Assoc. Am., Washington, DC, 1983.
[Thu1] A. Thue, *Uber die dichteste Zusammenstellung von kongruenten Kreisen in einer Ebene*, Norske Vid. Selsk. Skr., No. 1, 1910, pp. 1-9 [1].
[Tie1] A. Tietäväinen, *Bounds for binary codes just outside the Plotkin range*, IC **47** (1980), 85-93 [9].
R. Tijdeman: see C. Pomerance.
H. C. A. van Tilborg: see E. R. Berlekamp.
[Tit1] J. Tits, *Sur certaines classes d'espaces homogenes de groupes de Lie*, Mémoires Acad. Roy. Belg., Cl. Sci. **39** (1955), fasc. 3 [9].
[Tit2] — *Sur les systemes de Steiner associes aux trois "grands" groupes de Mathieu*, Rend. Math. e Appl. **23** (1964), 166-184 [11].
[Tit3] — *Classification of algebraic semisimple groups*, PSPM **9** (1966), 33-62 [10].
[Tit4] — *Le groupe de Janko d'ordre* 604,800, in [Bra2], 1969, pp. 91-95 [10].
[Tit5] — *Groupes finis simples sporadiques*, Sém. Bourbaki, **375**, 1970 [10].
[Tit6] — *Four presentations of Leech's lattice*, in [Col2], 1980, pp. 303-307 [4, 8, 10, 24].
[Tit7] — *Quaternions over* $\mathbf{Q}[\sqrt{5}]$, *Leech's lattice and the sporadic group of Hall-Janko*, J. Alg. **63** (1980), 56-75 [4, 8, 10, 24].
[Tit8] — *Le Monstre* (*d'après R. Griess, B. Fischer et al.*), Sém. Bourbaki **620**, Astérisque **121** (1985), 105-122 [10, 29].
[Tit9] — *On R. Griess' "Friendly Giant"*, Invent. Math. **78** (1984), 491-499 [10, 29].
J. A. Todd: see also H. S. M. Coxeter, G. C. Shephard.
[Tod1] — *The invariants of a finite collineation group in five dimensions*, PCPS **46** (1950), 73-90 [4].
[Tod2] — *The characters of a collineation group in five dimensions*, PRS **A200** (1950), 320-336 [4].
[Tod3] — *A representation of the Mathieu group* M_{24} *as a collineation group*, Ann. Mat. Pura Appl. **71** (1966), 199-238 [10].
N. Tokura: see T. Kasami.
[Tol1] R. Tolimieri, *The algebra of the finite Fourier transform and coding theory*, TAMS **287** (1985), 253-273 [7].
[Töl1] J. Tölke and J. M. Wills, editors, *Contributions to Geometry*, Birkhäuser, Boston, 1979 [B].
V. D. Tonchev: see also V. Pless.
[Ton1] — *Hadamard-type block designs and self-dual codes*, PPI **19** (No. 4, 1983),

25-30 = PIT **19** (1983), 270-274 [3].
[Ton2] — *Inequivalence of certain extremal self-dual codes*, Compt. Rend. Acad. Bulg. Sci. **36** (1983), 181-184 [7].
[Ton3] — *A characterization of designs related to the Witt system* $S(5, 8, 24)$, Math. Z. **191** (1986), 225-230 [11].
[Ton4] — and R. V. Raev, *Cyclic* 2-$(17, 8, 7)$ *designs and related doubly even codes*, Compt. Rend. Acad. Bulg. Sci. **35** (No. 10, 1982) [3].
E. M. Tory: see W. S. Jodrey.
[Tra1] D. Travis, *Spherical functions on finite groups*, J. Alg. **29** (1974), 65-76 [9].
D. W. Trenerry: see E. S. Barnes.
J. M. Tribolet: see J. L. Flanagan.
M. A. Tsfasman: see also G. L. Katsman, S. N. Litsyn, S. G. Vladuts.
[Tsf1] — *Goppa codes that are better than the Varshamov-Gilbert bound*, PPI **18** (No. 3, 1982), 3-6 = PIT **18** (1982), 163-165 [3].
[Tsf2] — S. G. Vladuts and T. Zink, *Modular curves, Shimura curves, and Goppa codes better than the Varshamov-Gilbert bound*, Math. Nach. **104** (1982), 13-28 [3, 9].
A. I. Tsitkin: see P. M. Gudivok.
A. W. Tucker: see R. J. Duffin.
[Tur1] R. Turyn, *Hadamard matrices, Baumert-Hall units, four-symbol sequences, pulse compression, and surface wave encodings*, JCT **A16** (1974), 313-333 [3].
W. T. Tutte: see J. J. Seidel.

U

[Ung1] G. Ungerboeck, *Channel coding with multilevel/phase signals*, PGIT **28** (1982), 55-67 [3].
[Ura1] H. Urakawa, *On the least positive eigenvalue of the Laplacian for the compact quotient of a certain Riemannian symmetric space*, Nagoya Math. J. **78** (1980), 137-152 [1, 9].

V

[Val1] A. Z. Val'fis, *On the representation of numbers by sums of squares – asymptotic formulae*, Usp. Mat. Nauk **7** (No. 6, 1952), 97-178 = Amer. Math. Soc. Transl. (2) **3** (1956), 163-248 [4].
[Val2] — *Uber Gitterpunkte in mehrdimensionalen Ellipsoiden*, Math. Z. **19** (1924), 300-307 [4].
[Val3] — *Gitterpunkte in Mehrdimensionalen Kugeln*, Państ. Wyd. Nauk., Warsaw, 1957 [4].
S. A. Vanstone: see D. M. Jackson.
F. D. Veldkamp: see M. Hazewinkel.
[Ven0] B. A. Venkov, *On a class of Euclidean polytopes*, Vestnik Leningrad. Univ., Ser. Mat. Fiz. Him. **9** (1954), 11-31 [21].
[Ven1] B. B. Venkov, *The classification of integral even unimodular* 24*-dimensional quadratic forms*, TMIS **148** (1978), 65-76 = PSIM (No. 4, 1980), 63-74 ≅ Chap. 18 of this book [18].
[Ven2] — *On odd unimodular lattices*, LOMI **86** (1979), 40-48 = JSM **17** (1981), 1967-1974 [17].
[Ven3] — *On even unimodular Euclidean lattices of dimension* 32, LOMI **116**

(1982), 44-45 and 161-162 = JSM **26** (1984), 1860-1867 [7].

[Ven4] — *Unimodular lattices and strongly regular graphs*, LOMI **129** (1983), 30-38 = JSM **29** (1985), 1121-1127.

[Ven5] — *Voronoi parallelohedra for certain unimodular lattices*, LOMI **132** (1983), 57-61 = JSM **30** (1985), 1833-1836 [2, 21].

[Ven6] — *On even unimodular Euclidean lattices of dimension* 32, II, LOMI **134** (1984), 34-58 = JSM **36** (1987), 21-38 [7].

[Ven7] — *Even unimodular extremal lattices*, TMIS **165** (1984), 43-48 = PSIM **165** (1984), 47-52 [7].

[Ver1] T. Verhoeff, *An updated table of minimum-distance bounds for binary linear codes*, PGIT **33** (1987), **33** (1987), 665-680 [3].

[Vet1] N. M. Vetchinkin, *The packings of uniform n-dimensional balls that are constructed from error-correcting codes*, Ivanov. Gos. Univ. Uc en. Zap. **89** (1974), 87-91 [5].

[Vet2] — *Uniqueness of the classes of positive quadratic forms on which the values of Hermite constants are attained for* $6 \leq n \leq 8$, TMIS **152** (1980) 34-86 = PSIM (1982, No. 3), pp. 37-95 [1, 4, 6].

[Vig1] M. F. Vigneràs, *Methodes analytiques*, in [Orz1], 1977, pp. 340-360 [2].

[Vil1] N. J. Vilenkin, *Special Functions and the Theory of Group Representations*, Nauka, Nauka, Moscow, 1965; English translation, Amer. Math. Soc., Providence RI, 1968 [9].

[Vin1] E. B. Vinberg, *Discrete groups generated by reflections in Lobachevskii spaces*, Mat. Sb **72** (No. 3, 1967), 471-488; **73** (1967), 303 = Sb. **1** (No. 3, 1967), 429-444 [4, 27, 28].

[Vin2] — *Some examples of crystallographic groups in Lobachevskii spaces*, Mat. Sb. **78** (1969), 633-639 = Sb. **7** (1969), 617-622 [4, 27, 28].

[Vin3] — *Discrete linear groups generated by reflections*, IANS **35** (1971), 1072-1112 = Izv. **5** (1971), 1083-1119 [4, 27, 28].

[Vin4] — *On the groups of units of certain quadratic forms*, Mat. Sb. **87** (No. 1, 1972), 18-36 = Sb. **16** (No. 1, 1972), 17-35 [4, 27, 28].

[Vin5] — *On unimodular integral quadratic forms*, Funkt. Analiz. Prilozen **6** (No. 2, 1972), 24-31 = Funct. Anal. Appl. **6** (1972), 105-111 [4, 16].

[Vin6] — *On the Schönflies-Bieberbach theorem*, DAN **221** (1975), 1013-1015 = SMD **16** (1975), 440-442 [4].

[Vin7] — *Some arithmetical discrete groups in Lobacevskii spaces*, in *Discrete Subgroups of Lie Groups and Applications to Moduli*, Oxford Univ. Press, 1975, pp. 323-348 [4, 27, 28, 30].

[Vin8] — *The nonexistence of crystallographic reflection groups in Lobachevskii spaces of large dimension*, Funkt. Analiz. Prilozen **15** (No. 2, 1981), 67-68 = Funct. Anal. Applic. **15** (1981), 128-130 [4, 27].

[Vin9] — *Discrete reflection groups in Lobachevskii spaces of large dimension*, LOMI **132** (1982), 62-68 = JSM **30** (1985), 1837-1841 [4, 27].

[Vin11] — *Discrete reflection groups in Lobachevskii spaces*, in *Proc. Intern. Congr. Math. Warsaw 1983*, North-Holland, Amsterdam, 1984, pp. 593-601 [4, 27].

[Vin12] — *On reflective hyperbolic lattices*, DAN **272** (No. 6, 1983), 1298-1301 = SMD **28** (1983), 517-520 [4, 27].

[Vin13] — *Absence of crystallographic reflection groups in Lobachevskii spaces of high dimension*, Trudy Moskov. Mat. Obshch. **47** (1984), 68-102 and 246 [4, 27] = Trans. Moscow Math. Soc. **47** (1985), 75-112.

[Vin14] — *Hyperbolic reflection groups*, Usp. Mat. Nauk. **40** (1985), 29-66 and 255 = Russ. Math. Surv. **40** (1985), 31-75 [4, 27].

[Vin15] — and I. M. Kaplinskaja, *On the groups* $O_{18,1}(Z)$ *and* $O_{19,1}(Z)$, DAN **238** (No. 6, 1978) 1273-1275 = SMD **19** (No. 1, 1978), 194-197 [4, 27, 28].

[Vit1] A. J. Viterbi and J. K. O'Mura, *Principles of Digital Communication and Coding*, McGraw-Hill, NY, 1979 [3].

S. G. Vladuts: see also G. L. Katsman, M. A. Tsfasman.

[Vla1] — *On the exhaustion boundary for algebraic-geometric "modular" codes*, PPI **23** (1987), 28-41 = PIT **23** (1987) 22-34 [3].

[Vla2] — G. L. Katsman and M. A. Tsfasman, *Modular curves and codes with a polynomial complexity of construction*, PPI **20** (No. 1, 1984), 47-55 = PIT **20** (1984), 35-42 [3].

[Vla3] — and Y. I. Manin, *Linear codes and modular curves*, Itogi Nauk. Tekh. **25** (1984), 209-257 = JSM **30** (No. 6, 1985), 2611-2643 [3].

[Vor1] G. F. Voronoi, *Nouvelles applications des paramètres continus à la théorie des formes quadratiques*, JRAM **133** (1908), 97-178; **134** (1908), 198-287; **136** (1909), 67-181 [2, 21].

[Vos1] H. Voskuil, *A special basis for the Leech lattice*, KNAW **A90** (1987), 73-86 [4].

C. de Vroedt: see C. Roos.

W

B. L. van der Waerden: see also W. Habicht, K. Schütte.

[Wae1] — *Punkte auf der Kugel. Drei Zusätze*, Math. Ann. **123** (1952), 213-222 [1].

[Wae2] — *Moderne Algebra*, Springer, Berlin, 2 vols., 1950; English translation, Ungar, NY, 1953 [8].

[Wae3] — *Die Reduktionstheorie der positiven quadratischen Formen*, Acta Math. **96** (1956), 263-309. Reprinted in part in [Wae5] [2, 15].

[Wae4] — *Pollenkörner, Punktverteilungen auf der Kugel und Informationstheorie*, Die Naturwissenschaften **7** (1961), 189-192.

[Wae5] — and H. Gross, editors, *Studien zur Theorie der Quadratischen Formen*, Birkhaüser, Basel, 1968 [2, 15, B].

S. L. Wagner: see K. T. Smith.

D. B. Wales: see also A. R. Calderbank, J. H. Conway, M. Hall, Jr.

[Wal1] — *Uniqueness of the graph of a rank three group*, PJM **30** (1969), 271-276 [10].

A. Z. Walfisz: see A. Z. Val'fis.

G. E. Wall: see E. S. Barnes.

J. S. Wallis: see W. D. Wallis.

W. D. Wallis: see also A. Hedayat.

[Wal2] — A. P. Street and J. S. Wallis, *Combinatorics: Room Squares, Sum-Free Sets, Hadamard Matrices*, LNM **292**, 1972 [3].

[Wan1] C.-S. Wang and A. Gersho, *A bound on the number of nearest neighbors of a Euclidean code*, in *Abstracts of Papers, IEEE Inter. Symp. Info. Theory, St. Jovite 1983*, IEEE Press, NY, 1983, p. 43 [2].

[Wan2] H.-C. Wang, *Two-point homogeneous spaces*, Ann. Math. **55** (1952), 177-191 [9].

Y. Wang: see L. K. Hua.

H. N. Ward: see also F. J. MacWilliams, V. Pless.

[War1] — *A form for* M_{11}, J. Alg. **37** (1975), 340-351 [11].

[War2] — *A restriction on the weight enumerator of a self-dual code*, JCT **A21**

(1976), 253-255 [7, 19].
[War3] G. Warner, *Harmonic Analysis on Semi-Simple Lie Groups*, Springer-Verlag, 2 vols., 1972 [9].
[War4] N. P. Warner, *The symmetry groups of the regular tessellations of S^2 and S^3*, PRS **A383** (1982), 379-398 [1].
S. Washburn: see E. Y. Miller.
[Was1] A. J. Wasserman, *The thirteen spheres problem*, Eureka **39** (1978), 46-49 [1].
[Wat1] D. F. Watson, *Computing the n-dimensional Delaunay tessellation with application to Voronoi polytopes*, Computer J. **24** (1981), 167-172 [2].
G. L. Watson: see also H. Davenport.
[Wat2] — *The covering of space by spheres*, Rend. Circ. Mat. Palermo **5** (1956), 93-100 [2, 21].
[Wat3] — *Integral Quadratic Forms*, Camb. Univ. Press, 1960 [15].
[Wat4] — *Transformations of a quadratic form which do not increase the class-number*, PLMS **12** (1962), 577-587 [15].
[Wat5] — *The class-number of a positive quadratic form*, PLMS **13** (1963), 549-576 [15].
[Wat6] — *Positive quadratic forms with small class-numbers*, PLMS **13** (1963), 577-592 [15].
[Wat7] — *One-class genera of positive quadratic forms*, JLMS **38** (1963), 387-392 [15].
[Wat8] — *On the minimum of a positive quadratic form in $n(\le 8)$ variables. Verification of Blichfeldt's calculations*, PCPS **62** (1966), 719 [1, 15].
[Wat9] — *On the minimal points of perfect septenary quadratic forms*, Math. **16** (1969), 170-177 [1, 15].
[Wat10] — *The number of minimum points of a positive quadratic form*, Dissertationes Math. **84** (1971), 42 pages [1, 5, 15].
[Wat11] — *On the minimum points of a positive quadratic form*, Math. **18** (1971), 60-70 [1, 15].
[Wat12] — *The least common denominator of the coefficients of a perfect quadratic form*, Acta Arith. **18** (1971), 29-36 [15].
[Wat13] — *The number of minimum points of a positive quadratic form having no perfect binary section with the same minimum*, PLMS **24** (1972), 625-646 [1, 15].
[Wat14] — *One-class genera of positive quaternary quadratic forms*, Acta Arith. **24** (1973), 461-475 [15].
[Wat15] — *One-class genera of positive quadratic forms in at least five variables*, Acta Arith. **26** (1975), 309-327 [15].
[Wat16] — *One-class genera of positive ternary quadratic forms*, Math. **19** (1972), 96-104; **22** (1975), 1-11 [15].
[Wat17] — *Transformations of a quadratic form which do not increase the class number* II, Acta Arith. **27** (1975), 171-189 [15].
[Wat18] — *The 2-adic density of a quadratic form*, Math. **23** (1976), 94-106 [15].
[Wat19] — *One-class genera of positive quadratic forms in nine and ten variables*, Math. **25** (1978), 57-67 [1, 15].
[Wat20] — *Determination of a binary quadratic form by its values at integer points*, Math. **26** (1979), 72-75; **27** (1980), 188 [2, 15].
[Wat21] — *One-class genera of positive quadratic forms in eight variables*, JLMS **26** (1982), 227-244 [15].
[Wat22] — *One-class genera of positive quadratic forms in seven variables*, PLMS **48** (1984), 175-192 [15].

G. N. Watson: see also E. T. Whittaker.

[Wat23] — *The problem of the square pyramid*, Messenger Math. **48** (1919), 1-22 [26].

[Wat24] G. S. Watson, *Statistics on Spheres*, Wiley, NY, 1983 [1].

[Weg1] G. Wegner, *Uber endliche Kreispackungen in der Ebene*, Studia Sci. Math. Hungar. **21** (1986), 1-28 [1].

[Wei1] L.-F. Wei, *Trellis-coded modulation with multi-dimensional constellations*, PGIT **33** (1987), 483-501 [3].

V. K. Wei: see A. R. Calderbank.

S. B. Weinstein: see G. J. Foschini.

[Wei2] E. Weiss, *Algebraic Number Theory*, McGraw-Hill, NY, 1963 [8].

G. L. Weiss: see W. M. Boothby, R. R. Coifman.

[Wei3] R. Weiss, *A geometric construction of Janko's group* J_3, Math. Z. **179** (1982), 91-95 [10].

[Wei4] — *On the geometry of Janko's group* J_3, Archiv Math. **38** (1982), 410-419 [10].

[Wei5] — *A characterization and another construction of Janko's group J_3*, TAMS **298** (1986), 621-633 [10].

A. J. Welch: see C. E. Briant.

L. R. Welch: see R. J. McEliece.

E. J. Weldon, Jr.: see also W. W. Peterson.

[Wel1] — *Justesen's construction — the low-rate case*, PGIT **19** (1973), 711-713 [3].

[Wel2] A. F. Wells, *Models in Structural Inorganic Chemistry*, Oxford Univ. Press, 1970 [1, 4].

[Wel3] — *Three-Dimensional Nets and Polyhedra*, Wiley, NY, 1977 [1, 4].

[Wel4] — *Structural Inorganic Chemistry*, Oxford Univ. Press, 5th ed., 1984 [1, 2, 4].

[Wel5] — *Survey of octahedral structures AX_n and A_2X_n*, Phil. Trans. Roy. Soc. London **A312** (1981), 553-600 [1].

[Wel6] G. R. Welti, *PCM/FDMA satellite telephony with 4-dimensionally-coded quadrature amplitude modulation*, COMSTAT Tech. Rev. **6** (1976), 323-338 [3].

[Wel7] — and S. L. Lee, *Digital transmission with coherent four-dimensional modulation*, PGIT **20** (1974), 497-502 [3].

[Wen1] M. J. Wenninger, *Polyhedron Models*, Camb. Univ. Press, 1971 [2, 21].

[Wen2] — *Spherical Models*, Camb. Univ. Press, 1979 [21].

A. T. White: see N. L. Biggs.

J. W. White: see C. E. Briant.

D. S. Whitehead: see N. J. A. Sloane.

[Whi1] T. A. Whitelaw, *On the Mathieu group of degree twelve*, PCPS **62** (1966), 351-364 [11].

[Whi2] — *Janko's group as a collineation group in $PG(6, 11)$*, PCPS **63** (1967), 663-677 [10].

G. J. Whitrow: see H. S. M. Coxeter.

[Wht1] E. T. Whittaker and G. N. Watson, *A Course of Modern Analysis*, Camb. Univ. Press, 4th ed., 1963 [4, 21].

[Why1] L. L. Whyte, *Unique arrangements of points on a sphere*, AMM **59** (1952), 606-611 [1].

M. M. Wiener: see L. Flatto.

[Wig1] E. Wigner and F. Seitz, *On the constitution of metallic sodium*, Phys. Rev. **43** (1933), 804-810 [2].

D. E. Williams: see Y. C. Lin.

J. M. Wills: see also G. Fejes Tóth, P. Gritzmann, P. M. Gruber, J. Tölke.

[Wil0] — *On the density of finite packings*, AMAH **46** (1985), 205-210 [1].

J. Wilson: see R. Askey.

R. A. Wilson: see also J. H. Conway, P. B. Kleidman, S. P. Norton.

[Wil1] — *The quaternionic lattice for* $2G_2(4)$ *and its maximal subgroups*, J. Alg. **77** (1982), 449-466 [8, 10].

[Wil2] — *The maximal subgroups of Conway's group* $\cdot 2$, J. Alg. **84** (1983), 107-114 [10].

[Wil3] — *The complex Leech lattice and maximal subgroups of the Suzuki group*, J. Alg. **84** (1983), 151-188 [7, 10].

[Wil4] — *The maximal subgroups of Conway's group* Co_1, J. Alg. **85** (1983), 144-165 [10].

[Wil5] — *The geometry and maximal subgroups of the simple groups of A. Rudvalis and J. Tits*, PLMS **48** (1984), 533-563 [10].

[Wil6] — *The subgroup structure of the Lyons group*, PCPS **95** (1984), 403-409 [10].

[Wil6a] — *On lexicographic codes of minimal distance* 4, Atti Sem. Mat. Fis. Univ. Modena **33** (1984), 125-135 [11].

[Wil7] — *The maximal subgroups of the Lyons group*, PCPS **97** (1985), 433-436 [10].

[Wil8] — *The maximal subgroups of the O'Nan group*, J. Alg. **97** (1985), 467-473 [10].

[Wil9] — *The geometry of the Hall-Janko group as a quaternionic reflection group*, Geom. Dedic. **20** (1986), 157-173 [8, 10].

[Wil10] — *On maximal subgroups of the Fischer group* Fi_{22}, PCPS **95** (1984), 197-222 [10].

[Wil11] — *Maximal subgroups of automorphism groups of simple groups*, JLMS **32** (1985), 460-466 [10].

[Wil12] — *Is* J_1 *a subgroup of the Monster*? BLMS **18** (1986), 349-350 [10].

[Wil13] — *Some subgroups of the Baby Monster*, Invent. Math. **89** (1987), 197-218 [10].

[Wil14] — *Some subgroups of the Thompson group*, JAMS **44** (1988), 17-32 [10].

[Wil15] — *The local subgroups of the Monster*, JAMS **44** (1988), 1-16 [10].

[Wil16] — *The subgroups of the Fischer groups*, JLMS **36** (1987), 77-94 [10].

[Wil17] — *Maximal subgroups of sporadic groups*, in *Groups-St. Andrews* 1985, ed. C. M. Campbell and E. F. Robertson, Camb. Univ. Press, 1986, pp. 352-358 [10].

R. M. Wilson: see also J. H. van Lint.

[Wil18] — *An existence theory for pairwise balanced designs* III: *proof of the existence conjectures*, JCT **A18** (1975), 71-79 [3].

[Wil19] — *On the theory of t-designs*, in [Jac1], 1984, pp. 19-49 [3].

[Wil20] S. G. Wilson, H. A. Sleeper and N. K. Smith, *Four-dimensional modulation and coding: an alternative to frequency-reuse*, in *Science, Systems and Services for Communications*, ed. P. Dewilde and C. A. May, IEEE, NY, 1984, pp. 919-923 [3].

G. L. Wise: see J. A. Bucklew.

H. S. Witsenhausen: see R. L. Graham.

[Wit1] E. Witt, *Theorie der quadratischen Formen in beliebigen Körpern*, JRAM

176 (1937), 31-44 [15].
[Wit2] — *Die 5-fach transitiven Gruppen von Mathieu*, ASUH **12** (1938), 256-264 [3, 10, 11].
[Wit3] — *Uber Steinersche Systeme*, ASUH **12** (1938), 265-275 [3, 10, 11, 12, 14].
[Wit4] — *Eine Identität zwischen Modulformen zweiten Grades*, ASUH **14** (1941), 323-337 [2, 7].
[Wit5] — *Spiegelungsgruppen und Aufzählung halbeinfacher Liescher Ringe*, ASUH **14** (1941), 289-322 [4].
[Wol0] A. Woldar, *On the maximal subgroups of Lyons' group*, Comm. in Alg. **15** (1987), 1195-1203 [10].
[Wol1] J. A. Wolf, *Spaces of Constant Curvature*, Publish or Perish, Boston MA, 5th ed., 1985 [9].
J. K. Wolf: see T. Berger.
P. M. de Wolff: see H. Burzlaff.
[Wol2] J. Wolfmann, *A permutation decoding of the* (24, 12, 8) *Golay code*, PGIT **29** (1983), 748-750 [20].
[Wol3] J. Wolfowitz, *The Coding Theorems of Information Theory*, Prentice-Hall, Englewood Cliffs NJ, 2nd ed., 1964 [3].
H. Wondratschek: see H. Brown, R. Bülow.
[Woo1] A. C. Woods, *Covering six-space with spheres*, JNT **4** (1972), 330-341 [2].
[Wor1] R. T. Worley, *The Voronoi region of* E_6^*, JAMS **A 48** (1987), 268-278 [4, 21].
[Wor2] — *The Voronoi region of* E_7^*, SIAD **1** (1988), 134-141 [4, 21].
[Woz1] J. M. Wozencraft and I. M. Jacobs, *Principles of Communication Engineering*, Wiley, NY, 1965 [3].
C. M. Wright: see E. L. Muetterties.
E. M. Wright: see G. H. Hardy.
[Wyc1] R. W. G. Wyckoff, *The Structure of Crystals*, Chemical Catalog Co., NY, 1924 [4].
[Wyc2] — *Crystal Structures*, Wiley, NY, 2nd ed., 6 vols., 1965 [4].
[Wyn1] A. D. Wyner, *Capabilities of bounded discrepancy decoding*, BSTJ **44** (1965), 1061-1122 [1].
[Wyn2] — *Random packings and coverings of the unit n-sphere*, BSTJ **46** (1967), 2111-2118 [2].
[Wyn3] — *Communication of analog data from a Gaussian source over a noisy channel*, BSTJ **47** (1968), 801-812 [3].
[Wyn4] — *On coding and information theory*, SIAM Rev. **11** (1969), 317-346 [3].
[Wyn5] — *Fundamental limits in information theory*, PIEEE **69** (1981), 239-251 [3].

Y

[Yag1] I. M. Yaglom, *Some results concerning distributions in n-dimensional space*, Appendix to the Russian edition of [Fej10], Moscow, 1958 [9].
[Yam1] Y. Yamada, S. Tazaki and R. M. Gray, *Asymptotic performance of block quantizers with difference distortion measures*, PGIT **26** (1980), 6-14 [2].
A. Yanushka: see E. Shult.
V. Y. Yorgov: see V. I. Iorgov.
[Yos1] S. Yoshiara, *The complex Leech lattice and the sporadic Suzuki group* (in Japanese), in *Topics in Finite Group Theory*, Kyoto Univ., 1982, pp. 26-46 [7].
[Yos2] — *The maximal subgroups of the sporadic simple group of O'Nan*, J. Fac. Sci. Univ. Tokyo **32** (1985), 105-141 [10].

Z

[Zad1] P. L. Zador, *Development and evaluation of procedures for quantizing multivariate distributions*, Ph.D. Dissertation, Stanford Univ., 1963 [2].

[Zad2] — *Asymptotic quantization error of continuous signals and their quantization dimension*, PGIT **28** (1982), 139-149 [2].

[Zag1] D. Zagier, *Zetafunktionen und Quadratische Korper*, Springer-Verlag, 1981 [15].

[Zar1] S. K. Zaremba, *La methode des "bon treillis" pour le calcul des integrales multiples*, in *Applications of Number Theory to Numerical Analysis*, ed. S. K. Zaremba, Ac. Press, NY, 1972, pp. 39-119 [1].

T. Zaslavsky: see P. D. Seymour.

H. J. Zassenhaus: see also H. Brown, A. G. Earnest, R. L. Graham.

[Zas1] — *Neuer Beweis der Endlichkeit der Klassenzahl bei unimodularer Äquivalenz endlicher ganzzahliger Substitutionsgruppen*, ASUH **12** (1938), 276-288 [14].

[Zas2] — *On the sphere packing problem*, Comm. Match in Math. Chem. **9** (1980), 127-135.

[Zet1] L. H. Zetterberg, *A class of codes for polyphase signals on a bandlimited Gaussian channel*, PGIT **11** (1965), 385-395 [3].

[Zet2] — and H. Brändström, *Codes for combined phase and amplitude modulated signals in a four-dimensional space*, COMM **25** (1977), 943-950 [3].

[Zim1] J. M. Ziman, *Models of Disorder*, Camb. Univ. Press, 1979 [1].

H. Zimmerman: see also H. Burzlaff.

[Zim2] — and H. Burzlaff, *DELOS-A computer program for the determination of a unique conventional cell*, Z. Krist. **170** (1985), 241-246 [2].

T. Zink: see M. A. Tsfasman.

V. A. Zinoviev: see also L. A. Bassalygo, N. V. Semakov.

[Zin1] — *A generalization of Johnson's bound for constant weight codes*, PPI **20** (No. 3, 1984), 105-108 [9].

[Zin2] — and S. N. Litsyn, *Shortening of codes*, PPI **20** (No. 1, 1984), 3-11 = PIT **20** (1984), 1-7 [3, 9].

[Zin3] — and S. N. Litsyn, *Codes that exceed the Gilbert bound*, PPI **21** (No. 1, 1985) 109-111 [3].

[Zin4] — — *A table of the best binary codes known*, preprint [3].

[Ziv1] J. Ziv, *On universal quantization*, PGIT **31** (1985), 344-347 [2].

[Ziv2] — and A. Lempel, *A universe algorithm for sequential data-compression*, PGIT **23** (1977), 337-343 [2].

G. Zolotareff: see A. Korkine.

I. J. Zucker: see also M. L. Glasser.

[Zuc1] — *New Jacobian theta functions and the evaluation of lattice sums*, J. Math. Phys. **16** (1975), 2189-2191; **17** (1976), 853 [2].

Supplementary Bibliography (1988-1998)

This supplementary bibliography covers the period 1988–1998, and also includes some earlier references that should have been included in the first edition. Besides the journal abbreviations used in the main bibliography, we also use

DCC	=	Designs, Codes and Cryptography
DCG	=	Discrete and Computational Geometry
EJC	=	European Journal of Combinatorics
JNB	=	Journal de Théorie des Nombres de Bordeaux (formerly Sém. Théor. Nombres Bordeaux)

Most (although not all) of these references are mentioned in the Preface to the Third Edition. (We apologize if we have overlooked any relevant papers, or, having listing them here, have failed to mention them in the Preface. No disrespect was intended.)

E. H. L. Aarts: see R. J. M. Vaessens.

[Abdu93] K. S. Abdukhalikov, *Integral invariant lattices in Lie algebras of type* A_{p^m-1} (in Russian), Mat. Sb. **184** (1993) = Russ. Acad. Sci. Sb. Math. **78** (1994), 447–478.

[AdM91] A. Adem, J. Maginnis and R. J. Milgram, *The geometry and cohomology of the Mathieu group* M_{12}, J. Alg. **139** (1991), 90–133.

[Adle81] A. Adler, *Some integral representations of* $PSL_2(\mathbb{F}_p)$ *and their applications*, J. Alg. **72** (1981), 115–145.

[AdB88] J.-P. Adoul and M. Barth, *Nearest neighbor decoding algorithms for spherical codes from the Leech lattice*, PGIT **34** (1988), 1188–1202.

D. Agrawal: see R. Urbanke.

[Agr96] E. Agrell, *Voronoi regions for binary linear block codes*, PGIT **42** (1996), 310–316.

[AgEr98] — and T. Eriksson, *Optimization of lattices for quantization*, PGIT **44** (1998), to appear.

[Ajt96] M. Ajtai, *Generating hard instances of lattice problems*, Electronic Colloq. Comp. Complexity, http://www.eccc.uni-trier.de/eccc/, Report TR96-007.

[Ajt97] — *The shortest vector problem in* L_2 *is NP-hard for randomized reductions*, Electronic Colloq. Comp. Complexity, http://www.eccc.uni-trier.de/eccc/, Report TR97-047.

[Allc96] D. Allcock, *Recognizing equivalence of vectors in the Leech lattice*, preprint, 1996.

[Allc97] — *New complex and quaternion-hyperbolic reflection groups*, preprint, 1997.

[Allc97a] — *The Leech lattice and complex hyperbolic reflections*, preprint, 1997.

[Alon97] N. Alon, *Packings with large minimum kissing numbers*, Discr. Math. **175** (1997), 249–251.

[AlBN92] — J. Bruck, J. Naor, M. Naor and R. M. Roth, *Construction of asymptotically good low-rate error-correcting codes through pseudo-random graphs*, PGIT **38** (1992), 509–516.

[Amr93] O. Amrani, Y. Be'ery and A. Vardy, *Bounded-distance decoding of the Leech lattice and the Golay code*, in *Algebraic Coding*, Lect. Notes Comp. Sci. **781** (1994), 236–248.

[Amr94] — — — F.-W. Sun and H. C. A. van Tilborg, *The Leech lattice and the Golay code: bounded-distance decoding and multilevel constructions*, PGIT **40** (1994), 1030–1043.

[Ang90] W. S. Anglin, *The square pyramid puzzle*, AMM **97** (1990), 102–124.

[Anz91] M. M. Anzin, *On variations of positive quadratic forms (with applications to the study of perfect forms)*, TMIS **196** (1991), 3–10.

[ArJ79] R. F. Arenstorf and D. Johnson, *Uniform distribution of integral points on 3–dimensional spheres via modular forms*, JNT **11** (1979), 218–238.

[ABB96] V. Arhelger, U. Betke and K. Böröczky, Jr., *Large finite lattice packings in* E^3, preprint, 1996.

[Asch94] M. Aschbacher, *Sporadic Groups*, Cambridge, 1994.

[AsL96] A. E. Ashikhmin and S. N. Litsyn, *Fast decoding algorithms for first order Reed-Muller and related codes*, DCC **7** (1996), 187–214.

[Atk82] K. Atkinson, *Numerical integration on the sphere*, JAMS **B 23** (1982), 332–347.

[AuC91] L. Auslander and M. Cook, *An algebraic classification of the three-dimensional crystallographic groups*, Advances Appl. Math. **12** (1991), 1–21.

[BaaGr97] M. Baake and U. Grimm, *Coordination sequences for root lattices and related graphs*, Zeit. Krist. **212** (1997), 253–256.

[BaaJK90] — D. Joseph, P. Kramer and M. Schlottmann, *Root lattices and quasicrystals*, J. Phys. **A 23** (1990), L1037–L1041.

[BaaPl95] — and P. A. B. Pleasants, *Algebraic solution of the coincidence problem in two and three dimensions*, Z. Naturforschung **50A** (1995) 711–717.

[BacSh96] E. Bach and J. O. Shallit, *Algorithmic Number Theory I: Efficient Algorithms*, MIT Press, Cambridge, MA, 1996.

[Bace94] R. Bacher, *Unimodular lattices without nontrivial automorphisms*, Int. Math. Res. Notes **2** (1994), 91–95.

[Bace97] — *Tables de réseaux entiers unimodulaires construits comme k-voisins de* $\mathbb{Z}^n$, JNB **9** (1997), 479–49.

[Bace96] — *Dense lattices in dimensions 28 and 29*, Invent. Math. **130** (1997), 153–158.

[BattVe98] — P. de la Harpe and B. B. Venkov, Séries de croissance et polynômes d'Ehrhart associés aux réseaux de racines, preprint.

[BaVe95] — and B. B. Venkov, *Lattices and association schemes: a unimodular example without roots*, Annales Inst. Fourier **45** (1995), 1163–1176.

[BaVe96] — — *Réseaux entiers unimodulaires sans racine en dimension 27 et 28*, Preprint No. 332, Inst. Fourier, Grenoble, 1996.

C. Bachoc: see also A. Bonnecaze.

[Baco95] — *Voisinages au sens de Kneser pour les réseaux quaternioniens*, Comm. Math. Helv. **70** (1995), 350–374.

[Baco97] — *Applications of coding theory to the construction of modular lattices*, JCT **A 78** (1997), 92–119.

[BacoB92] — and C. Batut, *Etude algorithmique de réseaux construits avec la forme trace*, Exper. Math. **1** (1992), 183–190.

[BacoN96] — and G. Nebe, *Classification of two genera of 32-lattices over the Hurwitz order*, Exper. Math. **6** (1997), 151–162.

[BacoN98] — — *Extremal lattices of minimum 8 related to the Mathieu group M_{22}*, JRAM **494** (1998), 155–171.

[Bag05] G. Bagnera, *I gruppi finiti di transforminazioni lineari dello spazio che contengono omologie*, Rend. Circ. Math. Palermo **19** (1905), 1–56.

[Baj91] B. Bajnok, *Construction of spherical 4- and 5-designs*, Graphs Combin. **7** (1991), 219–233.

[Baj91a] — *Construction of designs on the 2-sphere*, EJC **12** (1991), 377–382.

[Baj91b] — *Chebyshev-type quadrature formulas on the sphere*, Congr. Numer. **85** (1991), 214–218.

[Baj92] — *Construction of spherical t-designs*, Geom. Dedicata **43** (1992), 167–179.

[Baj96] — *Constructions of spherical 3-designs*, Graphs Combin. **14** (1998), 97–107.

[BanBl96] A. H. Banihashemi and I. F. Blake, *Trellis complexity and minimal trellis diagrams of lattices*, preprint, 1996.

[BanKh96] — and A. K. Khandani, *An inequality on the coding gain of densest lattices in successive dimensions*, preprint.

[BanKh97] — — *On the complexity of decoding lattices using the Korkin-Zolotarev reduced basis*, PGIT, to appear.

[BanDHO97] Eiichi Bannai, S. T. Dougherty, M. Harada and M. Oura, *Type II codes, even unimodular lattices and invariant rings*, preprint, June 1997.

[BanMO96] — S. Minashima and M. Ozeki, *On Jacobi forms of weight 4*, Kyushu J. Math. **50** (1996), 335–370.

[BanO96] — and M. Ozeki, *Construction of Jacobi forms from certain combinatorial polynomials*, Proc. Japan Acad. **A 72** (1996), 359–363.

[Bann90] Etsuko Bannai, *Positive Definite Unimodular Lattices with Trivial Automorphism Group*, Memoirs Amer. Math. Soc. **429** (1990), 1–70.

[Bara80] E. P. Baranovskii, *The Selling reduction domain of positive quadratic forms in five variables* (in Russian), TMIS **152** (1980) 5–33 = PSIM **3** (1982), 5–35.

[Bara91] — *Partitioning of Euclidean space into L-polytopes of some perfect lattices*, TMIS **196** (1991) 27–26 = PSIM (1992, No. 4), 29–51.

[Bara94] — *The perfect lattices $\Gamma(\mathcal{A}^n)$, and the covering density of $\Gamma(\mathcal{A}^9)$*, EJC **15** (1994), 317–323.

[Bart88] K. Bartels, *Zur Klassifikation quadratischer Gitter über diskreten Bewertungsringen*, Ph. Dissertation, Göttingen, 1988.

M. Barth: see J.-P. Adoul.

[Barv90] A. I. Barvinok, *The method of Newton sums in problems of combinatorial optimization* (in Russian), Disk. Mat. **2** (1990), 3–15.

[Barv91] — *Problems of combinatorial optimization, statistical sums and representations of the general linear group* (in Russian), Mat. Zamet. **49** (1991), 3–11 = Math. Notes **49** (1991), 3–9.

[Barv92] — *The method of statistical sums in combinatorial optimization problems*, Leningrad Math. J. **2** (1991), 987–1002.

[Barv92a] — *Optimization problems on lattices and theta functions*, preprint.

[Barv97] — *Lattice points and lattice polytopes*, Chap. 7 of [GoO'R97].

C. Batut: see also C. Bachoc.

[BatB91] — D. Bernardi, H. Cohen and M. Olivier, *User's Guide to PARI-GP*, 1991.

[Bat98] — *Classification of quintic eutactic forms*, Math. Comp., to appear.

[BatMa94] — and J. Martinet, *Radiographie des réseaux parfaits*, Exper. Math. **3** (1994), 39–49.

[BatQS95] — H.-G. Quebbemann and R. Scharlau, *Computations of cyclotomic lattices*, Exper. Math. **4** (1995), 175–179.

[Bay89] E. Bayer-Fluckiger, *Réseaux unimodulaires*, JNB **1** (1989), 535–589.

[Bay94] — *Galois cohomology and the trace form*, Jber. Deutsche Math.–Verein, **96** (1994), 35–55.

[Bay97] — *Lattices with automorphisms of given characteristic polynomial*, preprint.

[BayFa96] — and L. Fainsilber, *Non-unimodular Hermitian forms*, Invent. Math. **123** (1996), 233–240.

[BayM94] — and J. Martinet, *Formes quadratiques liées aux algèbres semi-simples*, JRAM **451** (1994), 51–69.

[BeaH89] S. Beale and D. K. Harrison, *A computation of the Witt index for rational quadratic forms*, Aeq. Math. **38** (1989), 86–98.

Y. Be'ery: see also O. Amrani, J. Snyders, A. Vardy.

[BeeSh91] — and B. Shahar, *VLSI architectures for soft decoding of the Leech lattice and the Golay codes*, Proc. IEEE Int. Workshop on Microelectronics in Comm., Interlaken, Switzerland, March 1991.

[BeeSh92] — — and J. Snyders, *Fast decoding of the Leech lattice*, IEEE J. Select. Areas Comm. **7** (1989), 959–967.

J. C. Belfiore: see J. Boutros.

[BenEHH] J. W. Benham, A. G. Earnest, J. S. Hsia and D. C. Hung, *Spinor regular positive definite ternary quadratic forms*, JLMS **42** (1990), 1–10.

[BenCa87] D. Benson and J. Carlson, *Diagrammatic methods for modular representations and cohomology*, Comm. Alg. **15** (1987), 53–121.

[Ber93] A.-M. Bergé, *Minimal vectors of a pair of dual lattices*, JNT **52** (1995), 284–298.

[BerM85] — and J. Martinet, *Sur la constante d'Hermite (étude historique)*, JNB **2** (No. 8, 1985), 1–15.

[BerM89] — — *Sur un problème de dualité lié aux sphères en géométrie des nombres*, JNT **32** (1989), 14–42.

[BerM89a] — — *Réseaux extrêmes pour un groupe d'automorphismes*, Astérisque **198–200** (1992), 41–46.

[BerM95] — — *Densité dans des familles de réseaux: application aux réseaux isoduaux*, L'Enseign. Math. **41** (1995), 335–365.

[BerM96] — — *Sur la classification des réseaux eutactiques*, JLMS **53** (1996), 417–432.

[BerMS92] — — and F. Sigrist, *Une généralisation de l'algorithme de Voronoi pour les formes quadratiques*, Astérisque **209** (1992), 137–158.

D. Bernardi: see C. Batut.

[BerSl97] M. Bernstein and N. J. A. Sloane, *Some lattices obtained from Riemann surfaces*, in *Extremal Riemann Surfaces (Contemporary Math. Vol. 201)*, J. R. Quine and P. Sarnak, Eds., Amer. Math. Soc., Providence, RI, 1997, pp. 29–32.

[BerSl97a] — — and P. E. Wright, *On sublattices of the hexagonal lattice*, Discr. Math. **170** (1997), 29–39.

D. Bessis: see H. J. Herrmann.

U. Betke: See also V. Arhelger.

[BetB96] — and K. Böröczky, Jr., *Asymptotic formulae for the lattice point enumerator*, preprint, 1996.

[BetB97] — — *Finite lattice packings and the Wulff-shape*, preprint, 1997.

[BetG84] — and P. Gritzmann, *Ueber L. Fejes Tóth's Wurstvermutung in kleinen Dimensionen*, AMAH **43** (1984), 299–307.

[BetG86] — — *An application of valuation theory to two problems in discrete geometry*, Discr. Math. **58** (1986), 81–85.

[BetGW82] — — and J. M. Wills, *Slices of L. Fejes Tóth's sausage conjecture*, Math. **29** (1982), 194–201.

[BetH98] — and M. Henk, *Finite packings of spheres*, DCG **19** (1998), 197–227.

[BetHW94] — — and J. M. Wills, *Finite and infinite packings*, JRAM **453** (1994), 165–191.

[BetHW95] — — — *Sausages are good packings*, DCG **13** (1995), 297–311.

[BetHW95a] — — — *A new approach to covering*, Math. **42** (1995), 251–263.

[BezBeCo] A. Bezdek, K. Bezdek and R. Connelly, *Finite and uniform stability of sphere packings*, preprint.

[BezK90] — and W. Kuperberg, *Maximum density space packings with congruent circular cylinders of infinite length*, Math. **37** (1990), 74–80.

[BezKu91] — — *Packing Euclidean space with congruent cylinders and with congruent ellipsoids*, in *Victor Klee Festschrift*, ed. P. Gritzmann and B. Sturmfels, DIMACS Series Vol. 4, Amer. Math. Soc., 1991, pp. 71–80.

[BezKM91] — — and E. Makai, Jr., *Maximum density space packing with parallel strings of spheres*, DCG **6** (1991), 277–283.

K. Bezdek: See also A. Bezdek.

[Bez97] — *Isoperimetric inequalities and the dodecahedral conjecture*, IJM **8** (1997), 759–780.

[BezCK85] — R. Connelly and G. Kertész, *On the average number of neighbors in a spherical packing of congruent circles*, in *Intuitive Geometry, Siofok 1985*, North-Holland, Amsterdam, 1987, pp. 37–52.

[BiEd98] J. Bierbrauer and Y. Edel, *Dense sphere packings from new codes*, preprint, 1998.

E. Biglieri: see also E. Viterbo.

[BDMS] — D. Divsalar, P. J. McLane and M. K. Simon, *Introduction to Trellis-Coded Modulation with Applications*, Macmillan, New York, 1991.

[Bla91] R. E. Blahut, *The Gleason-Prange theorem*, PGIT **37** (1991), 1269–1273.

I. F. Blake: see also A. H. Banihashemi, V. Tarokh.

[BlTa96] — and Tarokh, *On the trellis complexity of the densest lattices in* $\mathbb{R}^N$, SIAM J. Discr. Math. **9** (1996), 597–601.

[Blij59] F. van der Blij, *An invariant of quadratic forms mod 8*, Indag. Math. **21** (1959), 291–293.

[BocS91] S. Böcherer and R. Schulze-Pillot, *Siegel modular forms and theta series attached to quaternion algebras*, Nagoya Math. J. **121** (1991), 35–96.

[BocS97] — — *Siegel modular forms and theta series attached to quaternion algebras II*, Nagoya Math. J. **147** (1997), 71–106.

[BoHW72] J. Bokowski, H. Hadwiger and J. M. Wills, *Eine Ungleichung zwischen Volumen, Oberfläche und Gitterpunktanzahl konvexer Körper im n-dimensionalen euklidischen Raum*, Math. Zeit. **127** (1972), 363–364.

[BKT87] A. I. Bondal, A. I. Kostrikin and P. H. Tiep, *Invariant lattices, the Leech lattice and its even unimodular analogs in the Lie algebras* A_{p-1}, Mat. Sb. **130** (1986), 435–464 = Sb. **58** (1987), 435–465.

[Bonn90] P. G. Bonneau, *Weight distribution of translates of MDS codes*, Combinatorica **10** (1990), 103–105.

A. Bonnecaze: see also C. Bachoc.

[BonCS95] — A. R. Calderbank and P. Solé, *Quaternary quadratic residue codes and unimodular lattices*, PGIT **41** (1995), 366–377.

[BonGHKS] — P. Gaborit, M. Harada, M. Kitazume and P. Solé, *Niemeier lattices and Type II codes over* $\mathbb{Z}_4$, preprint.

[BonMS97] — B. Mourrain and P. Solé, *Jacobi polynomials, Type II codes, and designs*, DCC, submitted.

[BonS94] — and P. Solé, *Quaternary constructions of formally self-dual binary codes and unimodular lattices*, in *Algebraic Coding*, Lect. Notes Comp. Sci. **781** (1994), 194–205.

[BonSBM] — — C. Bachoc and B. Mourrain, *Type II codes over* $\mathbb{Z}_4$, *PGIT* **43** (1997), 969–976. PGIT **43** (1997), 969–976.

[Borch86] R. E. Borcherds, *Vertex algebras, Kac-Moody algebras, and the Monster*, PNAS **83** (1986), 3068–3071.

[Borch87] — *Automorphism groups of Lorentzian lattices*, J. Alg. **111** (1987), 133–153.

[Borch88] — *Generalized Kac-Moody algebras*, J. Alg. **115** (1988), 501–512.

[Borch90] — *Lattices like the Leech lattice*, J. Alg. **130** (1990), 219–234.

[Borch90a] — *The monster Lie algebra*, Adv. Math. **83** (1990), 30–47.

[Borch91] — *Central extensions of generalized Kac-Moody algebras*, J. Alg. **140** (1991), 330–335.

[Borch92] — *Introduction to the Monster Lie algebra*, in *Groups, Combinatorics and Geometry*, ed. M. W. Liebeck and J. Saxl, Camb. Univ. Press, 1992, pp. 99–107.

[Borch92a] — *Monstrous moonshine and monstrous Lie superalgebras*, Invent. Math. **109** (1992), 405–444.

[Borch94] — *Sporadic groups and string theory*, in *Proc. First European Congress of Math. (Paris, July 1992)*, ed. A. Joseph et al., Birkhauser, Vol. 1 (1994), pp. 411–421.

[Borch95] — *A characterization of generalized Kac-Moody algebras*, J. Alg. **174** (1995), 1073–1079.

[Borch95a] — *Automorphic forms on* $O_{s+2,2}(\mathbb{R})$ *and infinite products*, Invent. Math. **120** (1995), 161–213.

[Borch95b] — *Automorphic forms on* $O_{s+2,2}(\mathbb{R})^+$ *and generalized Kac-Moody algebras*, Proc. Internat. Congress Math. (Zürich, 1994), Vol II, pp. 744–742.

[Borch96] — *The moduli space of Enriques surfaces and the fake Monster Lie superalgebra*, Topology **35** (1996), 699–710.

[Borch96a] — *Modular moonshine III*, DMJ, to appear, 1998.

[Borch96b] — *Automorphic forms and Lie algebras*, in *Current Developments in Mathematics 1996*, to appear.

[BorchHP] — *Home page*: http://www.dpmms.cam.ac.uk/~reb/.

[BorchF98] — E. Freitag and R. Weissauer, *A Siegel cusp form of degree 12 and weight 12*, JRAM **494** (1998), 141-153.

[BorchR96] — and A. J. E. Ryba, *Modular moonshine II*, DMJ **83** (1996), 435–459.

K. Böröczky, Jr., see also V. Arhelger, U. Betke.

[BorSch97] — and U. Schnell, *Asymptotic shape of finite packings*, CJM, to appear.

[BorSch98] — — *Quasicrystals and the Wulff-shape*, DCG, to appear.

[BorSch98a] — — *Wulff-shape for non-peridic arrangements*, Lett. Math. Phys., to appear.

[BorWi96] — and J. M. Wills, *Finite sphere packing and critical radii*, Contributions to Algebra and Geometry **38** (1997), 193–211.

[BosC97] W. Bosma and J. Cannon, *Handbook of Magma Functions*, Sydney, May 22, 1995.

[BosCM94] — — and G. Mathews, *Programming with algebraic structures: Design of the Magma language*, in *Proceedings of the 1994 International Symposium on Symbolic and Algebraic Computation*, M. Giesbrecht, Ed., Association for Computing Machinery, 1994, 52–57.

[BosCP97] — — and C. Playoust, *The Magma algebra system I: The user language*, J. Symb. Comp. **24** (1997), 235–265.

[BoVRB] J. Boutros, E. Viterbo, C. Rastello and J. C. Belfiore, *Good lattice constellations for both Rayleigh fading and Gaussian channels*, PGIT **42** (1996), 502–518.

[BoV98] — — *A power and bandwidth efficient diversity technique for the Rayleigh fading channel*, PGIT to appear.

[Boy93] P. Boyvalenkov, *Nonexistence of certain symmetric spherical codes*, DCC **3** (1993), 69–74.

[Boy94] — *On the extremality of the polynomials used for obtaining the best known upper bounds for the kissing numbers*, J. Geom. **49** (1994), 67–71.

[Boy94a] — *Small improvements for the upper bounds for the kissing numbers in dimensions 19, 21 and 23*, Atti Sem. Mat. Fis. Univ. Modena **LXII** (1994), 159–163.

[Boy95] — *Computing distance distributions of spherical designs*, Lin. Alg. Applications **226–228** (1995), 277–286.

[Boy95a] — *Extremal polynomials for obtaining bounds for spherical codes and designs*, DCG **14** (1995), 167–183.

[BoyD95] — and D. Danev *Classification of spherical codes attaining the even Levenshtein bound*, in *Proc. Workshop on Optimal Codes and Related Topics*, Sozopol, Bulgaria, May 1995, pp. 21–24.

[BoyD95a] — — *On linear programming bounds for codes in polynomial metric spaces*, PPI, to appear.

[BoyD97] — — *On maximal codes in polynomial metric spaces*, Lect. Notes Comp. Sci. **1255** (1997), 29–38.

[BoyDB96] — — and S. P. Bumova, *Upper bounds on the minimum distance of spherical codes*, PGIT **42** (1996), 1576–1581.

[BoyDL97] — — and I. Landgev, *On maximal spherical codes II*, J. Combin. Designs, to appear.

[BoyDM] — — and M. Mitradjieva, *Linear programming bounds for codes in infinite projective spaces*, preprint.

[BoyDN] — — and S. Nikova, *Nonexistence of certain spherical designs of odd strengths and cardinalities*, DCG to appear.

[BoyD93] — and S. M. Dodunekov, *Some new bounds on the kissing numbers*, in *Proc. Sixth Joint Swedish-Russian Internat. Workshop on Information Theory*, Möele, 1993, pp. 389–393.

[BoyK94] — and P. Kazakov, *Nonexistence of certain spherical codes*, C. R. Bulgarian Acad. Sci. **47** (1994), 37–40.

[BoyK95] — — *New upper bounds for some spherical codes*, Serdica Math. J. **21** (1995), 231–238.

[BoyL95] — and I. Landgev, *On maximal spherical codes I*, in *Proc. XI AAECC Symposium*, Paris, 1995, Lect. Notes Comp. Sci. **948**, pp. 158–168.

[BoyN94] — and S. Nikova, *New lower bounds for some spherical designs*, in *Algebraic Coding*, Lect. Notes Comp. Sci. **781** (1994), 207–216.

[BoyN95] — — *Improvements of lower bounds for some spherical designs*, Math. Balcanika, to appear.

[BoyN97] — — *On lower bounds on the size of designs in compact symmetric spaces of rank 1*, Archiv. Math. **68** (1997), 81–88.

[Bro92] A. E. Brouwer, personal communication, 1992.

[Bro95] — *The linear programming bound for binary linear codes*, PGIT **39** (1995), 677–680.

[Bro98] — *Tables of bounds for linear codes*, to appear in [PHB].

[BrCN89] — A. M. Cohen and A. Neumaier, *Distance-Regular Graphs*, Springer-Verlag, NY, 1989.

[BrHOS] — H. O. Hämäläinen, P. R. J. Östergård, and N. J. A. Sloane, *Bounds on mixed binary ternary codes*, PGIT **44** (1998), 140–161.

[BrSSS] — J. B. Shearer, N. J. A. Sloane and W. D. Smith, *A new table of constant weight codes*, PGIT **36** (1990), 1334–1380.

[BrVe93] — and T. Verhoeff, *An updated table of minimum-distance bounds for binary linear codes*, PGIT **39** (1993), 662–677.

R. A. Brualdi: see also V. Pless.

[BrLP98] — S. Litsyn and V. S. Pless, *Covering radius*, in [PHB], to appear.

[BrP91] — and V. S. Pless, *Weight enumerators of self-dual codes*, PGIT **37** (1991), 1222–1225.

[Brau40] H. Braun, *Geschlecter quadratischer Formen*, JRAM **182** (1940), 32–49.

J. Bruck: see N. Alon.

[Bru86] N. D. de Bruijn, *Quasicrystals and their Fourier transform*, KNAW **48** (1986), 123–152.

G. O. Brunner: see also R. W. Grosse-Kunstleve.

[Brunn79] — *The properties of coordination sequences and conclusions regarding the lowest possible density of zeolites*, J. Solid State Chem. **29** (1979), 41–45.

[BrLa71] — and F. Laves, *Zum problem der koordinationszahl*, Wiss. Z. Techn. Univ. Dresden **20** (1971), 387–390.

[Bud89] R. de Buda, *Some optimal codes have structure*, IEEE J. Select. Areas Commun. **7** (1989), 893–899.

[Bue89] D. A. Buell, *Binary Quadratic Forms*, Springer-Verlag, NY, 1989.

[Bul73] R. Bülow, *Ueber Dadegruppen in* $GL(5, \mathbb{Z})$, Dissertation, Rheinisch-Westfälische Tech. Hochschule, Aachen, 1973.

S. P. Bumova: See P. G. Boyvalenkov.

[BuCD94] P. Buser, J. H. Conway, P. G. Doyle and K.-D. Semmler, *Some planar isospectral domains*, Inter. Math. Res. Notes **9** (1994), 391–400.

[BuSa94] — and P. Sarnak, *On the period matrix of a Riemann surface of large genus*, Invent. Math. **117** (1994), 27–56.

[BuTo89] F. C. Bussemaker and V. D. Tonchev, *New extremal doubly-even codes of length 56 derived from Hadamard matrices of order 28*, Discr. Math. **76** (1989), 45–49.

[BuTo90] — — *Extremal doubly-even codes of length 40 derived from Hadamard matrices*, Discr. Math. **82** (1990), 317–321.

A. R. Calderbank: see also A. Bonnecaze, G. D. Forney, Jr., A. R. Hammons, Jr.

[Cal91] — *The mathematics of modems*, Math. Intell. **13** (1991), 56–65.

[CaD92] — and P. Delsarte, *On error-correcting codes and invariant linear forms*, SIAM J. Discr. Math. **6** (1993), 1–23.

[CaD92a] — — *Extending the* t*-design concept*, TAMS, to appear.

[CaDS91] — — and N. J. A. Sloane, *A strengthening of the Assmus-Mattson theorem*, PGIT **37** (1991), 1261–1268.

[CaFR95] — P. C. Fishburn and A. Rabinovich, *Covering properties of convolutional codes and associated lattices*, PGIT **41** (1995), 732–746.

[CaFV98] — G. D. Forney, Jr. and A. Vardy, *Minimal tail-biting trellises: the Golay code and more*, PGIT, to appear.

[CHRSS] — R. H. Hardin, E. M. Rains, P. W. Shor and N. J. A. Sloane, *A group-theoretic framework for the construction of packings in Grassmannian space*, J. Algeb. Combin. (1998), to appear.

[CaMc97] — and G. McGuire, *Construction of a* $(64, 2^{37}, 12)$ *code via Galois rings*, DCC **10** (1997), 157–165.

[CaO90] — and L. H. Ozarow, *Nonequiprobable signaling on the Gaussian channel*, PGIT **36** (1990), 726–740.

[CaSl95] — and N. J. A. Sloane, *Modular and p-adic cyclic codes*, DCC **6** (1995), 21–35.

[CaSl97] — — *Double circulant codes over* $\mathbb{Z}_4$ *and even unimodular lattices*, J. Algeb. Combin. **6** (1997), 119–131.

J. Carlson: see D. Benson.

[ChaKR] W. Chan, M. H. Kim and S. Ragavan, *Ternary universal integral quadratic forms over real quadratic fields*, Indian J. Math., to appear.

[Cha82] G. R. Chapman, *Generators and relations for the cohomology ring of Janko's first group in the first twenty-one dimensions*, in *Groups – St. Andrews 1981*, London Math. Soc. Lect. Notes **71**, Cambridge Univ. Press, 1982.

[Chap96] R. Chapman, *Quadratic residue codes and lattices*, preprint.

[Chap97] — *Higher power residue codes*, Finite Flds. Applic. **3** (1997), 353–369.

[ChS96] — and P. Solé, *Universal codes and unimodular lattices*, JNB **8** (1996), 369–376.

[Chen91] C. L. Chen, *Construction of some binary linear codes of minimum distance five*, PGIT **37** (1991), 1429–1432.

W. Chen: see D. Li.

[Chow92] T. Y. Chow, *Penny-packings with minimal second moments*, Combinatorica **15** (1995), 151–158.

[Chu84] F. R. K. Chung, *The number of different distances determined by n points in the plane*, JCT **A 36** (1984), 342–354.

A. M. Cohen: see also A. E. Brouwer.

[Coh91] — *Presentations for certain finite quaternionic reflection groups*, in *Advances in Finite Geometries and Designs*, ed. J. W. P. Hirschfeld et al., Oxford, 1991, pp. 69–79.

[CNP96] — G. Nebe and W. Plesken, *Cayley orders*, Composit. Math. **103** (1996), 63–74.

[CHLL] G. D. Cohen, I. Honkala, S. N. Litsyn and A. C. Lobstein, *Covering Codes*, North-Holland, Amsterdam, 1997.

[CHLS] — — — and P. Solé, *Long packing and covering codes*, PGIT **43** (1997), 1617–1619.

[CLLM97] — S. N. Litsyn, A. C. Lobstein and H. F. Mattson, Jr., *Covering radius 1985–1994*, Applicable Algebra in Engineering, Communication and Computing, **8** (1997), 173–239.

H. Cohen: see also C. Batut.

[CohCNT] — *A Course in Computational Number Theory*, Springer-Verlag, NY, 1993.

[Coh92] S. D. Cohen, *The explicit construction of irreducible polynomials over finite fields*, DCC **2** (1992), 169–174.

R. Connelly: see A. Bezdek, K. Bezdek.

J. H. Conway: see also P. Buser, N. J. A. Sloane.

[Con92] — *The orbifold notation for surface groups*, in *Groups, Combinatorics and Geometry*, ed. M. W. Liebeck and J. Saxl, Camb. Univ. Press, 1992, pp. 438–447.

[Con93] — *From hyperbolic reflections to finite groups*, in *Groups and Computation*, DIMACS Series in Discrete Math., Vol. 11, Amer. Math. Soc., 1993, pp. 41–51.

[CoFu97] — assisted by P. Y. C. Fung, *The Sensual (Quadratic) Form*, Math. Assoc. Amer. 1997.

[CoHMS] — T. C. Hales, D. J. Muder and N. J. A. Sloane, *On the Kepler conjecture*, Math. Intelligencer **16** (No. 2, 1994), 5.

[CoHS96] — R. H. Hardin, and N. J. A. Sloane, *Packing lines, planes, etc.: packings in Grassmannian space*, Exper. Math. **5** (1996), 139–159.

[CoPS92] — V. Pless and N. J. A. Sloane, *The binary self-dual codes of length up to 32: A revised enumeration*, JCT **A 60** (1992), 183–195.

[CoSch98] — and W. A. Schneeberger, in preparation.

[CSLDL1] — and N. J. A. Sloane, *Low-dimensional lattices I: Quadratic forms of small determinant*, PRS **A 418** (1988), 17–41. (For errata, see end of Preface to 3rd Edition.)

[CSLDL2] — — *Low-dimensional lattices II: Subgroups of $GL(n, \mathbb{Z})$*, PRS **A 419** (1988), 29–68. (For errata, see end of Preface to 3rd Edition.)

[CSLDL3] — — *Low-dimensional lattices III: Perfect forms*, PRS **A 418B** (1988), 43–80. (For errata, see end of Preface to 3rd Edition.)

[CSLDL4] — — *Low-dimensional lattices IV: The mass formula*, PRS **A 419** (1988), 259–286. (For errata, see end of Preface to 3rd Edition.)

[CSLDL5] — — *Low-dimensional lattices V: Integral coordinates for integral lattices*, PRS **A 426** (1989), 211–232.

[CSLDL6] — — *Low-dimensional lattices VI: Voronoi reduction of three-dimensional lattices*, PRS **A 436** (1991), 55–68.

[CSLDL7] — — *Low-dimensional lattices VII: Coordination sequences*, PRS **A 453** (1997), 2369–2389.

[CSLDL8] — — *Low-dimensional lattices VIII: The 52 four-dimensional parallelotopes*, in preparation.

[CoSl90] — — *A new upper bound for the minimum of an integral lattice of determinant one*, BAMS **23** (1990), 383–387; **24** (1991), 479.

[CoSl90a] — — *A new upper bound for the minimal distance of self-dual codes*, PGIT **36** (1990), 1319–1333.

[CoSl90b] — — *Orbit and coset analysis of the Golay and related codes*, PGIT **36** (1990), 1038–1050.

[CoSl91] — — *Lattices with few distances*, JNT **39** (1991), 75–90.

[CoSl91a] — — *The cell structures of certain lattices*, in *Miscellanea mathematica*, P. Hilton, F. Hirzebruch and R. Remmert, editors, Springer-Verlag, NY, 1991, pp. 71–107.

[CoSl92] — — *On the covering multiplicity of lattices*, DCG **8** (1992), 109–130.

[CoSl92a] — — *Four-dimensional lattices with the same theta series*, DMJ **6** (Inter. Math. Res. Notices 4) 1992, 93–96.

[CoSl93] — — *Self-dual codes over the integers modulo 4*, JCT **A 62** (1993), 30–45.

[CoSl94] — — *On lattices equivalent to their duals*, JNT **48** (1994), 373–382.

[CoSl94a] — — *D_4, E_8, Leech and certain other lattices are symplectic*. Appendix to P. Buser and P. Sarnak [BuSa94].

[CoSl94b] — — *Quaternary constructions for the binary single-error-correcting codes of Julin, Best and others*, DCC **4** (1994), 31–42.

[CoSl95] — — *A lattice without a basis of minimal vectors*, Math. **42** (1995), 175–177.

[CoSl95a] — — *What are all the best sphere packings in low dimensions?*, DCG **13** (1995), 383–403.

[CoSl96] — — *The antipode construction for sphere packings*, Invent. Math. **123** (1996), 309–313.

[CoSl98] — — *A note on unimodular lattices*, JNT (submitted).

[CoSl98a] — — *Codes and lattices*, PGIT, to appear.

M. Cook: see L. Auslander.

[CoRa93] R. Cools and P. Rabinowitz, *Monomial cubature rules since "Stroud": a compilation*, J. Comput. Appl. Math. **48** (1993), 309–326.

[Cogn94] R. Coulangeon, *Réseaux quaternioniens et invariant de Venkov*, Manuscripta Math. **82** (1994), 41–50.

[Cogn95] — *Réseaux unimodulaires quaternioniens en dimension inférieure ou égale à 32*, Acta Arith. **71** (1995), 9–24.

[Cogn96] — *Réseaux k-extrêmes*, PLMS **73** (1996), 555–574.

[Cogn97] — *Minimal vectors in the second exterior power of a lattice*, J. Alg. **194** (1997), 467–476.

[Cogn98] — *Tensor products of Hermitian lattices*, preprint.

[Coul91] D. Coulson, *The dual lattice of an extreme six-dimensional lattice*, JAMS **A 50** (1991), 373–383.

[Coxe95] H. S. M. Coxeter, *Kaleidoscopes: Selected Writings*, ed. F. A. Sherk et al., Wiley, NY, 1995.

[CrL90] J. Cremona and S. Landau, *Shrinking lattice polyhedra*, SIAM J. Discr. Math. **3** (1990), 338–348.

[Cumm] C. J. Cummins, *Moonshine bibliography*, published electronically at http://www-cicma.concordia.ca/faculty/cummins/moonshine.refs.html.

[Cur89] R. T. Curtis, *Further elementary techniques using the Miracle Octad Generator*, Proc. Edin. Math. Soc. **32** (1989), 345–353.

[Cur89a] — *Natural constructions of the Mathieu groups*, PCPS **106** (1989), 423–429.

[Cur90] — *Geometric interpretations of the 'natural' generators of the Mathieu groups*, PCPS **107** (1990), 19–26.

D. Danev: see P. Boyvalenkov.

[Danz89] L. Danzer, *Three-dimensional analogues of the planar Penrose tilings and quasicrystals*, Discr. Math. **76** (1989), 1–7.

[DaZ87] M. H. Dauenhauer and H. J. Zassenhaus, *Local optimality of the critical lattice sphere-packing of regular tetrahedra*, Discr. Math. **64** (1987), 129–146.

[DaDr94] A. A. Davydov and A. Yu. Brozhzhina-Labinskaya, *Constructions, families and tables of binary linear covering codes*, PGIT **40** (1994), 1270–1279.

[Del29] B. N. Delone, *Sur la partition régulière de l'espace à 4–dimensions*, Izv. Akad. Nauk SSSR Otdel. Fiz.-Mat. Nauk (1929) pp. 79–110 and 145–164.

[Del37] — *Geometry of positive quadratic forms*, Uspehi Mat. Nauk **3** (1937), 16–62; **4** (1938), 102–164.

P. Delsarte: see A. R. Calderbank.

[DezG96] M. Deza and V. Grishukhin, *Bounds on the covering radius of a lattice*, Math. **43** (1996), 159–164. [But see the review: *Math. Reviews* MR98b: 11074.]

[Ding43] A. Dinghas, *Uber einen geometrischen Satz von Wulff über die Gleichgewichtsform von Kristallen*, Z. Krist. **105** (1043), 304–414.

D. Divsalar: see E. Biglieri.

S. M. Dodunekov: see also P. Boyvalenkov.

[DEZ91] — T. Ericson and V. A. Zinoviev, *Concatenation methods for construction of spherical codes in n-dimensional Euclidean space*, PPI **27** (1991), 34–38.

[DGM90] L. Dolan, P. Goddard and P. S. Montague, *Twisted conformal field theory*, in *Trieste Conf. Recent Developments in Conformal Field Theories*, edited S. Randjbar-Daemi et al., World Scientific, Singapore, 1990, pp. 39–54.

[DGM90a] — — — *Conformal field theory of twisted vertex operations*, Nuclear Phys. **B 338** (1990), 529–601.

[DGM90b] — — — *Conformal field theory, triality and the Monster group*, Phys. Lett. **B 236** (1990), 165–72.

[DoMa96] C. Dong and G. Mason, editors, *Moonshine, the Monster, and Related Topics*, Contemp. Math **193**, Amer. Math. Soc., 1996.

[Dou95] S. T. Dougherty, *Shadow codes and weight enumerators*, PGIT **41** (1995), 762–768.

[DoGH97] — T. A. Gulliver and M. Harada, *Type II codes over finite rings and even unimodular lattices*, J. Alg. Combin., to appear.

[DoGH97a] — — — *Extremal binary self-dual codes*, PGIT **43** (1997), 2036–2046.

[DoHa96] — and M. Harada, *Shadow optimal self-dual codes*, Kyushu J. Math., to appear.

[DoHa97] — — *New extremal self-dual codes of length 68*, preprint.

[DoHa97a] — — *Self-dual codes constructed from Hadamard matrices and symmetric designs*, preprint.

[DoHa09] — — and M. Oura, *Formally self-dual codes*, preprint.

[DoHaS97] — — and P. Solé, *Shadow codes over $\mathbb{Z}_4$*, preprint, July 1997.

[DowFV] R. G. Downey, M. R. Fellows, A. Vardy and G. Whittle, *The parameterized complexity of some fundamental problems in coding theory*, SIAM J. Computing, to appear.

P. G. Doyle: see P. Buser.

D. B. Drajić: see D. E. Lazić.

[DrS96] T. Drisch and P. Sonneborn, *The optimal results from linear programming for kissing numbers*, preprint.

A. Yu Drozhzhina-Labinskaya: see A. A. Davydov.

E. Dubois: see A. K. Khandani.

T. S. Duff: see N. J. A. Sloane.

[Duke93] W. Duke, *On codes and Siegel modular forms*, Intern. Math. Res. Notices **5** (1993), 125–126.

[DuS90] — and R. Schulze-Pillot, *Representations of integers by positive ternary quadratic forms and equidistribution of lattice points on ellipsoids*, Invent. Math. **99** (1990), 49–57.

[Dum94] N. Dummigan, *Mordell-Weil lattices and Shafarevich-Tate groups*, preprint, 1994.

[Dum95] — *The determinants of certain Mordel-Weil lattices*, Amer. J. Math. **117** (995), 1409–1429.

[Dum96] — *Symplectic group lattices as Mordell-Weil sublattices*, JNT **61** (1996), 365–387.

[Dum97] — *Algebraic cycles and even unimodular lattices*, preprint.

[Dye91] M. Dyer, *On counting lattice points in polyhedra*, SIAM J. Comput. **20** (1991), 695–707.

[DyKa98] — and R. Kannan, *On Barvinok's algorithm for counting lattice points in fixed dimension*, preprint, 1998.

A. G. Earnest: see also J. W. Benham.

[Earn88] — *Minimal discriminants of indefinite ternary quadratic forms having specified class number*, Math. **35** (1988), 95–100.

[Earn89] — *Binary quadratic forms over rings of algebraic integers: A survey of recent results*, in *Théorie des Nombres*, ed. J.-M. de Koninck and C. Levesque, Gruyter, Berlin, 1989, pp. 133–159.

[Earn90] — *Discriminants and class numbers of indefinite integral quadratic forms*, in *Number Theory*, ed. R A. Mollin, Gruyter, Berlin, 1990, pp. 115–123.

[Earn91] — *Genera of rationally equivalent integral binary quadratic forms*, Proc. Roy. Soc. Ed. **A 119** (1991), 27–30.

[Earn94] — *The representation of binary quadratic forms by positive definite quaternary quadratic forms*, TAMS **345** (1994), 853–863.

[EaH91] — and J. S. Hsia, *One-class spinor genera of positive quadratic forms*, Acta Arith. **58** (1991), 133–139.

[EaK97] — and A. Khosravani, *Universal binary Hermitian forms*, MTAC **66** (1997), 1161–1168.

[EaK97a] — — *Representations of integers by positive definite binary Hermitian lattices over imaginary quadratic fields*, JNT **62** (1997), 368–374.

[EaK97b] — — *Universal positive quaternary quadratic lattices over totally real number fields*, Math., to appear.

[EaN91] — and G. Nipp, *On the theta series of positive quaternary quadratic forms*, C. R. Math. Rep. Acad. Sci. Canada **13** (No. 1, Feb. 1991), 33–38.

[EaN97] — — *On the classification of positive quaternary quadratic forms by theta series*, preprint.

[Ebe87] W. Eberling, *The Monodromy Groups of Isolated Singularities of Complete Intersections*, LNM **1293** (1987).

[Ebe94] — *Lattices and Codes*, Vieweg, Wiesbaden, 1994.

Y. Edel: see also J. Bierbrauer.

[EdRS98] — E. M. Rains and N. J. A. Sloane, *On kissing numbers in dimensions 32 to 128*, preprint 1998.

[Ehr60] E. Ehrhart, *Sur les polyèdres rationnels et les systèmes diophantiens linéaires*, CR **250** (1960), 959–961.

[Ehr67] — *Démonstration de la loi réciprocité du polyèdre rationnel*, CR **265** (1967), 91–94.

[Ehr73] — *Une extension de la loi de réciprocité des polyèdres rationnels*, CR **277** (1973), 575–577.

[Ehr77] — *Polynômes Arithmétiques et Méthode des Polyèdres en Combinatoire*, Internat. Series Numer. Math., Vol. **35**, Birkhauser, Basel, 1977.

[EiZa85] M. Eichler and D. Zagier, *The Theory of Jacobi Forms*, Birkhäuser, Boston, 1985.

[Elki] N. D. Elkies, personal communication.

[Elki94] — *Mordell-Weil lattices in characteristic 2: I. Construction and first properties*, Internat. Mathematics Research Notices (No. 8, 1994), 353–361.

[Elki95] — *A characterization of the $\mathbb{Z}^n$ lattice*, Math. Research Letters **2** (1995), 321–326.

[Elki95a] — *Lattices and codes with long shadows*, Math. Research Letters **2** (1995), 643–651.

[Elki96] — *Mock-laminated lattices*, preprint, 1996.

[Elki97] — *Mordell-Weil lattices in characteristic 2: II. The Leech lattice as a Mordell-Weil lattice*, Invent. Math. **128** (1996), 1–8.

[Elki97a] — *Explicit modular towers*, in *Proc. Allerton Conf.*, 1997.

[ElkGr96] — and B. H. Gross, *The exceptional cone and the Leech lattice*, Internat. Math. Research Notices **14** (1996), 665–698.

[ElkOR91] — A. M. Odlyzko and J. A. Rush, *On the packing density of superballs and other bodies*, Ivent. Math. **105** (1991), 613–639.

[Eng86] P. Engel, *Geometric Crystallography*, Reidel, Dordrecht, Holland, 1986.

[Eng93] — *Geometric crystallography*, Chap. 3.7 of [GruWi93].

R. M. Erdahl: see also S. S. Ryskov.

[ErR87] — and S. S. Ryskov, *The empty sphere*, CJM **39** (1987), 794–824.

[Erd46] P. Erdős, *On sets of distances of n points*, AMM **53** (1946), 248–250.

[ErGH89] — P. M. Gruber and J. Hammer, *Lattice Points*, Longman Scientific, 1989.

T. Ericson: see also S. M. Dodunekov.

[ErZi95] — and V. Zinoviev, *Spherical codes generated by binary partitions of symmetric pointsets*, PGIT **41** (1995), 107–129.

T. Eriksson: see E. Agrell.

[ERS91] A. Eskin, Z. Rudnick and P. Sarnak, *A proof of Siegel's weight formula*, Internat. Math. Research Notes (Duke Math. J.) **5** (1991), 65–69.

M. Esmaeili: see A. K. Khandani.

[Ess90] F. Esselmann, *Ueber die maximale Dimension von Lorentz-Gittern mit coendlicher Spiegelungsgruppe*, JNT **61** (1996), 103–144.

[EsPa70] D. Estes and G. Pall, *The definite octonary quadratic forms of determinant 1*, IJM **14** (1970), 159–163.

[EtGr93] T. Etzion and G. Greenberg, *Constructions for perfect mixed codes and other covering codes*, PGIT **39** (1993), 209–213.

[EtGh93] — — and I. S. Honkala, *Normal and abnormal codes*, PGIT **39** (1993), 1453–1456.

[EtWZ95] — V. Wei and Z. Zhang, *Bounds on the sizes of constant weight covering codes*, DCC **5** (1995), 217–239.

L. Fainsilber: see E. Bayer-Fluckiger.

[FaL95] G. Fazekas and V. I. Levenshtein, *On upper bounds for code distance and covering radius of designs in polynomial metric spaces*, JCT **A 70** (1995), 267–288.

M. Feder: see R. Zamir.

[Fed85] E. S. Fedorov, *Elements of the study of figures*, Zap. Mineralog. Obsc. (2) **2** (1885), 1–279. Reprinted by Izdat. Akad. Nauk SSSR, Moscow, 1953.

[Fed91] — *The symmetry of regular systems of figures*, Zap. Mineralog. Obsc. (2) **28** (1891), 1–146. English translation in *Symmetry of Crystals*, ACA Monograph no. 7, Amer. Crystallographic Assoc., New York, 1971, pp. 50–131.

[FeFMMV] J. Feigenbaum, G. D. Forney, Jr., B. H. Marcus, R. J. McEliece and A. Vardy, editors, *Special Issue on Codes and Complexity*, PGIT **42** (No. 6, 1996), 1649–2064.

[Fej95] G. Fejes Tóth, *Review of [Hsi93]*, Math. Review **95g** #52032, 1995.

[Fej97] — *Packing and covering*, Chap. 2 of [GoO'R97].

[FGW90] — P. Gritzmann and J. M. Wills, *On finite multiple packings*, Archiv. Math. **55** (1990), 407–411.

[FGW91] — — — *Finite sphere packing and sphere covering*, DCG **4** (1989), 19–40.

[FeK93] — and W. Kuperberg, *Packing and covering with convex sets*, Chap. 3.3 of [GruWi93]

[FeK93a] — — *Blichfeldt's density bound revisited*, Math. Ann. **295** (1993), 721–727.

[Fel92] A. vom Felde, *A new presentation of Cheng-Sloane's [32,17,8] code*, Archiv. Math. **60** (1993), 508–511.

M. R. Fellows: see R. G. Downey.

[FeRa94] G.-L. Feng and T. R. N. Rao, *Decoding algebraic-geometric codes up to the designed minimum distance*, PGIT **39** (1994), 37–45.

[Ferg97] S. P. Ferguson, *Sphere packings V*, Abstract 926-51-196, Abstracts Amer. Math. Soc. **18** (1997).

[FeHa97] — and T. C. Hales, *A formulation of the Kepler conjecture*, preprint.

[FiP96] C. Fieker and M. E. Pohst, *On lattices over number fields*, Lect. Notes Comp. Sci. **1122** (1996), 133-139.

[Fiel80] K. L. Fields, *The fragile lattice packings of spheres in three-dimensional space*, Acta Cryst. **A 36** (1980), 194–197.

[FiNi80] — and M. J. Nicolich, *The fragile lattice packings of spheres in four-dimensional space*, Acta Cryst. **A 36** (1980), 588–591.

P. C. Fishburn: see A. R. Calderbank.

O. M. Fomenko: see E. P. Golubeva.

G. D. Forney, Jr.: see also A. R. Calderbank, J. Feigenbaum.

[Forn88] — *Coset codes I: Introduction and geometrical classification*, PGIT **34** (1988), 1066–1070.

[Forn88a] — *Coset codes II: Binary lattices and related codes*, PGIT **34** (1988), 1152–1187.

[Forn89] — *A bounded-distance decoding algorithm for the Leech lattice, with generalizations*, PGIT **35** (1989), 906–909.

[Forn89a] — *Multidimensional constellations II: Voronoi constellations*, IEEE J. Select. Areas Commun. **7** (1989), 941–958.

[Forn91] — *Geometrically uniform codes*, PGIT **37** (1991), 1241–1260.

[Forn93] — *Progress in geometrically uniform codes*, in *Proc. Sixth Swedish-Russian Joint Workshop on Information Theory*, Mölle, Sweden, 1993, pp. 16–20.

[Forn94] — *Dimension/length profiles and trellis complexity of linear block codes*, PGIT **40** (1994), 1741–1752.

[Forn94a] — *Dimension/length profiles and trellis complexity of lattices*, loc. cit., 1753–1791.

[Forn97] — *Approaching AWGN channel capacity with coset codes and multilevel coset codes*, preprint, 1997.

[FoCa89] — and A. R. Calderbank, *Coset codes for partial response channels; or, coset codes with spectral nulls*, PGIT **35** (1989), 925–943.

[FoST93] — N. J. A. Sloane and M. D. Trott, *The Nordstrom-Robinson code is the binary image of the octacode*, in *Coding and Quantization: DIMACS/IEEE Workshop October 19–21, 1992*, R. Calderbank, G. D. Forney, Jr., and N. Moayeri, Eds., Amer. Math. Soc., 1993, pp. 19–26.

[FoTr93] — and M. D. Trott, *The dynamics of group codes: state spaces, trellis diagrams, and canonical encoders*, PGIT **39** (1993), 1491–1513.

[FoVa96] — and A. Vardy, *Generalized minimum distance decoding of Euclidean-space codes and lattices*, PGIT **42** (1996), 1992–2026.

[FoWe89] — and L. F. Wei, *Multidimensional signal constellations I: Introduction, figures of merit, and generalized cross constellations*, IEEE J. Select. Areas Commun. **7** (1989), 877–892.

[Fort97] S. Fortune, *Voronoi diagrams and Delaunay triangulations*, Chap. 20 of [GrO'R97].

E. Freitag: see R. E. Borcherds.

[Fri89] S. Friedland, *Normal forms for definite integer unimodular quadratic forms*, PAMS **106** (1989), 917–921.

[FrTh94] J. Fröhlich and E. Thiran, *Integral quadratic forms, Kac-Moody algebras, and fractional quantum Hall effects. An ADE-O classification*, J. Statist. Phys. **76** (1994), 209–283.

F. Y. C. Fung: see J. H. Conway.

P. Gaborit: see A. Bonnecaze.

[GaWi92] P. M. Gandini and J. M. Wills, *On finite sphere-packings*, Math. Pannonica **3** (1992), 19–29.

[GaZu92] — and A. Zucco, *On the sausage catastrophe in 4-space*, Math. **39** (1992), 274–278.

[GaL90] T. Gannon and C. S. Lam, *Construction of four-dimensional strings*, Phys. Rev. **D 41** (1990), 492–506.

[GaL91] — — *Gluing and shifting lattice constructions and rational equivalence*, Reviews in Mathematical Physics **3** (1991), 331–369.

[GaL92] — — *Lattices and theta-functions identities I: Theta constants*, J. Math. Phys. **33** (1992), 854–870.

[GaL92a] — — *Lattices and theta-function identities II: Theta series*, J. Math. Phys. **33** (1992), 871–887.

[GaSt95] A. Garcia and H. Stichtenoth, *A tower of Artin-Schreier extensions of function fields attaining the Drinfeld-Valdut bound*, Invent. Math. **121** (1995), 211–222.

Z. Gáspár: see T. Tarnai.

A. Gersho: see K. Zeger.

[Gers72] L. J. Gerstein, *The growth of class numbers of quadratic forms*, AJM **94** (1972), 221–236.

[Gers91] — *Stretching and welding indecomposable quadratic forms*, JNT **37** (1991), 146–151.

[Gers95] — *Nearly unimodular quadratic forms*, Annals Math. **142** (1995), 597–610.

[Gers96] — *Characteristic elements of unimodular lattices*, preprint, 1996.

A. Gervois: see also L. Oger.

[GLOG] — M. Lichtenberg, L. Oger and E. Guyon, *Coordination number of disordered packings of identical spheres*, J. Phys. **A 22** (1989), 2119–2131.

P. Goddard: see also L. Dolan.

[Godd89] — *Meromorphic conformal field theory*, in *Infinite-Dimensional Lie Algebras and Lie Groups*, edited by V. G. Kac, World Scientific, Singapore, 1989, pp. 556–587.

[Golb70] M. Goldberg, *The packing of equal circles in a square*, Math. Mag. **43** (1970), 24–30.

[GoF87] E. P. Golubeva and O. M. Fomenko, *Asymptotic equidistribution of integral points on the three-dimensional sphere*, LOMI **160** (1987), 54–71.

[GoO'R97] J. E. Goodman and J. O'Rourke, editors, *Handbook of Discrete and Computational Geometry*, CRC Press, Boca Raton, 1997.

[Gop88] V. D. Goppa, *Gometry and Codes*, Kluwer, Dordrecht, 1988.

[GWW92] C. S. Gordon, D. L. Webb and S. A. Wolpert, *One cannot hear the shape of a drum*, BAMS **27** (1992), 134–138.

[Gow89] R. Gow, *Even unimodular lattices associated with the Weil representations of the finite symplectic groups*, J. Alg. **22** (1989), 510–519.

[Gow89a] — *Unimodular integral lattices associated with the basic spin representations of $2A_n$ and $2S_n$*, BLMS **21** (1989), 257–262.

[GraL95] R. L. Graham and B. D. Lubachevsky, *Dense packings of equal disks in an equilateral triangle: from 22 to 34 and beyond*, Electron. J. Combin. **2** (1995), #A1, 33 pp.

[GraLNO] — — K. J. Nuermela and P. R. J. Östergård, *Dense packings of congruent circles in a circle*, Discr. Math. **181** (1998), 139–154.

G. Greenberg: see T. Etzion.

[Grie88] R. L. Griess, *Code loops and a large finite group containing triality for* D_4, Rend. Circ. Mat. Palermo **19** (1988), 79–98.

[Grie90] — *A Moufang loop, the exceptional Jordan algebra, and a cubic form in 27 variables*, J. Alg. **131** (1990), 281–293.

U. Grimm: see M. Baake.

V. Grishukhin: see M. Deza.

P. Gritzmann: see also U. Betke, G. Fejes Tóth.

[Gri87] — *Ueber die j-ten Ueberdeckungsdichten konvexer Körper*, Monat. Math. **103** (1987), 207–220.

[GrW85] — and J. M. Wills, *On two finite covering problems of Bambah, Rogers, Woods and Zassenhaus*, Monat. Math. **99** (1985), 279–296.

[GriW93a] — — *Lattice points*, Chap. 3.2 of [GruWi93].

[GriW93b] — — *Finite packing and covering*, Chap. 3.4 of [GruWi93].

[GrG94] B. Groneick and S. Grosse, *New binary codes*, PGIT **40** (1994), 510–512.

C. de Groot: see R. Piekert.

B. H. Gross: see also N. D. Elkies.

[Gro90] — *Group representations and lattices*, J. Amer. Math. Soc. **3** (1990), 929–960.

[Gro96] — *Groups over* $\mathbb{Z}$, Invent. Math. **124** (1996), 263–279.

S. Grosse: see B. Groneick.

[GrBS] R. W. Grosse-Kunstleve, G. O. Brunner and N. J. A. Sloane, *Algebraic description of coordination sequences and exact topological densities of zeollites*, Acta Cryst. **A 52** (1996), 879–889.

P. M. Gruber: see also P. Erdős.

[GruWi93] — and J. M. Wills, *Handbook of Convex Geometry*, North-Holland, Amsterdam, 2 vols., 1993.

[GruS85] B. Grünbaum and G. C. Shephard, *Patterns of circular disks on a 2–sphere*, in *Diskrete Geometrie 3 Kolloq., Salzburg 1985*, North-Holland, 1985, 243–251.

T. A. Gulliver: see also S. T. Dougherty.

[Gull95] — *New optimal ternary linear codes*, PGIT **41** (1995), 1182–185.

E. Guyon: see A. Gervois, L. Oger.

[Habd95] B. Habdank-Eichelsbacher, *Unimodulare Gitter über Reel-quadratischen Zahlkörpern*, Report 95-005, Ergänzungsreihe des Sonderforschungsbereichs 343, Bielefield, 1994.

[Habs94] L. Habsieger, *Lower bounds for q-ary covering codes*, JCT **A 67** (1994), 199–222.

[Habs96] — *Some new lower bounds for ternary covering codes*, Elect. J. Combin. **3** (1996), #23.

[Habs96a] — *A new lower bound for the football pool problem for 7 matches*, JNB **8** (1996), 481–484.

[Habs97] — *Binary codes with covering radius one: some new lower bounds*, Discr. Math. **176** (1997), 115–130.

H. Hadwiger: see J. Bokowski.

[Hahn90] A. J. Hahn, *The coset lattices of E. S. Barnes and G. E. Wall*, JAMS **A 49** (1990), 418–433.

T. C. Hales: see also J. H. Conway, S. P. Ferguson.

[Hal92] — *The sphere packing problem*, J. Computational Applied Math. **44** (1992), 41–76.

[Hal93] — *Remarks on the density of sphere packings in three dimensions*, Combinatorica **13** (1993), 181–187.

[Hal94] — *The status of the Kepler conjecture*, Math. Intelligencer **16** (No. 3, 1994), 47–58.

[Hal97] — *Sphere packings I*, DCG **17** (1997), 1–51.

[Hal97a] — *Sphere packings II*, DCG **18** (1997), 135–149.

[Hal97b] — *Sphere packings III*, preprint.

[Hal98] — *The Kepler conjecture*, preprint.

H. Hämäläinen: see also A. E. Brouwer.

[Ham88] — *Two new binary codes with minimum distance three*, PGIT **34** (1988), 875.

[HaZe97] J. Hamkins and K. Zeger, *Asymptotically dense spherical codes — Part I: wrapped spherical codes*, PGIT **43** (1997), 1774–1785.

[HaZe97a] — — *Asymptotically dense spherical codes — Part II: laminated spherical codes*, PGIT **43** (1997), 1786–1798.

J. Hammer: see P. Erdős.

[HaKCSS] A. R. Hammons, Jr., P. V. Kumar, A. R. Calderbank, N. J. A. Sloane and P. Solé, *The $\mathbb{Z}_4$-linearity of Kerdock, Preparata, Goethals and related codes*, PGIT **40** (1994), 301–319.

[HaLa89] K. Harada and M. L. Lang, *Some elliptic curves arising from the Leech lattice*, J. Alg. **125** (1989), 298–310.

[HaLa90] — *On some sublattices of the Leech lattice*, Hokkaido Math. J. **19** (1990), 435–446.

[HaLaM94] — — and M. Miyamoto, *Sequential construction of Niemeier lattices and their uniqueness*, JNT **47** (1994), 198–223.

M. Harada: see also Eiichi Bannai, A. Bonnecaze, S. T. Dougherty.

[Hara96] — *Existence of new extremal doubly-even codes and extremal singly-even codes*, DCC **8** (1996), 273–284.

[Hara97] — *The existence of a self-dual* [70, 35, 12] *code and formally self-dual codes*, Finite Fields Applic. **3** (1997), 131–139.

[Hara98] — *New extremal ternary self-dual codes*, Australias. J. Combin., to appear.

R. H. Hardin: see also A. R. Calderbank, J. H. Conway, N. J. A. Sloane.

[HaSl92] — and N. J. A. Sloane, *New spherical 4 designs*, Discr. Math. **106/107** (1992), 255–264.

[HaSl93] — — *A new approach to the construction of optimal designs*, J. Stat. Planning Inference **37** (1993), 339–369.

[HaSl95] — — *Codes (spherical) and designs (experimental)*, in *Different Aspects of Coding Theory*, ed. A. R. Calderbank, AMS Series Proceedings Symposia Applied Math., Vol. 50, 1995, pp. 179–206.

[HaSl96] — — *McLaren's improved snub cube and other new spherical designs in three dimensions*, DCG **15** (1996), 429–441.

[HaSl96a] — — *Operating Manual for Gosset: A General Purpose Program for Designing Experiments (Second Edition)*, Statistics Research Report No. 106, AT&T Bell Laboratories, Murray Hill, NJ, Oct. 1996.

[HSS] — — and W. D. Smith, *Spherical Codes*, book in preparation.

D. K. Harrison: see S. Beale.

[Has89] Z. Hasan, *Intersection of the Steiner systems of* M_{24}, Discr. Math. **78** (1989), 267–289.

[HaK86] K. Hashimoto and H. Koseki, *Class numbers of positive definite binary and ternary unimodular hermitian forms*, Proc. Japan Acad. **A 62** (1986), 323–326.

[Has88] J. Hastad, *Dual vectors and lower bounds for the nearest lattice point problem*, Combinatorica **8** (1988), 75–81.

[HaL90] — and J. C. Lagarias, *Simultaneously good bases of a lattice and its reciprocal lattice*, Math. Ann. **287** (1990), 163–174.

A. Havemose: see J. Justesen.

T. Helleseth: see also A. G. Shanbhag.

[HeK95] — and P. V. Kumar, *The algebraic decoding of the* $\mathbb{Z}_4$*-linear Goethals code*, PGIT **41** (1995), 2040–2048.

M. Henk: see U. Betke.

[Hen86] C. L. Henley, *Sphere packings and local environments in Penrose tilings*, Phys. Rev. **34** (1986), 797–816.

[HMB90] H. J. Herrmann, G. Mantica and D. Bessis, *Space-filling bearings*, Phys. Rev. Lett. **65** (1990), 3223–3226.

[HiHu93] R. J. Higgs and J. F. Humphreys, *Decoding the ternary Golay code*, PGIT **39** (1993), 1043–1046.

[Hig69] G. Higman and J. McKay, Errata to [Hig5] in following issue (p. 219).

[HiN88] R. Hill and D. E. Newton, *Some optimal ternary linear codes*, Ars Comb. **A 25** (1988), 61–72.

[HiN92] — — *Optimal ternary linear codes*, DCC **2** (1992), 137–157.

T.-L. Ho: see D. A. Rabson.

[Hof91] D. W. Hoffmann, *On positive definite Hermitian forms*, Manuscripta Math. **71** (1991), 399–429.

[Hoehn95] G. Höhn, *Selbstduale Vertexoperatorsuperalgebren und das Babymonster*, Ph.D. Dissertation, Univ. Bonn, 1995. Also Bonner Mathematische Schriften **286** (1996).

[Hoehn96] — *Self-dual codes over the Kleinian four group*, preprint, August 1996.

T. Høholdt: see J. Justesen.

[HoPl89] D. F. Holt and W. Plesken, *Perfect Groups*, Oxford, 1989.

I. S. Honkala: see G. D. Cohen, T. Etzion.

[Hou96] X.-D. Hou, On the covering radius of $R(1, m)$ in $R(3, m)$, PGIT **42** (1996), 1035–1037.

J. S. Hsia: see also J. W. Benham, A. G. Earnest.

[Hsia87] — *On primitive spinor exceptional representations*, JNT **26** (1987), 38–45.

[Hsia89] — *Even positive definite unimodular quadratic forms over real quadratic fields*, Rocky Mtn. J. Math. **19** (1989), 725–733.

[Hsia91] — *Representations by spinor genera II*, preprint.

[HsH89] — and D. C. Hung, *Even unimodular 8–dimensional quadratic forms over* $\mathbb{Q}(\sqrt{2})$, Math. Ann. **283** (1989), 367–374.

[HsIc97] — and M. I. Icaza, *Effective version of Tartakowsky's theorem*, preprint.

[HsJ97] — and M. Jöchner, *Almost strong approximations for definite quadratic spaces*, Ivent. Math. **129** (1997), 471–487.

[HsPC] — and J. P. Prieto-Cox, *Representations of positive definite Hermitian forms with approximation and primitive properties*, JNT **47** (1994), 175–189.

[HJS] — — and Y.-Y. Shao, *A structure theorem for a pair of quadratic forms*, PAMS **119** (1993), 731–734.

[Hsi92] W.-Y. Hsiang, *A simple proof of a theorem of Thue on the maximal density of circle packings in* E^2, L'Enseign. Math. **38** (1992), 125–131.

[Hsi93] — *On the sphere packing problem and the proof of Kepler's conjecture*, Internat. J. Math. **93** (1993), 739–831.

[Hsi93a] — *On the sphere packing problem and the proof of Kepler's conjecture*, in *Differential Geometry and Topology (Alghero, 1992)*, Word Scientific, River Edge, NJ, 1993, pp. 117–127.

[Hsi93b] — *The geometry of spheres*, in *Differential Geometry (Shanghai, 1991)*, Word Scientific, River Edge, NJ, 1993, pp. 92–107.

[Hsi95] — *A rejoinder to T. C. Hales's article "The states of the Kepler conjecture"*, Math. Intelligencer **17** (No. 1, 1995), 35–42.

W. C. Huffman: see also V. Pless

[Huff90] — *On extremal self-dual quaternary codes of lengths 18 to 28*, PGIT **36** (1990), 651–660 and **37** (1991), 1206–1216.

[Huff91] — *On extremal self-dual ternary codes of lengths 28 to 40*, PGIT, submitted.

[Huff98] — *Codes and groups*, in [PHB].

[Huff98a] — *Decompositions and extremal Type II codes over* $\mathbb{Z}_4$, PGIT **44** (1998), to appear.

[HuTo95] — and V. D. Tonchev, *The existence of extremal self-dual* $[50, 25, 10]$ *codes and quasi-symmetric* $2-(49, 9, 6)$ *designs*, DCC **6** (1995), 97–105.

[Hump90] J. E. Humphreys, *Reflection Groups and Coxeter Groups*, Camb. Univ. Press, 1990.

J. F. Humphreys: see R. H. Higgs.

D. C. Hung: see also J. W. Benham, J. S. Hsia.

[Hun91] — *Even positive definite unimodular quadratic forms over* $\mathbb{Q}(\sqrt{3})$, MTAC **57** (1991), 351–368.

M. I. Icaza: see J. S. Hsia.

V. I. Iorgov: see V. Y. Yorgov.

[Iva90] A. A. Ivanov, *Geometric presentations of groups with an application to the Monster*, in Proc. Internat. Congress Math., Kyoto, 1990, Vol. II, pp. 1443–1453.

[Iva92] — *A geometric characterization of Fischer's Baby Monster*, J. Alg. Combin. **1** (1992), 45–69.

[Iva92a] — *A geometric characterization of the Monster*, in *Groups, Combinatorics and Geometry*, ed. M. W. Liebeck and J. Saxl, Camb. Univ. Press, 1992, pp. 46–62.

[IvSz96] G. Ivanyos and A. Szántó, *Lattice basis reduction for indefinite forms and an application*, Discr. Math. **153** (1996), 177–188.

[JKS97] W. C. Jagy, I. Kaplansky and A. Schiemann, *There are 913 regular ternary forms*, Math. **44** (1997), 332–341.

[Jame93] D. G. James, *Primitive representations by unimodular quadratic forms*, JNT **44** (1993), 356–366.

[JaLPW] C. Jansen, K. Lux, R. A. Parker and R. A. Wilson, *An Atlas of Brauer Characters*, Oxford, 1995.

[Jaq90] D.-O. Jaquet, *Domaines de Voronoi et algorithme de réduction des formes quadratiques définies positives*, JNB **2** (1990), 163–215.

[Jaq91] D.-O. Jaquet-Chiffelle, *Classification des réseaux dans* $\mathbb{R}^7$ *(à l'aide de la notion de formes parfaites)*, Astérisque **198–200** (1992), 7–8 and 177–185.

[Jaq92a] — *Description des voisines de* E_7, D_7, D_8 *et* D_9, JNB **4** (1992), 273–374.

[Jaq93] — *Enumération complète des classes de formes parfaites en dimension 7*, Ann. Inst. Fourier **43** (1993), 21–55.

[JaSi89] — and F. Sigrist, *Formes quadratiques contiguës à* D_7, CR **309** (1989), 641–644.

[JaSi94] — — *Classification des formes quadratiques réelles: un contre-example à la finitude*, Acta Arith. **68** (1994), 291–294.

H. E. Jensen: see J. Justesen.

[Jen93] J. N. Jensen, *On the construction of some very long cyclic codes*, PGIT **39** (1993), 1093–1094.

[Jen94] — *A class of constacyclic codes*, PGIT **40** (1994), 951–954.

A. Jessner: see W. Neutsch.

M. Jöchner: see J. S. Hsia.

[John91] R. W. Johnson, *Unimodularly invariant forms and the Bravais lattices*, Advances Appl. Math. **12** (1991), 22–56.

D. Joseph: see M. Baake.

[JuL88] D. Jungnickel and M. Leclerc, *A class of lattices*, Ars Comb. **26** (1988), 243–248.

J. Justesen: see also G. Lachaud.

[JuL89] — K. J. Larsen, A. Havemose, H. E. Jensen and T. Høholdt, *Construction and decoding of a class of algebraic geometry codes*, PGIT **35** (1989), 811–821.

[JuL92] — — H. E. Jensen and T. Høholdt, *Fast decoding of codes from algebraic curves*, PGIT **38** (1992), 111–119.

P. Kabal: see A. K. Khandani.

[Kac66] M. Kac, *Can one hear the shape of a drum?*, AMM **73** (1966), 1–23.

M. K. Kaikkonen: see also P. R. J. Östergård.

[Kaik98] — *Codes from affine permutation groups*, DCC, to appear.

[KaL95] G. Kalai and N. Linial, *On the distance distribution of codes*, PGIT **41** (1995), 1467–1472.

[KaNe90] J. Kallrath and W. Neutsch, *Lattice integration on the 7–sphere*, J. Comput. Appl. Math. **29** (1990), 9–14.

R. Kannan: see M. Dyer.

[Kant96] W. M. Kantor, *Note on Lie algebras, finite groups and finite geometries*, in *Groups, Difference Sets and the Monster*, eds. K. T. Arasu et al., de Gruyter, Berlin, 1996, pp. 73–81.

M. R. Kantorovitz: see K. Zeger.

I. Kaplansky: see also W. C. Jagy.

[Kapl95] — *Ternary positive quadratic forms that represent all odd positive integers*, Acta Arith. **70** (1995), 209–214.

[KaT90] S. N. Kapralov and V. D. Tonchev, *Extremal doubly-even codes of length 64 derived from symmetric designs*, Discr. Math. **83** (1990), 285–289.

[Kar93] J. K. Karlof, *Describing spherical codes for the Gaussian channel*, PGIT **39** (1993), 60–65.

P. Kazakov: see P. Boyvalenkov.

P. Keast: see also J. N. Lyness.

[Kea87] — *Cubature formulas for the surface of the sphere*, J. Comput. Appl. Math. **17** (1987), 151–172.

G. Kertész: see K. Bezdek.

[Kerv94] M. Kervaire, *Unimodular lattices with a complete root system*, L'Enseign. Math. **40** (1994), 59–104.

A. K. Khandani: see also A. H. Banihashemi.

[KhEs97] — and M. Esmaelili, *Successive minimization of the state complexity of self-dual lattices using Korkin-Zolotarev reduced bases*, preprint, 1997.

[KKD95] — P. Kabal and E. Dubois, *Computing the weight distribution of a set of points obtained by scaling, shifting and truncating a lattice*, PGIT **41** (1995), 1480–1482.

A. Khosravani: see A. G. Earnest.

M. H. Kim: see W. Chan.

[Kita93] Y. Kitaoka, *Arithmetic of Quadratic Forms*, Canbridge, 1993.

M. Kitazume: see also A. Bonnecaze.

[KiKM] — T. Kondo and I. Miyamoto, *Even lattices and doubly even codes*, J. Math. Soc. Japan **43** (1991), 67–87.

[KlL88] P. B. Kleidman and M. W. Liebeck, *The Subgroup Structure of the Finite Classical Groups*, London Math. Soc. Lect. Note Series **129**, Camb. Univ. Press, 1988.

[KlPW89] — R. A. Parker and R. A. Wilson, *The maximal subgroups of the Fischer group* Fi_{23}, JLMS **39** (1989), 89–101.

[KlW87] — and R. A. Wilson, *The maximal subgroups of* Fi_{22}, PCPS **102** (1987), 17–23; **103** (1988), 383.

[KlW90] — — $J_3 \subset E_6(4)$ *and* $M_{12} \subset E_6(5)$, JLMS **42** (1990), 555–561.

[KlW90a] — — *The maximal subgroups of* $E_6(2)$ *and* $Aut(E_6(2))$, PLMS **60** (1990), 266–294.

[KLV95] Y. Klein, S. Litsyn and A. Vardy, *Two new bounds on the size of binary codes with a minimum distance of three*, DCG **6** (1995), 219–227.

M. Kléman: see Z. Olamy.

[Klu89] P. Kluitmann, *Addendum zu der Arbeit "Ausgezeichnete Basen von Milnorgittern einfacher Singularitäten" von E. Voigt*, AMSH **59** (1989), 123–124.

[Kne54] M. Kneser, *Zur Theorie der Kristallgitter*, Math. Ann. **127** (1954), 105–106.

[Kne74] — *Quadratische Formen*, Lecture notes, Göttingen, 1974.

[Knill96] O. Knill, *Maximizing the packing density on a class of almost periodic sphere packings*, Expos. Math. **14** (1996), 227–246.

[Koch86] H. Koch, *Unimodular lattices and self-dual codes*, in *Proc. Intern. Congress Math., Berkeley 1986*, Amer. Math. Soc. Providence RI **1** (1987), pp. 457–465.

[Koch90] — *On self-dual doubly-even extremal codes*, Discr. Math. **83** (1990), 291–300.

[KoNe93] — and G. Nebe, *Extremal unimodular lattices of rank 32 and related codes*, Math. Nach. **161** (1993), 309–319.

[KoVe89] — and B. B. Venkov, *Ueber ganzzahlige unimodulare euklidische Gitter*, JRAM **398** (1989), 144–168.

[KoVe91] — — (with an appendix by G. Nebe), *Ueber gerade unimodulare Gitter der Dimension 32, III*, Math. Nach. **152** (1991), 191–213.

[Koik86] M. Koike, *Moonshines of* $PSL_2(\mathbb{F}_q)$ *and the automorphism group of the Leech lattice*, Japan J. Math. **12** (1986), 283–323.

[KolY97] A. V. Kolushov and V. A. Yudin, *Extremal dispositions of points on the sphere*, Analysis Math. **23** (1997), 25–34.

[Kon97] S. Kondo, *The automorphism group of a generic Jacobian Kummer surface*, J. Alg. Geom. **7** (1998), 589–609.

[Kon98] — *Niemeier lattices, Mathieu groups and finite groups of symplectic automorphisms of K9 surfaces*, Duke Math. J. **92** (1998), 593–598.

T. Kondo: see also M. Kitazume.

[KoTa87] — and T. Tasaka, *The theta functions of sublattices of the Leech lattice II*, J. Fac. Sci. Univ. Tokyo Sect. IA, Math. **34** (1987), 545–572

[KoMe93] J. Korevaar and J. L. H. Meyers, *Spherical Faraday cage for the case of equal point changes and Chebyshev-type quadrature on the sphere*, Integral Transforms and Special Functions **1** (1993), 105–117.

[KoMe94] — — *Chebyshev-type quadrature on multidimensional domains*, J. Approx. Theory **79** (1994), 144–164.

H. Koseki: see K. Hashimoto.

A. I. Kostrikin: see also A. I. Bondal.

[KoTi94] — and P. H. Tiep, *Orthogonal Decompositions and Integral Lattices*, De Gruyter, Berlin, 1994.

[Kott91] D. A. Kottwitz, *The densest packings of equal circles on a sphere*, Acta Cryst. **A 47** (1991), 158–165.

[KrM74] E. S. Kramer and D. M. Mesner, *Intersections among Steiner systems*, JCA **A 16** (1974), 273–285.

[Kram88] J. Kramer, *On the linear independence of certain theta-series*, Math. Ann. **281** (1988), 219–228.

P. Kramer: see M. Baake.

[KrLi97] I. Krasikov and S. Litsyn, *Linear programming bounds for doubly-even self-dual codes*, PGIT **43** (1997), 1238–1244.

[KrLi97a] — — *Linear programming bounds for codes of small size*, EJC **18** (1997), 647–656.

[Kra88] E. Krätzel, *Lattice Points*, Kluwer, Dordrecht, Holland, 1988.

F. R. Kschischang: see also A. Vardy.

[KsP92] — and S. Pasupathy, *Some ternary and quaternary codes and associated sphere packings*, PGIT **38** (1992) 227–246.

[Kuhn96] S. Kühnlein, *Partial solution of a conjecture of Schmutz*, Archiv. Math. **67** (1996), 164–172.

[Kuhn97] — *Multiplicities of non-arithmetic ternary quadratic forms and elliptic curves of positive rank*, preprint.

P. V. Kumar: see A. R. Hammons, Jr., T. Helleseth, O. Moreno, A. G. Shanbhag.

W. Kuperberg: see A. Bezdek, G. Fejes Tóth.

[Kuzj95] N. N. Kuzjurin, *On the difference between asymptotically good packings and coverings*, EJM **16** (1995), 35–40.

[LaS92] G. Lachaud and J. Stern, *Polynomial-time construction of codes I: linear codes with almost equal weights*, Proc. AAECC, vol. 3 (1992), pp. 151–161.

[LaS94] — — *Polynomial-time construction of codes II: spherical codes and the kissing number of spheres*, PGIT **40** (1994), 1140–1146.

[Lac95] — M. A. Tsfasman, J. Justesen and V. K.-W. Wei, editors, *Special Issue on Algebraic Geometry Codes*, PGIT **41** (No. 6, 1996), 1545–1772.

[LaVa95] A. Lafourcade and A. Vardy, *Asymptotically good codes have infinite trellis complexity*, PGIT **41** (1995), 555–559.

[LaVa95a] — — *Lower bounds on trellis complexity of block codes*, PGIT **41** (1995), 1938–1954.

[LaVa96] — — *Optimal sectionalization of a trellis*, PGIT **42** (1996), 689–703.

J. C. Lagarias: see also J. Hastad.

[Laga96] — *Point lattices*, in *Handbook of Combinatorics*, ed. R. L. Graham et al., North-Holland, Amsterdam, 1996, pp. 919–966.

J. Lahtonen: see T. Mittelholzer.

[LaiL98] T. Laihonen and S. Litsyn, *On upper bounds for minimum distance and covering radius of nonbinary codes*, DCC, to appear.

C. S. Lam: see T. Gannon.

[LmP90] C. W. H. Lam and V. Pless, *There is no (24,12,10) self-dual quaternary code*, PGIT **36** (1990), 1153–1156.

S. Landau: see J. Cremona.

I. Landgev: see P. Boyvalenkov.

[LaM92] P. Landrock and O. Manz, *Classical codes as ideals in group algebras*, DCC **2** (1992), 273–285.

[LaLo89] G. R. Lang and F. M. Longstaff, *A Leech lattice modem*, IEEE J. Select. Areas Comm. **7** (1989), 968–973.

M. L. Lang: see K. Harada.

K. J. Larsen: see J. Justesen.

[Laue43] M. v. Laue, *Der Wuffsche Satz für die Gleichgewichtsform von Kristallen*, Z. Krist. **105** (1943), 124–133.

F. Laves: see G. O. Brunner.

[LDS86] D. E. Lazić, D. B. Drajić and V. Senk, *A table of some small-size three-dimensional best spherical codes*, preprint.

[LDS87] — V. Senk and I. Seskar, *Arrangements of points on a sphere which maximize the least distance*, Bull. Appl. Math. **47** (No. 479, 1987), 5–21.

M. Leclerc: see D. Jungnickel.

[LePR93] H. Lefmann, K. T. Phelps and V. Rödl, *Rigid linear binary codes*, JCT **A** **63** (1993), 110–128.

P. Lemke: see S. S. Skiena.

[Lemm1] P. W. H. Lemmens, *On the general structure of holes in integral lattices*, preprint.

[Lemm2] — *A note on lattices and linear subspaces in* $\mathbb{R}^n$, preprint.

[Lemm3] — *On intersections and sums of integral lattices in* $\mathbb{R}^n$, preprint.

[Len92] H. W. Lenstra, Jr., *Algorithms in algebraic number theory*, BAMS **26** (1992), 211–244.

J. Leon: see V. Pless.

[Lep88] J. Lepovsky, *Perspectives on vertex operators and the Monster*, PSPM **48** (1988), 181–197.

[Lep91] — *Remarks on vertex operator algebras and moonshine*, in *Proc 20th Internat. Conf. Differential Geometric Methods in Theoretical Physics, New York, 1991*, edited S. Catto and A. Rocha, World Scientific, Singapore, 1992.

V. I. Levenshtein: see also G. Fazekas.

[Lev87] — *Packing of polynomial metric spaces*, Proc. Third. Intern. Workshop on Information Theory, Sochi, 1987, pp. 271–274.

[Lev91] — *On perfect codes in the metric of deletions and insertions* (in Russian), Discr. Mat. **3** (1991), 3–20. English translation in Discrete Math. and Applic. **2** (No. 3, 1992).

[Lev92] — *Packing and decomposition problems for polynomial association schemes*, EJC **14** (1993), 461–477.

[Lev92a] — *Bounds for self-complementary codes and their applications*, in Proc. Eurocode '92.

[Lev92b] — *Designs as maximum codes in polynomial metric spaces*, Acta Applicandae Math. **29** (1992), 1–82.

[Lev93] — *Bounds for codes as solutions of extremum problems for systems of orthogonal polynomials*, in *Proc. 10th Int. Sympos. Applied Algebra etc.*, Lect. Notes Comp. Sci. **673**, 1993, 25–42.

[Lev95] — *Krawtchouk polynomials and universal bounds for codes and designs in Hamming space*, PGIT **41** (1995), 1303–1321.

[Levi87] B. M. Levitan, *Asymptotic formulae for the number of lattice points in Euclidean and Lobachevskii spaces*, Russian Math. Surveys **42** (1987), 13–42.

[LeLi96] F. Levy-dit-Vehel and S. Litsyn, *More on the covering radius of BCH codes*, PGIT **42** (1996), 1023–1028.

[LiCh94] D. Li and W. Chen, *New lower bounds for binary covering codes*, PGIT **40** (1994), 1122–1129.

M. Lichtenberg: see A. Gervois, L. Oger.

M. W. Liebeck: see also P. B. Kleidman.

[LPS90] — C. E. Praeger and J. Saxl, *The Maximal Factorizations of the Finite Simple Groups and Their Automorphism Groups*, Memoirs Amer. Math. Soc. **432** (1990).

[LiSZ93] T. Linder, C. Schlegel and K. Zeger, *Corrected proof of de Buda's theorem*, PGIT **39** (1993), 1735–1737.

[Lin88] J. H. Lindsey II, *A new lattice for the Hall-Janko group*, PAMS **103** (1988), 703–709.

N. Linial: see G. Kalia.

J. H. van Lint: see R. J. M. Vaessens.

[LiW91] S. A. Linton and R. A. Wilson, *The maximal subgroups of the Fischer groups* Fi_{24} *and* Fi'_{24}, PLMS **63** (1991), 113–164.

S. Litsyn: see also A. E. Ashikhmin, R. A. Brualdi, G. D. Cohen, Y. Klein, I. Krasikov, F. Levy-dit-Vehel, T. Laihonen.

[Lits98] — *An updated table of the best binary codes known*, to appear in [PHB].

[LiV93] — and A. Vardy, *Two new upper bounds for codes of distance 3*, in *Algebraic Coding*, Lect. Notes Comp. Sci. **781** (1994), 253–262.

[LiV94] — — *The uniqueness of the Best code*, PGIT **40** (1994), 1693–1698.

A. C. Lobstein: see G. D. Cohen.

[Loel97] H.-A. Loeliger, *Averaging bounds for lattices and linear codes*, PGIT **43** (1997), 1767–1773.

[Loma94] G. A. Lomadze, *On the number of ways of writing a number as the sum of nine squares* (in Russian), Acta Arith. **68** (1994), 245–253.

F. M. Longstaff: see G. R. Lang.

P. Loyer: see also P. Solé.

[LoSo93] — and P. Solé, *Quantizing and decoding for the usual lattices in the* L_p-*metric*, in *Algebraic Coding (Paris, 1993)*, Lect. Notes Comp. Sci **781** (1994), 225–235.

[LoSo94] — and P. Solé, *Les réseaux BW32 et U32 sont équivalents*, JNB **6** (1994) 359–362.

B. D. Lubachevsky: see R. L. Graham.

K. Lux: see C. Jansen.

[LySK91] J. N. Lyness, T. Sørevik and P. Keast, *Notes on integration and integer sublattices*, MTAC **56** (1991), 243–255.

J. Maginnis: see A. Adem.

[Mak65] V. S. Makarov, *A class of partitions of a Lobachevskii space*, SMD **6** (1965), 400–401.

[Mak66] — *A class of discrete groups in Lobachevskii space having an infinite fundamental domain of finite measure*, SMD **7** (1966), 328–331.

[MaV85] Y. I. Manin and S. G. Vladuts, *Linear codes and modular curves*, JSM **30** (1985), 2611–2643.

G. Mantica: see H. J. Herrmann.
O. Manz: see P. Landrock.
B. H. Marcus: see J. Feigenbaum.
J. Martinet: see also C. Batut, E. Bayer-Fluckiger, A.-M. Bergé.

[Mar95] — *Structure algébriques sur les réseaux*, in *Number Theory (Seminaire de Théorie des Nombres de Paris, 1992–93)*, Cambridge Univ. Press, 1995, pp. 167–186.

[Mar96] — *Les Réseaux Parfaits des Espaces Euclidiens*, Masson, Paris, 1996.

[Mar97] — *Une famille de réseaux dual-extrêmes*, JNB **9** (1997), 169–181.

[Mar98] — *Perfect and eutactic lattices: extensions of Voronoi's theory*, preprint.
G. Mason: see C. Dong.
H. F. Mattson, Jr.: see G. D. Cohen.

[Maye93] A. J. Mayer, *A Tiling Property of the Minkowski Fundamental Domain*, Ph. D. Disseration, Princeton Univ., 1993.

[Maye95] — *Low dimensional lattices have a strict Voronoi base*, Math. **42** (1995), 229–238.

[MaO90] J. E. Mazo and A. M. Odlyzko, *Lattice points in high-dimensional spheres*, Monat. Math. **110** (1990), 47–61.
R. J. McEliece: see J. Feigenbaum.
J. McKay: see G. Higman.
P. J. McLane: see E. Biglieri.

[MeMo79] W. M. Meier and H. J. Moeck, *The topology of three-dimensional 4-connected nets: classification of zeolite framework types using coordination sequences*, J. Solid State Chem. **27** (1979), 349–355.

[Mel94] J. B. M. Melissen, *Densest packing of eleven congruent circles in a circle*, Geomet. Dedicata **50** (1994), 15–25.

[Mel97] — *Packing and Covering with Circles*, Ph.D. Dissertation, Univ. Utrecht, Dec. 1997.
N. D. Mermin: see also D. A. Rabson, D. S. Rokhsar.

[MRW87] — D. A. Rabson and D. C. Wright, *Beware of 46-fold symmetry: the classification of two-dimensional crystallographic lattices*, Phys. Rev. Lett. **58** (1987), 2099–2101.
D. M. Mesner: see E. S. Kramer.
J. L. H. Meyers: see J. Korevaar.

[MiRS95] L. Michel, S. S. Ryshkov and M. Senechal, *An extension of Voronoi's theorem on primitive parallelotopes*, EJC **16** (1995), 59–63.
R. J. Milgram: see A. Adem.

[Mil64] J. Milnor, *Eigenvalues of the Laplace operator on certain manifolds*, PNAS **51** (1964), 542.

[Mimu90] Y. Mimura, *Explicit examples of unimodular lattices with the trivial automorphism group*, in *Algebra and Topology*, Korea Adv. Inst. Sci. Tech., Taejon, 1990, pp. 91–95.
S. Minashima: see Eichii Bannai.

[Misch94] M. Mischler, *La formule de Minkowski-Siegel pour les formes bilinéaires symmétriques dégénérées et définies positives*, Théorie des nombres 1992/93 – 1993/94, Univ. Franche-Comté, Besançon, 1994.
M. Mitradjieva: see P. Boyvalenkov.

[MiL96] T. Mittelholzer and J. Lahtonen, *Group codes generated by finite reflection groups*, PGIT **42** (1996), 519–528.

[Miya89] T. Miyake, *Modular Forms*, Springer-Verlag, NY, 1989.
I. Miyamoto: see M. Kitazume.
M. Miyamoto: see also K. Harada.

[Miya95] — *21 involutions acting on the Moonshine module*, J. Alg. **175** (1995), 941–965.

[Miya96] — *Binary codes and vertex operator (super) algebras*, J. Alg. **181** (1996), 207–222.

[Miya98] — *Hamming code VOA and construction of VPOA's*, preprint.

H. J. Moeck: see W. M. Meier.

[MolP90] M. Mollard and C. Payan, *Some progress in the packing of equal circles in a square*, Discr. Math. **84** (1990), 303–307.

M. Monagan: see R. Peikert.

P. S. Montague: see also L. Dolan.

[Mont94] — *A new construction of lattices from codes over $GF(3)$*, Discr. Math. **135** (1994), 193–223.

[MoP92] R. V. Moody and J. Patera, *Voronoi and Delaunay cells of root lattices: classification of their faces and facets by Coxeter-Dynkin diagrams*, J. Phys. **A. 25** (1992), 5089–5134.

[MoP92a] — — *Voronoi domains and dual cells in the generalized kaleidescope with applications to root and weight lattices*, CJM **47** (1995), 573–605.

[MoKu93] O. Moreno and P. V. Kumar, *Minimum distance bounds for cyclic codes and Deligne's theorem*, PGIT **39** (1993), 1524–1534.

[MoPa93] W. O. J. Moser and J. Pach, *Research Problems in Discrete Geometry: Packing and Covering*, DIMACS Technical Report 93–32, DIMACS Center, Rutgers Univ., 1993.

B. Mourrain: see C. Bachoc, A. Bonnecaze.

D. J. Muder: see also J. H. Conway.

[Mud90] — *How big is an n-sided Voronoi polygon?*, PLMS **61** (1990), 91–108.

[Mude93] — *A new bound on the local density of sphere packings*, DCG **10** (1993), 351–375.

[Mus97] C. Musès, *The dimensional family approach in (hyper)sphere packing: a typological study of new patterns, structures, and interdimensional functions*, Applied Math. Computation **88** (1997), 1–26 (see Corrigenda in same issue).

J. Naor: see N. Alon.

M. Naor: see N. Alon.

[Napa94] H. Napias, *Sur quelques réseaux contenus dans les réseaux de Leech et de Quebbemann*, CR **319** (1994), 653–658.

[Napa96] — *A generalization of the LLL-algorithm over Euclidean rings or orders*, JNB **8** (1996), 387–396.

G. Nebe: see also C. Bachoc, A. Cohen, H. Koch.

[Nebe96] — *Finite subgroups of $GL_{24}(\mathbb{Q})$*, Exper. Math. **5** (1996), 163–195.

[Nebe96a] — *Finite subgroups of $GL_n(\mathbb{Q})$ for $25 \leq n \leq 31$*, Comm. Alg. **24** (1996), 2341–2397.

[Nebe97] — *The normaliser action and strongly modular lattices*, L'Enseign. Math. **43** (1997), 67–76.

[Nebe98] — *Some cyclo-quaternionic lattices*, J. Alg. **199** (1998), 472–498.

[Nebe98a] — *Finite quaternionic matrix groups*, Electron. J. Representation Theory, to appear.

[Nebe98b] — *A method of computing the minimum of certain lattices*, preprint.

[NePl95] — and W. Plesken, *Finite Rational Matrix Groups*, Memoirs A.M.S., vol. **116**, No. 556, 1995.

[NeSl] — and N. J. A. Sloane, *A Catalogue of Lattices*, published electronically at http://www.research.att.com/~njas/lattices/.

[NeVe96] — and B. B. Venkov, *Nonexistence of extremal lattices in certain genera of modular lattices*, JNT **60** (1996), 310–317.

A. Neumaier: see also A. E. Brouwer.

[NeS88] — — *Discrete measures for spherical designs, eutactic stars and lattices*, KNAW **50** (1988), 321–334.

[NeS92] — and J. J. Seidel, *Measures of strength 2e, and optimal designs of degree e*, Sankhyā **54** (1992), 299–309.

W. Neutsch: see also J. Kallrath.

[Neut83] — *Optimal spherical designs and numerical integration on the sphere*, J. Comput. Phys. **51** (1983), 313–325.

[NeSJ85] — E. Schrüfer and A. Jessner, *Efficient integration on the hypersphere*, J. Comput. Phys. **59** (1985), 167–175.

[Newm94] M. Newman, *Tridiagonal matrices*, Lin Alg. Appl. **201** (1994), 51–55.

D. E. Newton: see R. Hill.

M. J. Nicolich: see K. L. Fields.

S. Nikova: see P. Boyvalenkov.

G. Nipp: see also A. G. Earnest.

[Nip91] — *Quaternary Quadratic Forms: Computer Generated Tables*, Springer-Verlag, NY, 1991.

[Nip91a] — *Computer-Generated Tables of Quinary Quadratic Forms*, MTAC, Unpublished Mathematical Tables Collection, 1991.

[Nor90] S. P. Norton, *Presenting the Monster?*. Bull. Soc. Math. Belgium **62** (1990), 595–605.

[Nor92] — *Constructing the Monster*, in *Groups, Combinatorics and Geometry*, ed. M. W. Liebeck and J. Saxl, Camb. Univ. Press, 1992, pp. 63–76.

[NoW89] — and R. A. Wilson, *The maximal subgroups of* $F_4(2)$ *and its automorphism group*, Comm. Alg. **17** (1989), 2809–2824.

K. J. Nurmela: see also R. L. Graham.

[NuOs97] — and P. R. J. Östergård, *Packing up to 50 equal circles in a square*, DCG **18** (1997), 111–120.

A. M. Odlyzko: see N. D. Elkies, J. E. Mazo.

[Oes90] J. Oesterlé, *Empilements de sphères*, Sém. Bourbaki **42** (No. 727, 1990).

L. Oger: see also A. Gervois.

[OLGG] — M. Lichtenberg, A. Gervois and E. Guyon, *Determination of the coordination number in disordered packings of equal spheres*, J. Microscopy-Oxford **156** (1989), 65–78.

[OgS91] K. Oguiso and T. Shioda, *The Mordell-Weil lattice of a rational elliptic surface*, Comment. Math. Univ. St. Pauli **40** (1991), 83–99.

[O'Ke91] M. O'Keeffe, *N-dimensional diamond, sodalite and rare sphere packings*, Acta Cryst. **A 47** (1991), 748–753.

[O'Ke95] — *Coordination sequences for lattices*, Zeit. F. Krist. **210** (1995), 905–908.

[OlK89] Z. Olamy and M. Kléman, *A two-dimensional aperiodic dense tiling*, J. Phys. France **50** (1989), 19–37.

M. Olivier: see C. Batut.

[OrPh92] H. Oral and K. T. Phelps, *Almost all self-dual codes are rigid*, JCT **A 60** (1992), 264–276.

J. O'Rourke: see J. E. Goodman.

P. R. J. Östergård: see also A. E. Brouwer, R. L. Graham, K. J. Nurmela.

[OsK96] — and M. K. Kaikkonen, *New single-error-correcting codes*, PGIT **42** (1996), 1261–1262.

M. Oura: see Eiichi Bannai, S. T. Dougherty.

L. H. Ozarow: see A. R. Calderbank.

M. Ozeki: see also Eiichi Bannai.

[Oze76] — *On the basis problem for Siegel modular forms of degree 2*, Acta Arith. **31** (1976), 17–30.

[Oze87] — *Ternary code construction of even unimodular lattices*, in *Theorie des Nombres, Quebec 1987*, Gruyter, Berlin, 1989, pp. 772–784.

[Oze89a] — *On the structure of even unimodular extremal lattices of rank 40*, Rocky Mtn. J. Math. **19** (1989), 847–862.

[Oze89b] — *On a class of self-dual ternary codes*, Science Reports Hirosaki Univ. **36** (1989), 184–191.

[Oze91] — *Quinary code construction of the Leech lattice*, Nihonkai Math. J. **2** (1991), 155–167.

[Oze97] — *On the notion of Jacobi polynomials for codes*, Math. Proc. Camb. Phil. Soc. **121** (1997), 15–30.

J. Pach: see W. O. J. Moser.

[PacA95] — and P. K.Agarwal, *Combinatorial Geometry*, Wiley, New York, 1995.

G. Pall: see D. Estes.

R. A. Parker: see also C. Jansen, P. B. Kleidman.

[PaW90] — and R. A. Wilson, *The computer construction of matrix representations of finite groups over finite fields*, J. Symb. Comp. **9** (1990), 583–590.

[Pas81] G. Pasquier, *A binary extremal doubly even self-dual code (64,32,12) obtained from an extended Reed-Solomon code over* F_{16}, PGIT **27** (1981), 807–808.

S. Pasupathy: see F. R. Kschischang.

J. Patera: see R. V. Moody.

C. Payan: see M. Mollard.

[PaS87] A. Paz and C. P. Schnorr, *Approximating integer lattices by lattices with cyclic factor groups*, in *Automata, Languages and Programming, Proc. 14th Internat. Colloq.*, edited T. Ottmann, Lect. Notes Comp. Sci. **267** (1987), 386–393.

[PeWMG] R. Peikert, D. Würtz, M. Monagan and C. de Groot, *Packing circles in a square: A review and new results*, Maple Technical Newsletter **6** (1991), 28–34.

[Pel89] R. Pellikaan, *On a decoding algorithm for codes on maximal curves*, PGIT **35** (1989), 1228–1232.

[Per90] E. Pervin, *The Voronoi cells of the* E_6^* *and* E_7^* *lattices*, in D. Jungnickel et al., editors, *Coding Theory, Design Theory, Group Theory*, Wiley, NY, 1993, pp. 243–248.

[Pet80] M. Peters, *Darstellungen durch definite ternäre quadratische Formen*, Acta Arith **34** (1977), 57–80.

[Pet89] — *Jacobi theta series*, Rocky Mtn. J. Math. **19** (1989), 863–870.

[Pet90] — *Siegel theta series of degree 2 of extremal lattices*, JNT **35** (1990), 58–61.

K. T. Phelps: see H. Lefmann, H. Oral.

P. A. B. Pleasants: see M. Baake.

W. Plesken: see also A. M. Cohen, D. F. Holt, G. Nebe.

[Plesk91] — *Some applications of representation theory*, in *Representation Theory of Finite Groups and Finite-Dimensional Algebras (Bielefeld, 1991)*, Birkhauser, Basel, 1991, pp. 477–496.

[Plesk93] — *Kristallographische Gruppen*, in *Group Theory, Algebra and Number Theory (Saarbrücken, 1993)*, De Gruyter, Berlin, 1996, pp. 75–96.

[Plesk94] — *Additively indecomposable positive integral quadratic forms*, JNT **47** (1994), 273–283.

[Plesk96] — *Finite rational matrix groups — a survey*, in *Proc. Conf. "The ATLAS: Ten Years After" (1995)*, to appear.

[Plesk92] — and M. Pohst, *Constructing integral lattices with prescribed minimum II*, MTAC **60** (1993), 817–826.

[PlSo96] — and B. Souvignier, *Computing rational representations of finite groups*, Exper. Math. **5** (1996), 39–47.

V. Pless: see also R. A. Brualdi, J. H. Conway, C. W. H. Lam.

[Pless89] — *Extremal codes are homogeneous*, PGIT **35** (1989), 1329–1330.

[PHB] — W. C. Huffman and R. A. Brualdi, editors, *Handbook of Coding Theory*, Elsevier, Amsterdam, 1998, to appear.

[PlQ96] — and Z. Qian, *Cyclic codes and quadratic residue codes over* $\mathbb{Z}_4$, PGIT **24** (1996), 1594–1600.

[PTL92] — V. Tonchev and J. Leon, *On the existence of a certain (64,32,12) extremal code*, PGIT **39** (1993), 214–215.

M. Pohst: see also C. Fieker, W. Plesken.

[PoZ89] — and H. Zassenhaus, *Algorithmic Number Theory*, Cambridge Univ. Press, 1989.

[Polt94] G. Poltyrev, *On coding without restrictions for the AWGN channel*, PGIT **40** (1994), 409–417.

C. E. Praeger: see also M. W. Liebeck.

[Pra89] — *Finite primitive permutation groups: A survey*, LNM **1456** (1989), 63–84.

J. P. Prieto-Cox: see J. S. Hsia.

[Pro87] M. N. Prokhorov, *The absence of discrete reflection groups with noncompact fundamental polyhedron of finite volume in Lobachevsky space of large dimension*, Izv. **28** (1987), 401–411.

[Prop88] J. G. Propp, *Kepler's spheres and Rubik's cube*, Math Mag. **61** (1988), 231–239.

[Pul82] C. L. M. van Pul, *On Bounds on Codes*, Master's Dissertation, Eindhoven University of Technology, Netherlands, 99 pages, 1982.

Z. Qian: see V. Pless.

H.-G. Quebbemann: see also C. Batut.

[Queb88] — *Cyclotomic Goppa codes*, PGIT **34** (1988), 1317–1320.

[Queb88a] — *On even codes*, Discr. Math. **98** (1991), 29–34.

[Queb89] — *Lattices from curves over finite fields*, preprint, 1989.

[Queb91] — *A note on lattices of rational functions*, Math. **38** (1991), 10–13.

[Queb91a] — *Unimodular lattices with isometries of large prime order II*, Math. Nach. **156** (1992), 219–224.

[Queb95] — *Modular lattices in Euclidean spaces*, JNT **54** (1995), 190–202.

[Queb97] — *Atkin-Lehner eigenforms and strongly modular lattices*, L'Enseign. Math. **43** (1997), 55–65.

[Queb98] — *A shadow identity and an application to isoduality*, preprint.

[Quin95] J. R. Quine, *Jacobian of the Picard curve*, Contemp. Math. **201** (1997), 33–41.

[QuZh95] — and P. L. Zhang, *Extremal symplectic lattices*, preprint, 1995.

[RaBa91] P. Rabau and B. Bajnok, *Bounds for the number of nodes in Chebyshev type quadrature formulas*, J. Approx. Theory **67** (1991), 199–214.

A. Rabinovich: see A. R. Calderbank.

P. Rabinowitz: see R. Cools.

[RHM89] D. A. Rabson, T.-L. Ho and N. D. Mermin, *Space groups of quasicrystallographic tilings*, Acta Cryst. **A** (1989), 538–547.

S. Raghavan: see W. Chan.

E. M. Rains: see also A. R. Calderbank, Y. Edel.

[Rain98] — *Shadow bounds for self-dual codes*, PGIT **44** (1998), 134–139.

[Rain98a] — *Optimal self-dual codes over* $\mathbb{Z}_4$, preprint.

[RaSl98] — and N. J. A. Sloane, *Self-dual codes*, to appear in [PHB].

[RaSl98a] — — *The shadow theory of modular and unimodular lattices*, JNT (submitted).

[RaSi94] B. S. Rajan and M. U. Siddiqi, *A generalized DFT for abelian codes over* $\mathbb{Z}_m$, PGIT **40** (1994), 2082–2090.

[RaSh96] D. S. Rajan and A. M. Shende, *A characterization of root lattices*, Discr. Math. **161** (1996), 309–314.

[RaSn93] M. Ran and J. Snyders, *On maximum likelihood soft decoding of binary self-dual codes*, IEEE Trans. Commun. **41** (1993), 439–443.

[RaSn95] — — *Constrained designs for maximum likelihood soft decoding of $RM(2, m)$ and the extended Golay codes*, IEEE Trans. Commun. **43** (1995), 812–820.

[RaSn96] — — *A cyclic* $[6, 3, 4]$ *group code and the hexacode over* $GF(4)$, PGIT **42** (1996), 1250–1253.

[RaSn98] — — *Efficient decoding of the Gosset, Coxeter-Todd and the Barnes-Wall lattices* , preprint, 1998.

T. R. N. Rao: see G.-L. Feng.

C. Rastello: see J. Boutros.

[Rezn95] B. Reznick, *Some constructions of spherical 5-designs*, Lin. Alg. Appl. **226–228** (1995), 163–196.

[RiRo92] G. Riera and R. E. Rodríguez, *The period matrix of Bring's curve*, PJM **154** (1992), 179–200.

B. Rimoldi: see R. Urbanke.

V. Rödl: see H. Lefmann.

R. E. Rodríguez: see G. Riera.

D. S. Rokhsar: see also N. D. Mermin.

[RMW87] — N. D. Mermin and D. C. Wright, *Rudimentary quasicrystallography: the icosahedral and decagonal reciprocal lattices*, Phys. Rev. **B 35** (1987), 5487–5495.

[RWM88] — D. C. Wright and N. D. Mermin, *The two-dimensional quasicrystallographic space groups with rotational symmetries less than 23-fold*, Acta Cryst. **A 44** (1988), 197–211.

[RoT90] M. Y. Rosenbloom and M. A. Tsfasman, *Multiplicative lattices in global fields*, Invent. Math. **101** (1990), 687–696.

R. M. Roth: see N. Alon.

Z. Rudnick: see A. Eskin.

[Rung93] B. Runge, *On Siegel modular forms I*, JRAM **436** (1993), 57–85.

[Rung95] — *On Siegel modular forms II*, Nagoya Math. J. **138** (1995), 179–197.

[Rung95a] — *Theta functions and Siegel-Jacobi forms*, Acta Math. **175** (1995), 165–196.

[Rung96] — *Codes and Siegel modular forms*, Discr. Math. **148** (1996), 175–204.

J. A. Rush: see also N. D. Elkies.

[Rus89] — *A lower bound on packing density*, Invent. Math. **98** (1989), 499–509.

[Rus91] — *Constructive packings of cross polytopes*, Math. **38** (1991), 376–380.

[Rus92] — *Thin lattice coverings*, JLMS **45** (1992), 193–200.

[Rus93] — *A bound, and a conjecture, on the maximum lattice-packing density of a superball*, Math. **40** (1993), 137–143.

[Rus94] — *An indexed set of density bounds on lattice packings*, Geom. Dedicata **53** (1994), 217–221.

[Rus96] — *Lattice packing of nearly Euclidean balls in spaces of even dimension*, Proc. Edin. Math. Soc. **39** (1996), 163–169.

A. J. E. Ryba: see R. E. Borcherds.

S. S. Ryskov: see also R. M. Erdahl, L. Michel.

[RyE88] — and R. M. Erdahl, *The empty sphere, Part II*, CJM **40** (1988), 1058–1073.

[RyE89] — — *Stepwise construction of L-bodies of lattices*, Uspekhi Mat. Nauk **44** (1989), 241–242 = Russian Math. Surveys **44** (1989), 293–294.

[Sal94] A. Sali, *On the rigidity of spherical t-designs that are orbits of finite reflection groups*, DCC **4** (1994), 157–170.

[Samo98] A. Samorodnitsky, *On the optimum of Delsarte's linear program*, JCT, to appear.

P. Sarnak: see also P. Buser, A. Eskin.

[Sar90] — *Some Applications of Modular Forms*, Camb. Univ. Press, 1990.

[Sar95] — *Extremal geometries*, Contemp. Math. **201** (1997), 1–7.

J. Saxl: see M. W. Liebeck.

[Scha65] J. Schaer, *The densest packing of 9 circles in a square*, CMB **8** (1965), 273–277.

[Scha71] — *On the packing of ten equal circles in a square*, Math. Mag. **44** (1971), 139–140.

R. Scharlau: see also C. Batut.

[Scha94] — *Unimodular lattices over real quadratic fields*, Math. Zeit. **216** (1994), 437–452.

[SchaB96] — and B. Blaschke, *Reflective integral lattices*, J. Alg. **181** (1996), 934–961.

[SchaS98] — and R. Schulze-Pillot, *Extremal lattices*, preprint, 1998.

[SchaT96] — and P. H. Tiep, *Symplectic groups, symplectic spreads, codes and unimodular lattices*, Inst. Exper. Math., Essen, preprint No. 16, 1996.

[SchaT96a] — — *Symplectic group lattices*, preprint, 1996.

[SchaV94] — and B. B. Venkov, *The genus of the Barnes-Wall lattice*, Comm. Math. Helv. **69** (1994), 322–333.

[SchaV96] — — *The genus of the Coxeter-Todd lattice*, preprint, 1995.

[SchaW92] — and C. Walhorn, *Integral lattices and hyperbolic reflection groups*, Astérisque **209** (1992), 279–291.

[SchaW92a] — — *Tables of hyperbolic reflection groups*, Ergänzungsreihe der Sonderforschungsbereichs **343** der Univ. Bielefeld, preprint 92–001, April 1992.

[ScharS] W. Scharlau and D. Schomaker, personal communication, April 1991.

A. Schiemann: see also W. C. Jagy.

[Schi90] — *Ein Beispiel positiv definiter quadratischer Formen der Dimension 4 mit gleichen Darstellungszahlen*, Arch. Math. **54** (1990), 372–375.

[Schi97] — *Ternary positive definite quadratic forms are determined by their theta series*, Math. Ann. **308** (1997), 507–517.

M. Schlottmann: see M. Baake.

C. Schlegel: see T. Linder.

[Schma] B. Schmalz, *The t-designs with prescribed automorphism group, new simple 6-designs*, J. Combin. Designs **1** (1993), 125–170.

[Schm90] J. Gf. von Schmettow, *A practical view of KANT*, DMV-Seminar "Konstruktive Zahlentheorie", Düsseldorf, August, 1990.

[Schm91] — *KANT, a tool for computations in algebraic number fields*, in *Computational Number Theory*, A. Pethö et al., editors, Gruyter, 1991.

[Schmu93] P. Schmutz, *Riemann surfaces with shortest geodesics of maximal length*, Geomet. Functional Anal. **3** (1993), 564–631.

[Schmu95] — *Arithmetic groups and the length spectrum of Riemann surfaces*, DMJ, to appear.

[Schmu95a] P. Schmutz Schaller, *The modular torus has maximal length spectrum*, preprint.

W. A. Schneeberger: see also J. H. Conway.

[Schnee97] — Ph. D. Dissertation, Princeton Univ., 1997.

U. Schnell: see also K. Böröczky, Jr.

[Schn96] — Review of [Hsi93a], Math. Review **96**f: 52028.

[Schn98] — *On the Wulff-shape of periodic sphere packings*, preprint.

C. P. Schnorr: see also A. Paz.

[Schn87] — *A hierarchy of polynomial time basis reduction algorithms*, Theoret. Comp. Sci. **53** (1987), 201–224.

D. Schomaker: see also W. Scharlau.

[SchW92] — and M. Wirtz, *On binary cyclic codes of odd lengths from 101 to 127*, PGIT **38** (1992), 516–518.

E. Schrüfer: see W. Neutsch.

R. Schulze-Pillot: see also S. Böcherer, W. Duke, R. Scharlau.

[ScP91] — *An algorithm for computing genera of ternary and quaternary quadratic forms*, in *Proc. Intern. Symp. Symbolic Algebraic Computation (Bonn, 1991)*, pp. 134–143.

[ScP93] — *Quadratic residue codes and cyclotomic lattices*, Archiv. Math. **60** (1993), 40–45.

J. J. Seidel: see also A. Neumaier.

[Sei90] — *Designs and approximation*, Contemp. Math. **111** (1990), 179–186.

[Sei90a] — *More about two-graphs*, in *Fourth Czechoslovak Symp. Combinatorics, Prachatice 1990*, Ann. Discr. Math. **51** (1992), 297–308.

[Sei90b] — *Integral lattices, in particular those of Witt and of Leech*, in *Proceedings Seminar 1986—1987, Mathematical Structures in Field Theories*, Math. Centrum, Amsterdam, 1990, pp. 37–53.

[Sei91] — *Geometry and Combinatorics Selected Works*, eds. D. G. Corneil and R. Mathon, Academic Press, San Diego, 1991.

[Sei95] — *Spherical designs and tensors*, in *Progress in Algebraic Combinatorics (Fukuoka 1993)*, Adv. Studies Pure Math., **24** (1996), 309–321.

[Sel74] E. Selling, *Ueber die binären und ternären quadratischen Formen*, JRAM **77** (1874), 143–229.

K.-D. Semmler: see P. Buser.

M. Senechal: see L. Michel.

V. Senk: see D. E. Lazić.

I. Seskar: see D. E. Lazić.

B. Shahar: see Y. Be'ery.

J. O. Shallit: see E. Bach.

[ShKH96] A. G. Shanbhag, P. V. Kumar and T. Helleseth, *Improved binary codes and sequence families from $\mathbb{Z}_4$-linear codes*, PGIT **42** (1996), 1582–1587.

Y.-Y. Shao: see J. S. Hsia, F. Z. Zhu.

J. B. Shearer: see also A. E. Brouwer.

[Shea88] — personal communication, June, 1988.

A. M. Shende: see D. S. Rajan.

G. C. Shephard: see B. Grünbaum.

T. Shioda: see also K. Oguiso.

[Shiod88] — *Arithmetic and geometry of Fermat curves*, in *Algebraic Geometry Seminar, Singapore 1987*, World Scientific, Singapore, 1988, pp. 95–102.

[Shiod89] — *The Galois representation of type E_8 arising from certain Mordell-Weil groups*, Proc. Japan Acad. **A 65** (1989) 195–197.

[Shiod89a] — *Mordell-Weil lattices and Galois representation*, Proc. Japan Acad. **A 65** (1989), 268–271, 296–299, 300–303.

[Shiod90] — *Construction of elliptic curves over $\mathbb{Q}(t)$ with high rank: A preview*, Proc. Japan Acad. **A 66** (1990), 57–60.

[Shiod90a] — *On the Mordell-Weil lattices*, Comment. Math. Univ. St. Pauli **39** (1990), 211–240.

[Shi90b] — *The theory of Mordell-Weil lattices*, Proc. Inter. Congress Math., Tokyo, 1990, Vol. I, pp. 473–489.

[Shiod91] — *Mordell-Weil lattices and sphere packings*, AJM **113** (1991), 931–948.

[Shiod91a] — *Mordell-Weil lattices of type E_8 and deformation of singularities*, in *Prospects in Complex Geometry*, LNM **1468** (1991), 177–202.

[Shiod91b] — *Construction of elliptic curves with high rank via the invariants of the Weyl groups*, J. Math. Soc. Japan **43** (1991), 673–719.

[Shiod91c] — *An infinite family of elliptic curves over $\mathbb{Q}$ with large rank via Néron's method*, Invent. Math. **106** (1991), 109–119.

[Shiod91d] — *A Collection: Mordell-Weil Lattices*, Max-Planck Inst. Math., Bonn, 1991.

[Shiod91e] — *Mordell-Weil lattice theory and its applications – a point of contact of algebra, geometry and computers* (in Japanese), Sugaku **43** (No. 2, 1991), 97–114.

P. W. Shor: see also A. R. Calderbank.

[ShS98] — and N. J. A. Sloane, *A family of optimal packings in Grassmannian manifolds*, J. Alg. Combin., 1998, to appear.

M. U. Siddiqi: see B. S. Rajan.

F. Sigrist: see also A.-M. Bergé, D.-O. Jaquet.

[Sig90] — *Formes quadratiques encapsulées*, JNB **2** (1990), 425–429.

M. K. Simon: see E. Biglieri.

[Sim94] J. Simonis, *Restrictions on the weight distribution of binary linear codes imposed by the structure of Reed-Muller codes*, PGIT **40** (1994), 194–196.

[SiV91] — and C. de Vroedt, *A simple proof of the Delsarte inequalities*, DCC **1** (1991), 77–82.

[SipS96] M. Sipser and D. A. Spielman, *Expander codes*, PGIT **42** (1996), 1710–1722.

[SkSL] S. S. Skiena, W. D. Smith and P. Lemke, *Reconstructing sets from interpoint distances*, 6th ACM Symp. Comp. Geom., June 1990, pp. 332–339.

[SkV90] A. N. Skorobogatov and S. G. Vladuts, *On the decoding of algebraic-geometric codes*, PGIT **36** (1990), 1051–1061.

N. J. A. Sloane: see also M. Bernstein, A. E. Brouwer, A. R. Calderbank, J. H. Conway, Y. Edel, G. D. Forney, Jr., R. W. Grosse-Kunstleve, A. R. Hammons, Jr., R. H. Hardin, G. Nebe, E. M. Rains, P. W. Shor.

[SloEIS] — *The On-Line Encyclopedia of Integer Sequences*, published electronically at http://www.research.att.com/~njas/sequences/.

[SloHP] — *Home page*: http://www.research.att.com/~njas/.

[SlHDC] — R. H. Hardin, T. S. Duff and J. H. Conway, *Minimal-energy clusters of hard spheres*, DCG **14** (1995), 237–259.

[Slod80] P. Slodowy, *Simple Singularities and Simple Algebraic Groups*, LNM **815** (1980).

[Smith76] P. E. Smith, *A simple subgroup of M? and* $E_8(3)$, BLMS **8** (1976), 161–165.

W. D. Smith: see also A. E. Brouwer, R. H. Hardin, S. S. Skiena.

[Smi88] — *Studies in Computational Geometry Motivated by Mesh Generation*, Ph. D. Dissertation, Dept. of Applied Mathematics, Princeton Univ., Sept. 1988.

[Smi91] — *Few-distance sets and the second Erdős number*, in preparation.

J. Snyders: see also Y. Be'ery, M. Ran.

[SnBe89] — and Y. Be'ery, *Maximum likelihood soft decoding of binary block codes and decoders for the Golay codes*, PGIT **35** (1989), 963–975.

[Soi87] L. H. Soicher, *Presentations of some finite groups with applications to the O'Nan simple group*, J. Alg. **108** (1987), 310–316.

[Soi90] — *A new existence and uniqueness proof for the O'Nan group*, BLMS **22** (1990), 148–152.

[Soi91] — *More on the group* Y_{555} *and the projective plane of order 3*, J. Alg. **36** (1991), 168–174.

P. Solé: see also C. Bachoc, A. Bonnecaze, R. Chapman, G. D. Cohen, S. T. Dougherty, A. R. Hammons, Jr., P. Loyer.

[Sole95] — *D_4, E_6, E_8 and the AGM*, Lect. Notes Comp. Sci. **948** (1995), 448–455.

[SolL98] — and P. Loyer, *U_n lattices, Construction B and AGM iterations*, EJC **19** (1998), 227–236.

P. Sonneborn: see T. Drisch.

T. Sørevik: see J. N. Lyness.

B. Souvignier: see also W. Plesken.

[Sou94] — *Irreducible finite integral matrix groups of degree 8 and 10*, MTAC **63** (1994), 335–350.

D. A. Spielman; See also M. Sipser.

[Spiel96] — *Linear-time encodable and decodable error-correcting codes*, PGIT **42** (1996), 1723–1731.

[Stan80] R. P. Stanley, *Decompositions of rational convex polytopes*, Ann. Discrete Math. **6** (1980), 333–342.

[Stan86] — *Enumerative Combinatorics*, Vol. 1. Wadsworth, Monterey, CA, 1986.

J. Stern: see G. Lachaud.

H. Stichtenoth: see also A. Garcia.

[Stich93] — *Algebraic Function Fields and Codes*, Springer-Verlag, NY, 1993.

[StTs92] — and M. A. Tsfasman, editors, *Coding Theory and Algebraic Geometry*, LNM **1518**, Springer-Verlag, NY, 1992.

[Sto73] M. I. Stogrin, *Regular Dirichlet-Voronoi partitions for the second triclinic group*, TMIS **123** (1973) = PSIM **123** (1973).

[Stru94] R. Struik, *An improvement of the Van Wee bound for binary linear covering codes*, PGIT **40** (1994), 1280–1284.

[Stru94a] — *On the structure of linear codes with covering radius two and three*, PGIT **40** (1994), 1417–1424.

[Suda96] M. Sudan, *Maximum likelihood decoding of Reed-Solomon codes*, Proc. 37th FOCS (1996), 164–172.

[Suda97] — *Decoding of Reed-Solomon codes beyond the error-correction bound*, J. Compl. **13** (1997), 180–193.

[SuW95] I. A. I. Suleiman and R. A. Wilson, *Standard generators for J_3*, Exper. Math. **5** (1995).

[Sul90] J. M. Sullivan, *A Crystalline Approximation Theorem for Hypersurfaces*, Ph. D. Dissertation, Princeton Univ., 1990.

F.-W. Sun: see also O. Amrani.

[SuTi95] — and H. C. A. van Tilborg, *The Leech lattice, the octacode and decoding algorithms*, PGIT **41** (1995), 1097–1106.

A. Szántó: see G. Ivanyos.

[Tak85] I. Takada, *On the classification of definite unimodular lattices over the ring of integers in $\mathbb{Q}(\sqrt{2})$*, Math. Japan. **30** (1985), 423–433.

[Tak90] — *On the genera and the class number of unimodular lattices over the ring of integers of real quadratic fields*, JNT **36** (1990), 373–379.

[Tan91] L.-Z. Tang, *Rational double points on a normal octic K3 surface*, Nagoya Math. J. **129** (1993), 147–177.

[TaGa91] T. Tarnai and Z. Gáspár, *Covering a sphere by equal circles, and the rigidity of its graph*, PCPS **110** (1991), 71–89.

[TaGa91a] — — *Arrangement of 23 points on a sphere (on a conjecture of R. M. Robinson)*, PRS **A 433** (1991), 257–267.

V. Tarokh: see also I. F. Blake.

[TaBl96] — and I. F. Blake, *Trellis complexity versus the coding gain of lattices I*, PGIT **42** (1996), 1796–1807.

[TaBl96a] — — *Trellis complexity versus the coding gain of lattices II*, PGIT **42** (1996), 1808–1816.

[TaVa97] — and A. Vardy, *Upper bounds on trellis complexity of lattices*, PGIT **43** (1997), 1294–1300.

[TaVZ] — — and K. Zeger, *Universal bound on the performance of lattice codes*, PGIT, to appear.

T. Tasaka: see T. Kondo.

E. Thiran: see J. Fröhlich.

[Thomp76] J. G. Thompson, *A conjugacy theorem for E_8*, J. Alg. **38** (1976), 525–530.

P. H. Tiep: see also A. I. Bondal, A. I. Kostrikin, R. Scharlau.

[Tiep91] — *Weil representations of finite symplectic groups, and Gow lattices* (in Russian), Mat. Sb. **182** (1991), 1177–1199 = Math. USSR-Sb. **73** (1992), 535–555.

[Tiep92] — *Automorphism groups of some Mordell-Weil lattices* (in Russian), Izv. Ross. Akad. Nauk Ser. Mat. **56** (1992), 509–537 = Russian Acad. Sci. Izv. Math. **40** (1993), 477–501.

[Tiep92a] — *Basic spin representations of alternating groups, Gow lattices, and Barnes-Wall lattices* (in Russian), Mat. Sb. **183** (1992), 99–116 = Russian Acad. Sci. Sb. Math. **77** (1994), 351–365.

[Tiep94] — *Globally irreducible representations of the finite symplectic group $Sp_4(q)$*, Commun. Alg. **22** (1994), 6439–6457.

[Tiep95] — *Basic spin representations of $2\mathbb{S}_n$ and $2\mathbb{A}_n$ as globally irreducible representations*, Archiv. Math. **64** (1995), 103–112.

[Tiep97] — *Weil representations as globally irreducible representations*, Math. Nach. **184** (1997), 313–327.

[Tiep97a] — *Globally irreducible representations of finite groups and integral lattices*, Geom. Dedicata **64** (1997), 85–123.

[Tiep97b] — *Globally irreducible representations of $SL_3(q)$ and $SU_3(q)$*, preprint.

[Tiet91] A. Tietäväinen, *Covering radius and dual distance*, DCG **1** (1991), 31–46.

H. C. A. van Tilborg: see O. Amrani, F.-W. Sun.

V. D. Tonchev: see also F. C. Bussemaker, W. C. Huffman, S. N. Kapralov, V. Pless.

[Ton88] — *Combinatorial Configurations*, Longman, London, 1988.

[Ton89] — *Self-orthogonal designs and extremal doubly even codes*, JCT **A 52** (1989), 197–205.

[ToYo96] — and V. Y. Yorgov, *The existence of certain extremal [54, 27, 10] self-dual codes*, PGIT **42** (1996), 1628–1631.

[TrTr84] C. L. Tretkoff and M. D. Tretkoff, *Combinatorial group theory, Riemann surfaces, and differential equations*, Contemp. Math. **33** (1984), 467–519.

M. D. Tretkoff: see C. L. Tretkoff.

M. D. Trott: see G. D. Forney, Jr.

[Tsa91] H.-P. Tsai, *Existence of certain extremal self-dual codes*, PGIT **38** (1992), 501–504.

M. A. Tsfasman: see also G. Lachaud, M. Y. Rosenbloom, H. Stichtenoth.

[Tsf91] — *Algebraic-geometric codes and asymptotic problems*, Discr. Appl. Math. **33** (1991), 241–256.

[Tsf91a] — *Global fields, codes, and sphere packings*, Astérisque **198–200** (1992), 373–396.

[Tsf96] — *Algebraic curves and sphere packing*, in *Arithmetic, Geometry and Coding Theory*, ed. R. Pellikaan et al., de Gruyter, Berlin, 1996, pp. 225–251.

[TsV91] — and S. G. Vladuts, *Algebraic-Geometric Codes*, Kluwer, Dordrecht, 1991.

[Ura87] T. Urabe, *Elementary transformations of Dynkin graphs and singularities on quartic surfaces*, Invent. Math. **87** (1987), 549–572.

[Ura89] — *Dynkin graphs and combinations of singularities on plane sextic curves*, Contemp. Math. **90** (1989), 295–316.

[Ura90] — *Tie transformations of Dynkin graphs and singularities on quartic surfaces*, Invent. Math. **100** (1990), 207–230.

[Ura93] — *Dynkin Graphs and Quadrilateral Singularities*, LNM **1548**, 1993.

[UrA95] R. Urbanke and D. Agrawal, *A counter example to a Voronoi region conjecture*, PGIT **41** (1995), 1195–1199.

[UrB98] — and B. Rimoldi, *Lattice codes can achieve capacity on the AWGN channel*, PGIT **44** (1998), 273–278.

[VAL93] R. J. M. Vaessens, F. H. L. Aarts and J. H. van Lint, *Genetic algorithms in coding theory – a table for $A_3(n, d)$*, Discr. Appl. Math. **45** (1993), 71-87.

[Val90] B. Vallée, *A central problem in the algorithmic geometry of numbers: Lattice reduction*, CWI Quarterly **3** (1990), 95–120.

[Vall89] G. Vallette, *A better packing of ten equal circles in a square*, Discr. Math. **76** (1989), 57–59.

A. Vardy: see also O. Amrani, A. R. Calderbank, R. G. Downey, J. Feigenbaum, G. D. Forney, Jr., Y. Klein, A. Lafourcade, S. Litsyn, V. Tarokh.

[Vard94] — *The Nordstrom-Robinson code: representation over $GF(4)$ and efficient decoding*, PGIT **40** (1994), 1686–1693.

[Vard95] — *A new sphere packing in 20 dimensions*, Invent. Math. **121** (1995), 119–133.

[Vard95a] — *Even more efficient bounded-distance decoding of the hexacode, the Golay code, and the Leech lattice*, PGIT **41** (1995), 1495–1499.

[Vard97] — *The intractability of computing the minimum distance of a code*, PGIT **43** (1997), 1757–1766.

[Vard98] — *Density doubling, double-circulants and new sphere packings*, TAMS, to appear.

[Vard98a] — *Trellis structure of codes*, to appear in [PHB].

[VaBe91] — and Y. Be'ery, *More efficient soft-decision decoding of the Golay codes*, PGIT **37** (1991), 667–672.

[VaBe93] — — *Maximum-likelihood decoding of the Leech lattice*, PGIT **39** (1993), 1435–1444.

[VaKs96] — and F.R. Kschischang, *Proof of a conjecture of McEliece regarding the expansion index of the minimal trellis*, PGIT **42** (1996), 2027–2034.

[VaWi94] S. Vassallo and J. M. Wills, *On mixed sphere packings*, Contributions to Algebra and Geometry **35** (1994), 67–71.

B. B. Venkov: see R. Bacher, H. Koch, G. Nebe.

T. Verhoeff: see A. E. Brouwer.

E. Viterbo: see also J. Boutros.

[ViBi96] — and E. Biglieri, *Computing the Voronoi cell of a lattice: the diamond-cutting algorithm*, PGIT **42** (1996), 161–171.

S. G. Vladuts: see also Y. U. Manin, A. N. Skorobogatov, M. A. Tsfasman.

[Vlad90] — *On the decoding of algebraic-geometric codes over $\mathbb{F}_q$ for $q \geq 16$*, PGIT **36** (1990), 1461–1463.

[Voi85] E. Voigt, *Ausgezeichnete Basen von Milnorgittern einfacher Singularitäten*, AMSH **55** (1985), 183–190.

C. de Vroedt: see J. Simonis.

[Wan96] Z. Wan, On the uniqueness of the Leech lattice, EJC **18** (1997), 455–459.

[Wan91] M. Wang, *Rational double points on sextic K3 surfaces*, preprint.

[War92] H. H. Ward, *A bound for divisible codes*, PGIT **38** (1992), 191–194.
D. L. Webb: see C. S. Gordon.
[Wee93] G. J. M. van Wee, *Some new lower bounds for binary and ternary covering codes*, PGIT **39** (1993), 1422–1424.
L. F. Wei: see G. D. Forney, Jr.
V. K.-W. Wei: see T. Etzion, G. Lachaud.
R. Weissauer: see R. E. Borcherds.
G. Whittle: see R. G. Downey.
[Wille96] L. T. Wille, *New binary covering codes obtained by simulated annealing*, PGIT **46** (1996), 300–302.
[Will87] D. E. G. Williams, *Close packing of spheres*, J. Chem. Phys. **87** (1987), 4207–4210.
J. M. Wills: see also U. Betke, J. Bokowski, K. Böröczky, Jr., G. Fejes Tóth, P. M. Gandini, P. Gritzmann, P. M. Gruber, S. Vassallo.
[Wills83] — *Research Problem 35*, Period. Math. Hungar. **14** (1983), 312–314.
[Wills90] — *Kugellagerungen und Konvexgeometrie*, Jber. Dt. Math.-Verein. **92** (1990), 21–46.
[Wills90a] — *Dense sphere packings in cylindrical Penrose tilings*, Phys. Rev. **B 42** (1990), 4610–4612.
[Wills90b] — *A quasi-crystalline sphere-packing with unexpected high density*, J. Phys. France **51** (1990), 1061–1064.
[Wills91] — *An ellipsoid packing in E^3 of unexpected high density*, Math. **38** (1991), 318–320.
[Wills93] — *Finite sphere packings and sphere coverings*, Rend. Semin. Mat. Messina (II) **2** (1993), 91–97.
[Wills95] — *Sphere packings, densities and crystals*, in *Intuitive Geometry (Budapest 1995)*, Bolyai Society Math. Studies **6** (1997), pp. 223–234.
[Wills96] — *Sphere packings and the concept of density*, Rend. Circ. Mat. Palermo **41** (1996), 245–252.
[Wills96a] — *Lattice packings of spheres and the Wulff-shape*, Math. **43** (1996), 229–236.
[Wills97] — *On large lattice packings of spheres*, Geom. Dedicata **65** (1997), 117–126.
[Wills97a] — *Finite sphere packings and the methods of Blichfeldt and Rankin*, Acta Math. Hungar. **75** (1997), 337–342.
[Wills97b] — *Parametric density, online-packings and crystal growth*, Rend. Circ. Mat. Palermo **50** (1997), 413–424.
[Wills98] — *Crystals and quasicrystals, sphere packings and Wulff shape*, preprint.
[Wills98a] — *Spheres and sausages, crystals and catastrophes, and a joint packing theory*, Math. Intelligencer **20** (1998), 16–21.
R. A. Wilson: see also C. Jansen, P. B. Kleidman, S. A. Linton, S. P. Norton, R. A. Parker, A. I. A. Suleiman.
[Wil88] — *On the 3–local subgroups of Conway's group* Co_1, J. Alg. **113** (1988), 261–262.
[Wil89] — *Vector stabilizers and subgroups of Leech lattice groups*, J. Alg. **127** (1989), 387–408.
[Wil90] — *The 2– and 3-modular characters of* J_3, *its covering group and automorphism group*, J. Symb. Comp. **10** (1990), 647–656.
M. Wirtz: see D. Schomaker.
S. A. Wolpert: see C. S. Gordon.
D. C. Wright: see N. D. Mermin, D. S. Rokhsar.
P. E. Wright: see M. Bernstein.
D. Würtz: see R. Peikert.

[Xu1] X. Xu, *Untwisted and twisted gluing techniques for constructing self-dual lattices*, preprint.

[Yan94] J.-G. Yang, *Rational double points on a normal quintic K3 surface*, Acta Math. Sinica **10** (1994), 348–361.

V. Y. Yorgov: see also V. Tonchev.

[Yor89] — *On the extremal doubly-even codes of length 32*, Proc. Fourth Joint Swedish-Soviet International Workshop on Information Theory, Gotland, Sweden, 1989, pp. 275–279.

V. A. Yudin: see also A. V. Kolushov.

[Yud91] — *Sphere-packing in Euclidean space and extremal problems for trigonometrical polynomials*, Discr. Appl. Math. **1** (1991), 69–72.

[Yud97] — *Lower bounds for spherical designs*, Izv. Math. **61** (1997), 673–683.

D. Zagier: see M. Eichler.

[ZaF96] R. Zamir and M. Feder, *On lattice quantization noise*, PGIT **42** (1996), 1152–1159.

H. J. Zassenhaus: see also M. H. Dauenhauer, M. Pohst.

[Zas3] — *A new reduction method for positive definite quadratic forms*, preprint.

K. Zeger: see also J. Hamkins, C. Linder, V. Tarokh.

[ZeG94] — and A. Gersho, *Number of nearest neighbors in a Euclidean code*, PGIT **40** (1994), 1647–1649.

[ZeK93] — and M. R. Kantorovitz, *Average number of facets per cell in tree-structured vector quantizer partitions*, PGIT **39** (1993), 1053–1056.

P. L. Zhang: see J. R. Quine.

Z. Zhang: see T. Etzion.

[Zhu88] F. Z. Zhu, *On nondecomposability and indecomposability of quadratic forms*, Sci. Sinica **A 31** (1988), 265–273.

[Zhu91] — *Classification of positive definite Hermitian forms over $Q(\sqrt{-11})$*, J. Math (P.R.C.) **11** (1991), 188–195.

[Zhu91a] — *On the classification of positive definite unimodular Hermitian forms*, Chinese Sci. Bull. **36** (1991), 1506–1511.

[Zhu93] — *Classification of positive definite Hermitian forms over $Q(\sqrt{-19})$*, Northeastern Math. J. **9** (1993), 89–97.

[Zhu94] — *Construction of indecomposable definite Hermitian forms*, Chin. Ann. Math. **15 B** (1994), 349–360.

[Zhu95] — *On the construction of positive definite indecomposable unimodular even Hermitian forms*, JNT **50** (1995), 318–322.

[Zhu95a] — *On the construction of indecomposable positive definite Hermitian forms over $\mathbb{Z}[\sqrt{-m}]$*, Algebra Colloq. **2** (1995), 221–230.

[Zhu95b] — *Construction of indecomposable definite Hermitian forms*, Chinese Science Bull. **40** (1995), 177–180.

[Zhu88a] — and Y.-Y. Shao, *On the construction of indecomposable positive definite forms over $\mathbb{Z}$*, Chinese Ann. Math. **B 9** (1988), 79–94.

V. A. Zinoviev: see S. M. Dodunekov.

[Zo96] Ch. Zong, *Strange Phenomena in Convex and Discrete Geometry*, Springer-Verlag, NY, 1996.

A. Zurco: see P. M. Gandini.

Index

The extensive bibliography on pages 574–679 is intended to serve as a more complete author index; only a selection of names appears here. See also the list of symbols on pages xxx–xxxii.

Grundlehren der mathematischen Wissenschaften

A Series of Comprehensive Studies in Mathematics

A Selection

200. Doid: Lectures on Algebraic Topology
201. Beck: Continuous Flows in the Plane
202. Schmetterer: Introduction to Mathematical Statistics
203. Schoeneberg: Elliptic Modular Functions
204. Popov: Hyperstability of Control Systems
205. Nikol`skiĭ: Approximation of Functions of Several Variables and Imbedding Theorems
206. André: Homologie des Algèbres Commutatives
207. Donoghue: Monotone Matrix Functions and Analytic Continuation
208. Lacey: The Isometric Theory of Classical Banach Spaces
209. Ringel: Map Color Theorem
210. Gihman/Skorohod: The Theory of Stochastic Processes I
211. Comfort/Negrepontis: The Theory of Ultrafilters
212. Switzer: Algebraic Topology—Homotopy and Homology
213. Shafarevich: Basic Algebraic Geometry
214. van der Waerden: Group Theory and Quantum Mechanics
215. Schaefer: Banach Lattices and Positive Operators
216. Pólya/Szegö: Problems and Theorems in Analysis II
217. Stenström: Rings of Quotients
218. Gihman/Skorohod: The Theory of Stochastic Processes II
219. Duvant/Lions: Inequalities in Mechanics and Physics
220. Kirikov: Elements of the Theory of Representations
221. Mumford: Algebraic Geometry I: Complex Projective Varieties
222. Lang: Introduction to Modular Forms
223. Bergh/Löfström: Interpolation Spaces. An Introduction
224. Gilbarg/Trudinger: Elliptic Partial Differential Equations of Second Order
225. Schütte: Proof Theory
226. Karoubi: K-Theory. An Introduction
227. Grauert/Remmert: Theorie der Steinschen Räume
228. Segal/Kunze: Integrals and Operators
229. Hasse: Number Theory
230. Klingenberg: Lectures on Closed Geodesics
231. Lang: Elliptic Curves: Diophantine Analysis
232. Gihman/Skorohod: The Theory of Stochastic Processes III
233. Stroock/Varadhan: Multidimensional Diffusion Processes
234. Aigner: Combinatorial Theory
235. Dynkin/Yushkevich: Controlled Markov Processes
236. Grauert/Remmert: Theory of Stein Spaces
237. Köthe: Topological Vector Spaces II
238. Graham/McGehee: Essays in Commutative Harmonic Analysis
239. Elliott: Probabilistic Number Theory I
240. Elliott: Probabilistic Number Theory II
241. Rudin: Function Theory in the Unit Ball of C^R
242. Huppert/Blackburn: Finite Groups II
243. Huppert/Blackburn: Finite Groups III
244. Kubert/Lang: Modular Units
245. Cornfeld/Fomin/Sinai: Ergodic Theory
246. Naimark/Stern: Theory of Group Representations
247. Suzuki: Group Theory I
248. Suzuki: Group Theory II
249. Chung: Lectures from Markov Processes to Brownian Motion
250. Arnold: Geometrical Methods in the Theory of Ordinary Differential Equations
251. Chow/Hale: Methods of Bifurcation Theory
252. Aubin: Nonlinear Analysis on Manifolds. Monge-Ampère Equations
253. Dwork: Lectures on ρ-adic Differential Equations
254. Freitag: Siegelsche Modulfunktionen
255. Lang: Complex Multiplication
256. Hörmander: The Analysis of Linear Partial Differential Operators I
257. Hörmander: The Analysis of Linear Partial Differential Operators II

(continued after index)

258. Smoller: Shock Waves and Reaction–Diffusion Equations, 2nd Edition
259. Duren: Univalent Functions
260. Freidlin/Wentzell: Random Perturbations of Dynamical Systems, Second Edition
261. Bosch/Güntzer/Remmert: Non Archimedian Analysis—A System Approach to Rigid Analytic Geometry
262. Doob: Classical Potential Theory and Its Probabilistic Counterpart
263. Krasnosel'skiĭ/Zabreĭko: Geometrical Methods of Nonlinear Analysis
264. Aubin/Cellina: Differential Inclusions
265. Grauert/Remmert: Coherent Analytic Sheaves
266. de Rham: Differentiable Manifolds
267. Arbarello/Cornalba/Griffiths/Harris: Geometry of Algebraic Curves, Vol. I
268. Arbarello/Cornalba/Griffiths/Harris: Geometry of Algebraic Curves, Vol. II
269. Schapira: Microdifferential Systems in the Complex Domain
270. Scharlau: Quadratic and Hermitian Forms
271. Ellis: Entropy, Large Deviations, and Statistical Mechanics
272. Elliott: Arithmetic Functions and Integer Products
273. Nikol'skiĭ: Treatise on the Shift Operator
274. Hörmander: The Analysis of Linear Partial Differential Operators III
275. Hörmander: The Analysis of Linear Partial Differential Operators IV
276. Liggett: Interacting Particle Systems
277. Fulton/Lang: Riemann-Roch Algebra
278. Barr/Wells: Toposes, Triples and Theories
279. Bishop/Bridges: Constructive Analysis
280. Neukirch: Class Field Theory
281. Chandrasekharan: Elliptic Functions
282. Lelong/Gruman: Entire Functions of Several Complex Variables
283. Kodaira: Complex Manifolds and Deformation of Complex Structures
284. Finn: Equilibrium Capillary Surfaces
285. Burago/Zalgaller: Geometric Inequalities
286. Andrianov: Quadratic Forms and Hecke Operators
287. Maskit: Kleinian Groups
288. Jacod/Shiryaev: Limit Theorems for Stochastic Processes
289. Manin: Gauge Field Theory and Complex Geometry
290. Conway/Sloane: Sphere Packings, Lattices and Groups, 3rd Edition
291. Hahn/O'Meara: The Classical Groups and K-Theory
292. Kashiwara/Schapira: Sheaves on Manifolds
293. Revuz/Yor: Continuous Martingales and Brownian Motion
294. Knus: Quadratic and Hermitian Forms over Rings
295. Dierkes/Hildebrandt/Küster/Wohlrab: Minimal Surfaces I
296. Dierkes/Hildebrandt/Küster/Wohlrab: Minimal Surfaces II
297. Pastur/Figotin: Spectra of Random and Almost-Periodic Operators
298. Berline/Getzler/Vergne: Heat Kernels and Dirac Operators
299. Pommerenke: Boundary Behaviour of Conformal Maps
300. Orlik/Terao: Arrangements of Hyperplanes
301. Loday: Cyclic Homology
302. Lange/Birkenhake: Complex Abelian Varieties
303. DeVore/Lorentz: Constructive Approximation
304. Lorentz/v. Golitschek/Makovoz: Constructive Approximation. Advanced Problems
305. Hiriart-Urruty/Lemaréchal: Convex Analysis and Minimization Algorithms I. Fundamentals
306. Hiriart-Urruty/Lemaréchal: Convex Analysis and Minimization Algorithms II. Advanced Theory and Bundle Methods
307. Schwarz: Quantum Field Theory and Topology
308. Schwarz: Topology for Physicists
309. Adem/Milgram: Cohomology of Finite Groups
310. Giaquinta/Hildebrandt: Calculus of Variations I: The Lagrangian Formalism
311. Giaquinta/Hildebrandt: Calculus of Variations II: The Hamiltonian Formalism

312. Chung/Zhao: From Brownian Motion to Schrödinger's Equation
313. Malliavin: Stochastic Analysis
314. Adams/Hedberg: Function Spaces and Potential Theory
315. Bürgisser/Clausen/Shokrollahi: Algebraic Complexity Theory
316. Saff/Totik: Logarithmic Potentials with External Fields
317. Rockafellar/Wets: Variational Analysis
318. Kobayashi: Hyperbolic Complex Spaces